SAFETY SYMBOLS	HAZARD	EXAMPLES		
DISPOSAL 	Special disposal procedures need to be followed.	certain chemicals, living organisms	Do not dispose of these materials in the sink or trash can.	Dispose of wastes as directed by your teacher.
BIOLOGICAL 	Organisms or other biological materials that might be harmful to humans	bacteria, fungi, blood, unpreserved tissues, plant materials	Avoid skin contact with these materials. Wear mask or gloves.	Notify your teacher if you suspect contact with material. Wash hands thoroughly.
EXTREME TEMPERATURE 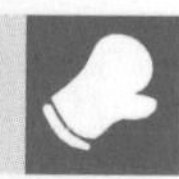	Objects that can burn skin by being too cold or too hot	boiling liquids, hot plates, dry ice, liquid nitrogen	Use proper protection when handling.	Go to your teacher for first aid.
SHARP OBJECT 	Use of tools or glassware that can easily puncture or slice skin	razor blades, pins, scalpels, pointed tools, dissecting probes, broken glass	Practice common-sense behavior and follow guidelines for use of the tool.	Go to your teacher for first aid.
FUME 	Possible danger to respiratory tract from fumes	ammonia, acetone, nail polish remover, heated sulfur, moth balls	Make sure there is good ventilation. Never smell fumes directly. Wear a mask.	Leave foul area and notify your teacher immediately.
ELECTRICAL 	Possible danger from electrical shock or burn	improper grounding, liquid spills, short circuits, exposed wires	Double-check setup with teacher. Check condition of wires and apparatus.	Do not attempt to fix electrical problems. Notify your teacher immediately.
IRRITANT 	Substances that can irritate the skin or mucous membranes of the respiratory tract	pollen, moth balls, steel wool, fiberglass, potassium permanganate	Wear dust mask and gloves. Practice extra care when handling these materials.	Go to your teacher for first aid.
CHEMICAL 	Chemicals that can react with and destroy tissue and other materials	bleaches such as hydrogen peroxide; acids such as sulfuric acid, hydrochloric acid; bases such as ammonia, sodium hydroxide	Wear goggles, gloves, and an apron.	Immediately flush the affected area with water and notify your teacher.
TOXIC 	Substance may be poisonous if touched, inhaled, or swallowed	mercury, many metal compounds, iodine, poinsettia plant parts	Follow your teacher's instructions.	Always wash hands thoroughly after use. Go to your teacher for first aid.
OPEN FLAME 	Open flame may ignite flammable chemicals, loose clothing, or hair	alcohol, kerosene, potassium permanganate, hair, clothing	Tie back hair. Avoid wearing loose clothing. Avoid open flames when using flammable chemicals. Be aware of locations of fire safety equipment.	Notify your teacher immediately. Use fire safety equipment if applicable.

Eye Safety
Proper eye protection should be worn at all times by anyone performing or observing science activities.

Clothing Protection
This symbol appears when substances could stain or burn clothing.

Animal Safety
This symbol appears when safety of animals and students must be ensured.

Radioactivity
This symbol appears when radioactive materials are used.

New York, New York Columbus, Ohio Woodland Hills, California Peoria, Illinois

Glencoe Science

INTEGRATED PHYSICS AND CHEMISTRY

Student Edition
Teacher Wraparound Edition
Interactive Teacher Edition CD-ROM
Interactive Lesson Planner CD-ROM
Texas Lesson Plans
Content Outline for Teaching
Dinah Zike's Teaching Science with Foldables
Directed Reading for Content Mastery
Foldables: Reading and Study Skills
Assessment
- *Chapter Review*
- *Chapter Tests*
- *ExamView Pro Test Bank Software*
- *Assessment Transparencies*
- *Performance Assessment in the Science Classroom*
- *The Princeton Review TAAS II Practice Booklet*

Directed Reading for Content Mastery in Spanish
Spanish Resources
English/Spanish Guided Reading Audio Program
Reinforcement
Enrichment
Activity Worksheets
Section Focus Transparencies
Teaching Transparencies
Laboratory Activities
Science Inquiry Labs
Critical Thinking/Problem Solving
Reading and Writing Skill Activities
Mathematics Skill Activities
Cultural Diversity
Texas Laboratory Management and Safety in the Science Classroom
Mindjogger Videoquizzes and Teacher Guide
Interactive Explorations and Quizzes CD-ROM
Vocabulary Puzzlemaker Software
Texas Cooperative Learning in the Science Classroom
Environmental Issues in the Science Classroom
Texas Home and Community Involvement
Using the Internet in the Science Classroom

"Study Tip," "Test-Taking Tip," and the "TAAS Practice" features in this book were written by The Princeton Review, the nation's leader in test preparation. Through its association with McGraw-Hill, The Princeton Review offers the best way to help students excel on standardized assessments.

The Princeton Review is not affiliated with Princeton University or Educational Testing Service.

Glencoe/McGraw-Hill

A Division of The **McGraw-Hill** *Companies*

Cover Images: Lead iodide, a bright-yellow solid, forms as potassium iodide is added to lead nitrate. The inset image is a circuit board.

Send all inquiries to:
Glencoe/McGraw-Hill
8787 Orion Place
Columbus, OH 43240

ISBN 0-07-823141-8
Printed in the United States of America.
2 3 4 5 6 7 8 9 10 071/055 06 05 04 03 02 01

Authors

National Geographic Society
Education Division
Washington, D.C.

Marilyn Thompson, EdD
Assistant Professor, College of Education
Arizona State University
Tempe, Arizona

Charles William McLaughlin, PhD
Senior Lecturer
University of Nebraska
Lincoln, Nebraska

Dinah Zike
Educational Consultant
Dinah-Might Activities, Inc.
San Antonio, Texas

Contributing Authors

Nancy Ross-Flanigan
Science Writer
Detroit, Michigan

Margaret K. Zorn
Science Writer
Yorktown, Virginia

Texas Science Consultants

José Luis Alvarez, PhD
Math/Science Mentor Teacher
TEKS for Leaders Trainer
Ysleta ISD
El Paso, Texas

José Alberto Marquez
TEKS for Leaders Trainer
Ysleta ISD
El Paso, Texas

Sandra West, PhD
Associate Professor of Biology
Southwest Texas State University
San Marcos, Texas

Content Consultants

Alan Bross, PhD
High Energy Physicist
Fermilab
Batavia, Illinois

Teresa Anne McCowen
Chemistry Instructor
Heartland Community College
Normal, Illnois

Jack Cooper
Adjunct Faculty Math and Science
Navarro College
Corsicana, Texas

Madelaine Meek
Physics Consultant
Editor
Lebanon, Ohio

Michael A. Hoggarth, PhD
Department of Life and Earth Sciences
Otterbein College
Westerville, Ohio

Carl Zorn, PhD
Staff Scientist
Jefferson Laboratory
Newport News, Virginia

Series Safety Consultants

Malcolm Cheney, PhD
OSHA Chemical Safety Officer
Hall High School
West Hartford, Connecticut

Sandra West, PhD
Associate Professor of Biology
Southwest Texas State University
San Marcos, Texas

Aileen Duc, PhD
Science II Teacher
Hendrick Middle School
Plano, Texas

Series Math Consultants

Michael Hopper, D.Eng
Manager of Aircraft Certification
Raytheon Company
Greenville, Texas

Teri Willard, EdD
Department of Mathematics
Montana State University
Belgrade, Montana

Series Reading Consultants

Elizabeth Babich
Special Education Teacher
Mashpee Public Schools
Mashpee, Massachusetts

Barry Barto
Special Education Teacher
John F. Kennedy Elementary School
Manistee, Michigan

Carol A. Senf, PhD
Associate Professor of English
Georgia Institute of Technology
Atlanta, Georgia

Rachel Swaters
Science Teacher
Rolla Middle Schools
Rolla, Missouri

Nancy Woodson, PhD
Professor of English
Otterbein College
Westerville, Ohio

Series Activity Testers

José Luis Alvarez, PhD
Math/Science Mentor Teacher
Ysleta ISD
El Paso, Texas

Nerma Coats Henderson
Teacher
Pickerington Jr. High School
Pickerington, Ohio

Mary Helen Mariscal-Cholka
Science Teacher
William D. Slider Middle School
Socorro ISD
El Paso, Texas

José Alberto Marquez
TEKS for Leaders Trainer
Ysleta ISD
El Paso, Texas

Science Kit and Boreal Laboratories
Tonawanda, New York

Reviewers

Sharla Adams
McKinney High School North
McKinney ISD
McKinney, Texas

Desiree Bishop
Baker High School
Mobile, Alabama

Nora M. Prestinari Burchett
Saint Luke School
McLean, Virginia

Mary Helen Mariscal-Cholka
William D. Slider Middle School
Socorro ISD
El Paso, Texas

Patricia Croft
Westside High School
Martinez, Georgia

Anthony DiSipio
Octorana Middle School
Atglen, Pennsylvania

George Gabb
Great Bridge Middle School
Chesapeake, Virginia

Maria Kelly
St. Leo School
Fairfax, Virginia

Eddie K. Lindsay
Vansant Middle School
Grundy, Virginia

H. Keith Lucas
Stewart Middle School
Fort Defiance, Virginia

Thomas E. Lynch Jr.
Northport High School
East Northport, New York

Linda Melcher
Woodmont Middle School
Piedmont, South Carolina

Annette Parrott
Lakeside High School
Atlanta, Georgia

Meredith Pickett
Memorial Middle School
Spring Branch ISD
Houston, Texas

Michelle Punch
Northwood Middle School
North Forest ISD
Houston, Texas

Pam Starnes
North Richland Middle School
Birdville ISD
Fort Worth, Texas

Clabe Webb
Sterling City High School
Sterling City ISD
Sterling City, Texas

Alison Welch
William D. Slider Middle School
Socorro ISD
El Paso, Texas

Kim Wimpey
North Gwinnett High School
Suwanee, Georgia

Contents in Brief

Integrated Physics & Chemistry TEKS*

***For a complete listing of all TEKS for Integrated Physics and Chemistry, see pages TX 3 and TX 4 at the end of the text.**

UNIT 1 Energy and Motion — 2

Contents

Energy — 98

CHAPTER 5

Work and Machines — 124

Thermal Energy — 156

UNIT 2 Electricity and Energy Resources — 190

CHAPTER 7 Electricity — 192

CHAPTER 8 Magnetism and Its Uses — 224

CHAPTER 9 Radioactivity and Nuclear Reactions — 256

CONTENTS

Energy Sources — 288

UNIT 3 Energy on the Move — 322

Waves — 324

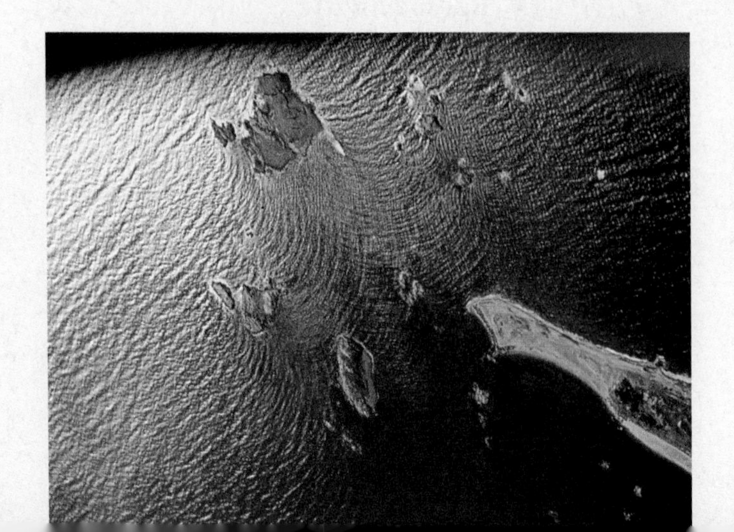

Sound — 356

Electromagnetic Waves — 388

CHAPTER 14

Light — 418

Mirrors and Lenses — 450

UNIT 4 The Nature of Matter — 484

Solids, Liquids, and Gases — 486

Classification of Matter — 516

CHAPTER 18 Properties of Atoms and the Periodic Table — 542

CHAPTER 19 Chemical Bonds — 572

UNIT 5 Diversity of Matter — 604

CHAPTER 20 Elements and Their Properties — 606

CONTENTS

Chemical Reactions — 736

Acids, Bases, and Salts — 764

Skill Handbooks — 798

Reference Handbooks — 826

English Glossary — 830

Spanish Glossary — 844

Index — 862

Handbook for Texas Students and Parents — TX 1–TX32

Interdisciplinary Connections

NATIONAL GEOGRAPHIC Unit Openers

NATIONAL GEOGRAPHIC VISUALIZING

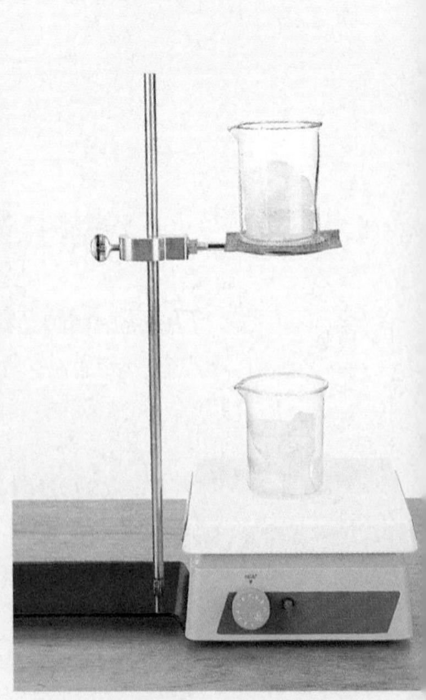

TIME

Science and Society

Science and History

Oops! Accidents in SCIENCE

Science and Language Arts

Science Stats

Activities

Feature Contents

Full Period Labs

Mini LAB

Activities

Feature Contents

EXPLORE ACTIVITY

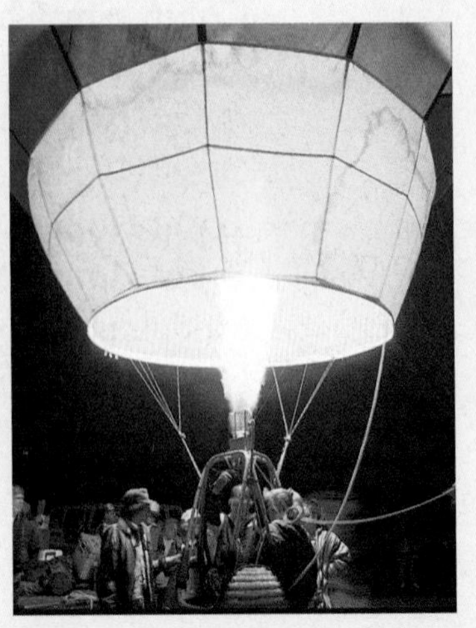

Problem-Solving Activities

Math Skills Activities

Activities

Skill Builder Activities

Science:

Math:

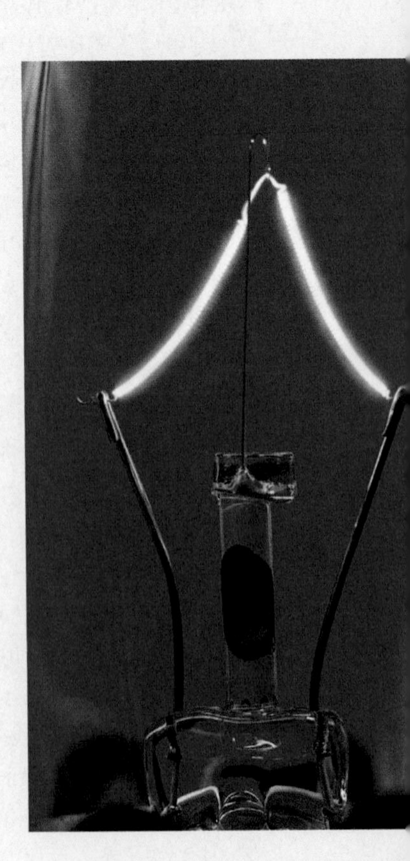

Technology:

Science INTEGRATION

SCIENCE *Online*

How Are Waffles & Running Shoes Connected?

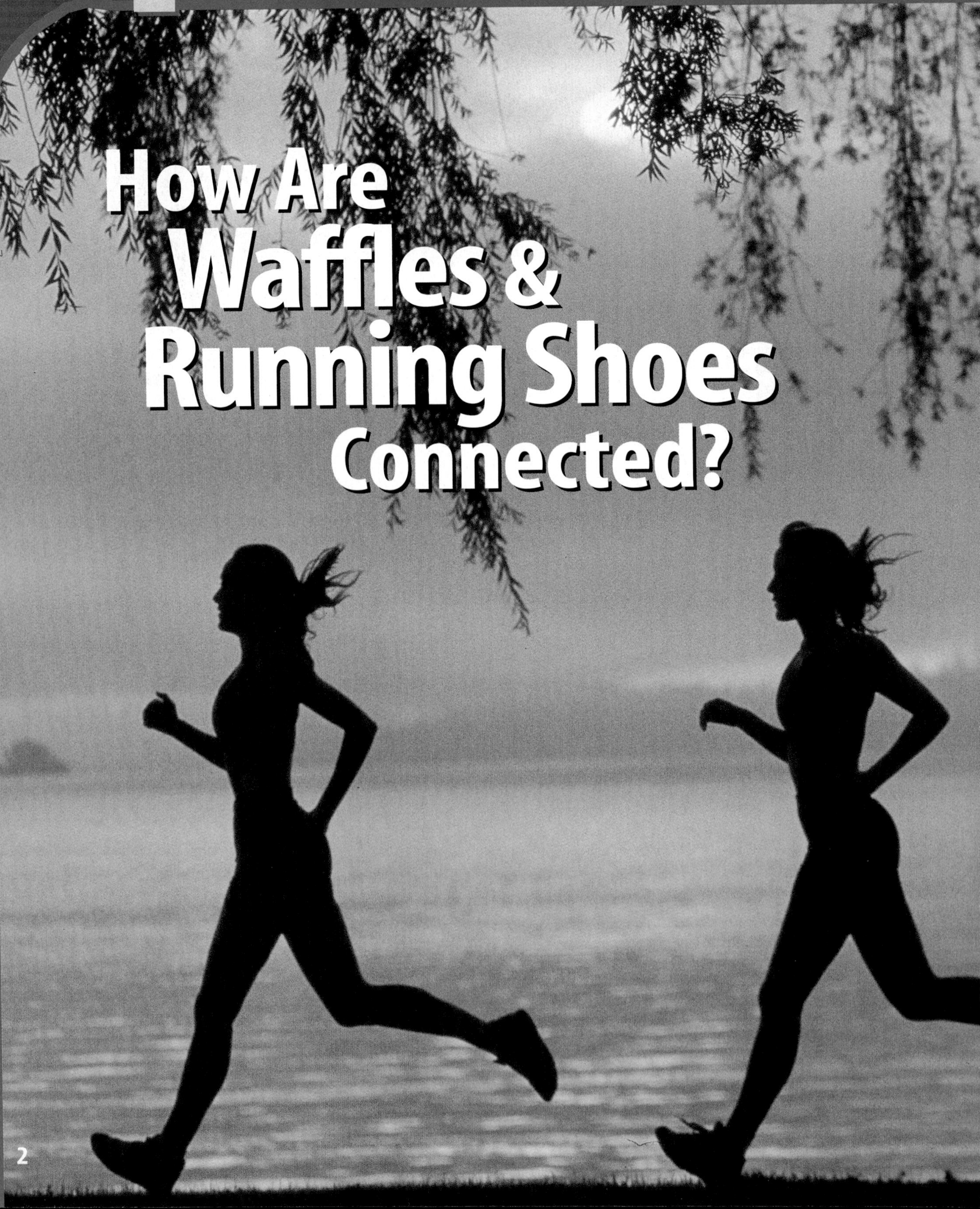

For centuries, shoes were made mostly of leather, cloth, or wood. These shoes helped protect feet, but they didn't provide much traction on slippery surfaces. In the early twentieth century, manufacturers began putting rubber on the bottom of canvas shoes, creating the first "sneakers." Sneakers provided good traction, but the rubber soles could be heavy—especially for athletes. One morning in the 1970s, an athletic coach stared into the waffles on the breakfast table and had an idea for a rubber sole that would be lighter in weight but would still provide traction. That's how the first waffle soles were born. Waffle soles soon became a world standard for running shoes.

SCIENCE CONNECTION

FRICTION Waffle soles improve traction by increasing friction—the force that opposes motion between two surfaces. At school or in a bicycle shop, examine and sketch the treads on different kinds of bicycle tires. Draw conclusions about how each type of tread increases or decreases friction between the tire and the ground. Then create a poster showing three different tire treads with an explanation about how each type of tread might suit a tire to particular riding conditions.

CHAPTER 1

The Nature of Science

Stacy Dragila of the United States won the women's pole vault event at the 2000 summer Olympics. The winner was decided through careful measurement of the jumps. In this chapter, you will learn how measurements are important to scientists. You also will learn about the methods scientists use to conduct their studies and how they communicate findings with graphs.

What do you think?

Science Journal Look at the picture below with a classmate. Discuss what this might be. Here's a hint: *Shipping authorities rely on these numbers for safety reasons.* Write your answer or best guess in your Science Journal.

During a track meet, one athlete ran 1 mile in 5 min and another athlete ran 5,000 m in 280 s. The two runners used different units to describe their races, so how can you compare them? Do the following activity to explore how choosing different units can make it difficult to compare measurements.

Discover how long a foot is

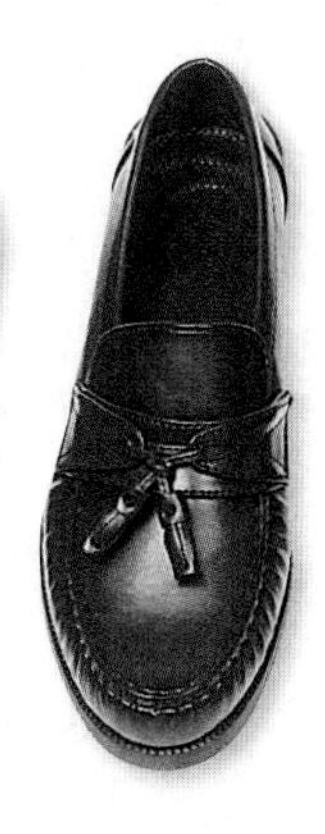
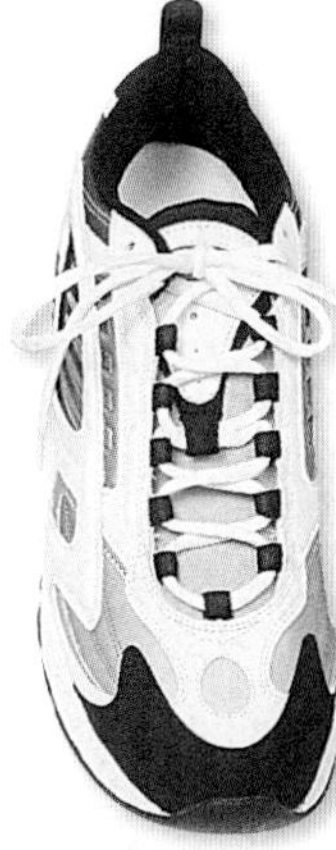

1. Measure the distance across your classroom using your foot as a measuring device.
2. Record your measurement and name your measuring unit.
3. Now, have your partner measure the same distance using his or her foot as the measuring device. Record this measurement and make up a different name for the unit.

Observe

In your Science Journal, explain why you think it might be important to have standard, well-defined units to make measurements.

Before You Read

Making a Question Study Fold **Asking yourself questions helps you stay focused and better understand scientific processes when you are reading the chapter.**

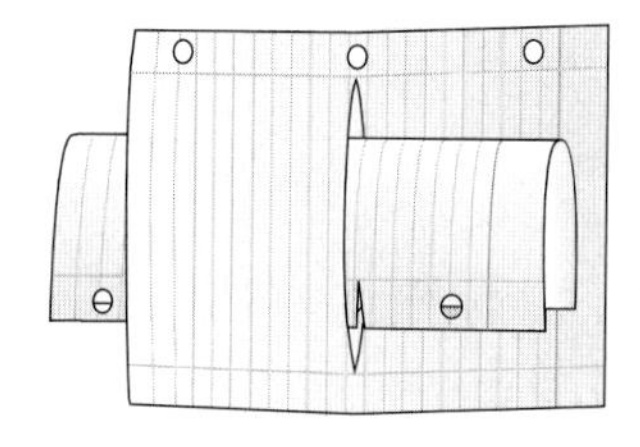

1. Stack two sheets of notebook paper in front of you so the long sides are at the top. Fold both in half from the left side to the right side. Unfold and separate.
2. Cut one sheet along the fold line, from one margin line to the other.
3. Place the second sheet in front of you so the long side is at the top. Cut along the fold line from the bottom of the paper to the margin line and then from the top of the paper to the margin line.
4. Insert the second sheet of paper into the cut of the first paper. Unfold it and align the cuts along the folds.
5. Title your book *Scientific Processes.* Before you read the chapter, write a question about something in your daily life on each page.

SECTION 1

The Methods of Science

As You Read

What You'll Learn

- **Identify** the steps scientists often use to solve problems.
- **Describe** why scientists use variables.
- **Compare and contrast** science and technology.

Vocabulary

scientific method
hypothesis
experiment
variable
dependent variable
independent variable
constant
control
bias
model
theory
scientific law
technology

Why It's Important

Using scientific methods will help you solve problems.

What is science?

Science is not just a subject in school. It is a method for studying the natural world. After all, science comes from the Latin word *scientia*, which means "knowledge." Science is a process that uses observation and investigation to gain knowledge about events in nature.

Nature follows a set of rules. Many rules, such as those concerning how the human body works, are complex. Other rules, such as the fact that Earth rotates about once every 24 h, are simpler. When you study these natural patterns, you are using science.

Major Categories of Science Science covers many different topics that can be classified according to three main categories. (1) Life science deals with living things. (2) Earth science investigates Earth and space. (3) Physical science deals with matter and energy. In this textbook, you will study mainly physical science. Sometimes, though, a scientific study will overlap the categories. One scientist, for example, might study the motions of the human body to understand how to build better artificial limbs. Is this scientist studying energy and matter or how muscles operate? She is studying both life science and physical science. It is not always clear what kind of science you are using, as shown in **Figure 1.**

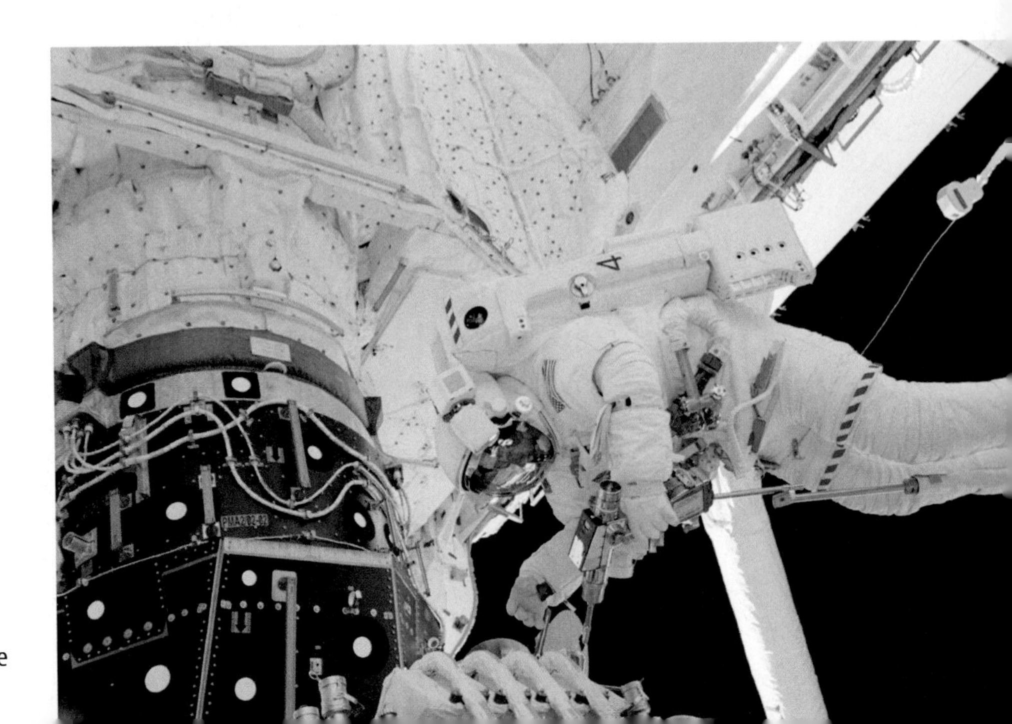

Figure 1
Astronaut Michael Lopez-Alegria uses a pistol grip tool on the *International Space Station.* *What evidence do you see of the three main branches of science in the photograph?*

Science Explains Nature Scientific explanations help you understand the natural world. Sometimes these explanations must be modified. As more is learned about the natural world, some of the earlier explanations might be found to be incomplete or new technology might provide more accurate answers.

For example, look at **Figure 2.** In the late eighteenth century, most scientists thought that heat was an invisible fluid with no mass. Heat seems to flow like a fluid. It also moves away from a warm body in all directions, just as a fluid moves outward when you spill it on the floor.

However, the heat fluid idea did not explain everything. If heat were an actual fluid, an iron bar that had a temperature of 1,000°C should have more mass than it did at 100°C because it would have more of the heat fluid in it. The eighteenth-century scientists thought they just were not able to measure the small mass of the heat fluid on the balances they had. When additional investigations showed no difference in mass was detected, scientists had to change the explanation.

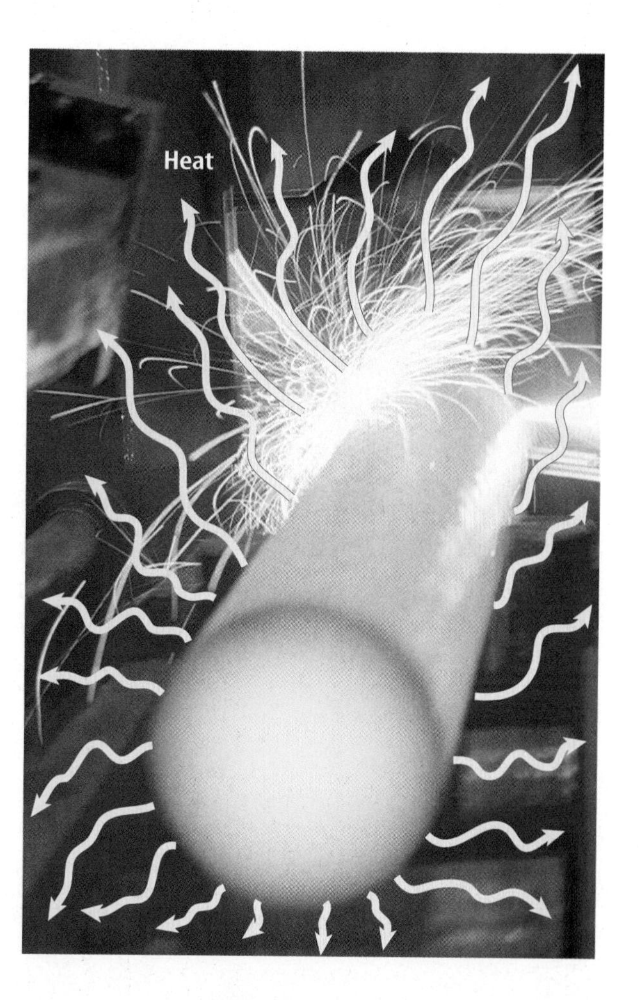

Figure 2
Many years ago, scientists thought that heat, such as in this metal rod, was a fluid. *How does heat act like a fluid?*

Investigations How do scientists learn more about the natural world? Scientists learn new information by performing investigations, which can be done many different ways. Some investigations involve simply observing something that occurs and recording the observations, perhaps in a journal. Other investigations involve setting up experiments that test the effect of one thing on another. Some investigations involve building a model that resembles something in the natural world and then testing the model to see how it acts. Often, a scientist will use something from all three ways when attempting to learn about the natural world.

✔ Reading Check *Why do scientific explanations change?*

Scientific Methods

Although scientists do not always follow a rigid set of steps, investigations often follow a general pattern. An organized set of investigation procedures is called a **scientific method.** Six steps often found in scientific methods are shown in **Figure 3.** A scientist might add new steps, repeat some steps many times, or skip other steps altogether when doing an investigation.

Research Visit the Glencoe Science Web site at **science.glencoe.com** for information about why leaves change color in the autumn. Make a prediction about why this occurs. Support your answer with evidence.

Figure 3
The series of procedures shown here is one way to use scientific methods to solve a problem.

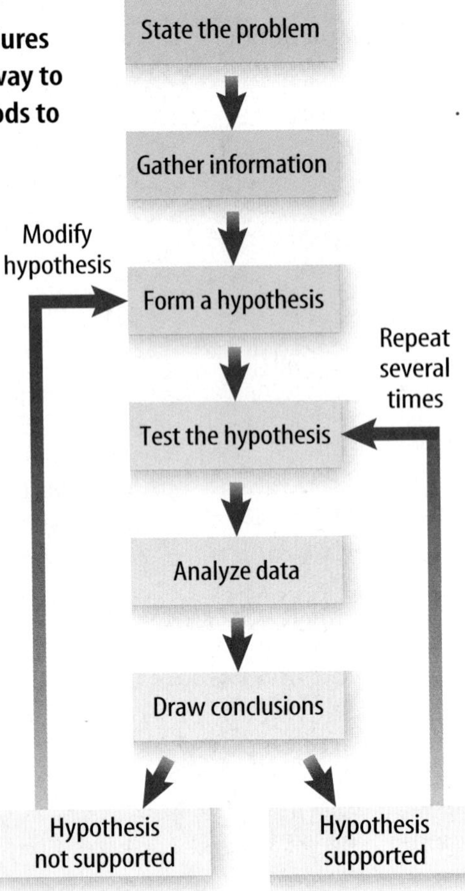

Stating a Problem Many scientific investigations begin when someone observes that an event in nature repeats itself and wonders why this is true. Then the question of "why" is the problem. Sometimes a statement of a problem arises from an activity that is not working. Some early work on guided missiles showed that the instruments in the nose of the missiles did not always work. The problem statement involved finding a material to protect the instruments from the harsh conditions of flight.

Later, National Aeronautics and Space Administration (NASA) scientists made a similar problem statement. They wanted to build a new vehicle—the space shuttle—that could carry people to outer space and back again. Guided missiles did not have this capability. NASA needed to find a material for the outer skin of the space shuttle that could withstand the heat and forces of reentry into Earth's atmosphere.

Researching and Gathering Information Before testing a hypothesis, it is useful to learn as much as possible about the background of the problem. Have others found information that will help determine what tests to do and what tests will not be helpful? The NASA scientists gathered information about melting points and other properties of the various materials that might be used. In many cases, tests had to be performed to learn the properties of new, recently created materials.

Forming a Hypothesis A **hypothesis** is a possible explanation for a problem using what you know and what you observe. NASA scientists knew that a ceramic coating had been found to solve the guided missile problem. They hypothesized that a ceramic material also might work on the space shuttle.

Testing a Hypothesis Some hypotheses can be tested by making observations. Others can be tested by building a model and relating it to real-life situations. One common way to test a hypothesis is to perform an experiment. An **experiment** tests the effect of one thing on another using controlled conditions.

Variables An experiment usually contains at least two variables. A **variable** is a quantity that can have more than a single value. You might set up an experiment to determine which of three fertilizers helps plants to grow the biggest. Before you begin your tests, you would need to think of all the factors that might cause the plants to grow bigger. Possible factors include plant type, amount of sunlight, water used, room temperature, type of soil, and type of fertilizer.

In this experiment, the amount of growth is the **dependent variable** because its value changes according to the changes in the other variables. The variable you change to see how it will affect the dependent variable is called the **independent variable.**

Constants and Controls To be sure you are testing to see how fertilizer affects growth, you must keep the other possible factors the same for each test, or trial. A factor that does not change when other variables change is called a **constant.** You might set up four trials, using the same soil and type of plant. Each plant is given the same amount of sunlight and water and is kept at the same temperature. These are constants. Three of the plants receive a different type of fertilizer. Fertilizer is the independent variable.

The fourth plant is not fertilized. This plant is a control. A **control** is the standard by which the test results can be compared. Suppose that after several days, the three fertilized plants grow between 2 and 3 cm. If the unfertilized plant grows 1.5 cm, you might infer that the growth of the fertilized plants was due to the fertilizers.

How might the NASA scientists set up an experiment to solve the problem of the damaged tiles shown in **Figure 4?** What are possible variables, constants, and controls?

✔ Reading Check *Why is a control used in an experiment?*

Life Science INTEGRATION

Through observations of living organisms, scientists have designed a classification system. The system groups organisms according to variables such as habits and physical and chemical features. As new organisms are discovered, scientific methods are used to determine their classification.

Figure 4
NASA has had an ongoing mission to improve the space shuttle. A technician is replacing tiles damaged upon reentry into Earth's atmosphere.

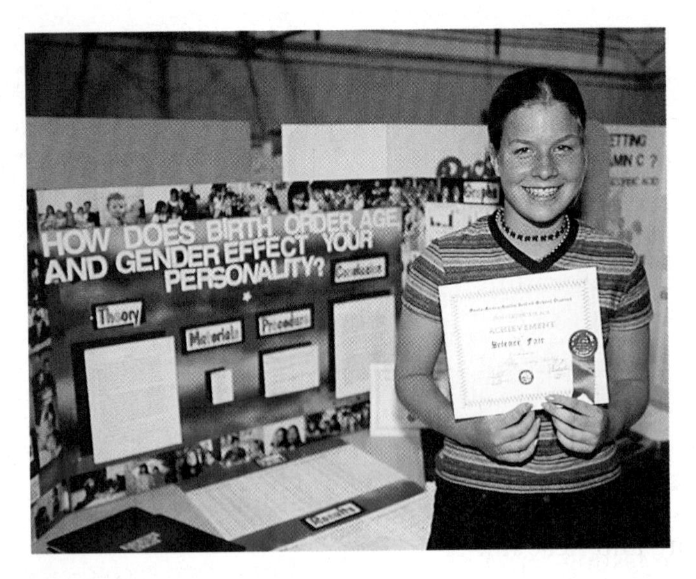

Figure 5
An exciting and important part of investigating something is sharing your ideas with others, as this student is doing at a science fair.

Analyzing the Data A part of an experiment includes recording observations and organizing the test data into easy-to-read tables and graphs. Later in this chapter you will study ways to display data. When you are making and recording observations, you should include all results, even unexpected ones. Many important discoveries have been made from unexpected occurrences.

Interpreting the data and analyzing the observations is an important step. If the data are not organized in a logical manner, wrong conclusions can be drawn. No matter how well a scientist communicates and shares that data, someone else might not agree with the data. Scientists share their data through reports and conferences. In **Figure 5** a student is displaying her data.

Drawing Conclusions Based on the analysis of your data, you decide whether or not your hypothesis is supported. When lives are at stake, such as with the space shuttle, you must be very sure of your results. For the hypothesis to be considered valid and widely accepted, the experiment must result in the exact same data every time it is repeated. If your experiment does not support your hypothesis, you must reconsider the hypothesis. Perhaps it needs to be revised or your experiment needs to be conducted differently.

Being Objective Scientists also should be careful to reduce bias in their experiments. A **bias** occurs when what the scientist expects changes how the results are viewed. This expectation might cause a scientist to select a result from one trial over those from other trials. Bias also might be found if the advantages of a product being tested are used in a promotion and the drawbacks are not presented.

Scientists can lessen bias by running as many trials as possible and by keeping accurate notes of each observation made. Valid experiments also must have data that are measurable. For example, a scientist performing a global warming study must base his or her data on accurate measures of global temperature. This allows others to compare the results to data they obtain from a similar experiment. Most importantly, the experiment must be repeatable. Findings are supportable when other scientists perform the same experiment and get the same results.

What is bias in science?

Visualizing with Models

Sometimes, scientists cannot see everything that they are testing. They might be observing something that is too large, too small, or takes too much time to see completely. In these cases, scientists use models. A **model** represents an idea, event, or object to help people better understand it.

Models in History Models have been used throughout history. One scientist, Lord Kelvin, who lived in England in the 1800s, was famous for making models. To model his idea of how light moves through space, he put balls into a bowl of jelly and encouraged people to move the balls around with their hands. Kelvin's work to explain the nature of temperature and heat still is used today.

High-Tech Models Scientific models don't always have to be something you can touch. Today, many scientists use computers to build models. NASA experiments involving space flight would not be practical without computers. The complex equations would take far too long to calculate by hand, and errors could be introduced much too easily.

Another type of model is a simulator, like the one shown in **Figure 6.** An airplane simulator enables pilots to practice problem solving with various situations and conditions they might encounter when in the air. This model will react the way a plane does when it flies. It gives pilots a safe way to test different reactions and to practice certain procedures before they fly a real plane.

Meteorology has changed greatly due to computer modeling. Using special computer programs, meteorologists now are able to more accurately predict disastrous weather. In your Science Journal, describe how computer models might help save lives.

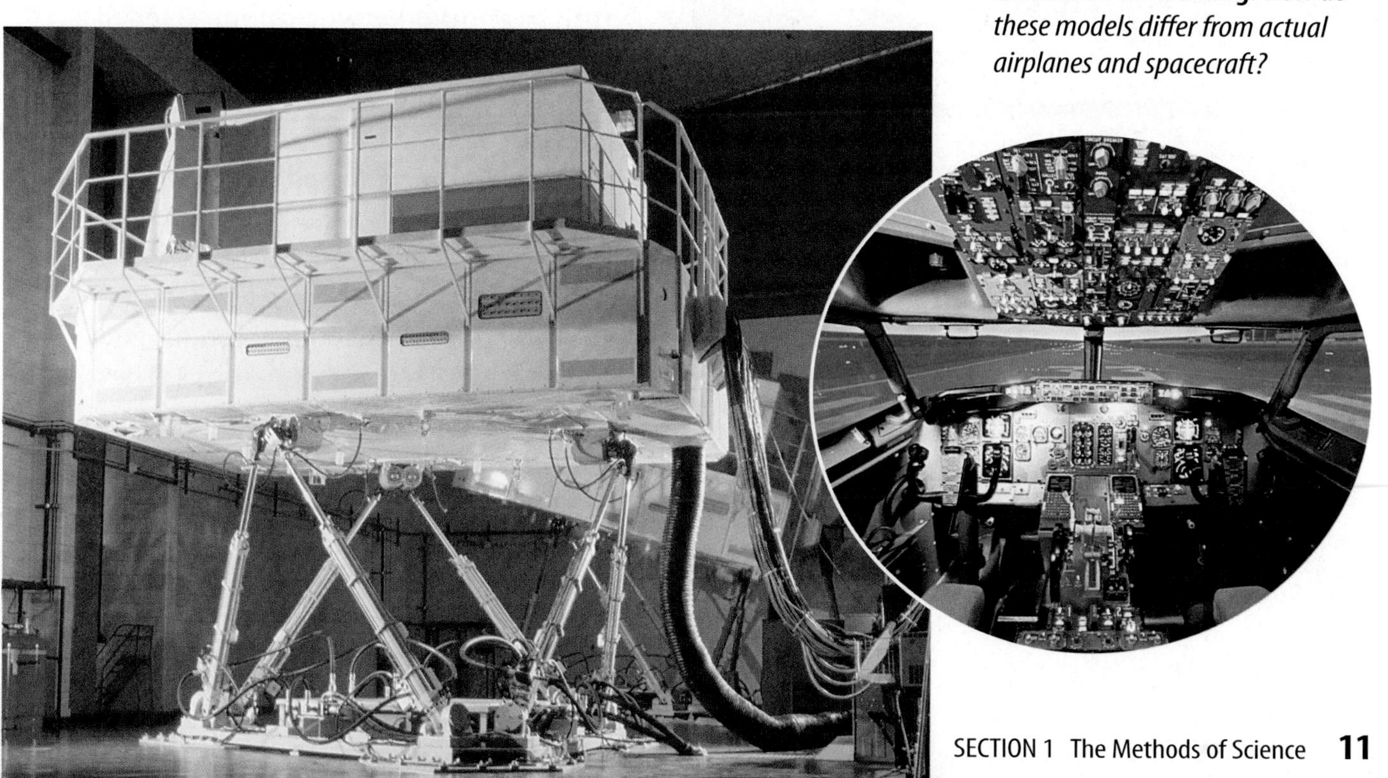

Figure 6
Pilots and astronauts use flight simulators for training. *How do these models differ from actual airplanes and spacecraft?*

SCIENCE *Online*

Research Visit the Glencoe Science Web site at **science.glencoe.com** to find out about Archimedes' principle. Would you classify this principle as a scientific theory or scientific law? Communicate to your class what you learn.

Scientific Theories and Laws

A scientific **theory** is an explanation of things or events based on knowledge gained from many observations and investigations. It is not a guess. If scientists repeat an investigation and the results always support the hypothesis, the hypothesis can be called a theory. Just because a scientific theory has data supporting it does not mean it will never change. Recall that the theory about heat being a fluid was discarded after further experiments. A theory accepted today might at some time in the future also be discarded.

A **scientific law** is a statement about what happens in nature and that seems to be true all the time. Laws tell you what will happen under certain conditions, but they don't explain why or how something happens. Gravity is an example of a scientific law. The law of gravity says that any one mass will attract another mass. To date, no experiments have been performed that disprove the law of gravity.

A theory can be used to explain a law. For example, many theories have been proposed to explain how the law of gravity works. Even so, there are few theories in science and even fewer laws.

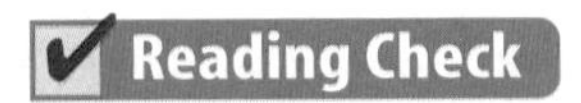

What is the difference between a scientific theory and a scientific law?

Figure 7
Science can't answer all questions. *Can anyone prove that you like this artwork? Explain.*

The Limitations of Science

Science can help you explain many things about the world, but science cannot explain or solve everything. Although it's the scientist's job to make guesses, the scientist also has to make sure his or her guesses can be tested and verified. But how do you prove that people will like a play or a piece of music? You cannot and science cannot.

Most questions about emotions and values are not scientific questions. They cannot be tested. You might take a survey to get people's opinions about such questions, but that would not prove that the opinions are true for everyone. A survey might predict that you will like the art in **Figure 7,** but science cannot prove that you or others will.

Using Science—Technology

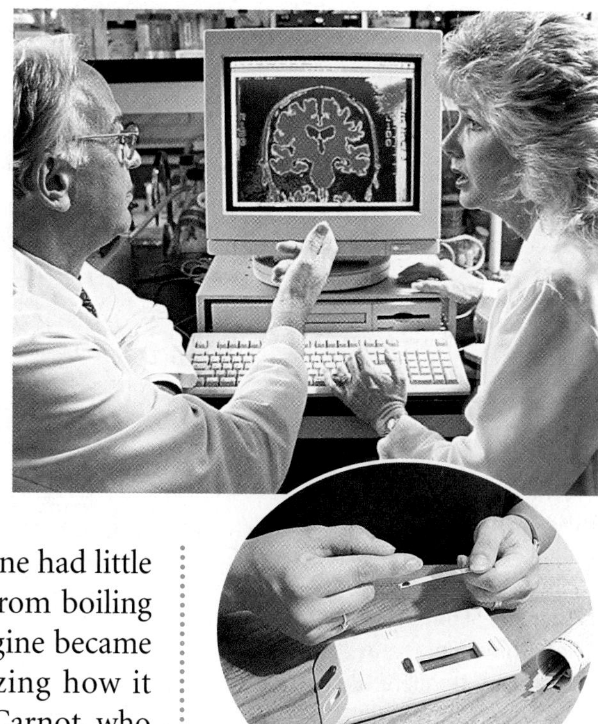

Many people use the terms *science* and *technology* interchangeably, but they are not the same. **Technology** is the application of science to help people. For example, when a chemist develops a new, lightweight material that can withstand great amounts of heat, this is science. When that material is used on the space shuttle, it is technology. **Figure 8** shows other examples of technology.

Technology doesn't always follow science, however. Sometimes the process of discovery can be reversed. One important historic example of science following technology is the development of the steam engine. The inventors of the steam engine had little idea of how it worked. They just knew that steam from boiling water could move the engine. Because the steam engine became so important to industry, scientists began analyzing how it worked. Lord Kelvin, James Prescott Joule and Sadi Carnot, who lived in the 1800s, learned so much from the steam engine that they developed revolutionary ideas about the nature of heat.

Do science and technology always produce positive results? Some people don't think so. The benefits of some technological advances, such as nuclear technology and genetic engineering, are subjects of debate. Being more knowledgeable about science can help society address these issues as they arise.

Figure 8
Technology is the application of science. *What type of science (life, Earth, or physical) is applied in these examples of technology?*

Section  Assessment

1. What is the first step a scientist usually takes to solve a problem?
2. What is the dependent variable in an experiment that shows how the volume of gas changes with changes in temperature?
3. Explain why a control is needed in a valid experiment.
4. How is science different from technology?
5. **Think Critically** You water your houseplant every Saturday. On Wednesday you notice its leaves are drooping. You give it some water, and the leaves perk up. You conclude you need to water twice a week. Was this a valid experiment? Explain.

Skill Builder Activities

6. **Forming Hypotheses** You don't have enough money to buy the music CD you want. Form a hypothesis about what you could do to solve the problem. How could you test your hypothesis before putting your plan in action? **For more help, refer to the Science Skill Handbook.**
7. **Communicating** You need to design a container to hold a new irregularly shaped device. Write the steps of the method you plan to use to help your team find the solution. **For more help, refer to the Science Skill Handbook.**

SECTION 2 Standards of Measurement

As You Read

What You'll Learn

- **Name** the prefixes used in SI and indicate what multiple of ten each one represents.
- **Identify** SI units and symbols for length, volume, mass, density, time, and temperature.
- **Convert** related SI units.

Vocabulary

standard, SI, volume, density, mass

Why It's Important

By using uniform standards, nations can exchange goods and compare information easily.

Units and Standards

Accurate measurement is needed in a valid experiment. Accuracy depends upon standards. A **standard** is an exact quantity that people agree to use for comparison. Measurements made using the same standard can be compared to each other.

Look at **Figure 9.** Suppose you and a friend want to make some measurements to find out whether a desk will fit through a doorway. You have no ruler, so you decide to use your hands as measuring tools. Using the width of his hands, your friend measures the doorway and says it is 8 hands wide. Using the width of your hands, you measure the desk and find it is 7¾ hands wide. Will the desk fit through the doorway? You can't be sure. What if your hands are wider than your friend's hands? The distance equal to 7¾ of your hands might be greater than the distance equal to 8 of your friend's hands.

What went wrong? Even though you both used hands to measure, you didn't check to see whether your hands are the same width as your friend's. In other words, you didn't use a measurement standard, so you can't compare the measurements.

Measurement Systems

Suppose the label on a ball of string indicates that the length of the string is 150. What is the length of the string? It could be 150 feet, 150 m, or 150 cm. In order for a measurement to make sense, it must include a number and a unit.

Your family might buy lumber by the foot, milk by the gallon, and potatoes by the pound. These measurement units are part of the English system of measurement, which is most commonly used in the United States. Most other nations use the metric system, which is a system of measurement based on multiples of ten. The metric system was devised by a group of scientists in the late 1700s.

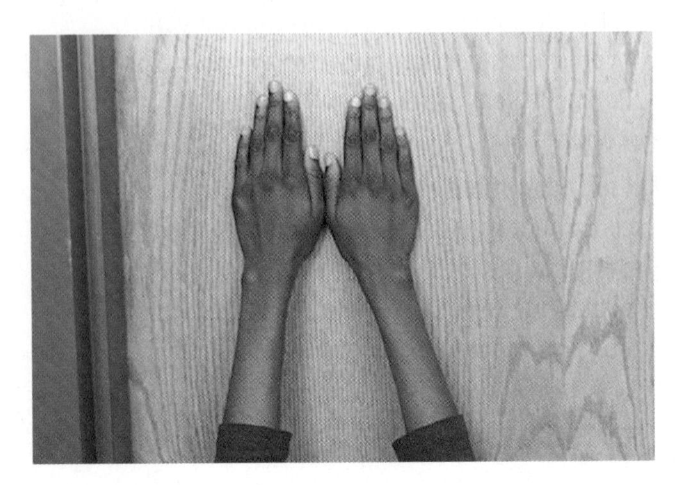

Figure 9
Hands are a convenient measuring tool, but using them can lead to misunderstanding.

International System of Units In 1960, an improved version of the metric system was devised. Known as the International System of Units, this system is often abbreviated SI, from the French *Le Systeme Internationale d'Unites.* All **SI** standards are universally accepted and understood by scientists throughout the world. The standard kilogram, which is kept in Sèvres, France, is shown in **Figure 10.** All kilograms used throughout the world must be exactly the same as the kilogram kept in France.

Each type of SI measurement has a base unit. The meter is the base unit of length. Every type of quantity measured in SI has a symbol for that unit. These names and symbols for the seven base units are shown in **Table 1.** All other SI units are obtained from these seven units.

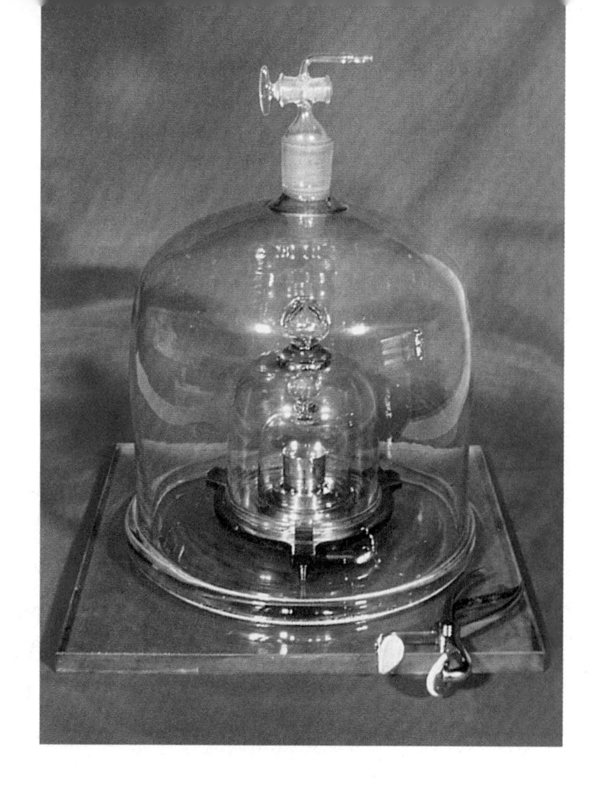

Figure 10
The standard for mass, the kilogram, and other standards are kept at the International Bureau of Weights and Measures in Sèvres, France. *What is the purpose of a standard?*

SI Prefixes The SI system is easy to use because it is based on multiples of ten. Prefixes are used with the names of the units to indicate what multiple of ten should be used with the units. For example, the prefix *kilo-* means "1,000." That means that one kilometer equals 1,000 meters. Likewise, one kilogram equals 1,000 grams. Because *deci-* means "one-tenth," one decimeter equals one tenth of a meter. A decigram equals one tenth of a gram. The most frequently used prefixes are shown in **Table 2.**

✔ Reading Check *How many meters is 1 km? How many grams is 1 dg?*

Table 1 SI Base Units

Quantity Measured	Unit	Symbol
Length	meter	m
Mass	kilogram	kg
Time	second	s
Electric Current	ampere	A
Temperature	kelvin	K
Amount of Substance	mole	mol
Intensity of Light	candela	cd

Table 2 Common SI Prefixes

Prefix	Symbol	Multiplying Factor
Kilo-	k	1,000
Deci-	d	0.1
Centi-	c	0.01
Milli-	m	0.001
Micro-	μ	0.000 001
Nano-	n	0.000 000 001

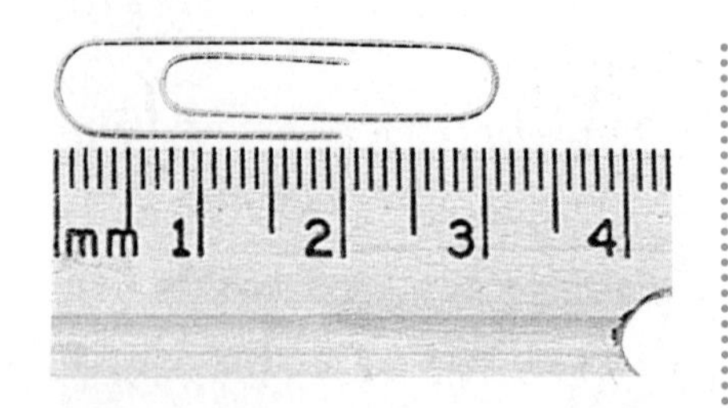

Figure 11
One centimeter contains 10 mm. *How many millimeters long is the paper clip?*

Converting Between SI Units Sometimes quantities are measured using different units. Conversion factors are used to change one unit to another. A conversion factor is a ratio that is equal to one. For example, there are 1,000 mL in 1 L, so 1,000 mL = 1 L. If both sides in this equation are divided by 1 L, the equation becomes:

$$\frac{1{,}000 \text{ mL}}{1 \text{ L}} = 1$$

The left side of this equation is a ratio equal to one and, therefore, is a conversion factor. You can make another conversion factor by placing 1L in the numerator and 1,000 mL in the denominator. The ratio still is equal to one.

To convert units, you multiply by the appropriate conversion factor. For example, to convert 1.255 L to mL, multiply 1.255 L by a conversion factor. Use the conversion factor with new units (mL) in the numerator and the old units (L) in the denominator.

$$1.255 \text{ L} \times \frac{1{,}000 \text{ mL}}{1 \text{ L}} = 1.255 \text{ mL}$$

The unit L cancels in this equation, just as if it were a number.

Math Skills Activity

Converting Units of Measure

Example Problem

You have a length of rope that measures 3,075 mm. How long is it in centimeters?

1. *This is what you know:* 1 m = 10 dm = 100 cm = 1,000 mm
2. *This is what you need to know:* 3,075 mm = ? cm
3. *This is the equation you need to use:* $? \text{ cm} = 3{,}075 \text{ mm} \times \frac{100 \text{ cm}}{1{,}000 \text{ mm}}$
4. *Cancel units and multiply:* $3{,}075 \cancel{\text{mm}} \times \frac{1\cancel{00} \text{ cm}}{10\cancel{00} \cancel{\text{mm}}} = 307.5 \text{ cm}$

Check your answer by multiplying your answer by $\frac{1{,}000 \text{ mm}}{100 \text{ cm}}$. *Do you calculate the original length in millimeters?*

Practice Problem

Your pencil is 11 cm long. How long is it in millimeters?

For more help, refer to the Math Skill Handbook.

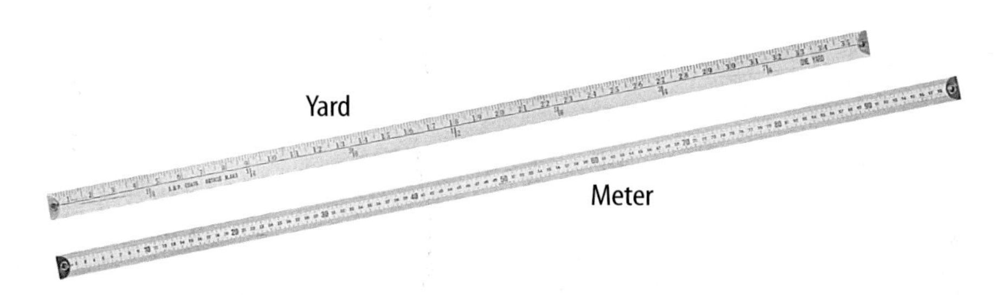

Figure 12
One meter is slightly longer than 1 yard and 100 m is slightly longer than a football field. *Would you expect your time for a 100-m dash to be slightly more or less than your time for a 100-yard dash?*

Measuring Distance

The word *length* is used in many different ways. For example, the length of a novel is the number of pages or words it contains. In scientific measurement, however, length is the distance between two points. That distance might be the diameter of a hair or the distance from Earth to the Moon. The SI base unit of length is the meter, m. A baseball bat is about 1 m long. Metric rulers and metersticks are used to measure length. **Figure 12** compares a meter and a yard.

Choosing a Unit of Length As shown in **Figure 13,** the size of the unit you measure with will depend on the size of the object being measured. For example, the diameter of a shirt button is about 1 cm. You probably also would use the centimeter to measure the length of your pencil and the meter to measure the length of your classroom. What unit would you use to measure the distance from your home to school? You probably would want to use a unit larger than a meter. The kilometer, km, which is 1,000 m, is used to measure these kinds of distances.

By choosing an appropriate unit, you avoid large-digit numbers and numbers with many decimal places. Twenty-one kilometers is easier to deal with than 21,000 m. And 13 mm is easier to use than 0.013 m.

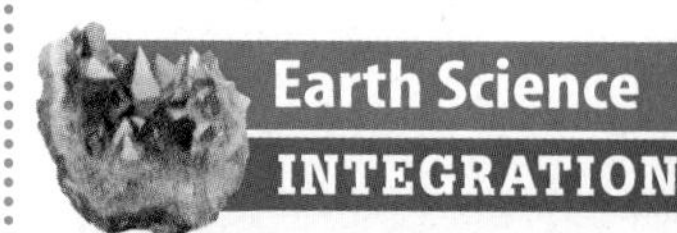

The standard measurement for the distance from Earth to the Sun is called the astronomical unit, AU. The distance is about 150 billion (1.5×10^{11}) m. In your Science Journal, calculate what 1 AU would equal in km.

Figure 13
The size of the object being measured determines which unit you will measure in. A tape measure measures in meters. The micrometer, shown at the left, measures in small lengths. *What unit do you think it measures in?*

Research Visit the Glencoe Science Web site at **science.glencoe.com** for more information on finding the volume of irregular objects using water displacement. Communicate to your class what you learn.

Measuring Volume

The amount of space occupied by an object is called its **volume.** If you want to know the volume of a solid rectangle, such as a brick, you measure its length, width, and height and multiply the three numbers and their units together ($V = l \times w \times h$). For a brick, your measurements probably would be in centimeters. The volume would then be expressed in cubic centimeters, cm^3. To find out how much a moving van can carry, your measurements probably would be in meters, and the volume would be expressed in cubic meters, m^3, because when you multiply you add exponents.

Measuring Liquid Volume How do you measure the volume of a liquid? A liquid has no sides to measure. In measuring a liquid's volume, you are indicating the capacity of the container that holds that amount of liquid. The most common units for expressing liquid volumes are liters and milliliters. These are measurements used in canned and bottled foods. A liter occupies the same volume as a cubic decimeter, dm^3. A cubic decimeter is a cube that is 1 dm, or 10 cm, on each side, as in **Figure 14.**

Look at **Figure 14.** One liter is equal to 1,000 mL. A cubic decimeter, dm^3, is equal to 1,000 cm^3. Because $1 \text{ L} = 1 \text{ dm}^3$, it follows that:

$$1 \text{ mL} = 1 \text{ cm}^3$$

Sometimes, liquid volumes such as doses of medicine are expressed in cubic centimeters.

Suppose you wanted to convert a measurement in liters to cubic centimeters. You use conversion factors to convert L to mL and then mL to cm^3.

$$1.5 \cancel{\text{L}} \times \frac{1{,}000 \cancel{\text{mL}}}{1 \cancel{\text{L}}} \times \frac{1 \text{ cm}^3}{1 \cancel{\text{mL}}} = 1{,}500 \text{ cm}^3$$

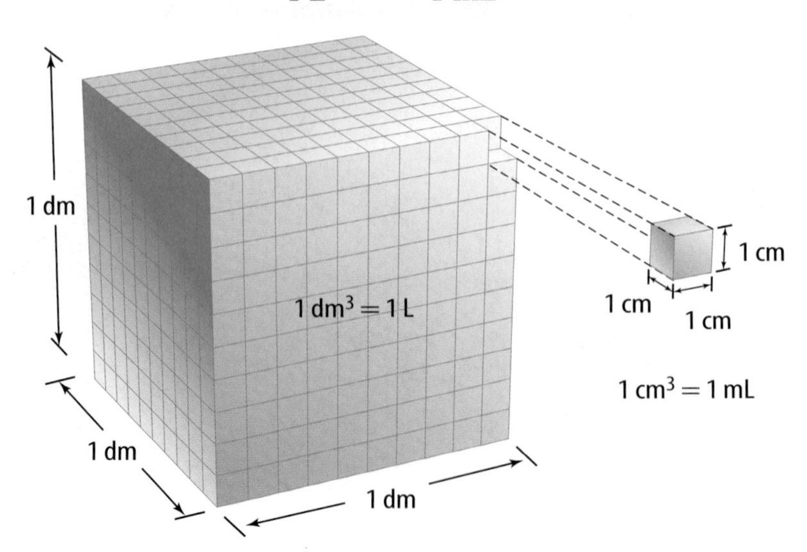

Figure 14
The large cube has a volume of 1 dm^3, which is equivalent to 1 L. *How many cubic centimeters (cm^3) are in the large cube?*

Table 3 Densities of Some Materials at 20°C

Material	Density (g/cm^3)	Material	Density (g/cm^3)
hydrogen	0.000 09	aluminum	2.7
oxygen	0.001 4	iron	7.9
water	1.0	gold	19.3

Measuring Matter

A table-tennis ball and a golf ball have about the same volume. But if you pick them up, you notice a difference. The golf ball has more mass. **Mass** is a measurement of the quantity of matter in an object. The mass of the golf ball, which is about 45 g, is almost 18 times the mass of the table-tennis ball, which is about 2.5 g. A bowling ball has a mass of about 5,000 g. This makes its mass roughly 100 times greater than the mass of the golf ball and 2,000 times greater than the table-tennis ball's mass. To visualize SI units, see **Figure 15** on the following page.

Density A cube of polished aluminum and a cube of silver that are the same size not only look similar but also have the same volume. The mass and volume of an object can be used to find the density of the material the object is made of. **Density** is the mass per unit volume of a material. You find density by dividing an object's mass by the object's volume. For example, the density of an object having a mass of 10 g and a volume of 2 cm^3 is 5 g/cm^3. **Table 3** lists the densities of some familiar materials.

Derived Units The measurement unit for density, g/cm^3, is a combination of SI units. A unit obtained by combining different SI units is called a derived unit. An SI unit multiplied by itself also is a derived unit. Thus the liter, which is based on the cubic decimeter, is a derived unit. A meter cubed, expressed with an exponent—m^3—is a derived unit.

Measuring Time and Temperature

It is often necessary to keep track of how long it takes for something to happen, or whether something heats up or cools down. These measurements involve time and temperature.

Time is the interval between two events. The SI unit for time is the second. In the laboratory, you will use a stopwatch or a clock with a second hand to measure time.

Mini LAB

Determining the Density of a Pencil

Procedure

1. Measure the mass of a **pencil** (unsharpened) in grams.
2. Put 90 mL of **water** into a 100-mL **graduated cylinder.** Lower the pencil, eraser end down, into the cylinder. Push the pencil until it is just submerged. This is known as water displacement. Hold it there and read the new volume to the nearest tenth of a milliliter.

Analysis

1. Calculate the pencil's density by dividing its mass by the change in volume of the water level.
2. Is the density of the pencil greater than or less than the density of water? How do you know?

NATIONAL GEOGRAPHIC VISUALIZING SI DIMENSIONS

Figure 15

The characteristics of most of these everyday objects are measured using an international system known as SI dimensions. These dimensions measure length, volume, mass, density, and time. Celsius is not an SI unit but is widely used in scientific work.

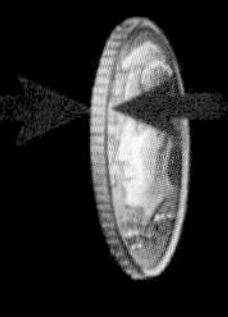

MILLIMETERS A dime is about 1 mm thick.

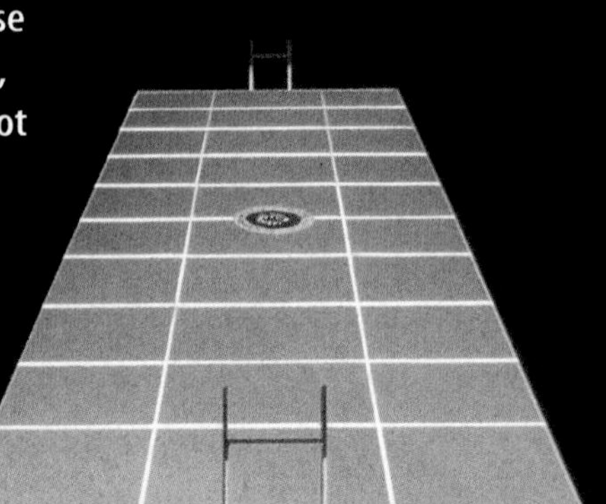

METERS A football field is about 91 m long.

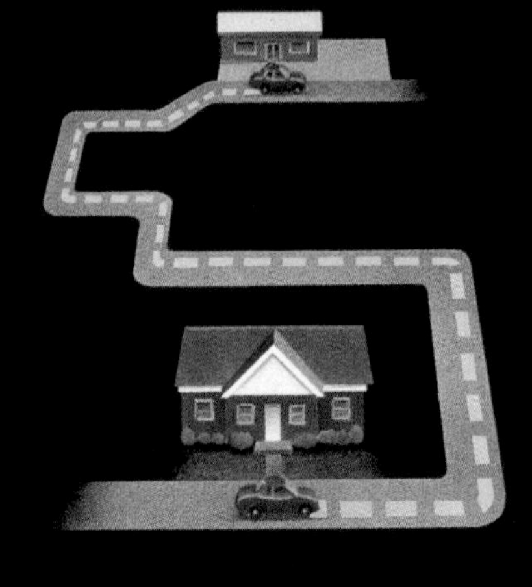

KILOMETERS The distance from your house to a store can be measured in kilometers.

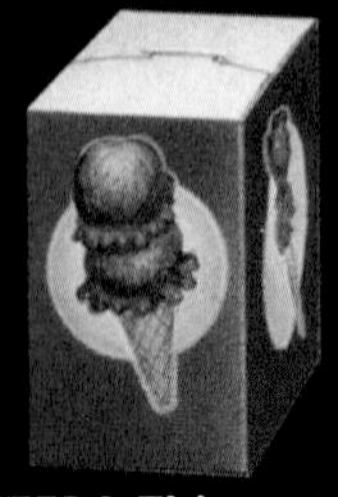

LITERS This carton holds 1.98 L of frozen yogurt.

MILLILITERS A teaspoonful of medicine is about 5 mL.

GRAMS/METER This stone sinks because it is denser—has more grams per cubic meter—than water.

GRAMS The mass of a thumbtack and the mass of a textbook can be expressed in grams.

METERS/SECOND The speed of a roller-coaster car can be measured in meters per second.

CELSIUS Water boils at 100°C and freezes at 0°C.

What's Hot and What's Not You will learn the scientific meaning of the word *temperature* in a later chapter. For now, think of temperature as a measure of how hot or how cold something is.

Look at **Figure 16.** For most scientific work, temperature is measured on the Celsius (C) scale. On this scale, the freezing point of water is 0°C, and the boiling point of water is 100°C. Between these points, the scale is divided into 100 equal divisions. Each one represents 1°C. On the Celsius scale, average human body temperature is 37°C, and a typical room temperature is between 20°C and 25°C.

Figure 16
These three thermometers illustrate the scales of temperature between the freezing and boiling points of water. *How do the scales compare at the boiling point?*

Kelvin and Fahrenheit The SI unit of temperature is the kelvin (K). Zero on the Kelvin scale (0 K) is the coldest possible temperature, also known as absolute zero. That is equal to −273°C, which is 273° below the freezing point of water.

Most laboratory thermometers are marked only with the Celsius scale. Because the divisions on the two scales are the same size, the Kelvin temperature can be found by adding 273 to the Celsius reading. So, on the Kelvin scale, water freezes at 273 K and boils at 373 K. Notice that degree symbols are not used with the Kelvin scale.

The temperature measurement you are probably most familiar with is the Fahrenheit scale, which was based roughly on the temperature of the human body.

Reading Check ***What is the relationship between the Celsius scale and the Kelvin scale?***

Section  Assessment

1. Why is it important to have standards of measurement that are exact?
2. What are the SI prefixes for 0.001, 1,000, 0.1, and 0.01?
3. Make the following conversions—100 cm to meters, 27°C to Kelvin, 20 dg to milligrams, and 3 m to decimeters.
4. Explain why density is a derived unit.
5. **Think Critically** What is the density of an unknown metal that has a mass of 158 g and a volume of 20 mL? Use **Table 3** to identify this metal.

Skill Builder Activities

6. **Concept Mapping** Make a network-tree concept map displaying the SI base units used to measure quantities of length, mass, time, and temperature. **For more help, refer to the Science Skill Handbook.**
7. **Solving One-Step Equations** Use a metric ruler to measure a shoe box and a pad of paper. Find the volume of each ($V = w \times l \times h$) in cubic centimeters. Then convert the units to mL. **For more help, refer to the Math Skill Handbook.**

Communicating with Graphs

As You Read

What You'll Learn

- **Identify** three types of graphs and explain the ways they are used.
- **Distinguish** between dependent and independent variables.
- **Analyze** data using the various types of graphs.

Vocabulary
graph

Why It's Important
Graphs are a quick way to communicate a lot of information in a small amount of space.

A Visual Display

Scientists often graph the results of their experiments because they can detect patterns in the data easier in a graph than in a table. A **graph** is a visual display of information or data. **Figure 17** is a graph that shows a girl walking her dog. The horizontal axis, or the *x*-axis, measures time. Time is the independent variable because as it changes, it affects the measure of another variable. The distance from home that the girl and the dog walk is the other variable. It is the dependent variable and is measured on the vertical axis, or *y*-axis.

Graphs are useful for displaying numerical information in business, science, sports, advertising, and many everyday situations. Different kinds of graphs—line, bar, and circle—are appropriate for displaying different types of information.

Reading Check *What are three common types of graphs?*

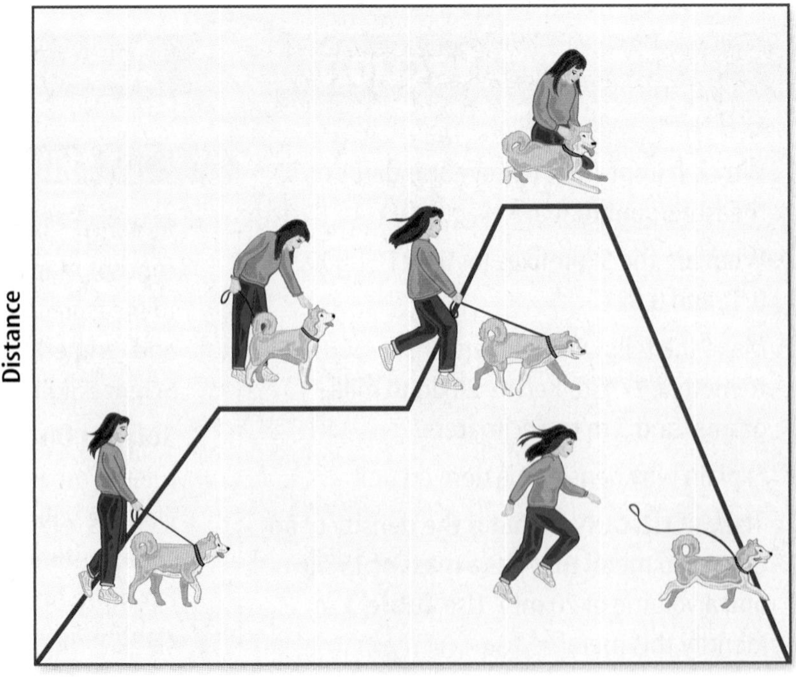

Figure 17
This graph tells the story of the motion that takes place when a girl takes her dog for an 8-min walk.

Line Graphs

A line graph can show any relationship where the dependent variable changes due to a change in the independent variable. Line graphs often show how a relationship between variables changes over time. You can use a line graph to track many things, such as how certain stocks perform or how the population changes over any period of time—a month, a week, or a year.

You can show more than one event on the same graph as long as the relationship between the variables is identical. Suppose a builder had three choices of thermostats for a new school. He wanted to test them to know which was the best brand to install throughout the building. He installed a different thermostat in classrooms A, B, and C. He set each thermostat at 20°C. He turned the furnace on and checked the temperatures in the three rooms every 5 min for 25 min. He recorded his data in **Table 4.**

Table 4 Room Temperature

Time*	Classroom Temperature (°C) A	B	C
0	16	16	16
5	17	17	16.5
10	19	19	17
15	20	21	17.5
20	20	23	18
25	20	25	18.5

*minutes after turning on heat

The builder then plotted the data on a graph. He could see from the table that the data did not vary much for the three classrooms. So he chose small intervals for the y-axis and left part of the scale out (the part between 0° and 15°). See **Figure 18.** This allowed him to spread out the area on the graph where the data points lie. You can see easily the contrast in the colors of the three lines and their relationship to the black horizontal line. The black line represents the thermostat setting and is the control. The control is what the resulting room temperature of the classrooms should be if the thermostats are working efficiently.

Figure 18
The room temperatures of classrooms A, B, and C are shown in contrast to the thermostat setting of 20°C. *Which classroom had the most efficient thermostat?*

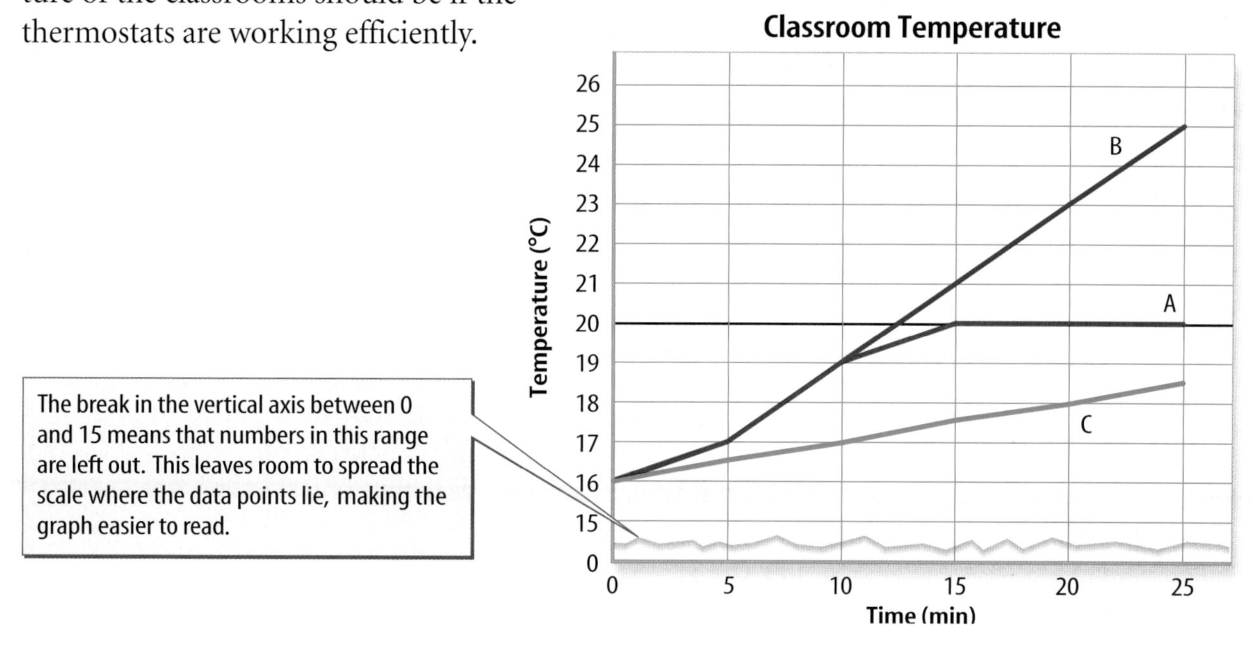

Figure 19
Graphing calculators are valuable tools for making graphs.

Building Line Graphs Besides choosing a scale that makes a graph readable, as illustrated in **Figure 18,** other factors are involved in building useful graphs. The most important factor in making a line graph is always using the *x*-axis for the independent variable. The *y*-axis always is used for the dependent variable. Because the points in a line graph are related, you connect the points.

Another factor in building a graph involves units of measurement. For example, you might use a Celsius thermometer for one part of your experiment and a Fahrenheit thermometer for another. But you must first convert your temperature readings to the same unit of measurement before you make your graph.

In the past, graphs had to be made by hand, with each point plotted individually. Today, scientists use a variety of tools, such as computers and graphing calculators like the one shown in **Figure 19,** to help them draw graphs.

Math Skills Activity

Line Graphing

Example Problem

In an experiment, you check the air temperature at certain hours of the day. At 8 A.M., the temperature is 27°C; at noon, the temperature is 32°C; and at 4 P.M., the temperature is 30°C. Graph the results of your experiment.

Solution

1. *This is what you know:* independent variable = time
   dependent variable = temperature
2. *Set up your graph:* Graph time on the *x*-axis and temperature on the *y*-axis. Mark the increments on the graph to include all measurements.
3. *Graph:* Plot each point on the graph by finding the time on the *x*-axis and moving up until you find the recorded temperature on the *y*-axis. Place a point there. Then connect the points from left to right.

Practice Problem

As you train for a marathon, you compare your previous times. In year one, you ran it in 5.2 h; in year two, you ran it in 5 h; in year three, you ran it in 4.8 h; in year four, you ran it in 4.3 h; and in year five, you ran it in 4 h. Graph the results of your marathon races.

For more help, refer to the Math Skill Handbook.

Bar Graphs

A bar graph is useful for comparing information collected by counting. For example, suppose you counted the number of students in every classroom in your school on a particular day and organized your data as in **Table 5.** You could show these data in a bar graph like the one shown in **Figure 20.** Uses for bar graphs include comparisons of oil, or crop productions, costs, or as data in promotional materials. Each bar represents a quantity counted at a particular time, which should be stated on the graph. As on a line graph, the independent variable is shown on the x-axis and the dependent variable is plotted on the y-axis.

Recall that you might need to place a break in the scale of the graph to better illustrate your results. For example, if your data were 1,002, 1,010, 1,030, and 1,040 and the intervals on the scale were every 100 units, you might not be able to see the difference from one bar to another. If you had a break in the scale and started your data range at 1,000 with intervals of ten units, you could make your comparison more accurately.

Reading Check *Describe possible data where using a bar graph would be better than using a line graph.*

TRY AT HOME Mini LAB

Observing Change Through Graphing

Procedure

1. Place a **thermometer** in a **plastic foam cup** of hot, but not boiling, **water.**
2. Measure and record the temperature every 30 s for 5 min.
3. Repeat the experiment with freshly heated water. This time, cover the cup with a **plastic lid.**

Analysis

1. Make a line graph of the changing temperature from step 2, showing time on the *x*-axis and temperature on the *y*-axis. Then plot the changing temperature from step 3 on the same graph.
2. Use the graph to describe the cooling process in each of the trials.

Table 5 Classroom Size

Number of Students	Number of Classrooms
20	1
21	3
22	3
23	2
24	3
25	5
26	5
27	3

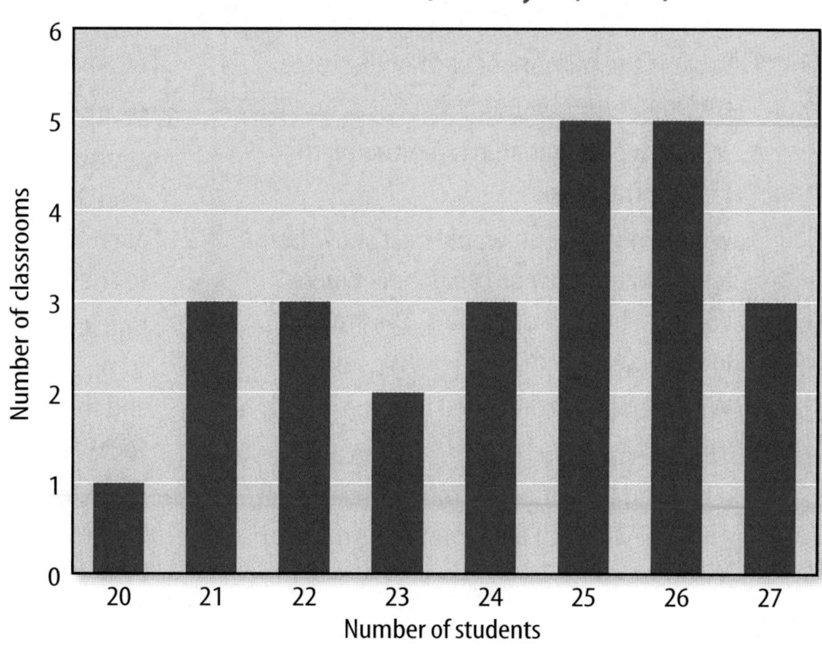

Figure 20
The height of each bar corresponds to the number of classrooms having a particular number of students.

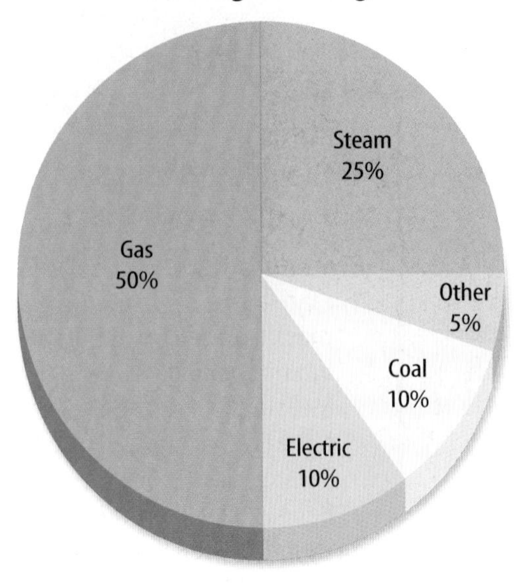

Figure 21
A circle graph shows the different parts of a whole quantity.

Circle Graphs

A circle graph, or pie graph, is used to show how some fixed quantity is broken down into parts. The circular pie represents the total. The slices represent the parts and usually are represented as percentages of the total.

Figure 21 illustrates how a circle graph could be used to show the percentage of buildings in a neighborhood using each of a variety of heating fuels. You easily can see that more buildings use gas heat than any other kind of system. What else does the graph tell you?

To create a circle graph, you start with the total of what you are analyzing. **Figure 21** starts with 72 buildings in the neighborhood. For each type of heating fuel, you divide the number of buildings using each type of fuel by the total (72). You then multiply that fraction (percent) by 360° to determine the angle that the fraction makes in the circle. Eighteen buildings use steam. Therefore, $18 \div 72 \times 360° = 90°$ on the circle graph. You then would measure 90° on the circle with your protractor to show 25 percent.

When you use graphs, think carefully about the conclusions you can draw from them. You want to make sure your conclusions are based on accurate information and that you use scales that help make your graph easy to read.

Section  Assessment

1. What is the purpose of each of the three common types of graphs?
2. Which type of variable is plotted on the *x*-axis? The *y*-axis?
3. What kind of graph would best show the results of a survey of 144 people where 75 ride a bus, 45 drive cars, 15 carpool, and 9 walk to work?
4. Why are points connected in a line graph?
5. **Think Critically** Describe one way that a bar graph is different from a circle graph and one way that a bar graph is similar to a circle graph.

Skill Builder Activities

6. **Making and Using Graphs** Find a graph in a newspaper or magazine. Identify the kind of graph you found and write an explanation of what the graph shows. **For more help, refer to the Science Skill Handbook.**
7. **Using an Electronic Spreadsheet** Some computer programs make creating data tables and making graphs an easier task. Use a spreadsheet and a graphing program to make a data table and a line graph of the data you collected in the Try at Home MiniLAB. **For more help, refer to the Technology Skill Handbook.**

Activity

Converting Kitchen Measurements

Look through a recipe book. Are any of the amounts of ingredients stated in SI units? Chances are, English measurements are used. How can you convert English measurements to SI measurements?

What You'll Investigate

How do kitchen measurements compare with SI measurements?

Safety Precautions

Materials

balance
100-mL graduated cylinder
measuring cup
measuring teaspoon
measuring tablespoon
corn meal
dried beans
dried rice
potato flakes
water
vinegar
salad oil

Goals

- **Determine** a relationship between two systems of measurements.
- **Calculate** the conversion factors for converting English units to SI units.

Procedure

1. Copy the data table into your Science Journal.
2. Use the appropriate English measuring cup or spoon to measure the amounts of each ingredient shown in the table.
3. Use a balance to measure the mass in grams of each dry ingredient. Use a graduated cylinder to measure the volume in milliliters of each liquid ingredient.
4. Record each SI equivalent in your data table.

English to SI Conversions

Ingredient	English Measure	SI Measure
Water	$^{1}/_{2}$ cup	
Corn Meal	2 cups	
Salad Oil	4 tablespoons	
Dried Rice	$^{1}/_{2}$ cup	
Potato Flakes	3 cups	
Vinegar	1 teaspoon	
Dried Beans	3 cups	

Conclude and Apply

1. **Calculate** the number of grams in one cup of each dry ingredient. Calculate the number of milliliters in one cup, one teaspoon, and one tablespoon of each liquid ingredient.
2. **Write** conversion factors that will convert each English unit to an SI unit for each ingredient.
3. **Calculate** how many milliliters you would measure if a recipe called for three tablespoons of salad oil.
4. **Compare and Contrast** your conversion factors for the dry ingredients and your conversion factors for the liquid ingredients.
5. **Explain** the benefits and problems of changing all recipes to SI units.

Write a recipe used in your home converting all the English units to SI units.

Activity Design Your Own Experiment

Setting High Standards for Measurement

To develop the International System of Units, people had to agree on set standards and basic definitions of scale. If you had to develop a new measurement system, people would have to agree with your new standards and definitions. In this activity, your team will use string to devise and test its own SI (String International) system for measuring length.

Recognize the Problem

What are the requirements for designing a new measurement system using string?

Form a Hypothesis

Based on your knowledge of measurement standards and systems, state a hypothesis about how exact units help to keep measuring consistent.

Possible Materials

string
scissors
marking pen
masking tape
miscellaneous objects for standards

Safety Precautions

Goals

- **Design** an experiment that involves devising and testing your own measurement system for length.
- **Measure** various objects with the string measurement system.

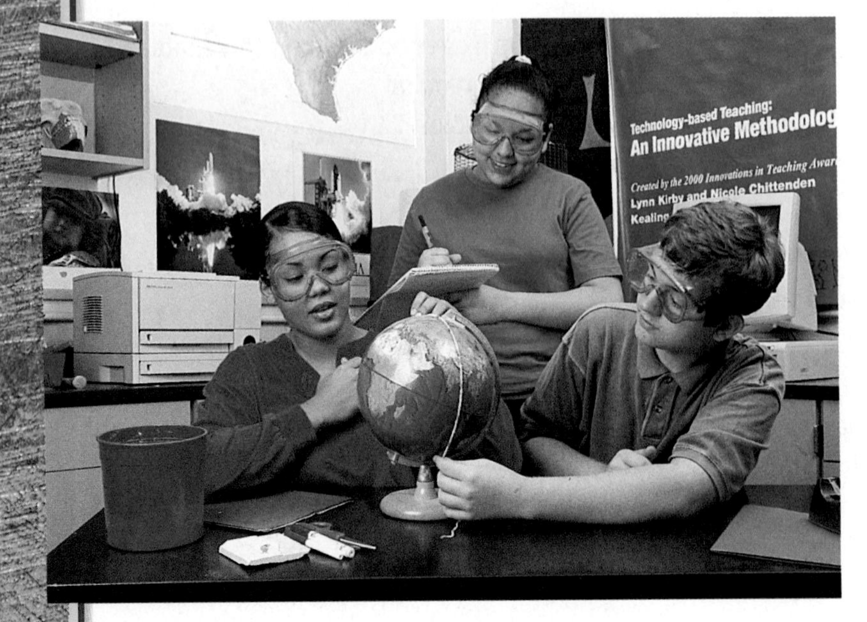

Test Your Hypothesis

Plan

1. As a group, agree upon and write out the hypothesis statement.
2. As a group, list the steps that you need to take to test your hypothesis. Be specific, describing exactly what you will do at each step.
3. Make a list of the materials that you will need.
4. **Design** a data table in your Science Journal so it is ready to use as your group collects data.
5. As you read over your plan, be sure you have chosen an object in your classroom to serve as a standard. It should be in the same size range as what you will measure.
6. Consider how you will mark scale divisions on your string. Plan to use different pieces of string to try different-sized scale divisions.
7. What is your new unit of measurement called? Come up with an abbreviation for your unit. Will you name the smaller scale divisions?
8. What objects will you measure with your new unit? Be sure to include objects longer and shorter than your string. Will you measure each object more than once to test consistency? Will you measure the same object as another group and compare your findings?

Do

1. Make sure your teacher approves your plan before you start.
2. Carry out the experiment as it has been planned.
3. **Record** observations that you make and complete the data table in your Science Journal.

Analyze Your Data

1. Which of your string scale systems will provide the most accurate measurement of small objects? Explain.
2. How did you record measurements that were between two whole numbers of your units?

Draw Conclusions

1. When sharing your results with other groups, why is it important for them to know what you used as a standard?
2. **Infer** how it is possible for different numbers to represent the same length of an object.

Compare your conclusions with other students' conclusions. **For more help, refer to the Science Skill Handbook.**

Thinking in Pictures: and other reports from my life with autism[1]

By Temple Grandin

Respond to the Reading

1. How do people with autism think differently than other people?
2. What did the author use to see from a cow's point of view?
3. What did the author use for models to design things when she was a child?

Temple Grandin is an animal scientist and writer who also happens to be autistic. People with autism are said to think in pictures. For instance, an autistic person might think of a "dog" by visualizing a specific dog that he or she has seen rather than the word "dog."

I think in pictures. Words are like a second language to me. I translate both spoken and written words into full-color movies, complete with sound, which run like a VCR tape in my head. When somebody speaks to me, his words are instantly translated into pictures. Language-based thinkers often find this phenomenon difficult to understand, but in my job as equipment designer for the livestock industry, visual thinking is a tremendous advantage.

... I credit my visualization abilities with helping me understand the animals I work with. Early in my career I used a camera to help give me the animals' perspective as they walked through a chute for their veterinary treatment. I would kneel down and take pictures through the chute from the cow's eye level. Using the photos, I was able to figure out which things scared the cattle.

Every design problem I've ever solved started with my ability to visualize and see the world in pictures. I started designing things as a child, when I was always experimenting with new kinds of kites and model airplanes.

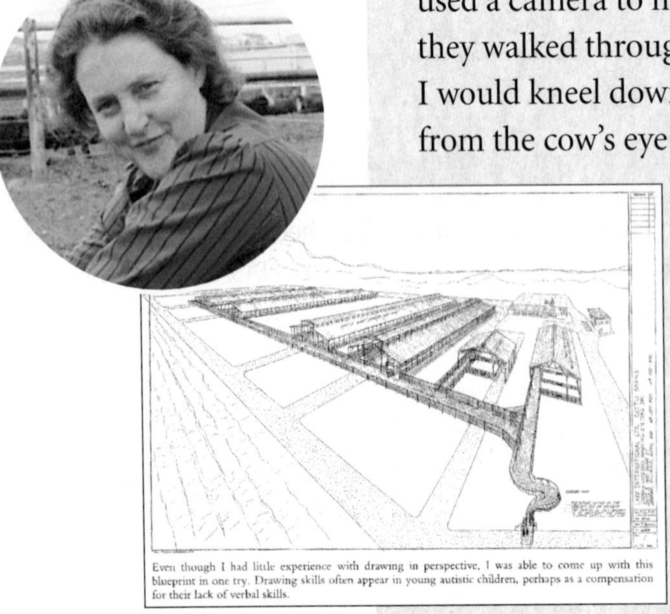

Even though I had little experience with drawing in perspective, I was able to come up with this blueprint in one try. Drawing skills often appear in young autistic children, perhaps as a compensation for their lack of verbal skills.

The author drew this blueprint of a cattle barn in one try.

[1] Autism is a complex developmental disability that usually appears during the first three years of life. Children and adults with autism typically have difficulties in communicating with others and relating to the outside world.

Understanding Literature

Identifying the Main Idea The most important idea expressed in a paragraph or essay is the main idea. The main idea in a reading might be clearly stated, but sometimes the main idea is implied. In other words, sometimes the reader has to summarize the contents of a reading in order to determine its main idea. What do you think is the main idea of the passage? Look closely at the contents of the first and third paragraphs to help you summarize and determine the main idea.

Science Connection In this chapter you learned that models are important tools for scientists. Models enable scientists to see things that are too big, too small, or take too much time to see completely. Scientists might build models of DNA, atoms, airplanes, or other equipment. Temple Grandin excels at building models because she is a visual thinker. Her visual thinking and ability to make models enables her to predict how things will work when they are put together.

Linking Science and Writing

Summarizing Research a magazine or newspaper article about the use of a scientific model. The model can be a blueprint design like that of Temple Grandin's. Write a paragraph that summarizes what you learned. Your summary should organize the information you learned by stating the main ideas and listing supporting details.

Career Connection

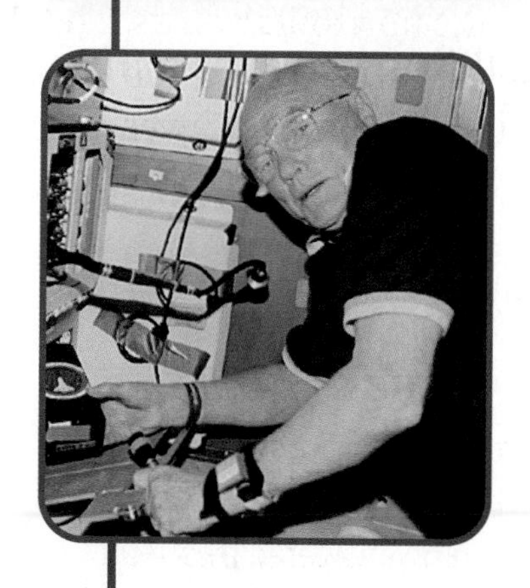

Astronaut

John Glenn, Jr. made history twice. In 1962 he became the first American to orbit Earth. In 1998, at age 77, he became the oldest person to fly in space. During Glenn's first mission, NASA learned how to place a craft into Earth's orbit and track its location. This mission also taught NASA the basics about how the human body reacts to weightlessness. On his second mission, Glenn studied the similarities between the effects of space flight and the effects of aging. After his retirement from the Marine Corps in 1965, Glenn went on to serve four terms as U.S. Senator from Ohio.

SCIENCE*Online* To learn more about a career as an astronaut, visit the Glencoe Science Web site at **science.glencoe.com**

Chapter 1 Study Guide

Reviewing Main Ideas

Section 1 The Methods of Science

1. Science is a way of learning about the natural world through investigation.
2. Scientific investigations can involve making observations, testing models, or conducting experiments. *What do you observe about hurricanes from the photo?*

3. Scientific experiments investigate the effect of one variable on another. All other variables are kept constant.
4. Scientific laws are repeated patterns in nature. Theories attempt to explain how and why these patterns develop.

Section 2 Standards of Measurement

1. A standard of measurement is an exact quantity that people agree to use as a basis of comparison.
2. When a standard of measurement is established, all measurements are compared to the same exact quantity—the standard. Therefore, all measurements can be compared with one another.
3. The most commonly used SI units include: length—meter, volume—liter, mass—kilogram, and time—second.
4. In SI, prefixes are used to make the base units larger or smaller by multiples of ten. *The Petronas Twin Towers in Malaysia is the world's tallest building, standing at 45,190 cm. Use a conversion factor to find out how many meters this is. How many kilometers?*

5. Any SI unit can be converted to any other related SI unit by multiplying by the appropriate conversion factor.

Section 3 Communicating With Graphs

1. Line graphs show continuous changes among related variables. Bar graphs are used to show data collected by counting. Circle graphs show how a fixed quantity can be broken into parts.
2. In a line graph, the independent variable is always plotted on the horizontal *x*-axis. The dependent variable is always plotted on the vertical *y*-axis.

After You Read

FOLDABLES Reading & Study Skills

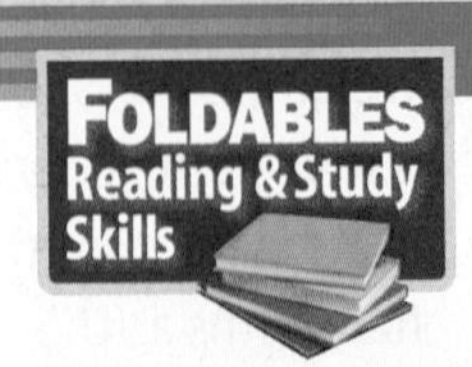

Using the information in this chapter, determine if the questions on your Question Study Fold can be answered by scientific processes. Then review your questions on the foldable and write the answers.

Chapter 1 Study Guide

Visualizing Main Ideas

Complete the following concept map on scientific methods.

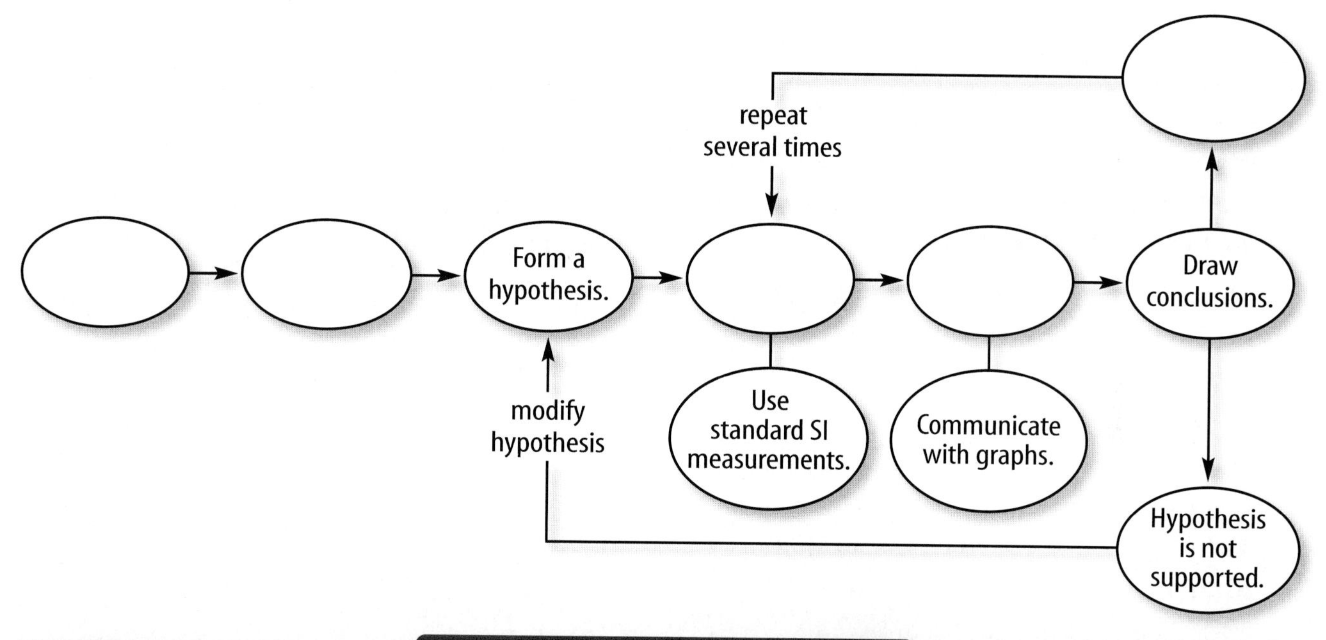

Vocabulary Review

Vocabulary Words

a. bias
b. constant
c. control
d. density
e. dependent variable
f. experiment
g. graph
h. hypothesis
i. independent variable
j. mass
k. model
l. scientific law
m. scientific method
n. SI
o. standard
p. technology
q. theory
r. variable
s. volume

Study Tip

If you're not sure of the relationship between terms in a question, try making a concept map of the terms and see how they fit together.

Using Vocabulary

Match each phrase with the correct term from the list of vocabulary words.

1. the modern version of the metric system
2. the amount of space occupied by an object
3. an agreed-upon quantity used for comparison
4. the amount of matter in an object
5. a variable that changes as another variable changes
6. a visual display of data
7. a test set up under controlled conditions
8. a variable that does NOT change as another variable changes
9. mass per unit volume
10. an educated guess using what you know and observe

Chapter 1 Assessment

Checking Concepts

Choose the word or phrase that best answers each question.

1. Which of the following questions CANNOT be answered by science?
A) How do birds fly?
B) Is this a good song?
C) What is an atom?
D) How does a clock work?

2. Which is an example of an SI unit?
A) foot
B) second
C) pound
D) gallon

3. Which system of measurement is used by scientists around the world?
A) SI
B) Standard system
C) English system
D) Kelvin system

4. Which of the following is SI based on?
A) inches
B) powers of five
C) English units
D) powers of ten

5. One one-thousandth is expressed by which prefix?
A) kilo-
B) nano-
C) centi-
D) milli-

6. What is the symbol for deciliter?
A) dL
B) dcL
C) dkL
D) Ld

7. What does the symbol μg stand for?
A) nanogram
B) kilogram
C) microgram
D) milligram

8. Which is the distance between two points?
A) volume
B) length
C) mass
D) density

9. Which of the following is NOT a derived unit?
A) dm^3
B) m
C) cm^3
D) g/ml

10. Which of the following is NOT equal to 1,000 mL?
A) 1 L
B) 100 cL
C) 1 dm^3
D) 1 cm^3

Thinking Critically

11. Make the following conversions.
a. 1,500 mL to L
b. 2 km to cm
c. 5.8 dg to mg
d. 22°C to K

12. Standards of measurement used during the Middle Ages often were based on such things as the length of the king's arm. What would you say to convince people to use a different system of standard units?

13. List the SI units of length you would use to express the following. Refer to **Table 2** in Section 2.
a. diameter of a hair
b. width of your classroom
c. width of a pencil lead
d. length of a sheet of paper

14. Suppose you set a glass of water in direct sunlight for 2 h and measure its temperature every 10 min. What type of graph would you use to display your data? What would the dependent variable be? The independent variable?

15. What are some advantages and disadvantages of adopting SI in the United States?

Developing Skills

16. Comparing and Contrasting Compare and contrast the ease with which conversions can be made among SI units versus conversions among units in the English system.

17. Forming Hypotheses A metal sphere is found to have a density of 5.2 g/cm^3 at 25°C and a density of 5.1 g/cm^3 at 50°C. Propose a hypothesis to explain this observation. How could you test your hypothesis?

18. Measuring in SI Not all objects have a volume that is measured easily. If you were to determine the mass, volume, and density of your textbook, a container of milk, and an air-filled balloon, how would you do it?

19. Interpreting Scientific Illustrations The illustrations show the items needed for an investigation. Which item is the independent variable? Which items are the constants? What might a dependent variable be?

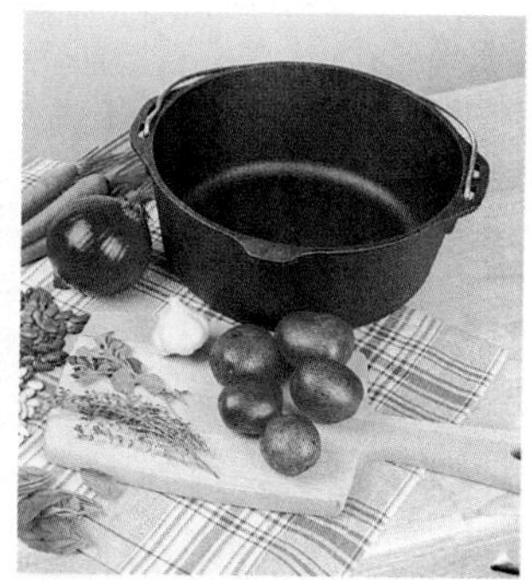

Performance Assessment

20. Making Observations and Inferences In the one-page activity, suppose you needed to measure five times the quantities. Compare the difficulty of multiplying the English measurements five times with multiplying the SI measurements five times.

Technology

Go to the Glencoe Science Web site at **science.glencoe.com** or use the **Glencoe Science CD-ROM** for additional chapter assessment.

Test Practice

A student made measurements of several items using units from the English system of measurement and the SI system of measurement to determine how units of the two systems were related. Using these data, the student determined unit conversion factors that allow one unit to be converted into another. These conversion factors and their related units are shown below.

Measurement Units

Measurement	English System Unit	SI Unit	Conversion Factor
Length	Foot	Meter	1 foot = 0.305 m
Mass	Slug	Kilogram	1 slug = 14.6 kg
Volume	Gallon	Liter	1 gallon = 3.78 L
Time	Second	Second	N/A

Study the table and answer the following questions.

1. According to this table, about how many liters are in a gallon?

A) ¼ **C)** 3/2
B) 3¾ **D)** ⅔

2. How many kilograms are equal to two slugs?

F) 42.8 **H)** 30.4
G) 11.6 **J)** 29.2

3. Which unit is the same in the English and SI systems of measurement?

A) meter **C)** gallon
B) slug **D)** second

CHAPTER 2

Motion and Speed

When you think of amusement parks, do you automatically think about roller coasters? Do you remember your last roller coaster ride? Do you recall the fast speeds, sharp turns, and plunging hills that cause your senses and balance to be in a state of total confusion. At this moment, it is doubtful that you are thinking about motion, the laws of gravity, or how to describe motion. In this chapter, you will learn about motion and speed—what they are and how to describe them.

What do you think?

Science Journal Look at the picture below with a classmate. Discuss what you think this is or what is happening. Here's a hint: *Without these on your car, it would slide on a dry road.* Write your answer or best guess in your Science Journal.

A cheetah can run at a speed of almost 120 km/h and is the fastest runner in the world. A horse can reach a speed of 64 km/h; an elephant's top speed is about 40 km/h, and the fastest snake slithers at a speed of about 3 km/h. The speed of an object is calculated by dividing the distance the object travels by the time it takes it to move that distance. How does the speed of a human compare to these animals?

Calculate your speed

1. Use a meterstick to mark off a 10-m distance.
2. Have your partner use a stopwatch to determine how fast you run 10 m.
3. Divide 10 m by your time in seconds to calculate your speed in m/s.
4. Multiply your answer by 3.6 to determine your speed in km/h.

Observe

Compare your speed with the maximum speed of a cheetah, horse, elephant, and snake. Could you win a race with any of them?

Before You Read

Making a Question Study Fold **Asking yourself questions helps you stay focused and better understand motion and speed when you are reading the chapter.**

1. Place a sheet of paper in front of you so the short side is at the top. Fold the paper in half from the left side to the right side.
2. Now fold the paper in half from top to bottom. Then fold it in half again top to bottom. Unfold the last two folds that you did.
3. Label the four sections *What motion?, How far?, How fast?,* and *In what direction?* as shown.
4. Through one thickness of paper, cut along each of the fold lines to form four tabs as shown.
5. Before you read the chapter, select a motion you can observe and write it on the front of the top tab. As you read the chapter, write answers to the other questions under the correct tab.

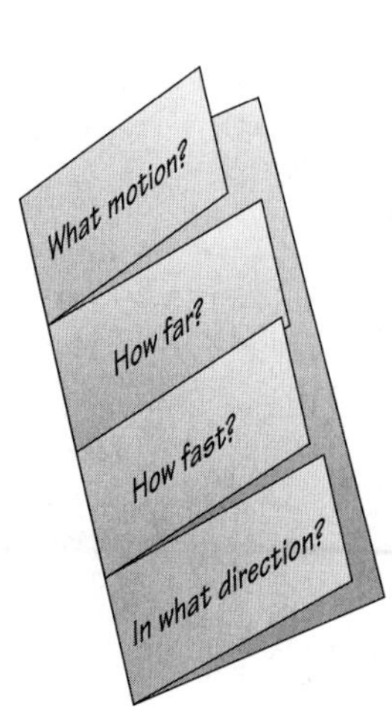

Describing Motion

As You Read

What You'll Learn

- **Distinguish** between distance and displacement.
- **Explain** the difference between speed and velocity.
- **Interpret** motion graphs.

Vocabulary

distance
displacement
speed
average speed
instantaneous speed
velocity

Why It's Important

Understanding the nature of motion and how to describe it helps you understand why motion occurs.

Motion

Are distance and time important in describing running events at the track and field meets in the Olympics? Would the winners of the 5-km race and the 10-km race complete the run in the same length of time?

Distance and time are important. In order to win a race, you must cover the distance in the shortest amount of time. The time required to run the 10-km race should be longer than the time needed to complete the 5-km race because the first distance is longer. How would you describe the motion of the runners in the two races?

Motion and Position You don't always need to see something move to know that motion has taken place. For example, suppose you look out a window and see a mail truck stopped next to a mailbox. One minute later, you look out again and see the same truck stopped farther down the street. Although you didn't see the truck move, you know it moved because its position relative to the mailbox changed.

Motion occurs when an object changes its position. To know whether the position of something has changed, you need a reference point such as the mailbox in **Figure 1.** A reference point also helps you determine how far the truck moved.

Figure 1
This mail truck is in motion.
How do you know the mail truck has moved?

Relative Motion Not all motion is as obvious as that of a truck that has changed its position. Even if you are sitting in a chair reading this book, you are moving. You are not moving relative to your desk or your school building, but you are moving relative to the other planets in the solar system and the Sun.

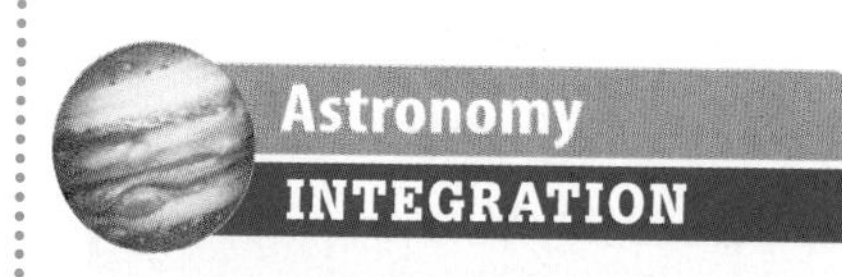

Using the Sun as your reference point, you are moving about 30 km through space every second. How many meters are in 30 km? What is this speed in meters per second? Record your answers in your Science Journal.

Distance In track and field events, have you ever run a 50-m dash? A distance of 50 m was marked on the track or athletic field to show you how far to run. An important part of describing the motion of an object is to describe how far it has moved, which is **distance.** The SI unit of length or distance is the meter (m). Longer distances are measured in kilometers (km). One kilometer is equal to 1,000 m. Shorter distances are measured in centimeters (cm). One meter is equal to 100 centimeters.

Displacement Suppose a runner jogs to the 50-m mark and then turns around and runs back to the 20-m mark, as shown in **Figure 2.** The runner travels 50 m in the original direction (north) plus 30 m in the opposite direction (south), so the total distance she ran is 80 m. How far is she from the starting line? The answer is 20 m. Sometimes you may want to know not only your distance but also your direction from a reference point, such as from the starting point. **Displacement** is the distance and direction of an object's change in position from the starting point. The runner's displacement in **Figure 2** is 20 m north.

The size of the runner's displacement and the distance traveled would be the same if the runner's motion was in a single direction. If the runner ran from the starting point to the finish line in a straight line, then the distance traveled would be 50 m and the displacement would be 50 m north.

Reading Check *How do distance and displacement differ?*

Speed

Think back to the example of the mail truck's motion in **Figure 1.** We now could describe the movement by the distance traveled and by the displacement from the starting point. You also might want to describe how fast it is moving. To do this, you need to know how far it travels in a given amount of time. **Speed** is the distance an object travels per unit of time.

Figure 2
Distance and displacement are not the same. The runner's displacement is 20 m north of the starting line. However, the total distance traveled is 80 m.

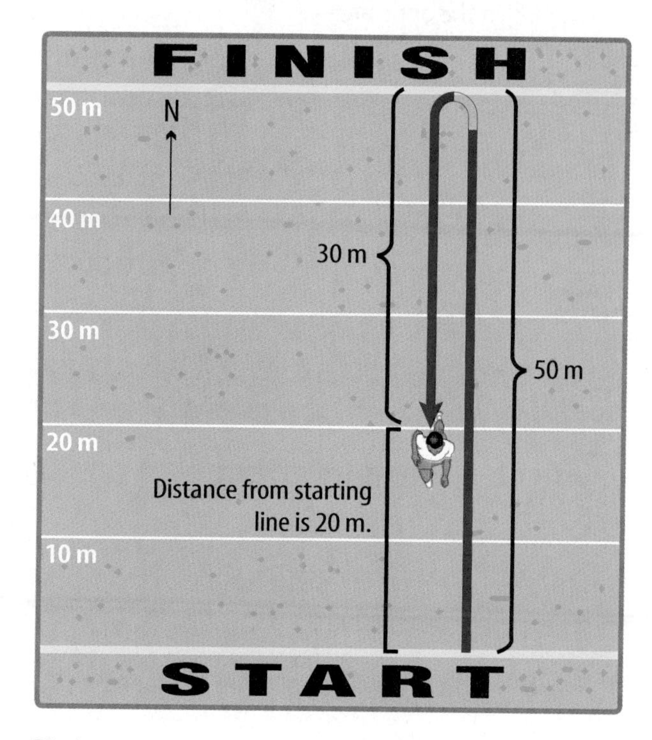

Displacement = 20 m north of starting line
Distance traveled = 50 m + 30 m = 80 m

Mini LAB

Describing the Motion of a Car

Procedure

1. Mark your starting point on the floor with **tape.**
2. At the starting line, give your **toy car** a gentle push forward. At the same time, start your **stopwatch.**
3. Stop timing when the car comes to a complete stop. Mark the spot at the front of the car with another **pencil.** Record the time for the entire trip.
4. Use a **meterstick** to measure the distance to the nearest tenth of a centimeter and convert it to meters.

Analysis

Calculate the speed. How would the speed differ if you repeated your experiment in exactly the same way but the car traveled in the opposite direction?

Rate Any change over time is called a rate. For example, your rate of growth is how much your height changes over a certain period of time, such as a year. If you think of distance as the change in position, then speed is the rate at which distance is traveled or the rate of change in position.

Calculating Speed Speed is related to the distance traveled and the time needed to travel the distance as follows:

$$\text{speed} = \frac{\text{distance}}{\text{time}}$$

If s = speed, d = distance, and t = time, this relationship can be written as follows:

$$s = \frac{d}{t}$$

Suppose you ran 2 km in 10 min. Your speed, or rate of change of position, would be found using the following equation:

$$s = \frac{d}{t} = \frac{2\text{ km}}{10\text{ min}} = 0.2\text{ km/min}$$

Because speed is calculated as distance divided by time, the units in which speed is measured always include a distance unit over a time unit. The SI unit for distance is the meter and the SI unit of time is the second (s), so in SI, units of speed are measured in meters per second (m/s). Speed also can be expressed in other units of distance and time, such as kilometers per hour (km/h) or centimeters per second (cm/s). **Table 1** shows some rates that show the range in which motion can occur. What units would you use to describe your rate of growth?

Motion with Constant Speed Suppose you are in a car traveling on a nearly empty freeway. You look at the speedometer and see that the car's speed hardly changes. If the car neither slows down nor speeds up, the car is traveling at a constant speed. Can you think of other examples of something moving at constant speed? If you are traveling at a constant speed, you can measure your speed over any distance interval from millimeters to light years.

Table 1 Examples of Units of Speed

Unit of Speed	Examples of Uses	Approximate Speed
km/s	rocket escaping Earth's atmosphere	11.2 km/s
km/h	car traveling at highway speed	100 km/h
cm/yr	geological plate movements	2 cm/yr–17 cm/yr

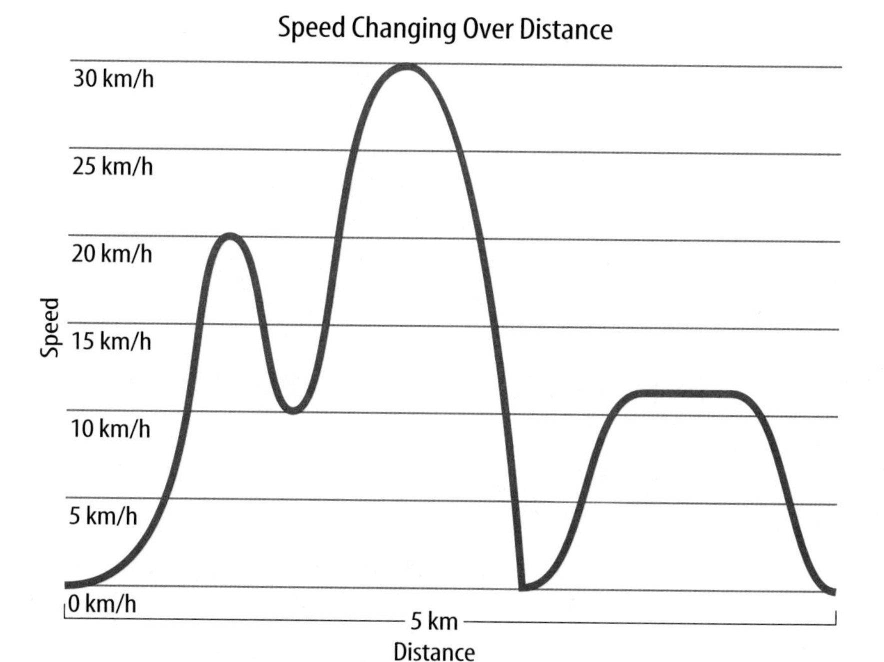

Figure 3
The cyclist is undergoing speed changes.
How do you describe the speed of an object when the speed is changing?

Changing Speed Much of the time, the speeds you experience are not constant. Think about riding a bicycle for a distance of 5 km as in **Figure 3.** As you start out, your speed increases from 0 km/h to, say, 20 km/h. You slow down to 10 km/h as you pedal up a steep hill and speed up to 30 km/h going down the other side of the hill. You stop for a red light, speed up again, and move at a constant speed for a while. As you near the end of the trip, you slow down and then stop. Checking your watch, you find that the trip took 15 min, or one-quarter of an hour. How would you express your speed on such a trip? Would you use your fastest speed, your slowest speed, or some speed between the two?

Research Visit the Glencoe Science Web site at **science.glencoe.com** for interesting facts about running speeds. In your Science Journal describe how running fast benefits the survival of animals in the wild.

Average Speed Average speed describes speed of motion when speed is changing. **Average speed** is the total distance traveled divided by the total time of travel. It can be calculated using the relationship among speed, distance, and time. For the bicycle trip just described, the total distance traveled was 5 km and the total time was 1/4 h, or 0.25 h. The average speed was:

$$s = \frac{d}{t} = \frac{5 \text{ km}}{0.25 \text{ h}} = 20 \text{ km/h}$$

Instantaneous Speed Suppose you watch a car's speedometer, like the one in **Figure 4,** go from 0 km/h to 60 km/h. A speedometer shows how fast a car is going at one point in time or at one instant. The speed shown on a speedometer is the instantaneous speed. **Instantaneous speed** is the speed at a given point in time.

Figure 4
The speed shown on the speedometer gives the instantaneous speed—the speed at one instant in time.

Changing Instantaneous Speed When something is speeding up or slowing down, its instantaneous speed is changing. The speed is different at every point in time. If an object is moving with constant speed, the instantaneous speed doesn't change. The speed is the same at every point in time.

Reading Check ***What are two examples of motion in which the instantaneous speed changes?***

Math Skills Activity

Calculating Time from Speed

Example Problem

Sound travels at a speed of 330 m/s. If a lightning bolt strikes the ground 1 km away from you, how long will it take for the sound to reach you?

Solution

1. *This is what you know:*

   distance: d = 1 km or 1,000 m

   speed: s = 330 m/s

2. *This is the equation you need to use to find time,* t:

$$s = \frac{d}{t}$$

3. *To find* t, *multiply both sides of this equation by* t *and divide both sides by* s:

$$t = \frac{d}{s}$$

4. *Substitute the known values, and then solve the equation for* t:

$$t = \frac{1\text{ km}}{330\text{ m/s}} = \frac{1{,}000\text{ m}}{330\text{ m/s}} = 3.03\text{ s}$$

Practice Problems

1. A passenger elevator operates at an average speed of 8 m/s. If the 60th floor is 219 m above the first floor, how long does it take the elevator to go from the first floor to the 60th floor?

2. A motorcyclist travels an average speed of 20 km/h. If the cyclist is going to a friend's house 5 km away, how long does it take the cyclist to make the trip?

For more help, refer to the Math Skill Handbook.

Graphing Motion

A distance-time graph makes it possible to display the motion of an object over a period of time. For example, the graph in **Figure 5** shows the motion of three swimmers during a 30-min workout. The straight, red line represents the motion of a swimmer who swam 800 m during each 10-min period. Her speed was constant at 80 m/min.

The straight blue line represents the motion of a swimmer who swam with a constant speed of 60 m/min. Notice that the line representing the motion of the faster swimmer is steeper. The steepness of a line on a graph is called the slope. On a distance-time graph, the slope of the line representing the motion of an object is the speed. Because the first swimmer has a greater speed, her line has a larger slope.

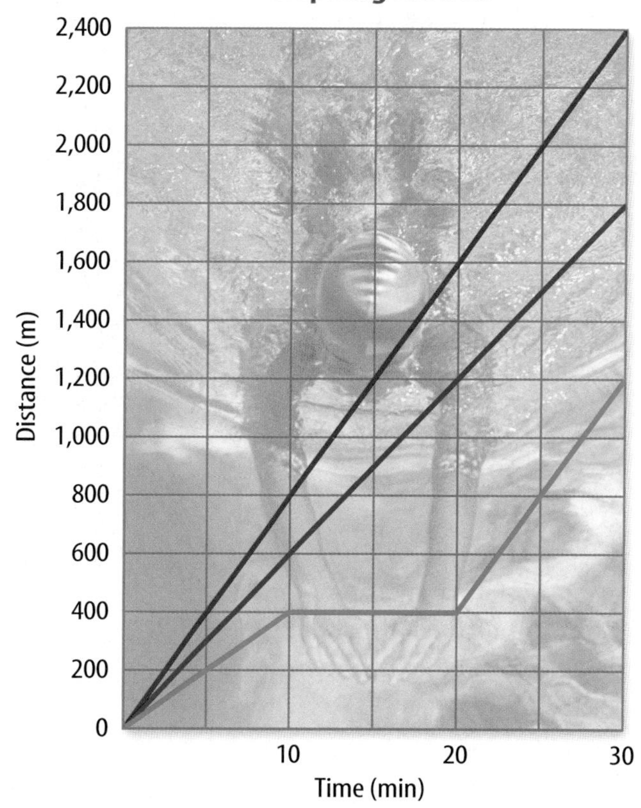

Figure 5
The slope of a distance-time graph gives the velocity of the object in motion.

Changing Speed The green line represents the motion of a third swimmer, who did not swim at a constant speed. She covered 400 m during the first 10 min at a constant speed, rested for the next 10 min, and covered 800 m during the final 10 min. During the first 10 min, she swam a shorter distance than the other two swimmers, so her line has a smaller slope. During the middle period her speed is zero, so her line over this interval is horizontal and has zero slope. During the last time interval, she swam as fast as the first swimmer so that part of her line has the same slope.

Reading Check ***What was the average speed of each swimmer over the 30-min period?***

Plotting a Distance-Time Graph Plotting a distance-time graph is simple. The distance is plotted on the vertical axis and the time on the horizontal axis. Each axis must have a scale that covers the range of numbers you are working with. For instance, the total distance that the swimmers traveled was 2,400 m. The scale for distance must range from 0 to 2,400 m. The total time the swimmers worked out was 30 min. Therefore, the time scale must range from 0 to 30 min. Then each axis must be divided into equal intervals to represent the data correctly. Once the scales for each axis are in place, the data points can be plotted. After plotting the data points, draw a line connecting the points.

Research Visit the Glencoe Science Web site at **science.glencoe.com** for information about the speed of Olympic swimmers for the past 60 years. Communicate to your class what you learn.

Figure 6
The speed of a storm is not enough information to plot the path. The direction the storm is moving must be known, too.

Velocity

You turn on the radio and hear the tail end of a news story about a hurricane, like the one in **Figure 6,** that is approaching land. The storm, traveling at a speed of 20 km/h, is located 100 km east of your location. Should you be worried?

Unfortunately, you don't have enough information to figure out the answer. Knowing only the speed of the storm isn't much help. Speed describes only how fast something is moving. To decide whether you need to move to a safer area, you also need to know the direction that the storm is moving. In other words, you need to know the velocity of the storm. **Velocity** includes the speed of an object and the direction of its motion.

Escalators like the one shown in **Figure 7A** are found in shopping malls and airports. The two sets of passengers are moving at constant speed, but in opposite directions. The speeds of the passengers are the same, but their velocities are different because the passengers are moving in different directions.

Because velocity depends on direction as well as speed, the velocity of an object can change even if the speed of the object remains constant. For example, look at **Figure 7B.** A race car has a constant speed of 100 km/h and is going around an oval track. Even though the speed remains constant, the velocity changes because the direction of the car's motion is changing constantly.

Reading Check *How are velocity and speed different?*

Figure 7
For an object to have constant velocity, speed and direction must not be changing.

A These two escalators have the same speed. However, their velocities are different because they are traveling in opposite directions.

B The speed of this car might be constant, but its velocity is not constant because the direction of motion is always changing.

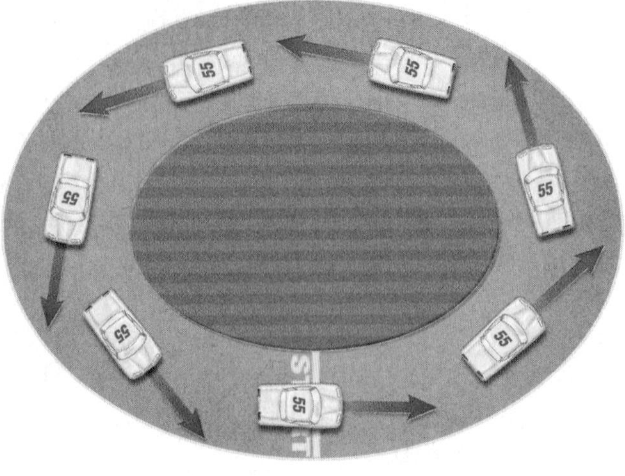

Motion of Earth's Crust

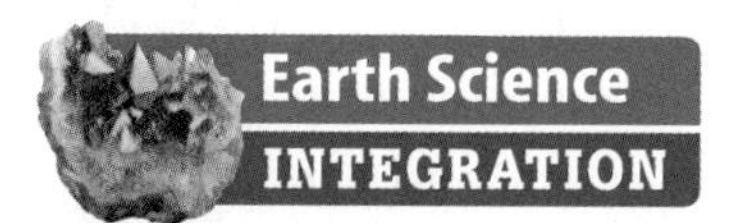

Can you think of something that is moving so slowly you cannot detect its motion, yet you can see evidence of its motion over long periods of time? As you look around the surface of Earth from year to year, the basic structure of the planet seems the same. Mountains, plains, lakes, and oceans seem to remain unchanged over hundreds of years. Yet if you examined geological evidence of what Earth's surface looked like over the past 250 million years, you would see that large changes have occurred. **Figure 8** shows how, according to the theory of plate tectonics, the positions of landmasses have changed during this time. Changes in the landscape occur constantly as continents drift slowly over Earth's surface. However, these changes are so gradual that you do not notice them.

Figure 8
Geological evidence suggests that continents have moved slowly over time.

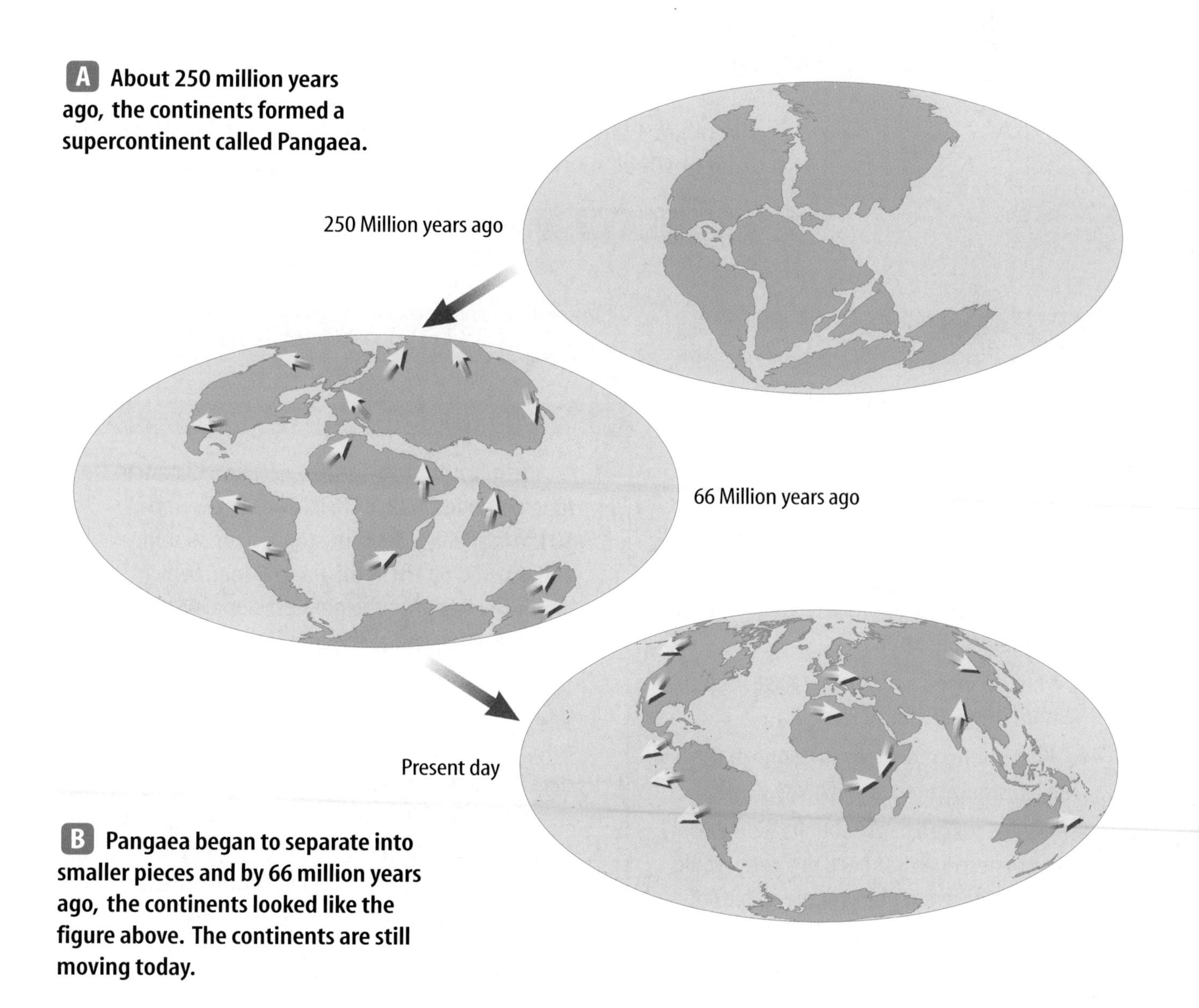

A About 250 million years ago, the continents formed a supercontinent called Pangaea.

B Pangaea began to separate into smaller pieces and by 66 million years ago, the continents looked like the figure above. The continents are still moving today.

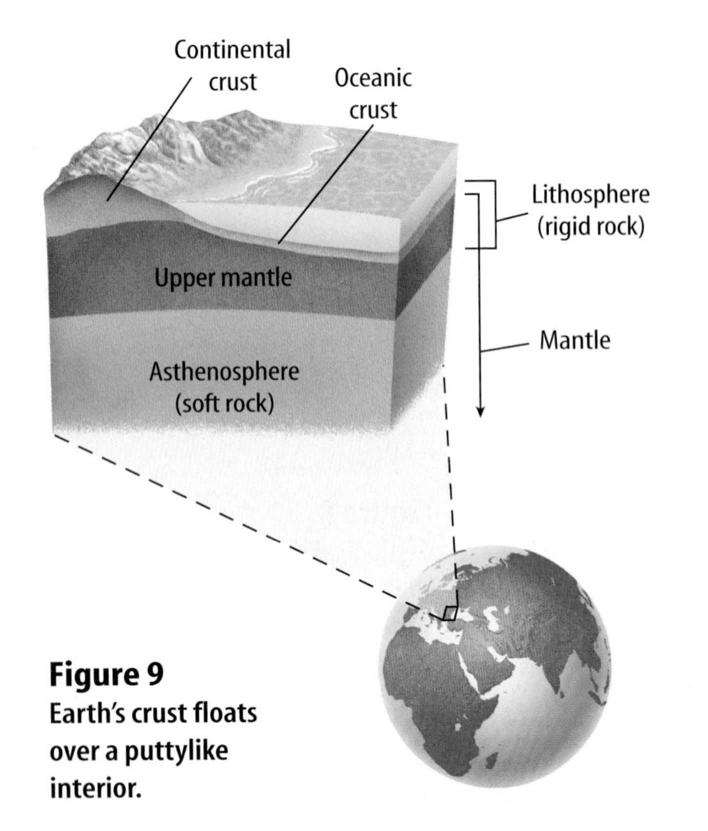

Figure 9
Earth's crust floats over a puttylike interior.

Moving Continents How can continents move around on the surface of Earth? Earth is made of layers, as shown in **Figure 9.** The outer layer is the crust, and the layer just below the crust is called the upper mantle. Together the crust and the top part of the upper mantle are called the lithosphere. The lithosphere is broken into huge sections called plates that slide slowly on the puttylike layers just below. If you compare Earth to an egg, these plates are about as thick as the eggshell. These moving plates cause geological changes such as the formation of mountain ranges, earthquakes, and volcanic eruptions.

Plates move so slowly that their speeds are given in units of centimeters per year. In California, two plates slide past each other along the San Andreas Fault with an average relative speed of about 1 cm per year. The Australian Plate's movement is one of the fastest, pushing Australia north at an average speed of about 17 cm per year.

How do the continents drift?

Section  Assessment

1. How does displacement differ from distance? Give an example of displacement and distance.
2. You bike from your house to school, covering a distance of 3 km in 15 min. What is your average speed? Give your answer in kilometers per hour.
3. What do speed and velocity have in common? How do they differ?
4. What information does the slope of the line in a distance-time graph give?
5. **Think Critically** What units would you use to describe the speed of a car? Would you use different units for the speeds of runners in a school race? Explain.

Skill Builder Activities

6. **Making and Using Graphs** Make a distance-time graph for a 2-h car trip. The car covered 50 km in the first 30 min, stopped for 30 min, and covered 60 km in the final 60 min. Which graph segment has the greatest slope? What was the car's average speed? **For more help, refer to the Science Skill Handbook.**
7. **Using an Electronic Spreadsheet** Use a computer to construct a data table. Using the data in problem 6, list the distance and time measurements at 5-min intervals. Use a spreadsheet program to make a data table and, if possible, re-create the graphs. **For more help, refer to the Technology Skill Handbook.**

SECTION

Acceleration

Acceleration, Speed, and Velocity

You're sitting in a car at a stoplight when the light turns green. The driver steps on the gas pedal and the car starts moving faster and faster. Just as speed is the rate of change of position, **acceleration** is the rate of change of velocity. When the velocity of an object changes, the object is accelerating.

Remember that velocity includes the speed and direction of an object. Therefore, a change in velocity can be either a change in how fast something is moving or a change in the direction it is moving. Acceleration occurs when an object changes its speed, its direction, or both.

Speeding Up and Slowing Down When you think of acceleration, you probably think of something speeding up. However, an object that is slowing down also is accelerating.

Imagine a car traveling through a city. If the speed is increasing, the car has positive acceleration. When the car slows down its speed is decreasing and the car has negative acceleration. In both cases the car is accelerating because its speed is changing.

An acceleration has a direction, just as a velocity does. If the acceleration is in the same direction as the velocity, as in **Figure 10A,** the speed increases and the acceleration is positive. If the speed decreases, the acceleration is in the opposite direction from the velocity, and the acceleration is negative for the car shown in **Figure 10B.**

As You Read

What You'll Learn

- **Identify** how acceleration, time, and velocity are related.
- **Explain** how positive and negative acceleration affect motion.
- **Describe** how to calculate the acceleration of an object.

Vocabulary
acceleration

Why It's Important

Acceleration occurs all around you as objects speed up, slow down, or change direction.

Figure 10
These cars are accelerating because their speed is changing.

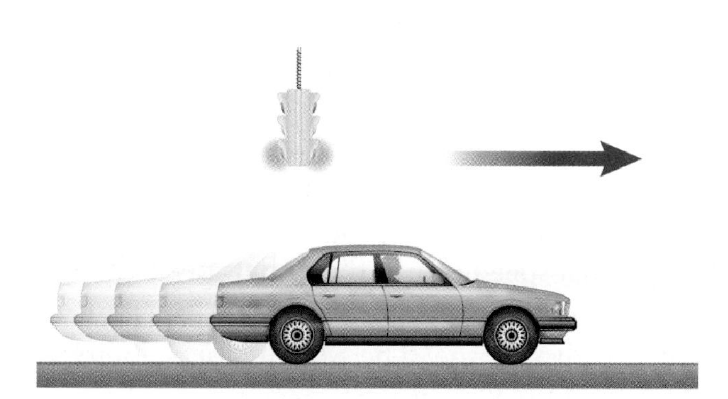

A **The speed of this car is increasing. The car has positive acceleration.**

B **The speed of this car is decreasing. The car has negative acceleration.**

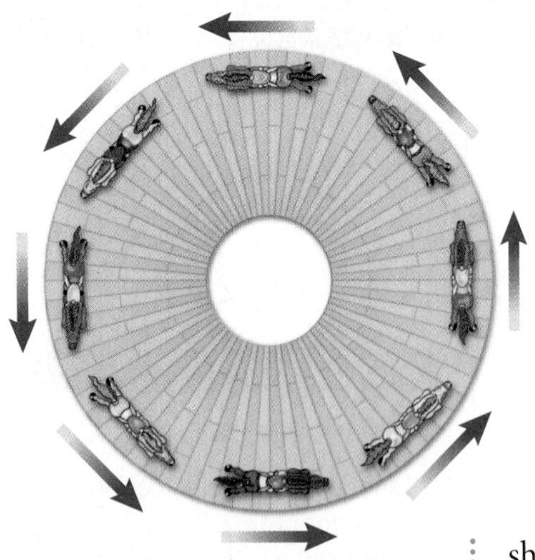

Figure 11
The speed of the horses in this carousel is constant, but the horses are accelerating because their direction is changing constantly.

Changing Direction A change in velocity can be either a change in how fast something is moving or a change in the direction of movement. Any time a moving object changes direction, its velocity changes and it is accelerating. Think about a horse on a carousel. Although the horse's speed remains constant, the horse is accelerating because it is changing direction constantly as it travels in a circular path, as shown in **Figure 11.** In the same way, Earth is accelerating constantly as it orbits the Sun in a nearly circular path.

Graphs of speed versus time can provide information about accelerated motion. The shape of the plotted curve shows when an object is speeding up or slowing down. **Figure 12** describes how motion graphs are constructed.

Calculating Acceleration

Remember that acceleration is the rate of change in velocity. To calculate the acceleration of an object, the change in velocity or speed is divided by the length of the time interval over which the change occurred. Another way to write this relationship is as follows:

$$\text{Acceleration} = \frac{\text{change in velocity}}{\text{time}}$$

How is the change in velocity calculated? Always subtract the initial velocity—the velocity at the beginning of the time interval—from the final velocity—the velocity at the end of the time interval. Let v_i stand for the initial velocity and v_f stand for the final velocity. The change in velocity is as follows:

$$\text{Change in velocity} = \text{final velocity} - \text{initial velocity}$$
$$= v_f - v_i$$

Then the relationship between acceleration, velocity, and time is as follows:

$$a = \frac{(v_f - v_i)}{t}$$

If the motion is in a single direction or a straight line, the change in speed can be used to calculate the change in velocity. The change in speed is the final speed minus the initial speed.

In the equation above, the unit of acceleration is a unit of velocity divided by a unit of time. The SI unit for velocity is meters/second (m/s), and the SI unit for time is seconds (s). So, the unit for acceleration is meters/second/second. This unit is written as m/s^2 and is read "meters per second squared."

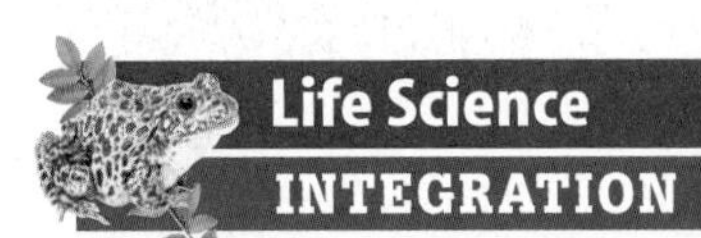

Your body is sensitive to acceleration. Much of the thrill of riding a roller coaster is due to the way your body feels while accelerating in different ways. Write a paragraph describing three situations in which you can feel accelerations while riding in a car.

NATIONAL GEOGRAPHIC VISUALIZING ACCELERATION

Figure 12

Acceleration can be positive, negative, or zero depending on whether an object is speeding up, slowing down, or moving at a constant speed. If the speed of an object is plotted on a graph, with time along the horizontal axis, the slope of the line is related to the acceleration.

A The car in the photograph on the right is maintaining a constant speed of about 90 km/h. Because the speed is constant, the car's acceleration is zero. A graph of the car's speed with time is a horizontal line.

B The green graph shows how the speed of a bouncing ball changes with time as it falls from the top of a bounce. The ball speeds up as gravity pulls the ball downward, so the acceleration is positive. For positive acceleration, the plotted line slopes upward to the right.

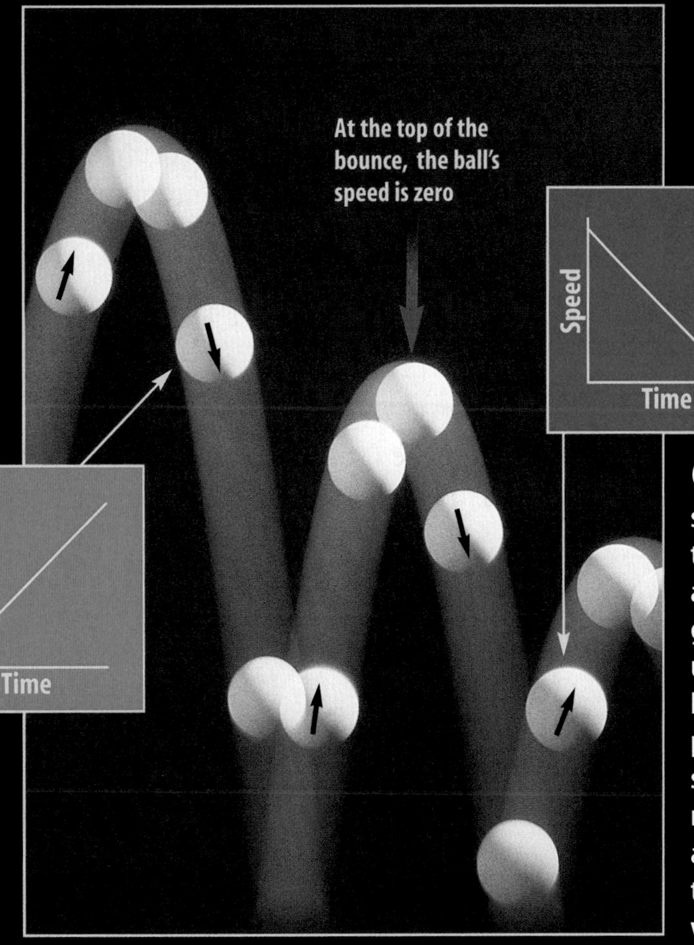

C The blue graph shows the change with time in the speed of a ball after it hits the ground and bounces upward. The climbing ball slows as gravity pulls it downward, so the acceleration is negative. For negative acceleration, the plotted line slopes downward to the right.

Calculating Positive Acceleration How is the acceleration for an object that is speeding up different from that of an object that is slowing down? Suppose the jet airliner in **Figure 13A** starts at rest at the end of a runway and reaches a speed of 80 m/s in 20 s. The airliner is traveling in a straight line down the runway, so its speed and velocity are the same. Because it started from rest, its initial speed was zero. Its acceleration can be calculated as follows:

$$a = \frac{(v_f - v_i)}{t} = \frac{(80\ \text{m/s} - 0\ \text{m/s})}{20\ \text{s}} = 4\ \text{m/s}^2$$

The airliner is speeding up, so the final speed is greater than the initial speed and the acceleration is positive.

Calculating Negative Acceleration Now imagine that the skateboarder in **Figure 13B** is moving in a straight line at a speed of 3 m/s and comes to a stop in 2 s. The final speed is zero and the initial speed was 3 m/s. The skateboarder's acceleration is calculated as follows:

$$a = \frac{(v_f - v_i)}{t} = \frac{(0\ \text{m/s} - 3\ \text{m/s})}{2\ \text{s}} = -1.5\ \text{m/s}^2$$

The skateboarder is slowing down, so the final speed is less than the initial speed and the acceleration is negative. The acceleration always will be positive if an object is speeding up and negative if the object is slowing down. What is the acceleration if an object moves with constant velocity?

Figure 13
A speed-time graph tells you if acceleration is a positive or negative number.

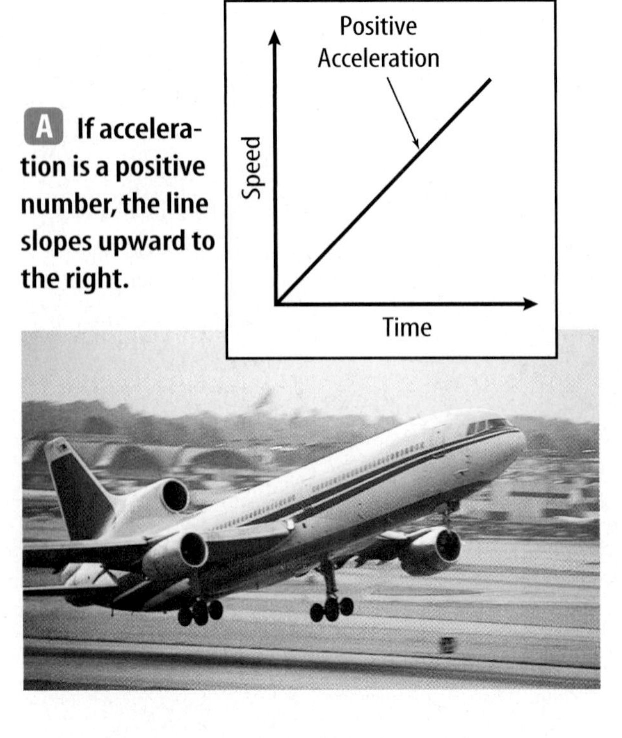

A **If acceleration is a positive number, the line slopes upward to the right.**

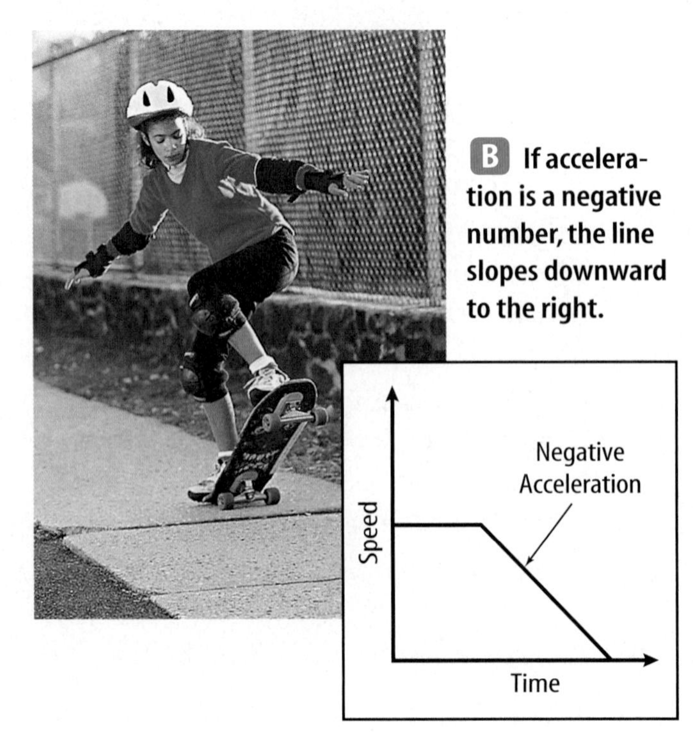

B **If acceleration is a negative number, the line slopes downward to the right.**

Amusement Park Acceleration

Riding roller coasters in amusement parks can give you the feeling of danger—but these rides are designed to be safe. Engineers use the laws of physics to provide a thrilling but harmless ride. Roller coasters are constructed of wood or steel. Wooden roller coasters do not have the high velocities and accelerations that steel roller coasters do. Wood is not as rigid as steel. Therefore, these roller coasters do not have steep hills or inversion loops that propel the rider at high speeds. However, wooden roller coasters can have a swaying movement that steel roller coasters do not have. This swaying motion can give the rider a different type of thrill.

Steel roller coasters can offer multiple steep drops and inversion loops, which give the rider large accelerations. As the rider moves down a steep hill or an inversion loop, he or she will accelerate toward the ground at $9.8\ m/s^2$ due to gravity. When riders go around a sharp turn, they also are accelerated. This acceleration makes them feel as if a force is pushing them toward the side of the car. The table in **Figure 14** shows the four fastest roller coasters in the United States. The fastest roller coaster goes from 0 to 160.9 km/h in 7 s.

Reading Check *What material are roller coasters with the steepest hills made from?*

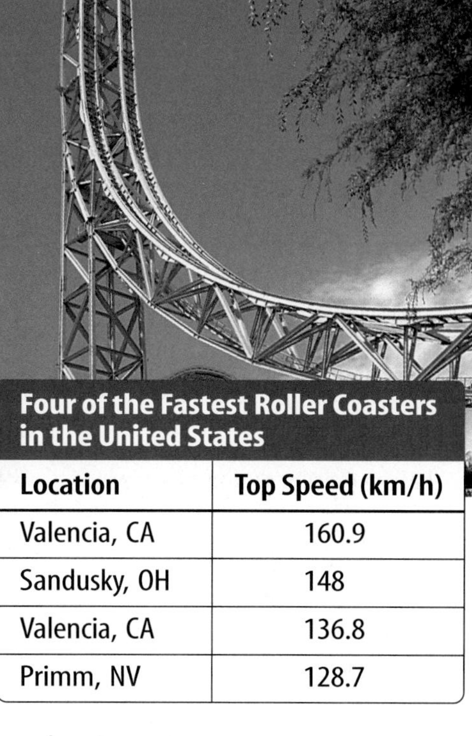

Four of the Fastest Roller Coasters in the United States

Location	Top Speed (km/h)
Valencia, CA	160.9
Sandusky, OH	148
Valencia, CA	136.8
Primm, NV	128.7

Figure 14
This roller coaster is the fastest roller coaster in the world. Riders reach a speed of 160.9 km/h.

Section 2 Assessment

1. How are velocity, time, and acceleration related mathematically?
2. A swimmer speeds up from 1.1 m/s to 1.3 m/s during the last 20 s of a workout. What is the swimmer's acceleration during this time interval?
3. While walking to school, you approach an intersection and slow down from 2 m/s to a stop in 3 s. What was your acceleration during this time interval?
4. Explain the term *negative acceleration.*
5. **Think Critically** Describe three ways to change your velocity while riding a bicycle.

Skill Builder Activities

6. **Making and Using Graphs** In the graph shown in **Figure 13A,** is the speed increasing or decreasing? Explain. Is the speed increasing or decreasing in **Figure 13B?** Explain. Describe how the graph would look for a jet airplane cruising at a constant speed. **For more help, refer to the Science Skill Handbook.**
7. **Communicating** In your Science Journal, explain why streets and highways have speed limits rather than velocity limits. Where might a velocity limit be used? **For more help, refer to the Science Skill Handbook.**

SECTION

Motion and Forces

As You Read

What You'll Learn

- **Explain** how force and velocity are related.
- **Describe** what inertia is and how it is related to Newton's first law of motion.
- **Identify** the forces and motion that are present during a car crash.

Vocabulary

force
net force
balanced force
inertia

Why It's Important

Force and motion are directly linked—without force, you cannot have motion.

What is force?

Passing a basketball to a team member or kicking a soccer ball into the goal are examples of applying force to an object. A **force** is a push or pull that one body exerts on another. In both examples, the applied force results in the movement of the ball. Sometimes it is obvious that a force has been applied. But other forces aren't as noticeable. For instance, are you conscious of the force the floor exerts on your feet? Can you feel the force of the atmosphere pushing against your body or gravity pulling on your body?

Think about all the forces you exert in a day. Every push, pull, stretch, or bend results in a force being applied to an object.

Changing Motion What happens to the motion of an object when you exert a force on it? A force can cause the motion of an object to change. Think of hitting a ball with a racket, as in **Figure 15.** The racket strikes the ball with a force that causes the ball to stop and then move in the opposite direction. If you have played billiards, you know that you can force a ball at rest to roll into a pocket by striking it with another ball. The force of the moving ball causes the ball at rest to move in the direction of the force. In these cases, the velocities of the ball and the billiard ball were changed by a force.

Figure 15
This ball is hit with a force. The racket strikes the ball with a force in the opposite direction of its motion. As a result, the ball changes the direction it is moving.

Figure 16
Forces can be balanced and unbalanced.

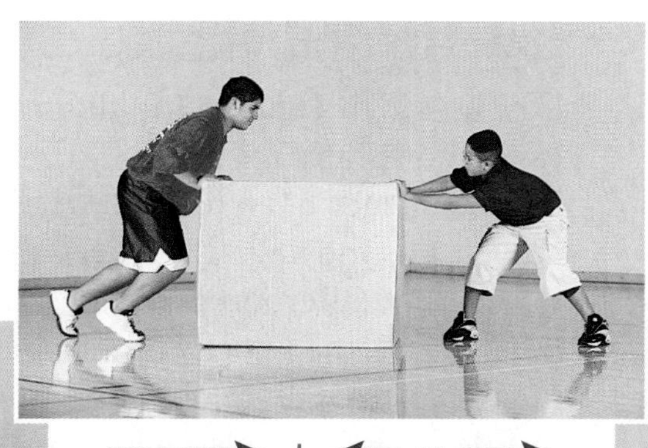

Net Force = →

B These students are pushing on the box with unequal forces in opposite directions. The box will be moved in the direction of the larger force.

Net Force = 0

A These students are pushing on the box with an equal force but in opposite directions. Because the forces are balanced, the box does not move.

Net Force = →

C These students are pushing on the box in the same direction. The combined forces will cause the box to move.

Balanced Forces Force does not always change velocity. In **Figure 16A,** two students are pushing on opposite sides of a box. Both students are pushing with an equal force but in opposite directions. When two or more forces act on an object at the same time, the forces combine to form the **net force.** The net force on the box in **Figure 16A** is zero because the two forces cancel each other. Forces on an object that are equal in size and opposite in direction are called **balanced forces.**

Unbalanced Forces Another example of how forces combine is shown in **Figure 16B.** When two students are pushing with unequal forces in opposite directions, a net force occurs in the direction of the larger force. In other words, the student who pushes with a greater force will cause the box to move in the direction of the force. The net force that moves the box will be the difference between the two forces because they are in opposite directions or they are considered to be unbalanced forces.

In **Figure 16C,** the students are pushing on the box in the same direction. These forces are combined or added together because they are exerted on the box in the same direction. The net force that acts on this box is found by adding the two forces together.

✔ Reading Check *What is an unbalanced force?*

SCIENCE *Online*

Research Visit the Glencoe Science Web site at **science.glencoe.com** for information about unbalanced forces along fault lines in the Earth's crust. In your Science Journal describe the significant activity that occurs along these fault lines.

Observing Inertia

Procedure

1. Create an inclined plane between 25° and 50° using a **board** and **textbooks.** Place a **stop block** (brick or other heavy object) at the end of the plane.
2. Place a **small object** in a **cart** and allow both to roll down the plane. Record the results in your journal.
3. Secure the object in the cart with **rubber bands** (seat belts). Allow both to roll down the plane again. Record the results.

Analysis

1. Identify the forces acting on the object in both runs.
2. Explain why it is important to wear seat belts in a car.

Inertia and Mass

The car in **Figure 17** is sliding on an icy road. This sliding car demonstrates the property of inertia. **Inertia** (ihn UR shuh) is the tendency of an object to resist any change in its motion. If an object is moving, it will keep moving at the same speed and in the same direction unless an unbalanced force acts on it. In other words, the velocity of the object remains constant unless a force changes it. If an object is at rest, it tends to remain at rest. Its velocity is zero unless a force makes it move.

Does a bowling ball have the same inertia as a table-tennis ball? Why is there a difference? You couldn't change the motion of a bowling ball much by swatting it with a table-tennis paddle. However, you easily could change the motion of the table-tennis ball. A greater force would be needed to change the motion of the bowling ball because it has greater inertia. Why is this? Recall that mass is the amount of matter in an object, and a bowling ball has more mass than a table-tennis ball does. The inertia of an object is related to its mass. The greater the mass of an object is, the greater its inertia.

Newton's First Law of Motion Forces change the motion of an object in specific ways. The British scientist Sir Isaac Newton (1642–1727) was able to state rules that describe the effects of forces on the motion of objects. These rules are known as Newton's laws of motion. They apply to the motion of all objects you encounter every day such as cars and bicycles, as well as the motion of planets around the Sun.

Figure 17
On an icy road, it is hard to turn or stop a car because the car has no traction. Because of its inertia, the car tends to move in a straight line with constant speed.

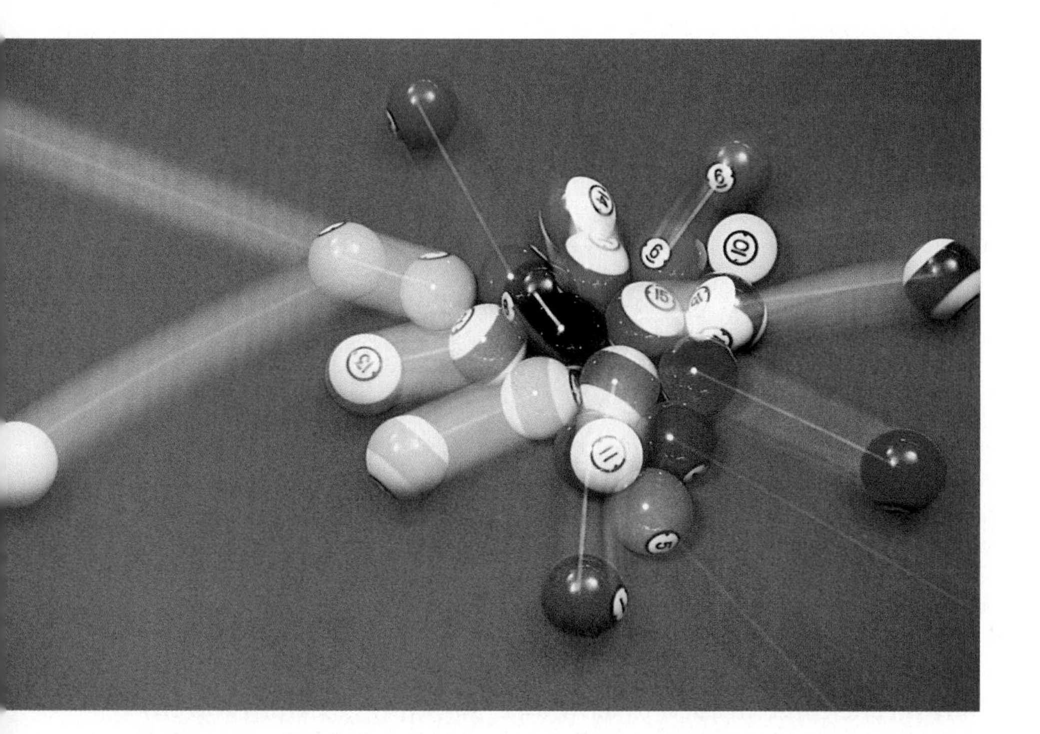

Figure 18
The inertia of the billiard balls causes them to remain at rest until a force is exerted on them by the cue ball.

The Law of Inertia According to Newton's first law of motion, an object moving at a constant velocity keeps moving at that velocity unless a net force acts on it. If an object is at rest, it stays at rest unless a net force acts on it. Does this sound familiar? It is the same as the earlier discussion of inertia. This law is sometimes called the law of inertia. You probably have seen and felt this law at work without even knowing it. **Figure 18** shows a billiard ball striking the other balls in the opening shot. What are the forces involved when the cue ball strikes the other balls? Are the forces balanced or unbalanced? How does this demonstrate the law of inertia?

✔ Reading Check *What is Newton's first law?*

What happens in a crash?

The law of inertia can explain what happens in a car crash. When a car traveling about 50 km/h collides head-on with something solid, the car crumples, slows down, and stops within approximately 0.1 s. Any passenger not wearing a seat belt continues to move forward at the same speed the car was traveling. Within about 0.02 s (1/50 of a second) after the car stops, unbelted passengers slam into the dashboard, steering wheel, windshield, or the backs of the front seats, as in **Figure 19.** They are traveling at the car's original speed of 50 km/h—about the same speed they would reach falling from a three-story building.

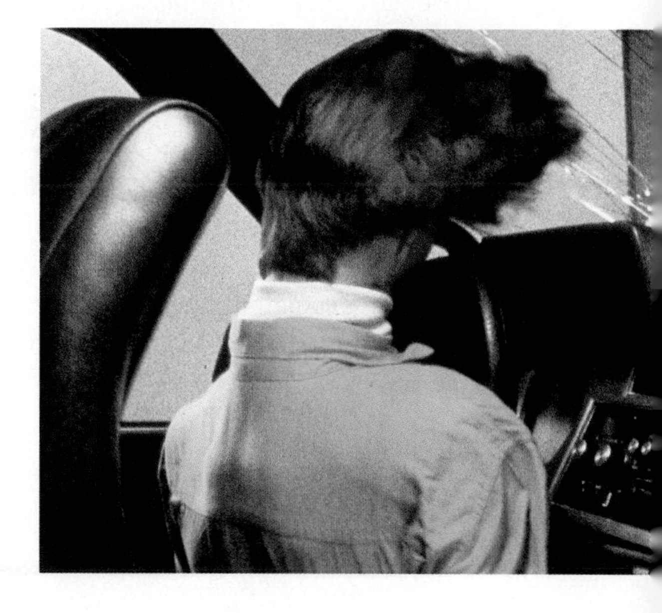

Figure 19
The crash dummy is not restrained in this low-speed crash. Inertia causes the dummy to slam into the steeting wheel.

Figure 20
These crash dummies were restrained safely with seat belts in this low-speed crash. Usually humans would have fewer injuries if they were restrained safely during an accident.

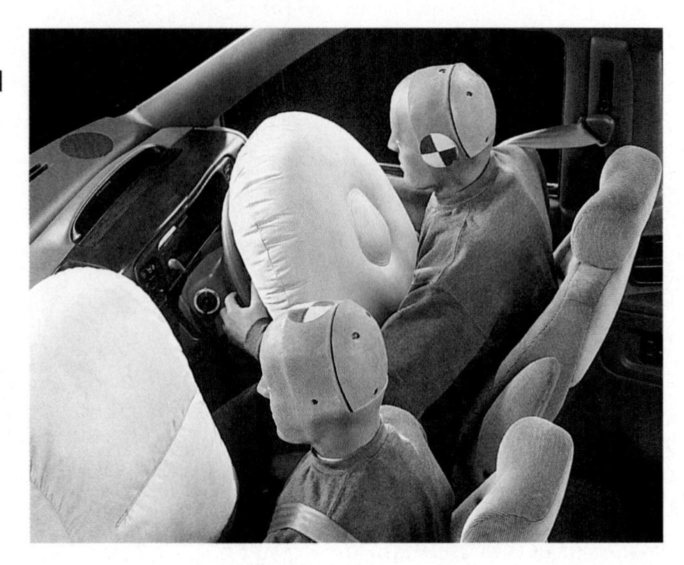

Seat belts The crash dummy wearing a seat belt in **Figure 20** will be attached to the car and will slow down as the car slows down. The force needed to slow a person from 50 km/h to zero in 0.1 s is equal to 14 times the force that gravity exerts on the person. The belt loosens a little as it restrains the person, increasing the time it takes to slow the person down. This reduces the force exerted on the person. The seat belt also prevents the person from being thrown out of the car. Car-safety experts say that about half the people who die in car crashes would survive if they wore seat belts. Thousands of others would suffer fewer serious injuries.

Section  Assessment

1. When a soccer player kicks a ball, the ball accelerates. Explain what causes this acceleration in terms of forces.
2. Explain which has greater inertia—a speeding car or a jet airplane sitting on a runway.
3. Do forces always cause motion? Explain.
4. While trying to explain a physics concept, a student said, "Stuff keeps doing what it's doing unless something messes with it." What law was this student summarizing? Explain your answer.
5. **Think Critically** Describe three examples from sports in which a force changes the velocity of an object or a person.

Skill Builder Activities

6. **Researching Information** Many states have passed seat belt laws requiring all passengers in cars to wear seat belts. Research whether your state has such a law and when it became a law. Record your answer in your Science Journal. **For more help, refer to the** Science Skill Handbook.
7. **Communicating** Inertia plays an important role in most sports. In your Science Journal, write a paragraph describing the role of inertia in your favorite sport. Write another paragraph describing how the sport would be different without inertia. **For more help, refer to the** Science Skill Handbook.

Activity

Force and Acceleration

If you stand at a stoplight, you will see cars stopping for red lights and then taking off when the light turns green. What makes the cars slow down? What makes them speed up? The cars accelerate because an unbalanced force is acting on them.

What You'll Investigate

How does an unbalanced force on a book affect its motion?

Materials

tape
paper clip
10-N spring scale
large book
this science book
triple beam balance
*electronic balance

*Alternate materials

Goals

- **Observe** the effect of force on the acceleration of an object.
- **Interpret** the data collected for each trial.

Safety Precautions

Proper eye protection should be worn at all times while performing this lab.

Procedure

1. With a piece of tape, attach the paper clip to your textbook so that the paper clip is just over the edge of the book.
2. Prepare a data table with the following headings: Force, Mass.
3. If available, use a large balance to find the mass of this science book.
4. Place the book on the floor or on the surface of a long table. Use the paper clip to hook the spring scale to the book.
5. Pull the book across the floor or table at a slow but constant velocity. While pulling, read the force you are pulling with on the spring scale and record it in your table.
6. Repeat step 5 two more times, once accelerating slowly and once accelerating quickly. Be careful not to pull too hard. Your spring scale will read only up to 10 N.
7. Place a second book on top of the first book and repeat steps 3 through 6.

Conclude and Apply

1. **Organize** the pulling forces from greatest to least for each set of trials. Do you see a relationship between force and acceleration? Explain your answer.
2. How did adding the second book change the results? Explain your answer.

Compare your conclusions with those of other students in your class. **For more help, refer to the Science Skill Handbook.**

Comparing Motion from Different Forces

Think about a small ball. How many ways could you exert a force on the ball to make it move? You could throw it, kick it, roll it down a ramp, blow it with a large fan, etc. Do you think the distance and speed of the ball's motion will be the same for all of these forces? Do you think the acceleration of the ball would be the same for all of these types of forces?

Recognize the Problem

How will the motion of a small toy car vary when different forces are applied to it?

Form a Hypothesis

Based on your reading and observations, state a hypothesis about how a force can be applied that will cause the toy car to go fastest.

Possible Materials

small toy car
ramps or boards of different lengths
springs or rubber bands
string
stopwatch
meterstick or tape measure
graph paper

Safety Precautions

Goals

- **Identify** several forces that you can use to propel a small toy car across the floor.
- **Demonstrate** the motion of the toy car using each of the forces.
- **Graph** the position versus time for each force.
- **Compare** the motion of the toy car resulting from each force.

Test Your Hypothesis

Plan

1. As a group, agree upon the hypothesis and decide how you will test it. Identify which results will confirm the hypothesis that you have written.
2. **List** the steps you will need to test your hypothesis. Be sure to include a control run. Be specific. Describe exactly what you will do in each step. List your materials.
3. Prepare a data table in your Science Journal to record your observations.
4. **Read** the entire experiment to make sure all steps are in logical order and will lead to a conclusion.
5. **Identify** all constants, variables, and controls of the experiment. Keep in mind that you will need to have measurements at multiple points. These points are needed to graph your results. You should make sure to have several data points taken after you stop applying the force and before the car starts to slow down. It might be useful to have several students taking measurements, making each responsible for one or two points.

Do

1. Make sure your teacher approves your plan before you start.
2. Carry out the experiment as planned.
3. While doing the experiment, record your observations and complete the data tables in your Science Journal.

Analyze Your Data

1. **Graph** the position of the car versus time for each of the forces you applied. How can you use the graphs to compare the speeds of the toy car?
2. **Calculate** the speed of the toy car over the same time interval for each of the forces that you applied. How do the speeds compare?

Draw Conclusions

1. Did the speed of the toy car vary depending upon the force applied to it?
2. For any particular force, did the speed of the toy car change over time? If so, how did the speed change? Describe how you can use your graphs to answer these questions.
3. Did your results support your hypothesis? Why or why not?

Compare your data to those of other students. **Discuss** how the forces you applied might be different from those others applied and how that affected your results.

A Brave and Startling Truth

by Maya Angelou

Respond to the Reading

1. What adjectives does the poet use to describe Earth?
2. What wonders of the world does the poet name?
3. What does the poet believe are the true wonders of the world?

We, this people, on a small and lonely planet
Traveling through casual space
Past aloof stars, across the way of indifferent suns
To a destination where all signs tell us
It is possible and imperative that we learn
A brave and startling truth...

When we come to it
Then we will confess that not the Pyramids
With their stones set in mysterious perfection
Nor the Gardens of Babylon
Hanging as eternal beauty
In our collective memory
Not the Grand Canyon
Kindled into delicious color
By Western sunsets
These are not the only wonders of the world...

When we come to it
We, this people, on this minuscule and kithless[1] globe...
We this people on this mote[2] of matter

When we come to it
We, this people, on this wayward[3], floating body
Created on this earth, of this earth
Have the power to fashion for this earth
A climate where every man and every woman
Can live freely without sanctimonious piety[4]
Without crippling fear

When we come to it
We must confess that we are the possible
We are the miraculous, the true wonder of the world
That is when, and only when
We come to it.

[1] to be without friends or neighbors
[2] small particle
[3] wanting one's own way in spite of the advice or wishes of another
[4] a self-important show of being religious

Understanding Literature

Descriptive Writing This poem is full of images of Earth moving through space. But the adjectives the author uses to describe Earth are from the perspective of the universe. This description from the point of view of the universe gives the impression that Earth is small and insignificant.

The poet also names some special places on Earth. These places, although marvelous, fall short of being really wonderful. Angelou contrasts Earth's position within the universe to emphasize the importance of people. The power that people have to make changes for better lives is more significant than the universe and the special places people have built on Earth.

Science Connection Sometimes a person doesn't need to see movement to know that something has moved. Even though we don't necessarily see Earth's movement, we know Earth moves because of reference points such as the Sun. We know Earth moves because the Sun appears to change its position in the sky. The poem describes Earth's movement from the point of view of the universe. The universe serves as a reference point for detecting Earth's movement.

Linking Science and Writing

Write a Poem In the poem you just read, Maya Angelou describes Earth's movement from the point of view of the universe. Write a six-line poem that describes Earth's movement from the point of view of the Moon. How might the Moon's point of view toward Earth be different from that of remote stars and suns?

Career Connection

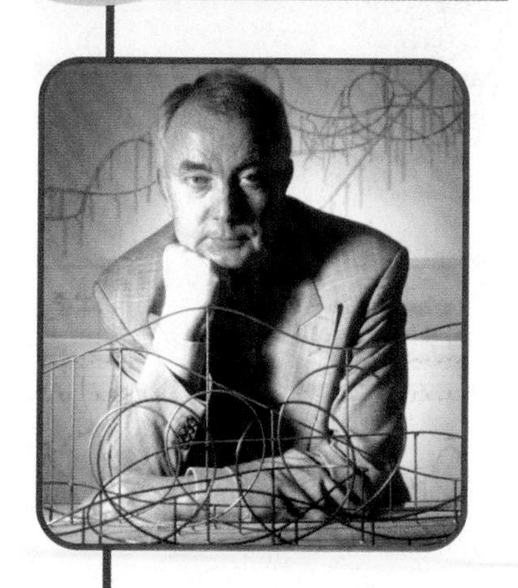

Roller Coaster Designer and Engineer

Ancient engineers designed the pyramids. Today, engineers design everything from can openers to cars. Werner Stengel and his team have designed and engineered more than 200 roller coasters. They compute the forces that react on roller coaster passengers. They also analyze the kind and amount of stress the roller-coaster structure will have to bear. Stengel has worked with safety committees and research groups to make sure that amusement park rides are safe for their riders.

SCIENCE*Online* To learn more about careers in engineering, visit the Glencoe Science Web site at **science.glencoe.com.**

Chapter 2 Study Guide

Reviewing Main Ideas

Section 1 Describing Motion

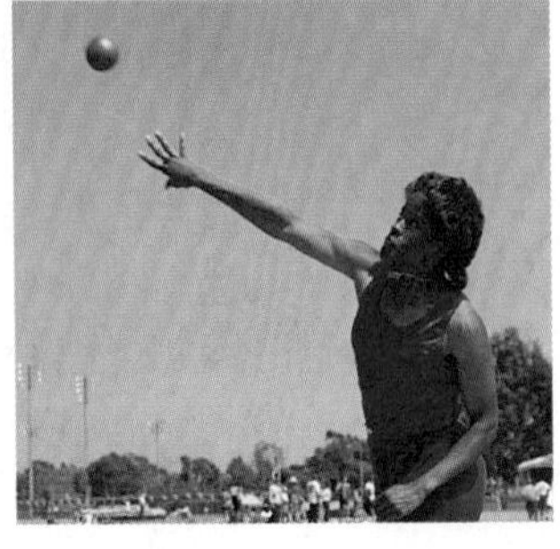

1. Motion is a change of position of a body. Distance is the measure of how far an object moved. Displacement is the distance and direction of an object's change in position from the starting point. *In the figure above, how can the motion of this shot put be described correctly?*
2. Average speed is the total distance traveled divided by the total time of travel.
3. Instantaneous speed is the speed at a given instant of time.
4. Velocity describes the speed and direction of a moving object.

Section 2 Acceleration

1. Acceleration is the rate of change of velocity for any object.
2. Any time the velocity of an object changes, the object must be accelerated. *In the figure below, describe the type of acceleration that occurs when the car stops for the red light and when it moves again for a green light.*

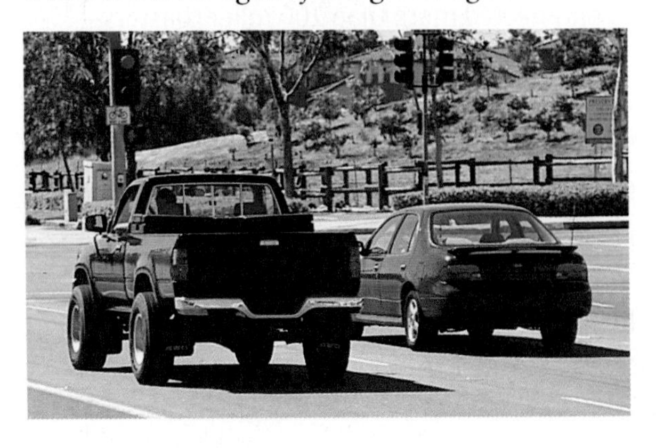

3. Acceleration occurs if an object speeds up, slows down, or changes direction.

Section 3 Motion and Forces

1. A force is a push or a pull one body exerts on another.
2. Balanced forces acting on a body do not change the motion of the body. Unbalanced forces result in a net force, which always changes the motion of a body.
3. Inertia is the resistance of an object to a change in its motion.
4. Newton's first law says an object's motion will not change unless a net force acts on it. *In the figure below, describe the forces that are acting on the car.*

After You Read

FOLDABLES Reading & Study Skills

To help you review the characteristics of motion, use the Foldable you made at the beginning of this chapter. Use the Foldable to review for quizzes, chapter tests, and semester exams.

Visualizing Main Ideas

Complete the following concept map about motion.

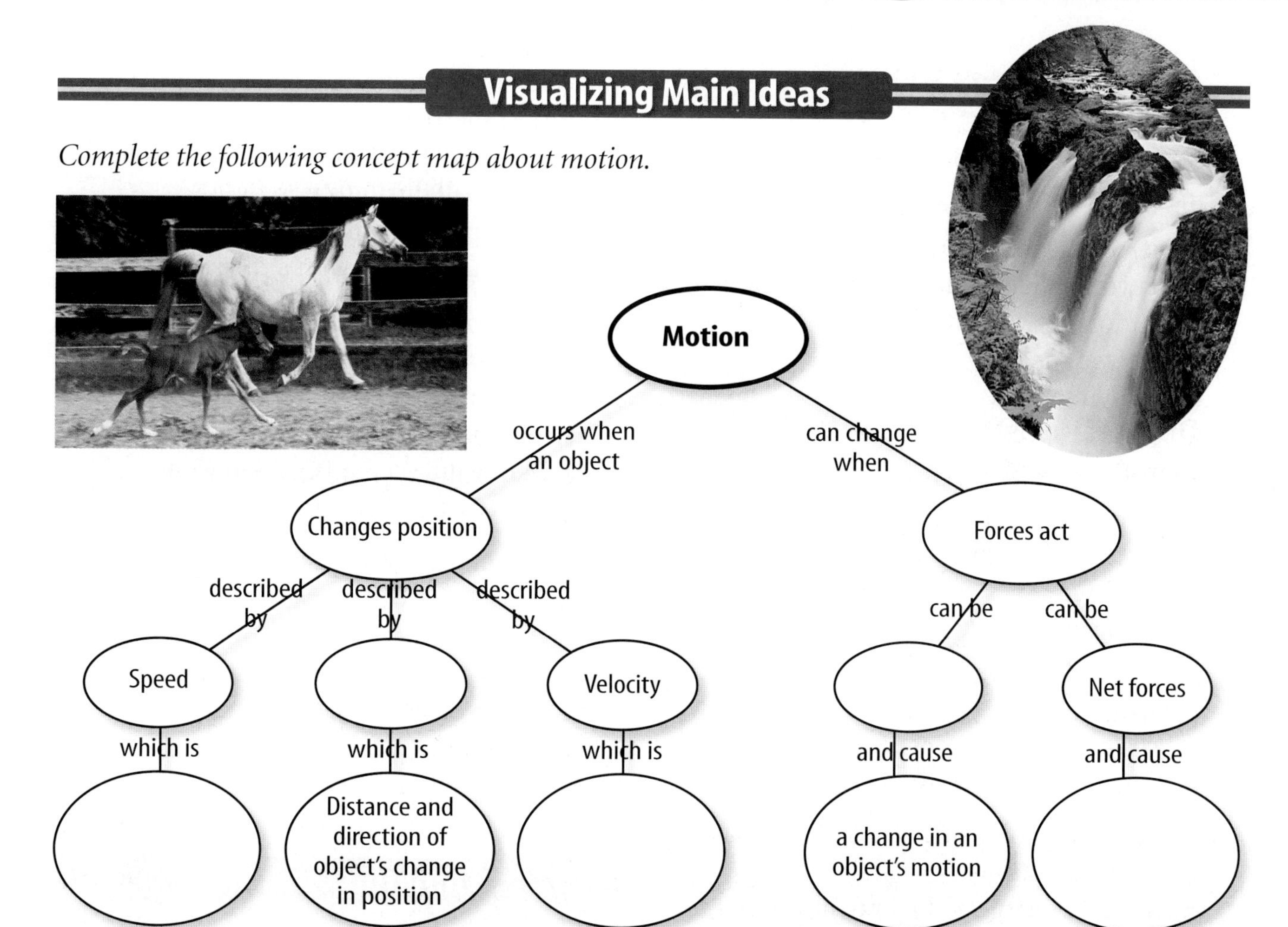

Vocabulary Review

Vocabulary Words

a. acceleration
b. average speed
c. balanced force
d. displacement
e. distance
f. force
g. inertia
h. instantaneous speed
i. net force
j. speed
k. velocity

Study Tip

To understand the information on a graph, write a sentence about the relationship between the x-axis and y-axis in the graph.

Using Vocabulary

Compare and contrast the following pairs of vocabulary words.

1. speed, velocity
2. distance, displacement
3. average speed, instantaneous speed
4. balanced force, net force
5. force, inertia
6. acceleration, velocity
7. velocity, instantaneous speed
8. force, net force
9. force, acceleration

Chapter 2 Assessment

Checking Concepts

Choose the word or phrase that best answers the question.

1. Which of the following do you calculate when you divide the total distance traveled by the total travel time?
A) average speed
B) constant speed
C) variable speed
D) instantaneous speed

2. What is the tendency for an object to resist any change in its motion called?
A) net force
B) acceleration
C) balanced force
D) inertia

3. Which of the following is a proper unit of acceleration?
A) s/km^2
B) km/h
C) m/s^2
D) cm/s

4. Which of the following is not used in calculating acceleration?
A) initial velocity
B) average speed
C) time interval
D) final velocity

5. In which of the following conditions does the car NOT accelerate?
A) A car moves at 80 km/h on a flat, straight highway.
B) The car slows from 80 km/h to 35 km/h.
C) The car turns a corner.
D) The car speeds up from 35 km/h to 80 km/h.

6. Which term below best describes the forces on an object with a net force of zero?
A) inertia
B) balanced forces
C) acceleration
D) unbalanced forces

7. How can speed be defined?
A) acceleration/time
B) change in velocity/time
C) distance/time
D) displacement/time

8. Which of the following objects has the greatest inertia?
A) a car parked on the side of the road
B) a baseball during a pop fly
C) a computer sitting on a desk
D) a woman running on a track

9. A man drives 3 km east from home to the store and then 2 km west to a friend's house. What is his displacement from his starting point at home?
A) 1 km west
B) 1 km east
C) 5 km west
D) 5 km east

10. Which answer best describes why a passenger who is not wearing a seat belt will likely hit the windshield in a head-on collision?
A) forces acting on the windshield
B) inertia of the unbelted person
C) acceleration of the car
D) gravity taking over

Thinking Critically

11. A cyclist must travel 800 km. How many days will the trip take if the cyclist travels 8 h/day at an average speed of 16 km/h?

12. A satellite's speed is 10,000 m/s. After 1 min, it is 5,000 m/s. What is the satellite's acceleration?

13. A cyclist leaves home and rides due east for a distance of 45 km. She returns home on the same bike path. If the entire trip takes 4 h, what is her average speed? What is her displacement?

14. The return trip of the cyclist in question 13 took 30 min longer than her trip east, although her total time was still 4 h. What was her velocity in each direction?

Chapter 2 Assessment

Developing Skills

15. Measuring in SI Which of the following represents the greatest speed: 20 m/s, 200 cm/s, or 0.2 km/s? Here's a hint: *Express all three in m/s and then compare.*

16. Recognizing Cause and Effect Acceleration can occur when a car is moving at constant speed. What must cause this acceleration?

17. Making and Using Graphs The following data were obtained for two runners. Make a distance-time graph that shows the motion of both runners. What is the average speed of each runner? Which runner stops briefly? During what time interval do Sally and Alonzo run at the same speed?

Distance-Time for Runners

Time (s)	1	2	3	4
Sally's Distance (m)	2	4	6	8
Alonzo's Distance (m)	1	2	2	4

Performance Assessment

18. Poster Research the current safety features available in cars, including seat belts, improved door locks, collapsible steering columns, and air bags. Use this information, along with the statistics you found in the Science Online Feature regarding safety in cars, to make a poster on the benefits of using seat belts.

Technology

Go to the Glencoe Science Web site at **science.glencoe.com** or use the **Glencoe Science CD-ROM** for additional chapter assessment.

Test Practice

Four runners ran for 40 min. The following table represents the distance each runner covered in that time:

Runners' Distances

Runner Name	Distance Covered (km)
Rosemarie (R)	12.5
Sam (S)	8.9
Jake (J)	10.5
Theresa (T)	7.8

Study the table and answer the following questions.

1. Which of these graphs best represents these data?

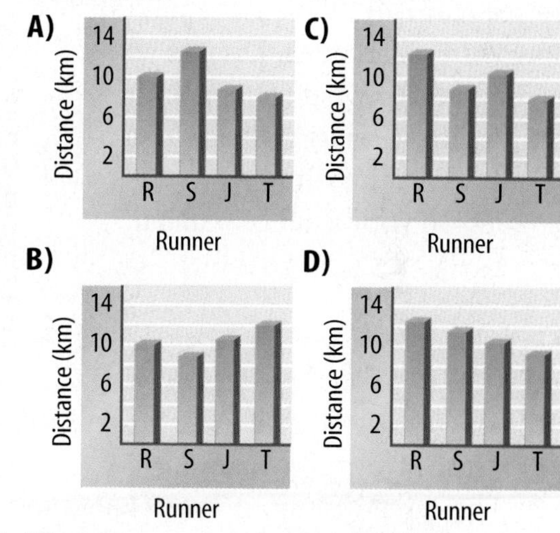

2. The average speed of each runner can be determined by dividing the total distance by the total travel time. Which runner has the fastest average speed?

F) Rosemarie
G) Sam
H) Jake
J) Theresa

CHAPTER 3

Forces

Crash! What brought this vehicle to its sudden stop? How did the impact affect the car and its driver? Automobile manufacturers consider factors like mass and acceleration to determine the forces on a driver during a crash. This helps them predict whether the driver would survive or be injured. In this chapter, you will learn how forces such as gravity and friction affect motion.

What do you think?

Science Journal Look at the picture below with a classmate. Discuss what this might be or what is happening. Here's a hint: *Gravity gives this object its structure.* Write your answer or best guess in your Science Journal.

What holds you to Earth, pulls footballs back to the ground, and keeps the Moon in orbit? The force of gravity, of course. But did you know that all objects near Earth's surface would fall with the same acceleration due to gravity? If this is true, why do bowling balls fall faster than feathers? Explore free-falling objects in this activity.

Observe free-falling objects

1. Drop a softball from a height of 2.5 m and use a stopwatch to measure the time it takes for the softball to fall the given distance.
2. Repeat the procedure using a tennis ball, a piece of paper crumpled into a ball, and a flat sheet of paper.

Observe

In your Science Journal, write a paragraph comparing the time it took for the four items to drop the 2.5 m. Infer why the crumpled paper fell faster than the flat sheet of paper even though they have the same mass.

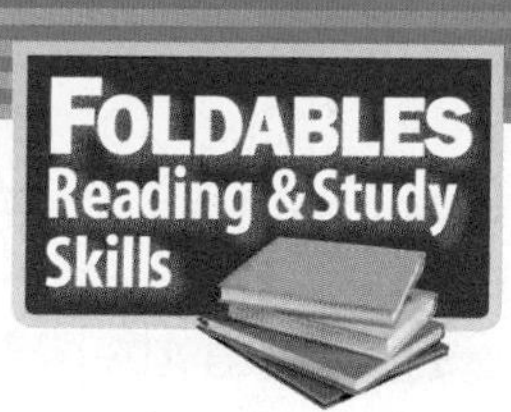

Before You Read

Making a Compare-and-Contrast Study Fold **Make the following Foldable to help you see how the three types of friction are similar and different.**

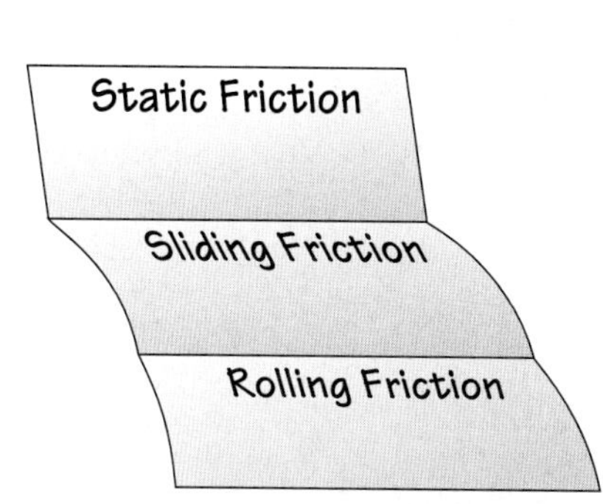

1. Place a sheet of paper in front of you so the short side is at the top. Fold the top of the paper down and the bottom up.
2. Open the paper and label the three rows *Static Friction, Sliding Friction,* and *Rolling Friction.*
3. Before you read the chapter, write the definition of each type of friction next to it. As you read the chapter, write more information about each type.

SECTION

Newton's Second Law

As You Read

What You'll Learn

- **Explain** how force, mass, and acceleration are related.
- **Describe** the three different types of friction.
- **Observe** the effects of air resistance on falling objects.

Vocabulary

Newton's second law of motion
friction

Why It's Important

Newton's second law explains why some objects move and some objects don't.

Force, Mass, and Acceleration

Newton's first law of motion states that the motion of an object, such as the volleyball in **Figure 1**, changes only if an unbalanced force acts on it. Force and motion are connected. How does force cause motion to change?

Force and Acceleration What's different about throwing a ball as hard as you can and tossing it gently? When you throw hard, you exert a much greater force on the ball. How is the motion of the ball different in each case?

In both cases, the ball was at rest in your hand before it began to move. However, when you throw hard, the ball has a greater velocity when it leaves your hand than it does when you throw gently. Thus the hard-thrown ball has a greater change in velocity, and the change occurs over a shorter period of time. Recall that acceleration is the change in velocity divided by the time it takes for the change to occur. So, a hard-thrown ball has a greater acceleration than a gently thrown ball.

For any object, the greater the force is that's applied to it, the greater its acceleration will be. This is true for anything from the blood cells swirling through your body to the galaxies swirling through outer space.

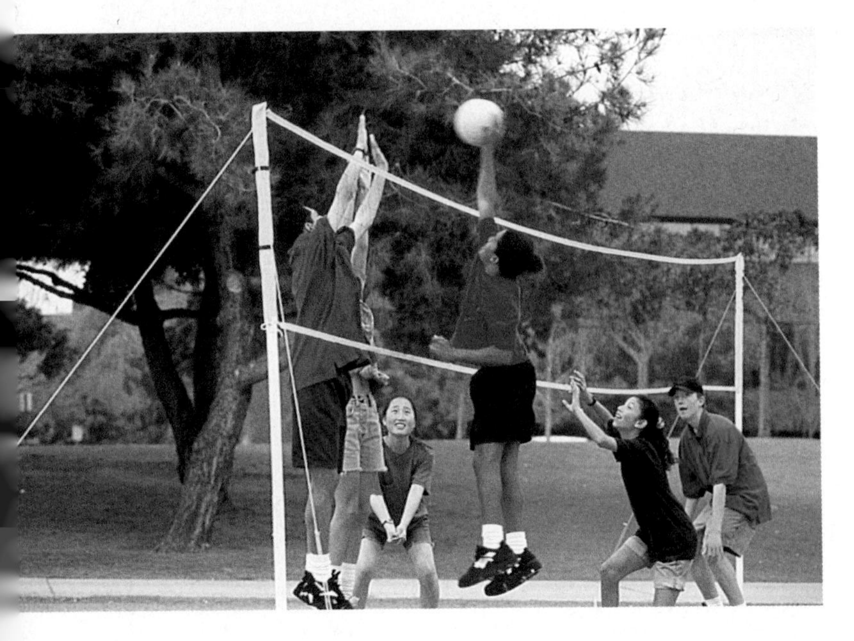

Figure 1
The motion of the volleyball changes when an unbalanced force is applied to the ball.

Force and Mass If you throw a softball and a baseball as hard as you can, why don't they have the same speed? The difference is due to their masses. The softball has a mass of about 0.20 kg, but a baseball's mass is about 0.14 kg. The softball has less velocity after it leaves your hand than the baseball does, even though you exerted the same force. If it took the same amount of time to throw both balls, the softball has less acceleration. The acceleration of an object depends on its mass as well as the force exerted on it. Force, mass, and acceleration are connected.

Newton's Second Law

Newton's second law of motion describes how force, mass, and acceleration are connected. Recall that if more than one force acts on an object, the forces combine to form a net force. According to **Newton's second law of motion,** the net force acting on an object causes the object to accelerate in the direction of the net force. The acceleration of an object is determined by the size of the net force and the mass of the object according to the equation:

$$\text{acceleration} = \frac{\text{net force}}{\text{mass}}$$

If a stands for the acceleration, F for the net force, and m for the mass, Newton's second law can be written as follows:

$$a = \frac{F}{m}$$

In SI units, the unit of mass is the kilogram (kg), and the unit of acceleration is meters per second squared (m/s^2). So, according to the second law of motion, force has the units $kg \times m/s^2$. The unit $kg \times m/s^2$ is called the newton (N).

Research Visit the Glencoe Science Web site at **science.glencoe.com** for information about how athletic trainers analyze the motions of athletes so they make the best use of Newton's second law. Select a sport and report to your class about some of the findings in that sport.

Math Skills Activity

Calculating Acceleration

You are pushing a friend on a sled. You push with a force of 40 N. Your friend and the sled together have a mass of 80 kg. Ignoring friction, what is the acceleration of your friend on the sled?

1. *This is what you know:* force: $F = 40$ N
   mass: $m = 80$ kg
2. *This is what you need to know:* acceleration: a
3. *This is the equation you need to use:* $a = F/m$
4. *Substitute the known values:* $a = 40 \text{ N}/80 \text{ kg} = 0.5 \text{ m/s}^2$

   Check your answer by multiplying it by the mass. Do you calculate the same force that was given?

Practice Problem

A student pedaling a bicycle applies a net force of 200 N. The mass of the rider and the bicycle is 50 kg. What is the acceleration of the bicycle and the rider?

For more help, refer to the Math Skill Handbook.

Figure 2
The force exerted on the ball by the tennis racket may be more than 500 times greater than the force needed to pick the ball up.

Using Newton's Second Law The second law can be used to calculate the net force on an object if the accleration and mass are known. Suppose a tennis ball like the one in **Figure 2** is accelerated by a tennis racket for five thousandths of a second—the time the racket is in contact with the ball. Because this time period is so short, a ball that leaves the racket at a speed of 100 km/h would have undergone an acceleration of about 5,500 m/s^2. How much force would the tennis racket have to exert to give the ball this acceleration? The ball has a mass of 0.06 kg, so by Newton's second law the force would have to be:

$$F = ma = (0.06 \text{ kg})(5{,}500 \text{ m/s}^2) = 330 \text{ N}$$

Friction

Suppose you give a skateboard a push with your hand. According to Newton's first law of motion, if no forces are acting on a moving object, it continues to move in a straight line with constant speed. What happens to the motion of the skateboard after it leaves your hand? Does it continue to move in a straight line with constant speed?

You know the answer. The skateboard gradually slows down and finally stops. Recall that when an object slows down, its velocity changes. If its velocity changes, it is accelerating. And if an object is accelerating, a net force must be acting on it.

The force that slows the skateboard and brings it to a stop is friction. **Friction** is the force that opposes motion between two surfaces that are touching each other. The amount of friction between two surfaces depends on two factors—the kinds of surfaces and the force pressing the surfaces together.

Reading Check ***The amount of friction between two objects depends on what two factors?***

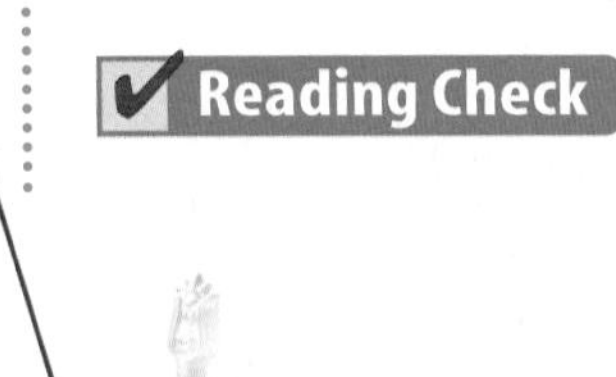

Figure 3
While surfaces might look and even feel smooth, they can be rough at the microscopic level.

What causes friction? Would you believe the surface of a highly polished piece of metal is rough? **Figure 3** shows a microscopic view of the dips and bumps on the surface of a polished silver teapot. If two surfaces, such as two pieces of silver, are pressed tightly together, welding, or sticking, occurs in those areas where the highest bumps come into contact with each other. These areas where the bumps stick together are called microwelds and are the source of friction.

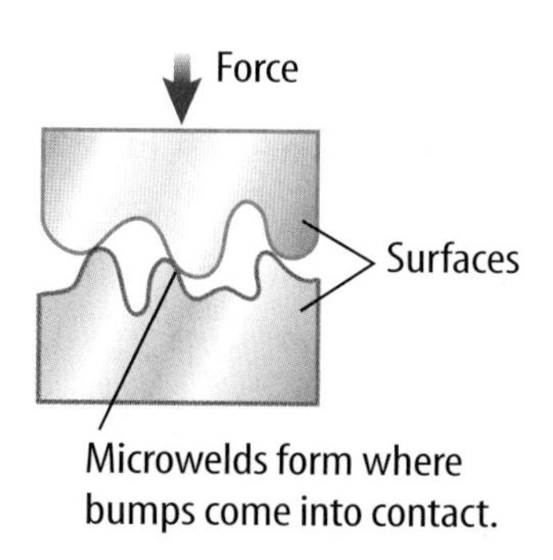

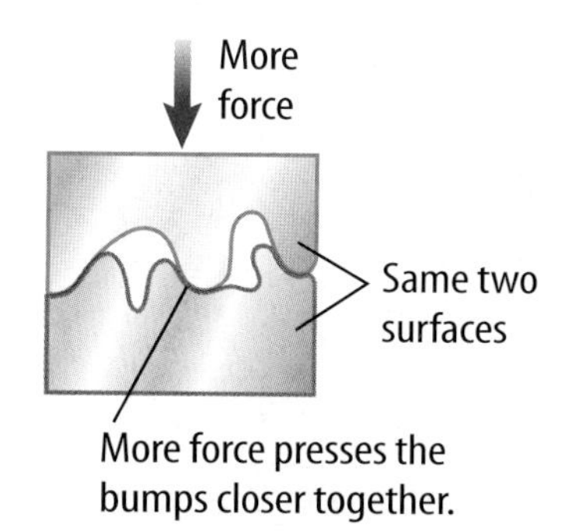

Figure 4
Friction is due to microwelds formed between two surfaces. The larger the force pushing the two surfaces together is, the stronger the microwelds will be.

Sticking Together The stronger the force pushing the two surfaces together is, the stronger these microwelds will be, because more of the surface bumps will come into contact, as shown in **Figure 4.** To break these microwelds and move one surface over the other, a force must be applied.

Static Friction Suppose you have filled a cardboard box, like the one in **Figure 5,** with books and want to move it. It's too heavy to lift, so you start pushing on it, but it doesn't budge. Is that because the mass of the box is too large? If the box doesn't move, then it has zero acceleration. According to Newton's second law, if the acceleration is zero, then the net force on the box is zero. Another force that cancels your push must be acting on the box. That force is friction due to the microwelds that have formed between the bottom of the box and the floor. This type of friction is called static friction. Static friction is the friction between two surfaces that are not moving past each other. In this case, your push is not large enough to break the microwelds, and the box remains stuck to the floor.

Mini LAB

Comparing Friction

Procedure

1. Position an **ice cube, rock, eraser, wood block,** and square of **aluminum foil** at one end of a **metal** or **plastic tray.**
2. Slowly lift the end of the tray with the items.
3. Have a partner use a **metric ruler** to measure the height at which each object slides to the other end of the tray. Record the heights in your **Science Journal.**

Analysis

1. List the height at which each object slid off the tray.
2. Why did the objects slide off at different heights?
3. What type of friction acted on each object?

Figure 5
The box doesn't move because static friction cancels the applied force.

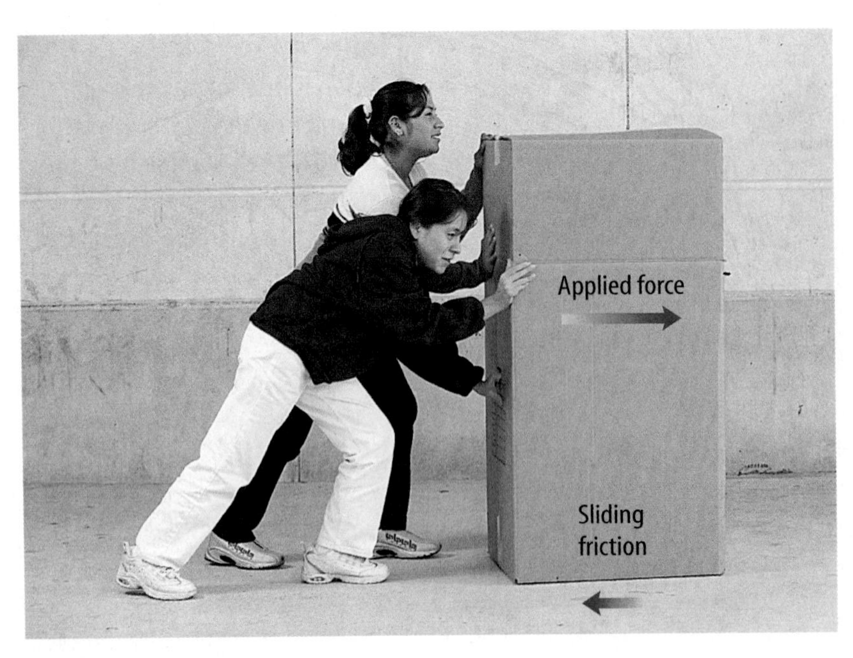

Figure 6
Sliding friction acts in the direction opposite the motion of the sliding box.

Sliding Friction To help you move the box, you ask a friend to push with you, as in **Figure 6.** Pushing together, the box starts to move, but it doesn't move easily. Also, if you stop pushing, it quickly comes to a stop. It might even seem as if someone is pushing on it from the opposite direction. But, by exerting enough force, you have broken the microwelds and started the box moving. However, as the box slides across the floor, another force—sliding friction—opposes the motion of the box. Sliding friction is the force that opposes the motion of two surfaces sliding past each other. Sliding friction is caused by microwelds constantly breaking and then forming again as the box slides along the floor. To keep the box moving, you must continually apply a force to overcome sliding friction.

✔ **Reading Check** ***What's the difference between sliding friction and static friction?***

Rolling Friction You might have watched a car stuck in snow, ice, or mud spinning its wheels. The driver steps on the gas, but the wheels just spin without the car moving. The car doesn't move when the wheels are on the slippery surface because there is not enough friction between the tires and the snow, ice, or mud. One way to make the car move might be to spread sand or gravel under the wheels. By spreading sand or gravel, the friction between the tires and the surface is increased. The friction between a rolling object and the surface it rolls on is rolling friction. As you can see in **Figure 7,** because of rolling friction, the wheels of the train rotate when they come in contact with the track rather than sliding over it. Rolling friction is due partly to the microwelds between a wheel and the surface it rolls over. Microwelds break and then reform as the wheel rolls over the surface. Rolling friction is usually much less than static or sliding friction. This is why it's easier to pull a load in a wagon rather than dragging it along the ground.

Figure 7
Rolling friction is what makes a train's wheels turn on the tracks or a car's wheels turn on the road.

Air Resistance

When an object falls toward Earth, it is pulled downward by the force of gravity. However, another force called air resistance acts on objects that fall through the air. Imagine dropping two identical plastic bags except that one is crumpled into a ball and one is spread out. The crumpled bag falls faster than the spread-out bag. So, the acceleration of the spread-out bag must be less than that of the crumpled bag. Therefore, the net force acting on the spread-out bag is less. This is because the force of air resistance is in the opposite direction to the force of gravity for both bags and is greater on the spread-out bag.

Air resistance affects anything that moves in Earth's atmosphere. Like friction, air resistance acts in the direction opposite to that of the object's motion. In the case of the two falling bags, air resistance is pushing up as gravity is pulling down, as shown in **Figure 8.**

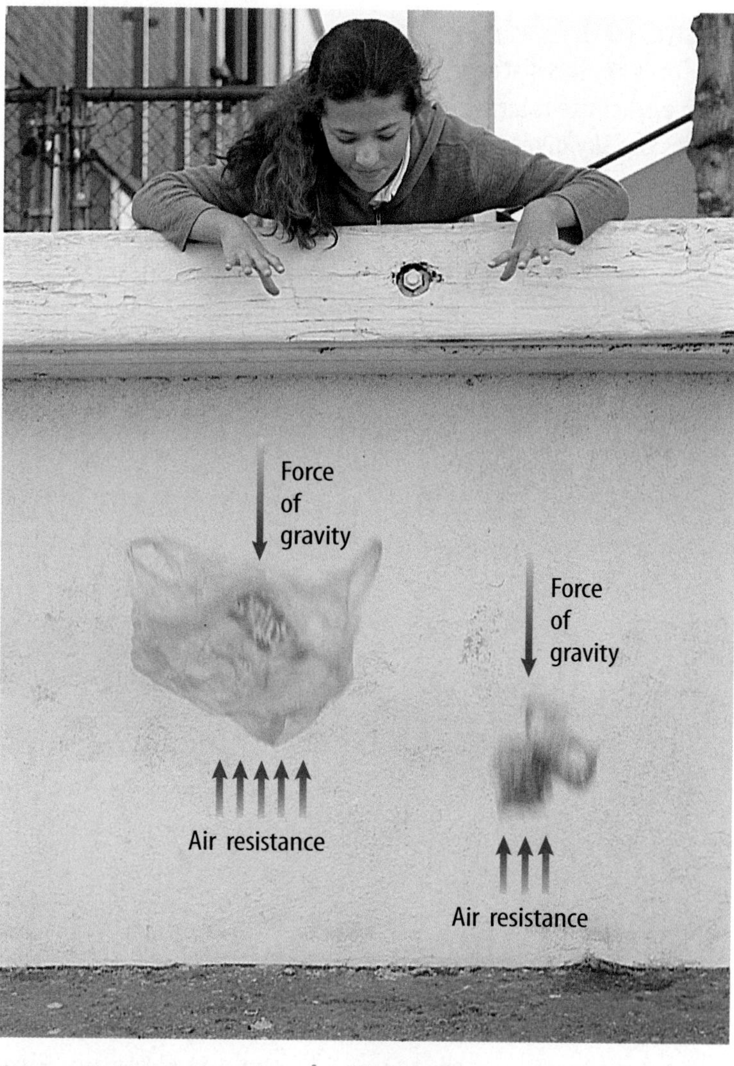

Figure 8
Because of its greater surface area, the spread-out bag has much more air resistance acting on it as it falls.

The amount of air resistance on an object depends on the speed, size, and shape of the object. Air resistance, not the amount of mass in the object, is why feathers, leaves, and pieces of paper fall more slowly than pennies, acorns, and apples. If no air resistance is present, then a feather and an apple fall at the same rate, as shown in **Figure 9.**

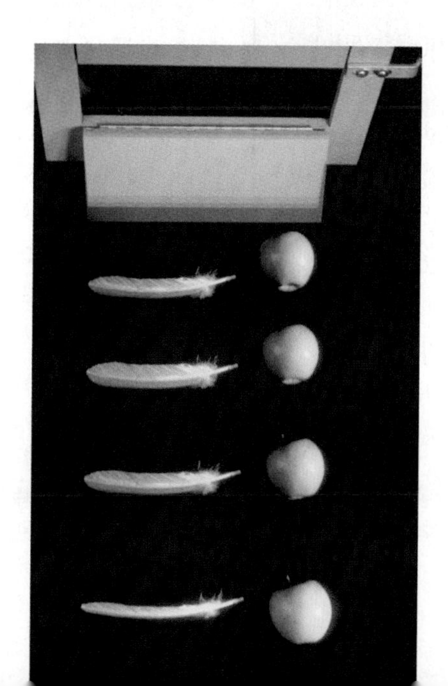

Figure 9
The apple and feather are falling in a vacuum. Because there is no air resistance, they both fall at the same rate.

Figure 10
The force of air resistance on an open parachute is large. This causes the sky diver to fall slowly.

Terminal Velocity The force of air resistance increases with speed. As an object falls, it accelerates and its speed increases. So the force of air resistance increases until it becomes large enough to cancel the force of gravity. Then the forces on the falling object are balanced, and the object no longer accelerates. It then falls with a constant speed called the terminal velocity. This terminal velocity is the highest velocity that a falling object will reach. A low terminal velocity enables a sky diver, such as the one shown in **Figure 10,** to land safely on the ground.

Section  Assessment

1. What is Newton's second law of motion?
2. The same force acts on two objects with different masses. Why does the object with less mass have a larger acceleration?
3. What is the difference between static friction and sliding friction?
4. A squirrel runs across a branch on an oak tree and knocks an acorn and a leaf loose. Why does the acorn hit the ground first?
5. **Think Critically** To reduce the friction in a metal door hinge, you might try coating it with oil. Why does oil reduce friction?

Skill Builder Activities

6. **Drawing Conclusions** Boxes of equal size are resting on the floor. Applying the same force on each box, you find that you can accelerate the first one 1 m/s^2, the second 4 m/s^2, and the third 6 m/s^2. What can you infer about the total mass of each box? What other factors might have influenced your results? **For more help, refer to the Science Skill Handbook.**
7. **Communicating** Describe activities in which friction would be useful. **For more help, refer to the Science Skill Handbook.**

SECTION 2 Gravity

The Law of Gravitation

There's a lot about you that's attractive. At this moment, you are exerting an attractive force on everything around you—your desk, your classmates, even the planet Jupiter millions of kilometers away. It's the attractive force of gravity.

Anything that has mass is attracted by the force of gravity. According to the **law of gravitation,** any two masses exert an attractive force on each other. The attractive force depends on the mass of the two objects and the distance between them. This force increases as the mass of either object increases, as shown in **Figure 11A.** Also, **Figure 11B** shows that the force of gravity increases as the objects move closer.

You can't feel any gravitational attraction between you and this book because the force is weak. Only Earth is close enough and has a large enough mass that you can feel its gravitational attraction. While the Sun has much more mass than Earth, the Sun is too far away to exert a noticeable gravitational attraction on you. And while this book is close, it doesn't have enough mass to exert an attraction you can feel.

Gravity—A Basic Force Gravity is one of the four basic forces. The other basic forces are the electromagnetic force, the strong nuclear force, and the weak nuclear force. The nuclear forces only act on particles in the nuclei of atoms. Electricity and magnetism are caused by the electromagnetic force. Chemical interactions between atoms and molecules also are due to the electromagnetic force.

As You Read

What You'll Learn

- **Describe** gravitational force.
- **Distinguish** between mass and weight.
- **Explain** why objects that are thrown or shot will follow a curved path.
- **Compare** motion in a straight line with circular motion.

Vocabulary

law of gravitation
weight
centripetal acceleration
centripetal force

Why It's Important

No matter where you might be in the universe, gravity will affect you.

A If the mass of either of the objects increases, the gravitational force between them increases.

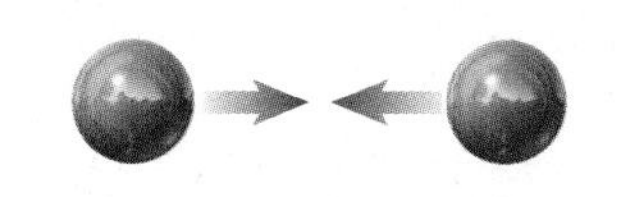

B If the objects are closer together, the gravitational force between them increases.

Figure 11
The gravitational force between two objects depends on their masses and the distance between them.

Figure 12
Gravity is at work in the formation of galaxies like this spiral galaxy.

The Range of Gravity You might think that a star in another galaxy is too far away to exert a gravitational force on you, but you'd be wrong. Despite the distance between two objects, the gravitational attraction between them never disappears. Gravity is a long-range force. All the stars in a galaxy, such as the one shown in **Figure 12,** exert a gravitational force on each other. These forces help give the galaxy its shape. In fact, a gravitational force exists between all matter in the universe. Gravity is the force that gives the universe its structure.

Data Update Visit the Glencoe Science Web site at **science.glencoe.com** for data comparing the gravitational accelerations objects would experience on different planets. Discuss with your class why the gravitational acceleration is greater on some planets than on others.

Gravitational Acceleration

Near Earth's surface, the gravitational attraction of Earth causes all falling objects to have an acceleration of 9.8 m/s^2. By Newton's second law, the net force, mass, and acceleration are related according to the following formula:

$$F = ma$$

According to the second law, the force on an object that has an acceleration of 9.8 m/s^2 is as follows:

$$F = m \times 9.8 \text{ m/s}^2$$

This is the force of gravity on an object near Earth's surface. This force depends only on the object's mass. A force has a direction associated with it. The force of Earth's gravity is always downward.

When an object is influenced only by the force of gravity, it is said to be in free fall. Suppose you were to drop a bowling ball and a marble from a bridge at the same time. Which would hit the water below first? Would it be the bowling ball because it has more mass?

Inertia and Gravity It's true that the force of gravity would be greater on the bowling ball because of its larger mass. But the larger mass also means the bowling ball has more inertia, so more force is needed to change its velocity. The gravitational force on the marble is smaller because it has less mass, but the inertia of the marble is smaller, too, so less force is needed to change its velocity. As a result, all objects fall with the same acceleration, no matter how large or small their mass is. Although the blue ball in **Figure 13** is more massive than the green one, they fall at the same rate.

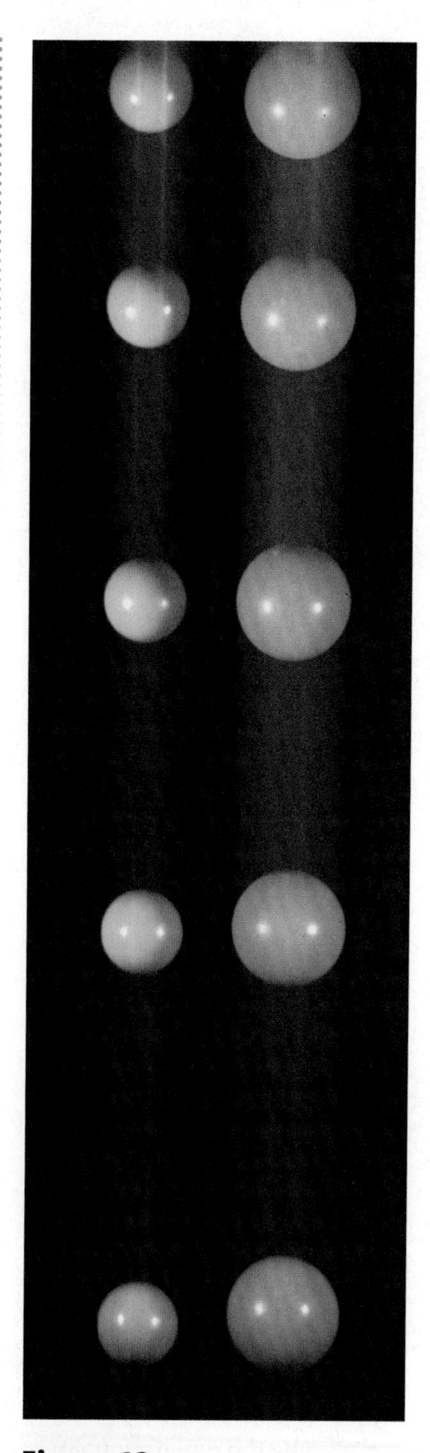

Figure 13
High-speed photography shows that two balls of different masses fall at the same rate.

Weight

If you are standing on the floor of your classroom, your acceleration is zero. According to Newton's second law, if your acceleration is zero, the net force on you must be zero. Does this mean Earth's gravitational attraction for you has disappeared? No. Earth still pulls you downward, but the floor also exerts an upward force that keeps you from falling.

Whether you are standing, jumping, or falling, Earth exerts a gravitational force on you. The gravitational force exerted on an object is called the object's **weight.** The symbol *W* stands for the weight. You can find gravitational force, or weight, using Newton's second law, as follows:

$$\text{gravitational force} = \text{mass} \times \begin{bmatrix}\text{acceleration} \\ \text{due to gravity}\end{bmatrix}$$

Because the gravitational force is the same as the weight and the acceleration due to gravity on Earth is 9.8 m/s^2, this equation can be written as follows:

$$W = m \times 9.8 \text{ m/s}^2$$

In other words, a mass of 1 kg weighs 1 kg $\times$ 9.8 m/s^2, or 9.8 N. You could calculate your weight in newtons if you knew your mass. For example, a person with a mass of 50 kg would have a weight of 490 N. On Earth, a cassette tape weighs about 0.5 N, a backpack full of books weighs about 40 N, and a jumbo jet weighs about 3.4 million N.

✔ Reading Check *How much does a person with a mass of 70 kg weigh on Earth?*

Losing Weight What would happen to your weight if you were far from Earth? Recall that the gravitational attraction between two objects becomes weaker as they move farther apart. So if you were to travel away from Earth, your weight would decrease.

Weight and Mass Weight and mass are not the same. Weight is a force, and mass is a measure of the amount of matter an object contains. However, weight and mass are related. The greater an object's mass is, the stronger the gravitational force between the object and Earth is. So the more mass an object has, the more it will weigh at the same location.

The weight of an object usually means the gravitational force between the object and Earth. But objects can have different weights, depending on what's pulling on them. For example, a person weighing about 480 N on Earth would weigh only about 80 N on the Moon. Does this mean the astronaut in **Figure 14** would have less mass on the Moon than on Earth? The answer is no, the mass of the astronaut would be unchanged. His weight is less than on Earth because the Moon has less mass and exerts a weaker gravitational force. **Table 1** shows how an object's weight depends upon the object's location.

Figure 14
The astronaut was able to take longer steps on the Moon because the gravitational attraction on him there is less than on Earth.

Reading Check *How are weight and mass related?*

Weightlessness and Free Fall

You've probably seen pictures of astronauts and equipment floating inside the space shuttle. Any item that is not fastened down in the shuttle floats throughout the cabin. They are said to be experiencing the sensation of weightlessness.

To be nearly weightless, the astronauts would have to be far from Earth to be significantly free from the effects of its gravity. Even while orbiting 400 km above Earth, the force of gravity pulling on the shuttle is still about 90 percent as strong as it is at Earth's surface, so they are not weightless.

Table 1 Weight Comparison Table

Weight on Earth (N)	Weight on Other Bodies in the Solar System (N)				
	Moon	Venus	Mars	Jupiter	Saturn
75	12	68	28	190	87
100	17	90	38	254	116
150	25	135	57	381	174
500	84	450	190	1,270	580
2,000	333	1,800	760	5,080	2,320

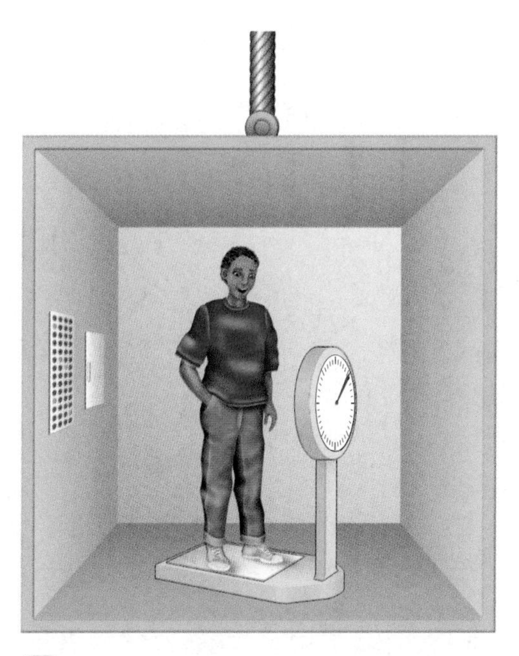

A When the elevator is stationary, the scale shows the boy's weight.

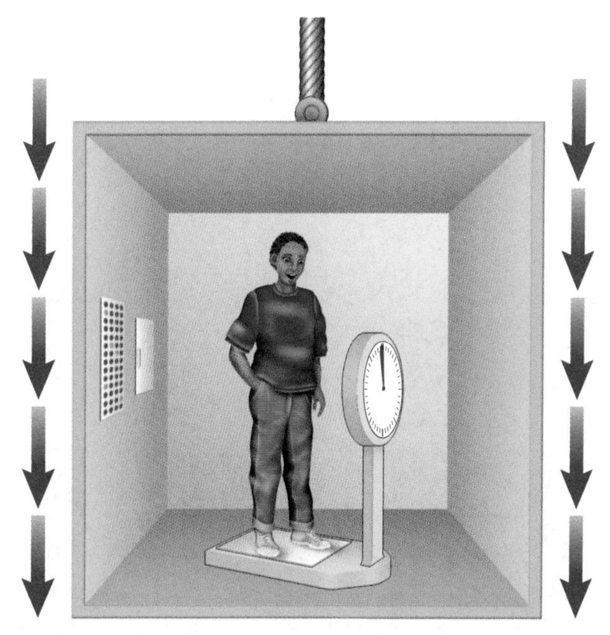

B If the elevator were falling, the scale would show a smaller weight.

Figure 15
The boy pushes down on the scale with less force when he and the scale are falling at the same rate.

Floating in Space So what does it mean to say that something is weightless in orbit? Think about how you measure your weight. When you stand on a scale, as in **Figure 15A,** you are at rest and the net force on you is zero. So the scale supports you and balances your weight by exerting an upward force. The dial on a scale shows the upward force exerted by the scale, which is your weight. Now suppose you stand on a scale in an elevator that is falling, as in **Figure 15B.** If you and the scale were in free fall, then you no longer would push down on the scale at all. The scale dial would say you have zero weight, even though the force of gravity on you hasn't changed.

Everything in the orbiting space shuttle is falling downward toward Earth at the same rate, in the same way you and the scale were falling in the elevator. Because objects in the shuttle have no force supporting them, they seem to be floating.

Projectile Motion

If you've tossed a ball to someone, you've probably noticed that thrown objects don't always travel in straight lines. They tend to curve downward. That's why quarterbacks, dart players, and archers aim above their targets. Anything that's thrown or shot through the air is called a projectile. Because of Earth's gravitational pull and their own inertia, projectiles follow a curved path. This is because they have horizontal and vertical velocities.

Apart from simply keeping your feet on the ground, gravity is important for life on Earth for other reasons, too. Because Earth has a sufficient gravitational pull, for example, it can hold around it the oxygen/nitrogen atmosphere necessary for sustaining life. Research other ways in which gravity has played a role in the formation of Earth.

Figure 16
The pitcher gives the ball a horizontal motion. Gravity, however, is pulling the ball down. The combination of these two motions causes the ball to move in a curved path.

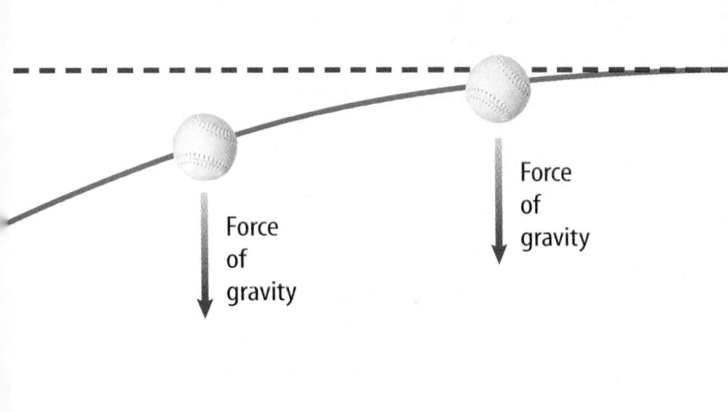

Figure 17
Time-lapse photography shows that each ball has the same acceleration downward, whether it's thrown or dropped.

Horizontal and Vertical Motions When you throw a ball, like the pitcher in **Figure 16,** the force from your hand makes the ball move forward. It gives the ball horizontal motion or motion parallel to Earth's surface. After you let go of the ball, no other force accelerates it forward, so its horizontal velocity is constant, if you ignore air resistance.

However, when you let go of the ball, something else happens. Gravity can now pull it downward, giving it vertical motion, or motion perpendicular to Earth's surface. Now the ball has constant horizontal velocity but increasing vertical velocity. Gravity exerts an unbalanced force on the ball, changing the direction of its path from only forward to forward and downward. The result of these two motions is that the ball appears to travel in a curve, even though its horizontal and vertical motions are completely independent of each other.

If you were to throw a ball as hard as you could from shoulder height in a perfectly horizontal direction, would it take longer to reach the ground than if you dropped a ball from the same height? Surprisingly, it won't. A thrown ball and one dropped will hit the ground at the same time. If you have a hard time believing this, **Figure 17** might help. The two balls have the same acceleration due to gravity—9.8 m/s^2 downward. How would a thrown ball's path look on the Moon?

Centripetal Force

Recall that acceleration is the rate of change of velocity due to a change in speed, direction, or both. Now, look at the path the ball follows as it travels through the pipe maze in **Figure 18.** The ball may accelerate in the straight sections of the pipe maze if it speeds up or slows down. However, when the ball enters a curve, even if its speed does not change, it is accelerating because its direction is changing. When the ball goes around a curve, the change in the direction of the velocity is toward the center of the curve. Acceleration toward the center of a curved or circular path is called **centripetal acceleration.** The word *centripetal* means to "move toward the center."

For the ball to be accelerating toward the center, an unbalanced force, called **centripetal force,** must be acting on it in a direction toward the center. The centripetal force acting on the ball running through the maze is exerted by the outside wall pushing against it and keeping it from going straight.

When a car rounds a sharp curve on a highway, the centripetal force is the friction between the tires and the road surface. If the road is icy or wet and the tires lose their grip, the centripetal force might not be enough to overcome the car's inertia. Then the car would keep moving in a straight line in the direction that it was traveling at the spot where it lost traction. Anything that moves in a circle, such as the people on the amusement park ride in **Figure 19,** is doing so because a centripetal force is accelerating it toward the center.

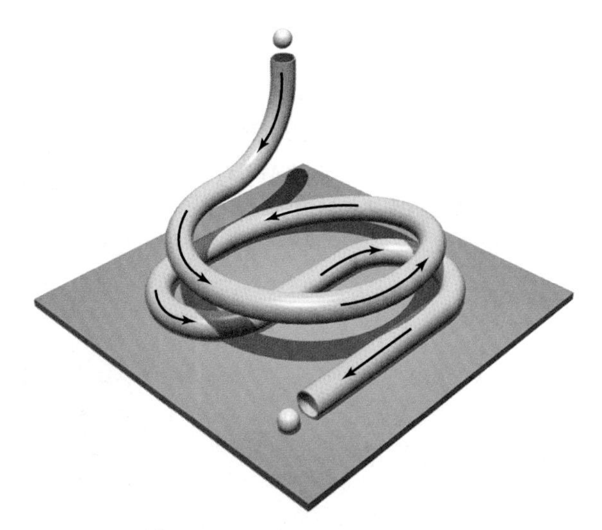

Figure 18
When the ball moves through the circular portions of the maze, it is accelerating because its velocity is changing. *Would you expect the ball to be traveling faster or slower if more curves were in the maze?*

Figure 19
Centripetal force keeps these riders moving in a circle.

Observing Centripetal Force

Procedure

1. Fill a **bucket** that has a secure handle with **water** to a level of about 3 cm.
2. Go outside and stand several meters away from any person or object.
3. Swing the bucket quickly in a circle. It should be upside down for just an instant.

Analysis

1. Why didn't you get wet?
2. What force did the bottom of the bucket exert on the water when you swung the bucket above you?
3. What would happen if you swung the bucket slowly?

Figure 20
The Moon would move in a straight line except that Earth's gravity keeps pulling it toward Earth. This gives the Moon its circular orbit.

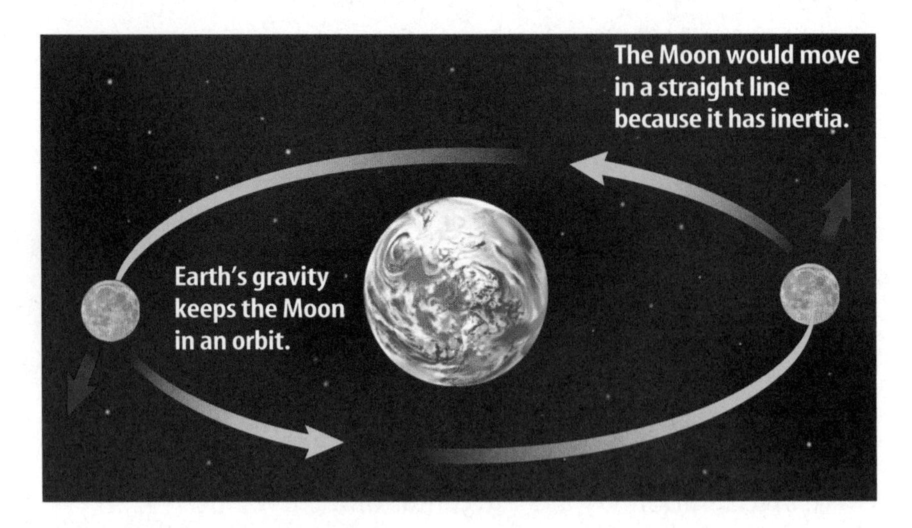

The Moon Is Falling A satellite is anything that moves around another body in a generally circular path called an orbit. Earth and the other planets are satellites because they orbit the Sun. The Moon is a natural satellite of Earth. The *International Space Station* is another of Earth's satellites. Because it was made here on Earth, it is an artificial satellite. Why do satellites move the way they do?

Imagine whirling an object tied to a string above your head. The string exerts a centripetal force on the object that keeps it moving in a circular path. In the same way, Earth's gravity exerts a centripetal force on the Moon that keeps it moving in a circular orbit, as shown in **Figure 20.**

Section  Assessment

1. What is gravity, and how does the size of two objects and the distance between them affect the gravitational force?
2. How is an object's mass different from its weight? Do objects always weigh the same? Explain.
3. What two motions contribute to the path of a projectile?
4. How do the planets stay in orbit around the Sun?
5. **Think Critically** Suppose that on a planet you weigh two-thirds as much as you do on Earth. Is the planet's mass greater than or less than Earth's? Explain.

Skill Builder Activities

6. **Using a Word Processor** Use a computer to make a table showing important characteristics of projectile motion, circular motion, and free fall. Table headings should include: *Kind of Motion, Shape of Path,* and *Laws or Forces Involved.* You can add other headings. **For more help, refer to the Technology Skill Handbook.**
7. **Communicating** Write a paragraph describing a situation in which you experienced something close to free fall or a feeling of weightlessness. Think about amusement park rides, elevators, athletic events, or even movie scenes. **For more help, refer to the Science Skill Handbook.**

SECTION 3

The Third Law of Motion

Newton's Third Law

Push against a wall and what happens? If the wall is sturdy enough, usually nothing happens. Now, think about what would happen if you pushed against a wall while wearing roller skates. You would go rolling backwards, of course. The harder you pushed, the more you would roll backwards. Your action on the wall produced a reaction—movement backwards. This is a demonstration of Newton's third law of motion.

Newton's third law of motion describes action-reaction pairs this way: When one object exerts a force on a second object, the second one exerts a force on the first that is equal in size and opposite in direction. Another way to say this is "to every action force there is an equal and opposite reaction force."

Action and Reaction When a force is applied in nature, a reaction to it occurs. When you jump on a trampoline, for example, you exert a downward force on the trampoline. The trampoline then exerts an equal force upward, sending you high into the air.

Action and reaction forces are acting on the two skaters in **Figure 21.** The male skater is pulling upward on the female skater, while the female skater is pulling downward on the male skater. The two forces are equal, but in opposite directions.

As You Read

What You'll Learn

- **Identify** when action and reaction forces occur.
- **Calculate** momentum.
- **Demonstrate** how momentum is conserved.

Vocabulary

Newton's third law of motion
momentum

Why It's Important

From walking across the floor to a rocket speeding through space, all motion occurs because every action has a reaction.

Figure 21
According to Newton's third law of motion, the two skaters exert forces on each other. The two forces are equal, but in opposite directions.

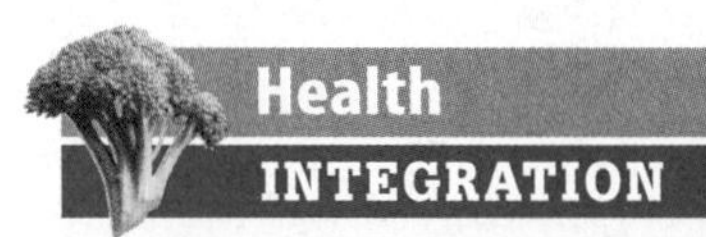

Astronauts who stay in outer space for extended periods of time may develop health problems. Their muscles, for example, may begin to weaken because they don't have to exert as much force to get the same reaction as they do on Earth. A branch of medicine called space medicine deals with the possible health problems that astronauts may experience. Research some other health risks that are involved in going into outer space. Do trips into outer space have any positive health benefits?

How You Move If action and reaction forces are equal, you might wonder how some things ever happen. For example, how does a swimmer move through the water in a pool if each time she pushes on the water, the water pushes back on her? An important point to remember when dealing with Newton's third law is that *action-reaction forces are acting on different objects.* Thus, even though the forces are equal, they are not balanced because they act on different objects. In the case of the swimmer, as she "acts" on the water, the "reaction" of the water pushes her forward. Thus, a net force, or unbalanced force, acts on her so a change in her motion occurs. Why is it harder for a swimmer to swim against a tide?

✔ Reading Check ***How is a swimmer able to move in the water?***

Rocket Propulsion Suppose you were standing on skates holding a softball. You exert a force on the softball when you throw the softball. According to Newton's third law, the softball exerts a force on you. This force pushes you in the direction opposite the softball's motion. Rockets use the same principle to move even in the vacuum of outer space. In the rocket engine, burning fuel produces hot gases. The rocket engine exerts a force on these gases and causes them to escape out the back of the rocket. By Newton's third law, the gases exert a force on the rocket and push it forward. The car in **Figure 22** uses a rocket engine to propel it forward. **Figure 23** shows how rockets move through space.

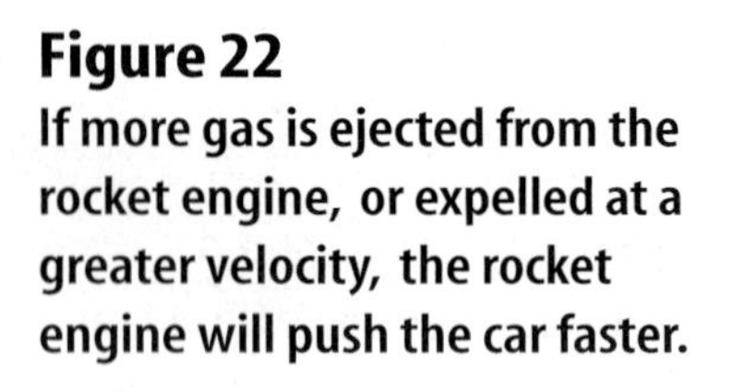

Figure 22
If more gas is ejected from the rocket engine, or expelled at a greater velocity, the rocket engine will push the car faster.

NATIONAL GEOGRAPHIC

VISUALIZING ROCKET MOTION

Figure 23

On the afternoon of July 16, 1969, *Apollo 11* lifted off from Cape Kennedy, Florida, bound for the Moon. Eight days later, the spacecraft returned to Earth, splashing down safely in the Pacific Ocean. The motion of the spacecraft to the Moon and back is governed by Newton's laws of motion.

▲ As *Apollo* rises, it burns fuel and ejects its rocket booster engines. This decreases its mass, and helps *Apollo* move faster. This is Newton's second law in action: As mass decreases, acceleration can increase.

◀ *Apollo 11* roars toward the Moon. At launch, a rocket's engines must produce enough force and acceleration to overcome the pull of Earth's gravity. A rocket's liftoff is an illustration of Newton's third law: For every action there is an equal and opposite reaction.

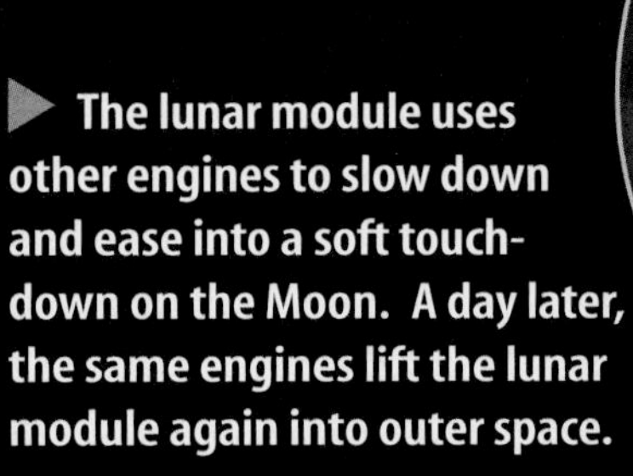

▶ The lunar module uses other engines to slow down and ease into a soft touchdown on the Moon. A day later, the same engines lift the lunar module again into outer space.

▲ After the lunar module returns to *Apollo,* the rocket fires its engines to set it into motion toward Earth. The rocket then shuts off its engines, moving according to Newton's first law. As it nears Earth, the rocket accelerates at an increasing rate because of Earth's gravity.

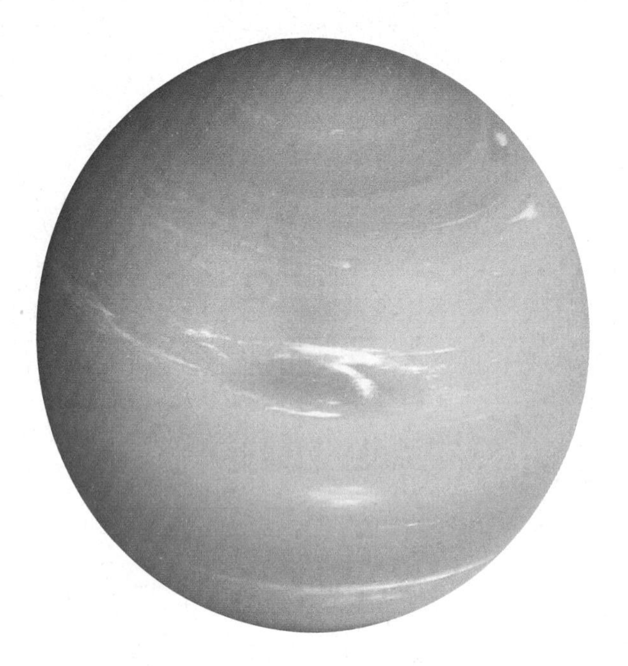

Finding Planets with Newton's Laws

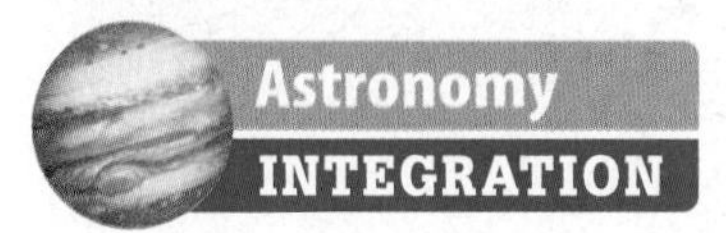

The gravitational force between Earth and the Sun causes Earth to orbit the Sun. However, Earth's orbit is also affected by the gravitational pulls of the other planets in the solar system. Each planet pulls on Earth with a force determined by its mass and its distance from Earth. In the same way, the orbit of every planet in the solar system is affected by the gravitational pulls from all the other planets.

In the 1840s, the most distant planet known was Uranus. Astronomers noticed that its orbit couldn't be explained by the forces exerted by the Sun and the other known planets. They concluded that there must be another planet affecting the orbit of Uranus that hadn't been discovered. Using Newton's laws of motion, Urbain Jean Leverrier and John Adams independently calculated where it must be located. The planet, shown in **Figure 24,** was found in 1846 where Leverrier and Adams said it would be, and named it Neptune.

Figure 24
The location of the planet Neptune was predicted correctly using Newton's laws.

Momentum

You know that a slow-moving bicycle is easier to stop than a fast-moving one. Also, a slow-moving bicycle is easier to stop than a car traveling at the same speed. Increasing either the speed or mass of an object makes it harder to stop.

A moving object has a property called momentum that is related to how much force is needed to change its motion. The **momentum** of an object is the product of its mass and velocity. Momentum is given the symbol p and can be calculated with this equation:

$$\text{momentum} = \text{mass} \times \text{velocity}$$

$$p = m \times v$$

The unit for momentum is kg m/s. Notice that momentum has a direction because velocity has a direction.

The two trucks in **Figure 25** might have the same velocity, but the bigger truck has more momentum because of its greater mass. An archer's arrow can have a large momentum because of its high velocity, even though its mass is small. A walking elephant may have a low velocity, but because of its large mass, it has a large momentum.

Force and Changing Momentum If you catch a fast-moving baseball, your hand might sting, even if you use a baseball glove. Your hand stings because the baseball exerted a force on your hand when it came to a stop, and its momentum changed.

Recall that acceleration is the difference between the initial and final velocity, divided by the time. Also, from Newton's second law, the net force on an object equals its mass times its acceleration. By combining these two relationships, Newton's second law can be written in this way:

$$F = (mv_f - mv_i)/t$$

In this equation mv_f is the final momentum and mv_i is the initial momentum. So the equation says that the net force exerted on an object can be calculated by dividing its change in momentum by the time over which the change occurs. When you catch a ball, your hand exerts a force on the ball that stops it. Here the final velocity is zero. The force depends on the mass and speed of the ball and how long it takes to come to a stop.

Law of Conservation of Momentum The momentum of an object doesn't change unless its mass, velocity, or both change. Momentum, however, can be transferred from one object to another. Consider the game of pool shown in **Figure 26.** Before the game starts, all the balls are motionless. The total momentum of the balls is, therefore, zero.

What happens when the cue ball hits the group of balls that are motionless? The cue ball slows down and the rest of the balls begin to move. The total momentum of all the balls just before and after the collision would be the same. The momentum the group of balls gained is equal to the momentum that the cue ball lost. If no other forces act on the balls, their total momentum is conserved—it isn't lost or created. This is the law of conservation of momentum—if a group of objects exerts forces only on each other, their total momentum doesn't change.

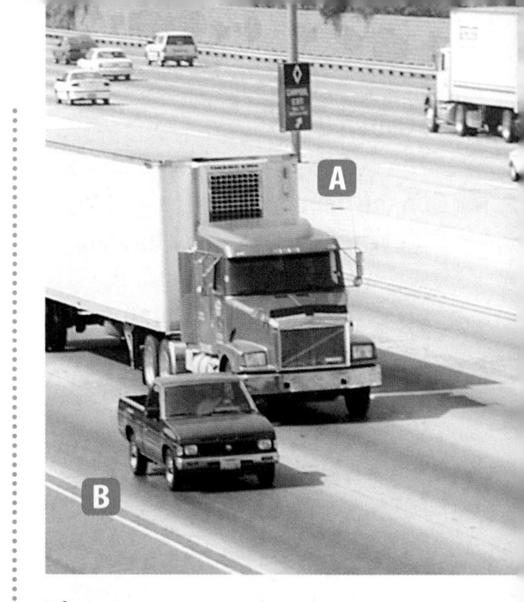

Figure 25
Suppose both trucks have the same speed. Truck A has more momentum than the smaller truck B because the larger truck has more mass.
Under what conditions would the smaller truck have a momentum greater than the big truck?

Figure 26
Momentum is transferred in collisions.
A At the start, the cue ball has all the momentum. The other balls have no momentum because they are not moving.
B When the cue ball strikes the other balls, it transfers some of its momentum to them.

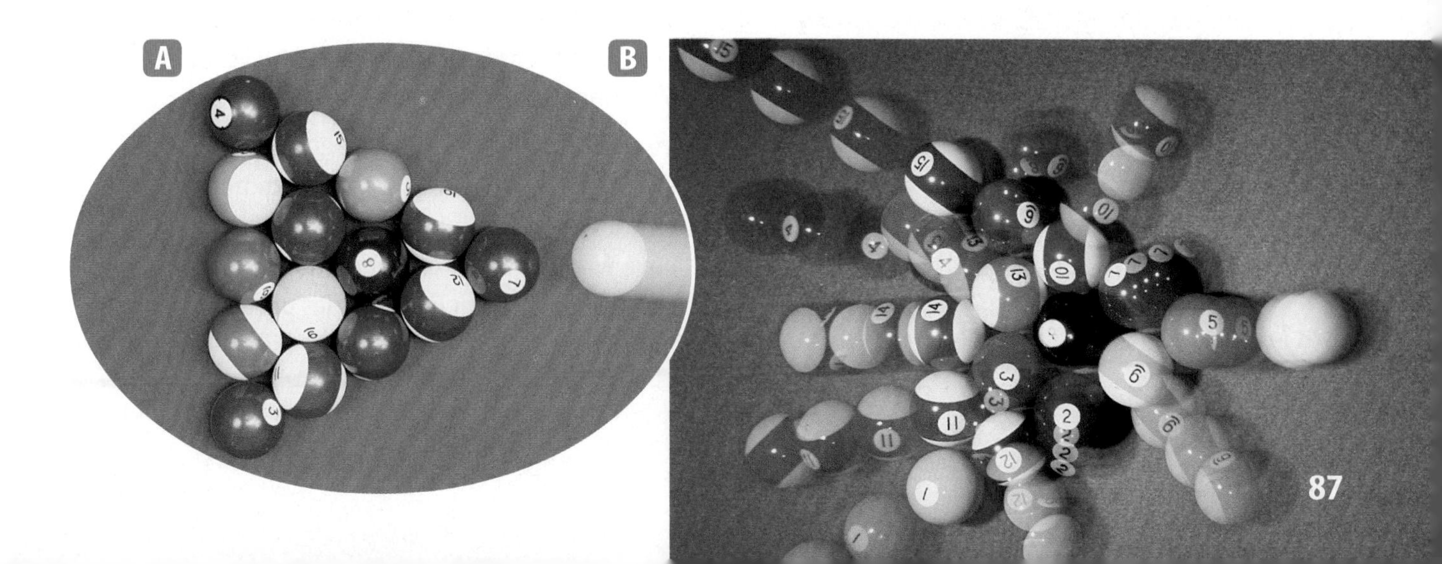

Figure 27
The results of a collision depend on the momentum of each object.
A When the first puck hits the second puck from behind, it gives the second puck momentum in the same direction.
B If the pucks are speeding toward each other with the same speed, the total momentum is zero. *How will they move after they collide?*

A

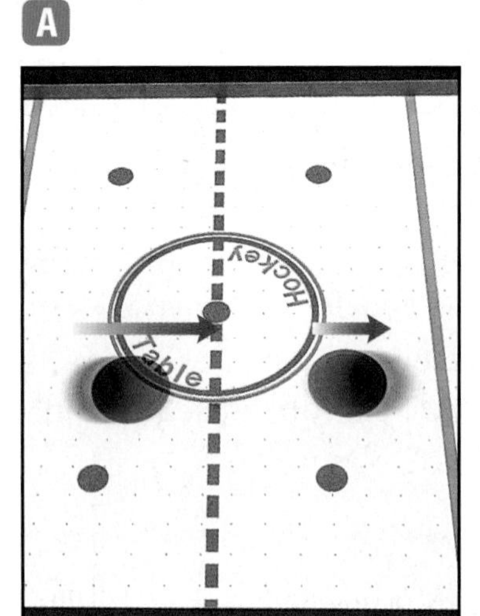

B

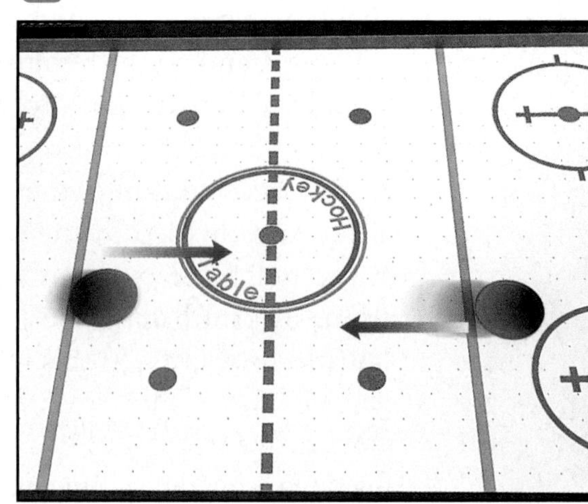

When Objects Collide Look at the pictures of the air-hockey table in **Figure 27.** Suppose one of the pucks was moving along the table in one direction and another struck it from behind. The puck that was struck would continue to move in the same direction but more quickly. The second puck has given it additional momentum in the same direction. What if the two pucks had the same mass and were moving toward each other with the same speed? Each would have the same momentum, but in opposite directions. So the total momentum would be zero. After the pucks collided, they would reverse direction, and move with the same speed. The total momentum would again be zero.

Section  Assessment

1. What is Newton's third law of motion?
2. How can a rocket move through outer space where no matter exists for it to push on?
3. Compare the momentum of a 50-kg dolphin swimming 10.4 m/s and a 6,300-kg elephant walking 0.11 m/s.
4. When two pool balls collide, what happens to the momentum of each?
5. **Think Critically** Some ballet directors assign larger dancers to perform slow, graceful steps and smaller dancers to perform quick movements. Does this plan make sense? Why?

Skill Builder Activities

6. **Predicting** You are a crane operator using a wrecking ball to demolish an old building. You can choose to use a 100-kg ball or a 150-kg ball. Which ball would knock the walls down faster? Which ball would be easier for you to control? Explain. **For more help, refer to the** Science Skill Handbook.
7. **Communicating** In your Science Journal, use the law of conservation of momentum to explain the results of a particular collision you have witnessed. For example, think of games, sports, or amusement park rides or contests. **For more help, refer to the** Science Skill Handbook.

Activity

Measuring the Effects of Air Resistance

If you dropped a bowling ball and a feather from the same height on the Moon, they would both hit the surface at the same time. All objects dropped on Earth are attracted to the ground with the same acceleration. But on Earth, a bowling ball and feather will not hit the ground at the same time. Air resistance slows the feather down.

What You'll Investigate

How does air resistance affect the acceleration of falling objects?

Materials

paper (4 sheets of equal size)
scissors
meterstick
stopwatch
masking tape

Goals

- **Measure** the effect of air resistance on sheets of paper with different shapes.
- **Design** and create a shape from a piece of paper that maximizes air resistance.

Safety Precautions 

Procedure

1. Copy the data table above in your Science Journal, or create it on a computer.
2. Measure a height of 2.5 m on the wall and mark the height with a piece of masking tape.
3. Have one group member drop the flat sheet of paper from the 2.5 m mark. Use the stopwatch to time how long it takes for the paper to reach the ground. Record your time in your data table.
4. Crumple a sheet of paper into a loose ball and repeat step 3.

Effects of Air Resistance

Paper Type	Time
Flat paper	
Loosely crumpled paper	
Tightly crumpled paper	
Your paper design	

5. Crumple a sheet of paper into a tight ball and repeat step 3.
6. Use scissors to shape a piece of paper so that it will fall slowly. You may cut, tear, or fold your paper into any design you choose.

Conclude and Apply

1. **Compare** the falling times of the different sheets of paper.
2. **Infer** the relationship between the falling time and the acceleration of each sheet of paper.
3. **Explain** why the different-shaped papers fell at different speeds.
4. **Explain** how your design maximized the effect of air resistance on your paper's gravitational acceleration.
5. **Infer** why a sky diver will fall in a spread-eagle position before opening her parachute.

Communicating Your Data

Compare your paper design with the designs created by your classmates. As a class, compile a list of characteristics that increase air resistance.

The Momentum of Colliding Objects

Many scientists hypothesize that dinosaurs became extinct 65 million years ago when an asteroid slammed into Earth. The asteroid's diameter was probably no more than 10 km. Earth's diameter is more than 12,700 km. How could an object that size change Earth's climate enough to cause the extinction of animals that had dominated life on Earth for 140 million years? The asteroid could because it may have been traveling at a velocity of 50 m/s, and had a huge amount of momentum. The combination of an object's velocity and mass will determine how much force it can exert. Explore how mass and velocity determine an object's momentum during this activity.

What You'll Investigate

How do the mass and velocity of a moving object affect its momentum?

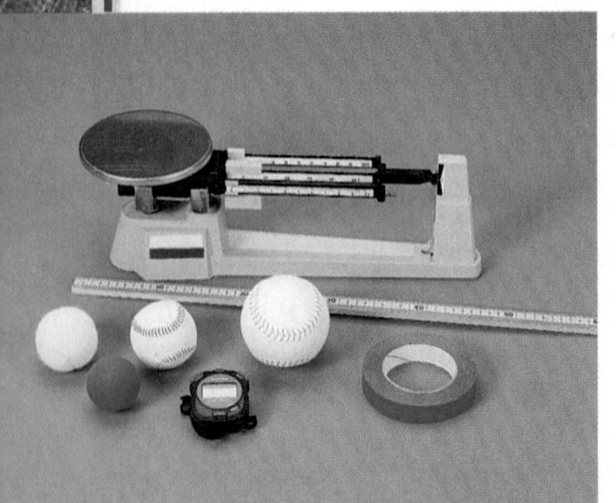

Materials

meterstick
softball
racquetball
tennis ball
baseball
stopwatch
masking tape
balance

Goals

- **Observe and calculate** the momentum of different balls.
- **Compare** results of collisions involving different amounts of momentum.

Safety Precautions 

Momentum of Colliding Balls

Action	Time	Velocity	Mass	Momentum	Distance ball moved softball
Racquetball rolled slowly					
Racquetball rolled quickly					
Tennis ball rolled slowly					
Tennis ball rolled quickly					
Baseball rolled slowly					
Baseball rolled quickly					

Procedure

1. Copy the data table on the previous page in your Science Journal.
2. Use the balance to measure the mass of the racquetball, tennis ball, and baseball. Record these masses in your data table.
3. Use your meterstick to measure a 2-m distance on the floor. Mark this distance with two pieces of masking tape.
4. Place the softball on one piece of tape. Starting from the other piece of tape, slowly roll the racquetball the 2-m distance so that it hits the softball squarely.
5. Use a stopwatch to time how long it takes the racquetball to roll the 2-m distance and hit the softball. Record this time in your data table.
6. Measure the distance the racquetball moved the softball. Record this distance in your data table.
7. Repeat steps 4-6, rolling the racquetball quickly.
8. Repeat steps 4-6, rolling the tennis ball quickly and then slowly.
9. Repeat steps 4-6, rolling the baseball quickly and then slowly.

Conclude and Apply

1. Using the formula $p = mv$, calculate the momentum for each type of ball and action. Record your calculations in the data table.
2. **Compare** the momentums you calculated. Which action had the greatest momentum? Which had the smallest momentum?
3. **Infer** the relationship between the momentum of each ball and the distance the softball was moved.
4. **Explain** why the baseball will have a greater momentum than the tennis ball even if both are traveling with an equal velocity.
5. **Explain** how you observed Newton's third law of motion occurring during this activity.

Use what you have learned about momentum to discuss the differences between the sports of softball and baseball.

Science Stats

Moving and Forcing

Did you know...

Nolan Ryan

...The fastest baseball pitch on record was thrown at 162.3 km/h. This superfast pitch was thrown by the California Angels' Nolan Ryan during a major league game in 1974.

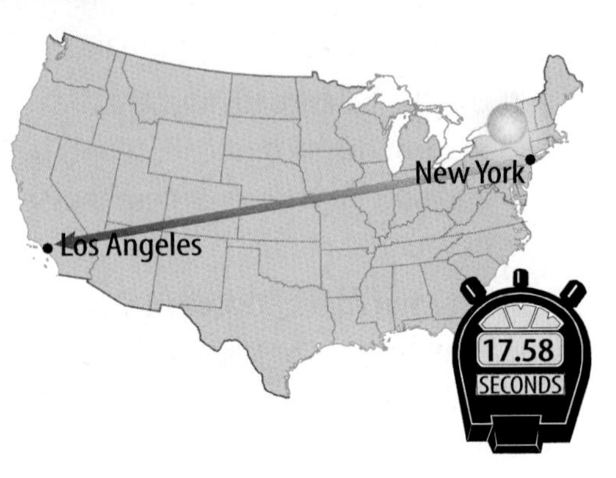

...The Sun moves in a circular path around the center of the galaxy at about 250 km/s. At this rate, the Sun could make a trip from New York to Los Angeles in about 18 s.

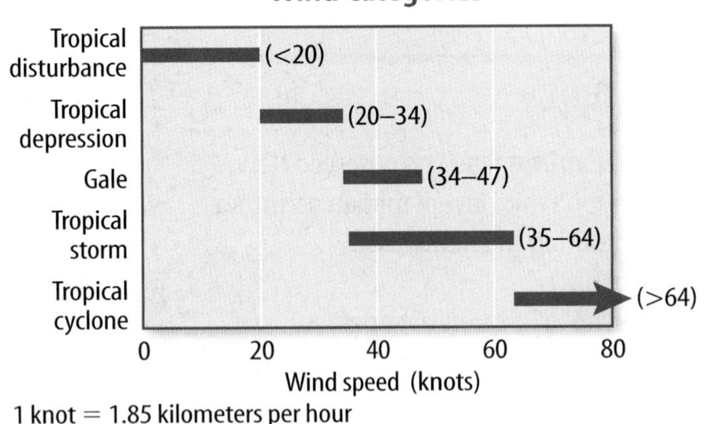

...Ocean waves are powerful. The Atlantic hurls an average of 10,000 kg of water per square meter at the shore in winter. During one storm, the waves ripped away a 1,496,880-kg steel-and-concrete portion of a breakwater. During a later storm, its 2,358,720-kg replacement met with the same fate.

Connecting To Math

Space shuttle

...The force exerted by the space shuttle's solid rocket booster engines when lifting off from Earth equals the force exerted by the engines of 35 jumbo jets. To escape Earth's gravity, a spacecraft must reach a speed through the atmosphere of more than 40,320 km/h. This is about 455 times faster than a typical highway speed limit of 88.5 km/h.

747 jumbo jet

...The force needed to stop a jumbo jet is equal to the frictional force applied by 1 million automobile brakes. Even so, the airplane has to travel a distance of almost 1 km on the ground before it stops.

Do the Math

1. What is the distance from New York to Los Angeles?
2. How far does the Sun travel in 5 min?
3. The following is a list of the masses of several types of balls: volleyball, 280 g; tennis ball, 60 g; baseball, 150 g; football, 425 g; basketball, 650 g; and softball, 200 g. Make a bar graph that compares the masses of these balls.

Go Further

Write Newton's first, second, and third laws of motion on three separate sheets of paper. Under each, write a paragraph or make a sketch that shows that law at work in some common situation.

Chapter 3 Study Guide

Reviewing Main Ideas

Section 1 Newton's Second Law

1. Newton's second law of motion states that a net force causes an object to accelerate in the direction of the net force, with an acceleration equal to the net force divided by the mass.

2. Friction is caused by the microwelds that develop between the microscopic bumps on two surfaces. The three types of friction are static, sliding, and rolling. *What type of friction is at work in this picture?*

3. All objects are attracted to Earth with the same acceleration. Air resistance exerts an upward force on objects falling through the atmosphere.

Section 2 Gravity

1. Gravity is the force of attraction that exists between any two objects having mass. The size of the gravitational force is determined by the mass of the objects and their distance from each other.

2. Projectiles have a horizontal and a vertical motion that makes them travel in a curved path. Circular motion is caused by a centripetal, or center-seeking force. *What exerts the center-seeking force in the photo?*

3. Weight is the measure of the gravitational force exerted on an object by Earth. Weight is expressed in newtons, N. You use the following equation to calculate weight

$$W = m \times 9.8 \text{ m/s}^2$$

where m is the mass of the object.

Section 3 The Third Law of Motion

1. Newton's third law of motion states that for every action there is an equal and opposite reaction.

2. The momentum of an object can be calculated by the equation $p = mv$. *If the objects in this photo are moving, which probably has the greatest momentum?*

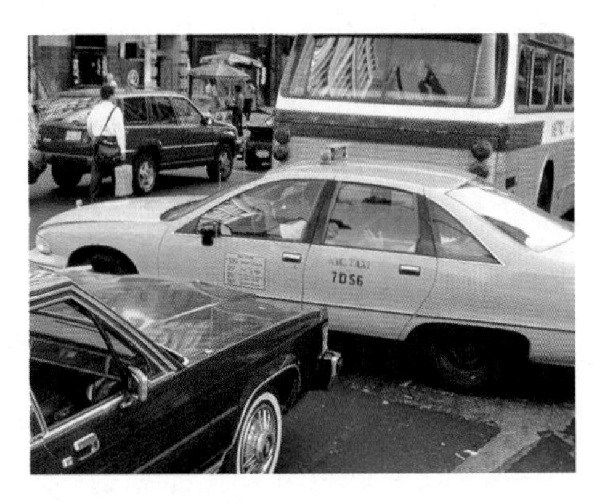

3. When two objects collide, momentum can be conserved. Some of the momentum from one object is transferred to the other.

After You Read

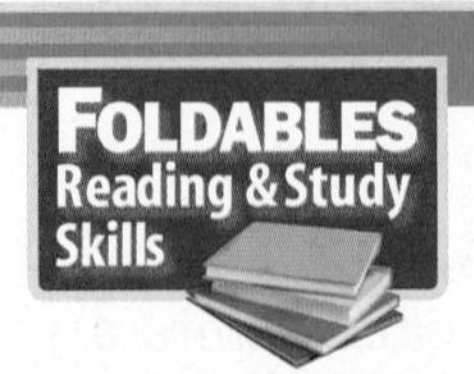

FOLDABLES Reading & Study Skills

Use the information on your Foldable to compare and contrast the types of friction. Write similarities and differences on the back of your Foldable.

Chapter 3 Study Guide

Visualizing Main Ideas

Complete the following concept map on forces.

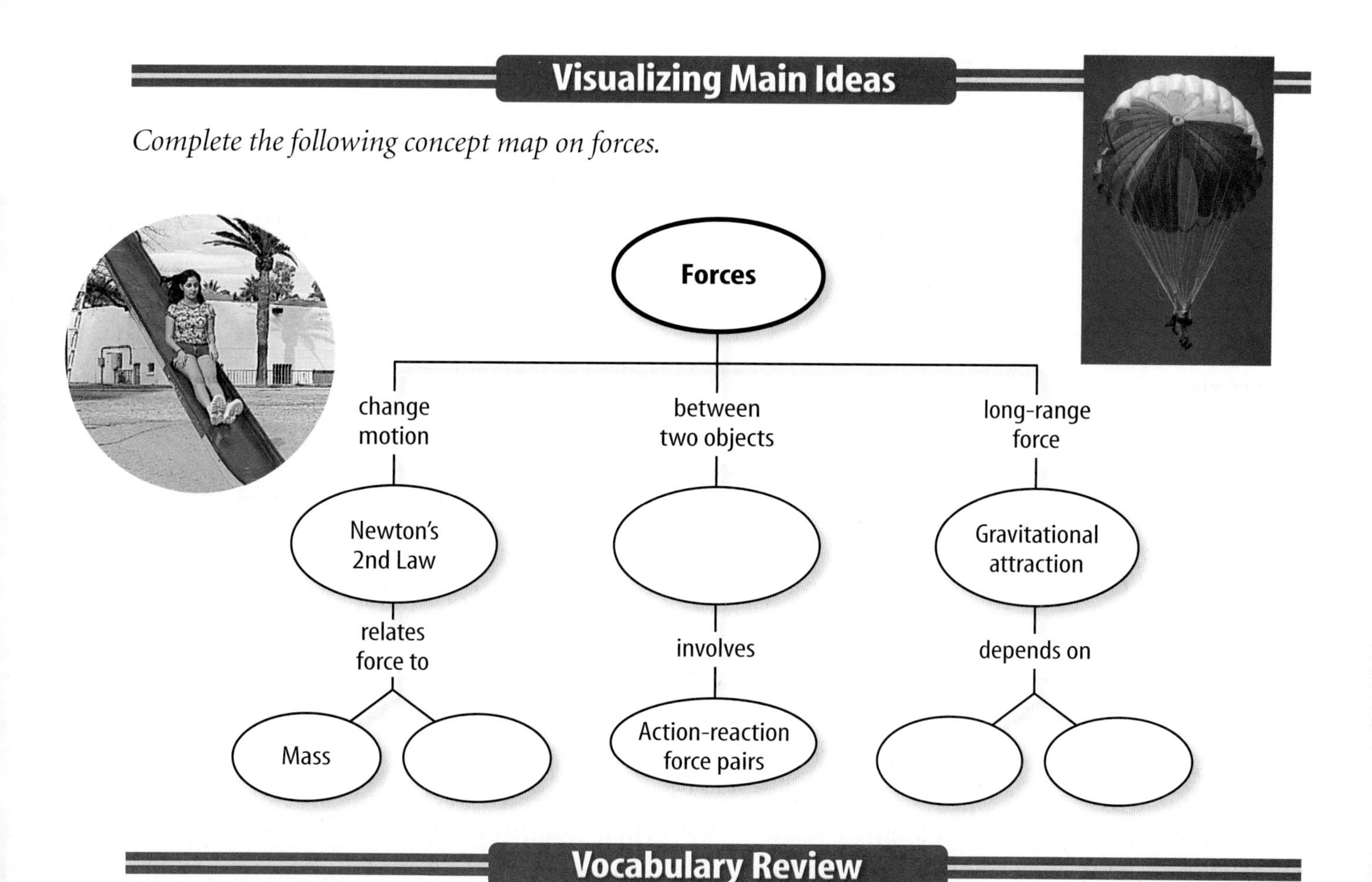

Vocabulary Review

Vocabulary Words

a. centripetal acceleration
b. centripetal force
c. friction
d. law of gravitation
e. momentum
f. Newton's second law of motion
g. Newton's third law of motion
h. weight

Study Tip

Make a note of anything you can't understand so that you'll remember to ask your teacher about it.

Using Vocabulary

Using the vocabulary words list, replace the underlined words with the correct words.

1. The microwelds that form between the surfaces of two objects sometimes make objects hard to move.
2. The Moon's acceleration toward the center of its circular path is caused by Earth's gravity.
3. For every action, there is an equal and opposite reaction.
4. The force of gravity exerted on an object is different on different planets.
5. The combined mass and velocity of the runaway train made it dangerous.
6. The acceleration of an object equals the net force divided by the mass.

Chapter 3 Assessment

Checking Concepts

Choose the word or phrase that best answers the question.

1. What will happen to an object when a net force acts on it?
A) fall
B) stop
C) accelerate
D) go in a circle

2. Which is Newton's second law?
A) $F = 1/2ma^2$
B) $F = 2\ ma$
C) $p = mv$
D) $a = F/m$

3. What is the force of gravity on an object known as?
A) centripetal force
B) friction
C) momentum
D) weight

4. Which of the following is NOT a type of friction?
A) static
B) sliding
C) centripetal
D) rolling

5. What's true about an object falling toward Earth?
A) It falls faster the heavier it is.
B) It falls faster the lighter it is.
C) Earth pulls on it, and it pulls on Earth.
D) It has no weight.

6. Why do projectiles follow a curved path?
A) They have a horizontal and a vertical motion.
B) They have centripetal force.
C) They have momentum.
D) They have inertia.

7. What is the product of mass and velocity known as?
A) gravity
B) momentum
C) friction
D) weight

8. Which body exerts the weakest gravitational force on Earth?
A) the Moon
B) Mars
C) Pluto
D) Venus

9. When a leaf falls, what force opposes gravity?
A) air resistance
B) terminal velocity
C) friction
D) weight

10. In circular motion, the centripetal force is in what direction?
A) forward
B) backward
C) toward the center
D) toward the side

Thinking Critically

11. What is the weight on Earth of a person with a mass of 65 kg?

12. Some people put chains on their tires in the winter. Why?

13. List some ways an astronaut could keep her supplies from floating away from her while she is in orbit around Earth.

14. As you in-line skate around the block, what action and reaction forces keep you moving?

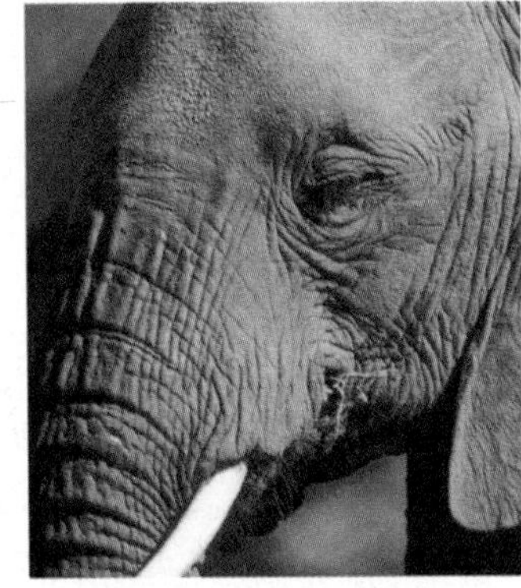

15. Which one of the following would have the most momentum—a charging elephant, a jumbo jet sitting on the runway, or a baseball traveling at 100 km/h? Explain.

Developing Skills

16. Classifying Classify the following as examples of static, sliding, or rolling friction: sledding down a hill, sitting in a chair, pushing a grocery cart, standing on a steep slope, and rowing a boat.

17. **Drawing Conclusions** A race car is moving in a circle at a constant speed around a track. Does a centripetal foce act on the car?

18. **Recognizing Cause and Effect** Suppose you stand on a scale next to a sink. What happens to the reading on the scale if you push down on the sink?

19. **Interpreting Data** The following table contains data about four objects that were dropped to Earth at the same time.
    a. Which object fell fastest? Slowest?
    b. Which object has the greatest weight?
    c. Is air resistance stronger on A or B?
    d. Why are the times different?

Time of Fall for Dropped Objects

Object	Mass (g)	Time of Fall (s)
A	5.0	2.0
B	5.0	1.0
C	30.0	0.5
D	35.0	1.5

Performance Assessment

20. **Poem** Write a poem about a falling leaf. Include the terms *gravity, free fall, air resistance,* and *terminal velocity.*

21. **Oral Presentation** Prepare a presentation to explain Newton's third law of motion to a group of first-grade students.

Technology

Go to the Glencoe Science Web site at **science.glencoe.com** or use the **Glencoe Science CD-ROM** for additional chapter assessment.

Test Practice

Tiffany learned that the acceleration of a free-falling body is 9.8 m/s^2. She wanted to find out what speed a sky diver reaches after several seconds. Her calculations are shown in the table below.

Speed of a Falling Sky Diver

Time (s)	Speed (m/s)
0	0
1	9.8
2	19.6
3	29.4
4	39.2
5	?

Study the table and answer the following questions.

1. According to these data, about how fast will the speed of a falling sky diver be after 5 s?
   A) 39.8 m/s
   B) 44.2 m/s
   C) 49.0 m/s
   D) 54.0 m/s

2. Which of these causes falling sky divers to accelerate?
   F) gravity
   G) inertia
   H) rotation of Earth on its axis
   J) the tilt of Earth on its axis

3. If Tiffany extended her table, what would the sky diver's speed be after 14 s?
   A) 107.8 m/s
   B) 78.4 m/s
   C) 147.0 m/s
   D) 137.2 m/s

CHAPTER 4

Energy

Snowboarding down the side of a mountain is an exhilarating experience. With little effort, you can easily reach speeds well over 50 km/h using nothing other than a snowboard, the slope, and a lot of snow. How do snowboarders achieve such high speeds? What supplies the energy to move them so fast? The answers to these questions can be found by studying energy and energy conservation.

What do you think?

Science Journal Look at the picture below with a classmate. Discuss what you think this might be or what is happening. Here's a hint: *It's used every day and allows people to work and play at any hour.* Write your answer or best guess in your Science Journal.

One of the most useful inventions of the nineteenth century was the electric lightbulb. Being able to light up the dark allows for extended work and recreation. A lightbulb uses electricity to produce light, but heat also is produced. To observe the conversion of electricity to light and heat, do the following activity.

Model how a lightbulb works 

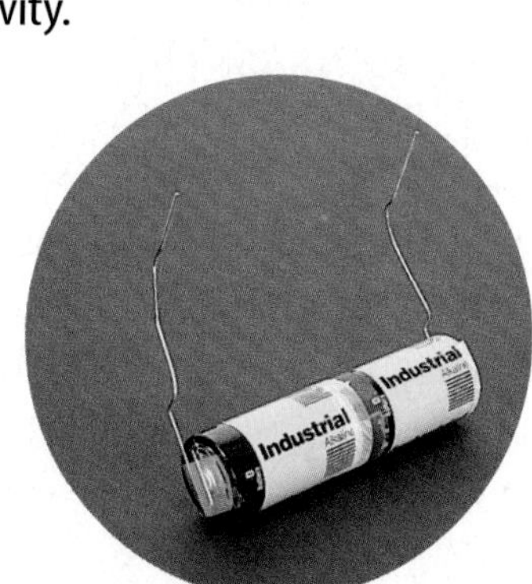

1. Obtain two D-cell batteries, two non-coated paper clips, tape, metal tongs and some steel wool. Separate the steel wool into thin strands and straighten the paper clips.
2. Tape the batteries together and then tape one end of each paper clip to the battery terminals as shown in the photograph.
3. While holding the strands of steel wool with the tongs, briefly complete the circuit by placing the steel wool in contact with both paper clip ends.

WARNING: *Steel wool can become hot—connect to battery only for a brief time.*

Observe

Describe in your Science Journal what you saw. Touch the steel wool. What changes are you observing?

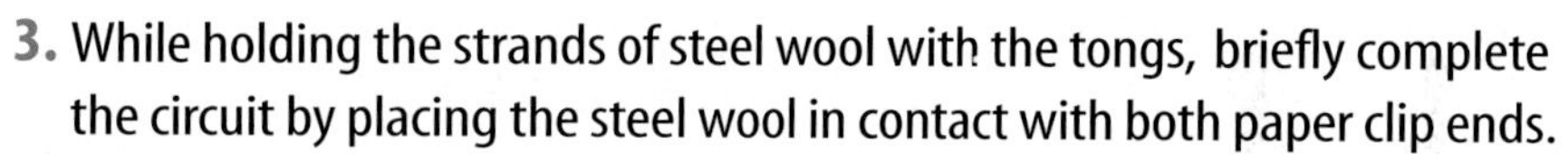

Before You Read

Making a Know-Want-Learn Study Fold **Make the following Foldable to help identify what you already know and what you want to know about energy.**

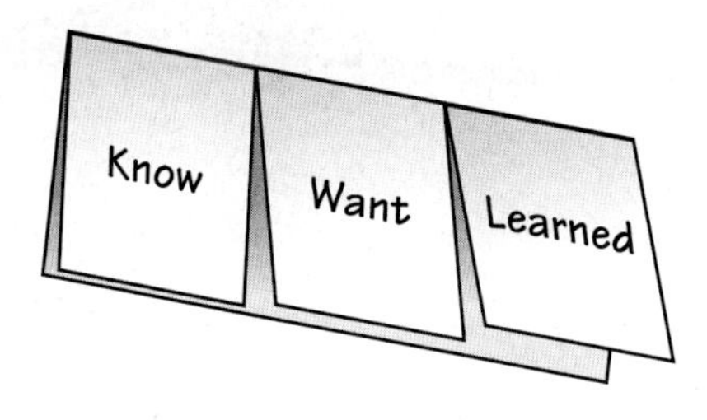

1. Place a sheet of paper in front of you so the long side is at the top. Fold the paper in half from top to bottom.
2. Fold both sides in to divide the paper into thirds. Unfold the paper so three sections show.
3. Through the top thickness of paper, cut along each of the fold lines to the top fold, forming three tabs. Label the tabs *Know, Want,* and *Learned,* as shown.
4. Before you read the chapter, write what you know about energy under the left tab and what you want to know under the middle tab.
5. As you read the chapter, write what you learn about energy under the right tab.

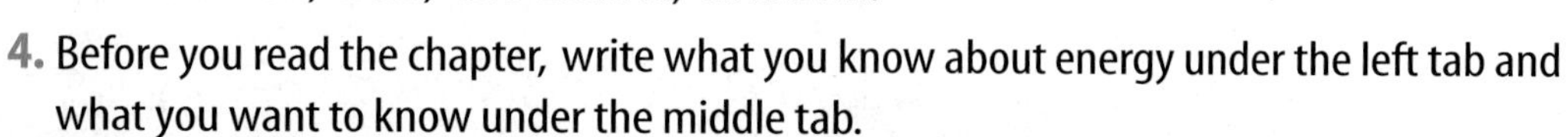

SECTION 1

The Nature of Energy

As You Read

What You'll Learn

- **Distinguish** between kinetic and potential energy.
- **Recognize** different ways that energy can be stored.

Vocabulary

kinetic energy
joule
potential energy
elastic potential energy
chemical potential energy
gravitational potential energy

Why It's Important

Understanding energy helps you understand how your environment is changing.

What is energy?

Wherever you are sitting as you read this, changes are taking place—lightbulbs are heating the air around them, the wind might be rustling leaves, or sunlight might be glaring off a nearby window. Even you are changing as you breathe, blink, or shift position in your seat.

Every change that occurs—large or small—involves energy. Imagine a baseball flying through the air. It hits a window, causing the glass to break as shown in **Figure 1.** The window changed from a solid sheet of glass to a number of broken pieces. The moving baseball caused this change—a moving baseball has energy. Even when you comb your hair or walk from class to class, energy is involved.

Change Requires Energy When something is able to change its environment or itself, it has energy. Energy is the ability to cause change. The moving baseball had energy. It certainly caused the window to change. Anything that causes change must have energy. You use energy to arrange your hair to look the way you want it to. You also use energy when you walk down the halls of your school between classes or eat your lunch. You even need energy to yawn, open a book, and write with a pen.

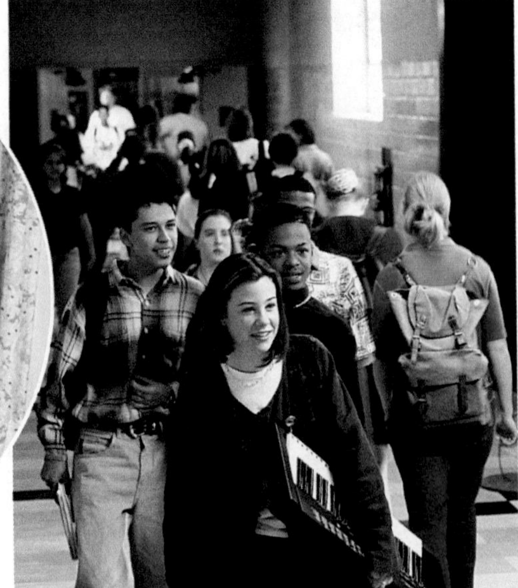

Figure 1
Each photo shows changes occurring.
Describe the changes that are occurring.

Figure 2
Energy can be stored in fuels, or it can travel through the environment. *Which objects are storing energy? Where is movement of energy occurring?*

Different Forms of Energy Turn on an electric light, and a dark room becomes bright. Turn on your CD player, and sound comes through your headphones. In both situations, energy moves from one place to another. These changes seem to differ from each other and differ from a baseball shattering a window. This is because energy has several different forms, such as electrical, chemical, and thermal.

Figure 2 shows some examples of everyday situations in which you might notice energy. Is the chemical energy stored in food the same as the energy that comes from the Sun or the energy stored in gasoline? Thermal energy from the Sun travels a vast distance through space to Earth, warming the planet and providing energy that enables green plants to grow. When you make toast in the morning, you are using electrical energy. In short, energy plays a role in every activity that you do.

An Energy Analogy Money can be used in an analogy to help you understand energy. If you have $100, you could store it in a variety of forms—cash in your wallet, a bank account, travelers' checks, or gold or silver coins. You could transfer that money to different forms. You could deposit your cash into a bank account or trade the cash for gold. Regardless of its form, money is money. The same is true for energy. Energy from the Sun that warms you and energy from the food that you eat are only different forms of the same thing.

Reading Check *How is energy like money?*

Kinetic Energy

Usually, when you think of energy, you think of action—or some sort of motion taking place. **Kinetic energy** is energy in the form of motion. A spinning bicycle wheel, a sprinting runner, and a football passing through the goalposts all have kinetic energy, but the amounts depend on two quantities—the mass of the moving object and its velocity.

The more mass a moving object has, the more kinetic energy it has. Similarly, the greater an object's velocity is, the more kinetic energy it has. **Figure 3** shows a truck and a motorcycle that are moving at 100 km/h. Which vehicle has more kinetic energy? Although they have the same velocity, the truck has more kinetic energy because it has a greater mass than the motorcycle. **Figure 3** also shows two motorcycles—one moving at 100 km/h and one moving at 80 km/h. Which motorcycle has more kinetic energy? Assuming that the motorcycles have the same mass, the one moving at 100 km/h has greater kinetic energy than the one moving at 80 km/h.

SCIENCE Online

Collect Data Visit the Glencoe Science Web site at **science.glencoe.com** for more information about the kinetic energy of various animals. Communicate to your class what you learn.

Calculating Kinetic Energy The kinetic energy of an object can be calculated using the following relationship.

$$\text{kinetic energy} = \frac{1}{2}\text{mass} \times \text{velocity}^2$$

$$KE\ (\text{J}) = \frac{1}{2}m\ (\text{kg}) \times v^2\ (\text{m}^2/\text{s}^2)$$

The **joule** (JEWL) is the SI unit of energy. It is named after the nineteenth-century British scientist James Prescott Joule. To calculate kinetic energy in joules (J), mass is measured in kilograms, and velocity is measured in meters per second.

Because velocity is squared in the equation for kinetic energy, increasing the velocity of an object can produce a large change in its kinetic energy. Without changing the mass of an object, doubling its velocity will quadruple its kinetic energy.

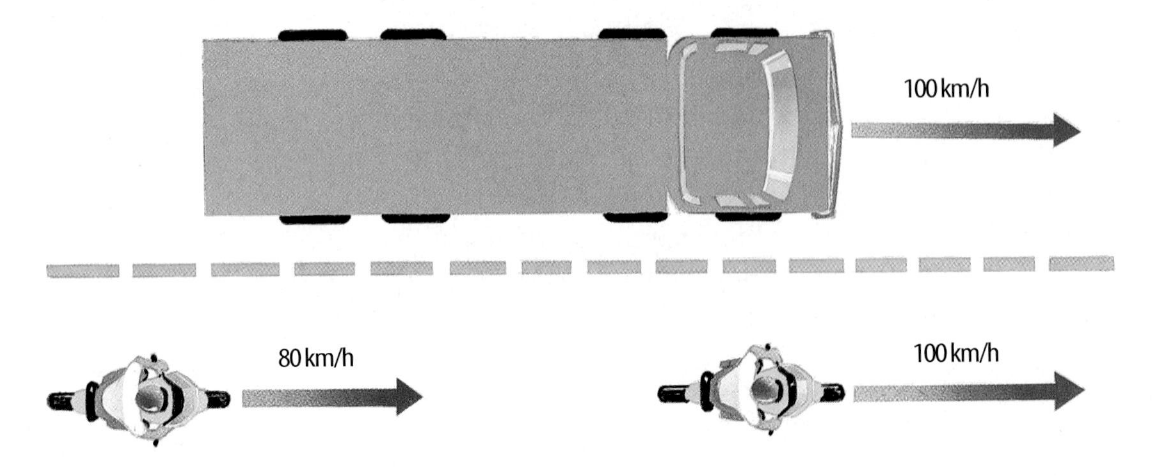

Figure 3
The kinetic energy of each vehicle is different because kinetic energy depends on an object's mass and its velocity.

Figure 4
As natural gas burns, it combines with oxygen to form carbon dioxide and water. In this chemical reaction, chemical potential energy is released.

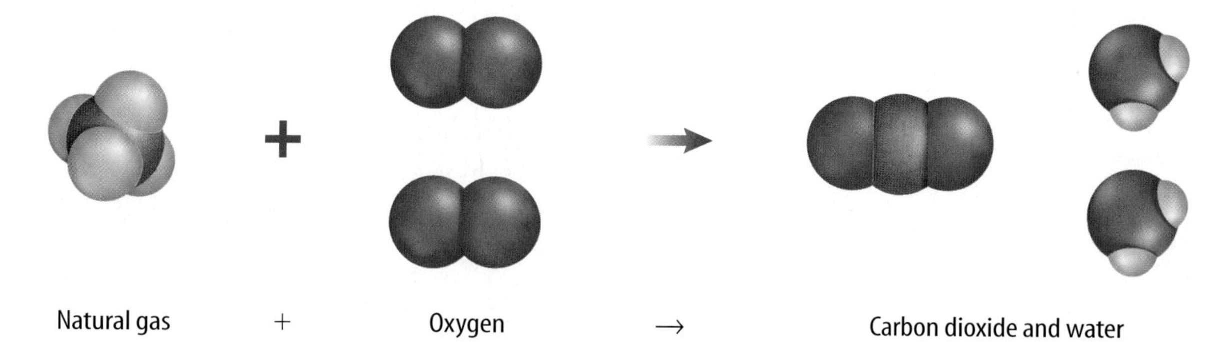

Potential Energy

Energy doesn't have to involve motion. Even motionless objects can have energy. This energy is stored in the object. Therefore, the object has potential to cause change. A hanging apple in a tree has stored energy. When the apple falls to the ground, a change occurs. Because the apple has the ability to cause change, it has energy. The hanging apple has energy because of its position above Earth's surface. Stored energy due to position is called **potential energy.** If the apple stays in the tree, it will keep the stored energy due to its height above the ground. If it falls, that stored energy of position is converted to energy of motion.

Elastic Potential Energy Energy can be stored in other ways, too. If you stretch a rubber band and let it go, it sails across the room. As it flies through the air, it has kinetic energy due to its motion. Where did this kinetic energy come from? Just as the apple hanging in the tree had potential energy, the stretched rubber band had energy stored as elastic potential energy. **Elastic potential energy** is energy stored by something that can stretch or compress, such as a rubber band or spring.

Chemical Potential Energy The cereal you eat for breakfast and the sandwich you eat at lunch also contain stored energy. Gasoline stores energy in the same way as food stores energy—in the chemical bonds between atoms. Energy stored in chemical bonds is **chemical potential energy. Figure 4** shows a molecule of natural gas. Energy is stored in the bonds that hold the carbon and hydrogen atoms together and is released when the gas is burned.

How is elastic potential energy different from chemical potential energy?

Interpreting Data from a Slingshot

Procedure

1. Using two fingers, carefully stretch a **rubber band** on a table until it has no slack.
2. Place a **nickel** on the table, slightly touching the midpoint of the rubber band.
3. Push the nickel back 0.5 cm and release. Measure the distance the nickel travels.
4. Repeat step 3, each time pushing the nickel back an additional 0.5 cm.

Analysis

1. How did the takeoff speed of the nickel seem to change relative to the distance that you stretched the rubber band?
2. What does this imply about the kinetic energy of the nickel?

Fast-flowing rivers and slow-moving glaciers have kinetic energy. A rock balanced on a hill contains potential energy. What are some other examples of kinetic and potential energy in nature?

Gravitational Potential Energy Gravity caused the apple to fall from the tree. Anything that can fall has stored energy called gravitational potential energy. **Gravitational potential energy** (GPE) is energy stored by objects that are above Earth's surface. The amount of gravitational potential energy depends on three things—the mass of the object, the acceleration due to gravity, and the height above the ground. The acceleration of gravity on Earth is 9.8 m/s^2, and the height is measured in meters.

The amount of gravitational potential energy an object has can be calculated using the following equation.

$$GPE = \text{mass} \times 9.8\ \text{m/s}^2 \times \text{height}$$

$$GPE\ (\text{J}) = m\ (\text{kg}) \times 9.8\ \text{m/s}^2 \times h\ (\text{m})$$

Just like kinetic energy, gravitational potential energy is measured in joules. All energy, no matter which form it is in, can be measured in joules.

Math Skills Activity

Calculating Gravitational Potential Energy

Example Problem

A 0.06-kg tennis ball starts to fall from a height of 2.9 m. How much gravitational potential energy does the ball have at that height?

Solution

1. *This is what you know:* mass of tennis ball: $m = 0.06$ kg; height of the tennis ball: $h = 2.9$ m; acceleration of gravity: 9.8 m/s^2
2. *This is what you want to find:* gravitational potential energy: GPE
3. *This is the equation you use:* $GPE = m \times 9.8\ \text{m/s}^2 \times h$
4. *Solve the equation by substituting the known values:* $GPE = 0.06\ \text{kg} \times 9.8\ \text{m/s}^2 \times 2.9\ \text{m} = 1.7\ \text{J}$

Check your answer by substituting it and the known values into the original equation. Do you get the same result?

Practice Problem

Bjorn is holding a tennis ball outside a second-floor window (3.5 m from the ground) and Billie Jean is holding one outside a third-floor window (6.25 m from the ground). How much more gravitational potential energy does Billie Jean's tennis ball have? (Each tennis ball has a mass of 0.06 kg.)

For more help, refer to the Math Skill Handbook.

Changing GPE Look at the objects on the bookshelf in **Figure 5.** Which of these objects has the most gravitational potential energy? According to the equation for gravitational potential energy, the GPE of an object can be increased by increasing its height above the ground. If two objects are at the same height, then the object with the larger mass has more gravitational potential energy.

In **Figure 5,** suppose the green vase on the lower shelf and the blue vase on the upper shelf have the same mass. Then the vase that is on the upper shelf has more gravitational potential energy because it is higher above the ground.

Imagine what would happen if the two vases were to fall. As they fall and begin moving, they have kinetic energy as well as gravitational potential energy. As the vases get closer to the ground, their gravitational potential energy decreases. At the same time they are moving faster, so their kinetic energy increases. The vase that initially had more gravitational potential energy will be moving faster when it hits the floor.

Do the books on the shelf above the green vase have more gravitational potential energy than the vase? That depends on the mass of the books. Even though they are twice as high as the vase, their gravitational potential energy also depends on their mass. If the mass of the books is less than half the mass of the vase, then the books have less gravitational potential energy, even though they are at a greater height.

✔ Reading Check *What does GPE depend on?*

Figure 5
An object's gravitational potential energy increases with increased height. *Which vase has more gravitational potential energy? Which one will have more kinetic energy when it strikes the ground?*

Section  Assessment

1. Two books with different masses fall off the same bookshelf. As they fall, which has more kinetic energy and why?
2. How can the gravitational potential energy of an object be changed?
3. How is energy stored in food? Is the form of energy stored in food different from the form stored in gasoline? Explain.
4. Contrast potential and kinetic energy.
5. **Think Critically** The food you eat supplies energy for your body. Suggest ways your body might make use of this energy.

Skill Builder Activities

6. **Comparing and Contrasting** Compare and contrast elastic potential energy, chemical potential energy, and gravitational potential energy. **For more help, refer to the Science Skill Handbook.**
7. **Solving One-Step Equations** An 80-kg diver jumps off a 10-m platform. Calculate how much gravitational potential energy the diver has at the top of the platform and halfway down. **For more help, refer to the Math Skill Handbook.**

Activity

Bouncing Balls

What happens when you drop a ball onto a hard, flat surface? It starts with potential energy. It bounces up and down until it finally comes to a rest. Where did the energy go?

What You'll Investigate

How do balls differ in their bouncing behavior?

Materials

tennis ball
rubber ball
balance
meterstick
masking tape
cardboard box
**shoe box*
**Alternate materials*

Goals

- **Identify** the forms of energy observed in a bouncing ball.
- **Infer** why the ball stops bouncing.

Safety Precautions 

Procedure

1. **Measure** the mass of the two balls.
2. Have a friend drop one ball from 1 m. Measure how high the ball bounced. Repeat this two more times so that you can calculate an average bounce height. Record your values on the data table.
3. Repeat step 2 for the other ball.
4. **Predict** whether the balls would bounce higher or lower if they were dropped onto the cardboard box. Design an experiment to measure how high the balls would bounce off the surface of a cardboard box.

Conclude and Apply

1. **Calculate** the gravitational potential energy of each ball before dropping them.
2. As the balls fall, what happens to their gravitational potential energy and their kinetic energy? What happens to their kinetic energy when they hit the floor?
3. **Calculate** the average bounce height for the three trials under each condition. Describe your observations.
4. How did the bounce heights compare when dropped on a cardboard box instead of the floor? Why? Hint: *Did you observe any movement of the box when the balls bounced?*
5. Use elastic potential energy to explain why the balls bounced to different heights.

Bounce Height

Type of Ball	Surface	Trial	Height (cm)
Tennis	Floor	1	
Tennis	Floor	2	
Tennis	Floor	3	
Rubber	Floor	1	
Rubber	Floor	2	
Rubber	Floor	3	
Tennis	Box	1	

Communicating Your Data

Meet with three other lab teams and compare average bounce heights for the tennis ball on the floor. Discuss why your results might differ. **For more help, refer to the Science Skill Handbook.**

SECTION 2

Conservation of Energy

Changing Forms of Energy

Unless you were talking about potential energy, you probably wouldn't think of the book on top of a bookshelf as having much to do with energy—until it fell. You'd be more likely to think of energy as race cars roar past or as your body uses energy from food to help it move, or as the Sun warms your skin on a summer day. You might be thankful for electrical energy as you play a favorite CD. These situations involve energy changing from one form to another form. Energy is most noticeable as it transforms from one type to another.

Transforming Electrical Energy You use many devices every day that convert one form of energy to other forms. For example, you might be reading this page in a room lit by lightbulbs. The lightbulbs transform electrical energy into light so you can see. The warmth you feel around the bulb is evidence that some of that electrical energy is turned into thermal energy, as illustrated in **Figure 6.** What other devices have you used today that make use of electrical energy? You might have been awakened by an alarm clock, styled your hair, made toast, listened to music, or played a video game. What form or forms of energy is electrical energy converted to in these examples?

As You Read

What You'll Learn

- **Describe** how energy is conserved when changing from one form to another.
- **Apply** the law of conservation of energy to familiar situations.

Vocabulary

mechanical energy
law of conservation of energy

Why It's Important

Conservation of energy is a universal principle that can explain how energy changes occur.

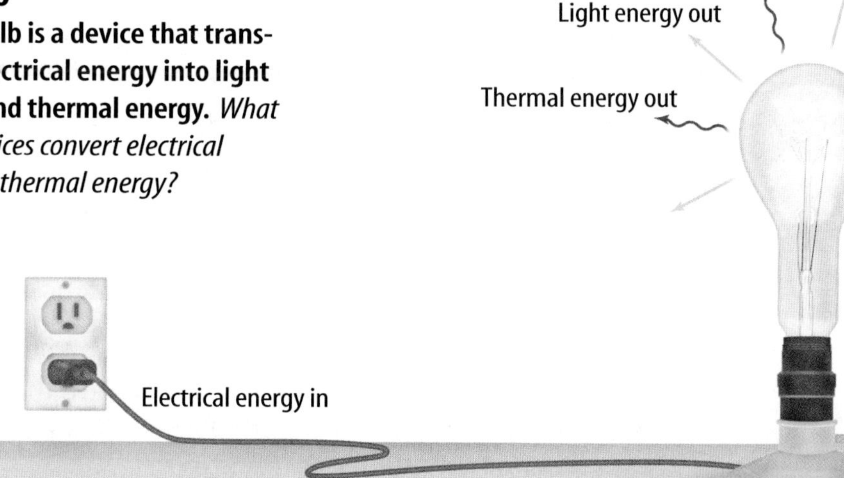

Figure 6
A lightbulb is a device that transforms electrical energy into light energy and thermal energy. *What other devices convert electrical energy to thermal energy?*

Figure 7

A In a car, a spark plug fires, initiating the conversion of chemical potential energy into thermal energy.

B As the hot gases expand, thermal energy is converted into kinetic energy.

Transforming Chemical Energy Fuel stores energy in the form of chemical potential energy. For example, the car or bus that might have brought you to school this morning probably runs on gasoline. The engine transforms the chemical potential energy stored in gasoline into the kinetic energy of a moving car or bus. Several energy conversions occur in this process, as shown in **Figure 7.** An electrical spark ignites a small amount of fuel. The burning fuel produces thermal energy. So chemical energy is changed to thermal energy. The thermal energy causes gases to expand and move parts of the car, producing kinetic energy.

Some energy transformations are less obvious because they do not result in visible motion, sound, heat, or light. Every green plant you see converts light energy from the Sun into energy stored in chemical bonds in the plant. If you eat an ear of corn, the chemical potential energy in the corn is transformed yet again when it is in your body.

Conversions Between Kinetic and Potential Energy

You have experienced many situations that involve conversions between potential and kinetic energy. Bicycles, roller coasters, and swings can be described in terms of potential and kinetic energy. Even something as simple as launching a rubber band or using a bow and arrow involves energy conversions. To understand the energy conversions in these activities, it is helpful to identify the mechanical energy of a system. **Mechanical energy** is the total amount of potential and kinetic energy in a system and can be expressed by this equation.

$$\text{mechanical energy} = \text{potential energy} + \text{kinetic energy}$$

In other words, mechanical energy is energy due to the position and the motion of an object. What happens to the mechanical energy of an object as potential and kinetic energy are converted into each other?

Falling Objects Standing under an apple tree can be hazardous. Here is an explanation of why this is true using energy terms. An apple on a tree, like the one in **Figure 8,** has gravitational potential energy due to Earth pulling down on it. The apple does not have kinetic energy while it hangs from the tree. However, the instant the apple comes loose from the tree, it accelerates due to gravity. As it falls, it loses height so its gravitational potential energy decreases. This potential energy is not lost. Rather, it is transformed into kinetic energy as the velocity of the apple increases.

Look back at the equation for mechanical energy. If the potential energy is being converted into kinetic energy, then the mechanical energy of the apple doesn't change as it falls. The potential energy that the apple loses is gained back as kinetic energy. The form of energy changes, but the total amount of energy remains the same.

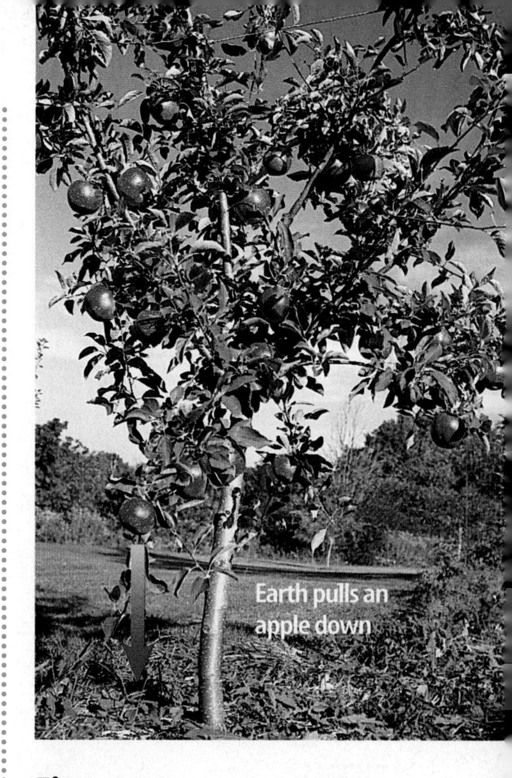

Figure 8
Objects that can fall have gravitational potential energy.
What objects around you have gravitational potential energy?

Reading Check ***What happens to the mechanical energy of the apple as it falls from the tree?***

Energy transformation also occurs when a baseball is hit into the air. Look at **Figure 9.** When the ball leaves the bat, it has mostly kinetic energy. As the ball rises, its velocity decreases, so its kinetic energy must decrease, too. However, the ball's gravitational potential energy increases as it goes higher. At its highest point, the baseball has the maximum amount of gravitational potential energy. The only kinetic energy it has at this point is due to its forward motion. Then, as the baseball falls, gravitational potential energy decreases while kinetic energy increases as the ball moves faster. Once again, the mechanical energy of the ball remains constant as it rises and falls.

Figure 9
A ball hit into the air illustrates how kinetic energy and gravitational potential energy are converted into each other.

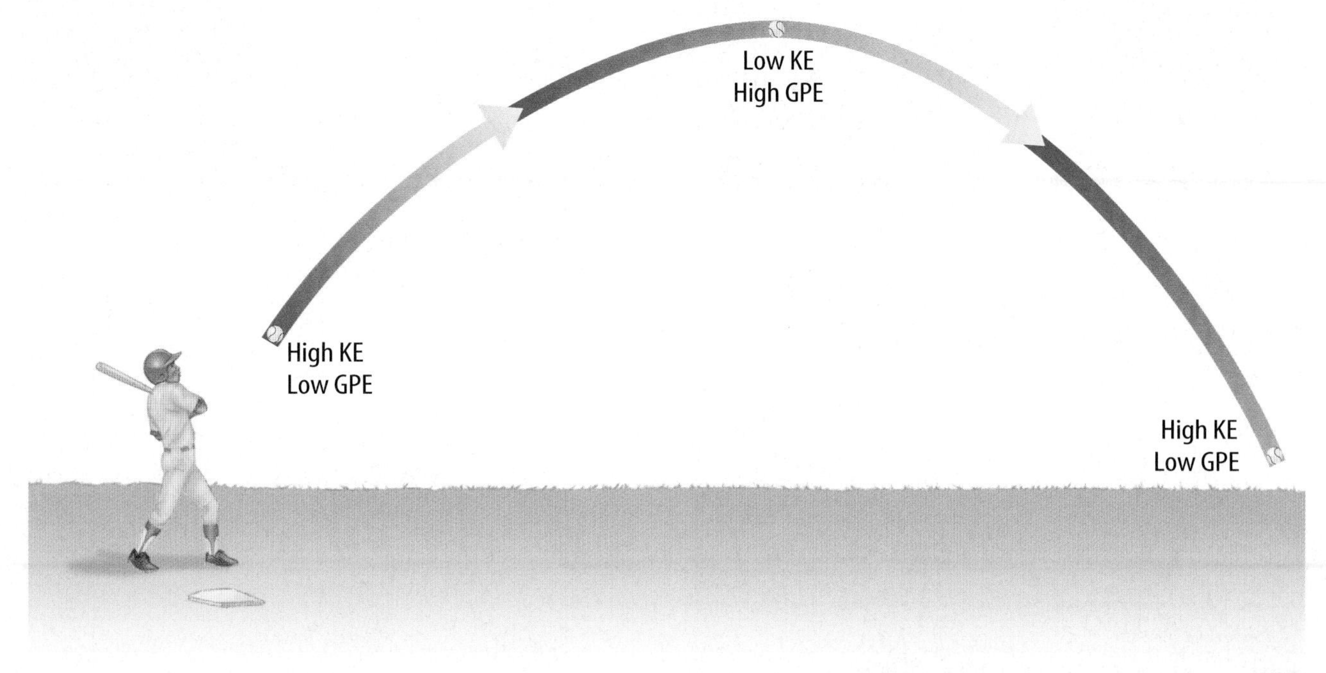

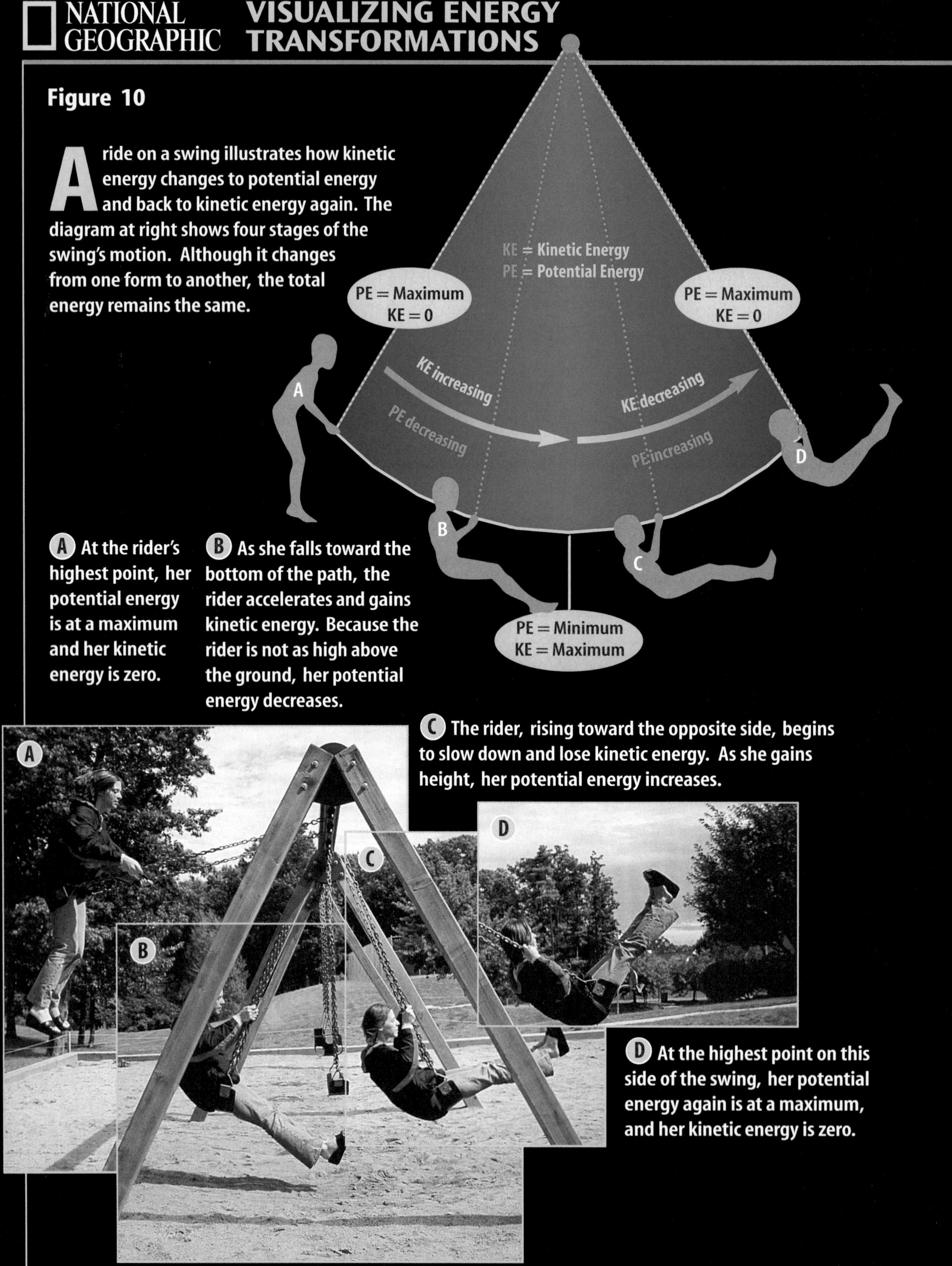

Figure 10

A ride on a swing illustrates how kinetic energy changes to potential energy and back to kinetic energy again. The diagram at right shows four stages of the swing's motion. Although it changes from one form to another, the total energy remains the same.

A At the rider's highest point, her potential energy is at a maximum and her kinetic energy is zero.

B As she falls toward the bottom of the path, the rider accelerates and gains kinetic energy. Because the rider is not as high above the ground, her potential energy decreases.

C The rider, rising toward the opposite side, begins to slow down and lose kinetic energy. As she gains height, her potential energy increases.

D At the highest point on this side of the swing, her potential energy again is at a maximum, and her kinetic energy is zero.

Swinging Along When you ride on a swing, like the one shown in **Figure 10,** part of the fun is the feeling of almost falling as you drop from the highest point to the lowest point of the swing's path. Think about energy conservation to analyze such a ride.

The ride starts with a push that gets you moving, giving you some kinetic energy. As the swing rises, you lose speed but gain height. In energy terms, kinetic energy changes to gravitational potential energy. At the top of your path, potential energy is at its greatest. Then, as the swing accelerates downward, potential energy changes to kinetic energy. At the bottom of each swing, the kinetic energy is at its greatest and the potential energy is at its minimum. As you swing back and forth, energy continually converts from kinetic to potential and back to kinetic. What happens to your mechanical energy as you swing?

Environmental Science INTEGRATION

One way energy enters ecosystems is when green plants transform radiant energy from the Sun into chemical potential energy in the form of food. Energy moves through the food chain as animals that eat plants are eaten by other animals. Some energy leaves the food chain, such as when living organisms release thermal energy to the environment. Diagram a simple biological food chain showing energy conservation.

The Law of Conservation of Energy

When a ball is thrown into the air or a swing moves back and forth, kinetic and potential energy are constantly changing as the object speeds up and slows down. However, mechanical energy stays constant. Kinetic and potential energy simply change forms and no energy is destroyed.

This is always true. Energy can change from one form to another, but the total amount of energy never changes. Even when energy changes form from electrical to thermal and other energy forms as in the hair dryer shown in **Figure 11,** energy is never destroyed. Another way to say this is that energy is conserved. This principle is recognized as a law of nature. The **law of conservation of energy** states that energy cannot be created or destroyed. On a large scale, this law means that the total amount of energy in the universe remains constant.

Reading Check *What law states that the total amount of energy never changes?*

You might have heard about energy conservation or been told to conserve energy. These ideas are related to reducing the demand for electricity and gasoline, which lowers the consumption of energy resources such as coal and fuel oil. The law of conservation of energy, on the other hand, is a universal principle that describes what happens to energy as it is transferred from one object to another or as it is transformed.

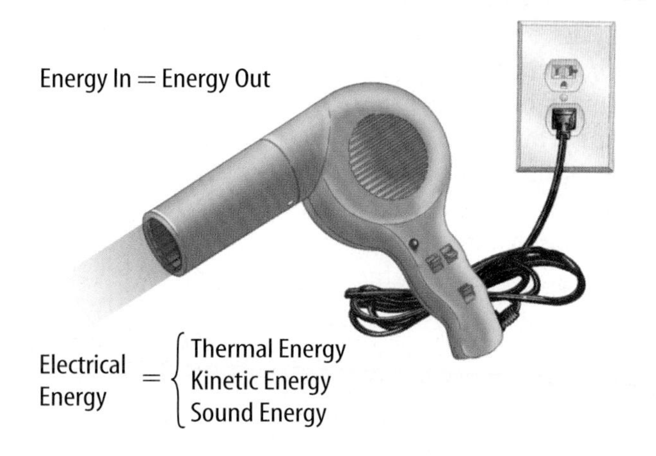

Figure 11
The law of conservation of energy requires that the total amount of energy going into a hair dryer must equal the total amount of energy coming out of the hair dryer.

Figure 12
In a swing, mechanical energy is transformed into thermal energy because of friction and air resistance.

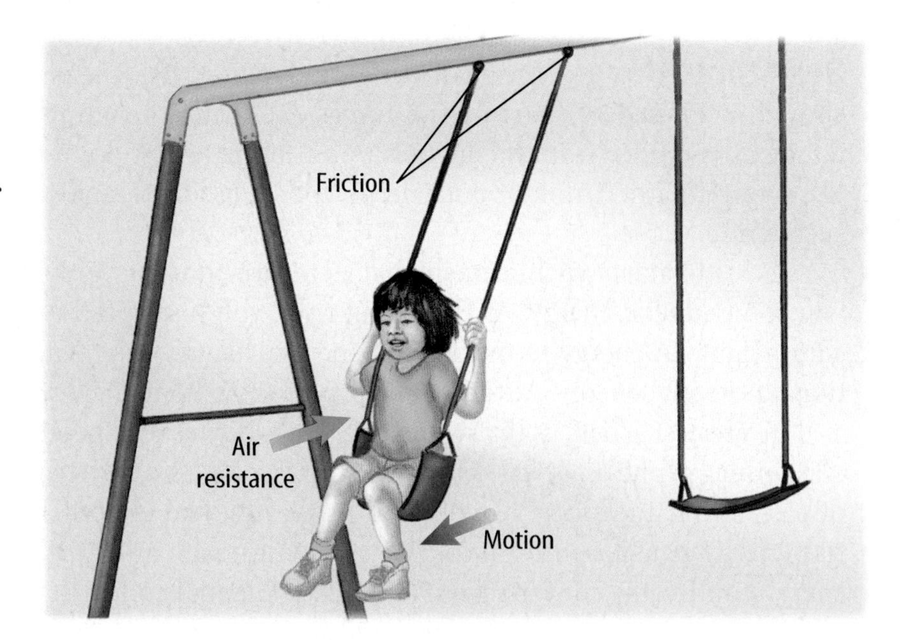

Transforming Energy Using a Paper Clip

Procedure

1. Straighten a **paper clip.** While holding the ends, touch the paper clip to the skin just below your lower lip. Note whether the paper clip feels warm, cool, or about room temperature.
2. Quickly bend the paper clip back and forth five times. Touch it below your lower lip again. Note whether the paper clip feels warmer or cooler than before.

Analysis

1. What happened to the temperature of the paper clip? Why?
2. As you bend the paper clip, explain all the energy conversions that take place.

Friction and the Law of Conservation of Energy You might be able to think of situations where it seems as though energy is not conserved. For example, while coasting along a flat road on a bicycle, you know that you will eventually stop if you don't pedal. If energy is conserved, why wouldn't your kinetic energy stay constant so that you would coast forever? In many situations, it might seem that energy is destroyed or created. Sometimes it is hard to see the law of conservation of energy at work.

Following Energy's Trail You know from experience that if you don't continue to pump a swing or be pushed by somebody else, your arcs will become lower and you eventually will stop swinging. In other words, the mechanical (kinetic and potential) energy of the swing seems to decrease, as if the energy were being destroyed. Is this a violation of the law of conservation of energy?

It can't be—it's the law! If the energy of the swing decreases, then the energy of some other object must increase by an equal amount to keep the total amount of energy the same. What could this other object be that experiences an energy increase? To answer this, you need to think about friction. With every movement, the swing's ropes or chains rub on their hooks and air pushes on the rider, as illustrated in **Figure 12.** Friction and air resistance cause some of the mechanical energy of the swing to change to thermal energy. With every pass of the swing, the temperature of the hooks and the air increases a little, so the mechanical energy of the swing is not destroyed. Rather, it is transformed into thermal energy. The total amount of energy always stays the same.

Converting Mass into Energy You might have wondered how the Sun unleashes enough energy to light and warm Earth from so far away. A special kind of energy conversion—nuclear fusion—takes place in the Sun and other stars. During this process a small amount of mass is transformed into a tremendous amount of energy. An example of a nuclear fusion reaction involving hydrogen nuclei is shown in **Figure 13A.**

Nuclear Fission Another process involving the nuclei of atoms, called nuclear fission, converts a small amount of mass into enormous quantities of energy. In this process, nuclei do not fuse—they are broken apart, as shown in **Figure 13B.** In either process, fusion or fission, mass is converted to energy. You must think of mass as energy when applying the law of conservation of energy to processes involving nuclear reactions. Here, as in all cases, the total amount of mass and energy is conserved. The process of nuclear fission is used by nuclear power plants to generate electrical energy. The fission process occurs in a nuclear reactor, and the heat released changes water to steam. The steam is used to spin an electric generator.

Research Visit the Glencoe Science Web site at **science.glencoe.com** for more information about the role of friction in the design of automobiles. Communicate to your class what you learn.

Figure 13
Mass is converted to energy in the processes of fusion and fission.

Nuclear fusion

H + H → He, Radiant energy

Mass **H** + Mass **H** > Mass **He** +Mass **neutron**

A **In this fusion reaction, the combined mass of the two hydrogen nuclei, H, is greater than the mass of the helium nucleus, He, and the neutron.**

Nuclear fission

+ U → Xe, Sr, Radiant energy

Mass **U** + Mass **neutron** > Mass **Xe** + Mass **Sr** + Mass **neutrons**

B **In nuclear fission, the mass of the large nucleus on the left is greater than the combined mass of the other two nuclei and the neutrons.** *Why aren't the masses equal?*

The Human Body—Balancing the Energy Equation

What forms of energy discussed in this chapter can you find in the human body? With your right hand, reach up and feel your left shoulder. With that simple action, stored potential energy within your body was converted to the kinetic energy of your moving arm. Did your shoulder feel warm to your hand? Some of the potential energy stored in your body is used to maintain a nearly constant internal temperature. A portion of this energy also is converted to the excess heat that your body gives off to its surroundings. Even the people shown standing in **Figure 14** require energy conversions to stand still.

Energy Conversions in Your Body The complex chemical and physical processes going on in your body also obey the law of conservation of energy. Your body stores energy in the form of fat and other chemical compounds. This chemical potential energy is used to fuel the processes that keep you alive, such as making your heart beat and digesting the food you eat. Your body also converts this energy to heat that is transferred to your surroundings, and you use this energy to make your body move. **Table 1** shows the amount of energy used in doing various activities. To maintain a healthy weight, you must have a proper balance between energy contained in the food you eat and the energy your body uses.

Figure 14
The runners convert the energy stored in their bodies more rapidly than the spectators do.
*Use **Table 1** to calculate how long a person would need to stand to burn as much energy as a runner burns in 1 h.*

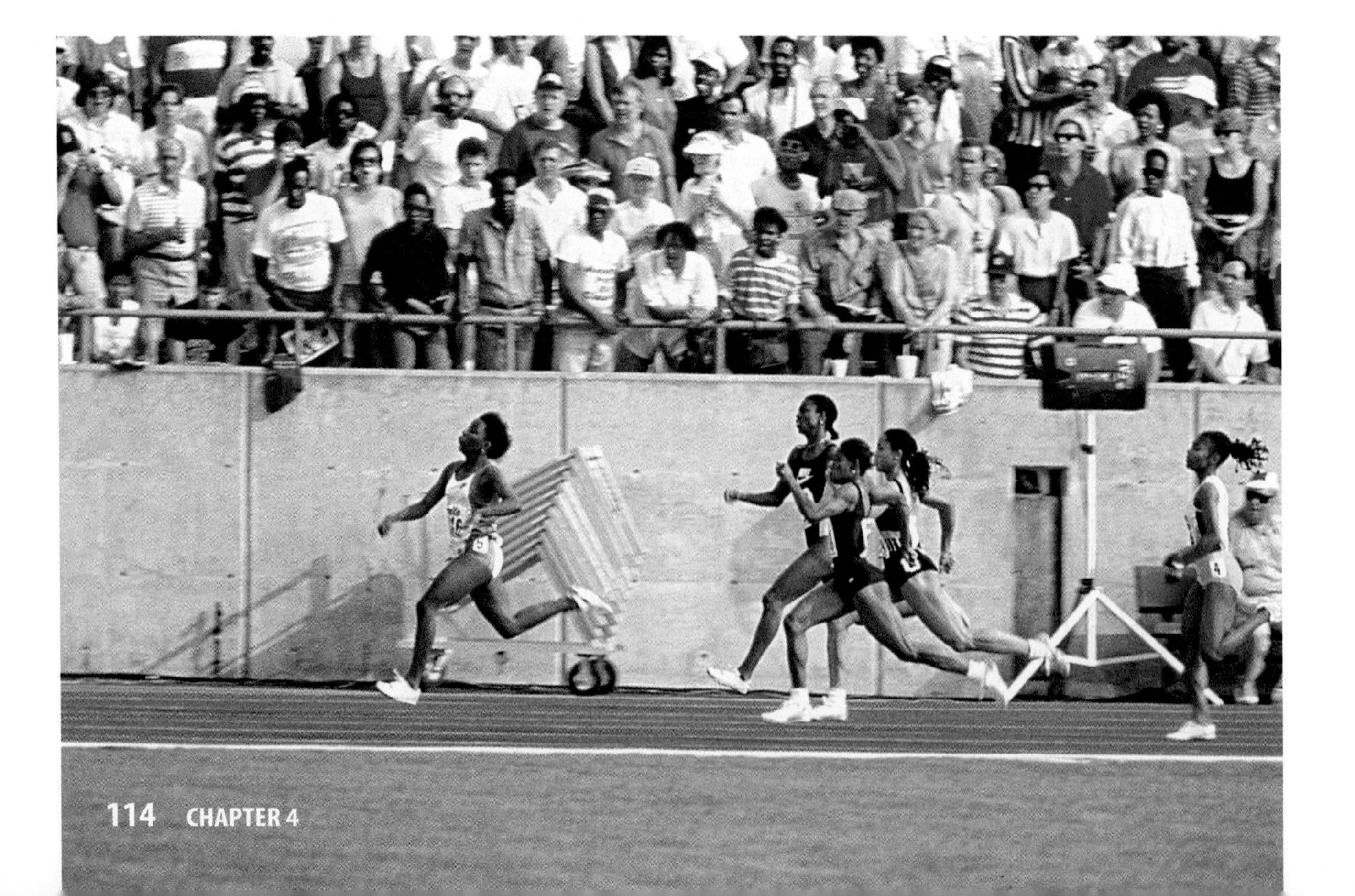

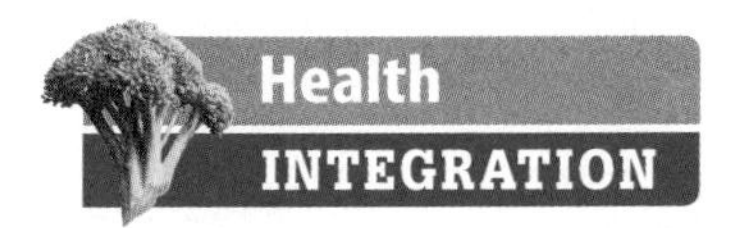

Food—Your Chemical Potential Energy Your body has been busy breaking down your breakfast into molecules that can be used as fuel. These fuel molecules, such as sugar, supply all the cells in your body with the energy they need to function. If you did not eat breakfast this morning, your body will convert energy stored in fat for its immediate needs until you eat again. The food Calorie (C) is a unit used by nutritionists to measure how much energy you get from various foods—1 C is equivalent to about 4,184 J. Every gram of fat a person consumes can supply 9 C of energy. Carbohydrates and proteins each supply about 4 C of energy per gram. The next time you go grocery shopping, look on the packages and notice how much energy each product eventually will supply to you.

Table 1 Calories Used in 1 h

Type of Activity	Body Frames		
	Small	Medium	Large
Sleeping	48	56	64
Sitting	72	84	96
Eating	84	98	112
Standing	96	112	123
Walking	180	210	240
Playing tennis	380	420	460
Bicycling (fast)	500	600	700
Running	700	850	1,000

Section  Assessment

1. Define the term *mechanical energy.* Describe the mechanical energy of a roller-coaster car immediately before it begins traveling down a long track.
2. What is the law of conservation of energy?
3. Applying bicycle brakes as you ride down a long hill causes the brake pads and the wheel rims to feel warm. Explain.
4. What is the source of the large amounts of energy released in nuclear reactors and in the Sun? Are the same processes occurring in the Sun and in reactors? Explain.
5. **Think Critically** Much discussion has focused on the need to drive more efficient cars and use less electricity. If the law of conservation of energy is true, why are people concerned about energy usage?

Skill Builder Activities

6. **Communicating** Suppose you drop a tennis ball out of a second-floor window. The first bounce will be the highest. Each bounce after that will be lower until the ball stops bouncing. Write a description of the energy conversions that take place, starting with dropping the ball. Accompany your description with an appropriate illustration. **For more help, refer to the Science Skill Handbook.**
7. **Communicating** Your body used energy as you walked into your school today. Where did this energy come from? In your Science Journal, write a paragraph describing where you acquired this energy. Trace it back through as many transformations as you can. **For more help, refer to the Science Skill Handbook.**

Activity Design Your Own Experiment

Swinging Energy

Imagine yourself swinging on a swing. What would happen if a friend grabbed the swing's chains as you passed the lowest point? Would you come to a complete stop or continue rising to your previous maximum height?

Recognize the Problem

How does the motion and maximum height reached by a swing change if the swing is interrupted?

Form a Hypothesis

Examine the diagram on this page. How is it similar to the situation in the introductory paragraph? An object that is suspended so that it can swing back and forth also is called a pendulum. Hypothesize what will happen to the pendulum's motion and final height if its swing is interrupted.

Goals

- **Construct** a pendulum to compare the exchange of potential and kinetic energy when a swing is interrupted.
- **Measure** the starting and ending heights of the pendulum.

Possible Materials

test-tube clamp
ring stand
support-rod clamp, right angle
30-cm support rod
2-hole, medium rubber stopper
string (1 m)
metersticks
graph paper

Safety Precautions

Be sure the base is heavy enough or well anchored so that the apparatus will not tip over.

Test Your Hypothesis

Plan

1. As a group, write your hypothesis and list the steps that you will take to test it. Be specific. Also list the materials you will need.
2. **Design** a data table and place it in your Science Journal.
3. Set up an apparatus similar to the one shown in the diagram.
4. **Devise** a way to measure the starting and ending heights of the stopper. Record your starting and ending heights in a data table. This will be your control.
5. **Decide** how to release the stopper from the same height each time.
6. Be sure you test your swing, starting it above and below the height of the cross arm. How many times should you repeat each starting point?

Do

1. Make sure your teacher approves your plan before you start.
2. Carry out the approved experiment as planned.
3. While the experiment is going on, write any observations that you make and complete the data table in your Science Journal.

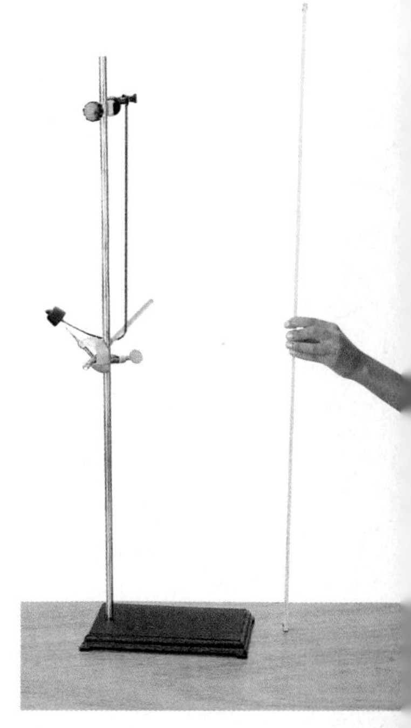

Analyze Your Data

1. When the stopper is released from the same height as the cross arm, is the ending height of the stopper exactly the same as its starting height? Use your data to support your answer.
2. **Analyze** the energy transfers. At what point along a single swing does the stopper have the greatest kinetic energy? The greatest potential energy?

Draw Conclusions

1. Do the results support your hypothesis? Explain.
2. **Compare** the starting heights to the ending heights of the stopper. Is there a pattern? Can you account for the observed behavior?
3. Do your results support the law of conservation of energy? Why or why not?
4. What happens if the mass of the stopper is increased? Test it.

Compare your conclusions with those of the other lab teams in your class. **For more help, refer to the Science Skill Handbook.**

The Impossible Dream

A machine that keeps on going? It has been tried for hundreds of years.

Science is a bit like an easygoing parent—it doesn't lay down a lot of laws. Those laws that do exist, however, are hard to get around. Just ask the hundreds of people who have tried throughout history—and failed—to build perpetual-motion machines.

In theory, a perpetual-motion machine would run forever and do work without a continual source of energy. You can think of it as a car that you could fill up once with gas, and with that single tankful, the car would run forever. Sound impossible? It is!

Artist M.C. Escher drew this never-ending staircase. You could walk on it perpetually!

Visitors look at the Keely Motor, the most famous perpetual-motion machine fraud of the late 1800s.

Science Puts Its Foot Down

For hundreds of years, people have tried to create perpetual-motion machines. But these machines won't work because they violate two of nature's laws. The first law is the law of conservation of energy, which states that energy cannot be created or destroyed. It can change form—say, from mechanical energy to electrical energy—but you always end up with the same amount of energy that you started with.

How does that apply to perpetual-motion machines? When a machine does work on an object, the machine transfers energy to the object. Unless that machine gets more energy from somewhere else, it can't keep doing work. If it did, it would be creating energy.

The second law states that heat by itself always flows from a warm object to a cold object. Heat will only flow from a cold object to a warm object if work is done. In the process, some heat always escapes. The hood of a car, for instance, feels hot when the car is running. That's because heat, or thermal energy has escaped from the car's engine.

To make up for these energy losses, energy constantly needs to be transferred to the machine. Otherwise, it stops. No perpetual motion. No free electricity. No devices that generate more energy than they use. No engine motors that run forever without fuel. No lights that shine or ships that sail without a continual source of energy. Some laws just can't be broken.

Losing Battles

For more than 300 years, people have tried to build a perpetual-motion machine that works. Nobody has ever succeeded. In fact, the U.S. Patent Office, which studies inventions to see if they work before they grant a patent, refuses to even look at machines that their inventors claim are perpetual-motion machines.

Some laws just can't be broken.

CONNECTIONS Analyze **Using your school or public library resources, locate a picture or diagram of a perpetual-motion machine. Figure out why it won't run forever. Explain to the class what the problem is.**

For more information, visit science.glencoe.com

Chapter 4 Study Guide

Reviewing Main Ideas

Section 1 The Nature of Energy

1. Energy is the ability of something to cause change.
2. Energy can take a variety of different forms. *What are some of the forms of energy illustrated in the photograph below?*

3. Moving objects have kinetic energy that depends on the mass of the object and the velocity of the object.
4. Potential energy is stored energy. Objects that can fall have gravitational potential energy—the amount depends on their weight and height above the ground. *If Balanced Rock, shown in the photograph below, fell, how would its gravitational potential energy change?*

Section 2 Conservation of Energy

1. Energy can change from one form to another. You observe many energy transformations every day.
2. The law of conservation of energy states that energy never can be created or destroyed. Because of friction, energy might seem to be lost, but it has changed into thermal energy.
3. Falling, flying, and swinging objects all involve transformations between potential and kinetic energy. The total amount of potential and kinetic energy is called mechanical energy. *What energy transformations are taking place in the figure below? Is energy being conserved?*

4. Mass can be converted into energy in nuclear fusion and fission reactions. Fusion and fission involve atomic nuclei and release tremendous amounts of energy.

After You Read

FOLDABLES Reading & Study Skills

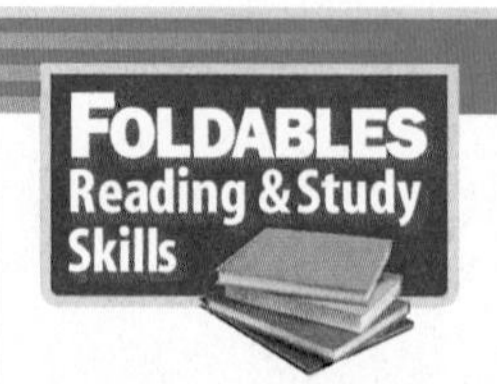

To help review what you've learned about energy and energy conservation, use the Foldable you made at the beginning of the chapter.

Visualizing Main Ideas

Complete the concept map below using the following terms: nuclear fusion/fission, potential energy, friction, joules, kinetic energy.

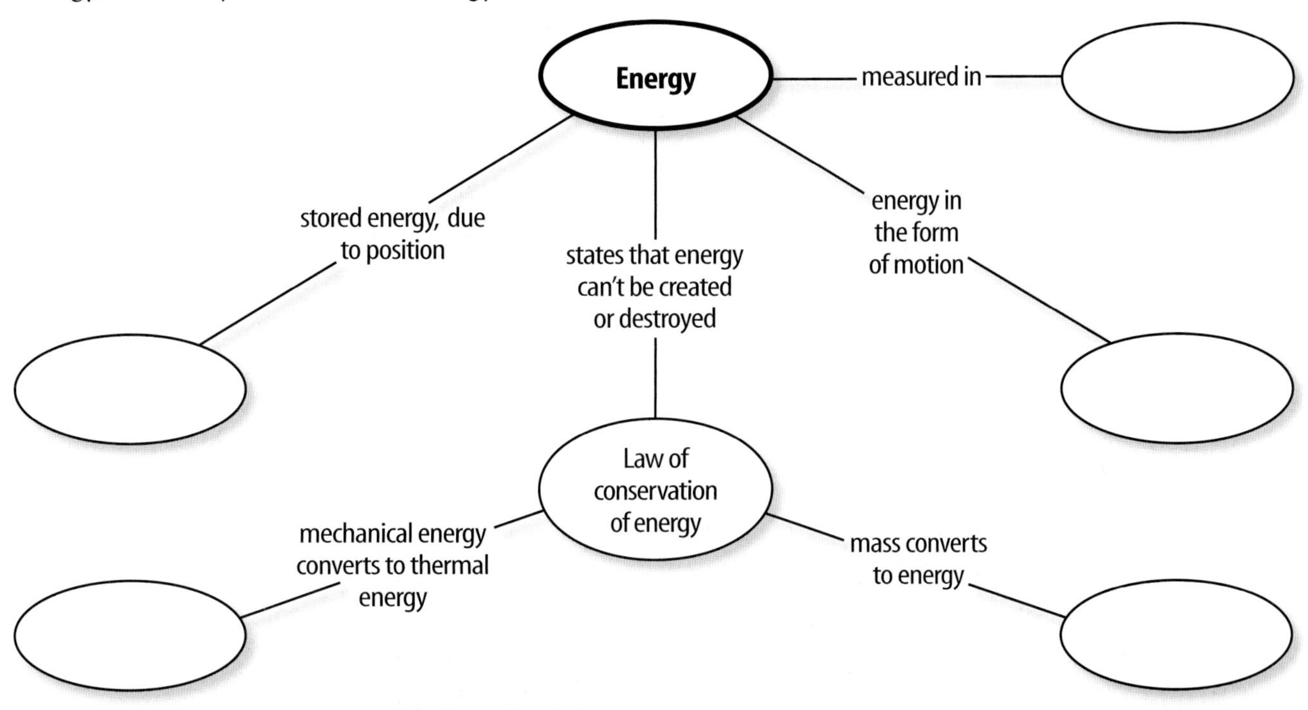

Vocabulary Review

Vocabulary Words

a. chemical potential energy
b. elastic potential energy
c. gravitational potential energy
d. joule
e. kinetic energy
f. law of conservation of energy
g. mechanical energy
h. potential energy

Study Tip

Be a teacher! Get together a group of friends and assign each person a section of the chapter to teach. When you teach you remember and understand information thoroughly.

Using Vocabulary

Using the list of vocabulary words, replace the underlined words with the correct science term.

1. In describing the energy changes of a bouncing ball, the total amount of kinetic and potential energy must be conserved.
2. The energy due to stretching of a bow can be used to propel an arrow forward.
3. Snow on the side of a mountain has energy due to Earth pulling down on it.
4. The fact that energy cannot be created means that dynamite does not create energy when it explodes.
5. The muscles of a runner transform chemical potential energy into energy of motion.

Chapter 4 Assessment

Checking Concepts

Choose the word or phrase that best answers the question.

1. When energy is transferred from one object to another, what must occur?
A) an explosion
B) a chemical reaction
C) nuclear fusion
D) a change

2. Which has more kinetic energy, a large dog sitting on a sidewalk or a small cat running down the street?
A) the large dog
B) the small cat
C) Both have the same kinetic energy.
D) need more information to answer

3. What energy transformations occur when a lump of clay is dropped?
A) *GPE* → *KE* → thermal energy
B) *KE* → chem *PE* → thermal energy
C) *KE* → *GPE* → thermal energy
D) *GPE* → *KE* → elastic *PE*

4. Suppose a juggler is juggling oranges. At an orange's highest point, what form of energy does it have?
A) mostly potential energy
B) mostly kinetic energy
C) no potential or kinetic energy
D) equal amounts of both

5. The gravitational potential energy of an object depends on which of the following?
A) velocity and height
B) velocity and weight
C) weight and height
D) weight and acceleration

6. Which idea is central to the law of conservation of energy?
A) Friction produces thermal energy.
B) People must conserve energy.
C) The total amount of energy is constant.
D) Energy is the ability to cause change.

7. To what property of an object is kinetic energy directly related?
A) volume
B) height
C) position
D) mass

8. Friction frequently causes some of an object's mechanical energy to be changed to which of the following forms?
A) thermal energy
B) nuclear energy
C) gravitational potential energy
D) chemical potential energy

9. What is the process of breaking apart large atomic nuclei into smaller nuclei called?
A) nuclear fusion
B) nuclear fission
C) atomic fracture
D) transformation

10. Green plants store energy from the Sun in what form?
A) light energy
B) chemical potential energy
C) gravitational potential energy
D) electrical energy

Thinking Critically

11. Briefly describe the energy changes in a swinging pendulum. Explain how energy is conserved, even as a pendulum slows.

12. How is energy transformed when one end of a stretched spring is released?

13. Describe the energy conversions that take place during a roller-coaster ride.

14. Explain why the law of conservation of energy must include mass.

15. A 15-kg bicycle carrying a 50-kg boy is traveling at a speed of 5 m/s. What is the kinetic energy of the bicycle (including the boy)?

Developing Skills

16. **Making and Using Tables** Make a table that reports the kinetic energy of a 1-kg object moving at various speeds. Compute the kinetic energy at speed increments of 1 m/s, from 0 m/s to 10 m/s.

Energy of a Moving Object

Mass (kg)	Speed (m/s)	KE (J)
1	0	
1	1	
1	2	
1	10	

17. **Making and Using Graphs** Make a graph to show how kinetic energy changes as speed increases. Using the data from question 16, plot speed on the *x*-axis and kinetic energy on the *y*-axis. Describe the shape of the plotted line. How is the shape of this line different from a straight line?

Performance Assessment

18. **Poster** Make an educational poster that highlights the law of conservation of energy. Include examples of energy conservation in your daily life.

19. **Oral Presentation** Research one type of alternative fuel. Find out the advantages and disadvantages of using the fuel. Present your findings to your class.

TECHNOLOGY

Go to the Glencoe Science Web site at **science.glencoe.com** or use the **Glencoe Science CD-ROM** for additional chapter assessment.

Test Practice

A team of students conducted an experiment in energy conversion and collected their data in a table.

Trial	Mass of Toy Car (kg)	Distance Traveled (m)	Time to Travel Ramp (s)	Speed (m/s)
1	0.05	1	3.1	0.32
	0.05	2	4.4	0.45
2	0.05	0.5	2.2	0.22
	0.05	1	3.1	0.32

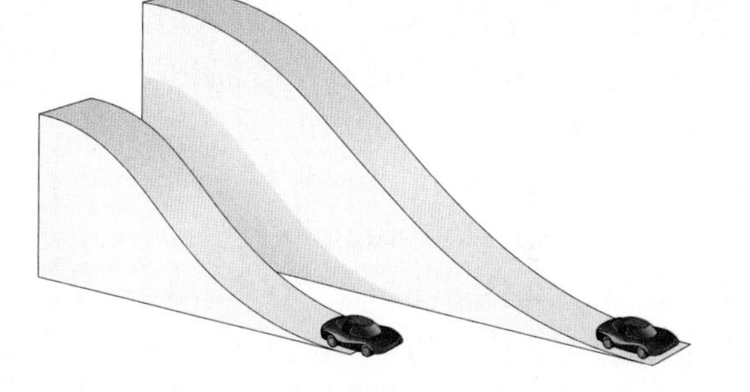

Study the experiment and the table above and answer the following questions.

1. The illustration above shows how these data were collected. Recording which one of the following would improve the experiment?
   A) owner of the toy cars
   B) weather conditions
   C) height of the ramps
   D) brand of the toy cars

2. At the beginning of the experiment, a toy car is at rest at the top of a ramp. In this position, what form of energy is stored in the car?
   F) kinetic energy
   G) fission
   H) potential energy
   J) mass

Work and Machines

Paola Pezzo couldn't have won a gold medal in the 2000 Summer Olympic Games without a machine—her mountain bike. Can you imagine your life without machines? Think of all the machines you use every day—in-line skates, staplers, pencil sharpeners. Machines make work easier. Many machines are simple. Others, such as mountain bikes and automobiles, are combinations of many simple machines. What kinds of machines are in a mountain bike? In this chapter, you will learn about simple and compound machines and how they change forces to make work easier.

What do you think?

Science Journal Look at the picture below with a classmate. Discuss what you think is happening. Here's a hint: *This makes a carpenter's work easier.* Write your answer or best guess in your Science Journal.

Before the hydraulic lift, mechanics used a pulley to raise cars off the ground. The pulley had many grooved wheels and a long chain threaded through them. The mechanic had to pull several meters of chain just to raise the car a few centimeters. In this activity, make your own pulley and experience the advantage of using simple machines.

Construct a Pulley

1. Tie a rope several meters in length to the center of a broom handle. Have one student hold both ends of the handle.
2. Have another student hold the ends of a second broom handle and face the first student. The two handles should be parallel, a meter apart.
3. Have a third student loop the free end of the rope around the second handle. Continue wrapping, making six or seven loops.
4. The third student should stand to the side of one of the handles and pull on the free end of the rope. The two students holding the broom handles should prevent the handles from coming together.

Observe

Write a paragraph in your Science Journal describing what happened when the rope was pulled. How far did the rope have to be pulled to bring the handles together?

Before You Read

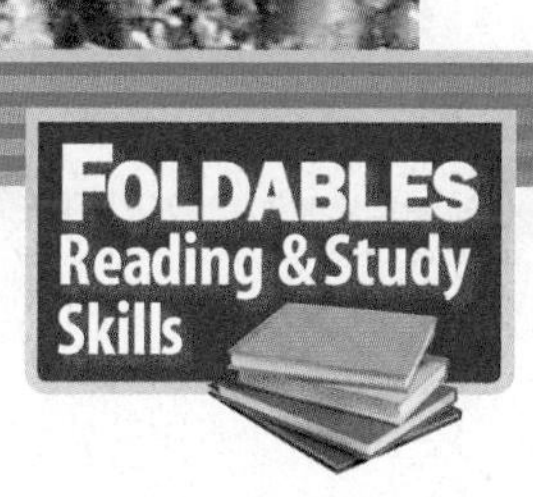

Making a Compare and Contrast Study Fold **Make the following Foldable to see how work and machines are similar and different.**

1. Place a sheet of paper in front of you so the short side is at the top. Fold the paper in half from the left side to the right side and unfold.
2. Through the top thickness of paper, cut along the middle fold line to form two tabs. Label them *Work without Machines* and *Work with Machines*.
3. List examples of work you do without machines under its tab. As you read the chapter, rate the work you did without machines on a scale of 1 (little force) to 10 (great force). Write it next to the work.

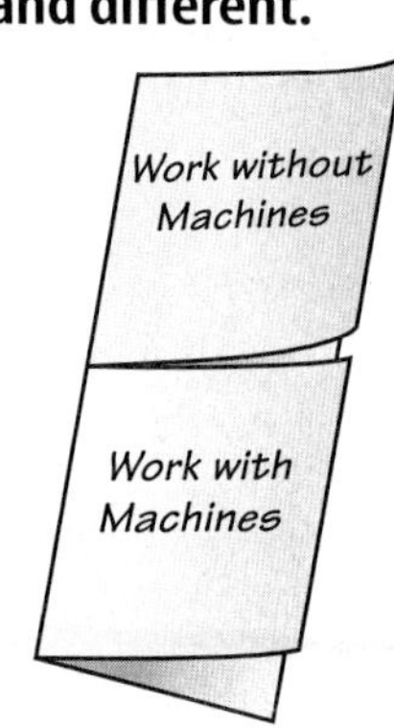

Work

As You Read

What You'll Learn

- **Explain** the meaning of work.
- **Explain** how work and energy are related.
- **Calculate** work.
- **Calculate** power.

Vocabulary
work
power

Why It's Important

Learning the scientific meaning of work is a key to understanding how machines make life easier.

What is work?

Press your hand against the surface of your desk as hard as you can. Although your muscles might start to feel tired, you haven't done any work. Most people feel that they have done work if they push or pull something. However, the scientific meaning of work is more specific. **Work** is the transfer of energy that occurs when a force makes an object move. Recall that a force is a push or a pull. For work to be done, a force must make something move. If you push against the desk and nothing moves, you haven't done any work.

Doing Work Two conditions have to be satisfied for work to be done on an object. One is that the object has to move, and the other is that the motion of the object must be in the same direction as the applied force. For example, if you pick up a pile of books from the floor as in **Figure 1,** you do work on the books. The books move upward in the direction of the force you are applying. If you hold the books in your arms without moving, you are not doing work on the books. You're still applying an upward force to keep the books from falling, but no movement is taking place.

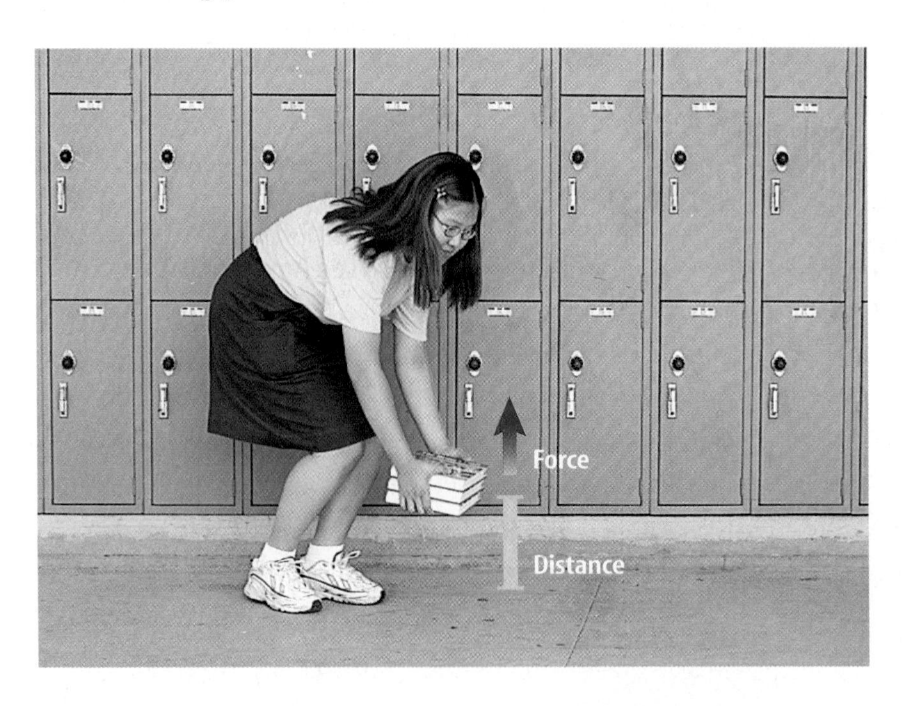

Figure 1
When you lift a stack of books, your arms apply a force upward and the books move upward. Because the force and distance are in the same direction, your arms have done work on the books.

Direction of Motion Now suppose you start walking as in **Figure 2.** The books are moving horizontally, but your arms still do no work on the books. The force exerted by your arms is still upward, and is at right angles to the direction the books are moving. It is your legs that are exerting the force that causes you and the books to move forward. It is your legs, not your arms, that cause work to be done on the books.

Reading Check *What must you ask to determine if work is being done?*

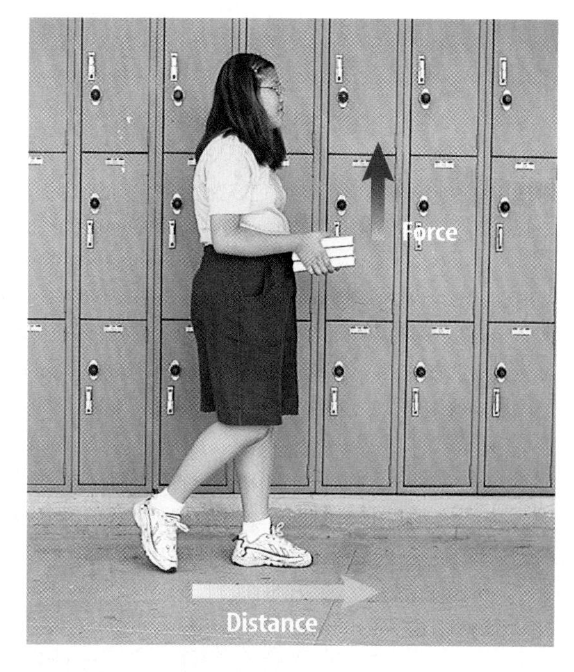

Figure 2
If you hold a stack of books and walk forward, your arms are exerting a force upward. However, the distance the books move is horizontal. Therefore your arms are not doing work on the books. *Does this mean no work is done on the books? Explain.*

Work and Energy

How are work and energy related? When work is done, a transfer of energy always occurs. This is easy to understand when you think about how you feel after carrying a heavy box up a flight of stairs. Remember that when the height of an object above Earth's surface increases, the potential energy of the object increases. You transferred energy from your moving muscles to the box and increased its potential energy by increasing its height.

You may recall that energy is the ability to cause change. Another way to think of energy is that energy is the ability to do work. If something has energy, it can transfer energy to another object by doing work on that object. When you do work on an object, you increase its energy. If you do work, such as the person carrying the box in **Figure 3,** your energy decreases. Energy is always transferred from the object that is doing the work to the object on which the work is done.

Figure 3
By carrying a box up the stairs, you are doing work. You transfer some of your energy to the box. *How is work done on the box?*

Calculating Work

Which of these tasks would involve more work—lifting a pack of gum or a pile of books from the floor to waist level? Would you do more work if you lifted the books from the floor to your waist or over your head? You probably can guess the answers to these questions. You do more work when you exert more force and when you move an object a greater distance. In fact, the amount of work done depends on two things: the amount of force exerted and the distance over which the force is applied.

When a force is exerted and an object moves in the direction of the force, the amount of work done can be calculated as follows.

$$Work = force \times distance$$
$$W = F \times d$$

In this equation, force is measured in newtons (N) and distance is measured in meters. Work, like energy, is measured in joules. One joule is about the amount of work required to lift a baseball a vertical distance of 0.7 m.

Math Skills Activity

Calculating Work Given Force and Distance

Example Problem

You move a 75-kg refrigerator 35 m. This requires a force of 90 N. How much work, in joules, was done while moving the refrigerator?

Solution

1. *This is what you know:* force: $F = 90$ N; distance: $d = 35$ m; 1 newton-meter (N·m) = 1 joule (J)
2. *This is what you need to find:* Work W
3. *This is the equation you need to use:* $W = F \times d$
4. *Substitute the known values:* $W = (90 \text{ N}) \times (35 \text{ m}) = 3{,}150 \text{ N·m} = 3{,}150 \text{ J}$

   Check your answer by dividing the work you calculated by the given distance. Did you calculate the force that was given?

Practice Problem

When you and a friend move a 45-kg couch to another room, you exert a force of 75 N over 5 m. How much work, in joules, did you do?

For more help, refer to the Math Skill Handbook.

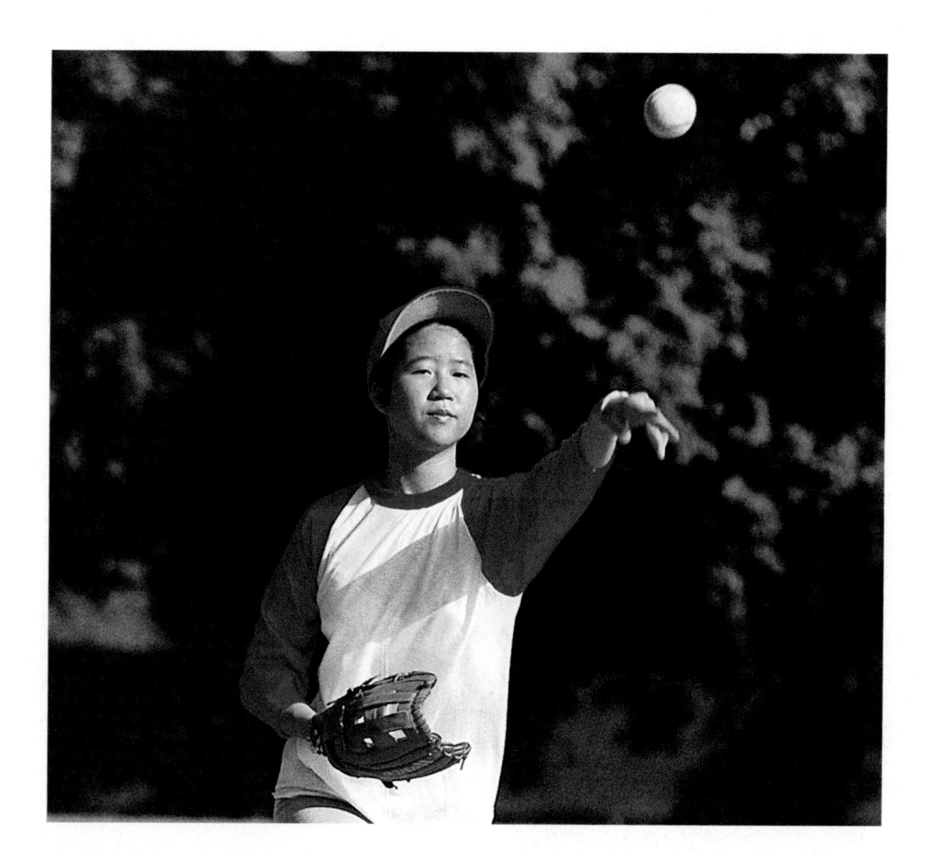

Figure 4
A softball pitcher exerts a force on the ball to throw it to the catcher. After the ball leaves her hand, she no longer is exerting any force on the ball. The only work she does on the ball is moving it from the back of the pitch to the release.

When is work done? Suppose you give a book a push and it slides along a table for a distance of 1 m before it comes to a stop. The distance you use to calculate the work you did is how far the object moves while the force is being applied. Even though the book moved 1 m, you do work on the book only while your hand is in contact with it. The distance in the formula for work is the distance the book moved while your hand was pushing on the book. As **Figure 4** shows, work is done on an object only when a force is being applied to the object.

Power

Suppose you and another student are pushing boxes of books up a ramp to load them into a truck. To make the job more fun, you make a game of it, racing to see who can push a box up the ramp faster. The boxes weigh the same, but your friend is able to push a box a little faster than you can. She moves a box up the ramp in 30 s. It takes you 45 s. You both do the same amount of work on the books because the boxes weigh the same and are moved the same distance. The only difference is the time it takes to do the work.

In this game, your friend has more power than you do. **Power** is the amount of work done in a certain amount of time. It is a rate—the rate at which work is done.

Reading Check *How is power related to work?*

Mini LAB

Calculating Your Work and Power

Procedure

1. Find a set of **stairs** that you can safely walk and run up. Measure the total height of the stairs in meters.
2. Record how many seconds it takes you to walk and run up the stairs.
3. Calculate the work you did in walking and running up the stairs using $W = F \times d$. For force, use your weight in newtons (your weight in pounds $\times$ 4.5).
4. Use the formula $P = W/t$ to calculate the power you needed to walk and run up the stairs.

Analysis

1. Is the work you did walking and running the steps the same?
2. Which required more power—walking or running up the steps? Why?

Research Visit the Glencoe Science Web site at **science.glencoe.com** for more information about energy-efficient devices. Communicate to your class what you learn.

Calculating Power To determine the power you deliver by pushing a box up the ramp, you need a way to calculate power. To calculate power, divide the work done by the time that is required to do the work.

$$\text{Power} = \text{work}/\text{time}$$

$$P = W/t$$

Power is measured in watts, named for James Watt, who helped develop the steam engine in the eighteenth century. A watt (W) is 1 J/s. A watt is fairly small—about equal to the power needed to raise a glass of water from a table to your mouth in 1 s. Because the watt is such a small unit, large amounts of power often are expressed in kilowatts. One kilowatt (kW) equals 1,000 W. If you were to run up a flight of steps in about 1.5 s, it would take about 1 kW of power.

Math Skills Activity

Calculating Power Given Work and Time

Example Problem

It took five minutes to move a refrigerator. You did 3,150 joules of work in the process. How much power was required to move the refrigerator?

Solution

1. *This is what you know:* Work: $W = 3{,}150$ J; Time: $t = 5$ min $= 300$ s; 1 J/s $=$ 1 W
2. *This is what you need to find:* Power (P)
3. *This is the equation you need to use:* $P = W/t$
4. *Substitute the known values:* $P = 3{,}150\text{ J}/300\text{ s} = 10.5\text{ J/s} = 10.5\text{ W}$

Check your answer by multiplying the power you calculated by the time given in the problem. Did you calculate the same work that was given?

Practice Problem

How much power is required to push a car for 10 seconds if the amount of work done during that time is 5,500 joules?

For more help, refer to the Math Skill Handbook.

Power and Energy Doing work is a way of transferring energy from one object to another. Remember that energy can be transferred in other ways that don't involve doing work. For example, a lightbulb like the one in **Figure 5** uses electrical energy to produce heat and light, but no work is done. Power is produced or used any time energy is transferred from one object to another. Power is the energy transferred divided by the time needed for the transfer to occur. Anytime energy is transferred from one object to another, power can be calculated from the following equation.

$$\text{Power} = \text{energy}/\text{time}$$
$$P = E/t$$

Figure 5
This 100 W lightbulb uses energy at 100 J/s, converting electrical energy into light and heat. *Even though energy is being transferred, why is no work done?*

Changing Energy by Doing Work What happens to the energy of a book when you lift it off your desk? You changed the height of the book, so its potential energy increased. Where did this increase in energy come from? You transferred energy to the book by doing work on the book when you lifted it. You can also increase the kinetic energy of an object by doing work on it, as when you push furniture from one place to another. In another example, think about using sandpaper on a piece of wood. Feel the wood and you will notice that it is warm. The energy of the wood has increased in the form of heat from friction. Anytime you do work on an object, you cause its energy to increase.

Section  Assessment

1. Explain how the scientific definition of work is different from the everyday meaning.
2. How are work and energy related?
3. A person pushed a bowling ball 20 m. The amount of work done was 1,470 J. How much force did the person exert?
4. How are power, work, and time related?
5. **Think Critically** In which of the following situations is work being done? Explain.
   - A person shovels snow off a sidewalk.
   - A worker lifts bricks, one at a time, from the ground to the back of a truck.
   - A roofer's assistant carries a bundle of shingles across a construction site.

Skill Builder Activities

6. **Solving One-Step Equations** A passenger weighing 500 N is inside an elevator weighing 24,500 N that rises 30 m in 1 min. How much power is needed for the elevator's trip? **For more help, refer to the Math Skill Handbook.**
7. **Communicating** In your Science Journal, write down everything you did today that would be considered work in the everyday sense. From this list, choose one task that also fits the scientific description of work. Write a paragraph explaining how the task fits the scientific and the everyday descriptions. **For more help, refer to the Science Skill Handbook.**

SECTION 2

Using Machines

As You Read

What You'll Learn

- **Explain** how machines make work easier.
- **Calculate** mechanical advantage.
- **Calculate** efficiency.

Vocabulary

machine
effort force
resistance force
mechanical advantage
efficiency

Why It's Important

Complex devices are made up of simpler parts called simple machines. Even your body contains simple machines.

What is a machine?

How many machines did you use today? Did you cut your food with a knife? Or maybe you used a pair of scissors? If you did, you used a machine. A **machine** is a device that makes doing work easier. When you think of a machine you may picture a device with an engine and many moving parts. But not all machines are complicated or powered by engines or electric motors. Machines can be simple, and can be powered by a force applied by a person. Some, like knives, scissors, and doorknobs, are used everyday to make doing work easier.

Making Work Easier

Machines can make work easier by increasing the force that can be applied to an object. A bottle opener increases the force you can apply to a bottle cap, causing the cap to bend. A car jack enables you to lift a heavy automobile. A second way that machines can make work easier is by increasing the distance over which a force can be applied. A leaf rake is an example of this type of machine. Machines can also make work easier by changing the direction of an applied force. When you open window blinds by pulling on a cord, the downward force on the cord is changed to an upward force that opens the blinds.

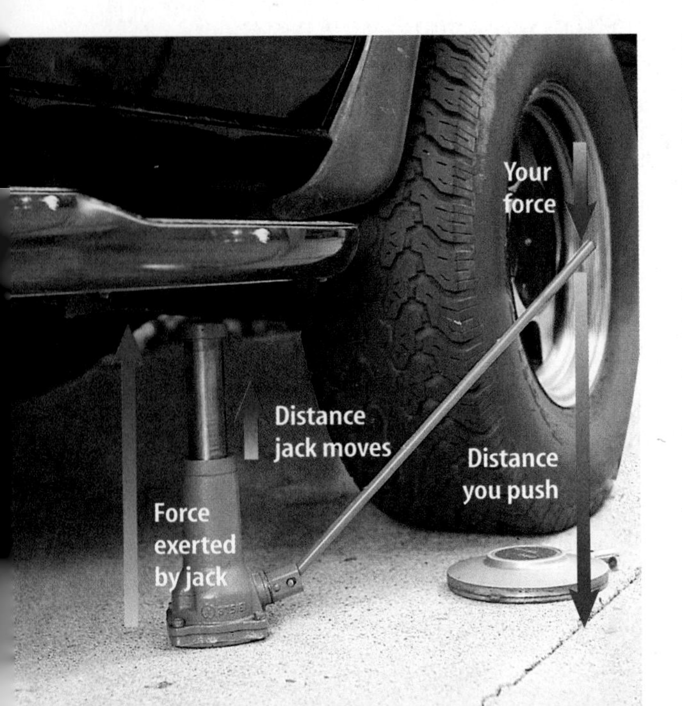

Figure 6
A car jack works by increasing your force.

Increasing Force Nobody could lift a car to change a flat tire without help. A car jack, like the one in **Figure 6,** is an example of a machine that multiplies your force.

Remember that work is the product of force and distance. You can do the same amount of work by applying a small force over a long distance as you can by applying a large force over a short distance. For example, the distance you push the handle of a car jack downward is longer than the distance the car moves upward, and the upward force exerted by the jack is greater than the downward force you exert on the handle. Can you think of other machines that multiply force?

Figure 7
Whether the mover slides the chair up the ramp or lifts it directly into the truck, she will do the same amount of work. Doing the work over a longer distance allows her to use less force.

Increasing Distance Why does the mover in **Figure 7** push the heavy furniture up the ramp instead of lifting it directly into the truck? It is easier for her because less force is needed to move the furniture.

The work done in lifting an object depends on the change in height of the object. The same amount of work is done whether the mover pushes the furniture up the long ramp or lifts it straight up. If she uses a ramp to lift the furniture, she moves the furniture a longer distance than if she just raised it straight up. If work stays the same and the distance is increased, then less force will be needed to do the work.

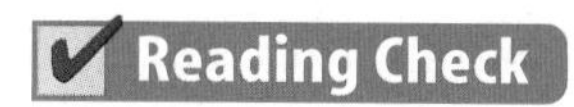

How does a ramp make lifting an object easier?

Changing Direction Some machines change the direction of the force you apply. When you use the car jack, you are exerting a force downward on the jack handle. The force exerted by the jack on the car is upward. The direction of the force you applied is changed from downward to upward. Some machines change the direction of the force that is applied to them in another way. The wedge-shaped blade of an ax is one example. When you use an ax to split wood, you exert a downward force as you swing the ax toward the wood. As **Figure 8** shows, the blade changes the downward force into a horizontal force that splits the wood apart.

Figure 8
An ax blade changes the direction of the force from vertical to horizontal.

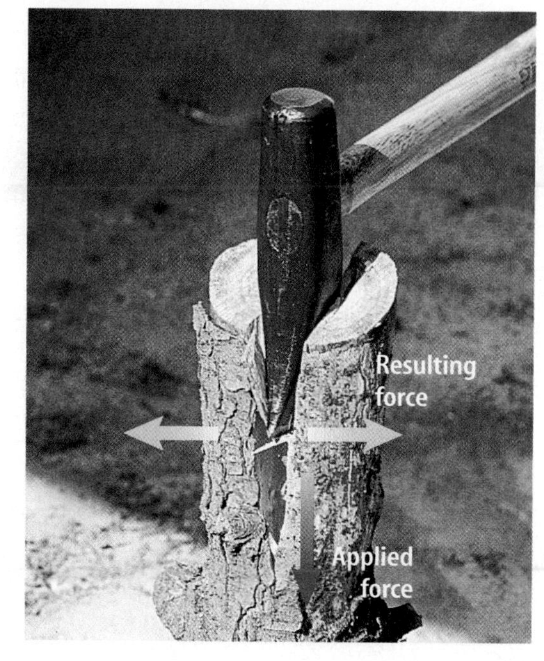

Figure 9
A crowbar increases the force you apply and changes its direction.

TRY AT HOME Mini LAB

Machines Multiplying Force

Procedure

1. Open a **can of food** using a **manual can opener**. **WARNING:** Do not touch can opener's cutting blades or cut edges of the can's lid.
2. Use a **metric ruler** to measure the diameter of the cutting blade of the can opener.
3. Measure the length of the handle you turn.

Analysis

1. Compare how difficult it is to open the can using the can opener with how difficult it would have been to open the can by turning the cutting blade with a smaller handle.
2. Compare the diameter of the cutting blade with the diameter of the circle formed by turning the can opener's handle.
3. Infer why a can opener makes it easier to open a metal can.

The Work Done by Machines

To pry the lid off a wooden crate with a crowbar, you'd slip the end of the crowbar under the edge of the crate lid and push down on the handle. By moving the handle downward, you do work on the crowbar. As the crowbar moves, it does work on the lid, lifting it up. **Figure 9** shows how the crowbar increases the amount of force being applied and changes the direction of the force.

When you use a machine such as a crowbar, you are trying to move something that resists being moved. For example, if you use a crowbar to pry the lid off a crate, you are working against the friction between the nails in the lid and the crate. You also could use a crowbar to move a large rock. In this case, you would be working against gravity—the weight of the rock.

Effort and Resistance Forces Two forces are involved when a machine is used to do work. You exert a force on the machine, such as a bottle opener, and the machine then exerts a force on the object you are trying to move, such as the bottle cap. The force applied to the machine is called the **effort force.** F_e stands for the effort force. The force applied by the machine to overcome resistance is called the **resistance force,** symbolized by F_r. When you try to pull a nail out with a hammer as in **Figure 10,** you apply the effort force on the handle. The resistance force is the force the claw applies to the nail.

Two kinds of work need to be considered when you use a machine—the work done by you on the machine and the work done by the machine. When you use a crowbar, you do work when you apply force to the crowbar handle and make it move. The work done by you on a machine is called the input work and is symbolized by W_{in}. The work done by the machine is called the output work and is abbreviated W_{out}.

Conserving Energy Remember that energy is always conserved. When you do work on the machine, you transfer energy to the machine. When the machine does work on an object, energy is transferred from the machine to the object. Because energy cannot be created or destroyed, the amount of energy the machine transfers to the object cannot be greater than the amount of energy you transfer to the machine. A machine cannot create energy, so W_{out} is never greater than W_{in}.

However, the machine does not transfer all of the energy it receives to the object. In fact, when a machine is used, some of the energy transferred changes to heat due to friction. The energy that changes to heat cannot be used to do work, so W_{out} is always smaller than W_{in}.

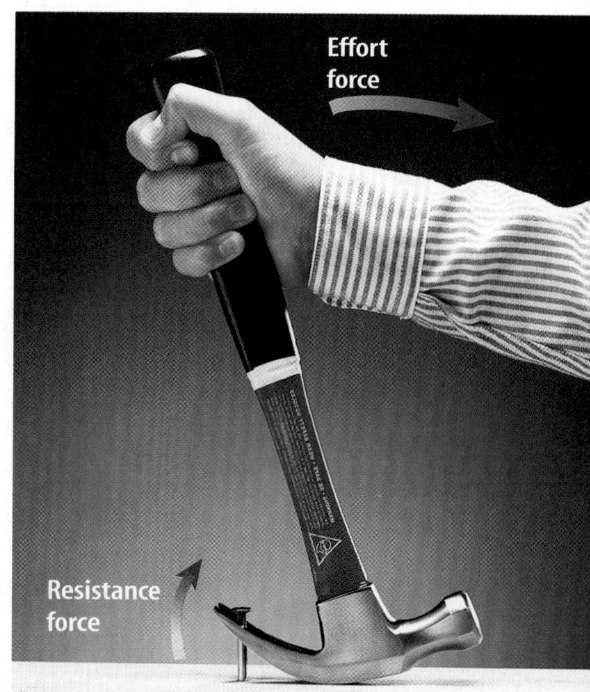

Figure 10
When prying a nail out of a piece of wood with a claw hammer, you exert the effort force on the handle of the hammer, and the claw exerts the resistance force.

Ideal Machines Remember that work is calculated by multiplying force by distance. The input work is the product of the effort force and the distance over which the effort force is exerted. The output work is the product of the resistance force and the distance that force moves the object.

Suppose a perfect machine could be built in which there was no friction. None of the input work or output work would be converted to heat. For such an ideal machine, the input work equals the output work. So for an ideal machine,

$$W_{in} = W_{out}$$

Suppose the ideal machine increases the force applied to it. This means that the resistance force, F_r, is greater than the effort force F_e. Recall that work is equal to force times distance. If F_r is greater than F_e, then W_{in} and W_{out} can be equal only if the effort force is applied over a greater distance than the resistance force is exerted over.

For example, suppose the hammer claw in **Figure 10** moves a distance of 0.10 m to remove a nail. If a resistance force of 1,500 N is exerted by the claw of the hammer, and you move the handle of the hammer 0.5 m, you can find the effort force as follows.

$$W_{in} = W_{out}$$
$$F_e \times d_e = F_r \times d_r$$
$$F_e \times (0.5\text{ m}) = (1{,}500\text{ N}) \times (0.1\text{ m})$$
$$F_e \times (0.5\text{ m}) = 150\text{ N}\cdot\text{m}$$
$$F_e = 300\text{ N}$$

Because the distance you move the hammer is longer than the distance the hammer moves the nail, the effort force is less than the resistance force.

Figure 11
Mini blinds are a familiar example of a simple machine that changes the direction of a force. When you pull down on the cord, the direction of your force is changed to upward. Because the effort force equals the resistance force, the MA of the mini blinds is 1.

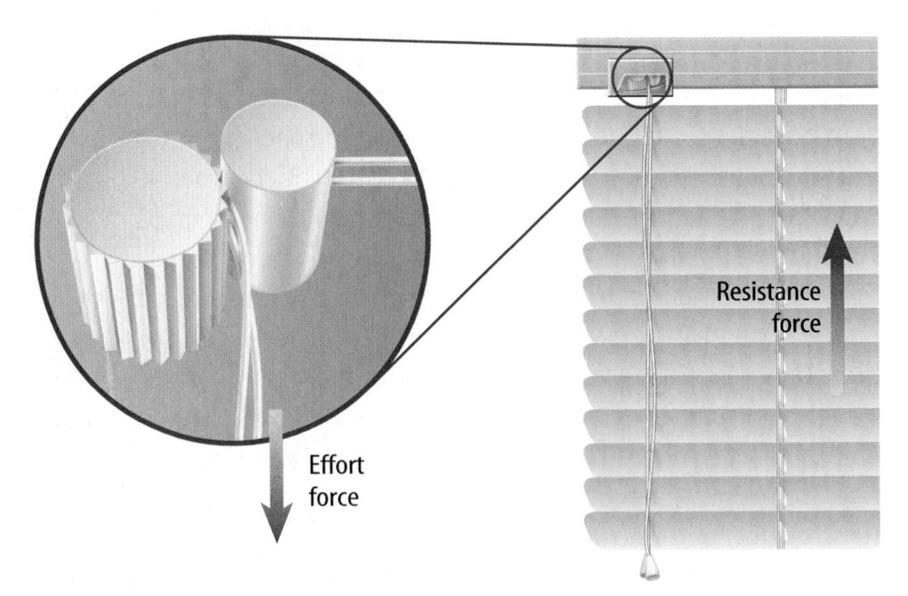

Mechanical Advantage

Think again about the crowbar example. The crowbar increases the force you apply, so the force exerted on the crate lid is greater than the force you exert on the handle. In other words, the resistance force is greater than the effort force. However, just as for the hammer, the effort distance you move the crowbar handle is greater than the distance the crowbar moves the lid. The machine multiplies your effort, but you must move the handle a greater distance.

The number of times a machine multiplies the effort force is the **mechanical advantage** (MA) of the machine. To calculate mechanical advantage, you divide the resistance force by the effort force. Some machines simply change the direction of the effort force, such as the window blinds in **Figure 11.** When only the direction of the force changes, the effort force and resistance force are equal, so the mechanical advantage is 1.

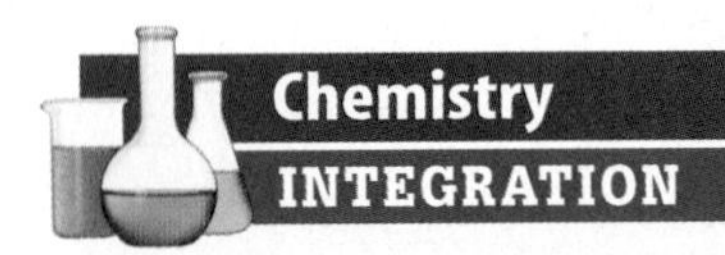

A material called graphite is sometimes used as a lubricant to increase the efficiency of machines. Find out what element graphite is made of and infer why graphite eases the movement of machines.

Efficiency

When you use a hammer to pull a nail out of a piece of wood, the friction between the wood and the nail causes the nail to get warm as it's pulled out. For real machines, some of the energy put into a machine is always lost as heat produced by friction. For that reason, the output work of a machine is always less than the work put into the machine. Machines that lose less energy to friction are said to be more efficient.

Efficiency is a measure of how much of the work put into a machine is changed into useful output work by the machine. A machine with high efficiency produces less heat from friction so more of the input work is changed to useful output work.

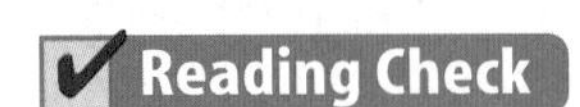

What is the efficiency of a machine?

Calculating Efficiency To calculate the efficiency of a machine, the output work is divided by the input work. Efficiency is usually expressed as a percentage by this equation:

$$\text{efficiency} = \left(\frac{W_{out}}{W_{in}}\right) \times 100\%$$

Because friction causes the output work to always be less than the input work, the efficiency of a real machine is always less than 100 percent. Because no machine is 100 percent efficient, the actual mechanical advantage is always less than the mechanical advantage of an ideal machine.

Machines can be made more efficient by reducing friction. This usually is done by adding a lubricant, such as oil or grease, to surfaces that rub together, as shown in **Figure 12.** When a lubricant is used in a machine, it fills in the gaps between the surfaces. This allows the two surfaces to slide more easily across each other, reducing friction and increasing efficiency. Sometimes, dirt will build up on the lubricant, and it will lose its effectiveness. The dirty lubricant should be wiped off and replaced with clean oil or grease.

You might have heard some household appliances or automobiles described as being energy efficient. By using less energy to do work, these machines cost less to operate. They also help conserve resources, such as coal, oil, and natural gas, that are used to produce electricity.

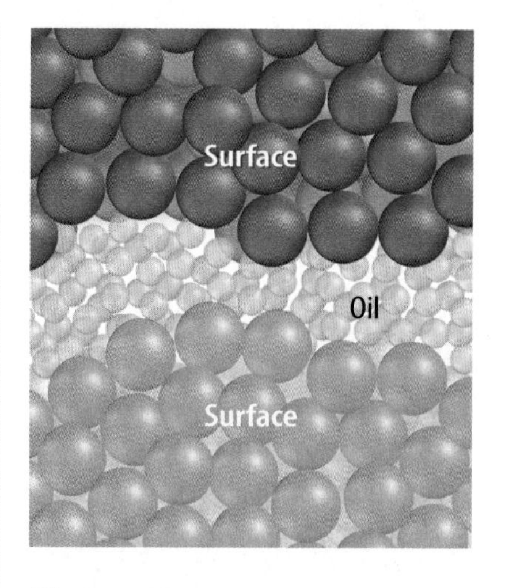

Figure 12
Oil reduces the friction between two surfaces. Oil fills the space between the surfaces so high spots don't rub against each other.

Section  Assessment

1. Explain how machines can make work easier without violating the law of conservation of energy.
2. A claw hammer is used to pull a nail from a board. If the claw exerts a resistance of 2,500 N to the applied effort force of 125 N, what is the MA of the hammer?
3. Explain why W_{out} is always less than W_{in}.
4. How would you calculate the efficiency of a machine?
5. **Think Critically** Give an example of a machine you've used recently. How did you apply effort force? How did the machine apply resistance force?

Skill Builder Activities

6. **Recognizing Cause and Effect** When you operate a machine, it's often easy to observe cause and effect. For example, when you turn a doorknob, the latch in the door moves. Give five examples of machines and describe one cause-and-effect pair in the action of each machine. **For more help, refer to the Science Skill Handbook.**
7. **Solving One-Step Equations** Suppose you want to use a machine to lift a 6,000-N log. What effort force will you need if your machine has a mechanical advantage of 25? 15? 1? Show your calculations. **For more help, refer to the Math Skill Handbook.**

SECTION

Simple Machines

As You Read

What You'll Learn

- **Describe** the six types of simple machines.
- **Calculate** the ideal mechanical advantage for different types of simple machines.

Vocabulary

simple machine
lever
pulley
wheel and axle
inclined plane
screw
wedge
compound machine

Why It's Important

If you know the principles behind simple machines, you'll make better use of everyday tools.

Types of Simple Machines

Without realizing it, you use many simple machines every day. A **simple machine** is a machine that does work with only one movement. The six types of simple machines are: lever, pulley, wheel and axle, inclined plane, screw, and wedge. As you'll see, the pulley and wheel and axle are modified forms of the lever, and the screw and wedge are modified forms of the inclined plane.

Reading Check *All simple machines are variations of which two basic machines?*

Levers

You probably won't try to pry the cap off a soft drink bottle with your fingers. Instead you would use a bottle opener to remove the cap. A bottle opener like the one in **Figure 13** is an example of a lever. A **lever** is a bar that is free to pivot, or turn, about a fixed point. The fixed point on the lever is called the fulcrum. The distance from the fulcrum to where the effort force is applied is called the effort arm. The distance from the fulcrum to where the resistance force is applied is called the resistance arm.

There are three different classes of levers. The differences are based upon the positions of the effort force, resistance force, and fulcrum.

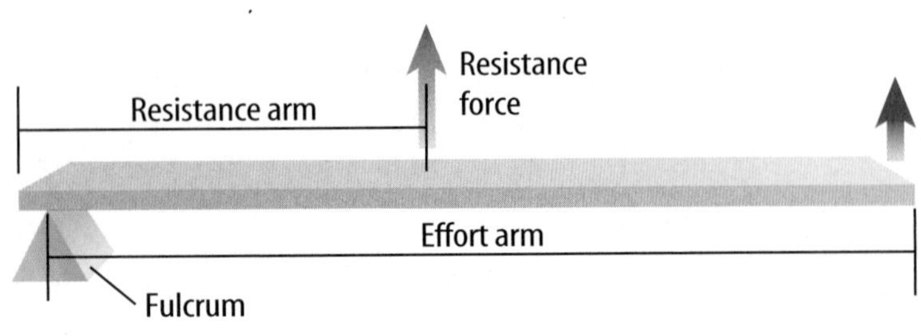

Figure 13
When you push up on the bottle opener (effort force), the opener bends the cap up (resistance force). *What acts as the fulcrum in a bottle opener?*

Figure 14
Levers are classified by the location of the effort force, resistance force, and the fulcrum.

A For a first-class lever, the fulcrum is between the effort force and resistance force.

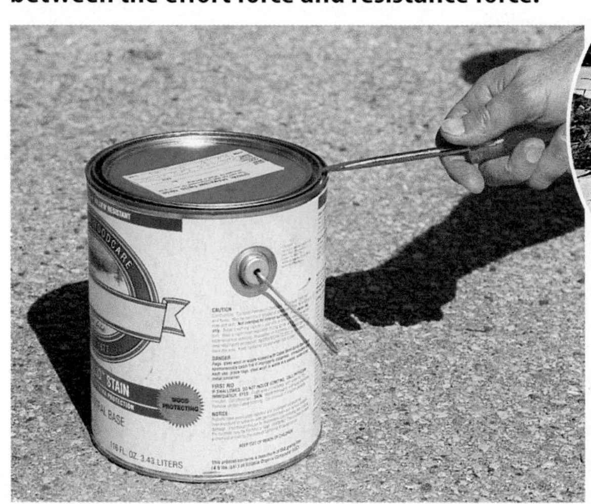

B For a second-class lever, the resistance force is between the effort force and the fulcrum.

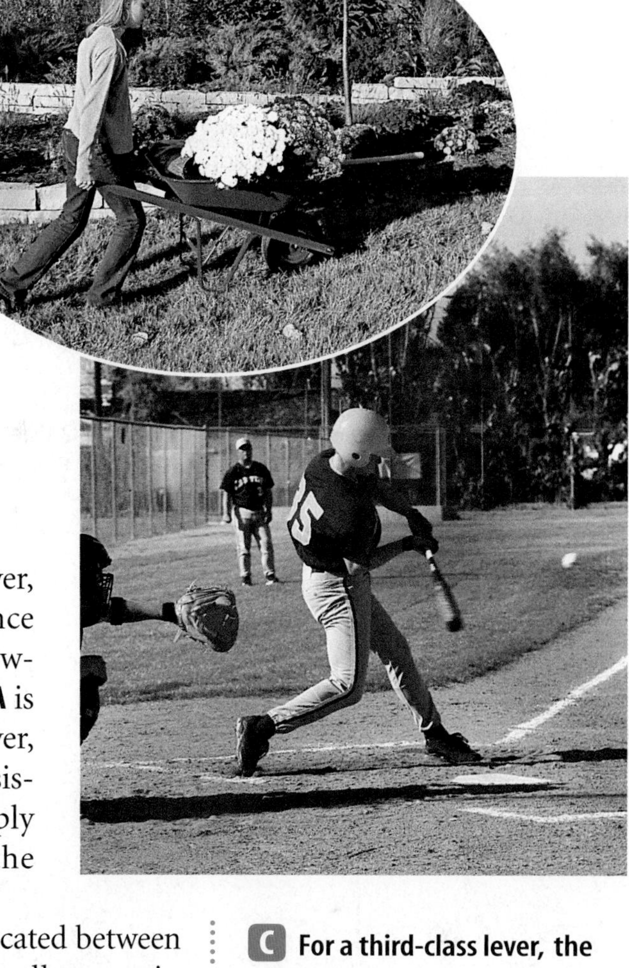

C For a third-class lever, the effort force is between the resistance force and the fulcrum.

Types of Levers To determine the class of a lever, you need to know where the effort and resistance forces are exerted relative to the fulcrum. The screwdriver used to open the can of paint in **Figure 14A** is an example of a first-class lever. In a first-class lever, the fulcrum is located between the effort and resistance forces. A first-class lever is used to multiply force, and it always changes the direction of the applied force.

In a second-class lever, the resistance force is located between the effort force and fulcrum. Look at the wheelbarrow in **Figure 14B.** You provide the effort force at the end, and the wheel acts as the fulcrum. The load you are lifting, the resistance force, is located in between. Second-class levers always multiply force.

Many pieces of sports equipment are examples of third-class levers. In a third-class lever, the effort force is located between the resistance force and fulcrum. Think about swinging a baseball bat like the one in **Figure 14C.** If you are right-handed, you hold the base of the bat with your left hand—the fulcrum. You use your right hand to apply the effort force and swing the bat. The resistance force is provided by the baseball when it hits the bat. Third-class levers cannot multiply force because the effort arm is always smaller than the resistance arm. Instead they increase the distance over which the resistance force is applied.

Every lever can be placed into one of these classes. Each class can be found in your body, as shown in **Figure 15** on the next page.

NATIONAL GEOGRAPHIC

VISUALIZING LEVERS IN THE HUMAN BODY

Figure 15

All three types of levers—first-class, second-class, and third-class—are found in the human body. The forces exerted by muscles in your body can be increased by first-class and second-class levers, while third-class levers increase the range of movement of a body part. Examples of how the body uses levers to help it move are shown here.

▲ Fulcrum
Effort force
Resistance force

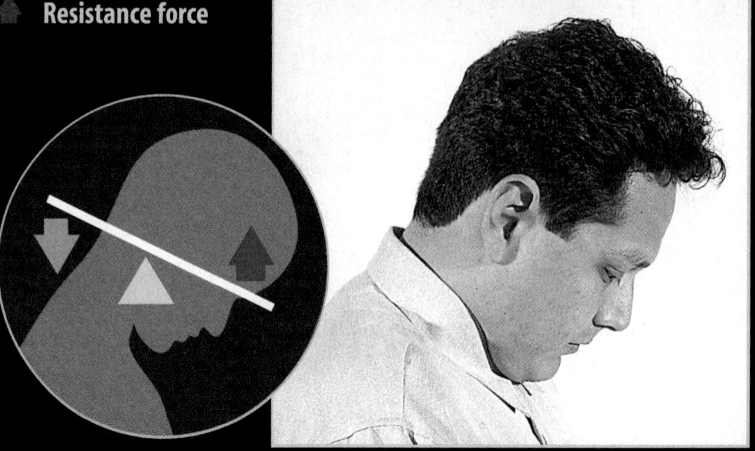

▲ FIRST-CLASS LEVER The fulcrum lies between the effort force and the resistance force. Your head acts like a first-class lever. Your neck muscles provide the effort force to support the weight of your head.

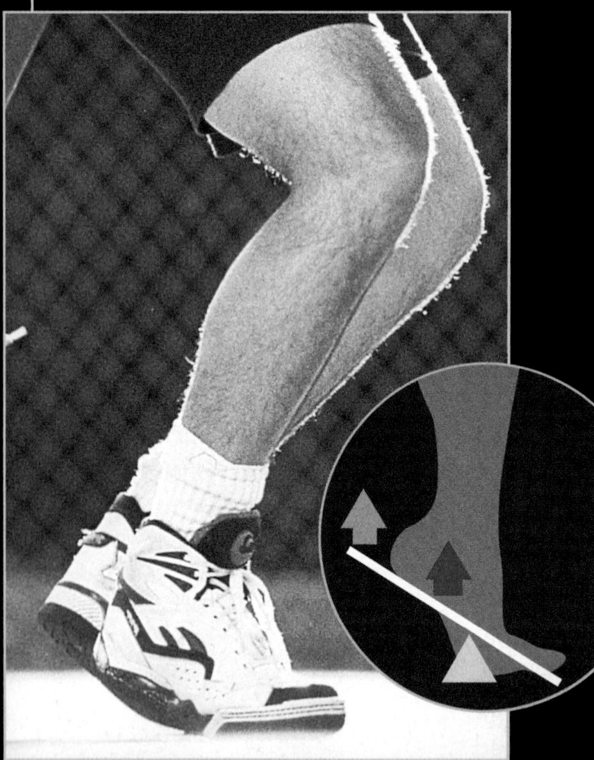

◀ SECOND-CLASS LEVER The resistance force is between the fulcrum and the effort force. Your foot becomes a second-class lever when you stand on your toes.

▶ THIRD-CLASS LEVER The effort force is between the fulcrum and the resistance force. A third class lever increases the range of motion of the resistance force. When you do a curl with a dumbbell, your forearm is a third-class lever.

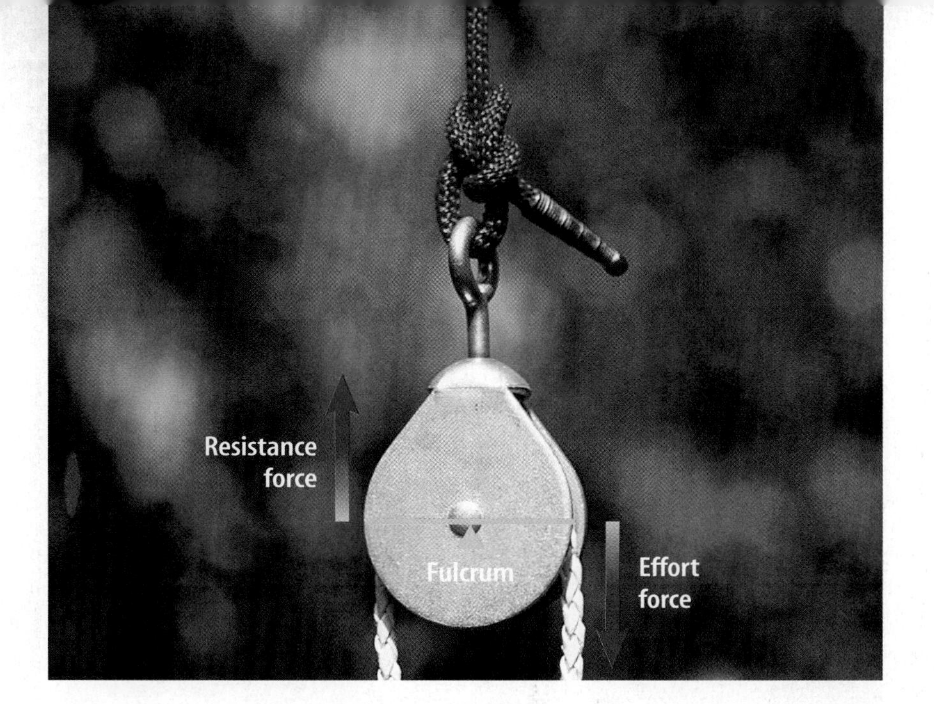

Figure 16
A fixed pulley is another form of the lever. *What are the lengths of the effort arm and resistance arm in a pulley?*

Mechanical Advantage of a Lever The bottle opener makes work easier by multiplying your effort force and changing the direction of your force. To calculate the ideal mechanical advantage (IMA) of the lever, you can use the lengths of the arms of the lever. The distance from the fulcrum to the place you exert your force is the effort arm. The resistance arm is the distance from the fulcrum to the point the resistance force is applied. Assuming no friction, the IMA is as follows:

$$\text{IMA} = \frac{\text{length of effort arm}}{\text{length of resistance arm}} = \frac{L_e}{L_r}$$

Making the effort arm longer increases the ideal mechanical advantage.

Pulleys

What causes an elevator to rise? A cable attached to the elevator is wrapped around a pulley, allowing the elevator to be raised and lowered. A **pulley** is a grooved wheel with a rope, chain, or cable running along the groove. A fixed pulley is a modified first-class lever, as shown in **Figure 16.** The axle of the pulley acts as the fulcrum. The two sides of the pulley are the effort arm and resistance arm. A pulley can multiply the effort force, but all pulleys can change the direction of the effort force.

Fixed Pulleys The cable attached to an elevator passes over a fixed pulley at the top of the elevator shaft. A fixed pulley, such as the one in **Figure 17,** is attached to something that doesn't move, such as a ceiling or wall. Because a fixed pulley changes only the direction of force, the effort force is not multiplied and the IMA is 1.

Figure 17
A fixed pulley changes only the direction of your force. You still need to apply 4 N of force to lift the weight.

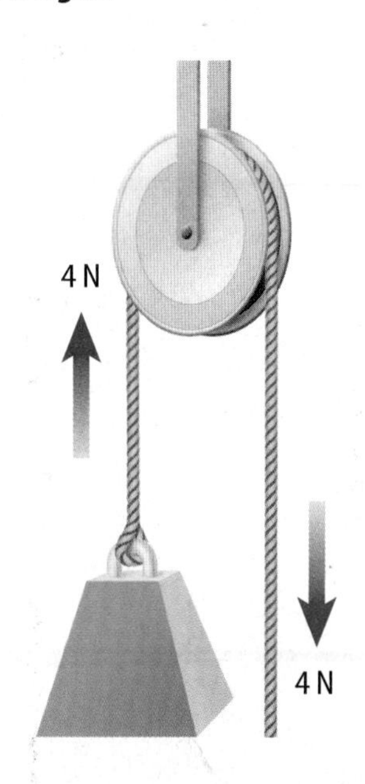

Figure 18
The fixed pulley does not multiply force, while a movable pulley and block and tackle do.

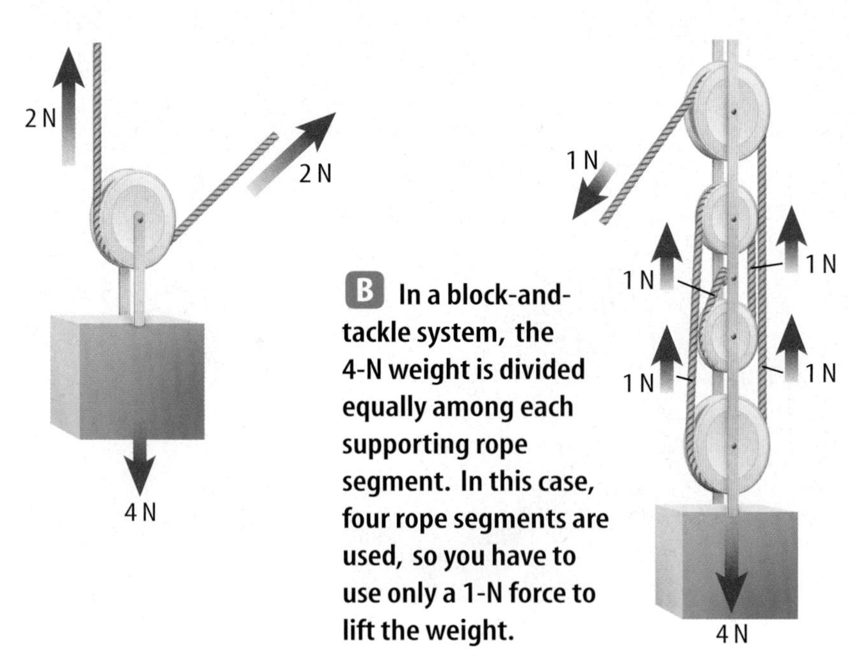

A With a movable pulley, the attached side of the rope supports half of the 4-N weight.

B In a block-and-tackle system, the 4-N weight is divided equally among each supporting rope segment. In this case, four rope segments are used, so you have to use only a 1-N force to lift the weight.

Movable Pulleys A pulley in which one end of the rope is fixed and the wheel is free to move is called a movable pulley. Unlike a fixed pulley, a movable pulley does multiply force. Suppose a 4-N weight is hung from the movable pulley in **Figure 18A.** The ceiling acts like someone helping you to lift the weight. The string attached to the ceiling will support half of the weight—2 N. You need to exert only the other half of the weight—2 N—in order to support and lift the weight. Since the resistance force is 4 N and your effort force is 2 N, the IMA of the movable pulley will be 2.

Because the movable pulley increases your effort force, the distance must increase to conserve energy. The IMA is 2, so the distance you pull must be twice as large as the resistance distance.

Reading Check *How does a movable pulley multiply the effort force?*

The Block and Tackle A system of pulleys consisting of fixed and movable pulleys is called a block and tackle. **Figure 18B** shows a block and tackle made up of two fixed pulleys and two movable pulleys. If a 4-N weight is suspended from the movable pulley, each rope segment supports one fourth of the weight, reducing the effort force to 1 N. The IMA of a pulley system is equal to the number of rope segments that support the resistance weight. A block and tackle can have a large mechanical advantage. When designing a block and tackle you must keep in mind that the more pulleys that are involved, the effects of friction are greater, which will reduce the overall mechanical advantage.

Figure 19
If the handle on the pencil sharpener were removed, you would be unable to sharpen your pencil.

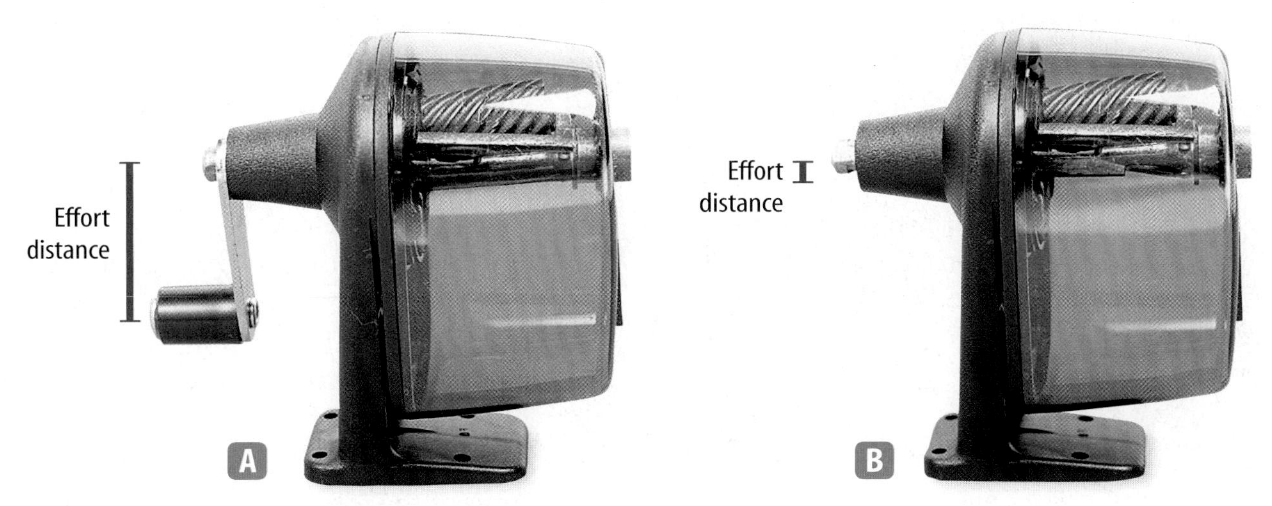

Wheel and Axle

Could you use the pencil sharpener in **Figure 19** if the handle weren't attached? The handle on the pencil sharpener is an example of a wheel and axle. A **wheel and axle** is a machine consisting of two wheels of different sizes that rotate together.

When you think of a wheel and axle, you might picture something like a bicycle tire. Both parts of a wheel and axle move in a circle. Even though the handle of a pencil sharpener doesn't roll, it moves in a circular path just as the bicycle wheel does. Usually, effort force is exerted on the larger wheel. The smaller wheel, the axle, usually exerts the resistance force. Doorknobs and faucet handles are machines that use a wheel and axle.

The wheel and axle is another modified form of a lever. On the pencil sharpener, the point where the handle connects to the sharpening mechanism acts as the fulcrum. The length of the handle is the effort arm, or the wheel. The radius of the wheel inside is the resistance arm, or the axle.

Mechanical Advantage of the Wheel and Axle

Remember that the IMA of a lever can be calculated by dividing the effort arm by the resistance arm. In a wheel and axle, each travels in a circular path, so the effort arm is the same as the radius of the wheel. Likewise, the resistance arm is the same as the radius of the axle. Thus, the IMA can be calculated as follows.

$$\text{IMA} = \frac{\text{radius of wheel}}{\text{radius of axle}} = \frac{r_w}{r_a}$$

The mechanical advantage of a wheel and axle can be increased by making the radius of the wheel larger.

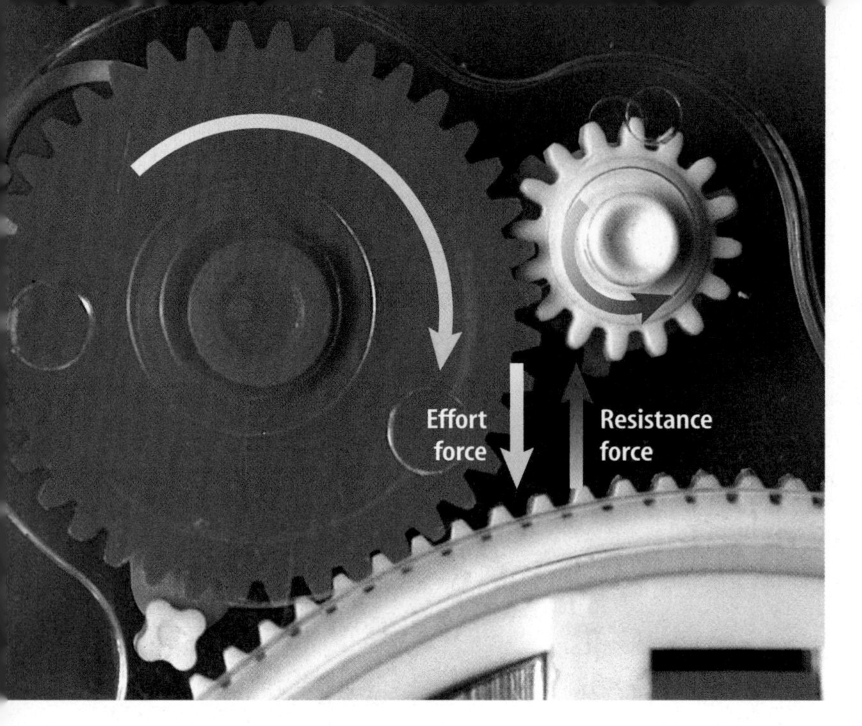

Gears A modified form of the wheel and axle that you may be familiar with is a gear. A gear usually consists of two wheels of different sizes with interlocking teeth along their circumferences. When one of the wheels is turned, the teeth force the other wheel to turn.

If the larger wheel is the effort gear one turn of the effort gear can result in many turns of the resistance gear, because it has more teeth than the smaller wheel. A system of gears in which the effort gear is larger reduces the effort force.

Gears also may change the direction of the force. Notice that when the effort gear in **Figure 20** is rotated clockwise, the resistance gear rotates counterclockwise. You may have noticed this when you use a can opener. As you twist the handle in one direction, the can revolves in the opposite direction.

Figure 20
When one gear turns, the teeth that are interlocked with the other gear make it turn.

Inclined Planes

Why do the roads and paths on mountains zigzag? Would it be easier to climb directly up a steep incline or walk a longer path gently sloped around the mountain? You have learned that it takes less force to lift something if you use a ramp. A sloping surface, such as a ramp that reduces the amount of force required to do work, is called an **inclined plane**.

Mechanical Advantage of an Inclined Plane You do the same work by lifting a box straight up or sliding it up a ramp. But the ramp, an inclined plane, reduces the amount of force required by increasing the distance over which the force is applied. You can calculate the IMA of an inclined plane by dividing the length of the ramp by the height of the ramp.

$$\text{IMA} = \frac{\text{effort distance}}{\text{resistance distance}} = \frac{\text{length of slope}}{\text{height of slope}} = \frac{l}{h}$$

As the ramp is made longer and less steep, less force is required to push the object up the ramp.

When you think of an inclined plane, you normally think of moving an object up a ramp—you move and the inclined plane remains stationary. The screw and the wedge, however, are variations of the inclined plane in which the inclined plane moves and the object remains stationary.

Figure 21
A screw has an inclined plane that wraps around the post of the screw.

A The thread gets thinner farther from the post. This helps the screw force its way into materials.

B Many lids, such as those on peanut butter jars, also contain threads.

The Screw

You normally think of a screw as a carpenter's tool like the one in **Figure 21A.** A **screw** is an inclined plane wrapped in a spiral around a cylindrical post. If you look closely at the screw in **Figure 21A,** you'll see that the threads form a tiny ramp that runs upward from its tip. As you turn the screw, the threads seem to pull the screw into the wood. The wood seems to slide up the inclined plane. Actually, the plane slides through the wood.

There are many other examples of the screw that you encounter every day. How do you remove the lid off a jar of peanut butter, like in **Figure 21B?** If you look closely, you see the threads similar to the ones in **Figure 21A.** Where else can you find examples of a screw?

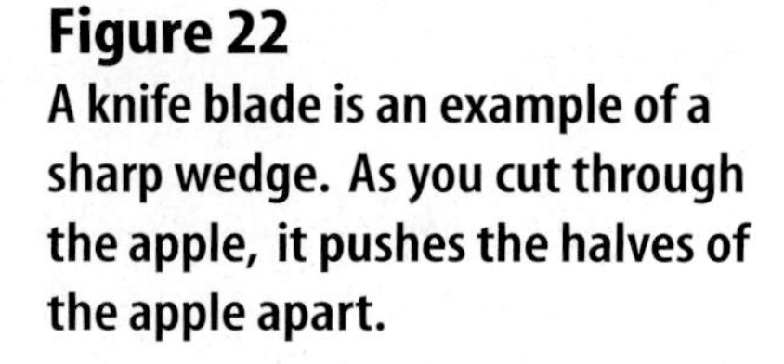

Figure 22
A knife blade is an example of a sharp wedge. As you cut through the apple, it pushes the halves of the apple apart.

The Wedge

Like the screw, the wedge is also a simple machine where the inclined plane moves through an object or material. A **wedge** is an inclined plane with one or two sloping sides. It changes the direction of the effort force.

Look closely at the knife in **Figure 22.** One edge is extremely sharp, and it slopes outward at both sides, forming an inclined plane. As it moves through the apple in a downward motion, the force is changed to a horizontal motion, forcing the apple apart.

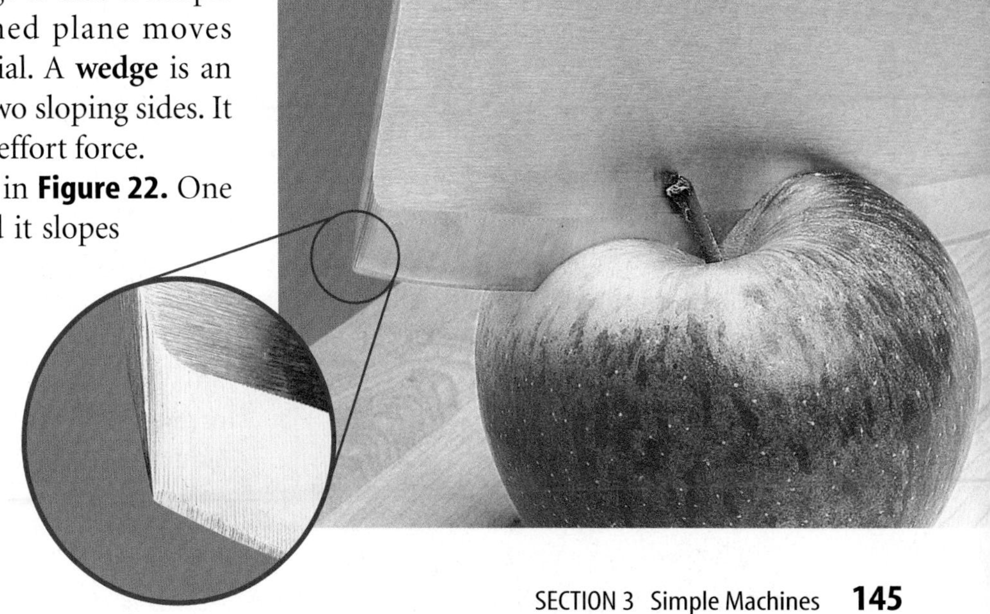

Figure 23
A compound machine, such as a can opener, is made up of simple machines.

Compound Machines

Some of the machines you use every day are made up of several simple machines. When two or more simple machines are used together, it is called a **compound machine.**

Look at the can opener in **Figure 23.** To open the can you first squeeze the handles together. The handles act as a lever and increase the force applied on a wedge, which then pierces the can. You then turn the handle, a wheel and axle, to open the can. The overall mechanical advantage of a compound machine is related to the mechanical advantages of all the machines involved.

A car is also a compound machine. Burning fuel in the cylinders of the engine causes the pistons to move up and down. This up-and-down motion makes the crankshaft rotate. The force exerted by the rotating crankshaft is transmitted to the wheels through other parts of the car, such as the transmission and the differential. Both of these parts contain gears, which are simple machines. When a large and a small gear are in contact, the larger gear rotates a shorter distance, but the force it exerts is increased. In this way, these gears can change the rate at which the wheels rotate, the force exerted by the wheels, and even reverse the direction of rotation.

Section  Assessment

1. Give one example of each kind of simple machine. Use examples different from the ones in the text.
2. Explain why the six kinds of simple machines are variations of two basic machines.
3. Suppose you are using a screwdriver to pry the lid off a paint can. Identify the fulcrum, the effort arm, and the resistance arm.
4. A 6-m ramp runs from a ground-level sidewalk to a porch. The porch is 2 m off the ground. What is the ideal mechanical advantage of the ramp?
5. **Think Critically** When would the friction of an inclined plane be useful?

Skill Builder Activities

6. **Making and Using Tables** Organize information about the six kinds of simple machines into a table. Include the type of machine, an example of each type, and a brief description of how it works. You may include other information, such as mechanical advantage, if you wish. **For more help, refer to the** Science Skill Handbook.
7. **Using a Word Processor** Using a word processor, write a separate paragraph about each class of lever. Describe at least two examples of each class, identifying the effort force, resistance force, and fulcrum for each. Use **Figure 15** for reference. **For more help, refer to the** Technology Skill Handbook.

Activity

Levers

Did you ever play on a seesaw? Wasn't it much easier to balance your friend on the other end if you both weighed the same? If your friend was lighter, you had to move toward the fulcrum to balance the seesaw. In this activity, you will use a lever to determine the mass of a coin.

What You'll Investigate

How can a lever measure mass?

Materials

$8^1/_2$ " × 11" sheet of paper
coins (one quarter, one dime, one nickel)
balance
metric ruler

Goals

- **Measure** effort arm and resistance arm.
- **Observe** how mass can affect the fulcrum.

Safety Precautions

Procedure

1. Make a lever by folding the paper into a strip 3 cm wide by 28 cm long.
2. Mark a line 2 cm from one end of the paper strip. Label this line Resistance.
3. Slide the other end of the paper strip over the edge of a table until the strip begins to tip. Mark a line across the paper at the table edge and label this line Effort.
4. **Measure** the mass of the paper to the nearest 0.1 g. Write this mass on the Effort line.
5. Center a dime on the Resistance line. Locate the fulcrum by sliding the paper strip until it begins to tip. Mark the balance line. Label it Fulcrum #1.

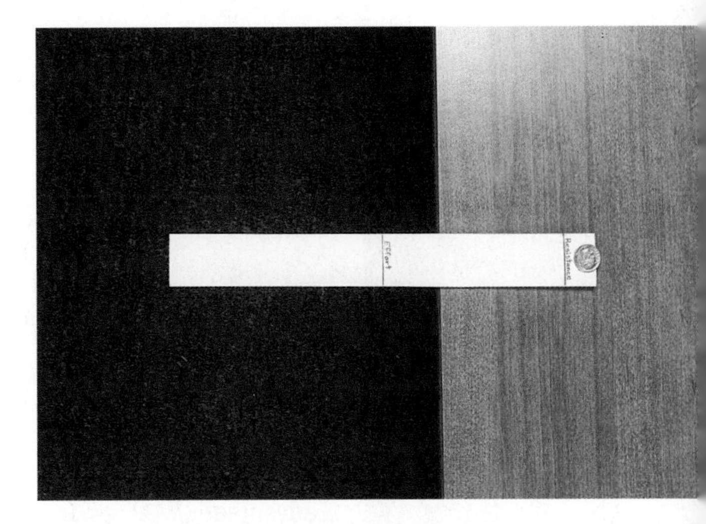

6. **Measure** the lengths of the resistance and effort arms to the nearest 0.1 cm.
7. Calculate the IMA of the lever. Multiply the IMA by the mass of the lever to find the approximate mass of the coin.
8. Repeat steps 5 through 7 with the nickel and the quarter. Mark the fulcrum line #2 for the nickel and #3 for the quarter.

Conclude and Apply

1. What provides the effort force?
2. What does it mean if the IMA is less than 1.0?
3. The calculations are done as if the entire weight of the paper is located at what point?
4. **Infer** why mass units can be used in place of force units in this kind of problem.

Communicating Your Data

Compare your results with those of other students in your class. **For more help, refer to the Science Skill Handbook.**

Activity Model and Invent

Work Smarter

You are the contractor on a one-story building with a large air-conditioner. The lower the force, the easier the job for your crew. What ways can you think of to get the air conditioner to the roof?

Recognize the Problem

How can you minimize the force needed to lift an object? What machines could you use?

Thinking Critically

Is a lever practical for this job? Why? Consider a fixed pulley with ideal mechanical advantage (IMA) $=$ 1, a movable pulley with $IMA = 2$, a block and tackle with one fixed double pulley and one movable double pulley with $IMA = 4$, and an inclined plane with $IMA =$ slope / height $=$ 4. The latter two machines may differ in efficiency. How can you find the efficiency of machines?

Goals

- **Model** lifting devices based on a block and tackle and on an inclined plane.
- **Calculate** the output work that will be accomplished.
- **Measure** the force needed by each machine to lift a weight.
- **Calculate** the input work and efficiency for each model machine.
- **Select** the best machine for your job based on force required.

Possible Materials

Spring scale, 0 – 10 N range
9.8 N weight (1 kg mass)
Two double pulleys
String for pulleys
Stand or support for the pulleys
Wooden board, 40 cm long
Support for board, 10 cm high

Problem Data

	Control	Inclined Plane	Block and Tackle
Ideal Mechanical Advantage, IMA	1	4	4
Effort Force, F_e, N			
Effort Distance, d_e, m	0.10		
Resistance Force, F_r, N	9.8	9.8	9.8
Resistance Distance, d_r, m	0.10	0.10	0.10
$Work_{in} = F_e \times d_e$, Joules			
$Work_{out} = F_r \times d_r$, Joules	0.98	0.98	0.98
% Efficiency, $(Work_{out} / Work_{in}) \times 100$			

Planning the Model

1. Work in teams of at least two. **Collect** all the needed equipment.
2. Sketch a model for each lifting machine. **Model** the inclined plane with a board 40 cm long and raised 10 cm at one end. Include a control in which the weight is lifted while being suspended directly from the spring scale.
3. Make a table for data.

Check the Model Plans

1. Is the pulley support high enough that the block and tackle can lift a weight 10 cm?
2. Obtain your teacher's approval of your sketches and data table before proceeding.

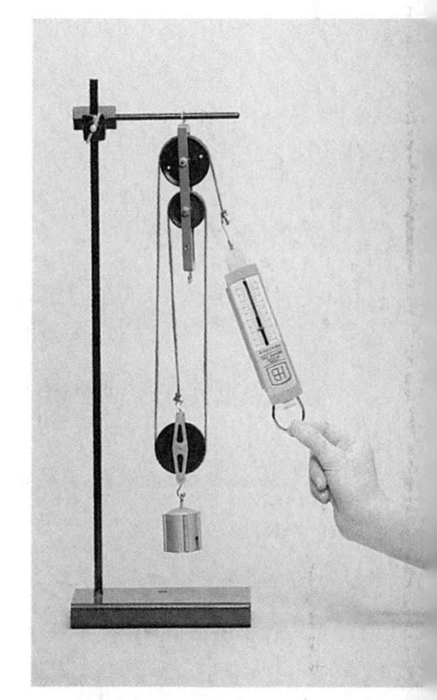

Making the Model

1. Tie the weight to the spring scale and measure the force required to lift it. Record the effort force in your data table under Control, along with the 10-cm effort distance.
2. Assemble the inclined plane so that the weight can be pulled up the ramp at a constant rate. The 40-cm board should be supported so that one end is 10 cm higher.
3. Tie the string to the spring scale and measure the force required to move the weight up the ramp at a constant speed. Record this effort force under inclined plane in your data table. Record 40 cm as the effort distance for the inclined plane.
4. Assemble the block and tackle using one fixed double pulley and one movable double pulley.
5. Tie the weight to the lower pulley and tie the spring scale to the string at the top of the upper pulley.
6. **Measure** the force required to lift the weight with the block and tackle. Record this effort force.
7. **Measure** the length of string that must be pulled to raise the weight 10 cm. Record this effort distance.

Analyzing and Applying Results

1. **Calculate** the output work for all three methods of lifting the 9.8 N weight 10 cm.
2. **Calculate** the input work and the efficiency for the control, the inclined plane, and the block and tackle.
3. Which machine used the lowest force to raise the weight? How do you account for the observed differences in efficiencies? How could you improve the efficiency for each?

Make a poster showing how the best machine would be used to lift the air conditioner to the roof of your building.

THE SCIENCE

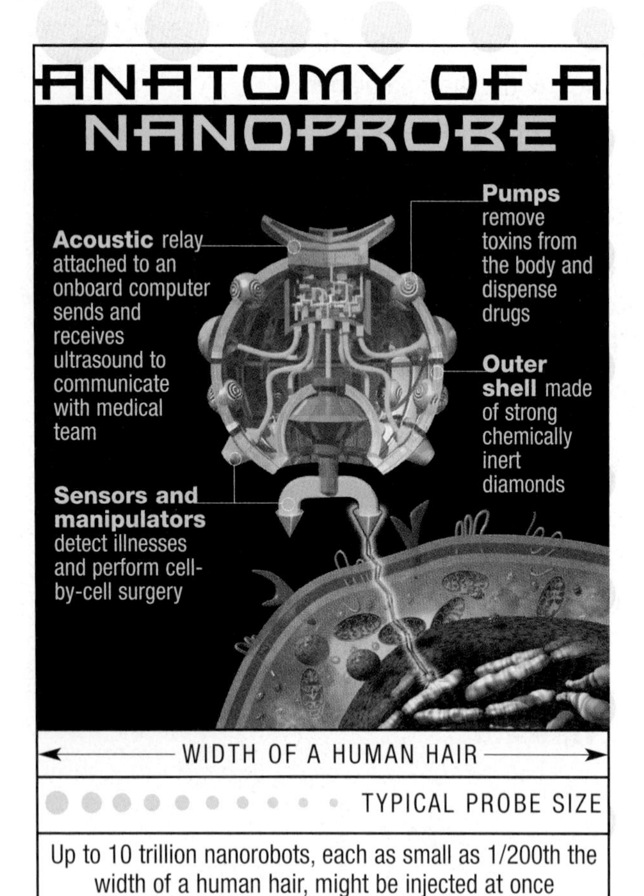

Up to 10 trillion nanorobots, each as small as 1/200th the width of a human hair, might be injected at once

Imagine an army of tiny robots, each no bigger than a bacterium, swimming through your bloodstream. One type of robot takes continuous readings of blood pressure in different parts of your body. Another type monitors cholesterol. Still others measure your blood sugar, the beginnings of possible blockages in arteries leading to your heart, and your general health.

Welcome to the world of nanotechnology—the science of creating molecular-sized machines. The machines, called nanobots, are each about one one billionth of a meter in size. The prefix *nano* refers to a billionth part of a unit, and comes from *nanos*, the Greek word for "dwarf." These machines are so small they can skillfully control matter one atom at a time.

Nanotechnology uses microscopic machines to do microscopic work. By combining engineering and biology, scientists might actually be able to reorganize atoms and molecules to create new objects. For example, nanobots could create diamonds from coal—all they'd need to do is rearrange a few atoms.

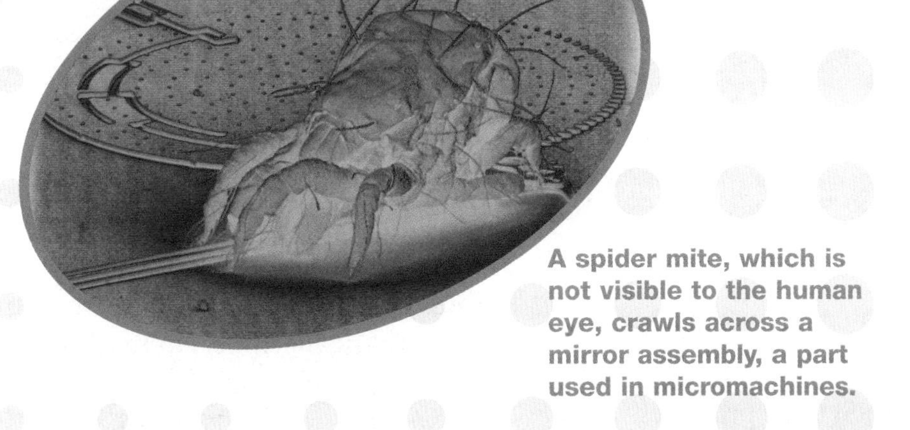

A spider mite, which is not visible to the human eye, crawls across a mirror assembly, a part used in micromachines.

OF VERY, VERY SMALL

One day you might have robots swimming through your bloodstream

Nanobots also could be used to clean up oil spills and toxic waste sites. Toxic wastes usually are made up of atoms that are arranged into noxious molecules. Nanobots will be able to break down these poisonous molecules, converting dangerous waste into harmless forms. Nanobots will have all sorts of uses in nature, in the body, and in the workplace.

Small, Smaller, Smallest

Nanotechnologists are predicting that within a few decades they will be creating machines that can do just about anything, as long as it's small. Already, nanotechnologists have built gears 10,000 times thinner than a human hair. They've also built tiny molecular "motors" only 50 atoms long. At Cornell University, nanotechnologists created the world's smallest guitar. It is appoximately the size of a white blood cell and it even has six strings.

This is the smallest guitar in the world. It is about as big as a human white blood cell. Each of its six silicon strings is 100 atoms wide. You can see the guitar only with an electron microscope.

Of course, the tiny guitar isn't meant to be played—only to illustrate the reality of the "science of the small," and to give us a glimpse into what lies ahead.

And getting back to those nanobots in your body—in the future, they might transmit your internal vital signs to a nanocomputer, which might be implanted under your skin. There the data could be analyzed for signs of disease.

Nanomachines then could be sent to scrub your arteries clean of dangerous blockages, or mop up cancer cells, or even vaporize blood clots with tiny lasers. These are just some of the possibilities in the imaginations of those studying the new science of nanotechnology.

CONNECTIONS Design Think up a very small simple or complex machine that could go inside the body and do something. What would the machine do? Where would it go? Share your diagram or design with your classmates.

SCIENCE *Online*

For more information, visit science.glencoe.com

Chapter 5 Study Guide

Reviewing Main Ideas

Section 1 Work

1. Work is the transfer of energy when a force makes an object move.
2. Work is done only when force produces motion in the direction of the force. *Does this forklift do work by lifting these crates? Explain.*

3. Power is the amount of work, or the amount of energy transferred, in a certain amount of time.

Section 2 Using Machines

1. A machine makes work easier by changing the size of the force applied, by increasing the distance an object is moved, or by changing the direction of the applied force. *How is work made easier for the person pulling the nail from a board?*

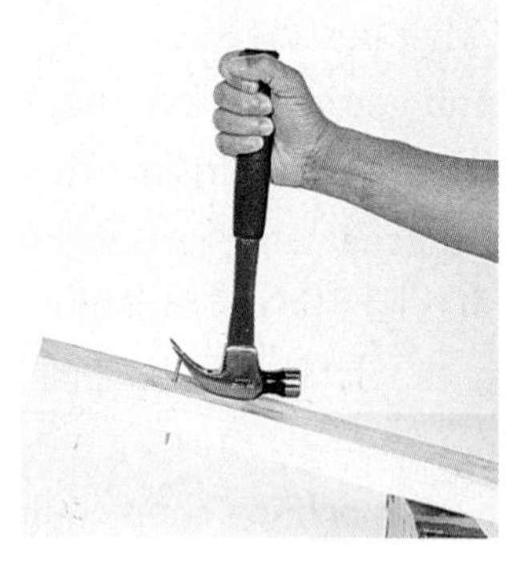

2. The number of times a machine multiplies the force applied to it is the mechanical advantage of the machine. The actual mechanical advantage is always less than the ideal mechanical advantage.
3. The efficiency of any machine is a ratio of the work done by the machine to the work put into the machine. No machine can be 100 percent efficient.

Section 3 Simple Machines

1. A simple machine is a machine that can do work with a single movement.
2. A simple machine can increase an applied force, change its direction, or both.
3. A lever is a bar that is free to pivot about a fixed point called a fulcrum. A pulley is a grooved wheel with a rope running along the groove. A wheel and axle consists of two different-sized wheels that rotate together. An inclined plane is a sloping surface used to raise objects. The screw and wedge are special types of inclined planes.
4. A combination of two or more simple machines is called a compound machine. *What simple machines make up a pair of scissors?*

After You Read

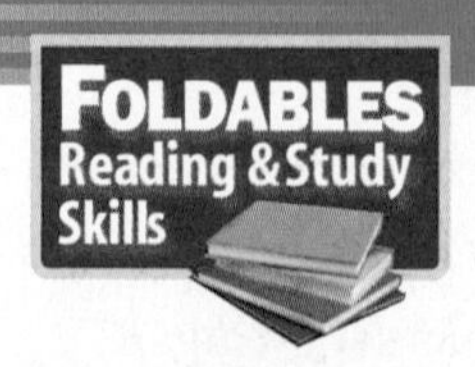

On your Foldable, list work you would do with machines under its tab. Now compare and contrast work with and without machines.

Visualizing Main Ideas

Complete the concept map for simple machines using the following terms: inclined plane, lever, lever types, pulley, screw, wedge, and wheel and axle.

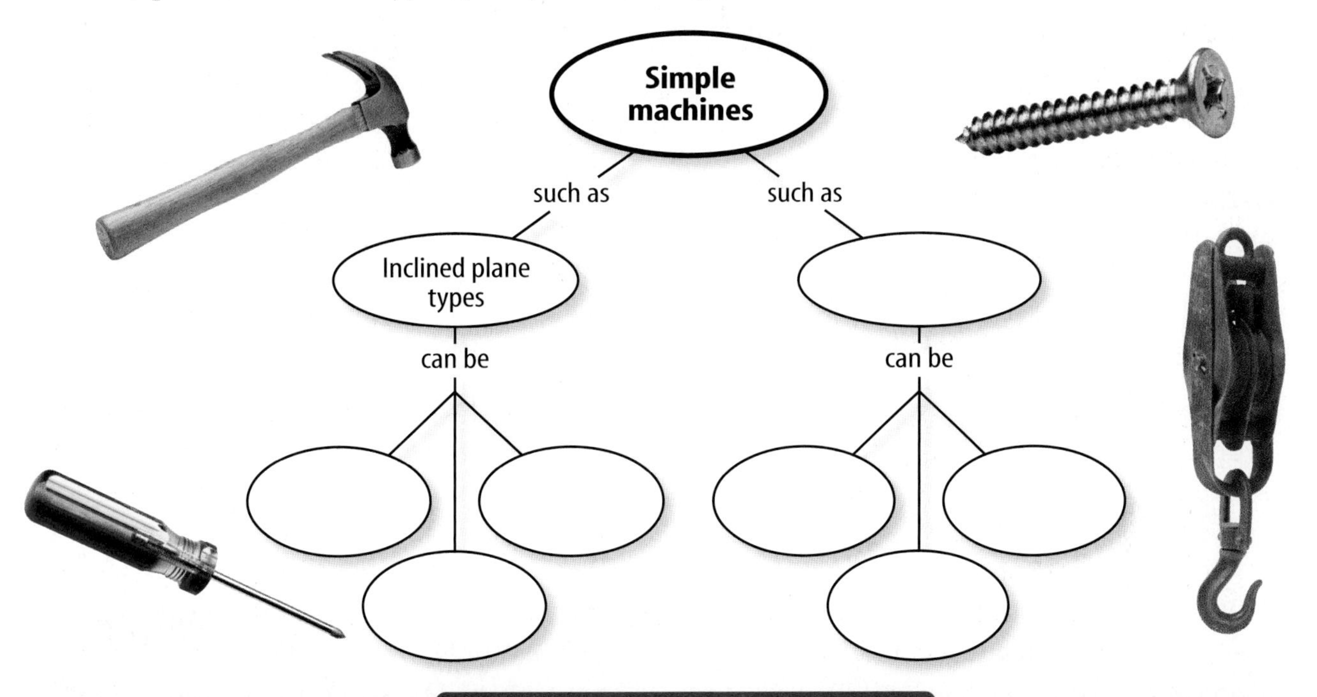

Vocabulary Review

Vocabulary Words

- **a.** compound machine
- **b.** efficiency
- **c.** effort force
- **d.** inclined plane
- **e.** lever
- **f.** machine
- **g.** mechanical advantage
- **h.** power
- **i.** pulley
- **j.** resistance force
- **k.** screw
- **l.** simple machine
- **m.** wedge
- **n.** wheel and axle
- **o.** work

Study Tip

When you encounter new vocabulary, write it down in a sentence. This will help you understand, remember, and use new vocabulary words.

Using Vocabulary

Replace the underlined words with the correct vocabulary word(s).

1. A combination of two or more simple machines is a(n) <u>ideal machine</u>.
2. A wedge is another form of a <u>wheel and axle</u>.
3. The amount by which a machine multiplies your force is called <u>efficiency</u>.
4. The force that you exert on a lever is called the <u>resistance force</u>.
5. A(n) <u>inclined plane</u> is a grooved wheel with a rope or chain running through it.
6. <u>Efficiency</u> is the rate at which work is done.
7. <u>Power</u> is when a force causes an object to move in the direction of the force.

Chapter 5 Assessment

Checking Concepts

Choose the word or phrase that best answers the question.

1. Using the scientific definition, which of the following is true of work?
   A) It must be difficult.
   B) It must involve levers.
   C) It must involve the transfer of energy.
   D) It must be done with a machine.

2. How many types of simple machines exist?
   A) three **C)** eight
   B) six **D)** ten

3. In an ideal machine, which of the following is true?
   A) Work input is equal to work output.
   B) Work input is greater than work output.
   C) Work input is less than work output.
   D) Work input is independent of work output.

4. Which of these cannot be done by a machine?
   A) multiply force
   B) multiply energy
   C) change direction of a force
   D) work

5. What term indicates the number of times a machine multiplies the effort force?
   A) efficiency
   B) power
   C) mechanical advantage
   D) resistance

6. How could you increase the IMA of an inclined plane?
   A) increase its length
   B) increase its height
   C) decrease its length
   D) make its surface smoother

7. How far must the effort rope of a single fixed pulley move to raise a resistance 4 m?
   A) 1 m **C)** 4 m
   B) 2 m **D)** 8 m

8. In a wheel and axle, which of the following usually exerts the resistance force?
   A) the axle **C)** the gear ratio
   B) the larger wheel **D)** the pedals

9. What is the IMA of an inclined plane 8 m long and 2 m high?
   A) 2 **C)** 8
   B) 4 **D)** 16

10. Which of the following increases as the efficiency of a machine increases?
    A) work input **C)** friction
    B) work output **D)** IMA

Thinking Critically

11. An adult and a small child get on a seesaw that has a movable fulcrum. When the fulcrum is in the middle, the child can't lift the adult. How should the fulcrum be moved so the two can seesaw? Explain.

12. Using a ramp 6 m long, workers apply an effort force of 1,250 N to move a 2,000-N crate onto a platform 2 m high. What is the efficiency of the ramp?

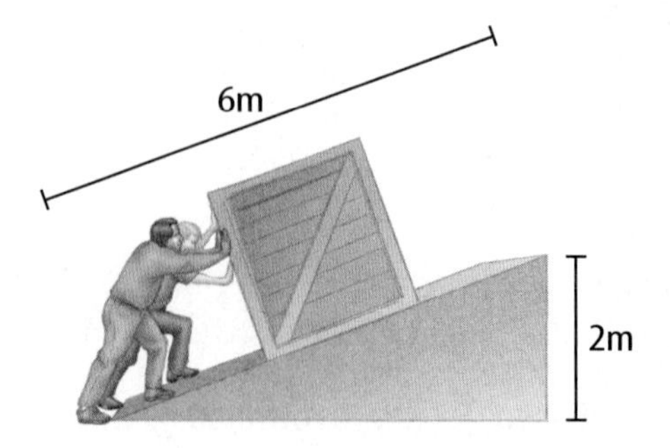

13. How much power does a person weighing 500 N need to climb a 3-m ladder in 5 s?

14. You have two screwdrivers. One is long with a thin handle, and the other is short with a fat handle. Which would you use to drive a screw into a board? Explain.

Developing Skills

15. Concept Mapping Complete the concept map of simple machines using the following terms: *compound machines, mechanical advantage, resistance force, work.*

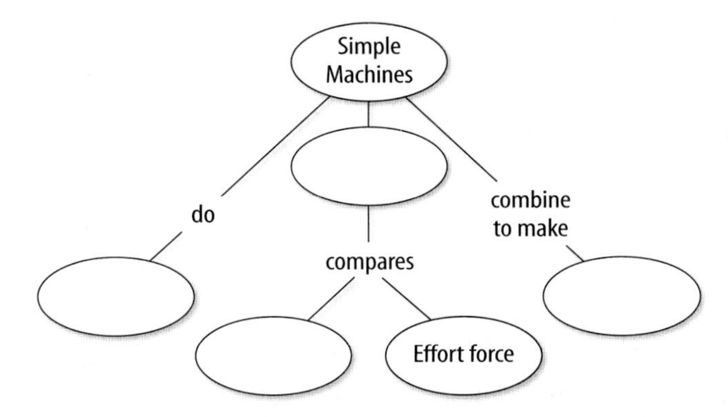

16. Communicating A pair of scissors is a compound machine. Draw a diagram of a pair of scissors and label the simple machines you can identify. Explain to a classmate the purpose of each simple machine in this device.

Performance Assessment

17. Invention Design a human-powered machine of some kind. Describe the simple machines used in your design, and tell what each of these machines does.

Technology

Go to the Glencoe Science Web site at **science.glencoe.com** or use the **Glencoe Science CD-ROM** for additional chapter assessment.

Test Practice

Maria went to the library to do some research on inventions that made it easier to perform everyday tasks. Some of the pictures that she photocopied are shown in the diagram below.

Study the diagram and answer the following questions.

1. What do these simple machines have in common?
- **A)** They are all wedges.
- **B)** They are all levers.
- **C)** All were invented in the United States.
- **D)** They are all recent inventions.

2. The larger scissors can cut through a thick object because _____.
- **F)** levers with longer effort arms are easier to hold
- **G)** levers with longer effort arms multiply force more
- **H)** levers with longer effort arms multiply her force less
- **J)** levers with longer effort arms are made of stronger materials

CHAPTER

Thermal Energy

You probably couldn't shape a piece of cold steel unless it was very thin like a wire. But if the steel is heated enough, it melts, and the liquid steel can be poured into molds, as shown in this picture. In the mold, the steel cools and again becomes a solid, this time in the desired shape. In this chapter you will learn how heat and temperature are related, and how thermal energy is transferred. You will also learn how the flow of heat can be controlled.

What do you think?

Look at the picture below with a classmate. Discuss what this might be or what is happening. Here's a hint: *You can do this, but a dog can't.* Write your answer or your best guess in your Science Journal.

Why does hot water burn your skin but warm water does not? Molecules move faster and have more energy at a higher temperature than at a lower temperature. The energy of moving molecules is called kinetic energy. When fast-moving molecules of hot water touch your skin, they trigger nerve cells to send pain signals to your brain. Warm water molecules have less energy and cause no pain. In this activity, observe and compare other effects of fast-moving and slow-moving water molecules.

Observe the effects of molecules at different temperatures

1. Pour 200 mL of room-temperature water into a beaker.
2. Pour 200 mL of water into a beaker and add some ice.
3. Put one drop of food coloring into each beaker.
4. Compare how quickly the food coloring causes the color of the water to change in each beaker.

Observe

Write a paragraph in your Science Journal describing the results of your experiment. Infer why the food coloring spread throughout the water in the two beakers at different rates.

Before You Read

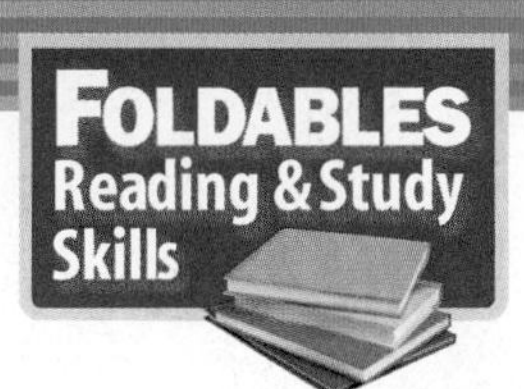

Making a Compare and Contrast Study Fold Make the following Foldable to help you see how temperature and heat are similar and different.

1. Place a sheet of paper in front of you so the long side is at the top. Fold the paper in half from top to bottom. Fold from the left side to the right side and crease. Then unfold.

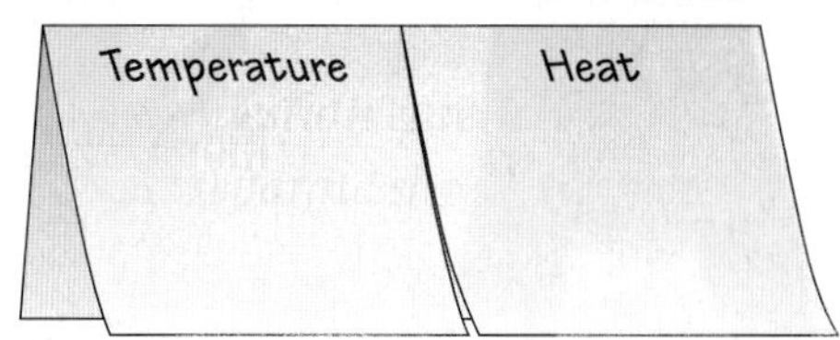

2. Through the top thickness of paper, cut along the middle fold line to form two tabs. Label the tabs *Temperature* and *Heat.*
3. Before you read the chapter, write what you know about temperature and heat under the tabs. As you read the chapter, add to and correct what you have written.

Temperature and Heat

As You Read

What You'll Learn

- **Explain** the difference between heat and temperature.
- **Define** thermal energy.
- **Explain** the meaning of specific heat.

Vocabulary

temperature
heat
thermal energy
specific heat

Why It's Important

If you know the difference between temperature and heat, you can understand why heat flows.

Temperature

The words hot and cold are commonly used to describe the temperature of a material. Although the terms *hot* and *cold* are not very precise, they still are useful. Everyone understands that hot indicates high temperature and that cold indicates low temperature. But what is temperature and how is temperature related to heat?

Matter in Motion All matter is made of tiny particles—atoms and molecules. Molecules are made of atoms held together by chemical bonds. Atoms and molecules are so small that a speck of dust has trillions of them. However, in all materials—solids, liquids, or gases—these particles are in constant motion.

Like all objects that are moving, these moving particles have kinetic energy. The faster these particles move, the more kinetic energy they have. **Figure 1** shows how molecules are moving in hot and cool objects.

Figure 1
The atoms in an object are in constant motion.

A When the horseshoe is hot, the particles in it move very quickly.

B When the horseshoe has cooled, its particles are moving more slowly.

Temperature Why do some objects feel hot and others feel cold? The **temperature** of an object is related to the average kinetic energy of the atoms or molecules. The faster these particles are moving, the more kinetic energy they have, and the higher the temperature of the object is. Think about a cup of hot tea and a glass of iced tea. The temperature of the hot tea is higher because the molecules in the hot tea are moving faster than those in the iced tea. In SI units, temperature is measured in kelvins (K), and a change in temperature of one kelvin is the same as a change of one degree Celsius.

Thermal Energy

If you let cold butter sit at room temperature for a while, it warms and becomes softer. Because the air in the room is at a higher temperature than the butter, molecules in air have more kinetic energy than butter molecules. Collisions between molecules in butter and molecules in air transfer energy from the faster-moving molecules in air to the slower-moving butter molecules. The butter molecules then move faster and the temperature of the butter increases.

Molecules in the butter can exert attractive forces on each other. Recall that Earth exerts an attractive gravitational force on a ball. When the ball is above the ground, the ball and Earth are separated, and the ball has potential energy. In the same way, atoms and molecules that exert attractive forces on each other have potential energy when they are separated. The sum of the kinetic and potential energy of all the molecules in an object is the **thermal energy** of the object. Because the kinetic energy of the butter molecules increased as it warmed, the thermal energy of the butter increased.

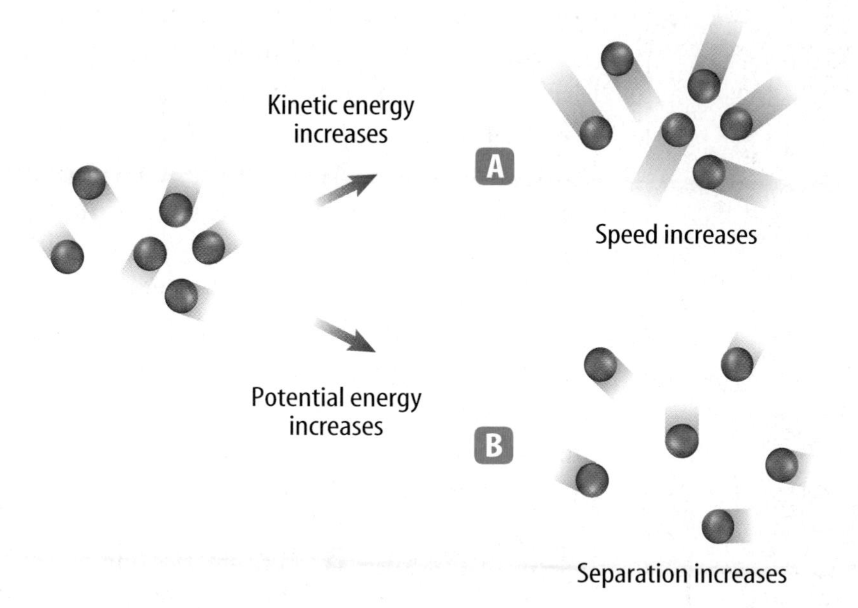

Figure 2

The thermal energy of a substance is the sum of the kinetic and potential energy of its molecules. A The kinetic energy increases as the molecules move faster. B The potential energy increases as the molecules move farther apart.

SCIENCE Online

Research Visit the Glencoe Science Web site at **science.glencoe.com** for information about how weather satellites use thermal energy. Communicate to your class what you learn.

Thermal Energy and Temperature Thermal energy and temperature are related. When the temperature of an object increases, the average kinetic energy of the molecules in the object increases. Because thermal energy is the total kinetic and potential energy of all the molecules in an object, the thermal energy of the object increases when the average kinetic energy of its molecules increases. Therefore, the thermal energy of an object increases as its temperature increases.

Thermal Energy and Mass Suppose you have a glass and a beaker of water that are at the same temperature. The beaker contains twice as much water as the glass. The water in both containers is at the same temperature, so the average kinetic energy of the water molecules is the same in both containers. But there are twice as many water molecules in the beaker as there are in the glass. So the total kinetic energy of all the molecules is twice as large for the water in the beaker. As a result, even though they are at the same temperature, the water in the beaker has twice as much thermal energy as the water in the glass does. If the temperature doesn't change, the thermal energy in an object increases if the mass of the object increases.

Heat

Can you tell if someone has been sitting in your chair? Perhaps you've noticed that your chair feels warm, and maybe you concluded that someone has been sitting in it recently. The chair feels warmer because thermal energy from the person's body flowed to the chair and increased its temperature.

Heat is thermal energy that flows from something at a higher temperature to something at a lower temperature. Heat is a form of energy, so it is measured in joules—the same units that energy is measured in. Heat always flows from warmer to cooler materials. How did the ice cream in **Figure 3** become cold? Heat flowed from the warmer liquid ingredients to the cooler ice-and-salt mixture. The liquid ingredients lost enough thermal energy to become cold enough to form solid ice cream. Meanwhile, the ice-and-salt solution gained thermal energy, causing some of the ice to melt.

Reading Check *How are heat and thermal energy related?*

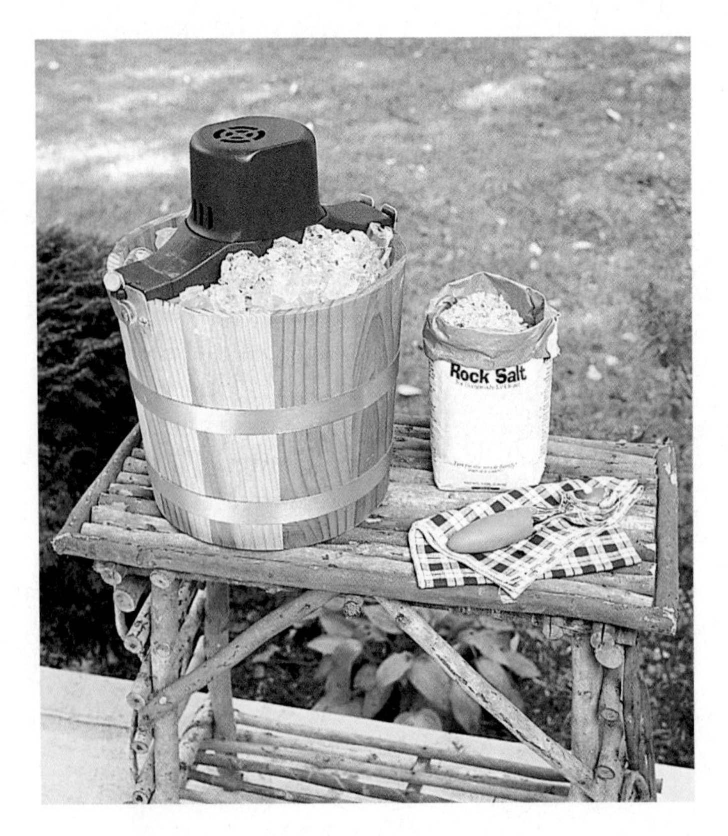

Figure 3
Heat flows from the warmer ingredients inside the container to the ice-and-salt mixture.

Specific Heat

If you are at the beach in the summertime, you might notice that the ocean seems much cooler than the air or sand. Even though energy from the Sun is falling on the air, sand, and water at the same rate, the temperature of the water has changed less than the temperature of the air or sand has.

As a substance absorbs heat, its temperature change depends on the nature of the substance, as well as the amount of heat that is added. For example, compared to 1 kg of sand, the amount of heat that is needed to raise the temperature of 1 kg of water by 1°C is about six times greater. So the ocean water at the beach would have to absorb six times as much heat as the sand to be at the same temperature. The amount of heat that is needed to raise the temperature of 1 kg of some material by 1°C or 1 K is called the **specific heat** of the material. Specific heat is measured in joules per kilogram kelvin [J/(kg K)]. **Table 1** shows the specific heats of some familiar materials.

Table 1 Specific Heat of Some Common Materials

Substance	Specific Heat [J/(kg K)]
Water	4,184
Wood	1,760
Carbon (graphite)	710
Glass	664
Iron	450

Water as a Coolant Compared with the other common materials in **Table 1,** water has the highest specific heat, as shown in **Figure 4.** Because water can absorb heat without a large change in temperature, it is useful as a coolant. A coolant is a subtance that is used to absorb heat. For example, water is used as the coolant in the cooling systems of automobile engines. As long as the water temperature is lower than the engine temperature, heat will flow from the engine to the water. Compared to other materials, water can absorb more heat from the engine before its temperature rises. Because it takes water longer to heat up compared with other materials, it also takes water longer to cool down.

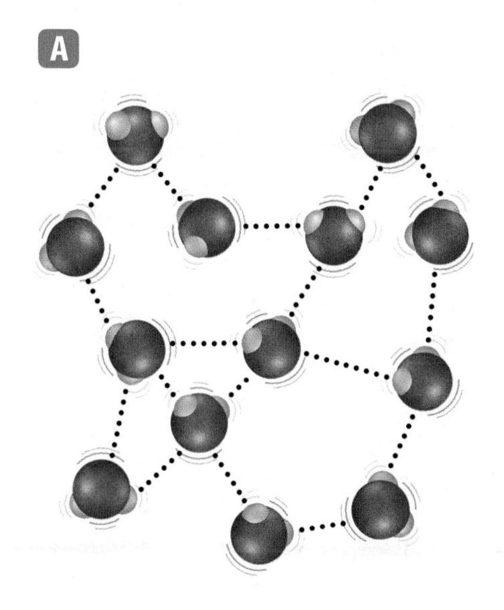

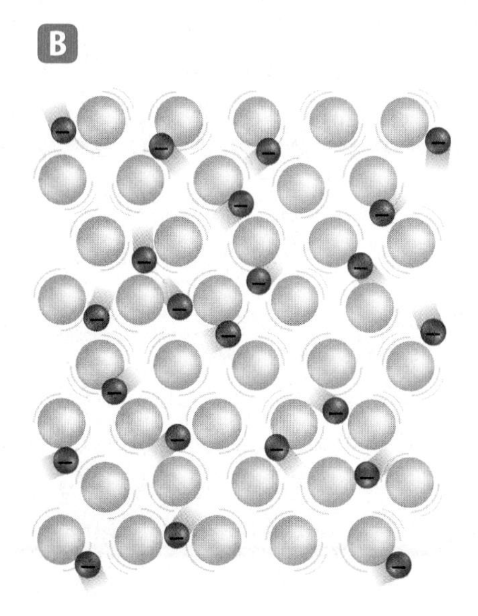

Figure 4
The specific heat of water is high because water molecules form strong bonds with each other.
A When heat is added, some of the added heat has to break some of these bonds before the molecules can start moving faster.
B In metals, electrons can move freely. When heat is added, no strong bonds have to be broken before the electrons can start moving faster.

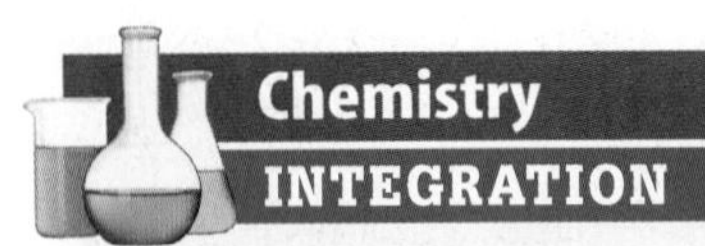

While Karen was blow-drying her hair, she noticed that her silver earring had become uncomfortably hot. If her earring was about the same mass as her earlobe, infer which of the two had the higher specific heat. Assuming a human earlobe is composed largely of water, find data from **Table 1** to support your inference and record it in your Science Journal.

Calculating Changes in Thermal Energy

The thermal energy of an object changes when heat flows into or out of it. The change in thermal energy is related to the mass of the object, its specific heat, and its change in temperature in this way:

$$\text{Change in thermal energy} = \text{mass} \times \text{change in temperature} \times \text{specific heat}$$

The change in temperature is calculated by subtracting the initial temperature from the final temperature in this way:

$$\text{Change in temperature} = T_{\text{final}} - T_{\text{initial}}$$

If Q is the change in thermal energy, m is the mass, and C is the specific heat, then the change in thermal energy can be calculated from this formula:

$$Q = m \times (T_{\text{final}} - T_{\text{initial}}) \times C$$

Math Skills Activity

Calculating Changes in Thermal Energy

Example Problem

The temperature of a 32-g silver spoon increases from 20°C to 60°C. If silver has a specific heat of 235 J/(kg K) what is the change in the thermal energy of the spoon?

Solution

1. *This is what you know:* mass of spoon, m = 32 g = 0.032 kg
specific heat of silver, C = 235 J/(kg K)
initial temperature, T_{initial} = 20°C
final temperature, T_{final} = 60°C
2. *This is what you want to find:* change in thermal energy, Q
3. *This is the equation you need to use:* $Q = m \times (T_{\text{final}} - T_{\text{initial}}) \times C$
4. *Solve the equation for Q:* Q = 0.032 kg × (60°C − 20°C) × 235 J/(kg K)
Q = 301 J

Check your answer by dividing it by the change in temperature then by the specific heat of silver. Did you get the original mass of the spoon?

Practice Problem

1. A 45-kg brass sculpture gains 203,000 J of thermal energy as its temperature increases from 28°C to 40°C. What is the specific heat of brass?

For more help, refer to the Math Skill Handbook.

Thermal Energy and Temperature Changes

When the temperature of an object changes, the amount of thermal energy it contains changes also. When heat flows into an object its temperature usually increases. So T_{final} is greater than $T_{initial}$ and the change in temperature is a positive number. According to the formula for the change in thermal energy, if the change in temperature is a positive number, then Q is a positive number. Therefore, when the temperature increases, heat flows into an object, and Q is positive.

In the same way, when heat flows out of an object its temperature usually decreases. Then the temperature of an object decreases, the change in temperature is a negative number, and Q is also negative.

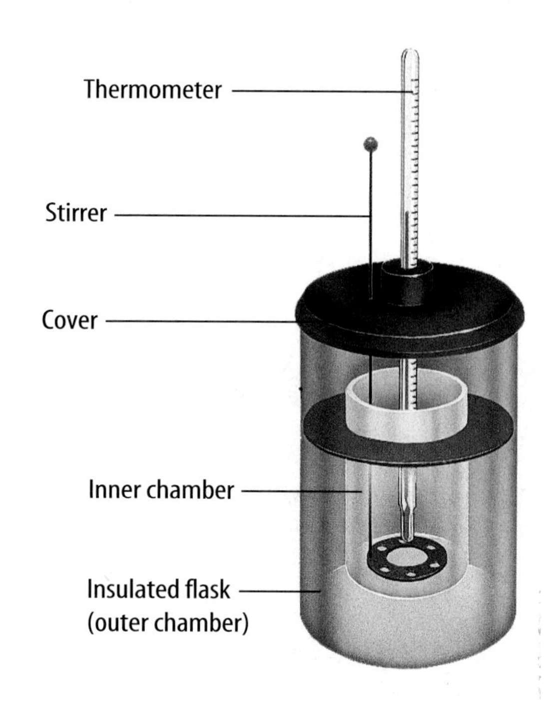

Figure 5
A calorimeter can be used to measure the specific heat of materials. The sample is placed in the inner chamber.

Measuring Specific Heat The specific heat can be measured using a device called a calorimeter, shown in **Figure 5.** To do this, the mass of the water in the calorimeter and its temperature are measured. Then a sample of a material of known mass is heated and its temperature measured. The heated sample is then placed in the calorimeter. Heat flows from the sample to the water until they both reach the same temperature. Then the increase in thermal energy of the water can be calculated. This is equal to the thermal energy lost by the sample. Because the mass of the sample, its change in temperature, and change in thermal energy are known, the sample's specific heat can be calculated.

Section  Assessment

1. Explain the difference between heat and temperature.
2. What is thermal energy and what causes it to change?
3. How is heat transferred between two objects?
4. What is specific heat?
5. **Think Critically** Using your knowledge of heat, explain what happens when you heat a pan of soup on the stove, and then put some of the leftover warm soup in the refrigerator.

Skill Builder Activities

6. **Interpreting Data** Equal amounts of iron, water, and glass with the same initial temperature were placed in an oven and heated briefly. Use the data in **Table 1** to match each final temperature with the appropriate material: 51.5°C, 25°C, and 66°C. **For more help, refer to the Science Skill Handbook.**
7. **Solving One-Step Equations** Calculate the change in thermal energy of 30 g of wood that cools from 100°C to 30°C. **For more help, refer to the Math Skill Handbook.**

SECTION

Transferring Thermal Energy

As You Read

***What* You'll Learn**

- **Compare and contrast** thermal energy transfer by conduction, convection, and radiation.
- **Compare and contrast** conductors and insulators.
- **Explain** how insulation affects the transfer of energy.

Vocabulary

conduction
convection
radiation
insulator

***Why* It's Important**

Understanding how thermal energy is transferred can help you use energy more efficiently.

Conduction

Thermal energy travels as heat from a material at a higher temperature to a material at a lower temperature. When you pick up a handful of snow to make a snowball, thermal energy from your hand is transferred to the snow, causing the snow to begin melting and your hand to get a little colder. When you come back indoors, the opposite happens if you hold a cup of hot chocolate. The thermal energy from the cup moves to your hand, making it warmer and the cup cooler. Direct contact is one way heat can travel from one place to another. The transfer of thermal energy through matter by the direct contact of particles is called **conduction.** Conduction occurs because all matter is made of atoms and molecules that are in constant motion.

Transfer by Collisions Think about what happens when you grasp a handful of snow. The slower-moving molecules in the snow come into contact with the faster-moving molecules in your warm hand. These particles then collide with one another, and some of the kinetic energy from the faster-moving particles is transferred to the slower-moving particles. This causes the slower-moving particles to speed up and the faster-moving particles to slow down. As the collisions continue, thermal energy gets transferred from your bare hand to the snow. As a result, your hand gets colder and the snow melts.

Heat can be transferred by conduction from one material to another or through one material. What happens to a metal ladle that's used to stir a pot of simmering soup? The hot soup transfers thermal energy to the part of the ladle sitting in the soup. At first, this end of the ladle is hotter than the rest. Eventually, however, the entire ladle becomes hot. Heat was transferred from the soup to the ladle—from one material to another—and through the length of the ladle—through one material. **Figure 6** shows how heat is transferred by conduction.

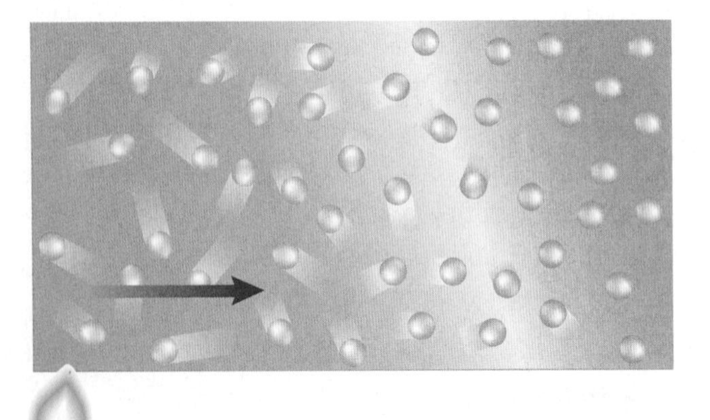

Figure 6
Conduction transfers heat from the hotter part of the material to the cooler part.

Heat Conductors Although conduction can occur in solids, liquids, and gases, solids usually conduct heat much more effectively. The particles in a solid are usually much closer together than they are in liquids and gases, so they collide with one another more often. The loosely held electrons in a metal make them especially good heat conductors. In a metal, electrons can move easily, and they readily transfer kinetic energy to other nearby particles. Silver, copper, and aluminum are among the best heat conductors. Wood, plastic, glass, and fiberglass are poor conductors of heat. Why do you think most cooking pots are made of metal, but the handles usually are not?

Convection

One way liquids and gases differ from solids is that they can flow. Any material that flows is a fluid. The ability to flow allows fluids to transfer heat in another way—convection. **Convection** is the transfer of energy in a fluid by the movement of the heated particles. In conduction, heated particles collide with each other and transfer their energy. In convection, however, the more energetic fluid particles move from one location to another, and carry their energy along with them.

As the particles move faster, they tend to be farther apart. In other words, the fluid expands as its temperature increases. Recall that density is the mass of a material divided by its volume. When a fluid expands, its volume increases, but its mass doesn't change. As a result, its density decreases. The same is true for parts of a fluid that have been heated. The density of the warmer fluid, therefore, is less than that of the surrounding cooler fluid.

Heat Transfer by Currents How does convection occur? Look at the lamp shown in **Figure 7.** Some of these lamps contain oil and alcohol. When the oil is cool, its density is greater than the alcohol, and it sits at the bottom of the lamp. When the two liquids are heated, the oil becomes less dense than the alcohol. Because it is less dense than the alcohol, it rises to the top of the lamp. As it rises, it loses heat by conduction to the cooler fluid around it. When the oil reaches the top of the lamp, it has become cool enough that it is denser than the alcohol, and it sinks. This rising-and-sinking action is a convection current. Convection currents transfer heat from warmer to cooler parts of the fluid. In a convection current, both conduction and convection transfer thermal energy.

✔ Reading Check *How are conduction and convection different?*

Figure 7
The heat from the light at the bottom of the lamp causes one fluid to expand more than the other. This creates convection currents in the lamp.

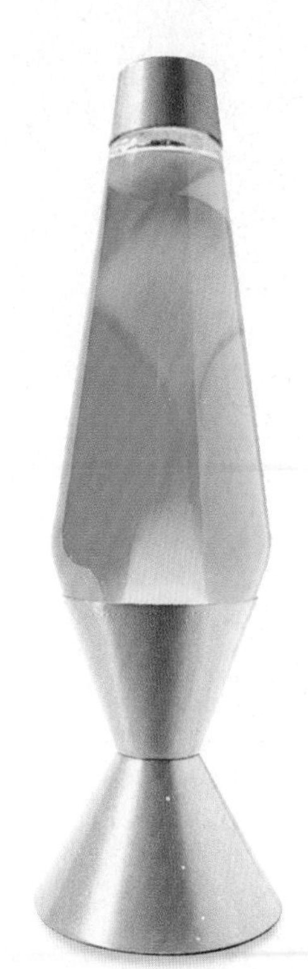

Figure 8

When the Sun beats down on the equator, warm, moist air begins to rise. As it rises, the air cools and loses its moisture as rain that sustains rain forests near the equator. Convection currents carry the now dry air farther north and south. Some of this dry air descends at the tropics, where it creates a zone of deserts.

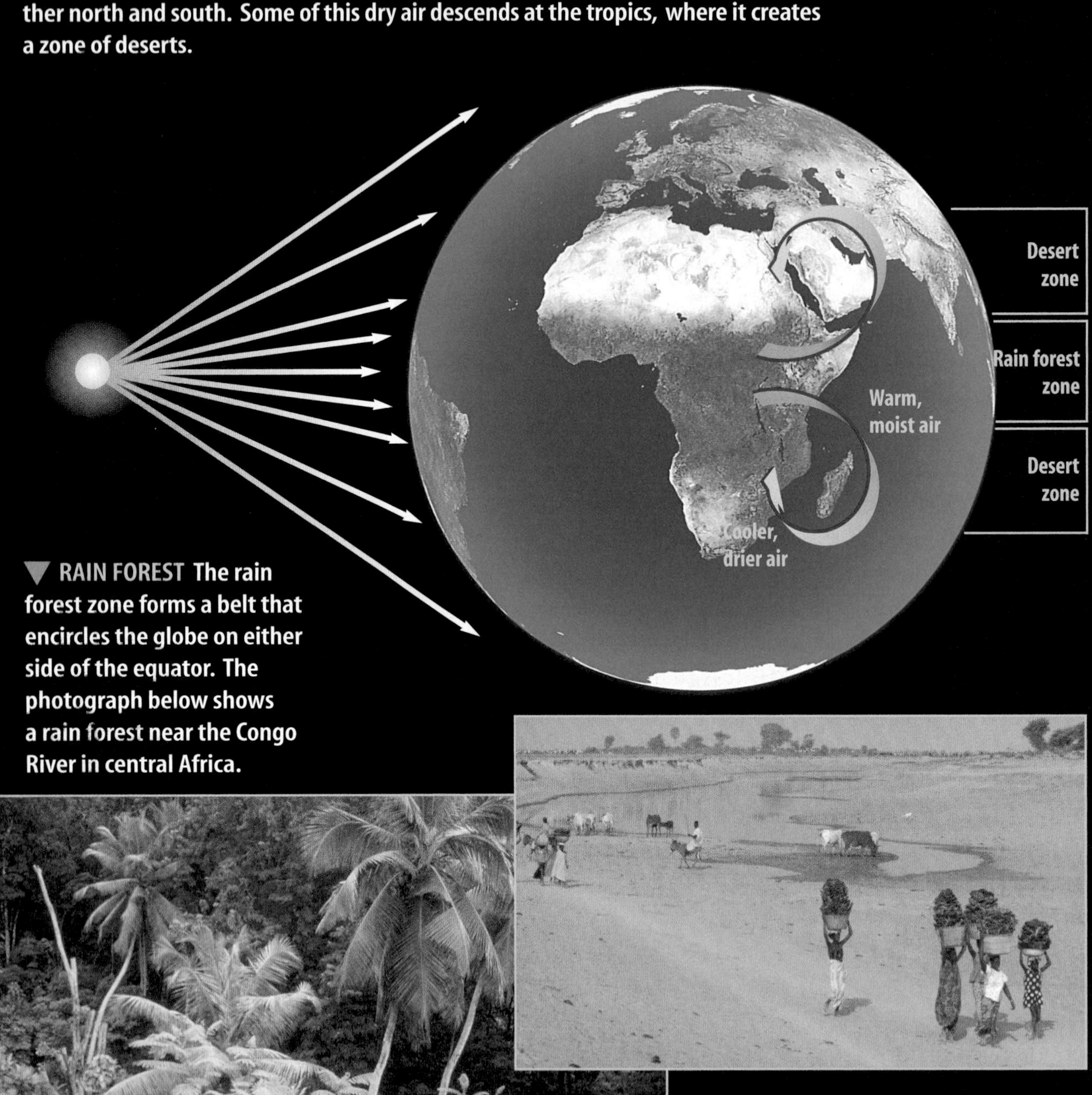

▼ RAIN FOREST The rain forest zone forms a belt that encircles the globe on either side of the equator. The photograph below shows a rain forest near the Congo River in central Africa.

▲ DESERT Like many of the great desert regions of the world, the Sahara, in northern Africa, is largely a result of atmospheric convection currents. Here, a group of nomads gather near a dried-up river in Mali.

Desert and Rain Forests Earth's atmosphere is made of various gases and is a fluid. The atmosphere is warmer at the equator than it is at the north and south poles. Also, the atmosphere is warmer at Earth's surface than it is at higher altitudes. These temperature differences create convection currents that carry heat to cooler regions. **Figure 8** shows how these convection currents create rain forests and deserts over different regions of Earth's surface.

Radiation

Earth gets heat from the Sun, but how does that heat travel through space? Almost no matter exists in the space between Earth and the Sun, so heat cannot be transferred by conduction or convection. Instead, the Sun's heat reaches Earth by radiation.

Radiation is the transfer of energy by electromagnetic waves. These waves can travel through space even when no matter is present. Energy that is transferred by radiation often is called radiant energy. When you stand near a fire and warm your hands, much of the warmth you feel has been transferred from the fire to your hands by radiation.

Radiant Energy and Matter When radiation strikes a material, some of the energy is absorbed, some is reflected, and some may be transmitted through the material. **Figure 9** shows what happens to radiant energy from the Sun as it reaches Earth. The amount of energy absorbed, reflected, and transmitted depends on the type of material. Materials that are light-colored reflect more radiant energy, while dark-colored materials absorb more radiant energy. When radiant energy is absorbed by a material, the thermal energy of the material increases.

For example, when a car sits outside in the Sun, some of the radiation from the Sun passes through the transparent car windows. Materials inside the car absorb some of this radiation and become hot. Radiation sometimes can pass through solids and liquids, as well as gases.

Radiation in Solids, Liquids, and Gases The transfer of energy by radiation is most important in gases. In a solid, liquid or gas, radiant energy can travel through the space between molecules. Molecules can absorb this radiation and re-emit some of the energy they absorbed. This energy then travels through the space between molecules, and is absorbed and re-emitted by other molecules. Because molecules are much farther apart in gases than in solids or liquids, radiation usually passes more easily through gases than through solids or liquids.

Figure 9
Not all of the Sun's radiation reaches Earth. Some of it is reflected by the atmosphere. Some of the radiation that does reach the surface is also reflected.

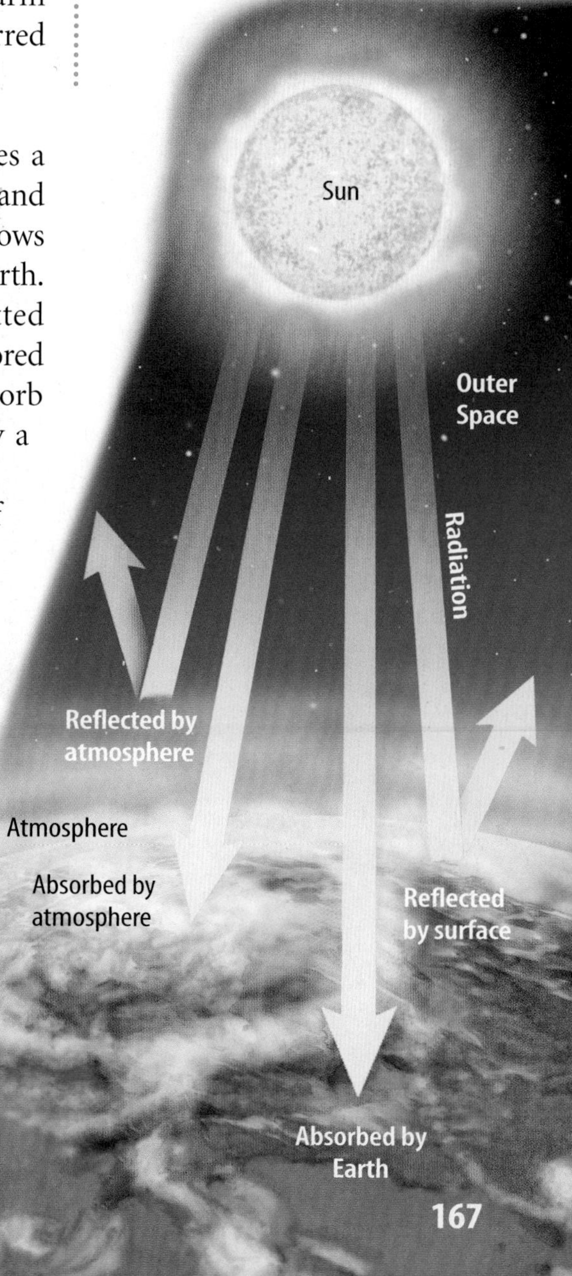

Observing Heat Transfer by Radiation

Procedure

1. On a sunny day, go outside and place the back of your hand in **direct sunlight** for two min.
2. Go inside and find a **window exposed to direct sunlight.**
3. Place the back of your hand in the sunlight that has passed through the window for two min.

Analysis

1. Explain how heat was transferred from the Sun to your skin when you were outside.
2. Compare how warm your skin felt inside and outside.
3. Was thermal energy transferred through the glass in the window? Explain.

Controlling Heat Flow

You might not realize it, but you probably do a number of things every day to control the flow of heat. For example, when its cold outside, you put on a coat or a jacket before you leave your home. When you reach into an oven to pull out a hot dish, you might put a thick, cloth mitten over your hand to keep from being burned. In both cases, you used various materials to help control the flow of heat. Your jacket kept you from getting cold by reducing the flow of heat from your body to the surrounding air. And the oven mitten kept your hand from being burned by reducing the flow of heat from the hot dish.

As shown in **Figure 10,** almost all living things have special features that help them control the flow of heat. For example, the antarctic fur seal's thick coat and the emperor penguin's thick layer of blubber help keep them from losing heat. This helps them survive in a climate in which the temperature is often below freezing. In the desert, however, the scaly skin of the desert spiny lizard has just the opposite effect. It reflects the Sun's rays and keeps the animal from becoming too hot. An animal's color also can play a role in keeping it warm or cool. The black feathers on the penguin's back, for example, allow it to absorb as much radiant energy as possible. Can you think of any other animals that have special adaptations for cold or hot climates?

Reading Check *What are two animal adaptations that control the flow of heat?*

Figure 10
Animals have different features that help them control heat flow.

A The antarctic fur seal grows a coat that can be as much as 10 cm thick.

B The emperor penguin has a thick layer of blubber and thick, closely grown feathers, which help reduce the loss of body heat.

C The scaly skin of the desert spiny lizard reflects the rays of the Sun. This prevents it from losing water by evaporation in an environment where water is scarce.

Figure 11
The tiny pockets of air in fleece make it a good insulator. They help prevent the jogger's body heat from escaping.

Insulators

A material that doesn't allow heat to flow through it easily is called an **insulator.** A material that is a good conductor of heat, such as a metal, is a poor insulator. Materials such as wood, plastic, and fiberglass are good insulators and, therefore, are poor conductors of heat.

Gases, such as air, are usually much better insulators than solids or liquids. Some types of insulators contain many pockets of trapped air. These air pockets conduct heat poorly and also keep convection currents from forming. Fleece jackets, like the one shown in **Figure 11,** work in the same way. When you put the jacket on, the fibers in the fleece trap air and hold this air next to you. This air slows down the flow of your body heat to the colder air outside the jacket. Gradually, the air trapped by the fleece is warmed by your body heat, and underneath the jacket you are wrapped in a blanket of warm air.

Why does trapped air make a material like fleece a good insulator?

Insulating Buildings Insulation, or materials that are insulators, helps keep warm air from flowing out of buildings in cold weather and from flowing into buildings in warm weather. Building insulation is usually made of some fluffy material, such as fiberglass, that contains pockets of trapped air. The insulation is packed into a building's outer walls and attic, where it reduces the flow of heat between the building and the surrounding air.

Insulation helps furnaces and air conditioners work more effectively, saving energy. In the United States, about 55 percent of the energy used in homes is used for heating and cooling.

Mini LAB

Comparing Thermal Conductors

Procedure

1. Obtain a **plastic spoon,** a **metal spoon,** and a **wooden spoon** with similar lengths.
2. Stick a small **plastic bead** to the handle of each spoon with a dab of **butter or wax.** Each bead should be the same distance from the tip of the spoon.
3. Stand the spoons in a beaker, with the beads hanging over the edge of the beaker.
4. Carefully pour about 5 cm of boiling water in the beaker holding the spoons.

Analysis

1. In what order did the beads fall from the spoons?
2. Describe how heat was transferred from the water to the beads.
3. Rank the spoons in their ability to conduct heat.

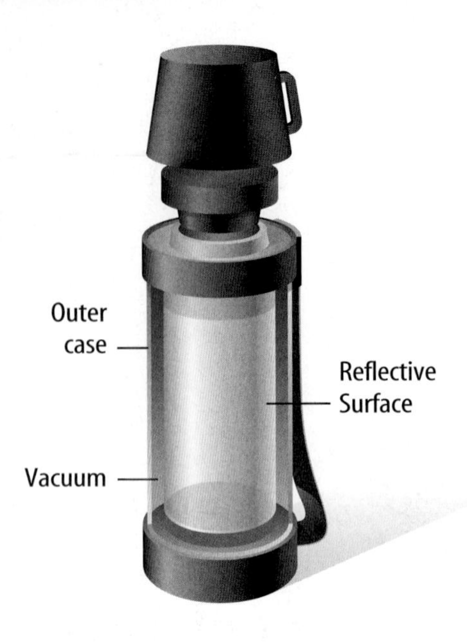

Figure 12
The vacuum layer of the thermos bottle is a very poor conductor of heat.

Using Insulators You might have used a thermos bottle, like the one in **Figure 12,** to carry hot soup or iced tea. The vacuum layer in the thermos helps keep your soup hot by reducing the heat flow due to conduction, convection, and radiation. Thermos bottles don't use a thick layer of material to reduce heat flow, instead they use nothing. A thermos bottle has two glass layers with a vacuum between the layers. The vacuum between the two layers contains almost no matter, and so prevents heat transfer either by conduction or by convection.

To reduce heat transfer by radiation, the inside and outside glass surface of a thermos bottle is coated with aluminum to make each surface highly reflective. This causes the electromagnetic waves that carry heat energy by radiation to be reflected at each surface. The inner coating prevents electromagnetic waves from leaving the bottle, and the outer coating prevents them from entering the bottle.

Think about the things you do to stay warm or cool. Sitting under a shady umbrella reduces the heat energy transferred to you by radiation. Also, wearing light-colored or dark-colored clothing changes the amount of heat you absorb due to radiation. To change the amount of heat transferred by convection, you can open and close windows. Putting on a sweater reduces the heat transferred from your body by conduction and convection. In what other ways do you control the flow of heat?

Section  Assessment

1. Describe the three ways thermal energy can be transferred.
2. Why are materials that are good conductors of heat also poor insulators?
3. How is heat transfer by radiation different from conduction and convection?
4. How could insulating a home save money?
5. **Think Critically** Several days after a snowfall, the roofs of some houses on a street have almost no snow on them, while the roofs of other houses are still snow-covered. Which houses are better insulated? Explain.

Skill Builder Activities

6. **Testing a Hypothesis** Design an experiment to find out which material makes the best insulation: plastic foam pellets, shredded newspaper, or crumpled plastic bags. Remember to state your hypothesis and indicate what factors must be held constant. **For more help, refer to the Science Skill Handbook.**
7. **Concept Mapping** Make a concept map showing the three types of thermal energy transfer and ways you can control the flow of heat in each type. **For more help, refer to the Science Skill Handbook.**

Activity

Convection in Gases and Liquids

A hawk gliding through the sky will rarely flap its wings. Hawks and some other birds conserve energy by gliding on columns of warm air rising up from the ground. These convection currents form when gases or liquids are heated unevenly, and the warmer, less dense fluid is forced upward. In this activity, you will create and observe your own convection currents.

What You'll Investigate

How can convection currents be modeled and observed?

Materials

burner or hot plate
water
candle
500-mL beaker
black pepper

Safety Precautions 

WARNING: *Use care when working with hot materials. Remember that hot and cold glass appear the same.*

Goals

- **Model** the formation of convection currents in water.
- **Observe** convection currents formed in water.
- **Observe** convection currents formed in air.

Procedure

1. Pour about 450 mL of water into a 500-mL beaker.
2. Use a balance to measure 1 g of black pepper.
3. Sprinkle the pepper into the beaker of water.
4. Let the pepper settle to the bottom of the beaker.
5. Heat the bottom of the beaker using the burner or by placing it on the hotplate.
6. **Observe** how the particles of pepper move as the water is heated, and make a drawing showing their motion in your Science Journal.
7. Turn off the hot plate or burner. Light the candle and let it burn for a few minutes.
8. Blow out the candle, and observe the motion of the smoke.
9. Make a drawing of the movement of the smoke in your Science Journal.

Conclude and Apply

1. **Describe** how the particles of pepper moved as the water became hotter.
2. How is the motion of the pepper particles related to the motion of the water?
3. **Explain** how a convection current formed in the beaker.
4. **Explain** why the motion of the pepper changed when the heat was turned off.
5. **Predict** how the pepper would move if the water were heated from the top.
6. **Describe** how the smoke particles moved when the candle was blown out.
7. **Explain** why the smoke moved as it did.

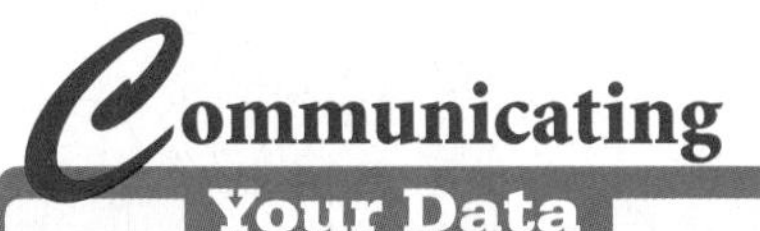

Compare your conclusions with other students in your class. **For more help, refer to the** Science Skill Handbook.

SECTION

Using Heat

As You Read

What You'll Learn

- **Compare and Contrast** three types of conventional heating systems.
- **Distinguish** between passive and active solar heating systems.
- **Describe** how internal combustion engines work.
- **Explain** how a heat mover can transfer thermal energy in a direction opposite to that of its natural movement.

Vocabulary

solar energy
solar collector
heat engine
internal combustion engine
heat mover

Why It's Important

Heat is used to heat homes and operate vehicles.

Heating Systems

Almost everywhere in the United States air temperatures at some time become cold enough that a source of heat is needed. As a result, most homes and public buildings contain some type of heating system. No one heating system is best in all conditions. The best heating system for any home or building depends on the local climate and how the building is constructed.

All heating systems require some source of energy. In the simplest and oldest heating system, wood or coal is burned in a stove. The heat that is produced by the burning fuel is transferred from the stove to the surrounding air by conduction, convection, and radiation. One disadvantage of this system is that heat transfer from the room in which the stove is located to other rooms in the building can be slow.

Forced-Air Systems The most common type of heating system in use today is the forced-air system, shown in **Figure 13.** In this system, fuel is burned in a furnace and heats a volume of air. A fan then blows the warm air through a series of large pipes called ducts. The ducts lead to openings called vents in each room. Cool air returns to the furnace through additional vents, where it is reheated.

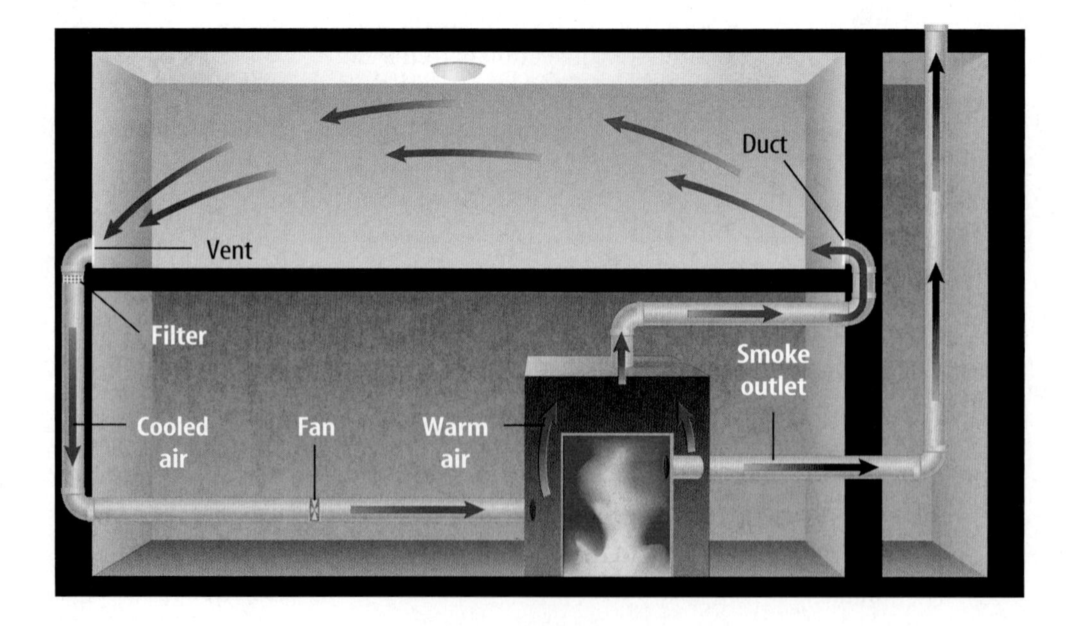

Figure 13
In forced-air systems, air heated by the furnace gets blown through ducts that usually lead to every room.

Radiator Systems Before forced-air systems were widely used, many homes and buildings were heated by a system of radiators. A radiator is a closed metal container that contains hot water or steam. The thermal energy contained in the hot water or steam is transferred to the air surrounding the radiator by conduction. This warm air then moves through the room by convection.

In a radiator heating system fuel is burned in a central furnace and heats a tank of water. A system of pipes carries the hot water to radiators that are located in the rooms of the building. Usually each room has one radiator, although large rooms may have several. When the water cools, pipes take it back down to the water tank, where it is reheated. In some radiator systems water is heated to produce steam that flows to the radiators. As the steam cools, it condenses into water and flows back to the tank.

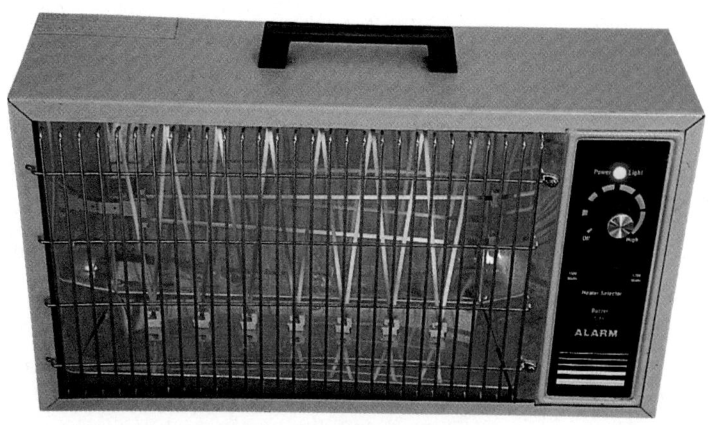

Figure 14
Electric radiators like this one convert electric energy to thermal energy.

You might have seen some electric radiators like the one in **Figure 14** that are not connected to a central furnace. These radiators contain metal coils that are heated when an electric current passes through them. The hot coils then transfer thermal energy to the room, mainly by radiation. This type of radiator often is used to provide heat in rooms that do not receive enough heat from the central heating system.

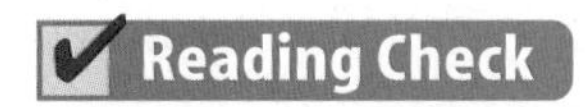
What are two ways that heat is transferred from a radiator?

Electric Heating Systems An electric heating system has no central furnace as forced-air and radiator systems do. Instead, it uses electrically heated coils placed in ceilings and floors to heat the surrounding air by conduction. Convection then distributes the heated air through the room. Electric heating systems are not as widely used as forced-air systems. In many areas, electric heating systems cost more to use. However, the walls and floors of some buildings might not be thick enough to include the pipes and ducts that forced-air and radiator systems require. Then an electric heating system might be the only practical way to provide heat.

Electric heating systems may seem to be a pollution-free way to provide heat. However, most power plants that produce electric energy burn fossil fuels, producing various pollutants. Also, it is much more efficient to burn fuels to produce heat rather than generate electricity. As a result, less fuel is burned to heat a house using a conventional furnace.

Solar Heating

Radiant energy from the Sun can make a greenhouse warm. The windows of the greenhouse allow the Sun's radiant energy to be transferred inside the greenhouse where materials absorb the radiant energy and become warmer. As they become warmer, the greenhouse heats up. The windows now keep the thermal energy from being transferred to the air outside, so the greenhouse stays warm.

The energy from the Sun is called **solar energy.** Solar energy is not only free, it is also available in a seemingly endless supply. Just as solar energy can heat a greenhouse, it also can be used to help heat homes and buildings.

Passive Solar Heating Two types of solar heating systems—passive and active—are available. In passive solar heating systems, solar energy heats rooms inside a building, but no mechanical devices are used to move heat from one area to another. Just as in a greenhouse, materials such as water or concrete inside a building absorb radiant energy from the Sun during the day and heat up. At night when the building begins to cool, the thermal energy absorbed by these materials helps keep the room warm.

Figure 15 shows a room in a house that uses passive solar heating. The south side of buildings usually receives the most solar energy. Homes that are heated by passive solar systems often have a wall of windows on the south side of the house. Walls of windows receive the maximum amount of sunlight possible during the day. The other walls are heavily insulated and have no windows to help reduce the loss of heat.

Figure 15
Passive solar heating systems make use of many materials' ability to hold heat. *In what climates would these systems work well?*

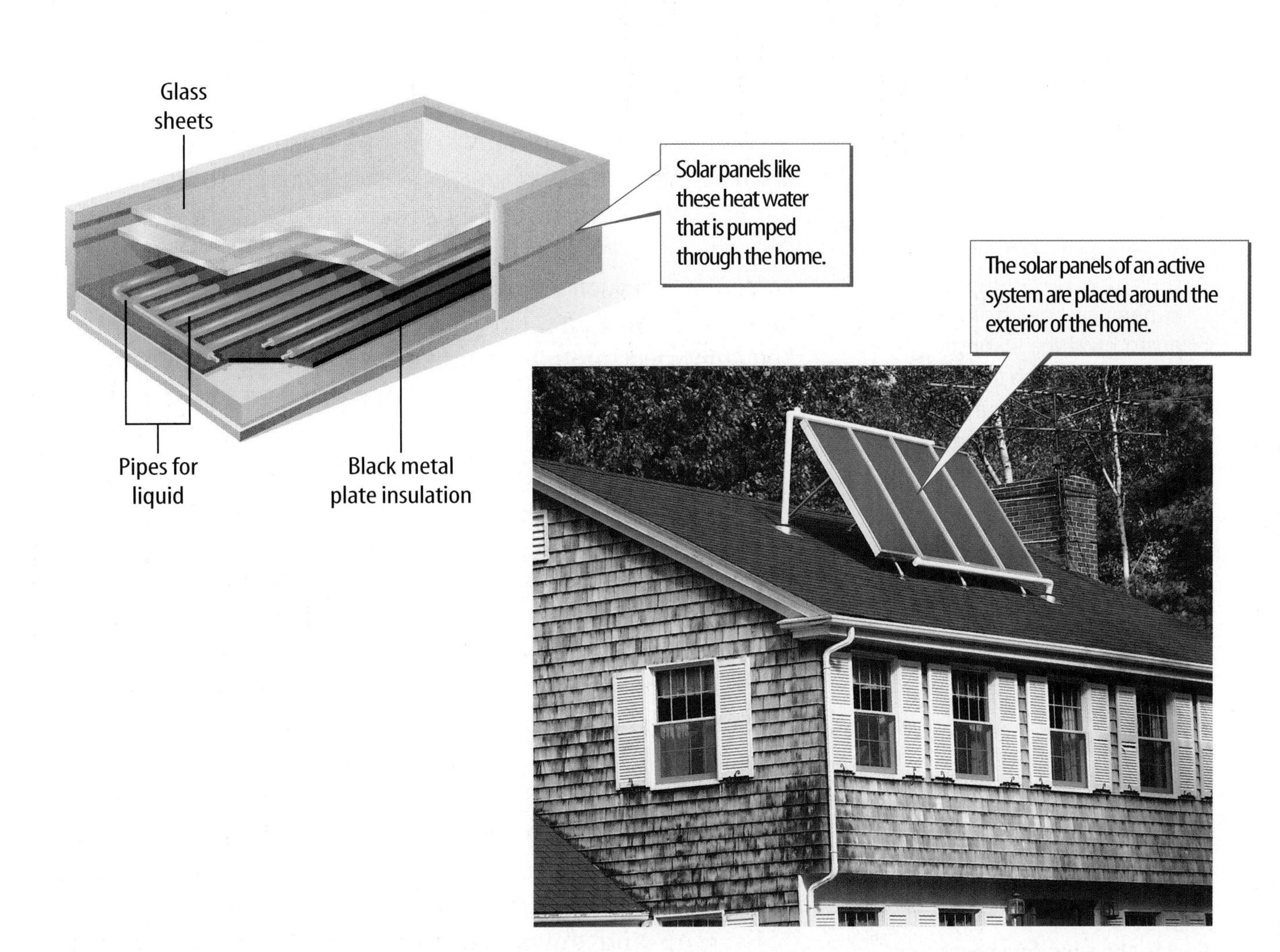

Active Solar Heating Active solar heating systems use devices called **solar collectors** that absorb radiant energy from the Sun. The collectors usually are installed on the roof or south side of a building where exposure to the Sun is greatest. Radiant energy from the Sun heats air or water in the solar collectors. This hot air or water is then circulated through the house.

Figure 16 shows a home with one type of active solar collector. The metal plate absorbs radiant energy from the Sun. The glass sheets reduce energy loss due to convection and conduction. Water-filled pipes are located just below the metal plate. The absorbed radiant energy heats the water in the pipes. A pump circulates the warm water to radiators in the rooms of the house. The cooled water is pumped back to the collector to be reheated. Some systems have large, insulated tanks for storing heated water. The water can then be used as needed to provide heat. Some homes use active solar heating to provide hot water for cleaning and bathing.

If your were asked to design a building, what kind of heating system would you install? How would the climate of your area affect the choice you make? What other factors would you need to consider?

Figure 16
In an active solar heating system, solar energy is absorbed and can be used to heat different rooms in the house.

Data Update Visit the Glencoe Science Web site at **science.glencoe.com** for information about improvements in solar heating designs.

Hurricanes are severe tropical storms that form over the ocean in regions of low pressure. Because hurricanes use heat from warm air to produce strong winds, they are sometimes called nature's heat engines. Research hurricanes and draw a diagram showing how they behave like a heat engine.

Using Heat to Do Work

Thermal energy not only keeps you warm, it also is the form of energy that enables cars and other vehicles to operate. For example, a car's engine converts the chemical energy in gasoline to thermal energy by a process called combustion. During combustion, a fuel such as gasoline combines with oxygen and produces heat and light. The thermal energy that is produced is then converted into mechanical energy. An engine that converts thermal energy into mechanical energy is called a **heat engine.**

Internal Combustion Engines A heat engine, like those usually used in cars, is called an **internal combustion engine** because fuel is burned inside the engine in chambers, or cylinders. Automobile engines usually have four, six, or eight cylinders. The more cylinders an engine has, the more power it can produce. Each cylinder contains a piston, which can move up and down. **Figure 17** shows how the motion of pistons is transmitted to the wheels of a car. Each piston is connected to a metal shaft called the crankshaft. The up-and-down motion of the pistons causes the crankshaft to spin. The motion of the crankshaft is transmitted through the transmission and the differential to the tires of the car. As the tires turn, rolling friction between the tires and the road causes the car to move.

Each up-and-down movement of the piston is called a stroke. In a gasoline or diesel engine there are four different strokes, so these engines are called four-stroke engines. A sequence of the four strokes is called a cycle. When the engine is running, many cycles are repeated for each piston every second. **Figure 18** shows the four-stroke cycle in an automobile engine.

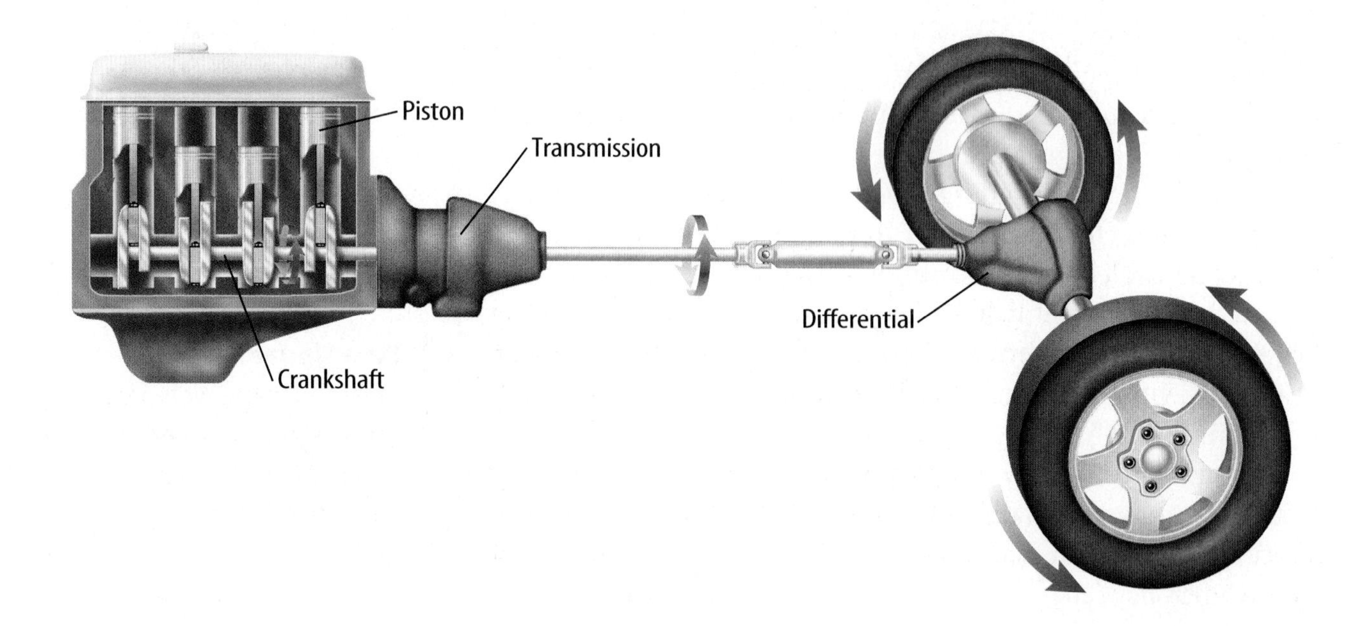

Figure 17
Fuel in the cylinders burns and pushes the pistons downward. A system of gears then translates the up and down motion of the pistons into motions that can turn the wheels.

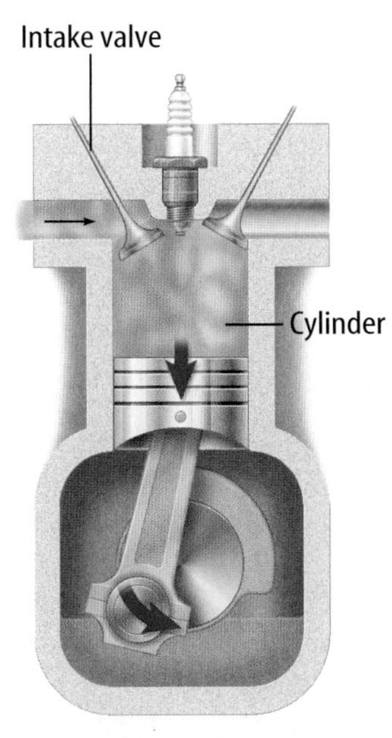

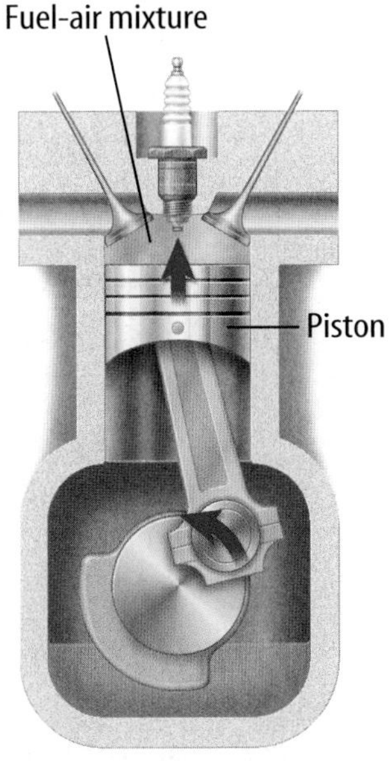

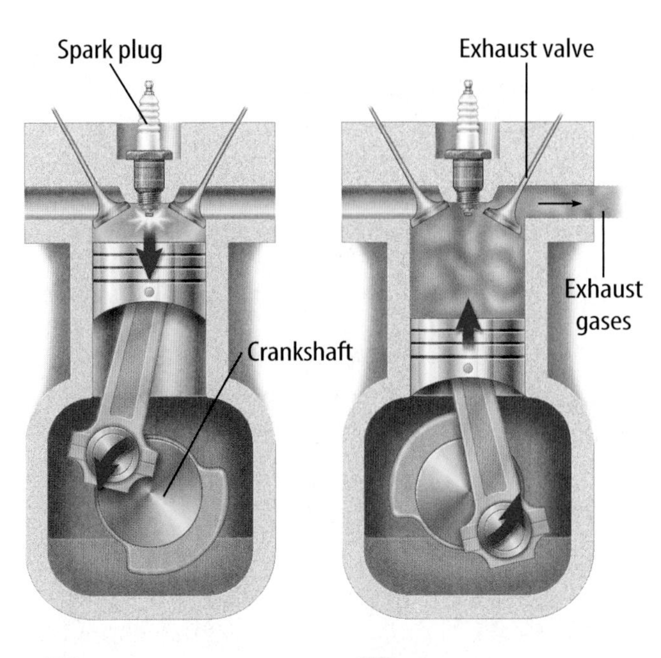

A Intake stroke
The intake valve opens as the piston moves downward, drawing a mixture of gasoline and air into the cylinder.

B Compression stroke
The intake valve closes as the piston moves upward, compressing the fuel-air mixture.

C Power stroke
The spark from a spark plug ignites the fuel-air mixture. As the mixture burns, hot gases expand, pushing the piston down.

D Exhaust stroke
As the piston moves up, the exhaust valve opens, and the waste gases from burning the fuel-air mixture are pushed out of the cylinder.

Figure 18
Automobile engines are usually four-stroke engines. Each four-stroke cycle converts thermal energy to mechanical energy.

Why are engines hot? If you've opened the hood of a car after it has been driven, you know the engine is too hot to touch. In an internal combustion engine, only part of the thermal energy that is produced by the burning fuel is converted into mechanical energy. Some of the thermal energy from the burning fuel has made the engine and other car parts hotter. So much heat is transferred to the engine that a cooling system is needed to keep the engine from becoming too hot to run properly. Gasoline automobile engines convert only about 26 percent of the chemical energy in the fuel to mechanical energy.

Heat Movers

How can the inside of a refrigerator stay cold? Heat should flow from the warmer room into the refrigerator until the room and refrigerator are at the same temperature. Instead, a refrigerator moves thermal energy out of the refrigerator into the warmer room. Heat can be made to flow from the cool refrigerator to the warm room only if work is done. The energy to do the work comes from the electricity that powers the refrigerator. A refrigerator is an example of a heat mover. A **heat mover** is a device that removes thermal energy from one location and transfers it to another location at a different temperature.

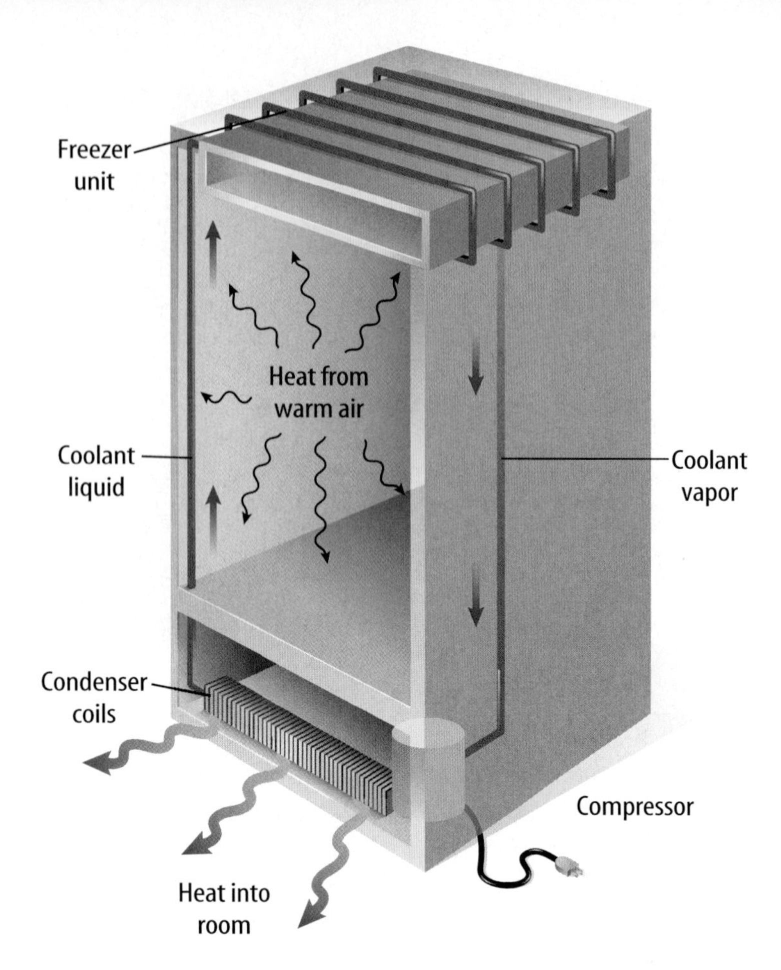

Figure 19
In refrigeration systems, the coolant that flows through the pipes absorbs heat from the food that is stored inside. It then condenses and releases the heat into the room.

Refrigerators A refrigerator contains a coolant that is pumped through pipes on the inside and outside of the refrigerator. The coolant is a special substance that evaporates at a low temperature. **Figure 19** shows how a refrigerator operates. Liquid coolant is pumped through a device where it changes into a gas. When the coolant changes to a gas, it cools. The cold gas is pumped through pipes inside the refrigerator, where it absorbs thermal energy. As a result, the inside of the refrigerator cools.

The gas then is pumped to a compressor that does work by compressing the gas. This makes the gas warmer than the temperature of the room. The warm gas is pumped through the condenser coils. Because the gas is warmer than the room, thermal energy flows from the gas to the room. Some of this heat is the thermal energy that the coolant gas absorbed from the inside of the refrigerator. As the gas gives off heat, it cools and changes to a liquid. The liquid coolant then is changed back to a gas, and the cycle is repeated.

What happens to a coolant when it is compressed?

Air Conditioners and Heat Pumps An air conditioner is another type of heat mover. It operates like a refrigerator, except that warm air from the room is forced to pass over tubes containing the coolant. The warm air is cooled and is forced back into the room. The thermal energy that is absorbed by the coolant is transferred to the air outdoors. Refrigerators and air conditioners are heat engines working in reverse—they use mechanical energy supplied by the compressor motor to move thermal energy from cooler to warmer areas.

A heat pump is a two-way heat mover. In warm weather, it operates as an air conditioner. In cold weather, a heat pump operates like an air conditioner in reverse. The coolant gas is cooled and is pumped through pipes outside the house. There, the coolant absorbs heat from the outside air. The coolant is then compressed and pumped back inside the house, where it releases heat.

The Human Coolant After exercising on a warm day, you might feel hot and be drenched with sweat. But if you were to take your temperature, you would probably find that it's close to your normal body temperature of 37°C. How does your body stay cool in hot weather?

Your body uses evaporation to keep its internal temperature constant. When a liquid changes to a gas, energy is absorbed from the liquid's surroundings. As you exercise, your body generates sweat from tiny glands within your skin. As the sweat evaporates, it carries away heat, making you cooler, as shown in **Figure 20.** The thermal energy that is lost by your body becomes part of the thermal energy of your evaporated sweat.

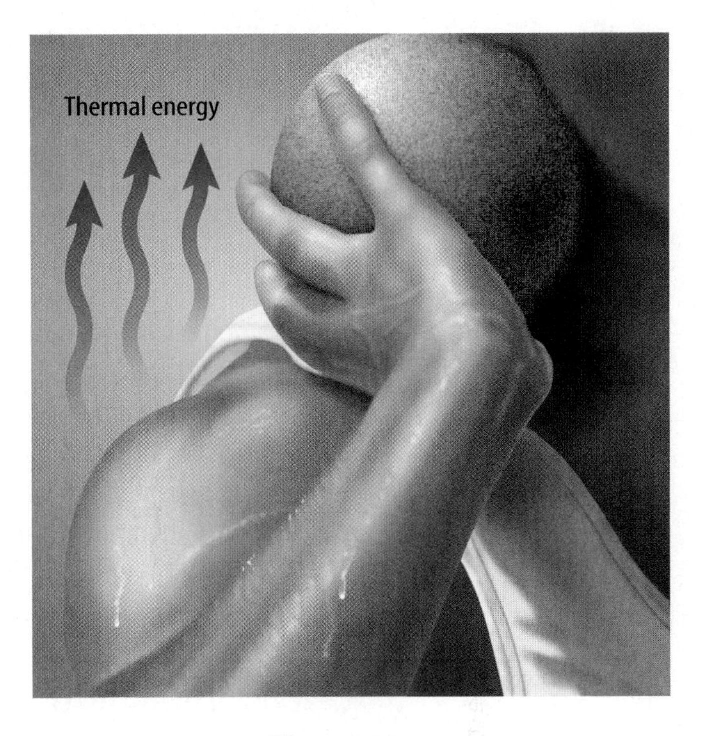

Figure 20
As perspiration evaporates from your skin, it carries heat away, cooling your body.

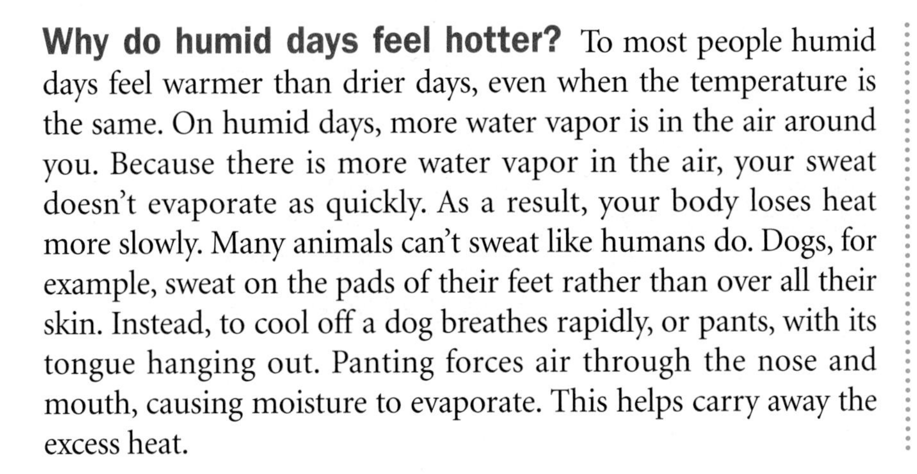

Why do humid days feel hotter? To most people humid days feel warmer than drier days, even when the temperature is the same. On humid days, more water vapor is in the air around you. Because there is more water vapor in the air, your sweat doesn't evaporate as quickly. As a result, your body loses heat more slowly. Many animals can't sweat like humans do. Dogs, for example, sweat on the pads of their feet rather than over all their skin. Instead, to cool off a dog breathes rapidly, or pants, with its tongue hanging out. Panting forces air through the nose and mouth, causing moisture to evaporate. This helps carry away the excess heat.

Section  Assessment

1. Compare and contrast electrical, radiator, and forced-air heating systems.
2. Compare and contrast active and passive solar heating systems.
3. Describe how internal combustion engines work.
4. What is a heat mover?
5. **Think Critically** How does a heat mover absorb heat from cold air?

Skill Builder Activities

6. **Making and Using Tables** In a table, organize information about heating systems. **For more help, refer to the Science Skill Handbook.**
7. **Communicating** Only a small part of the chemical energy of gasoline is used to move your car. In your Science Journal, describe ways to improve an engine's efficiency. **For more help, refer to the Science Skill Handbook.**

Activity

Conduction in Gases

Does smog occur where you live? If so, you may have experienced a temperature inversion. Usually the Sun warms the ground, and the air above it. When the air near the ground is warmer than the air above, convection occurs. This convection also carries smoke and other gases emitted by cars, chimneys, and smokestacks upward into the atmosphere. If the air near the ground is colder than the air above, convection does not occur. Then smoke and other pollutants can be trapped near the ground, sometimes forming smog. In this activity you will use a temperature inversion to investigate the conduction of heat in air.

What You'll Investigate

How is heat transferred by conduction in gases?

Goals

- **Measure** temperature changes in air near a heat source.
- **Observe** conduction of heat in air.

Materials

Thermometers (3)
2 foam cups
400-mL beakers (2)
burner or hot plate
paring knife
thermal mitts (2)

Safety Precautions

WARNING: *Use care when handling hot water. Pour hot water using both hands.*

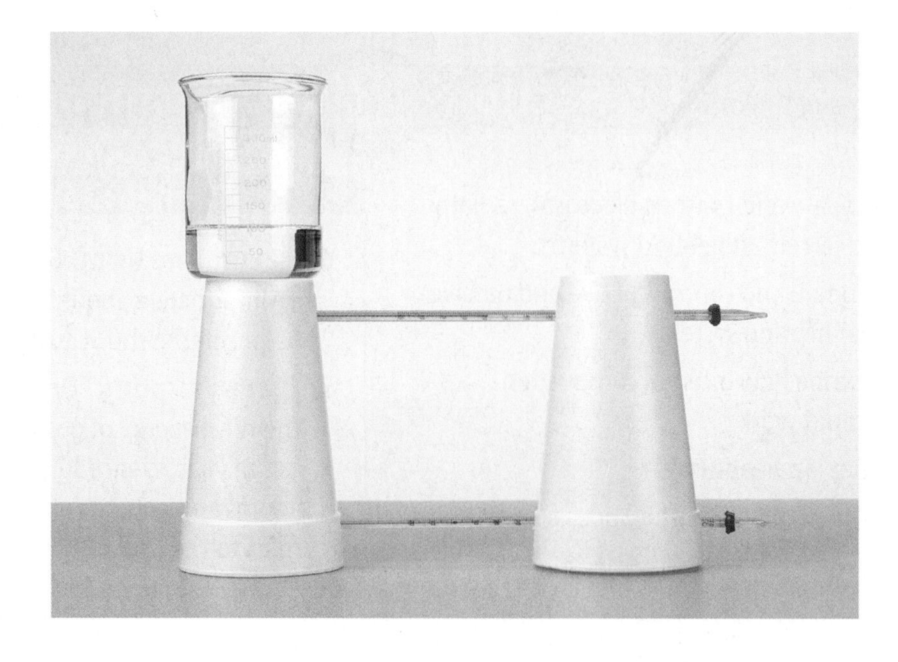

Procedure

1. Using the paring knife, carefully cut the bottom from one foam cup.
2. Use a pencil or pen to poke holes about 2 cm from the top and bottom of each foam cup.
3. Turn both cups upside down, and poke the ends of the thermometers through the upper holes and lower holes, so both thermometers are supported horizontally. The bulb end of both thermometers should extend into the middle of the bottomless cup.
4. Heat about 350 mL of water to about 80°C in one of the beakers.
5. Place an empty 400-mL beaker on top of the bottomless cup. Record the temperature of the two thermometers in your data table.
6. Add about 100 mL of hot water to the empty beaker. After one minute, record the temperatures of the thermometers in a data table like the one shown here.
7. Continue to record the temperatures every minute for 10 min. Add hot water as needed to keep the temperature of the water at about 80°C.

Air Temperatures in Foam Cup

Time (min)	Upper Thermometer (°C)	Lower Thermometer (°C)
0		
1		
2		
3		

Conclude and Apply

1. **Explain** whether convection can occur in the foam cup if it's being heated from the top.
2. **Describe** how heat was transferred through the air in the foam cup.
3. **Graph** the temperatures measured by the upper and lower thermometers, with time on the horizontal axis.
4. **Explain** why the temperature of the two thermometers changed differently.

Science Stats

Surprising Thermal Energy

Did you know...

. . . The average amount of solar energy that reaches the United States each year is about 600 times greater than the nation's annual energy demands.

. . . Energy travels from the Sun to Earth at the speed of light—299,274 km/s. This is fast enough to travel around Earth's equator almost eight times in a second.

. . . A lightning bolt heats the air in its path to temperatures of about 25,000°C. That's about 4 times hotter than the average temperature on the surface of the Sun.

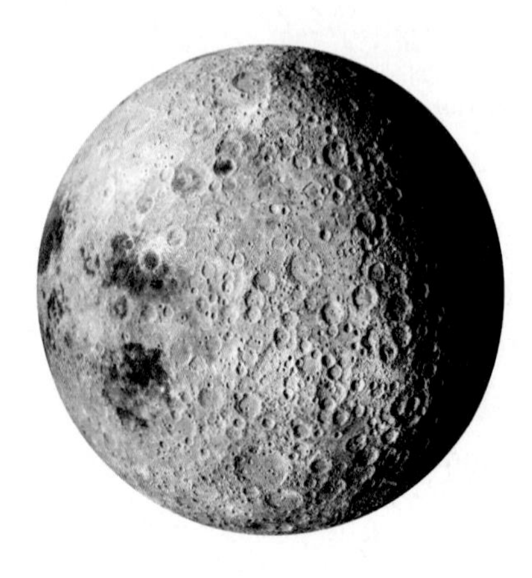

. . . Without energy from the Sun, Earth's average temperature would be −240°C. Earth would be like the side of the Moon that never sees the Sun. There could be no life at such frigid temperatures. The lowest recorded temperature on Earth is −89°C.

Connecting To Math

. . . More than 70 percent of the energy used by the average home is for temperature control. This includes your refrigerator, heating, air-conditioning, and hot-water heater.

Energy Efficiency

Incandescent bulb
Internal combustion engine (gasoline)
Human body
Fluorescent bulb
Steam engine

0 10 20 30 40 50 60 70 80 90 100
Percent of total energy used

Energy that does work
Energy wasted as heat

. . . When a space shuttle reenters Earth's atmosphere at more than 28,000 km/h, its outer surface is heated by friction to nearly 1,650°C. This temperature is high enough to melt steel.

Do the Math

1. The highest recorded temperature on Earth is 58°C and the lowest is −89°C. What is the range between the highest and lowest recorded temperatures?
2. What is the average temperature of the surface of the Sun? Draw a bar graph comparing the temperature of a lightning bolt to the temperature of the surface of the Sun.
3. The Sun is almost 150 million km from Earth. How long does it take solar energy to reach Earth?

Go Further

Use the library and the Glencoe Science Web site at **science.glencoe.com** to research how much energy is used in the United States. How much of the energy used comes from the Sun?

Chapter 6 Study Guide

Reviewing Main Ideas

Section 1 Temperature and Heat

1. The temperature of a material is a measure of the average kinetic energy of the molecules in the material.
2. Heat is thermal energy that flows from a higher to a lower temperature. *How is heat flowing in the photo at the right?*

3. The thermal energy of an object is the total kinetic and potential energy of the molecules in the object.
4. The specific heat is the amount of heat needed to raise the temperature of 1 kg of a substance by 1 K.

Section 2 Transferring Thermal Energy

1. Thermal energy is transferred by conduction, convection, and radiation.
2. Conduction and convection can occur only when matter is present.
3. Heat can flow easily through materials that are conductors. Heat flows more slowly through insulators. *What materials below are conductors, and which are insulators?*

Section 3 Using Heat

1. Conventional heating systems use air, hot water, and steam to transfer thermal energy through a building.
2. A solar heating system converts radiant energy from the Sun to thermal energy. Active solar systems use solar collectors to absorb the thermal radiant energy.
3. Heat engines convert thermal energy produced by burning fuel into mechanical energy. In an internal combustion engine, fuel is burned inside the engine in cylinders.
4. Heat movers, like refrigerators and air conditioners, move thermal energy from one place and release it somewhere else.
5. Sweating helps humans cool their bodies through evaporation. *Why is this person shivering after getting out of a pool, even though it is a warm day?*

After You Read

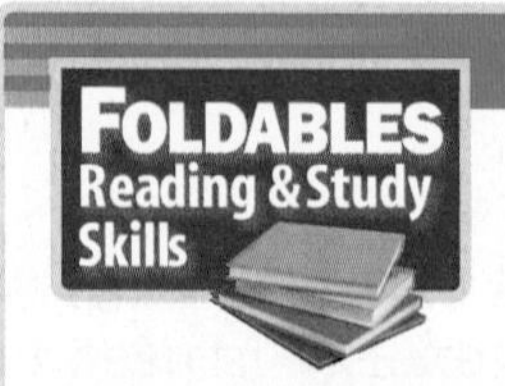

Use what you learned and write how temperature and heat are related but different on the front of your Foldable.

Chapter 6 Study Guide

Visualizing Main Ideas

Complete the following concept map of thermal energy transfer.

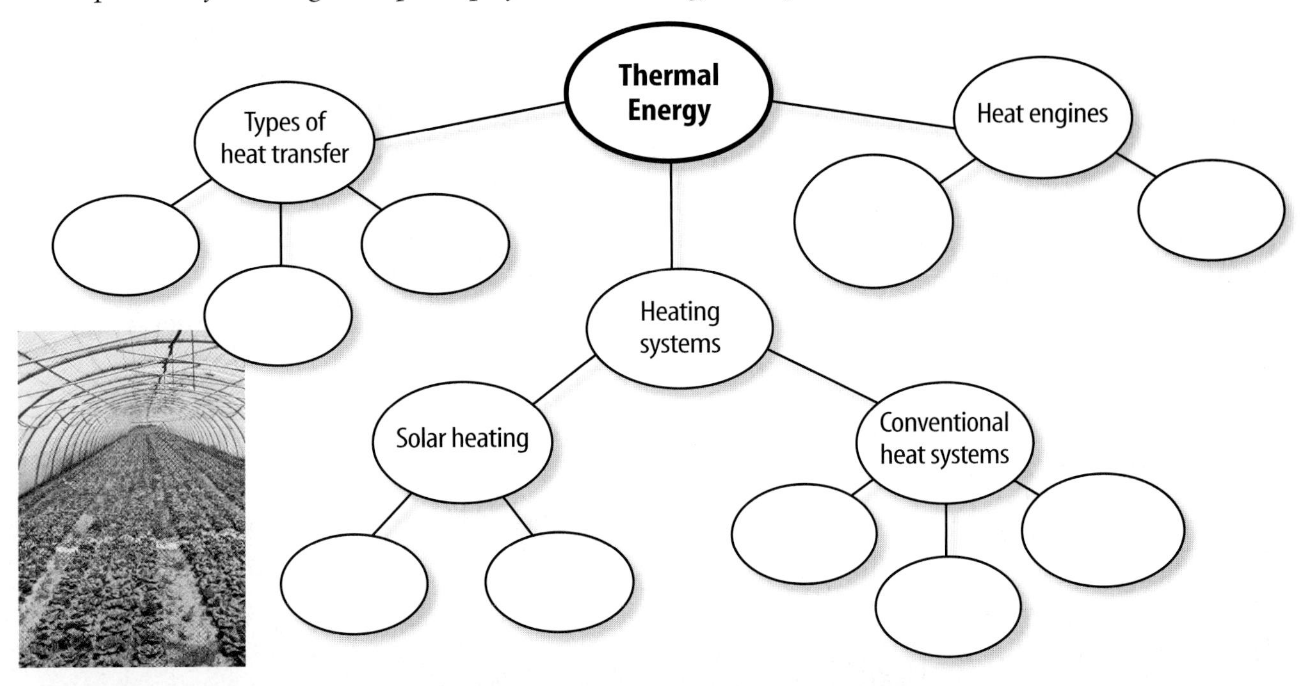

Vocabulary Review

Vocabulary Words

a. conduction
b. convection
c. heat
d. heat engine
e. heat mover
f. insulator
g. internal combustion engine
h. radiation
i. specific heat
j. solar collector
k. solar energy
l. temperature
m. thermal energy

Study Tip

Make a plan. Before you start your homework, write out a checklist of what you need to do for each subject. As you finish each item, check it off.

Using Vocabulary

Using the vocabulary words, change the incorrect terms to make the sentences read correctly.

1. A heat mover is a device that converts thermal energy into mechanical energy.

2. Solar energy is energy that is transferred from warmer to cooler materials.

3. A heat engine is a device that absorbs the Sun's radiant energy.

4. The energy required to raise the temperature of 1 kg of a material 1 K is a material's thermal energy.

5. Heat is a measure of the average kinetic energy of the particles in a material.

6. Convection is energy transfer by electromagnetic waves.

Chapter 6 Assessment

Checking Concepts

Choose the word or phrase that best answers the question.

1. Which is NOT a method of heat transfer?
A) conduction **C)** radiation
B) specific heat **D)** convection

2. In which of the following devices is fuel burned inside chambers called cylinders?
A) internal combustion engines
B) radiators
C) heat pumps
D) air conditioners

3. During which phase of a four-stroke engine are waste gases removed?
A) power stroke
B) intake stroke
C) compression stroke
D) exhaust stroke

4. Which material is a poor insulator of heat?
A) iron **C)** air
B) feathers **D)** plastic

5. Which of the following devices is an example of a heat mover?
A) solar panel
B) refrigerator
C) internal combustion engine
D) diesel engine

6. Which term describes the measure of the average kinetic energy of the particles in an object?
A) potential energy **C)** temperature
B) thermal energy **D)** specific heat

7. Which of these is NOT used to calculate change in thermal energy?
A) volume
B) temperature change
C) specific heat
D) mass

8. Which of these does NOT require the presence of particles of matter?
A) radiation **C)** convection
B) conduction **D)** combustion

9. Which of the following is the name for thermal energy that is transferred only from a higher temperature to a lower temperature?
A) potential energy **C)** heat
B) kinetic energy **D)** solar energy

10. Which device changes thermal energy into mechanical energy?
A) conductor **C)** solar collector
B) refrigerator **D)** heat engine

Thinking Critically

11. On a hot day a friend suggests that you can make your kitchen cooler by leavng the refrigerator door open. Explain whether leaving the refrigerator door open would or would not cause the air temperature in the kitchen to decrease.

12. A copper bowl and a silver bowl of equal mass were heated from 27°C to 100°C. Which required more heat? Explain.

Specific Heat [J/(kg K)]	
Copper	385
Silver	235

13. Explain whether or not the following statement is true: If the thermal energy of an object increases, the temperature of the object must also increase.

14. Which has the greater amount of thermal energy, one liter of water at 50°C or two liters of water at 50°C? Explain.

15. Suppose a beaker of water is heated from the top. Will heat transfer by convection occur in the water? Explain.

Developing Skills

16. **Classifying** Order the events that occur in the removal of heat from an object by a refrigerator. Show the complete cycle, from the placing of a warm object in the refrigerator to the changes in the coolant.

17. **Recognizing Cause and Effect** Describe what might happen if the following occur in an internal combustion engine, and indicate the engine stroke that will be affected: exhaust valve stuck closed, bad spark plug, and intake valve will not close.

18. **Concept Mapping** Complete the following events chain to show how an active solar heating system works.

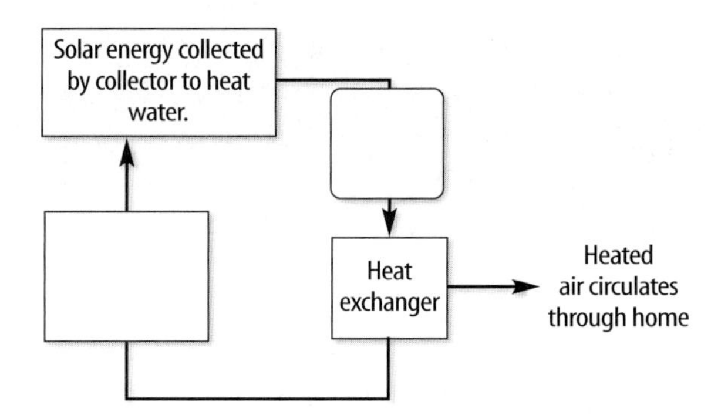

Performance Assessment

19. **Making Observations and Inferences** Identify the building materials that are used in or on the walls, ceilings, or attic space in your home and school. Which ones are good heat insulators?

Technology

Go to the Glencoe Science Web site at **science.glencoe.com** or use the **Glencoe Science CD-ROM** for additional chapter assessment.

Test Practice

Pierre's science class did an experiment involving heat. The setup for the experiment is shown in the diagram below. Pierre took temperature readings every 2 min for 20 min. For the first 10 min, the light was turned on. For the last 10 min, the light was turned off.

Study the diagram and answer the following questions.

1. What is probably being measured in the setup in the diagram?
   - **A)** the rate at which electricity affects the temperature of soil and water
   - **B)** the rate at which soil and water evaporate
   - **C)** the rate at which soil and water absorb heat energy
   - **D)** the effect of light on plant growth in soil and water

2. How could Pierre improve his experiment?
   - **F)** by putting soil in both containers
   - **G)** by taking temperature readings every 5 min
   - **H)** by using more or less water
   - **J)** by moving the lamp so that it heats both beakers equally

Standardized Test Practice

Reading Comprehension

Read the passage. Then read each question that follows the passage. Decide which is the best answer to each question.

Bouncing Back

Have you ever noticed that the balls you use for different sports bounce differently? If you played baseball with a tennis ball, the ball would probably fly way out into the outfield without much effort when you hit it with your bat. In contrast, if you used a baseball in a tennis match, the ball probably would not bounce high enough for your opponent to hit it very well. The difference in the way balls bounce depends upon the materials that make up the balls and the way in which the balls are constructed.

A ball drops to the floor as a result of gravity. As the ball drops, it gathers speed. When the ball hits the floor, the energy that it has gained goes into deforming the ball, changing it from its round shape. As the ball changes shape, the molecules within it stretch farther apart in some places and squeeze closer together in other places. The strength of the bonds between molecules determines how much they stretch apart and squeeze together. This depends on the chemical composition of the materials in the ball.

Most balls are made of rubber. Rubber is elastic, which means that it returns to its original shape after it's been deformed. Rubber is made of molecules called polymers that are long chains. Normally these chains are coiled up, but when the rubber is stretched, the chains straighten out. Then when the stretching force is removed, the chains coil up again. How high a ball bounces depends on the type of polymer molecules the rubber is made of. Bouncing balls sometimes feel warm because some of the ball's kinetic energy is converted into thermal energy.

Test-Taking Tip Make a list of the important details in the passage.

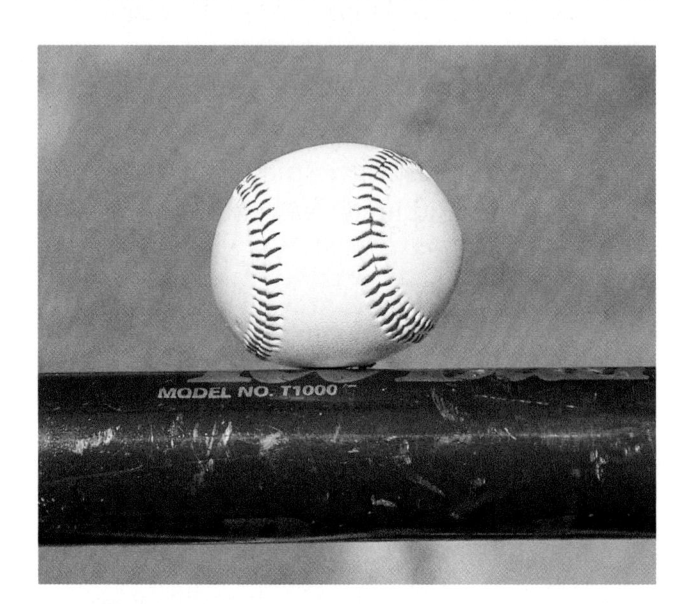

The force applied by a bat caused the shape of a baseball to change.

1. According to information in the passage, it is probably accurate to conclude that _____.
A) all rubber balls bounce the same, no matter what they are made of
B) the way rubber balls bounce depends upon the polymers that they are made of
C) baseballs are better for playing tennis than tennis balls are for playing baseball
D) the higher a ball bounces, the more thermal energy is produced

2. In the context of this passage, the word elastic means _____.
F) able to retain its shape
G) inflexible
H) tightly linked
J) deformed

Standardized Test Practice

Reasoning and Skills

Read each question and choose the best answer.

1. Latifah wanted to figure out which race car had the most kinetic energy during a competition. She researched the different race cars to find out their masses. Her experiment could be improved by _____.

A) writing down a list of observations during the competition
B) finding out the velocity of each race car during the competition
C) weighing the cars after the competition
D) researching motorcycles

Test-Taking Tip Consider how mass and velocity affect an object's kinetic energy.

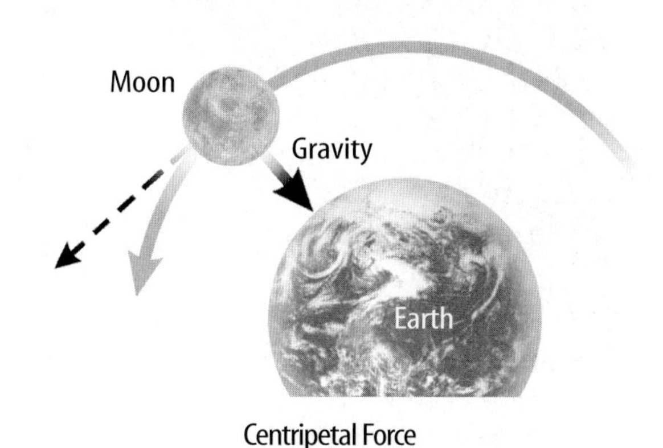

2. Which kind of scientist would most likely use this picture?

F) botanist
G) physicist
H) chemist
J) zoologist

Test-Taking Tip Consider the type of information presented by the picture.

Training record	
End of Month	**Maximum Speed (km/h)**
1	7.0
2	7.5
3	8.0
4	?

3. These data were collected while an athlete trained for a marathon. If the trend continues, what will be the maximum speed of the athlete at the end of the fourth month?

A) 6.5 km/h
B) 8.0 km/h
C) 9.0 km/h
D) 8.5 km/h

Test-Taking Tip Carefully consider the information in both the chart and the question in order to identify the trend.

Consider this question carefully before writing your answer on a separate sheet of paper.

4. Explain the similarities and differences between simple and complex machines. Give some examples of simple and complex machines you see or use everyday.

Test-Taking Tip Recall examples of simple and complex machines from everyday life.

UNIT 2

Electricity and Energy Resources

How Are Clouds & Toasters Connected?

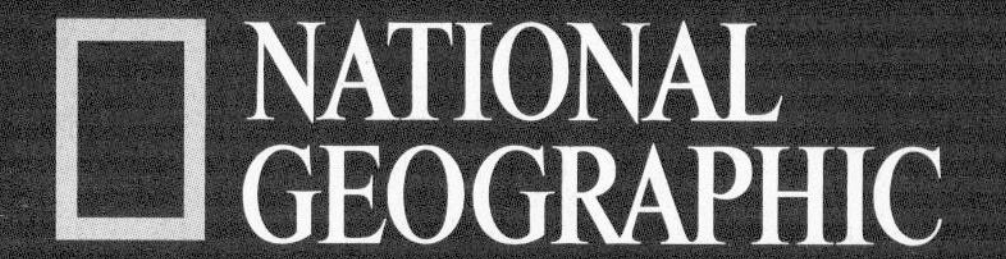

In the late 1800s, a mysterious form of radiation called X rays was discovered. One French physicist wondered whether uranium would give off X rays after being exposed to sunlight. He figured that if X rays were emitted, they would make a bright spot on a wrapped photographic plate. But the weather turned cloudy, so the physicist placed the uranium and the photographic plate together in a drawer. Later, on a hunch, he developed the plate and found that the uranium had made a bright spot anyway. The uranium was giving off some kind of radiation even without being exposed to sunlight! Scientists soon determined that the atoms of uranium are radioactive—that is, they give off particles and energy from their nuclei. In today's nuclear power plants, this energy is harnessed and converted into electricity. This electricity provides some of the power used in homes to operate everything from lamps to toasters.

SCIENCE CONNECTION

NUCLEAR ENERGY Does some of the electricity you use every day come from a nuclear power plant? Contact your local electric company to find out how much of the electricity produced in your area comes from nuclear, hydroelectric, and fossil-fuel-burning power plants. Make a graph of your research results. As a class, investigate and debate the advantages and disadvantages of using nuclear reactors to generate electricity.

Electricity

A city at night. Lights of every color and size reflect from windows and the river's glassy surface. What makes the lights so bright? Electricity. It not only provides us with light, but also heat, refrigeration, and power to run the countless electrical devices we use every day. Where does electricity come from? How does it get into our homes, schools, and offices? And how can you control it by flicking a switch or pushing a button? In this chapter, you will learn the answers to these questions.

What do you think?

Science Journal Look at the picture below with a classmate. Discuss what this might be or what is happening. Here's a hint: *What force might cause her hair to behave this way?* Write your answer or best guess in your Science Journal.

Imagine life before electricity. CD players, refrigerators, dishwashers, TVs, and dozens of other things that make your life comfortable and enjoyable would be impossible without electricity. You wouldn't even have lightbulbs and would have to use candles or oil lamps to provide light at night. So how do these electrical devices work? Explore how electric lights work during this activity.

How many ways?

1. Obtain a battery, a flashlight bulb, and some wire.
2. Arrange the materials so that the lightbulb lights.
3. Record all the ways that you were able to light the bulb.
4. Record a few of the ways that didn't work.
5. Can you light the bulb using only one wire and one battery?

Observe

In your Science Journal, write a paragraph describing the requirements to light the bulb. Write out a procedure for lighting the bulb and have a classmate follow your procedure.

Before You Read

Making a Know-Want-Learn Study Fold **Make the following Foldable to help identify what you already know and what you want to know.**

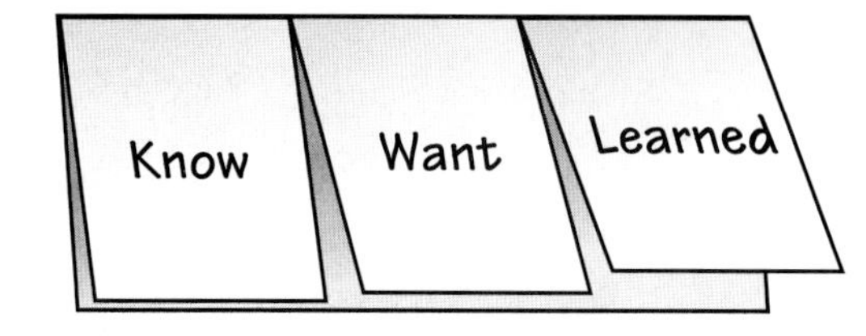

1. Place a sheet of paper in front of you so the long side is at the top. Fold the paper in half from top to bottom.
2. Fold both sides in to divide the paper into thirds. Unfold the paper so three sections show.
3. Through the top thickness of paper, cut along each of the fold lines to the top fold, forming three tabs. Label the tabs *Know, Want,* and *Learned,* as shown.
4. Before you read the chapter, write what you know under the left tab and what you want to know under the middle tab.
5. As you read the chapter, add to or correct what you have written under the tabs.

SECTION

Electric Charge

As You Read

What You'll Learn

- **Describe** the properties of static electricity.
- **Distinguish** between conductors and insulators.
- **Recognize** the presence of charge in an electroscope.

Vocabulary

static electricity
law of conservation of charge
conductor
insulator
charging by contact
charging by induction

Why It's Important

Static electricity can damage electrical equipment and be lethal to humans.

Static Electricity

You know from experience that walking across a carpeted floor and touching something can often result in a shocking experience. What causes this startling and sometimes painful phenomenon?

Electric Charges When your shoes rub on the carpet, some of the atoms in the carpet are disturbed. Atoms contain particles called protons, neutrons, and electrons, as shown in **Figure 1.** Protons and electrons have a property called electric charge, and neutrons have no electric charge.

No one knows exactly what electric charge is, but there are two different types of electric charge. Protons have positive electric charge, and electrons have negative electric charge. The amount of positive charge on a proton is exactly equal to the amount of negative charge on an electron. An atom contains equal numbers of protons and electrons, so the positive and negative charges cancel out and an atom has no net electric charge. Objects such as shoes and carpets are made of atoms and usually have no net electric charge. Objects with no net charge are said to be neutral.

Building Up Charge Some atoms hold their electrons more tightly than other atoms. For example, atoms in your shoes hold their electrons more tightly than atoms in the carpet. As you walk on carpet, some electrons that are loosely held by the atoms in the carpet are transferred to your shoes. Your shoes gain electrons and then have more electrons than protons. Because your shoes have an excess of negative charge, your shoes are said to be negatively charged. The carpet loses electrons and has more protons than electrons. Because the carpet has an excess of positive charge, it is positively charged. This is an example of static electricity. **Static electricity** is the accumulation of excess electric charges on an object. Can you think of other examples of static electricity?

Figure 1
The center of an atom contains protons (orange) and neutrons (blue). Electrons (red) swarm around the atom's center.

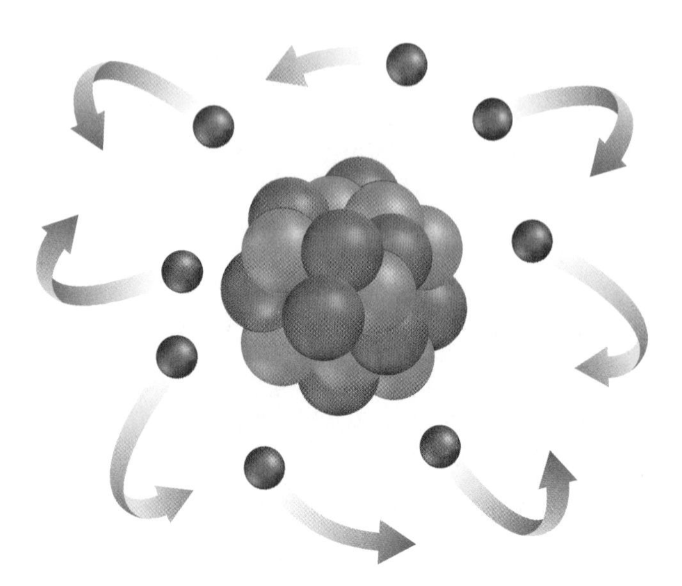

A Before rubbing

B After rubbing

Figure 2
A Before the shoes scuff against the carpet, the shoes and the carpet have equal numbers of electrons and protons. This balance means the shoes and the carpet have no net charge. B Later, as electrons move from the carpet to the shoes, the shoes become negatively charged and the carpet becomes positively charged.

Conservation of Charge It is important to remember that when an object becomes charged, charge is neither created nor destroyed. Electrons simply have moved from one object to another. For example, before you rub your shoes against the carpet, your shoes and the carpet have equal numbers of electrons and protons. Your shoes and the carpet have no net charge and are electrically neutral, as shown in **Figure 2A.** After rubbing the carpet, your shoes gain electrons from the carpet and become negatively charged. The carpet, which now has more protons than electrons, is positively charged, as shown in **Figure 2B.** According to the **law of conservation of charge,** charge can be transferred from object to object, but it cannot be created or destroyed. In every case, when an object becomes charged, electric charges have moved.

How does an object acquire charge?

Opposites Attract As shown in **Figure 3,** electrically charged objects obey two rules—opposite charges attract, and like charges repel. Have you noticed how clothes sometimes cling together when removed from the dryer? While the clothes tumble around, electrons are transferred from fabrics that hold their electrons loosely to those that hold their electrons tightly. Clothes that gain electrons become negatively charged and cling to clothes that have lost electrons and are positively charged. Clothes that are oppositely charged attract each other and stick together.

Figure 3
The only two kinds of electric charge are positive and negative. *How do charged objects interact with each other?*

Opposite charges attract

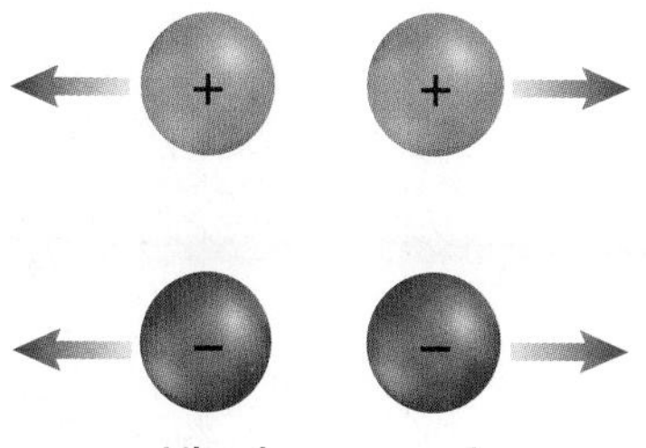

Like charges repel

Figure 4
Surrounding every charge is an electric field. Through the electric field, a charge is able to push or pull on another charge.

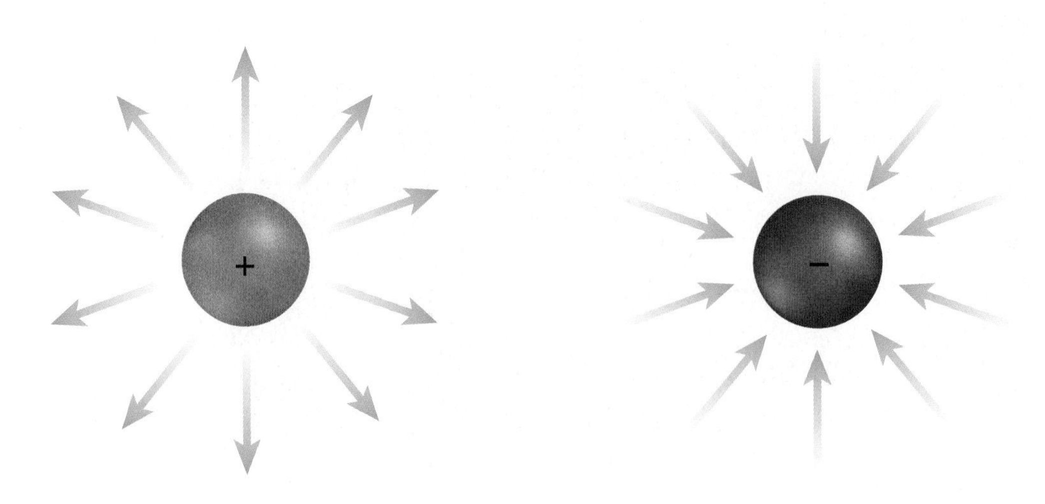

Force at a Distance You might have seen bits of tissue paper fly up and stick to a charged balloon. The balloon didn't even have to touch the bits of paper. The force of gravity behaves in the same way. A football in the air does not need to touch Earth for gravity to pull it downward. Likewise, the bits of paper do not need to touch the charged balloon for an electric force to act on them. If the balloon and the paper are not touching, what causes the paper to move?

An electric field surrounds every electric charge, as shown in **Figure 4.** The electric field exerts the force that causes other electric charges to move. Any charge that is placed in an electric field will be pushed or pulled by the field. Electric fields are represented by lines with arrows drawn away from positive charges and toward negative charges. The arrows show how the electric field would make a positive charge move.

Figure 5
As you walk across a carpeted floor, your body builds up a static charge. When you reach for a metal doorknob, the charges flow between your hand and the doorknob and you feel a shock.

Conductors and Insulators

If you reach for a metal doorknob after walking across a carpet, you could feel a shock and see a spark. The spark is caused by excess electrons moving from your hand to the doorknob. Excess electrons were transferred from the carpet to your shoes. How did excess electrons move from your shoes to your hand?

Conductors Look at **Figure 5.** An excess of electrons can move more easily through some materials, called **conductors,** than through others. Electrons on your shoes repel each other and some are pushed onto your skin. Because your skin is a conductor, the electrons spread out over your skin and onto your hand. The metal doorknob is also a conductor.

Metallic Conductors Metals are excellent conductors of electricity. The atoms in metals have electrons that are able to move easily through the material. For this reason, electric wires usually are made of metals, such as copper, which is one of the best conductors. Gold and silver wire are also excellent conductors of electricity but are much more expensive than copper.

Insulators Wires in cords attached to telephones and other household appliances are coated with some type of insulating material. An **insulator** is a material that doesn't allow electrons to move through it easily. Most plastics are insulators. Electrons are held strongly to atoms in insulating materials. The plastic coating around electric wires prevents a dangerous electrical shock when you touch the wire, as shown in **Figure 6.** In addition to plastic, other good insulators are wood, rubber, and glass.

Reading Check *What is an insulator?*

Transferring Electric Charge

Objects can become charged in several ways. Perhaps the most familiar situations are a static charge resulting from contact. For example, you probably have felt a charge on your own body after combing your hair on a dry day. You might have noticed socks being attracted to each other after they have been rubbed together in a clothes dryer. Rubbing two materials together can result in a transfer of electrons between the objects. Then one object is left with a positive charge and the other with an equal amount of negative charge. The process of transferring charge by touching or rubbing is called **charging by contact.**

Figure 6
The plastic coating around wires is an insulator. A damaged electrical cord is hazardous when the conducting wire is exposed.

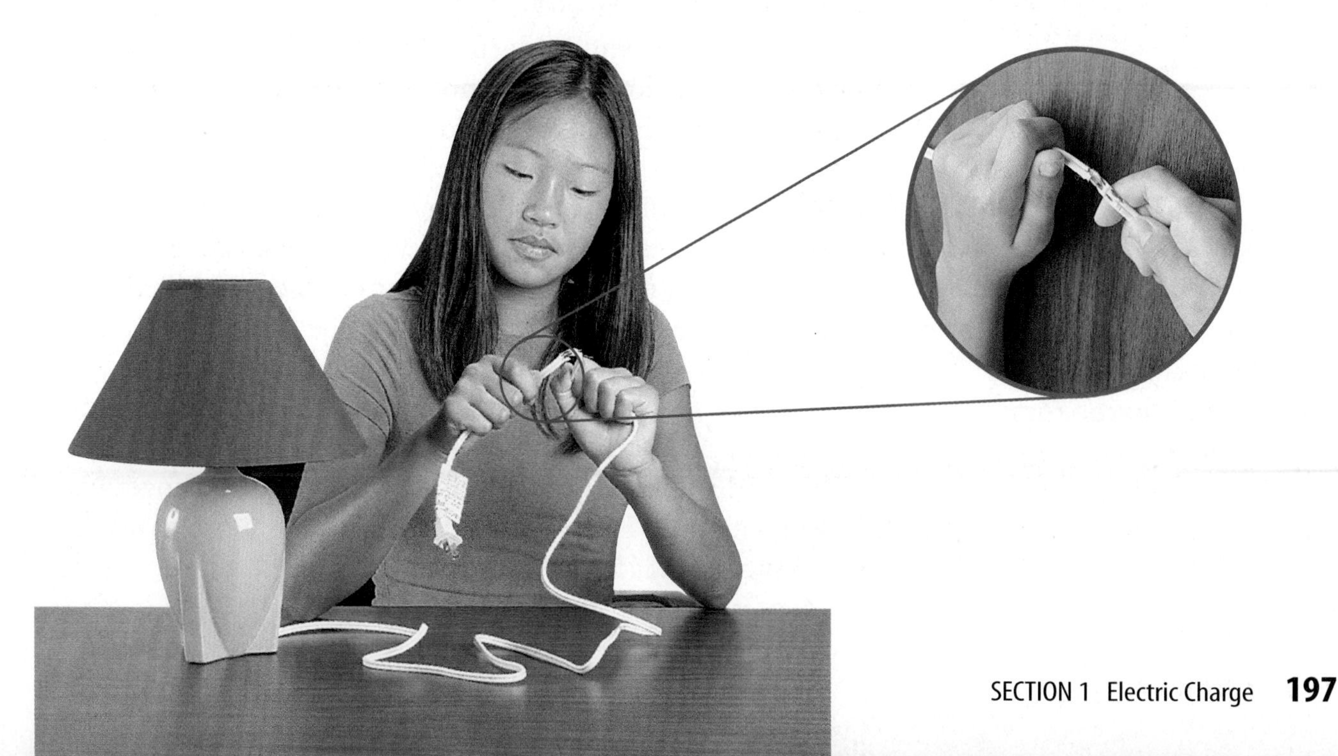

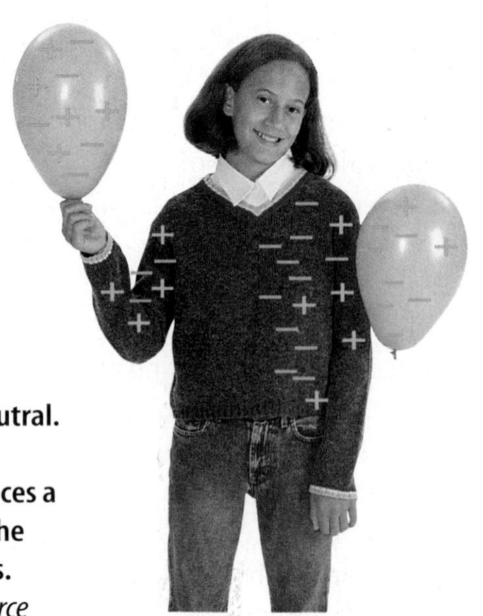

Figure 7
The balloon on the left is neutral. The balloon on the right is negatively charged. It produces a positively charged area on the sleeve by repelling electrons.
What is the direction of the force acting on the balloon?

Charging at a Distance

Because electrical forces act at a distance, charged objects brought near a neutral object will cause electrons to rearrange their positions on the neutral object. Suppose you charge a balloon by rubbing it with a cloth. If you bring the negatively charged balloon near your sleeve, the extra electrons on the balloon repel the electrons in the sleeve. The electrons near the sleeve's surface move away from the balloon, leaving a positively charged area on the surface of the sleeve, as shown in **Figure 7.** As a result, the negatively charged balloon attracts the positively charged area of the sleeve. The rearrangement of electrons on a neutral object caused by a nearby charged object is called **charging by induction.** The sweater was charged by induction. The balloon will now cling to the sweater, being held there by an electrical force.

Lightning Have you ever seen lightning strike Earth? Lightning is a large static discharge. A static discharge is a transfer of charge through the air between two objects because of a buildup of static electricity. A thundercloud is a mighty generator of static electricity. As air masses move and swirl in the cloud, areas of positive and negative charge build up. Eventually, enough charge builds up to cause a static discharge between the cloud and the ground. As the electric charges move through air, they collide with atoms and molecules. These collisions cause the atoms and molecules in air to emit light. You see this light as a spark, as shown in **Figure 8.**

Thunder Not only does lightning produce a brilliant flash of light, it also generates powerful sound waves. The electrical energy in a lightning bolt rips electrons off atoms in the atmosphere and produces great amounts of heat, warming the surrounding air to temperatures of about 25,000°C—several times hotter than the Sun's surface. The heat causes air in the bolt's path to expand rapidly, producing sound waves that you hear as thunder.

The sudden discharge of so much energy can be dangerous. It is estimated that Earth is struck by lightning more than 100 times every second. It can cause power outages, injury, loss of life, and fires.

Collect Data Visit the Glencoe Science Web site at **science.glencoe.com** for an online update of lightning strikes each day. Communicate to your class what you learn.

Figure 8

Storm clouds can form when humid, sun-warmed air rises to meet a colder air layer. As these air masses churn together, the stage is set for the explosive electrical display we call lightning. Lightning strikes when negative charges at the bottom of a storm cloud are attracted to positive charges on the ground.

A Convection currents in the storm cloud cause charge separation. The top of the cloud becomes positively charged, the bottom negatively charged.

B Negative charges on the bottom of the cloud induce a positive charge on the ground below the cloud by repelling negative charges in the ground.

C When the bottom of the cloud has accumulated enough negative charges, the attraction of the positive charges below causes electrons in the bottom of the cloud to move toward the ground.

D When the electrons get close to the ground, they attract positive charges that surge upward, completing the connection between cloud and ground. This is the spark you see as a lightning flash.

INTRA-CLOUD LIGHTNING can occur ten times more often in a storm than cloud-to-ground lightning and never strikes Earth.

Figure 9
A lightning rod directs the charge from a lightning bolt safely to the ground.

Grounding The sensitive electronics in a computer can be harmed by large static discharges. A discharge can occur any time that charge builds up in one area. Providing a path for charge to reach Earth prevents any charge from building up. Earth is a large, neutral object that is also a conductor of charge. Any object connected to Earth by a good conductor will transfer any excess electric charge. Connecting an object to Earth with a conductor is called grounding. For example, buildings often have a metal lightning rod that provides a conducting path from the highest point on the building to the ground to prevent damage by lightning, as shown in **Figure 9.**

Plumbing fixtures, such as metal faucets, sinks, and pipes, often provide a convenient ground connection. Look around. Do you see anything that might act as a path to the ground?

Detecting Electric Charge

The presence of electric charges can be detected by an electroscope. One kind of electroscope is made of two thin, metal leaves attached to a metal rod with a knob at the top. The leaves are allowed to swing freely from the metal rod. When the device is not charged, the leaves hang straight down, as shown in **Figure 10A.**

Suppose a negatively charged balloon touches the knob. Because the metal is a good conductor, electrons travel down the rod into the leaves. Both leaves become negatively charged as they gain electrons, as shown in **Figure 10B.** Because the leaves have similar charges, they repel each other.

If a glass rod is rubbed with silk, electrons are pulled from the atoms in the glass rod and build up on the silk. The glass rod becomes positively charged.

Investigating Charged Objects

Procedure

1. Fold over about 1 cm on the end of a **roll of tape** to make a handle. Tear off a strip of tape about 10 cm long.
2. Stick the strip on a clean, dry, smooth surface, such as a countertop. Make another identical strip and stick it directly on top of the first.
3. Pull both pieces off the counter together and pull them apart. Then bring the nonsticky sides of both tapes together. What happens?
4. Now stick the two strips of tape side by side on the smooth surface. Pull them off and bring the nonsticky sides near each other again.

Analysis

1. What happened when you first brought the pieces close together? Were they charged alike or opposite? What might have caused this?
2. What did you observe when you brought the pieces together the second time? How were they charged? What did you do differently that might have changed the behavior?

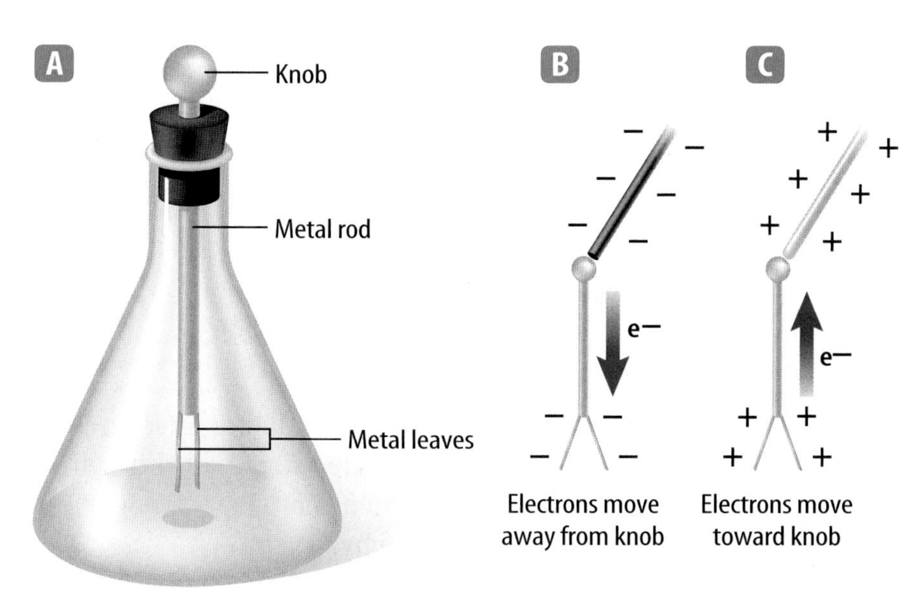

Figure 10
Notice the position of the leaves on the electroscope when they are A uncharged, B negatively charged, and C positively charged. *How can you tell whether an electroscope is positively or negatively charged?*

When the positively charged glass rod is brought into contact with the metal knob of an uncharged electroscope, electrons flow out of the metal leaves and onto the rod. The leaves repel each other because each leaf becomes positively charged as it loses electrons, as shown in **Figure 10C.**

Think of any other examples of static electricity that you have seen. Can you explain them in terms of like or opposite charges? How do objects become charged, and what happens when they discharge?

Section  Assessment

1. What is static electricity?
2. Distinguish between electrical conductors and insulators and give an example of each.
3. How is lightning produced?
4. How do like charges affect each other? Unlike charges? What could you use to detect the presence of electric charges?
5. **Think Critically** Assume you have already charged an electroscope with a positively charged glass rod. Hypothesize what would happen if you touched the knob again with another positively charged object.

Skill Builder Activities

6. **Drawing Conclusions** Suppose you observe that the individual hairs on your arm rise up when a balloon is placed near them. Using the concept of induction and the rules of static electricity, what could you conclude about the cause of this phenomenon? **For more help, refer to the Science Skill Handbook.**
7. **Communicating** Moist air is a better conductor than dry air. It is more difficult to observe events related to static electricity, such as clothes clinging or hair standing out, on humid days when the air is moist. In your Science Journal, explain how humidity affects static electricity. **For more help, refer to the Science Skill Handbook.**

SECTION

Electric Current

As You Read

What You'll Learn

- **Describe** how electric current is different from static electricity.
- **Explain** how a dry cell provides a source of voltage difference.
- **Describe** the relationship among voltage difference, resistance, and current.

Vocabulary

voltage difference
circuit
electric current
resistance
Ohm's law

Why It's Important

Without electric current, many devices would not exist, including telephones, personal computers, and lighting.

Charge on the Move

You just learned that if you touch a conductor after building up a negative charge on your body, electrons will move from you to the conductor. You also learned that charges flow easily through a conductor. Do you know why?

Electrical Pressure To understand this question, think about water. Why does water flow? To answer, you might develop a rule—water will flow from a place where the pressure is high to a place where the pressure is low, as in **Figure 11A.** Air also follows this rule, as wind always blows from a high-pressure region to a lower-pressure region. For water or air to flow, a pressure difference must exist. To explain why charges flow, a similar rule can be developed. Charges flow from high-voltage areas to low-voltage areas. Voltage is like an electrical pressure that pushes charge. Just as water or air must have a pressure difference to flow, a voltage difference must be present for electric charges to flow, as shown in **Figure 11B.** A **voltage difference** is the push that causes charges to move and is measured in volts (V).

Reading Check *What do electric charges need in order to flow?*

Look at **Figure 12A.** The water flows because the pump increases the pressure to push water through the loop. The turbine makes use of the flowing water to do work. The turbine might turn gears in a machine or connect to a generator to produce electricity.

Figure 11
Water pressure and voltage are similar.

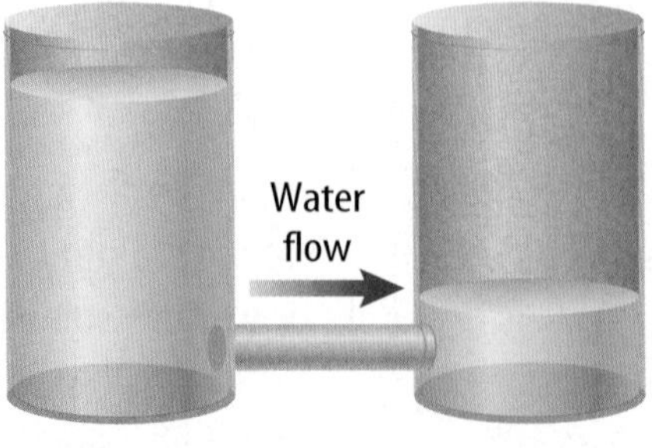

A **A pressure difference causes water to flow.**

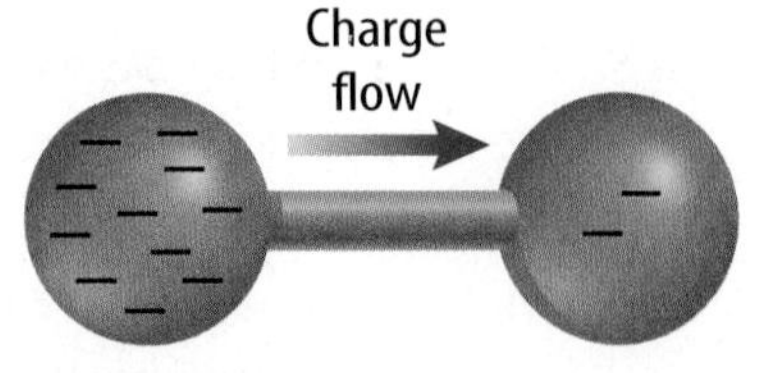

B **A voltage difference causes charge to flow.**

Figure 12
There are some similarities between the flow of water in a pipe and the flow of electric current through a circuit.

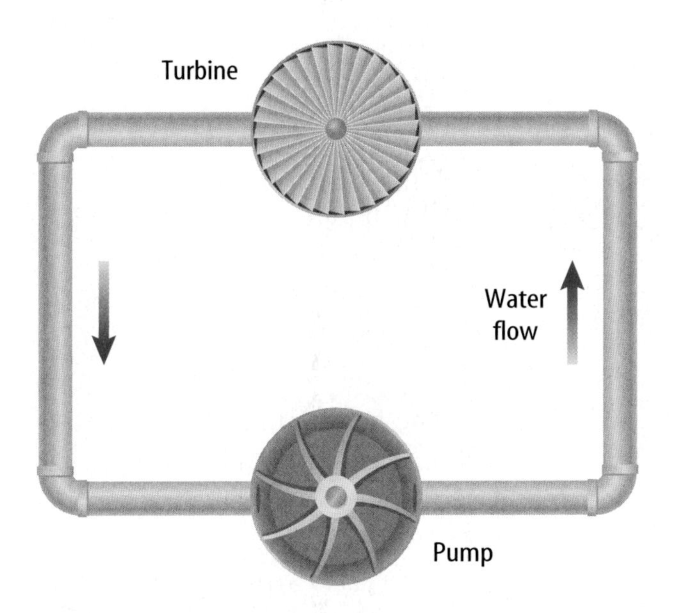

A Water flows only when the pipe makes a closed loop.

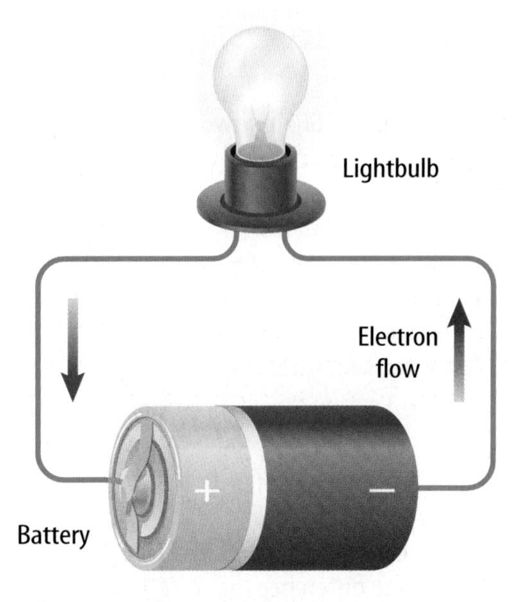

B Electric charge flows only when the wire makes a closed loop.

Closed Circuits In **Figure 12A,** what would happen if the pipe broke? Rather quickly the water would stop flowing. For water to continue to flow, the pipe must always make a closed loop. The same is true for charges flowing in a wire. For charges to flow, the wire must always be connected in a closed loop, or circuit. A **circuit** is a closed, conducting path, as shown in **Figure 12B.**

The flow of charges through a wire or any conductor is called **electric current.** The electric current in a circuit is measured in amperes (A). Current is almost always the flow of electrons. Protons have charge, but they are locked deep within the center of atoms and do not move. Only the outer, loosely held electrons are free to move.

Batteries

In a static discharge, charges move from one place to another in a short period of time. In order to keep the current moving continuously through a circuit, a device must be present that maintains a voltage difference. One common source of a voltage difference is a battery. Portable radios and flashlights use the voltage difference provided by batteries for power. Batteries also are used to provide the energy needed to start a car. How do batteries cause an electric current to flow?

Mini LAB

Investigating Battery Addition

Procedure

1. Make a circuit by linking two **bulbs** and one **D-cell battery** in a loop. Observe the brightness of the bulbs.
2. Assemble a new circuit by linking two bulbs and two D-cell batteries in a loop. Observe the brightness of the bulbs.

Analysis

1. What is the voltage difference of each D cell? Add them together to find the total voltage difference for the circuit you tested in step 2.
2. Assuming that a brighter bulb indicates a greater current, what can you conclude about the relationship between the voltage difference and current?

Dry-Cell Batteries The individual batteries you are most familiar with are dry cells. Look at the dry cell shown in **Figure 13A.** The zinc container of the dry cell surrounds a moist chemical paste with a solid carbon rod suspended in the middle. Can you locate the positive and negative terminals of the dry cell in the diagram? A dry cell produces a voltage difference between the positive and negative terminals. What causes this voltage difference?

When the two terminals of a standard D-cell battery are connected in a circuit, such as in a flashlight, a reaction involving zinc and several chemicals in the paste occurs. The carbon rod does not take part in the reaction. Instead, it serves as a conductor to transfer electrons. Electrons are transferred between some of the compounds in this chemical reaction. As a result, the carbon rod becomes positive, forming the positive (+) terminal. Electrons accumulate on the zinc, making it the negative (−) terminal.

The voltage difference between these two terminals causes current to flow through a closed circuit, such as when you turn on a portable CD player. You make a battery when you connect two or more cells together to produce a higher voltage difference. Can you think of a device in your home or school that requires more than one battery to operate?

Wet-Cell Batteries Another type of battery that is used commonly is the wet-cell battery. A wet cell, like the one shown in **Figure 13B,** contains two connected plates made of different metals or metallic compounds in a conducting solution.

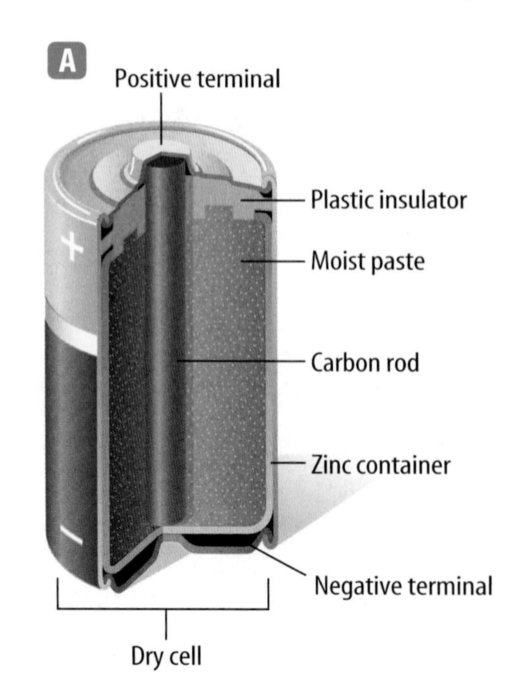

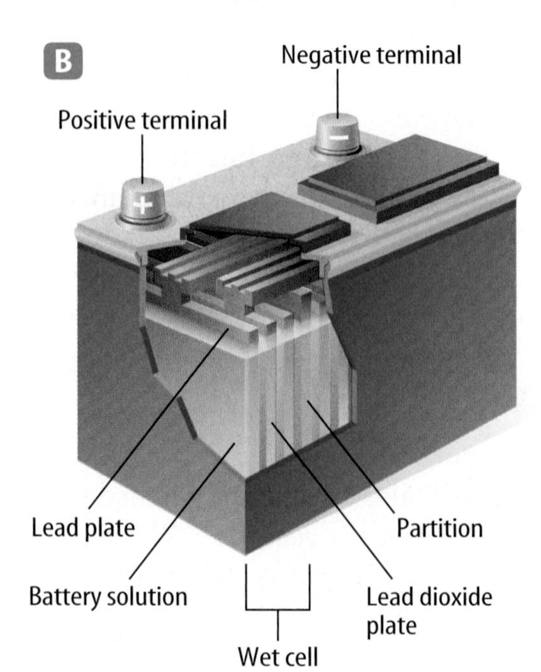

Figure 13 Batteries produce a voltage difference between the positive and negative terminals. **A** In this dry cell, chemical reactions in the moist paste transfer electrons to the zinc container. **B** In this wet cell, chemical reactions transfer electrons from the lead plates to the lead dioxide plates.

Providing Voltage Most car batteries, also called lead-storage batteries, contain a series of six wet cells made up of lead and lead dioxide plates in a sulfuric acid solution. The chemical reaction in each cell provides a voltage difference of about 2 V, giving a total voltage difference of 12 V. As a car is driven, the alternator recharges the battery by sending current through the battery in the opposite direction to reverse the chemical reaction.

In addition to batteries, a voltage difference is provided at electrical outlets, such as a wall socket. Most types of household devices are designed to use the voltage difference supplied by a wall socket. In the United States, the voltage difference across the two holes in a wall socket is usually 120 V. Some wall sockets supply 240 V, which is required by electric ranges and clothes dryers.

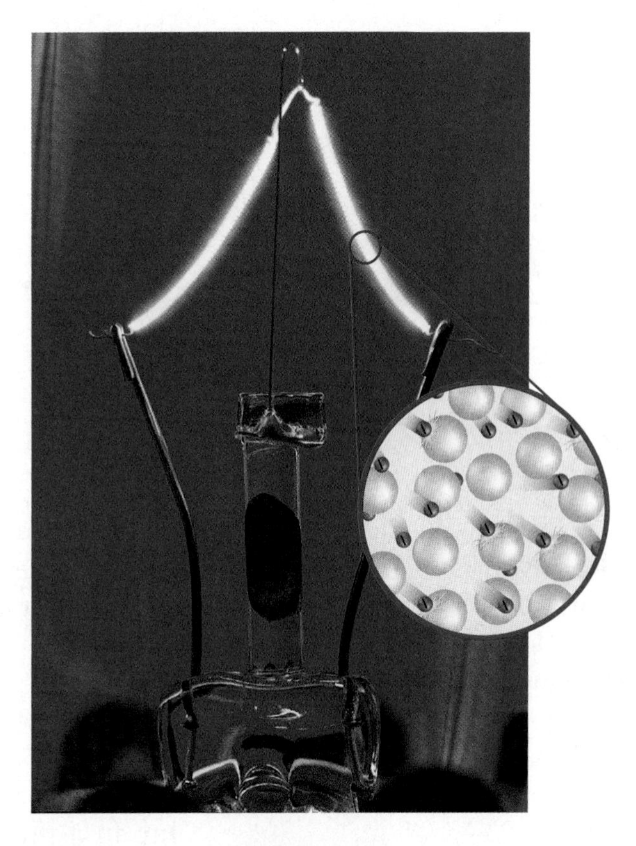

Figure 14
As electrons move through the filament in a lightbulb, they bump into metal atoms. Due to the collisions, the metal heats up and starts to glow.

Resistance

Flashlights use dry-cell batteries to provide the electric current that lights a lightbulb. What makes a lightbulb glow? Look at the lightbulb in **Figure 14.** Part of the circuit through the bulb contains a thin wire called a filament. As the electrons flow through the filament, they bump into the metal atoms that make up the filament. As these collisions occur, some of the electrical energy of the electrons is converted into thermal energy. Eventually, the metal filament becomes hot enough to glow, producing radiant energy that can light up a dark room.

Oppose the Flow Electric current loses energy as it moves through the filament because the filament resists the flow of electrons. **Resistance** is the tendency for a material to oppose the flow of electrons, changing electrical energy into thermal energy and light. With the exception of a few substances that become superconductors at low temperatures, all materials have some electrical resistance. Electrical conductors have much less resistance than insulators. Resistance is measured in ohms (Ω).

Copper is an excellent conductor and has low resistance to the flow of electrons. Copper is used in household wiring because little electrical energy is converted to thermal energy as current passes through the wires. In contrast, tungsten wire glows white-hot as current passes through it. Tungsten's high resistance to current makes it suitable for use as filaments in lightbulbs but not for carrying current through a house.

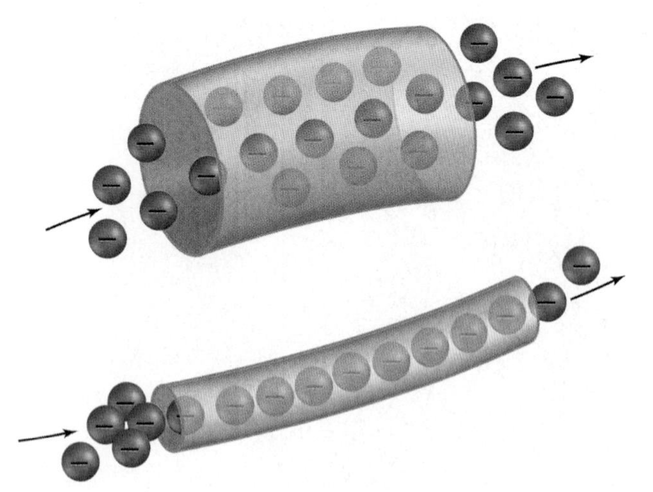

Figure 15
The resistance of a short, thick piece of wire is less than the resistance of a long, thin piece of wire.

Resistance in Wires The size of a wire also affects its resistance. **Figure 15** illustrates how electrons move more efficiently through thick wires than thin wires. In wires of the same length and material, thinner wires have greater resistance to electron flow. Making a wire longer causes the resistance to increase because more collisions occur as electrons flow through the longer wire. In most conductors, the resistance also increases as the temperature increases.

Reading Check *Why might you want to use a material with high resistance?*

Control the Flow

So far, you have learned two ways that the flow of charges, or current, in a circuit can be changed. A voltage difference causes the charges to flow, and an electrical resistance restricts the movement of charges. To help visualize this, think of water flowing in a pipe. If you increase the water pressure, the water flows faster and the water current increases. On the other hand, if you place obstructions in the pipe, the water current decreases.

Figure 16
The amount of current flowing through a circuit is related to the amount of resistance in the circuit.

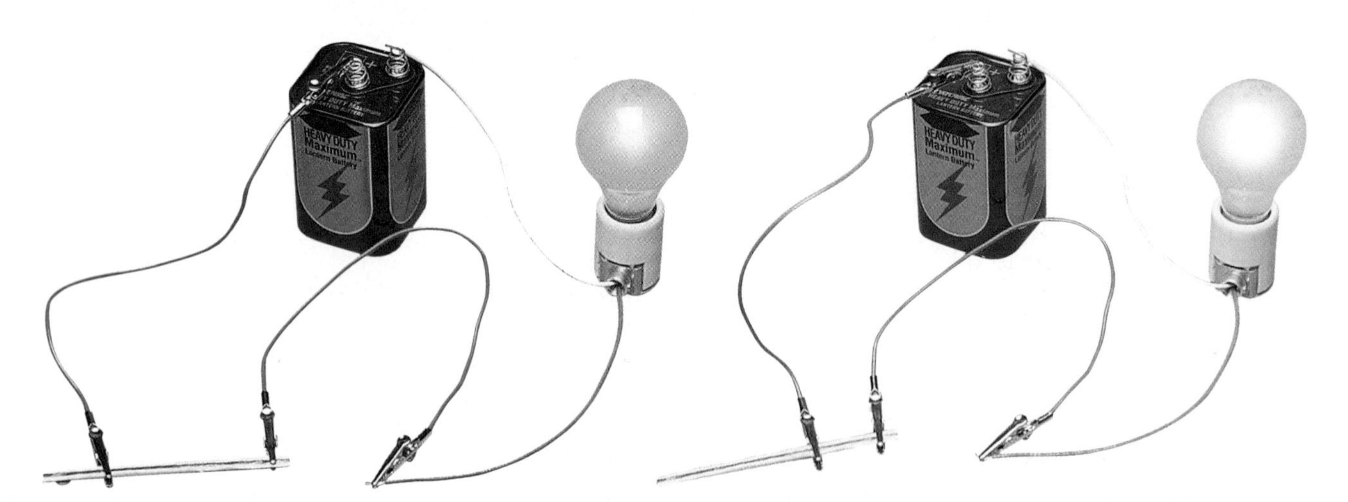

A By changing the length of the graphite rod that the current must pass through, the resistance of the circuit can be changed. Recall that longer wires of a given material have higher resistances than shorter wires. *How does changing the resistance affect the voltage difference of the circuit?*

B Notice that here the current flows through a shorter section of the graphite rod. This decreases the total resistance of the circuit, while the voltage difference produced by the battery remains the same. *How does the brightness of the bulb compare to the brightness of the bulb in the first photo?*

Ohm's Law For water flowing in a pipe, increasing the resistance causes the current to decrease, while increasing the pressure causes the current to increase. A similar relationship is true for electric current and is called Ohm's law.

$$\text{current} = \frac{\text{voltage difference}}{\text{resistance}}$$

$$I\,(\text{A}) = \frac{V\,(\text{V})}{R\,(\Omega)}$$

According to **Ohm's law,** the current in a circuit equals the voltage difference divided by the resistance. Consequently, as the resistance in a circuit increases, the current decreases. This relationship is shown in **Figure 16.** The graphite rod resists the flow of current in the circuit. As the length of the graphite rod increases, the resistance in the circuit increases, and the current through a lightbulb decreases. As a result, the bulb becomes less bright. On the other hand, from Ohm's law the current increases if the voltage difference increases.

By multiplying both sides of the above equation by the resistance, R, Ohm's law also can be written as follows.

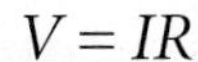

$$V = IR$$

Health INTEGRATION

Harm from electricity is due to high current. Wet skin has a much lower resistance than dry skin. According to Ohm's law, low resistance means high current. Research the effects of high current on the human body.

Section  Assessment

1. How does a current traveling through a circuit differ from a static discharge?
2. Briefly describe how a carbon-zinc dry cell supplies electric current for your CD player.
3. Describe three factors that affect the resistance of a copper wire.
4. Compare and contrast the flow of water through a pipe and the flow of electrons through a wire.
5. **Think Critically** Calculate the voltage difference across a 25-Ω resistor if a 0.3-A current is flowing through it. What happens to the voltage difference if the current is doubled?

Skill Builder Activities

6. **Interpreting Data** Suppose you connect three copper wires of unequal length to a 1.5-V dry cell. The following currents flow in the wires: wire 1, 1.2 A; wire 2, 1.4 A; wire 3, 1.1 A. Use Ohm's law to calculate the resistance of each wire. Make a graph of current versus resistance. Describe the shape of the line on your graph. **For more help, refer to the Science Skill Handbook.**
7. **Using Fractions** Suppose you place a bulb with a resistance of 60 Ω in a circuit with a 12-V battery. What is the current through this circuit? How does the current change if you add one more bulb? Two more bulbs? **For more help, refer to the Math Skill Handbook.**

Activity

Identifying Conductors and Insulators

Have you ever had a flashlight that you couldn't seem to make work any longer? You replaced the batteries and put in a new bulb, yet the flashlight still wouldn't light. The most likely cause for such a broken flashlight is a break in the circuit. If you could find the break and then repair it, you could fix the flashlight.

What You'll Investigate

Compare the ability of different materials to conduct a current.

Materials

battery
flashlight bulb
bulb holder
insulated wire

Goals

- **Identify** conductors and insulators.
- **Describe** the common characteristics of conductors and insulators.

Procedure

1. Set up an incomplete circuit as pictured in the photograph.
2. Touch the free bare ends of the wires to various objects around the room. Test at least 12 items.
3. In a table like the one below, record which materials make the lightbulb light and which don't.

Material Tested with Lightbulb Circuit	
Lightbulb Lights	**Lightbulb Stays Out**

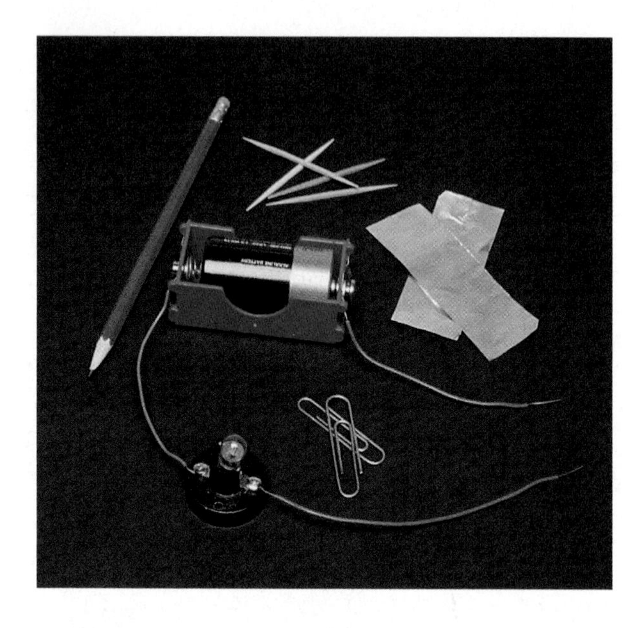

Conclude and Apply

1. Is there a pattern to your data?
2. Do all or most of the materials that light the lightbulb have something in common?
3. Do all or most of the materials that don't light the lightbulb have something in common?
4. **Explain** why a material may allow the lightbulb to light and what will prevent the lightbulb from lighting.
5. **Predict** what other materials will allow the lightbulb to light and what will prevent the lightbulb from lighting.
6. **Classify** all the materials you have tested as conductors or insulators.

Compare your conclusions with those of other students in your class. **For more help, refer to the Science Skill Handbook.**

Electrical Energy

Electric Circuits

Look around. How many electrical devices such as lights, clocks, stereos, and televisions do you see that are plugged into wall outlets? These devices rely on a source of electrical energy and wires to complete an electric circuit. Circuits typically include a voltage source, a conductor such as a wire, and one or more devices that use the electrical energy to do work.

Consider, for example, a circuit that includes an electric hair dryer. The dryer must be plugged into a wall outlet to operate. A generator at a power plant produces a voltage difference across the outlet, causing charges to move when the circuit is complete. The dryer and the circuit in the house contain conducting wires to carry current. The hair dryer turns the electrical energy into thermal energy and mechanical energy. When you unplug the hair dryer or turn off its switch, you open the circuit and break the path of the current. To use electrical energy, a complete circuit must be made. Several kinds of circuits exist.

Series Circuits One kind of circuit is called a series circuit. In a **series circuit,** the current has only one loop to flow through, as shown in **Figure 17.** Series circuits are used in flashlights and some holiday lights.

Reading Check *How many loops are in a series circuit?*

As You Read

What You'll Learn

- **Describe** the difference between series and parallel circuits.
- **Recognize** the function of circuit breakers and fuses.
- **Explain and calculate** electrical power.

Vocabulary

series circuit
parallel circuit
electrical power
kilowatt-hour

Why It's Important

The convenience and safety of household electricity depend on how the electric circuits in your home are designed.

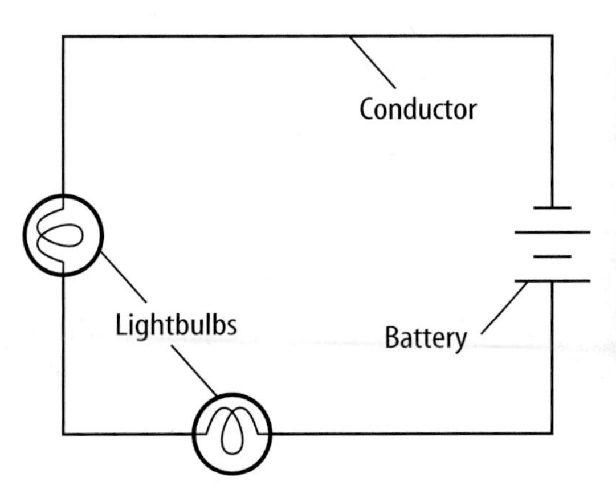

Figure 17
A series circuit provides only one path for the current to follow.
What happens to the brightness of each bulb as more bulbs are added?

Rivers sometimes form different branches that separate and then rejoin, making an island. Write a paragraph describing which kind of circuit this is most like and why.

Open Circuit If you have ever decorated a window or a tree with a string of lights, you might have had the frustrating experience of trying to find one burned-out bulb. How can one faulty bulb cause the whole string to go out? Because the parts of a series circuit are wired one after another, the amount of current is the same through every part. When any part of a series circuit is disconnected, no current flows through the circuit. This is called an open circuit. The burned-out bulb causes an open circuit in the string of lights.

Parallel Circuits What would happen if your home were wired in a series circuit and you turned off one light? This would cause an open circuit, and all the other lights and appliances in your home would go out, too. This is why houses are wired with parallel circuits. **Parallel circuits** contain two or more branches for current to move through. Look at the parallel circuit in **Figure 18.** The current splits up to flow through the different branches. Because all branches connect the same two points of the circuit, the voltage difference is the same in each branch. Then, according to Ohm's law, more current flows through the branches that have lower resistance.

Parallel circuits have several advantages. When one branch of the circuit is opened, such as when you turn a light off, the current continues to flow through the other branches. Houses, automobiles, and most electrical systems use parallel wiring so individual parts can be turned off without affecting the entire circuit.

Figure 18
In parallel circuits, the current follows more than one path. *How will the voltage difference compare in each branch?*

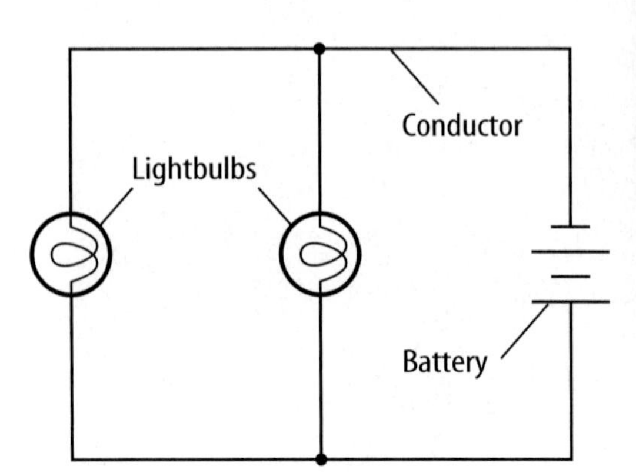

Figure 19
The wiring in a house must allow for the individual use of various appliances and fixtures.
What type of circuit is most common in household wiring?

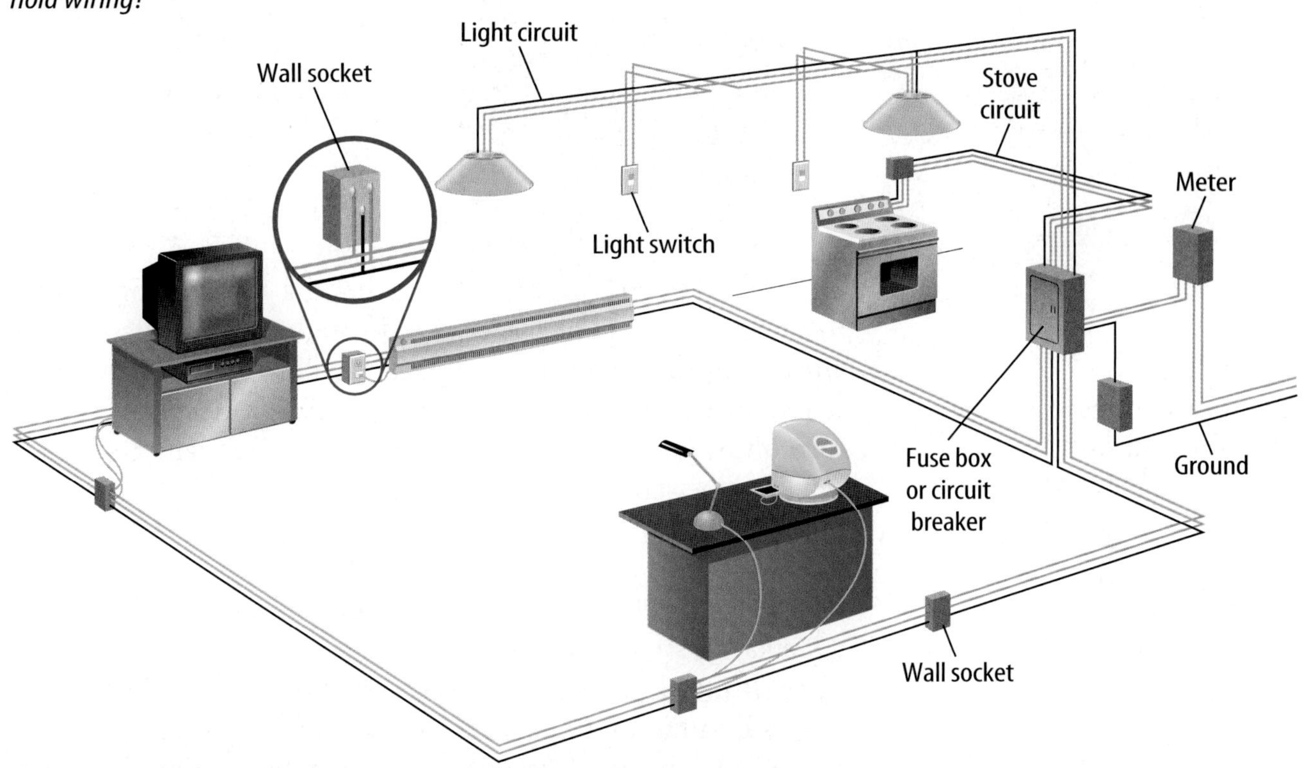

Household Circuits

Count how many different things in your home require electrical energy. You don't see the wires because most of them are hidden behind the walls, ceilings, and floors. This wiring is made up mostly of a combination of parallel circuits connected in an organized and logical network. **Figure 19** shows how electrical energy enters a home and is distributed. Each branch receives the standard voltage difference from the electric company, which is 120 V in the United States. The main switch and circuit breaker or fuse box serve as an electrical headquarters for your home. Parallel circuits branch out from the breaker or fuse box to wall sockets, major appliances, and lights.

In a house, many appliances draw current from the same circuit. If more appliances are connected, more current will flow through the wires. As the amount of current increases, so does the amount of heating in the wires. If the wires get too hot, the insulation can melt and the bare wires can cause a fire. To protect against overheating of the wires, all household circuits contain either a fuse or a circuit breaker.

Fuses When you hear that somebody has "blown a fuse," it means that the person has lost his or her temper. This expression comes from the function of an electrical fuse, **Figure 20A,** which contains a small piece of metal that melts if the current becomes too high. When it melts, it causes a break in the circuit, stopping the flow of current through the overloaded circuit. To fix this, you must replace the blown fuse with a new one. However, before you replace the blown fuse, you should turn off or unplug some of the appliances. Too many appliances in use at the same time is the most likely cause for the overheating of the circuit.

Figure 20
Two useful devices to prevent electric circuits from overheating are A fuses and B circuit breakers. *Which device, a fuse or a circuit breaker, seems more convenient to have in the home?*

Circuit Breaker A circuit breaker, **Figure 20B,** is another guard against overheating a wire. A circuit breaker contains a piece of metal that bends when it gets hot. The bending causes a switch to flip and open the circuit, stopping the flow of current. Circuit breakers usually can be reset by moving the switch to its "on" position. Again, before you reset a circuit breaker, you should turn off or unplug some of the appliances from the overloaded circuit.

What is the purpose of fuses and circuit breakers in household circuits?

Electrical Power

The reason that electricity is so important to your everyday life is that electrical energy is converted easily to other types of energy. For example, electrical energy is converted to mechanical energy as the blades of a fan rotate to cool you. Electrical energy is converted to light energy in lightbulbs. A hair dryer changes electrical energy into thermal energy. The rate at which electrical energy is converted to another form of energy is called **electrical power.**

The electrical power used by appliances varies. Appliances often are labeled with a power rating that describes how much power the appliance uses. Appliances that have electric heating elements, such as ovens and hair dryers, have a large power rating. Why might ovens and hair dryers require a high power rating?

Calculating Power Appliances with high power ratings can be supplied with the electrical power they need by increasing the amount of charge flowing into the appliance or increasing the electrical pressure on the charge that is flowing already. The relationship among power, voltage, and current can be expressed as follows.

$$\text{power} = \text{current} \times \text{voltage difference}$$
$$P\,(\text{watts}) = I\,(\text{amperes}) \times V\,(\text{volts})$$

Electrical power is expressed in watts (W). For example, a hair dryer might draw 10 A of current at a voltage difference of 120 V. The power rating of the hair dryer is then 10 A times 120 V, or 1,200 W.

Power Rating Every electrical appliance comes with a label that shows how much power it uses. **Figure 21** shows the power-rating label for a typical hair dryer, and **Table 1** lists the power requirements of some appliances. Which appliance requires the most electrical power to operate? You can tell by looking at the number of watts listed for that appliance under the Power Rating column.

Figure 21
All appliances come with a power rating. *Why is a power rating important?*

Table 1 Power and Energy Used by Home Appliances

Appliance	Time of Usage (h/day)	Power Rating (W)	Energy Usage (kWh/day)
Hair dryer	0.25	1,000	0.25
Microwave oven	0.5	700	0.35
Stereo	2.5	109	0.27
Range (oven)	1	2,600	2.60
Refrigerator/freezer (15 ft^3, frostless)	10	615	6.15
Television (color)	3.25	200	0.65
Electric toothbrush	0.08	7	0.0006
100-W lightbulb	6	100	0.60
40-W fluorescent lightbulb	1	40	0.04

SCIENCE Online

Data Update For an online update about the current energy costs in different communities across the United States, visit the Glencoe Science Web site at **science. glencoe.com** and select the appropriate chapter.

Electrical Energy

Do you leave the light on or the stereo playing in your room when you aren't there? Consider that any electrical energy you use costs money. Furthermore, most electrical energy is produced from natural resources, such as oil and coal, which are in limited supply.

The amount of electrical energy you use depends on two things. One is the power required by appliances in your home, and the other is how long they are used. Many appliances with high power ratings, such as hair dryers, are used for such a short amount of time that the total amount of electrical energy they require in a given month is small. Appliances that run constantly, such as refrigerators, usually use more total energy. The last column of **Table 1** shows typical energy usage per day for various household appliances.

Math Skills Activity

Calculating Energy

Example Problem

You use your fan for 3 h each day. It has a power rating of 50 W. How much energy does it use in one day? Express your answer in kilowatt-hours.

Solution

1. *This is what you know:* time: $t = 3$ h; power: $P = 50$ W
2. *This is what you need to find:* energy: E
3. *This is the equation you need to use:* $E = P \times t$
4. *To calculate energy, the unit of power must be kW. So convert* P *from W to kW by dividing by 1,000:* $P = \frac{50\ \text{W}}{1{,}000\ \text{W/kW}} = 0.05\ \text{kW}$
5. *Substitute the known values:* $E = 0.05\ \text{kW} \times 3\ \text{h}$; $E = 0.15\ \text{kWh}$

Check your answer by solving the original equation, E = P × t, *for* t. *Then substitute* E *and* P. *Do you calculate the same time that was given?*

Practice Problems

1. A 100-W lightbulb has a power rating of 100 W. How much energy in kWh is used when you leave it on for 5 h?
2. Find the power rating for a hair drier on **Table 1.** How much energy is used if you run it for 12 min (0.20 h)?

For more help, refer to the Math Skill Handbook.

Calculating Energy You can calculate the amount of energy an appliance uses in a day by multiplying the power required by the amount of time it uses that power.

$$\text{energy} = \text{power} \times \text{time}$$
$$E\,(\text{kWh}) = P\,(\text{kW}) \times t\,(\text{h})$$

Notice that to calculate energy, power is expressed in kilowatts. One kilowatt is 1,000 W. The unit of electrical energy is the **kilowatt-hour** (kWh). One kilowatt-hour is 1,000 W of power used for 1 h. The electric company charges you periodically for each kilowatt-hour you use. You can figure your electric bill by multiplying the energy used by the cost per kilowatt-hour. **Table 2** shows some sample costs of running electrical appliances. For example, to determine the cost of using a 100-W lightbulb for 20 h, the following calculation is made.

$$\text{cost} = 0.1\text{ kW} \times 20\text{ h} \times \$0.09/\text{kWh} = \$0.18$$

Table 2 Cost of Using Home Appliances

Appliance	Hair Dryer	Stereo	Color Television
Average power in watts	1,000	109	200
Hours used daily	0.25	2.5	2.5
Hours used monthly	7.5	75.0	75.0
Monthly watt-hours	7,500	8,175	15,000
kWh used monthly	7.5	8.175	15.000
Rate charge	$0.09	$0.09	$0.09
Monthly cost	$0.68	$0.74	$1.35

Section  Assessment

1. What is electrical power? What is electrical energy? How are the two related?
2. Compare and contrast fuses and circuit breakers. Which is easier to use? Why?
3. Do appliances with the highest power ratings always use the most energy per month? Use examples to explain why or why not.
4. How does a series circuit differ from a parallel circuit? Sketch an example of each.
5. **Think Critically** How much energy would be needed for brushing your teeth with an electric toothbrush daily for the month of May? How much would it cost at $0.09 kWh?

Skill Builder Activities

6. **Concept Mapping** Prepare a concept map that shows the steps that are followed in calculating the energy used in operating an electrical device with known voltage difference and current for a known amount of time. **For more help, refer to the Science Skill Handbook.**
7. **Using an Electronic Spreadsheet** On a spreadsheet, list the appliances your family uses daily, the estimated hours per day, and, from **Table 1,** the power usage. Multiply the power usage by the hours per day to find the electrical energy used daily for each appliance. Which appliance uses the most energy? **For more help, refer to the Technology Skill Handbook.**

Activity Design Your Own Experiment

Comparing Series and Parallel Circuits

Imagine what a bedroom might be like if it were wired in series. For an alarm clock to keep time and wake you in the morning, your lights and anything else that uses electricity would have to be on. Fortunately, most outlets in homes are wired on separate branches of the main circuit. Can you design simple circuits that have specific behaviors and uses?

Recognize the Problem

How do the behaviors of series and parallel circuits compare?

Form a Hypothesis

Predict what will happen to the other bulbs when one bulb is unscrewed from a series circuit and from a parallel circuit. Also, write a hypothesis predicting in which circuit the lights shine the brightest.

Possible Materials

6-V dry-cell battery
small lights with sockets (3)
aluminum foil
paper clips
tape
scissors
paper

Goals

- **Design and construct** series and parallel circuits.
- **Compare and contrast** the behaviors of series and parallel circuits.

Safety Precautions 

Some parts of circuits can become hot. Do not leave the battery connected or the circuit closed for more than a few seconds at a time. Never connect the positive and negative terminals of the dry-cell battery directly without including at least one bulb in the circuit.

Test Your Hypothesis

Plan

1. As a group, agree upon and write the hypothesis statement.
2. Work together determining and writing the steps you will take to test your hypothesis. Include a list of the materials you will need.
3. How will your circuits be arranged? On a piece of paper, draw a large parallel circuit of three lights and the dry-cell battery as shown. On the other side, draw another circuit with the three bulbs arranged in series.
4. Make conducting wires by taping a 30-cm piece of transparent tape to a sheet of aluminum foil and folding the foil over twice to cover the tape. Cut these to any length that works in your design.

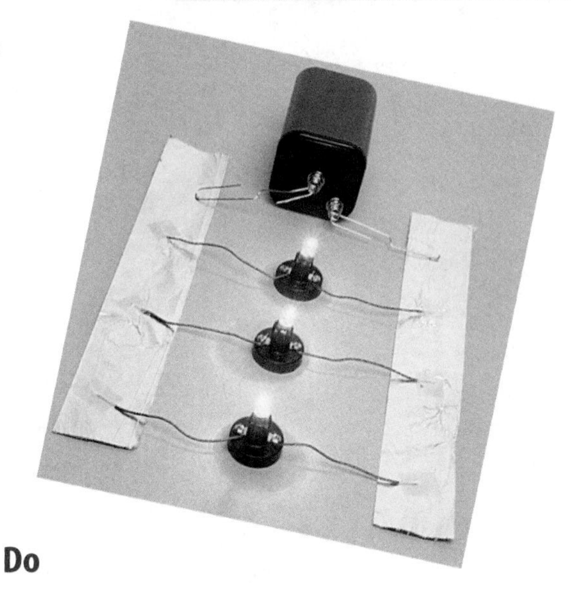

Do

1. Make sure your teacher approves your plan before you start.
2. Carry out the experiment. **WARNING:** *Leave the circuit on for only a few seconds at a time to avoid overheating.*
3. As you do the experiment, record your predictions and your observations in your Science Journal.

Analyze Your Data

1. **Predict** what will happen in the series circuit when a bulb is unscrewed at one end. What will happen in the parallel circuit?
2. **Compare** the brightness of the lights in the different circuits. Explain.
3. **Predict** what happens to the brightness of the bulbs in the series circuit if you complete it with two bulbs instead of three bulbs. Test it. How does this demonstrate Ohm's law?

Draw Conclusions

1. Did the results support your hypothesis? Explain by using your observations.
2. Where in the parallel circuit would you place a switch to control all three lights? Where would you place a switch to control only one light? Test it.

Communicating Your Data

Prepare a poster to highlight the differences between a parallel and a series circuit. Include possible practical applications of both types of circuits. **For more help, refer to the Science Skill Handbook.**

The Invisible Man

by Ralph Ellison

Respond to the Reading

1. Sometimes you can figure out the meaning of words by their contexts. A word's context refers to the other words in a sentence or phrase that shed light on that word's meaning. Can you guess the meaning of the words *ectoplasm* and *epidermis* by their contexts?
2. What clues does the narrator give that he is not really invisible?
3. Why does the narrator believe he is in the "great American tradition of tinkers"?

I am an invisible man. No, I am not a spook like those who haunted Edgar Allan Poe; nor am I one of your Hollywood-movie ectoplasms.[1] I am a man of substance, of flesh and bone, fiber and liquids—and I might even be said to possess a mind. I am invisible, understand, simply because people refuse to see me. . . . Nor is my invisibility exactly a matter of biochemical accident to my epidermis.[2] That invisibility to which I refer occurs because of a peculiar disposition of the eyes of those with whom I come in contact. A matter of the construction of their *inner* eyes . . .

. . . Now don't jump to the conclusion that because I call my home a "hole" it is damp and cold like a grave. . . . Mine is a warm hole.

My hole is warm and full of light. Yes, *full* of light. I doubt if there is a brighter spot in all New York than this hole of mine, and I do not exclude Broadway. . . . Perhaps you'll think it strange that an invisible man should need light, desire light, love light. Because maybe it is exactly because I *am invisible.* Light confirms my reality, gives birth to my form. . . . I myself, after existing some twenty years, did not become alive until I discovered my invisibility.

. . . In my hole in the basement there are exactly 1,369 lights. I've wired the entire ceiling, every inch of it. . . . Though invisible, I am in the great American tradition of tinkers. That makes me kin to Ford, Edison and Franklin.

[1]The outer layer of a part of the cell.
[2]The outer layer of skin.

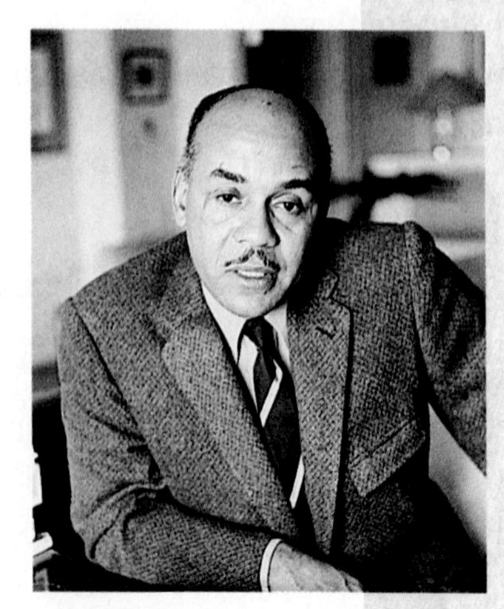

Ralph Ellison

Understanding Literature

Prologue The passage you have just read is a prologue to a novel. A prologue is an introduction to a novel, play, or other work of literature. Often a prologue contains useful information about events to come in the story. In a prologue to a play, an actor addresses the audience directly and tells them what the play will be about or describes the setting of the play.

Foreshadowing is the use of clues by the author to prepare readers for events that will happen.

The prologue of *The Invisible Man*, likewise, sets the stage for the reader by foreshadowing two themes that will reoccur in the novel: invisibility and light.

Science Connection The narrator of *The Invisible Man* says that he has strung 1,369 lights in his basement room. How were this many bulbs wired together? If all the bulbs were all wired together in a series circuit, the electrical resistance in the circuit would be high. By Ohm's law, the current in the circuit would be low and the bulbs wouldn't glow. If all the bulbs were wired in a parallel circuit, so much current would flow in the circuit that the connecting wires would melt. For the bulbs to light, the narrator must have wired them in many independent circuits.

Linking Science and Writing

Prologue Write a prologue to a make-believe book describing Edison's invention of the light bulb. Recall that a prologue is not a summary of the book. The prologue can state general themes that the work of literature will address, or it can set the stage or describe the setting of the story. You might want to discuss what was happening in the world during Edison's time or foreshadow the character and personality traits that enabled him to be a great inventor.

Career Connection

Electrical Engineer

At the age of 10, **Hans Moravec** wired up a tin can man and started building animate things out of inanimate objects. Since 1980, he has been a Principal Research Scientist at the Robotics Institute of Carnegie Mellon University in Pittsburgh. He and his team have built several mobile robots—the latest one is being designed to "see" in three dimensions and to move in crowded spaces without going bump in the night. By 2010, he expects robots, with the improved computers of that time, to be doing many simple tasks.

SCIENCE *Online* To learn more about careers in electrical engineering, visit the Glencoe Science Web site at **science.glencoe.com**.

Chapter 7 Study Guide

Reviewing Main Ideas

Section 1 Electric Charge

1. There are two types of electric charge—positive charge and negative charge. Like charges repel and unlike charges attract. *Why are the towels clinging together?*

2. Electric charge is conserved. Charges cannot be created or destroyed.
3. An electrical conductor allows electrons to move through it easily. An electrical insulator doesn't allow electrons to move through it easily.
4. Objects can be charged by contact or by induction. Charging by induction occurs when a charged object is brought near an electrically neutral object.

Section 2 Electric Current

1. Charges flow through a conductor due to a voltage difference. *How is the flow of water similar to the flow of electric charge?*

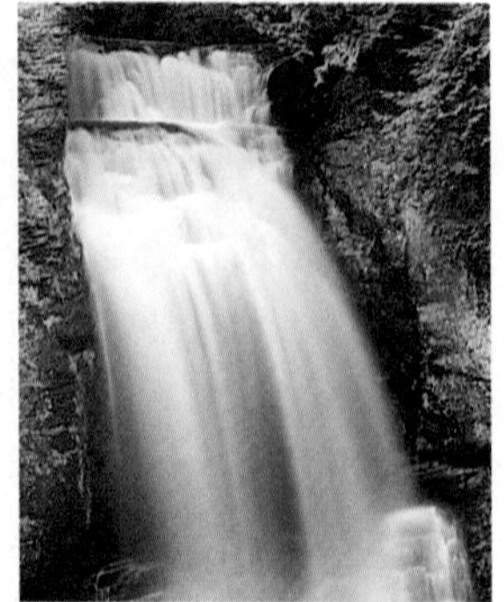

2. Electric current is the movement of electric charges. A circuit is a closed conducting loop through which electric charges can move.
3. A battery establishes a voltage difference in a circuit by separating positive and negative charges.
4. In an electric circuit, increasing the voltage difference increases the current. Increasing the resistance decreases the current. These relations are known as Ohm's law.

Section 3 Electrical Energy

1. Current has only one path in a series circuit and more than one path in a parallel circuit.
2. Circuit breakers and fuses are safety devices that prevent excessive current from flowing in a circuit.
3. Electrical power is the rate at which electrical energy is used.
4. Utility companies sell electrical energy by the kilowatt-hour, which is 1,000 W of power used for 1 h. *How would your life change if electrical energy were no longer available?*

After You Read

FOLDABLES Reading & Study Skills

Reflect on what you have learned about electricity in this chapter. Record your thoughts under the Learned tab of your Know-Want-Learn Study Fold.

Visualizing Main Ideas

Use the following terms to complete the concept map: voltage difference, attract, conductor, repel, law of conservation of charge, insulator.

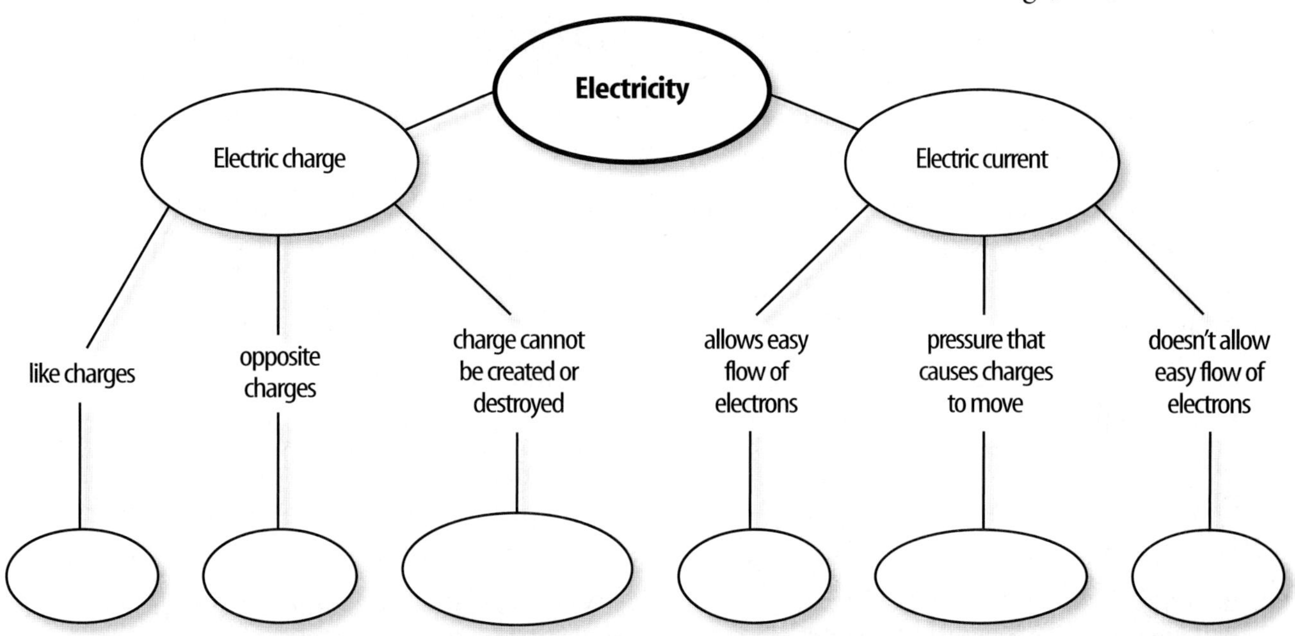

Vocabulary Review

Vocabulary Words

a. charging by contact
b. charging by induction
c. circuit
d. conductor
e. electric current
f. electrical power
g. insulator
h. kilowatt-hour
i. law of conservation of charge
j. Ohm's law
k. parallel circuit
l. resistance
m. series circuit
n. static electricity
o. voltage difference

Study Tip

Find a quiet place to study, whether you are at home or at school. Turn off the television or radio and give your full attention to your lessons.

Using Vocabulary

Distinguish between the terms in each of the following groups of words.

1. law of conservation of charge, Ohm's law
2. electric current, static electricity
3. voltage difference, electrical power
4. series circuit, parallel circuit
5. conductor, insulator
6. electrical power, kilowatt-hour
7. charging by contact, charging by induction
8. circuit, conductor
9. voltage difference, resistance
10. voltage difference, electric current

Chapter 7 Assessment

Checking Concepts

Choose the word or phrase that best answers the question.

1. An object becomes positively charged when which of the following occurs?
A) loses electrons
B) loses protons
C) gains electrons
D) gains neutrons

2. How do two negative charges interact when they are brought close together?
A) repel
B) attract
C) no interaction
D) ground

3. What is a common source for a voltage difference?
A) circuits
B) batteries
C) wires
D) lightning

4. Which of the following is an insulator?
A) copper
B) silver
C) wood
D) salt water

5. What is the process of connecting an object to Earth with a conductor called?
A) charging
B) grounding
C) draining
D) induction

6. What is the SI unit used to measure the difference in voltage between two places?
A) amperes
B) coulombs
C) ohms
D) volts

7. What is the rate at which appliances consume energy?
A) kilowatt-hour
B) resistance
C) current
D) power

8. Resistance in wires causes electrical energy to be converted to what energy form?
A) chemical energy
B) nuclear energy
C) thermal energy
D) sound

9. Which of the following wires would tend to have the least amount of resistance?
A) long
B) fiberglass
C) hot
D) thick

10. What SI unit measures electrical energy?
A) volts
B) newtons
C) kilowatts
D) kilowatt-hours

Thinking Critically

11. How do lightning rods protect buildings from lightning?

12. Explain how an electroscope could be used to detect a negatively charged object.

13. A toy car with a resistance of 2 Ω is connected to a 3 V battery. How much current flows through the car?

14. The current flowing through an appliance connected to a 120-V source is 2 A. How many kilowatt-hours of electrical energy does the appliance use in 4 h?

15. You are asked to connect a stereo, a television, a VCR, and a lamp in a single, complete circuit. Would you connect these appliances in parallel or in series? How would you prevent an electrical fire? Draw a diagram of your circuit.

Developing Skills

16. Making and Using Graphs The resistance in a 1-cm length of copper wire at different temperatures is shown below. One microohm equals one millionth of an ohm. Construct a line graph for the data. Is copper a better conductor on a cold day or a hot day?

Copper Wire Resistance

Resistance in Microohms	Temperature (°C)
2	50
3	200
5	475

17. **Concept Mapping** Make a network concept map sequencing the events that occur when an electroscope is brought near a positively charged object and a negatively charged object. Indicate which way electrons flow and the charge and responses of the leaves.

18. **Identifying and Manipulating Variables and Controls** Design an experiment to test the effect on current and voltage differences in a circuit when two identical batteries are connected in series. What is your hypothesis? What are the variables and control?

19. **Interpreting Scientific Illustrations** The diagram below shows a series circuit containing a lamp connected to a standard wall outlet. Using the information in the diagram, compute the current in the circuit shown.

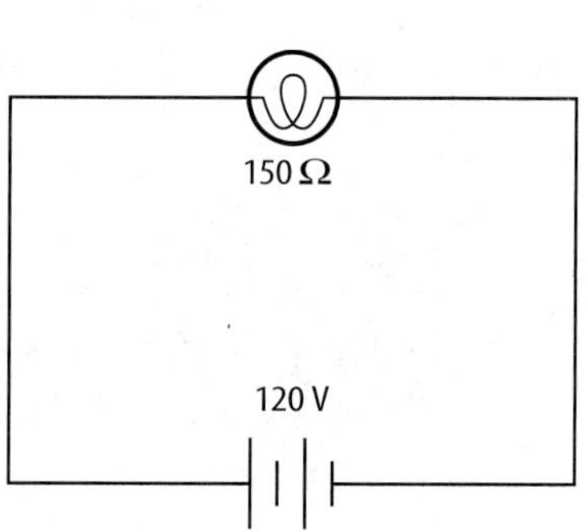

Performance Assessment

20. **Poster** You probably have seen warnings about contacting overhead power lines. However, birds can perch safely on power lines. Find out how this is possible. Share the information you learn by making a poster for your classroom.

Technology

Go to the Glencoe Science Web site at **science.glencoe.com** or use the **Glencoe Science CD-ROM** for additional chapter assessment.

Test Practice

The local electric company did a study of the power ratings of some common appliances. The results of this study are shown in the chart below.

Power Used by Common Appliances

Appliance	Power (watts)	Appliance	Power (watts)
Clock	3	Stove/oven	2,600
Microwave oven	1,450	Dishwasher	2,300
Clothes dryer	4,000	Refrigerator/freezer	600
Radio	100	Hair dryer	1,000
Color TV	300	Toaster	700

Study the chart and answer the following questions.

1. According the chart, which appliance requires the least amount of power?
   A) toaster
   B) clock
   C) radio
   D) color television

2. According to this information, which appliance requires more than 3,000 W of power?
   F) dishwasher
   G) microwave oven
   H) clothes drier
   J) stove/oven

CHAPTER 8

Magnetism and Its Uses

A giant solar flare erupts from the Sun, spewing high energy particles and other forms of radiation toward Earth. Fortunately, Earth's magnetic field deflects most of these particles so they don't damage you and other living creatures. In this chapter, you will learn how magnetism and electricity are related, and how some common devices use magnetism.

What do you think?

Look at the picture below with a classmate. Discuss what this might be or what is happening. Here's a hint: *The force holding up this cube also spins electric motors.* Write your answer or your best guess in your Science Journal.

Magnets can do more than hold papers on a refrigerator door. Did you know that they are used in TVs, computers, stereo speakers, and electric motors? Magnets play an important role in making the electricity you use at home. Magnetism also is used to make images of the organs and tissues inside the human body. What properties of magnets make them so useful? This activity will help you find out.

Observe the strength of a magnet

1. Hold a bar magnet horizontally and suspend a paper clip from one end of it. Continue adding paper clips to make a chain until the magnet will hold no more. Record the number of paper clips the magnet held.
2. Repeat step 1 three times. First, suspend the paper clips about 2 cm from the end of the magnet, then near the center of the magnet, and finally at the other end of the magnet.

Observe

In your Science Journal, compare the number of clips suspended from each point on the magnet. Infer which part of the magnet has the strongest attraction for the paper clips.

Before You Read

Making a Question Study Fold **Asking yourself questions helps you to stay focused and better understand magnets when you are reading the chapter.**

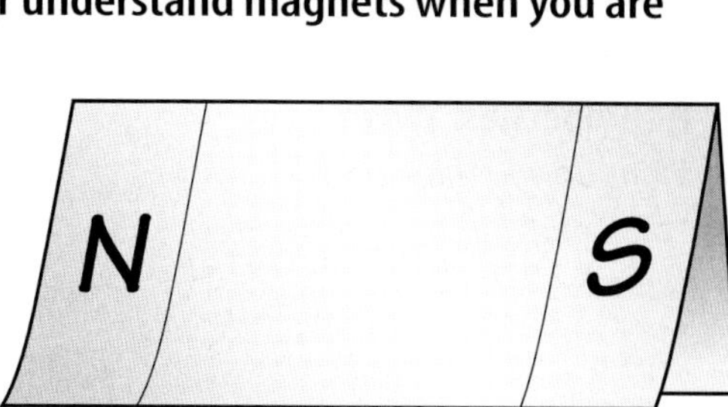

1. Place a sheet of paper in front of you so the long side is at the top. Fold the paper in half from top to bottom.
2. Make the front and back look like a magnet by writing *N* for north on the left side and *S* for south on the right side as shown.
3. Before you read the chapter, write two questions about magnets inside.
4. As you read the chapter, write answers to your questions.

SECTION

Magnetism

As You Read

What You'll Learn

- **Describe** the properties of temporary and permanent magnets.
- **Explain** how a magnet exerts a force on an object.
- **Explain** why some materials are magnetic and others are not.
- **Model** magnetic behavior using magnetic domains.

Vocabulary

magnetism
magnetic pole
magnetic domain

Why It's Important

Without magnets, you could not use computers, CD players, or even the lights in your home.

Magnets

You may be familiar with magnets because they help display artwork on refrigerators, but magnets also fascinated early Greek and Chinese cultures long before refrigerators were invented. The Greeks discovered a mineral, shown in **Figure 1,** that was a natural magnet. They found the mineral in a region called Magnesia, so the Greeks called the mineral magnetic. More than 2,000 years later, magnets play an important role in business, medicine, transportation, and science. Today, the word **magnetism** refers to the properties and interactions of magnets.

✔ Reading Check *Why did the Greeks use the term magnetic?*

Magnetic Force You probably have played with magnets to attract a metal object. You might have noticed that two magnets also exert a force on each other. Depending on which ends of the magnets are close together, the magnets either repel or attract each other. You probably noticed that the interaction between two magnets can be felt even before the magnets touch. This interaction is called magnetic force. Its strength increases as magnets move closer together and decreases as the distance between the magnets increases.

Figure 1
The Greeks found a mineral, now called magnetite, with natural magnetic properties. *What explanations do you think they gave for the behavior of magnetite?*

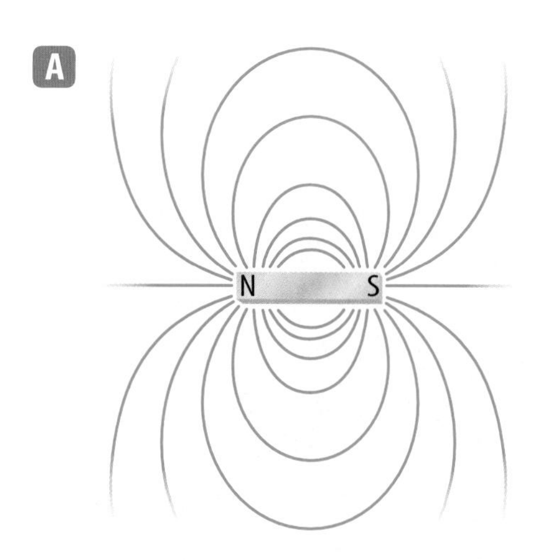

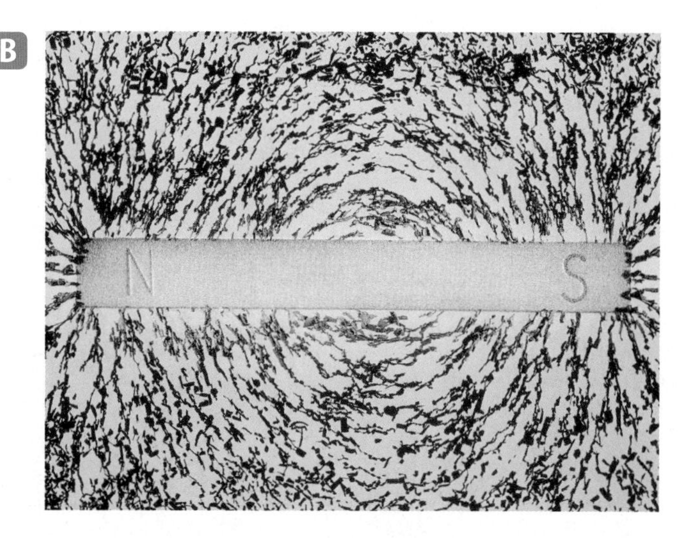

Figure 2
A magnet is surrounded by a magnetic field. A A magnet's magnetic field is represented by magnetic field lines. B Iron filings sprinkled around a magnet line up along the magnetic field lines.

Magnetic Field A magnet is surrounded by a magnetic field that exerts the magnetic force. When objects made of iron or another magnet is placed in this magnetic field, it reacts to the magnetic force. The magnetic field is strongest close to the magnet and weakest far away The magnetic field can be represented by lines of force, or magnetic field lines. **Figure 2** shows the magnetic field lines surrounding a bar magnet.

Magnetic Poles Look again at **Figure 2.** Do you notice that the magnetic field lines are closest together at the ends of the bar magnet? These regions, called the **magnetic poles,** are where the magnetic force exerted by the magnet is strongest. All magnets have a north pole and a south pole. For a bar magnet, the north and south poles are at the opposite ends. If a bar magnet is suspended so it turns freely, the north pole of the magnet will point north. Even magnets with more complicated shapes have north and south poles, as **Figure 3** shows. The two ends of a horseshoe-shaped magnet are the north and south poles. A magnet shaped like a disk has opposite poles on the top and bottom of the disk. Magnetic field lines always connect the north pole and the south pole of a magnet.

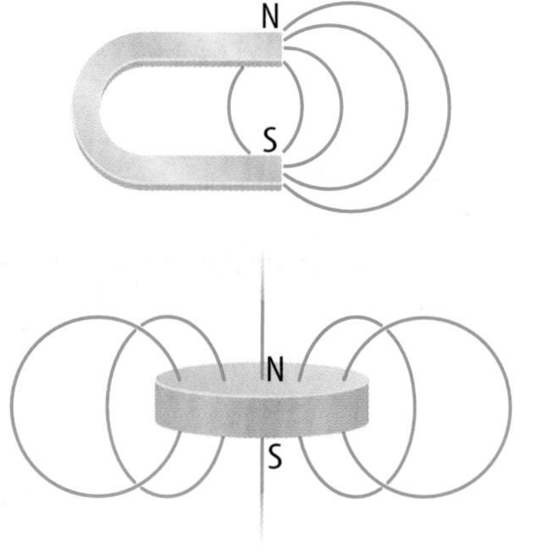

Figure 3
The magnetic field lines around horseshoe and disk magnets are closest together at the magnets' poles. ***Where would a horseshoe magnet have the weakest attraction for metal objects?***

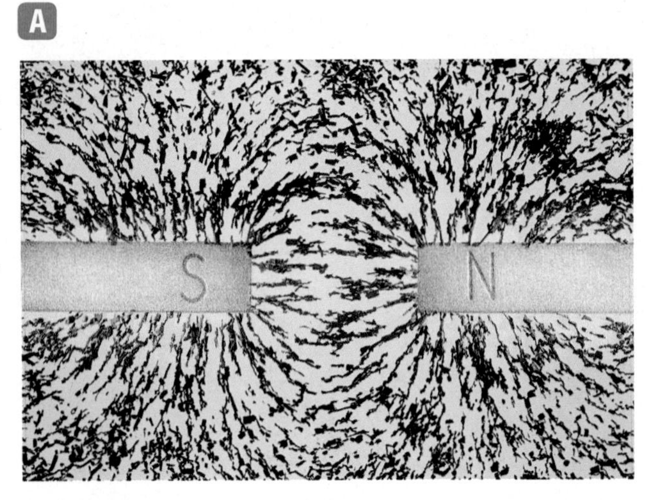

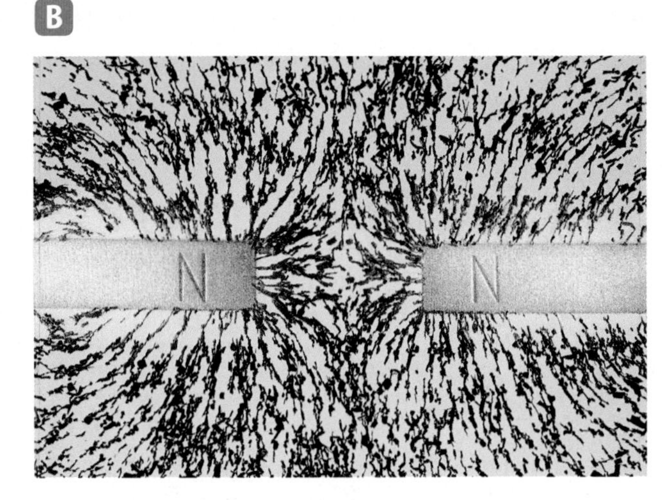

Figure 4
Magnets can attract or repel each other. A Unlike poles attract. When unlike poles are brought together, their magnetic field lines seem to connect with each other. B Like poles repel. When like poles are brought together, their magnetic field lines seem to push away from each other. *How would two horseshoe magnets interact?*

How Magnets Interact Two magnets can either attract or repel each other. If you try to bring the two north poles or the two south poles of two magnets close to each other, you can feel a force preventing the magnets from touching. However, north poles always attract south poles. When two magnets are brought close to each other, their magnetic fields can combine to produce a new magnetic field. **Figure 4** shows the magnetic field that results when like poles and unlike poles of bar magnets are brought close to each other.

Reading Check *How do magnetic poles interact with each other?*

A Compass Needle A magnet that is free to rotate can turn when it is placed in a magnetic field. A compass contains a needle, a small bar magnet, that can freely rotate. If you place a small compass near a bar magnet, the compass needle will turn so that the north pole of the needle points toward the south pole of the bar magnet. The compass needle also lines up along the magnetic field lines that pass near it. **Figure 5** shows how compass needles placed at several positions around a bar magnet are aligned along the magnetic field lines.

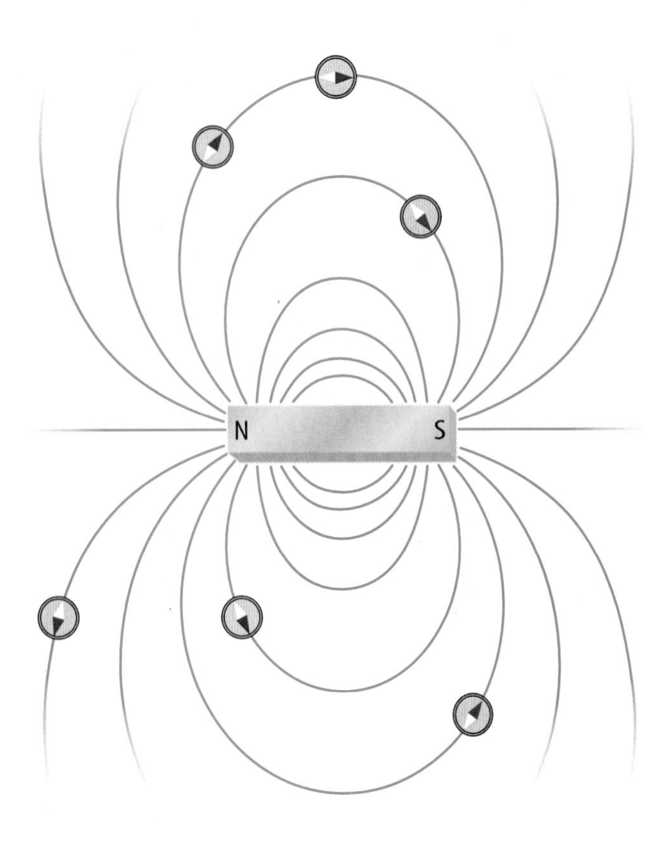

Figure 5
Compass needles rotate to line up with the magnetic field lines of a bar magnet.

Earth's Magnetic Field Imagine you are in a boat in the middle of an ocean. Without landmarks nearby, how could you tell in which direction you were traveling? A compass would help determine your direction because the north pole of the compass needle always points north. This is because Earth acts like a giant bar magnet and is surrounded by a magnetic field that extends into space. Just as with a bar magnet, the compass needle aligns with Earth's magnetic field lines, as shown in **Figure 6.**

Earth's Magnetic Poles The north pole of a magnet is defined as the end of the magnet that points toward the geographic north. Sometimes the north pole and south pole of magnets are called the north-seeking pole and the south-seeking pole. Because opposite magnetic poles attract, the north pole of a compass is being attracted by a south magnetic pole. So Earth is like a bar magnet with its south magnetic pole near its geographic north pole.

The location of Earth's south magnetic pole currently is in northern Canada about 1,500 km from the geographic north pole. So if you were north of the south magnetic pole, your compass needle would point south, away from the geographic north pole.

No one is sure what produces Earth's magnetic field. Earth's core is made of a solid ball of iron and nickel, surround by a liquid layer of molten iron and nickel. According to one theory, circulation of the molten iron and nickel caused by heat produces Earth's magnetic field.

Mini LAB

Observing Magnetic Interference

Procedure

1. Clamp a **bar magnet** to a **ring stand.** Tie a **thread** around one end of a **paper clip** and stick the paper clip to one pole of the magnet.
2. Anchor the other end of the thread under a **book** on the table. Slowly pull the thread until the paper clip is suspended below the magnet but not touching the magnet.
3. Without touching the paper clip, slip a piece of paper between the magnet and the paper clip. Does the paper clip fall?
4. Try other materials, such as **aluminum foil, fabric,** or a **butter knife.**

Analysis

1. Which materials caused the paper clip to fall? Why do you think these materials interfered with the magnetic field?
2. Which materials did not cause the paper clip to fall? Why do you think these materials did not interfere with the magnetic field?

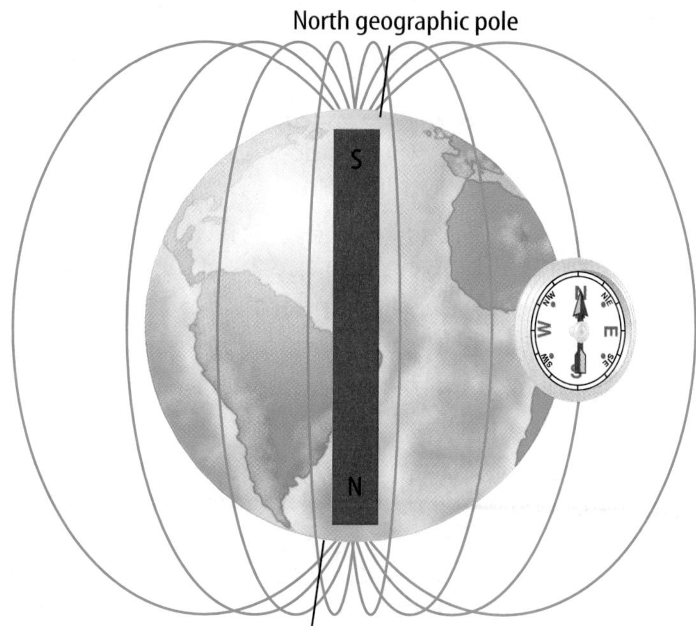

Figure 6
A compass needle aligns with the magnetic field lines of Earth's magnetic field. *Which way would a compass needle point if you held it while you stood directly over the south magnetic pole?*

Some animals may use Earth's magnetic field to help find their way around. Some species of birds, insects, and bacteria have been shown to contain small amounts of the mineral magnetite. Research how one species uses Earth's magnetic field, and report your findings to your class.

Magnetic Materials

You might have noticed that a magnet will not attract all metal objects. For example, a magnet will not attract pieces of aluminum foil. Only a few metals such as iron, cobalt, or nickel are attracted to magnets or can be made into permanent magnets. What makes these elements magnetic? Remember that every atom contains electrons. Electrons have magnetic properties. In the atoms of most elements, the magnetic properties of the electrons cancel out. But in the atoms of iron, cobalt, and nickel, these magnetic properties don't cancel out. Each atom in these elements behaves like a small magnet and has its own magnetic field.

Even though these atoms have their own magnetic fields, objects made from these metals are not always magnets. For example, if you hold an iron nail close to a refrigerator door and let go, it falls to the floor. However, you can make the nail behave like a magnet temporarily.

Problem-Solving Activity

How can magnetic parts of a junk car be salvaged?

Every year, over 10 million cars are scrapped. Magnets are often used to help retrieve valuable materials from these cars for recycling. Once the junk car has been fed into a shredder, big magnets can easily separate many of its metal parts from its nonmetal parts. How much of the car does a magnet actually help separate? Use your ability to interpret a circle graph to find out.

Percentage weight of materials in a car

Other 10%
Plastic, glass, and rubber 15%
Nonmagnetic metals 10%
Magnetic metals 65%

Identifying the Problem

The graph at the right shows the average percent by weight of the different materials in a car. Included in the magnetic metals are steel and iron. The nonmagnetic metals refer to aluminum, copper, lead, zinc, and magnesium. According to the chart, how much of the car can a magnet separate for recycling?

Solving the Problem

1. What percent of the car's weight will a magnet recover? Explain your answer.
2. Plastics are replacing steel in many new cars. How might this affect the future of car recycling?

Magnetic Domains—A Model for Magnetism In iron, cobalt, nickel, and other magnetic materials, the magnetic field created by each atom exerts a force on the other nearby atoms. Because of these forces, groups of atoms align their magnetic poles so that all like poles point in the same direction. The groups of atoms with aligned magnetic poles are called **magnetic domains.** Each domain contains billions of atoms, yet the domains are too small to be seen with your naked eye. Because the magnetic poles of the individual atoms in a domain are aligned, the domain itself behaves like a magnet with a north pole and a south pole.

Lining Up Domains An iron nail contains an enormous number of these magnetic domains, so why doesn't the nail behave like a magnet? Even though each domain behaves like a magnet, the poles of the domains are arranged randomly and point in different directions, as shown in **Figure 7A.** As a result, the magnetic fields from all the domains cancel each other out.

If you place a magnet against the same nail, the atoms in the domains orient themselves in the direction of the nearby magnetic field, as shown in **Figure 7B.** The like poles of all the domains point in the same direction and no longer cancel each other out. The nail now acts as a magnet itself. But when the external magnetic field is removed, the constant motion and vibration of the atoms bump the magnetic domains out of their alignment. The magnetic domains in the nail return to random arrangement. For this reason, the nail is a temporary magnet. Paper clips and other objects containing iron also can become temporary magnets.

TRY AT HOME

Mini LAB

Making Your Own Compass

Procedure

WARNING: *Use care when handling sharp objects.*

1. Cut off the bottom of a **plastic foam cup** to make a polystyrene disk.
2. Magnetize a **sewing needle** by continuously stroking the needle in the same direction with a magnet for 1 min.
3. **Tape** the needle to the center of the foam disk.
4. Fill a **plate** with **water** and float the disk, needle side up, in the water.
5. Bring the magnet close to the foam disk.

Analysis

1. What happened to the needle and disk when you placed them in the water? Explain.
2. How did the needle and disk move when the magnet was brought near it? Explain.

Figure 7
Magnetic materials contain magnetic domains.

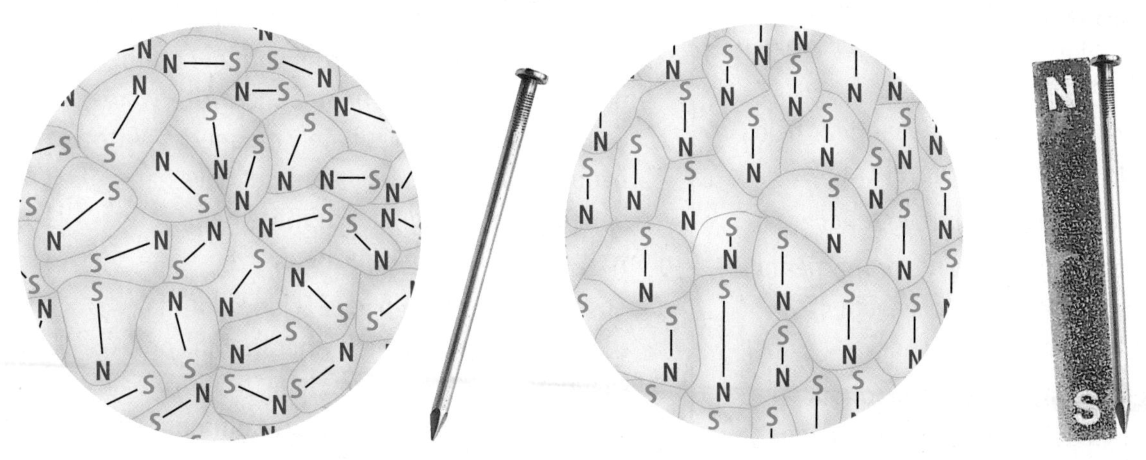

A A normal iron nail is made up of billions of domains that are arranged randomly.

B The domains will align themselves along the magnetic field lines of a nearby magnet.

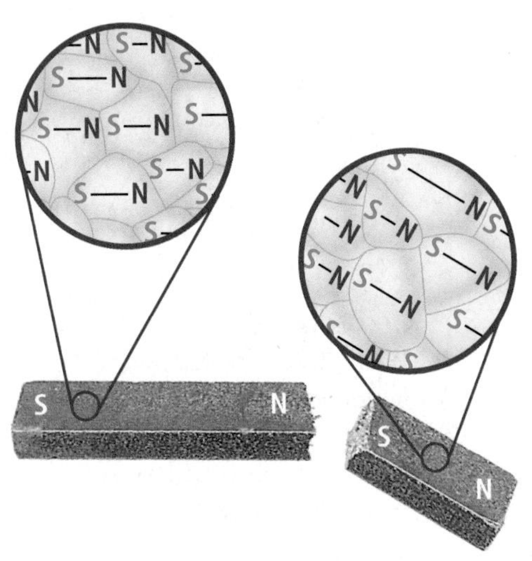

Figure 8
Each piece of a broken magnet still has a north and a south pole.

Permanent Magnets A permanent magnet can be made by placing a piece of magnetic material, such as iron, cobalt, or nickel, in a strong magnetic field. The strong magnetic field causes a large number of the magnetic domains in the material to line up. The magnetic fields of these aligned domains add together and create a magnetic field inside the material that may be several thousand times larger than the magnetic field outside the material. This then prevents the constant motion of the atoms from bumping all the domains out of alignment. The material is then a permanent magnet, and it can retain its magnetic properties for a long time.

But even permanent magnets can lose their magnetic behavior if they are heated or dropped. Heating causes atoms to move faster, so they can jostle magnetic domains out of alignment. If the material is heated enough, its atoms may be moving fast enough to jostle all the domains out of alignment. Then the material is no longer a magnet.

Can a pole be isolated? What happens when a magnet is broken in two? Can one piece be a north pole and one piece be a south pole? Look at the domain model of the broken magnet in **Figure 8.** Recall that even individual atoms of magnetic materials act as tiny magnets. Because every magnet is made of many aligned smaller magnets, even the smallest pieces have a north pole and a south pole. As a result, a magnetic pole cannot be isolated.

Section  Assessment

1. Describe what happens when you bring two like magnetic poles together. Draw a picture to illustrate your answer.
2. If a compass is placed in a magnetic field, how does the compass needle move?
3. Why aren't all materials magnetic?
4. What would happen to the properties of a bar magnet if it were broken in half? In thirds? Explain your answer.
5. **Think Critically** Use the magnetic domain model to explain why a magnet sticks to a refrigerator door.

Skill Builder Activities

6. **Forming Hypotheses** Your younger brother or sister played with a bar magnet. Afterward, you noticed that it was barely magnetic. Write a hypothesis to explain what might have happened to your magnet. **For more help, refer to the Science Skill Handbook.**
7. **Communicating** In your Science Journal, make a list of all the uses you can think of for magnets. Write a paragraph describing what these magnets seem to have in common. **For more help, refer to the Science Skill Handbook.**

SECTION 2

Electricity and Magnetism

Electric Current and Magnetism

Even in science, it can help to be lucky. In 1820, Hans Christian Oersted, a Danish physics teacher, found that electricity and magnetism are related. While demonstrating the operation of electric circuits to his class, he happened to have a compass near a piece of wire. When current flowed through the wire, he noticed that the compass needle was turned, or deflected. When the current was reversed, he saw that the compass needle was deflected in the opposite direction. The compass needle returned to its normal position when he stopped the current in the wire. Oersted hypothesized that the electric current must produce a magnetic field around the wire, and the direction of the field changes with the direction of the current.

Moving Charges and Magnetic Fields It is now known that moving charges, like those in an electric current, produce magnetic fields. Oersted's hypothesis that passing electric current through a wire creates a magnetic field was correct. The magnetic field around a current-carrying wire forms a circular pattern about the wire, as shown in **Figure 9.** The direction of the field depends on the direction of the current. The strength of the magnetic field depends on the amount of current flowing in the wire. When no current flows in a wire, the magnetic field disappears. This discovery of the connection between electricity and magnetism has led to many useful devices.

As You Read

What You'll Learn

- **Understand** the relationship between electric current and magnetism.
- **Explain** how electromagnets are constructed.
- **Describe** how electromagnets are used.
- **Describe** how an electric motor operates.

Vocabulary

electromagnet
galvanometer
electric motor

Why It's Important

Many of the devices you use every day use the relationship between electricity and magnetism to operate.

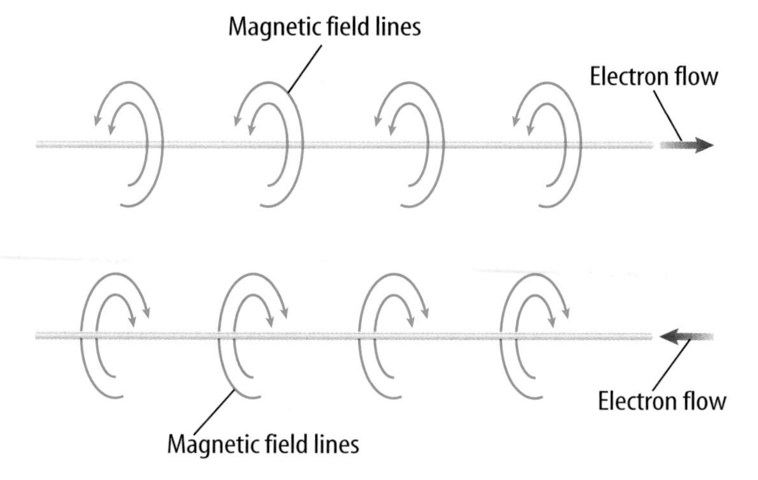

Figure 9
When electric current flows through a wire, a magnetic field forms around the wire. The direction of the magnetic field depends on the direction of the current in the wire.

Electromagnets

One of the most important devices that uses the connection between electricity and magnetism is the electromagnet (ih lek troh MAG nut). An **electromagnet** is a temporary magnet made by placing a piece of iron inside a current-carrying coil of wire. When a current flows through a circular loop of wire, magnetic-field lines form all around the wire as shown in **Figure 10A.** If more loops of wire are added to make a coil, the magnetic-field lines formed around each loop will overlap and add together, as shown in **Figure 10B.** As a result, the magnetic field inside the coil is made stronger. If an iron core is inserted into the coil as in **Figure 10C,** the magnetic field inside the coil causes the iron core to become magnetized. When a magnetized iron core and a coil are combined this way, the magnetic field comes mostly from the iron core.

Properties of Electromagnets Electromagnets are temporary magnets because the magnetic field is present only when current is flowing in the wire coil. The strength of the magnetic field can be increased by adding more turns to the wire coil or by increasing the current passing through the wire.

An electromagnet behaves like any other magnet when current flows through the wire coil. One end of the electromagnet is a north pole and the other end is a south pole. If placed in a magnetic field, an electromagnet will align itself along the magnetic field lines, just as a compass needle will. An electromagnet also will attract magnetic materials and be attracted or repelled by other magnets. What makes electromagnets so useful is that their magnetic properties can be controlled by changing the electric current flowing through the wire coil.

When current flows through the electromagnet and it moves toward or away from another magnet, it converts electric energy into mechanical energy to do work. Electromagnets do work in various devices such as stereo speakers and electric motors. They also can lift large metal objects.

Figure 10
An electromagnet is made from a current-carrying wire.

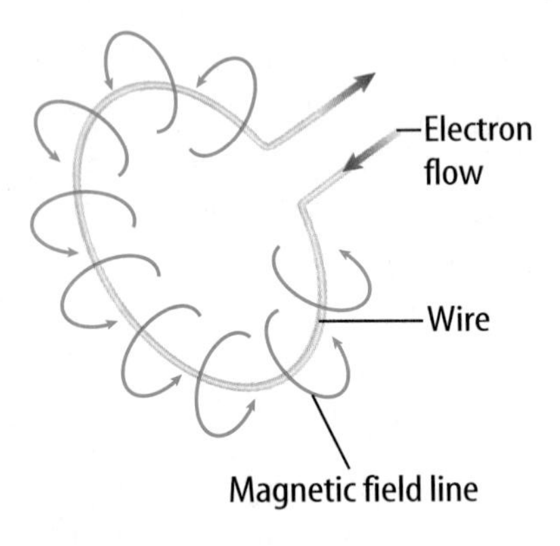

A Magnetic field lines circle around a loop of current-carrying wire.

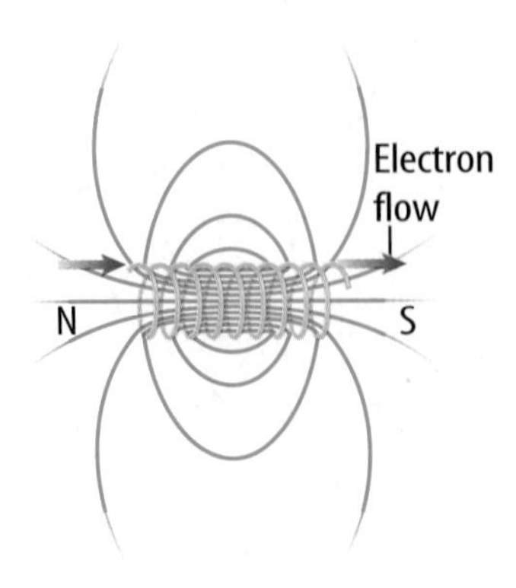

B When many loops of current-carrying wire are formed into a coil, the magnetic field is increased inside the coil. The coil has a north pole and a south pole. *What would happen if you switched the direction of the current in the coil?*

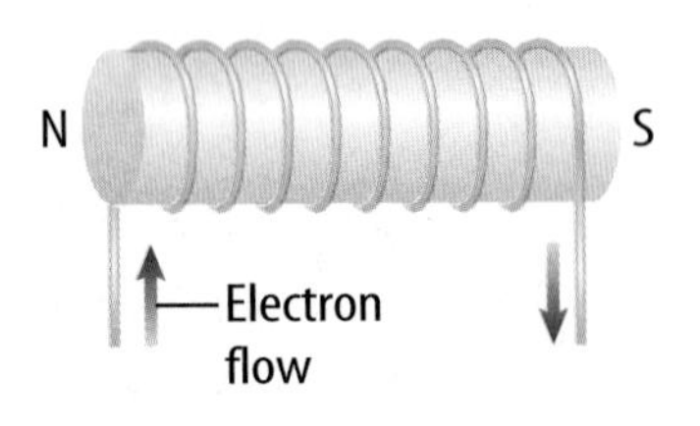

C An iron core inserted into the coil becomes a magnet.

Music to Your Ears—Stereo Speakers How does musical information stored on a CD become sound you can hear? The sound is produced by a stereo speaker that contains an electromagnet. The electromagnet changes electrical energy to mechanical energy that vibrates parts of the speaker to produce sound.

Reading Check *How does an electromagnet allow a stereo speaker to produce sound?*

When you listen to a CD, the CD player produces an electric current that changes according to the musical information on the CD. This varying electric current passes through a coiled wire inside the speaker that is part of an electromagnet, as in **Figure 11.** A magnetic field is generated in the electromagnet. This magnetic field changes depending on the varying characteristics of the electric current. The electromagnet is then attracted to or repelled by a permanent, fixed magnet, making the electromagnet move back and forth. This movement vibrates the speaker's flexible surface and produces sound. The vibration of the speaker cone reproduces the original musical information stored on the CD.

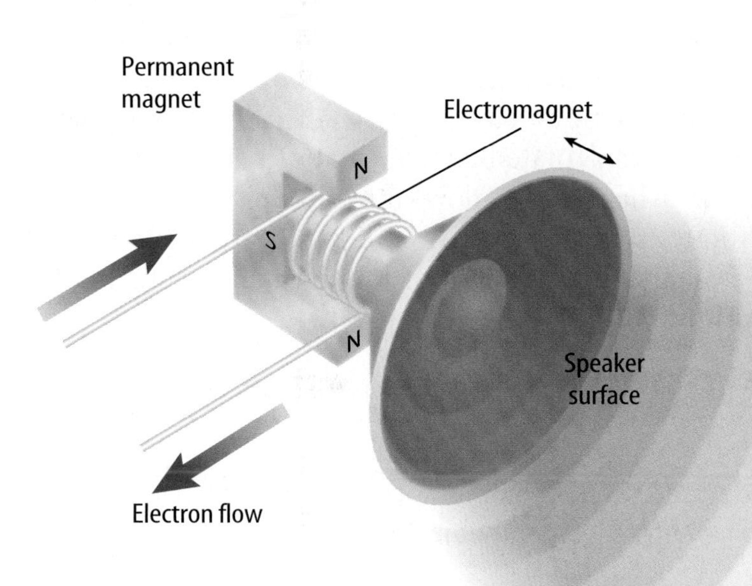

Figure 11
The electromagnet in a speaker turns electrical energy into mechanical energy to produce sound.

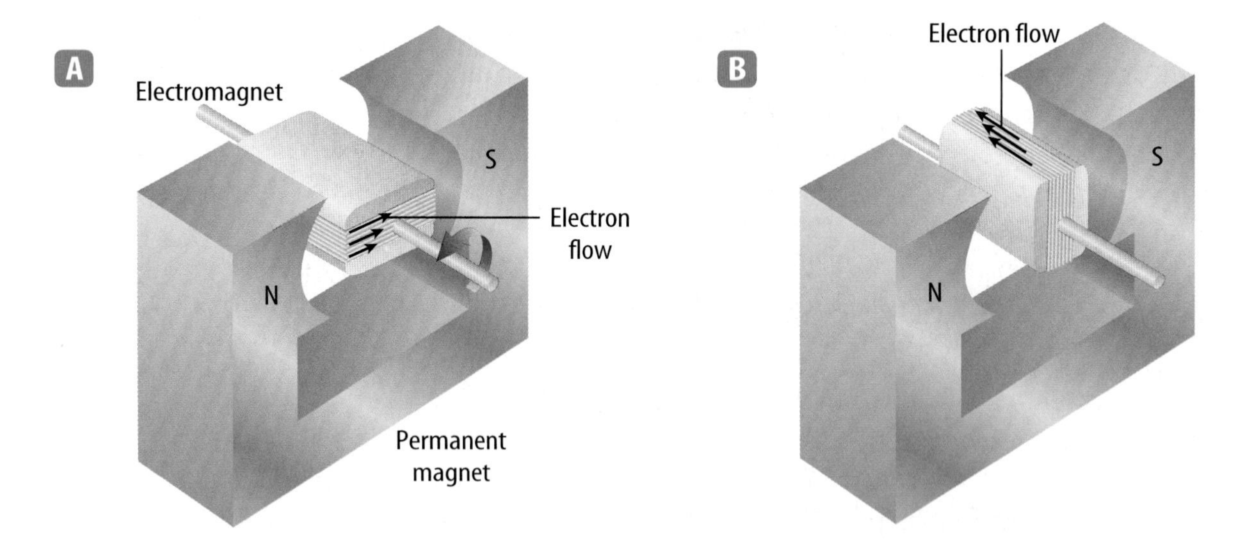

Figure 12
A When current flows through the coil, an electromagnet is formed that is attracted to and repelled by the poles of the permanent magnet. B The magnetic forces on the coil cause it to rotate, aligning it with the field of the permanent magnet.

Galvanometers You've probably noticed the gauges in the dashboard of a car. One gauge shows the amount of gasoline left, and another shows the engine temperature. How does a change in the amount of gasoline in a tank or the water temperature in the engine make a needle move in a gauge on the dashboard? These gauges are **galvanometers** (gal vuh NAHM ut urs), which are devices that use an electromagnet to measure electric current. For example, a temperature sensor in the engine sends an electric current to the temperature gauge. This current changes as the engine temperature changes. The needle of the temperature gauge is connected to an electromagnet. This electromagnet is suspended so it can rotate between the poles of a permanent, fixed magnet. When current flows through the coil, the electromagnet rotates so that its north and south poles are aligned along the magnetic-field lines of the permanent magnet, as shown in **Figure 12.**

✔ Reading Check *What is measured by a galvanometer?*

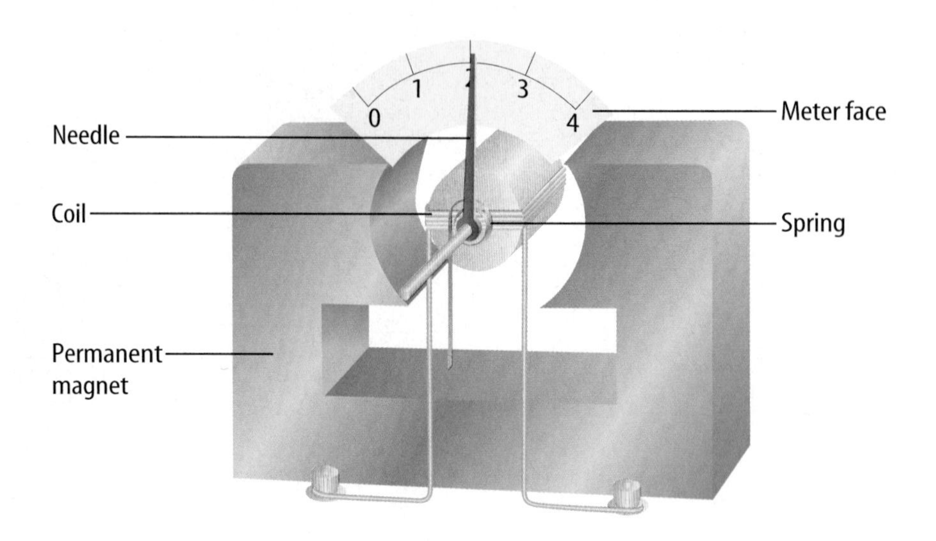

Figure 13
A galvanometer includes a permanent magnet, an electromagnet that rotates against a spring, and a scale that gives a measurement of the current.
Would it matter which way you hooked up the galvanometer to the two terminals of the circuit you were testing? Why?

Using Galvanometers If the coil is connected to a small spring, then the coil can act as a galvanometer. If the current through the coil is small, only a weak magnetic field is produced in the electromagnet. Then the magnetic force between the electromagnet and the permanent magnet is weak. The coil can rotate only a small amount against the resistance of the spring, and the needle moves by only a small amount. When a large current flows in the coil, the magnetic force between the electromagnet and the permanent magnet is stronger. The coil rotates further, and the needle moves further along the scale. To be used as a gauge, a galvanometer must be calibrated by sending a known current through the coil and seeing how much the needle is deflected. **Figure 13** shows an example of a galvanometer.

Research Visit the Glencoe Science Web site at **science.glencoe.com** for more information about how motors do work in electric devices. In your Science Journal, summarize the similarities among the motors you learn about.

Electric Motors

On sizzling summer days, do you ever use an electric fan to keep cool? A fan uses an **electric motor,** which is a device that changes electrical energy into mechanical energy. The motor in a fan turns the fan blades, moving air past your skin to make you feel cooler.

Like a galvanometer, an electric motor contains an electromagnet that is free to rotate between the poles of a permanent, fixed magnet. The coil in the electromagnet is connected to a source of electric current, such as a battery, as shown in **Figure 14.** When a current flows through the electromagnet, a magnetic field is produced in the coil.

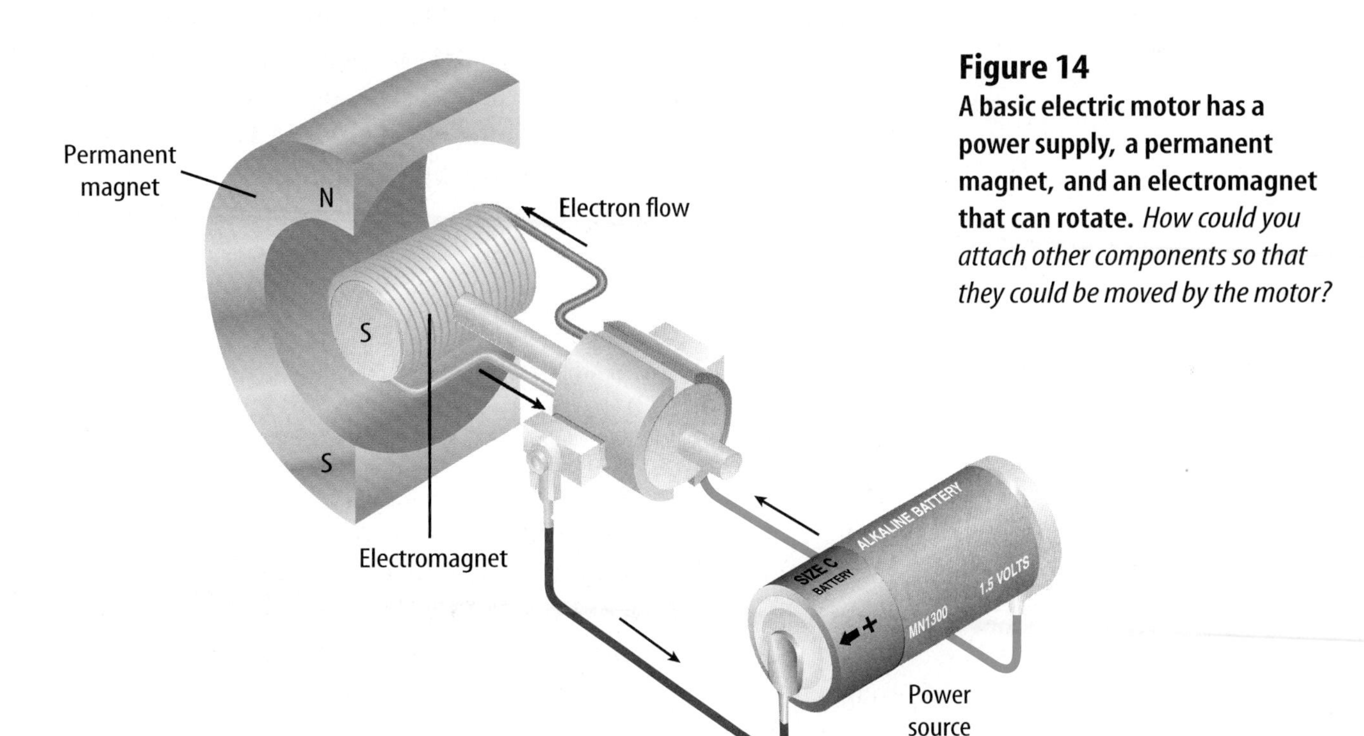

Figure 14
A basic electric motor has a power supply, a permanent magnet, and an electromagnet that can rotate. *How could you attach other components so that they could be moved by the motor?*

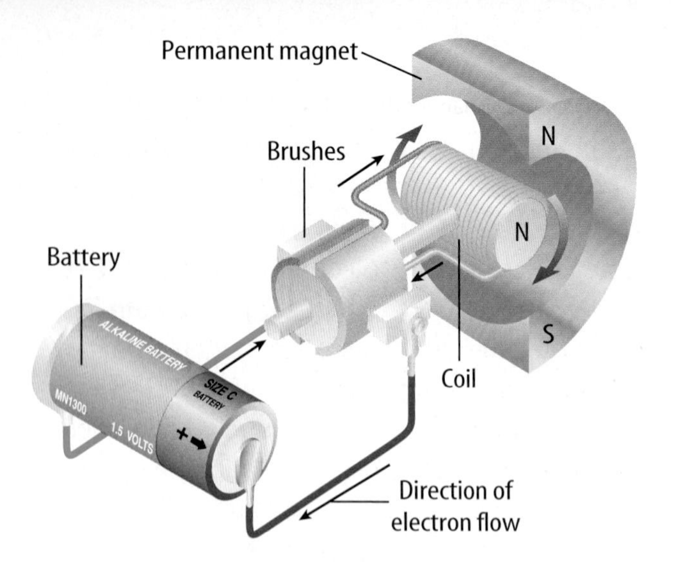

A **A battery causes an electric current to flow through the coil of the electromagnet.**

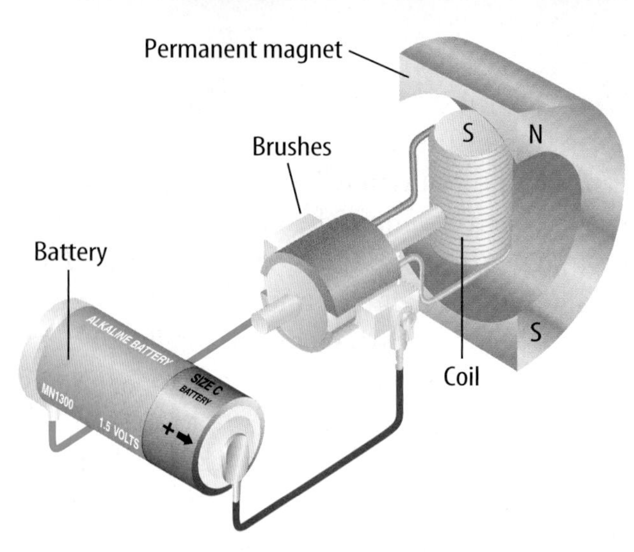

B **Unlike poles of the two magnets attract each other, and the like poles repel. This causes the coil to rotate until the opposite poles are next to each other.**

Figure 15
The shaft of an electric motor is made to rotate by the forces between magnets.

Switching Poles **Figures 15A** and **15B** show how the magnetic force between the electromagnet and the permanent magnet causes the coil to turn. Just as in a galvanometer, the coil in an electric motor turns so that its north and south poles are aligned along the magnetic-field lines of the permanent magnet.

However, once the coil is aligned, there is no longer a force that will keep the coil rotating. Now suppose that the magnetic field in the coil is flipped so the north and south poles switch ends. The direction of the coil's magnetic field can be flipped by reversing the direction of the electric current in the coil. Then the like poles of the coil and magnet will be next to each other.

After flipping the field, the coil will be repelled and will rotate further, as shown in **Figures 15C** and **15D**. The coil will then rotate until it is once again aligned along the field lines of the permanent magnet. Then the current is reversed again. In this way, the coil is kept rotating.

In some motors a switch called a commutator reverses the current in the coil. Other motors don't need a commutator because they use household alternating current, which reverses direction 120 times a second.

Controlling Electric Motors Electric motors can be more useful if their rotation speed can be controlled. One way to do this is to vary the amount of current flowing through the coil. Because the coil is an electromagnet, its magnetic field becomes stronger if more current flows through the coil. This causes the magnetic force between the coil and the permanent magnet to increase. As a result, the coil turns faster.

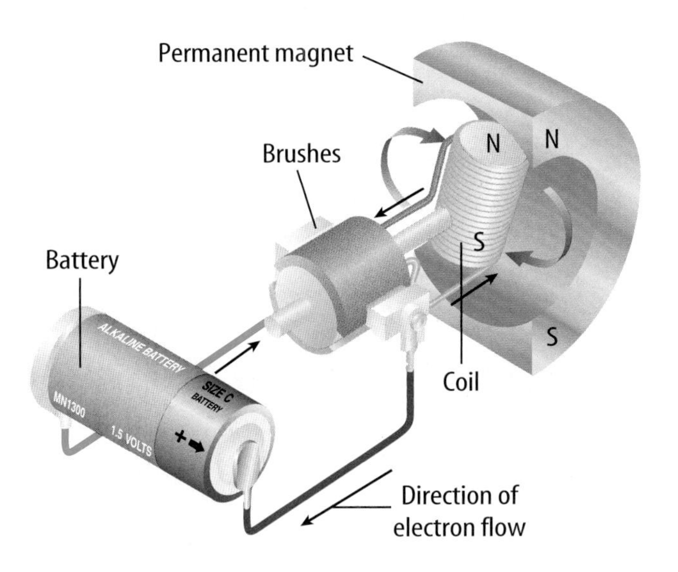

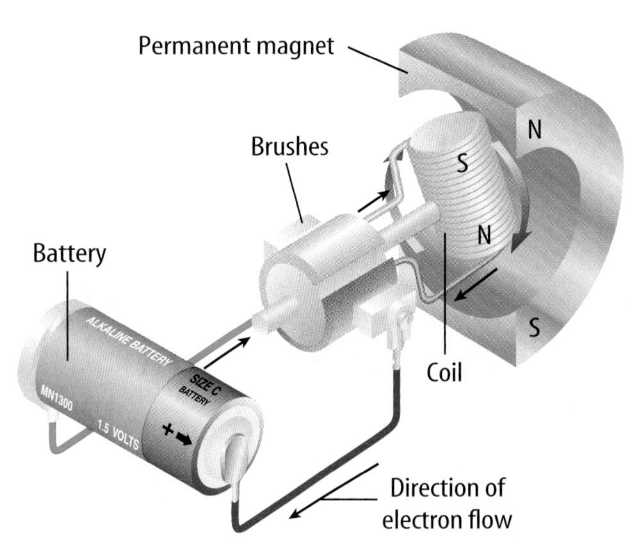

C If the current in the coil is switched, the direction of the coil's magnetic field also switches. The north and south poles of the magnet trade places.

D The coil is repelled by and attracted once again to the poles of the permanent magnet. The coil rotates until it is again aligned with the permanent magnet's field.

Using Electric Motors The first electric motor to be widely used was developed in 1873. This motor used direct current. The first motor to use alternating current was invented in 1888. Since that time many additional developments have made electric motors smaller, more powerful, and more efficient. Today electric motors are used everywhere. Almost every appliance with moving parts uses an electric motor. Can you find an electric motor in every room of your home?

Section  Assessment

1. Does a straight wire or a looped wire have a stronger magnetic field when both carry the same amount of current? Explain.
2. How is the magnetic field of an electromagnet controlled?
3. What are galvanometers used for?
4. How does an electric motor rotate once its electromagnet is aligned along the magnetic field of its permanent magnet?
5. **Think Critically** Could an electromagnet use a nickel core in the coil of wire instead of iron? Why or why not?

Skill Builder Activities

6. **Comparing and Contrasting** Compare and contrast galvanometers and electric motors. **For more help, refer to the Science Skill Handbook.**
7. **Using an Electronic Spreadsheet** Take an inventory of all the devices in your home or school that use an electric motor. Organize your inventory using a database or spreadsheet to indicate the following: *name of the device, the place you found it, the power source used, and which parts the motor causes to move.* **For more help, refer to the Technology Skill Handbook.**

Producing Electric Current

As You Read

What **You'll Learn**

- **Describe** how a generator produces an electric current.
- **Distinguish** between alternating current and direct current.
- **Explain** how a transformer can change the voltage of an alternating current.

Vocabulary

electromagnetic induction
generator
turbine
direct current (DC)
alternating current (AC)
transformer

Why **It's Important**

Power plants use electromagnetic induction to generate electricity for you to use at home and school.

From Mechanical to Electrical Energy

After it was discovered that an electric current could produce a magnetic field, some people wondered whether the opposite could happen: could a magnetic field produce an electric current? Working independently in 1831, Michael Faraday in Britain and Joseph Henry in the United States found that moving a loop of wire through a magnetic field caused an electric current to flow in the wire. They also found that moving a magnet through a loop of wire produces a current. In both cases the mechanical energy associated with the motion of the wire loop or the magnet is converted into electrical energy associated with the electrical current in the wire. Producing an electric current by moving a loop of wire through a magnetic field or moving a magnet through a wire loop is called **electromagnetic induction** (ihn DUK shun). The discovery of electromagnetic induction has led to many applications.

Generators How is the electricity that comes to your home and school produced? Most of the electricity you use each day is produced by generators using electromagnetic induction. A **generator** produces electric current by rotating a coil of wire in a magnetic field. Just as in a galvanometer or an electric motor, the wire coil is wrapped around an iron core and placed between the poles of a permanent magnet. The coil is rotated by an outside source of mechanical energy, as shown in **Figure 16.** As the coil turns within the magnetic field of the permanent magnet, an electric current flows through the coil.

Reading Check *How does a generator use electromagnetic induction?*

Permanent magnet
N
Electromagnet
Source of mechanical energy
S
Direction of electron flow

Figure 16
The electromagnet in a generator is rotated by some outside source of mechanical energy. In this setup, a student can rotate a crank to turn the electromagnet.

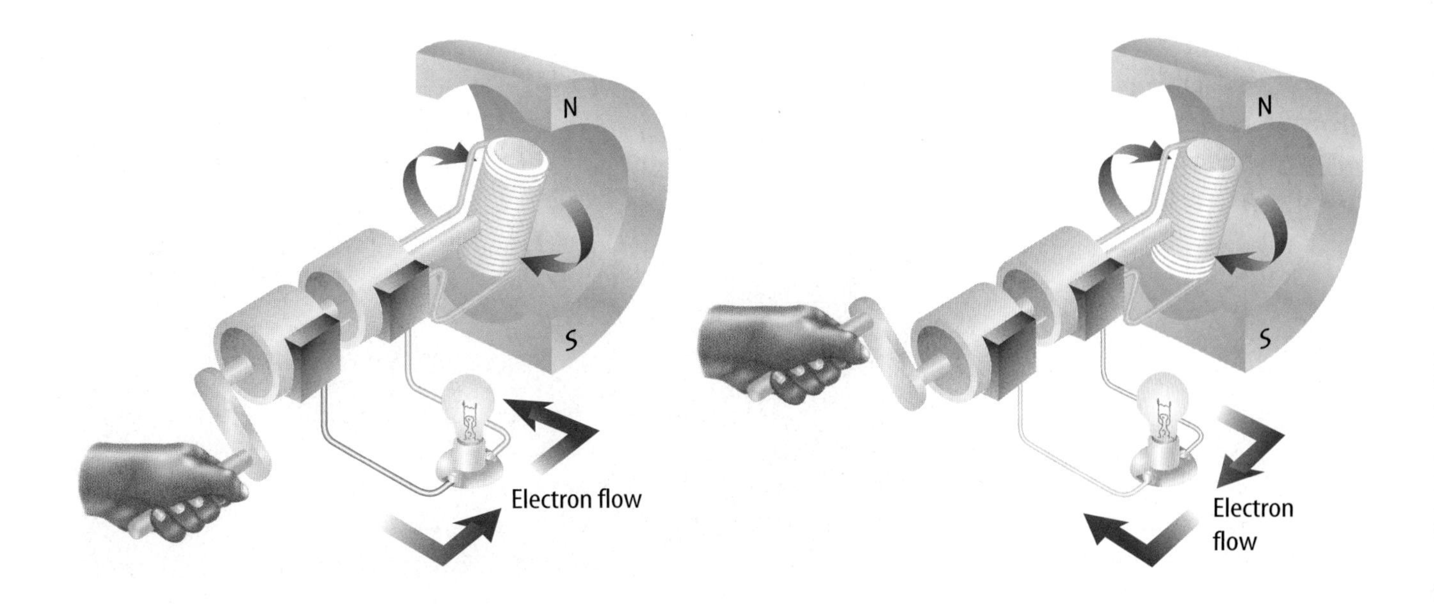

Figure 17
The direction that current flows in a wire coil depends on how the wire coil is aligned with the permanent magnet. *Would a generator still work if the electromagnet were held steady and the permanent magnet moved around it? Explain.*

Switching Direction As the generator's wire coil rotates through the magnetic field of the permanent magnet, current flows through the coil. After the wire coil makes one half of a revolution, the ends of the coil are moving past the opposite poles of the permanent magnet. This causes the current to change direction. Remember that the current flowing to a motor must switch directions periodically so the electromagnetic coil can keep turning. In a generator, as the electromagnetic coil continuously turns, the current that is produced periodically changes direction, as **Figure 17** shows. The direction of the current in the coil changes twice with each revolution. The frequency with which the current changes direction can be controlled by regulating the rotation rate of the generator. In the United States, current is produced by generators that rotate 60 times a second, or 3,600 revolutions per minute.

Using Electric Generators The type of generator shown in **Figure 17** is used in a car, where it is called an alternator. The alternator provides electrical energy to operate lights and other accessories. Spark plugs in the car's engine also use this electricity to ignite the fuel in the cylinders of the engine. Once the engine is running, it provides the mechanical energy that is used to turn the coil in the alternator.

Suppose instead of using mechanical energy to rotate the coil in a generator, the coil was fixed, and the permanent magnet rotated instead. In fact the current generated would be the same as when the coil rotates and the magnet doesn't move. The huge generators used in electric power plants are made this way. The current is produced in the stationary coil, and mechanical energy is used to rotate the magnet.

Figure 18
Electric power plants use huge generators such as the ones shown here to produce the electric current you use every day.

Environmental Science INTEGRATION

A dam can be placed on a river to create a lake. Water can then be released slowly from the dam to turn the turbines in a generator. Discuss with your classmates what effects a dam would have on the environment.

Generating Electricity for Your Home You probably do not have a generator in your home that supplies all the electricity you need to watch television or wash your clothes. Your electricity comes from a power plant with huge generators like the one in **Figure 18.** The electromagnets in these generators are made of many coils of wire wrapped around huge iron cores. The rotating magnets are connected to a **turbine** (TUR bine)—a large wheel that rotates when pushed by water, wind, or steam.

For example, some power plants first produce thermal energy by burning fossil fuels or using the heat produced by nuclear reactions. This thermal energy is used to heat water and produce steam. Thermal energy is then converted to mechanical energy as the steam pushes the turbine blades. The generator then changes the mechanical energy of the rotating turbine into an electric current that flows to your home. In some areas, fields of windmills like those in **Figure 19** can be used to capture the mechanical energy in wind to turn generators. Other power plants use the mechanical energy in falling water to drive the turbine. Look at **Figure 20** to compare and contrast the characteristics of generators and motors.

Figure 19
These windmills harness the energy in wind so it can be transformed into electrical energy by a generator. *What are some advantages and disadvantages of using windmills?*

NATIONAL GEOGRAPHIC VISUALIZING MOTORS AND GENERATORS

Figure 20

Electric motors power many everyday machines, from CD players to vacuum cleaners. Generators produce the electricity those motors need to run. Both motors and generators use electromagnets, but in different ways. The table below compares motors and generators.

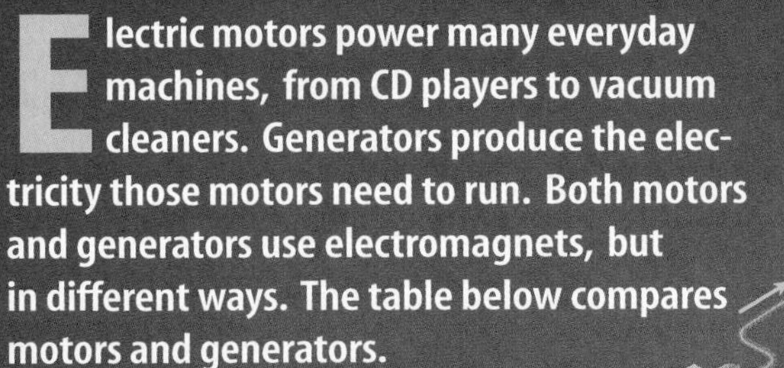

	Electric Motor	Generator
What does it do?	Changes electricity into movement	Changes movement into electricity 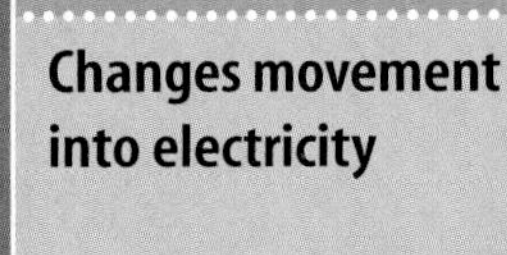
What makes its electromagnetic coil rotate?	Attractive and repulsive forces between the coil and the permanent magnet 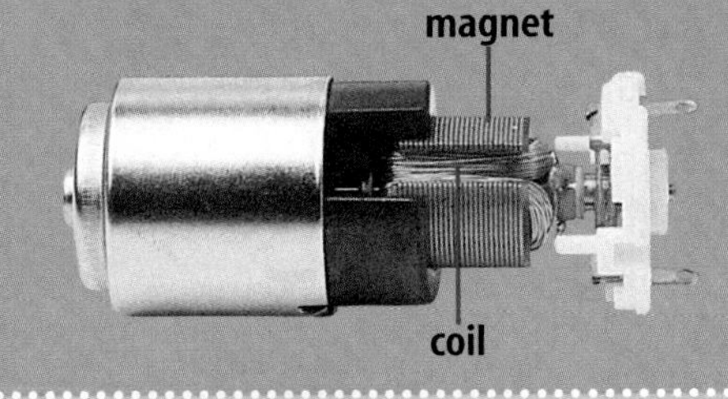	An outside source of mechanical energy 
What is the source of the current that flows in its coil?	An outside power source 	Electromagnetic induction from moving the coil through the field of the permanent magnet 
How often does the current in the coil change direction?	Twice during each rotation of the coil	Twice during each rotation of the coil

Research Visit the Glencoe Science Web site at **science.glencoe.com** for more information about how transformers are used in transmitting electric current. Communicate to your class what you learned.

Direct and Alternating Currents

Modern society relies heavily on electricity. Just how much you rely on electricity becomes obvious during a power outage. Out of habit you might walk into a room and flip on the light switch. You might try to turn on a radio or television or check the clock to see what time it is. Because power outages occur, some electrical devices, like the one in **Figure 21,** use batteries as a backup source of electrical energy. Is the current produced by a battery the same as the current from a generator? Both devices cause electrons to move through a wire. However, the currents produced by these electric sources are different from each other in an important way.

A battery produces a direct current. **Direct current** (DC) flows in only one direction through a wire. When you plug your CD player or any other appliance into a wall outlet, you are using alternating current. **Alternating current** (AC) reverses the direction of the current flow in a regular way. In North America, generators produce alternating current at a frequency of 60 cycles per second, or 60 Hz. The electric current produced by a generator changes direction twice during each cycle or each rotation of the coil. So a 60-Hz alternating current changes direction 120 times each second.

Transformers

The current that flows in an electric circuit carries electrical energy. This electrical energy is related to the voltage in the circuit. The alternating current traveling through power lines is at an extremely high voltage. Before alternating current from the power plant can enter your home safely, its voltage must be decreased. The voltage is decreased by passing the current through a transformer. A **transformer** is a device that increases or decreases the voltage of an alternating current.

A transformer is made of two coils of wire called the primary and secondary coils. These coils are wrapped around the same iron core. As an alternating current passes through the primary coil, the iron core becomes an electromagnet. The current changes direction many times each second, so the magnetic field of the iron core also changes direction. This changing magnetic field induces an alternating current in the secondary coil.

Figure 21
Some devices can use either direct or alternating current.
Why might it be a good idea to keep batteries in a clock or a VCR?

Stepping Up and Stepping Down If the secondary coil in a transformer has more turns of wire than the primary coil does, then the transformer increases, or steps up, voltage. For example, the secondary coil of the step-up transformer in **Figure 22A** has two times more turns than the primary coil has. This means than an input voltage in the primary coil of 60 V would increase by two times to 120 V in the secondary coil.

A transformer that reduces voltage is called a step-down transformer. **Figure 22B** shows how the output voltage of a transformer is decreased if the number of turns in the secondary coil is less than the number of turns in the primary coil. If the secondary coil of a transformer has half as many turns as the primary coil does, the output voltage will be half the input voltage.

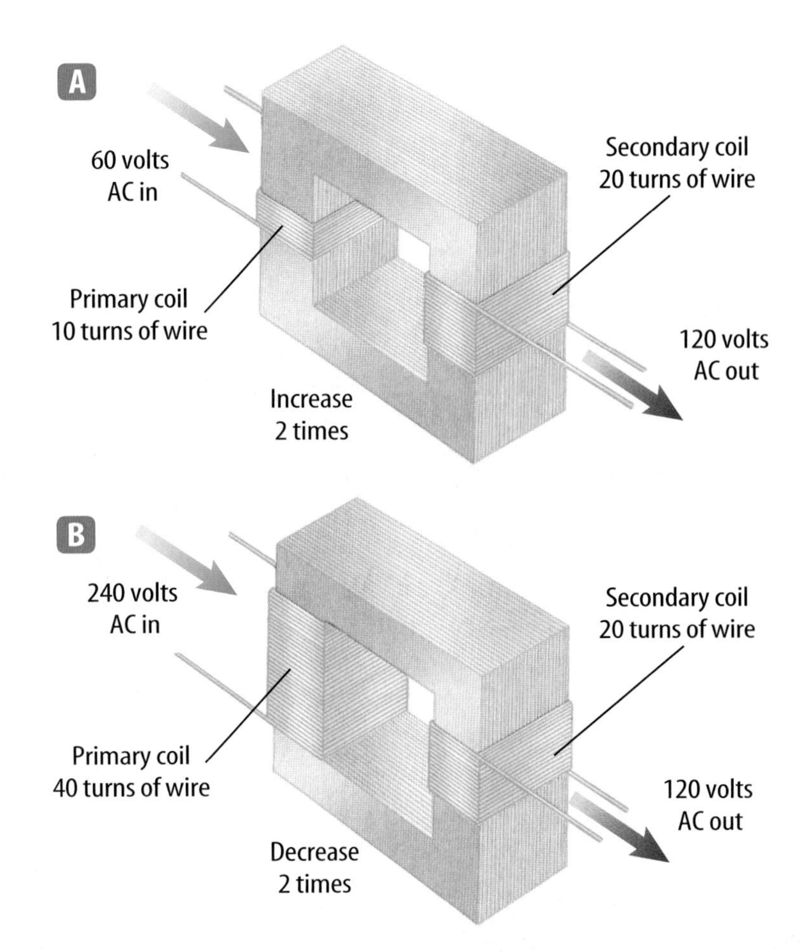

Figure 22
Transformers can increase or decrease voltage.
A A step-up transformer increases voltage. The secondary coil has more turns than the primary coil does.
B A step-down transformer decreases voltage. The secondary coil has fewer turns than the primary coil does.

✔ Reading Check ***What type of transformer has more turns of wire in the secondary coil than the primary coil?***

Transmitting Alternating Current When an electric current flows in a wire, some of the energy carried by the current is lost as heat. This heat loss is due to the resistance of the wire and increases as the wire is made longer. If the current produced by a power plant is transmitted over long distances, as much as ten percent of the electrical energy can be lost as heat. This energy loss can be reduced greatly by transmitting the power at high voltages. Power plants commonly produce alternating current because the voltage can be increased or decreased with transformers. In the United States, some power lines carry power at voltages as high as 750,000 V—high voltage indeed although most power lines you see carry lower voltages.

Such high voltage is dangerous and cannot be used in home appliances. Step-down transformers reduce the voltage of the alternating current to 120 V before it enters your home. You can then operate devices such as microwaves and hair dryers with 120-V household current.

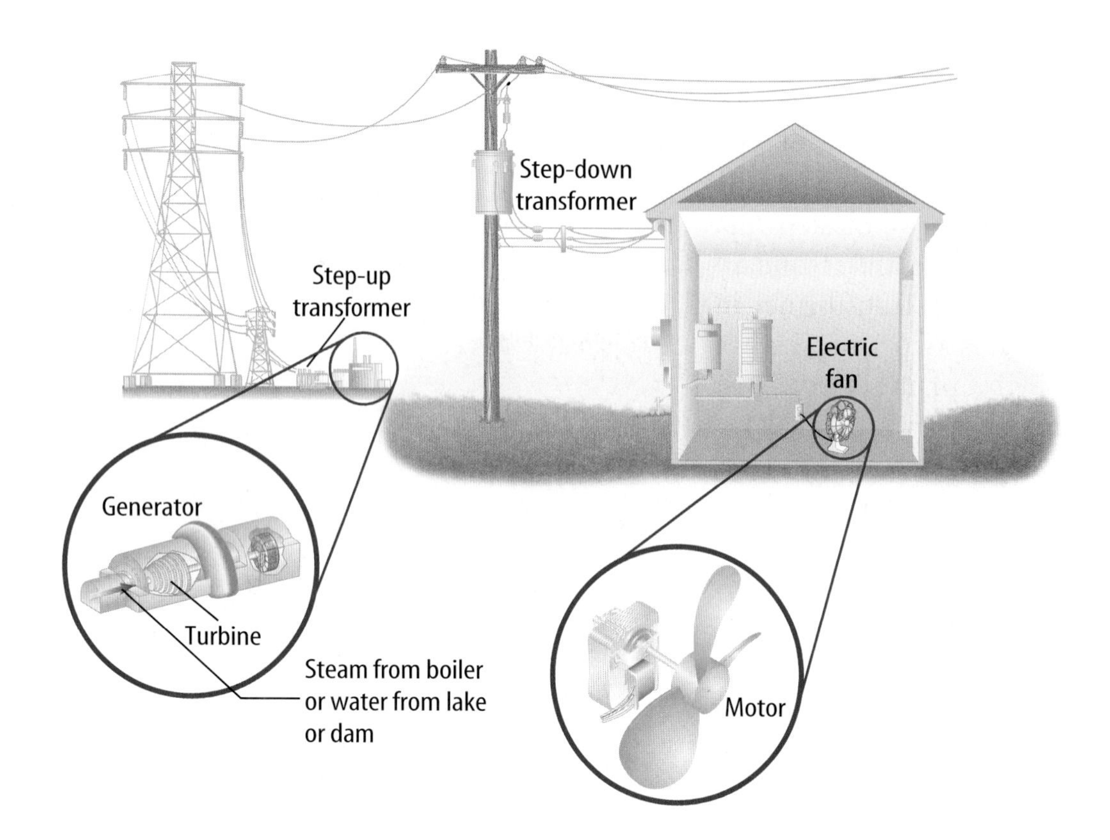

Figure 23
Many steps are involved in the creation, transportation, and use of the electric current in your home. *Which steps involve electromagnetic induction?*

Electric current in your home Think back over this section. You have learned how electromagnetic induction, generators, alternating current, and transformers all work together to make your electric fan operate. **Figure 23** illustrates the series of steps used in producing, transporting, and delivering alternating current to your home in a form that you can use safely.

Section  Assessment

1. How does a generator use electromagnetic induction to produce a current?
2. A transformer contains 20 turns in the primary coil and 80 turns in the secondary coil. Which is greater—the output voltage or the input voltage?
3. Contrast alternating and direct current.
4. What type of current do power plants produce? Why is it convenient?
5. **Think Critically** Why can't a transformer step up the voltage in a direct current?

Skill Builder Activities

6. **Concept Mapping** Prepare an events chain concept map to show how electricity is produced by a generator. **For more help, refer to the Science Skill Handbook.**
7. **Using Proportions** An alternating current in a wire has a voltage of 2,800 V. It needs to be reduced to 70 V. The wire makes 120 turns in the primary coil of a step-down transformer. How many turns of wire need to be in the secondary coil? **For more help, refer to the Math Skill Handbook.**

Activity

Electricity and Magnetism

Huge generators in power plants produce electricity by moving magnets past coils of wire. How can you use a magnet to make your own electric current?

What You'll Investigate

How can a magnet be used to create an electric current?

Materials

cardboard tube
scissors
bar magnet
thin, flexible, insulated wire
galvanometer or ammeter

Goals

- **Observe** how a magnet can produce an electric current in a wire.
- **Compare and contrast** the currents created by moving the magnet in different ways.

Safety Precautions

Be careful with scissors. Do not touch bare wires when current is running through them.

Procedure

1. Wrap the wire around a cardboard tube to make a coil of about 20 turns. Leave about 15 cm for a lead at each end of the wire.
2. Use the scissors to cut through the insulation 2 cm from each end of the wire. Pull the insulation off with your fingers. Remove the tube from the coil.
3. Connect the ends of the wire to a galvanometer or ammeter. Record the reading on your meter.
4. While closely watching the meter, insert one end of the bar magnet into the coil.
5. Pull the magnet out of the coil and repeat. Record the reading on the meter. Move the magnet at different speeds and record your measurements.
6. Watch the meter and move the bar magnet in different directions around the outside of the coil. Record your observations.

Conclude and Apply

1. Which circumstances that you tested generated the greatest current?
2. Does the current generated by moving the magnet always flow in the same direction? How do you know?
3. **Predict** what would happen if you tried the experiment with a coil made with fewer turns of wire.
4. **Infer** whether a current would have been generated if the cardboard tube were left in the coil. Why or why not? Try it.

Communicating Your Data

Compare the currents generated by different members of the class. **For more help, refer to the Science Skill Handbook.**

Activity *Design Your Own Experiment*

Putting Electromagnets to Work

You have learned that a current flowing through loops of wire around an iron core forms an electromagnet. You use electromagnets every day in electric motors, stereo speakers, power door locks and many other devices. To make electromagnets work in these devices, you must be able to control the strength of their magnetic fields. When might you want to make a magnet stronger? When would you want to make it weaker?

Recognize the Problem

How can you control the strength of an electromagnet?

Form a Hypothesis

Think about how an electromagnet is constructed. As a group, write down the components of an electromagnet which might affect the strength of its magnetic field. Which component could have the most effect on the strength of the electromagnet? Which could be easiest to control? Form a hypothesis about the best way to control an electromagnet's strength.

Goals

- **Make** electromagnets.
- **Measure** relative strengths of electromagnets.
- **Modify** electromagnets to change their strength.
- **Determine** which factors affect the strength of an electromagnet.
- **Determine** which factor has the most effect on its strength.
- **Describe** how you could control the strength of an electromagnet.

Possible Materials

22-gauge insulated wire
16-penny iron nail
Aluminum rod or nail
0-6 v DC power supply
Three 1.5 -V "D" cells
Steel paper clips
Magnetic compass
Duct tape (to hold "D" cells together)

Safety Precautions

Do not leave the electromagnet connected for a long time because the battery will run down. Magnets with only a few turns of wire will get hot. Use caution in handling them when current is flowing through the coil. Do not apply voltages higher than 6 V to your electromagnets.

Test Your Hypothesis

Plan

1. Write your hypothesis for the best way to control the strength of an electromagnet.
2. As a group, decide how you will assemble and test the electromagnets. Which features will you change to determine effect on the strength of the magnetic fields? How many changes will you need to try? How many electromagnets do you need to build?
3. Decide how you are going to test the strength of your electromagnets. Several ways are possible with the materials listed. Which way would be the most sensitive? Be prepared to change test methods if necessary.
4. Write your plan of investigation. Make sure your plan tests only one variable at a time.

Do

1. Before you begin to build and test the electromagnets, make sure your teacher approves of your plan.
2. Carry out your planned investigation. Record your results.

Analyze Your Data

1. **Make a table** showing how the strength of your electromagnet depends on changes you made in its construction or operation.
2. **Examine** the trends shown by your data. Are there any data points which seem out of line? How can you account for them?

Draw Conclusions

1. How did the strength of the electromagnet depend on its construction or operation?
2. Which feature of the electromagnet's construction had the greatest effect on its strength? Which do you think would be easiest to control?
3. How might you use your electromagnet to make a doorbell? Would it work with both AC and DC?
4. Did your results support your hypothesis? Why or why not?

Compare your group's results with those of other groups. Did any other group use a different method to test the strength of the magnet? Did you get the same results?

Body Art

The invention of a machine that uses magnetism means better lives for many

The year is 1975. A surgeon stands facing an exposed human brain. She has already removed part of the patient's skull and is looking for a growth on the brain. From the patient's symptoms, the surgeon can only infer where to find the tumor. But can she find it and remove it without causing more damage than the tumor was causing? "We're going in," the doctor says. She puts out her hand to the nurse. "Scalpel."

Flash forward to the present day.

The surgeon turns the computer screen so the patient can see it. Pointing to a dark area on a colorful image of the patient's brain, she reassures the worried patient. "This MRI shows exactly where your tumor is. We can remove it with very little danger to you." "Thank goodness for the MRI," the patient says in relief.

MRI for the Soft Stuff

MRI stands for "magnetic resonance imaging." It's a way to take 3-D pictures of the inside of your body. Before the 1980s, doctors could x-ray solid tissue like bones, but had no way to see soft tissue like the brain. Well, they had one way—surgery, which sometimes caused injury and infection, risking a patient's health.

MRIs were originally used to identify substances in chemistry and physics labs.

These are MRI scans of brains from different people. The colors help doctors read the scans more easily, and detect any problems.

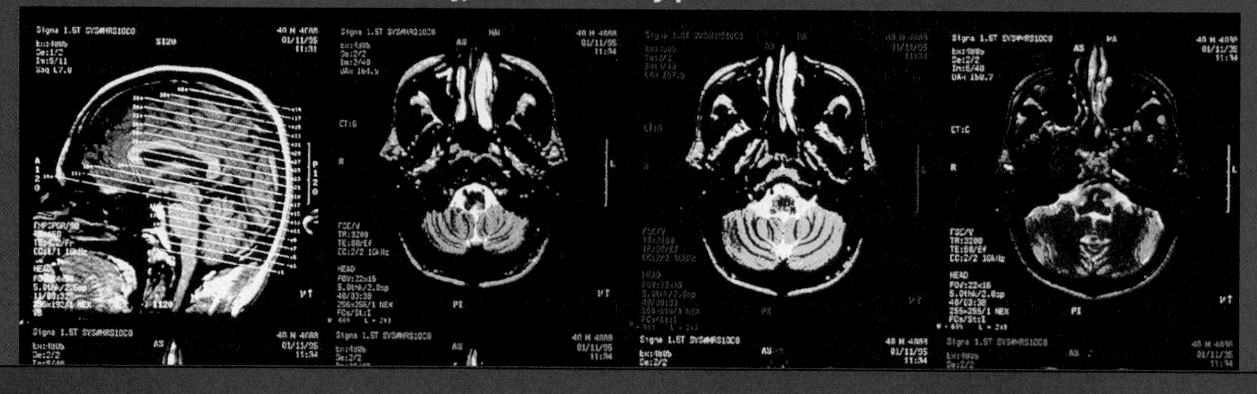

Then after research and modifications in the method, doctors began to use MRIs to make images of the tissues inside the human body.

MRI uses a strong magnet and radio waves. Tissues in your body contain water molecules that are made of oxygen and hydrogen atoms. The nucleus of a hydrogen atom is a proton, which behaves like a tiny magnet. A strong magnetic field inside the MRI tube makes these proton magnets line up in the direction of the field. Radio waves are then applied to the body. The protons absorb some of the radio-wave energy, and flip their direction.

When the radio waves are turned off, the protons realign themselves with the magnetic field and emit the energy they absorbed. Different tissues in the body absorb and emit different amounts of energy. The emitted energy is detected, and a computer uses this information to form images of the body.

A girl gets ready for an MRI scan. It doesn't hurt!

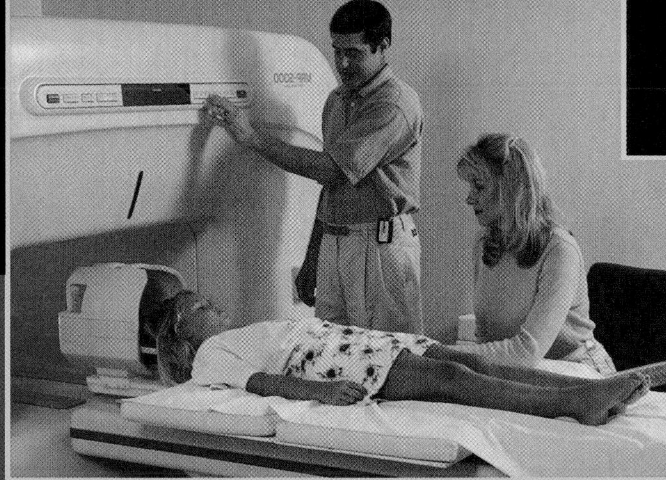

Your Brain Is Getting Bigger!

MRI has been most useful in finding and treating tumors. But it has also turned into an important research tool. For example, Elizabeth Sowell's research team at the University of California, Los Angeles, used MRIs to study the brain growth of middle school students. She and other researchers have found that the brain grows rapidly during adolescence. Before this groundbreaking research, people thought that the brain stopped growing in childhood. MRI has proved that adolescents are getting bigger brains all the time.

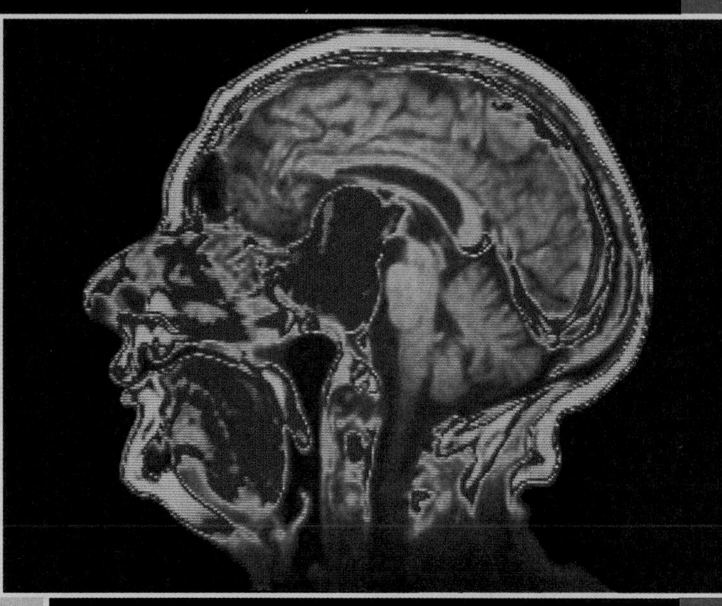

An MRI scan of a brain shows a tumor on the pituitary gland. The gland is the large pink area in the middle of the photo.

CONNECTIONS Interview **As an oral history project, interview a retired physician or surgeon. Ask him or her to discuss with you how tools such as the MRI changed during his or her career. Make a list of the tools and how they have helped improve medicine.**

SCIENCE *Online*
For more information, visit science.glencoe.com

Chapter 8 Study Guide

Reviewing Main Ideas

Section 1 Magnetism

1. A magnetic field surrounds a magnet and exerts a magnetic force. *Why is this magnet attracted to the refrigerator but not to the cupboard?*
2. All magnets have two poles: a south pole and a north pole.
3. Opposite poles of magnets attract; like poles repel.
4. Groups of atoms with aligned magnetic poles are called magnetic domains.

Section 2 Electricity and Magnetism

1. An electric current flowing through a wire produces a magnetic field.
2. An electric current passing through a coil of wire can produce a magnetic field inside the coil. The coil becomes an electromagnet. One end of the coil is the north pole, and the other end is the south pole.
3. The magnetic field produced by an electromagnet depends on the current and the number of coils. *How is an electromagnet used in this temperature gauge from a car?*

4. An electric motor contains a rotating electromagnet that converts electrical energy to mechanical energy.

Section 3 Producing Electric Current

1. By moving a magnet near a wire, you can create an electric current in the wire. This is called electromagnetic induction.
2. A generator produces electric current by rotating a coil of wire in a magnetic field. *How does the dam in the picture below help make electricity?*

3. Direct current flows in one direction through a wire. Alternating current reverses the direction of current flow in a regular way.
4. The number of turns of wire in the primary and secondary coils of a transformer determines whether it increases or decreases voltage.

After You Read

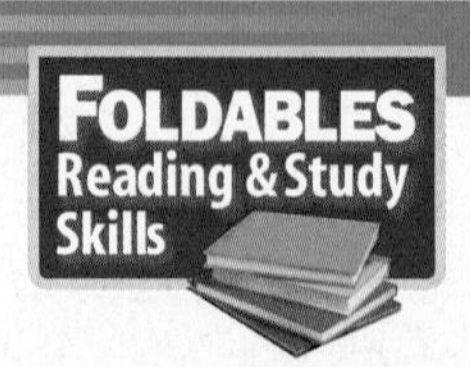

To help you review what you've learned about magnets, use the Question Study Fold you made at the beginning of this chapter.

Visualizing Main Ideas

Complete the following concept map on how an electric motor works.

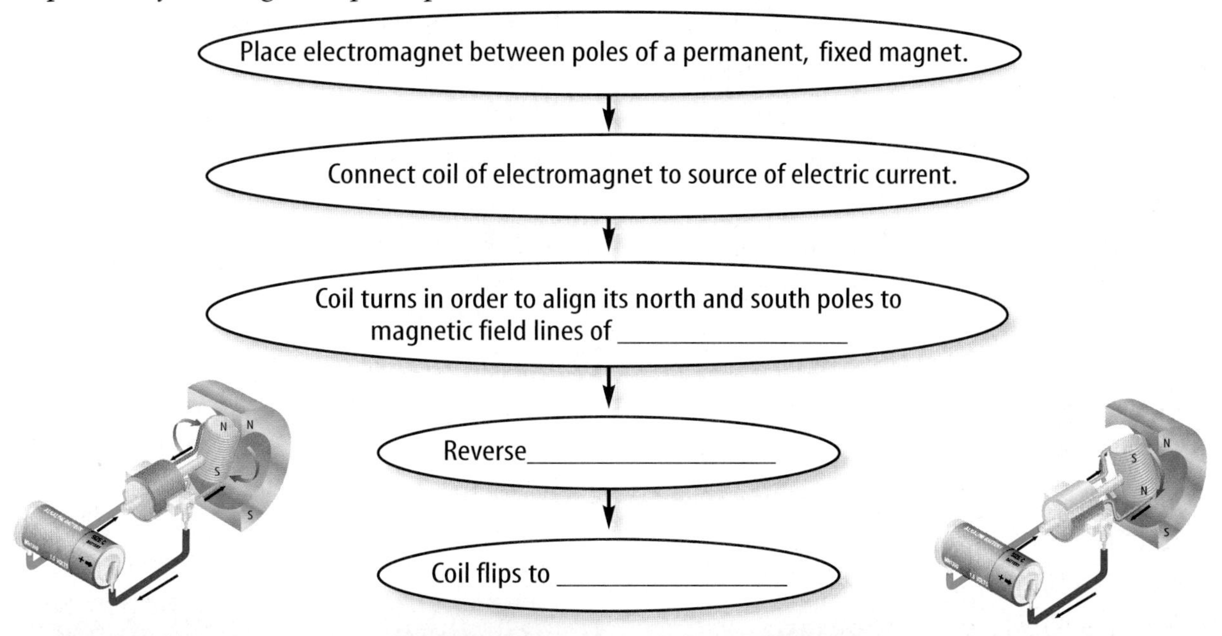

Vocabulary Review

Vocabulary Words

a. alternating current (AC)
b. direct current (DC)
c. electric motor
d. electromagnet
e. electromagnetic induction
f. galvanometer
g. generator
h. magnetic domain
i. magnetic pole
j. magnetism
k. transformer
l. turbine

Study Tip

Study the material you *don't* understand as well first! It's easy to review the material you know, but harder to force yourself to really go over the tough stuff.

Using Vocabulary

Each of the following sentences is false. Make the sentence true by replacing the underlined word with a vocabulary word.

1. An <u>electric motor</u> can be used to change the voltage of an alternating current.
2. <u>Flat current</u> does not change direction.
3. A <u>magnetic domain</u> is the region where the magnetic force of a magnet is strongest.
4. Current flows in the coil of a <u>galvanometer</u> because of electromagnetic induction.
5. The properties and interactions of magnets are called <u>electricity</u>.
6. A <u>transformer</u> can rotate in a magnetic field when a current passes through it.
7. A generator uses <u>alternating current</u> to produce an electric current.

Chapter 8 Assessment

Checking Concepts

Choose the word or phrase that best answers the question.

1. Where is the magnetic force exerted by a magnet strongest?
A) both poles **C)** north pole
B) south pole **D)** center

2. What happens to the magnetic force as the distance between two magnetic poles decreases?
A) stays constant **C)** increases
B) decreases sharply **D)** decreases slightly

3. What type of magnetic poles do the domains at the north pole of a magnet have?
A) north magnetic poles only
B) south magnetic poles only
C) no magnetic poles
D) north and south magnetic poles

4. Which of the following would not change the strength of an electromagnet?
A) increasing the amount of current
B) changing the current's direction
C) inserting an iron core inside the coil
D) increasing the number of loops

5. Which of the following would NOT be part of a generator?
A) turbine **C)** electromagnet
B) battery **D)** permanent magnet

6. Which conversion does an electric motor make?
A) electrical energy to mechanical energy
B) thermal energy to wind energy
C) mechanical energy to electrical energy
D) wind energy to electrical energy

7. Which of the following describes the direction of the electric current in AC?
A) remains constant **C)** changes regularly
B) is direct **D)** changes irregularly

8. Before current in power lines can enter your home, what must it pass through?
A) step-up transformer
B) step-down transformer
C) commutator
D) motor

9. A generator creates a 40-Hz alternating current. How many times does the current change direction every second?
A) 40 times **C)** 80 times
B) 60 times **D)** 20 times

10. When current flows through a wire, what is created around the wire?
A) an electromagnet **C)** a magnetic field
B) a galvanometer **D)** a direct current

Thinking Critically

11. How could you use a horseshoe magnet to find the direction north?

12. In Europe, generators produce alternating current at a frequency of 50 Hz. How is the frequency of this current changed by a step-down transformer? How is it changed by a step-up transformer?

13. Audiotapes, computer disks, and videotapes are recorded using magnets, and their information is coded magnetically. Why would it be harmful to a tape or computer disk, to expose it to a strong magnetic field?

14. A step-down transformer reduces a 1,200-V current to 120 V. If the primary coil has 100 turns, how many must its secondary coil have?

15. Suppose the magnetic fields of magnets were not strongest at the magnets' poles. Would motors, galvanometers, and generators still work? Why or why not?

16. **Comparing and Contrasting** Compare and contrast electric and magnetic forces.
17. **Forming Hypotheses** A compass needle will point north due to the magnetic field of Earth. When a bar magnet is brought near the compass, the needle is attracted or repelled by the bar magnet. Propose a hypothesis about the relative strengths of a bar magnet and Earth's magnetic field.
18. **Interpreting Scientific Illustrations** Review **Figure 14** and describe the function of each labeled part of the motor.
19. **Comparing and Contrasting** Compare and contrast electromagnetic induction and the formation of electromagnets.
20. **Concept Mapping** Complete the following Venn diagram of AC generators and DC motors.

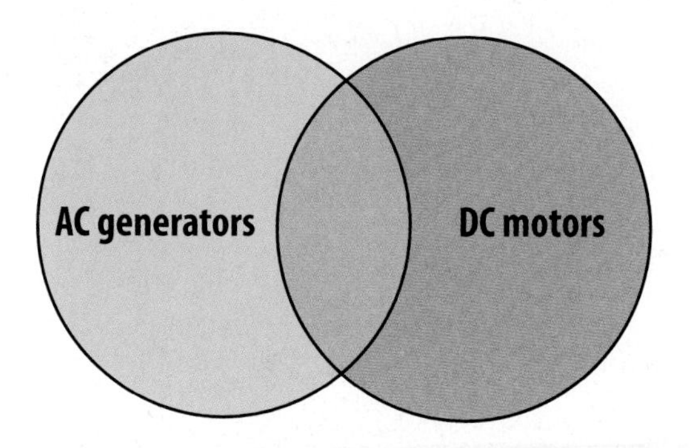

Performance Assessment

21. **Invention** Invent a device that uses an electric motor. Describe it to your class.

Technology

Go to the Glencoe Science Web site at **science.glencoe.com** or use the **Glencoe Science CD-ROM** for additional chapter assessment.

Test Practice

The diagram below is taken from the instructions that come with a blank videocassette.

Proper Storage

Avoid exposing cassettes to:

Direct sunlight and heat

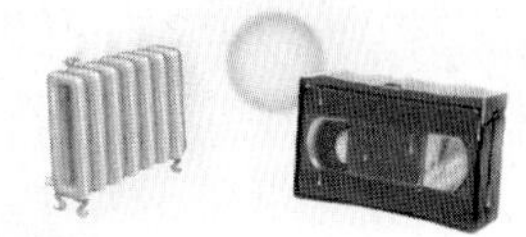

Dust

Strong magnetic fields

Humidity

Study the diagram and answer the following questions.

1. Why do videocassettes come with a warning to avoid strong magnetic fields?
   A) The videocassette will be turned into a powerful magnet.
   B) The information on the videocassette will be damaged.
   C) The videocassette's case will become electrically charged.
   D) The videocassette will melt.
2. According to this diagram, all of the following could be harmful to a videocassette **EXCEPT**______.
   F) air conditioning
   G) moisture
   H) direct sunlight and heat
   J) dust

CHAPTER 9

Radioactivity and Nuclear Reactions

The Sun gives off tremendous amounts of energy from day to day, year to year. Almost all of the Sun's energy comes from nuclear reactions in which the nuclei of atoms are fused together. In this chapter, you will learn about unstable nuclei and how they emit different types of radiation. You will also learn how this radiation can be used to determine the age of objects, produce energy, or treat diseases.

What do you think?

Science Journal Look at the picture below with a classmate. Discuss what you think this is or what is happening. Here's a hint: *There's probably one of these on the ceiling.* Write your answer or best guess in your Science Journal.

Do you realize you are made up mostly of empty space? Your body is made of atoms, and atoms are made of electrons whizzing around a small nucleus of protons and neutrons. The size of this region of space in which the electrons are moving is the same as the size of the atom. An atom is much larger than its nucleus. During this activity you will find out just how small a nucleus is.

Model the space inside an atom

1. Go outside and pour several grains of sugar onto a sheet of paper.
2. Choose a tiny grain of sugar with a diameter equal to the width of one of the lines on a ruler. This sugar grain represents the nucleus of an atom.
3. Brush the rest of the sugar off the paper and place the sugar grain in the center of the paper.

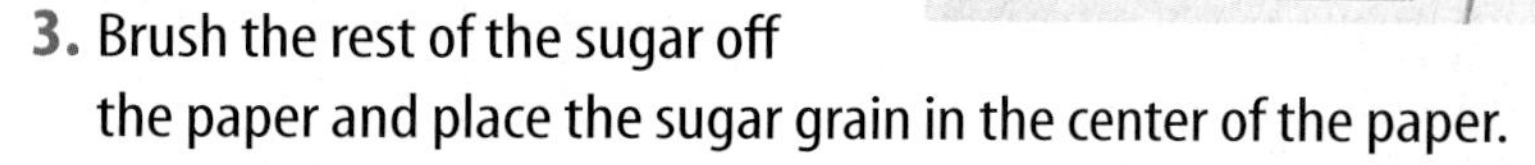

4. Use a meterstick to measure a 10 m distance away from the sugar grain.

Observe

In your Science Journal, explain why an atom contains mostly empty space. Use the fact that an electron is much smaller than the nucleus of an atom.

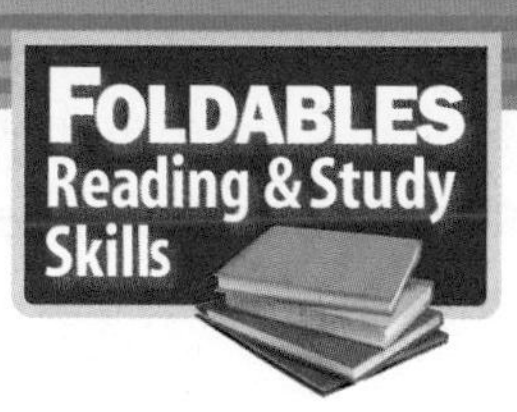

Before You Read

Making a Main Ideas Study Fold Make the following Foldable to help you identify the major topics about radioactivity and nuclear reactions.

1. Place a sheet of paper in front of you so the short side is at the top. Fold the paper in half from top to bottom and then unfold.
2. Fold in to the centerfold line to divide the paper into fourths.
3. Label the flaps *Radioactivity* and *Nuclear Reactions.*
4. As you read the chapter, write what you learn about radioactivity and nuclear reactions under the flaps.

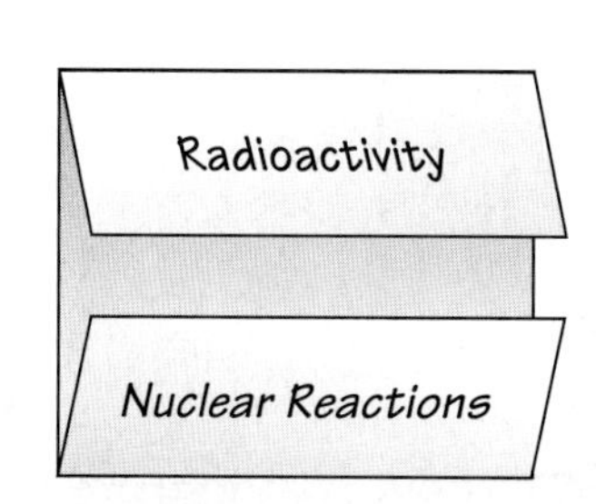

SECTION 1

Radioactivity

As You Read

***What* You'll Learn**

- **Describe** the structure of an atom and its nucleus.
- **Explain** what radioactivity is.
- **Contrast** properties of radioactive and stable nuclei.
- **Discuss** the discovery of radioactivity.

Vocabulary
strong force
radioactivity

***Why* It's Important**
The characteristics of atomic nuclei determine whether or not they will undergo radioactive decay.

The Nucleus

Every second you are being bombarded by energetic particles. Some of these particles come from unstable atoms in soil, rocks, and the atmosphere. What types of atoms are unstable? What type of particles do unstable atoms emit? The answers to these questions begin with the nucleus of an atom.

You remember that atoms are composed of protons, neutrons, and electrons. The nucleus of an atom contains the protons, which have a positive charge, and neutrons, which have no electric charge. The total amount of charge in a nucleus is determined by the number of protons, which also is called the atomic number. You might remember that an electron has a charge that is equal but opposite to a proton's charge. Atoms usually contain the same number of protons as electrons. Negatively charged electrons are electrically attracted to the positively charged nucleus and swarm around it.

Protons and Neutrons in the Nucleus Protons and neutrons are packed together tightly in a nucleus. The region outside the nucleus in which the electrons are located is large compared to the size of the nucleus. As **Figure 1** shows, the nucleus occupies only a tiny fraction of the space in the atom. If an atom were enlarged so that it was 1 km in diameter, its nucleus would have a diameter of only a few centimeters. But the nucleus contains almost all the mass of the atom, because the mass of one proton or neutron is almost 2,000 times greater than the mass of an electron.

Figure 1
The size of a nucleus in an atom can be compared to a marble sitting in the middle of an empty football stadium.

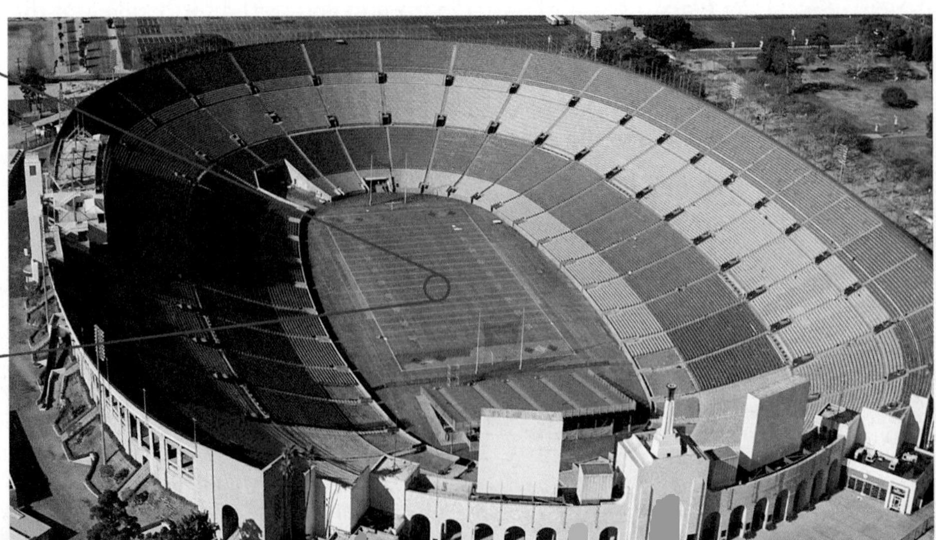

Figure 2
The particles in the nucleus are attracted to each other by the strong force.

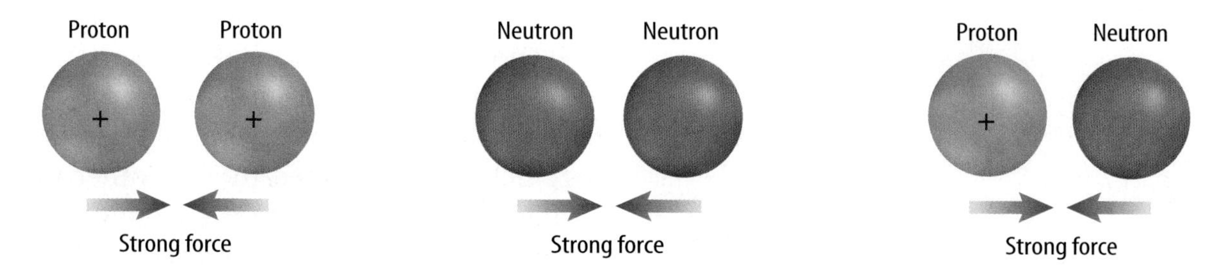

The Strong Force

How do you suppose protons and neutrons are held together so tightly in the nucleus? Positive electric charges repel each other, so why don't the protons in a nucleus push each other away? Another force, called the **strong force,** causes protons and neutrons to be attracted to each other, as shown in **Figure 2.**

The strong force is one of the four basic forces and is about 100 times stronger than the electric force. The attractive forces between all the protons and neutrons in a nucleus keep the nucleus together. However, protons and neutrons have to be close together, like they are in the nucleus, to be attracted by the strong force. The strong force is a short-range force that quickly becomes extremely weak as protons and neutrons get farther apart. The electric force is a long-range force, so protons that are far apart still are repelled by the electric force, as shown in **Figure 3.**

✔ Reading Check ***What causes the attraction between protons and neutrons?***

Figure 3
The total force between two protons depends on how far apart they are.

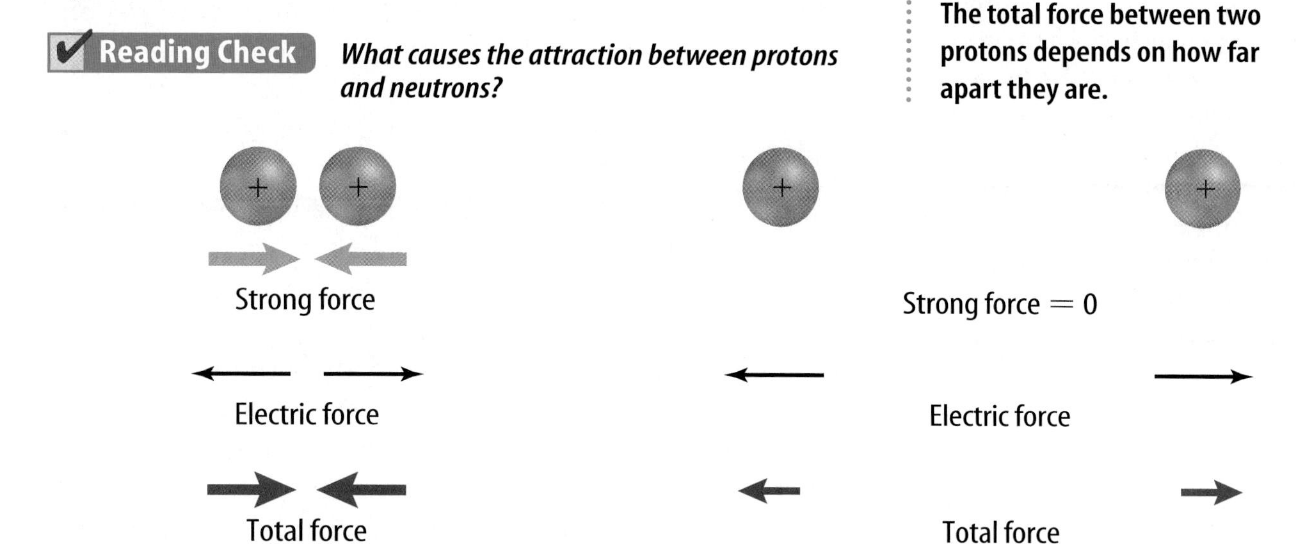

A **When protons are close together, they are attracted to each other. The attraction due to the short-range strong force is much stronger than the repulsion due to the long-range electric force.**

B **When protons are too far apart to be attracted by the strong force, they still are repelled by the electric force between them. Then the total force between them is repulsive.**

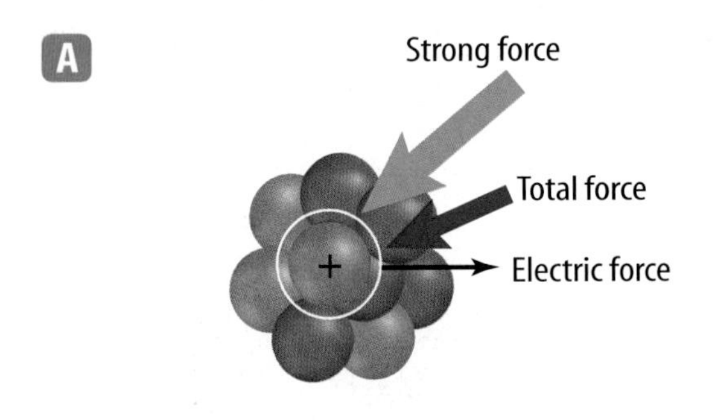

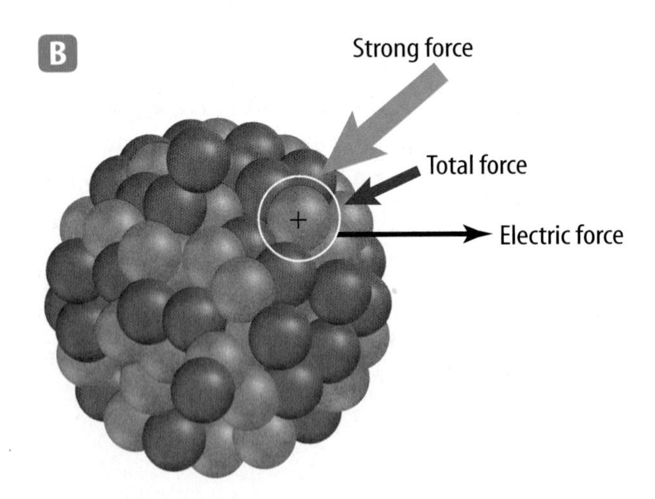

Attraction and Repulsion Some atoms, such as uranium, have many protons and neutrons in their nuclei. These nuclei are held together less tightly than nuclei containing only a few protons and neutrons. To understand this, look at **Figure 4A.** If a nucleus has only a few protons and neutrons, they are all close enough together to be attracted to each other by the strong force. Because only a few protons are in the nucleus, the total electric force causing protons to repel each other is small. As a result, the overall force between the protons and the neutrons attracts the particles to each other.

Forces in a Large Nucleus However, if nuclei have many protons and neutrons, each proton or neutron is attracted to only a few neighbors by the strong force, as shown in **Figure 4B.** The other protons and neutrons are too far away. Because only the closest protons and neutrons attract each other in a large nucleus, the strong force holding them together is about the same as in a small nucleus. However, all the protons in a large nucleus exert a repulsive electric force on each other. Thus, the electric repulsive force on a proton in a large nucleus is larger than it would be in a small nucleus. Because the repulsive force increases in a large nucleus while the attractive force on each proton or neutron remains about the same, protons and neutrons are held together less tightly in a large nucleus.

Figure 4
Protons and neutrons are held together less tightly in large nuclei. The circle shows the range of the attractive strong force. A Small nuclei have few protons, so the repulsive force on a proton due to the other protons is small. B In large nuclei, the attractive strong force is exerted only by the nearest neighbors, but all the protons exert repulsive forces. The total repulsive force is large.

Radioactivity

In many nuclei the strong force is able to keep the nucleus permanently together, and the nucleus is stable. When the strong force is not large enough to hold a nucleus together tightly, the nucleus can decay and give off matter and energy. This process of nuclear decay is called **radioactivity.**

Large nuclei tend to be unstable and can break apart or decay. In fact, all nuclei that contain more than 83 protons are radioactive. However, many other nuclei that contain fewer than 83 protons also are radioactive. Even some nuclei with only a few protons are radioactive.

Almost all elements with more than 92 protons don't exist naturally on Earth. They have been produced only in laboratories and are called synthetic elements. These synthetic elements are unstable, and decay soon after they are created.

Stable and Unstable Nuclei The atoms of an element all have the same number of protons in the nucleus. For example, the nucleus of all carbon atoms contains six protons. However, not all naturally occurring carbon nuclei have the same numbers of neutrons. Some carbon nuclei have six neutrons, some have seven, and some have eight neutrons. Nuclei that have the same number of protons but different numbers of neutrons are called isotopes. The element carbon has three isotopes that occur naturally. The atoms of all isotopes of an element have the same number of electrons, and have the same chemical properties.

Isotopes of elements differ in the ratio of neutrons to protons, as shown in **Figure 5.** This ratio is related to the stability of the nucleus. In less massive elements, an isotope is stable if the ratio is about 1 to 1. Isotopes of the heavier elements are stable when the ratio of neutrons to protons is about 3 to 2. However, the nuclei of any isotopes that differ much from these ratios are unstable, whether the elements are light or heavy. In other words, nuclei with too many or too few neutrons compared to the number of protons are radioactive. Radioactive isotopes are sometimes called radioisotopes.

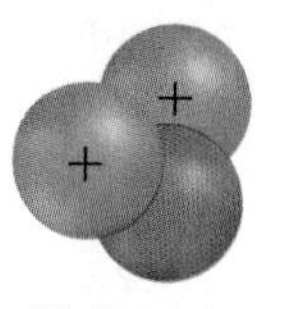

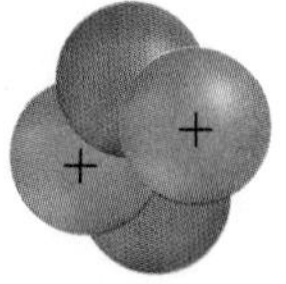

Figure 5
These two different isotopes of helium each have the same number of protons, even though they have different numbers of neutrons. *What is the ratio of protons to neutrons in each of these isotopes of helium?*

Nucleus Numbers A nucleus can be described by the number of protons and neutrons it contains. The number of protons in a nucleus is called the atomic number. Because the mass of all the protons and neutrons in a nucleus is nearly the same as the mass of the atom, the number of protons and neutrons is called the mass number.

A nucleus can be represented by a symbol that includes its atomic number, mass number, and the symbol of the element it belongs to, as shown in **Figure 6.** The symbol for the nucleus of the stable isotope of carbon is shown as an example.

$$\text{mass number} \rightarrow {}^{12}_{\;6}\text{C} \leftarrow \text{element symbol}$$
$$\text{atomic number} \rightarrow 6$$

This isotope is called carbon-12. The number of neutrons in the nucleus is the mass number minus the atomic number. So the number of neutrons in the carbon-12 nucleus is $12 - 6 = 6$. Carbon-12 has six protons and six neutrons. Now, compare the isotope carbon-12 to this radioactive isotope of carbon:

$$\text{mass number} \rightarrow {}^{14}_{\;6}\text{C} \leftarrow \text{element symbol}$$
$$\text{atomic number} \rightarrow 6$$

The radioactive isotope is carbon-14. How many neutrons does carbon-14 have?

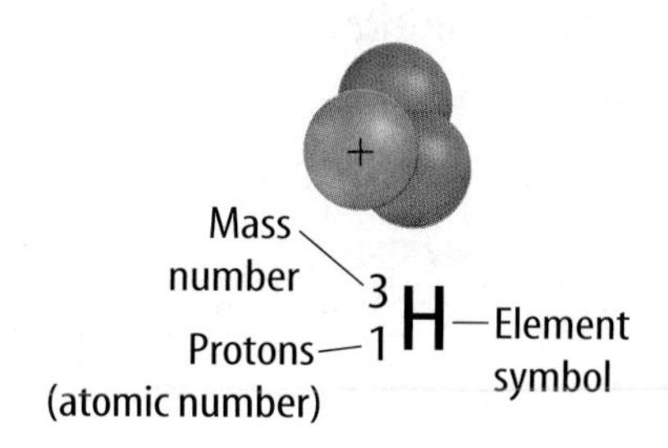

Figure 6
A simple way to indicate the atomic mass of an isotope is to use a symbol like this one for hydrogen. *How many neutrons are in this isotope?*

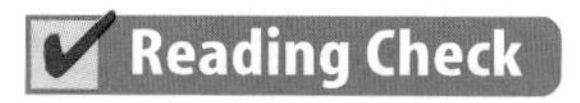

What is the atomic number of a nucleus?

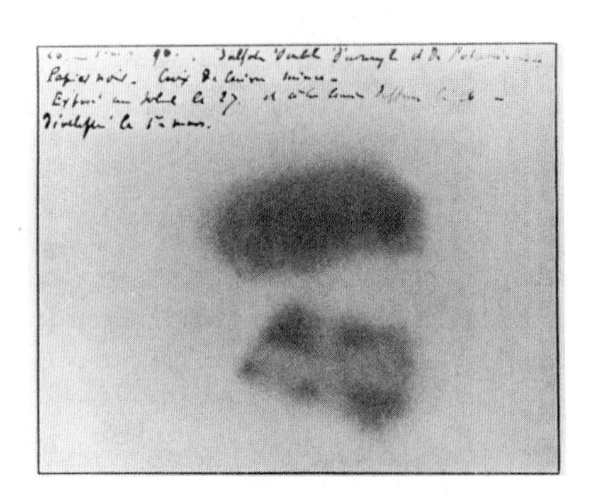

Figure 7
Henri Becquerel found outlines of uranium salt on a photographic plate.

Discovery of Radioactivity

Look around the room. Can you detect any evidence of radioactivity? You can't see, hear, taste, touch, or smell radioactivity. Do you realize that small amounts of radioactivity are all around you, even inside your body? How was radioactivity first discovered if it can't be detected by your senses?

Henri Becquerel accidentally discovered radioactivity in 1896 when he left uranium salt in a desk drawer with a photographic plate. Later, when he removed the plate and developed it, he found an outline of the clumps of the uranium salt like in **Figure 7.** He hypothesized that the uranium had given off some invisible energy, or radiation, and exposed the film. Two years after Becquerel's discovery, Marie and Pierre Curie discovered the elements polonium and radium. These elements are even more radioactive than uranium. What are the atomic numbers of these elements?

SCIENCE *Online*

Research Visit the Glencoe Science Web site at **science.glencoe.com** for more information about the scientists who discovered and developed applications for radioactivity. Create a time line that shows key events and people in the use of radioactivity.

Section 1 Assessment

1. What force keeps stable nuclei permanently together?
2. What is radioactivity?
3. Why are large nuclei unstable?
4. Identify the contributions of the three scientists who discovered the first radioactive elements.
5. **Think Critically** What is the ratio of protons to neutrons in lead-214? Explain whether you would expect this isotope to be radioactive or stable.

Skill Builder Activities

6. **Comparing and Contrasting** Compare and contrast stable and unstable nuclei. **For more help, refer to the Science Skill Handbook.**
7. **Communicating** In your Science Journal, make a list of the first things you think of when you hear the word *radioactivity.* Write one paragraph describing your positive thoughts about radioactivity and another describing your negative thoughts. **For more help, refer to the Science Skill Handbook.**

SECTION

Nuclear Decay

Nuclear Radiation

When an unstable nucleus decays, particles and energy are emitted from the decaying nucleus. These particles and energy are called nuclear radiation. The three types of nuclear radiation are alpha, beta (BAYT uh), and gamma radiation. Alpha and beta radiation are particles. Gamma radiation behaves like a wave that is similar to light but of much higher frequency.

Alpha Particles

When alpha radiation occurs, an alpha particle is emitted from the decaying nucleus. An **alpha particle** is made of two protons and two neutrons, as shown in **Table 1.** Notice that the alpha particle is the same as a helium nucleus. The symbol for an alpha particle is the same as for the helium nucleus, $^{4}_{2}He$. An alpha particle has an electric charge of +2 and an atomic mass of 4.

Reading Check *What does an alpha particle consist of?*

Damage from Alpha Particles Compared to beta and gamma radiation, alpha particles are much more massive. They also have the most electric charge. As a result, alpha particles lose energy more quickly when they interact with matter than the other types of nuclear radiation do. When alpha particles pass through matter, they exert an electric force on the electrons in atoms in their path. This force pulls electrons away from atoms and leaves behind charged ions. Alpha particles lose energy quickly during this process. As a result, alpha particles are the least penetrating form of nuclear radiation. Alpha particles cannot even pass through a sheet of paper.

However, alpha particles can be dangerous if they are released by radioactive atoms inside your body. Biological molecules inside your body are large and easily damaged. A single alpha particle can damage many fragile biological molecules. Damage from alpha particles can cause cells in your body to no longer function properly, leading to illness and disease.

As You Read

***What* You'll Learn**

- **Compare and contrast** alpha, beta, and gamma radiation.
- **Define** the half-life of a radioactive material.
- **Describe** the process of radioactive dating.

Vocabulary

alpha particle
transmutation
beta particle
gamma ray
half-life

***Why* It's Important**

Different types of nuclear radiation are used in medicine and for calculating the ages of artifacts.

Table 1 Alpha Particles

Symbol	$^{4}_{2}He$
Mass	4
Charge	+2

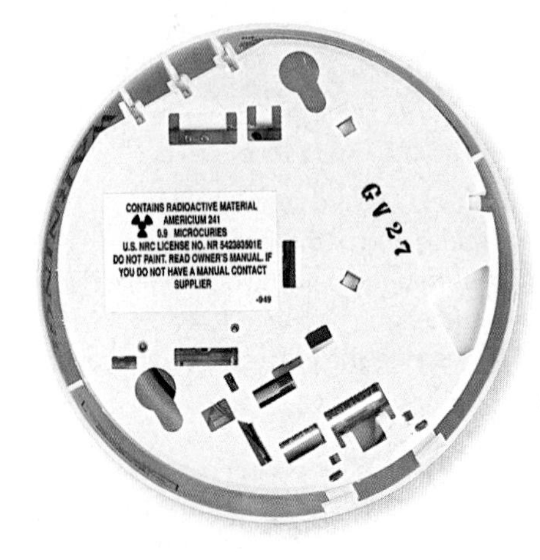

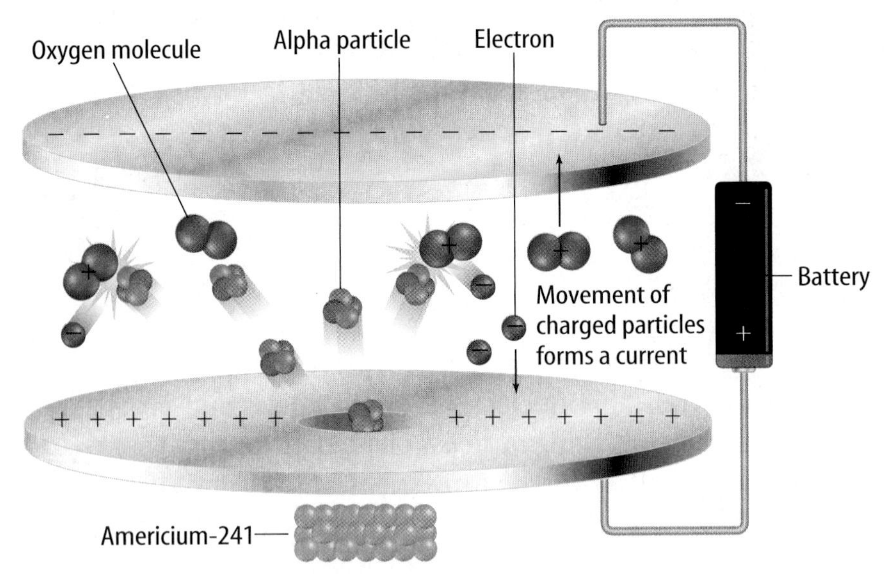

Figure 8
When alpha particles collide with molecules in the air, positively-charged ions and electrons result. The ions and electrons move toward charged plates, creating a current in the smoke detector.

Smoke Detectors Some smoke detectors give off alpha particles that ionize the surrounding air. Normally, an electric current flows through this ionized air to form a circuit, as in **Figure 8.** But if smoke particles enter the ionized air, they will absorb the ions and electrons. The circuit is broken and the alarm goes off.

Transmutation When an atom loses an alpha particle, it no longer has the same number of protons, so it no longer is the same element. **Transmutation** is the process of changing one element to another through nuclear decay. In alpha decay, two protons and two neutrons are lost from the nucleus, so the new element formed has an atomic number two less than that of the original element. The mass number of the new element is four less than the original element. The nuclear equation in **Figure 9** shows a nuclear transmutation caused by alpha decay. Notice in the equation that the charge of the original nucleus equals the sum of the charges of the nucleus and the alpha particle that are formed.

Figure 9
In this transmutation, polonium emits an alpha particle and changes into lead. *Do the charges and mass numbers of the products add up to the charge and mass number of the polonium nucleus?*

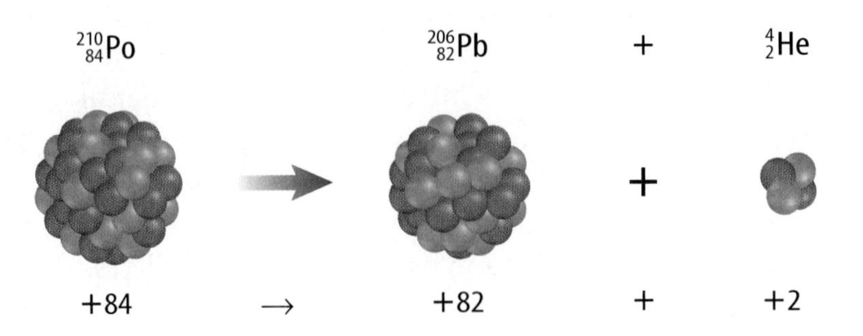

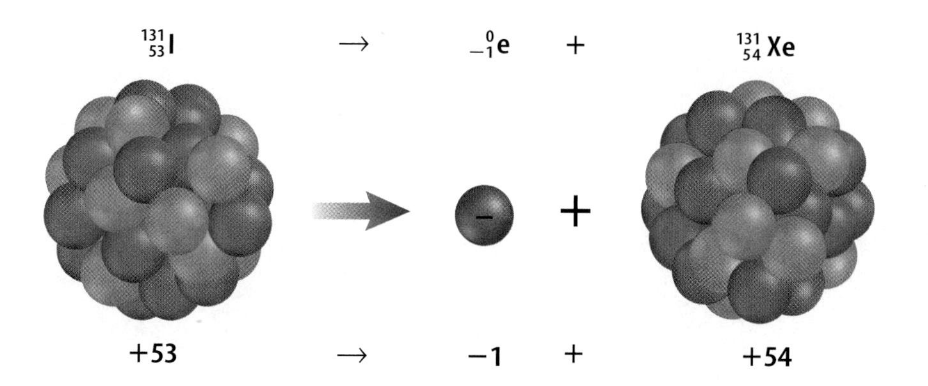

Figure 10
Nuclei that emit beta particles undergo transmutation. In this process, iodine changes to xenon. *Show how the charges and masses of the products add up to the charge and mass of the iodine nucleus.*

Beta Particles

A second type of radioactive decay is called beta decay, which is summarized in **Table 2.** Sometimes in an unstable nucleus a neutron decays into a proton and emits an electron. The electron is emitted from the nucleus and is called a **beta particle.** Beta decay is caused by another basic force called the weak force.

Because the atom now has one more proton, it becomes the element with an atomic number one greater than that of the original element. Atoms that lose beta particles undergo transmutation. However, because the total number of protons and neutrons does not change during beta decay, the atomic mass number of the new element is the same as that of the original element. **Figure 10** shows a transmutation caused by beta decay.

Table 2 Beta Particles

Symbol	β
Mass	0.0005
Charge	−1

Damage from Beta Particles Beta particles are much faster and more penetrating than alpha particles. They can pass through paper but are stopped by a sheet of aluminum foil. Just like alpha particles, beta particles can damage cells when they are emitted by radioactive nuclei inside the human body.

Gamma Rays

The most penetrating form of radiation is not made of protons, neutrons, or electrons. **Gamma rays** are a form of radiation called electromagnetic waves. Like water and sound waves, gamma rays carry energy. They have no mass and no charge, and they travel at the speed of light. They usually are released along with alpha or beta particles. The characteristics of gamma rays are summarized in **Table 3.**

Thick blocks of dense materials, such as lead and concrete, are required to stop gamma rays. However, gamma rays cause less damage to biological molecules as they pass through living tissue. Suppose an alpha particle and a gamma ray travel the same distance through matter. The gamma ray produces far fewer ions because it has no electric charge.

Table 3 Gamma Radiation

Symbol	γ
Mass	0
Charge	0

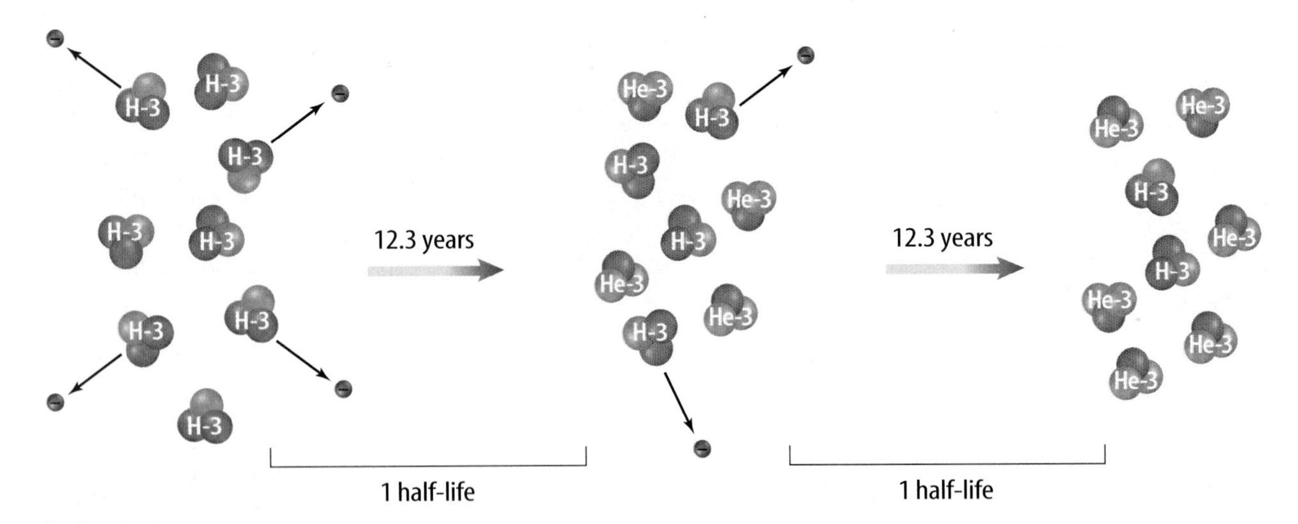

Figure 11
The half-life of $^{3}_{1}H$ is 12.3 years. During each half-life, half of the atoms in the sample decay into helium. *How many hydrogen atoms will be left in the sample after the next half-life?*

Radioactive Half-Life

If an element is radioactive, how can you tell when its atoms are going to decay? Some radioisotopes decay to stable atoms in less than a second. However, the nuclei of certain radioactive isotopes require millions of years to decay. A measure of the time required by the nuclei of an isotope to decay is called the half-life. The **half-life** of a radioactive isotope is the amount of time it takes for half the nuclei in a sample of the isotope to decay. The nucleus left after the isotope decays is called the daughter nucleus. **Figure 11** shows how the number of decaying nuclei decreases after each half-life.

Half-lives vary widely among the radioactive isotopes. For example, polonium-214 has a half-life of less than a thousandth of a second, but uranium-238 has a half-life of 4.5 billion years. The half-lives of some other radioactive elements are listed in **Table 4.**

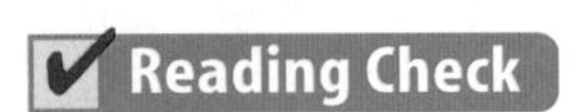
Reading Check *What is a daughter nucleus?*

Table 4 Sample Half-Lives

Isotope	Half-Life
$^{3}_{1}H$	12.3 years
$^{212}_{82}Pb$	10.6 hr
$^{14}_{6}C$	5,730 years
$^{211}_{84}Po$	0.5 s
$^{235}_{92}U$	7.04×10^{8} years
$^{131}_{53}I$	8.04 days

Radioactive Dating

Some geologists, biologists, and archaeologists, among others, are interested in the ages of rocks and fossils found on Earth. The ages of these materials can be determined using radioactive isotopes and their half-lives. First, the amounts of the radioactive isotope and its daughter nucleus in a sample of material are measured. Then, the number of half-lives that need to pass to give the measured amounts of the isotope and its daughter nucleus is calculated. The number of half-lives is the amount of time that has passed since the isotope began to decay. It is also usually the amount of time that has passed since the object was formed, or the age of the object. Different isotopes are useful in dating different types of materials.

Carbon Dating The radioactive isotope carbon-14 often is used to find the ages of objects that were once living. Carbon-14 is found in molecules throughout the environment, including some carbon dioxide molecules plants take in as they carry out photosynthesis. Carbon-14 atoms behave chemically just like nonradioactive carbon-12 atoms, so all living plants contain some carbon-14. When animals eat plants, they ingest some of the radioactive carbon-14.

An atom of carbon-14 eventually will decay into nitrogen-14. The half-life of carbon-14 is 5,730 years. The amount of carbon-14 in living plants and animals remains fairly constant as decaying carbon-14 is replaced constantly by new carbon-14 when an animal eats or a plant makes food. However, when an organism dies, its carbon-14 atoms decay without being replaced. By measuring the amount of carbon-14 in a sample and comparing it to the amount of carbon-12, scientists can determine the approximate age of the material. Only the remains of plants and animals that lived within the last 50,000 years contain enough carbon-14 to measure.

Uranium Dating Radioactive dating also can be used to estimate the ages of rocks. Some rocks contain uranium, which has two radioactive isotopes with long half-lives. Each of these uranium isotopes decays into a different isotope of lead. The amount of these uranium isotopes and their daughter nuclei are measured. From the ratios of these amounts, the time since the rock was formed can be calculated.

Modeling the Strong Force

Procedure

1. Gather together **15 yellow candies** to represent neutrons and **13 red** and **two green candies** to represent protons.
2. Model a small nucleus by arranging one red and one green proton and two neutrons in a tight group.
3. Model a larger nucleus by arranging the remaining candies in a tight group.

Analysis

1. Compare the number of protons and neutrons touching the green proton in both nuclei. From this, compare the strong force on a proton in both nuclei.
2. How would the strong force on a proton change if the number of protons and neutrons in a nucleus were much larger?

Section  Assessment

1. Describe three types of radiation.
2. Write a nuclear equation to show how radon-222 decays to give off an alpha particle and another element. What is the other element?
3. What is a half-life?
4. How is radioactivity useful in determining the age of material that was once part of a living organism?
5. **Think Critically** Is it possible for an isotope to decay to an element with a higher atomic number? Explain.

Skill Builder Activities

6. **Using an Electronic Spreadsheet** Write a short program that allows you to input the mass and the half-life of a radioactive sample and calculate what mass remains after a certain number of half-lives. **For more help, refer to the Technology Skill Handbook.**
7. **Using Fractions** The half-life of iodine-131 is about eight days. Calculate how much of a 40-g sample will be left after eight days, after 16 days, and after 32 days. **For more help, refer to the Math Skill Handbook.**

SECTION

Detecting Radioactivity

As You Read

What You'll Learn

- **Describe** how radioactivity can be detected in cloud and bubble chambers.
- **Explain** how an electroscope can be used to detect radiation.
- **Explain** how a Geiger counter can measure nuclear radiation.

Vocabulary

cloud chamber
bubble chamber
Geiger counter

Why It's Important

Devices to detect and measure radioactivity are needed to monitor exposure to humans.

Radiation Detectors

Because you can't see or feel alpha particles, beta particles, or gamma rays, you must use instruments to detect their presence. Some tools that are used to detect radioactivity rely on the fact that radiation forms ions in the matter it passes through. The tools detect these newly formed ions in several ways.

Cloud Chambers A **cloud chamber,** shown in **Figure 12,** can be used to detect alpha or beta particle radiation. A cloud chamber is filled with water or ethanol vapor. When a radioactive sample is placed in the cloud chamber, it gives off charged alpha or beta particles that travel through the water or ethanol vapor. As each charged particle travels through the chamber, it knocks electrons off the atoms in the air, creating ions. It leaves a trail of ions in the chamber. The water or ethanol vapor condenses around these ions, creating a visible path of droplets along the track of the particle. Beta particles leave long, thin trails, and alpha particles leave shorter, thicker trails.

Reading Check *Why are trails produced by alpha and beta particles seen in cloud chambers?*

Figure 12
If a sample of radioactive material is placed in a cloud chamber, a trail of condensed vapor will form along the paths of the emitted particles.

Bubble Chambers Another way to detect and monitor the paths of nuclear particles is by using a bubble chamber. A **bubble chamber** holds a superheated liquid, which doesn't boil because the pressure in the chamber is high. When a moving particle leaves ions behind, the liquid boils along the trail. The path shows up as tracks of bubbles, like the ones in **Figure 13.**

Figure 13
Particles of nuclear radiation can be detected as they leave trails of bubbles in a bubble chamber.

Electroscopes Do you remember how an electroscope can be used to detect electric charges? When an electroscope is given a negative charge, its leaves repel each other and spread apart, as in **Figure 14A.** They will remain apart until their extra electrons have somewhere to go and discharge the electroscope. The excess charge can be neutralized if it combines with positive charges. Nuclear radiation moving through the air can remove electrons from some molecules in air, as shown in **Figure 14B,** and cause other molecules in air to gain electrons. When this occurs near the leaves of the electroscope, some positively charged molecules in the air can come in contact with the electroscope and attract the electrons from the leaves, as **Figure 14C** shows. As these negatively charged leaves lose their charges, they move together. **Figure 14D** shows this last step in the process. The same process also will occur if the electroscope leaves are positively charged. Then the electrons move from negative ions in the air to the electroscope leaves.

Figure 14
Nuclear radiation can cause an electroscope to lose its charge.

A The electroscope leaves are charged with negative charge.

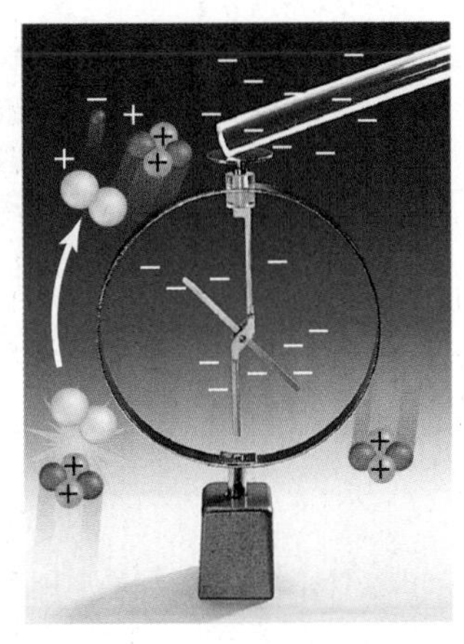

B Nuclear radiation, such as alpha particles, can create positive ions.

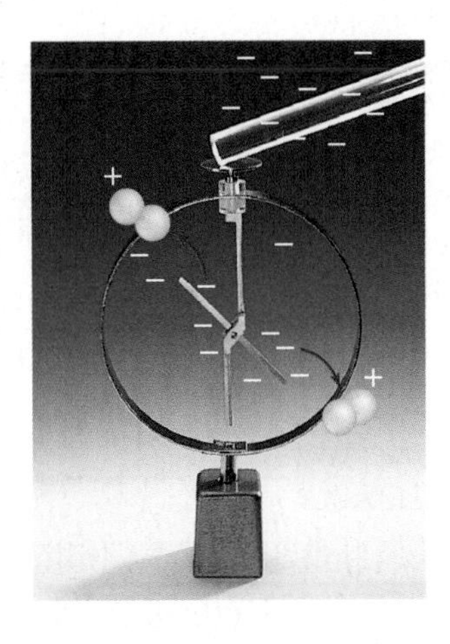

C Negative charges move from the leaves to positively charged ions.

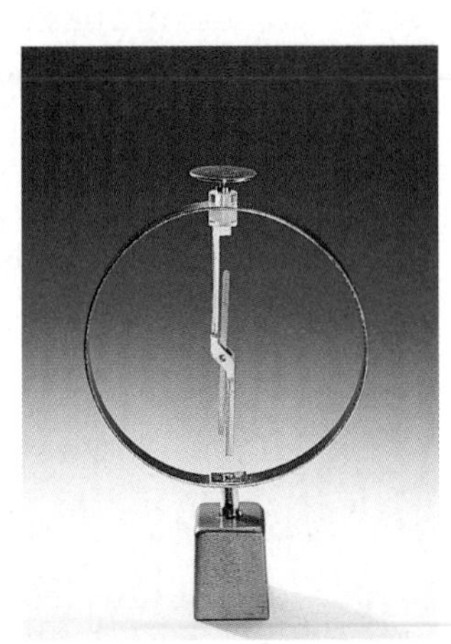

D The electroscope leaves lose their negative charge and come together.

Measuring Radiation

Large doses of radiation are harmful to living tissue. If you worked or lived in an environment that had potential for exposure to high levels of radiation—for example, a nuclear testing facility—you might want to know exactly how much radiation you were being exposed to. You could measure the radiation with a Geiger (GI gur) counter. A **Geiger counter** is a device that measures radioactivity by producing an electric current when radiation is present.

Math Skills Activity

How can radioactive half lives be used to measure geological time?

Example Problem

The time it takes for half of the atoms of one element in a piece of rock to change into another element is called its half-life. Scientists use the half-lives of radioactive isotopes to measure geological time. Potassium-40 has a half-life of 1.28 billion years before it produces the stable daughter product argon-40. If three-fourths of the potassium-40 atoms in a rock had changed into atoms of argon-40, how old would you predict the rock to be?

Solution

1. *This is what you know:*
   half-life of potassium-40 = 1.28 billion years
   75% of potassium-40 atoms have decayed
2. *This is what you want to find:*
   age of the rock
3. *Set up a pattern to help solve:*

Rate of Decay

Time	% Potassium-40	% Argon-40
1.28 billion years	50%	50%
2.56 billion years	25%	75%
3.84 billion years	12.5%	87.5%

The age of the rock would be 2.56 billion years old.

Practice Problem

Uranium-238 has a half-life of 4.5 billion years before half of the atoms change into lead-206. Determine the age of a rock in which approximately 94% of the atoms are lead-206.

For more help, refer to the Math Skill Handbook.

Figure 15
Electrons that are stripped off gas molecules in a Geiger counter move to a positively charged wire in the device. This causes current to flow in the wire. The current then is used to produce a click or a flash of light.

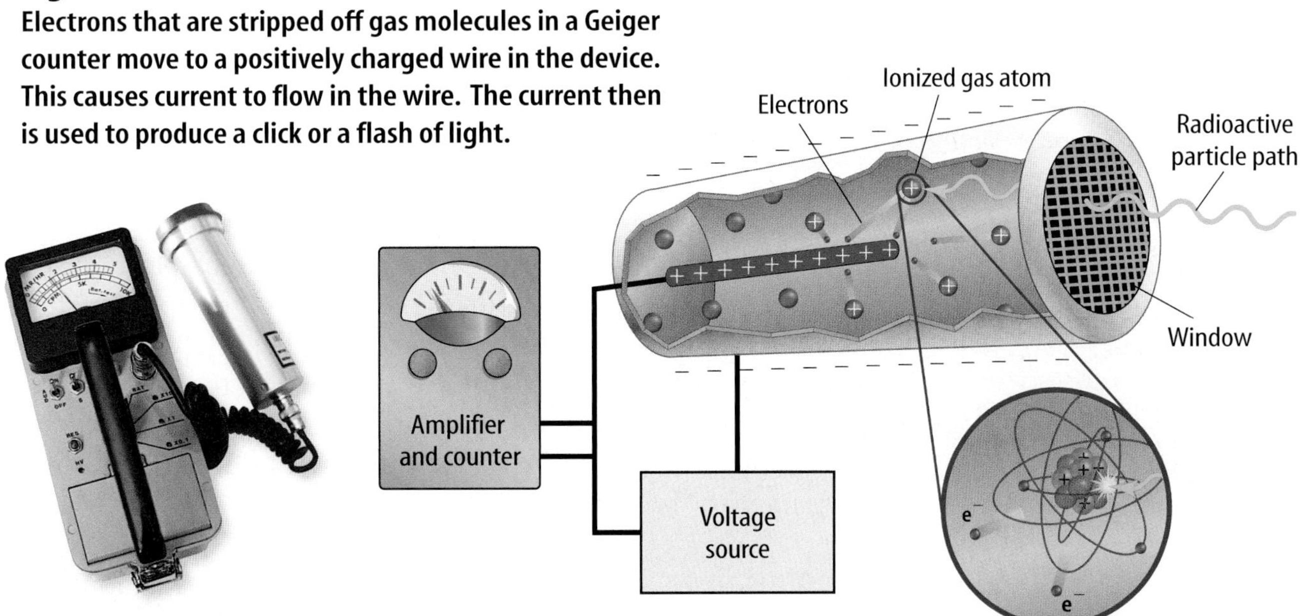

Geiger Counters **Figure 15** shows a Geiger counter. A Geiger counter has a tube with a positively charged wire running through the center of a negatively charged copper cylinder. This tube is filled with gas at a low pressure. When radiation enters the tube at one end, it knocks electrons from the atoms of the gas. These electrons then knock more electrons off other atoms in the gas, and an "electron avalanche" is produced. The free electrons are attracted to the positive wire in the tube. When a large number of electrons reaches the wire, a short, intense current is produced in the wire. This current is amplified to produce a clicking sound or flashing light. The intensity of radiation present is determined by the number of clicks or flashes of light each second.

How does a Geiger counter indicate that radiation is present?

Background Radiation

It might surprise you to know that you are bathed in radiation that comes from your environment. This radiation, called background radiation, is not produced by humans. Instead it is low-level radiation emitted mainly by naturally occurring radioactive isotopes found in Earth's rocks, soils, and atmosphere. Building materials such as bricks, wood, and stones contain traces of these radioactive materials. Traces of naturally occurring radioactive isotopes are found in the food, water, and air consumed by all animals and plants. As a result, animals and plants also contain small amounts of these isotopes.

Earth Science INTEGRATION

The formation of droplets in a cloud chamber is similar to the formation of rain drops in a cloud. Clouds contain droplets of very cold water. Rain forms when these droplets freeze around microscopic particles of dust and then melt as they fall through air warmer than freezing. Many attempts have been made to make rain fall from clouds. Research some attempts at artificial rainmaking and report your findings to your class.

Sources of Background Radiation

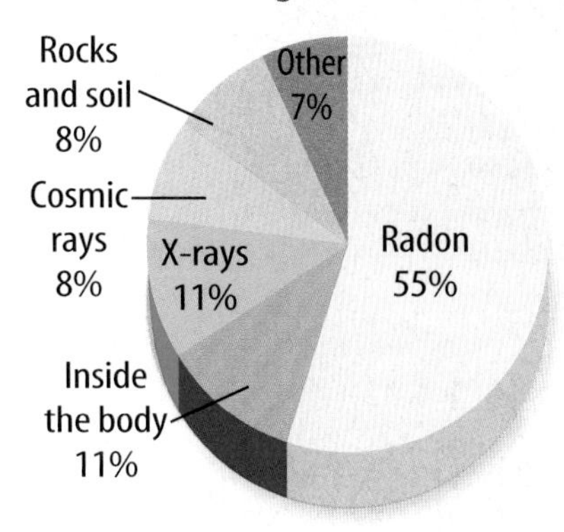

Figure 16
This circle graph shows the sources of background radiation received on average by a person living in the United States.

Sources of Background Radiation Background radiation comes from several sources, as shown in **Figure 16.** The largest source comes from the decay of radon gas. Radon is produced in Earth's crust by the decay of uranium-238 and emits an alpha particle when it decays. Radon gas can seep into houses and basements from the surrounding soil and rocks and can be inhaled.

Some background radiation comes from high-speed nuclei, called cosmic rays, that hit the top of Earth's atmosphere. They produce showers of particles, including alpha, beta, and gamma radiation. Most of this radiation is absorbed by the atmosphere. As you go higher, less atmosphere is above you to absorb this radiation. Therefore, the background radiation from cosmic rays increases with elevation.

Radiation in Your Body Naturally occurring radiation also is found inside your body. Some of the elements in your body that are essential for life have naturally occurring radioactive isotopes. For example, about one out of every trillion carbon atoms is carbon-14, which emits a beta particle when it decays. With each breath, you inhale about 3 million carbon-14 atoms.

The amount of background radiation a person receives can vary greatly. The amount depends on the type of rocks underground, the type of materials used to construct the person's home, and the elevation at which the person lives, among other things. However, because it comes from naturally occurring processes, background radiation never can be eliminated.

Section  Assessment

1. What are four ways that radioactivity can be detected?
2. How are cloud and bubble chambers similar? How are they different?
3. How can an electroscope be used to detect nuclear radiation?
4. Briefly explain how a Geiger counter operates.
5. **Think Critically** Which device would be used to check the amount of radiation present in your home? What are the possible sources of the radiation?

Skill Builder Activities

6. **Drawing Conclusions** You are observing the presence of nuclear radiation with a bubble chamber and see two kinds of trails. Some trails are short and thick, and others are long and thin. What type of nuclear radiation might have caused each trail? **For more help, refer to the Science Skill Handbook.**
7. **Drawing Conclusions** Explain why homes can contain radon gas even though radon-22 has a half-life of only four days. **For more help, refer to the Science Skill Handbook.**

Nuclear Reactions

Nuclear Fission

Do you know what the first step in a game of pool is? One player shoots the cue ball into a triangle of densely packed billiard balls. If the cue ball hits the triangle right on, the balls spread apart, or break. In 1938, two physicists named Otto Hahn and Fritz Strassmann found that a similar result occurs when a neutron is shot into the large nucleus of a uranium-235 atom. The nucleus is split.

Lise Meitner was the first to offer a theory to explain the splitting of a nucleus. She concluded that the neutron fired into the nucleus disturbs and distorts the uranium-235 nucleus. The nuclear strong force is no longer enough to overcome the electrical repulsion within the nucleus, causing it to split into two nuclei, as in **Figure 17.** The process of splitting a nucleus into two nuclei with smaller masses is called **nuclear fission.** The word *fission* means "to divide."

Reading Check ***What initiates nuclear fission of a uranium-235 nucleus?***

Only large nuclei, such as the nuclei of uranium and plutonium atoms, can undergo nuclear fission. The products of a fission reaction usually include several individual neutrons in addition to the smaller nuclei. The total mass of the products is slightly less than the mass of the original nucleus and the neutron. This small amount of missing mass is converted to a tremendous amount of energy during the fission reaction.

As You Read

What You'll Learn

- **Explain** nuclear fission and how it can begin a chain reaction.
- **Discuss** how nuclear fusion occurs in the Sun.
- **Describe** how radioactive tracers can be used to diagnose medical problems.
- **Discuss** how nuclear reactions can help treat cancer.

Vocabulary

nuclear fission
chain reaction
critical mass
nuclear fusion
tracer

Why It's Important

Radiation from nuclear reactions can be used to generate power and diagnose and treat medical problems.

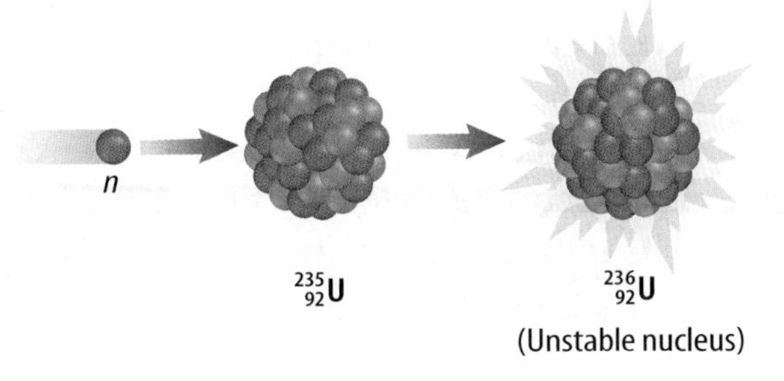

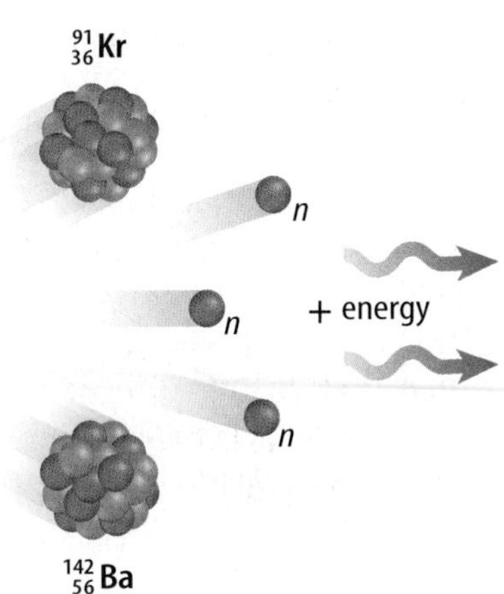

Figure 17
When a neutron hits a uranium-235 nucleus, the uranium nucleus splits into two smaller nuclei and two or three free neutrons. Energy also is released.

Figure 18
A chain reaction quickly grows larger and larger if there are enough nuclei to keep it going.

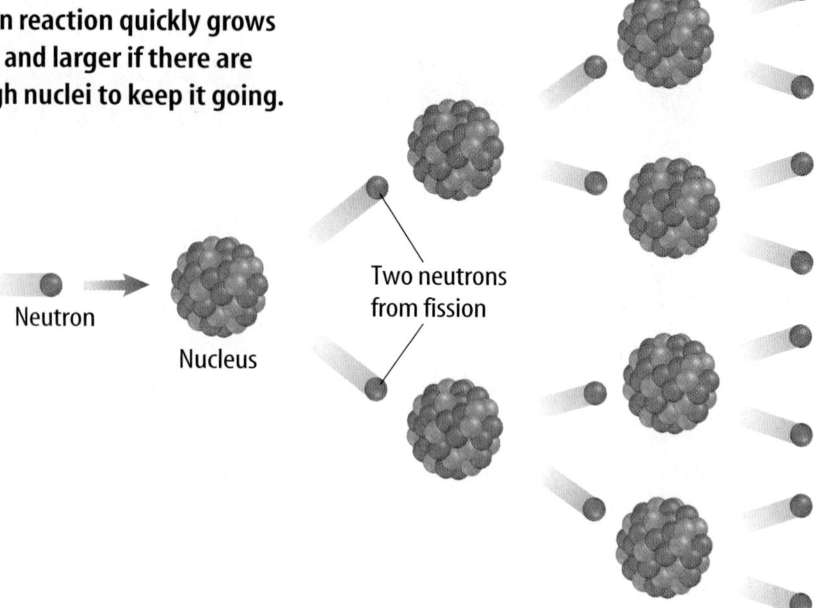

Modeling a Nuclear Reaction

Procedure

1. Put **32 marbles,** each with an attached lump of **clay,** into a large **beaker.** These marbles with clay represent unstable atoms.
2. During a 1-min period, remove half of the marbles and pull off the clay. Place the removed marbles into another beaker and place the lumps of clay into a pile. Marbles without clay represent stable atoms. The clay represents waste from the reaction—smaller atoms that still might decay and give off energy.
3. Repeat this procedure four more times.

Analysis

1. How does this model show one of the main problems that is associated with using nuclear power to make electricity?
2. Why is it difficult to find a place for waste products from nuclear reactions?

Chain Reactions The energy released in a nuclear fission reaction is much greater than the energy released by the natural, spontaneous decay of a radioactive isotope. The neutrons produced in the fission reaction can then bombard and split other nuclei in the sample. These reactions each release more neutrons. If some other material is not present to absorb some of these neutrons as they are released, an uncontrolled chain reaction can result. A **chain reaction,** represented in **Figure 18,** is an ongoing series of fission reactions. Billions of reactions can occur each second during a chain reaction, resulting in the release of tremendous amounts of energy.

When controlled, the large amounts of energy released in nuclear fission reactions can be used to generate electricity. Nuclear fission reactions also are used in nuclear weapons.

Critical Mass To be useful, chain reactions cannot grow out of control or die out. Especially when chain reactions are used to produce electricity, they should occur at a relatively constant rate. How can the rate of chain reactions be controlled when they involve so much energy? Chain reactions can be controlled if the number of neutrons that are available to start additional fission reactions is controlled carefully. This can be done by adding materials that absorb neutrons. If enough neutrons are absorbed, the chain reaction will continue at a constant rate. The **critical mass** is the amount of fissionable material required so that each fission reaction produces approximately one more fission reaction. If less than the critical mass of reaction material is present, a chain reaction will not occur.

Nuclear Fusion

Tremendous amounts of energy can be released in nuclear fission. In fact, splitting one uranium-235 nucleus produces about 30 million times more energy than chemically reacting one molecule of dynamite. Even more energy can be released in another type of nuclear reaction, called nuclear fusion. In **nuclear fusion,** two nuclei with low masses are combined to form one nucleus of larger mass. Fusion fuses atomic nuclei together, and fission splits nuclei apart.

Research Visit the Glencoe Science Web site at **science.glencoe.com** for more information about scientists' research involving nuclear fusion. Communicate to your class what you learn.

Temperature and Fusion For nuclear fusion to occur, positively charged nuclei must get close to each other. However, all nuclei repel each other because they have the same positive electric charge. If nuclei are moving fast, they can have enough kinetic energy to overcome the repulsive electrical force between them and get close to each other.

Remember that the kinetic energy of atoms or molecules increases as their temperature increases. Only at temperatures of millions of degrees Celsius are nuclei moving so fast that they can get close enough for fusion to occur. These extremely high temperatures are found in the center of stars, including the Sun.

Nuclear Fusion and the Sun The Sun is composed mainly of hydrogen. Most of the energy given off by the Sun is produced by a process involving the fusion of hydrogen nuclei. This process occurs in several stages, and one of the stages is shown in **Figure 19.** The net result of this process is that four hydrogen nuclei are converted into one helium nucleus. As this occurs, a small amount of mass is changed into an enormous amount of energy. Earth receives a small amount of this energy as heat and light.

As the Sun ages, the hydrogen nuclei are used up as they are converted into helium. So far, only about one percent of the Sun's mass has been converted into energy. Eventually, no hydrogen nuclei will be left, and the fusion reaction that changes hydrogen into helium will stop. However, it is estimated that the Sun has enough hydrogen to keep this reaction going for another 5 billion years.

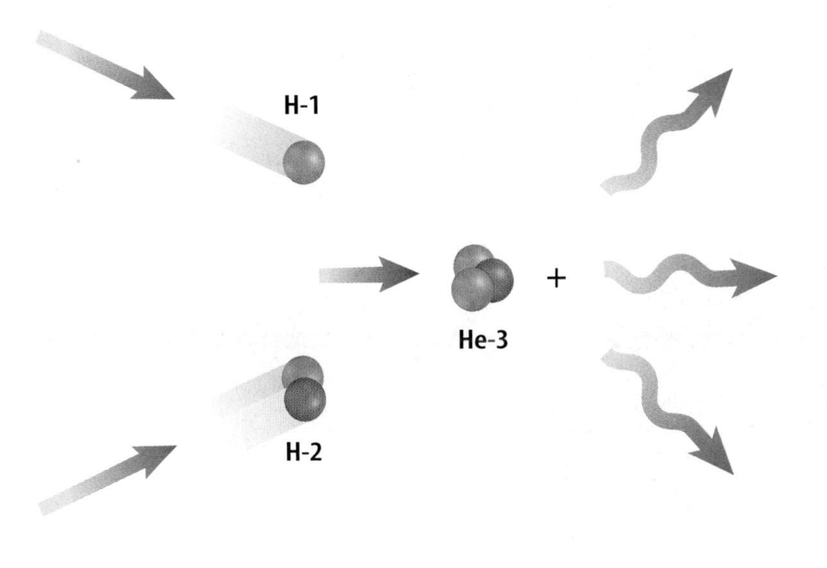

Figure 19
The fusion of hydrogen to form helium takes place in several stages in the Sun. One of these stages is shown here. An isotope of helium is produced when a proton and the hydrogen isotope H-2 undergo fusion.

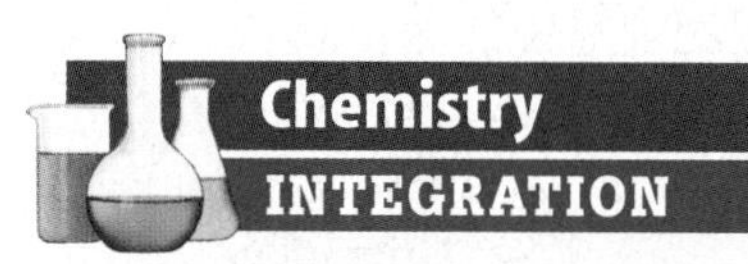

One way a uranium-235 atom can fission, or break apart, is into barium and krypton. Use a periodic table to find the atomic numbers of barium and krypton. What do they add up to? A uranium-235 atom can fission in several other ways such as producing neodymium and another element. What is the other element?

Using Nuclear Reactions in Medicine

If you were going to meet a friend in a crowded area, it would be easier to find her if your friend told you that she would be wearing a red hat. In a similar way, scientists can find one molecule in a large group of molecules if they know that it is "wearing" something unique. Although a molecule can't wear a red hat, if it has a radioactive atom in it, it can be found easily in a large group of molecules, or even a living organism. Radioactive isotopes can be located by detecting the radiation they emit.

When a radioisotope is used to find or keep track of molecules in an organism, it is called a **tracer.** Scientists can use tracers to follow where a particular molecule goes in your body or to study how a particular organ functions. Tracers also are used in agriculture to monitor the uptake of nutrients and fertilizers. Examples of tracers include carbon-11, iodine-131, and sodium-24. These three radioisotopes are useful tracers because they are important in certain body processes. As a result, they accumulate inside the organism being studied.

Reading Check *What use do tracers have in agriculture?*

Iodine Tracers in the Thyroid The thyroid gland is located in your neck and produces chemical compounds called hormones. These hormones help regulate several body processes, including growth. Because the element iodine accumulates in the thyroid, the radioisotope iodine-131 can be used to diagnose thyroid problems. As iodine-131 atoms are absorbed by the thyroid, their nuclei decay, emitting beta particles and gamma rays. The beta particles are absorbed by the surrounding tissues, but the gamma rays penetrate the skin. The emitted gamma rays can be detected and used to determine whether the thyroid is healthy, as shown in **Figure 20.** If the detected radiation is not intense, then the thyroid has not properly absorbed the iodine-131 and is not functioning properly. This could be due to the presence of a tumor. **Figure 21** shows how radioactive tracers are used to study the brain.

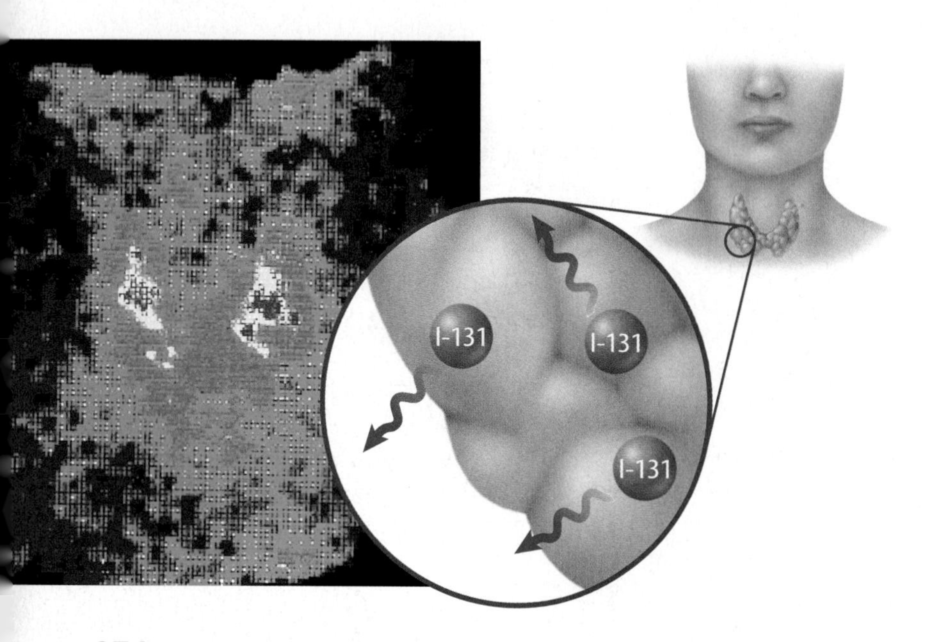

Figure 20
Radioactive iodine-131 accumulates in the thyroid gland and emits gamma rays, which can be detected to form an image of a patient's thyroid. *What are some advantages of being able to use iodine-131 to form an image of a thyroid?*

NATIONAL GEOGRAPHIC

VISUALIZING PET SCANS

Figure 21

The diagram below shows an imaging technique known as Positron Emission Tomography, or PET. Positrons are emitted from the nuclei of certain radioactive isotopes when a proton changes to a neutron. PET can form images that show the level of activity in different areas of the brain. These images can reveal tumors and regions of abnormal brain activity.

A When positrons are emitted from the nucleus of an atom, they can hit electrons from other atoms and become transformed into gamma rays.

B The radioactive isotope fluorine-18 emits positrons when it decays. Fluorine-18 atoms are chemically attached to molecules that are absorbed by brain tissue. These compounds are injected into the patient and carried by blood to the brain.

C Inside the patient's brain, the decay of the radioactive fluorine-18 nuclei emits positrons that collide with electrons. The gamma rays that are released are sensed by the detectors.

D A computer uses the information collected by the detectors to generate an image of the activity level in the brain. This image shows normal activity in the right side of the brain (red, yellow, green) but below-normal activity in the left (purple).

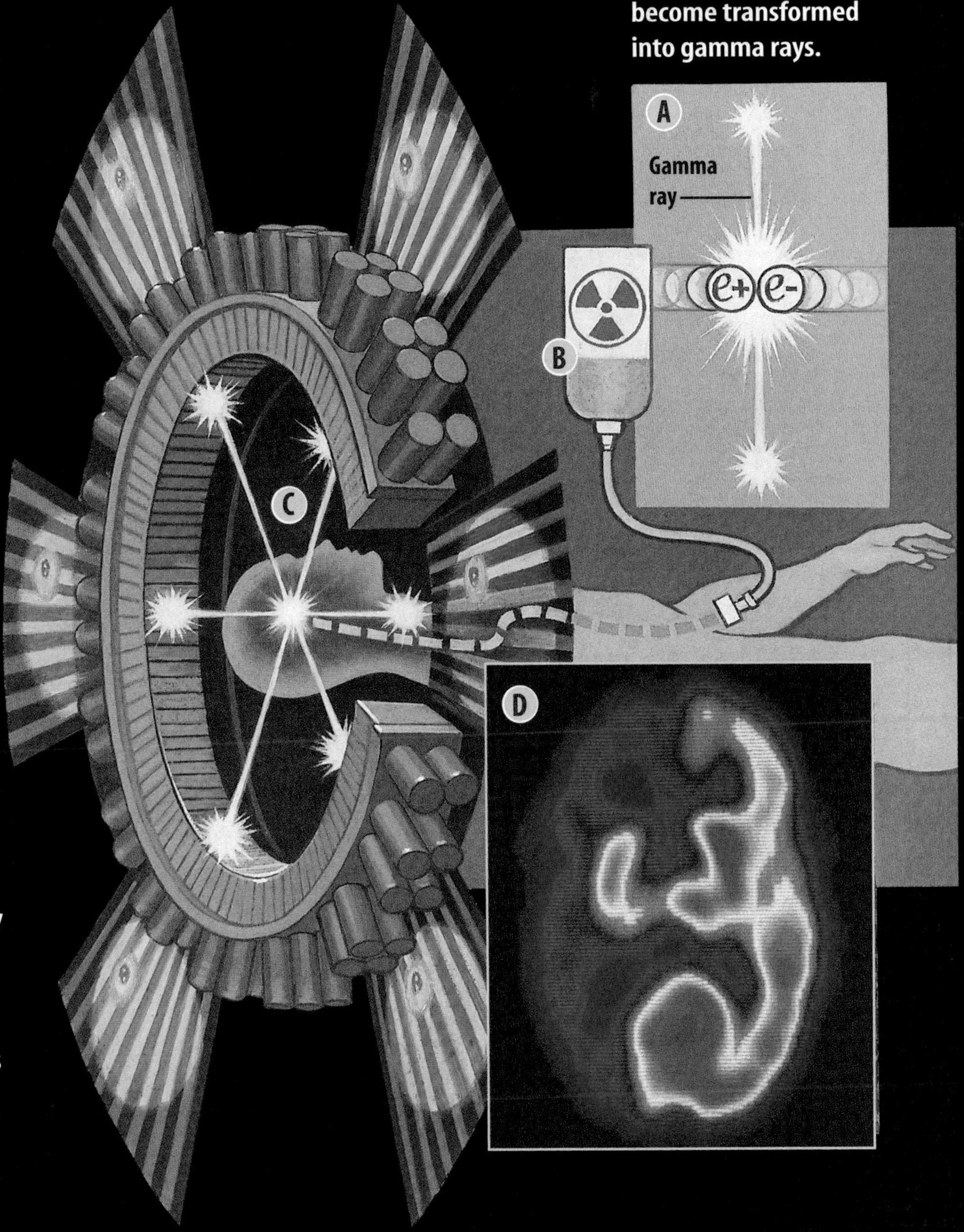

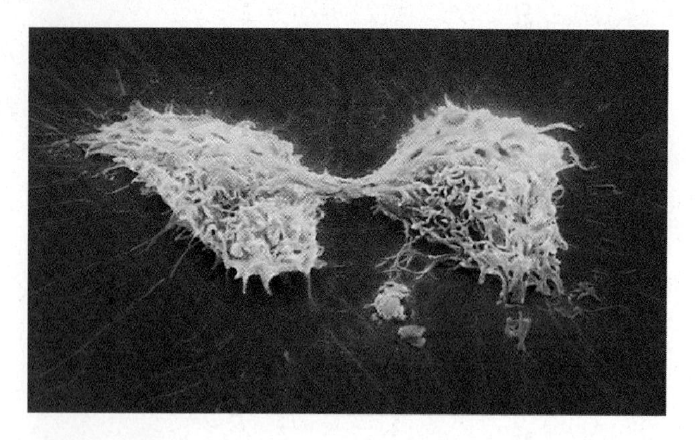

Figure 22
Cancer cells, such as the ones shown here, can be killed with carefully-measured doses of radiation.

Treating Cancer with Radioactivity

When a person has cancer, a group of cells in that person's body grows out of control and can form a tumor. Radiation can be used to stop some types of cancerous cells from growing. Remember that the radiation that is given off during nuclear decay is strong enough to ionize nearby atoms. If a source of radiation is placed near cancer cells, atoms in the cells can be ionized. If the ionized atoms are in a critical molecule, such as the DNA or RNA of a cancer cell, then the molecule might no longer function properly. The cell then could die or stop growing, as shown in **Figure 22.**

When possible, a radioactive isotope such as gold-198 or iridium-192 is implanted within or near the tumor. Other times, tumors are treated from outside the body. Typically, an intense beam of gamma rays from the decay of cobalt-60 is focused on the tumor for a short period of time. The gamma rays pass through the body and into the tumor. How can physicians be sure that only the cancer cells will absorb radiation? Because cancer cells grow quickly, they are more susceptible to absorbing radiation and being damaged than healthy cells are. However, other cells in the body that grow quickly also are damaged, which is why cancer patients who have radiation therapy sometimes experience severe side effects.

Section  Assessment

1. Why is critical mass important in some applications of nuclear fission?
2. Explain why it would be difficult to start a fusion reaction on Earth.
3. How might a tracer be used to diagnose a digestive problem?
4. Why does nuclear radiation cause damage to living cells?
5. **Think Critically** During nuclear fission, large nuclei with high masses are split into two nuclei with smaller masses. During nuclear fusion, two nuclei with low masses are combined to form one nucleus of larger mass. How are the two processes similar?

Skill Builder Activities

6. **Concept Mapping** Make a concept map to show how a chain reaction occurs when U-235 is bombarded with a neutron. Show how each of the three neutrons given as products begins another fission reaction. **For more help, refer to the Science Skill Handbook.**
7. **Identifying a Question** Suppose you had a medical problem and the doctor suggested a diagnostic test that involved a radioactive tracer. Using what you have learned about medical applications of radioactivity, write a set of questions you might ask your doctor. **For more help, refer to the Science Skill Handbook.**

Activity

Chain Reactions

In an uncontrolled nuclear chain reaction, the number of reactions increases as additional neutrons split more nuclei. In a controlled nuclear reaction, neutrons are absorbed, so the reaction continues at a constant rate. How could you model a controlled and an uncontrolled nuclear reaction in the classroom?

What You'll Investigate

How can you use dominoes to model chain reactions?

Materials

dominoes
stopwatch

Goals

- **Model** a controlled and uncontrolled chain reaction
- **Compare** the two types of chain reactions

Procedure

1. Set up a single line of dominoes standing on end so that when the first domino is pushed over, it will knock over the second and each domino will knock over the one following it.
2. Using the stopwatch, time how long it takes from the moment the first domino is pushed over until the last domino falls over. Record the time.
3. Using the same number of dominoes as in step 1, set up a series of dominoes in which at least one of the dominoes will knock down two others, so that two lines of dominoes will continue falling. In other words, the series should have at least one point that looks like the letter Y.
4. Repeat step 2.

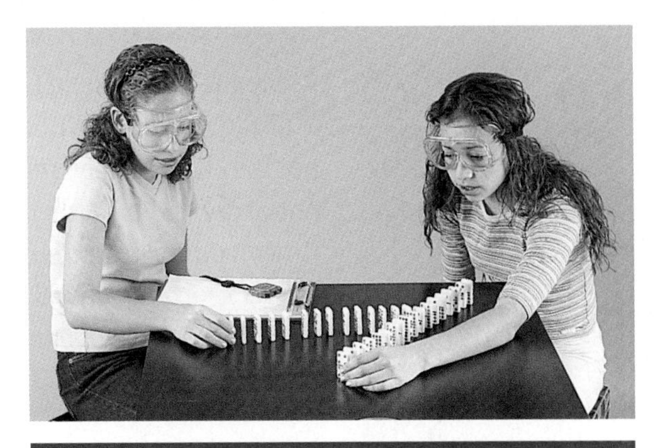

Conclude and Apply

1. **Compare** the amount of time it took for all of the dominoes to fall in each of your two arrangements.
2. Were the same number of dominoes falling at a particular time in both domino arrangements? Explain.
3. Which of your domino arrangements represented a controlled chain reaction? Which represented an uncontrolled chain reaction?
4. **Describe** how the concept of critical mass was represented in your model of a controlled chain reaction.
5. Assuming that they had equal amounts of material, which would finish faster—a controlled or an uncontrolled nuclear chain reaction? Explain.

Communicating Your Data

Explain to friends or members of your family how a controlled nuclear chain reaction can be used in nuclear power plants to generate electricity.

Activity Model and Invent

Modeling Transmutation

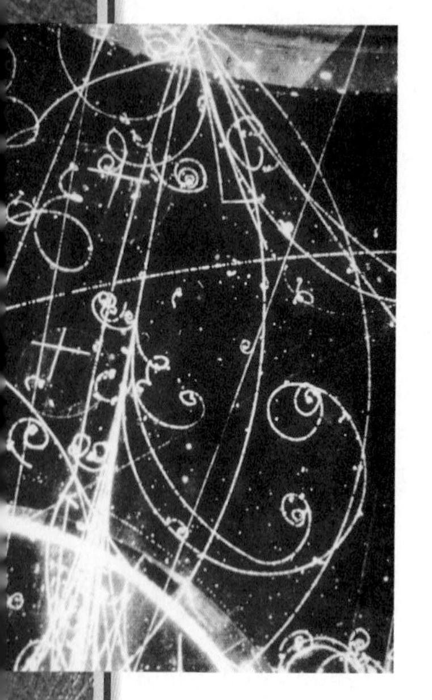

Imagine what would happen if the oxygen atoms around you began changing into nitrogen atoms. Without oxygen, most living organisms, including people, could not live. Fortunately, more than 99.9 percent of all oxygen atoms are stable and do not decay. Usually, when an unstable nucleus decays, an alpha or beta particle is thrown out of its nucleus, and the atom becomes a new element. A uranium-238 atom, for example, will undergo eight alpha decays and six beta decays to become lead. This process of one element changing into another element is called transmutation. You will model this decay process during this activity.

Recognize the Problem

How could you create a model of a uranium-238 atom and the decay process it undergoes during transmutation?

Thinking Critically

What types of materials could you use to represent the protons and neutrons in a U-238 nucleus? How could you use these materials to model transmutation?

Possible Materials

- brown rice
- white rice
- colored candies
- dried beans
- dried seeds
- glue
- poster board

Safety Precautions

Never eat foods used in the lab.

Data Source

Refer to your textbook for general information about transmutation.

Planning the Model

1. **Choose** two materials of different colors or shapes for the protons and neutrons of your nucleus model. Choose a material for the negatively charged beta particle.
2. **Decide** how to model the transmutation process. Will you create a new nucleus model for each new element? How will you model an alpha or beta particle leaving the nucleus?
3. **Create** a transmutation chart to show the results of each transmutation step of a uranium-238 atom with the identity, atomic number, and mass number of each new element formed and the type of radiation particle emitted at each step. A uranium-238 atom will undergo the following decay steps before transmuting into a lead-206 atom: alpha decay, beta decay, beta decay, alpha decay, alpha decay, alpha decay, alpha decay, alpha decay, beta decay, beta decay, alpha decay, beta decay, beta decay, alpha decay.

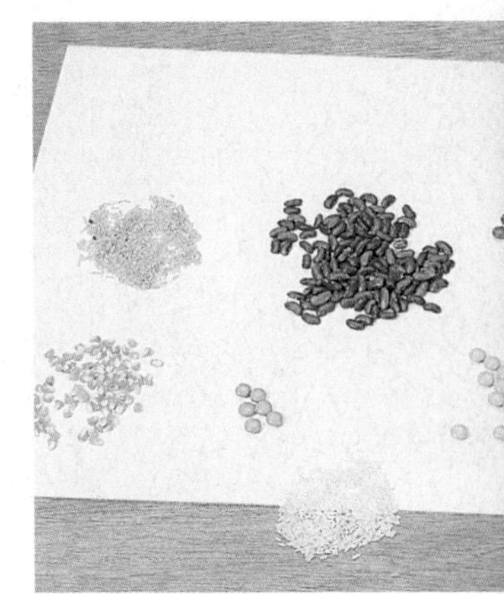

Check the Model Plans

1. **Describe** your model plan and transmutation chart to your teacher and ask how they can be improved.
2. **Present** your plan and chart to your class. Ask classmates to suggest improvements in both.

Making the Model

1. Construct your model of a uranium-238 nucleus showing the correct number of protons and neutrons.
2. Using your nucleus model, demonstrate the transmutation of a uranium-238 nucleus into a lead-206 nucleus by following the decay sequence outlined in the previous section.
3. Show the emission of an alpha particle or beta particle between each transmutation step.

Analyzing and Applying Results

1. **Compare** how alpha and beta decay change an atom's atomic number.
2. **Compare** how alpha and beta decay change the mass number of an atom.
3. **Calculate** the ratio of neutrons to protons in lead-206 and uranium-238. In which nucleus is the ratio closer to 1.5?
4. Alchemists living during the Middle Ages spent much time trying to turn lead into gold. Identify the decay processes needed to accomplish this task.

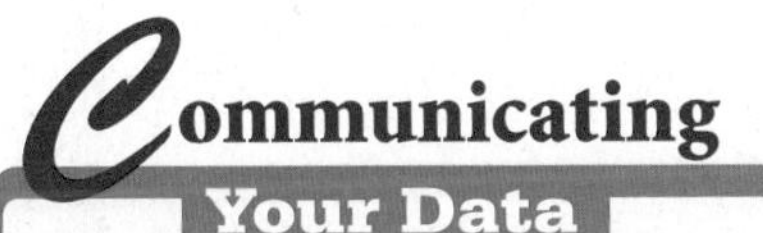

Show your model to the class and explain how your model represents the transmutation of U-238 into Pb-206.

Oops! Accidents in SCIENCE

SOMETIMES GREAT DISCOVERIES HAPPEN BY ACCIDENT!

X-Ray Surprise

The first X ray, taken in 1896 by Röntgen of a hand.

The image on the X-ray screen looked like a metal box with wires attached to it. The airport security guard blinked, but her expression didn't change. "Excuse me, sir, could you step over here?" The passenger cooperated and the mysterious metal box turned out to be a tackle box tangled up with wires from a miniature radio headset. "Next time you fly, please bring a plastic tackle box," the guard advised. "It won't set off any alarms!"

Today, X rays of people and objects are fairly routine. But back in 1895, in Germany, an X ray was unknown and about to be discovered. Physics professor Wilhelm Röntgen was experimenting with a glass tube from which most of the air had been removed.

He sent a jolt of electricity from one end of the tube to the other and tried to observe the results. "What's wrong?" he asked himself. There was too much light in the room to see what was happening in the tube. So Röntgen darkened the room, put black paper around the tube, and restarted the electricity. The tube glowed. And in the dark room, a screen also glowed. Röntgen knew that something besides light must be coming from the tube, but what? He hadn't a clue. So he called these mysterious, unknown rays, X rays.

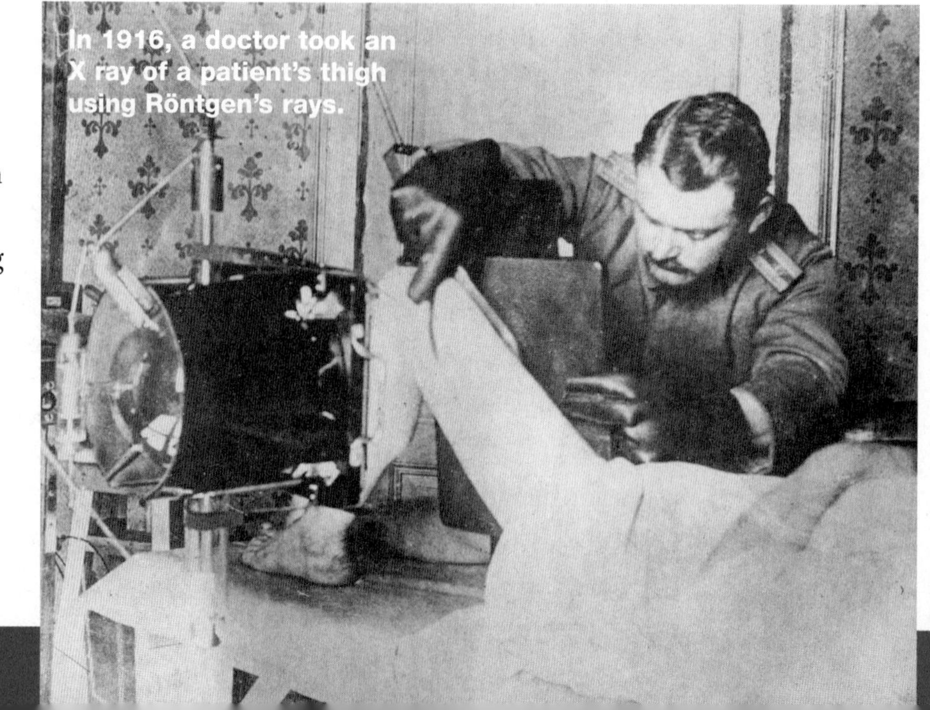

In 1916, a doctor took an X ray of a patient's thigh using Röntgen's rays.

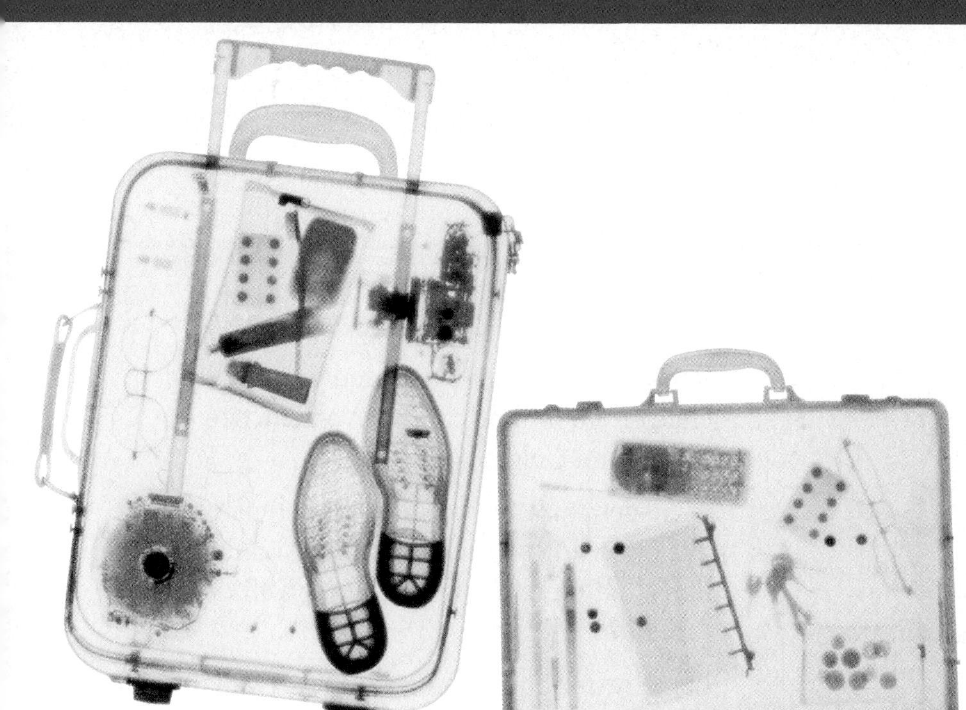

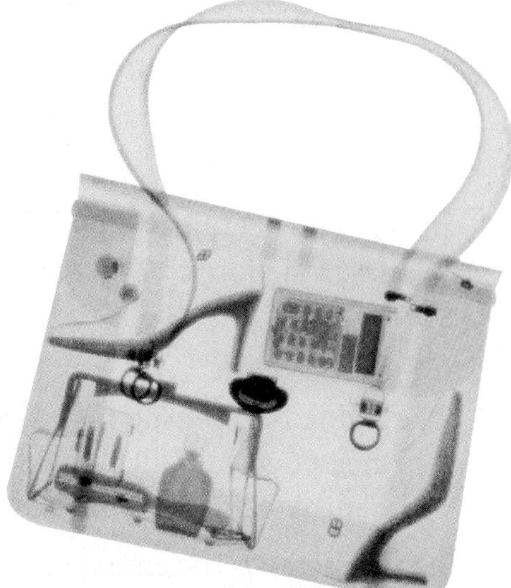

An X ray of a suitcase, a briefcase, and a handbag reveals their contents. What objects can you identify?

Medical Breakthrough

What Röntgen didn't know at the time is that X-ray radiation is a part of the electromagnetic spectrum with a shorter wavelength than light. X rays are created in an X-ray machine when a tungsten target is hit with electrons. The radiation produced passes easily through soft body tissues and materials, but is stopped by dense materials, such as metal and bone. X rays that pass through the soft material can be captured on film, leaving an outline of the dense material—an X ray.

Thanks to Röntgen's accidental discovery, doctors can look inside the human body. Fractured bones can be spotted, and certain diseases detected. However, too much exposure to X rays is dangerous to the body. But even this downside has been put to use, as radiation treatment to destroy cancerous cells.

An airport security check

X-Ray Visions

New uses for X rays continue to emerge. For example, one powerful type of X ray can look through vehicles and buildings to identify terrorists. Another new use of X rays is in identifying land mines so they can be safely removed from war zones. Other scientists are trying to focus X-ray radiation in the same way that light is focused in a laser beam. These "xasers" will allow biologists to study the structure of proteins. The most far out use is far out in space. Astronomers are using satellites to study sources of X rays from black holes deep in outer space.

CONNECTIONS **Research** Investigate the jobs of radiologists and radiology technicians. What training do they receive? How do they contribute to keeping people healthy?

SCIENCE *Online*

For more information, visit science.glencoe.com

Chapter 9 Study Guide

Reviewing Main Ideas

Section 1 Radioactivity

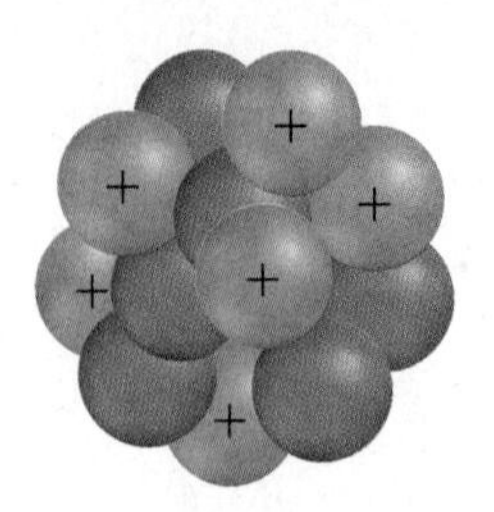

1. The protons and neutrons in an atomic nucleus are held together by the strong force. *What other force acts among particles in the nucleus?*
2. The ratio of protons to neutrons indicates whether a nucleus will be stable or unstable. Large nuclei tend to be unstable.
3. Radioactivity is the emission of energy or particles from an unstable nucleus.
4. Radioactivity was discovered accidentally by Henri Becquerel about 100 years ago.

Section 2 Nuclear Decay

1. The three common types of radiation emitted from a decaying nucleus are alpha particles, beta particles, and gamma rays. In alpha and beta decay, particles are given off. In gamma decay, energy is released.
2. Alpha and beta decay cause transmutation where the nucleus of one element changes into the nucleus of another element.
3. Half-life is the amount of time that it takes for half of the atoms in a radioactive sample to decay.
4. The half-lives of radioactive carbon and uranium isotopes can be used to calculate the ages of objects that contain these substances. *Would you use carbon-14 or uranium dating to find the age of a bone?*

Section 3 Detecting Radioactivity

1. Radioactivity can be detected with a cloud chamber, a bubble chamber, an electroscope, or a Geiger counter.
2. A Geiger counter indicates the intensity of radiation with a clicking sound or a flashing light that increases in frequency as more radiation is present.
3. Background radiation is low-level radiation emitted by naturally occurring isotopes found in Earth's rocks and soils, the atmosphere, and inside your body.

Section 4 Nuclear Reactions

1. Nuclei are split during fission and combined during fusion. In each reaction, large amounts of energy are released. *What are two ways that this energy can be used?*
2. Neutrons released from a nucleus during fission can split other nuclei, leading to a chain reaction.
3. Radioactive tracers can go to targeted areas of the body and then be detected to diagnose some health problems.
4. Some radiation can kill cancer cells.

After You Read

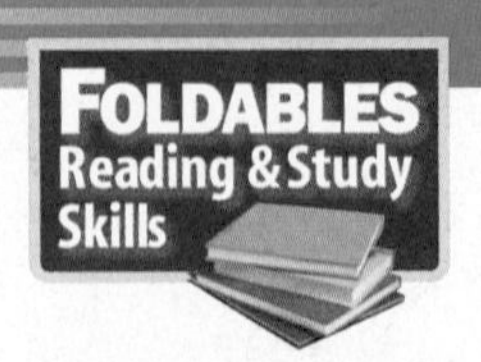

On the inside center section of your Main Ideas Study Fold, list positive and negative uses of radioactive materials and nuclear reactions.

Chapter 9 Study Guide

Visualizing Main Ideas

Complete the following concept map on radioactivity.

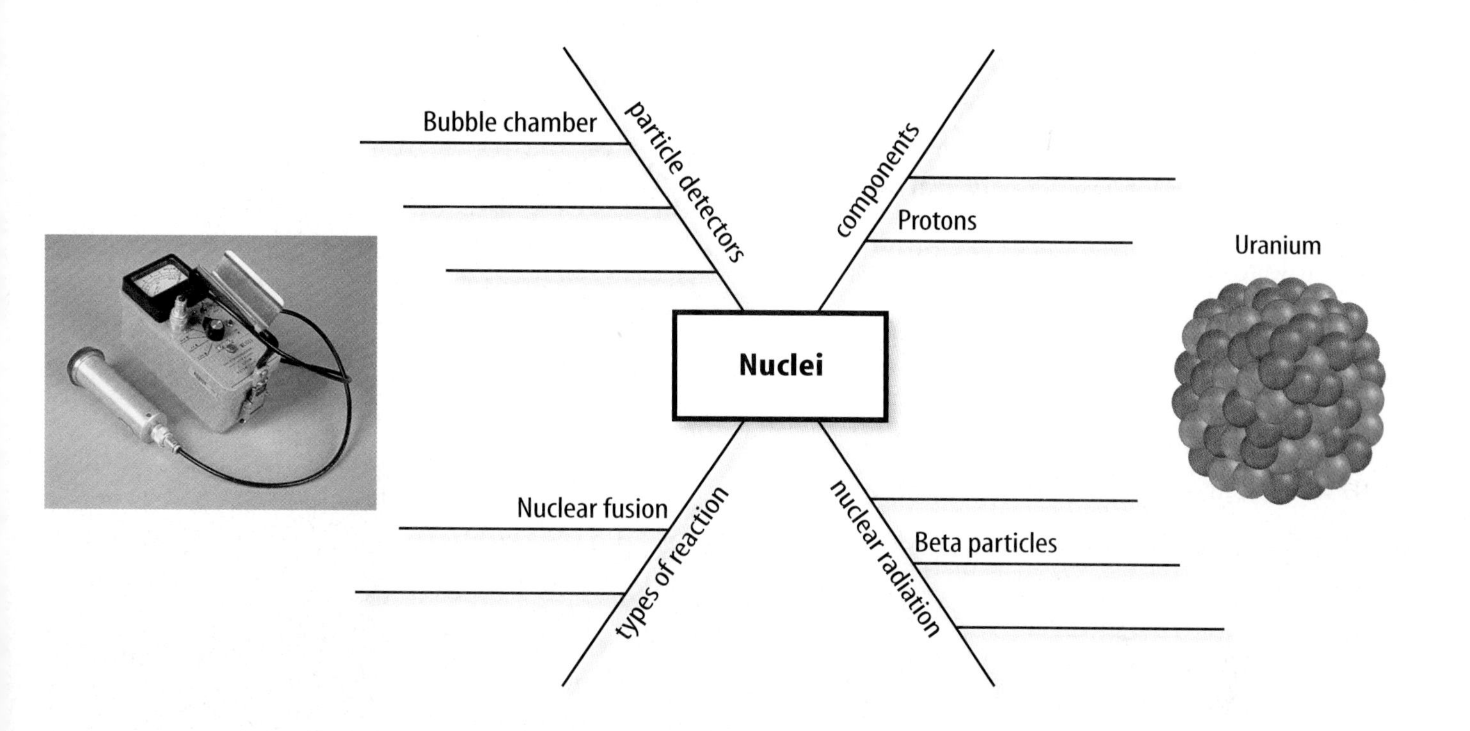

Vocabulary Review

Vocabulary Words

a. alpha particle
b. beta particle
c. bubble chamber
d. chain reaction
e. cloud chamber
f. critical mass
g. gamma ray
h. Geiger counter
i. half-life
j. nuclear fission
k. nuclear fusion
l. radioactivity
m. strong force
n. tracer
o. transmutation

Study Tip

Make a study schedule for yourself. If you have a planner, write down exactly which hours you plan to spend studying and stick to it.

Using Vocabulary

Use what you know about the vocabulary words to explain the differences in the following sets of words. Then explain how the words are related.

1. cloud chamber, bubble chamber
2. chain reaction, critical mass
3. nuclear fission, nuclear fusion
4. radioactivity, half-life
5. alpha particle, beta particle, gamma ray
6. Geiger counter, tracer
7. nuclear fission, transmutation
8. electroscope, Geiger counter
9. strong force, radioactivity

Chapter 9 Assessment

Checking Concepts

Choose the word or phrase that best answers the question.

1. What keeps particles in a nucleus together?
   A) strong force **C)** electrical force
   B) repulsion **D)** atomic glue

2. Which of the following describes all nuclei with more than 83 protons?
   A) radioactive **C)** synthetic
   B) repulsive **D)** stable

3. What is an electron that is produced when a neutron decays called?
   A) an alpha particle **C)** gamma radiation
   B) a beta particle **D)** a negatron

4. Which of the following describes an isotope's half-life?
   A) a constant time interval
   B) a varied time interval
   C) an increasing time interval
   D) a decreasing time interval

5. For which of the following could carbon-14 dating be used?
   A) a Roman scroll **C)** dinosaur fossils
   B) a marble column **D)** rocks

6. Which device would be most useful for measuring the amount of radiation in a nuclear laboratory?
   A) a cloud chamber **C)** an electroscope
   B) a Geiger counter **D)** a bubble chamber

7. Which term describes an ongoing series of fission reactions?
   A) chain reaction **C)** positron emission
   B) decay reaction **D)** fusion reaction

8. Which process is responsible for the tremendous energy released by the Sun?
   A) nuclear decay **C)** nuclear fusion
   B) nuclear fission **D)** combustion

9. Which radioisotope acts as an external source of ionizing radiation in the treatment of cancer?
   A) cobalt-60 **C)** gold-198
   B) carbon-14 **D)** technetium-99

10. Which of the following is a common medical application of radiation?
    A) assist breathing **C)** heal broken bones
    B) treat infections **D)** treat cancers

Thinking Critically

11. When a nucleus emits gamma radiation, what happens to the atomic number?

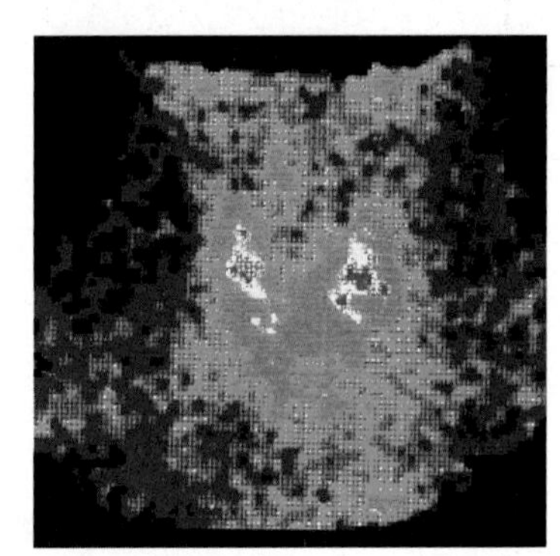

12. How do the properties of alpha particles make them harmful? Explain.

13. Why does the amount of background radiation a person receives vary greatly?

14. Explain how radioisotopes could be used to study how plants take up nutrients from the soil.

15. How can doctors tell if a patient's thyroid is not working correctly by using iodine tracers?

Developing Skills

16. **Communicating** Prepare a presentation to explain how a Geiger counter works.

17. **Making and Using Tables** Construct a table summarizing the characteristics of each of the three common types of radiation. Include the symbol for the radiation, what type of particle or energy it produces, and what it can penetrate.

18. Predicting Predict what type of radiation will be emitted during each of the following nuclear reactions:

a. uranium-238 decays into thorium-234

b. boron-12 decays into carbon-12

19. Interpreting Data Using the data below, construct a graph plotting the mass numbers versus. the half-lives of radioisotopes. Is it possible to use your graph to predict the half-life of a radioisotope given its mass number?

Isotope Half-Lives

Radioisotope	Mass Number	Half-Life
Radon	222	4 days
Thorium	234	24 days
Iodine	131	8 days
Bismuth	210	5 days
Polonium	210	138 days

20. Recognizing Cause and Effect Describe nuclear fission chain reactions. How are these chain reactions controlled? What might happen if they were not controlled?

Performance Assessment

21. Oral Presentation Research the causes and effects of radon pollution in homes. Report your findings to the class.

TECHNOLOGY

Go to the Glencoe Science Web site at **science.glencoe.com** or use the **Glencoe Science CD-ROM** for additional chapter assessment.

Test Practice

In 1903, Ernest Rutherford placed radioactive uranium ore in a lead box with a small pinhole. He aimed it so that radiation escaping from the hole passed through an electric field to reach a photographic plate.

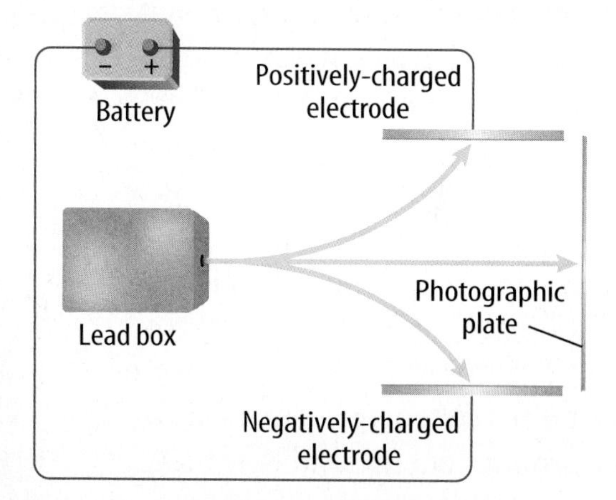

Study the diagram and answer the following questions.

1. In this experiment, beta particles were deflected toward the positive electrode while alpha particles were deflected toward the negative electrode because:

A) Beta particles are negatively charged and alpha particles are positively charged.

B) Beta particles have a negative charge and a larger mass than alpha particles.

C) Beta particles have a positive charge and alpha particles are negative.

D) Beta particles have a positive charge and a larger mass than alpha particles.

2. Why were gamma rays not deflected by the charged plates in this experiment?

F) Gamma rays have negative charge.

G) Gamma rays have positive charge.

H) Gamma rays have no charge.

J) Gamma rays travel too fast.

CHAPTER 10

Energy Sources

It takes energy to build a car. The welding torches use energy. The robots that operate the torches require energy. The assembly line runs on energy. And the car, when it is finished, will need energy to be driven. Energy heats and cools your home, refrigerates and cooks your food, pumps your water, and turns on your lights. Where does all that energy come from? How does it get to your home? Will we ever run out of energy? In this chapter, you will learn about different energy sources, how they produce energy, and how they affect the environment.

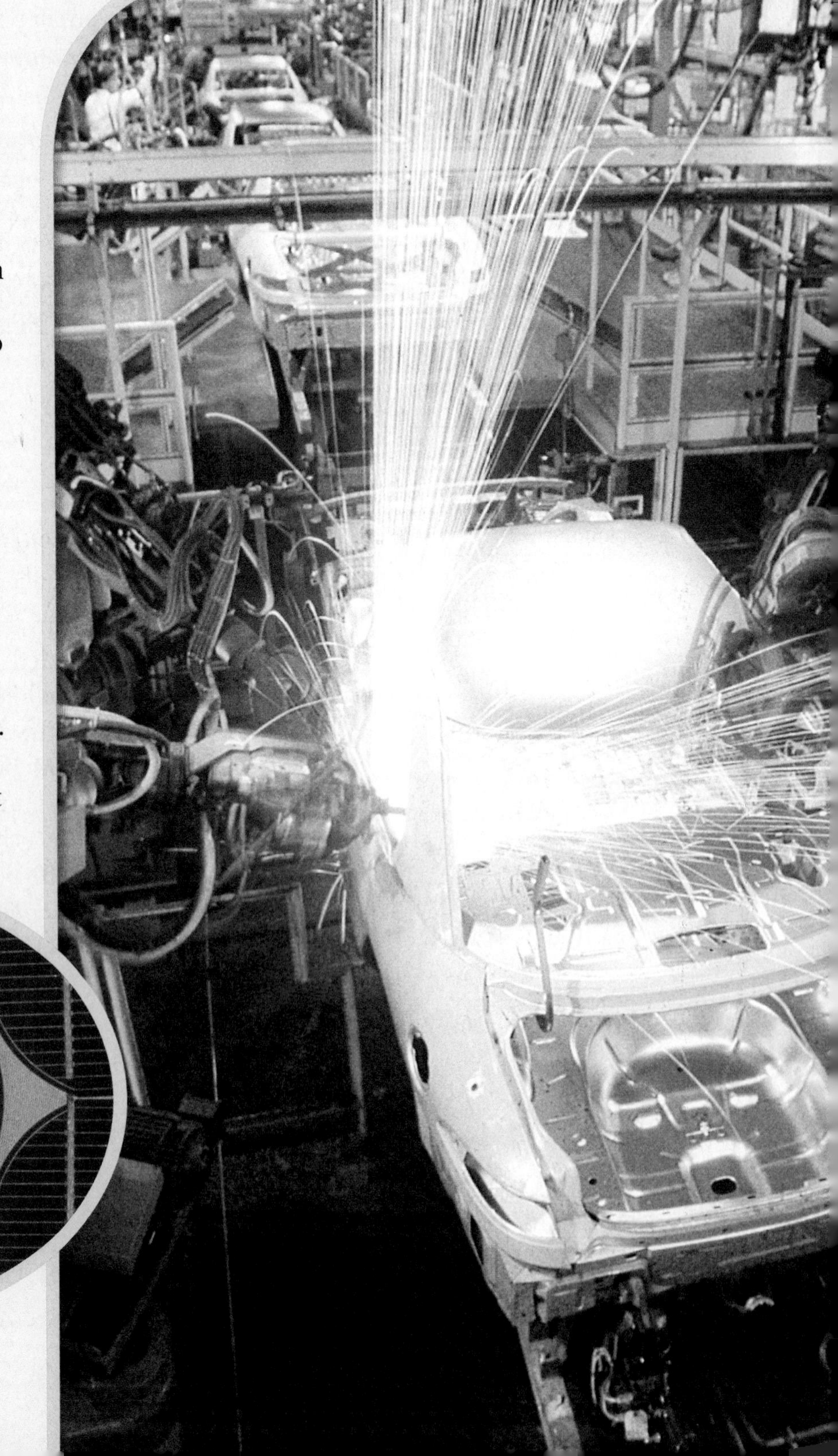

What do you think?

Science Journal Look at the picture below with a classmate. Discuss what you think this is or what is happening. Here's a hint: *It can be used to operate a calculator.* Write your answer or best guess in your Science Journal.

The Sun constantly bathes our planet with enormous amounts of energy, and this energy can be captured and used to make electricity, heat homes, and provide hot water. In this activity, you will explore a way to capture the Sun's energy to heat water.

Observe solar heating 

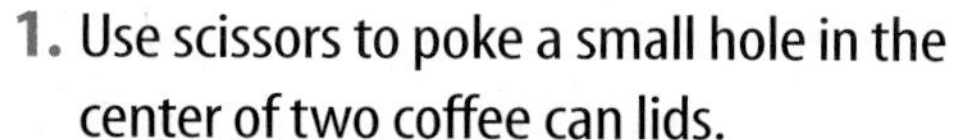

1. Use scissors to poke a small hole in the center of two coffee can lids.
2. Fill a coffee can that has been painted black with water at room temperature. Snap on the lid and push a thermometer through the hole in the lid. Record the temperature.
3. Repeat step 2 using the coffee can that has been painted white.
4. Place both cans in direct sunlight. After 15 min, record the temperature of the water in both cans again.

Observe

Write a paragraph in your Science Journal explaining why the temperature change differed between the two cans.

Before You Read

Making a Concept Map Study Fold Make the following Foldable to help you organize information by diagramming ideas about energy sources.

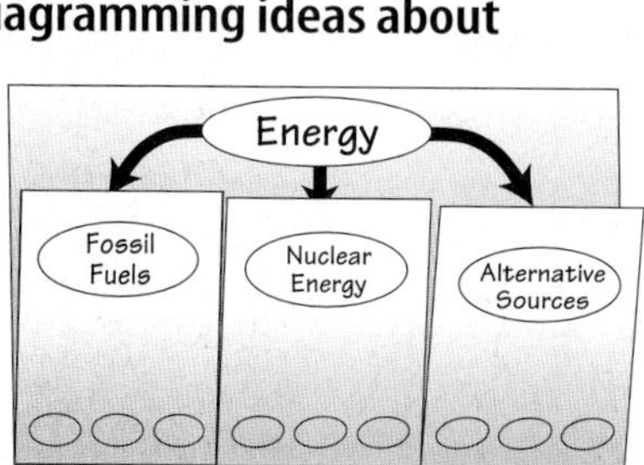

1. Place a sheet of paper in front of you so the long side is at the top. Fold the bottom of the paper up, stopping about four centimeters from the top.
2. Draw an oval above the fold. Write *Energy* in the oval.
3. Fold both sides in. Unfold. Through the top thickness of the paper, cut along each of the fold lines to form three tabs. Draw an oval on each tab and draw arrows from the large oval to the smaller ovals.
4. Write *Fossil Fuels, Nuclear Energy,* and *Alternative Sources* in the ovals. Draw three smaller ovals at the bottom of each tab, but don't write in them yet.
5. As you read the chapter, write about each source of energy under the tabs.

SECTION 1

Fossil Fuels

As You Read

What You'll Learn

- **Discuss** properties and uses of the three main types of fossil fuels.
- **Explain** how fossil fuels are formed.
- **Describe** how the chemical energy in fossil fuels is converted into electrical energy.

Vocabulary

fossil fuels
petroleum
nonrenewable resource

Why It's Important

Fossil fuels are widely used to generate electricity.

Using Energy

How many different ways have you used energy today? You can see energy being used in many ways, throughout the day, such as those shown in **Figure 1.** Furnaces and stoves use thermal energy to heat buildings and cook food. Air conditioners use electrical energy to move thermal energy outdoors. Cars and other vehicles use mechanical energy to carry people and materials from one part of the country to another.

Transforming Energy According to the law of conservation of energy, energy cannot be created or destroyed. Energy can only be transformed, or converted, from one form to another. To use energy means to transform one form of energy to another form of energy that can perform a useful function. For example, energy is used when the chemical energy in fuels is transformed into thermal energy that is used to heat your home.

Sometimes energy is transformed into a form that isn't useful. For example, when an electric current flows through power lines, about 10 percent of the electrical energy is changed to thermal energy. This reduces the amount of useful electrical energy that is delivered to homes, schools, and businesses.

Figure 1
Energy is used in many ways.

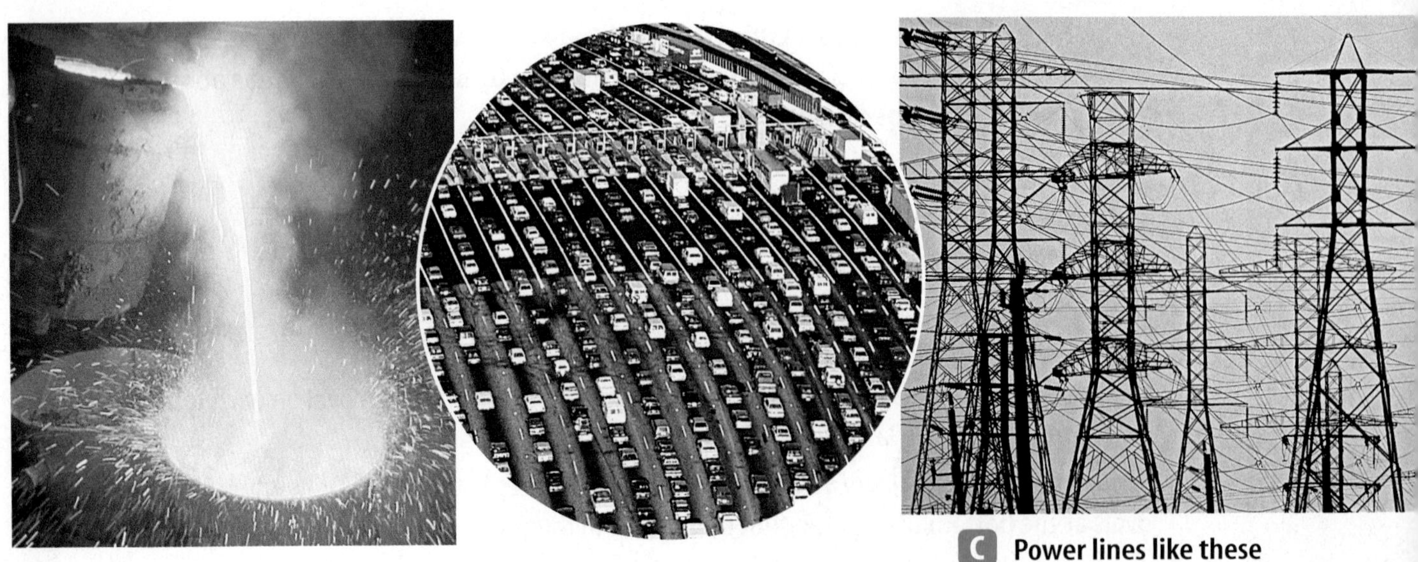

A A steel plant uses energy to make steel products.

B Automobiles burn gasoline to provide energy.

C Power lines like these carry the electrical energy you use every day.

Energy Use in the United States More energy is used in the United States than in any other country in the world. **Figure 2A** shows how energy is used in the United States. About 20 percent of the energy is used in homes for heating and cooling, to run appliances, and to provide lighting and hot water. About 27 percent is used for transportation to power vehicles such as cars, trucks, and aircraft. Another 16 percent is used by businesses to heat, cool, and light stores, shops, and office buildings. Finally, about 37 percent of this energy is used by industry and agriculture to manufacture products and produce food. **Figure 2B** shows the main sources of the energy used in the United States. Almost 85 percent of the energy used in the United States comes from burning petroleum, natural gas, and coal. Nuclear power plants provide about eight percent of the energy used in the United States.

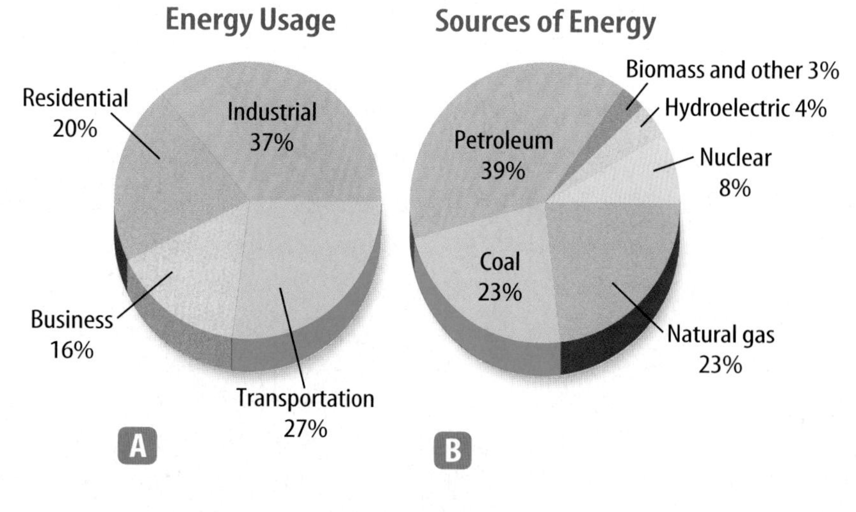

Figure 2
A This circle graph shows the percentages of energy in the United States used by homes, businesses, transportation and industry. B This circle graph shows the sources of the energy used in the United States.

Making Fossil Fuels

In one hour of freeway driving a car might use several gallons of gasoline. It may be hard to believe that it took millions of years to make the fuels that are used to produce electricity, provide heat, and transport people and materials. **Figure 4** on the next page shows how coal, petroleum, and natural gas are formed by the decay of ancient plants and animals. Fuels such as petroleum, or oil, natural gas, and coal are called **fossil fuels** because they are formed from the decaying remains of ancient plants and animals.

Concentrated Energy Sources When fossil fuels are burned, carbon and hydrogen atoms combine with oxygen molecules in the air to form carbon dioxide and water molecules. This process converts the chemical energy that is stored in the chemical bonds between atoms to heat and light. Compared to other fuels such as wood, the chemical energy that is stored in fossil fuels is more concentrated. For example, burning 1kg of coal releases two to three times as much energy as burning 1 kg of wood. **Figure 3** shows the amount of energy that is produced by burning different fossil fuels.

Figure 3
The bar graph shows the amount of energy released by burning one gram of four different fuels.

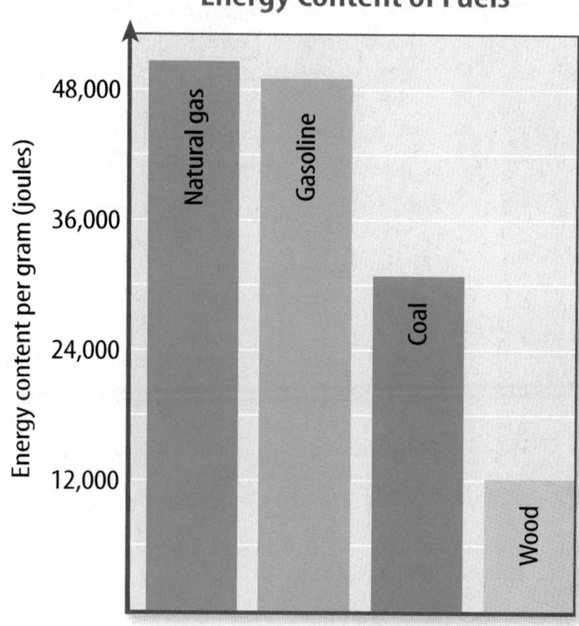

VISUALIZING THE FORMATION OF FOSSIL FUELS

Figure 4

Oil and natural gas form when organic matter on the ocean floor, gradually buried under additional layers of sediment, is chemically changed by heat and crushing pressure. The oil and gas may bubble to the surface or become trapped beneath a dense rock layer. Coal forms when peat—partially decomposed vegetation—is compressed by overlying sediments and transformed first into lignite (soft brown coal) and then into harder, bituminous (buh TYEW muh nus) coal. These two processes are shown below.

HOW OIL AND NATURAL GAS ARE FORMED

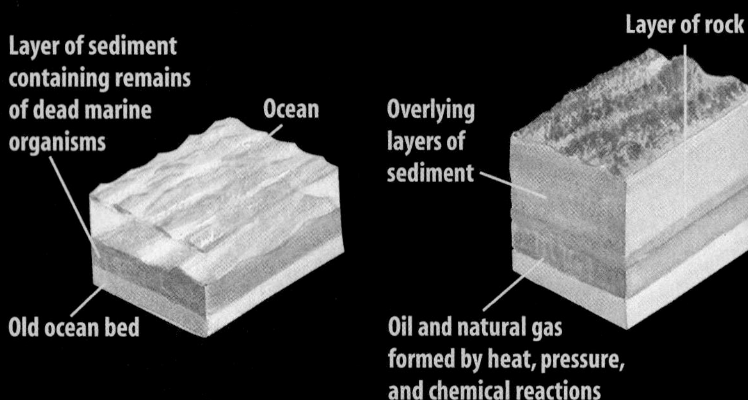

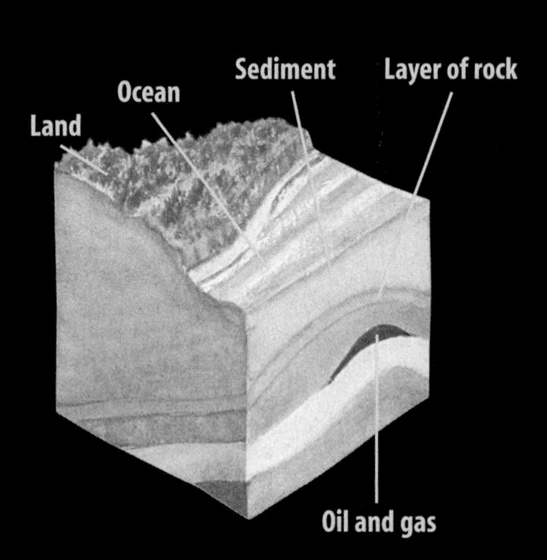

HOW COAL IS FORMED

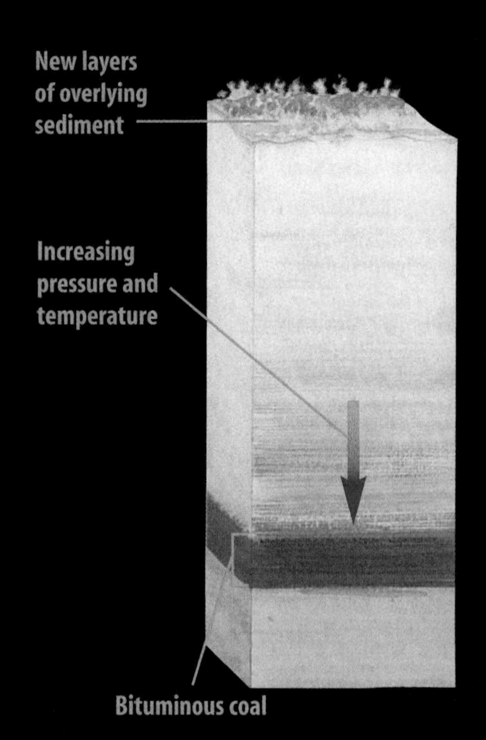

Petroleum

Millions of gallons of petroleum, or crude oil, are pumped every day from wells deep in Earth's crust. **Petroleum** is a highly flammable liquid formed by decayed ancient organisms, such as microscopic plankton and algae. Petroleum is a mixture of thousands of chemical compounds. Most of these compounds are hydrocarbons, which means they contain only carbon and hydrogen.

Separating Hydrocarbons The different hydrocarbon molecules found in petroleum have different numbers and arrangements of carbon and hydrogen atoms. The composition and structure of hydrocarbons determines their properties.

The many different compounds that are found in petroleum are separated in a process called fractional distillation. This separation occurs in the tall towers of oil-refinery plants. First, crude oil is pumped into the bottom of the tower and heated. The chemical compounds in the crude oil boil and vaporize according to their individual boiling points. Materials with the lowest boiling points rise to the top of the tower as vapor and are collected. Hydrocarbons with high boiling points, such as asphalt and some types of waxes, remain liquid and are drained off through the bottom of the tower.

✔ Reading Check *What is fractional distillation used for?*

Other Uses for Petroleum Not all of the products obtained from petroleum are burned to produce energy. About 15 percent of the petroleum-based substances that are used in the United States go toward nonfuel uses. Look around at the materials in your home or classroom. Do you see any plastics? In addition to fuels, plastics and synthetic fabrics are made from the hydrocarbons found in crude petroleum. Also, lubricants such as grease and motor oil, as well as the asphalt used in surfacing roads, are obtained from petroleum. Some synthetic materials produced from petroleum are shown in **Figure 5.**

Designing an Efficient Water Heater

Procedure

1. Measure and record the mass of a **candle.**
2. Measure 50 mL of **water** into a **beaker.** Record the temperature of the water.
3. Use the lighted candle to increase the temperature of the water by 10°C. Put out the candle and measure its mass again.
4. Repeat steps 1 to 3 with an **aluminum chimney** surrounding the candle to help direct the heat upward.

Analysis

1. Compare the mass change in the two trials. Does a smaller or larger mass change in the candle show greater efficiency?
2. Gas burners are used to heat hot-water tanks. What must be considered in the design of these heaters?

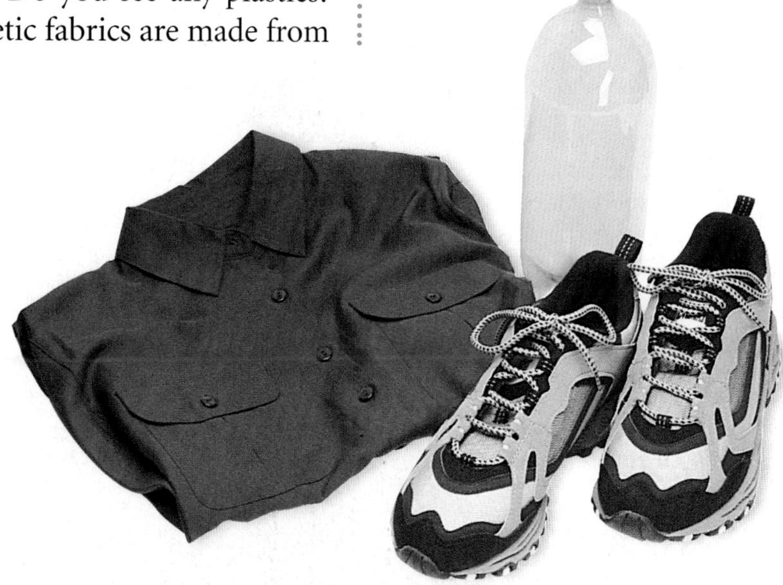

Figure 5
The objects shown here are made from chemical compounds found in petroleum.

Natural Gas

The chemical processes that produce petroleum as ancient organisms decay also produce gaseous compounds called natural gas. These compounds rise to the top of the petroleum deposit and are trapped there. Natural gas is composed mostly of methane, CH_4, but it also contains other hydrocarbon gases such as propane, C_3H_8, and butane, C_4H_{10}. Natural gas is burned to provide energy for cooking, heating, and manufacturing. About one fourth of the energy consumed in the United States comes from burning natural gas. There's a good chance that your home has a stove, furnace, hot-water heater, or clothes drier that uses natural gas.

Natural gas contains more energy per kilogram than petroleum or coal does. It also burns more cleanly than other fossil fuels, produces fewer pollutants, and leaves no residue such as ash.

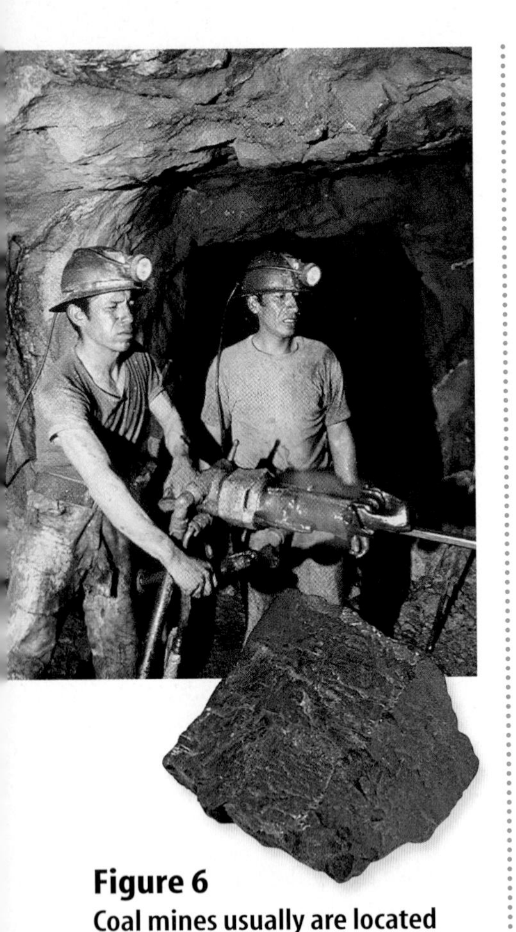

Figure 6
Coal mines usually are located deep underground.

Coal

Coal is a solid fossil fuel that is found in mines underground, such as the one shown in **Figure 6.** In the first half of the twentieth century, most houses in the United States were heated by burning coal. In fact, during this time, coal provided more than half of the energy that is used in the United States. Now almost two thirds of the energy used comes from petroleum and natural gas, and only about one fourth comes from coal. About 90 percent of all the coal that is used in the United States is burned by power plants to generate electricity.

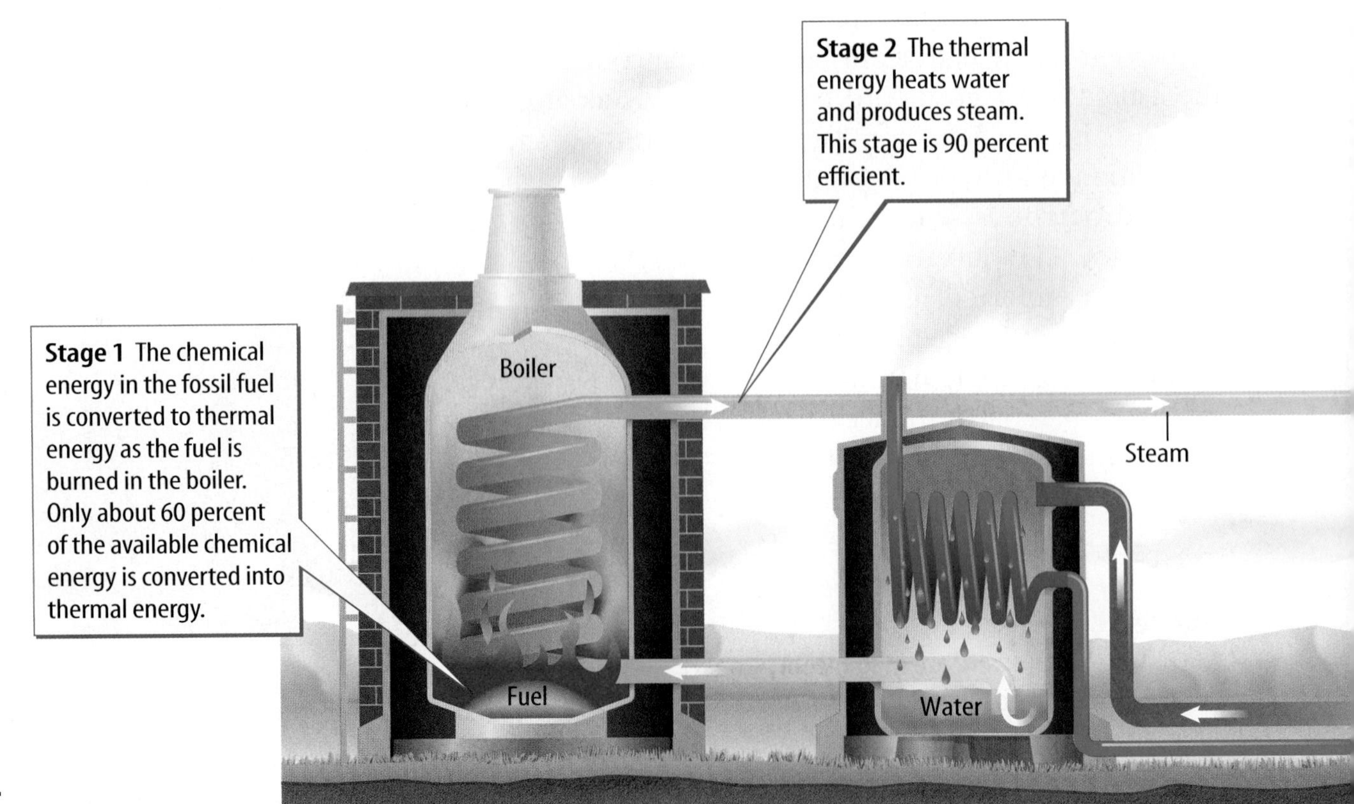

Origin of Coal Coal mines were once the site of ancient swamps where large, fernlike plants grew. Coal formed from this plant material. Worldwide, the amount of coal that is potentially available is estimated to be 20 to 40 times greater than the supply of petroleum.

Coal also is a complex mixture of hydrocarbons and other chemical compounds. Compared to petroleum and natural gas, coal contains more impurities, such as sulfur. As a result, more pollutants, such as sulfur dioxide, are produced when coal is burned.

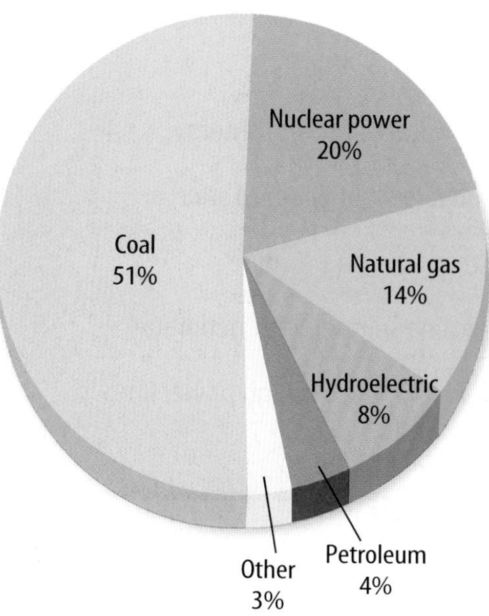

Figure 7
This circle graph shows the percentage of electricity generated in the United States that comes from various energy sources.

Generating Electricity

Figure 7 shows that almost 70 percent of the electrical energy used in the United States is produced by burning fossil fuels. How is the chemical energy contained in fossil fuels converted to electrical energy in an electric power station?

The process is shown in **Figure 8.** In the first stage, fuel is burned in a boiler or combustion chamber, and it releases thermal energy. In the second stage, this thermal energy heats water and produces steam under high pressure. In the third stage, the steam strikes the blades of a turbine, causing it to spin. The shaft of the turbine is connected to an electric generator. In the fourth stage, electric current is produced when the spinning turbine shaft rotates magnets inside the generator. In the final stage, the electric current is transmitted to homes, schools, and businesses through power lines.

Figure 8
Fossil fuels are burned to generate electricity in a power plant.

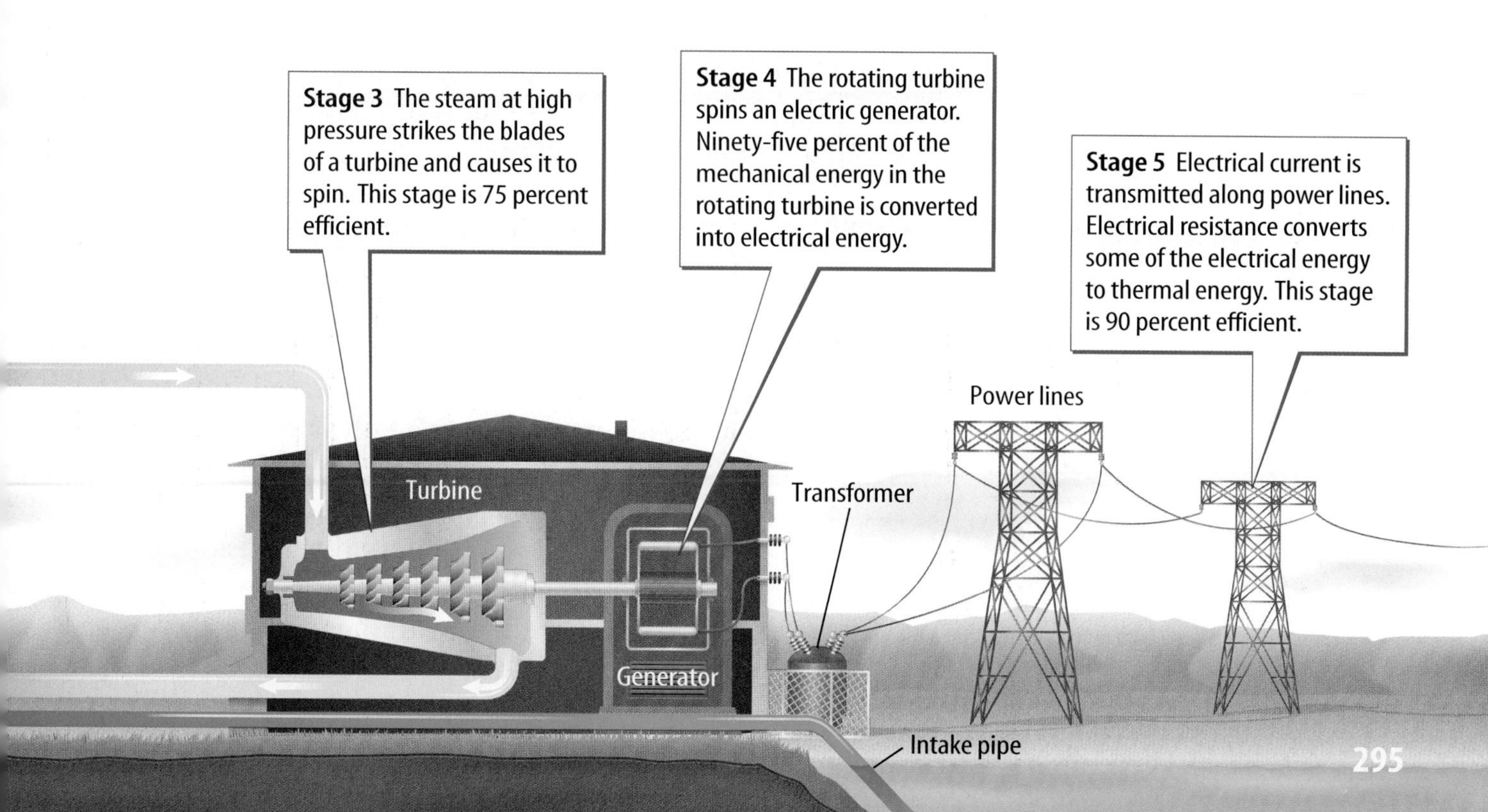

Process	Efficiency (%)
Chemical to thermal energy	60
Conversion of water to steam	90
Steam-turning turbine	75
Turbine spins electric generator	95
Transmission through power lines	90
Overall efficiency	35

Table 1 Efficiency of Fossil Fuel Conversion

Efficiency of Power Plants

When fossil fuels are burned to produce electricity, not all the chemical energy in the fuel is converted to electrical energy. Energy is lost in every stage of the process. No stage is 100 percent efficient.

The overall efficiency of the entire process is given by multiplying the efficiencies of each stage of the process shown in **Table 1**. If you were to do this, you'd find that the resulting overall efficiency is only about 35 percent. This means that only about 35 percent of the energy contained in the fossil fuels is delivered to homes, schools, and businesses as electrical energy. The other 65 percent is lost as the chemical energy in the fuel is transformed into electrical energy and delivered to energy users.

The Costs of Using Fossil Fuels

Although fossil fuels are a useful source of energy for generating electricity and providing the power for transportation, their use has some undesirable side effects. When petroleum products and coal are burned, smoke is given off that contains small particles called particulates. These particulates cause breathing problems for some people. Burning fossil fuels also releases carbon dioxide. **Figure 9** shows how the carbon dioxide concentration in the atmosphere has increased from 1960 to 1999. The increased concentration of carbon dioxide in the atmosphere might cause Earth's surface temperature to increase.

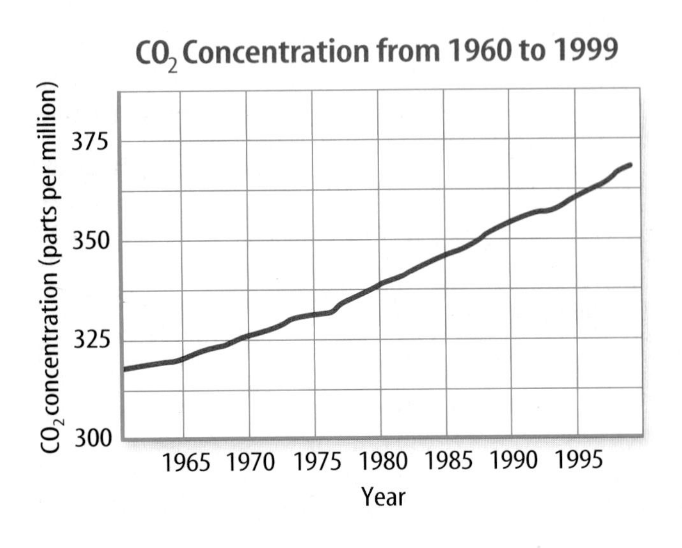

Figure 9
The carbon dioxide concentration in Earth's atmosphere has been measured at Mauna Loa in Hawaii. From 1960 to 1999, the carbon dioxide concentration has increased by about 16 percent.

Using Coal The most abundant fossil fuel is coal, but coal contains even more impurities than oil or natural gas. Many electric power plants that burn coal remove some of these pollutants before they are released into the atmosphere. Removing sulfur dioxide, for example, helps to prevent the formation of compounds that might cause acid rain. Mining coal also can be dangerous. Miners risk being killed or injured, and some suffer from lung diseases caused by breathing coal dust over long periods of time.

Nonrenewable Resources

It's a safe bet that almost any time you use an electrical appliance or ride in a car, some type of fossil fuel is the energy source that is being used. All fossil fuels are **nonrenewable resources,** which means they are resources that cannot be replaced by natural processes as quickly as they are used. Therefore, fossil fuel reserves are decreasing as population and industrial demands are increasing. **Figure 10** shows how the production of oil might decline over the next 50 years as oil reserves are used up. As the production of energy from fossil fuels continues, the remaining reserves of fuel will decrease. Fossil fuels will become more difficult to obtain, causing them to become more costly in the future.

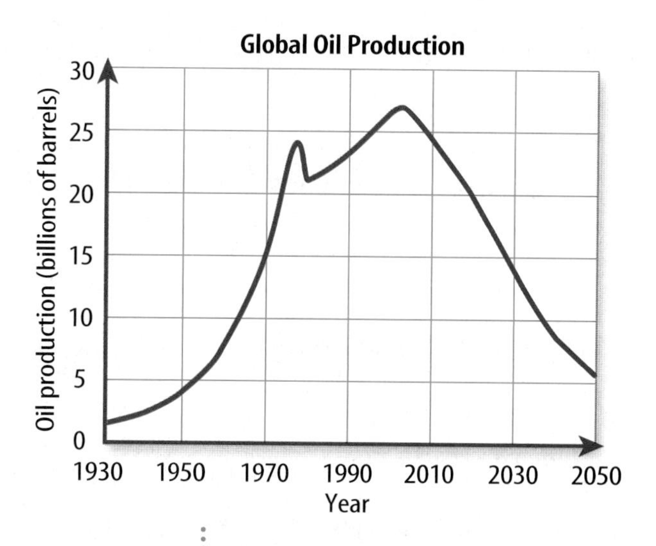

Figure 10
Some predictions show that worldwide oil production will peak by 2005 and then decline rapidly over the following 50 years.

Conserving Fossil Fuels

Even as reserves of fossil fuels decrease and they become more costly, the demand for energy continues to increase as the world's population increases. To meet these energy demands, the use of fossil fuels must be reduced and energy must be obtained from other sources. One way to reduce the use of fossil fuels is to make vehicles that are more fuel efficient. You can help reduce the demand for energy by not wasting energy in your daily activities.

Why is it important to conserve nonrenewable resources?

Section  Assessment

1. Describe the three main forms of fossil fuels.
2. What are the advantages and disadvantages of using coal to generate electricity?
3. How are the different chemicals in crude oil separated?
4. Give three examples of different products derived from chemicals in crude oil.
5. **Think Critically** If fossil fuels are still forming, why are they considered to be nonrenewable resources?

Skill Builder Activities

6. **Comparing and Contrasting** Compare and contrast the different fossil fuels. Include the advantages and disadvantages of using each as a source of energy. **For more help, refer to the Science Skill Handbook.**
7. **Communicating** In your Science Journal, make a list of areas in your school where energy use could be reduced. **For more help, refer to the Science Skill Handbook.**

Nuclear Energy

As You Read

What You'll Learn

- **Outline** the steps in the operation of a nuclear reactor.
- **Compare** the advantages and disadvantages of using nuclear energy to produce electricity.
- **Discuss** nuclear fusion as a possible energy source.

Vocabulary
nuclear reactor
nuclear waste

Why It's Important

Nuclear energy can help reduce the use of fossil fuels, but its use may cause environmental problems.

Using Nuclear Energy

Over the past several decades, electric power plants have been developed that generate electricity without burning fossil fuels. Some of these power plants, such as the one shown in **Figure 11,** convert nuclear energy to electrical energy. Energy is released when the nucleus of an atom breaks apart. In this process, called nuclear fission, an extremely small amount of mass is converted into an enormous amount of energy. Today almost 20 percent of all the electricity produced in the United States comes from nuclear power plants. Overall, nuclear power plants produce about eight percent of all the energy consumed in the United States. Currently, there are more than 100 nuclear power plants in the United States, with 6 more under construction.

Nuclear Reactors

A **nuclear reactor** uses the energy from controlled nuclear reactions to generate electricity. Although nuclear reactors vary in design, all have some parts in common, as shown in **Figure 12.** They contain a fuel that can be made to undergo nuclear fission; they contain control rods that are used to control the nuclear reactions; and they have a cooling system that keeps the reactor from being damaged by the heat produced. The actual fission of the radioactive fuel occurs in a relatively small part of the reactor known as the core.

Figure 11
A nuclear power plant generates electricity using the energy released in nuclear fission. The dome in the center of the photo contains the reactor core. A cooling tower is on the left.

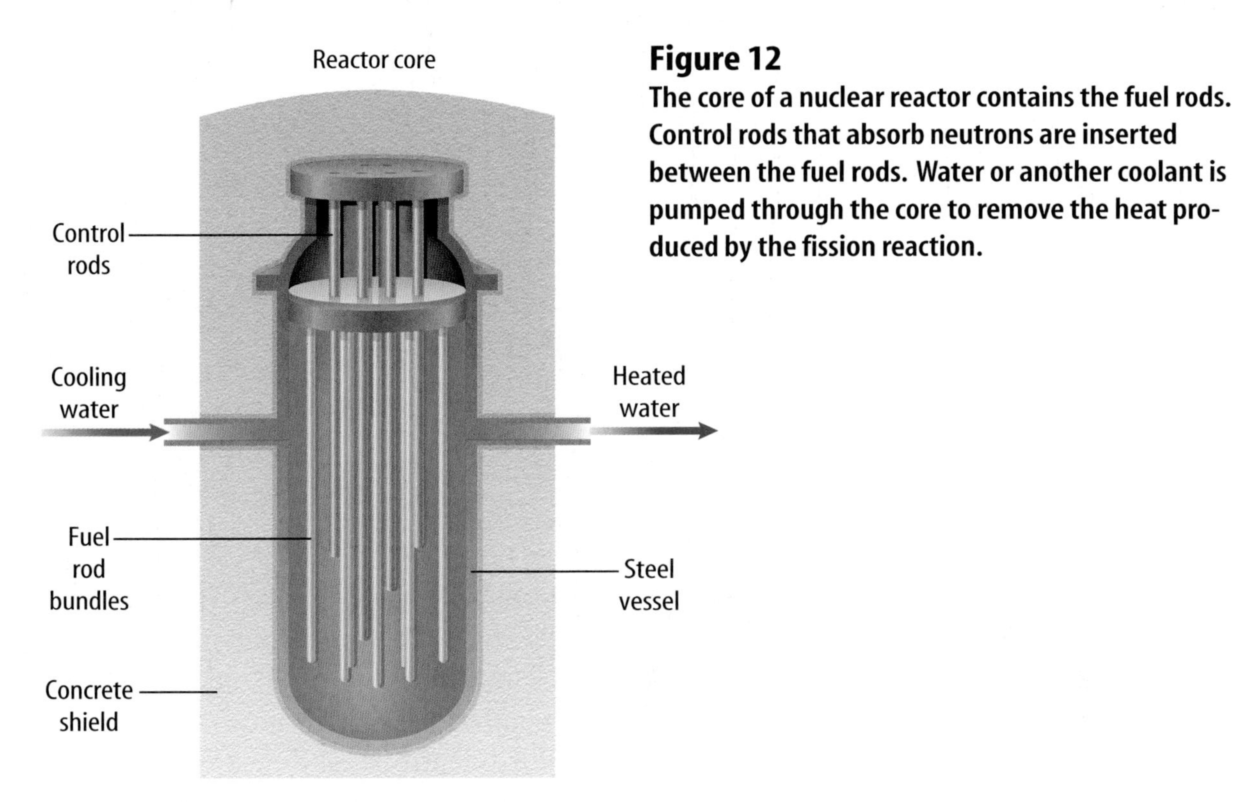

Figure 12
The core of a nuclear reactor contains the fuel rods. Control rods that absorb neutrons are inserted between the fuel rods. Water or another coolant is pumped through the core to remove the heat produced by the fission reaction.

Nuclear Fuel Only certain elements have nuclei that can undergo fission. Naturally occurring uranium contains an isotope, U-235, whose nucleus can split apart. As a result, the fuel that is used in a nuclear reactor is usually uranium dioxide. Naturally occurring uranium contains only about 0.7 percent of the U-235 isotope. In a reactor, the uranium usually is enriched so that it contains three percent to five percent U-235.

The Reactor Core The reactor core contains uranium dioxide fuel in the form of tiny pellets like the ones in **Figure 13.** The pellets are about the size of a pencil eraser and are placed end to end in a tube. The tubes are then bundled and covered with a metal alloy, as shown in **Figure 13.** The core of a typical reactor contains about a hundred thousand kilograms of uranium in hundreds of fuel rods. For every kilogram of uranium that undergoes fission in the core, 1 g of matter is converted into energy. The energy released by this gram of matter is equivalent to the energy released by burning more than 3 million kg of coal.

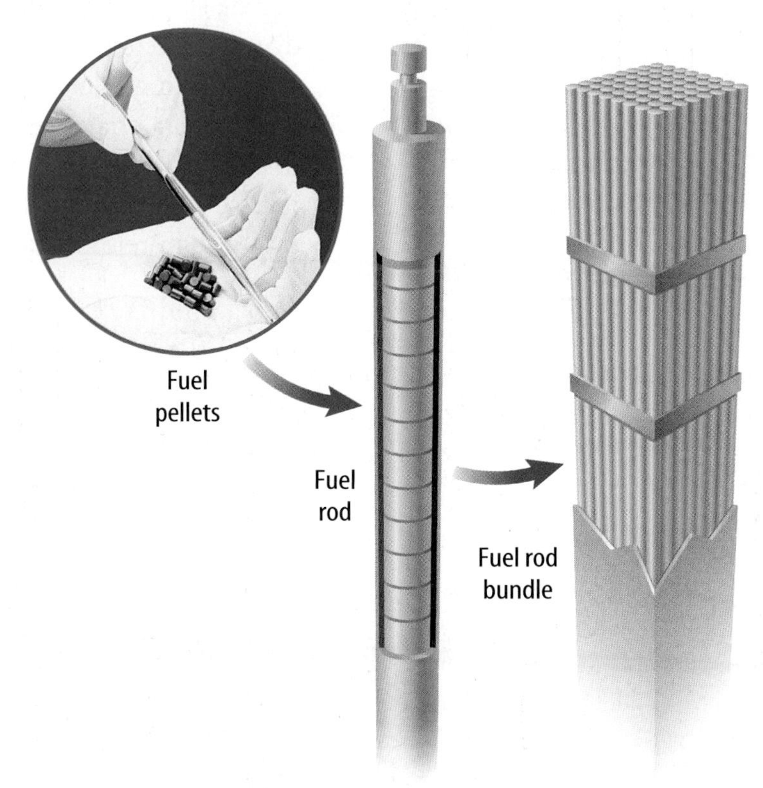

Figure 13
Nuclear fuel pellets are stacked together to form fuel rods. The fuel rods are bundled together, and the bundle is covered with a metal alloy.

Figure 14
When a neutron strikes the nucleus of a U-235 atom, the nucleus splits apart into two smaller nuclei. In the process two or three neutrons also are emitted. The smaller nuclei are called fission products.

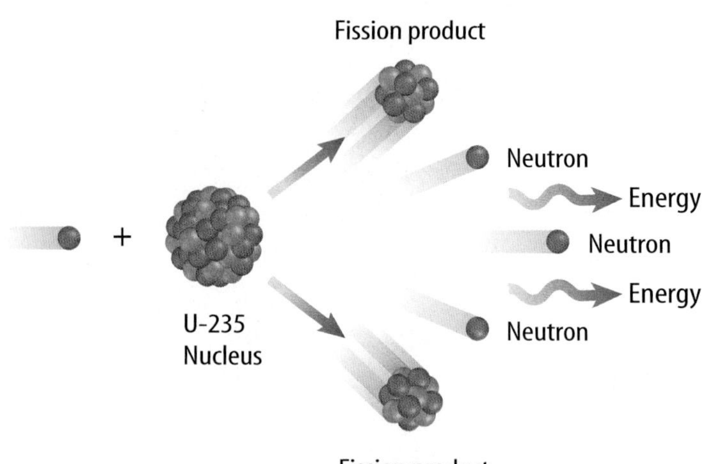

Nuclear Fission

Nuclear Fission How does the nuclear reaction proceed in the reactor core? Neutrons that are produced by the decay of U-235 nuclei are absorbed by other U-235 nuclei. When a U-235 nucleus absorbs a neutron, it splits into two smaller nuclei and two or three additional neutrons, as shown in **Figure 14.** These neutrons strike other U-235 nuclei, causing them to release two or three more neutrons each when they split apart.

Because every uranium atom that splits apart releases neutrons that cause other uranium atoms to split apart, this process is called a nuclear chain reaction. In the chain reaction involving the fission of uranium nuclei, the number of nuclei that are split can more than double at each stage of the process. As a result, an enormous number of nuclei can be split after only a small number of stages. For example, if the number of nuclei involved doubles at each stage, after only 50 stages more than a quadrillion nuclei might be split.

Nuclear chain reactions take place in a matter of milliseconds. If the process isn't controlled, the chain reaction will release an ever-increasing amount of energy each millisecond, rather than releasing energy at a constant rate.

Life Science INTEGRATION

Nuclear reactions can be harmful if they aren't carefully controlled. An accident at the Chernobyl nuclear power plant in the Ukraine in 1986 caused radiation sickness in many people and a worldwide concern about nuclear power. Research how the nuclear power industry has attempted to reduce concern about nuclear power. What type of promotional material exists for nuclear power?

Reading Check *What is a nuclear chain reaction?*

Controlling the Chain Reaction To control the chain reaction, some of the neutrons that are released when U-235 splits apart must be prevented from striking other U-235 nuclei. These neutrons are absorbed by rods containing boron or cadmium that are inserted into the reactor core. Moving these control rods deeper into the reactor allows them to absorb more neutrons and slow down the chain reaction. Eventually, only one neutron per fission reacts with a U-235 atom to produce another fission, and energy is released at a constant rate.

Nuclear Power Plants

Nuclear fission reactors produce electricity in much the same way that conventional power plants do. **Figure 15** shows how a nuclear reactor produces electricity. The thermal energy released in nuclear fission is used to heat water and produce steam. This steam then is used to drive a turbine that rotates an electric generator. To transfer thermal energy from the reactor core to heat water and produce steam, the core is immersed in a fluid coolant. The coolant absorbs heat from the core and is pumped through a heat exchanger. There thermal energy is transferred from the coolant and boils water to produce steam. The overall efficiency of nuclear power plants is about 35 percent, similar to that of fossil fuel power plants.

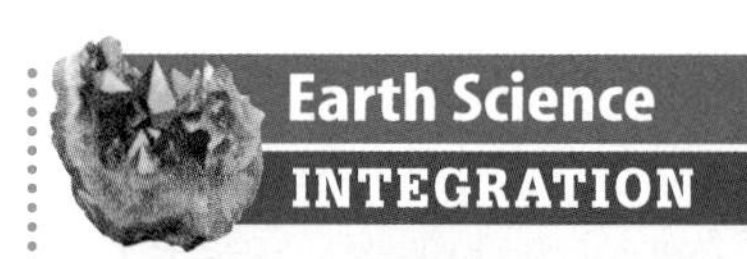

Uranium is used to determine the age of rocks in which it occurs. As uranium decays into lead at a constant rate, the age of a rock can be found by comparing the amount of uranium to the amount of lead produced. Uranium-lead dating is used by scientists to date rocks as old as 4.6 billion years. Research other methods used to determine the age of rocks.

The Risks of Nuclear Power

Producing energy from nuclear fission has some environmental advantages over burning fossil fuels. Nuclear power plants do not produce the air pollutants that are released by fossil-fuel burning power plants. Also, nuclear power plants don't produce carbon dioxide.

The nuclear generation of electricity, however, has its problems. The mining of the uranium can cause environmental damage. Water that is used as a coolant in the reactor core must cool before it is released into streams and rivers. Otherwise, the excess heat could harm fish and other animals and plants in the water.

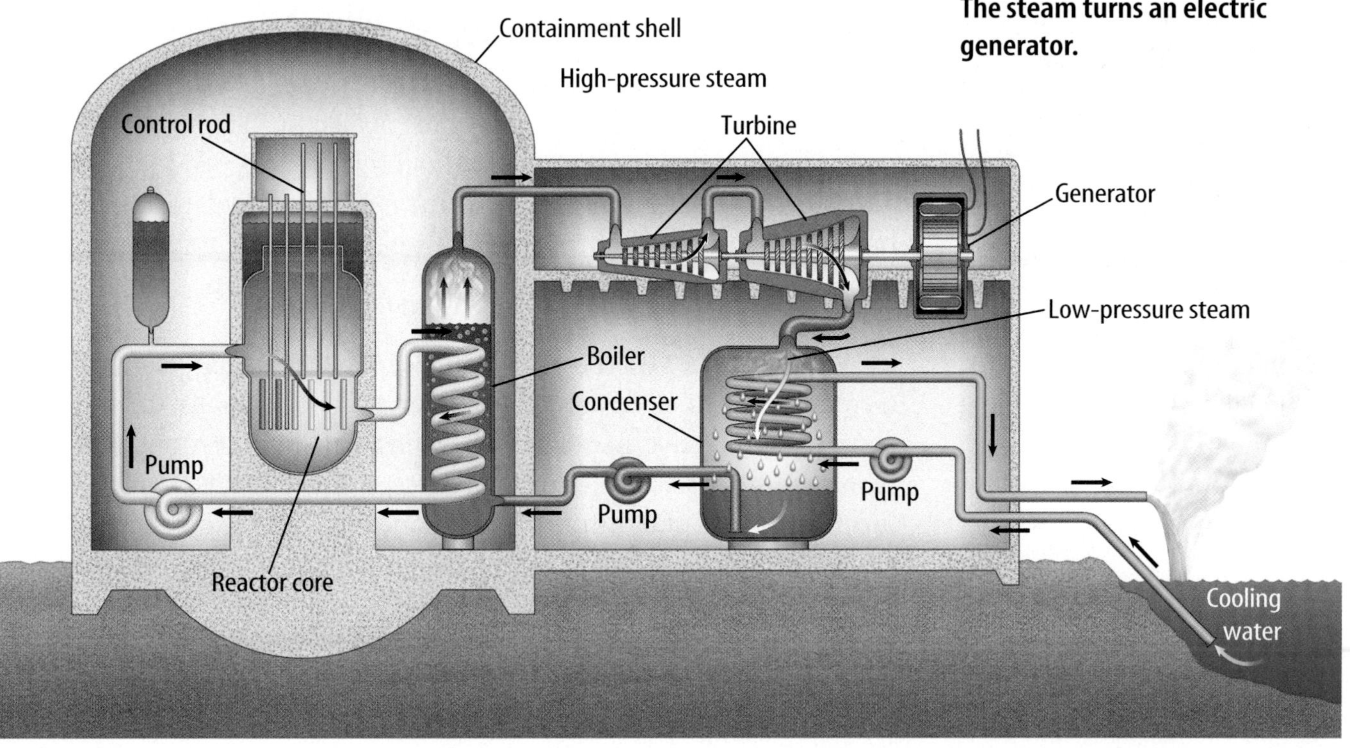

Figure 15
A nuclear power plant uses the heat produced by nuclear fission in its core to produce steam. The steam turns an electric generator.

Figure 16
An explosion occurred at the Chernobyl reactor in the Ukraine after graphite control rods caught fire. The explosion shattered the reactor's roof.

The Release of Radioactivity

One of the most serious risks of nuclear power is the escape of harmful radiation from power plants. The fuel rods contain radioactive elements with various half-lives. Some of these elements could cause damage to living organisms if they were released from the reactor core. Nuclear reactors have elaborate systems of safeguards, strict safety precautions, and highly trained workers in order to prevent accidents. In spite of this, accidents have occurred.

For example, in 1986 in Chernobyl, Ukraine, an accident occurred when a reactor core overheated during a safety test. Materials in the core caught fire and caused a chemical explosion that blew a hole in the reactor, as shown in **Figure 16.** This resulted in the release of radioactive materials that were carried by winds and deposited over a large area. As a result of the accident, 28 people died of acute radiation sickness. It is possible that 260,000 people might have been exposed to levels of radiation that could affect their health.

In the United States, power plants are designed to prevent accidents such as the one that occurred at Chernobyl. But many people still are concerned that similar accidents are possible.

The Disposal of Nuclear Waste

After about three years, not enough fissionable U-235 is left in the fuel pellets in the reactor core to sustain the chain reaction. The spent fuel contains radioactive fission products in addition to the remaining uranium. **Nuclear waste** is any radioactive by-product that results when radioactive materials are used.

Research Visit the Glencoe Science Web site at **science.glencoe.com** for a link to more information about storing nuclear wastes. Communicate to your class what you learn.

Low-Level Waste Low-level nuclear wastes usually contain a small amount of radioactive material. They usually do not contain radioactive materials with long half-lives. Products of some medical and industrial processes are low-level wastes, including items of clothing used in handling radioactive materials. Low-level wastes also include used air filters from nuclear power plants and discarded smoke detectors. Low-level wastes usually are sealed in containers and buried in trenches 30 m deep at special locations. When dilute enough, low-level waste sometimes is released into the air or water.

High-Level Waste High-level nuclear waste is generated in nuclear power plants and by nuclear weapons programs. After spent fuel is removed from a reactor, it is stored in a deep pool of water, as shown in **Figure 17.** Many of the radioactive materials in high-level nuclear waste have short half-lives. However, the spent fuel also contains materials that will remain radioactive for tens of thousands of years. For this reason, the waste must be disposed of in extremely durable and stable containers.

Figure 17
Spent nuclear fuel rods are radioactive and are placed underwater after they are removed from the reactor core. The water absorbs the nuclear radiation and prevents it from escaping into the environment.

✔ Reading Check ***What is the difference between low-level and high-level nuclear wastes?***

One method proposed for the disposal of high-level waste is to seal the waste in ceramic glass, which is placed in protective metal-alloy containers. The containers then are buried hundreds of meters below ground in stable rock formations or salt deposits. It is hoped that this will keep the material from contaminating the environment for thousands of years.

Problem-Solving Activity

Can a contaminated radioactive site be reclaimed?

In the early 1900s with the discovery of radium, extensive mining for the element began in the Denver, Colorado, area. Radium is a radioactive element that was used to make watch dials and instrument panels that glowed in the dark. After World War I, the radium industry collapsed. The area was left contaminated with 97,000 tons of radioactive soil and debris containing heavy metals and radium, which is now known to cause cancer. The soil was used as fill, foundation material, left in place, or mishandled.

Identifying the Problem

In the 1980s one area became known as the Denver Radium Superfund Site and was cleaned up by the Environmental Protection Agency. The land then was reclaimed by a local commercial establishment.

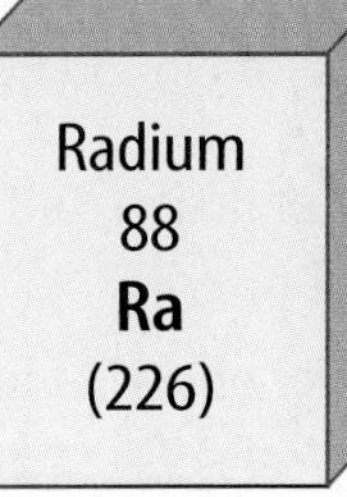

Solving the Problem

1. The contaminated soil was placed in one area and a protective cap was placed over it. This area also was restricted from being used for residential homes. Explain why it is important for the protective cap to be maintained and why homes could not be built in this area.
2. The advantages of cleaning up this site are economical, environmental, and social. Give an example of each.

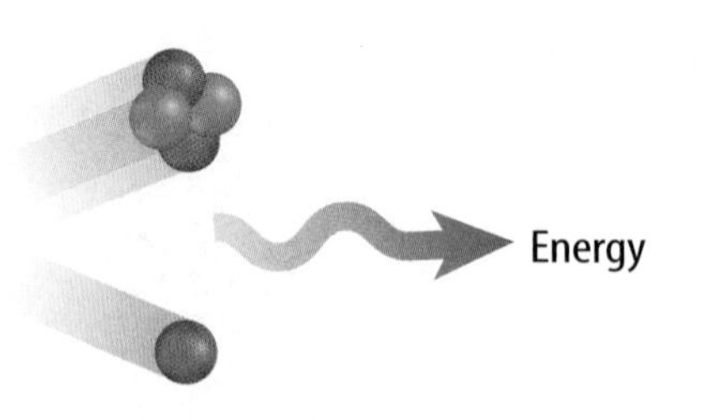

Figure 18
In nuclear fusion, two smaller nuclei join together to form a larger nucleus. Energy is released in the process. In the reaction shown here, two isotopes of hydrogen come together to form a helium nucleus.

Nuclear Fusion

Imagine the amount of energy the Sun must give off to heat Earth from 150 million kilometers away. It gets this energy from thermonuclear fusion. Thermonuclear fusion is the joining together of small nuclei at high temperatures, as shown in **Figure 18.** In this process, a small amount of mass is converted into energy. Fusion is the most concentrated energy source known.

An advantage of producing energy from fusion is that the fuel it uses, hydrogen, is the most abundant element in the universe. Hydrogen can be obtained from water in the oceans in practically limitless amounts. Another advantage is that the fusion reaction produces helium gas, which is not radioactive and is chemically nonreactive.

However, fusion reactions occur only at temperatures of millions of degrees Celsius. The biggest challenge lies in creating and maintaining the high temperatures fusion requires. To do this requires enormous amounts of energy in order to start the fusion reaction. Because of these and other problems, the use of fusion as an energy source remains in the future.

Section  Assessment

1. How is the rate of fission controlled in a nuclear reactor?
2. Explain the major obstacles in controlling nuclear fusion.
3. What types of environmental hazards do nuclear reactors produce?
4. In what way are nuclear power plants similar to those that burn fossil fuels?
5. **Think Critically** In a research project, a scientist has generated a 10-g sample of nuclear waste. The materials have a short half-life and are not present in a significant amount. How would you classify this waste, and how will it likely be disposed of?

Skill Builder Activities

6. **Concept Mapping** Using a computer, design an events-chain concept map for the generation of electricity in a nuclear fission reactor. Begin with the bombarding neutron and end with electricity in overhead lines. **For more help, refer to the Science Skill Handbook.**
7. **Using a Word Processor** Use a word processor to create a table with two divisions for the advantages and disadvantages of nuclear power. Type as many entries under each as you can. Do you think nuclear power is or isn't a good idea? Explain. **For more help, refer to the Technology Skill Handbook.**

Renewable Energy Sources

Energy Options

The demand for energy increases continually, but supplies of fossil fuels are decreasing. Using more nuclear reactors to produce electricity will produce more high-level nuclear waste that has to be disposed of safely. As a result, sources of energy that can meet Earth's increasing demands and not damage the environment are being developed. A number of possible energy sources are renewable resources. A **renewable resource** is an energy source that is replaced nearly as quickly as it is used. Although no completely adequate energy source has been found, some promising options exist.

As You Read

What **You'll Learn**

- **Analyze** the need for alternate energy sources.
- **Describe** alternate methods of generating electricity.
- **Compare** the advantages and disadvantages of various alternate energy sources.

Vocabulary

renewable resource
photovoltaic cell
hydroelectricity
geothermal energy
biomass

Why **It's Important**

Our primary energy sources are nonrenewable, so alternative energy sources need to be explored.

Energy from the Sun

The amount of solar energy that falls on the United States in one day, on average, is more than the total amount of energy used in the United States in one year. Because the Sun is expected to continue to supply energy for several billion years, solar energy cannot be used up like fossil fuels. Solar energy is a renewable resource that can provide a source of energy for the foreseeable future. Even if a small fraction of this solar energy could be used, it could significantly reduce the consumption of fossil fuels.

Solar Cells

The radiant energy from the Sun can be used to heat homes and provide hot water. This energy also can be converted directly into electricity. A device that is used to convert solar energy into electricity is the **photovoltaic cell,** which also is called a solar cell. Do you own a solar-powered calculator, like the one shown in **Figure 19?** It contains solar cells.

Reading Check *What does a photovoltaic cell do?*

Figure 19
This calculator uses a solar cell to produce the electricity it needs to operate.

Figure 20
Solar cells convert radiant energy from the Sun to electricity.

When sunlight strikes a solar cell, electrons are ejected from the electron-rich semiconductor. These electrons can travel in a closed circuit back to the electron-poor semiconductor.

Sunlight
Antireflective coating
Glass cover
Metal contact
Current

A solar cell is made of two layers of semiconductor material.

Electron-rich semiconductor
Electron-poor semiconductor
Metal contact
Current

Using Solar Power at Home

Procedure

1. Cut a piece of **cloth** into four equal sized pieces.
2. Wet the pieces and wring them out so they are the same dampness.
3. Spread the pieces out to dry—two pieces inside and two pieces outdoors. One piece of each set should be in direct sunlight and one piece should be in the shade.
4. In your **Science Journal,** record the time it takes for each cloth piece to dry.

Analysis

1. How long did it take for each cloth piece to dry?
2. What conditions determined how quickly the cloth dried?
3. Infer how you can use solar energy in your home to conserve electricity.

Making Electricity Solar cells are made of two layers of semiconductor materials sandwiched between two layers of conducting metal, as shown in **Figure 20.** One layer of semiconductor is rich in electrons, while the other layer is electron poor. When sunlight strikes the surface of the solar cell, electrons flow through an electrical circuit from the electron-rich semiconductor to the electron-poor material. This process of converting radiant energy from the Sun directly to electricity is only about 7 percent to 11 percent efficient.

Using Solar Energy Producing electricity using solar cells is, however, more expensive on a large scale than the use of nonrenewable fuels is. The cost of electricity produced by a conventional fossil-fuel power plant is about 8 cents to 10 cents per kilowatt-hour. In 1998, the cost of electricity generated by solar cells was about 28 cents per kilowatt-hour. However, in remote areas the difference in cost drops if the cost of building transmission lines to those areas is considered.

One disadvantage of using solar cells to generate electricity is that the Sun's rays do not strike any place on Earth every hour of every day. Therefore, the electricity generated by solar cells when the Sun is shining must be stored in batteries to be used when the Sun isn't out. However, a large amount of energy is needed to manufacture batteries, and large batteries contain heavy metals such as lead that are environmental hazards. In spite of this disadvantage, solar energy is an energy resource that is becoming more economical as solar technology improves. One of the world's largest and most productive solar energy plants is located in California.

Energy from Water

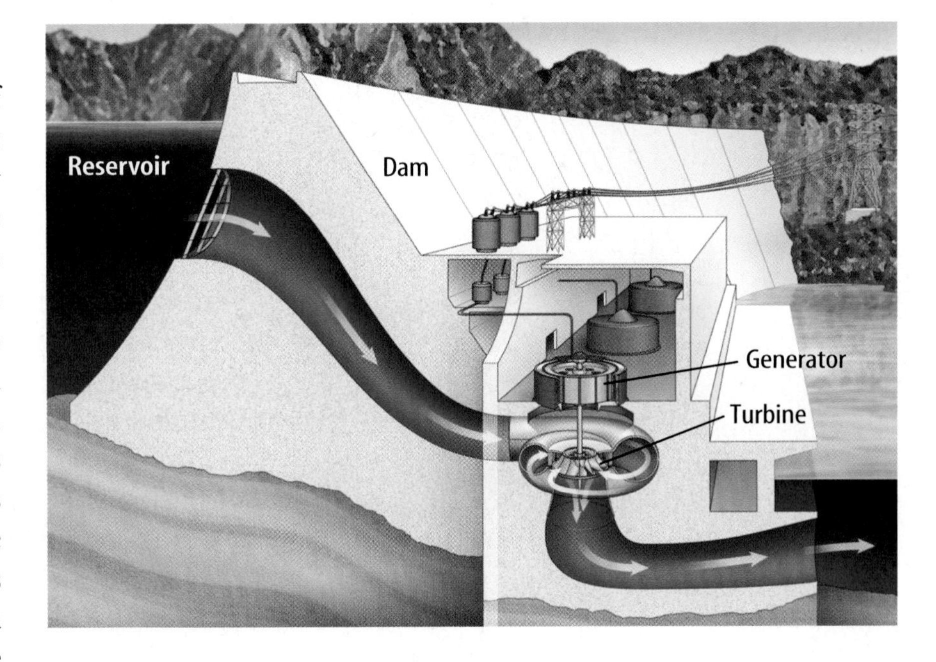

Figure 21
The potential energy contained in water stored behind the dam is converted to electrical energy in a hydroelectric power plant.

Just as the expansion of steam can turn an electric generator, a river's rapidly moving water can as well. The energy carried by water can be increased if the water is retained by a high dam. This increases the gravitational potential energy of the water. This potential energy is released when the water flows through tunnels near the base of the dam. **Figure 21** shows how the rushing water spins a turbine, which rotates the shaft of an electric generator to produce electricity. Dams built for this purpose are called hydroelectric dams.

Using Hydroelectricity Electricity produced from the energy of moving water is called **hydroelectricity.** Currently about 8 percent of the electrical energy used in the United States is produced by hydroelectric power plants. Hydroelectric power plants are an efficient way to produce electricity with almost no pollution. Because no exchange of heat is involved in producing steam to spin a turbine, hydroelectric power plants are almost twice as efficient as fossil fuel or nuclear power plants.

Another advantage is that the bodies of water held back by dams can form lakes that can provide water for drinking and crop irrigation. These lakes also can be used for boating and swimming. Also, after the initial cost of building a dam and a power plant, the electricity is relatively cheap.

However, artificial dams can disturb the balance of natural ecosystems. Some species of fish that live in the ocean migrate back to the rivers in which they were hatched to breed. This migration can be blocked by dams, which causes a decline in the fish population. Fish ladders, such as those shown in **Figure 22,** have been designed to enable fish to migrate upstream past some dams. Also, some water sources suitable for a hydroelectric power plant are not near the people needing the power.

Figure 22
Fish ladders enable fish to migrate upstream past dams.

✔ Reading Check *Could your area use a hydroelectric power plant? Explain.*

Energy from the Tides

The gravity of the Moon and Sun causes bulges in Earth's oceans. As Earth rotates, the two bulges of ocean water move westward. Each day, the level of the ocean on a coast rises and falls continually. Hydroelectric power can be generated by these ocean tides. As the tide comes in, the moving water spins a turbine that generates electricity. The water is then trapped behind a dam. At low tide the water behind the dam flows back out to the ocean, spinning the turbines and generating electric power.

Tidal energy is nearly pollution free. The efficiency of a tidal power plant is similar to that of a conventional hydroelectric power plant. However, only a few places on Earth have large enough differences between high and low tides for tidal energy to be a useful energy source. The only tidal power station in use in North America is at Annapolis Royal, Nova Scotia, shown in **Figure 23.** Tidal energy probably will be a limited, but useful, source of energy in the future.

Figure 23
This tidal energy plant at Annapolis Royal, Nova Scotia, generates 20 megawatts of electric power.

Harnessing the Wind

You might have seen a windmill on a farm or pictures of windmills in a book. These windmills use the energy of the wind to pump water. Windmills also can use the energy of the wind to generate electricity. Wind spins a propeller that is connected to an electric generator. Usually, areas that make use of wind power have several hundred windmills working together, as shown in **Figure 24.**

Figure 24
Wind energy is converted to electricity as the spinning propeller turns a generator.

However, only a few places on Earth consistently have enough wind to rely on wind power to meet energy needs. Also, windmills are only about 20 percent efficient on average. Research is underway to improve the design of wind generators and increase their efficiency. Wind generators do not consume any resources. They do not pollute the atmosphere or water. However, they can be noisy and do change the appearance of a landscape. Also, they can disrupt the migration patterns of some birds. Still, wind energy can be a useful source of energy in some areas.

Energy from Inside Earth

Earth is not completely solid. Heat is generated within Earth by the decay of radioactive elements. This heat is called geothermal heat. Geothermal heat causes the rock beneath Earth's crust to soften and melt. This hot molten rock is called magma. The thermal energy that is contained in hot magma is called **geothermal energy.**

In some places, Earth's crust has cracks or thin spots that allow magma to rise near the surface. Active volcanoes, for example, permit hot gases and magma from deep within Earth to escape. Perhaps you have seen a geyser, like Old Faithful in Yellowstone National Park, shooting steam and hot water. The water that shoots from the geyser is heated by magma close to Earth's surface. In some areas, this hot water can be pumped into houses to provide heat.

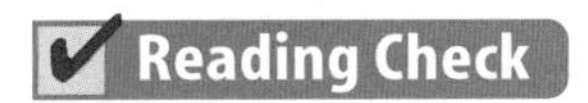

What two natural phenomena are caused by geothermal heat?

Research Visit the Glencoe Science Web site at **science.glencoe.com** for a link to more information about geothermal energy. Communicate to your class what you learn.

Geothermal Power Plants

Geothermal energy also can be used to generate electricity, as shown in **Figure 25.** Where magma is close to the surface, the surrounding rocks are also hot. A well is drilled and water is pumped into the ground, where it makes contact with the hot rock and changes into steam. The steam then returns to the surface, where it is used to rotate turbines that spin electric generators.

The efficiency of geothermal power plants is about 16 percent. Although geothermal power plants can release some gases containing sulfur compounds, pumping the water created by the condensed steam back into Earth can help reduce this pollution. However, the use of geothermal energy is limited to areas where magma is relatively close to the surface.

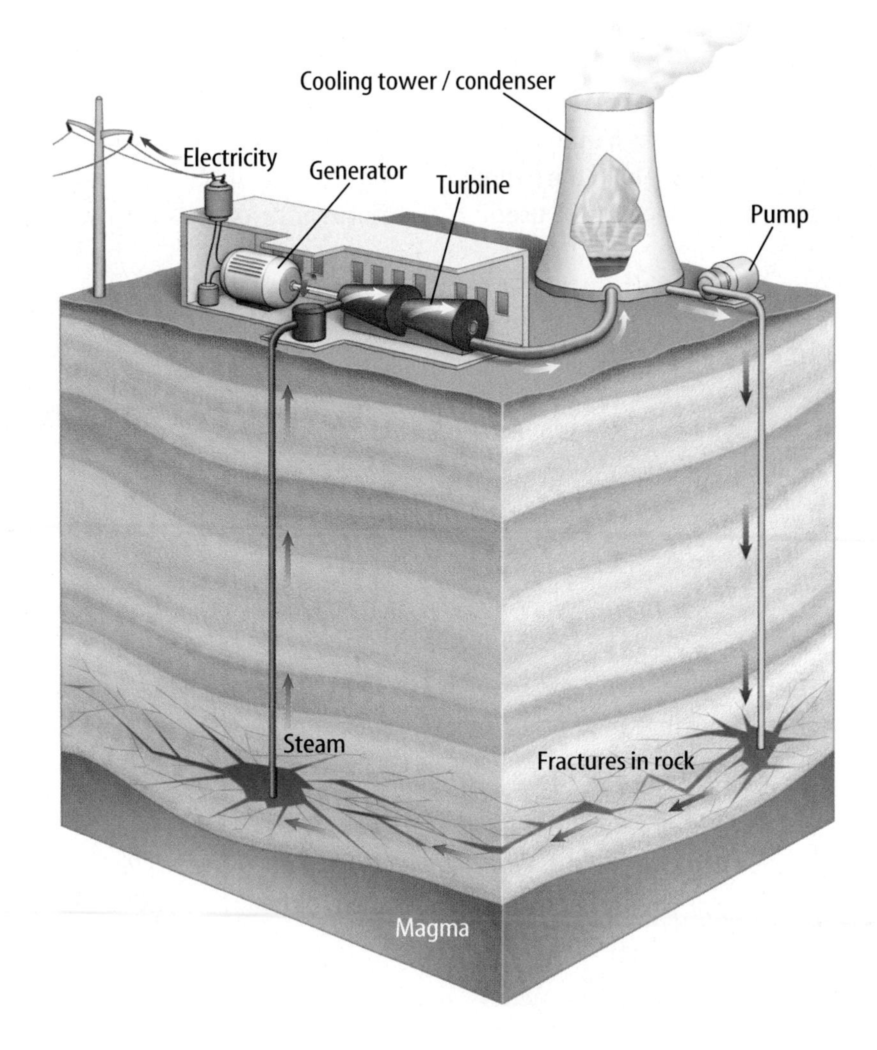

Figure 25
A geothermal power plant converts geothermal energy to electrical energy. Water is changed to steam by the hot rock, and the steam is pumped to the surface where it turns a turbine attached to an electric generator.

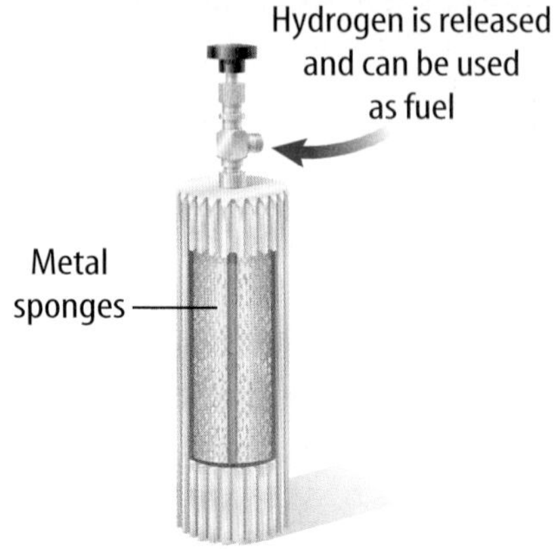

Figure 26
The hydrogen car might one day replace gasoline automobiles. One possible way of storing hydrogen is to use a metal sponge that will hold the hydrogen until it is ready to be used.

Alternative Fuels

More than two thirds of the petroleum used in the United States powers cars and other vehicles. The use of fossil fuels would be greatly reduced if cars could run on other fuels or sources of energy. For example, cars have been developed that use electrical energy supplied by batteries as a power source. Other designs use both electric motors and gasoline engines.

Hydrogen gas is another possible alternative fuel. It produces only water vapor when it burns and creates no pollution. The oceans contain an almost limitless supply of hydrogen that is combined with oxygen in water molecules. The hydrogen in water can be released by passing an electric current through the water. Producing the electric current, however, requires more energy than can be generated by burning the hydrogen gas that is produced. Other ways of producing hydrogen are being studied and might be useful in the future. **Figure 26** shows one possible way to store the fuel.

Biomass Fuels Fossil fuels and nuclear fission produce electricity by heating water. Could any other materials be used to heat water and produce electricity? **Biomass,** for example, is renewable organic matter, such as wood, sugarcane fibers, rice hulls, and animal manure. It, too, can be burned in the presence of oxygen to convert the stored chemical energy to thermal energy. In fact, burning biomass is probably the oldest use of natural resources for meeting human energy needs.

Section  Assessment

1. Why do humans need to develop and use alternative energy sources?
2. Describe three ways that solar energy can be used.
3. How is the generation of electricity by hydroelectric, tidal, and wind sources similar to each other? How is it similar to fossil fuel and nuclear power?
4. Why is geothermal energy unlikely to become a major energy resource?
5. **Think Critically** What single resource do most energy alternatives depend on, either directly or indirectly?

Skill Builder Activities

6. **Classifying** On a computer, draw a chart of the energy sources in this section. List the advantages and disadvantages of each. Which source do you feel is most promising? Why? **For more help, refer to the Science Skill Handbook.**
7. **Using Percentages** U.S. sources of energy follow: 39 percent petroleum, 23 percent natural gas, 23 percent coal, 8 percent nuclear, and 7 percent other. If the percent of nuclear energy was shown with a 1-m strip of paper, how long would the other strips be? **For more help, refer to the Math Skill Handbook.**

Activity

Solar Heating

Energy from the Sun is absorbed by Earth and makes its temperature warmer. In a similar way, solar energy also is absorbed by solar collectors to heat water and buildings. Does the rate at which an object absorbs solar energy depend on the color of the object?

What You'll Investigate

How does color affect the amount of heat absorbed from the Sun?

Materials

small cardboard boxes
black, white, and colored paper
tape or glue
thermometer
watch with a second hand

Goals

- **Demonstrate** solar heating.
- **Compare** the effectiveness of heating items of different colors.
- **Graph** your results.

Procedure

1. Cover at least three small boxes with colored paper. The colors should include black and white as well as at least one other color.
2. Copy the following data table into your Science Journal. Replace "Other color" with whatever color you are using.
3. Place the three objects on a windowsill or other sunny spot and note the starting time.
4. **Measure and record** the temperature inside each box at 2-min intervals for at least 10 min.

Conclude and Apply

1. **Graph** your data using a line graph.
2. **Describe** the shapes of the lines on your graph. What color heated up the fastest? Which heated up the slowest?
3. **Explain** why the colored boxes heated at different rates.
4. Suppose you wanted to heat a tub of water using solar energy. Based on what you discovered in this activity, what color would you want the tub to be? Explain.
5. **Explain** why you might want to wear a white or light-colored shirt on a hot, sunny, summer day.

Communicating Your Data

Compare your results with those of other students in your class. Discuss any differences found in your graphs, particularly if different colors were used by different groups.

Temperature Due to Different Colors

Color	Minute 2	Minute 4	Minute 6	Minute 8	Minute 10
Black					
White					
Other color					

Activity Use the Internet

How much does energy really cost?

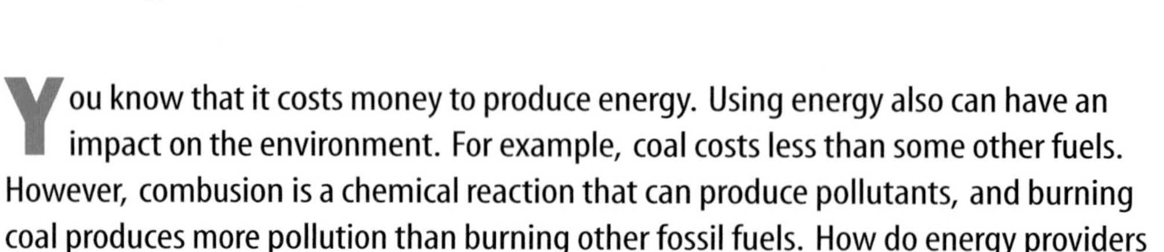

You know that it costs money to produce energy. Using energy also can have an impact on the environment. For example, coal costs less than some other fuels. However, combusion is a chemical reaction that can produce pollutants, and burning coal produces more pollution than burning other fossil fuels. How do energy providers convince consumers that their service is the most cost-efficient and the least polluting?

Recognize the Problem

What are the costs and environmental impacts of various energy-producing sources?

Form a Hypothesis

Form a hypothesis about which energy source you think will have the lowest cost and which will have the least impact on the environment.

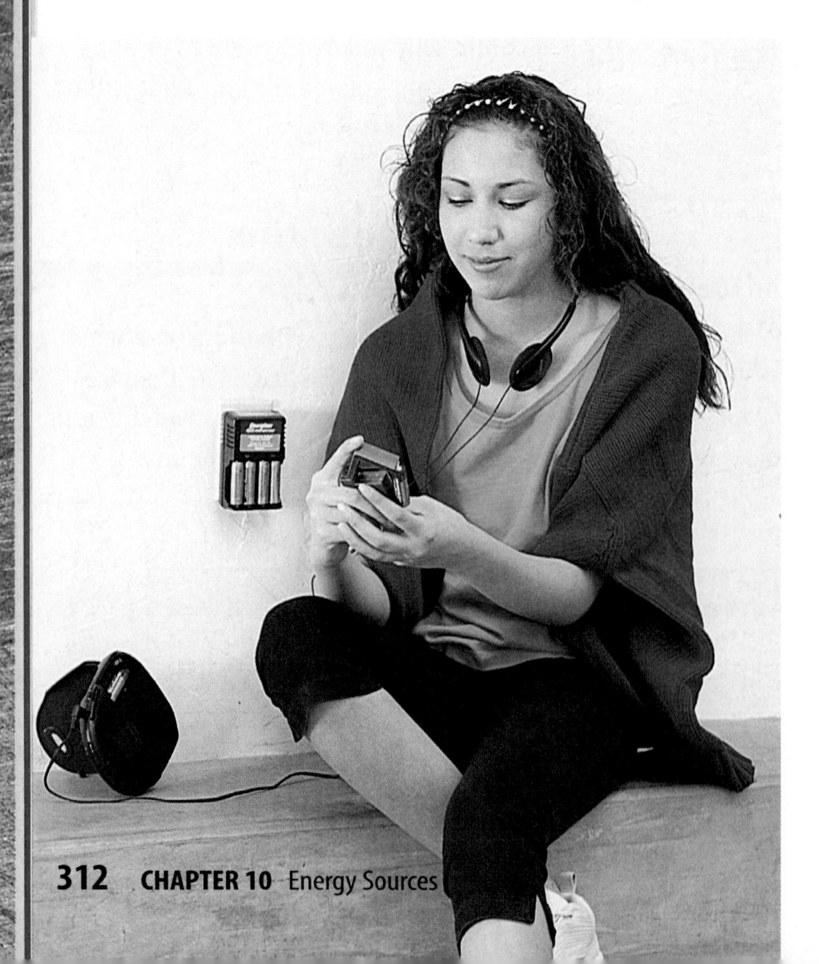

Goals

- **Identify** three energy sources that people use.
- **Determine** the cost of each source.
- **Describe** the environmental impact of each source.
- **Describe** which energy source is most cost-efficient as well as the one which causes the least environmental impact.

Data Source

SCIENCE*Online* Go to the Glencoe Science Web site at **science.glencoe.com** to get more information about various energy sources and services and for data collected by other students.

Test Your Hypothesis

Plan

1. Think about the various sources of power used in different areas of the United States and make a list of possible energy sources to investigate.
2. Find the cost of 1 kWh of electric energy generated by three of these energy sources.
3. Determine whether your sources have a negative impact on the environment. Which sources are renewable?
4. Use your data to create a table of energy sources, costs, and impacts.
5. Write a summary describing which of your three energy sources is the best for producing 1 kWh of electric energy. Consider the cost of the energy and the environmental impact. Provide facts from your research to support your conclusions.

Energy Sources

Energy Source	Cost per kWh	Environmental Impacts
Energy source 1		
Energy source 2		
Energy source 3		

Do

1. Make sure your teacher approves your plan before you start.
2. Go to the Glencoe Science Web site at **science.glencoe.com** to post your data.

Analyze Your Data

1. Of the energy sources you investigated, which is the most expensive to use? The least expensive?
2. Which energy source do you think has the most impact on the environment? The least impact?

Draw Conclusions

1. Find this *Use the Internet* activity on the Glencoe Science Web site at **science.glencoe.com.** Post your data in the table provided. Compare your data to that of other students.
2. Of the energy sources you investigated, which is the least expensive energy source? Which is the best choice to use? Why or why not?
3. Of the energy sources you investigated, how did the environmental impact of that power influence your choice of the best energy solution?
4. Which data support your decision?

Make a poster of magazine pictures to illustrate impact on the environment for each of the three energy sources.

Reacting to

Solar panels help power homes in this development.

The power of the wind is harnessed by these wind turbines in California.

Most people agree that thanks to energy sources, we have many things that make our quality of life better. Energy runs our cars, lights our homes, and powers our appliances. What many people don't agree on is where that energy should come from. Nuclear energy is a topic that stirs up strong opinions in people. As you read the summaries of the issues given here, think about your own opinions.

A Question of the Environment

Almost all of the world's electric energy is produced by thermal power plants. Most of these plants burn fossil fuels—such as coal, oil, and natural gas—to produce energy. Nuclear energy is produced by fission, which is the splitting of an atom's nucleus. People in favor of nuclear energy argue that, unlike fossil fuels, nuclear energy is nonpolluting. When coal is burned, it releases large amounts of sulfur and other pollutants into the air. These pollutants contribute to serious environmental problems such as smog and acid rain. Uranium, the key fuel for nuclear reactors, releases no chemical or solid pollutants into the air during use.

Opponents counter, though, that the poisonous radioactive waste created in nuclear reactors qualifies as pollution—and will be lingering in the ground and water for hundreds of thousands of years.

Supporters of nuclear energy also cite the spectacular efficiency of nuclear energy—one metric ton of nuclear fuel produces the same amount of energy as up to 3 million tons of coal. Opponents point out that uranium is in very short supply and, like fossil fuels, is likely to run out in the next 100 years.

Nuclear Energy

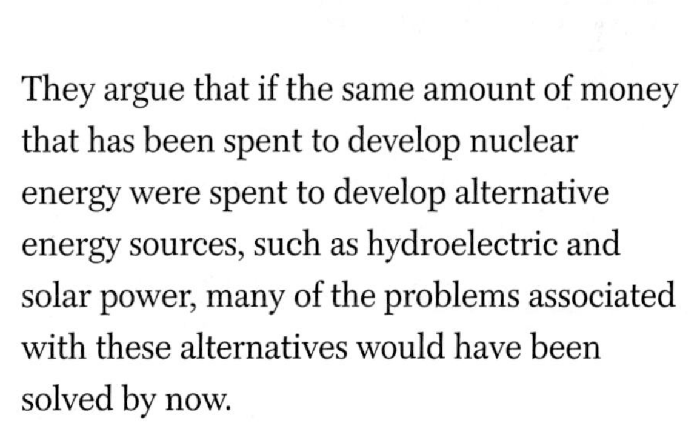

Radioactivity caused by an accident at this Ukraine nuclear power plant has given this baby physical problems.

A Question of Health & Safety

Opponents of nuclear energy point out the health dangers associated with mining and processing nuclear fuel. *Radiation sickness* is the term for a variety of symptoms that result when a person is exposed to damaging amounts of radiation. Exposure to high levels of radiation can cause lasting illness or even death. Opponents worry that as utilities come under less government regulation, safety standards will be ignored in the interest of profit. This could result in more accidents like the one that occurred at Chernobyl in the Ukraine. There, an explosion in the reactor core released radiation over a wide area.

Supporters counter that it will never be in the best interests of those running nuclear plants to relax safety standards since those safety standards are the best safeguard of workers' health. They cite the overall good safety record of nuclear power plants.

A Third Side of the Coin

Others argue that the solution to energy woes lies elsewhere. They say nuclear energy and fossil fuels are both non-renewable and produce dangerous by-products—and that investments should be made in sources of energy that are renewable and safe. They argue that if the same amount of money that has been spent to develop nuclear energy were spent to develop alternative energy sources, such as hydroelectric and solar power, many of the problems associated with these alternatives would have been solved by now.

This view is challenged by those who say that some alternative energy sources are not always available to people. For example, tidal energy isn't available everywhere, and solar power will not work well in areas that receive little sunlight.

CONNECTIONS Debate Form three teams and have each team defend one of the views presented here. If you need more information, go to the Glencoe Science Web site. "Debrief" after the debate. Did the arguments change your understanding of the issues?

SCIENCE *Online*

For more information, visit science.glencoe.com

Chapter 10 Study Guide

Reviewing Main Ideas

Section 1 Fossil Fuels

1. Fossil fuels include oil, natural gas, and coal. They formed from the buried remains of plants and animals.

2. Fossil fuels can be burned to supply energy for generating electricity. Petroleum has other uses, as well. *What materials in this picture might have been made from petroleum?*

3. Fossil fuels are nonrenewable energy resources. They can be replaced but it takes millions of years.

Section 2 Nuclear Energy

1. A nuclear reactor uses the energy from a controlled nuclear chain reaction to generate electricity.

2. Nuclear wastes must be contained and disposed of carefully so radiation from nuclear decay will not leak into the environment. *What are the differences between low-level wastes, shown below, and high-level wastes?*

3. Nuclear fusion releases energy when two nuclei combine. Fusion only occurs at high temperatures that are difficult to produce in a laboratory.

Section 3 Renewable Energy Sources

1. Alternate energy resources can be used to supplement or replace nonrenewable energy resources.

2. Other sources of energy for generating electricity include hydroelectricity and solar, wind, tidal, and geothermal energy. Each source has its advantages and disadvantages. Also, some of these sources can damage the environment.

3. Although some alternate energy sources produce less pollution than fossil fuels do and are renewable, their use often is limited to the regions where the energy source is available. *What type of alternate energy might provide power for the area in this photo?*

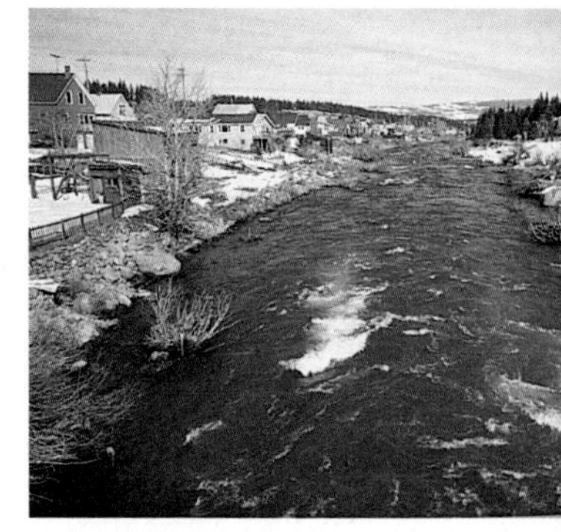

4. It's possible that humans might one day drive hydrogen-powered cars. Biomass is an alternate fuel that has been used for thousands of years.

After You Read

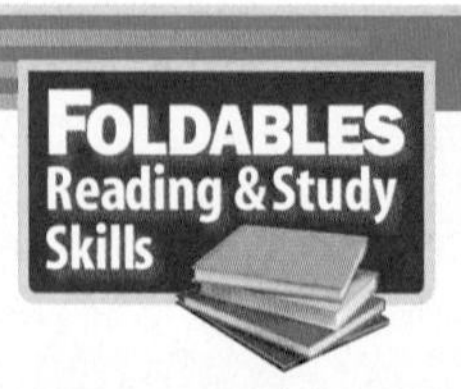

Using information from the chapter, list an example of each energy source in the smaller ovals on the front of your Concept Map Study Fold.

Visualizing Main Ideas

Complete the following concept map on energy sources.

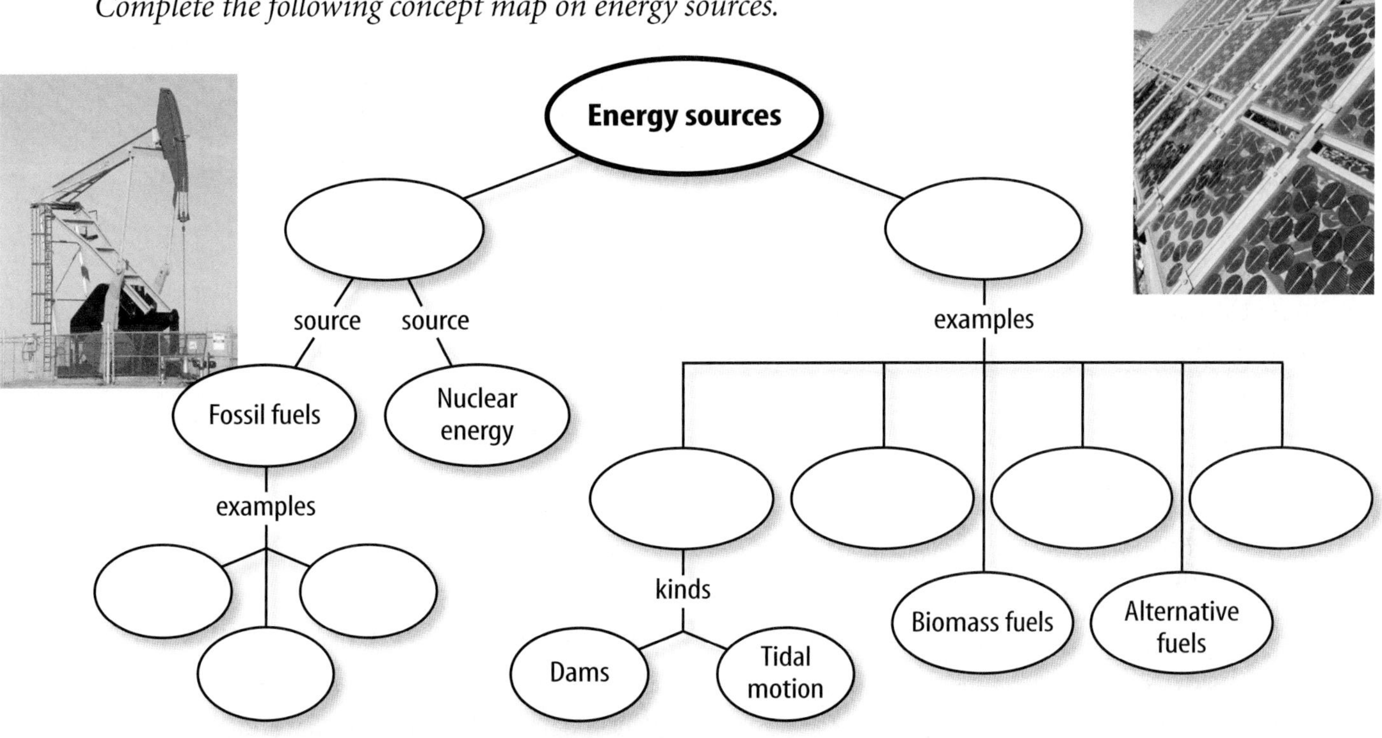

Vocabulary Review

Vocabulary Words

a. biomass
b. fossil fuel
c. geothermal energy
d. hydroelectricity
e. nonrenewable resource
f. nuclear reactor
g. nuclear waste
h. petroleum
i. photovoltaic cell
j. renewable resource

Using Vocabulary

Change the incorrect terms so that the sentences read correctly. Underline your changes.

1. A nuclear reactor uses the Sun to generate electricity.

2. Hydroelectricity makes use of thermal energy inside Earth.

3. Energy produced by the rise and fall of ocean levels is a nonrenewable resource.

4. Petroleum includes the following: oil, natural gas, and coal.

5. Fossil fuels are a renewable resource because they are being used up faster than they are being made.

6. Special caution should be taken in disposing of a photovoltaic cell.

Study Tip

Listening is a learning tool, too. Record a reading of your notes on tape and listen to it several times each week.

Chapter 10 Assessment

Checking Concepts

Choose the word or phrase that best answers the question.

1. How much energy in the United States comes from burning petroleum, natural gas, and coal?
   A) 85% **C)** 65%
   B) 35% **D)** 25%

2. What do hydrocarbons react with when fossil fuels are burned?
   A) carbon dioxide **C)** oxygen
   B) carbon monoxide **D)** water

3. Why are fossil fuels considered to be nonrenewable resources?
   A) They are no longer being produced.
   B) They are in short supply.
   C) They are not being produced as fast as they're being used.
   D) They contain hydrocarbons.

4. To generate electricity, nuclear power plants produce which of the following?
   A) steam **C)** plutonium
   B) carbon dioxide **D)** water

5. What is a major disadvantage of using nuclear fusion reactors?
   A) use of hydrogen as fuel
   B) less radioactivity produced
   C) extremely high temperatures required
   D) use of only small nuclei

6. Which is NOT a source of nuclear waste?
   A) products of fission reactors
   B) materials with short half-lives
   C) some medical and industrial products
   D) products of coal-burning power plants

7. To what can most of Earth's energy resources ultimately be traced?
   A) plants **C)** magma
   B) the Sun **D)** fossil fuels

8. How are spent nuclear fuel rods usually disposed of?
   A) releasing them into a river
   B) storing them in a deep pool of water
   C) burying them at the reactor site
   D) releasing them into the air

9. What characteristic would enable photovoltaic cells to be used more widely?
   A) pollution free **C)** less expensive
   B) nonrenewable **D)** larger

10. What energy source uses water that is heated naturally by Earth's internal heat?
    A) hydroelectricity **C)** tidal energy
    B) nuclear fission **D)** geothermal energy

Thinking Critically

11. Why aren't alternate energy resources more widely used?

12. Match each of the energy resources described in the chapter with the proper type of energy conversion listed below.
    a. kinetic energy to electricity
    b. thermal energy to electricity
    c. nuclear energy to electricity
    d. chemical energy to electricity
    e. light energy to electricity

13. Why isn't fusion currently used as a source of energy?

14. Classify the energy resources discussed in this chapter and in the photo as renewable or nonrenewable.

15. Suppose new reserves of fossil fuels and a way to burn them cleanly were found. Why would it still be a good idea to decrease use of them as a source of energy?

Developing Skills

16. Communicating Describe the steps that must occur before you can use the Sun's energy to power a car.

17. Drawing Conclusions Discuss whether or not alternate energy sources could have negative effects on the environment.

18. Recognizing Cause and Effect Complete the table describing possible effects of changes in the normal operation of a nuclear reactor.

Reactor Problems

Cause	Effect
The cooling water is released hot.	
The control rods are removed.	
	The reactor core overheats and meltdown occurs.

19. Using Percentages What is the overall efficiency of a power plant whose stages have efficiencies of 65 percent, 75 percent, 90 percent, and 70 percent?

Performance Assessment

20. Newspaper Article Write a newspaper article to raise public awareness of current energy problems and solutions. In your article, discuss the economic and environmental costs of various energy sources.

TECHNOLOGY

Go to the Glencoe Science Web site at **science.glencoe.com** or use the **Glencoe Science CD-ROM** for additional chapter assessment.

Test Practice

Concerned about air pollution, Andrew gathered information about compressed natural gas-powered cars. He made a graph comparing emissions from a compressed natural gas (CNG)-powered car to its gasoline-powered counterpart.

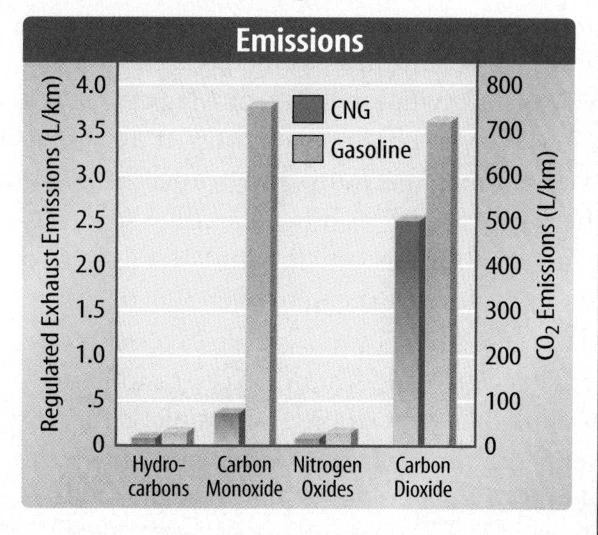

Study the graph and answer the following questions.

1. According to the graph, which substance is emitted into the air in the greatest amount?
   A) hydrocarbons
   B) carbon monoxide
   C) nitrogen oxides
   D) carbon dioxide

2. Based on the data in the graph, which is the greatest benefit of using cars that are powered by compressed natural gas?
   F) better gas mileage when traveling on highways
   G) fewer carbon monoxide emissions
   H) increased carbon dioxide emissions
   J) use of noncombustible fuel reclaimed

Standardized Test Practice

Reading Comprehension

Read the passage. Then read each question that follows the passage. Decide which is the best answer to each question.

Magnetic Levitation Train

One of the first things people learn about magnets is that like magnetic poles repel each other. This is the basic principle behind the Magnetic Levitation Train, or Maglev.

Maglev is a high-speed train. It uses high-strength magnets to lift and propel the train to incredible speeds as it hovers only a few centimeters above the track. A full-size Maglev in Japan achieved a speed of over 500 km/h! Its electromagnetic motor can be precisely controlled to provide smooth acceleration and braking between stops. The magnetic field prevents the vehicle from drifting away from the center of the guideway.

Because there is no friction between wheels and rails, Maglevs eliminate the principal limitation of <u>conventional</u> trains, which is the high cost of maintaining the tracks to avoid excessive vibration and wear that can cause dangerous derailments. Critics point out that Maglevs require enormous amounts of energy. However, studies have shown that Maglevs use 30 percent less energy than other high-speed trains traveling at the same speed. Others worry about the dangers from magnetic fields; however, measurements show that humans are exposed to magnetic fields no stronger than those from toasters or hair dryers.

In Japan, a series of Maglev vehicles are slated to begin tests later this year on a 43-km demonstration line. In Germany, a 290-km Maglev line between Berlin and Hamburg is scheduled to go into service in 2005. Perhaps, in the not-too-distant future, Maglev trains also will transport commuters to and from work or school here in the United States.

Test-Taking Tip After you read the passage, write a one-sentence summary of the main idea for each paragraph.

This is a Maglev train test in Japan.

1. Which of these is the best summary of this passage?
   A) Maglev transportation is currently in use in Germany and Japan.
   B) Maglev might be a high-speed transport system of the future.
   C) Maglevs use more energy than conventional high-speed trains.
   D) Maglevs expose passengers to strong magnetic fields.

2. In this passage, the word <u>conventional</u> means _____.
   F) customary
   G) innovative
   H) political
   J) unusual

Standardized Test Practice

Reasoning and Skills

Read each question and choose the best answer.

1. Voltage gets stepped up when the secondary coil in a transformer has more turns of wire h of the follow- st?

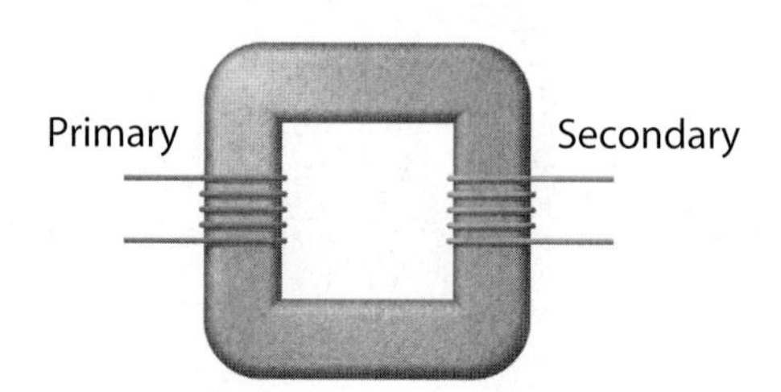

B)

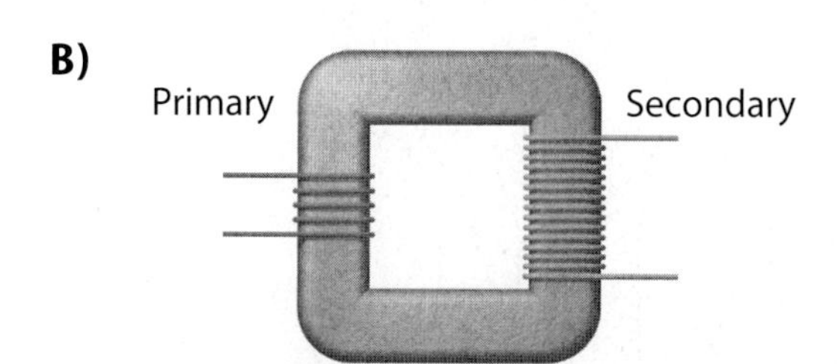

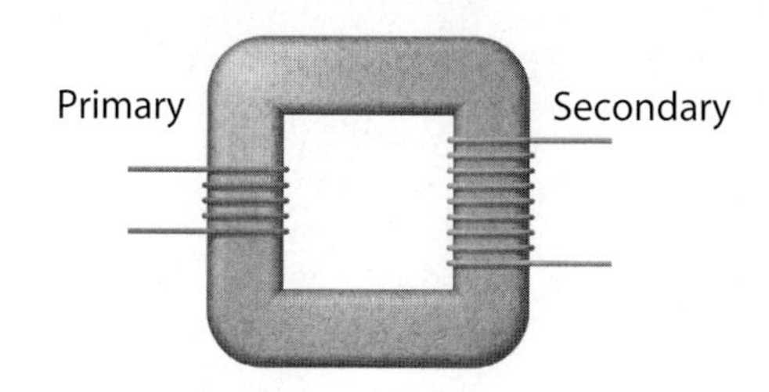

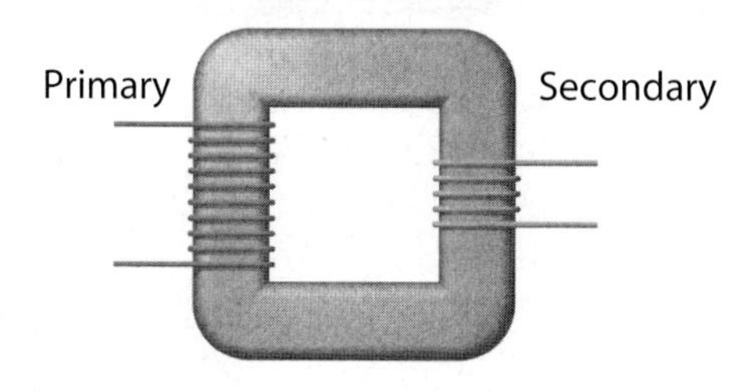

Test-Taking Tip Use the information provided in the question to closely consider each answer choice.

2. Nuclear decay produces radiation, which can ionize nearby atoms. How could this radiation benefit human health?
   F) Absorbing excess hormones produced by the thyroid.
   G) Increasing an organ's absorption of radioactive isotopes.
   H) Destroying cells in cancerous tumors.
   J) Boosting the immune system.

Test-Taking Tip Review information about cancer treatments that use radiation.

3. Shahid wanted to pick up pieces of metal with a magnet. Which observation would mean that the magnet would NOT allow Shahid to pick up the pieces of metal?
   A) The metal pieces were small and far away from the magnet.
   B) The magnet was brand new.
   C) The metal pieces were made out of aluminum foil.
   D) The metal pieces and the magnet have the same magnetic poles.

Test-Taking Tip Review what you know about magnetic materials.

Consider this question carefully before writing your answer on a separate sheet of paper.

4. Recall what you know about the production of current. Explain the similarities and differences between direct current (DC) and alternating current (AC).

Test-Taking Tip Use the clues *direct* and *alternating* to guide your answer.

UNIT 3 Energy on the Move

How Are Glassblowing & X Rays Connected?

Glassblowing (far left) is an art in which air is blown through a tube to shape melted glass. In the mid-1800s, a glassblower created a glass tube, sealed metal electrodes into the ends, and removed most of the air from inside. When electricity was passed through the tube, it glowed. The glow aroused the curiosity of scientists, who began experimenting with similar tubes. In order to observe the glow more closely, one physicist surrounded a tube with black cardboard and darkened the laboratory. When the electric current was turned on, the tube glowed—but so did an object across the room! Apparently the tube was emitting some kind of radiation that could pass through cardboard. The mysterious radiation became known as X rays. Scientists eventually learned that X rays are a form of electromagnetic radiation, similar to visible light but with shorter wavelengths and higher energy. Since X rays pass through many substances, they have become important in medicine and science, making it possible to "see" structures inside the bodies of people—and also fish.

SCIENCE CONNECTION

X RAYS AND BODY STRUCTURES X rays are used routinely by doctors to examine bones and some other structures inside the body. On a piece of drawing paper, use a pencil to trace around one of your hands, including all the fingers. Inside the outline, sketch what you think an X-ray image of your hand might look like. Compare your drawing to a real X ray of a human hand. What types of body structures are most visible in X rays? What types of body structures are hard to see?

CHAPTER 11

Waves

The lights flash, guitar strings vibrate, keyboards wail, and the beat of the drums makes you want to get up and dance. All the sights and sounds of this concert are brought to you by waves. Waves are all around you. Some, like water and light waves, you can see. Others, like sound and radio waves, you cannot see. In this chapter, you will learn what waves are and how they travel. You will learn about the different kinds of waves and the properties all waves have in common. You also will find out how waves interact to transform energy into bright lights and spectacular sound.

What do you think?

Science Journal Look at the picture below with a classmate. Discuss what this might be. Here's a hint: *They make movies out of events like this.* Write your answer or best guess in your Science Journal.

Light enters your eyes and sound strikes your ears, enabling you to sense the world around you. Light and sound are waves that carry energy from one place to another. What else gets transferred from place to place when a wave carries energy? Does a wave transfer matter as well as energy? In this activity you'll observe one way that waves can transfer energy.

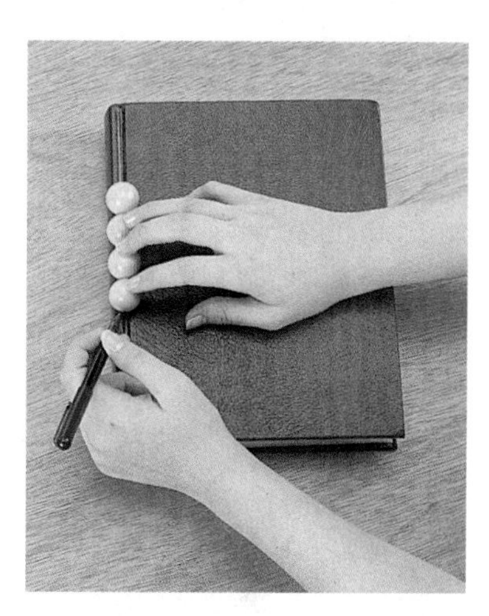

Demonstrating energy transfer

1. Line up four marbles on the groove formed by the spine of your textbook so that the marbles are touching each other.
2. Hold the first three marbles in place using three fingers of one hand.
3. Use your other hand to tap the first marble with a pen or pencil.
4. Observe what happens to the fourth marble.

Observe

Write a paragraph in your Science Journal explaining how the fourth marble reacted to the pen tap. Draw a diagram showing the energy transfer through the marbles.

Before You Read

Making a Venn Diagram Study Fold Make the following Foldable to compare and contrast two types of waves.

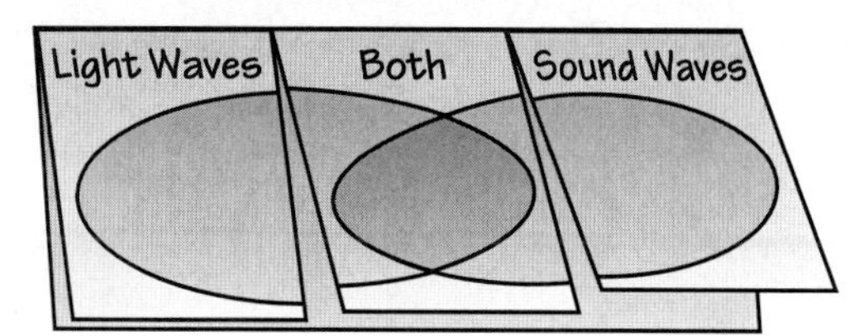

1. Place a sheet of paper in front of you so the long side is at the top. Fold the paper in half from top to bottom.
2. Fold both sides in. Unfold the paper so three sections show.
3. Through the top thickness of paper, cut along each of the fold lines to the top fold, forming three tabs. Label the tabs *Light Waves, Sound Waves,* and *Both* and draw ovals across the front of the foldable, as shown.
4. As you read the chapter, write characteristics of light and sound waves under the left and right tabs. Under the middle tab, write what light and sound waves have in common.

SECTION

The Nature of Waves

As You Read

What You'll Learn

- **Recognize** that waves carry energy but not matter.
- **Define** mechanical waves.
- **Distinguish** between transverse waves and compressional waves.

Vocabulary

wave
medium
transverse wave
compressional wave

Why It's Important

You hear and see the world around you because of the energy carried by waves.

What's in a wave?

A surfer bobs in the ocean waiting for the perfect wave, microwaves warm up your leftover pizza, and sound waves from your CD player bring music to your ears. Do these and other types of waves have anything in common with one another?

A **wave** is a repeating disturbance or movement that transfers energy through matter or space. For example, ocean waves disturb the water and transfer energy through it. During earthquakes, energy is transferred in powerful waves that travel through Earth. Light is a type of wave that can travel through empty space to transfer energy from one place to another, such as from the Sun to Earth.

Waves and Energy

Kerplop! A pebble falls into a pool of water and ripples form. As **Figure 1** shows, the pebble causes a disturbance that moves outward in the form of a wave. Because it is moving, the falling pebble has energy. As it splashes into the pool, the pebble transfers some of its energy to nearby water molecules, causing them to move. Those molecules then pass the energy along to neighboring water molecules, which, in turn, transfer it to their neighbors. The energy moves farther and farther from the source of the disturbance. What you see is energy traveling in the form of a wave on the surface of the water.

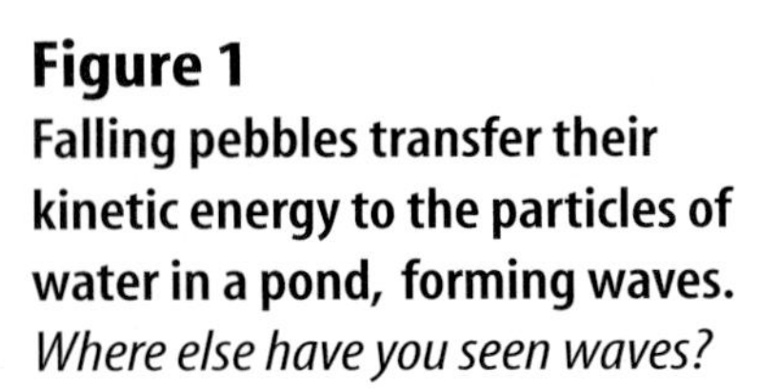

Figure 1
Falling pebbles transfer their kinetic energy to the particles of water in a pond, forming waves.
Where else have you seen waves?

Waves and Matter Imagine you're in a boat on a lake. Approaching waves bump against your boat, but they don't carry it along with them as they pass. The boat does move up and down and maybe even a short distance back and forth because the waves transfer some of their energy to it. But after the waves have moved on, the boat is still in nearly the same place. The waves don't even carry the water along with them. Only the energy carried by the waves moves forward. All waves have this property—they carry energy without transporting matter from place to place.

Reading Check *What do waves carry?*

Making Waves A wave will travel only as long as it has energy to carry. For example, when you drop a pebble into a puddle, the ripples soon die out and the surface of the water becomes still again.

Suppose you are holding a rope at one end, and you give it a shake. You would create a pulse that would travel along the rope to the other end, and then the rope would be still again, as **Figure 2** shows. Now suppose you shake your end of the rope up and down for a while. You would make a wave that would travel along the rope. When you stop shaking your hand up and down, the rope will be still again. It is the up-and-down motion of your hand that creates the wave.

Anything that moves up and down or back and forth in a rhythmic way is vibrating. The vibrating movement of your hand at the end of the rope created the wave. In fact, all waves are produced by something that vibrates.

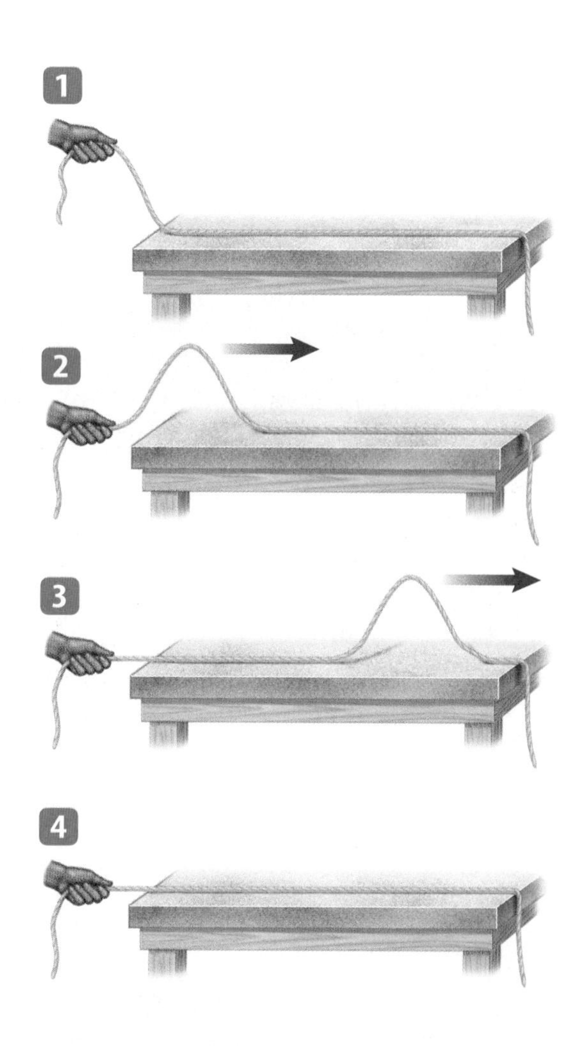

Figure 2
A wave will exist only as long as it has energy to carry. *What happened to the energy that was carried by the wave in this rope?*

Mechanical Waves

Sound waves travel through the air to reach your ears. Ocean waves move through water to reach the shore. In both cases, the matter the waves travel through is called a **medium.** The medium can be a solid, a liquid, a gas, or a combination of these. For sound waves the medium is air, and for ocean waves the medium is water. Not all waves need a medium. Some waves, such as light and radio waves, can travel through space. Waves that can travel only through matter are called mechanical waves. The two types of mechanical waves are transverse waves and compressional waves.

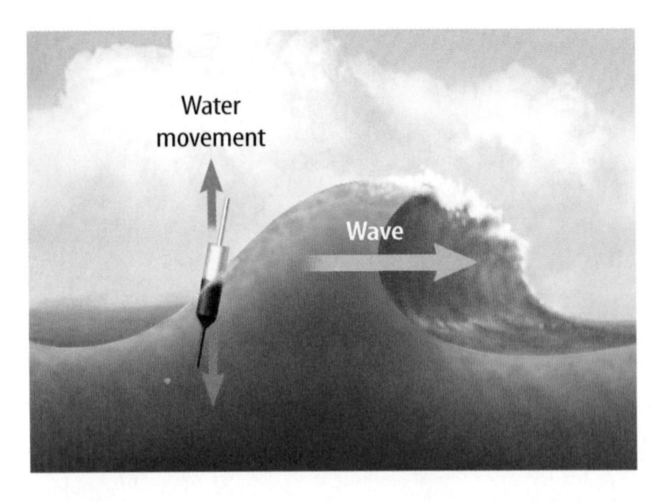

Figure 3
A water wave travels horizontally as the water moves vertically up and down.

Transverse Waves In a **transverse wave,** matter in the medium moves back and forth at right angles to the direction that the wave travels. For example, **Figure 3** shows how a wave in the ocean moves horizontally, but the water that the wave passes through moves up and down. When you shake one end of a rope while your friend holds the other end, you are making transverse waves. The wave and its energy travel from you to your friend as the rope moves up and down.

Compressional Waves In a **compressional wave,** matter in the medium moves back and forth in the same direction that the wave travels. You can model compressional waves with a coiled spring toy, as shown in **Figure 4.** Squeeze several coils together at one end of the spring. Then let go of the coils, still holding onto coils at both ends of the spring. A wave will travel along the spring. As the wave moves, it looks as if the whole spring is moving toward one end. Suppose you watched the coil with yarn tied to it as in **Figure 4.** You would see that it moves back and forth as the wave passes, and then stops moving after the wave has passed. The wave carries energy, but not matter, forward along the spring.

Sound Waves Sound waves are compressional waves. When a noise is made, such as when a locker door slams shut and vibrates, nearby air molecules are pushed together by the vibrations. The air molecules are squeezed together like the coils in a coiled spring toy are when you make a compressional wave with it. The compressions travel through the air to make a wave.

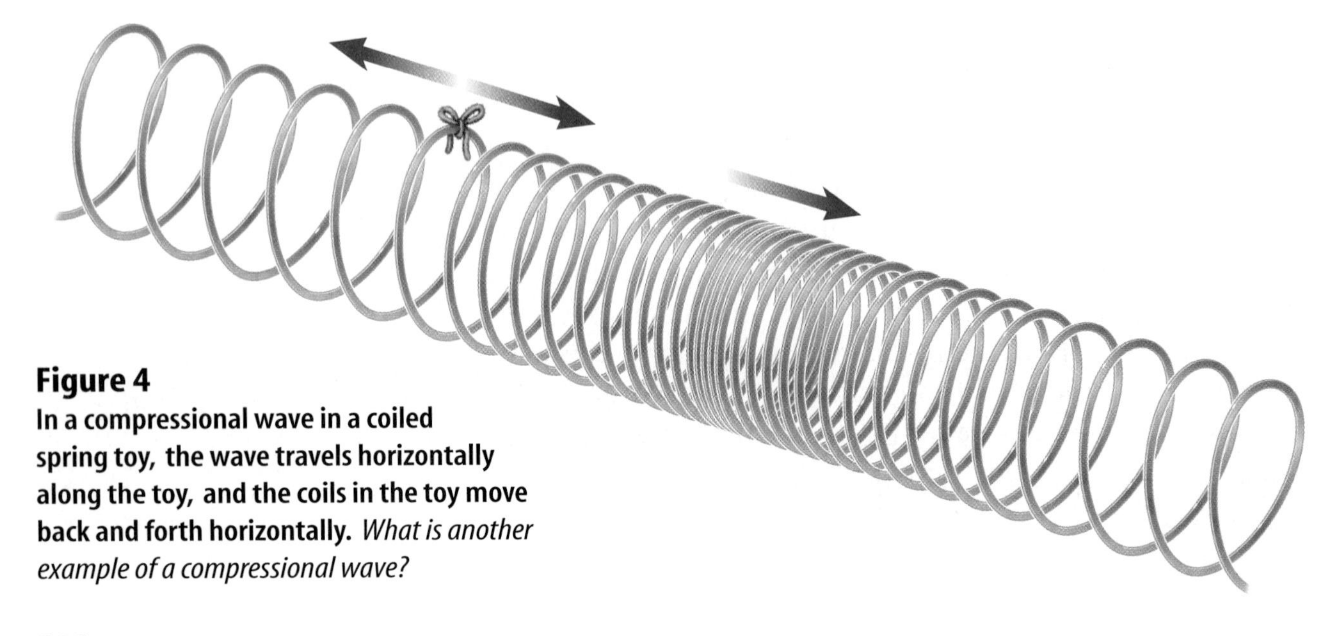

Figure 4
In a compressional wave in a coiled spring toy, the wave travels horizontally along the toy, and the coils in the toy move back and forth horizontally. *What is another example of a compressional wave?*

Sound in Other Materials Sound waves also can travel through other mediums, such as water and wood. Particles in these mediums also are pushed together and move apart as the sound waves travel through them. When a sound wave reaches your ear, it causes your eardrum to vibrate. Your inner ear then sends signals to your brain, and your brain interprets the signals as sound.

✔ Reading Check *How do sound waves travel in solids?*

Water Waves Water waves are not purely transverse waves. The surface of the water moves up and down as the waves go by. But the water also moves a short distance back and forth. This movement happens because the low part of the wave can be formed only by pushing water forward or backward toward the high part of the wave, as in **Figure 5A.** Then as the wave passes, the water that was pushed aside moves back to its initial position, as in **Figure 5B.** In fact, if you looked closely, you would see that the combination of this up-and-down and back-and-forth motion causes water to move in circles. Anything floating on the surface of the water absorbs some of the waves' energy and bobs in a circular motion.

Ocean waves are formed most often by wind blowing across the ocean surface. As the wind blows faster and slower, the changing wind speed is like a vibration. The size of the waves that are formed depends on the wind speed, the distance over which the wind blows, and how long the wind blows. **Figure 6** on the next page shows this process.

Figure 5
When a wave passes, the surface of the water doesn't just move up and down.

A **The low point of a water wave is formed when water is pushed aside and up to the high point of the wave.**

B **The water that is pushed aside returns to its initial position.**

NATIONAL GEOGRAPHIC

VISUALIZING FORMATION OF OCEAN WAVES

Figure 6

When wind blows across an ocean, friction between the moving air and the water causes the water to move. As a result, energy is transferred from the wind to the surface of the water. The waves that are produced depend on the length of time and the distance over which the wind blows, as well as the wind speed.

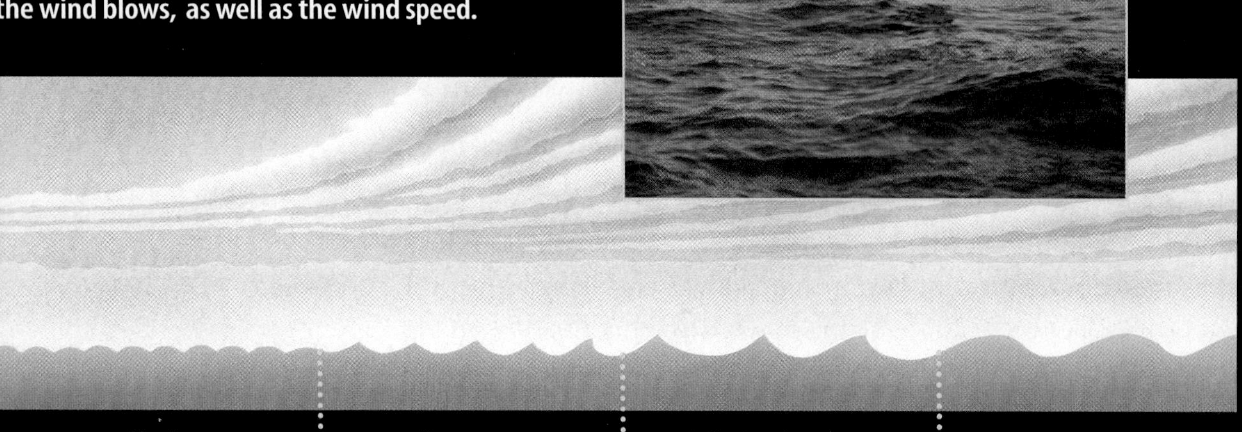

Ripples | Choppy seas | Fully developed seas | Swells

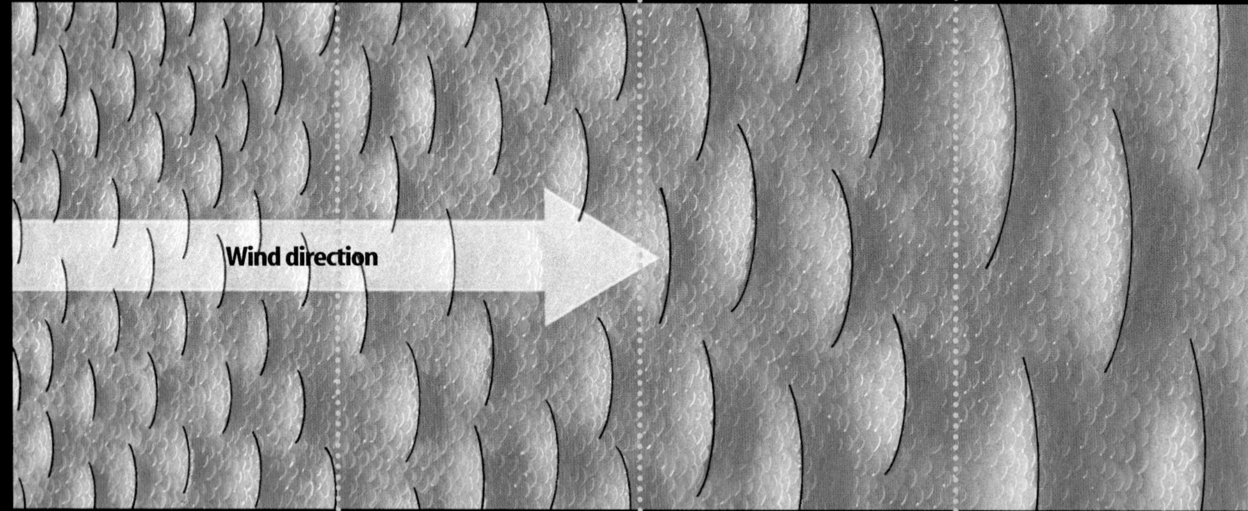

▲ Wind causes ripples to form on the surface of the water. As ripples form, they provide an even larger surface area for the wind to strike, and the ripples increase in size.

▲ Waves that are higher and have longer wavelengths grow faster as the wind continues to blow, but the steepest waves break up, forming whitecaps. The surface is said to be choppy.

▲ The shortest-wavelength waves break up, while the longest-wavelength waves continue to grow. When these waves have reached their maximum height, they form fully developed seas.

▲ After the wind dies down, the waves lose energy and become lower and smoother. These smooth, long-wavelength ocean waves are called swells.

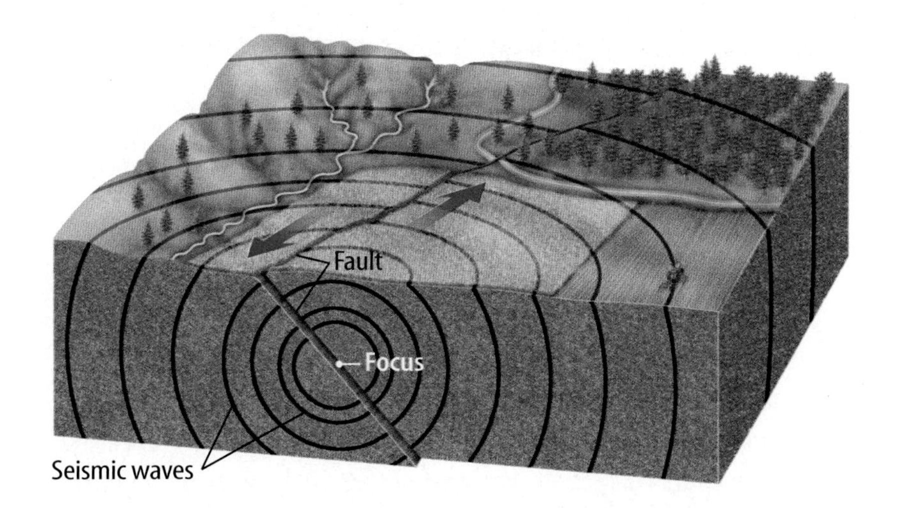

Figure 7
When Earth's crust breaks, the energy that is released is transmitted outward, causing an earthquake. *Why are earthquakes mechanical waves?*

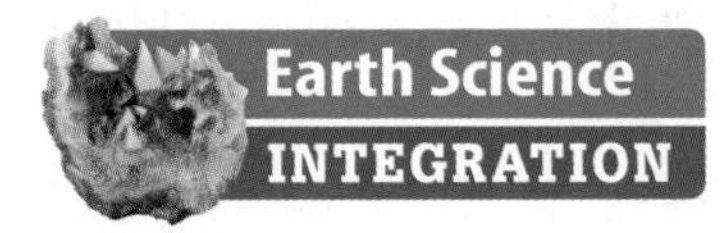

Seismic Waves If you pulled too hard on a guitar string, the string would break and you would hear a noise. The noise occurs because the string vibrates for a short time after it breaks, and creates a sound wave. In a similar way, forces in Earth's crust can cause regions of the crust to shift, bend, or even break. The breaking crust vibrates, creating seismic (SIZE mihk) waves that carry energy outward, as shown in **Figure 7.** Seismic waves are a combination of compressional and transverse waves. They can travel through Earth and along Earth's surface. When objects on Earth's surface absorb some of the energy carried by seismic waves, they move and shake. The more the crust moves during an earthquake, the more energy is released.

SCIENCE Online

Research Seismic waves generated by earthquakes are used to map the interior of Earth. Visit the Glencoe Science Web site at **science.glencoe.com** to find out more about interpreting seismic waves. Write a summary of what you learn.

Section 1 Assessment

1. Give one example of a transverse wave and one example of a compressional wave.
2. Why doesn't a boat on a lake move forward when a water wave passes? Describe the boat's motion.
3. Describe how to model compressional waves using a coiled spring toy.
4. What is a mechanical wave?
5. **Think Critically** If ocean waves do not carry matter forward, why do boats need anchors?

Skill Builder Activities

6. **Comparing and Contrasting** Compare and contrast transverse and compressional waves. What does each type of wave carry? How does matter in the medium move? **For more help, refer to the Science Skill Handbook.**
7. **Communicating** In your Science Journal, describe waves you have observed. Have you ever observed the effects of a wave without being able to see it? Explain. **For more help, refer to the Science Skill Handbook.**

SECTION

Wave Properties

As You Read

What You'll Learn

- **Compare and contrast** transverse and compressional waves.
- **Describe** the relationship between frequency and wavelength.
- **Explain** how a wave's amplitude is related to the wave's energy.
- **Calculate** wave speed.

Vocabulary

crest
trough
rarefaction
wavelength
frequency
amplitude

Why It's Important

Changing the properties of waves enables them to be used in many ways.

The Parts of a Wave

Besides the fact that sound waves, water waves, and seismic waves travel in different mediums, what makes these waves different from each other? Waves can differ in how much energy they carry and in how fast they travel. Waves also have other characteristics that make them different from each other. These characteristics can be used to describe waves.

Suppose you shake the end of a rope and make a transverse wave. The transverse wave has alternating high points and low points. **Figure 8A** shows that the highest points of a transverse wave are called the **crests,** and the lowest points are called the **troughs.**

Reading Check *What are the highest and lowest points of a transverse wave?*

On the other hand, a compressional wave has no crests and troughs. When a compressional wave passes through a medium, it creates a region where the medium becomes crowded together and more dense, as in **Figure 8B.** This region is called the compression. When you make compressional waves in a coiled spring, the compression is the region where the coils are close together. **Figure 8B** also shows that the coils in the region next to a compression are spread apart, or less dense. This less-dense region of a compressional wave is called a **rarefaction** (rar uh FAK shun).

Figure 8
Transverse and compressional waves have different characteristics.

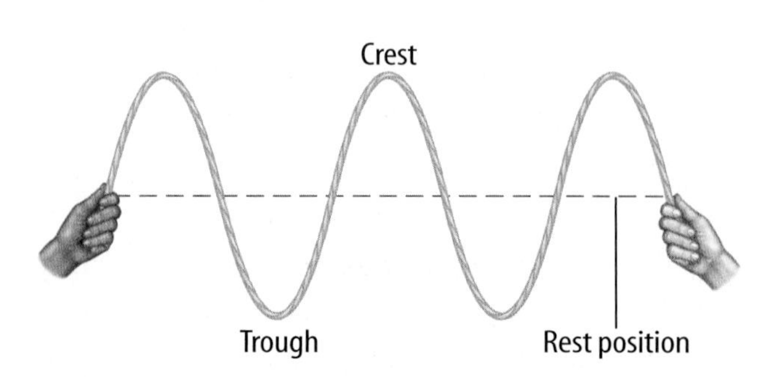

A **The highest point of a transverse wave is a crest. The lowest point is a trough.**

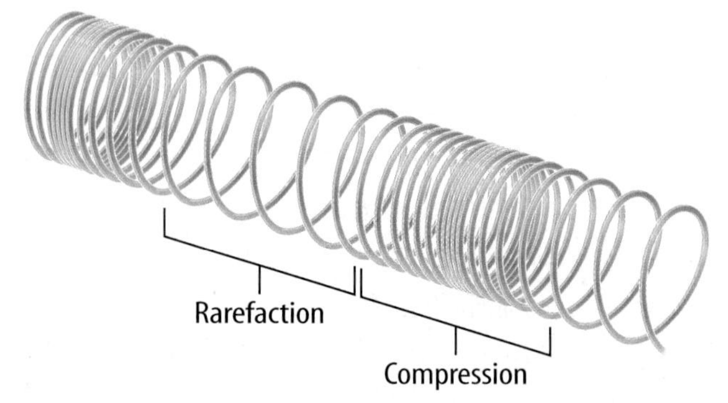

B **The densest parts of a compressional wave are compressions. The least dense parts are rarefactions.**

Figure 9
One wavelength starts at any point on a wave and ends at the nearest point just like it.

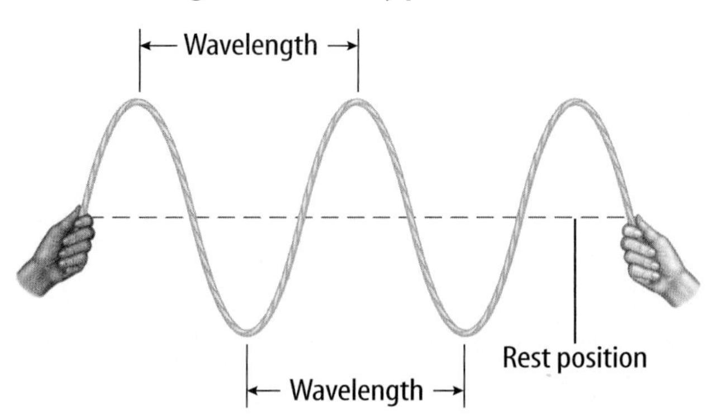

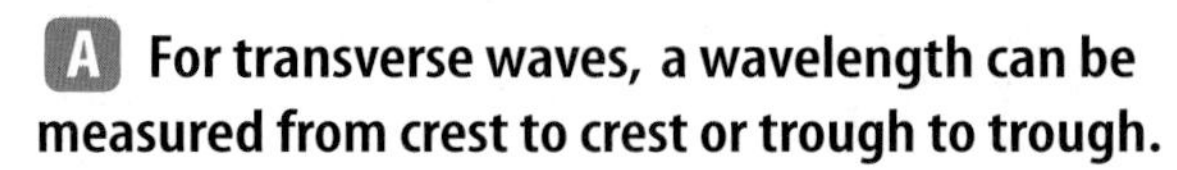

A For transverse waves, a wavelength can be measured from crest to crest or trough to trough.

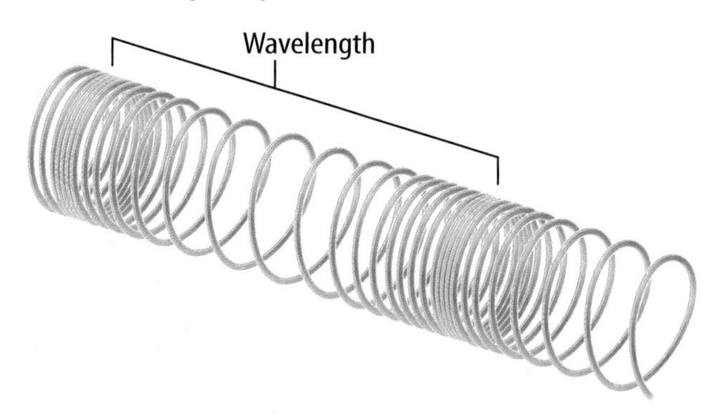

B The wavelength of a compressional wave can be measured from the start of one compression to the start of the next or the start of one rarefaction to the start of the next. *How many compressions are in each wavelength?*

Wavelength

Waves also have a property called wavelength. A **wavelength** is the distance between one point on a wave and the nearest point just like it. For example, in transverse waves you can measure wavelength from crest to crest or from trough to trough, as shown in **Figure 9A.**

A wavelength in a compressional wave is the distance between two neighboring compressions or two neighboring rarefactions, as shown in **Figure 9B.** You can measure from the start of one compression to the start of the next compression or from the start of one rarefaction to the start of the next rarefaction. The wavelengths of sound waves that you can hear range from a few centimeters for the highest-pitched sounds to about 15 m for the deepest sounds.

Frequency

What is your favorite radio station? When you tune your radio to a station, you are choosing radio waves of a certain frequency. The **frequency** of a wave is the number of wavelengths that pass a fixed point each second. You can find the frequency of a transverse wave by counting the number of crests or troughs that pass by a point each second. The frequency of a compressional wave is the number of compressions or rarefactions that pass a point every second. Frequency is expressed in hertz (Hz). A frequency of 1 Hz means that one wavelength passes by in 1 s. In SI units, 1 Hz is the same as 1/s.

Reading Check *What does a frequency of 7 Hz mean?*

Observing Wavelength

1. Fill a pie plate or other wide pan with about 2 cm of water.
2. Lightly tap your finger once per second on the surface of the water and observe the spacing of the water waves.
3. Increase the rate of your tapping, and observe the spacing of the water waves.

Analysis

1. How is the spacing of the water waves related to their wavelength?
2. How does the spacing of the water waves change when the rate of tapping increases?

Wavelength Is Related to Frequency If you make transverse waves with a rope, you increase the frequency by moving the rope up and down faster. Moving the rope faster also makes the wavelength shorter. This relationship is always true—as frequency increases, wavelength decreases. **Figure 10** compares the wavelengths and frequencies of two different waves.

The frequency of a wave is always equal to the rate of vibration of the source that creates it. If you move the rope up, down, and back up in 1 s, the frequency of the wave you generate is 1 Hz. If you move the rope up, down, and back up five times in 1 s, the resulting wave has a frequency of 5 Hz.

Wave Speed

You're at a large stadium watching a baseball game, but you're high up in the bleachers, far away from the action. The batter swings and you see the ball rising in the air. An instant later you hear the crack of the bat hitting the ball. You see the impact before you hear it because all waves do not travel at the same speed. Light waves travel much faster than sound waves do. Therefore, the light waves reflected from the flying ball reach your eyes before the sound waves created by the crack of the bat reach your ears.

The speed of a wave depends on the properties of the medium it is traveling through. For example, sound waves usually travel faster in liquids and solids than they do in gases. On the other hand, light waves travel more slowly in liquids and solids than they do in gases or in empty space. Also, sound waves usually travel faster in a material if the temperature of the material is increased. For example, sound waves travel faster in air at 20°C than in air at 0°C.

Figure 10
The wavelength of a wave decreases as the frequency increases.

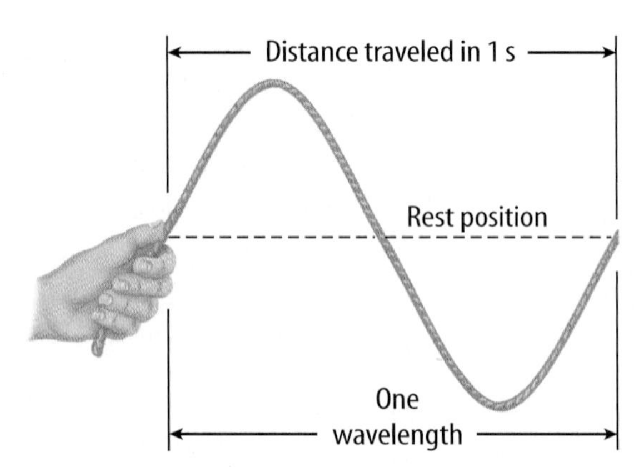

A The rope is moved down, up, and down again one time in 1 s. One wavelength is created on the rope.

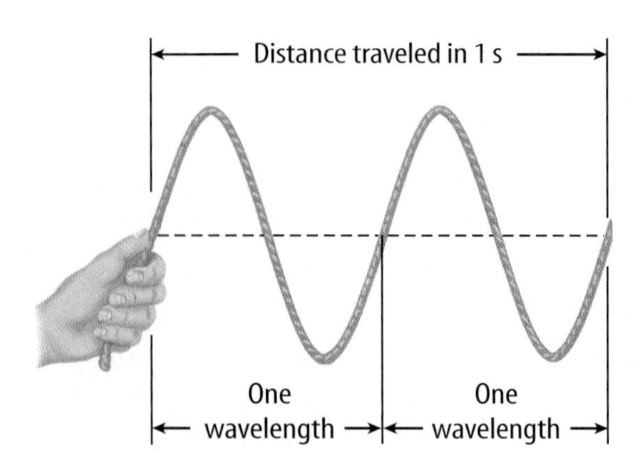

B The rope is shaken down, up, and down again twice in 1 s. Two wavelengths are created on the rope.

Calculating Wave Speed Sometimes people want to know how fast a wave is traveling. For example, earthquakes beneath the ocean floor can produce giant water waves called tsunamis. Knowing how fast the wave is moving helps determine when the wave will reach land. Wave velocity (v) describes how fast the wave moves forward. You can calculate the velocity of a wave by multiplying its frequency times its wavelength. Wavelength is represented by the Greek letter lambda (λ) and frequency is represented by f.

$$\text{velocity} = \text{wavelength} \times \text{frequency}$$
$$v = \lambda \times f$$

For example, what is the speed of a wave with a wavelength of 2 m and a frequency of 3 Hz? Because 3 Hz equals 3 wavelengths/second or 3 × 1/s, the wave's speed is:

$$v = \lambda \times f = 2\text{ m} \times 3\text{ Hz} = 2\text{ m} \times 3/\text{s} = 6\text{ m/s}$$

Tsunamis can cause serious damage when they hit land. These waves can be up to 30 m tall and can travel more than 700 km/h. Research areas of the world where tsunamis are most likely to occur.

Math Skills Activity

Calculating Wave Speed

Example Problem

A wave is traveling at a velocity of 12 m/s and its wavelength is 3 m. Calculate the wave frequency.

Solution

1. *This is what you know:* velocity (v) = 12 m/s; wavelength (λ) = 3 m
2. *This is what you want to find:* wave frequency (f)
3. *This is the equation you need to use:* $v = \lambda \times f$
4. *Solve for* f *and then substitute the known values in the equation.* $f = v/\lambda$; $f = 12\text{ m/s} / 3\text{ m} = 4 \times 1/\text{s} = 4\text{ Hz}$

Check your answer by substituting the frequency and given wavelength into the original equation. Do you calculate the velocity that was given?

Practice Problem

1. A wave is traveling at a speed of 18 m/s with a frequency of 3 Hz. A second wave is traveling at a speed of 16 m/s with a frequency of 4 Hz. What is the difference between these two wavelengths?

For more help, refer to the Math Skill Handbook.

Amplitude and Energy

Why do some earthquakes cause terrible damage, while others are hardly felt? This is because the amount of energy a wave carries can vary. **Amplitude** is related to the energy carried by a wave. The greater the wave's amplitude is, the more energy the wave carries. Amplitude is measured differently for compressional and transverse waves.

Amplitude of Compressional Waves The amplitude of a compressional wave is related to how tightly the medium is pushed together at the compressions. The denser the medium is at the compressions, the larger its amplitude is and the more energy the wave carries. For example, it takes more energy to push the coils in a coiled spring toy tightly together than to barely move them. The closer the coils are in a compression, the farther apart they are in a rarefaction. So the less dense the medium is at the rarefactions, the more energy the wave carries. **Figure 11** shows compressional waves with different amplitudes.

Figure 11
The amplitude of a compressional wave depends on how dense its medium is at each compression. A This coiled spring has the greater amplitude. B This coiled spring has the smaller amplitude.

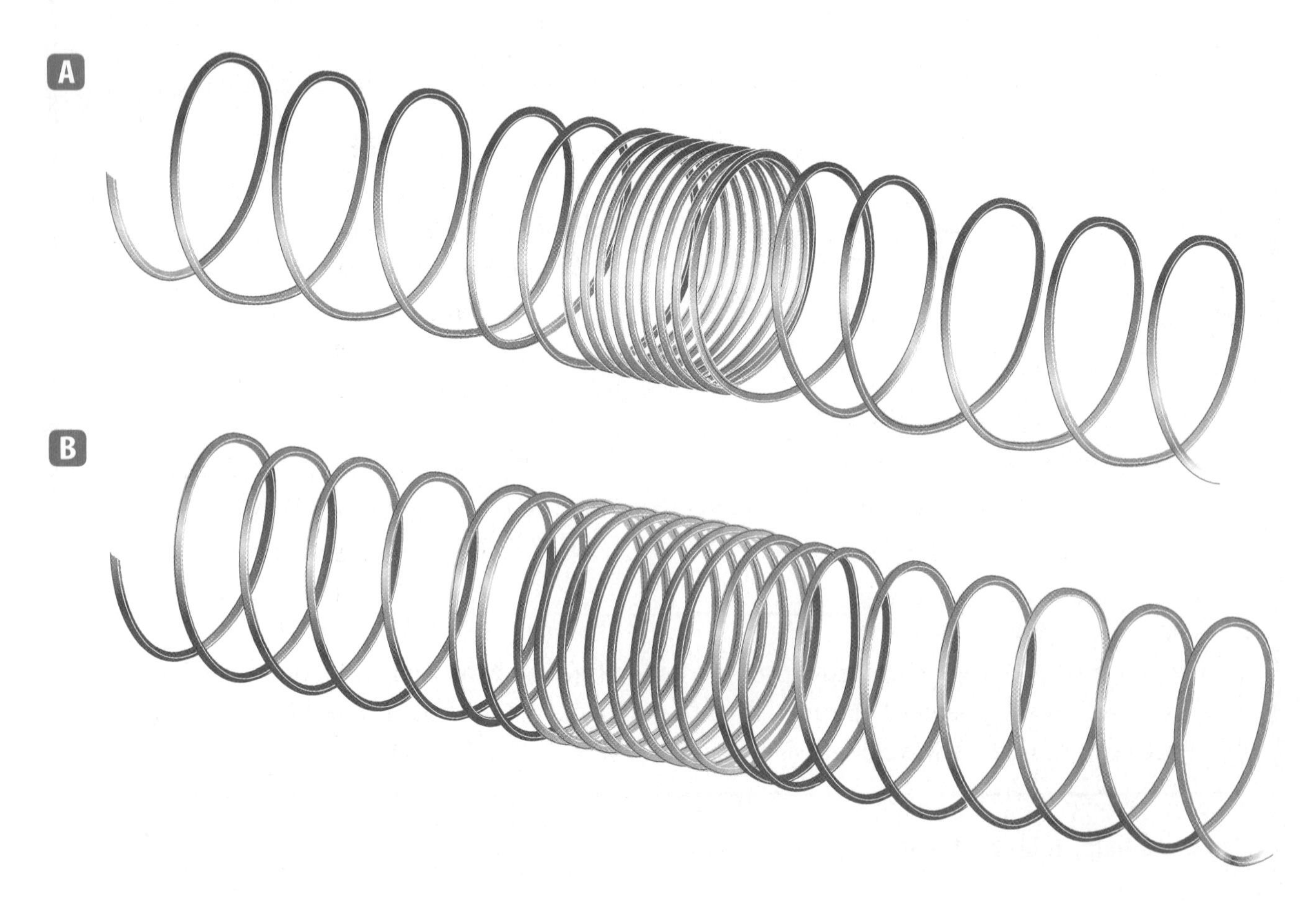

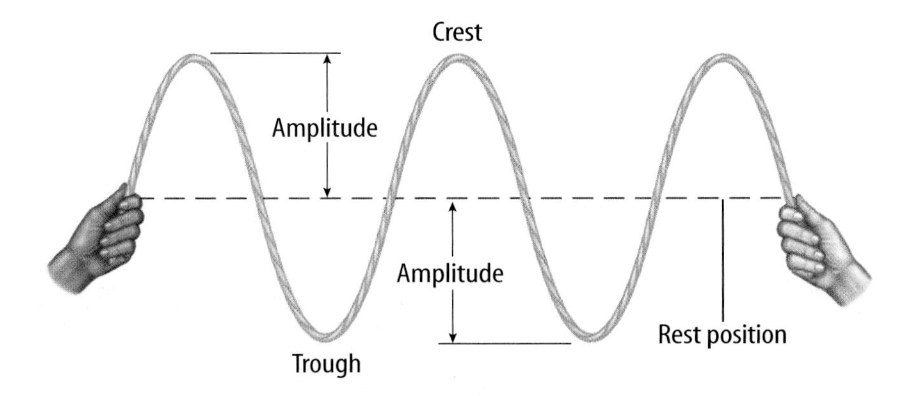

Figure 12
The amplitude of a transverse wave is the distance between a crest or a trough and the position of the medium at rest. *How could you create waves with different amplitudes in a piece of rope?*

Amplitude of Transverse Waves How can you tell the difference between a transverse wave that carries a lot of energy from one that carries little energy? If you've ever been knocked over by an ocean wave, you know that the higher the wave, the more energy it carries. Remember that the amplitude of a wave increases as the energy carried by the wave increases. So a tall ocean wave has a greater amplitude than a short ocean wave does. The amplitude of any transverse wave is the distance from the crest or trough of the wave to the rest position of the medium, as shown in **Figure 12.**

Section  Assessment

1. Sketch a transverse wave and label the crest, trough, wavelength, rest position, and amplitude.
2. How does the wavelength of a wave change when frequency decreases? When frequency increases?
3. How is the density at a compression in a compressional wave like the height of a transverse wave?
4. A wave travels at a speed of 4.0 m/s and has a frequency of 3.5 Hz. What is the wavelength?
5. **Think Critically** Remember that sound waves are compressional waves. Why do you think sound waves travel faster in solids than in gases?

Skill Builder Activities

6. **Concept Mapping** Create a concept map that shows the relationships among the following: *crest, trough, compression, rarefaction, wavelength, wave frequency, amplitude,* and *wave speed.* **For more help, refer to the Science Skill Handbook.**
7. **Drawing Conclusions** The unit *megahertz (MHz)* means "1 million Hertz." Your favorite FM radio station broadcasts at a frequency of 104.1 MHz, or 104.1 million Hz. Your friend prefers a station at 101.9 MHz. If the radio waves from both stations travel at the same speed, which station uses longer wavelengths? Explain. **For more help, refer to the Science Skill Handbook.**

Activity

Waves in Different Mediums

Have you ever swum underwater? If so, even with your head underwater, you probably still heard some sounds. Sound waves can travel through more than one medium, including air and water. The noises probably sounded different underwater than they do in air. How do waves change when they pass through different mediums?

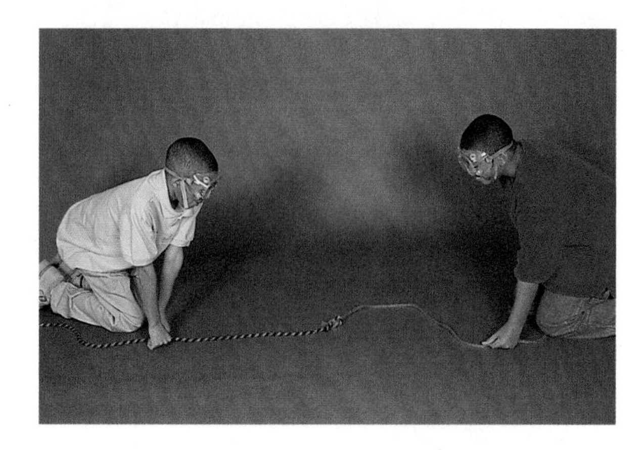

What You'll Investigate

How is the speed of a wave affected by the type of material it is traveling through?

Possible Materials

small coiled spring toys (made out of metal and plastic)
rope, both heavy and light
string
long rubber band, such as those used for exercising
strip of heavy cloth, such as a towel
strip of light cloth, such as nylon panty hose
stopwatch

Goals

- **Demonstrate** transverse and compressional waves.
- **Compare** the speed of waves traveling through different mediums.

Safety Precautions

Procedure

1. Use pieces of each material that are about the same length. For each material, have a partner hold one end of the material still while you shake the material back and forth between two set points to make a wave. Identify the points by placing markers or chairs on the floor. Shake each material in the same way.
2. Have someone time how long a pulse takes to reach the opposite end of the material.
3. Tie two different types of rope together or tie a heavy piece of cloth to a lighter piece. **Observe** how the wave changes when it moves from one material to the other.
4. **Observe** compressional waves using coiled spring toys. You can connect two different types of coiled spring toys together to see how a compressional wave changes in different mediums.

Conclude and Apply

1. Did the waves traveling through the different mediums have the same amplitude? Explain why or why not.
2. Did the waves travel at the same speed through the different mediums? **Explain.**
3. **Explain** how the waves changed when they moved from one material to another.
4. Waves carry energy. Where did the waves created in this activity get their energy?

SECTION

The Behavior of Waves

Reflection

If you are one of the last people to leave your school building at the end of the day, you'll probably find the hallways quiet and empty. When you close your locker door, the sound echoes down the empty hall. Your footsteps also make a hollow sound. Thinking you're all alone, you may be startled by your own reflection in a classroom window. The echoes and your image looking back at you from the window are caused by wave reflection.

Reflection occurs when a wave strikes an object and bounces off of it. All types of waves—including sound, water, and light waves—can be reflected. How does the reflection of light allow the boy in **Figure 13** to see himself in the mirror? It happens in two steps. First, light strikes his face and bounces off. Then, the light reflected off his face strikes the mirror and is reflected into his eyes.

A similar thing happens to sound waves when your footsteps echo. Sound waves form when your foot hits the floor and the waves travel through the air to both your ears and other objects. Sometimes when the sound waves hit another object, they reflect off it and come back to you. Your ears hear the sound again, a few seconds after you first heard your footstep.

Bats and dolphins use echoes to learn about their surroundings. A dolphin makes a clicking sound and listens to the echoes. These echoes enable the dolphin to locate nearby objects.

As You Read

What You'll Learn

- **Identify** the law of reflection.
- **Recognize** what makes waves bend.
- **Explain** how waves combine.

Vocabulary

refraction, diffraction, interference, standing wave, resonance

Why It's Important

You can check your reflection in a mirror, hear an echo, and see shadows because of how waves behave.

Figure 13
The light that strikes the boy's face is reflected into the mirror. The light then reflects off the mirror into his eyes. *What kinds of waves can be reflected?*

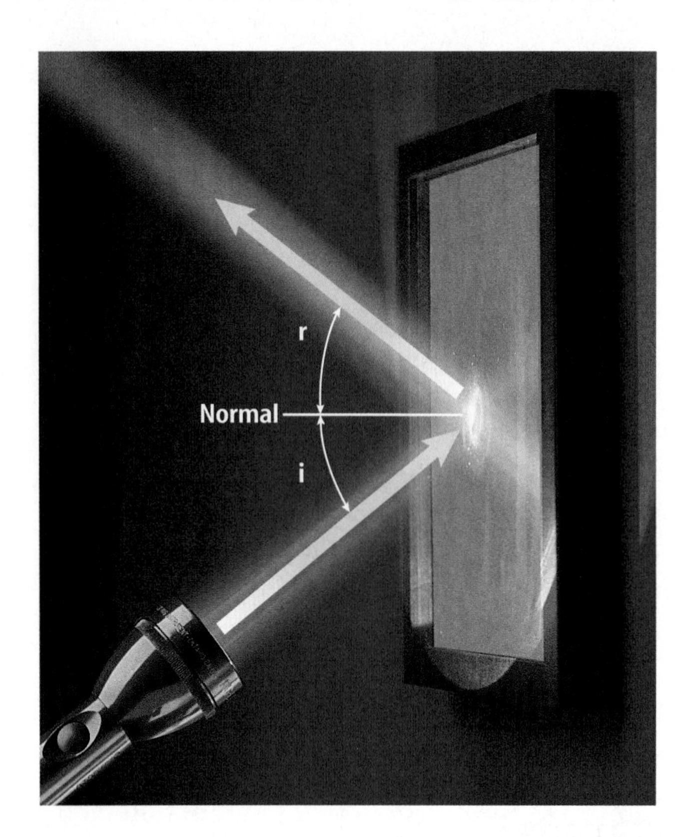

Figure 14
A flashlight beam is made of light waves. When any wave is reflected, the angle of incidence, *i*, equals the angle of reflection, *r*.

The Law of Reflection Look at the two light beams in **Figure 14.** The beam striking the mirror is called the incident beam. The beam that bounces off the mirror is called the reflected beam. The line drawn perpendicular to the surface of the mirror is called the normal. The angle formed by the incident beam and the normal is the angle of incidence, labeled *i.* The angle formed by the reflected beam and the normal is the angle of reflection, labeled *r.* According to the law of reflection, the angle of incidence is equal to the angle of reflection. All reflected waves obey this law. Objects that bounce from a surface sometimes behave like waves that are reflected from a surface. For example, suppose you throw a bounce pass while playing basketball. The angle between the ball's direction and the normal to the floor is the same before and after it bounces.

Refraction

Do you notice anything unusual in **Figure 15?** The pencil looks as if it is broken into two pieces. But if you pulled the pencil out of the water, you would see that it is unbroken. This illusion is caused by refraction. How does it work?

Remember that a wave's speed depends on the medium it is moving through. When a wave passes from one medium to another—such as when a light wave passes from air to water—it changes speed. If the wave is traveling at an angle when it passes from one medium to another, it changes direction, or bends, as it changes speed. **Refraction** is the bending of a wave caused by a change in its speed as it moves from one medium to another. The greater the change in speed is, the more the wave bends.

✔ Reading Check *When does refraction occur?*

Figure 16A on the next page shows what happens when a wave passes into a material in which it slows down. The wave is refracted (bent) toward the normal. **Figure 16B** shows what happens when a wave passes into a medium in which it speeds up. Then the wave is refracted away from the normal.

Figure 15
The pencil looks like it is broken at the surface of the water because of refraction. *Does light travel faster in water or air?*

Figure 16
Light travels slower in water than in air.

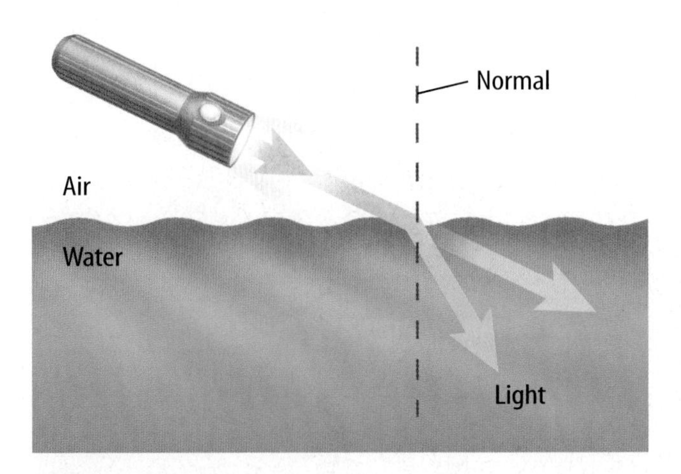

A **When light travels from air to water, it slows down and bends toward the normal.**

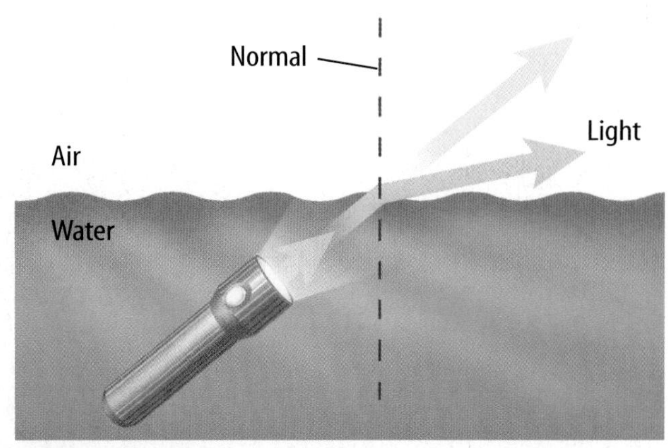

B **When light leaves water and travels to air, it speeds up and bends away from the normal.** *How would the beam bend if the speed were the same in both air and water?*

Refraction of Light in Water You may have noticed that objects that are underwater seem closer to the surface than they really are. **Figure 17** shows how refraction causes this illusion. In the figure, the light waves reflected from the swimmer's foot are refracted away from the normal and enter your eyes. However, your brain assumes that all light waves have traveled in a straight line. The light waves that enter your eyes seem to have come from a foot that was higher in the water. This is also why the pencil in **Figure 15** seems broken. The light waves coming from the part of the pencil that is underwater are refracted, but your brain interprets them as if they have traveled in a straight line. However, the light waves coming from the part of the pencil above the water are not refracted. So, the part of the pencil that is underwater looks as if it has shifted.

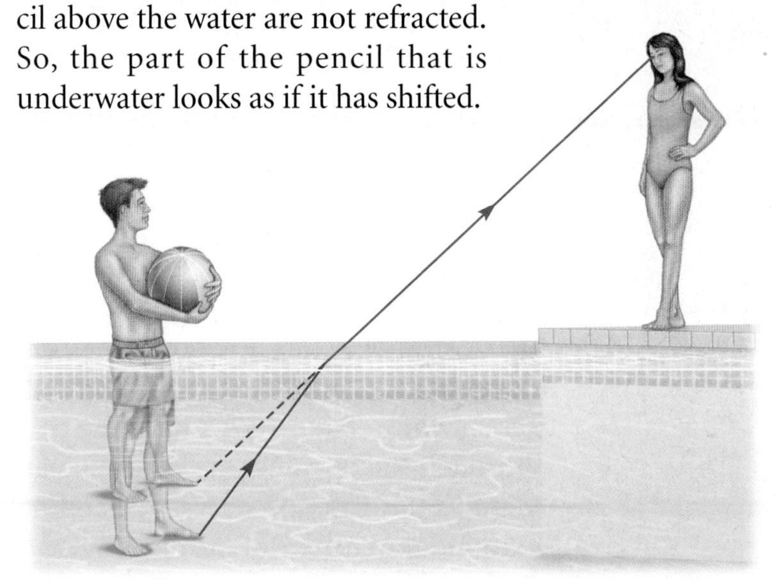

Figure 17
To the observer on the side of the pool, the swimmer's foot looks closer to the surface than it actually is. *When the boy looks down at his feet, will they seem closer to the surface than they really are?*

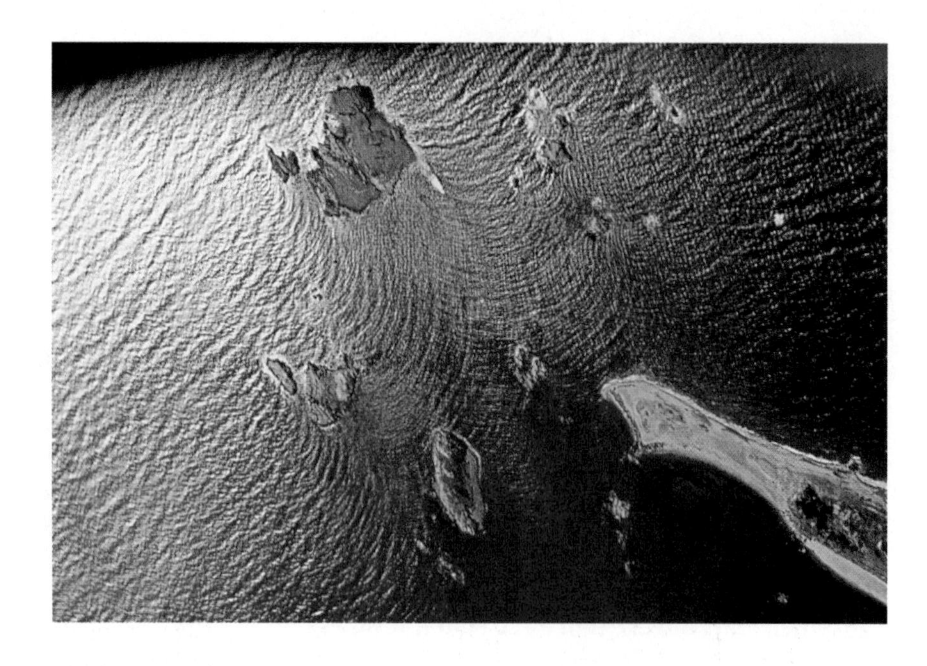

Figure 18
Ocean waves change direction as they pass a group of islands. *How are the waves different before and after they pass the islands?*

Diffraction

When waves strike an object, several things can happen. The waves can bounce off, or be reflected. If the object is transparent, light waves can be refracted as they pass through it. Sometimes the waves may be both reflected and refracted. If you look into a glass window, sometimes you can see your reflection in the window, as well as objects behind it. Light is passing through the window and is also being reflected at its surface.

Waves also can behave another way when they strike an object. The waves can bend around the object. **Figure 18** shows how ocean waves change direction and bend after they strike an island. **Diffraction** occurs when an object causes a wave to change direction and bend around it. Diffraction and refraction both cause waves to bend. The difference is that refraction occurs when waves pass through an object, while diffraction occurs when waves pass around an object.

Figure 19
When water waves pass through a small opening in a barrier, they diffract and spread out after they pass through the hole.

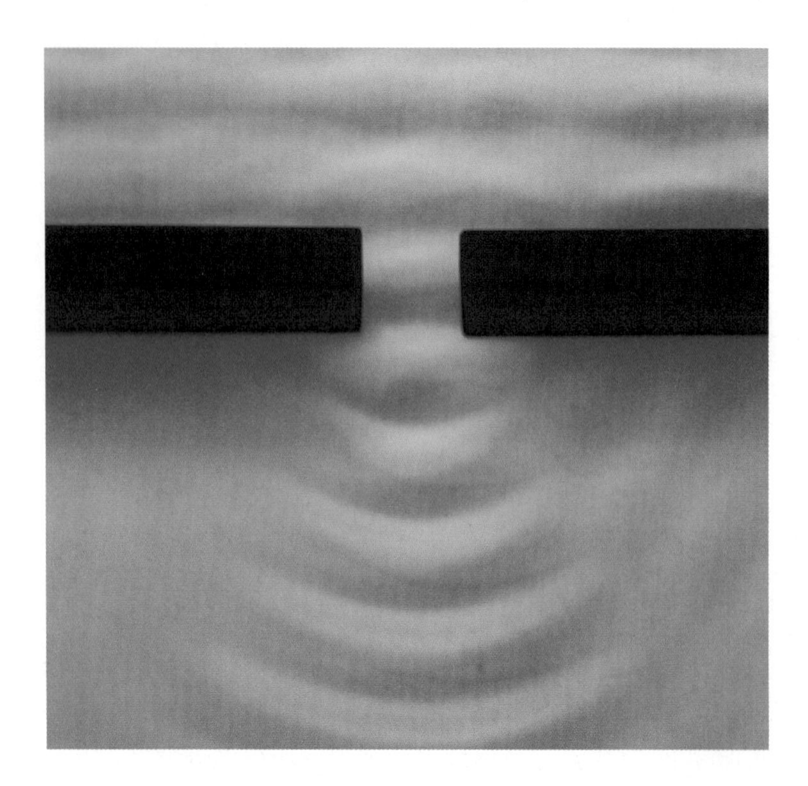

Reading Check *What is diffraction?*

Waves also can be diffracted when they pass through a narrow opening, as shown in **Figure 19.** After they pass through the opening, the waves spread out. In this case the waves are bending around the corners of the opening.

Diffraction and Wavelength How much does a wave bend when it strikes an object or an opening? The amount of diffraction that occurs depends on how big the obstacle or opening is compared to the wavelength. When an obstacle is smaller than the wavelength, the waves bend around it. But if the obstacle is larger than the wavelength, the waves do not diffract as much. In fact, if the obstacle is much larger than the wavelength, almost no diffraction occurs. The obstacle casts a shadow because almost no waves bend around it. The larger the obstacle is compared to the wavelength, the less the waves will diffract, as shown in **Figure 20.**

For example, you're walking down the hallway and you can hear sounds coming from the lunchroom before you reach the open lunchroom door. However, you can't see into the room until you reach the doorway. Why can you hear the sound waves but not see the light waves while you're still in the hallway? The wavelengths of sound waves are similar in size to a door opening. Sound waves diffract around the door and spread out down the hallway. Light waves have a much shorter wavelength. They are hardly diffracted at all by the door. The light waves from the lunchroom bend only slightly around the doorway, and you can't see into the room until you get close to the door.

Radio Waves Diffraction also affects your radio's reception. AM radio waves have longer wavelengths than FM radio waves do. Because of their longer wavelengths, AM radio waves diffract around obstacles like buildings and mountains. The FM waves with their short wavelengths do not diffract as much. As a result, AM radio reception is often better than FM reception around tall buildings and natural barriers.

SCIENCE Online

Research Visit the Glencoe Science Web site at **science.glencoe.com** for more information about diffraction. Communicate to your class what you learned.

Figure 20
The diffraction of waves around an obstacle depends on the wavelength and the size of the obstacle.

A Less diffraction occurs if the wavelength is smaller than the obstacle.

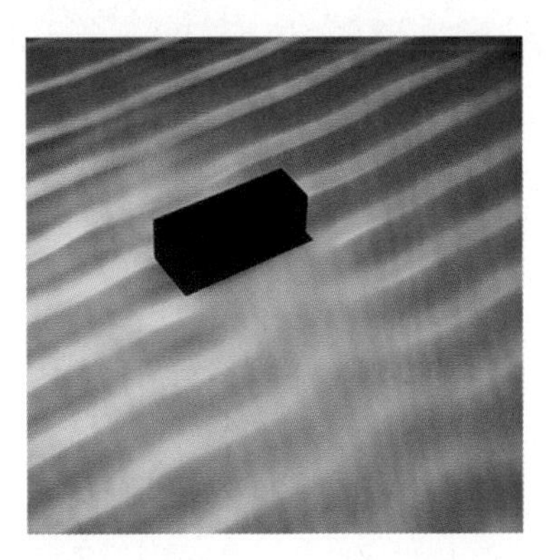

B More diffraction occurs if the wavelength is the same size as the obstacle.

Interference

Suppose two waves are traveling toward each other on a long rope as in **Figure 21A.** What will happen when the two waves meet? If you did this experiment, you would find that the two waves pass right through each other, and each one continues to travel in its original direction, as shown in **Figure 21B** and **Figure 21C.** If you look closely at the waves when they meet each other in **Figure 21B,** you see a wave that looks different than either of the two original waves. When the two waves arrive at the same place at the same time, they combine to form a new wave. When two or more waves overlap and combine to form a new wave, the process is called **interference.** This new wave exists only while the two original waves continue to overlap. The two ways that the waves can combine are called constructive interference and destructive interference.

Figure 21
At the ocean, when one wave retreats from shore, it can meet a new wave coming in. The two waves combine to form a new wave. The same thing happens with the waves on this rope.

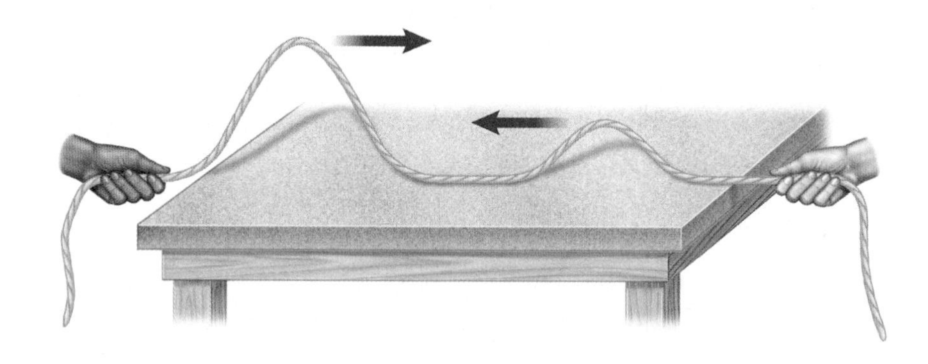

A **Two waves move toward each other on a rope.**

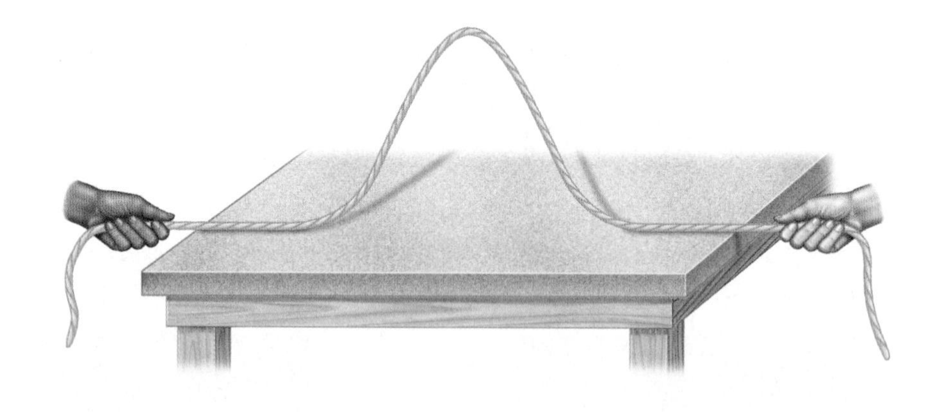

B **As the waves overlap, they interfere to form a new wave.**
What is the amplitude of the new wave?

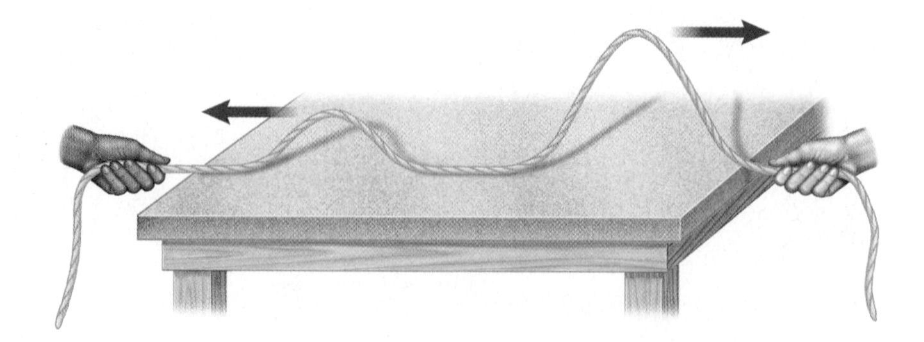

C **When the two waves overlap, they move right through each other. Afterward, they continue moving unchanged, as if they had never met.**

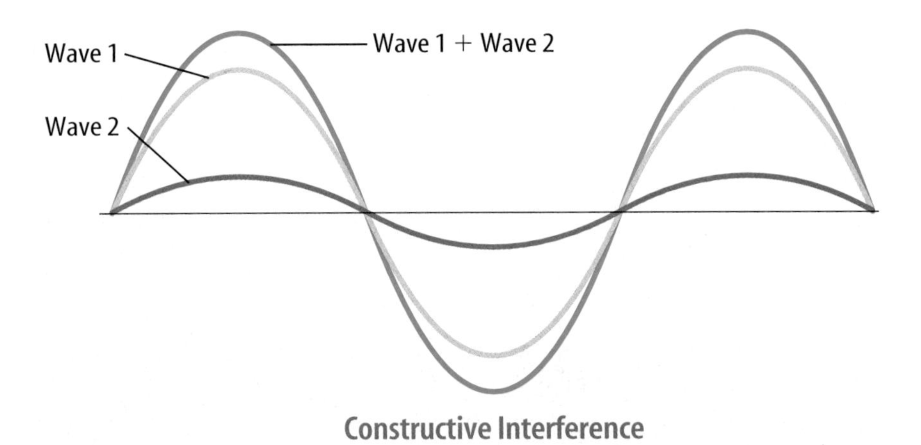

Wave 1

Wave 1 + Wave 2

Wave 2

Destructive Interference

Figure 22
Two types of interference are possible.

A If Wave 1 and Wave 2 were moving toward each other on a rope, they would constructively interfere and form the green wave. Wave 1 and Wave 2 are in phase.

B If Wave 1 and Wave 2 were traveling toward each other on a rope, they would destructively interfere and form the green wave. Wave 1 and Wave 2 are out of phase.

Constructive Interference In constructive interference, shown in **Figure 22A,** the waves add together. This happens when the crests of two or more transverse waves arrive at the same place at the same time and overlap. The amplitude of the new wave that forms is equal to the sum of the amplitudes of the original waves. Constructive interference also occurs when the compressions of different compressional waves overlap. If the waves are sound waves, for example, constructive interference produces a louder sound. Waves undergoing constructive interference are said to be in phase.

Destructive Interference In destructive interference, the waves subtract from each other as they overlap. This happens when the crests of one transverse wave meet the troughs of another transverse wave, as shown in **Figure 22B.** The amplitude of the new wave is the difference between the amplitudes of the waves that overlapped. With compressional waves, destructive interference occurs when the compression of one wave overlaps with the rarefaction of another wave. The compressions and rarefactions combine and form a wave with reduced amplitude. When destructive interference happens with sound waves, it causes a decrease in loudness. Waves undergoing destructive interference are said to be out of phase.

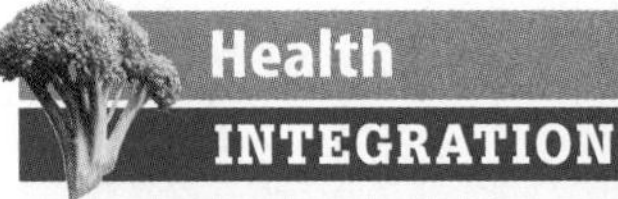

People who are exposed to constant loud noises, such as those made by airplane engines, can suffer hearing damage. Scientists have developed ways to reduce loud noises by using destructive interference. Special ear protectors use destructive interference to cancel damaging noise. With a classmate, list all the jobs you can think of that require ear protectors.

Figure 23
Standing waves seem to stay in the same place. *How do nodes form?*

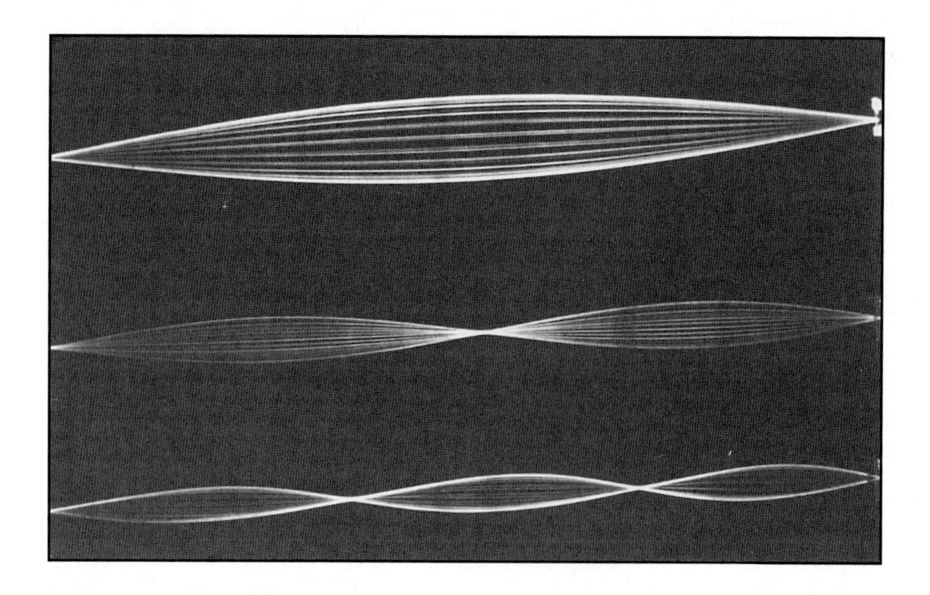

Standing Waves

When you make transverse waves with a rope, you might shake one end while your friend holds the other end still. What would happen if you both shook the rope continuously to create identical waves moving toward each other? As the two waves travel in opposite directions down the rope, they continually pass through each other. Interference takes place as the waves from each end overlap along the rope. At any point where a crest meets a crest, a new wave with a larger amplitude forms. But at points where crests meet troughs, the waves cancel each other and no motion occurs.

The interference of the two identical waves makes the rope vibrate in a special way, as shown in **Figure 23.** The waves create a pattern of crests and troughs that do not seem to be moving. Because the wave pattern stays in one place, it is called a standing wave. A **standing wave** is a special type of wave pattern that forms when waves equal in wavelength and amplitude, but traveling in opposite directions, continuously interfere with each other. The places where the two waves always cancel are called nodes. The nodes always stay in the same place on the rope. Meanwhile, the wave vibrates between the nodes.

Standing Waves in Music When the string of a violin is played with a bow, it vibrates and creates standing waves. The standing waves in the string help produce a rich, musical tone. Other instruments also rely on standing waves to produce music. Some instruments, like flutes, create standing waves in a column of air. In other instruments, like drums, a tightly stretched piece of material vibrates in a special way to create standing waves. As the material in a drum vibrates, nodes are created on the surface of the drum.

Resonance

You may have noticed that bells of different sizes and shapes create different notes. When you strike a bell, you cause it to vibrate to produce sound. The bell vibrates at a certain frequency called the natural frequency. The note you hear depends on the bell's natural frequency. The natural frequency of vibration depends on the bell's size, shape, and the material it is made from. Other objects, including windows, bridges, and columns of air, also vibrate at their own natural frequencies.

There is another way to make something vibrate at its natural frequency. Suppose you have a tuning fork that has a natural frequency of 440 Hz. Imagine that a sound wave of the same frequency strikes the tuning fork. Because the sound wave has the same frequency as the natural frequency of the tuning fork, the tuning fork will vibrate. The ability of an object to vibrate by absorbing energy at its natural frequency is called **resonance** (RE zun unts).

Sometimes resonance can cause an object to absorb a large amount of energy. What happens to the tuning fork if it continues to absorb energy from the sound wave? Remember that the amplitude of a wave increases as the energy it carries increases. In the same way, an object vibrates more strongly as it continues to absorb energy at its natural frequency. If enough energy is absorbed, the object can vibrate so strongly that it breaks apart.

Mini LAB

Experimenting with Resonance

Procedure

1. Strike a **tuning fork** with a **mallet.**
2. Hold the vibrating tuning fork near a **second tuning fork** that has the same frequency.
3. Strike the tuning fork again. Hold it near a **third tuning fork** that has a different frequency.

Analysis

What happened when you held the vibrating tuning fork near each of the other two? Explain.

Section

Assessment

1. Describe how the reflection of light rays allows you to see your image in a mirror.
2. Sketch a diagram showing what happens when a wave enters a medium and slows down. Also sketch a wave speeding up as it enters a new medium. In each diagram, label the normal, the angle of incidence, and the angle of refraction.
3. What happens when waves overlap?
4. What is resonance?
5. **Think Critically** Aluminum foil is shiny like a mirror, yet you can't see your reflection in a piece of crumpled aluminum foil. Explain.

Skill Builder Activities

6. **Recognizing Cause and Effect** Imagine you are on the shore of a large lake and see waves moving toward you from the center of the lake. However, before reaching shore, the waves pass by a boat dock. The waves then move toward you at a slightly different angle. What would you infer is happening? **For more help, refer to the Science Skill Handbook.**
7. **Using a Word Processor** Use a word processor to make an outline showing important points about constructive interference and destructive interference. **For more help, refer to the Technology Skill Handbook.**

Activity

Measuring Wave Properties

Some waves travel through space; others pass through a medium such as air, water, or earth. Each wave has a wavelength, speed, frequency, and amplitude. In this activity you will make waves in the classroom, and observe, measure, and change some of the properties of these waves.

What You'll Investigate

How can the speed of a wave be measured?
How can the wavelength be determined from the frequency?

Materials

long spring, rope, or hose
meterstick
stopwatch

Goals

- **Measure** the speed of a transverse wave.
- **Create** waves with different amplitudes.
- **Measure** the wavelength of a transverse wave.

Safety Precautions

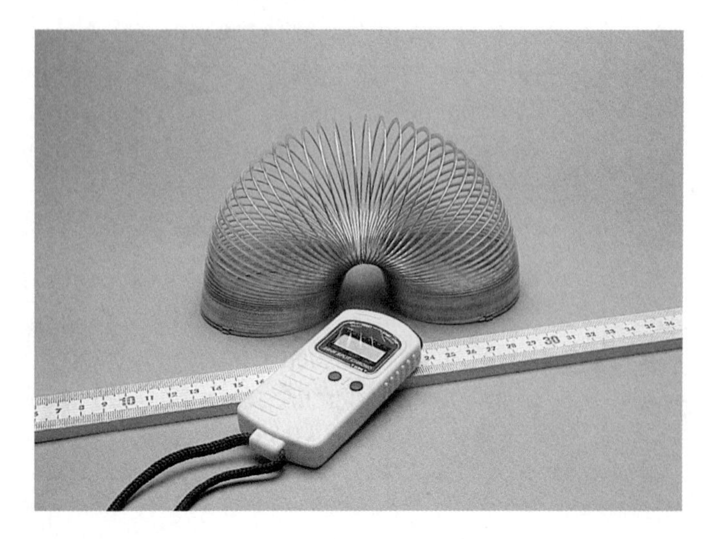

Procedure

1. With a partner, stretch your spring across an open floor and measure the length of the spring. Record this measurement in the data table. Make sure the spring is stretched to the same length for each step.
2. Have your partner hold one end of the spring. Create a single wave pulse by shaking the other end of the spring back and forth.
3. Have a third person use a stopwatch to measure the time needed for the pulse to travel the length of the spring. Record this measurement in the "Wave Time" column of your data table.
4. Repeat steps 2 and 3 two more times.
5. **Calculate** the speed of waves 1, 2, and 3 in your data table by using the formula:

   speed = distance / time

   Average the speeds of waves 1, 2, and 3 to find the speed of waves on your spring.

6. **Create** a wave with several wavelengths. You make one wavelength when your hand moves up, down, and up again. Count the number of wavelengths that you generate in 10 s. Record this measurement for wave 4 in the Wavelength Count column in your data table.
7. Repeat step 6 two more times. Each time, create a wave with a different wavelength by shaking the spring faster or slower.
8. **Calculate** the frequency of waves 4, 5, and 6 by dividing the number of wavelengths by 10 s.

9. Calculate the wavelength of waves 4, 5, and 6 using the formula

wavelength = wave speed / frequency

Use the average speed calculated in step 5 for the wave speed.

Wave Property Measurements

	Spring Length	Wave Time	Wave Speed
Wave 1			
Wave 2			
Wave 3			
	Wavelength Count	**Frequency**	**Wavelength**
Wave 4			
Wave 5			
Wave 6			

Conclude and Apply

1. Was the wave speed different for the three different pulses you created? Why or why not?
2. Why would you average the speeds of the three different pulses to calculate the speed of waves on your spring?
3. How did the wavelength of the waves you created depend on the frequency of the waves?

Communicating Your Data

Ask your teacher to set up a contest between the groups in your class. Have each group compete to determine who can create waves with the longest wavelength, the highest frequency, and the largest wave speed. Record the measurements of each group's efforts on the board. **For more help, refer to the Science Skill Handbook.**

Making Waves

Sonar Helps Create Deep-Sea Pictures and Save Lives

What is sonar?

Sonar is a device that uses sound waves to find the location and distance of underwater objects. Its name is a shortened version of SOund NAvigation and Ranging.

Sonar was used to find enemy subs during World War II.

How does sonar work?

There are two kinds of sonar—active and passive. Active sonar sends out a ping sound that is reflected back when it hits an underwater object. Since sound travels through water at a known speed (1,500 m/s), scientists use the time the sound takes to return to calculate the distance. Passive sonar only listens for sounds given off by underwater objects, such as the noise made by a submarine's engines or by torpedoes.

Why was it invented?

Sonar was first developed by scientists in the early twentieth century as a way to detect icebergs and prevent boating disasters. Its technical advancement was hurried by the Allies' need to detect German submarines in World War I. Before 1916, antisubmarine sonar was passive—a series of microphones towed underwater. By 1918, the United States and Britain had developed an active sonar system placed in submarines sent to attack other subs.

The range of early sonar was only 1.6 km. (Today it is more than 16 km.) Even so, in World War II, sonar allowed ships to defend themselves effectively from enemy subs.

Their strategy was to use sonar to find subs and then fire rocket-fueled depth charges from a safe distance. After the war, quieter nuclear submarines were developed. Sonar-absorbing hulls and quiet engines and machinery ensured that the subs could partly shield themselves from sonar.

Since the war, sonar has been used to help scientists find schools of fish. It also has been used by oceanographers to map ocean and lake floors. Most dramatically, sonar has been vital in the discovery of submerged wrecks, such as downed airplanes and ships, including the *Titanic*—the passenger liner that sank in 1912.

In 1985, a French and American team used a new type of sonar device called the side-scan sonar to locate the *Titanic*. This kind of sonar projects a tight beam of sound that can create accurate images of the sea bed. Members of the expedition towed this sonar device about 170 m above the seabed across a section of the Atlantic Ocean where the *Titanic* went down. Although the expedition's ship passed above the *Titanic* early on, the sonar readings were misinterpreted. Weeks later, video cameras finally spotted the wreck. In 1996, a French expedition to the *Titanic* used a special sonar device that produced 3D images of the wreck site. This sonar was also powerful enough to penetrate the 15 m of mud that covered the bow of the ship. It enabled researchers to see how the hull had been damaged when the ship had collided with an iceberg.

The *Titanic* was found thanks to sonar.

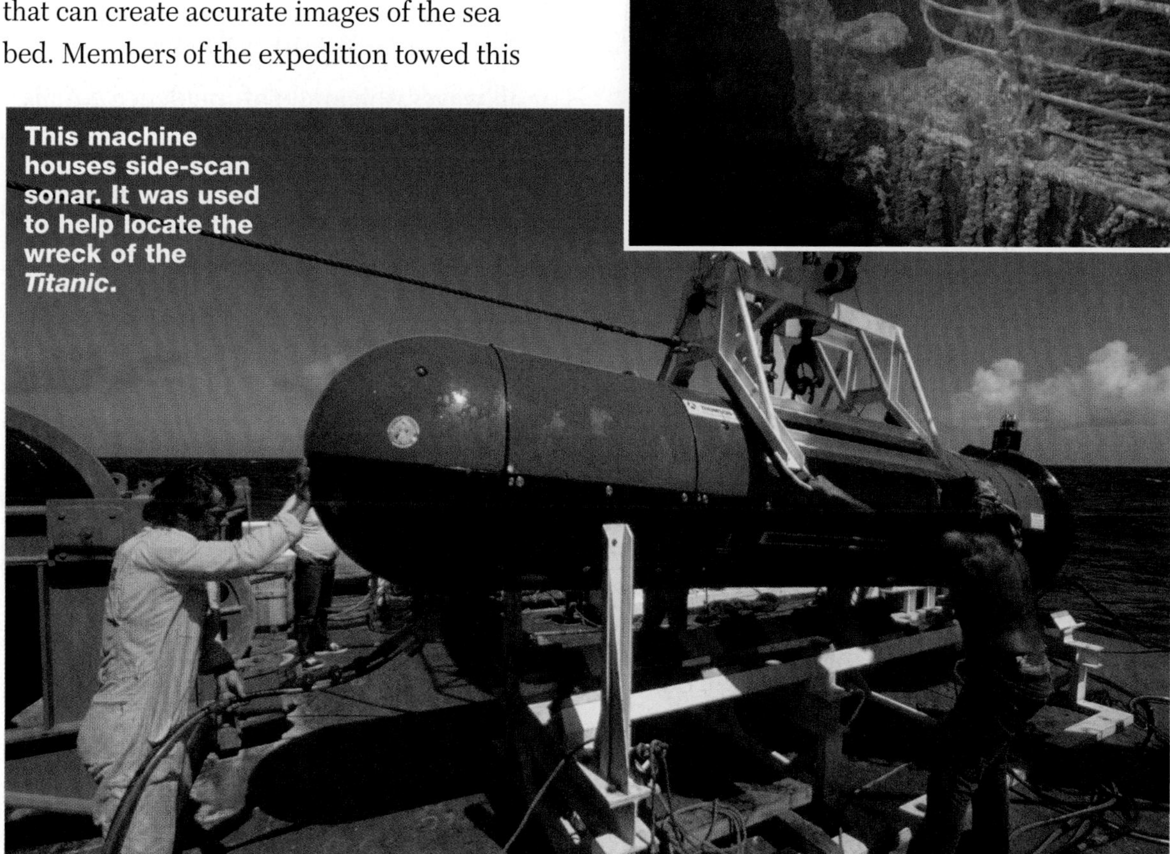

This machine houses side-scan sonar. It was used to help locate the wreck of the *Titanic*.

CONNECTIONS Report **Research how sonar was used by navies in World War I and World War II. Did sonar affect each war's outcome? How did it save lives? What uses can you think of for sonar if it could be used in everyday life?**

SCIENCE Online
For more information, visit science.glencoe.com

Chapter 11 Study Guide

Reviewing Main Ideas

Section 1 The Nature of Waves

1. Waves are rhythmic disturbances that transfer energy through matter or space.
2. Waves transfer only energy, not matter. *Is a human "wave" in a stadium really a wave? Explain.*

3. Mechanical waves need matter to travel through. Mechanical waves can be compressional or transverse.
4. When a transverse wave passes through a medium, matter in the medium moves at right angles to the direction the wave travels. For a compressional wave, matter moves back and forth in the same direction as the wave travels. Matters returns to its original position after the wave passes.

Section 2 Wave Properties

1. Transverse waves have high points (crests) and low points (troughs). Compressional waves have more dense areas (compressions) and less dense areas (rarefactions).
2. Transverse and compressional waves can be described by their wavelengths, frequencies, and amplitudes. As frequency increases, wavelength always decreases.
3. The greater a wave's amplitude is, the more energy it carries. *How would you measure the amplitude of these ocean waves?*

4. A wave's velocity can be calculated by multiplying its frequency times its wavelength.

Section 3 The Behavior of Waves

1. For all waves, the angle of incidence equals the angle of reflection.
2. A wave is bent, or refracted, when it changes speed as it enters a new medium. *How does refraction affect how this fisher aims with his spear?*

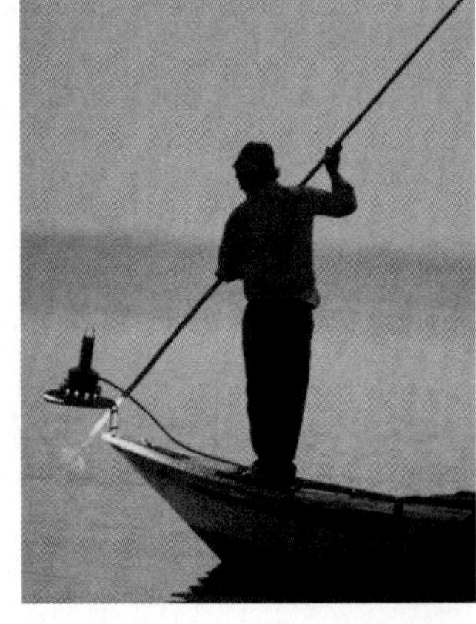

3. When two or more waves overlap, they combine to form a new wave. This process is called interference.

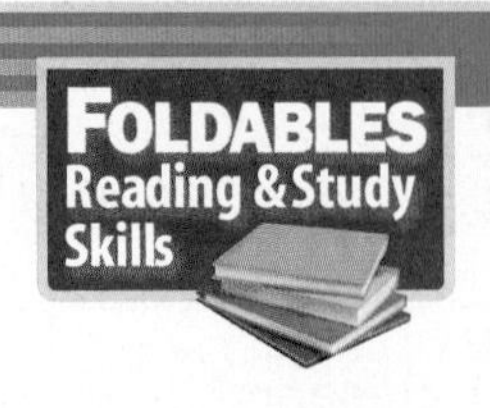

After You Read

Use your Foldable to help you review characteristics of light and sound waves.

Chapter 11 Study Guide

Visualizing Main Ideas

Complete the following concept map on waves.

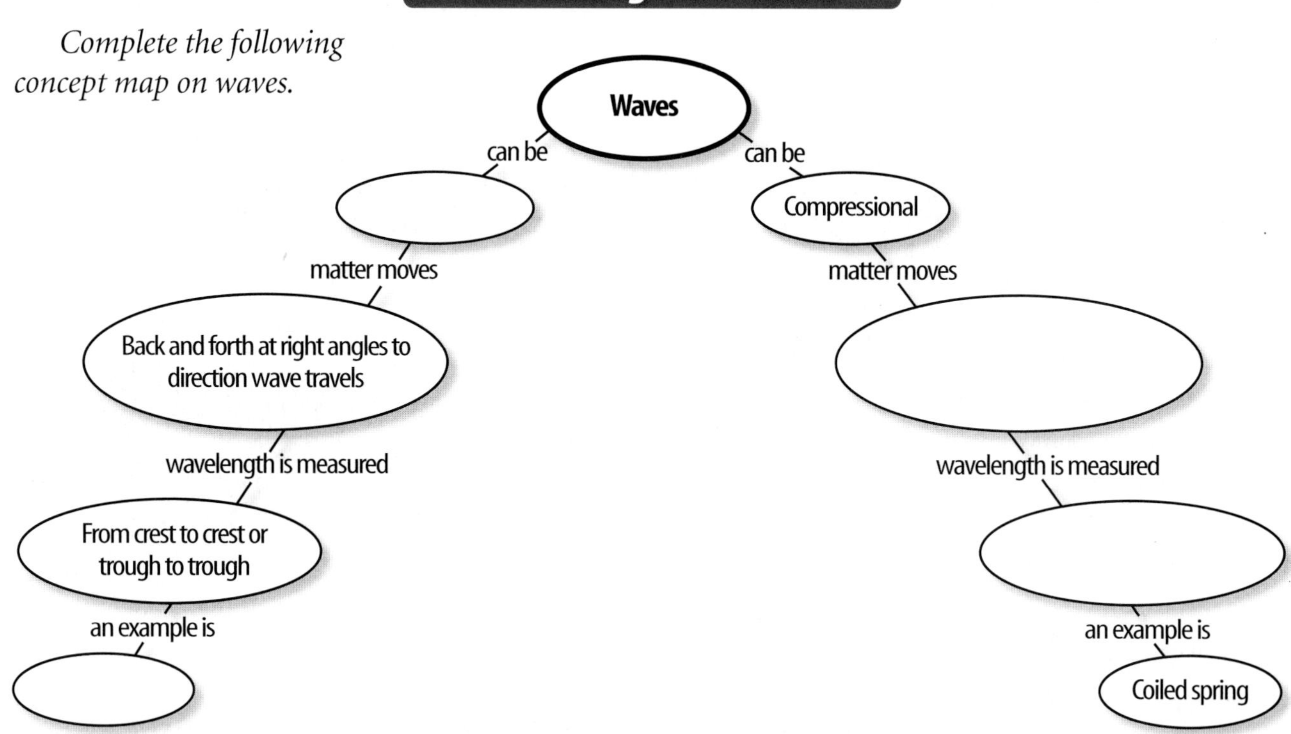

Vocabulary Review

Vocabulary Words

a. amplitude
b. compressional wave
c. crest
d. diffraction
e. frequency
f. interference
g. medium
h. rarefaction
i. refraction
j. resonance
k. standing wave
l. transverse wave
m. trough
n. wave
o. wavelength

Study Tip

Use word webs. Write down the main idea of the chapter on a piece of paper and circle it. Connect other related facts to it with lines and arrows.

Using Vocabulary

Answer the following questions using complete sentences.

1. Compare and contrast reflection and refraction.
2. Which type of wave has points called nodes that do not move?
3. Which part of a compressional wave has the lowest density?
4. Find two words in the vocabulary list that describe the bending of a wave.
5. Describe what happens when waves overlap.
6. What is the relationship among amplitude, crest, and trough?
7. What does frequency measure?
8. What does a mechanical wave always travel through?

Chapter 11 Assessment

Checking Concepts

Choose the word or phrase that best answers the question.

1. Which of the following do waves carry?
A) matter **C)** matter and energy
B) energy **D)** the medium

2. When a compressional wave travels through a medium, which way does matter in the medium move?
A) backward
B) all directions
C) at right angles to the direction the wave travels
D) in the same direction the wave travels

3. What is the formula for calculating wave speed?
A) $v = \lambda \times f$ **C)** $v = \lambda / f$
B) $v = f$ **D)** $v = \lambda + f$

4. What is the highest point of a transverse wave called?
A) crest **C)** wavelength
B) compression **D)** trough

5. As the frequency of a wave increases, what happens to the wavelength?
A) It moves forward. **C)** It vibrates.
B) It decreases. **D)** It increases.

6. What is the amplitude of a wave related to?
A) the wave's energy **C)** wave speed
B) frequency **D)** refraction

7. Which term describes the bending of a wave around a barrier?
A) resonance **C)** diffraction
B) interference **D)** reflection

8. What types of waves have nodes?
A) seismic waves **C)** water waves
B) radio waves **D)** standing waves

9. What is equal to the angle of reflection?
A) refraction angle **C)** bouncing angle
B) normal angle **D)** angle of incidence

10. When two or more waves arrive at the same place at the same time, what do they do?
A) turn around
B) bend toward the normal
C) stop
D) combine

Thinking Critically

11. An earthquake on the ocean floor produces a tsunami that hits a remote island. Is the water that hits the island the same water that was above the earthquake? Explain.

12. When a wave's amplitude increases, does its frequency change? Explain.

13. Use the law of reflection to explain why you see only a portion of the area behind you when you look in a mirror.

14. Explain why you can hear a fire engine coming around a street corner before you can see it.

15. Describe what vibrated to produce three of the sounds you've heard today.

Developing Skills

16. Forming Hypotheses In 1981, swaying dancers on the balconies of a Kansas City, Missouri, hotel caused the balconies to collapse. Use what have you learned about wave behavior to form a hypothesis that explains why this happened.

17. Comparing and Contrasting Compare and contrast diffraction and refraction.

18. Interpreting Data According to the data in the table below, approximately how many times faster does sound travel in steel than in air?

Sound Transmission

Substance	Speed of Sound at 25°C (m/s)
Air	347
Brick	3,650
Cork	500
Water	1,498
Steel	5,200

19. Making and Using Tables Find newspaper articles describing five recent earthquakes. Construct a table that shows for each earthquake the date, location, magnitude, and whether the damage caused by each earthquake was light, moderate, or heavy.

20. Concept Mapping Design a concept map that shows the characteristics of transverse waves. Include the terms *crest, trough, medium, wavelength, frequency,* and *amplitude.*

Performance Assessment

21. Oral Presentation A seismograph is an instrument that measures the magnitude of earthquakes. Research seismographs, and make an oral presentation explaining how they work.

Technology

Go to the Glencoe Science Web site at **science.glencoe.com** or use the **Glencoe Science CD-ROM** for additional chapter assessment.

Test Practice

A scientist is studying the formation of ocean waves during windy storms. Her observations are listed in the table below.

Ocean Wave Observations

Wind Conditions	Wind Speed (km/h)	Ocean Wave Height (m)	Description of Ocean Waves
Calm	1-5	0.05	Like a Small Lake
Light Breeze	6-11	0.10	Small Wavelets
Gentle Breeze	?	0.60	Small Waves
Mod. Breeze	20-28	1.00	Large Wavelets
Fresh Breeze	29-38	2.00	Mod. Waves
Strong Breeze	39-49	3.00	Large Waves
Gale	62-75	7.00	Breaking Waves

Study the chart and answer the following questions.

1. What wind speeds are considered a "gentle breeze?"
A) 9–17 (km/h)
B) 10–18 (km/h)
C) 12–19 (km/h)
D) 12–20 (km/h)

2. According to the table, a 33 km/h wind produced ______.
F) large wavelets
G) moderate waves
H) large waves
J) breaking waves

3. What is the height of a large wave?
A) 1.50 m
B) 0.25 m
C) 1.00 m
D) 0.10 m

CHAPTER 12

Sound

You might have tried the following experiment. With a friend, you go below the water surface in a swimming pool and scream a message. After you come up for air, your friend tries to guess what was said. Your words were hard to understand under water. These scuba divers use hand signals to communicate under water. In this chapter, learn how sound travels and how noise differs from music. Also, learn how musical instruments produce sound, and how sound is used in medicine.

What do you think?

Science Journal Look at the picture below with a classmate. Discuss what this might be or what is happening. Here's a hint: *It's music to your ears.* Write your answer or best guess in your Science Journal.

Think of all the different kinds of musical instruments you've seen and heard. Some have strings, some have hollow tubes, and others have keys or pedals. Musical instruments come in many shapes and sizes and are played with various techniques. The differences between the instruments give each one a unique sound. What would an instrument made out of a ruler sound like?

Create music with a ruler

1. Hold one end of a thin ruler firmly down on a desk, allowing the free end to extend beyond the edge of the desk.
2. Gently pull up on and release the end of the ruler. What do you see and hear?
3. Vary the length of the overhanging portion and repeat the experiment several times.

Observe

How does the length of the overhanging part of the ruler affect the sound you hear? In your Science Journal, write a paragraph about how you could write instructions for playing a song with the ruler.

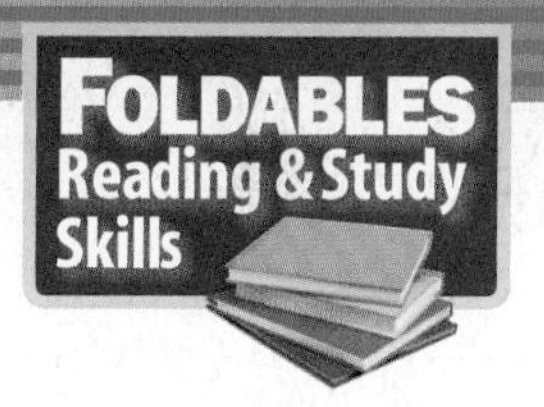

Before You Read

Making a Question Study Fold **Asking yourself questions helps you stay focused and better understand sound when you are reading the chapter.**

1. Place a sheet of paper in front of you so the short side is at the top. Fold the paper in half from the left side to the right side.
2. Fold in the top and the bottom. Unfold the paper so three equal sections show.
3. Through the top thickness of paper, cut along each of the fold lines to the left fold, forming three tabs.
4. Before you read the chapter, write these questions on the tabs: Can sound travel through solids? Can sound travel through liquids? Can sound travel through gases? Now answer each question on the front of the tab.
5. As you read the chapter, write what you learn about how sound travels under the tabs.

SECTION

The Nature of Sound

As You Read

What You'll Learn

- **Explain** how sound travels through different mediums.
- **Identify** what influences the speed of sound.
- **Describe** how the ear enables you to hear.

Vocabulary

eardrum
cochlea

Why It's Important

The nature of sound affects how you hear and interpret sounds.

What causes sound?

An amusement park can be a noisy place. With all the racket of carousel music and booming loudspeakers, it can be hard to hear what your friends say. These sounds are all different, but they do have something in common—each sound is produced by an object that vibrates. For example, your friends' voices are produced by the vibrations of their vocal cords, and music from a carousel and voices from a loudspeaker are produced by vibrating speakers. All sounds are created by something that vibrates.

Sound Waves

How does the sound made by a vibrating speaker get to your ears? When an object like a radio speaker vibrates, it collides with nearby molecules in the air, transferring some of its energy to them. These molecules then collide with other molecules in the air and pass the energy on to them. The energy originally transferred by the vibrating object continues to pass from one molecule to another. This process of collisions and energy transfer forms a sound wave. Eventually, the wave reaches your ears and you hear a sound.

Sound Is a Compressional Wave Sound waves are compressional waves. Remember that a compressional wave is made up of two types of regions called compressions and rarefactions. If you look at **Figure 1A,** you'll see that when a radio speaker vibrates outward, the nearby molecules in the air are pushed together to form compressions. As **Figure 1B** shows, when the speaker moves inward, the nearby molecules in the air have room to spread out, and a rarefaction forms. As long as the speaker continues to vibrate back and forth, compressions and rarefactions are formed.

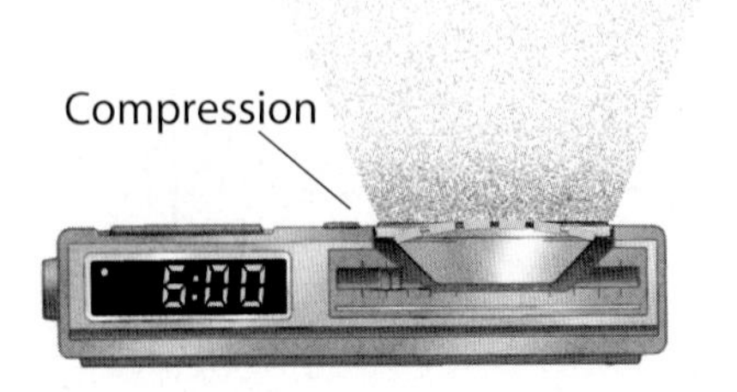

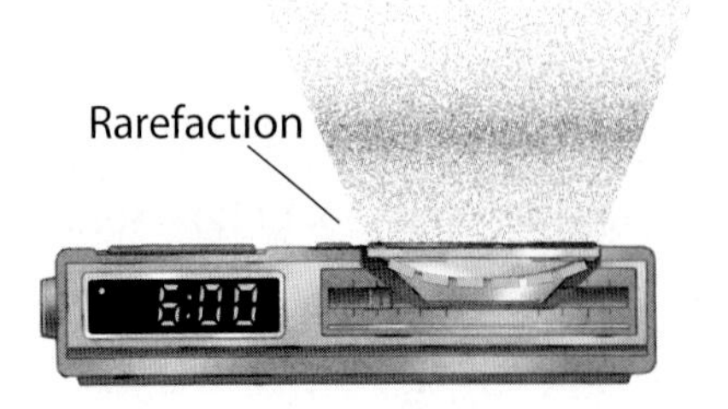

Figure 1
A When the speaker vibrates outward, molecules in the air next to it are pushed together to form a compression. B When the speaker vibrates inward, the molecules spread apart to form a rarefaction.

Traveling as a Wave Compressions and rarefactions move away from the speaker as molecules in the air collide with their neighbors. As the speaker continues to vibrate, more molecules in the air are alternately pushed together and spread apart. A series of compressions and rarefactions forms that travels from the speaker to your ear. This sound wave is what you hear.

Moving Through Mediums

Most sounds you hear travel through air to reach your ears. However, if you've ever been swimming underwater and heard garbled voices, you know that sound also travels through water. In fact, sound waves can travel through any type of matter—solid, liquid, or gas. The matter that a wave travels through is called a medium. Sound waves create compressions and rarefactions in any medium they travel through.

What would happen if no matter existed to form a medium? Could sound be transmitted without particles of matter to compress, expand, and collide? On the Moon, which has no atmosphere, the energy in sound waves cannot be transmitted from particle to particle because no particles exist. Sound waves cannot travel through empty space. Astronauts must talk to each other using electronic communication equipment.

The Speed of Sound Through Different Mediums

The speed of a sound wave through a medium depends on the substance the medium is made of and whether it is solid, liquid, or gas. For example, **Table 1** shows that at room temperature, sound travels at 347 m/s through air, at 1,498 m/s through water, and at 4,877 m/s through aluminum. In general, sound travels the slowest through gases, faster through liquids, and even faster through solids.

Reading Check *What are two things that affect the speed of sound?*

Sound travels faster in liquids and solids than in gases because the individual molecules in a liquid or solid are closer together than the molecules in a gas. When molecules are close together, they can transmit energy from one to another more rapidly. However, the speed of sound doesn't depend on the loudness of the sound. Loud sounds travel through a medium at the same speed as soft sounds.

Table 1 Speed of Sound in Different Mediums

Medium	Speed of Sound (in m/s)
Air	347
Cork	500
Water	1,498
Brick	3,650
Aluminum	4,877

Mini LAB

Observing Sound in Different Materials

Procedure 

1. Tie a piece of **string** onto a **metal object,** such as a wire hanger or a spoon, so that you have two long ends of the string.
2. Wrap each string around a finger on each hand.
3. Gently placing your fingers in your ears, let the object swing until it bumps against the edge of a **chair** or **table.** Listen to the sound.
4. Take your fingers out of your ears and listen to the sound made by the collisions.

Analysis

1. Compare and contrast the sounds you hear when your fingers are and are not in your ears.
2. Do sounds travel better through air or the string?

Figure 2
A line of people passing a bucket is a model for molecules transferring the energy of a sound wave.

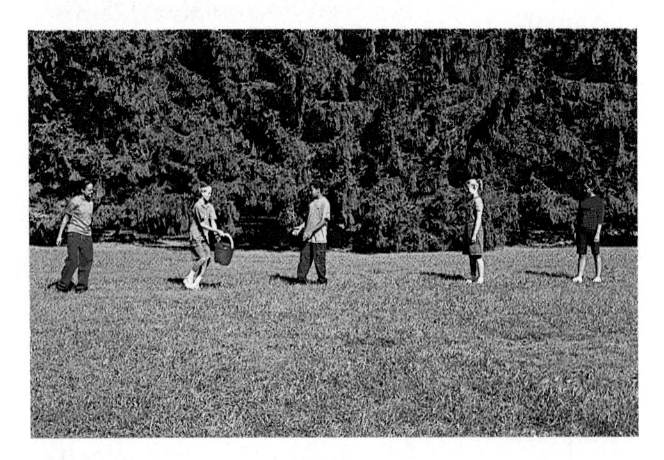

A When the people are far away from each other, like the molecules in a gas, it takes longer to transfer the bucket of water from person to person.

B The bucket travels quickly down the line when the people stand close together. *Why would sound travel more slowly in cork than in steel?*

A Model for Transmitting Sound You can understand why solids and liquids transmit sound well by picturing a large group of people standing in a line. Imagine that they are passing a bucket of water from person to person. If everyone stands far apart, each person has to walk a long distance to transfer the bucket, as in **Figure 2A.** However, if everyone stands close together, as in **Figure 2B,** the bucket quickly moves down the line.

The people standing close to each other are like particles in solids and liquids. Those standing far apart are like gas particles. The closer the particles, the faster they can transfer energy from particle to particle.

Temperature and the Speed of Sound The speed of sound waves also depends on the temperature of a medium. As the temperature of a substance increases, its molecules move faster. This makes them more likely to collide with each other. Remember that sound waves depend on the collisions of particles to transfer energy through a medium. If the particles in a medium are colliding with each other more often, more energy can be transferred in a shorter amount of time. Then sound waves move faster. For example, when the temperature is 0°C, sound travels through the air at only 331 m/s, but at a temperature of 20°C, it speeds up to 343 m/s.

Human Hearing

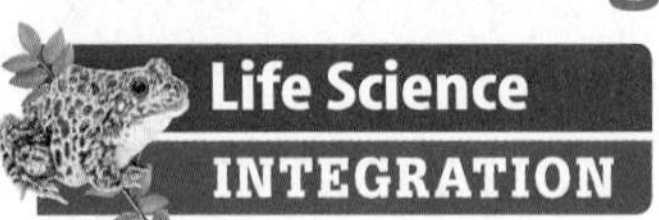

Think of the last conversation you had. Vocal cords and mouths move in many different ways to produce various kinds of compressional waves, but you were somehow able to make sense of these different sound waves. Your ears and brain work together to turn the compressional waves caused by speech, music, and other sources into something that has meaning. Making sense of these waves involves four stages. First, the ear gathers the compressional waves. Next, the ear amplifies the waves. In the ear, the amplified waves are converted to nerve impulses that travel to the brain. Finally, the brain decodes and interprets the nerve impulses.

Gathering Sound Waves—The Outer Ear When you think of your ear, you probably picture just the fleshy, visible, outer part. But, as shown in **Figure 3,** the human ear has three sections called the outer ear, the middle ear, and the inner ear.

The visible part of your ear, the ear canal, and the eardrum make up the outer ear. The outer ear is where sound waves are gathered. The gathering process starts with the outer part of your ear, which is shaped to help capture and direct sound waves into the ear canal. The ear canal is a passageway that is 2-cm to 3-cm long and is a little narrower than your index finger. The sound waves travel along this passageway, which leads to the eardrum. The **eardrum** is a tough membrane about 0.1 mm thick. When incoming sound waves reach the eardrum, they transfer their energy to it and it vibrates.

Reading Check *What makes the eardrum vibrate?*

Amplifying Sound Waves—The Middle Ear When the eardrum vibrates, it passes the sound vibrations into the middle ear, where three tiny bones start to vibrate. These bones are called the hammer, the anvil, and the stirrup. They make a lever system that multiplies the force and pressure exerted by the sound wave. The bones amplify the sound wave. The stirrup is connected to a membrane on a structure called the oval window, which vibrates as the stirrup vibrates.

Some types of hearing loss involve damage to the inner ear. The use of a hearing aid often can improve hearing. Research different advertisements for hearing aids. How would you evaluate different companies' claims that they are the best?

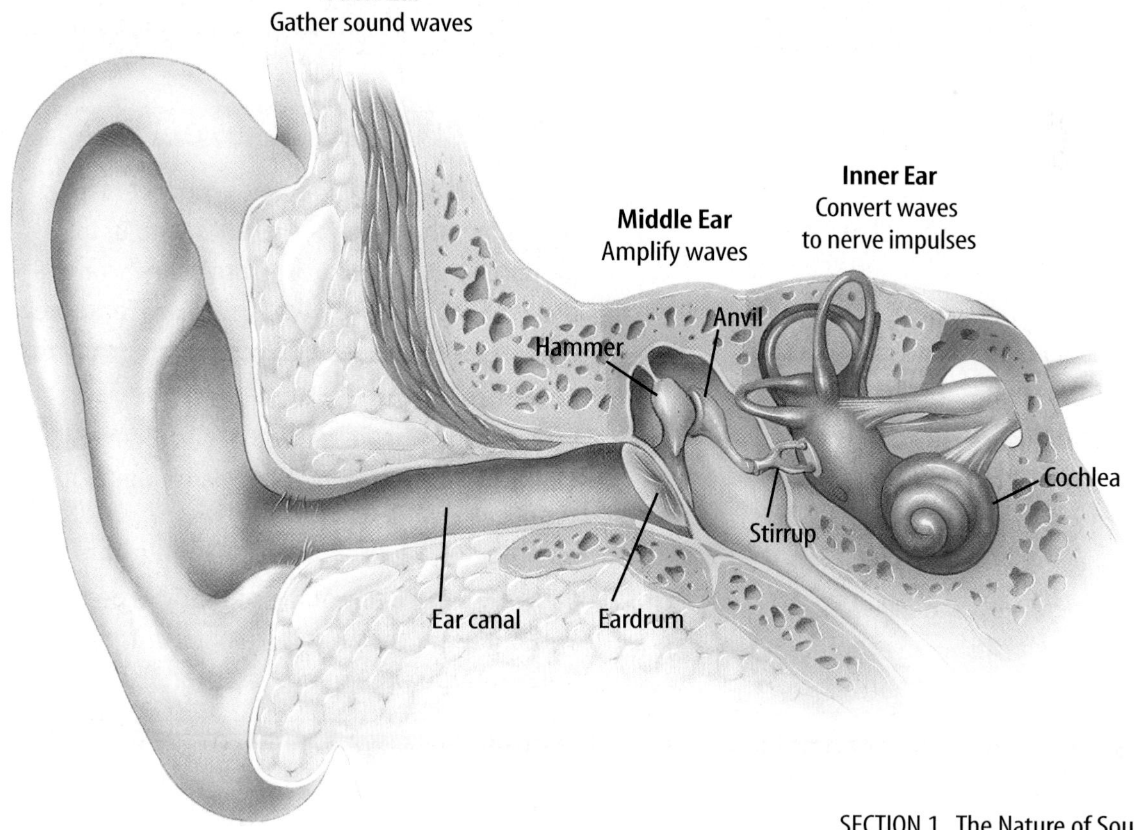

Figure 3
The ear has three regions and each performs specific functions in hearing.

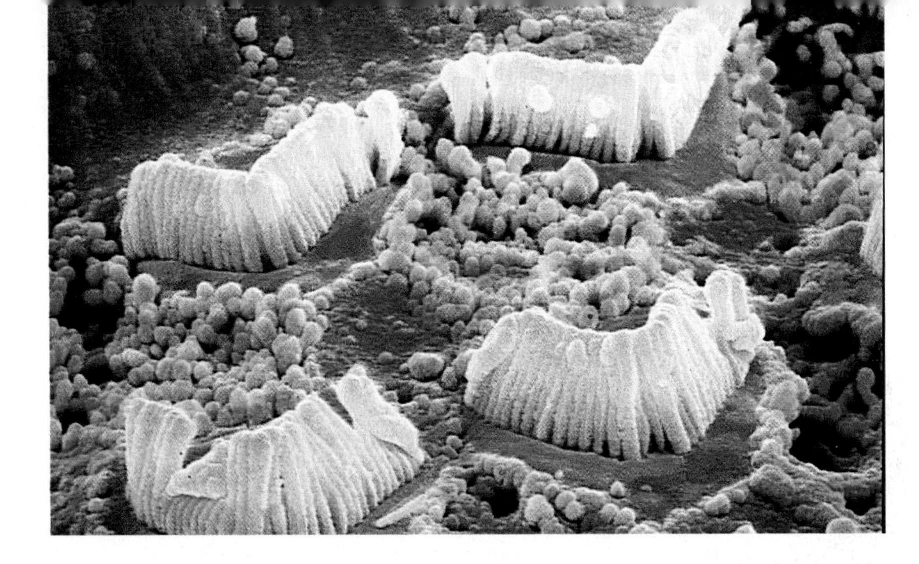

Figure 4
These hair cells in the human ear send nerve impulses to the brain when sound waves cause them to vibrate. In this photo the hair cells are magnified 3,900 times.

Converting Sound Waves—The Inner Ear When the membrane in the oval window vibrates, the sound vibrations are transmitted into the inner ear. The inner ear contains the **cochlea** (KOH klee uh), which is a spiral-shaped structure that is filled with liquid and contains tiny hair cells like those shown in **Figure 4.** When these tiny hair cells in the cochlea begin to vibrate, nerve impulses are sent through the auditory nerve to the brain. It is the cochlea that converts sound waves to nerve impulses.

When someone's hearing is damaged, it's usually because the tiny hair cells in the cochlea are damaged or destroyed, often by loud sounds. New research suggests that these hair cells may be able to repair themselves.

Section  Assessment

1. Explain how sound travels from your vocal cords to your friend's ears when you talk.
2. Compare the speed of sound waves through liquids and air.
3. Explain how the temperature and density of a medium affect the speed of sound waves traveling through it.
4. Describe each section of the human ear and its role in hearing.
5. **Think Critically** Some people hear ringing in their ears, called tinnitus, even when there are no sound waves vibrating their eardrums. Form a hypothesis to explain how this might occur.

Skill Builder Activities

6. **Concept Mapping** Prepare a concept map that shows the series of events that occur to produce sound. Include the terms *rarefaction, medium, vibration,* and *compression.* **For more help, refer to the Science Skill Handbook.**
7. **Communicating** Make a list of 10 different sounds you've heard today. For each sound, identify what was vibrating to cause the sound. Write a description of the vibration and tell whether you could see the vibration. Could you have sensed any of the sounds you heard without using your ears? Explain. **For more help, refer to the Science Skill Handbook.**

Properties of Sound

Intensity and Loudness

If the phone rings while you're listening to the radio, you might have to turn the radio down to be able to hear the person on the phone. What happens to the sound waves from your radio when you adjust the volume? The notes sound the same as when the volume was higher, but something about the sound changes. The difference is that quieter sound waves do not carry as much energy as louder sound waves do.

Recall that the amount of energy a wave carries corresponds to its amplitude. For a compressional wave, amplitude is related to the density of the particles in the compressions and rarefactions. Look at **Figure 5.** When an object vibrates strongly with a lot of energy, it makes sound waves with tight, dense compressions. When an object vibrates with low energy, it makes sound waves with loose, less dense compressions. The density of particles in the rarefactions behaves in the opposite way. In a loud sound wave with lots of energy, the particles in its rarefactions are far apart. In quiet sound waves with low energy, particles in the rarefactions are closer together.

As You Read

What You'll Learn

- **Recognize** how amplitude, intensity, and loudness are related.
- **Describe** how sound intensity is measured and what levels can damage hearing.
- **Explain** the relationship between frequency and pitch.
- **Discuss** the Doppler effect.

Vocabulary

intensity
loudness
decibel
pitch
ultrasonic
Doppler effect

Why It's Important

The properties of sound waves determine how things sound to you—from a blaring CD player to someone's whisper.

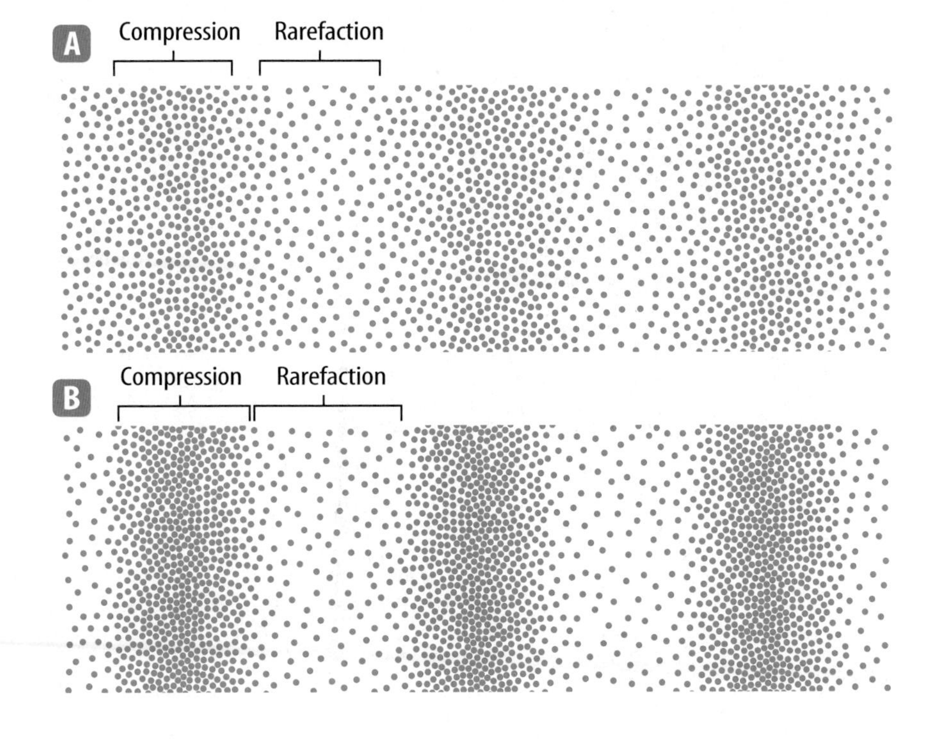

Figure 5
The amplitude of a sound wave depends on how tightly packed molecules are in the compressions and rarefactions. A This sound wave has low amplitude. B This sound wave has a greater amplitude. Molecules are more tightly packed together in compressions and are farther apart in rarefactions.

Intensity Imagine sound waves moving through the air from your radio to your ear. If you held a square loop between you and the radio, as in **Figure 6,** and could measure how much energy passed through the loop in 1 s, you would measure intensity. **Intensity** is the amount of energy that flows through a certain area in a specific amount of time. When you turn down the volume of your radio, you reduce the energy carried by the sound waves, so you also reduce their intensity.

Intensity influences how far away a sound can be heard. If you and a friend whisper a conversation, the sound waves you create have low intensity and do not travel far. You have to sit close together to hear each other. However, when you shout to each other, you can be much farther apart. The sound waves made by your shouts have high intensity and can travel far.

Intensity influences how far a wave will travel because some of a wave's energy is converted to other forms of energy when it is passed from particle to particle. Think about what happens when you drop a basketball. The ball has potential energy as you hold it above the ground. This potential energy is converted into energy of motion as the ball falls. When the ball hits the ground and bounces up, a small amount of that energy has been transferred to the ground. The ball no longer has enough energy to bounce back to the original level. The ball transfers a small amount of energy with each bounce, until finally the ball has no more energy. If you held the ball higher above the ground, it would have more energy and would bounce for a longer time before it came to a stop. In a similar way, a sound wave of low intensity loses its energy more quickly, and travels a shorter distance than a sound wave of higher intensity.

Figure 6
The intensity of the sound waves from the CD player is related to the amount of energy that passes through the loop in a certain amount of time. *How would the intensity change if the loop were 10 m away from the radio?*

Loudness When you hear different sounds, you do not need special equipment to know which sounds have greater intensity. Your ears and brain can tell the difference. **Loudness** is the human perception of sound intensity. Sound waves with high intensity carry more energy. When sound waves of high intensity reach your ear, they cause your eardrum to move back and forth a greater distance than sound waves of low intensity do. The bones of the middle ear convert the increased movement of the eardrum into increased movement of the hair cells in the inner ear. As a result, you hear a loud sound. As the intensity of a sound wave increases, the loudness of the sound you hear increases.

✔ Reading Check *How are intensity and loudness related?*

SCIENCE Online

Research Visit the Glencoe Science Web site at **science.glencoe.com** to find a longer list of sounds and their intensities in decibels. Make a table that lists sounds you heard today, from loudest to quietest. In another column, write the intensity level of each sound.

A Scale for Loudness It's hard to say how loud too loud is. Two people are unlikely to agree on what is too loud, because people vary in their perception of loudness. A sound that seems fine to you may seem earsplitting to your teacher. Even so, the intensity of sound can be described using a measurement scale. Each unit on the scale for sound intensity is called a **decibel** (DES uh bel), abbreviated dB. On this scale, the faintest sound that most people can hear is 0 dB. Sounds with intensity levels above 120 dB may cause pain and permanent hearing loss. During some rock concerts, sounds reach this damaging intensity level. Wearing ear protection, such as earplugs, around loud sounds can help protect against hearing loss. **Figure 7** shows some sounds and their intensity levels in decibels.

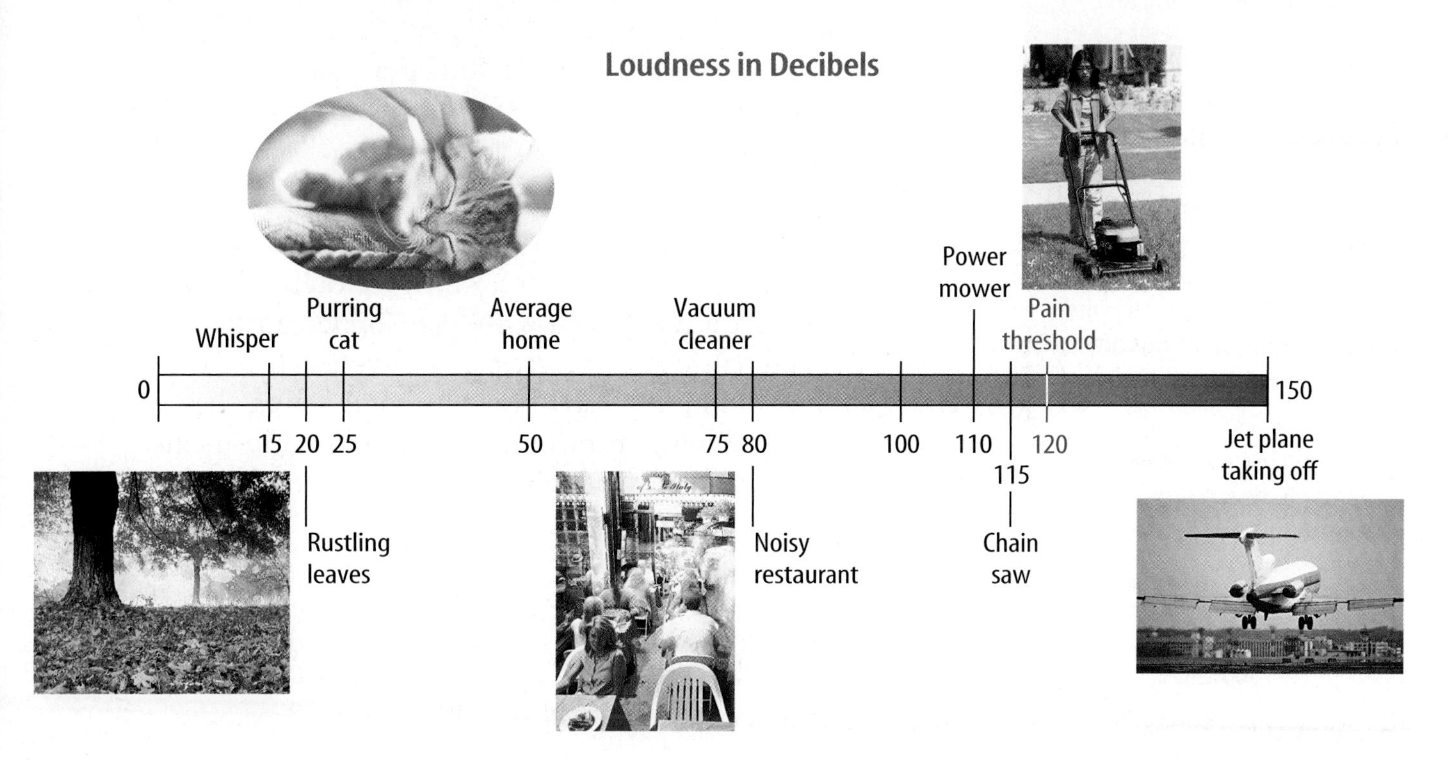

Figure 7
The decibel scale measures the intensity of sound. ***Where would a normal speaking voice fall on the scale?***

C	D	E	F	G	A	B	C
do	re	mi	fa	so	la	ti	do
262 Hz	294 Hz	330 Hz	349 Hz	393 Hz	440 Hz	494 Hz	524 Hz

Figure 8
Every note has a different frequency, which gives it a distinct pitch. *How does pitch change when frequency increases?*

Pitch

If you have ever taken a music class, you are probably familiar with the musical scale do, re, mi, fa, so, la, ti, do. If you were to sing this scale, your voice would start low and become higher with each note. You would hear a change in **pitch,** which is how high or low a sound seems to be. The pitch of a sound is related to the frequency of the sound waves.

Frequency and Pitch Frequency is a measure of how many wavelengths pass a particular point each second. For a compressional wave, such as sound, the frequency is the number of compressions or the number of rarefactions that pass by each second. Frequency is measured in hertz (Hz)—1 Hz means that one wavelength passes by in 1 s.

When a sound wave with high frequency hits your ear, many compressions hit your eardrum each second. The wave causes your eardrum and all the other parts of your ear to vibrate more quickly than if a sound wave with a low frequency hit your ear. Your brain interprets these fast vibrations caused by high-frequency waves as a sound with a high pitch. As the frequency of a sound wave decreases, the pitch becomes lower. **Figure 8** shows different notes and their frequencies. For example, a whistle with a frequency of 1,000 Hz has a high pitch, but low-pitched thunder has a frequency of less than 50 Hz.

A healthy human ear can hear sound waves with frequencies from about 20 Hz to 20,000 Hz. The human ear is most sensitive to sounds in the range of 440 Hz to about 7,000 Hz. In this range, most people can hear much fainter sounds than at higher or lower frequencies.

Making a Model for Hearing Loss

Procedure

1. To simulate a hearing loss, tune a **radio** to a news station. Turn the volume down to the lowest level you can hear and understand.
2. Turn the bass to maximum and the treble to minimum. If the radio does not have these controls, hold thick wads of **cloth** over your ears.
3. Observe which sounds are hardest and easiest to hear.

Analysis

1. Are high or low pitches harder to hear? Are vowel or consonant sounds harder to hear?
2. How could you help a person with hearing loss understand what you say?

Ultrasonic and Infrasonic Waves Most people can't hear sound frequencies above 20,000 Hz, which are called **ultrasonic** waves. Dogs can hear sounds with frequencies up to about 35,000 Hz, and bats can detect frequencies higher than 100,000 Hz. Even though humans can't hear ultrasonic waves, they use them for many things. Ultrasonic waves are used in medical diagnosis and treatment. They also are used to estimate the size, shape, and depth of underwater objects.

Infrasonic, or subsonic, waves have frequencies below 20 Hz—too low for most people to hear. These waves are produced by sources that vibrate slowly, such as wind, heavy machinery, and earthquakes. Although you can't hear infrasonic waves, you may feel them as a rumble inside your body.

The Doppler Effect

Imagine that you are standing at the side of a racetrack with race cars zooming past. As they move toward you, the different pitches of their engines become higher. As they move away from you, the pitches become lower. The change in pitch or wave frequency due to a moving wave source is called the **Doppler effect. Figure 9** shows how the Doppler effect occurs.

Reading Check *What is the Doppler effect?*

Moving Sound As a race car moves, it sends out sound waves in the form of compressions and rarefactions. In **Figure 9A,** the race car creates a compression, labeled A. Compression A moves through the air toward the flagger standing at the finish line. By the time compression B leaves the race car in **Figure 9B,** the car has moved forward. Because the car has moved since the time it created compression A, compressions A and B are closer together than they would be if the car had stayed still. Because the compressions are closer together, more compressions pass by the flagger each second than if the car were at rest. As a result, the flagger hears a higher pitch. You also can see from **Figure 9B** that the compressions behind the moving car are farther apart, resulting in a lower frequency and a lower pitch after the car passes and moves away from the flagger.

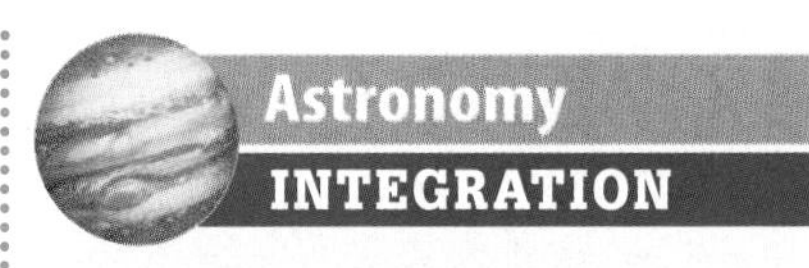

The Doppler effect can also be observed in light waves emanating from moving sources—although the sources must be moving at tremendous speeds. Astronomers have learned that the universe is expanding by observing the Doppler effect in light waves. Research the phenomenon known as red shift and explain in your Science Journal how it relates to the Doppler effect.

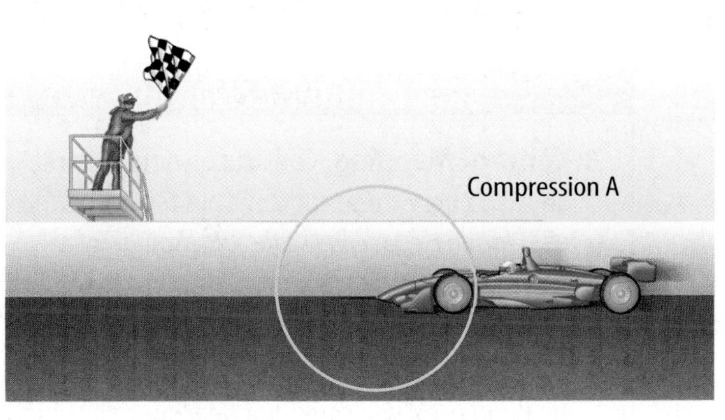

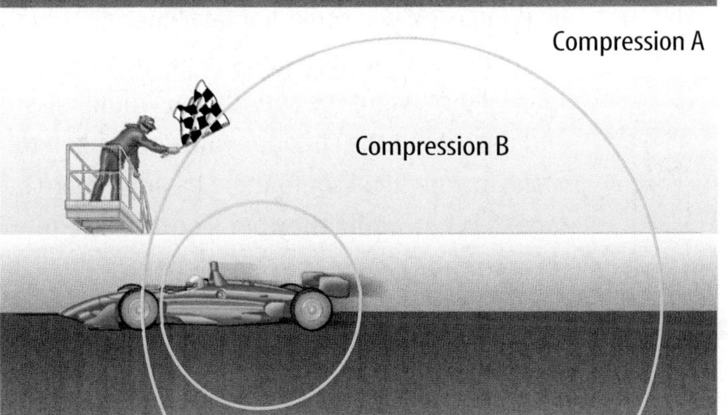

Figure 9
The Doppler effect occurs when the source of a sound wave is moving relative to a listener.

A The race car creates compression A.

B The car is closer to the flagger when it creates compression B. Compressions A and B are closer together in front of the car, so the flagger hears a higher-pitched sound.

A Moving Observer You also can observe the Doppler effect when you are moving past a sound source that is standing still. Suppose you were riding in a school bus and passed a building with a ringing bell. The pitch would sound higher as you approached the building and lower as you rode away from it. The Doppler effect happens any time the source of a sound is changing position compared with the observer. It occurs no matter whether it is the sound source or the observer that is moving. The faster the change in position, the greater the change in frequency and pitch.

Figure 10
Doppler radar can show the movement of winds in storms, and, in some cases, can detect the wind rotation that leads to the formation of tornadoes, like the one shown here. This can help provide early warning and reduce the injuries and loss of life caused by tornadoes.

Using the Doppler Effect The Doppler effect also occurs for other waves besides sound waves. For example, the frequency of electromagnetic waves, such as radar waves, changes if an observer and wave source are moving relative to each other. Radar guns use the Doppler effect to measure the speed of cars. The radar gun sends radar waves toward a moving car. The waves are reflected from the car and their frequency is shifted, depending on the speed and direction of the car. From the Doppler shift of the reflected waves, the radar gun determines the car's speed. Weather radar also uses the Doppler shift to show the movement of winds in storms, such as the tornado in **Figure 10.**

Section  Assessment

1. When you turn up the volume on a radio, which of the following change: velocity of sound, intensity, pitch, amplitude, frequency, wavelength, or loudness?
2. What does the decibel scale measure? What decibel levels can damage hearing?
3. Compare frequency and pitch.
4. Sketch a diagram that shows why the pitch of an ambulance siren seems to change as the ambulance passes you.
5. **Think Critically** Why is the Doppler effect more dramatic when a race car speeds past you than it is when a police car siren passes you on the street?

Skill Builder Activities

6. **Concept Mapping** Construct a network tree showing how these concepts are related: *intensity, loudness, decibel scale, ultrasonic, infrasonic, pitch,* and *frequency.* **For more help, refer to the Science Skill Handbook.**
7. **Communicating** Imagine how your life would be different if you could not hear sounds at high frequencies. Write a paragraph in your Science Journal discussing the sounds you would miss and describing the changes you would need to make in your life. Would there be any advantages? What would they be? **For more help, refer to the Science Skill Handbook.**

Music

What is music?

To someone else, your favorite music might sound like a jumble of noise. Music and noise are caused by vibrations—with some important differences, as shown in **Figure 11.** You can easily make noise—just tap a pencil on a desk. Noise has random patterns and pitches. **Music** is made of sounds that are deliberately used in a regular pattern.

Natural Frequencies Every material has a particular frequency at which it will vibrate, called its natural frequency. No matter how you pluck a guitar string, you hear the same pitch, because the string always vibrates at its natural frequency. The guitar string's natural frequency depends on the string's thickness and length and how tightly it is stretched across the guitar. Each string is tuned to a different natural frequency, which lets you play different notes and make music. Musical instruments contain strings, membranes or columns of air—something that vibrates at its natural frequency to create a pitch.

Resonance In wind instruments, the column of air inside vibrates. The air vibrates because of resonance—the ability of a medium to vibrate by absorbing energy at its own natural frequency. When air is blown into the instrument, the reed or the mouthpiece vibrates. The air column absorbs some of this energy and also starts to vibrate. The resonance of the air column amplifies the sound of the instrument. Resonance helps amplify the sound created in many musical instruments.

As You Read

What You'll Learn

- **Distinguish** between noise and music.
- **Describe** why different instruments have different sound qualities.
- **Explain** how string, wind, and percussion instruments produce music.
- **Describe** the formation of beats.

Vocabulary

music
quality
overtone
resonator

Why It's Important

Music enhances your enjoyment of life, but noise pollution can interfere with it.

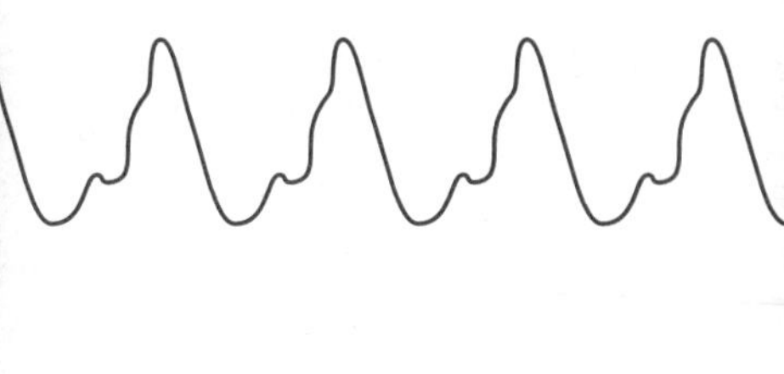

Piano

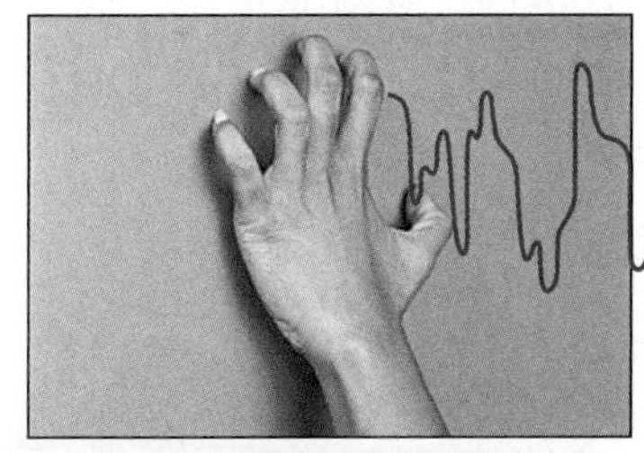

Scraping fingernails

Figure 11
These wave patterns represent the sound waves created by the piano and the scraping fingernails. ***Which of these has a regularly repeating pattern?***

SCIENCE *Online*

Research Visit the Glencoe Science Web site at **science.glencoe.com** for information about noise pollution. Identify sources of noise pollution that you can help eliminate.

Sound Quality

Suppose your classmate played a note on a flute and then a note of the same pitch and loudness on a piano. Even if you closed your eyes, you could tell the difference between the two instruments. Their sounds wouldn't be the same. Each of these instruments has a unique sound quality. **Quality** of sound describes the differences among sounds of the same pitch and loudness. Objects can be made to vibrate at other frequencies besides their natural frequency. This produces sound waves with more than one frequency. The specific combination of frequencies produced by a musical instrument is what gives it a distinctive quality of sound.

✔ **Reading Check** ***What does sound quality describe and how it is created?***

Overtones Even though an instrument vibrates at many different frequencies at once, you still hear just one note. All of the frequencies are not at the same intensity. The main tone that is played and heard is called the fundamental frequency. On a guitar, for example, the fundamental frequency is produced by the entire string vibrating back and forth, as in **Figure 12.** In addition to vibrating at the fundamental frequency, the string also vibrates to produce overtones. An **overtone** is a vibration whose frequency is a multiple of the fundamental frequency. The first two guitar string overtones also are shown in **Figure 12.** These overtones create the rich sounds of a guitar. The number and intensity of overtones vary with each instrument. These overtones produce an instrument's distinct sound quality.

Figure 12
A guitar string can vibrate at more than one frequency at the same time. Here the guitar string vibrations produce the fundamental frequency, and first and second overtones are shown.
How would the string vibrate to produce the third overtone?

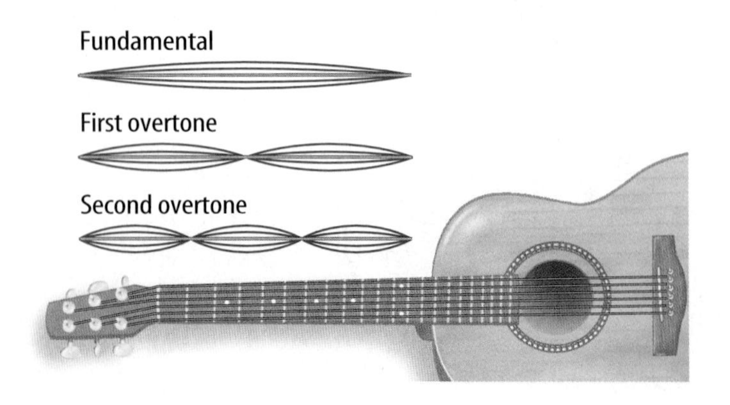

Musical Instruments

A musical instrument is any device used to produce a musical sound. Violins, cello, oboes, bassoons, horns, and kettledrums are musical instruments that you might have seen and heard in your school orchestra. These familiar examples are just a small sample of the diverse assortment of instruments people play throughout the world. For example, Australian Aborigines accompany their songs with a woodwind instrument called the didgeridoo (DIH juh ree dew). Caribbean musicians use rubber-tipped mallets to play steel drums, and a flutelike instrument called the nay is played throughout the Arab world.

Strings Soft violins, screaming electric guitars, and elegant harps are types of string instruments. In string instruments, sound is produced by plucking, striking, or drawing a bow across tightly stretched strings. Because the sound of a vibrating string is soft, string instruments usually have a resonator, like the violin in **Figure 13.** A **resonator** (RE zen ay tur) is a hollow chamber filled with air that amplifies sound when the air inside of it vibrates. For example, if you pluck a guitar string that is stretched tightly between two nails on a board, the sound is much quieter than if the string were on a guitar. When the string is attached to a guitar, the guitar frame and the air inside the instrument begin to vibrate as they absorb energy from the vibrating string. The vibration of the guitar body and the air inside the resonator makes the sound of the string louder and also affects the quality of the sound.

Figure 13
The air inside the violin's resonator vibrates when the string is played. The vibrating air amplifies the string's sound. *What causes the air to vibrate?*

Sound waves

Brass and Woodwinds Brass and woodwind instruments rely on the vibration of air to make music. The many different brass and wind instruments—such as horns, oboes, and flutes—use various methods to make air vibrate inside the instrument. For example, brass instruments have cone-shaped mouthpieces like the one in **Figure 14.** This mouthpiece is inserted into metal tubing, which is the resonator in a brass instrument. As the player blows into the instrument, his or her lips vibrate against the mouthpiece. The air in the resonator also starts to vibrate, producing a pitch. On the other hand, to play a flute, a musician blows a stream of air against the edge of the flute's mouth hole. This causes the air inside the flute to vibrate.

In brass and wind instruments, the length of the vibrating tube of air determines the pitch of the sound produced. For example, in flutes and trumpets, the musician changes the length of the resonator by opening and closing finger holes or valves. In a trombone, however, the tubing slides in and out to become shorter or longer.

Figure 14
When the trumpeter makes the mouthpiece vibrate, the air in the trumpet resonates to amplify the sound.

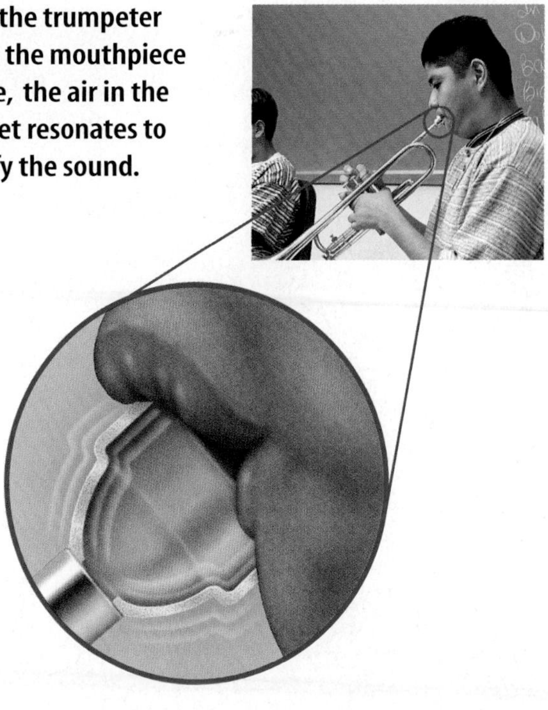

Figure 15
The air inside the resonator of the drum amplifies the sound created when the musician strikes the membrane's surface. *How does the natural frequency of the air in the drum affect the sound it creates?*

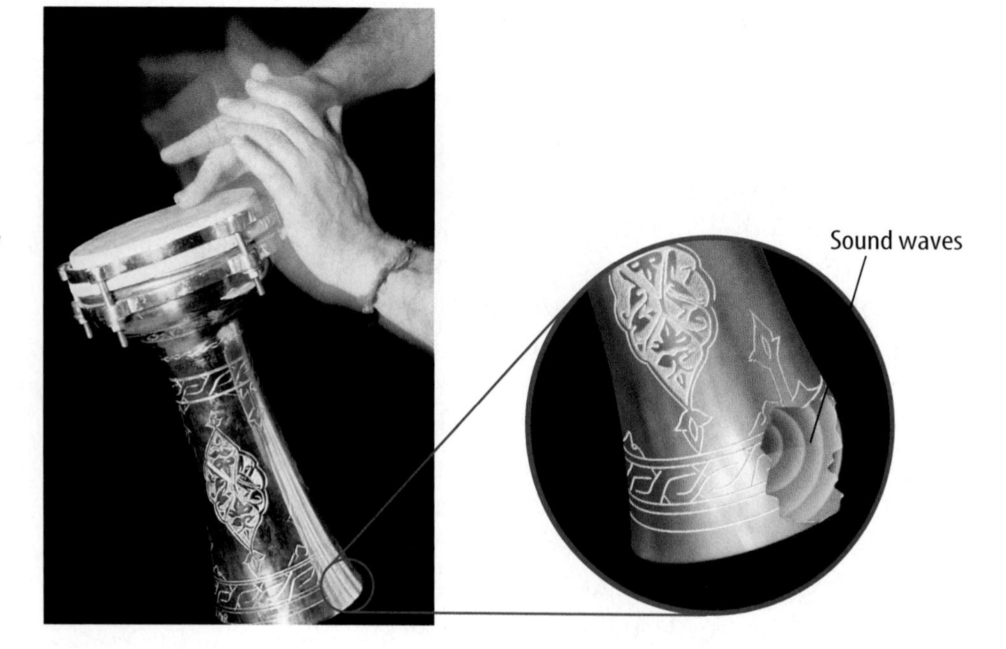

Percussion Does the sound of a bass drum make your heart start to pound? Since ancient times, people have used drums and other percussion instruments to send signals, accompany important rituals, and entertain one another. Percussion instruments are struck, shaken, rubbed, or brushed to produce sound. Some, such as kettledrums or the drum shown in **Figure 15,** have a membrane stretched over a resonator. When the drummer strikes the membrane with sticks or hands, the membrane vibrates and causes the air inside the resonator to vibrate. The resonator amplifies the sound made when the membrane is struck. Some drums have a fixed pitch, but others have a pitch that can be changed by tightening or loosening the membrane.

Figure 16
Xylophones are made with many wooden bars that each have their own resonator tubes. *Why are the resonators and bars on a xylophone different sizes?*

Caribbean steel drums were developed in the 1940s in Trinidad. As many as 32 different striking surfaces hammered from the ends of 55-gallon oil barrels create different pitches of sound. The side of a drum acts as the resonator.

Reading Check *How have people used drums?*

The xylophone shown in **Figure 16** is another type of percussion instrument. It has a series of wooden bars, each with its own tube-shaped resonator. The musician strikes the bars with mallets, and the type of mallet affects the sound quality. Hard mallets make crisp sounds, while softer rubber mallets make duller sounds. Other types of percussion instruments include cymbals, rattles, and even old-fashioned washboards.

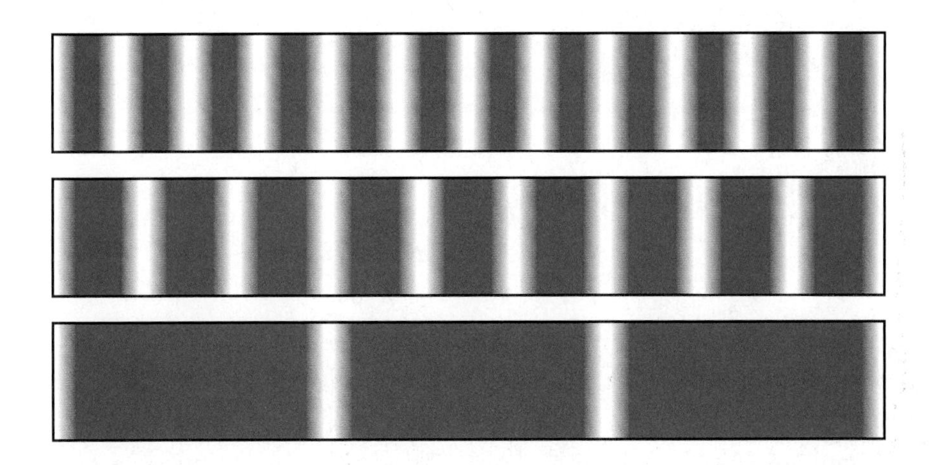

Figure 17
Beats can occur when sound waves of different frequencies, shown in the top two panels, combine. These sound waves interfere with each other, forming a wave with a lower frequency, shown in the bottom panel. This wave causes a listener to hear beats.

Beats

Have you ever heard two flutes play the same note when they weren't properly tuned? The sounds they produce have slightly different frequencies. You may have heard a pulsing variation in loudness called beats.

When two instruments play at the same time, the sound waves produced by each instrument interfere. The amplitudes of the waves add together when compressions overlap and rarefactions overlap, causing an increase in loudness. When compressions and rarefactions overlap each other, the loudness decreases. Look at **Figure 17.** If two waves of different frequencies interfere, a new wave is produced that has a different frequency. The frequency of this wave is actually the difference in the frequencies of the two waves. The frequency of the beats that you hear decreases as the two waves become closer in frequency. If two flutes that are in tune play the same note, no beats are heard.

Section  Assessment

1. Compare and contrast music and noise.
2. Explain how an instrument's overtones contribute to its sound quality.
3. Describe how a flute produces sound.
4. What happens when two instruments that are out of pitch play the same note?
5. **Think Critically** The bull roarer is a blade-shaped, wooden musical instrument. When it is whirled around on a string, it produces a rumbling sound. Explain how this sound might be produced.

Skill Builder Activities

6. **Recognizing Cause and Effect** What causes resonance? How does resonance affect the sound quality of a musical instrument? Describe how a resonator amplifies the sound of a musical instrument. **For more help, refer to the Science Skill Handbook.**
7. **Communicating** In your Science Journal, write a poem about music. Use the terms *resonance, overtone, quality,* and *beat*. **For more help, refer to the Science Skill Handbook.**

Activity

Making Music

There are many different types of musical instruments. You can also make music using every day objects that are not formal instruments, such as pots and pot lids, garbage can covers or boxes of matches. How can you create a musical instrument that requires air to be blown across it in order to make sound?

What You'll Investigate

How can you make different tones using only test tubes and water?

Materials

test tubes
test-tube rack

Goals

- **Demonstrate** how to make music using water and test tubes.
- **Predict** how the tones will change when there is more or less water in the test tube.

Procedure

1. Put different amounts of water into each of the test tubes.
2. **Predict** any differences you expect in how the tones from the different test tubes will sound.
3. Blow across the top of each test tube.
4. **Record** any differences that you noticed in the tones that you heard from each test tube .

Conclude and Apply

1. **Describe** how the tones changed depending on the amount of water in the test tube.
2. How did the pitch depend on the height of the water?
3. Why were the tones different from the different test tubes? Explain.
4. **Explain** how resonance amplifies the sound from a test tube.
5. **Explain** how the natural frequencies of the columns of air in each of the tubes differ.
6. **Compare and contrast** the way the test tubes make music with the way a flute makes music.

Communicating Your Data

When you are listening to music with family or friends, describe to them what you have learned about how musical instruments produce sound.

SECTION 4

Using Sound

The Uses of Sound

You already know that sound, in the form of music, can provide entertainment. Sounds such as sirens and fire alarms can be warning signals. The bell on a microwave timer tells you when your food is ready. People and animals use sound in a number of important ways.

As You Read

What **You'll Learn**

- **Recognize** some of the factors that determine how a concert hall or theater is designed.
- **Describe** how some animals use sound waves to hunt and navigate.
- **Discuss** the uses of sonar.
- **Explain** how ultrasound is useful in medicine.

Vocabulary

acoustics
sonar
echolocation

Why **It's Important**

Sound waves have many uses, from discovering sunken treasures to diagnosing and treating diseases.

Acoustics

Think of the last time you heard someone talk into a microphone. Did you hear echoes? When an orchestra is done playing, does it seem as if the sound of its music lingers for a couple of seconds? The sound waves produced by a speaker or an orchestra reflect off the walls and objects around you. The sounds and their reflections reach your ears at different times, so you hear echoes. This echoing effect produced by many reflections of sound is called reverberation (rih vur buh RAY shun).

If you're in the audience at an orchestra performance, reverberation can ruin the sound of the music. To prevent this problem, scientists and engineers who design concert halls must understand how the size, shape, and furnishings of the room affect the reflection of sound waves. These scientists and engineers specialize in **acoustics** (uh KEW stihks), which is the study of sound. They know, for example, that soft, porous materials can reduce excess reverberation, so they might recommend that the walls of concert halls be lined with carpets and draperies. **Figure 18** shows a concert hall that has been designed to create a good listening environment.

Figure 18
This concert hall uses cloth drapes to help reduce reverberations. *Do you think the drapes absorb or reflect sound waves?*

Echolocation

At night, bats swoop around in darkness without bumping into anything. They even manage to hunt insects and other prey in the dark. Their senses of sight and smell help them navigate, but many species of bats also depend on echolocation. **Echolocation** is the process of locating objects by emitting sounds and interpreting the sound waves that are reflected back. Look at **Figure 19** to learn how echolocation works.

NATIONAL GEOGRAPHIC **VISUALIZING BAT ECHOLOCATION**

Figure 19

Many bats emit ultrasonic—very high-frequency—sounds. The sound waves bounce off objects, and bats locate prey by using the returning echoes. Known as echolocation, this technique is also used by dolphins, which produce clicking sounds as they hunt. The diagrams below show how a bat uses echolocation to capture a flying insect.

A Sound waves of a bat's ultrasonic cries spread out in front of it.

B Some of the waves strike a moth and bounce back to the bat.

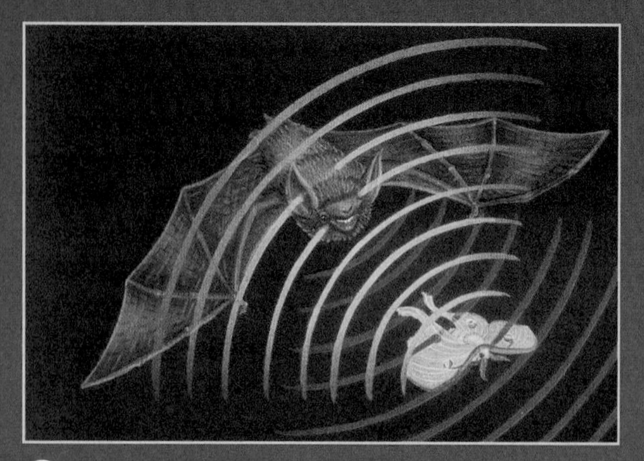

C The bat determines the moth's location by continuing to emit cries, then changes its course to catch the moth.

D By emitting a continuous stream of ultrasonic cries, the bat homes in on the moth and captures its prey.

Sonar

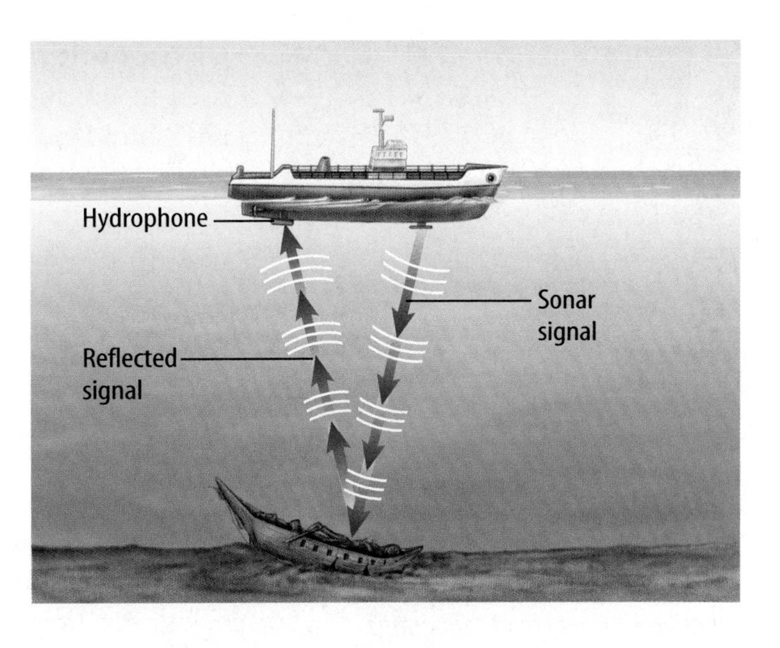

Figure 20
Sonar uses sound waves to find objects that are underwater.
How is sonar like echolocation?

More than 140 years ago, a ship named the *Central America* disappeared in a hurricane off the coast of South Carolina. In its hold lay 21 tons of newly minted gold coins and bars that would be worth $1 billion or more in today's market. When the shipwreck occurred, there was no way to search for the ship in the deep water where it sank. The *Central America* and its treasures lay at the bottom of the ocean until 1988, when crews used sonar to locate the wreck under 2,400 m of water. **Sonar** is a system that uses the reflection of underwater sound waves to detect objects. First, a sound pulse is emitted toward the bottom of the ocean. The sound travels through the water and is reflected when it hits something solid, as shown in **Figure 20.** A sensitive underwater microphone called a hydrophone picks up the reflected signal. Because the speed of sound in water is known, the distance to the object can be calculated by measuring how much time passes between emitting the sound pulse and receiving the reflected signal.

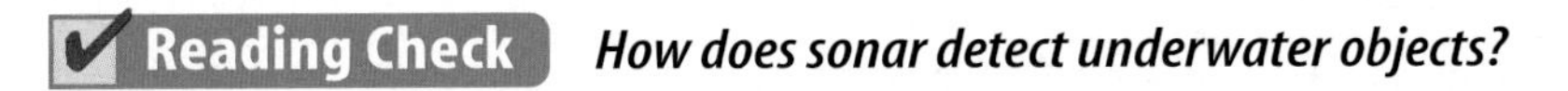
Reading Check *How does sonar detect underwater objects?*

The idea of using sonar to detect underwater objects was first suggested as a way of avoiding icebergs, but many other uses have been developed for it. Navy ships use sonar for detecting, identifying, and locating submarines. Fishing crews also use sonar to find schools of fish, and scientists use it to map the ocean floor. More detail can be revealed by using sound waves of high frequency. As a result, most sonar systems use ultrasonic frequencies.

Ultrasound in Medicine

High-frequency sound waves are used in more than just echolocation and sonar. Ultrasonic waves also are used to break up and remove dirt or buildup from jewelry. Chemists sometimes use ultrasonic waves to clean glassware. One of the main uses of ultrasonic waves, though, is in medicine. Using special instruments, medical professionals can send ultrasonic waves into a specific part of a patient's body. Reflected ultrasonic waves are used to detect and monitor conditions such as pregnancy, certain types of heart disease and cancer.

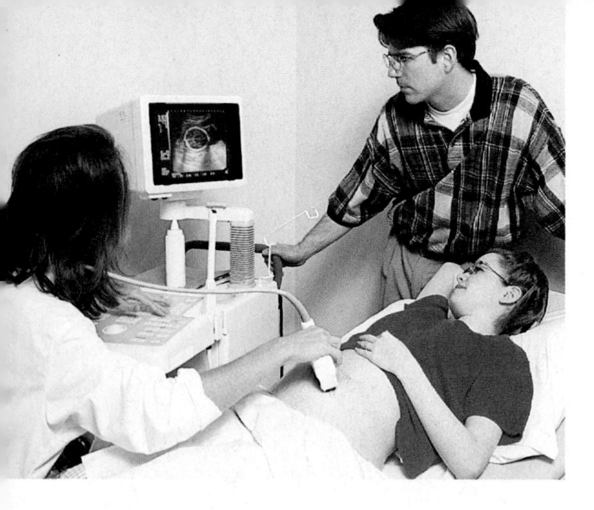

Figure 21
Ultrasonic waves are directed into a pregnant woman's uterus to form images of her fetus.
What benefits can fetal sonograms offer expectant parents?

Ultrasound Imaging Like X rays, ultrasound can be used to produce images of internal structures. A medical ultrasound technician directs the ultrasound waves toward a target area of a patient's body. The sound waves reflect off the targeted organs or tissues, and the reflected waves are used to produce electrical signals. A computer program converts these electrical signals into video images, called sonograms. Physicians trained to interpret sonograms can use them to detect a variety of medical problems.

Reading Check *How does ultrasound imaging use reflected waves?*

Medical professionals use ultrasound to examine many parts of the body, including the heart, liver, gallbladder, pancreas, spleen, kidneys, breast, and eye. Probably the best-known use of ultrasound in medicine is to monitor the development of a fetus in a pregnant woman's uterus, as shown in **Figure 21.** Ultrasound is better than X rays for producing images of soft tissue structures inside the body. However, it is not as good for examining bones and lungs, because hard tissues and air absorb the ultrasonic waves instead of reflecting them.

Math Skills Activity

Calculating Distance with Sonar

Example Problem

You send out a sonar pulse that is returned from a sunken pirate ship in 3 s. How far away is the ship? Hint: *The sonar pulse travels twice the distance from your ship to the sunken ship.*

1. *This is what you know:* speed of sound in water: $v = 1{,}439$ m/s; time: $t = 3$ s
2. *This is what you need to know:* distance: d
3. *This is the equation you need to use:* $d = vt/2$
4. *Substitute the known values:* $d = (1{,}439 \text{ m/s})\ (3 \text{ s}) / 2 = 2{,}158.5$ m

Check your answer by dividing the time and multiplying by 2. Do you calculate the same speed of sound that was given?

Practice Problem

The sonar pulse you emitted returns in 1 s when you are directly over the sunken treasure you were hoping to find. How far away is it?

For more help, refer to the Math Skill Handbook.

Figure 22
Ultrasonic waves can be used to break apart kidney stones.
What are the benefits of ultrasound therapy for kidney stones?

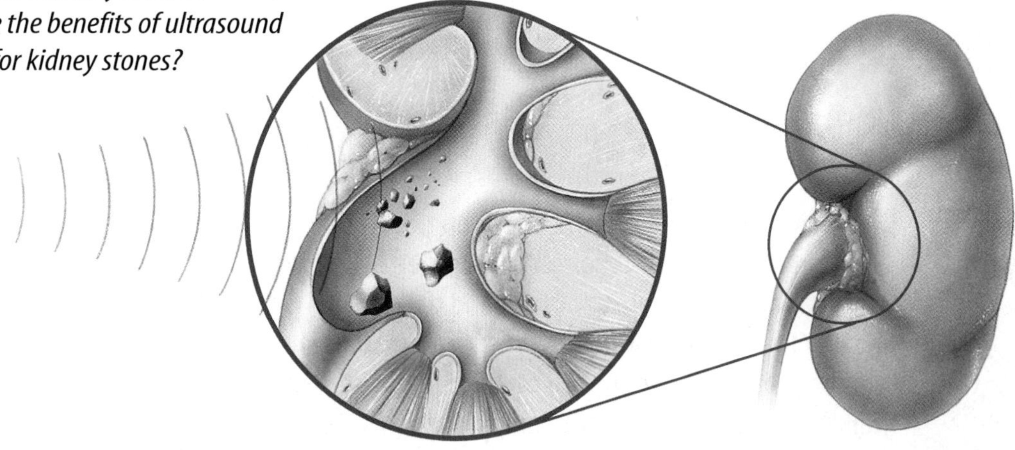

Treating with Ultrasound High-frequency sound waves can be used to treat certain medical problems. For example, sometimes small, hard deposits of calcium compounds or other minerals form in the kidneys, making kidney stones. In the past, physicians had to perform surgery to remove kidney stones. But now ultrasonic treatments are commonly used to break them up instead. Bursts of ultrasound create vibrations that cause the stones to break into small pieces, as shown in **Figure 22.** These fragments then pass out of the body with the urine. A similar treatment is available for gallstones. Patients who are treated successfully with ultrasound recover more quickly than those who have surgery.

Health INTEGRATION

Physicians can measure blood flow by studying the Doppler effect in ultrasonic waves. Brainstorm how the Doppler-shifted waves could help doctors diagnose diseases of the arteries and monitor their healing.

Section 4 Assessment

1. What are some differences between a gym and a concert hall that affect the amount of reverberation you hear in each place?
2. Explain how echolocation helps bats find food and avoid obstacles in the dark.
3. How can sound waves be used to find objects that are underwater?
4. Describe at least three uses of ultrasonic technology in medicine.
5. **Think Critically** How is sonar technology useful in locating deposits of oil and mineral resources?

Skill Builder Activities

6. **Drawing Conclusions** Imagine you are a physician looking at an ultrasound image and an X ray of the same area of a patient's body. A structure that you see on the X ray doesn't appear on the ultrasound image. What can you conclude about the structure? **For more help, refer to the Science Skill Handbook.**
7. **Solving One-Step Equations** Sound travels at about 1,500 m/s in water. How far will a sonar pulse travel in 45 s? **For more help, refer to the Math Skill Handbook.**

Activity

Blocking Noise Pollution

What loud noises do you enjoy, and which ones do you find annoying? Most people enjoy a music concert performed by their favorite artist, booming displays of fireworks on the Fourth of July, and the roar of a crowd when their team scores a goal or touchdown. Although these are loud noises, people enjoy them for short periods of time. Most people find certain loud noises, such as traffic, sirens, and loud talking, annoying. Constant, annoying noises are called noise pollution. What can be done to reduce noise pollution? What types of barriers best block out loud noises?

Recognize the Problem

What types of barriers will best block out noise pollution?

Form a Hypothesis

Based on your experiences with loud noises, form a hypothesis about how effectively different types of barriers block out noise pollution.

Goals

- **Design** an experiment that tests the effectiveness of various types of barriers and materials for blocking out noise pollution.
- **Test** different types of materials and barriers to determine the best noise blockers.

Possible Materials

radio, CD player, horn, drum, or other loud noise source
shrubs, trees, concrete walls, brick walls, stone walls, wooden fences, parked cars, or hanging laundry
sound meter
meterstick or metric tape measure

Test Hypothesis

Plan

1. **Decide** what type of barriers or materials you will test.
2. Describe exactly how you will use these materials.
3. **Identify** the controls and variables you will use in your experiment.
4. **List** the steps you will use and describe each step precisely.
5. **Prepare** a data table in your Science Journal to record your measurements.
6. **Organize** the steps of your experiment in logical order.

Do

1. Ask your teacher to examine your plan and data table before you start.
2. **Conduct** your experiment as planned.
3. **Test** each barrier two or three times.
4. **Record** your test results in your data table in your Science Journal.

Analyze Your Data

1. **Identify** the barriers that most effectively reduced noise pollution.
2. **Identify** the barriers that least effectively reduced noise pollution.
3. **Compare** the effective barriers and identify common characteristics that might explain why they reduced noise pollution.
4. **Compare** the natural barriers you tested with the artificial barriers. Which type of barrier best reduced noise pollution?
5. **Compare** the different types of materials the barriers were made of. Which type of material best reduced noise pollution?

Draw Conclusions

1. Did your results support your hypothesis?
2. **Predict** your results if you had used a louder source of noise such as a siren.
3. Use your results to infer how people living near a busy street could reduce noise pollution.
4. **Identify** major sources of noise pollution in or near your home. How could this be reduced?
5. **Research** how noise pollution can be unhealthy.

Draw a poster illustrating how builders and landscapers could use certain materials to better insulate a home or office from excess noise pollution.

TURNING UP THE VOLU

NOISE POLLUTION AND HEARING LOSS

Now hear this: More than 28 million Americans have hearing loss. Twelve million more people have a condition called tinnitus (TIN uh tus), or ringing in the ears. In at least 10 million of the 28 million cases mentioned above, hearing loss could have been prevented, because it was caused by noise pollution.

People take their music very seriously in the United States. And a lot of people like it loud. So loud, in fact, that it can damage their hearing. The medical term is *auditory over-stimulation*. You may have experienced a high-pitched ringing in your ears for days after standing too close to a loudspeaker at a concert. That's how hearing loss starts.

Music isn't the only cause of hearing loss caused by noise pollution. Other kinds of environmental noise can be strong enough to damage the ears, too. Intense, short-duration noise, like the sound of a gunshot or even a thunderclap, can cause some hearing loss. All of the structures of the inner ear can be damaged this way. A less intense but longer duration noise, like the sound of a lawn mower, a low-flying plane, or a drill blasting away at cement, can possibly damage the ear, as well.

All the Better to Hear You

There are 20,000 to 30,000 sensory receptors, or hair cells, located in the inner ear, or the cochlea. When vibrations reach these hair cells, electrical impulses are triggered.

The impulses send messages to the auditory center of the brain. But the human ear was not made to withstand all the very loud sounds of the modern world. Once hair cells are damaged, they don't grow back.

What can you do to avoid hearing damage? Well, the first thing is to turn the volume down on the stereo and TV. And keep the volume low when you've got your earphones on, no matter how tempting it is to blast it. Also, if you go to a rock concert, you can wear earplugs to muffle the sound. You'll still hear everything, but you won't damage your ears. And earplugs are now small enough that they're pretty much undetectable. So enjoy all that music and the street sounds of urban life, but mind your ears.

Did You Know?

Sounds can be described by their intensity. Sounds of high intensity carry more energy than sounds of low intensity. Sound intensity is usually measured in units of decibels (dB). Conversational speech is between 65 dB and 70 dB. A rock concert can measure up to 140 dB, which is considered beyond the threshold of pain for the ear.

CONNECTIONS Test **Put on a blindfold and have a friend test your hearing. Have your friend choose several noise-making objects. (Not so loud, please, and make sure you have your teacher's permission.) Guess what object is making each sound.**

Chapter 12 Study Guide

Reviewing Main Ideas

Section 1 The Nature of Sound

1. Sound is a compressional wave created by something that is vibrating.
2. Sound travels fastest through solids and slowest through gases. The speed of a sound wave also increases as the temperature of the medium increases. *Would the sound of thunder travel faster in a winter or summer storm?*

3. The human ear can be divided into three sections—the outer ear, the middle ear, and the inner ear. Each section plays a specific role in hearing.
4. Hearing involves four stages: gathering sound waves, amplifying them, converting them to nerve impulses, and interpreting the signals in the brain.

Section 2 Properties of Sound

1. Intensity is a measure of how much energy a wave carries. Humans interpret the intensity of sound waves as loudness.
2. The pitch of a sound becomes higher as the frequency increases.
3. The Doppler effect is a change in frequency that occurs when a source of sound is moving relative to a listener.

Section 3 Music

1. Music is made of sounds used deliberately in a regular pattern.
2. Instruments use a variety of methods to produce and amplify sound waves. *How is the sound produced by vibrating strings amplified by this cello?*

3. When sound waves of similar frequencies overlap, they interfere with each other to form beats.

Section 4 Using Sound

1. Acoustics is the study of sound. *Why would a gym be a poor place for a symphony concert?*

2. Some animals emit sounds and interpret the echoes to navigate and catch prey.
3. Sonar uses reflected sound waves to detect objects.
4. Ultrasound waves can be used for imaging body tissues or treating medical conditions.

After You Read

FOLDABLES Reading & Study Skills

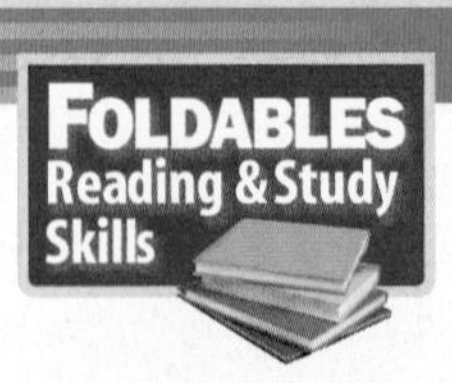

Use the information on your Question Study Fold to help explain how the temperature of a solid, liquid, or gas affects how sound moves.

Chapter 12 Study Guide

Visualizing Main Ideas

Complete the following table on musical instruments.

Characteristics of Musical Instruments

	Guitar	Flute	Bongo Drum
How Played	plucked	blown into	
Role of Resonator	amplifies sound		amplifies sound
Type of Instrument		wind	percussion

Vocabulary Review

Vocabulary Words

- **a.** acoustics
- **b.** cochlea
- **c.** decibel
- **d.** Doppler effect
- **e.** eardrum
- **f.** echolocation
- **g.** intensity
- **h.** loudness
- **i.** music
- **j.** overtone
- **k.** pitch
- **l.** quality
- **m.** resonator
- **n.** sonar
- **o.** ultrasonic

THE PRINCETON REVIEW Study Tip

Take good notes, even during lab. Lab experiments reinforce key concepts, and looking back at these notes can help you better understand what happened and why.

Using Vocabulary

Each of the following sentences is false. Make the sentence true by replacing the underlined word with the correct vocabulary word.

1. The eardrum is filled with fluid and contains tiny hair cells that vibrate.
2. Echolocation is the study of sound.
3. A change in pitch or wave frequency due to a moving wave source is called ultrasonic.
4. Overtone is a combination of sounds and pitches that follows a specified pattern.
5. Differences among sounds of the same pitch and loudness are described as the intensity of sound.
6. Decibel is how humans perceive the intensity of sound.

Chapter 12 Assessment

Checking Concepts

Choose the word or phrase that best answers the question.

1. For a sound with a low pitch, what else is always low?
A) amplitude **C)** wavelength
B) frequency **D)** wave velocity

2. Sound intensity decreases when which of the following decreases?
A) wave velocity **C)** quality
B) wavelength **D)** amplitude

3. When specific pitches and sounds are put together in a pattern, what are they called?
A) overtones **C)** white noise
B) music **D)** resonance

4. Sound can travel through all but which of the following?
A) solids **C)** gases
B) liquids **D)** empty space

5. What is the term for variations in the loudness of sound caused by wave interference?
A) beats **C)** pitch
B) standing waves **D)** forced vibrations

6. What does the outer ear do to sound waves?
A) scatter them **C)** gather them
B) amplify them **D)** convert them

7. Which of the following occurs when a sound source moves away from you?
A) The sound's velocity decreases.
B) The sound's loudness increases.
C) The sound's frequency decreases.
D) The sound's frequency increases.

8. Sounds with the same pitch and loudness traveling in the same medium may differ in which of these properties?
A) frequency **C)** quality
B) amplitude **D)** wavelength

9. What part of a musical instrument amplifies sound waves?
A) resonator **C)** mallet
B) string **D)** finger hole

10. What is the name of the method used to find objects that are underwater?
A) sonogram **C)** sonar
B) ultrasonic bath **D)** percussion

Thinking Critically

11. A car comes to a railroad crossing. The driver hears a train's whistle and then hears the whistle's pitch become lower. What can be assumed about how the train is moving?

12. A whistle is blown and a dog responds, even though the person can't hear the whistle. Explain.

13. Acoustic scientists sometimes do research in rooms that absorb all sound waves. How could such a room be used to study how bats find their food?

14. Explain why windows might begin to rattle when an airplane flies overhead.

15. The Indian sitar has extra strings that are tuned to certain pitches but are never played by the musician. Explain why these strings might be present.

Developing Skills

16. Forming Hypotheses Sound travels slower in air at high altitudes than at low altitudes. State a hypothesis to explain this.

17. Identifying a Question Some people say that an ultrasound can harm a fetus. Write a list of questions to ask a doctor about the safety of ultrasound imaging.

18. Making and Using Tables You use a lawn mower with a sound level of 100 dB for your lawn-mowing business. Using the table below, determine how many hours a day you can safely work mowing lawns.

Federally Recommended Noise Exposure Limits

Sound Level (dB)	Time Permitted (hours per day)
90	8
95	4
100	2
105	1
110	0.5

19. Communicating Some people enjoy using snowmobiles. Others object to the noise that they make. Write a proposal for a policy that seems fair to both groups for the use of snowmobiles in a state park.

20. Concept Mapping Design a concept map that shows how the characteristics of sound waves are related. Include these terms: *pitch, compression, medium, speed, loudness, frequency,* and *intensity.*

Performance Assessment

21. Project Using materials you have at home, make a musical instrument. Give a demonstration to your class. Explain how you change the pitch of your instrument.

TECHNOLOGY

Go to the Glencoe Science Web site at **science.glencoe.com** or use the **Glencoe Science CD-ROM** for additional chapter assessment.

Test Practice

Lightning and thunder are always associated with each other, but they are very different things. You see lightning, but you hear thunder. Lightning is a large discharge of static electricity. The electrical energy in a lightning bolt ionizes atoms in the atmosphere and produces great amounts of heat. The heat causes air in the bolt's path to expand rapidly producing compressional waves you hear as thunder.

A Rapid heating of air

B Lightning

C Rapid compression of air

D Rapid expansion of air

E Thunder

Study the events listed above and answer the following questions.

1. The different events lead to the creation of thunder. Which of the following shows the correct order of events?

A) D, A, C, B, E
B) B, A, D, C, E
C) A, D, B, C, E
D) C, A, D, B, E

2. Which of the events above is most closely accompanied by the creation of rarefactions?

F) B
G) A
H) C
J) E

CHAPTER

Electromagnetic Waves

What do cordless phones and microwave ovens have in common with the stars? Each emits electromagnetic waves. These waves also help your computer read a CD-ROM and they enable quick communication across the globe. The data and information explosion is made possible by manipulating electromagnetic waves. In this chapter, you will learn about the usefulness of the entire spectrum of electromagnetic waves.

What do you think?

Science Journal Look at the picture below with a classmate. Discuss what you think this is. Here's a hint: *Cell phones would be useless without them.* Write your answer or best guess in your Science Journal.

You often hear about the danger of too much exposure to the Sun's ultraviolet rays, which can damage the cells of your skin. When the exposure isn't too great, your cells can repair themselves, but too much exposure at one time can cause a painful sunburn. Repeated overexposure to the Sun over many years can damage cells and cause skin cancer. In the activity below, observe how energy carried by ultraviolet waves can cause changes in other materials.

Observe damage by ultraviolet waves

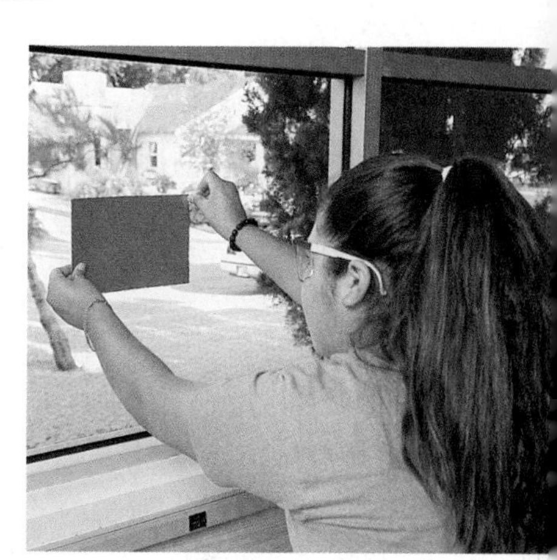

1. Cut a sheet of red construction paper in half.
2. Place one piece outside in direct sunlight. Place the other in a shaded location or behind a window.
3. Keep the construction paper in full sunlight for at least 45 min. If possible, allow it to stay there for 3 h or more before taking it down.

Observe

In your Science Journal, describe any differences you notice in the two pieces of construction paper. Comment on your results.

Before You Read

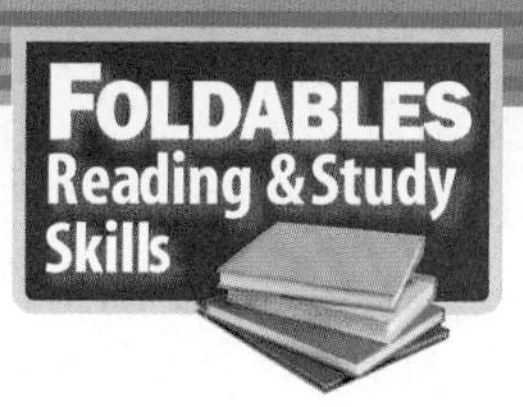

Making a Question Study Fold **Asking yourself questions helps you stay focused and better understand electromagnetic waves when you are reading the chapter.**

1. Place a sheet of paper in front of you so the long side is at the top. Fold the paper in half from the left side to the right side. Fold top to bottom and crease. Then unfold.
2. Through the top thickness of paper, cut along the middle fold line to form two tabs, as shown.
3. Write these questions on the tabs: *How do electromagnetic waves travel through space? How do electromagnetic waves transfer energy to matter?*
4. As you read the chapter, write answers to the questions under the tabs.

How do electromagnetic waves travel through space?
How do electromagnetic waves transfer energy to matter?

SECTION

What are electromagnetic waves?

As You Read

What You'll Learn

- **Explain** how vibrating charges produce electromagnetic waves.
- **Describe** properties of elecromagnetic waves.

Vocabulary

electromagnetic wave
radiant energy
frequency
photon

Why It's Important

Knowledge of electromagnetic waves helps you understand much of the technology around you.

Waves in Space

Stay calm. Do not panic. As you are reading this sentence, no matter where you are, you are surrounded by electromagnetic waves. Even though you can't feel them, some of these waves are traveling right through your body. They enable you to see. They make your skin feel warm. You use electromagnetic waves when you watch television, talk on a cordless phone, or prepare popcorn in a microwave oven.

Sound and Water Waves Waves are produced by something that vibrates, and they carry energy from one place to another. Look at the sound wave and the water wave in **Figure 1.** Both waves are moving through matter. The sound wave is moving through air and the water wave through water. These waves travel because energy is transferred from particle to particle. Without matter to transfer the energy, they cannot move.

Electromagnetic Waves However, electromagnetic waves do not require matter to transfer energy. **Electromagnetic waves** are made by vibrating electric charges and can travel through space where matter is not present. Instead of transferring energy from particle to particle, electromagnetic waves travel by transferring energy between vibrating electric and magnetic fields.

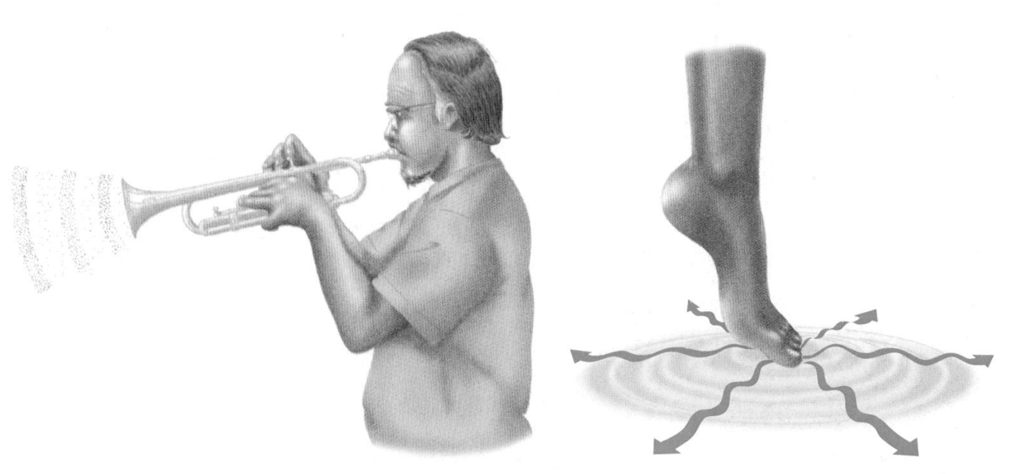

Figure 1
Water waves and sound waves require matter to move through. Energy is transferred from one particle to the next as the wave travels through the matter.

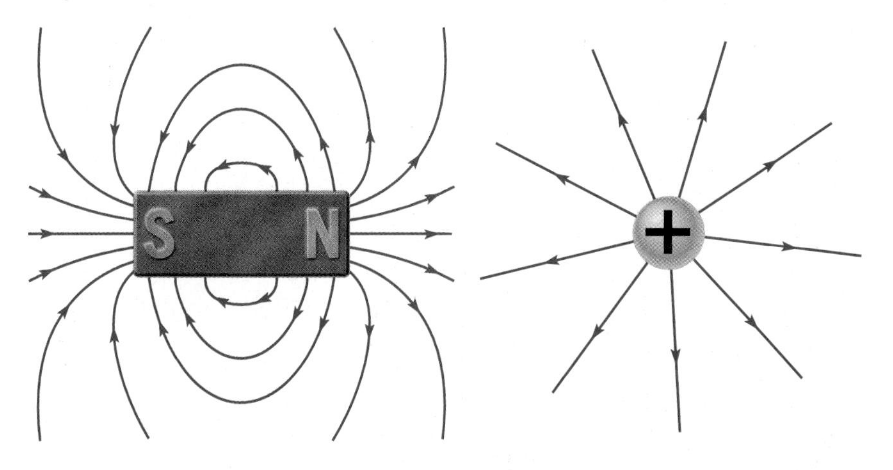

A A magnetic field surrounds all magnets.

B An electric field surrounds all charges.

Figure 2
Fields enable magnets and charges to exert forces at a distance. These fields extend throughout space. *How can you detect a magnetic field?*

Electric and Magnetic Fields

When you bring a magnet near a metal paper clip, the paper clip moves toward the magnet and sticks to it. The paper clip moved because the magnet exerted a force on it. The magnet exerted this force without having to touch the paper clip. The magnet exerts a force without touching the paper clip because all magnets are surrounded by a magnetic field, as shown in **Figure 2A.** Magnetic fields exist around magnets even if the space around the magnet contains no matter.

Just as magnets are surrounded by magnetic fields, electric charges are surrounded by electric fields, as shown in **Figure 2B.** An electric field enables charges to exert forces on each other even when they are far apart. Just as a magnetic field around a magnet can exist in empty space, an electric field exists around an electric charge even if the space around it contains no matter.

Magnetic Fields and Moving Charges Electricity and magnetism are related. An electric current flowing through a wire is surrounded by a magnetic field, as shown in **Figure 3.** An electric current is created by the movement of electrons in a wire. It is the motion of these electrons that creates the magnetic field around the wire. In fact, any moving electric charge is surrounded by a magnetic field, as well as an electric field.

Figure 3
Electrons moving in a wire are surrounded by a magnetic field. *How would you confirm that a magnetic field exists around a current-carrying wire?*

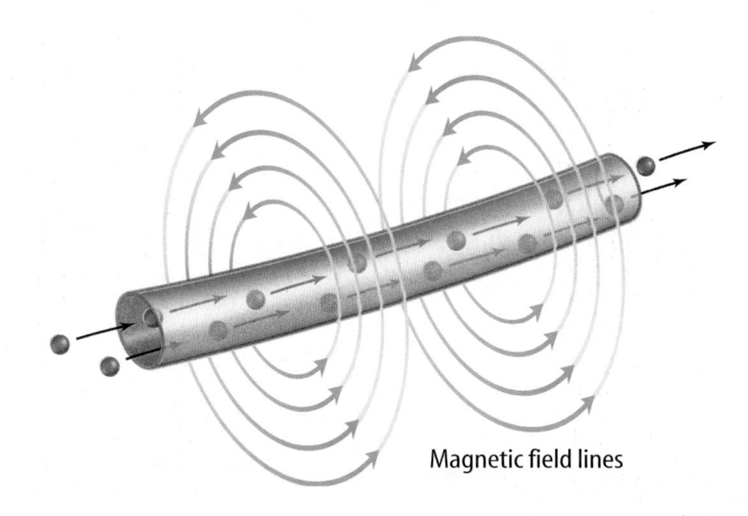

Investigating Electromagnetic Waves

Procedure

1. Point your **television remote control** in different directions and observe whether it will still control the **television.**
2. Place various materials in front of the infrared receiver on the television and observe whether the remote still will control the television. Some materials you might try are **glass, a book, your hand, paper,** and a **metal pan.**

Analysis

1. Was it necessary for the remote to be pointing exactly toward the receiver to control the television? Explain.
2. Did the remote continue to work when the various materials were placed between it and the receiver? Explain why or why not.

Changing Electric and Magnetic Fields The relationship between electricity and magnetism can explain the behavior of electric motors, generators, and transformers. This behavior is the result of the relationship between changing electric and magnetic fields. A changing magnetic field creates a changing electric field. The reverse is also true—a changing electric field creates a changing magnetic field.

Making Electromagnetic Waves

Waves such as sound waves are produced when something vibrates. Electromagnetic waves also are produced when something vibrates—an electric charge that moves back and forth.

Reading Check *What produces an electromagnetic wave?*

Vibrating Fields When an electric charge vibrates, the electric field around it vibrates. Because the electric charge is in motion, it also has a magnetic field around it. This magnetic field is changing as the charge moves back and forth. As a result, the vibrating electric charge is surrounded by vibrating electric and magnetic fields.

How do the vibrating electric and magnetic fields around the charge become a wave that travels through space? The changing electric field around the charge creates a changing magnetic field. This changing magnetic field then creates a changing electric field. This process continues, with the magnetic and electric fields continually creating each other. These vibrating electric and magnetic fields are perpendicular to each other and travel outward from the moving charge, as shown in **Figure 4.**

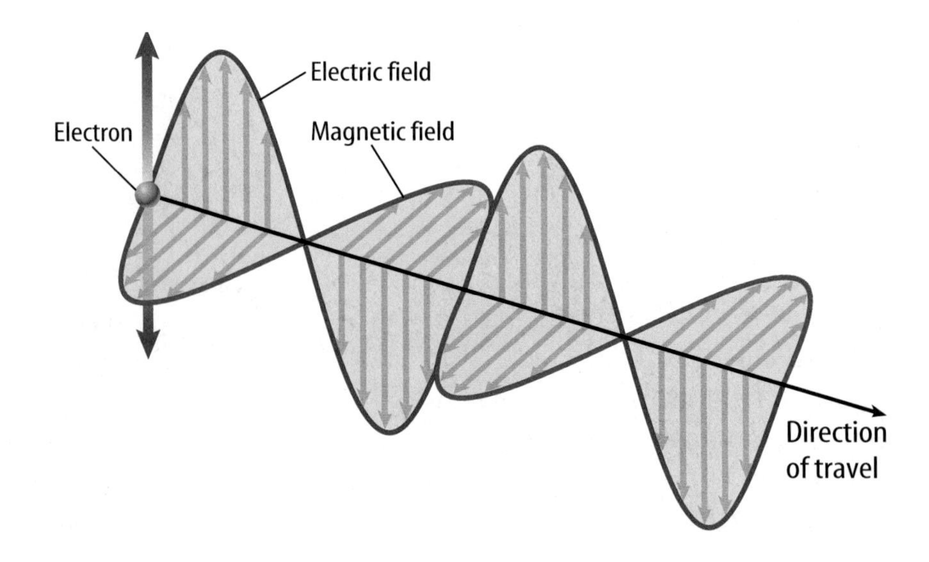

Figure 4
A vibrating electric charge creates an electromagnetic wave that travels outward in all directions from the charge. Only one direction is shown here.

Properties of Electromagnetic Waves

Electromagnetic waves travel outward from a vibrating charge in all directions. Recall that a wave transfers energy without transporting matter. How does an electromagnetic wave transfer energy? As an electromagnetic wave moves, its electric and magnetic fields encounter objects. These vibrating fields can exert forces on charged particles and magnetic materials, causing them to move. For example, electromagnetic waves from the Sun cause electrons in your skin to vibrate and gain energy, as shown in **Figure 5.** The energy carried by an electromagnetic wave is called **radiant energy.** Radiant energy makes a fire feel warm and enables you to see.

Figure 5
As an electromagnetic wave strikes your skin, electric charges in your skin gain energy from the vibrating electric and magnetic fields.

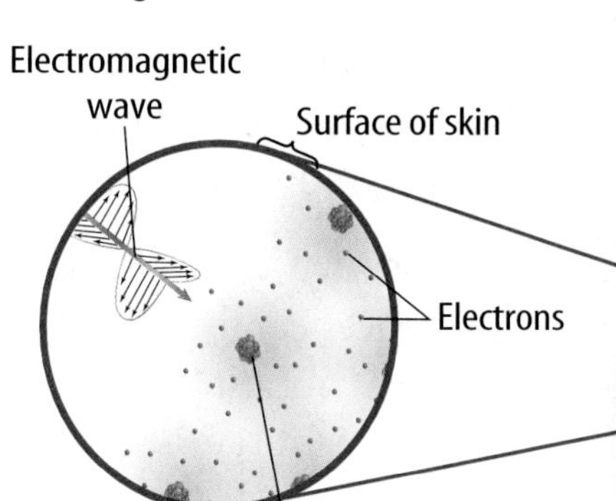

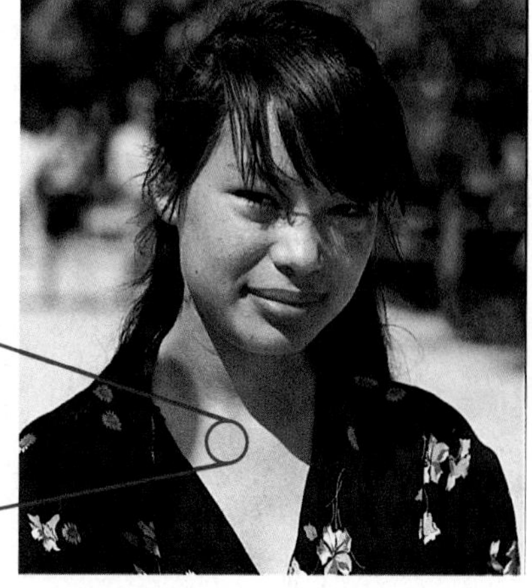

Problem-Solving Activity

What is scientific notation?

In science, numbers such as the speed of light (300,000,000 m/s) and the size of a gold atoms (0.000 000 000 288 m) are either too large or too small to use easily. By using scientific notation, numbers that are very large or very small can be written in a more compact way. For example, in scientific notation the speed of light is 3.00×10^8 m/s and the size of a gold atom is 2.88×10^{-10} m. Scientific notation follows the form $M \times 10^n$. M is a number with only one number to the left of the decimal point. The number of places the decimal point was moved is represented by n. If the original number is greater than 1, n is positive. If the original number is less than 1, n is negative.

Example Problem

Put the numbers 2,000 and 0.003 into scientific notation.

Solution

1. *For the number 2000, move the decimal point 3 places to the left.*
2. *Because you moved the decimal point 3 places and the number is greater than 1, n equals 3. In scientific notation the number is* 2×10^3.
3. *For the number 0.003, move the decimal point 3 places to the right.*
4. *Because you moved the decimal point 3 places and the number is less than 1, n equals* -3*. In scientific notation the number is* 3×10^{-3}

Practice Problems

1. Put the following numbers into scientific notation:
   40; 7,000; 100,000.
2. Put the following numbers into scientific notation:
   0.09; 0.000,6; 0.000,005.

Wave Speed Electromagnetic waves travel through space, which is empty, as well as through various materials—and they travel fast. In the time it takes you to blink your eyes, an electromagnetic wave can travel around the entire Earth. All electromagnetic waves travel at 300,000 km/s in the vacuum of space. Because light is an electromagnetic wave, the speed of electromagnetic waves in space is usually called the "speed of light." However, when electromagnetic waves travel through matter, they slow down. The speed of the wave depends upon the material they travel through. Electromagnetic waves usually travel the slowest in solids and the fastest in gases. **Table 1** lists the speed of visible light in various materials.

Table 1 Speed of Visible Light

Material	Speed (km/s)
Vacuum	300,000
Air	slightly less than 300,000
Water	226,000
Glass	200,000
Diamond	124,000

Frequency and Wavelength Like all waves, electromagnetic waves can be described by their frequency and their wavelength. A vibrating charge produces an electromagnetic wave. A charge can vibrate at different speeds, or frequencies. **Frequency** is the number of vibrations that occur in 1 s. Frequency is measured in hertz. One Hz is one vibration each second. For example, if you clap your hands four times each second, then you are clapping at a frequency of 4 Hz.

The wavelength of an electromagnetic wave is the distance from one crest to another, as shown in **Figure 6.** Wavelength is measured in meters. The frequency and wavelength of electromagnetic waves are related. As the frequency of the wave increases, the wavelength becomes smaller.

Reading Check *How are the wavelength and frequency of electromagnetic waves related?*

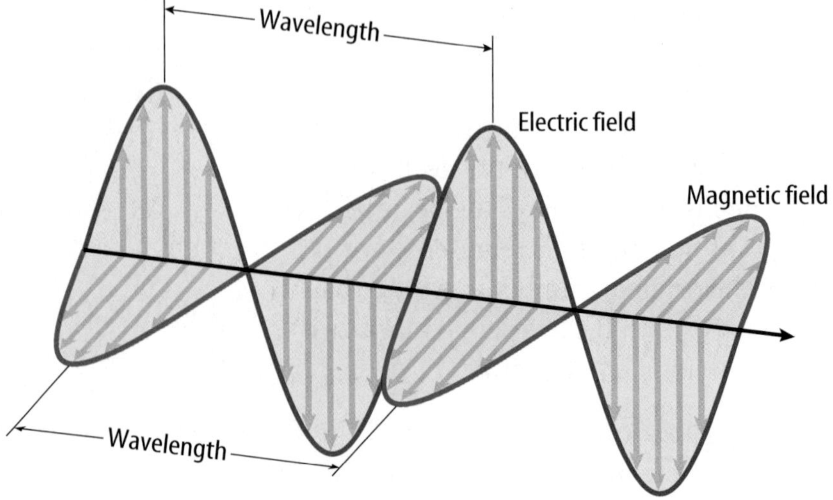

Figure 6
The wavelength of an electromagnetic wave is the distance from one crest to the next one. *What happens to the wavelength as the frequency decreases?*

Waves and Particles

The difference between a wave and a particle might seem obvious—a wave is a disturbance that carries energy, and a particle is a piece of matter. However, in reality the difference is not so clear.

Waves as Particles In 1887, Heinrich Hertz found that by shining light on a metal, electrons were ejected from the metal. Hertz found that whether or not electrons were ejected depended on the frequency of the light and not the amplitude. Because the energy carried by a wave depends on its amplitude and not its frequency, this result was mysterious. Years later, Albert Einstein provided an explanation—light can behave as a particle, called a **photon,** whose energy depends on the frequency of the light.

Particles as Waves Because light could behave as a particle, others wondered whether matter could behave as a wave. If a beam of electrons were sprayed at two tiny slits, you might expect that the electrons would strike only the area behind the slits, like the spray paint in **Figure 7A.** Instead, it was found that the electrons formed an interference pattern, as shown in **Figure 7B.** This type of pattern is produced by waves when they pass through two slits and interfere with each other, as the water waves do in **Figure 7C.** This experiment showed that electrons can behave like waves. It is now known that all particles, not only electrons, can behave like waves.

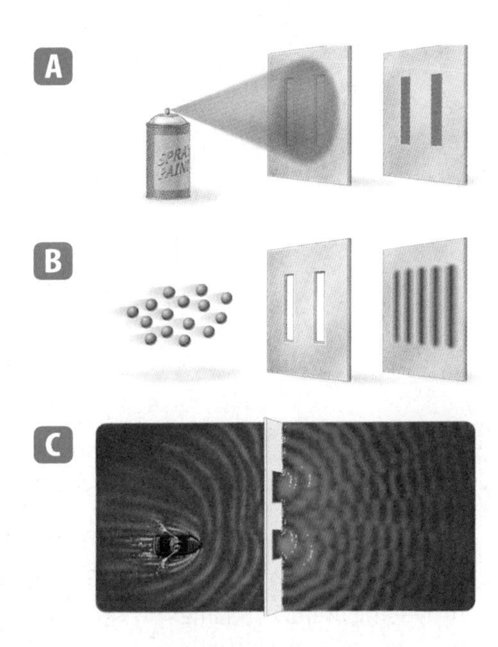

Figure 7
When electrons are sent through two narrow slits, they behave as a wave. A Particles of paint sprayed through two slits coat only the area behind the slits. B Electrons fired at two closely-spaced openings don't strike only the area behind the slits. Instead they form a wave-like interference pattern. C Water waves produce an interference pattern after passing through two openings.

Section  Assessment

1. What produces electromagnetic waves?
2. How is an electromagnetic wave similar to the wave created when a pebble is dropped into a pond?
3. What is the relationship between the frequency and wavelength of electromagnetic waves?
4. **Think Critically** Is it possible to have just an electric-field wave or just a magnetic-field wave? Explain.

Skill Builder Activities

5. **Testing a Hypothesis** Hypothesize that light behaves like water waves. Design an experiment to test your hypothesis. **For more help, refer to the Science Skill Handbook.**
6. **Solving One-Step Equations** When light travels in ethyl alcohol, its speed is about three-fourths its speed in air. What is the speed of light in ethyl alcohol? **For more help, refer to the Math Skill Handbook.**

SECTION 2

The Electromagnetic Spectrum

As You Read

What You'll Learn

- **Compare** the various types of electromagnetic waves.
- **Identify** some useful and some harmful properties of electromagnetic waves.

Vocabulary

radio wave
microwave
infrared wave
visible light
ultraviolet wave
X ray
gamma ray

Why It's Important

Different types of electromagnetic waves can be used in different ways.

A Range of Frequencies

Electromagnetic waves can have a wide variety of frequencies. They might vibrate once each second or trillions of times each second. The entire range of electromagnetic wave frequencies is known as the electromagnetic spectrum (SPEK trum), shown in **Figure 8.** Various portions of the electromagnetic spectrum interact with matter differently. As a result, they are given different names. The electromagnetic waves that humans can detect with their eyes, called visible light, are a small portion of the entire electromagnetic spectrum. However, devices have been built to detect the other frequencies. For example, the antenna of your radio detects radio waves.

Radio Waves

Stop reading for a moment and look around you. Everywhere you look, radio waves are traveling. You can't see or hear them, but they are there. Radio waves carry the signal from a radio station to your radio. It might seem that radio waves should be the same as sound waves. However, sound waves are compressions and expansions of groups of molecules, while radio waves shake electrons, not molecules of air. Therefore, you can't hear radio waves. You can hear sound waves because molecules bump against your eardrums.

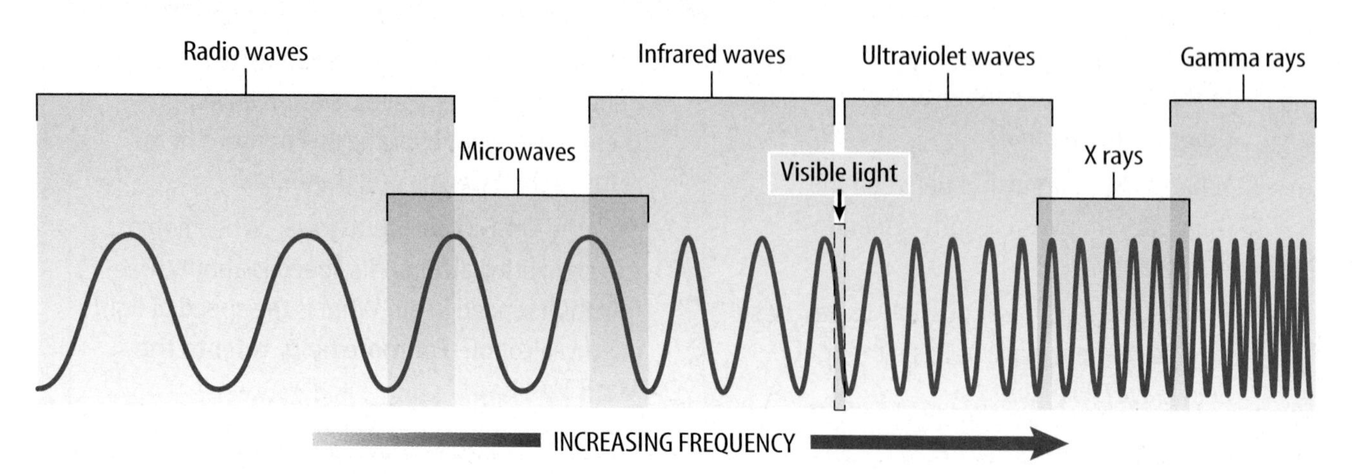

Figure 8
Electromagnetic waves are described by different names depending on their frequency and wavelength.

Figure 9
Microwave ovens use electromagnetic waves to heat food.

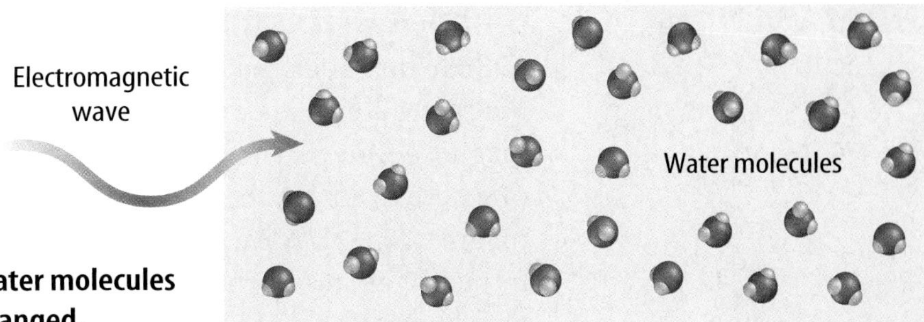

A Normally water molecules are randomly arranged.

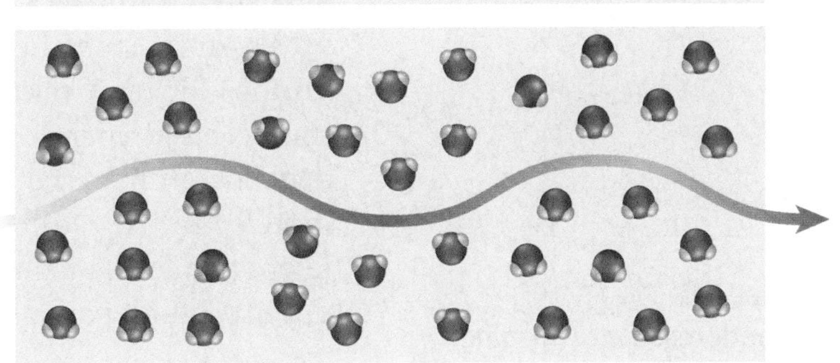

B The electric fields of microwaves cause the molecules to flip back and forth. This flipping motion causes the food to be heated.

Microwaves **Radio waves** are low-frequency electromagnetic waves with wavelengths greater than about 1 mm. Radio waves with wavelengths of less than 1 m are called **microwaves.** Microwaves with wavelengths of about 1 cm to 10 cm are widely used for communication, such as for cellular telephones and satellite signals. You are probably most familiar with microwaves because of their use in microwave ovens.

Reading Check *What is the difference between a microwave and a radio wave?*

Microwave ovens heat food when microwaves interact with water molecules in food, as shown in **Figure 9.** Each water molecule is positively charged on one side and negatively charged on the other side. The vibrating electric field in microwaves causes water molecules in food to flip direction billions of times each second. This motion causes the molecules to bump one another. Bumping causes friction between the molecules, changing the microwave's radiant energy into thermal energy. It is the thermal energy that cooks your food.

Radar Another use for radio waves is to find the position and movement of objects by a method called radar. Radar stands for **RA**dio **D**etecting **A**nd **R**anging. With radar, radio waves are transmitted toward an object. By measuring the time required for the waves to bounce off the object and return to a receiving antenna, the location of the object can be found. Law enforcement officers use radar to measure how fast a vehicle is moving. Radar also is used for tracking the movement of aircraft, watercraft, and spacecraft.

Mini LAB

Heating Food with Microwaves

Procedure

1. Obtain two small **beakers or baby food jars.** Place 50 mL of **dry sand** into each. In one of the jars, add 20 mL of **room-temperature water** and stir well.
2. Record the temperature of the sand in each jar.
3. Together, **microwave** both jars of sand for 10 s and immediately record the temperature again.

Analysis

1. Compare the initial and final temperatures of the wet and dry sand.
2. Infer why there was a difference.

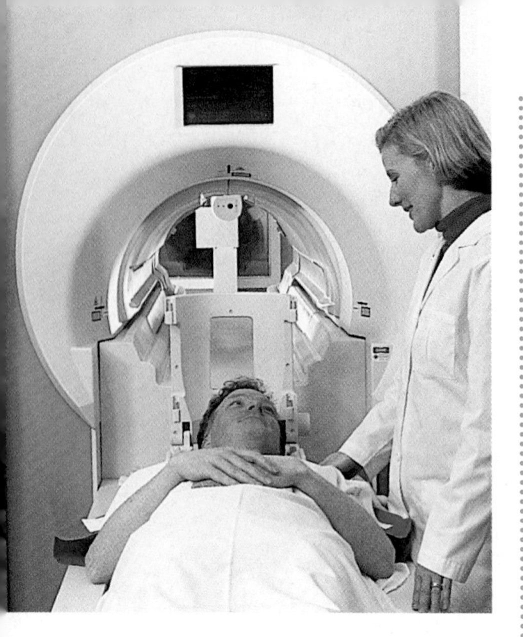

Figure 10
Magnetic resonance imaging technology uses radio waves as an alternative to X-ray imaging.

Magnetic Resonance Imaging (MRI) In the early 1980s, medical researchers developed a technique called Magnetic Resonance Imaging, which uses radio waves to help diagnose illness. The patient lies inside a large cylinder, like the one shown in **Figure 10.** Housed in the cylinder is a powerful magnet, a radio wave emitter, and a radio wave detector. Protons in hydrogen atoms in bones and soft tissue behave like magnets and align with the strong magnetic field. Energy from radio waves causes some of the protons to flip their alignment. As the protons flip, they release radiant energy. A radio receiver detects this released energy. The amount of energy a proton releases depends on the type of tissue it is part of. The released energy detected by the radio receiver is used to create a map of the different tissues. A picture of the inside of the patient's body is produced without pain or risk.

Infrared Waves

Most of the warm air in a fireplace moves up the chimney, yet when you stand in front of a fireplace, you feel the warmth of the blazing fire. Why do you feel the heat? The warmth you feel is thermal energy transmitted to you by **infrared waves,** which are a type of electromagnetic wave with wavelengths between about 1 mm and about 750 billionths of a meter.

You use infrared waves every day. A remote control emits infrared waves to communicate with your television. A computer uses infrared waves to read CD-ROMs. In fact, every object emits infrared waves. Hotter objects emit more than cooler objects do. Your world would look strange if you could see infrared waves. It is possible to take photographs called thermograms with a special film that is sensitive to infrared waves. These photographs show cool and warm areas in different colors. Infrared photography is used in many ways. **Figure 11** shows how cities appear different from surrounding vegetation in infrared imagery.

Figure 11
A This visible-light image of the region around San Francisco Bay in California was taken from an aircraft at an altitude of 20,000 m. **B** This infrared image of the same area was taken from a satellite. In this image, vegetation is red and buildings are gray.

Visible Light

Visible light is the range of electromagnetic waves that you can detect with your eyes. Light differs from radio waves and infrared waves only by its frequency and wavelength. Visible light has wavelengths around 400 billionths to 750 billionths of a meter. Your eyes contain substances that react differently to various wavelengths of visible light, so you see different colors. These colors range from short-wavelength blue to long-wavelength red. If all the colors are present, you see the light as white.

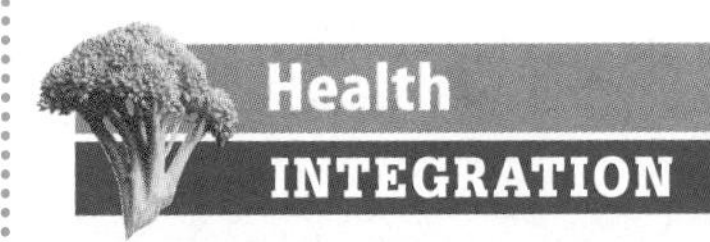

In certain situations, doctors will perform a CT scan on a patient instead of a traditional X ray. Research to find out more about CT scans. Compare and contrast the CT scan to X rays. What are the advantages and disadvantages of a CT scan? Write a paragraph about your findings in your Science Journal.

Ultraviolet Waves

Ultraviolet waves are electromagnetic waves with wavelengths from about 400 billionths to 10 billionths of a meter. Ultraviolet waves are energetic enough to enter skin cells. Overexposure to ultraviolet rays can cause skin damage and cancer. Most of the ultraviolet radiation that reaches Earth's surface are longer-wavelength UVA rays. The shorter-wavelength UVB rays cause sunburn, and both UVA and UVB rays can cause skin cancers and skin damage such as wrinkling. Although too much exposure to the Sun's ultraviolet waves is damaging, some exposure is healthy. Ultraviolet light striking the skin enables your body to make vitamin D, which is needed for healthy bones and teeth.

Useful UVs A useful property of ultraviolet waves is their ability to kill bacteria on objects such as food or medical supplies. When ultraviolet light enters a cell, it damages protein and DNA molecules. For some single-celled organisms, damage can mean death, which can be a benefit to health. Ultraviolet waves are also useful because they make some materials fluoresce (floor ES). Fluorescent materials absorb ultraviolet waves and reemit the energy as visible light. As shown in **Figure 12,** police detectives sometimes use fluorescent powder to show fingerprints when solving crimes.

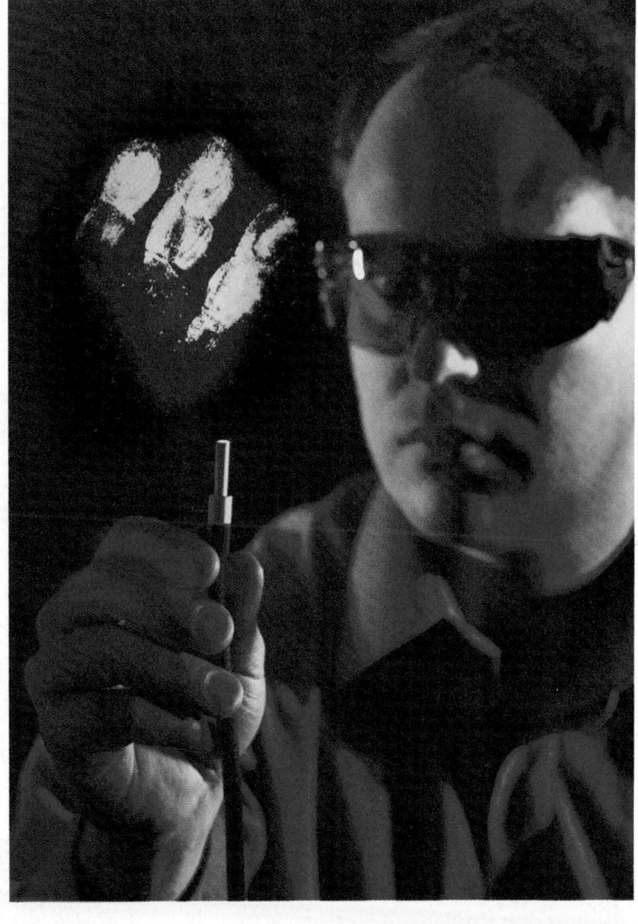

Figure 12
The police detective in this picture is shining ultraviolet light on a fingerprint dusted with fluorescent powder.

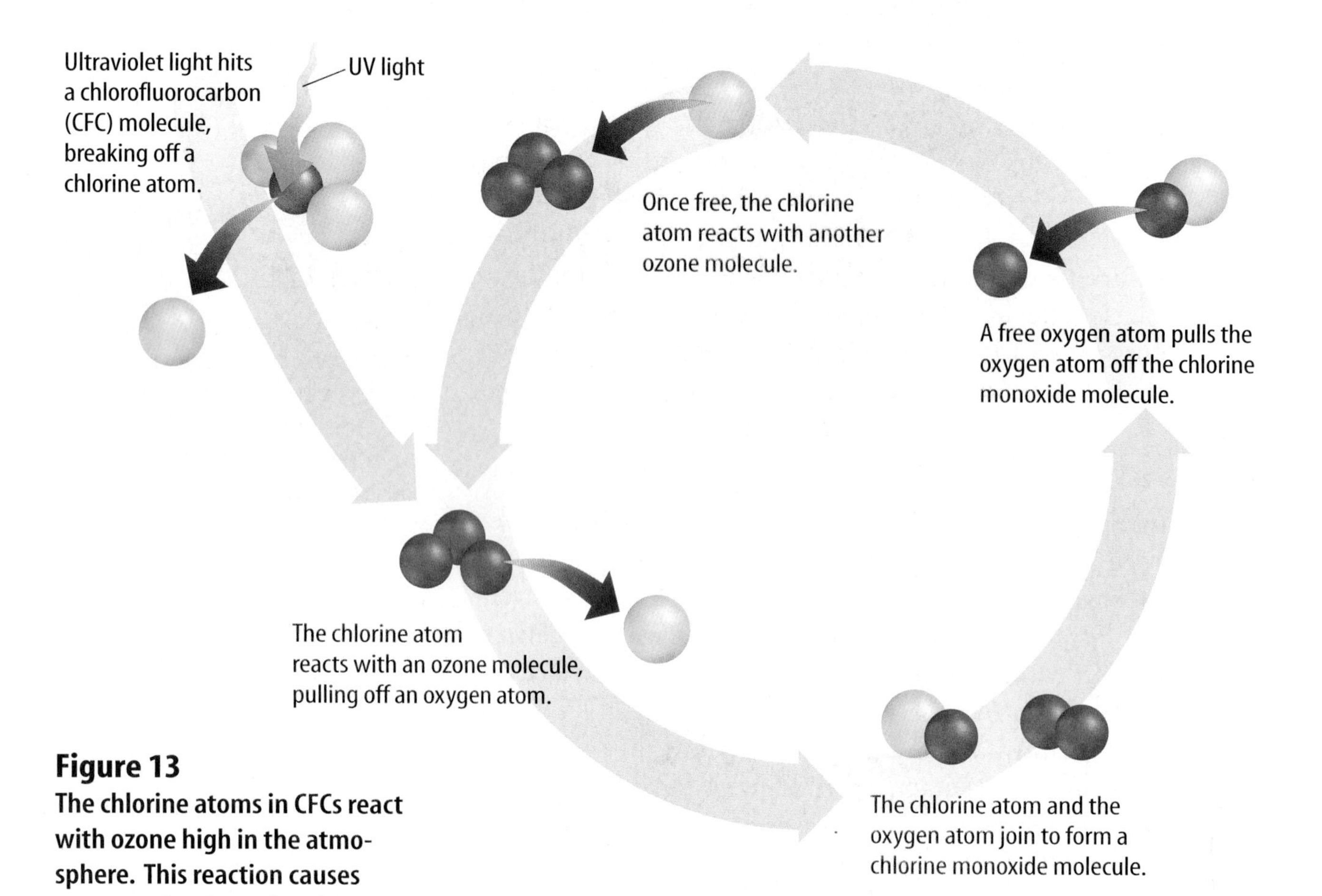

Figure 13
The chlorine atoms in CFCs react with ozone high in the atmosphere. This reaction causes ozone molecules to break apart.

The Ozone Layer About 20 to 50 km above Earth's surface is a region called the ozone layer. Ozone is a molecule composed of three oxygen atoms. It is continually being formed and destroyed high in the atmosphere. The ozone layer is vital to life on Earth because it absorbs most of the Sun's harmful ultraviolet waves. You might have heard of an ozone hole that forms over Antarctica. In fact, thinning of the ozone layer has occurred over areas of Earth but is greatest over Antarctica.

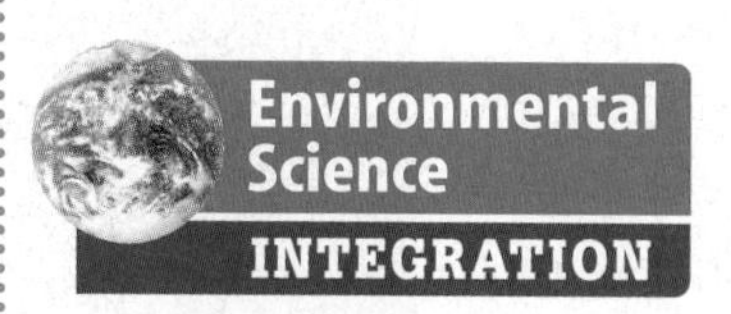

The greatest threat to the ozone layer is from ozone-depleting chemicals, such as nitric oxides and CFCs. CFCs, which stands for chlorofluorocarbons, are used in air conditioners, refrigerators, and as cleaning fluids. When these substances reach the ozone layer, they react chemically with ozone, breaking the ozone molecule apart, as shown in **Figure 13.** One chlorine atom of a CFC molecule will break apart many ozone molecules. To reduce the damage to the ozone layer, many countries in the world are reducing their use of ozone-depleting chemicals.

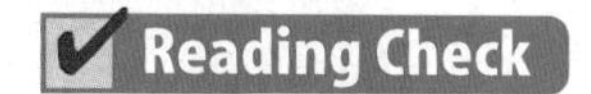

What chemicals can reduce the amount of ozone in the ozone layer?

X Rays and Gamma Rays

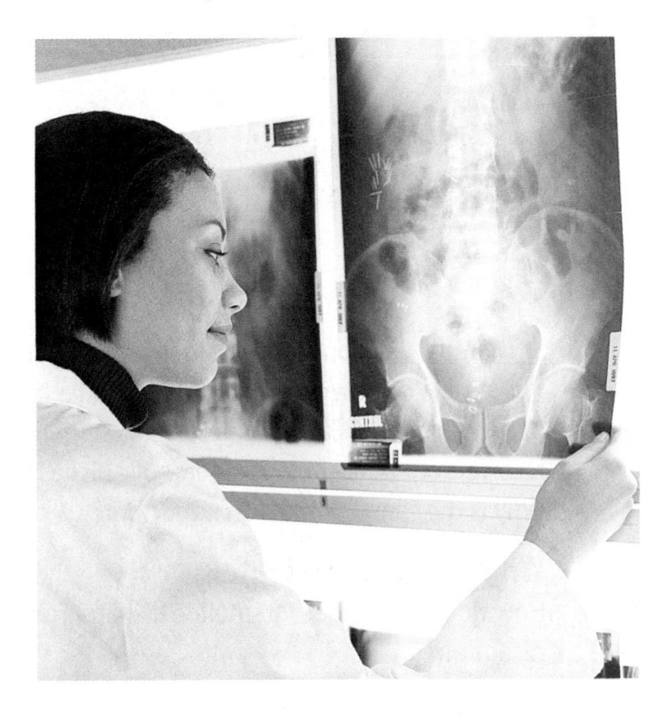

At the far end of the electromagnetic spectrum are **X rays** and **gamma rays.** These ultra-high-frequency electromagnetic waves are so energetic that they can travel through matter, breaking molecular bonds as they go. Doctors and dentists often send low doses of X rays through a patient's body onto photographic film. Dense parts of the body such as bones or teeth absorb more X rays than soft parts do. **Figure 14** shows the shadow image of bones produced by X rays. New techniques are being developed to use gamma rays for more precise medical imaging. X rays are used at airports to inspect luggage without opening it. X rays and gamma rays are used at low doses in industry to check metal objects for cracks and defects.

Radiation therapy is a technique used in medicine for exposing part of a patient's body to X rays or gamma rays to kill diseased cells. X rays and gamma rays have short wavelengths and are highly energetic. When X rays or gamma rays pass through matter, part of the energy damages molecules. This eventually kills cells. However, nearby healthy cells also are damaged by the radiation. By carefully controlling the amount of radiation, the damage to healthy cells is reduced.

Figure 14
Bones are more dense than surrounding tissues and absorb more X rays. The image of a bone on an X ray is the shadow cast by the bone as X rays pass through the soft tissue.

Section  Assessment

1. Explain how a microwave oven heats food. Draw a diagram to help your explanation.
2. Describe how light you see with your eyes differs from other forms of electromagnetic waves, such as X rays and radio waves.
3. Name some ways that ultraviolet waves are useful and some ways in which they are dangerous.
4. Describe the ozone layer and why damage to the ozone layer could be harmful.
5. **Think Critically** Why are ultraviolet waves, X rays, and gamma rays far more dangerous to humans than other forms of electromagnetic waves?

Skill Builder Activities

6. **Researching Information** Many scientists around the world are studying ozone depletion and how we can solve the problem. Learn about one of these scientists and write a paragraph about the work he or she is doing. **For more help, refer to the Science Skill Handbook.**
7. **Using Graphics Software** Use graphics software to create your own version of the electromagnetic spectrum. Be sure to include all of the forms of electromagnetic waves mentioned in this section. Use clip art to represent how each part of the spectrum is used. **For more help, refer to the Technology Skill Handbook.**

Activity

The Shape of Satellite Dishes

Communications satellites transmit signals with a narrow beam pointed toward a particular area of Earth. To detect this signal, receivers are typically large, parabolic dishes. Why is the shape of the dish important?

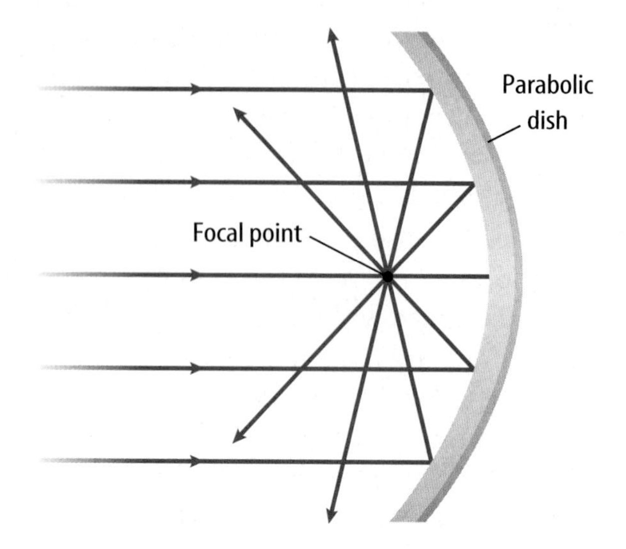

What You'll Investigate

How does the shape of a satellite dish improve reception?

Materials

flashlight
several books
aluminum foil
small, parabolically shaped bowl (such as a mortar bowl)
large, metal spoon
** Alternate materials*

Goals

- **Make** a model of a satellite reflecting dish.
- **Observe** how the shape of the dish affects reception.

Safety Precautions 

Procedure

1. Cover one side of a book with aluminum foil. Be careful not to wrinkle the foil.
2. Line the inside of the bowl with foil, also keeping it as smooth as possible.
3. Place some of the books on a table. Put the flashlight on top of the books so that its beam of light will shine several centimeters above and across the table.
4. Hold the foil-covered book on its side at a right angle to the top of the table. The foil-covered side should face the beam of light.
5. **Observe** the intensity of the light on the foil.
6. Repeat these procedures, replacing the foil-covered book with the bowl, keeping it at the same distance from the flashlight.
7. **Observe** how the intensity of light differs when the flat surface is used rather than the curved surface.

Conclude and Apply

1. **Compare** the difference in intensity of light when the two surfaces were used.
2. **Infer** what caused this difference.
3. **Explain** how these results relate to why parabolic dishes are used for satellite signal receivers.

Communicating Your Data

Compare your conclusions with those observed by other students in your class. **For more help, refer to the Science Skill Handbook.**

SECTION

Radio Communication

Radio Transmission

When you listen to the radio, you hear music and words that are produced at a distant location. The music and words are sent to your radio by radio waves. You can't actually hear the radio waves, because they are electromagnetic waves. Ears only detect sound waves. It is the metal antenna of your radio that detects radio waves. As the electromagnetic waves pass by your radio's antenna, the electrons in the metal vibrate, as shown in **Figure 15.** These vibrating electrons produce an electric signal that contains the information about the music and words. An amplifier boosts the signal and sends it to speakers, causing them to vibrate. The vibrating speakers create sound waves that travel to your ears.

Dividing the Radio Spectrum Many radio stations broadcast programs for you to listen to. What is it that allows you to listen to only one station at a time? Each station is assigned to broadcast at one particular radio frequency. Turning the tuning knob on your radio allows you to select a particular frequency to listen to. The specific frequency of the electromagnetic wave that a radio station is assigned is called the **carrier wave.**

The radio station must do more than simply transmit a carrier wave. The station has to send information about the sounds that you are to receive. This information is sent by modifying the carrier wave. The carrier wave is modified to carry information in one of two ways, as shown in **Figure 16.** One way is to vary the amplitude of the carrier wave. This method is called amplitude modulation, or AM. The other way is to vary the frequency of the carrier wave. This method is called frequency modulation, or FM.

As You Read

What **You'll Learn**

- **Explain** how modulating carrier waves make radio transmissions.
- **Distinguish** between AM and FM radio.
- **Identify** various ways of communicating using radio waves.

Vocabulary

carrier wave
cathode-ray tube
transceiver

Why **It's Important**

Every day you use radio waves to communicate.

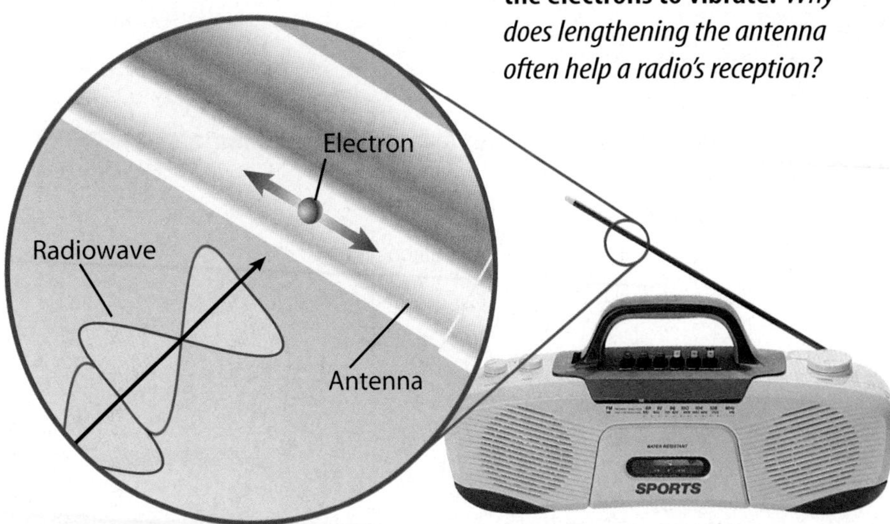

Figure 15
Radio waves exert a force on the electrons in an antenna, causing the electrons to vibrate. *Why does lengthening the antenna often help a radio's reception?*

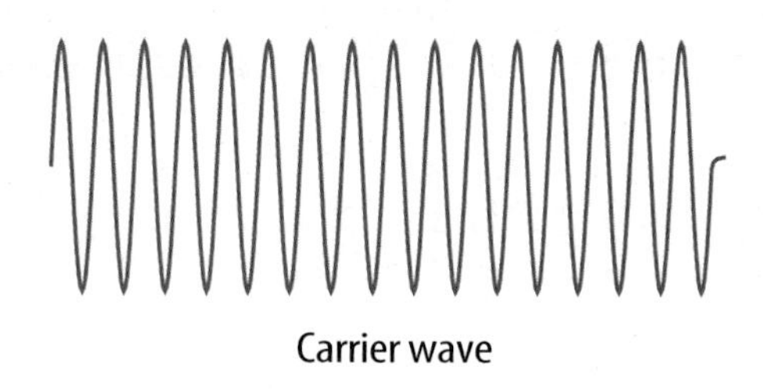

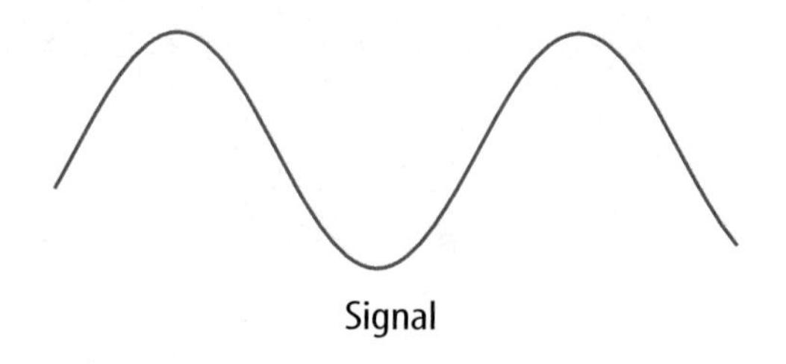

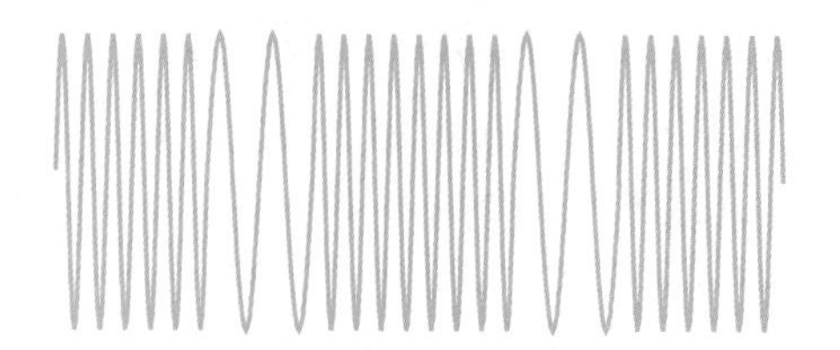

Figure 16
A carrier wave broadcast by a radio station can be altered in one of two ways to transmit a signal: amplitude modulation (AM), or frequency modulation (FM).

AM Radio An AM radio station broadcasts information by varying the amplitude of the carrier wave, as shown in **Figure 16.** Your radio detects the variations in amplitude of the carrier wave and makes an electronic signal from these slight variations. The electronic signal is then used to make the speaker vibrate. AM carrier wave frequencies range from 540,000 to 1,600,000 vibrations each second.

FM Radio Electronic signals are transmitted by FM radio stations by varying the frequency of the carrier wave, as in **Figure 16.** Your radio detects the changes in frequency of the carrier wave. Because the strength of the FM waves is kept fixed, FM signals tend to be more clear than AM signals. FM carrier frequencies range from 88 million to 108 million vibrations each second. This is much higher than AM frequencies, as shown in **Figure 17. Figure 18** shows how radio signals are broadcast.

Figure 17
Cellphones, TVs, and radios broadcast at frequencies that range from more than 500,000 Hz to almost 1 billion Hz.

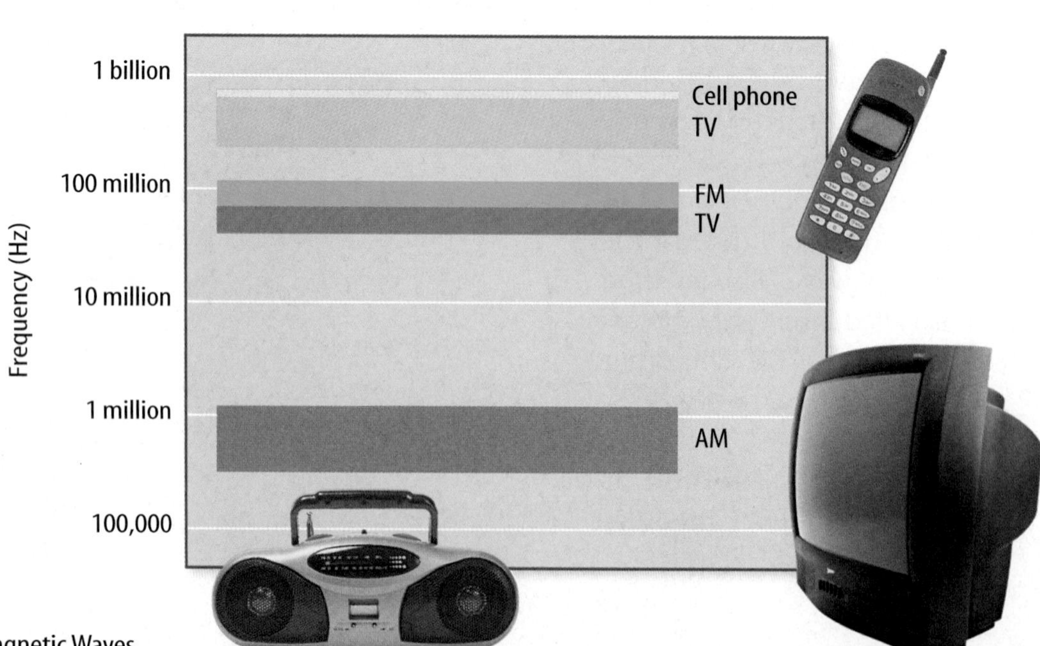

NATIONAL GEOGRAPHIC

VISUALIZING RADIO BROADCASTS

Figure 18

You flick a switch, turn the dial, and music from your favorite radio station fills the room. Although it seems like magic, sounds are transmitted over great distances by converting sound waves to electromagnetic waves and back again, as shown here.

A At the radio station, musical instruments and voices create sound waves by causing air molecules to vibrate. Microphones convert these sound waves to a varying electric current, or electronic signal.

B This signal then is added to the station's carrier wave. If the station is an AM station, the electronic signal modifies the amplitude of the carrier wave. If the station is a FM station, the electronic signal modifies the frequency of the carrier wave.

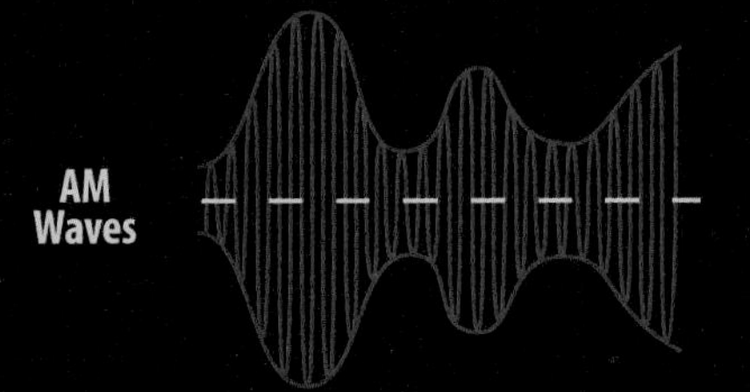

FM Waves

C The modified carrier wave is used to vibrate electrons in the station's antenna. These vibrating electrons create a radio wave that travels out in all directions at the speed of light.

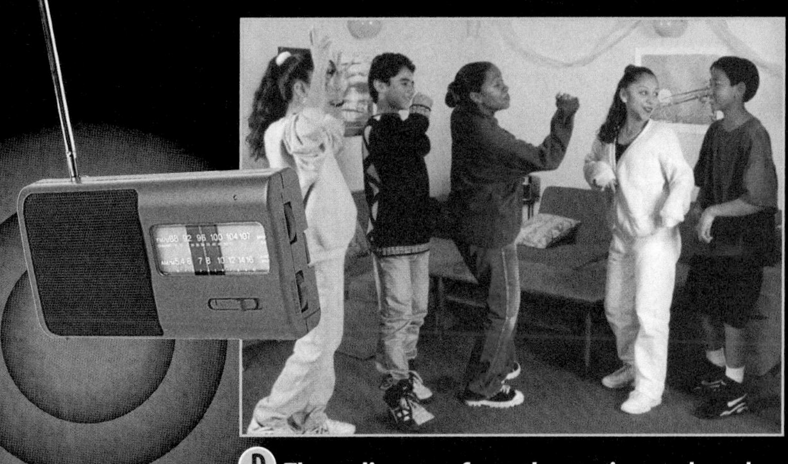

D The radio wave from the station makes electrons in your radio's antenna vibrate. This creates an electric current. If your radio is tuned to the station's frequency, the carrier wave is removed from the original electronic signal. This signal then makes the radio's speaker vibrate, creating sound waves that you hear as music.

Television

What would people hundreds of years ago have thought if they had seen a television? It might seem like magic, but it's not if you know how they work. Television and radio transmissions are similar. At the television station, sound and images are changed into electronic signals. These signals are broadcast by carrier waves. The audio part of television is sent by FM radio waves. Information about the color and brightness is sent at the same time by AM signals.

Astronomy INTEGRATION

Do you ever look up at the stars at night and wonder how they were formed? With so many stars and so many galaxies, life might be possible on other planets. Research ways that astronomers use electromagnetic waves to investigate the universe. Choose one project astronomers currently are working on, and write about it in your Science Journal.

Cathode-Ray Tubes In many television sets, images are displayed on a cathode-ray tube (CRT), as shown in **Figure 19.** A **cathode-ray tube** is a sealed vacuum tube in which one or more beams of electrons are produced. The CRT in a color TV produces three electron beams that are focused by a magnetic field and strike a coated screen. The screen is speckled with more than 100,000 rectangular spots that are of three types. One type glows red, another glows green, and the third type glows blue when electrons strike it. The spots are grouped together with a red, green, and blue spot in each group.

An image is created when the three electron beams of the CRT sweep back and forth across the screen. Each electron beam controls the brightness of each type of spot, according to the information in the video signal from the TV station. By varying the brightness of each spot in a group, the three spots together can form any color so that you see a full-color image.

Reading Check *What is a cathode-ray tube?*

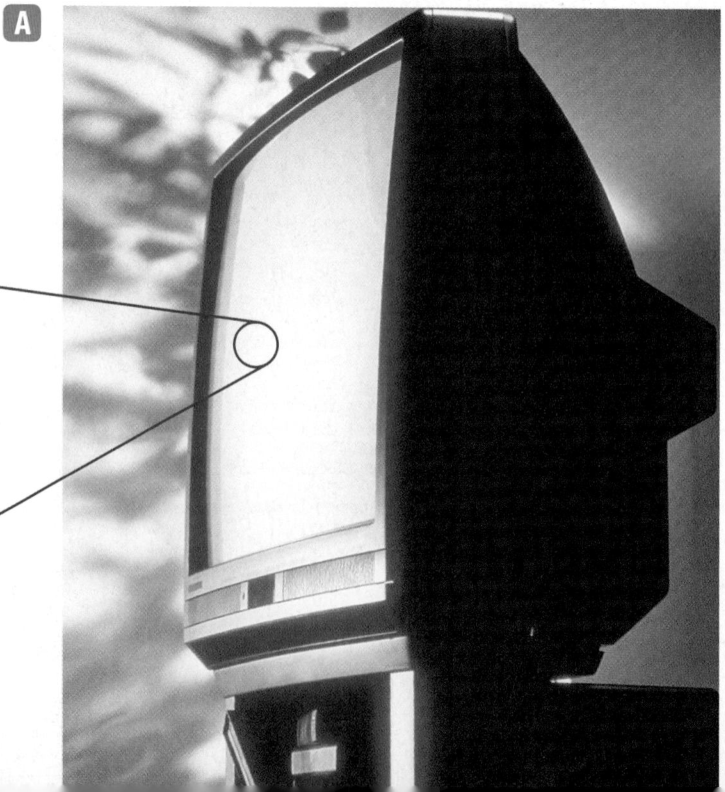

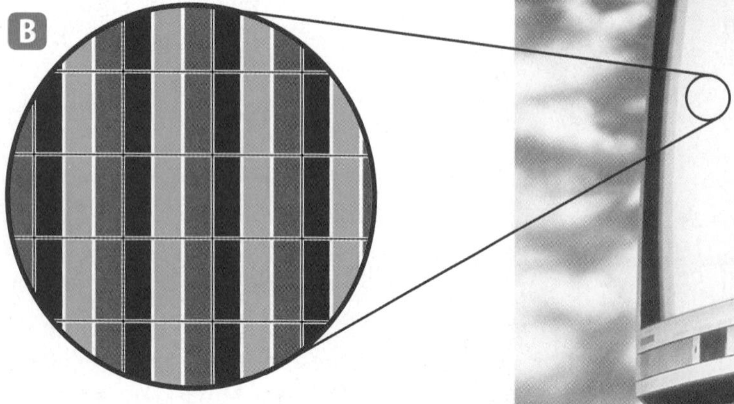

Figure 19
A Cathode-ray tubes produce the images you see on television.
B The inside surface of a television screen is covered by groups of spots that glow red, green, or blue when struck by an electron beam.

Telephones

Unitl about 1950, human operators were needed to connect many calls between people. Just 20 years ago you never would have seen someone walking down the street talking on a telephone. Today, cell phones are seen everywhere. When you speak into a telephone, a microphone converts sound waves into an electrical signal. In cell phones, this current is used to create radio waves that are transmitted to and from a microwave tower, as shown in **Figure 20.** A cell phone uses one radio signal for sending information to a tower at a base station. It uses another signal for receiving information from the base station. The base stations are several kilometers apart. The area each one covers is called a cell. If you move from one cell to another while using a cell phone, an automated control station transfers your signal to the new cell.

Figure 20
The antenna at the top of a microwave tower receives signals from nearby cell phones. *Are any microwave towers located near your school or home? Where?*

Cordless Telephones Like a cellular telephone, a cordless telephone is a transceiver. A **transceiver** transmits one radio signal and receives another radio signal from a base unit. Having two signals at different frequencies allows you to talk and listen at the same time. Cordless telephones work much like cell phones. With a cordless telephone, however, you must be close to the base unit. Another drawback is that when someone nearby is using a cordless telephone, you could hear that conversation on your phone if the frequencies match. For this reason, many cordless phones have a channel button. This allows you to switch your call to another frequency.

Pagers Another method of transmitting signals is a pager, which allows messages to be sent to a small radio receiver. A caller leaves a message at a central terminal by entering a callback number through a telephone keypad or by entering a text message from a computer. At the terminal, the message is changed into an electronic signal and transmitted by radio waves. Each pager is given a unique number for identification. This identification number is sent along with the message. Your pager receives all messages that are transmitted in the area at its assigned frequency. However, your pager responds only to messages with its particular identification number. Newer pagers can send data as well as receive them.

Research Visit the Glencoe Science Web site at **science.glencoe.com** for recent news or magazine articles about advances in radio wave technology. Communicate to your class what you learn.

How does a pager know when to beep you?

Figure 21
Currently, more than 2,000 satellites orbit Earth. *Other than communications, what might the satellites be used for?*

Communications Satellites

Since satellites were first developed, thousands have been launched into Earth's orbit. Many of these, like the one shown in **Figure 21,** are used for communication. A station broadcasts a high-frequency microwave signal to the satellite. The satellite receives the signal, amplifies it, and transmits it to a particular region on Earth. To avoid interference, the frequency broadcast by the satellite is different than the frequency broadcast from Earth.

Research Visit the Glencoe Science Web site at **science.glencoe.com** for more information about ways satellites are used for communication. Report to your class what you learn.

Satellite Telephone Systems If you have a mobile telephone, you can make a phone call when sailing across the ocean. To call on a mobile telephone, the telephone transmits radio waves directly to a satellite. The satellite relays the signal to a ground station, and the call is passed on to the telephone network. Satellite links work well for one-way transmissions, but two-way communications can have an annoying delay caused by the large distance the signals travel to and from the satellite.

Television Satellites The satellite-reception dishes that you sometimes see in yards or attached to houses are receivers for television satellite signals. Satellite television is used as an alternative to ground-based transmission. Communications satellites use microwaves rather than the longer-wavelength radio waves used for normal television broadcasts. Short-wavelength microwaves travel more easily through the atmosphere. The ground receiver dishes are rounded to help focus the microwaves onto an antenna.

The Global Positioning System

Getting lost while hiking is not uncommon, but if you are carrying a **Global Positioning System** (GPS) receiver, it is much less likely to happen. The GPS is a system of satellites, ground monitoring stations, and receivers that provide details about your exact location at or above Earth's surface. The 24 satellites necessary for 24-hour, around-the-world coverage became fully operational in 1995. GPS satellites are owned and operated by the United States Department of Defense, but the microwave signals they send out can be used by anyone. As shown in **Figure 22,** signals from four satellites are needed to determine the location of an object using a GPS receiver.

GPS receivers are used in airplanes, ships, cars, and even by hikers. Many police cars, fire trucks, and ambulances have GPS receivers. This allows the closest help to be sent in an emergency. Many automobile GPS receivers come with a high-resolution, color display screen that can show you a map of the area, display mileage to various locations, and provide information on the services provided at the next interstate exit. Can you think of other uses for the Global Positioning System?

Figure 22
A GPS receiver uses signals from orbiting satellites to determine the user's location. *How would having a GPS receiver in an automobile be useful?*

Section  Assessment

1. Explain the difference between AM and FM radio. Make a sketch of how a carrier wave is modulated in AM and FM radio.
2. What is a cathode-ray tube and how is it used in a television?
3. What happens if you are talking on a cell phone while riding in a car and you travel from one cell to another cell?
4. Explain some of the uses of a Global Positioning System. Why might emergency vehicles all be equipped with GPS receivers?
5. **Think Critically** Why do cordless telephones stop working if you move too far from the base unit?

Skill Builder Activities

6. **Researching Information** For a cellular phone system to work, microwave antennas must be spaced every few kilometers throughout the area. Look around your community to see where microwave antennas are located. Draw a map of the area and note where they are. **For more help, refer to the Science Skill Handbook.**
7. **Communicating** Technology using radio waves for communication is changing rapidly. In your Science Journal, write some ways that communication with radio waves might be different in the future. **For more help, refer to the Science Skill Handbook.**

Activity Use the Internet

Radio Frequencies

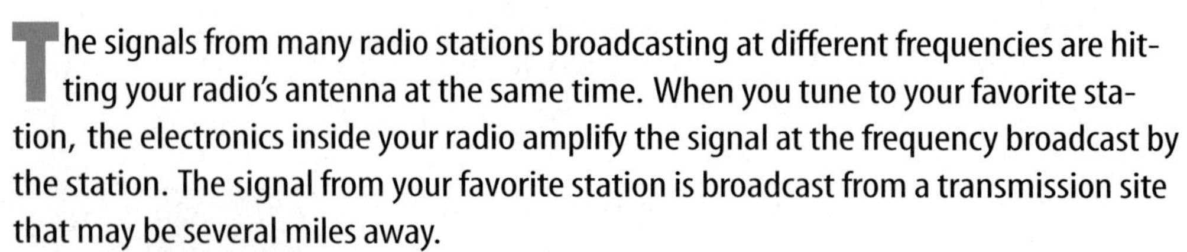

The signals from many radio stations broadcasting at different frequencies are hitting your radio's antenna at the same time. When you tune to your favorite station, the electronics inside your radio amplify the signal at the frequency broadcast by the station. The signal from your favorite station is broadcast from a transmission site that may be several miles away.

You may have noticed that if you're listening to a radio station while driving in a car, sometimes the station gets fuzzy and you'll hear another station at the same time. Sometimes you lose the station completely. How far can you drive before that happens? Does the distance vary depending on the station you listen to?

Recognize the Problem

What are the ranges of radio stations?

Form a Hypothesis

How far can a radio station transmit? Which type of signal, AM or FM, has a greater range? Form a hypothesis about the range of your favorite radio station.

Goals

- **Research** which frequencies are used by different radio stations.
- **Observe** the reception of your favorite radio station.
- **Make** a chart of your findings and communicate them to other students.

Data Source

SCIENCE*Online* Go to the Glencoe Science Web site at **science.glencoe.com** for more information on radio frequencies, different frequencies of radio stations around the country, and the ranges of AM and FM broadcasts.

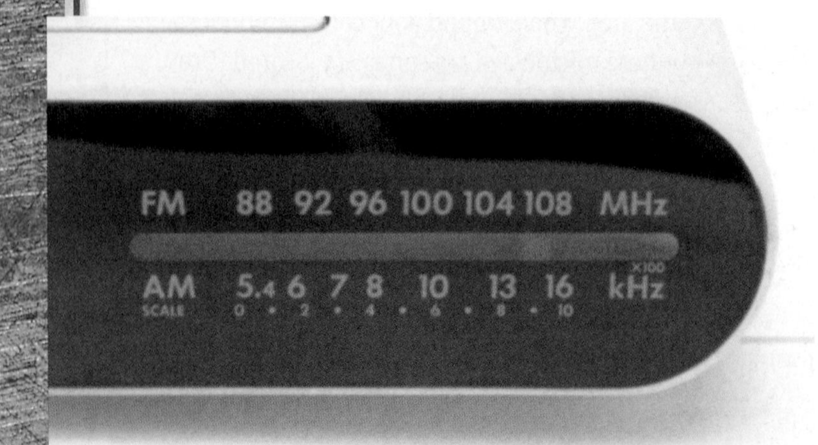

Test Your Hypothesis

Plan

1. **Research** what frequencies are used by AM and FM radio stations in your areas and other areas around the country.
2. **Determine** these stations' broadcast locations.
3. **Determine** the broadcast range of radio stations in your area.
4. **Observe** how frequencies differ. What is the maximum difference between frequencies for FM stations in your area? AM stations?

Do

1. Make sure your teacher approves your plan before you start.
2. Visit the Glencoe Science Web site for links to different radio stations.
3. **Compare** the different frequencies of the stations and the locations of the broadcasts.
4. **Determine** the range of radio stations in your area and the power of their broadcast signal in watts.
5. **Record** your data in your Science Journal.

Analyze Your Data

1. **Make** a map of the radio stations in your area. Do the ranges of AM stations differ from FM stations?
2. **Make** a map of different radio stations around the country. Do you see any patterns in the frequencies for stations that are located near each other?
3. **Write** a description that compares how close the frequencies of AM stations are and how close the frequencies of FM stations are. Also compare the power of their broadcast signals and their ranges.
4. **Share** your data by posting it on the Glencoe Science Web site.

Draw Conclusions

1. Compare your findings to those of your classmates and other data that was posted on the Glencoe Science Web site. Do all AM stations and FM stations have different ranges?
2. Look at your map of the country. How close can stations with similar frequencies be? Do AM and FM stations appear to be different in this respect?
3. The power of the broadcast signal also determines its range. How does the power (wattage) of the signals affect your analysis of your data?

Communicating Your Data

Find this *Use the Internet* activity on the Glencoe Science Web site at **science.glencoe.com**. Post your data in the table provided. Compare your data to that of other students. Then combine your data with theirs and make a map for your class that shows all of the data.

TIME

SCIENCE AND Society

SCIENCE ISSUES THAT AFFECT YOU!

Can phoning and driving go together safely?

Cell Phones

If you use a cell phone, you're not alone. More than 92 million Americans have them, and 30,000 more new cell phone users sign up each day. Although it seems that you can't eat in a restaurant or ride a train without hearing someone else's cell phone conversation, one of the most popular places for cell phone use is the car. And many people think that's a problem.

Danger, Danger

According to the National Highway Transportation Safety Board, driver inattention is a factor in half of all accidents. Many people believe that cell phone use distracts drivers, causing accidents. Drivers can get so excited or involved in a phone conversation that they forget they are behind the wheel. Drivers who hold phones (rather than use speaker phones) don't have complete control of the car. Dialing a phone number can make drivers take their eyes off the road.

One study found that people who talk on cell phones while driving are four times more likely to get into a car accident than people who do not. In Oklahoma, accident reports suggest that drivers with cell phones are more likely to speed and swerve between lanes. They are also involved in more fatal accidents. Because of findings such as these, many people think laws should restrict cell phone use by drivers. This is already the case in countries such as Brazil, Sweden, and Australia. Several communities in the United States have restricted cell phone use as well. In Suffolk County, New York, for example, lawmakers have passed a bill making it illegal to use a hand-held cellular phone while driving.

Cellulars Aren't All Bad

Regardless of this evidence, some people feel that singling out cell phones as the cause of accidents is unfair. They say that drivers are inattentive for many reasons. Changing CDs, eating, or looking at maps while driving, can take attention from the road. A driver looking at a digital map display takes his or her eyes off the road about 20 times in a short period, according to one study. This can spell danger in a car speeding down the road at 100 km/h. Yet there are no laws against looking at maps while driving.

Supporters of cell phone use in cars also point out that cellular phones are useful during emergencies. Many drivers have used these phones to report accidents or roadside injuries. These reports have helped people in trouble and have saved lives.

The best course may be to just learn to use car phones more carefully. For example, drivers should pull off the road to make calls. Drivers shouldn't use a hand-held cell phone but should use a speaker phone in order to keep both hands on the steering wheel. But even if people followed those two suggestions, the debate won't end. As cars become more loaded with gadgets that enable drivers to send faxes or even microwave snacks on the road, the question of whether drivers should do anything in the car other than drive will become more important.

CONNECTIONS Survey With a partner, write a questionnaire on car cell phone use for classmates to answer. Include questions on what they think of cell phone use by drivers, and whether adults in their families use a cell phone while driving. Tabulate the results and post them.

Chapter 13 Study Guide

Reviewing Main Ideas

Section 1 What are electromagnetic waves?

1. A vibrating charge creates electromagnetic waves. *In what ways are electromagnetic waves similar to and different from the waves in this picture?*

2. Electromagnetic waves have radiant energy and travel through a vacuum or through matter.
3. Electromagnetic waves sometimes can behave like particles. The particles are called photons.

Section 2 The Electromagnetic Spectrum

1. Electromagnetic waves with the lowest frequency are called radio waves. Infrared waves have frequencies between radio waves and visible light. *Why would being able to detect infrared waves help this pit viper catch its prey?*
2. Human eyes can see electromagnetic waves that span a wavelength range of 390 billionths to 770 billionths of a meter.
3. Ultraviolet waves have frequencies slightly higher than visible light. They have useful and harmful properties.
4. X rays and gamma rays are two types of ultrahigh-frequency electromagnetic waves that are energetic enough to travel through many materials and destroy living cells.

Section 3 Radio Communication

1. Modulated radio waves are used often for communication. AM and FM are two forms of carrier wave modulation. *What is the purpose of AM and FM modulation of carrier waves?*
2. Television signals are transmitted as a combination of AM and FM waves.
3. Cellular telephones, cordless telephones, and pagers all rely on radio waves for signal transmission. Communications satellites are used for telephone and television transmissions.
4. The Global Positioning System uses satellites to help people determine their exact location.

After You Read

FOLDABLES Reading & Study Skills

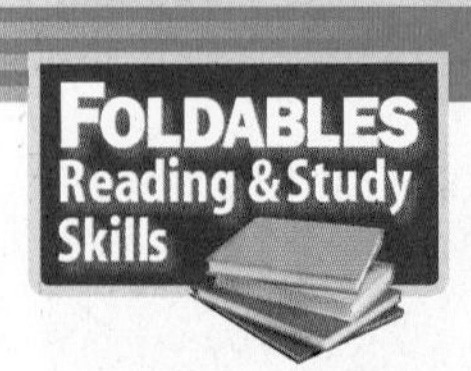

Make a list of different types of electromagnetic waves arranged in order of increasing frequency on the back of the tabs of your Foldable.

Chapter 13 Study Guide

Visualizing Main Ideas

Complete the following table about the electromagnetic spectrum.

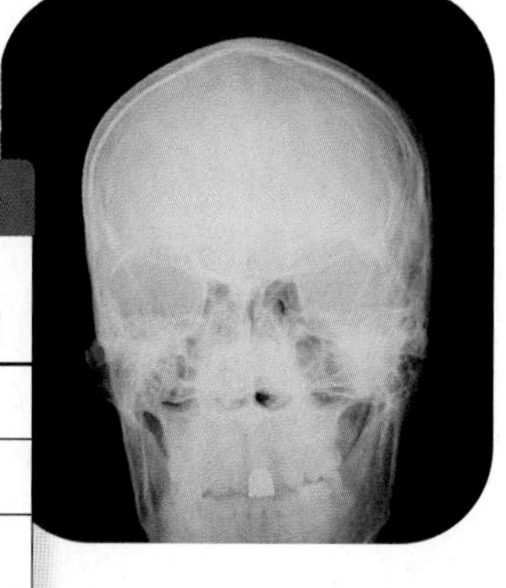

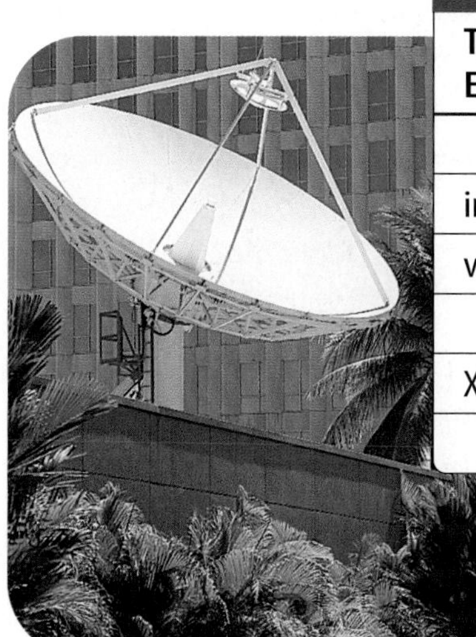

Uses of Electromagnetic Waves

Type of Electromagnetic Waves	Examples of How Electromagnetic Waves Are Used
	radio, TV transmission
infrared waves	
visible light	vision
X rays	
	destroying harmful cells

Vocabulary Review

Vocabulary Words

a. carrier wave
b. cathode-ray tube
c. electromagnetic wave
d. frequency
e. gamma ray
f. infrared wave
g. microwave
h. photon
i. radiant energy
j. radio wave
k. transceiver
l. ultraviolet wave
m. visible light
n. X ray

Study Tip

Keep all your homework assignments and reread them from time to time. Make sure you understand any problems you answered incorrectly.

Using Vocabulary

Replace the underlined words with the correct vocabulary word.

1. Gamma rays are a type of electromagnetic wave often used for communication.
2. Visible light and radio waves are used often for medical imaging.
3. A remote control is able to communicate with a television by sending X rays.
4. Electromagnetic waves are composed of massless particles called carrier waves.
5. If you stay outdoors too long, your skin might be burned by exposure to radio waves from the Sun.
6. Microwaves are waves of unique frequencies used by radio stations to broadcast information.

Chapter 13 Assessment

Checking Concepts

Choose the word or phrase that best answers the question.

1. Which type of electromagnetic wave is the most energetic?
A) gamma rays **C)** infrared waves
B) ultraviolet waves **D)** microwaves

2. Electromagnetic waves can behave like what type of particle?
A) electrons **C)** photons
B) molecules **D)** atoms

3. Which type of electromagnetic wave enables skin cells to produce vitamin D?
A) visible light **C)** infrared waves
B) ultraviolet waves **D)** X rays

4. Which of the following describes X rays?
A) short wavelength, high frequency
B) short wavelength, low frequency
C) long wavelength, high frequency
D) long wavelength, low frequency

5. Which of the following is changing in an AM radio wave?
A) speed **C)** amplitude
B) frequency **D)** wavelength

6. Which type of electromagnetic wave is used to produce a thermogram?
A) X rays **C)** infrared waves
B) ultraviolet waves **D)** gamma rays

7. Which type of electromagnetic wave has wavelengths greater than about 1 mm?
A) X rays **C)** gamma rays
B) radio waves **D)** ultraviolet waves

8. What is the name of the ability of some materials to absorb ultraviolet light and re-emit it as visible light?
A) modulation **C)** transmission
B) handoff **D)** fluorescence

9. Which of these colors of visible light has the shortest wavelength?
A) blue **C)** red
B) green **D)** white

10. Which type of electromagnetic wave has wavelengths slightly longer than humans can see?
A) X rays **C)** infrared waves
B) ultraviolet waves **D)** gamma rays

Thinking Critically

11. When you heat food in a microwave oven, the ceramic, glass, or plastic containers usually remain cool, even though the food can become hot. Why is this?

12. Doctors and dentists often use X rays for medical imaging. Why are X rays useful for this purpose?

13. Give one reason why communications satellites don't use ultraviolet waves to transmit and receive information.

14. Could an electromagnetic wave travel through space if its electric and magnetic fields were not changing with time? Explain.

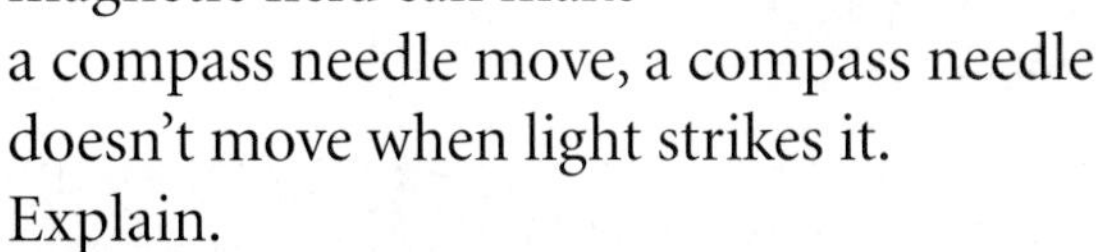

15. Electromagnetic waves consist of vibrating electric and magnetic fields. Even though a magnetic field can make a compass needle move, a compass needle doesn't move when light strikes it. Explain.

Developing Skills

16. Developing Multimedia Presentations Create a multimedia presentation on how radio waves are used for communication.

17. **Classifying** Look around your home, school, and community. Make a list of the many types of technology that use electromagnetic waves. Beside each item, write the type of electromagnetic wave it uses.

18. **Writing a Paper** Some people warn that microwaves from cell phones can be harmful. Research this problem and write a paper describing your opinion.

19. **Concept Mapping** Copy the diagram and fill in the missing events about ozone destruction.

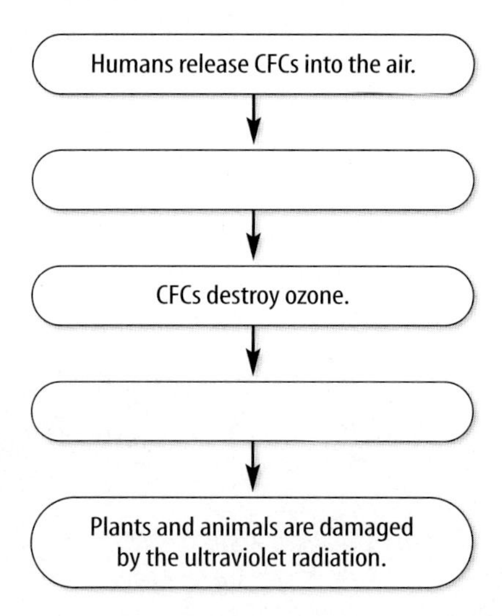

Performance Assessment

20. **Observe and Infer** Tune a radio to various AM and FM frequencies. Notice how strong or weak the signals are at various times of the day and night. Record your observations in your Science Journal.

Technology

Go to the Glencoe Science Web site at **science.glencoe.com** or use the **Glencoe Science CD-ROM** for additional chapter assessment.

Test Practice

Because radio waves can reflect off objects, they are used to detect an object's location. The two rescue ships below are using radio waves to search for a lifeboat.

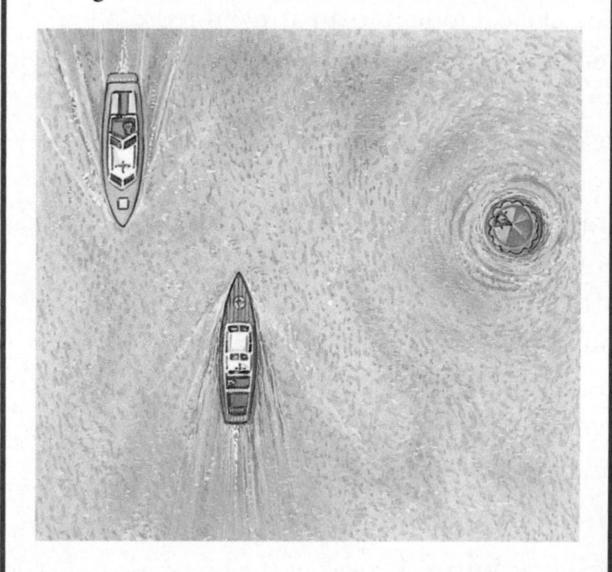

Study the picture and answer the following questions.

1. What do the rescuers need to know to help find the lifeboat?
   A) the strength of the carrier wave
   B) the amplitude and frequency of the radio waves that were sent out
   C) the frequency of the reflected radio waves
   D) the direction from which the reflected radio waves came

2. Why will the rescuers detect the reflected radio waves at different times?
   F) The speed of the radio waves leaving each ship is different.
   G) The life boat is a different distance from each ship.
   H) Only one ship is sending radio waves.
   J) The lifeboat is avoiding detection.

Light

A sunrise over the ocean is a spectacle of light and color. As the sun tops the horizon, sea and sky explode into an array of reds, pinks, and oranges. Why does the sky appear red? How are its colors reflected in the water? In this chapter, you will learn about light—how it travels, bends, reflects, and enables you to see different colors. You will also learn how different kinds of bulbs produce light and how light is used by lasers, optical scanners, and optical fibers.

What do you think?

Science Journal Look at the picture below with a classmate. Discuss what this might be or what is happening. Here's a hint: *It's full of air and doesn't last long.* Write your answer or best guess in your Science Journal.

Light passing through a prism can produce exciting patterns of color. Imagine what your surroundings would look like now if humans could see only shades of gray instead of distinct colors. The ability to see color depends on the cells in your eyes that are sensitive to different wavelengths of light. What color is the light produced by a flashlight or the sun?

Make your own rainbow

1. In a darkened room, shine a flashlight through a glass prism. Project the resulting colors onto a white wall or ceiling.
2. In a darkened room, shine a flashlight over the surface of some water with dishwashing liquid bubbles in it. What do you see?
3. Aim a flashlight at the surface of a compact disc.

Observe

How did your observations in each case differ? In your Science Journal, explain where you think the colors came from.

Before You Read

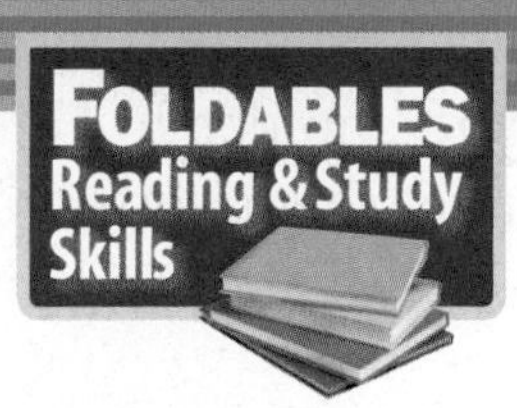

Making a Compare and Contrast Study Fold **Make the following Foldable to compare and contrast the characteristics of opaque, translucent, and transparent.**

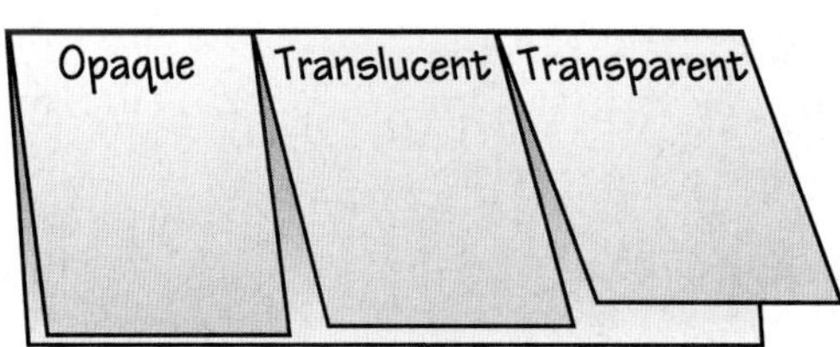

1. Place a sheet of paper in front of you so the short side is at the top. Fold the paper in half from top to bottom.
2. Fold both sides in to divide the paper into equal thirds. Unfold the paper so three sections show.
3. Through one thickness of paper, cut along each of the fold lines to the topfold, forming three tabs. Label each tab *Opaque, Translucent*, and *Transparent* as shown.
4. As you read the chapter, write characteristics of these materials under the tabs, using the words absorb, reflect, and transmit.

SECTION

The Behavior of Light

As You Read

What You'll Learn

- **Describe** the differences among opaque, transparent, and translucent materials.
- **Explain** how light is reflected.
- **Discuss** how refraction separates white light.

Vocabulary

opaque
translucent
transparent
index of refraction
mirage

Why It's Important

Knowing how light behaves will help you understand various sights, such as reflections in a store window, rainbows, and mirages.

Light and Matter

Look around your room after turning off the lights at night. At first you can't see anything, but as your eyes adjust to the darkness, you begin to recognize some familiar objects. You know that some of the objects are brightly colored, but they look gray or black in the dim light. Turn on the light, and you clearly can see all the objects in the room, including their colors. What you see depends on the amount of light in the room and the color of the objects. To see an object, it must reflect some light back to your eyes.

Opaque, Transparent, and Translucent Objects can absorb light, reflect light, and allow light to pass through them. The type of matter in an object determines the amount of light it absorbs, reflects, and transmits. For example, the **opaque** (oh PAYK) material in the candleholder in **Figure 1A,** only absorbs and reflects light—no light passes through it. As a result, you cannot see the candle inside.

Other materials allow some light to pass through, but you cannot see clearly through them. These are **translucent** (trans LEW sunt) materials, like the candleholder in **Figure 1B.**

Transparent materials like the candleholder in **Figure 1C** transmit almost all of the light that strikes them, so you can see objects clearly through them. Only a small amount of light is absorbed and reflected.

Figure 1
These candleholders have different light-transmitting properties.

A Opaque

B Translucent

C Transparent

Reflection of Light

Just before you left for school this morning, did you take one last glance in a mirror to check your appearance? To see your reflection in the mirror, light had to reflect off you, hit the mirror, and reflect off the mirror into your eye. Reflection occurs when a light wave strikes an object and bounces off.

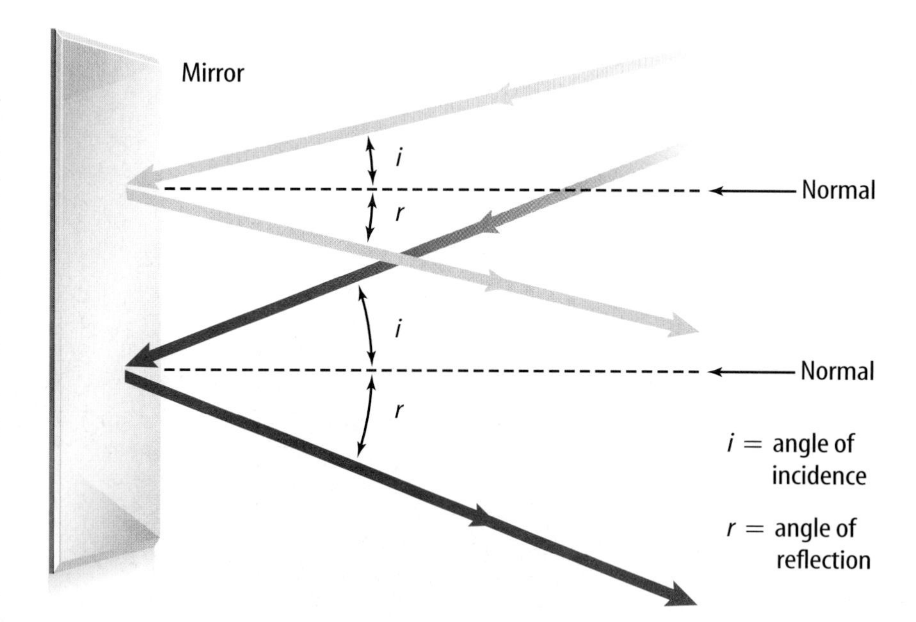

Figure 2
Light is reflected according to the law of reflection so that the angle of incidence always equals the angle of reflection.

The Law of Reflection

Because light behaves as a wave, it obeys the law of reflection, as shown in **Figure 2.** According to the law of reflection, the angle at which a light wave strikes a surface is the same as the angle at which it is reflected. Light reflected from any surface—a mirror or a sheet of paper—follows this law.

Regular and Diffuse Reflection Why can you see your reflection in a store window but not in a brick wall? The answer has to do with the smoothness of the surfaces. A smooth, even surface like that of a pane of glass produces a sharp image by reflecting parallel light waves in only one direction. Reflection of light waves from a smooth surface is regular reflection. A brick wall has an uneven surface that causes incoming parallel light waves to be reflected in many directions, as shown in **Figure 3.** Reflection of light from a rough surface is diffuse reflection.

✔ Reading Check ***What is diffuse reflection?***

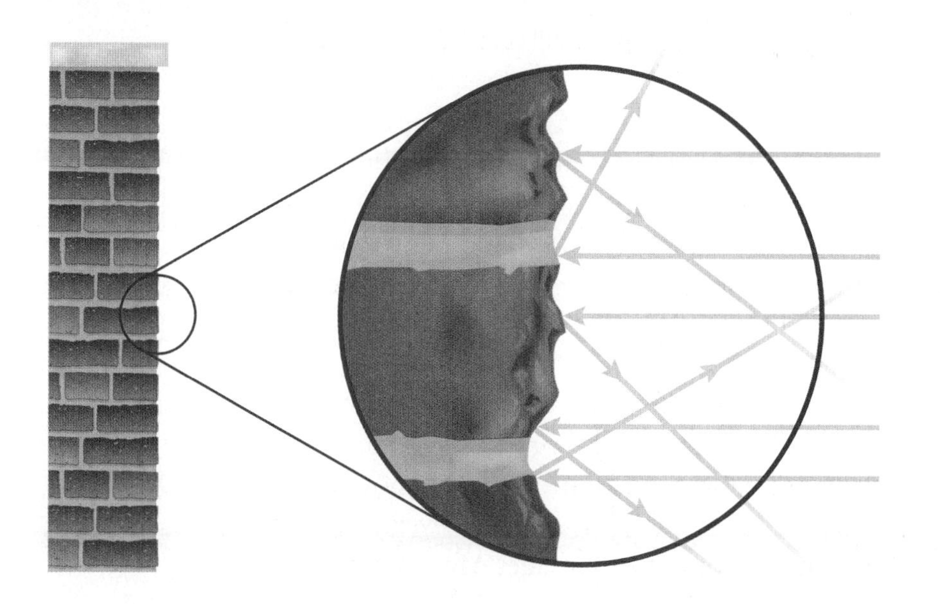

Figure 3
This brick wall has an uneven surface, so it produces a diffuse reflection.

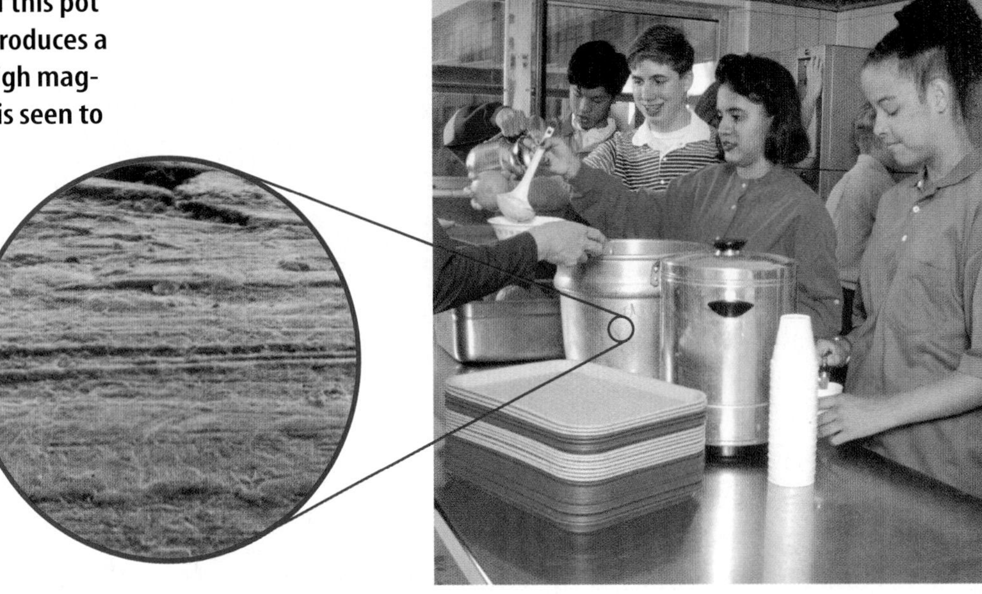

Figure 4
Although the surface of this pot may seem smooth, it produces a diffuse reflection. At high magnification, the surface is seen to be rough.

Roughness of Surfaces Even a surface that appears to be smooth can be rough enough to cause diffuse reflection. For example, a metal pot might seem smooth, but at high magnification, the surface shows rough spots, as shown **Figure 4.** To cause a regular reflection, the roughness of the surface must be less than the wavelengths it reflects.

Refraction of Light

What occurs when a light wave passes from one material to another—from air to water, for example? Refraction is caused by a change in the speed of a wave when it passes from one material to another. If the light wave is traveling at an angle and the speed that light travels is different in the two materials, the wave will be bent, or refracted.

Reading Check *How does refraction occur?*

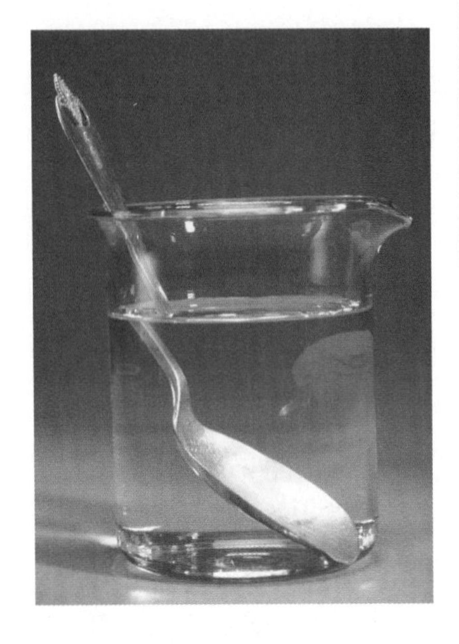

Figure 5
The spoon looks bent in the water because the light waves are refracted as they change speed when they pass from the water to the air.

The Index of Refraction The amount of bending that takes place depends on the speed of light in both materials. The greater the difference is, the more the light will be bent as it passes at an angle from one material to the other. **Figure 5** shows an example of refraction. Every material has an **index of refraction**—a property of the material that indicates how much it reduces the speed of light.

The larger the index of refraction, the more light is slowed down in the material. For example, because glass has a larger index of refraction than air, light moves more slowly in glass than air. Many useful devices like eyeglasses, binoculars, cameras, and microscopes form images using refraction.

Prisms A sparkling glass prism hangs in a sunny window, refracting the sunlight and projecting a colorful pattern onto the walls of the room. How does the bending of light create these colors? It occurs because the amount of bending usually depends on the wavelength of the light. Wavelengths of visible light range from the longer red waves to the shorter violet waves. White light, such as sunlight, is made up of this whole range of wavelengths.

Figure 6
Refraction causes a prism to separate a beam of white light into different colors.

Figure 6 shows what occurs when white light passes through a prism. The triangular prism refracts the light twice—once when it enters the prism and again when it leaves the prism and reenters the air. Because the longer wavelengths of light are refracted less than the shorter wavelengths are, red light is bent the least. As a result of these different amounts of bending, the different colors are separated when they emerge from the prism. Which color of light would you expect to bend the most?

Rainbows Does the light leaving the prism in **Figure 6** remind you of a rainbow? Like prisms, rain droplets also refract light. The refraction of the different wavelengths can cause white light from the Sun to separate into the individual colors of visible light, as shown in **Figure 7.** In a rainbow, the human eye usually can distinguish only about seven colors clearly. In order of decreasing wavelength, these colors are red, orange, yellow, green, blue, indigo, and violet.

Figure 7
As white light passes through the water droplet, different wavelengths are refracted by different amounts. This produces the separate colors seen in a rainbow.

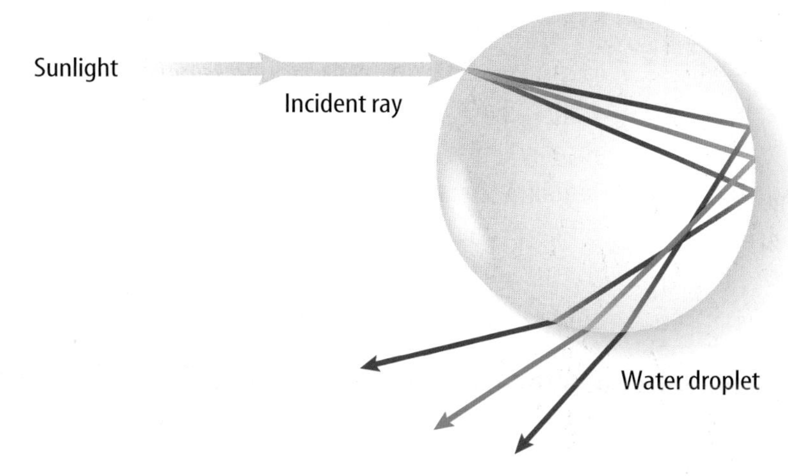

Observing Refraction in Water

Procedure

1. Place a **penny** at the bottom of a **short, opaque cup.** Set it on a **table** in front of you.
2. Have a partner slowly slide the cup away from you until you can't see the penny.
3. Without disturbing the penny or the cup and without moving your position, have your partner slowly pour **water** into the cup until you can see the penny.
4. Reverse roles and repeat the experiment.

Analysis

1. What did you observe? Explain how this is possible.
2. In your **Science Journal,** sketch the light path from the penny to your eye after the water was added.

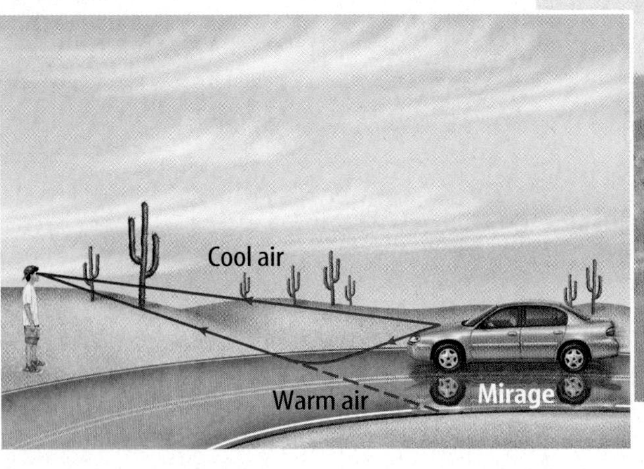

Figure 8
Mirages result when air near the ground is much warmer or cooler than the air above. This causes some lightwaves reflected from the object to refract, creating one or more additional images.

Mirages When you're riding in a car on a hot day, you might see something that looks like a shimmering pool of water on the road ahead. As you get closer, the water seems to disappear. What you saw was a mirage. A **mirage** is an image of a distant object produced by the refraction of light through air layers of different densities. The density of air increases as the air gets cooler. The greater the difference in densities is, the more the light is refracted. Mirages result when the air at ground level is much warmer or much cooler than the layers of air above it, as **Figure 8** shows. The image you see is always some distance away from the actual object. For example, the water you think you see on a hot road surface sometimes is an image of the sky.

Section  Assessment

1. Contrast opaque, transparent, and translucent materials. Give at least one example of each.
2. Explain why you can see your reflection in a smooth piece of aluminum foil but not in a crumpled ball of foil.
3. Why are you more likely to see a mirage on a hot day than on a mild day?
4. What happens to white light when it passes through a prism?
5. **Think Critically** Consider the following parts of your body: the lens of your eye, a fingernail, your skin, and a tooth. Decide whether each of these is opaque, transparent, or translucent. Explain.

Skill Builder Activities

6. **Making and Using Tables** Construct a table that shows the light-reflecting or light-absorbing properties of different materials. Use the words *opaque, transparent,* and *translucent.* **For more help, refer to the Science Skill Handbook.**
7. **Communicating** Walk around your classroom noting five reflecting objects. Which objects display diffuse reflection and which display regular reflection? How does the surface differ on each? For each object you note, list the colors the object is absorbing. **For more help, refer to the Science Skill Handbook.**

Light and Color

As You Read

What You'll Learn

- **Explain** how you see color.
- **Describe** the difference between light color and pigment color.
- **Predict** what happens when different colors are mixed.

Vocabulary
pigment

Why It's Important

From traffic lights to great works of art, color plays an important role in your world.

Colors

Why do some apples appear red, while others look green or yellow? An object's color depends on the wavelength of light it reflects. You know that white light is a blend of all colors of visible light. When a red apple is struck by white light, it reflects red light back to your eyes and absorbs all of the other colors. **Figure 9** shows white light striking a green leaf. Only the green light is reflected to your eyes.

Although some objects appear to be black, black isn't a color that is present in visible light. Objects that appear black absorb all colors of light and reflect little or no light back to your eye. White objects appear to be white because they reflect all colors of visible light.

Reading Check *Why does a white object appear white?*

Colored Filters Wearing tinted glasses changes the color of almost everything you look at. If the lenses are yellow, the world takes on a golden glow. If they are rose colored, everything looks rosy. Something similar would occur if you placed a colored, clear plastic sheet over this white page. The paper would appear to be the same color as the plastic. The plastic sheet and the tinted lenses are filters. A filter is a transparent material that transmits one or more colors of light but absorbs all others. The color of a filter is the color of the light that it transmits.

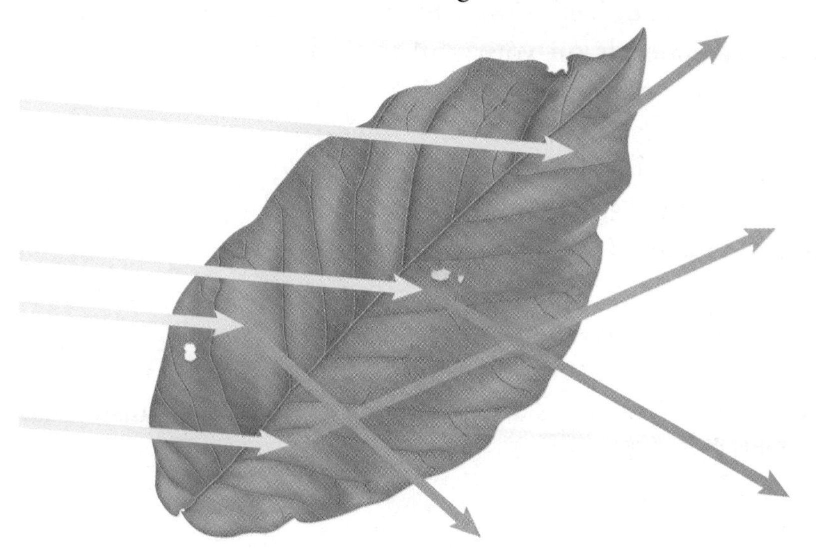

Figure 9
This green leaf absorbs all wavelengths of visible light except green.

Figure 10
The color of this cooler seems to change under different lighting conditions.

B The cooler appears blue when viewed through a blue filter.

A The blue cooler is shown in white light.

C The cooler appears black through a red filter.

Looking Through Colored Filters **Figure 10** shows what happens when you look at a colored object through various colored filters. In the white light in **Figure 10A,** a blue cooler looks blue because it reflects only the blue light in the white light striking it. It absorbs the light of all other colors. If you look at the cooler through a blue filter as in **Figure 10B,** the cooler still looks blue because the filter transmits the reflected blue light. **Figure 10C** shows how the cooler looks when you examine it through a red filter. Why does it appear to be black?

Seeing Color

As you approach a busy intersection, the color of the traffic light changes from green to yellow to red. On the cross street, the color changes from red to green. At a busy intersection, traffic safety depends on your ability to detect immediate color changes. How do you see colors?

Light and the Eye In a healthy eye, light enters and is focused on the retina, an area on the inside of your eyeball, as shown in **Figure 11A.** The retina is made up of two types of cells that absorb light, as shown in **Figure 11B.** When these cells absorb light energy, chemical reactions convert light energy into nerve impulses that are transmitted to the brain. One type of cell in the retina, called a cone, allows you to distinguish colors and detailed shapes of objects. Cones are most effective in daytime vision. Why?

Figure 11
Light enters the eye and focuses on the retina.

A The retina is at the back of your eye.

B The two types of nerve cells that make up the retina are called rods and cones. The rods are the thinner of the two.

Lens

Retina

Cones and Rods Your eyes have three types of cones, each of which responds to a different range of wavelengths. Red cones respond to mostly red and yellow, green cones respond to mostly yellow and green, and blue cones respond to mostly blue and violet. The second type of cell, called a rod, is sensitive to dim light and is useful for night vision.

Interpreting Color Why does a banana look yellow? The light reflected by the banana causes the cone cells that are sensitive to red and green light to send signals to your brain. Your brain would get the same signal if a mixture of red light and green light reached your eye. Again, your red and green cones would respond, and you would see yellow light because your brain can't perceive the difference between incoming yellow light and yellow light produced by combining red and green light. The next time you are at a play or a concert, look at the lighting above the stage. Watch how the colored lights combine to produce effects onstage.

Color Blindness If one or more of your sets of cones did not function properly, you would not be able to distinguish between certain colors. About eight percent of men and one-half percent of women have a form of color blindness. Most people who are said to be color blind are not truly blind to color, but they have difficulty distinguishing between a few colors, most commonly red and green. **Figure 12** shows a plate of a color blindness test. Because these two colors are used in traffic signals, drivers and pedestrians must be able to identify them.

Figure 12
Color blindness is an inherited sex-linked condition in which certain cones do not function properly. *What number do you see in the dots?*

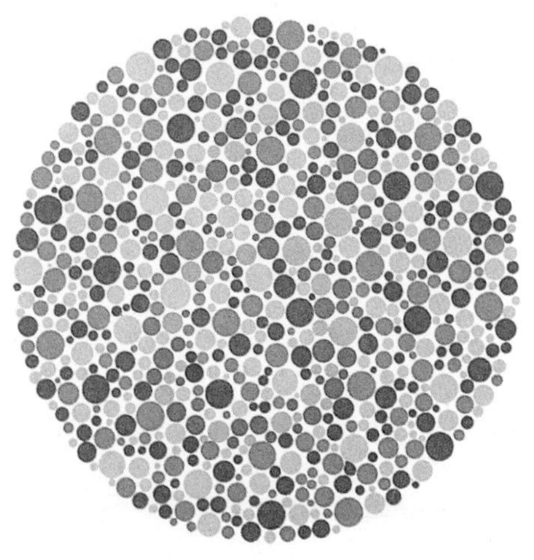

Mixing Colors

If you have ever browsed through a paint store, you have probably seen displays where customers can select paint samples of almost every imaginable color. The colors are a result of mixtures of pigments. For example, you might have mixed blue and yellow paint to produce green paint. A pigment is a colored material that absorbs some colors and reflects other colors. The color of a pigment results from the different wavelengths of light that the pigment reflects.

Life Science INTEGRATION

Plant pigments allow plants to select the wavelengths of light they use for photosynthesis. Leaves usually look green due to the pigment chlorophyll. Chlorophyll absorbs most wavelengths of visible light except green, which it reflects. But not all plants are green. Research different plant pigments to find how they allow plant species to survive in diverse habitats.

Mixing Colored Lights From the glowing orange of a sunset to the deep blue of a mountain lake, all the colors you see can be made by mixing of three colors of light. These three colors—red, green, and blue—are the primary colors of light. They correspond to the three different types of cones in the retina of your eye. When mixed together in equal amounts, they produce white light, as **Figure 13** shows. Mixing the primary colors in different proportions can produce the colors you see.

Reading Check *What are primary colors?*

Paint Pigments If you were to mix equal amounts of red, green, and blue paint, would you get white paint? If mixing colors of paint were like mixing colors of light, you would, but mixing paint is different. Paints are made with pigments. Paint pigments usually are made of chemical compounds such as titanium oxide, a bright white pigment, and lead chromate, which is used for painting yellow lines on highways.

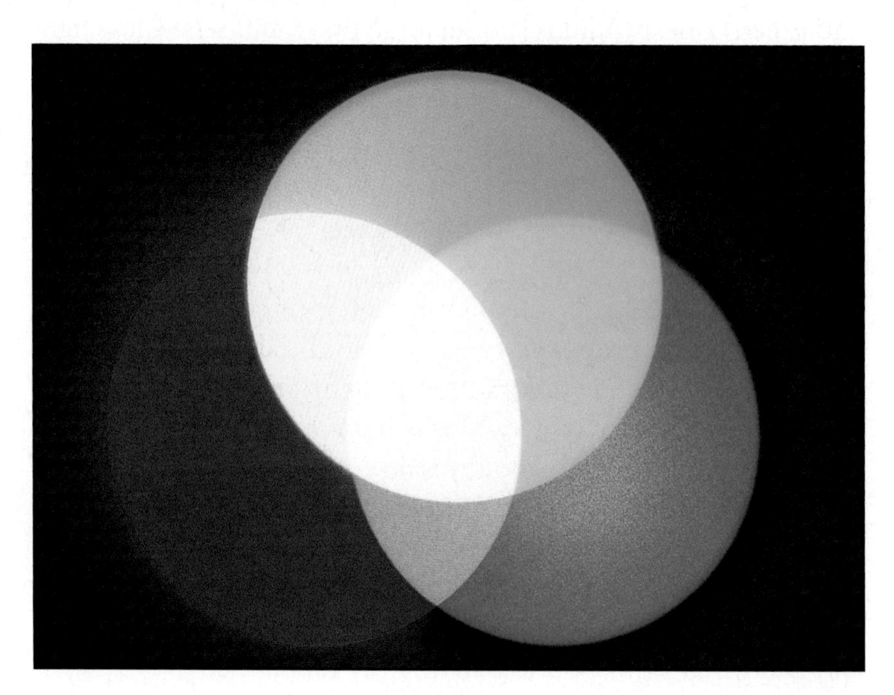

Figure 13
White light is produced when the three primary colors of light are mixed.

Mixing Pigments You can make any pigment color by mixing different amounts of the three primary pigments—magenta (bluish red), cyan (greenish blue), and yellow. In fact, color printers use those pigments to make full-color prints like the pages in this book. However, color printers also use black ink to produce a true black color. A primary pigment's color depends on the color of light it reflects. Actually, pigments both absorb and reflect a range of colors in sending a single color message to your eye. For example, in white light, the yellow pigment appears yellow because it reflects yellow, red, orange, and green light but absorbs blue and violet light. The color of a mixture of two primary pigments is determined by the primary colors of light that both pigments reflect.

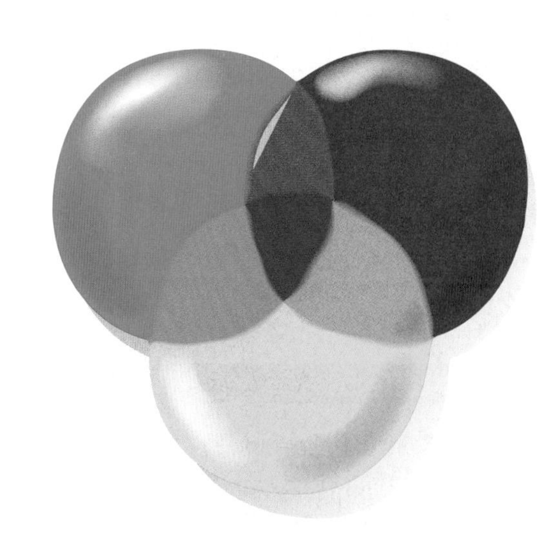

Figure 14
The three primary colors of pigment appear to be black when they are mixed.

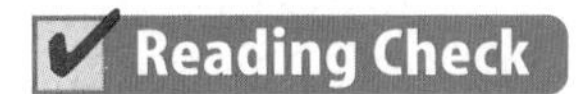

What colors are the three primary pigments?

Look at **Figure 14.** The area in the center where the colors all overlap appears to be black because the three blended primary pigments absorb all the primary colors of light. Recall that the primary colors of light combine to produce white light. They are called additive colors. However, the primary pigment colors combine to produce black. Because black results from the absence of reflected light, the primary pigments are called subtractive colors.

Section  Assessment

1. If a white light shines on a red shirt, what colors are reflected and what colors are absorbed?
2. How do the primary colors of light differ from the primary pigment colors?
3. Explain why a person with color blindness can distinguish among some colors but not others.
4. If all colors are present in white light, why does a white fence appear to be white instead of multicolored?
5. **Think Critically** If you had only magenta, cyan, and yellow paints, could you paint a picture of a zebra? Explain why this would or would not be possible.

Skill Builder Activities

6. **Concept Mapping** Design a concept map to show the chain of events that must happen for you to see a blue object. Work with a partner. **For more help, refer to the Science Skill Handbook.**
7. **Researching Information** Research the electromagnetic spectrum to find the wavelength and frequency range of visible light. Make a poster showing the wavelengths for the seven main colors in the visible light spectrum. Explain the units used, and the relationship between wavelength and wave frequency. **For more help, refer to the Science Skill Handbook.**

Producing Light

As You Read

What You'll Learn

- **Explain** how incandescent and fluorescent lightbulbs work.
- **Analyze** the advantages and disadvantages of different lighting devices.
- **Explain** how a laser produces coherent light.
- **Describe** various uses of lasers.

Vocabulary

incandescent light
fluorescent light
coherent light
incoherent light

Why It's Important

Knowing how different lighting devices work will help you choose the right one for your needs.

Incandescent Lights

It's only been 100 years or so since lightbulbs became common in households. Their shapes, sizes, and varieties today are a far cry from a single lightbulb swinging from an electric cord over a table. Most of the lightbulbs in your house produce incandescent light. Heating a piece of metal until it glows produces **incandescent light.** If you look into an unlit incandescent lightbulb, you will see a thin wire called the filament. It usually is made of the element tungsten. Turn on the light and electricity flows through the filament and causes it to heat up. When the tungsten filament gets hot, it gives off light. However, if you've ever accidentally touched a lit lightbulb, you know that it also produces a great deal of heat. More than 80 percent of the energy given off by incandescent bulbs is in the form of heat.

✔ Reading Check *Why does an incandescent lightbulb get hot?*

Fluorescent Lights

Your house also may have fluorescent (floo RE sunt) lights. A fluorescent bulb, like the one shown in **Figure 15,** is filled with a gas at low pressure. The inside of the bulb is coated with phosphors that emit visible light when they absorb ultraviolet radiation. The tube also contains electrodes at each end. Electrons are given off when the electrodes are connected in a circuit. When these electrons collide with the gas atoms, ultraviolet radiation is emitted. The phosphors on the inside of the bulb absorb this radiation and give off visible light.

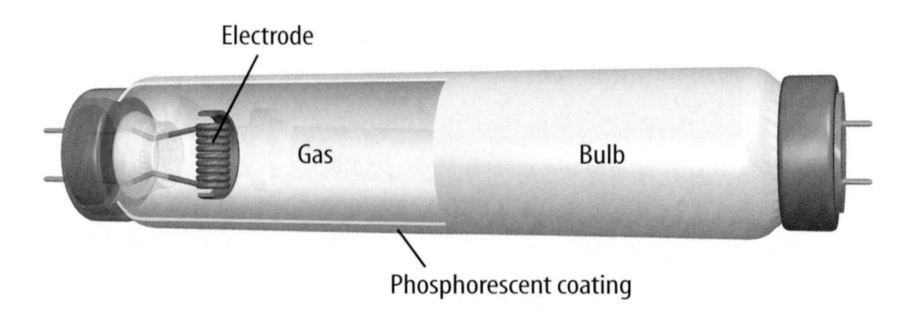

Figure 15
Fluorescent lightbulbs do not use filaments. *What property of phosphors makes them useful in fluorescent bulbs?*

Efficient Lighting A **fluorescent light** uses phosphors to convert ultraviolet radiation to visible light. Fluorescent lights use as little as one fifth the electrical energy to produce the same amount of light as incandescent bulbs. Fluorescent bulbs also last much longer than incandescent bulbs. This higher efficiency can mean lower energy costs over the life of the bulb. Reduced energy usage could reduce the amount of fossil fuels burned to generate electricity, which also decreases the amount of carbon dioxide and pollutants released into Earth's atmosphere.

Fluorescent bulbs are used widely in hospitals, office buildings, schools, and factories. Compact fluorescent bulbs, which can be screwed into traditional lightbulb sockets, have been developed that are more practical for use in homes.

Neon Lights

The vivid, glowing colors of neon lights, such as those shown in **Figure 16,** make them a popular choice for signs and eye-catching decorations on buildings. These lighting devices are glass tubes filled with gas, typically neon, and work similarly to fluorescent lights. When an electric current flows through the tube, electrons collide with the gas molecules. In this case, however, the collisions produce visible light. If the tube contains only neon, the light is bright red. Different colors can be produced by adding other gases to the tube.

Reading Check *What causes the color in a neon light?*

Discovering Energy Waste in Lightbulbs

Procedure

1. Obtain an **incandescent bulb** and a **fluorescent bulb** of identical wattage.
2. Make a heat collector by covering the top of a **foam cup** with a piece of **plastic food wrap** to make a window. Carefully make a small hole (diameter less than the thermometer's) in the side of the cup. Push a **thermometer** through the hole.
3. Measure the temperature of the air inside the cup. Then, hold the window of the tester 1 cm from one of the lights for 2 minutes and measure the temperature.
4. Cool the heat collector and thermometer. Repeat step 3 using the second bulb.

Analysis

1. What was the temperature for each bulb?
2. Which bulb appears to give off more heat? Explain why this occurs.

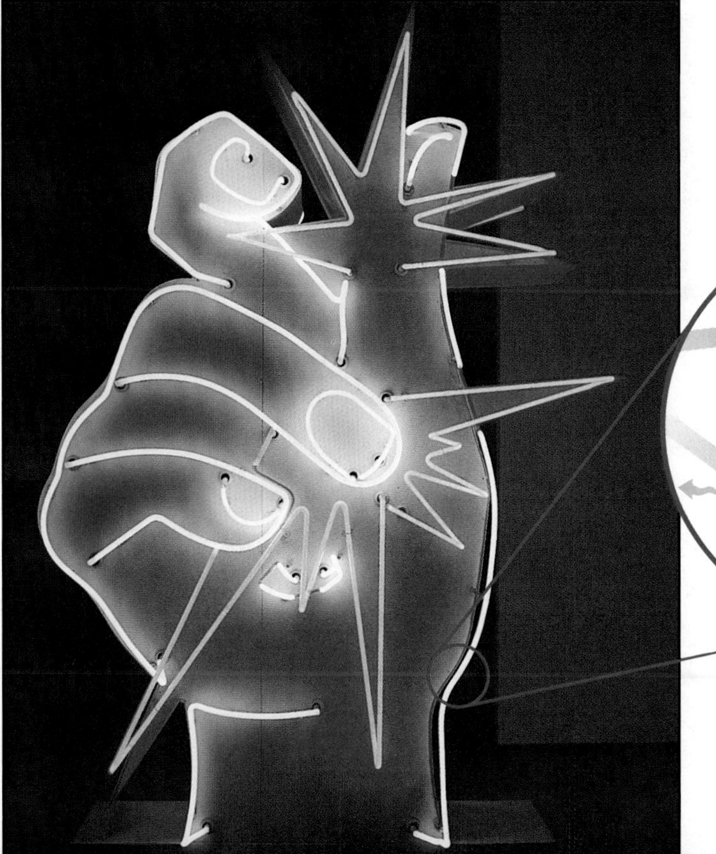

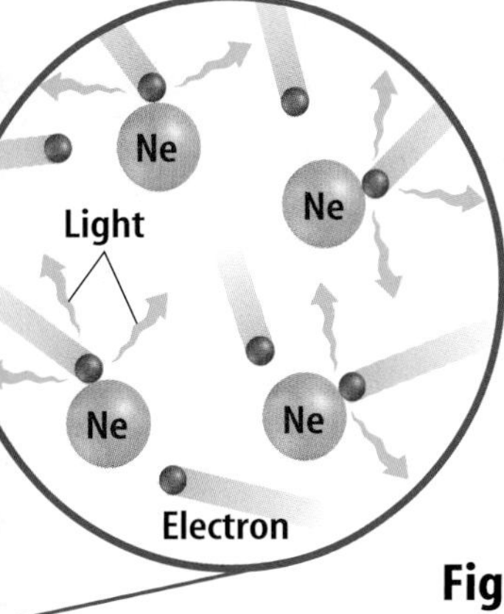

Figure 16
This neon light has vivid, glowing colors.

Sodium-Vapor Lights

Sodium-vapor lights often are used for streetlights and other outdoor lighting. Inside a sodium-vapor lamp is a tube that contains a mixture of neon gas, a small amount of argon gas, and a small amount of sodium metal. When the lamp is turned on, the gas mixture becomes hot. The hot gases cause the sodium metal to turn to vapor, and the hot sodium vapor emits a yellow-orange glow, as shown in **Figure 17.**

Figure 17
Sodium-vapor lights emit mostly yellow light. Half of this photo was taken under sunlight and half was taken under sodium-vapor lighting.

Tungsten-Halogen Lights

Tungsten-halogen lights sometimes are used to create intensely bright light. These lights have a tungsten filament inside a quartz bulb or tube. The tube is filled with a gas that contains one of the halogen elements such as fluorine or chlorine. The presence of this gas enables the filament to become much hotter than the filament in an ordinary incandescent bulb. As a result, the light is much brighter and also lasts longer. Tungsten-halogen lights sometimes are used on movie sets and in underwater photography.

Lasers

From laser surgery to a laser light show, lasers have become a large part of the world you live in. A laser's light begins when a number of light waves are emitted at the same time. To achieve this, a number of identical atoms each must be given the same amount of energy. When they release their energy, each atom sends off an identical light wave. This light wave is reflected between two facing mirrors at opposite ends of the laser. One of the mirrors is coated only partially with reflective material, so it reflects most light but allows some to get through. Some emitted light waves travel back and forth between the mirrors many times, stimulating other atoms to emit identical light waves also. **Figure 18** shows how this process produces a beam of laser light.

Research Visit the Glencoe Science Web site at **science.glencoe.com** for more information about sodium vapor lights and light pollution.

✔ Reading Check *How do mirrors help in creating lasers?*

Lasers can be made with many different materials, including gases, liquids, and solids. One of the most common is the helium-neon laser, which produces a beam of red light. A mixture of helium and neon gases sealed in a tube with mirrors at both ends is excited by a flashtube. The excited atoms then lose their excess energy by emitting coherent light waves.

Figure 18

Lasers produce light waves that have the same wavelength. Almost all of these waves travel in the same direction and are in phase. As a result, beams of laser light can be made more intense than ordinary light. In modern eye surgery, shown at the right, lasers are often used instead of a traditional scalpel.

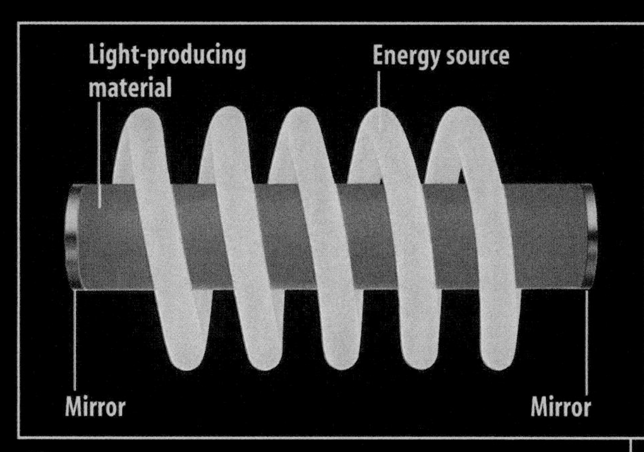

A The key parts of a laser include a material that can be stimulated to produce light, such as a ruby rod, and an energy source. In this example, the energy source is a lamp that spirals around the ruby rod and emits an intense light.

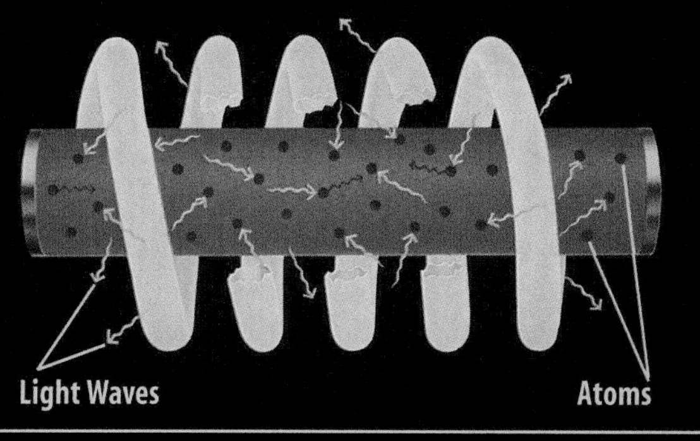

B When the lamp flashes, energy is absorbed by the atoms in the rod. These atoms then re-emit that energy as light waves that are in phase and have the same wavelength.

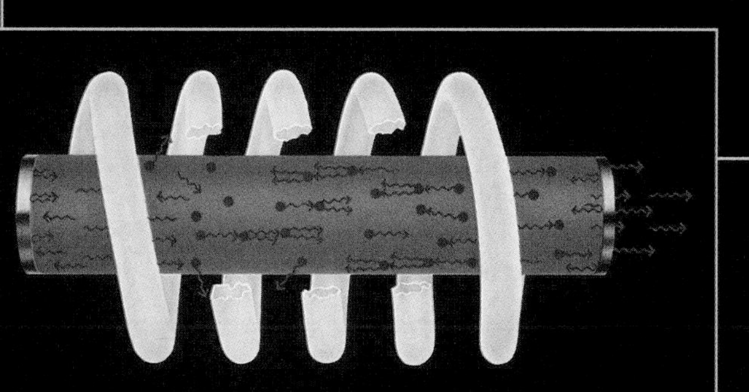

C Most of these waves are reflected between the mirrors located at each end of the laser. One of the mirrors, however, is only partially reflective, allowing one percent of the light waves to pass through it and form a beam.

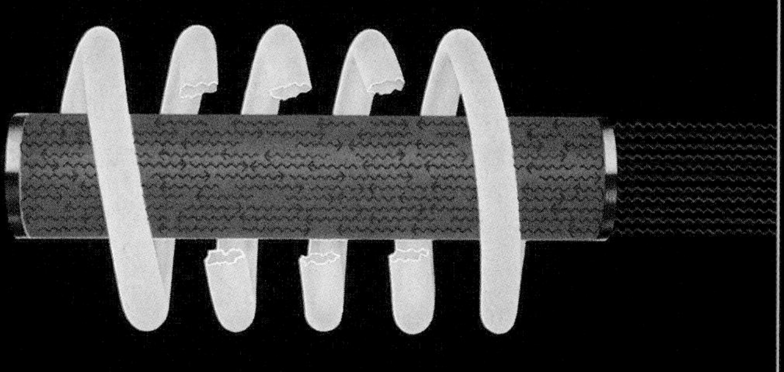

D As the waves travel back and forth between the mirrors, they stimulate other atoms in the ruby rod to emit light waves. In a fraction of a second, billions of identical waves are bouncing between the mirrors. The waves are emitted from the partially reflective mirror in a stream of laser light.

Figure 19
Light waves can be either coherent or incoherent.

A **These waves are coherent because they have the same wavelength and travel with their crests and troughs aligned.**

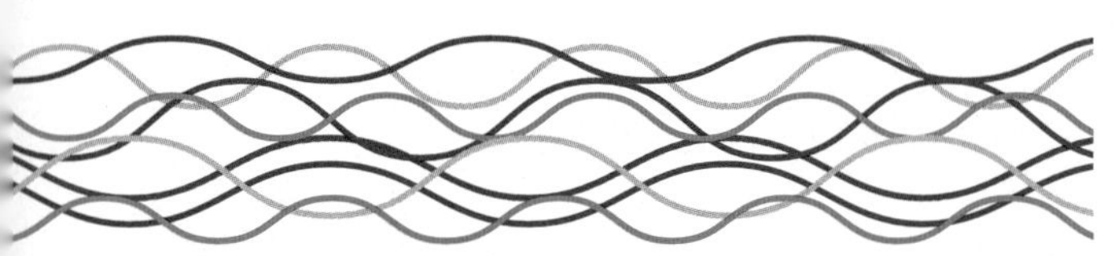

B **Incoherent waves such as these can contain more than one wavelength, and do not travel with their crests and troughs aligned.**

Coherent Light Lasers produce the narrow beams of light that zip across the stage and through the auditorium during some rock concerts. Beams of laser light do not spread out because laser light is coherent. **Coherent light** is light of only one wavelength that travels with its crests and troughs aligned. The beam does not spread out because all the waves travel in the same direction, as shown in **Figure 19A.** As a result, the energy carried by the beam remains concentrated over a small area.

Incoherent Light Light from an ordinary lightbulb is incoherent. **Incoherent light** can contain more than one wavelength, and its electromagnetic waves are not aligned, as in **Figure 19B.** The waves don't travel in the same direction, so the beam spreads out. The energy carried by the light waves is spread over a large area, so the intensity of the light is much less than that of the laser beam.

Using Lasers

Compact disc players, surgical tools, and many other useful devices take advantage of the unique properties of lasers. A laser beam is narrow and does not spread out as it travels over long distances. So lasers can apply large amounts of energy to small areas. In industry, powerful lasers are used for cutting and welding materials. Surveyors and builders use lasers for measuring and leveling. To measure the moon's orbit with great accuracy, scientists use laser light reflected from mirrors placed on the Moon's surface. Information also can be coded in pulses of light from lasers. This makes them useful for communications. In telephone systems, pulses of laser light transmit conversations through long glass fibers called optical fibers.

Lasers in Medicine Lasers are routinely used to remove cataracts, reshape the cornea, and repair the retina. In the eye and other parts of the body, surgeons can use lasers in place of scalpels to cut through body tissues. The energy from the laser seals off blood vessels in the incision and reduces bleeding. Because most lasers do not penetrate deeply through the skin, they can be used to remove small tumors or birthmarks on the surface without damaging deeper tissues. By sending laser light into the body through an optical fiber, physicians can also treat conditions such as blocked arteries.

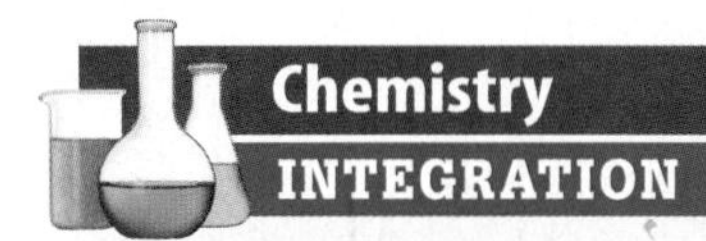

A particular helium-neon laser contains a mixture of 15 percent He and 85 percent Ne. Where are these gases located on the periodic table? Analyze their chemical characteristics. Would you be concerned that a chemical reaction might occur in the laser? Explain.

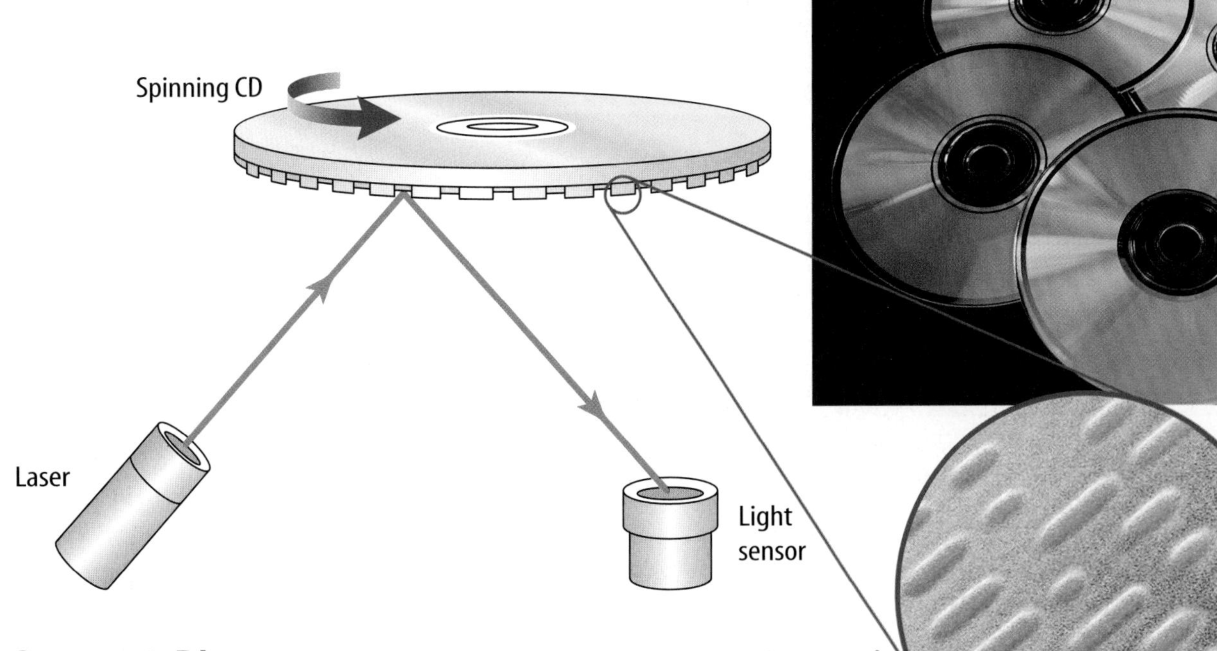

Figure 20
The blowup shows the pits (blue) on the bottom surface of a CD. A CD player uses a laser to convert the information on the CD to an electric signal.

Compact Discs Lasers play important roles in producing and using compact discs, which are plastic discs with reflective surfaces used to store sound, images, and text in digital form. When a CD is produced, the information is burned into the surface of the disc with a laser. The laser creates millions of tiny pits in a spiral pattern that starts at the center of the disc and moves out to the edge. A CD player, shown in **Figure 20,** also uses a laser to read the disc. The laser's beam is aimed at the surface of the disc. As the laser beam strikes a pit or flat spot, different amounts of light are reflected to a light sensor. The reflected light is converted to an electric signal used by the speakers to create sound.

Section 3 Assessment

1. Explain how light is produced in an ordinary incandescent bulb.
2. What are the advantages of using a fluorescent bulb instead of an incandescent bulb?
3. Describe the difference between coherent and incoherent light.
4. **Think Critically** Which type of lighting device would you use for each of the following needs: an economical light source in a manufacturing plant, an eye-catching sign that will be visible at night, and a baseball stadium? Explain.

Skill Builder Activities

5. **Concept Mapping** Make a concept map showing the sequence of events that occur in a laser to produce coherent light. Begin with the emission of lightwaves from atoms. **For more help, refer to the Science Skill Handbook.**
6. **Using an Electronic Spreadsheet** Create a spreadsheet listing the types of lighting devices described in this chapter. Compare and contrast their features in separate columns. Include a column for how each is used. **For more help, refer to the Technology Skill Handbook.**

SECTION

Using Light

As You Read

What You'll Learn

- **Describe** polarized light and the uses of polarizing filters.
- **Apply** the concept of total internal reflection to the uses of optical fibers.

Vocabulary
polarized light
holography
total internal reflection

Why It's Important
Light is used in entertainment, medicine, manufacturing, scientific research, communications, and just about every other facet of life.

Polarized Light

You may have a pair of sunglasses with a sticker on them that says polarized. Do you know what makes them different from other sunglasses? The difference has to do with the vibration of light waves that pass through the lenses. You can make transverse waves in a rope vibrate in any direction—horizontal, vertical, or anywhere in between. Light also is a transverse wave and can vibrate in any direction. In **polarized light,** however, the waves vibrate in only one direction.

Polarizing Filters If the light passes through a special polarizing filter, the light becomes polarized. A polarizing filter acts like a group of parallel slits. Only light waves vibrating in the same direction as the slits can pass through. If a second polarizing filter is lined up with its slits at right angles to those of the first filter, no light can pass through, as **Figure 21** shows.

Polarized lenses are useful for reducing glare without interfering with your ability to see clearly. When light is reflected from a horizontal surface, such as a lake or a shiny car hood, it becomes partially horizontally polarized. The lenses of polarizing sunglasses have vertically polarizing filters that block out the reflected light that has been polarized horizontally, while allowing vertically polarized light through.

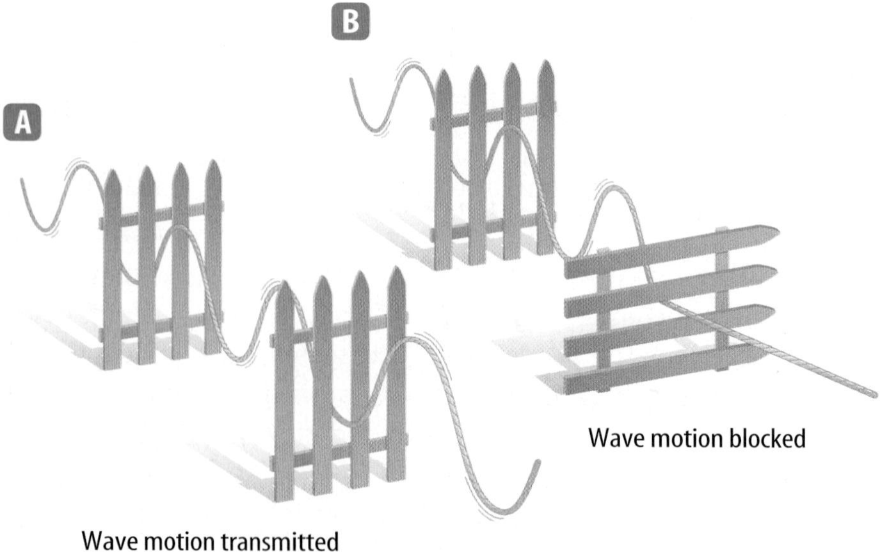

Figure 21
Slats in a fence behave like a polarizing filter for a transverse wave on a rope. A If the slats in the fence are in the same direction, the wave passes through. B If the slats are aligned at right angles to each other, the wave can't pass through.

Figure 22
Lasers can be used to make holograms like this one.

Holography

Science museums often have exhibits where a three-dimensional image seems to float in space, like the one shown in **Figure 22.** You can see the image from different angles, just as you would if you passed the real object. Three-dimensional images on credit cards are produced by holography. **Holography** (hoh LAH gruh fee) is a technique that produces a hologram—a complete photographic image of a three-dimensional object.

Making Holograms Illuminating objects with laser light produces holograms. Laser light reflects from the object onto photographic film. At the same time, a second beam split from the laser also is directed at the film. The light from the two beams creates an interference pattern on the film. The pattern looks nothing like the original object, but when laser light shines on the pattern on the film, a holographic image is produced.

Information in Light An ordinary photographic image captures only the brightness or intensity of light reflected from an object's surface, but a hologram records the intensity as well as the direction. As a result, it conveys more information to your eye than a conventional two-dimensional photograph does, but it also is more difficult to copy. Holographic images are used on credit cards, identification cards, and on the labels of some products to help prevent counterfeiting. Using X-ray lasers, scientists can produce holographic images of microscopic objects. It may be possible to create three-dimensional views of biological cells.

How are holographic images produced?

Research Visit the Glencoe Science Web site at **science.glencoe.com** for more information about holograms. Communicate to your class what you learned.

Optical Fibers

When laser light must travel long distances or be sent into hard-to-reach places, optical fibers often are used. These transparent glass fibers can transmit light from one place to another. A process called total internal reflection makes this possible.

Total Internal Reflection Remember what happens when light speeds up as it travels from one medium to another. For example, when light travels from water to air the direction of the light ray is bent away from the normal, as shown in **Figure 23.** If the underwater light ray makes a larger angle with the normal, the light ray in the air bends closer the surface of the water. At a certain angle, called the critical angle, the refracted ray has been bent so that it is traveling along the surface of the water, as shown in **Figure 23.** For a light ray traveling from water into air, the critical angle is about 49°.

Figure 23 shows what happens if the underwater light ray strikes the boundary between the air and water at an angle larger than the critical angle. There is no longer any refraction, and the light ray does not travel in the air. Instead, the light ray is reflected at the boundary, just as if a mirror were there. This behavior of light is called total internal reflection. **Total internal reflection** occurs when light traveling from one medium to another is completely reflected at the boundary between the two materials. Then the light ray obeys the law of reflection. For total internal reflection to occur, light must travel slower in the first medium, and must strike the boundary at an angle greater than the critical angle.

Reading Check *How does total internal reflection occur?*

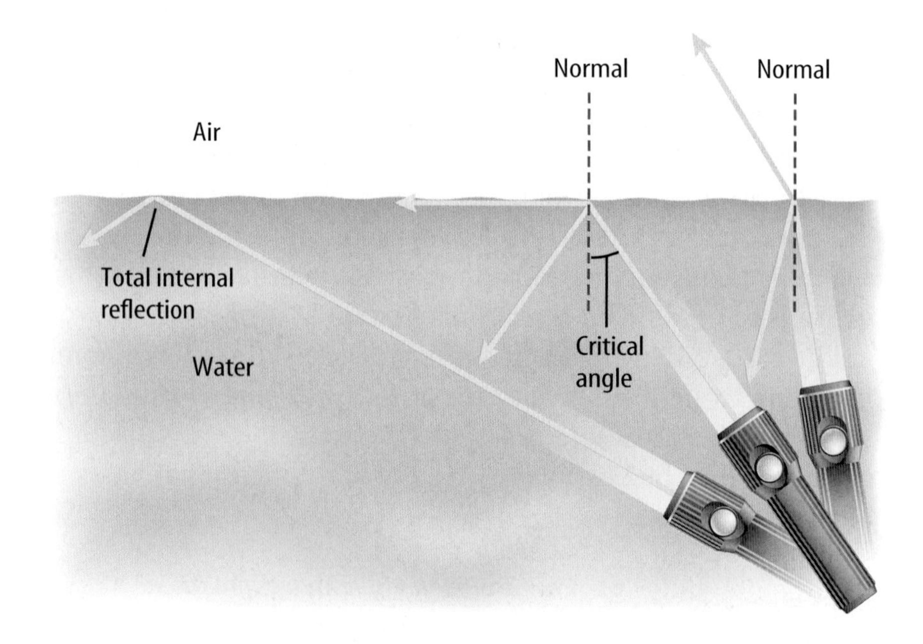

Figure 23
A light wave is bent away from the normal as it passes from water to air. At the critical angle, the refracted wave is traveling along the water surface. At angles greater than the critical angle, total internal reflection occurs.

Figure 24
Optical fibers make use of total internal reflection.

Light

Plastic covering

Glass core

Glass layer

Light rays

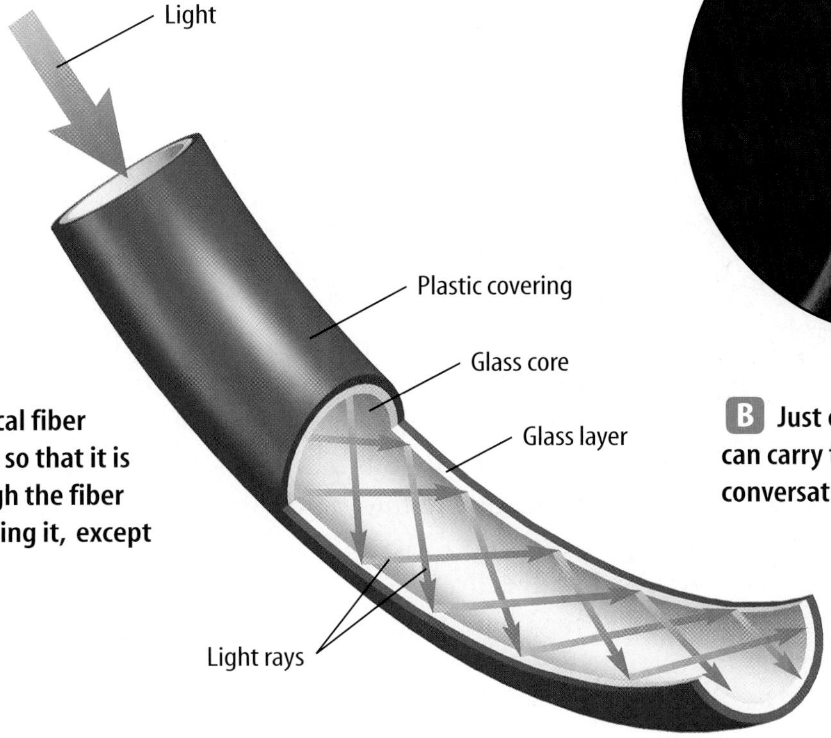

A An optical fiber reflects light so that it is piped through the fiber without leaving it, except at the ends.

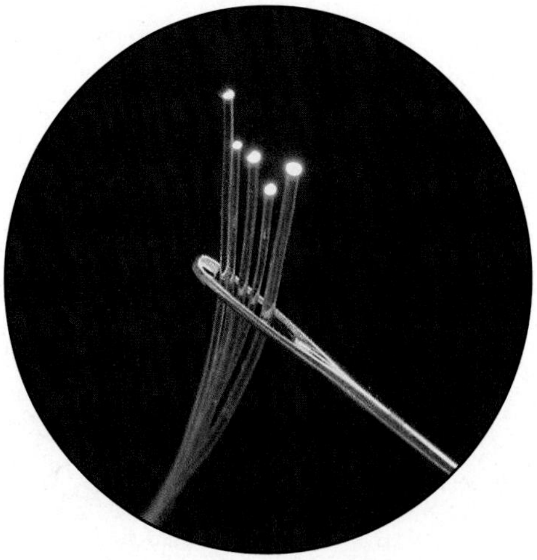

B Just one of these optical fibers can carry thousands of phone conversations at the same time.

Light Pipes Total internal reflection makes light transmission in optical fibers possible. As shown in **Figure 24A,** light entering one end of the fiber is reflected continuously from the sides of the fiber until it emerges from the other end. Like water moves through a pipe, almost no light is lost or absorbed in optical fibers.

Using Optical Fibers Optical fibers are most often used in communications. Telephone conversations, television programs, and computer data can be coded in light beams. Signals can't leak from one fiber to another and interfere with other messages, so the signal is transmitted clearly. To send telephone conversations through an optical fiber, sound is converted into digital signals consisting of pulses of light by a light-emitting diode or a laser. Some systems use multiple lasers, each with its own wavelength to fit multiple signals into the same fiber. You could send a million copies of the play *Romeo and Juliet* in one second on a single fiber. **Figure 24B** shows the size of typical optical fibers.

Optical fibers also are used to explore the inside of the human body. One bundle of fibers transmits light, while another carries the reflected light back to the doctor.

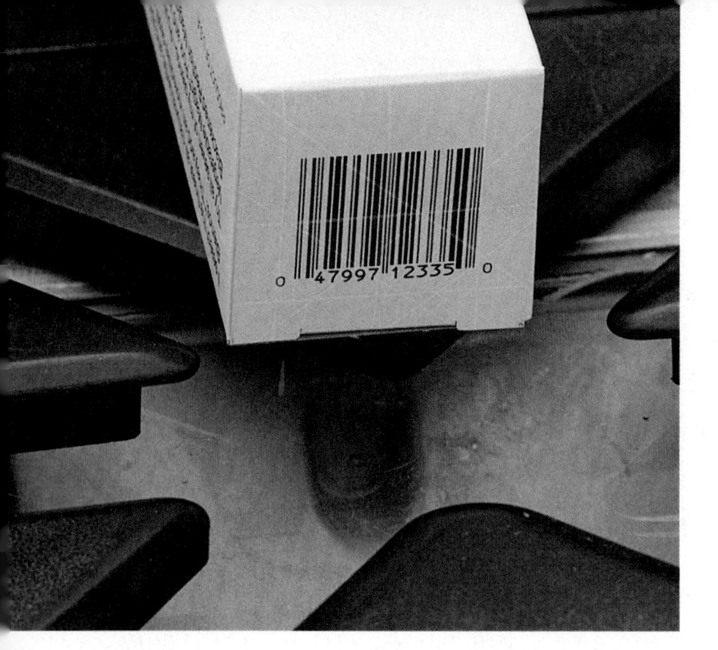

Figure 25
Optical scanners like this one use lasers to find the price of various products.

Optical Scanners

In supermarkets and many other kinds of stores, a cashier passes your purchases over a glass window in the checkout counter or holds a handheld device up to each item, like the one in **Figure 25.** In an instant, the optical scanner beeps and the price of the item appears on a screen. An optical scanner is a device that reads intensities of reflected light and converts the information to digital signals. You may have noticed that somewhere on each item the cashier scans is a pattern of thick and thin stripes called a bar code. An optical scanner detects the pattern and translates it into a digital signal, which goes to a computer. The computer searches its database for a matching item, finds its price, and sends the information to the cash register. The U.S. Postal Service also uses optical scanners to sort mail and keep track of mail delivery.

You may have used another type of optical scanner to convert pictures or text into forms you can use in computer programs. With a flatbed scanner, for example, you lay a document or picture facedown on a sheet of glass and close the cover. An optical scanner passes underneath the glass and reads the pattern of light and dark areas (or colors, if you are scanning a color picture). Some scanners are used with software that can read text on a page and convert the scanned information into a text file that you can work on. Photocopy machines and facsimile (fax) machines also use optical scanners.

Section  Assessment

1. What is polarized light?
2. Why is a holographic image three-dimensional?
3. What conditions are necessary for total internal reflection to occur?
4. Explain how an optical fiber is able to transmit light.
5. **Think Critically** Geologists and surveyors often use lasers for aligning equipment, measuring, and mapping. Explain why.

Skill Builder Activities

6. **Comparing and Contrasting** Make a table that compares and contrasts incoherent, coherent, and polarized light. **For more help, refer to the Science Skill Handbook.**
7. **Communicating** Many people wear polarized sunglasses while they are working. Write a list of jobs or occupations in which wearing polarized sunglasses is helpful. Explain why. **For more help, refer to the Science Skill Handbook.**

Activity

Make a Light Bender

From a hilltop you can see the reflection of pine trees and a cabin in the calm surface of a lake. This is possible because some of the light that reflects off these objects strikes the water's surface and reflects into your eyes. However, you don't see a clear, colorful image because much of the light enters the water rather than being reflected.

What You'll Investigate

How does water affect the viewer's image of an object that is above the water's surface?

Materials

light source
unsharpened pencil
clear rectangular container
water
clay

Goals

- Identify reflection of an image in water.
- Identify refraction of an image in water.

Safety Precautions

Procedure

1. Fill the container with water.
2. Place the container so that a light source—window or overhead light—reaches it.
3. Stand the pencil on end in the clay and place it by the container as shown in the figure above. The pencil must be taller than the level of the water. Also, place the pencil on the same side of the container as the light source.

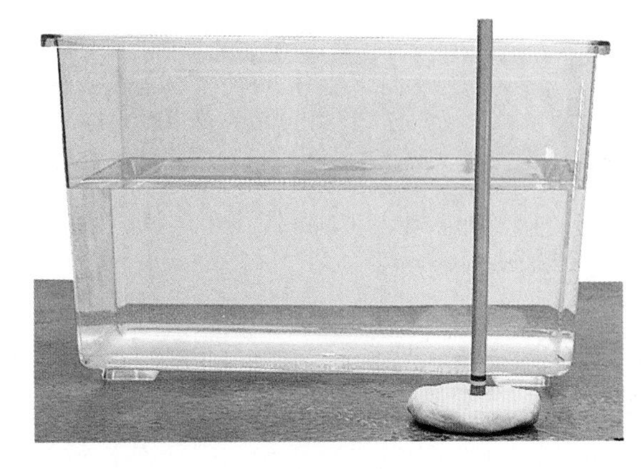

4. Looking down through the surface of the water from the side opposite the pencil, observe the reflection and refraction of the image of the pencil.
5. **Draw** a diagram of the image and label "reflection" and "refraction."
6. Repeat steps 4 and 5 two more times but position the pencil at two different angles.

Conclude and Apply

1. How would the image you see change or be different if the surface of the water were a mirror?
2. **Predict** how the angles of reflection or refraction would change if the surface of the container were curved. Explain.

Communicating Your Data

Make a poster of your diagrams and use it to explain reflection and refraction of light waves to your class. **For more help, refer to your Science Skill Handbook.**

Activity Design Your Own Experiment

Polarizing Filters

Polarizing filters cause light waves to vibrate only in one direction. Wearing polarized sunglasses can help reduce glare while allowing you to see clearly. If you have two polarizing filters on top of one another, when will light shine through and when will it not? What might happen if you added a third filter in between the first two?

Recognize the Problem

How can you demonstrate the effects of polarizing filters?

Form a Hypothesis

Form a hypothesis about how two polarizing filters that are placed on top of one another must be oriented for light to shine through and for no light to shine through.

Goals

- **Demonstrate** when light does and does not shine through a pair of polarizing filters.
- **Predict** what will happen when you add a third polarizing filter.

Possible Materials

polarizing filters (3)
lamp or flashlight

Safety Precautions

Never look directly at the sun, even with a polarizing filter.

Test your Hypothesis

Do

1. Using a pair of polarizing filters, choose at least three orientations of the filters to test your hypothesis.
2. When the two filters are oriented to allow the maximum amount of light to shine through, predict how a third filter placed between the two must be oriented for the same results.
3. Repeat step 2 but allow no light to shine through.
4. Make sure that your teacher approves your plan before you start.
5. Using an appropriate light source, test when light does and does not shine through a pair of polarizing filters. Test each of the orientations you planned in step 1. Record the results.
6. Test three orientations of the third filter for allowing the maximum amount of light to pass through. Record the results.
7. Repeat step 6 but allow no light to shine through. Record the results.

Analyze Your Data

1. **Describe** how the pair of polarizing filters were oriented when light did and did not shine through. In cases where light did shine through, was it always the same amount of light? Or did the amount of light change in different orientations?
2. In each case, describe what happened when you added a third filter between the two. How did the three orientations of the third filter change the amount of light that passed through? Explain.

Draw Conclusions

1. **Explain** why light did or did not shine through two polarizing filters in the various orientations.
2. Was your hypothesis supported? Why or why not?
3. **Explain** why light did or did not shine through various orientations of three polarizing filters.
4. Were your predictions correct? Why or why not?
5. If a polarizing filter reduces the brightness of light reflected from the surface of a lake, what can you conclude about the polarization of reflected light?

Communicating Your Data

The next time you see a family member or friend wearing sunglasses, **explain** to them how polarizing lenses can reduce problems of glare.

Science and Language Arts

A Haiku Garden:
The Four Seasons in Poems and Prints
by Stephen Addiss with Fumiko and Akira Yamamoto

Respond to the Reading

1. What do the words *one-color world* mean?
2. How do the illustrations help the reader better understand the poems?
3. What do you think is meant by the word *lingering* in the Haiku about spring sunlight?

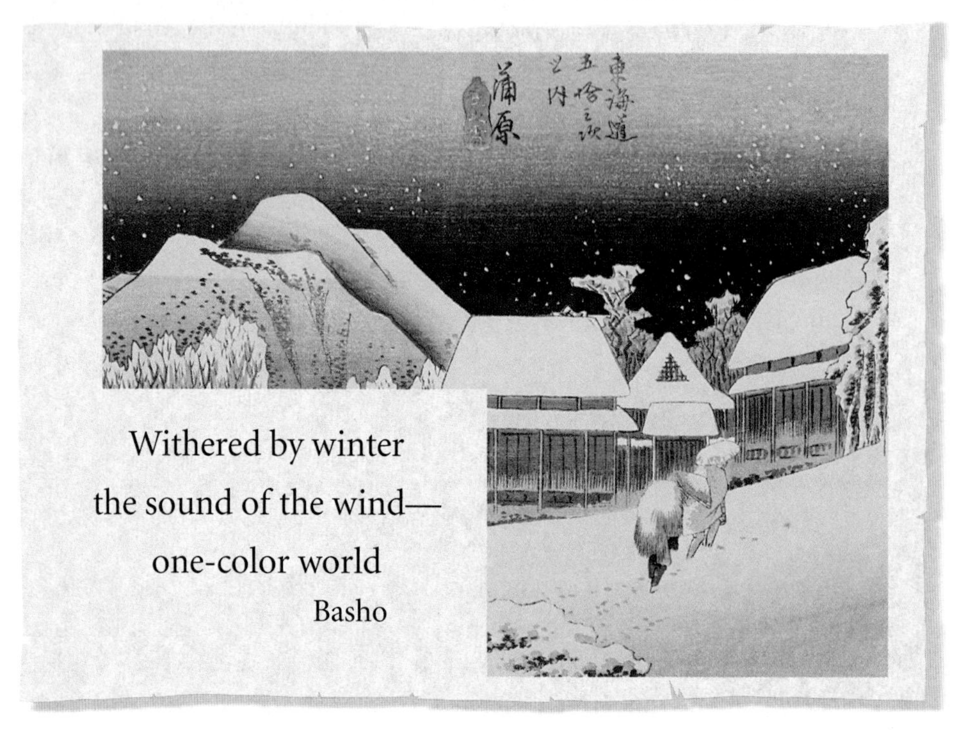

Withered by winter
the sound of the wind—
one-color world
Basho

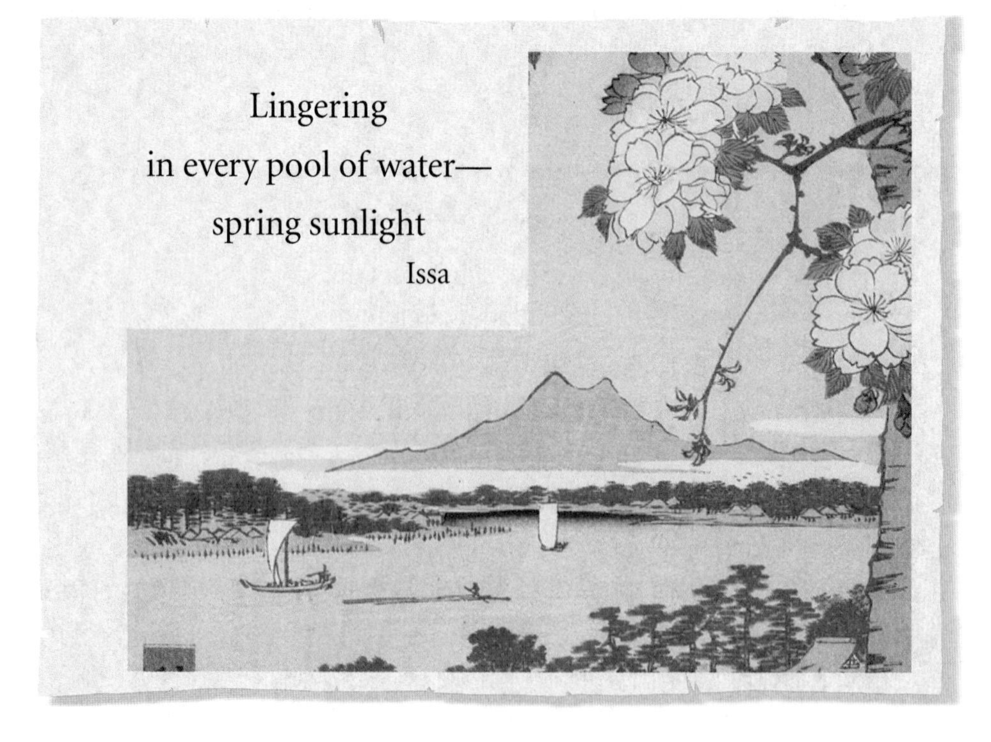

Lingering
in every pool of water—
spring sunlight
Issa

Understanding Literature

Japanese Haiku You have just read English translations of two poems from a book that combines Japanese haiku and art. A haiku is a verse that consists of three lines and 17 syllables in the Japanese language. The first and third lines have five syllables each, and the middle line has seven syllables. Using few words, a haiku allows the reader's imagination to complete the picture. For instance, the *sound of the wind* might make you think of the sound of winter wind whipping around the house while you sit snuggly inside.

Science Connection Research has determined that there is a connection between color and mood. For example, the color of a room can affect a person's feelings and behavior. Warm colors have longer wavelengths, and can be more stimulating. Cool colors, which have shorter wavelengths, tend to have a calming or soothing effect on people. Light and color have long been used as literary symbols. When the haiku is combined with Japanese prints, do you read it differently? Does the use of color change what you imagine when you read the haiku?

Linking Science and Writing

Writing and Illustrating Haiku Write two haiku poems—one about summer and one about fall. In one poem, use color to help you describe the season. In the other, use light or some property of light to help describe the season. When you have finished, exchange poems with one of your classmates. Read your classmate's haiku and create illustrations to accompany them.

Career Connection

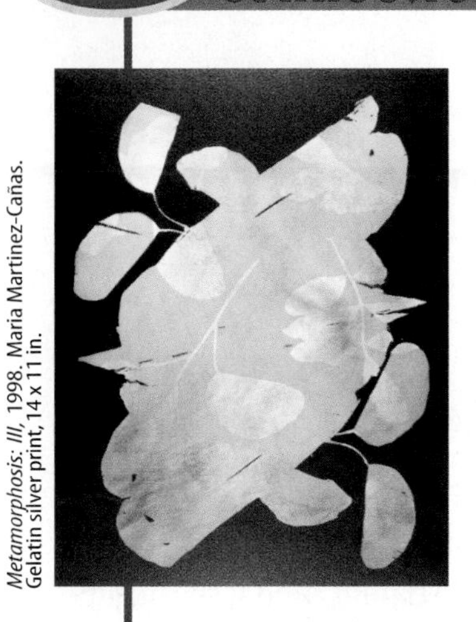

Metamorphosis: III, 1998, Maria Martinez-Cañas. Gelatin silver print, 14 x 11 in.

Photographer

Maria Martinez-Cañas uses light in an inventive and exciting way. Martinez-Cañas was born in Havana, Cuba and grew up in Puerto Rico. She went to art school in the United States, earning a master of fine arts degree from the School of Art Institute in Chicago. Her innovative technique involves using a type of photographic material that blocks light. She cuts the material into artistic shapes, then prints these shapes on white paper in a manner similar to using stencils. Martinez-Cañas lives in Miami and her photography can be seen in art exhibits all over the world.

SCIENCE*Online* To learn more about careers in photography, visit the Glencoe Science Web site at **science.glencoe.com**.

Chapter 14 Study Guide

Reviewing Main Ideas

Section 1 The Behavior of Light

1. You can't see through opaque materials. You can see hazily through translucent materials and clearly through transparent materials.
2. Light behaves as a wave, so it can be reflected. *Is this reflection from the lake regular or diffuse?*

3. Light waves can be refracted, or bent, when a wave changes speed as it travels from one material to another.

Section 2 Light and Color

1. You see color when light is reflected off objects and into your eyes.
2. Specialized cells in your eyes called cones allow you to distinguish colors and shapes of objects. Other cells, called rods, allow you to see in dim light.
3. The three primary colors of light can be mixed to form all other colors.
4. The colors of pigments are determined by the colors they reflect.

Section 3 Producing Light

1. Incandescent bulbs produce light by heating a tungsten filament until it glows brightly.
2. Fluorescent bulbs give off light when ultraviolet radiation produced inside the bulb causes the phosphor coating inside the bulb to glow.
3. Neon lights contain a gas that glows when electric current passes through it.
4. A laser produces coherent light by emitting a beam of light waves that have only one wavelength, with their crests and troughs aligned, and moving in a single direction.

Section 4 Using Light

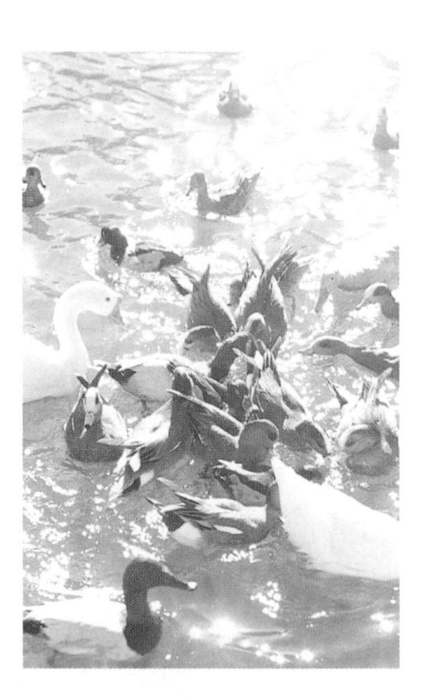

1. Polarized light consists of transverse waves that vibrate along only one plane. *What could be done to reduce the glare in this photo?*
2. Total internal reflection occurs when a light wave strikes the boundary between two materials at an angle greater than the critical angle.
3. Optical scanners sense reflected light and convert the information to digital signals.

After You Read

FOLDABLES Reading & Study Skills

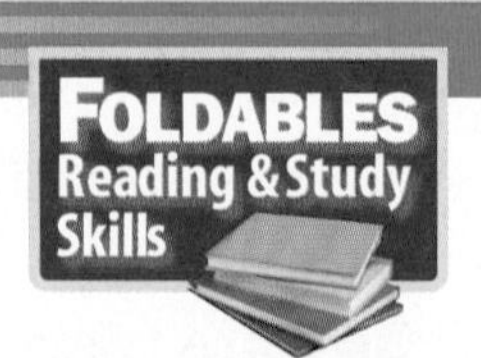

List examples of common materials that are opaque, transparent, and translucent on the front of your Foldable.

Chapter 14 Study Guide

Visualizing Main Ideas

Complete the following concept map about light.

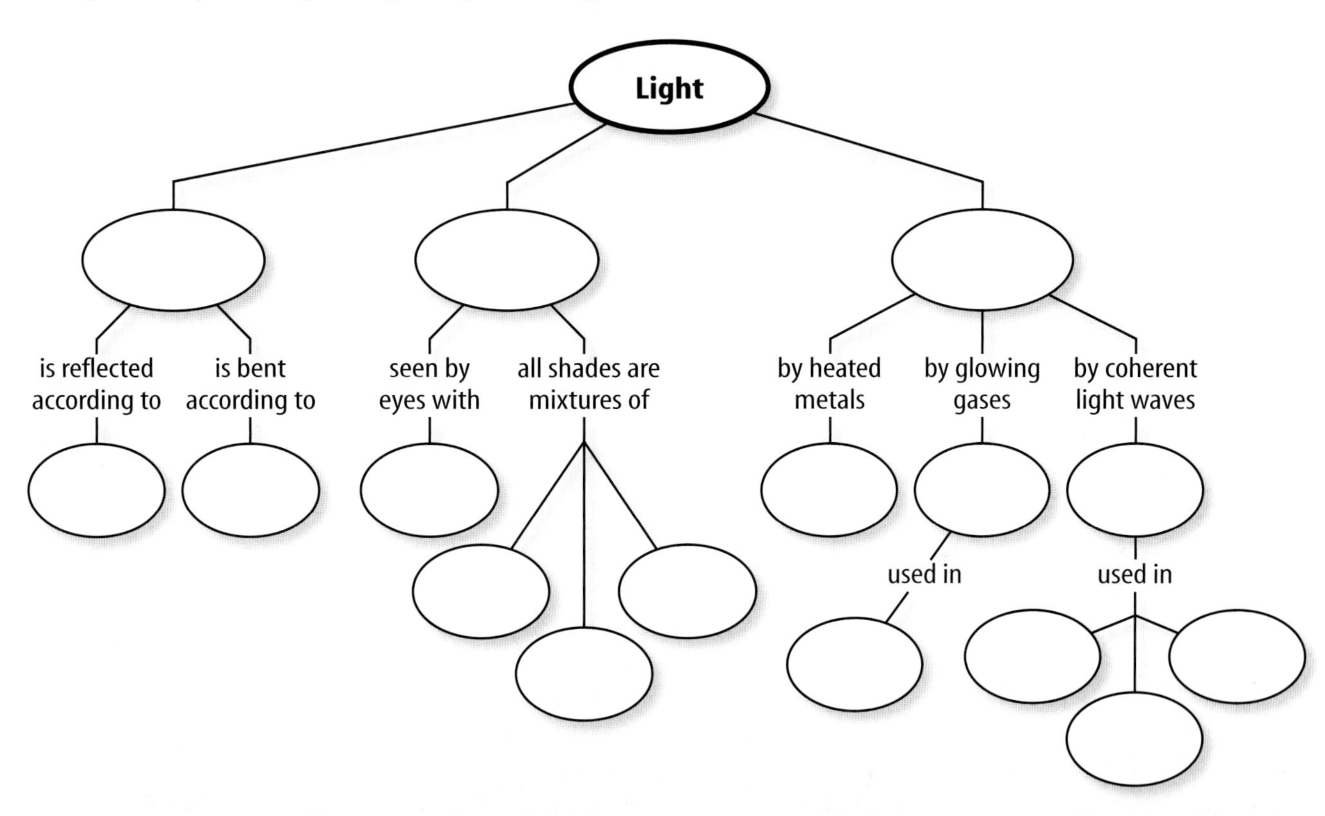

Checking Concepts

Vocabulary Words

a. coherent light
b. fluorescent light
c. holography
d. incandescent light
e. incoherent light
f. index of refraction
g. mirage
h. opaque
i. pigment
j. polarized light
k. total internal reflection
l. translucent
m. transparent

Study Tip

Get together with a friend. Quiz each other from textbook and class material.

Using Vocabulary

Answer the following questions using complete sentences.

1. What type of light does heating a filament until it glows produce?
2. What process would you use to produce a complete three-dimensional image of an object?
3. How would you describe an object that you can see through?
4. What process makes it possible for optical fibers to transmit telephone conversations over long distances?
5. What is a false image of a distant object?

Chapter 14 Assessment

Checking Concepts

Choose the word or phrase that best answers the question.

1. Which word describes materials that absorb or reflect all light?
   A) translucent **C)** ultraviolet
   B) opaque **D)** diffuse

2. What is the term for the property of a material that indicates how much light slows down when traveling in the material?
   A) pigment **C)** index of refraction
   B) filter **D)** mirage

3. Which of the following explains why a prism separates white light into the colors of the rainbow?
   A) interference **C)** diffraction
   B) fluorescence **D)** refraction

4. What do you see when noting the color of an object?
   A) the light it reflects
   B) the light it absorbs
   C) polarization
   D) diffuse reflection

5. What do the phosphors inside fluorescent bulbs absorb to create a glow?
   A) incandescent light
   B) ultraviolet radiation
   C) halogens
   D) argon

6. What term describes objects that allow some light, but not a clear image to pass through?
   A) translucent **C)** transparent
   B) reflective **D)** opaque

7. Which light waves are bent most when passing through a prism?
   A) red waves **C)** blue waves
   B) yellow waves **D)** violet waves

8. Which type of cells in your eyes allows you to see the color violet?
   A) red cones **C)** blue cones
   B) green cones **D)** rods

9. What color of light is produced when the three primary colors of light are combined in equal amounts?
   A) black **C)** white
   B) yellow **D)** cyan

10. Which term describes laser light?
    A) incoherent **C)** incandescent
    B) coherent **D)** fluorescent

Thinking Critically

11. Explain how light is produced by an incandescent bulb. What is a disadvantage of these bulbs?

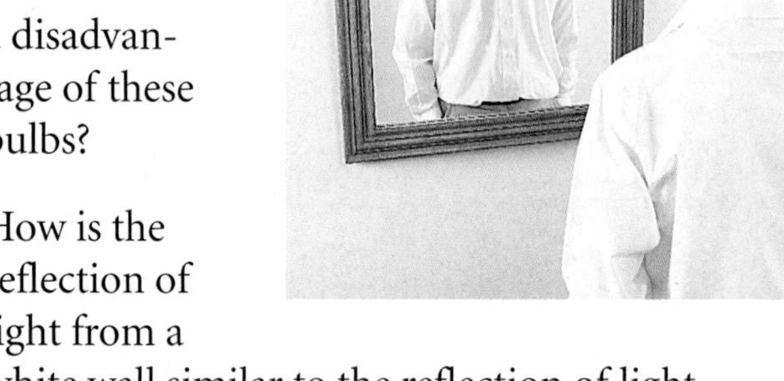

12. How is the reflection of light from a white wall similar to the reflection of light from a mirror? How is it different?

13. What color would a blue shirt appear to be if a blue filter were placed in front of it? A red filter? A green filter?

14. Which color of light changes speed the most when it passes through a prism? Explain.

Developing Skills

15. **Drawing Conclusions** Most mammals, such as dogs and cats, can't see colors. Infer how a cat's eye might be different from your eye.

16. Concept Mapping Use this blank concept map to show the steps in the production of fluorescent light.

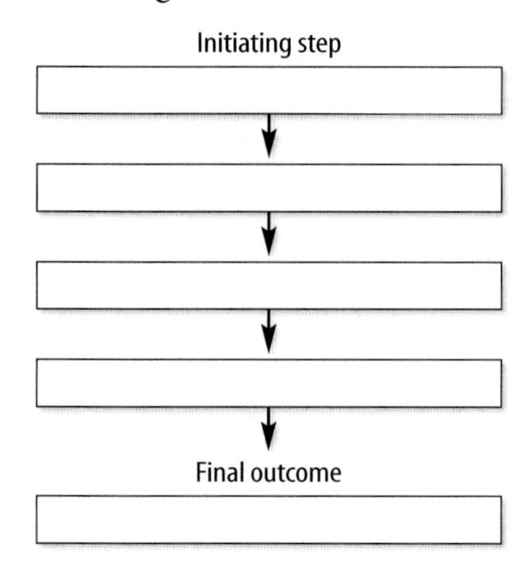

17. Interpreting Scientific Illustrations Make a drawing that shows how a fluorescent bulb produces light.

18. Drawing Conclusions Why is the inside of a camera painted black?

19. Making and Using Tables Construct a table to show the properties and applications of incandescent, fluorescent, neon, sodium-vapor, and tungsten-halogen lighting devices. For each type of device, include information on how light is produced, typical uses, and its advantages or disadvantages.

Performance Assessment

20. Poster Make a poster to show how the three primary pigments are combined to produce common colors such as blue, red, yellow, green, purple, brown, and black.

Technology

Go to the Glencoe Science Web site at **science.glencoe.com** or use the **Glencoe Science CD-ROM** for additional chapter assessment.

Test Practice

Judy and Markus are studying light. They have just read about an experiment that shows how light has certain properties and follows certain rules.

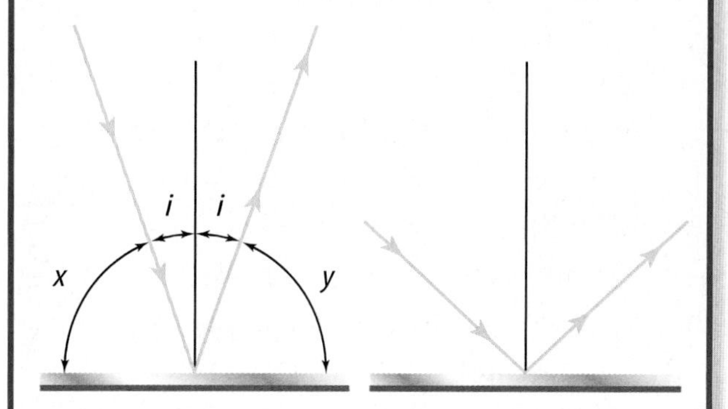

Study the pictures above and answer the following questions.

1. Angle *i* measures 20°. Using this information, what is angle *x*?
A) 50°
B) 60°
C) 70°
D) 80°

2. Which of the following is taking place in both of the experiments above?
F) refraction
G) reflection
H) diffusion
J) fluorescence

3. Which of the following statements is true about reflection?
A) $\angle i + \angle x = 100°$
B) $\angle i + \angle x = 60°$
C) $\angle i + \angle x = 80°$
D) $\angle i + \angle x = 90°$

Mirrors and Lenses

The dark-tinted glass panes of this office building act as mirrors, reflecting an image of the Iowa State Capitol Building. Why is the image distorted? Depending on whether it is flat or curved, a mirror can reflect an object's true size and shape, magnify the object, or reduce the object. In this chapter, you'll learn about mirrors and lenses and how they are used in cameras, telescopes, microscopes, and your own eyes. You also will learn how the shape of a lens determines the image you see.

What do you think?

Science Journal Look at the picture below with a classmate. Discuss what this might be or what is happening. Here's a hint: *If you could see like this, you would have eyes in the back of your head.* Write your answer or best guess in your Science Journal.

Have you ever used a magnifying glass, a camera, a microscope, or a telescope? If so, you were using a lens to create an image. A lens is a transparent material that bends rays of light and forms an image. In this activity, you will use water to create a lens.

Observe objects through a drop of water

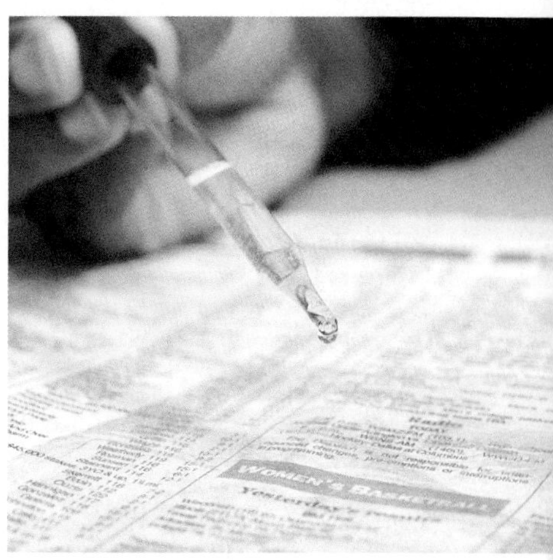

1. Cut a 10-cm × 10-cm piece of plastic wrap. Set it on a page of printed text.
2. Place a small water drop on the plastic. Look at the text through the drop. What do you observe?
3. Make your water drop larger and observe the text through it again.
4. Carefully lift the piece of plastic wrap a few centimeters above the text and look at the text through the water drop again.

Observe

In your Science Journal, describe how the text looked in steps 2, 3, and 4. Why do you think water affects the way the text looks? What other materials might you use to magnify the text?

Before You Read

Making a Cause and Effect Study Fold **Make the following Foldable to help you understand the cause and effect relationship of reflections.**

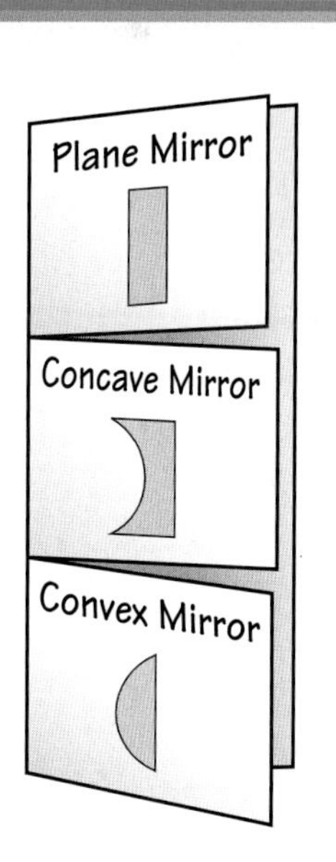

1. Place a sheet of paper in front of you so the short side is at the top. Fold the paper in half from the left side to the right side.
2. Fold in the top and the bottom. Unfold the paper so three sections show.
3. Through the top thickness of paper, cut along each of the fold lines to the left fold, forming three tabs.
4. Label the tabs *Plane Mirror*, *Concave Mirror*, and *Convex Mirror*. Draw examples of the three types of mirrors.
5. As you read the chapter, write under the tabs how each mirror works and how each reflects light differently.

Mirrors

As You Read

What You'll Learn

- **Describe** how an image is formed in three types of mirrors.
- **Explain** the difference between real and virtual images.
- **Identify** examples and uses of plane, concave, and convex mirrors.

Vocabulary

plane mirror
virtual image
concave mirror
optical axis
focal point
focal length
real image
convex mirror

Why It's Important

Mirrors allow you to check your appearance, use flashlights, and see your surroundings.

How do you use light to see?

Have you tried to read a book under the covers with only a small flashlight? Or have you ever tried to find an address number on a house or an apartment at night on a poorly lit street? It's harder to do those activities in the dark than it is when there is plenty of light. Your eyes see by detecting light, so anytime you see something, it is because light has come from that object to your eyes. Light can start from a light source, such as the Sun or a lightbulb, or it can reflect off of an object, such as the page of a book or someone's face. Either way, when light travels from an object to your eye, you see the object. Light can reflect more than once. For example, light can reflect off of an object into a mirror and then reflect into your eyes. When no light is available to reflect off of objects and into your eye, your eyes cannot see anything. This is why it is hard to read a book or see an address in the dark.

Light Rays Light sources send out light waves that travel in all directions. These waves spread out from the light source just as ripples on the surface of water spread out from the point of impact of a pebble.

You also could think of the light coming from the source as being many narrow beams of light. Each narrow beam of light travels in a straight line and is called a light ray. **Figure 1** shows how a light source, such as a candle, gives off light rays that travel away from the source in all directions. Even though light rays can change direction when they are reflected or refracted, your brain interprets images as if light rays travel in a single direction.

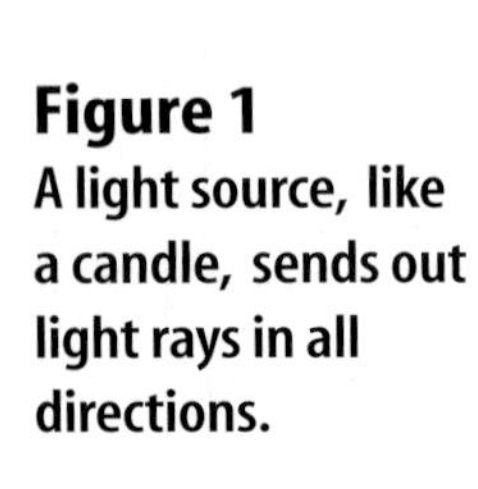

Figure 1
A light source, like a candle, sends out light rays in all directions.

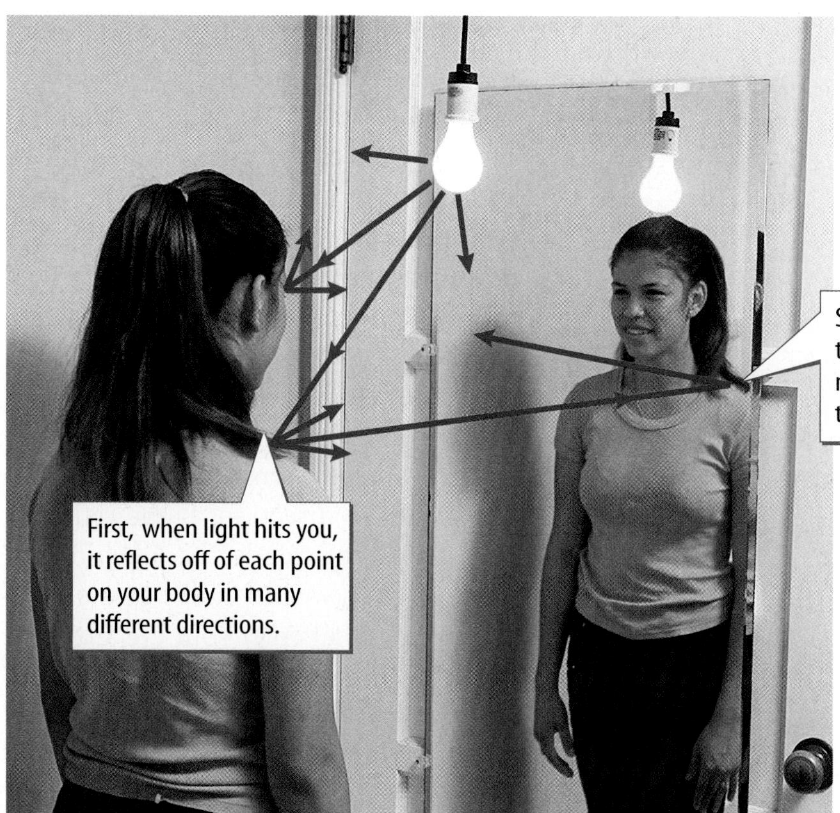

Figure 2
Seeing an image of yourself in a mirror involves two sets of reflections.

Seeing with Plane Mirrors

Greek mythology tells the story of a handsome young man named Narcissus who noticed his image in a pond and fell in love with himself. Like pools of water, mirrors are smooth surfaces that reflect light to form images. Just as Narcissus did, you can see yourself as you glance into a quiet pool of water or walk past a shop window. Most of the time, however, you probably look for your image in a flat, smooth mirror called a **plane mirror**.

What is a plane mirror?

Reflection from Plane Mirrors What do you see when you look into a plane mirror? Your reflection appears upright. If you were one meter from the mirror, your image would appear to be 1 m behind the mirror, or 2 m from you. In fact, your image is what someone standing 2 m from you would see. **Figure 2** shows how your image is formed by a plane mirror. First, light rays from a light source strike you. Every point that is struck by the light rays reflects these rays so they travel outward in all directions. If your friend were looking at you, these reflected light rays coming from you would enter her eyes so she could see you. However, if a mirror is placed between you and your friend, the light rays are reflected back to your eyes.

Life Science INTEGRATION

Your left hand and right hand are mirror images of each other. Some of the molecules in your body exist in two forms that are mirror images. However, your body uses some molecules only in the left-handed form and other molecules only in the right-handed form. Using different colors of gumdrops and toothpicks, make a model of a molecule that has a mirror image.

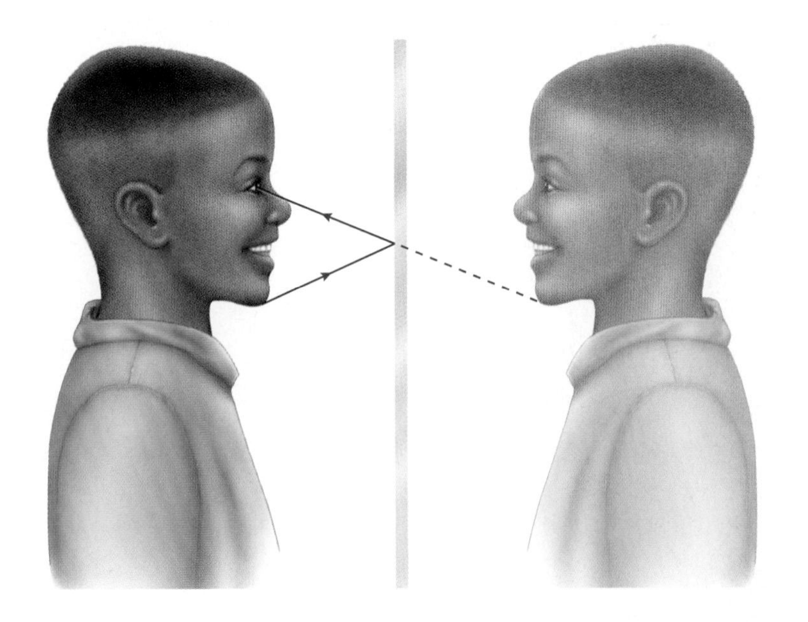

Figure 3
Your brain thinks that the light rays that reflect off of the mirror come from a point behind the mirror. *Why does your reflection seem to be as far behind the mirror as you are in front of it?*

Virtual Images You can understand how your brain interprets your reflection in a mirror by looking at **Figure 3.** The light waves that are reflected off of you travel in all directions. Light rays reflected from your chin strike the mirror at different places. Then, they reflect off of the mirror in different directions. Recall that your brain always interprets light rays as if they have traveled in a straight line. It doesn't realize that the light rays have been reflected and that they changed direction. If the reflected light rays were extended back behind the mirror, they would meet at a single point. Your brain thinks that the rays that enter your eye are coming from this point behind the mirror. You seem to see the reflected image of your chin at this point. An image like this, which your brain perceives even though no light rays actually pass through it, is called a **virtual image.** The virtual image formed by a plane mirror is always upright and appears to be as far behind the mirror as the object is in front of it.

Concave Mirrors

Not all mirrors are flat like plane mirrors are. If the surface of a mirror is curved inward, it is called a **concave mirror.** Concave mirrors, like plane mirrors, reflect light waves to form images. The difference is that the curved surface of a concave mirror reflects light in a unique way.

Features of Concave Mirrors A concave mirror has an optical axis. The **optical axis** is an imaginary straight line drawn perpendicular to the surface of the mirror at its center. There is a point on this optical axis that every light ray parallel to the optical axis is reflected through called the **focal point**. Using the focal point and the optical axis, you can diagram how some of the light rays that travel to a concave mirror are reflected, as shown in **Figure 4.** On the other hand, if a light ray passes through the focal point before it hits the mirror, it is reflected parallel to the optical axis. The distance from the center of the mirror to the focal point is called the **focal length.**

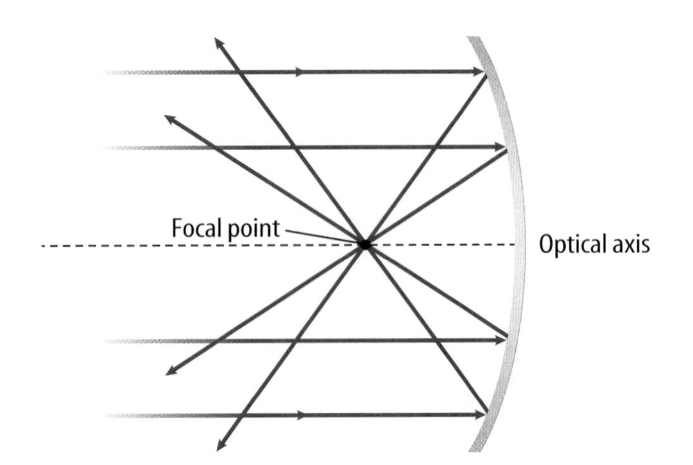

Figure 4
A concave mirror has an optical axis and a focal point. When light rays travel toward the mirror parallel to the optical axis, they reflect through the focal point.

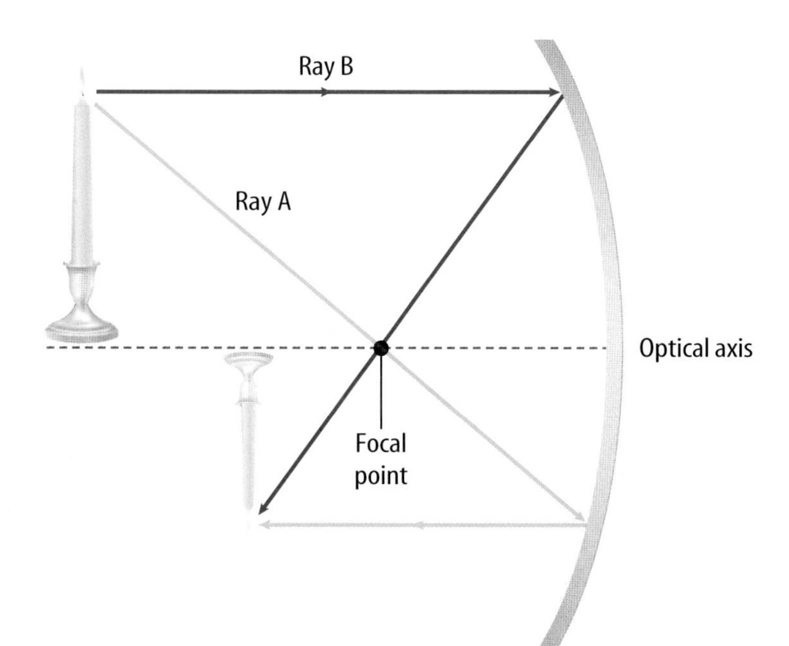

Figure 5
Rays A and B start from the same place on the candle, travel in different directions, and meet again on the reflected image. *How are the other points on the image of the candle formed?*

How a Concave Mirror Works The image that is formed by a concave mirror changes depending on where the object is located relative to the focal point of the mirror. You can diagram how an image is formed. For example, suppose that the distance between the object, such as the candle in **Figure 5,** and the mirror is a little greater than the focal length. Light rays bounce off of each point on the candle in all directions. One light ray, labeled Ray A, starts from a point on the flame of the candle and passes through the focal point on its way to the mirror. Ray A is then reflected so it travels parallel to the optical axis. Another ray, Ray B, starts from the same point on the candle's flame but travels parallel to the optical axis as it moves toward the mirror. When Ray B is reflected by the mirror, it passes through the focal point. The place where Ray A and Ray B meet after they are reflected forms a point on the flame of the reflected image.

More points on the reflected image can be located in this way. From each point on the candle, one ray can be drawn that passes through the focal point and is reflected parallel to the optical axis. Another ray can be drawn that travels parallel to the optical axis and passes through the focal point after it is reflected. The point where the two rays meet is on the reflected image.

Real Images Notice that the image that is formed is not virtual. Rays of light pass through the location of the image. A **real image** is formed when light rays converge to form the image. You could hold a sheet of paper at the location of the real image and see the image on the paper. When an object is farther from a concave mirror than twice the focal length, the image that is formed is real, smaller, and upside down, or inverted.

Observing Images

Procedure

1. Look at the inside of a shiny **spoon.** Move it close to your face and then far away. The place where your image changes is the focal point.
2. Hold the inside of the spoon facing a bright **light,** a little farther away than the focal length of the spoon.
3. Place a piece of **poster board** between the light and the spoon without blocking all of the light.
4. Move the poster board between the spoon and the light until you see the reflected light on it.

Analysis

Which of the images you observed were real and which were virtual?

Figure 6
A flashlight uses a concave mirror to create a beam of light. *Why are all of the reflected rays of light in the diagram parallel to each other?*

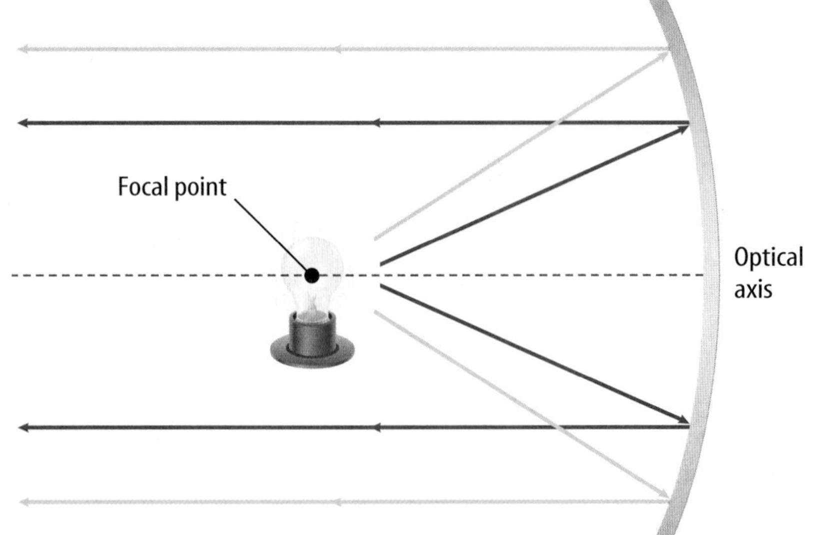

Creating Light Beams What happens if you place an object exactly at the focal point of the concave mirror? **Figure 6** shows that if the object is at the focal point, the mirror reflects all light rays parallel to the optical axis. No image forms because the rays never meet—not even if the rays are extended back behind the mirror. Therefore, a light placed at the focal point is reflected in a beam. Car headlights, flashlights, lighthouses, spotlights, and other devices use concave mirrors in this way to create concentrated light beams of nearly parallel rays.

Mirrors That Magnify The image formed by a concave mirror changes again when you place an object between it and its focal point. The location of the reflected image again can be found by drawing two rays from each point. **Figure 7** shows that in this case, these rays never meet after they reflect from the mirror. Instead, the reflected light rays spread apart, or diverge. Just as it does with a plane mirror, your brain interprets the light rays as if they came from one point behind the mirror. Because no light rays are behind the mirror where the image seems to be, the image formed is virtual. The image also is upright and enlarged.

Shaving mirrors and makeup mirrors are concave mirrors. They form an enlarged, upright image of a person's face so it's easier to see small details. The bowl of a shiny spoon also forms an enlarged, upright image of your face when it is placed close to your face.

Figure 7
If the candle is between the mirror and its focal point, the reflected image is enlarged and virtual. *Why couldn't this image be projected on a screen?*

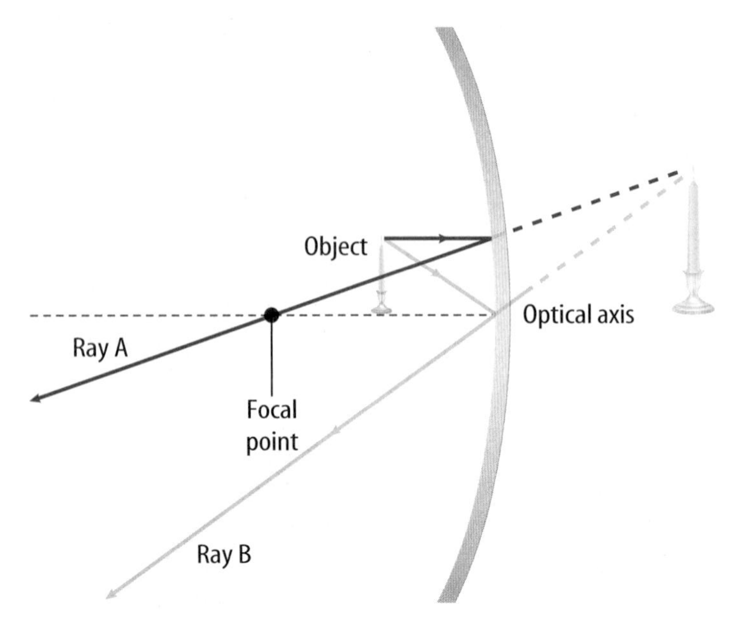

Convex Mirrors

Why do you think the security mirrors in banks and stores are shaped the way they are? The next time you are in a store, look up to one of the back corners or at the end of an aisle to see if a large, rounded mirror is mounted there. You can see a large area of the store in the mirror. A mirror that curves outward like the back of a spoon is called a **convex mirror.** Light rays that hit a convex mirror diverge, or spread apart, after they are reflected. Look at **Figure 8** to see how the rays from an object are reflected to form an image. The reflected rays diverge and never meet, so the image formed by a convex mirror is a virtual image. The image also is always upright and smaller than the actual object is.

Reading Check *What is a convex mirror, and what happens to light rays that hit it?*

Uses of Convex Mirrors Because convex mirrors spread out the reflected light, they allow large areas to be viewed. In addition to increasing the field of view in places like grocery stores and factories, convex mirrors can widen the view of traffic that can be seen in rearview or side-view mirrors of automobiles. However, because the image created by a convex mirror is smaller than the actual object, your perception of distance can be distorted. Objects look farther away than they truly are in a convex mirror. Distances and sizes seen in a convex mirror are not realistic, so most convex side mirrors carry a printed warning that says "Objects in mirror are closer than they appear."

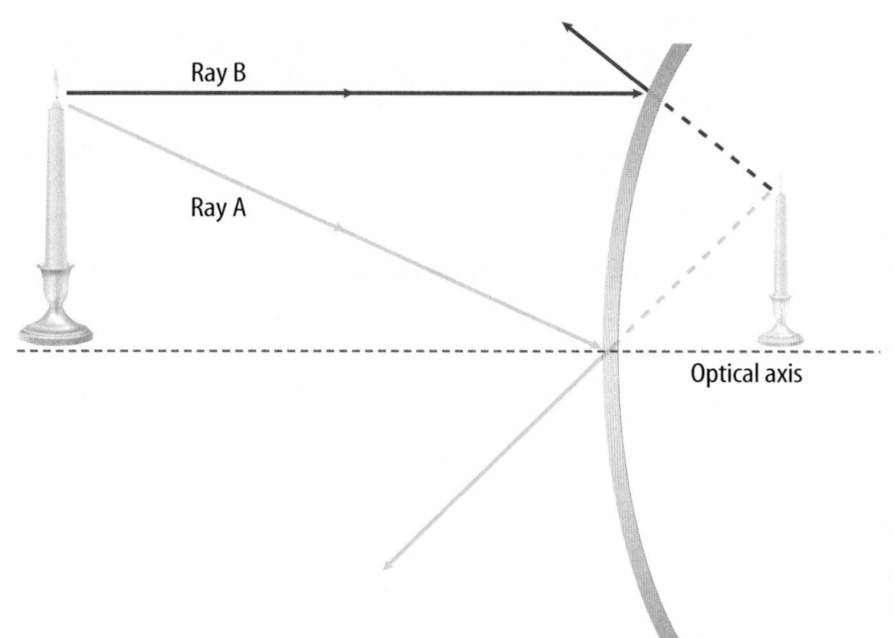

Figure 8
A convex mirror forms a reduced, upright, virtual image. *Why are convex mirrors useful?*

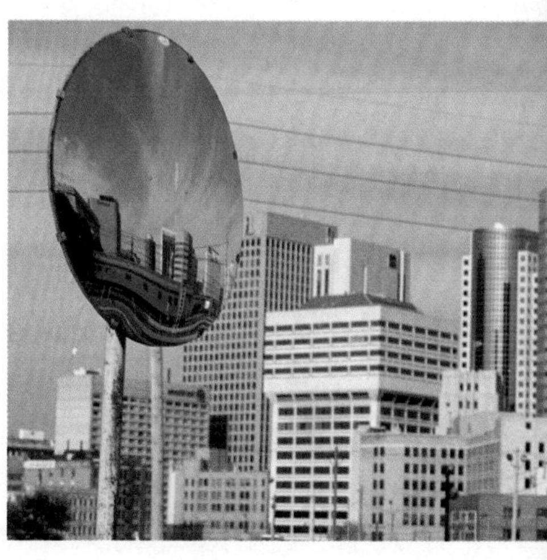

Table 1 Images Formed by Mirrors

Mirror Shape		Virtual/ Real	Image Created Upright /Upside Down	Size
Plane		virtual	upright	same as object
Concave	Object is positioned more than 2× focal length	real	upside down	smaller than object
	Object at focal point	none	none	none
	Object within focal length	virtual	upright	larger than object
Convex		virtual	upright	smaller than object

Mirror Images The different shapes of plane, concave, and convex mirrors cause them to reflect light in distinct ways. Each type of mirror has different uses. **Table 1** summarizes the images formed by plane, concave, and convex mirrors.

Section  Assessment

1. Contrast the differences between the surfaces of the following types of mirrors: plane, concave, and convex.
2. Diagram how light rays are reflected to form an image in a convex mirror.
3. What is the difference between a real and a virtual image?
4. How are concave mirrors used? Explain.
5. **Think Critically** What kind of mirror would you use to focus light entering a telescope? Explain.

Skill Builder Activities

6. **Recognizing Cause and Effect** You drop a flashlight that has a concave mirror in it. The light is then less intense than it was before you dropped the flashlight. Explain. **For more help, refer to the Science Skill Handbook.**
7. **Communicating** In your Science Journal, list all the mirrors you see during an average day. Describe how each is used, and identify each one as plane, concave, or convex. **For more help, refer to the Science Skill Handbook.**

Activity

Reflections of Reflections

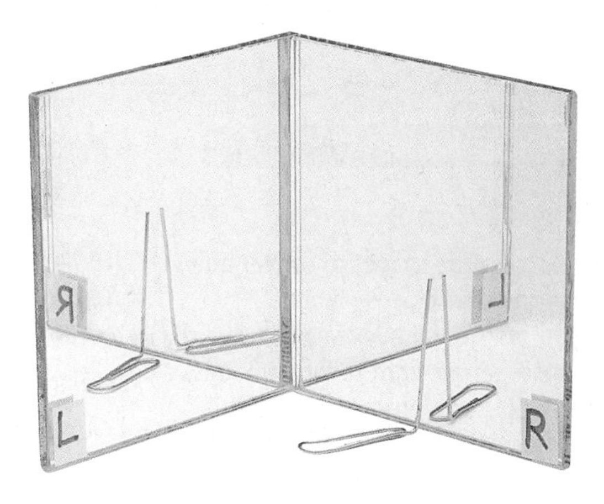

How can you see the back of your head? You can use two mirrors to see a reflection of the original reflection of the back of your head. How many reflections of your head can you create?

What You'll Investigate

How can you change the number of reflections of an object that is created in two mirrors?

Materials

plane mirrors (2)
masking tape
protractor
paper clip

Goals

- **Observe** multiple reflections of an object.
- **Infer** how many reflections will be made when mirrors are placed at a certain angle.

Safety Precautions

Handle glass mirrors and paper clips carefully.

Procedure

1. Lay one mirror on top of the other with the mirror surface inward. Tape them together so they will open and close. Use tape to label them *L* and *R*.
2. Stand the mirrors up on a sheet of paper. Using the protractor, close the mirrors to an angle of 72°.
3. Bend one leg of a paper clip up 90° and place it close to the front of the R mirror.
4. **Count** the number of images of the clip you see in the R and L mirrors. Record these numbers in the data table.
5. The mirror arrangement creates an image of a circle divided into wedges by the mirrors. Record the number of wedges.
6. Hold the R mirror still and slowly open the L mirror to 90°. Count and record the images of the clip and the wedges in the circle. Repeat, this time opening the mirrors to 120°.

Images and Wedges Seen in the Mirrors

Angle of Mirrors	Number of Paper Clip Images		Number of Wedges
	R	L	
72°			
90°			
120°			

Conclude and Apply

1. What is the relationship between the number of wedges and paper clip images you can see?
2. What angle would divide a circle into six wedges? Hypothesize how many images would be produced.

Communicating Your Data

Demonstrate for younger students the relationship between the angle of the mirrors and the number of reflections. **For more help, refer to the Science Skill Handbook.**

Lenses

As You Read

What You'll Learn

- **Describe** the shapes of convex and concave lenses.
- **Explain** how convex and concave lenses refract light to form images.
- **Explain** how lenses are used to correct vision.

Vocabulary

convex lens
concave lens
cornea
retina

Why It's Important

Even if you don't wear glasses or contacts, you rely on lenses to see.

What is a lens?

What do your eyes have in common with cameras, eyeglasses, and microscopes? Each of these things contains at least one lens. A lens is a transparent material with at least one curved surface that causes light rays to bend, or refract, as they pass through. The image that a lens forms depends on the shape of the lens. Like curved mirrors, a lens can be convex or concave.

Convex Lenses

A **convex lens** is thicker in the middle than at the edges. Its optical axis is an imaginary straight line that is perpendicular to the surface of the lens at its thickest point. When light rays approach a convex lens traveling parallel to its optical axis, the rays are refracted toward the center of the lens, as in **Figure 9A.** All light rays traveling parallel to the optical axis are refracted so they pass through a single point, which is the focal point of the lens. The focal length of the lens depends on the shape of the lens. If the sides of a convex lens are less curved, light rays are bent less. As a result, lenses with flatter sides have longer focal lengths. Light rays that travel along the optical axis are not bent at all, as shown in **Figure 9B.**

Figure 9
Convex lenses are thicker in the middle than at the edges.

A **A convex lens focuses light rays at a focal point.** *Why does the light bend as it goes through the lens?*

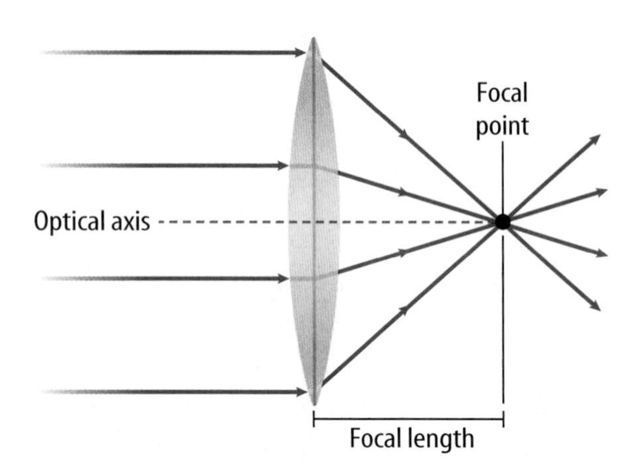

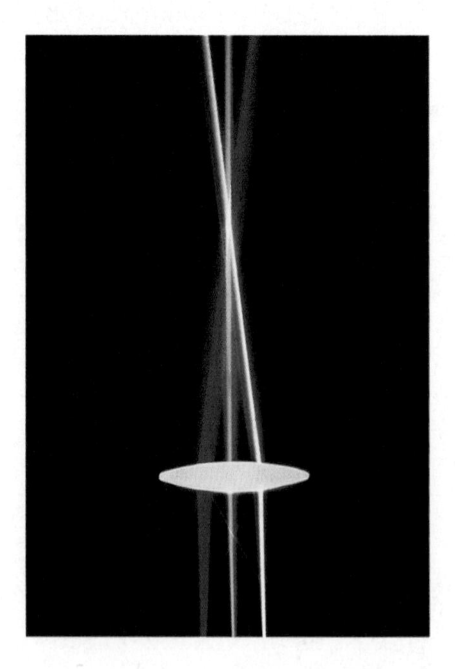

B **The green beam of light is not refracted at all as it passes through the center of the lens.**

Figure 10
The image formed by a convex lens changes depending on where the lens is compared to the object.

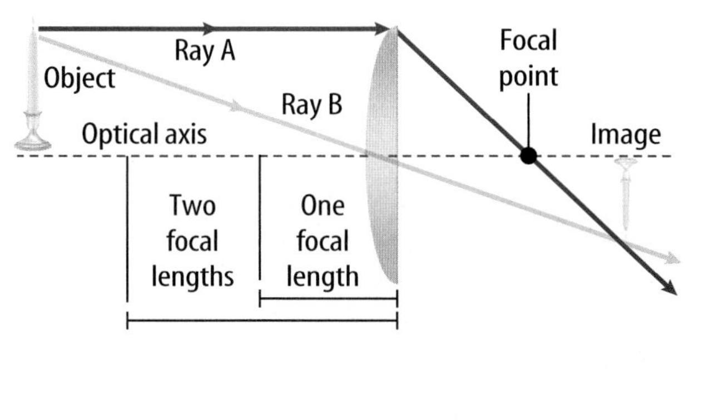

A When the candle is more than two focal lengths away from the lens, its image is real, reduced, and upside down.

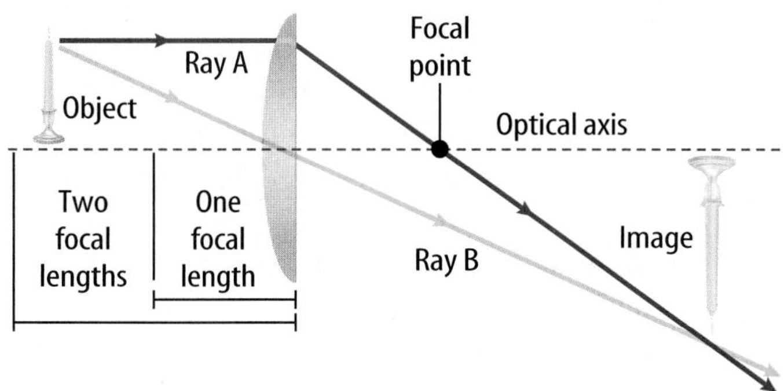

B When the candle is between one and two focal lengths from the lens, its image is real, enlarged, and upside down.

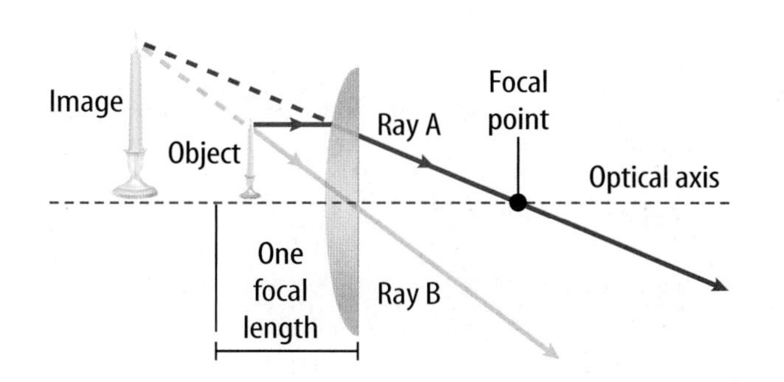

C When the candle is less than one focal length from the lens, its image is virtual, enlarged, and upright.

Forming Images with a Convex Lens The type of image a convex lens forms depends on where the object is relative to the focal point of the lens. If an object is more than two focal lengths from the lens, as in **Figure 10A,** the image is real, reduced, and inverted, and on the opposite side of the lens from the object. As the object moves closer to the lens, the image gets larger. **Figure 10B** shows how the image forms when the object is between one and two focal lengths from the lens. When the object is less than one focal length away from the lens, as in **Figure 10C,** the image seems to be on the same side of the lens as the object. When you use a magnifying glass, you place a convex lens less than one focal length away from an object. The image you see is upright and magnified and appears to be on the same side of the lens as the object you are inspecting.

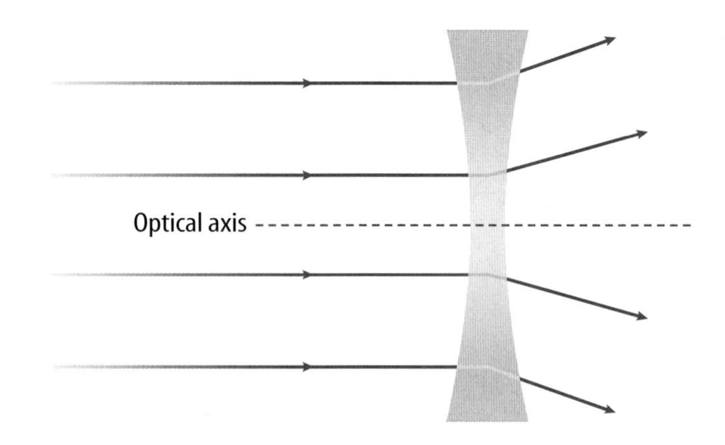

Figure 11
A concave lens refracts light rays so they spread out. *Is a concave lens most like a concave mirror or a convex mirror?*

Concave Lenses

A **concave lens** is thinner in the middle and thicker at the edges. As shown in **Figure 11,** light rays that pass through a concave lens bend outward away from the optical axis. The rays spread out and never meet at a focal point, so they never form a real image. The image is always virtual, upright, and smaller than the actual object is. Concave lenses are used in some types of eyeglasses and some telescopes. Concave lenses usually are used in combination with other lenses.

✔ Reading Check *What shape is a concave lens?*

Problem-Solving Activity

Comparing Object and Image Distances

The size and orientation of an image formed by a convex lens depends on the location of the object. What happens to the location of the image formed by a convex lens as the object moves closer to the lens or farther from the lens? The distance from the lens to the object is the object distance, and the distance from the lens to the image is the image distance. How are the focal length, object distance, and image distance related to each other?

Object and Image Distances

Focal Length	Object Distance	Image Distance
15.0 cm	45.0 cm	22.5 cm
15.0 cm	30.0 cm	30.0 cm
15.0 cm	20.0 cm	60.0 cm

Identifying the Problem

A 5-cm-tall object is placed at different lengths from a double convex lens with a focal length of 15 cm. The table above lists the different object and image distances. How are these two measurements related?

Solving the Problem

1. What is the relationship between the object distance and the image distance?
2. The lens equation relates the focal length and the image and object distances.
   1/focal length = 1/object distance + 1/image distance
   Using this equation, calculate the image distance of an object placed at a distance of 60.0 cm.

Table 2 Images Formed by Lenses

Lens Shape	Location of Object	Type of Image		
		Virtual/ Real	Upright/ Inverted	Size
Convex	Object beyond 2 focal lengths	real	inverted	smaller than object
	Object between 1 and 2 focal lengths	real	inverted	larger than object
	Object within 1 focal length	virtual	upright	larger than object
Concave	Object at any position	virtual	upright	smaller than object

Lenses and Eyesight

What determines how well you can see the words on this page? Perhaps you wear eyeglasses with lenses like those summarized in **Table 2.** If you don't need eyeglasses, the structure of your eye gives you the ability to focus on these words and other objects around you. Look at **Figure 12.** Light enters your eye through a transparent covering on your eyeball called the **cornea** (KOR nee uh). The light then passes through an opening called the pupil. Behind the pupil is a flexible convex lens. When light rays pass through it, they bend and form an inverted image on your retina. The **retina** (RET nuh) is the inner lining of your eye. It has cells that convert the light image into electrical signals, which are then carried along the optic nerve to your brain to be interpreted.

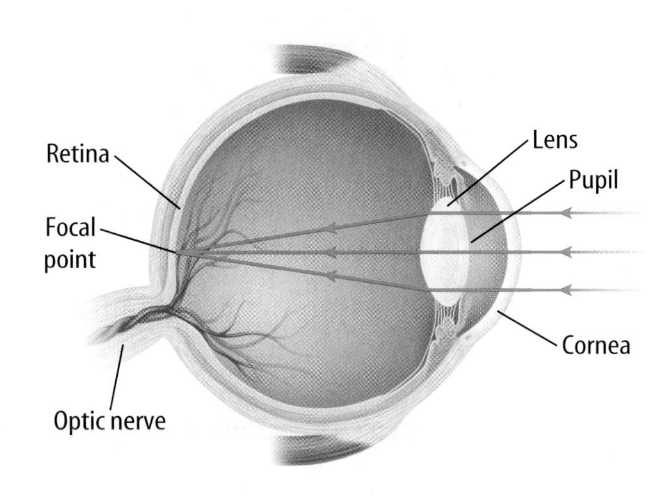

Figure 12
Light bends as it passes through the lens in your eye.
What would happen if the lens were a concave lens?

Focusing on Near and Far

Focusing on Near and Far How can your eyes focus on close objects, like the watch on your wrist, and distant objects, like a clock across the room? For you to see an object clearly, its image must be focused exactly on your retina. However, the retina is always a fixed distance from the lens. Remember that the location of an image formed by a convex lens depends on the focal length of the lens and the location of the object. For example, look back at **Figure 10.** As an object moves farther from a convex lens, the position of the image moves closer to the lens.

For an image to be formed on the retina, the focal length of the lens needs to be able to change as the distance of the object changes. The lens in your eye is flexible, and muscles attached to it change its shape and its focal length. This is why you can see objects that are near and far away.

Look at **Figure 13.** As an object gets farther from your eye, the focal length of the lens has to increase. The muscles around the lens stretch it so it has a less convex shape. But when you focus on a nearby object, these muscles make the lens more curved, causing the focal length to decrease.

Reading Check *How does the shape of the lens change when you focus on a nearby object?*

SCIENCE Online

Research Visit the Glencoe Science Web site at **science.glencoe.com** for information about diseases that affect the retina. Make a poster that shows what you learn.

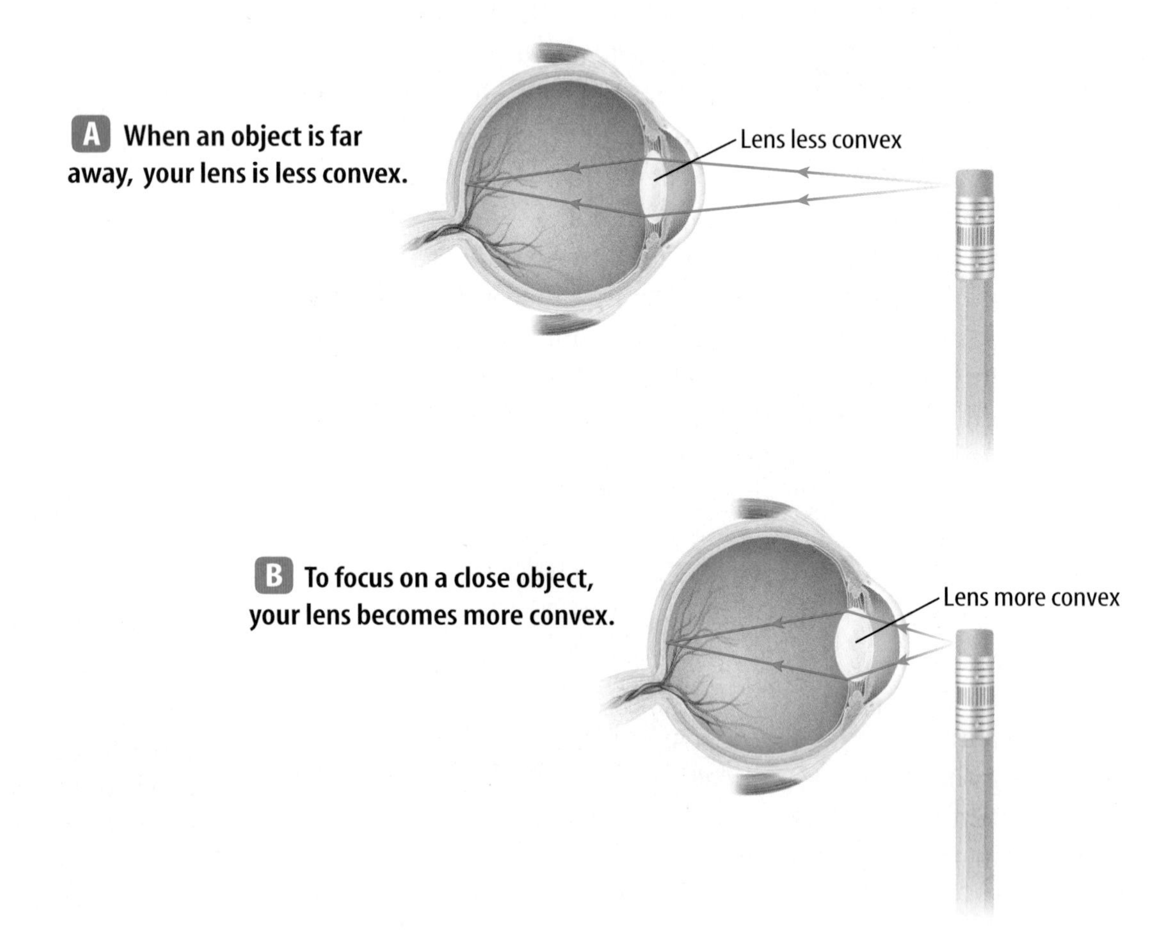

Figure 13
The lens in your eye changes shape so you can focus on objects at different distances.

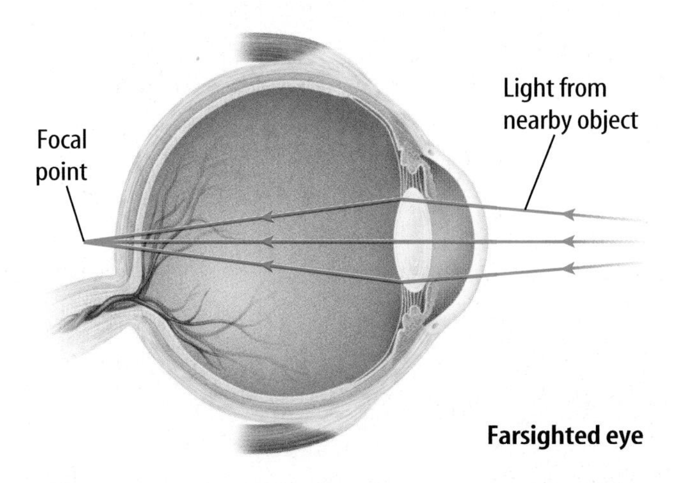

Focal
point
Light from
nearby object
Corrected
farsighted eye

A The lenses in farsighted people's eyes focus the image of a close object behind the retina.

B A convex lens in front of a farsighted eye will help converge the light rays on the retina.

Figure 14
Farsightedness can be corrected by convex lenses.

Vision Problems

If you have healthy vision, you should be able to see objects clearly when they are 25 cm or farther away from your eyes. The image of an object should be formed exactly on your retina. However, for many people, the image is blurry or formed in the wrong place, causing vision problems.

Farsightedness If you can see distant objects clearly but can't bring nearby objects into focus, then you are farsighted. The eyeball is too short or the lens isn't curved enough to form an image of close objects on the retina. The image of a close object would be focused behind the retina, as shown in **Figure 14A.** To correct the problem, convex lenses, as in **Figure 14B,** converge incoming light rays before they enter the eye.

As many people age, their eyes develop a condition that makes them unable to focus on close objects. The lenses in their eyes become less flexible. The muscles around the lenses still contract as they try to change the shape of the lens. However, the lenses have become more rigid, and cannot be made curved enough to form an image on the retina. People who are 40 years old might not be able to focus on objects that are closer than 1 m from their eyes. Some vision problems are caused by diseases of the retina. **Figure 15** shows how using new technology allows people with diseased retinas to recover some vision.

Astigmatism Another vision problem, called astigmatism (uh STIHG muh tih zum) occurs when the surface of the cornea is curved unevenly. When people have astigmatism, their corneas are more oval than round in shape. Astigmatism causes blurry vision at all distances. Corrective lenses also have an uneven curvature, canceling out the effect of an uneven cornea.

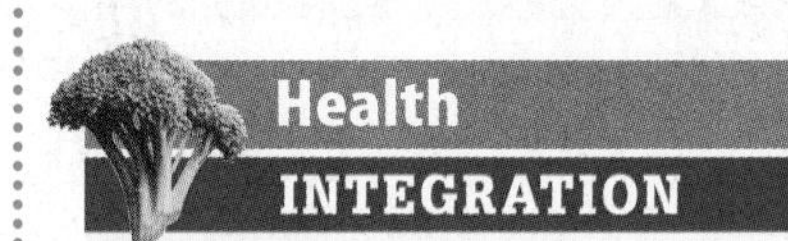

If you need eyeglasses, what do the letters and numbers on your eyeglass prescription mean? O.D. specifies right eye, and O.S. specifies left eye. The sphere number tells how nearsighted or farsighted you are. The cylinder and axis numbers are a measure of astigmatism. These numbers allow eyeglasses to be made specifically for your vision problems.

Figure 15

Millions of people worldwide suffer from vision problems associated with diseases of the retina. Until recently, such people had little hope of improving their eyesight. Now, however, scientists are developing specialized silicon chips that convert light into electrical pulses, mimicking the function of the retina. When implanted in the eye, these artificial silicon retinas may restore sight.

Viewed with normal vision

Viewed with retinitis pigmentosa

Viewed with macular degeneration

▲ These three photos show how normal vision can deteriorate as a result of diseases that attack the retina. Retinitis pigmentosa (ret uh NYE tis pig men TOE suh) causes a lack of peripheral vision. Macular degeneration can lead to total blindness.

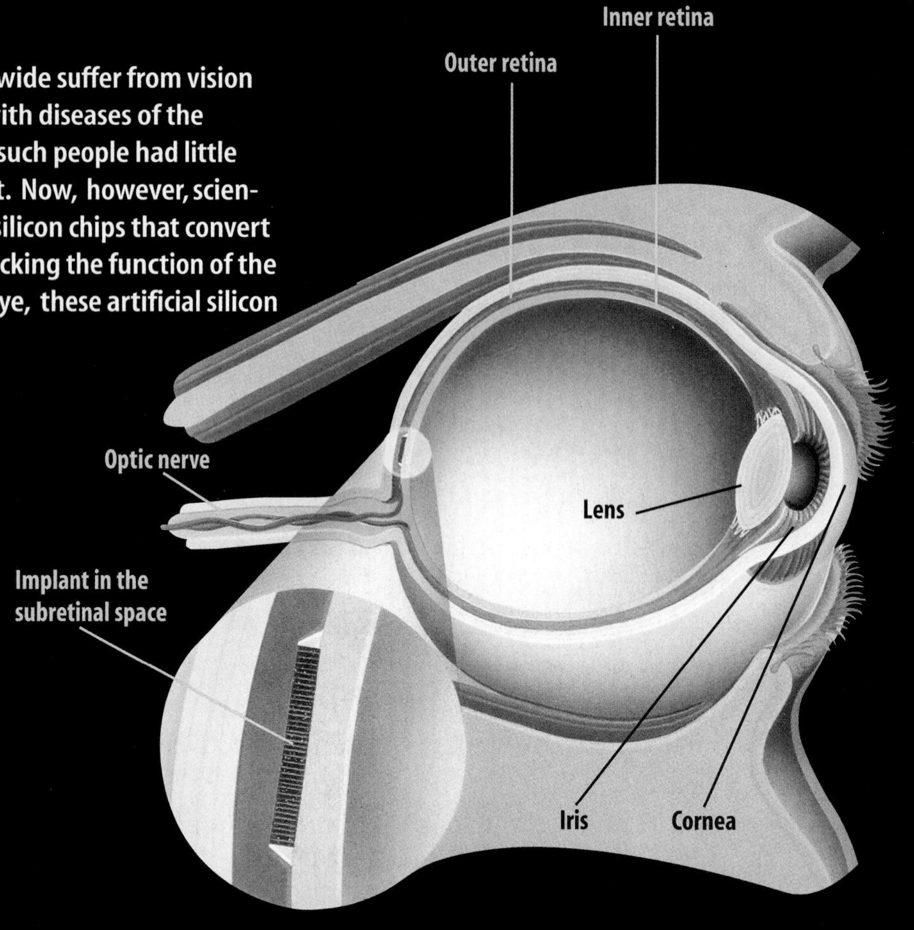

▲ After making a number of incisions, surgeons implant the artificial silicon retina between the outer and inner retinal layers. Then they reseal the retina over the silicon chip.

▲ The artificial silicon retina, above right, is thinner than a human hair and only 2 mm in diameter—the same diameter as the white dot on this penny.

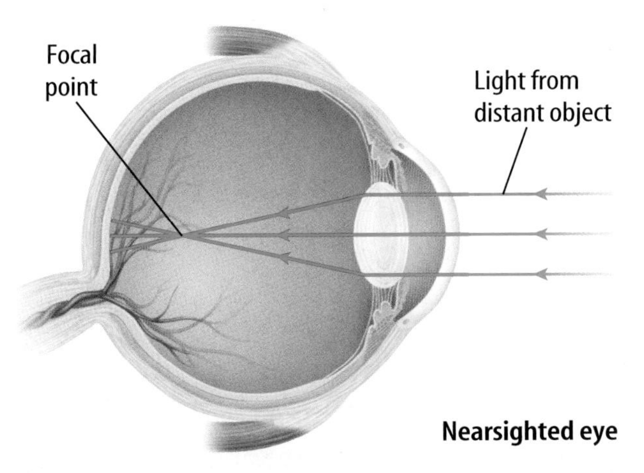

A **When a nearsighted person looks at distant objects, the light rays from the objects are focused in front of the retina.**

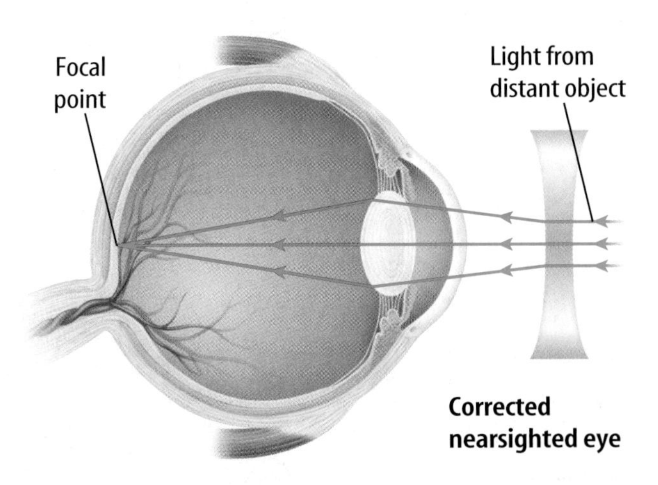

B **If the light rays pass through a concave lens on the way to the nearsighted eye, they will spread apart. The eye's lens then can form an image of the distant object on the retina.**

Figure 16
Nearsightedness can be corrected with concave lenses.

Nearsightedness If you have nearsighted friends, you know that they can see clearly only when objects are nearby. Their eyeballs may be too long or their lens cannot be made flat enough to form an image on the retina of an object that is far away. Instead, the image is formed in front of the retina, as in **Figure 16A.** To correct this problem, a nearsighted person can wear concave lenses. **Figure 16B** illustrates how the concave lenses will spread out incoming light rays before they reach the eye. Then when the diverging light rays reach the lens, they are refracted the right amount to form an image on the retina.

Section  Assessment

1. How can you tell the difference between a convex lens and a concave lens?
2. How is light refracted when it passes through a concave lens?
3. What type of lens would you use to examine a tiny spider on your desk?
4. If you have difficulty reading the chalkboard from the back row, what is most likely your vision problem? How can it be corrected?
5. **Think Critically** When using a slide projector, why do you insert the slides in the projector upside down?

Skill Builder Activities

6. **Concept Mapping** Mirrors and lenses are the simplest optical devices. Design a network tree concept map to show some uses for each shape of mirror and lens. **For more help, refer to the Science Skill Handbook.**
7. **Using an Electronic Spreadsheet** Prepare a spreadsheet that organizes the information you have acquired about vision problems. For nearsightedness, astigmatism, and farsightedness, display the following: *the symptom*, *the cause*, and *the method of correction*. **For more help, refer to the Technology Skill Handbook.**

SECTION 3 Optical Instruments

As You Read

What You'll Learn

- **Compare** refracting and reflecting telescopes.
- **Explain** why a telescope in space would be useful.
- **Describe** how a microscope uses lenses to magnify small objects.
- **Explain** how a camera creates an image.

Vocabulary

refracting telescope
reflecting telescope
microscope

Why It's Important

Optical instruments help the human eye see distant and small objects.

Telescopes

Imagine a clear evening when a full moon is just starting to rise. Even though the Moon might seem large and close, it is still too far away for you to see the details on its surface. You know from your own experience that it's hard to see faraway objects clearly. When you look at an object, only some of the light reflected from its surface enters your eye. Much of the light is reflected in other directions. As the object moves farther away, the amount of light entering your eye decreases, as shown in **Figure 17.** As a result, the object appears dimmer and less detailed.

A telescope uses a lens or a concave mirror that is much larger than your eye to gather more of the light from distant objects. The largest telescopes can gather more than a million times more light than the human eye. As a result, objects such as distant galaxies appear much brighter. Because the image formed by the telescope's lens or mirror is so much brighter, more detail can be seen when the image is magnified. If the light gathered by a telescope is focused on electronic sensors or photographic film for several hours, an even brighter image can be produced.

Figure 17
The amount of light that reaches the eye decreases as the distance between the eye and the object increases.

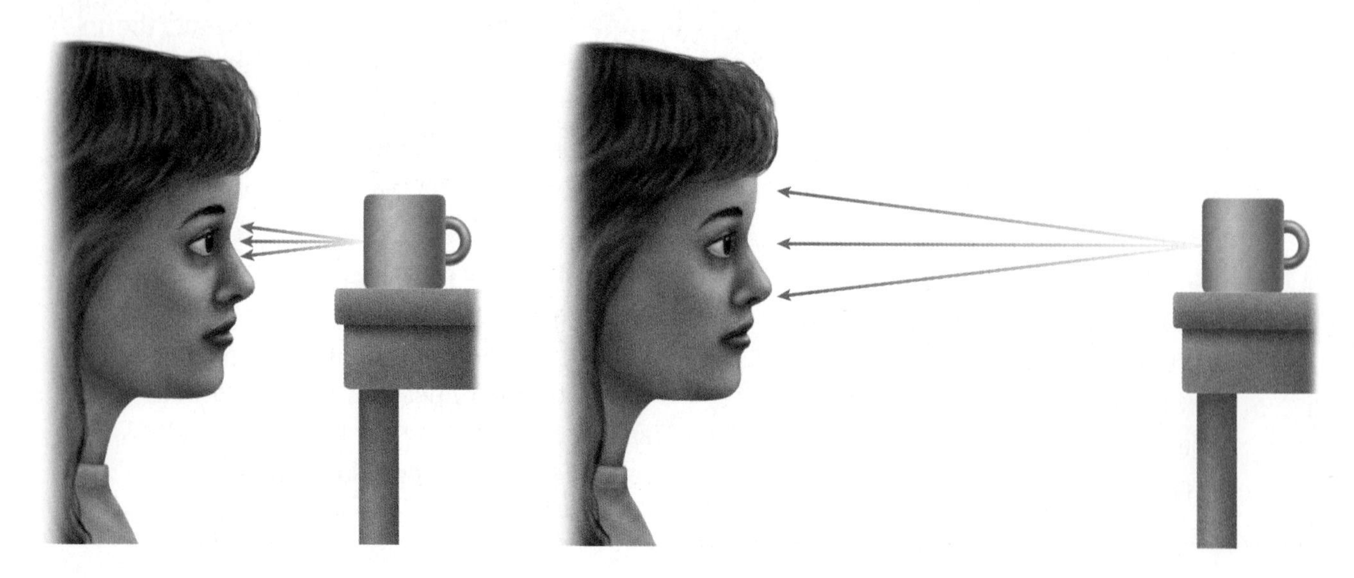

Refracting Telescopes One common telescope is the refracting telescope. A simple **refracting telescope,** shown in **Figure 18,** uses two convex lenses to gather and focus light from distant objects. Incoming light from distant objects passes through the first lens, called the objective lens. Because the objects are so far away, light rays from these objects are nearly parallel to the optical axis of the lens. As a result, the rays form a real image at the focal point of the lens, within the body of the telescope. The second convex lens, called the eyepiece lens, acts like a magnifying glass and magnifies this real image. When you look through the eyepiece lens, you see an enlarged, inverted, virtual image of the real image formed by the objective lens.

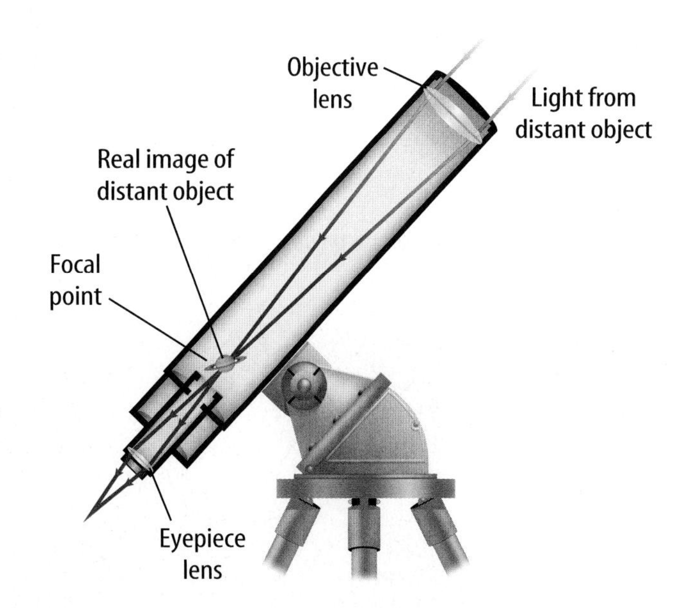

Figure 18
Light from a distant object refracts twice in a refracting telescope to create a large virtual image.

How do the two lenses in a refracting telescope work together?

Several problems are associated with refracting telescopes. In order to form a detailed image of distant objects, such as planets and galaxies, the objective lens must be as large as possible. A large lens is heavy and can be supported in the telescope tube only around its edge. The lens can sag or flex due to its own weight, distorting the image it forms. Also, these heavy glass lenses are costly and difficult to make.

Reflecting Telescopes Due to the problems with making large lenses, most large telescopes today are reflecting telescopes. A **reflecting telescope** uses a concave mirror, a plane mirror, and a convex lens to collect and focus light from distant objects. **Figure 19** shows a reflecting telescope. Light from a distant object enters one end of the telescope and strikes a concave mirror at the opposite end. The light reflects off of this mirror and converges. Before it converges at a focal point, the light hits a plane mirror that is placed at an angle within the telescope tube. The light is reflected from the plane mirror toward the telescope's eyepiece. The light rays converge at the focal point, creating a real image of the distant object. Just as in a refracting telescope, a convex lens in the eyepiece then magnifies this image.

Figure 19
Reflecting telescopes use two mirrors to create a real image, which then is magnified by a convex lens. *Is the final image you see real or virtual?*

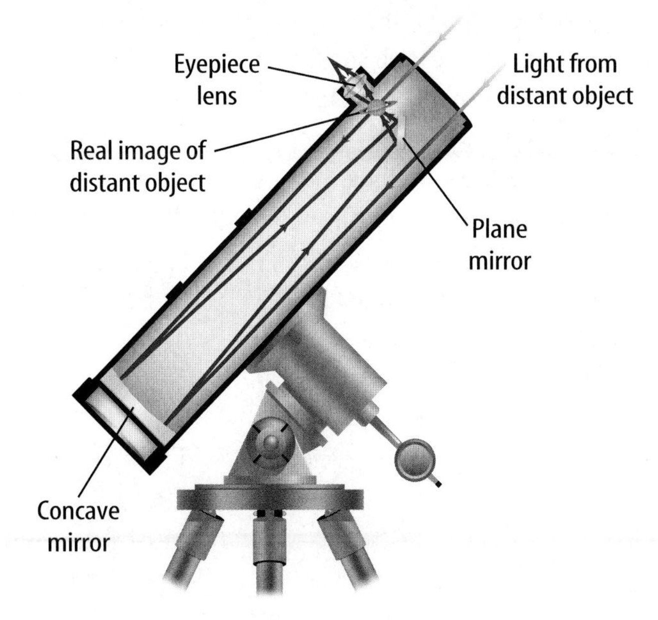

Figure 20
The view from telescopes on Earth is different from the view from telescopes in space.

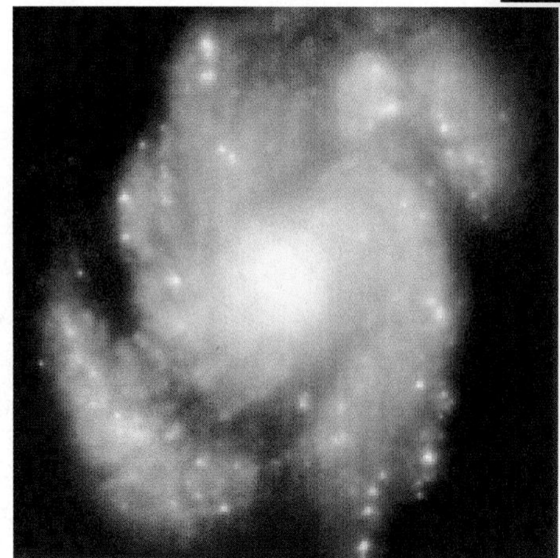

A Telescopes on Earth can form blurry images of objects in space due to Earth's atmosphere.

B The *Hubble Space Telescope* is above Earth's atmosphere and forms clearer images of objects in space.

Telescopes In Space Imagine being at the bottom of a swimming pool and trying to read a sign by the pool's edge. The water in the pool would distort your view of any object beyond the water's surface. In a similar way, Earth's atmosphere blurs your view of objects in outer space. To overcome the blurriness of humans' view into space, the National Aeronautics and Space Administration (NASA) built a telescope called the *Hubble Space Telescope* to be placed into space high above Earth's atmosphere. On April 25, 1990, NASA used the space shuttle Discovery to launch this telescope into an orbit about 600 km above Earth. The *Hubble Space Telescope* has produced images much sharper and more detailed than the largest telescopes on Earth can. **Figure 20** shows the difference in the images produced by telescopes on Earth and the *Hubble* telescope. With the *Hubble Space Telescope,* scientists can detect visible light—as well as other types of radiation—from the planets, stars, and distant galaxies that usually is blocked by Earth's atmosphere.

Reading Check *Why is the* **Hubble Space Telescope** *able to produce clearer images than telescopes on Earth?*

The *Hubble* telescope is a type of reflecting telescope that uses two mirrors to collect and focus light to form an image. The primary mirror in the telescope is 2.4 m across. When the *Hubble* was first launched, a defect in this primary mirror caused the telescope to create blurry images. The telescope was repaired by astronauts in December 1993.

Research Visit the Glencoe Science Web site at **science.glencoe.com** for data about the *Hubble Space Telescope.* Do you think the *Hubble Space Telescope* is worthwhile? Prepare a persuasive speech to defend your opinion on whether or not the *Hubble Space Telescope* is useful and important.

Microscopes

A telescope would be useless if you were trying to study the cells in a butterfly wing, a sample of pond scum, or the differences between a human hair and a horse hair. You would need a microscope to look at such small objects. A **microscope** uses two convex lenses with relatively short focal lengths to magnify small, close objects. A microscope, like a telescope, has an objective lens and an eyepiece lens. However, it is designed differently because the objects viewed are not far away.

Figure 21 shows a simple microscope. The object to be viewed is placed on a transparent slide and illuminated from below. The light passes by or through the object on the slide and then travels through the objective lens. The objective lens is a convex lens. It forms a real, enlarged image of the object, because the distance from the object to the lens is between one and two focal lengths. The real image is then magnified again by the eyepiece lens (another convex lens) to create a virtual, enlarged image. This final image can be hundreds of times larger than the actual object, depending on the focal lengths of the two lenses.

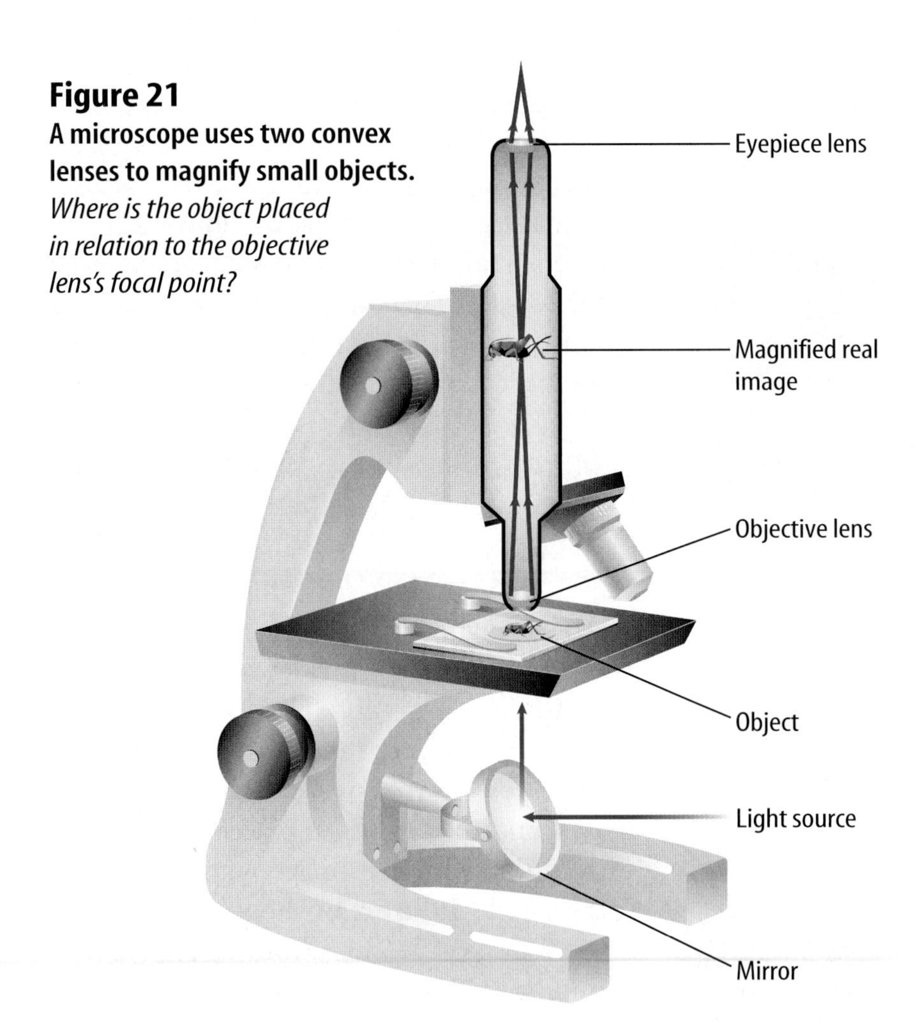

Figure 21
A microscope uses two convex lenses to magnify small objects.
Where is the object placed in relation to the objective lens's focal point?

Experimenting with Focal Lengths

Procedure

1. Fill a glass **test tube** with **water** and seal it with a **lid or stopper.**
2. Type or print the compound name SULFUR DIOXIDE in capital letters on a piece of **paper or a note card.**
3. Set the test tube horizontally over the words and observe them. What do you notice?
4. Hold the tube 1 cm over the words and observe them again. Record your observations. Repeat, holding the tube at several other heights above the words.

Analysis

1. What were your observations of the words at the different distances? How do you explain your observations?
2. Is the image you see at each height real or virtual?

Figure 22
A camera's lens focuses an image on photographic film.

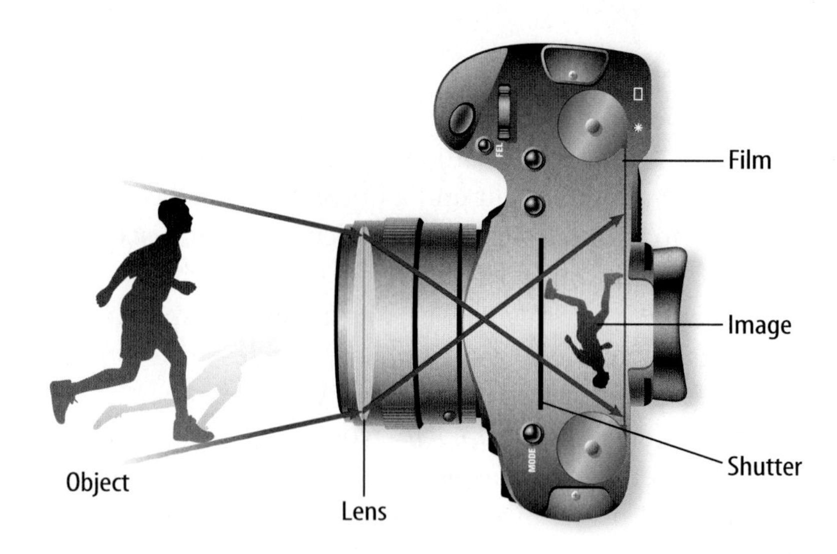

Cameras

Imagine swirls of lavender, gold, and magenta clouds sweeping across the sky at sunset. With the click of a button, you can capture the beautiful scene in a photo. How does a camera make a reduced image of a life-sized scene on film? A camera works by gathering and bending light with a lens. This lens then projects an image onto light-sensitive film to record a scene.

When you take a picture with a camera, a shutter opens to allow light to enter the camera for a specific length of time. The light reflected off your subject enters the camera through an opening called the aperture (AP uh choor). It passes through the camera lens, which focuses the image on the film, as in **Figure 22.** The image is real, inverted, and smaller than the actual object. The size of the image depends upon the focal length of the lens and how close the lens is to the film.

Figure 23
A Each object in the image produced by a wide-angle lens is small. **B** This allows more of the surroundings to be seen. *What situations would be good for using a wide-angle lens?*

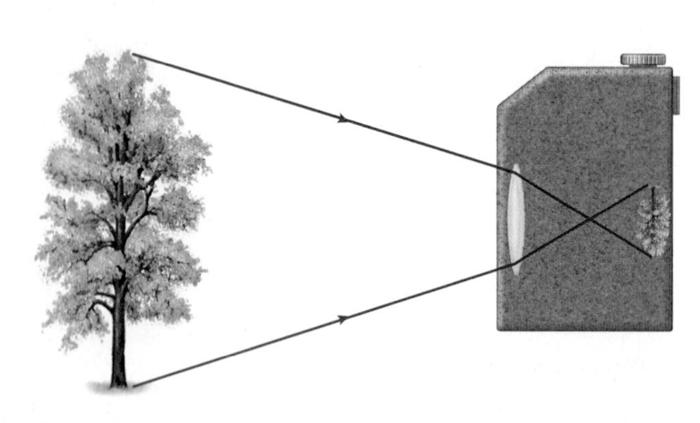

B

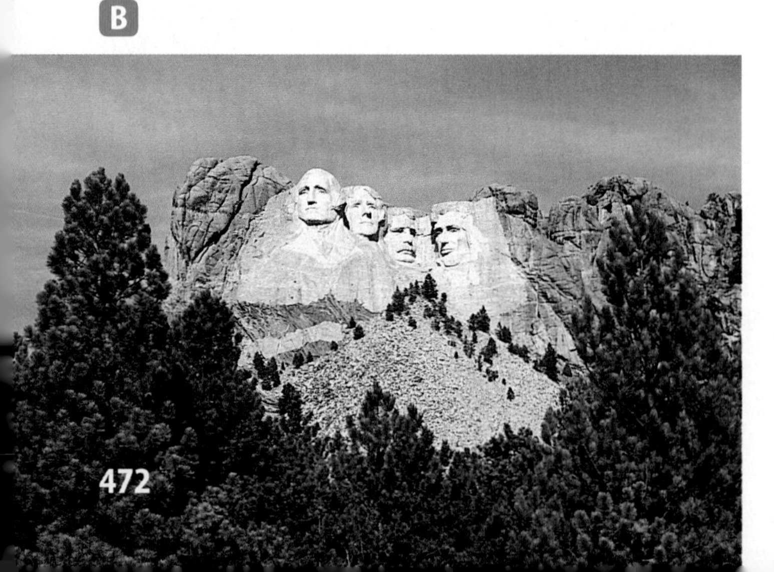

Wide-Angle Lenses Suppose you and a friend use two different cameras to photograph the same object at the same distance. If the cameras have different lenses, your pictures might look different. For example, some lenses have short focal lengths that produce a relatively small image of the object but include much of its surroundings. These lenses are called wide-angle lenses, and they must be placed close to the film to form a sharp image with their short focal length. **Figure 23A** shows how a wide-angle lens works. The photo in **Figure 23B** was taken with a wide-angle lens.

Figure 24
A telephoto lens creates a larger image of an object than a wide-angle lens does.

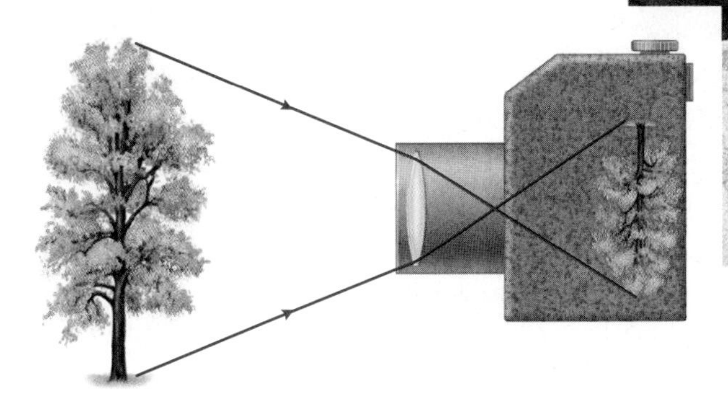

A The telephoto lens has a long focal length.

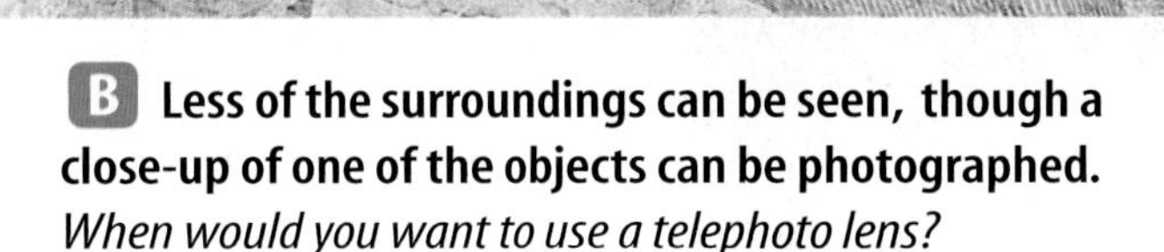

B Less of the surroundings can be seen, though a close-up of one of the objects can be photographed.
When would you want to use a telephoto lens?

Telephoto Lenses Telephoto lenses have longer focal lengths and are located farther from the film than wide-angle lenses are. **Figure 24A** shows how a telephoto lens forms an image. The image you see through a telephoto lens seems enlarged and the object seems closer than it actually is, as shown in **Figure 24B.** Telephoto lenses are easy to recognize because they are usually long. They protrude from the camera to increase the distance between the lens and the film.

Section  Assessment

1. Compare and contrast the components of reflecting and refracting telescopes and the images they form.
2. What advantages or disadvantages does the Hubble Space Telescope have compared to telescopes on Earth?
3. What does each lens in a microscope do?
4. If you wanted to photograph a single rose on a rosebush, what kind of lens would you use? Explain.
5. **Think Critically** Which optical instrument—a telescope, a microscope, or a camera—forms images in a way most like your eye? Explain.

Skill Builder Activities

6. **Forming Hypotheses** You notice that all the objects in a photograph you've taken are blurry. Use your knowledge of lenses and focal lengths to form a hypothesis that could explain why the photo was blurred and how to correct it. **For more help, refer to the Science Skill Handbook.**
7. **Solving One-Step Equations** Suppose the objective lens in a microscope forms an image that is 100 times the size of an object. The eyepiece lens magnifies this image ten times. What is the total magnification of the microscope? **For more help, refer to the Math Skill Handbook.**

Activity Model and Invent

Up Close and Personal

Galileo used the telescope to enhance his eyesight. It enabled him to see planets and stars beyond the range of his eyes alone. Today more powerful instruments, such as the *Hubble Space Telescope*, are used to learn more about our universe.

Recognize the Problem

How do the lenses in a simple telescope form an image?

Thinking Critically

By combining two lenses, distant objects can be magnified. A simple refracting telescope uses a small convex eyepiece lens and a larger convex objective lens at the other end. How do the lenses in a telescope change the light rays reflected from a distant object to make it appear closer? How do the focal lengths of the eyepiece and objective lenses determine the magnification of the telescope?

Goals

- **Build** a simple telescope.
- **Estimate** the magnification of the telescope.
- **Compare** convex and concave eyepieces.

Safety Precautions

WARNING: *Do not look directly at the Sun through a telescope. Permanent eye damage can result.*

Possible Materials

objective lens—convex, 25 cm to 30 cm focal length, about 4 cm diameter
eyepiece lenses—one each convex and concave, 2 cm to 3 cm focal length, about 2.5 cm to 3 cm diameter
cardboard tubes—one with inside diameter of about 4 cm; one with inside diameter of about 3 cm. (The smaller tube should slide inside the larger one with a snug fit.)
clay to hold the lenses in place
**cellophane tape or duct tape*
scissors

**Alternate materials*

Data Source

SCIENCE*Online* Go to the Glencoe Science Web site at **science.glencoe.com** for more information about telescopes.

Planning the Model

1. Hold the small, concave eyepiece lens near your eye. Hold the objective lens in front of it and move it away until a distant object appears focused. Note the approximate distance between the two lenses at this point. Subtract half the length of the larger-diameter cardboard tube from this measurement to get the length needed for the smaller tube.
2. **Decide** how you will attach each lens to the tube. You can hold it in place with clay or tape.

Check the Model

1. The lenses must be perpendicular to your line of sight to get the best results. How can you make sure the lenses are in the right position?
2. The smaller-diameter tube must have some room to slide in and out of the larger tube to focus properly on distant objects. Will your calculated length for the shorter tube allow this?
3. Make sure your teacher approves your plan before you start.

Making the Model

1. Cut the smaller-diameter cardboard tube to the calculated length. Make two pieces this size.
2. Attach the objective lens to the end of the larger tube.
3. Attach the convex eyepiece lens to one of the smaller tubes.
4. Slide the smaller tube into the larger one and look through the eyepiece.
5. Move the smaller tube in and out until a distant object is focused clearly.
6. **Estimate** how much larger the image seen through the eyepiece is than the image you see with your unaided eye. Note the appearance of the object.
7. Attach the concave eyepiece to the second smaller tube that you cut.
8. Repeat the observations using the concave eyepiece. Again, note the appearance of the object.

Analyzing and Applying Results

1. How did the image appear when using the convex and concave eyepieces? What happens when you turn the telescope around and look through the objective lens?
2. What was the estimated magnification of your telescope? How could you change its magnification?

Communicating Your Data

Compare your telescope and its operation with those of other members of your class. Try reading numbers or letters on a distant sign. Which telescope helps you see more detail?

TIME SCIENCE AND Society

SCIENCE ISSUES THAT AFFECT YOU!

Sight Lines

Lasers make it possible to throw away eyeglasses

Imagine seeing the world through the eyes of a hawk. With your crystal-clear vision, you could spot a ripple in the grass from hundreds of meters away. This kind of super-hero vision may be possible in the not-so-distant future. Already, scientists have developed ways to improve human eyesight beyond "perfect" 20/20 vision. With this technology, most of the 160 million Americans who wear eyeglasses or contact lenses can kiss them goodbye forever. Poor vision can be changed, permanently.

Laser Eye Surgery

Back in the 1970s, scientists developed a special kind of laser to make microscopic notches in computer chips. The laser is also perfect for eye surgery. It does not generate a lot of heat, so it doesn't damage the delicate tissues of the eye. Plus, the laser is very precise. Eye surgeons like this. They know that one tiny slip-up can cause a lifetime of vision problems.

In a Blink of an Eye . . .

Laser eye surgery only takes about 15 min. The patient is awake the entire time, staring at a red light to keep his or her eyes from moving while the doctor slices through his or her corneas.

Although some people experience complications, such as hazy vision, most couldn't be happier with the results. Soon after the surgery, one patient declared, "When I looked up at the sky I could see the stars clearly—just like that! It's a completely new life for me."

How It Works

A The eye is measured to determine the vision problem. These measurements are fed into the computer that controls the laser.

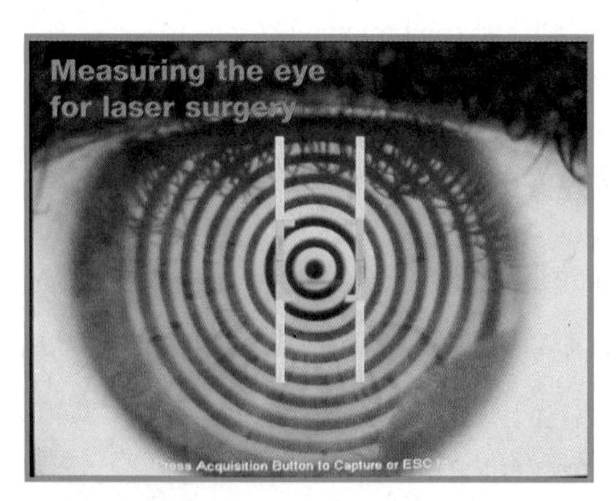

B The doctor numbs the eye with liquid drops, then props the eyelids open. The cornea, the transparent cover in front of the eye, is marked with a special ink before any cutting begins. These marks help the doctor put everything back in the right place.

The Cornea and Vision

The cornea is where light begins its journey into the eye. The transparent cornea is a lens that helps focus light that enters the eye. Most of the bending of these light rays occurs when they pass from air into the cornea. They then are bent more by the lens of the eye, which adjusts its shape to focus the light on the retina. Unlike the lens though, the eye doesn't adjust the shape of the cornea.

When someone is nearsighted, light rays are brought to a focus in front of the retina. To focus an image on the retina, the light rays must be bent less. This is usually done by placing a concave lens in front of the eye. However, another way would be to make the shape of the cornea flatter. Then light rays are bent less when they enter the eye.

If someone is farsighted, light rays aren't bent enough to form a focused image on the retina. A convex lens placed in front of the eye causes light rays to be bent before they enter the eye. Light rays also would be bent more if the shape of the cornea were thicker in the center, so it was less flat.

C A suction ring is attached to the eye to hold it steady during surgery.

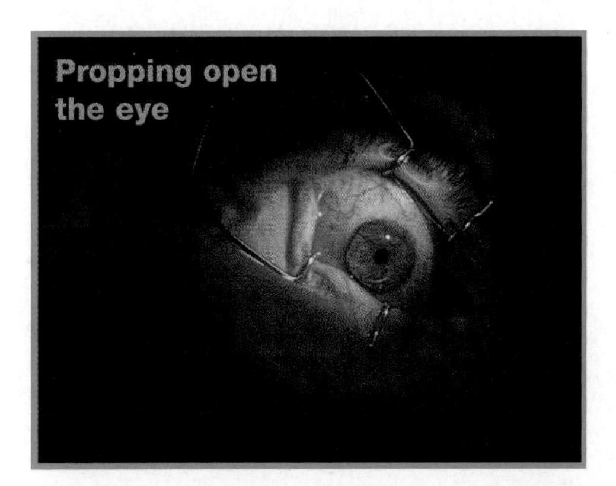
Propping open the eye

D The cutting instrument slices across a guided path through the outer layers of the eye. The blade leaves an uncut section that's lifted until the surgery is completed. Because the eye has been numbed, this procedure is not painful.

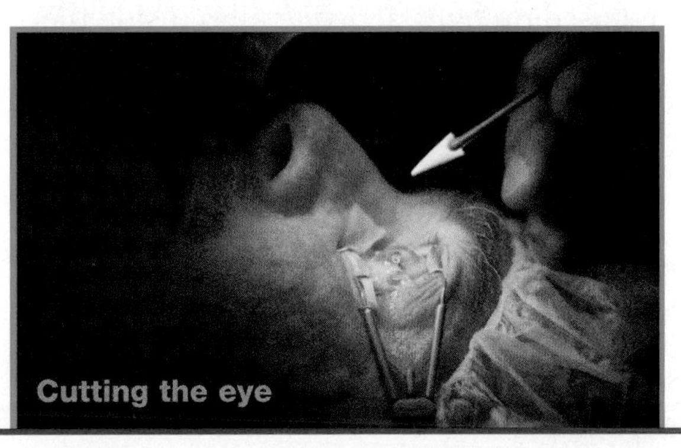
Cutting the eye

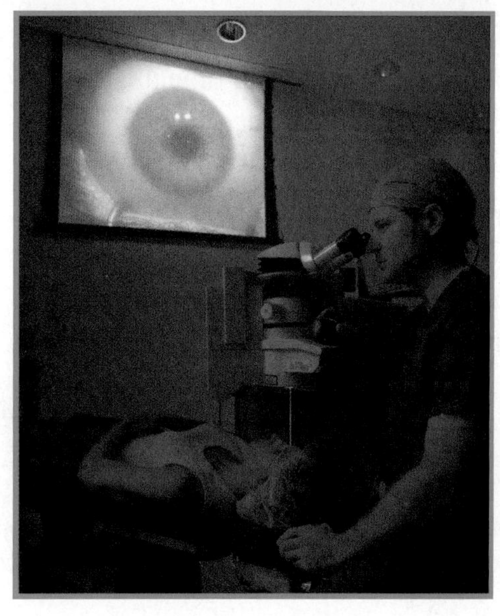
A screen helps this doctor keep tabs on his laser work.

E The blade is taken away and the laser goes to work. For people who are nearsighted—they see near objects clearly—the laser vaporizes the inner cornea, making it flatter. For people who are farsighted—they see far objects clearly—the laser removes a ring of tissue to make the cornea steeper.

CONNECTIONS Interview **Ophthamologists are medical doctors who specialize in healing eyes. Optometrists make glasses and check vision. Interview an optometrist or ophthamologist to find out how he or she detects eye problems and how these problems can be corrected.**

SCIENCE *Online*
For more information, visit science.glencoe.com

Chapter 15 Study Guide

Reviewing Main Ideas

Section 1 Mirrors

1. Plane mirrors reflect light to form upright, virtual images. *Why is the image you see in a plane mirror a virtual image?*

2. Concave mirrors can form various types of images, depending on where an object is relative to the focal point of the mirror. Concave mirrors can be used to magnify objects or create beams of light.
3. Convex mirrors spread out reflected light to form a reduced image. Convex mirrors allow you to see large areas.

Section 2 Lenses

1. Convex lenses converge light rays. Convex lenses can form real or virtual images, depending on the distance from the object to the lens.
2. Concave lenses diverge light rays to form virtual images. They often are used in combination with other lenses.
3. The human eye has a flexible lens that focuses an image on the retina. *Why does the lens in your eye need to change shape when you look at objects at different distances?*

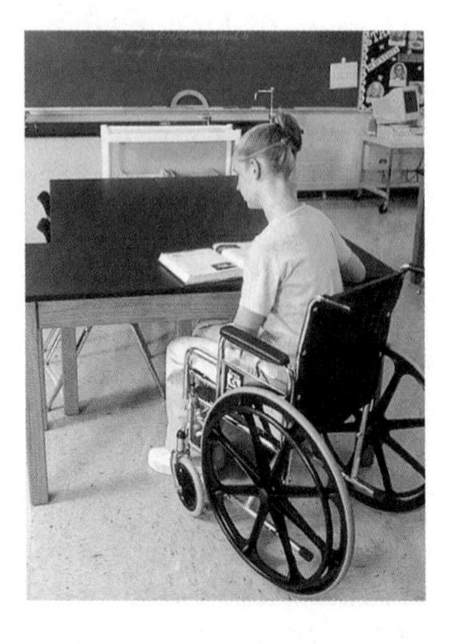

4. People with imperfect vision can use corrective lenses to improve their vision. Farsighted people wear convex lenses, and nearsighted people wear concave lenses.

Section 3 Optical Instruments

1. A refracting telescope uses convex lenses to magnify distant objects. A reflecting telescope uses concave and plane mirrors and a convex lens to magnify distant objects.
2. By avoiding atmospheric distortion, the *Hubble Space Telescope* produces sharper images than telescopes on Earth can.
3. Microscopes use two convex lenses to magnify small objects. *Is the image you see in a microscope real or virtual?*

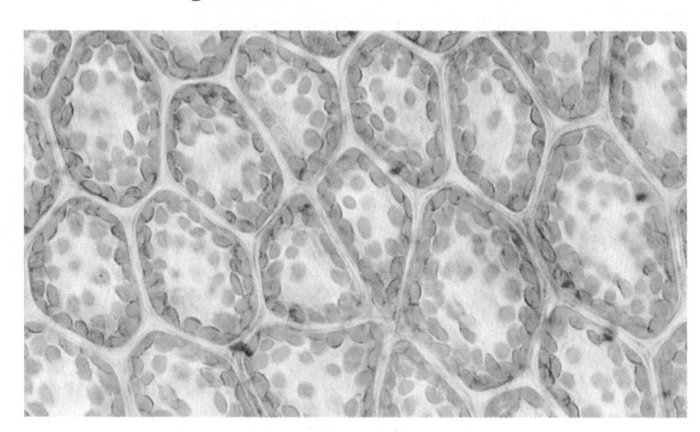

4. Light passing through the lens of a camera is focused on light-sensitive film inside the camera. The image on the film is real, inverted, and reduced.

After You Read

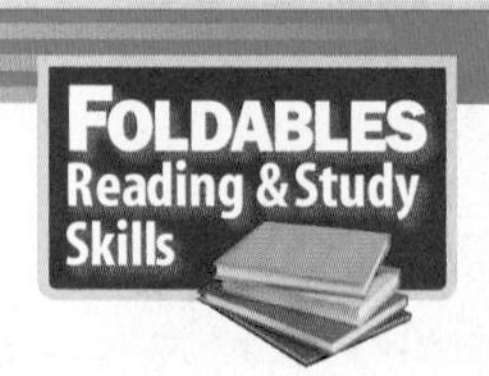

FOLDABLES Reading & Study Skills

On the back of the tabs of your Foldable explain the cause and effect of the reflections of the three types of mirrors.

Visualizing Main Ideas

Complete the following table about optical instruments.

Types of Optical Instruments		
Optical Instrument	**What is it?**	**How does it work?**
Convex Lens	transparent material that is thicker in the middle than at the edges	causes light rays to converge
Concave Lens	transparent material that is thicker at the edges than in the middle	causes light rays to diverge
Refracting Telescope	tube containing two convex lenses	
Reflecting Telescope	tube containing concave mirror, plane mirror, and convex lens	
Microscope		real image formed by first lens is magnified by second lens

Vocabulary Review

Vocabulary Words

a. concave lens
b. concave mirror
c. convex lens
d. convex mirror
e. cornea
f. focal length
g. focal point
h. microscope
i. optical axis
j. plane mirror
k. real image
l. reflecting telescope
m. refracting telescope
n. retina
o. virtual image

THE PRINCETON REVIEW

Study Tip

Write out the full questions and answers to end-of-chapter quizzes, not just the answers. This will help you form complete responses to important questions.

Using Vocabulary

Each of the following sentences is false. Make the sentence true by replacing the underlined word with the correct vocabulary word.

1. A flat, smooth surface that reflects light and forms an image is a convex mirror.
2. A reflecting telescope uses two convex lenses to magnify small, close objects.
3. Every light ray that travels parallel to the optical axis before hitting a concave mirror is reflected to pass through the retina.
4. A concave lens is thicker in the middle than at the edges.
5. The inner lining of the eye that converts light images into electrical signals is called the cornea.

Chapter 15 Assessment

Checking Concepts

Choose the word or phrase that best answers the question.

1. Which of the following types of images is not formed by plane mirrors?
A) upright
B) life-sized
C) enlarged
D) virtual

2. What object reflects light and curves inward?
A) plane mirror
B) concave mirror
C) convex mirror
D) concave lens

3. What type of mirror can be used to form a magnified image?
A) convex
B) plane
C) concave
D) transparent

4. What object is used in a headlight, flashlight, or spotlight to create a beam of light?
A) concave lens
B) convex lens
C) concave mirror
D) convex mirror

5. What do lenses do?
A) reflect light
B) refract light
C) diffract light
D) interfere with light

6. Which way does a concave lens bend light?
A) toward its optical axis
B) toward its center
C) toward its edges
D) toward its focal point

7. What kinds of lenses are most helpful for farsighted people?
A) flat lenses
B) convex lenses
C) concave lenses
D) plane lenses

8. Which object is not in a reflecting telescope?
A) plane mirror
B) concave mirror
C) convex lens
D) concave lens

9. Which of the following images do light rays never pass through?
A) real
B) virtual
C) enlarged
D) reduced

10. Which lens would you use to take a close-up photograph?
A) lens with a short focal length
B) lens with a long focal length
C) a small lens
D) wide-angle lens

Thinking Critically

11. Magicians often make objects disappear by using trick mirrors. How might a magician seem to make an object disappear by using a concave mirror?

12. A movie or a slide projector projects an image that is magnified and upright. Could such an image be formed by a single lens? Explain why or why not.

13. Which lens in a refracting telescope has a shorter focal length? Why?

14. Would a reflecting telescope work properly if a convex mirror replaced its concave mirror? Explain.

15. The magnification of a refracting telescope can be calculated by dividing the focal length of the objective lens by the focal length of the eyepiece lens. If an objective lens has a focal length of 1 m and the eyepiece has a focal length of 1 cm, what is the magnification of the telescope?

Developing Skills

16. Classifying Classify the different types of images formed by plane, concave, and convex mirrors.

17. Recognizing Cause and Effect Infer the effects of a hard, rigid eye lens on human vision. Would this make the eye more or less like a simple camera?

18. Comparing and Contrasting Compare and contrast a microscope and a refracting telescope.

19. Forming Hypotheses Suppose you use a magnifying glass underwater. Propose a hypothesis to explain how the magnification of images would be affected under water.

20. Interpreting Scientific Illustrations Describe the image of the candle seen through the lens.

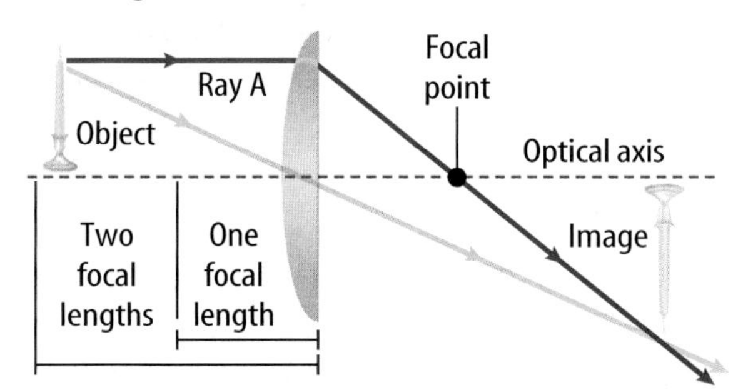

Performance Assessment

21. Writing in Science Write a report tracing the development of the telescope from the time of Galileo to the *Hubble Space Telescope.* Include illustrations with your report.

22. Oral Presentation Investigate the types of mirrors used in fun houses. Explain how these mirrors are formed and why they give distorted images. Demonstrate your findings to the class.

Technology

Go to the Glencoe Science Web site at **science.glencoe.com** or use the **Glencoe Science CD-ROM** for additional chapter assessment.

Test Practice

A student explored the properties of several converging lenses and created the table below.

Exploring Convex Lenses

Lens	Pencil's Distance to Lens	What Image Appears as
1		
2		
3		

Study the table and answer the following questions.

1. The data table could be improved by also recording the _____.
- **A)** pencil color
- **B)** lens height
- **C)** focal length of the lenses
- **D)** lens temperature

2. Each of the lenses above is able to make the image of the pencil larger. This occurs because _____.
- **F)** they bend light as light passes through them
- **G)** light bounces off their surfaces
- **H)** the lenses are larger than the pencil
- **J)** they produce light and send it off in beams to form images

Standardized Test Practice

Reading Comprehension

Read the passage. Then read each question that follows the passage. Decide which is the best answer to each question.

The History of the Telescope: An International Story

Roger Bacon, an English scientist, first wrote about the basic ideas behind the operation of a telescope in the 1200s. It was not until the early 1600s, however that Han Lippershey, a Dutchman who made spectacles for people with poor vision, made the first telescope. Lippershey noticed that objects appeared closer if he viewed them through a combination of a concave and a convex lens. He placed the lenses in a tube to hold them more easily. This was the world's first refracting telescope.

A few years later, an Italian scientist, Galileo, was the first to point a telescope toward the stars. Galileo first learned of the Dutch invention in 1609. At the time, it was mainly used to see objects on Earth, such as distant ships and enemy armies. This is why the telescope was first called a "spyglass." Galileo made his own telescope and began using it to view the sky. Before this, Galileo had not been particularly interested in astronomy. That quickly changed as he recorded observations of the Moon's surface, spots on the Sun, and four moons circling Jupiter.

Another advance in telescope technology occurred in 1663 when James Gregory, a Scottish scientist, designed the first reflecting telescope. Unfortunately, it would take twenty-five years until Isaac Newton would build the first reflecting telescope. The earliest, most valuable contribution to astronomy made by an American was the construction of the Hooker telescope, a reflecting telescope on Mount Wilson. Completed in 1917, its 100-inch reflecting concave mirror allowed astronomers to see other galaxies clearly for the first time.

Since then, scientists have continued to design and build larger and larger telescopes. The development of the modern telescope is the result of many years of work by many scientists across the world.

Test-Taking Tip As you read the passage, make a time line of the history of the telescope.

1. The telescope was first called a "spyglass" because it _____.

A) was helpful in observing the Moon and stars
B) was designed by Roger Bacon
C) could be used to watch other people
D) was first made by a Dutchman

2. According to the passage, scientists often _____.

F) use each other's work
G) are slow workers
H) aren't interested in many things
J) never read the work of other scientists

3. The earliest, most valuable contribution to astronomy made by an American was _____.

A) the first refracting telescope built in the 1600s
B) Roger Bacon's basic ideas about the operation of a telescope in the 1200s
C) the construction of the Hooker telescope on Mount Wilson which allowed astronomers to see other galaxies clearly for the first time
D) using a telescope to view the Moon's surface, spots on the Sun, and four moons circling Jupiter

Reasoning and Skills

Power of a Lens

Lens	Diopter	Focal length (m)
1	1/4	4
2	1/5	5
3	1/6	6
4	1/7	7
5	1/9	?

1. Diopters are one way to measure the strength of a lens. What is the focal length of lens 5?

A) 5
B) 8
C) 9
D) 10

Test-Taking Tip Study the values for the first four lenses and consider how the diopter value is related to the focal length.

Famous Telescopes

Observatory	Mirror's Diameter(m)	Location	Altitude(m)
Roque de los Muchachos	4.2	Spain	2,400
Mauna Kea	10	U.S.(HI)	4,200
Russian Academy	6	Russia	2,070
Cerro Tololo	4	Chile	2,200

2. Reflecting telescopes use large mirrors to view distant objects. To reduce the amount of atmosphere that interferes with observing the stars, they are built atop of mountains. According to the table above, which of these telescopes is at the lowest altitude?

F) Roque de los Muchachos
G) Mauna Kea
H) Russian Academy
J) Cerro Tololo

Test-Taking Tip Read the table's column headings carefully and then reread the question.

CORNEA
RETINA
EYE

3. Which of these belongs with the group above?

A) concave lens
B) convex lens
C) plane mirror
D) convex mirror

Test-Taking Tip In your mind, review the differences between lens and mirrors and between concave and convex.

4. In the last 15 years, the National Aeronautics and Space Administration (NASA) has launched several telescopes into orbit around Earth. From space, these telescopes transmit electronic signals with the images of distant stars, planets and other galaxies. What are the advantages of using a telescope in space?

Test-Taking Tip Think about what has to happen for the light from a distant star to reach a telescope on Earth. Compare that to what has to happen to that same light for it to reach a telescope in space.

UNIT 4

The Nature of Matter

How Are Playing Cards & the Periodic Table Connected?

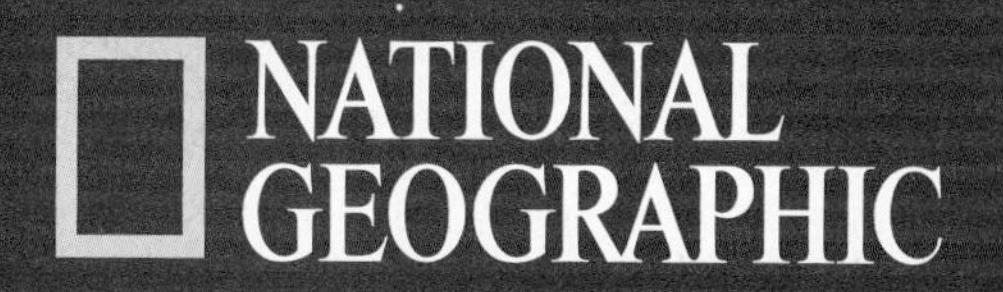

By 1860, scientists knew of about 60 elements. However, they had yet to clearly organize their knowledge. A Russian scientist named Dmitri Mendeleev changed that. Mendeleev loved to play solitaire, a type of card game in which playing cards are arranged into patterns according to their properties. One day, Mendeleev decided to make a set of cards on which he wrote the names and properties of the known elements. Then he began to arrange the cards into rows. The result was a table in which certain chemical properties could be seen to occur periodically—that is, to occur in a repeating pattern. In 1869, Mendeleev published his "periodic table" (seen here in a more advanced version). He left blank spaces in the table where the pattern seemed to call for elements that were not yet known. Over the next several decades, other scientists refined the table, and new elements were added. Modern periodic tables—like the one probably hanging in your classroom—still follow the basic pattern laid out by Mendeleev.

SCIENCE CONNECTION

ELEMENTS Working as a class, investigate all the elements that make up the modern periodic table. Divide up the task so that every student researches 3 to 5 different elements. On index cards, record the name, symbol, and atomic number of each element, as well as whether it is a metal, nonmetal, or metalloid. Also include at least two physical and two chemical properties of each element. Assemble the cards on a bulletin board to create an "enhanced" periodic table.

Solids, Liquids, and Gases

If you were traveling on this scenic highway, you couldn't help but notice the lake beside the road and the beautiful, snowcapped mountain in the distance. The clouds you see were formed when water in the air (a gas) came together to form droplets. In this chapter you will learn about the three states of matter—solid, liquid, and gas.

What do you think?

Science Journal Look at the picture below with a classmate. Discuss what you think this might be. Here's a hint: *It may be on your windows on a cold winter day.* Write your answer or best guess in your Science Journal.

Why does the mercury in a thermometer rise? Why do sidewalks have cracks? Many substances expand when heated and contract when cooled as you will see during this activity.

Safety Precautions Use caution when handling hot items.

Observe the expansion and contraction of air

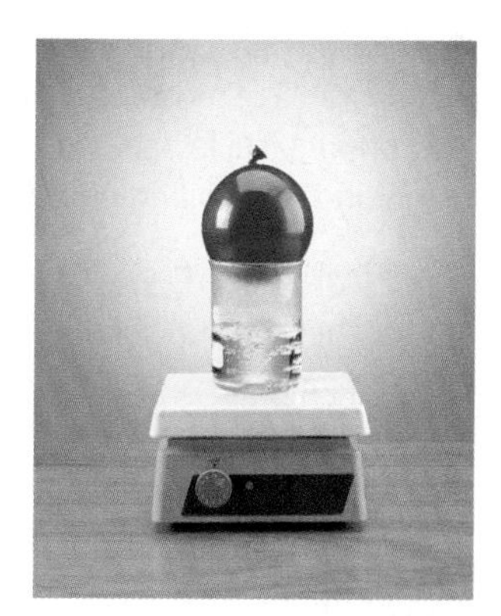

1. Blow up a balloon until it is half filled. Use a tape measure to measure the circumference of the balloon.
2. Pour water into a large beaker until it is half full. Place the beaker on a hotplate and wait for the water to boil.
3. Set the balloon on the mouth of beaker and observe for five minutes. Be careful not to allow the balloon to touch the hotplate. Measure the circumference of the balloon.

Observe

Write a paragraph in your Science Journal describing the changing size of the balloon's circumference. Infer why the balloon's circumference changed.

Before You Read

Making a Concept Map Study Fold **Make the following Foldable to help you organize information by diagramming ideas about solids, liquids, and gases.**

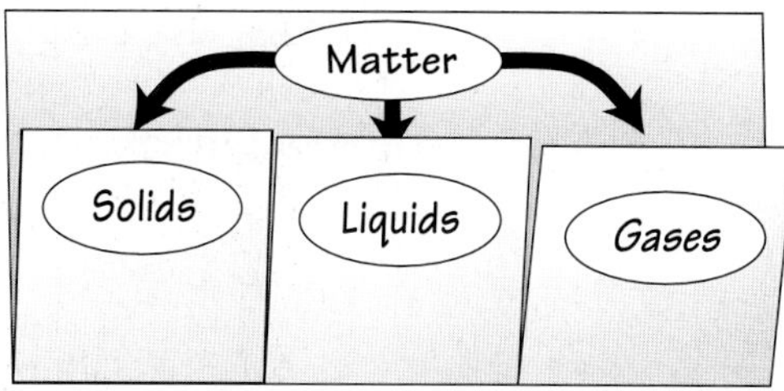

1. Place a sheet of paper in front of you so the long side is at the top. Fold the bottom of the paper to the top, stopping about 4 cm from the top.
2. Draw an oval above the fold. Write *Matter* inside the oval.
3. Fold both sides in and then unfold. Through the top thickness of the paper, cut along each of the fold lines to form three tabs.
4. Label the tabs *Solids, Liquids,* and *Gases* and draw an oval around each word. Draw arrows from the large oval to the smaller ovals.
5. Before you read, list examples of each you already know on the front of the tabs. As you read the chapter, add to your lists.

SECTION

Kinetic Theory

As You Read

***What* You'll Learn**

- **Explain** the kinetic theory of matter.
- **Describe** particle movement in the four states of matter.
- **Explain** particle behavior at the melting and boiling points.

Vocabulary

kinetic theory
melting point
heat of fusion
heat of vaporization
boiling point
diffusion
plasma
thermal expansion

***Why* It's Important**

You can use energy that is lost or gained when a substance changes from one state to another.

States of Matter

If you don't finish lunch quickly, you'll be late for practice. The soup is boiling on the stove. You hastily pour the soup into the bowl, but now it's too hot to eat. You add an ice cube and stir. The soup's temperature drops—now you can eat it without burning your tongue. Does this sound familiar? If you look closely at the situation, as shown in **Figure 1,** you can identify two states or phases of matter—solid and liquid. The boiling soup on the stove is in the liquid state. The steam directly above the boiling soup also is in the liquid state. The ice cube you dropped into your soup is in the solid state. Do these states have anything in common? How do they differ? Take a closer look at what is going on at the particle level in the three states of matter.

Kinetic Theory The **kinetic theory** is an explanation of how particles in matter behave. To explain the behavior of particles, it is necessary to make some basic assumptions. The three assumptions of the kinetic theory are as follows:

1. All matter is composed of small particles (atoms, molecules, and ions).
2. These particles are in constant, random motion.
3. These particles are colliding with each other and the walls of their container.

Particles lose some energy during collisions with other particles. But the amount of energy lost is very small and can be neglected in most cases.

To visualize the kinetic theory, think of each particle as a tiny table tennis ball in constant motion. These balls are bouncing and colliding with each other. Mentally visualizing matter in this way can help you understand the movement of particles in matter.

Figure 1
Two states of water are present in this photograph.
Can you identify the solid and liquid states?

Thermal Energy Think about the ice cube in the soup. Does the ice cube appear to be moving? How can a frozen, solid ice cube have motion? Remember to focus on the particles. Atoms in solids are held tightly in place by the attraction between the particles. This attraction between the particles gives solids a definite shape and volume. However, the thermal energy in the particles causes them to vibrate in place. Thermal energy is the total energy of a material's particles, including kinetic—vibrations and movement within and between the particles—and potential—resulting from forces that act within or between particles. When the temperature of the substance is lowered, the particles will have less thermal energy and will vibrate more slowly.

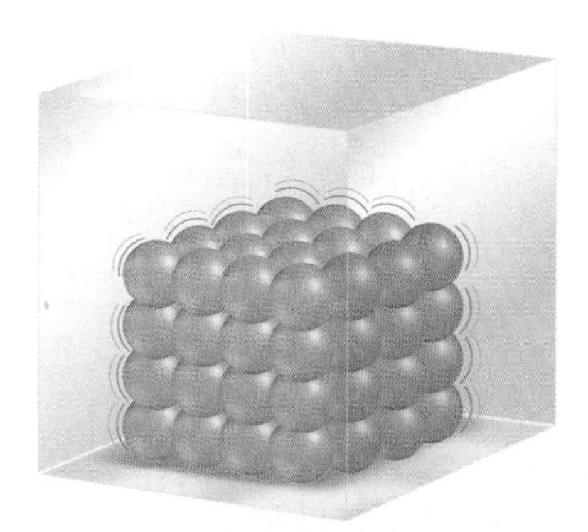

Figure 2
The particles in a solid are packed together tightly and are constantly vibrating in place.

 ***What is thermal energy?***

Average Kinetic Energy Temperature is the term used to explain how hot or cold an object is. In science, temperature means the average kinetic energy in the substance, or how fast the particles are moving. On average, molecules of frozen water at 0°C will move slower than molecules of water at 100°C. Therefore, water molecules at 0°C have lower average kinetic energy than the molecules at 100°C. Molecules will have some movement and kinetic energy at all temperatures, except at absolute zero. Scientists theorize that at absolute zero, or –273.15°C, particle motion is so slow that thermal energy is equal to zero.

 ***How are kinetic energy and temperature related?***

Solid State An ice cube is an example of a solid. The particles of a solid are closely packed together, as shown in **Figure 2.** Most solid materials have a specific type of geometric arrangement in which they form when cooled. The type of geometric arrangement formed by a solid is important. Chemical and physical properties of solids often can be attributed to the type of geometric arrangement that the solid forms. **Figure 3** shows the geometric arrangement of solid water. Notice that the hydrogen and oxygen atoms are alternately spaced in the arrangement.

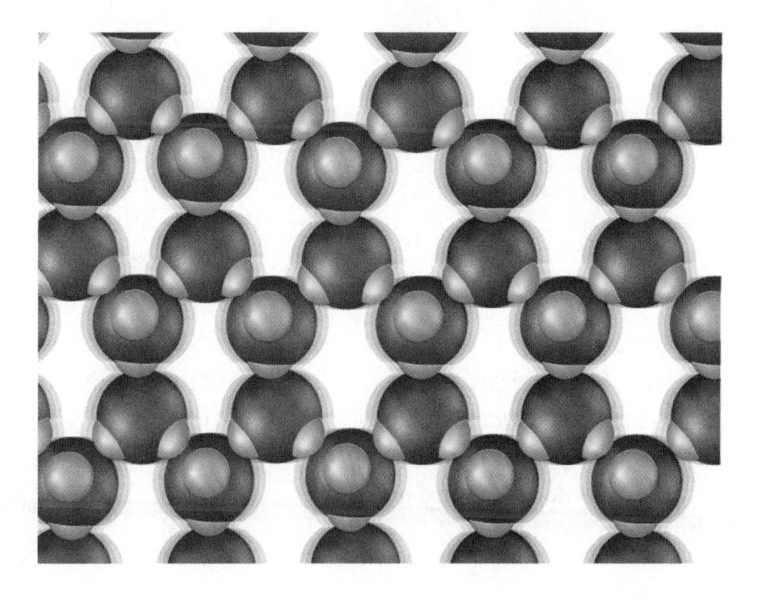

Figure 3
The particles in water align themselves in an ordered geometric pattern. Even though a solid ice cube doesn't look like it is moving, its molecules are vibrating in place.

Liquid

Figure 4
The particles in a liquid are moving more freely than the particles in a solid. They have enough kinetic energy to slip out of the ordered arrangement of a solid.

Liquid State What happens to a solid when thermal energy or heat is added to it? Think about the ice cube in the hot soup. The particles in the hot soup are moving fast and colliding with the vibrating particles in the ice cube. The collisions of the particles transfer energy from the soup to the ice cube. The particles on the surface of the ice cube vibrate faster. These particles collide with and transfer energy to other ice particles. Soon the particles of ice have enough kinetic energy to overcome the attractive forces. The particles of ice gain enough kinetic energy to slip out of their ordered arrangement and the ice melts. This is known as the **melting point,** or the temperature at which a solid begins to liquefy. Energy is required for the particles to slip out of the ordered arrangement. The amount of energy required to change a substance from the solid phase to the liquid phase at its melting point is known as the **heat of fusion.**

✔ Reading Check *What is heat of fusion?*

Liquids Flow Particles in a liquid, shown in **Figure 4,** have more kinetic energy than particles in a solid. This extra kinetic energy allows particles to partially overcome the attractions to other particles. Thus, the particles can slide past each other allowing liquids to flow and take the shape of their container. However, the particles in a liquid have not completely overcome the attractive forces between them. This causes the particles to cling together, giving liquids a definite volume.

✔ Reading Check *Why do liquids flow?*

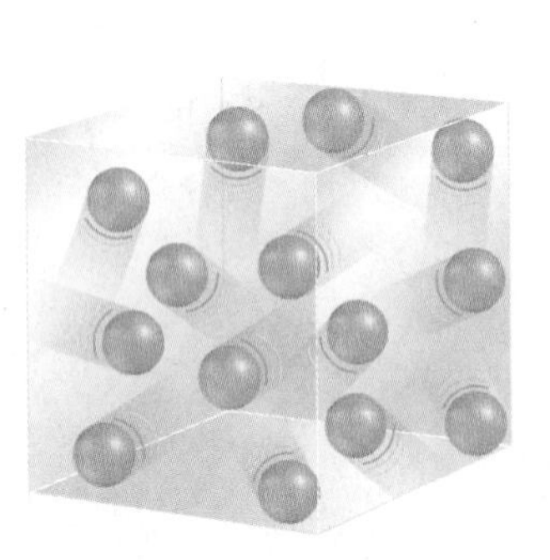

Gas

Figure 5
In gases, the particles are far apart and the attractive forces between the particles are overcome. Gases do not have a definite volume or shape.

Gaseous State Particles in the gas state are shown in **Figure 5.** Gas particles have enough kinetic energy to overcome the attractions between them. Gases do not have a fixed volume or shape. Therefore, they can spread far apart or contract to fill the container that they are in. How does a liquid become a gas? The particles in a liquid are constantly moving. Some particles are moving faster and have more kinetic energy than others. The particles that are moving fast enough can escape the attractive forces of other particles and enter the gas phase. This process is called vaporization. Vaporization can occur in two ways—evaporation and boiling. Evaporation is vaporization that occurs at the surface of a liquid and can occur at temperatures below the liquid's boiling point. To evaporate, particles must have enough kinetic energy to escape the attractive forces of the liquid. They must be at the liquid's surface and traveling away from the liquid.

✔ Reading Check *What occurs on a molecular level when a liquid begins to boil?*

Figure 6
Boiling occurs throughout a liquid when the pressure of the vapor in the liquid equals the pressure of the vapor on the surface of the liquid.

Boiling Point A second way that a liquid can vaporize is by boiling. Unlike evaporation, boiling occurs throughout a liquid at a specific temperature depending on the pressure on the surface of the liquid. Boiling is shown in **Figure 6.** The **boiling point** of a liquid is the temperature at which the pressure of the vapor in the liquid is equal to the external pressure acting on the surface of the liquid. This external pressure is a force pushing down upon a liquid, keeping particles from escaping. Particles require energy to overcome this force. **Heat of vaporization** is the amount of energy required for the liquid at its boiling point to become a gas.

✔ Reading Check *How does external pressure affect the boiling point of a liquid?*

Gases Fill Their Container What happens to the attractive forces between the particles in a gas? The gas particles are moving so quickly and are so far apart that they have overcome the attractive force between them. Because the attractive forces between them are overcome, gases do not have a definite shape or a definite volume. The movement of particles and the collisions between them cause gases to diffuse. **Diffusion** is the spreading of particles throughout a given volume until they are uniformly distributed. Diffusion occurs in solids and liquids but occurs most rapidly in gases. For example, if you spray air freshener in one corner of a room, it's not long before you smell the scent all over the room. The particles of gas have moved, collided, and "filled" their container—the room. The particles have diffused. Gases will fill the container that they are in even if the container is a room. The particles continue to move and collide in a random motion within their container.

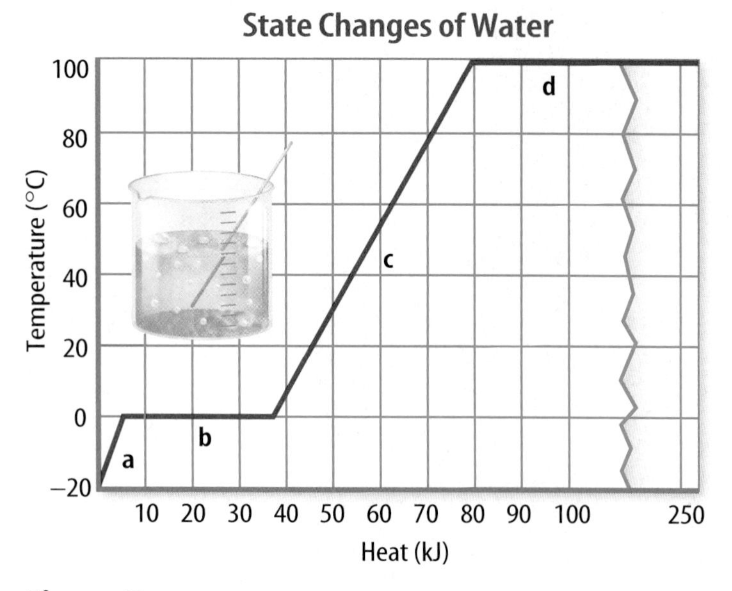

Figure 7
This graph shows the heating curve of water. At A and C the water is increasing in kinetic energy. At B and D the added energy is used to overcome the bonds between the particles.

Heating Curve of a Liquid A graph of water being heated from –20°C to 100°C is shown in **Figure 7.** This type of graph is called a heating curve because it shows the temperature change of water as thermal energy or heat is added. Notice the two areas on the graph where the temperature does not change. At 0°C, ice is melting. All of the energy put into the ice at this temperature is used to overcome the attractive forces between the particles in the solid. The temperature remains constant during melting. After the attractive forces are overcome, particles move more freely and their average kinetic energy or temperature increases. At 100°C, water is boiling or vaporizing and the temperature remains constant again. All of the energy that is put into the water goes to overcoming the remaining attractive forces between the water particles. When all of the attractive forces in the water are overcome, the energy goes to increasing the temperature of the particles.

✔ **Reading Check** *What is occurring at the two temperatures on the heat curve where the graph is a flat line?*

Plasma State So far, you've learned about the three familiar states of matter—solids, liquids, and gases. But none of these is the most common state of matter in the universe. Scientists estimate that much of the matter in the universe is plasma. **Plasma** is gas consisting of positively and negatively charge particles. Although this high-temperature gas contains positive and negative particles, the overall charge of the gas is neutral because equal numbers of both charges are present. Recall that on average, particles of matter move faster as the matter is heated to higher temperatures. The faster the particles move the greater the force is with which they collide. The forces produced from high-energy collisions are so great that electrons from the atom are stripped off. This state of matter is called plasma. All of the observed stars including the Sun, shown in **Figure 8,** consist of plasma. Plasma also is found in lightning bolts, neon and fluorescent tubes, and auroras.

Figure 8
Stars including the Sun contain matter that is in the plasma phase. Plasma exists where the temperature is extremely high.

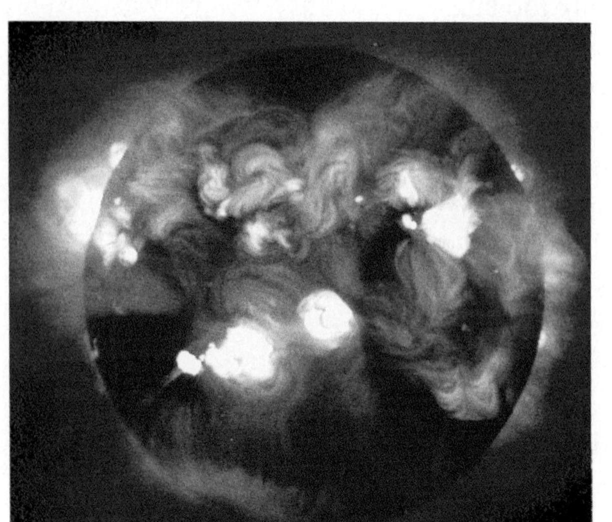

✔ **Reading Check** *What is plasma?*

Thermal Expansion

You have learned how the kinetic theory is used to explain the behavior of particles in different states of matter. The kinetic theory also explains other characteristics of matter in the world around you. Have you noticed the seams in a concrete driveway or sidewalk? A gap often is left between the sections to clearly separate them. These separation lines are called expansion joints. When concrete absorbs heat, it expands. Then when it cools, it contracts. If expansion joints are not used, the concrete will crack when the temperature changes.

Figure 9
As the thermometer is heated, the column of liquid in the thermometer expands. As the temperature cools, the liquid in the thermometer contracts.

Expansion of Matter The kinetic theory can be used to explain this behavior in concrete. Recall that particles move faster and separate as the temperature rises. This separation of particles results in an expansion of the entire object known as thermal expansion. **Thermal expansion** is an increase in the size of a substance when the temperature is increased. The kinetic theory can be used to explain the contraction in objects, too. When the temperature of an object is lowered, particles slow down. The attraction between the particles increases and the particles move closer together. The movements of the particles closer together result in an overall shrinking of the object known as contraction.

Expansion in Liquids Expansion and contraction occur in most solids, liquids, and gases. A common example of expansion in liquids occurs in thermometers, as shown in **Figure 9.** The addition of energy causes the particles of the liquid in the thermometer to move faster. The particles in the liquid in the narrow thermometer tube start to move farther apart as their motion increases. The liquid has to expand only slightly to show a large change on the temperature scale.

Expansion in Gases **Figure 10** is an example of thermal expansion in gases. Hot-air balloons are able to rise due to thermal expansion of air. The air in the balloon is heated, causing the distance between the particles in the air to expand. As the air in the hot-air balloon expands, the number of particles per cubic centimeter decreases. This expansion results in a decreased density of the hot air. Because the density of the air in the hot-air balloon is lower than the density of the cool air outside, the balloon will rise.

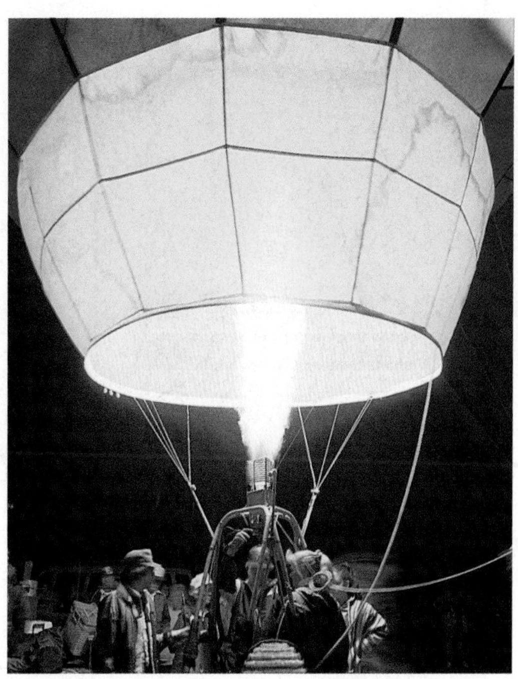

Figure 10
Heating the air in this hot-air balloon causes the particles in the air to separate, creating a lower density inside the balloon.

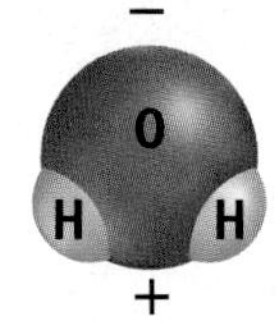

Figure 11
The positively and negatively charged regions on a water molecule interact to create empty spaces in the crystal lattice. These interactions cause water to expand when it is in the solid phase.

The Strange Behavior of Water Normally, substances expand as the temperature rises, because the particles move farther apart. An exception to this rule, however, is water. Water molecules are unusual in that they have highly positive and highly negative areas. **Figure 11** is a diagram of the water molecule showing these charged regions. These charged regions affect the behavior of water. As the temperature of water drops, the particles move closer together. The unlike charges will be attracted to each other and line up so that only positive and negative zones are near each other. Because the water molecules orient themselves according to charge, empty spaces occur in the structure. These empty spaces are larger in ice than in liquid water, so water expands when going from a liquid to a solid state. Solid ice is less dense than liquid water. That is why ice floats on the top of lakes in the winter.

Solid or a Liquid?

There are other substances that have unusual behavior when changing states. Amorphous solids and liquid crystals are two classes of materials that do not react as you would expect when they are changing states.

Amorphous Solids Ice melts at 0°C, gold melts at 1,064°C and lead melts at 327°C. But not all solids have a definite temperature at which they change from solid to liquid. Some solids merely soften and gradually turn into a liquid over a temperature range. There is not an exact temperature like a boiling point where the phase change occurs. These solids lack the highly ordered structure found in crystals. They are known as amorphous solids from the Greek word for "without form."

You are familiar with two amorphous solids—glass and plastics. The particles that make up amorphous solids are typically long, chainlike structures that can get jumbled and twisted instead of being neatly stacked into geometric arrangements. Interactions between the particles occur along the chain, which gives amorphous solids some properties that are very different from crystalline solids.

For example, glass appears to be a solid, but glass windows actually change over time. In old houses if you measure the thickness of the top and bottom of a windowpane, you will find that the top is thinner than the bottom. Because of the gravitational pull and the lack of crystalline structure, the glass in the windowpane will flow to the bottom over time.

Reading Check *What are two examples of amorphous solids?*

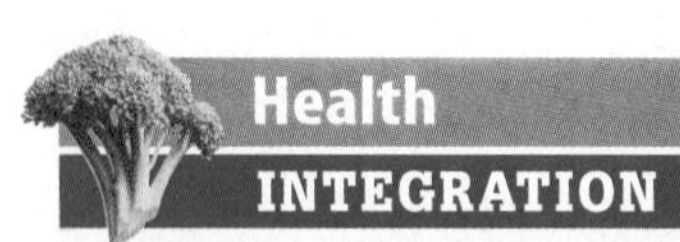

Some liquid crystals can form thin layers that are one molecule thick. These liquid crystals react to tiny temperature changes by changing color making them useful in determining temperature changes over the surface of the skin such as in a thermometer. In your Science Journal, identify other possible uses for this type of liquid crystal.

Liquid Crystals Liquid crystals are another group of materials that do not change states in the usual manner. Normally, the ordered geometric arrangement of a solid is lost when the substance goes from the solid state to the liquid state. Liquid crystals start to flow during the melting phase similar to a liquid, but they do not lose their ordered arrangement completely, as most substances do. Liquid crystals will retain their geometric order in specific directions.

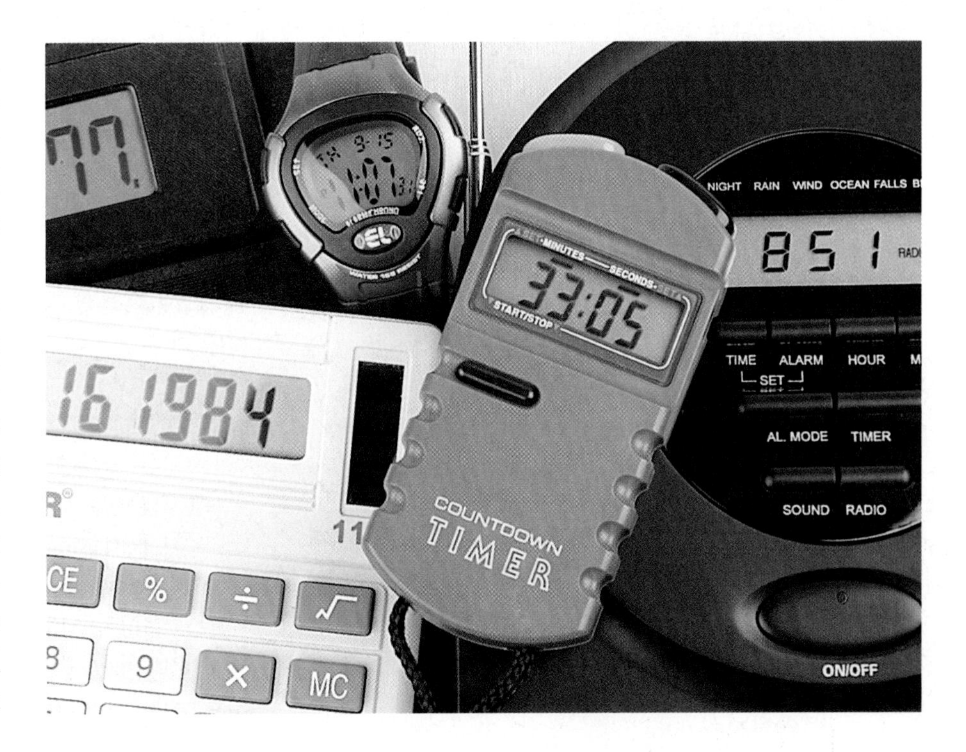

Figure 12
Liquid crystals are used in the displays of watches, clocks, calculators, and some notebook computers because they respond to electric fields.

Liquid crystals are placed in classes depending upon the type of order they maintain when they liquefy. They are highly responsive to temperature changes and electric fields. Scientists use these unique properties of liquid crystals to make liquid crystal displays (LCD) in the displays of watches, clocks, and calculators, as shown in **Figure 12.**

Reading Check ***What unusual property do liquid crystals have when they melt?***

Section 1 Assessment

1. What are the three basic assumptions of the kinetic theory?
2. Describe the movement of the particles in solids, liquids, and gases.
3. Describe the movement of the particles at the melting point of a substance.
4. Describe the movement of the particles at the boiling point of a substance.
5. **Think Critically** Would the boiling point of water be higher or lower on the top of a mountain peak? How would the boiling point be affected in a pressurized boiler system? Explain.

Skill Builder Activities

6. **Interpreting Data** Using the graph in **Figure 7,** describe the energy changes that are occurring when water goes from −15°C to 100°C. **For more help, refer to the Science Skill Handbook.**
7. **Making and Using Graphs** The melting point of acetic acid is 16.6°C and the boiling point is 117.9°C. Draw a graph showing the phase changes for acetic acid similar to the graph in **Figure 7.** Clearly mark the three phases, the boiling point, and the melting point on the graph. **For more help, refer to the Science Skill Handbook.**

Activity

How Thermal Energy Affects Matter

The states of matter and its characteristics change as its thermal energy changes.

What You'll Investigate

How does thermal energy affect the state of matter?

Materials

beakers (2)
ring clamp (1)
ring stand
wire mesh
hot plate
ice
thermometer

Goals

- **Explain** the thermal energy changes that occur as matter goes from the solid to gas state.

Safety Precautions

Procedure

1. Set up the equipment as pictured. Prepare a data table in your Science Journal.
2. Gently heat the ice in the lower beaker. Every 3 min record your observations and the temperature of the water in the bottom container. Do not touch the thermometer to the bottom or sides of the container.
3. After the ice in the beaker melts and the water begins to boil, observe the system for several more minutes and record your observations.
4. Turn off the heat and let your system completely cool before you clean up.

Conclude and Apply

1. **Draw** a picture of the system used in this lab in your Science Journal. Label the state the water started at in the lower beaker, the state it changed into in the lower beaker, the state above the lower beaker, and the state on the outside of the upper beaker.
2. Which location on the diagram has the greatest thermal energy and which has the least amount of thermal energy?
3. Make a time-temperature graph using your data for your Science Journal.

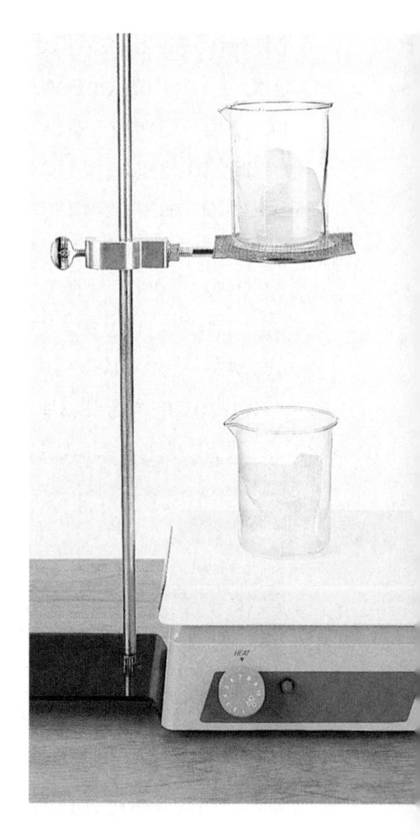

Communicating Your Data

Compare your results with other groups in the lab. **For more help, refer to the Science Skill Handbook.**

SECTION

Properties of Fluids

How do ships float?

Some ships are so huge that they are like floating cities. For example, aircraft carriers are large enough to allow airplanes to take off and land on their decks. Despite their weight, these ships are able to float. This is because a greater force pushing up on the ship opposes the weight—or force—of the ship pushing down. What is this force? This supporting force is called the buoyant force. **Buoyancy** is the ability of a fluid—a liquid or a gas—to exert an upward force on an object immersed in it. If the buoyant force is equal to the object's weight, the object will float. If the buoyant force is less than the object's weight, the object will sink.

Archimedes' Principle In the third century B.C., a Greek mathematician named Archimedes made a discovery about buoyancy. Archimedes found that the buoyant force on an object is equal to the weight of the fluid displaced by the object. For example, if you place a block of wood in water, it will push water out of the way as it begins to sink—but only until the weight of the water displaced equals the block's weight. When the weight of water displaced—the buoyant force—becomes equal to the weight of the block, it floats. If the weight of the water displaced is less than the weight of the block, the object sinks. **Figure 13** shows the forces that affect an object in a fluid.

As You Read

What You'll Learn

- **Explain** Archimedes' principle.
- **Explain** Pascal's principle.
- **Explain** Bernoulli's principle and explain how we use it.

Vocabulary

buoyancy
pressure
viscosity

Why It's Important

Properties of fluids determine the design of ships, airplanes, and hydraulic machines.

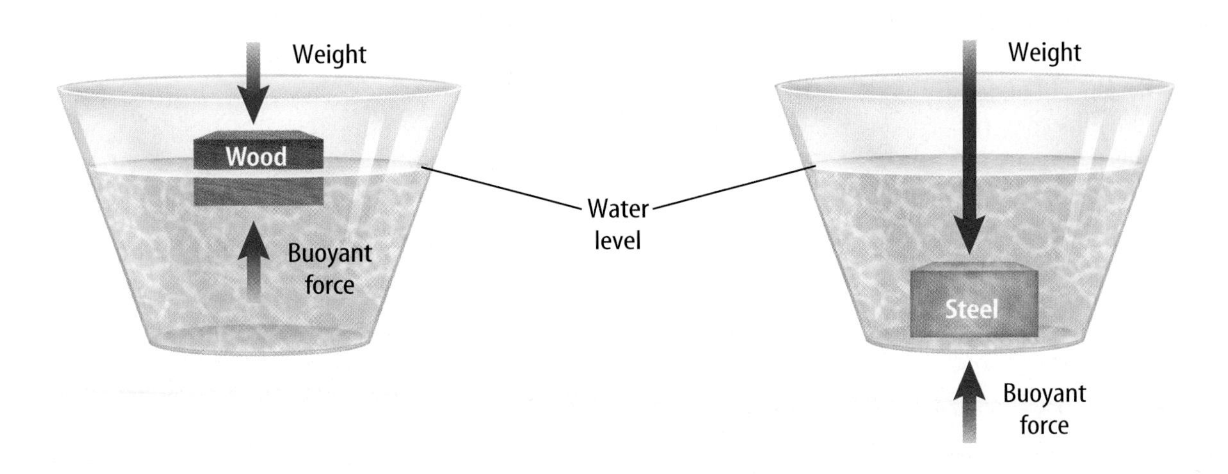

Figure 13
If the buoyant force of the fluid is equal to the weight of the object, the object floats. If the buoyant force of the fluid is less than the weight of the object, the object sinks.

Figure 14
An empty hull of a ship contains mostly air. Its density is much lower than the density of a solid-steel hull. The lower density of the steel and air combination is what allows the ship to float in water.

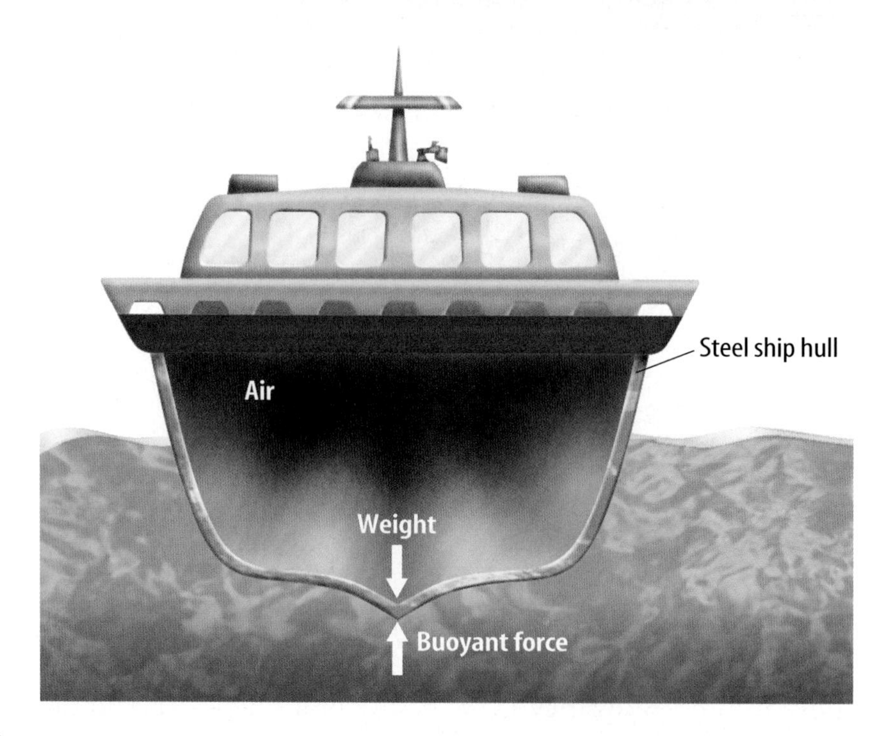

Observing Density and Buoyancy of Substances

Procedure

1. Pour 10 mL of **corn syrup** into a 100-mL beaker. In another **beaker,** add 3 to 4 drops of **food coloring** to 10 mL of **water.** Pour the dyed water into the 100-mL beaker containing corn syrup. Add 10 mL of **vegetable oil** to the beaker.
2. Drop a 0.5 cm square piece of **aluminum foil, a steel nut,** and a **whole peppercorn** into the 100-mL beaker.

Analysis

1. Using the concept of density, explain why the contents of the beaker separated into layers.
2. Using the concept of buoyancy, explain why the foil, steel nut, and peppercorn settled in their places.

Density Would a steel block the same size as a wood block float in water? They both displace the same volume and weight of water when submerged. Therefore, the buoyant force on the blocks is equal. Yet the steel block sinks and the wood block floats. What is different? The volume of the blocks and the volume of the water displaced each have different masses. If the three equal volumes have different masses, they must have different densities. Remember that density is mass per unit volume. The density of the steel block is greater than the density of water. The density of the wood block is less than the density of water. An object will float if its density is less than the density of the fluid it is placed in.

Suppose you formed the steel block into the shape of a hull filled with air as in **Figure 14.** Now the same mass takes up a larger volume. The overall density of the steel boat and air is less than the density of water. The boat will now float.

Pascal's Principle

If you are underwater, you can feel the pressure of the water all around you. **Pressure** is force exerted per unit area. Do you realize that Earth's atmosphere is a fluid? Earth's atmosphere exerts pressure all around you.

Blaise Pascal (1623–1662), a French scientist, discovered a useful property of fluids. According to Pascal's principle, pressure applied to a fluid is transmitted throughout the fluid. For example, when you squeeze one end of a balloon, the balloon expands out on the other end. When you squeeze one end of a toothpaste tube, toothpaste emerges from the other end. The pressure has been transmitted through the fluid toothpaste.

Applying the Principle Hydraulic machines are machines that move heavy loads in accordance with Pascal's principle. Maybe you've seen a car raised using a hydraulic lift in an auto repair shop. A pipe that is filled with fluid connects small and large cylinders as shown in **Figure 15.** Pressure applied to the small cylinder is transferred through the fluid to the large cylinder. Because pressure remains constant throughout the fluid, according to Pascal's principle, more force is available to lift a heavy load by increasing the surface area. With a hydraulic machine, you could use your weight to lift something much heavier than you are. Do the Math Skills Activity to see how force, pressure, and area are related.

Figure 15
The pressure remains the same throughout the fluid in a hydraulic lift.

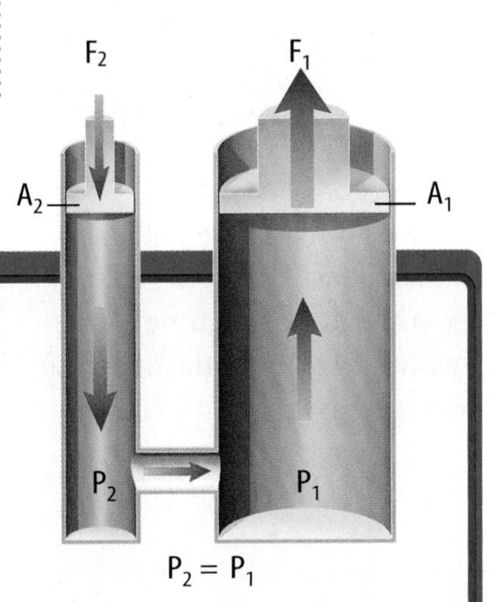

Math Skills Activity

Calculating Forces Using Pascal's Principle

Example Problem

A hydraulic lift is used to lift a heavy machine that is pushing down on a 2.8 m^2 piston (A_1) with a force (F_1) of 3,700 N. What force (F_2) needs to be exerted on a 0.072 m^2 piston (A_2) to lift the machine?

Solution

1 *This is what you know:*

$$\text{pressure} = \frac{\text{Force}}{\text{Area}} = \frac{F}{A}$$

$F_1 = 3{,}700$ N

$A_1 = 2.8\ m^2$

$A_2 = 0.072\ m^2$

2 *This is what you need to find:* Force needed: F_2

3 *Because $P_1 = P_2$, this is the equation that you need to use:*

$$\frac{F_1}{A_1} = \frac{F_2}{A_2}$$

4 *Solve the equation for F_2 and then substitute the known values:*

$$F_2 = \frac{F_1\,A_2}{A_1} = \frac{3{,}700\ \text{N} \times 0.072\ m^2}{2.8\ m^2} = 95\ N$$

Check your answer by substituting it and the known values back into the original equation.

Practice Problem

A heavy crate applied a force of 1,500 N on a 25-m^2 piston. What force needs to be exerted on the 0.80-m^2 piston to lift the crate?

For more help, refer to the Math Skill Handbook.

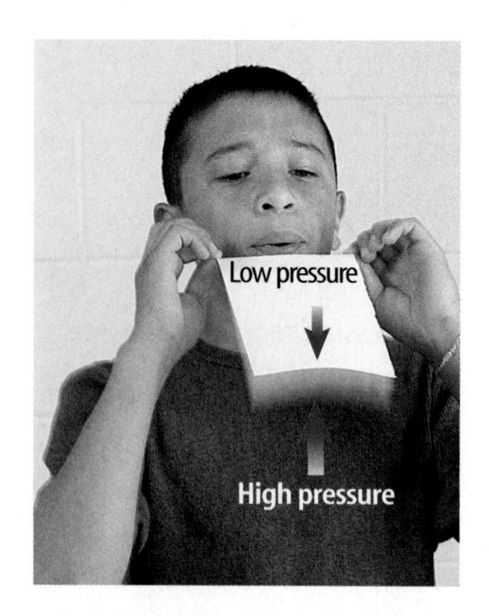

Figure 16
The air above the sheet of paper is moving faster than the air under the paper, creating a low-pressure area above the paper, so the paper rises.

Bernoulli's Principle

It took humans thousands of years to learn to do what birds do instinctively—fly, glide, and soar. It wasn't easy to build a machine that could lift itself off the ground and fly with people aboard. This ability is a property of fluids stated in Bernoulli's principle. Daniel Bernoulli (1700–1782) was a Swiss scientist who studied the properties of moving fluids such as water and air. He published his discovery in 1738. According to Bernoulli's principle, as the velocity of a fluid increases, the pressure exerted by the fluid decreases. One way to demonstrate Bernoulli's principle is to blow across the top surface of a sheet of paper, as in **Figure 16.** The paper will rise. The velocity of the air you blew over the top surface of the paper is greater than that of the quiet air below it. As a result, the air pressure pushing down on the top of the paper is lower than the air pressure pushing up on the paper. The net force below the paper pushes the paper upward.

Now, look at the curvature of the airplane wing in **Figure 17.** As the plane moves forward, the air passing over the top of the wing travels faster than the air passing below it. Thus, the pressure above the wing is less than the pressure below it. The result is a net upward force on the wing. This upward force contributes to the lift of an airplane wing.

Reading Check *How does pressure change as the velocity of a fluid increases?*

Notice that the airflow over the wing is a smooth path. If the wing encounters air that is rotating, the airplane might not have lift and the plane could crash. It is important for the pilot to be aware of any turbulent airflow, especially when an airplane is landing where there is little room for error.

Research Visit the Glencoe Science Web site at **science.glencoe.com** for information about flying machines. Communicate to your class what you learn.

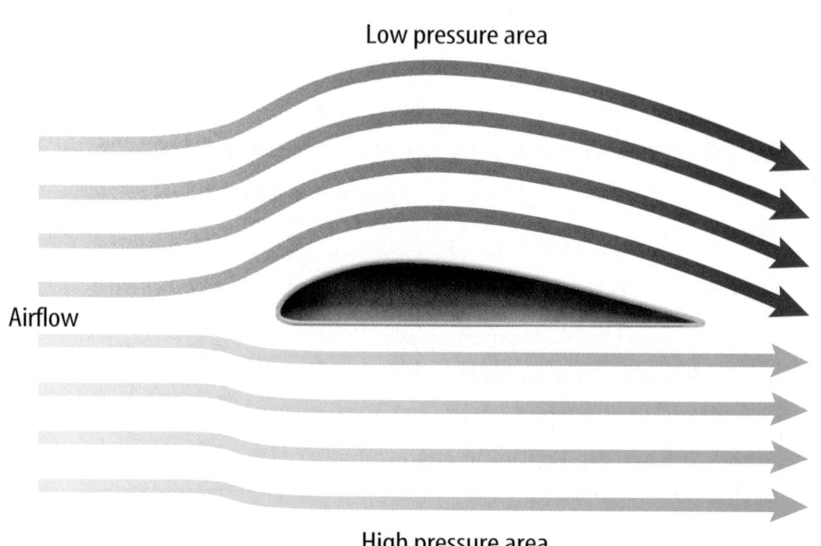

Figure 17
This is a side view of an airplane wing. The air above the wing travels faster over the wing than under it. This creates a low-pressure area above the wing.

Fluid Flow

Another property exhibited by fluid is its tendency to flow. A resistance to flow by a fluid is called **viscosity.** Fluids vary in their tendency to flow. For example, when you take syrup out of the refrigerator and pour it, the flow of syrup is slow. But if this syrup were heated, it would flow much faster. Water has a low viscosity because it flows easily. Cold syrup has a high viscosity because it flows slowly.

Fluids vary in their tendency to flow because their structures differ. When a container of liquid is tilted to allow flow to begin, the flowing particles will transfer energy to the particles that are stationary. In effect, the flowing particles are pulling the other particles, causing them to flow, too. If the flowing particles do not effectively pull the other particles into motion, then the liquid has a high viscosity, or a high resistance to flow. If the flowing particles pull the other particles in motion easily, then the liquid has low viscosity, or a low resistance to flow.

A rise in temperature increases the movement of particles in any substance. In substances that are heated, such as the syrup mentioned above, the particles move much faster. If the particles are moving faster, then energy transfer occurs much faster. Heating the syrup causes the particles to interact more, resulting in a faster energy transfer and a lower viscosity or a lower resistance to flow.

Reading Check *How does temperature affect viscosity?*

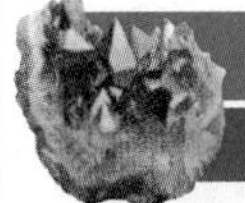
Earth Science INTEGRATION

Magma, or liquefied rock from a volcano, is an example of a liquid with varying viscosity. The viscosity of magma depends upon its composition. The viscosity of the magma flow determines the shape of the volcanic cone. In your Science Journal, infer the type of volcano cone that is created with high- and low-viscosity lava flows.

Section  Assessment

1. Explain what two opposing forces are acting on an object floating in water.
2. What is Archimedes' principle? Explain how it enables heavy ships to float.
3. What is Pascal's principle? Explain how it works in a plastic mustard bottle.
4. Using Bernoulli's principle, explain how roofs are lifted off building in tornados.
5. **Think Critically** If you fill a balloon with air, tie it off, and release it, it will fall to the floor. Why does it fall instead of float? What if the balloon contained helium?

Skill Builder Activities

6. **Measuring in SI** The density of water is 1.0 g/cm^3. How many kilograms of water does a submerged 120-cm^3 block displace? One kilogram weighs 9.8 N. What is the buoyant force on the block? **For more help, refer to the Science Skills Handbook.**
7. **Solving One-Step Equations** If you wanted to lift an object weighing 20,000 N, how much force would you need to exert on the small piston in **Figure 15? For more help, refer to the Math Skills handbook.**

Behavior of Gases

As You Read

What You'll Learn

- **Explain** how a gas exerts pressure on its container.
- **Explain** how a gas is affected when pressure, temperature, or volume is changed.

Vocabulary

pascal

Why It's Important

Being able to explain and to predict the behavior of gases is useful because you live in a sea of air.

Pressure

Relax and take a deep breath. If the air is clean and fresh, it is primarily a mixture of nitrogen, oxygen, argon, and carbon dioxide. Small amounts of hydrogen, water vapor, and a few other elements are present also. The atmosphere is held in place by the gravitational force on these tiny gas particles. Without the force of gravity acting on these particles, they would escape into space. More information about the atmosphere is on the next page in **Figure 19.**

Particle Collisions You learned from kinetic theory that gas particles are constantly moving and colliding with anything in their path. The collisions of these particles in the air result in atmospheric pressure. Pressure is the amount of force exerted per unit of area, or $P = F/A$. It is measured in units called **pascals** (Pa), the SI unit of pressure. Because pressure is the amount of force divided by area, one pascal of pressure is one Newton per square meter or 1 N/m^2. This is a small pressure unit, so most pressures are given in kilopascals (kPa) or 1,000 pascals. At sea level, atmospheric pressure is 101.3 kPa. This means that at Earth's surface, the atmosphere exerts a force of about 101,300 N on every square meter—about the weight of a large truck.

Often, gases are confined within containers. A balloon and a bicycle tire are considered to be containers. They remain inflated because of collisions the air particles have with the walls of their container, as shown in **Figure 18.** This collection of forces, caused by the collisions of the particles, pushes the walls of the container outward. If more air is pumped into the balloon, the number of air particles is increased. This causes more collisions with the walls of the container, which causes it to expand. Since the bicycle tire can't expand much, its pressure increases.

Reading Check *How are force, area, and pressure related?*

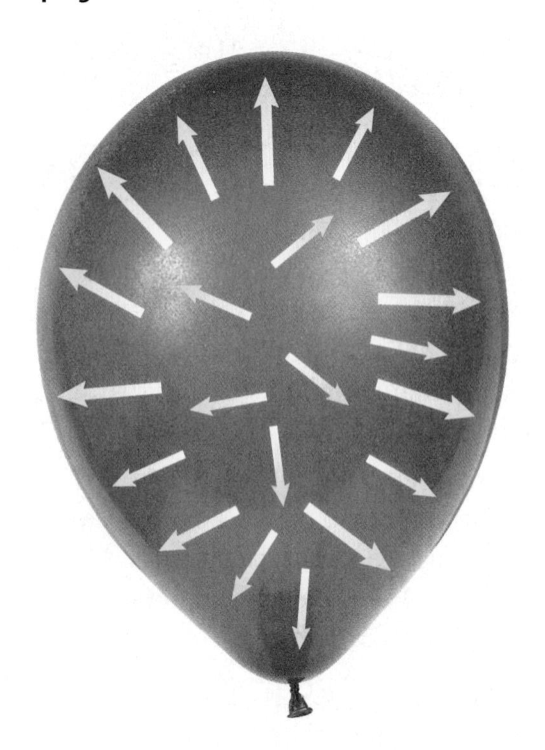

Figure 18
The force created by the many particles in air striking the balloon's walls forces them outward, keeping the balloon inflated.

NATIONAL GEOGRAPHIC

VISUALIZING ATMOSPHERIC LAYERS

Figure 19

Earth's atmosphere is divided into five layers. The air gets thinner as distance from Earth's surface increases. Temperature is variable, however, due to differences in the way the layers absorb incoming solar energy.

The Hubble Space Telescope

Exosphere (on average, 1,100°C; pressure negligible)

500 km

Gas molecules are sparse in the exosphere (beyond 500 km). The Landsat 7 satellite and the Hubble Space Telescope orbit in this layer, at an altitude of about 700 km and 600 km respectively. Beyond the exosphere there is nothing but the vacuum of interplanetary space.

The space shuttle crosses all the atmosphere's layers.

Thermosphere (−80°C to 1,000°C; pressure negligible)

Compared to the exosphere, gas molecules are slightly more concentrated in the thermosphere (85–500 km). Air pressure is still very low, however, and temperatures range widely. Light displays called auroras form in this layer over polar regions.

Auroras

The temperature drops dramatically in the mesosphere (50–85 km), the coldest layer. The stratosphere (10–50 km) contains a belt of ozone, a gas that absorbs most of the Sun's harmful ultraviolet rays. Clouds and weather systems form in the troposphere (1–10 km), the only layer in which air-breathing organisms typically can survive.

Meteors

Jets and weather balloons fly in the atmosphere's lowest layers.

85 km

Mesosphere (−80°C to −25°C; 0.3 to 0.01 kPa)

50 km

Ozone Layer

Stratosphere (−55°C to −20°C; 27 to 0.3 kPa)

10 km

0 km

Troposphere (−55°C to 15°C; 100 to 27 kPa)

Figure 20
Balloons are used to measure the weather conditions at high altitudes. These balloons expand as they rise due to decreased pressure.

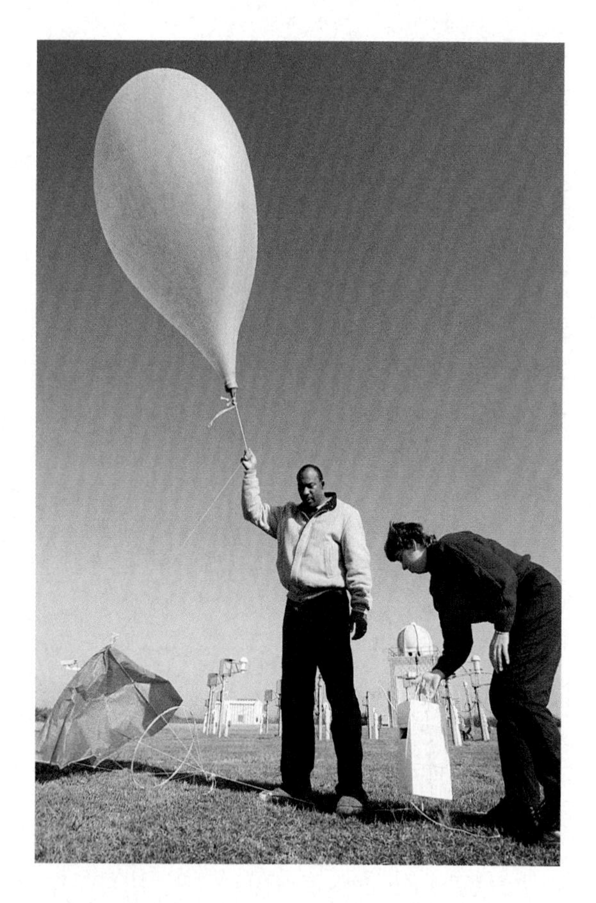

Boyle's Law

You now know how gas creates pressure in a container. What happens to the gas pressure if you decrease the size of the container? You know that the pressure of a gas depends on how often its particles strike the walls of the container. If you squeeze gas into a smaller space, its particles will strike the walls more often—giving an increased pressure. The opposite is true, too. If you give the gas particles more space, they will hit the walls less often—gas pressure will be reduced. Robert Boyle (1627–1691), a British scientist, described this property of gases. According to Boyle's law, if you decrease the volume of a container of gas and hold the temperature constant, the pressure of the gas will increase. An increase in the volume of the container causes the pressure to drop, if the temperature remains constant.

The behavior of weather balloons, as shown in **Figure 20,** can be explained using Boyle's law. Rubber or neoprene weather balloons are used to carry sensing instruments to high altitudes to detect weather information. The balloons are inflated near Earth's surface with a low-density gas. As the balloon rises, the atmospheric pressure decreases. The balloon gradually expands to a volume of 30 to 200 times its original size. At some point the expanding balloon ruptures. Boyle's law states that as pressure is decreased the volume increases, as demonstrated by the weather balloon. The opposite also is true, as shown by the graph in **Figure 21.** As the pressure is increased, the volume will decrease.

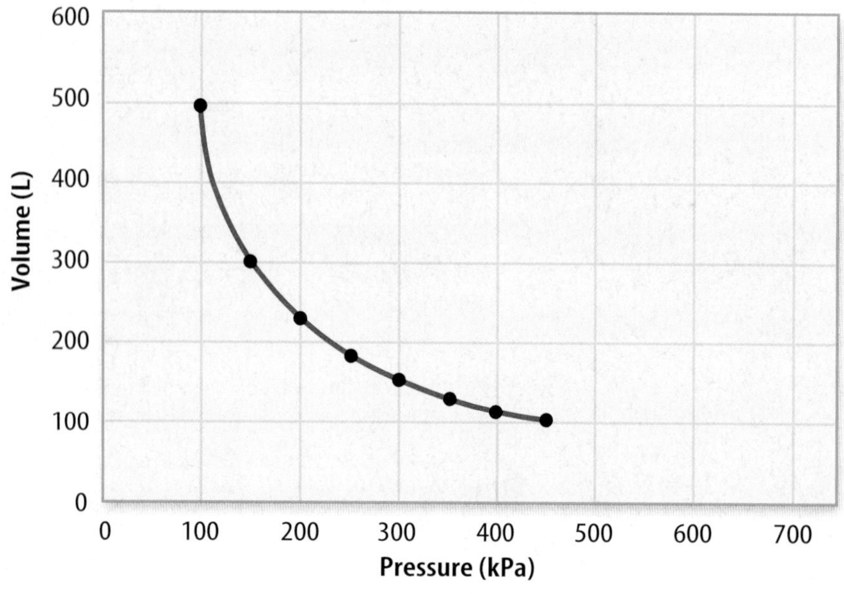

Figure 21
As you can see from the graph, as pressure increases, volume decreases; as pressure decreases, volume increases. *What is the volume of the gas at 100 kPa?*

Boyle's Law in Action When Boyle's law is applied to a real life situation, we find that the pressure multiplied by the volume is always equal to a constant if the temperature is constant. As the pressure and volume change indirectly, the constant will remain the same. You can use the equations P_1V_1 = constant = P_2V_2 to express this mathematically. This shows us that the product of the initial pressure and volume—designated with the subscript 1—is equal to the product of the final pressure and volume—designated with the subscript 2. Using this equation, you can find one unknown value, as shown in the example problem below.

✔ **Reading Check** *What is $P_1V_1 = P_2V_2$ known as?*

Research Visit the Glencoe Science Web site at **science.glencoe.com** for various industrial uses for compressed gases. Communicate to your class what you learn.

Math Skills Activity

Using Boyle's Law

Example Problem

A balloon has a volume of 10.0 L at a pressure of 101 kPa. What will be the new volume when the pressure drops to 43 kPa?

Solution:

1. *This is what you know:* initial pressure: $P_1 = 101$ kPa; initial volume: $V_1 = 10.0$ L; final pressure: $P_2 = 43$ kPa
2. *This is what you need to find:* final volume: V_2
3. *This is the equation you need to use:* $P_1V_1 = P_2V_2$
4. *Solve the equation for V_2:* $V_2 = \frac{P_1V_1}{P_2}$

Substitute the known values: $V_2 = \frac{(101\ \text{kPa})(10.0\ \text{L})}{43\ \text{kPa}} = 23\ \text{L}$

Check your answer by substituting 23 L back into the original equation and solving for P_1. Does the result reflect the theory of Boyle's law—volume increased as the pressure decreased?

Practice Problem

A volume of helium occupies 11.0 L at 98.0 kPa. What is the new volume if the pressure drops to 86.2 kPa?

For more help, refer to the Math Skill Handbook.

Observing Pressure

Safety Precautions

Procedure

1. Blow up a **balloon** to about half its maximum size.
2. Place the balloon on a **beaker** filled with **ice water.**

Analysis

1. Explain what happened to the balloon when you placed it on the beaker.
2. If the volume of the half-filled balloon was 0.5 L at a temperature of 298 K, what would the volume of the balloon be if the temperature increased to 358 K?

The Pressure-Temperature Law

Have you ever read the words "keep away from heat" on a pressurized spray canister? What happens if you heat an enclosed gas? The particles of gas will strike the walls of the canister more often. Because this canister is rigid, its volume cannot increase. Instead, its pressure increases. If the pressure becomes greater than the canister can hold, it will explode. After the container explodes, the pressure is released and the gas expands.

Reading Check *How does temperature affect pressure at constant volume?*

Charles's Law

If you've watched a hot air balloon being inflated, you know that gases expand when they are heated. Because particles in the hot air are further apart than particles in the cool air, the hot air is less dense than the cool air. This difference in density allows the hot air balloon to rise. Jacques Charles (1746–1823) was a French scientist who studied gases. According to Charles's law, the volume of a gas increases with increasing temperature, as long as pressure does not change. As with Boyle's law, the reverse is true, also. The volume of a gas shrinks with decreasing temperature, as shown in **Figure 22.**

Charles's law can be explained using the kinetic theory of matter. As a gas is heated, its particles move faster and faster and its temperature increases. Because the gas particles move faster, they begin to strike the walls of their container more often and with more force. In the hot air balloon, the walls have room to expand so instead of increased pressure, the volume increases.

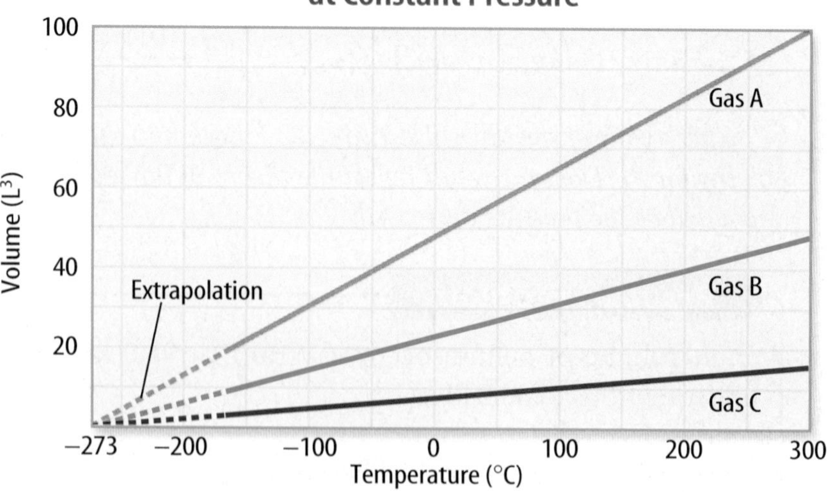

Figure 22
The volume of a gas increases when the temperature increases at constant pressure. Notice that when the graphs are extended to absolute zero, the volume is theoretically zero.

Using Charles's Law The formula that relates the variables of temperature to volume shows a direct relationship, $V_1/T_1 = V_2/T_2$, when temperature is given in Kelvin. When using Charles's law, the pressure must be kept constant. What would be the resulting volume of a 2.0-L balloon at 25.0°C that was placed in a container of ice water at 3.0°C, as shown in **Figure 23?**

$V_1 = 2.0 \text{ L} \qquad T_1 = 25.0°\text{C} + 273 = 298 \text{ K}$

$V_2 = ? \qquad T_2 = 3.0°\text{C} + 273 = 276 \text{ K}$

$$\frac{V_1}{T_1} = \frac{V_2}{T_2}$$

$$\frac{2.0 \text{ L}}{298 \text{ K}} = \frac{V_2}{276 \text{ K}}$$

$$V_2 = \frac{(2.0 \text{ L})(276 \text{ K})}{298 \text{ K}}$$

$$V_2 = 1.9 \text{ L}$$

As Charles's law predicts, the volume decreased as the temperature of the trapped gas decreased. This assumed no changes in pressure.

Figure 23
Charles's law states that as the temperature is lowered, the volume decreases. *If the balloon in the text was placed in a freezer at 5°C, what would be the new volume?*

According to Charles's law, what happens to the volume of a gas if the temperature increases?

Section  Assessment

1. Why does a gas have pressure?
2. What is the pressure of Earth's atmosphere at sea level?
3. Explain Boyle's law. Give an example of Boyle's law at work.
4. Explain Charles's law. Give an example of Charles's law at work.
5. **Think Critically** Labels on cylinders of compressed gases state the highest temperature to which the cylinder may be exposed. Give a reason for this warning.

Skill Builder Activities

6. **Forming Hypotheses** A bottle of ammonia begins to leak. An hour later, you can smell ammonia almost everywhere, especially near the bottle. State a hypothesis to explain your observations. **For more help, refer to the Science Skill Handbook.**
7. **Solving One-Step Equations** If a 5-L balloon at 25°C was gently heated to 30°C, what new volume would the balloon have? **For more help, refer to the Math Skill Handbook.**

Activity

Testing the Viscosity of Common Liquids

The resistance to flow of a liquid is called viscosity, and it can be measured and compared. One example of the importance of a liquid's viscosity is motor oil in car engines. The viscosity of motor oil in your family car is very important because it keeps the engine lubricated. It must cling to the moving parts and not run off leaving the parts dry and unlubricated. If the engine is not properly lubricated, it will be damaged eventually. The motor oil must maintain its viscosity in all types of weather from extreme heat in the summer to freezing cold in the winter.

What You'll Investigate

How can you compare the resistance to flow, or viscosity, of common household liquids?

Materials

room temperature household liquids such as:
- dish detergent
- corn syrup
- pancake syrup
- shampoo
- vegetable oil
- vinegar
- molasses
- water

spheres such as glass marbles or steel balls
100-mL graduated cylinders
150-mL beaker
ruler
stopwatch

Goals

- Observe and compare the viscosity of common liquids.

Safety Precautions

Dispose of wastes as directed by your teacher.

Procedure

1. Measure equal amounts of the liquids to be tested into the graduated cylinders.
2. Measure the depth of the liquid.
3. Copy the data chart into your Science Journal.
4. Place the sphere on the surface of the liquid. Using a stopwatch, measure and record how long it takes for it to travel to the bottom of the liquid.
5. Remove the sphere and repeat step 4 two more times for the same liquid.
6. Rinse and dry the sphere.
7. Repeat steps 4, 5, and 6 for two more liquids.

Viscosity of Common Liquids

Substance	Depth of Liquid (cm)	Time (s)	Velocity (cm/s)

Conclude and Apply

1. **Graph** the average speed of the sphere for each liquid on a bar graph.
2. In which liquid did the sphere move the fastest? Would that liquid have a high or low viscosity? Explain.
3. Would it matter if you dropped or threw the sphere into the liquid instead of placing it there? Explain your answer.
4. What effect does temperature play in the viscosity of a liquid? What would happen to the viscosity of your slowest liquid if you made it colder? Explain.

Compare your results with other groups and discuss differences noted. Why might these differences have occurred? **For more help, refer to the Science Skill Handbook.**

Science Stats

Hot and Cold

Did you know...

. . . The world's coldest substance, liquid helium, is about −269°C. It's used in cryogenics research, which is the study of extremely low temperatures. Cryogenics has enabled physicians to freeze and preserve body parts, such as corneas from human eyes. The freezing keeps cells alive until they are needed.

Cryogenics Laboratory

Hot Springs National Park, Arkansas

. . . Hot springs—also called thermal springs— are a popular tourist attraction. Thousands of people visit Hot Springs National Park in Arkansas each year. The average water temperature of the hot springs at the park is about 62°C.

Oxyacetylene Torch

. . . The hottest known flame is made by burning a mixture of oxygen and acetylene. The flame of an oxyacetylene torch can become as hot as 3,300°C. That's more than two times hotter than the melting point of steel.

. . . The hottest temperature in the universe occurs during a supernova—the explosion of a giant star. Temperatures can reach 3,500,000,000 Kelvin. The temperature of the surface of the Sun, by comparison, is about 5,600 K and the temperature of molten lava is 2,000 K.

Supernova

. . . Air becomes a liquid at −195°C. Scientists use liquid air to extract liquid oxygen, which is used in high-energy fuels for rocket engines.

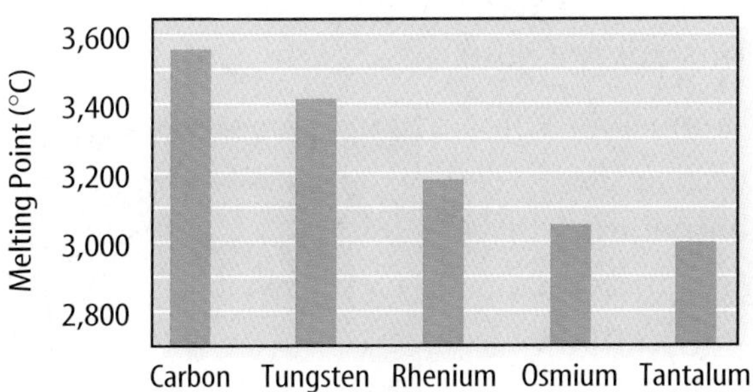

. . . The ideal serving temperature of ice cream is between −14°C and −12°C. In this temperature range, the ice cream is firm enough to hold its shape and deliver flavor.

Do the Math

1. In 1983, the temperature dropped to −89°C in Vostok, Antarctica. How many more degrees Celsius would the temperature need to drop for the air to become a liquid?
2. Make a bar graph that compares the temperature of liquid helium, liquid air, and ice cream at serving temperature.
3. Look at the graph above. List the elements that could be melted by an oxyacetylene torch.

Go Further

Go to **science.glencoe.com** to research the melting points of six metals. Make a bar graph showing the data you find.

Chapter 16 Study Guide

Reviewing Main Ideas

Section 1 Kinetic Theory

1. Four states of matter exist: solid, liquid, gas, and plasma.
2. According to the kinetic theory, all matter is made of constantly moving particles that collide without losing energy.
3. Most matter expands when heated and contracts when cooled. *Why is the expansion joint needed in this concrete bridge shown below?*

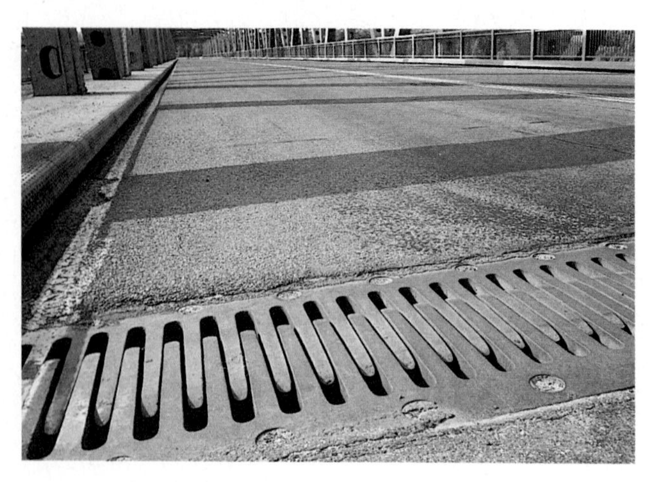

4. Changes of state can be interpreted in terms of the kinetic theory of matter.

Section 2 Properties of Fluids

1. Archimedes' principle states that the buoyant force of an object in a fluid is equal to the weight of the fluid displaced. *In the photograph on the right, explain why the penny sank in the beaker of water.*

2. Pascal's principle states that pressure applied to a fluid is transmitted unchanged throughout the fluid.
3. Bernoulli's principle states that the pressure exerted by a fluid decreases as its velocity increases.

Section 3 Behavior of Gases

1. Gas pressure results from moving particles colliding with the inside walls of the container.
2. Boyle's law states that the volume of a gas decreases when the pressure increases at constant temperature.
3. Charles's law states that the volume of a gas increases when the temperature increases at constant pressure.
4. At constant volume, as the temperature of a gas increases, so does the pressure of a gas. *What happens to the pressure in this cylinder as the temperature outdoors rises?*

After You Read

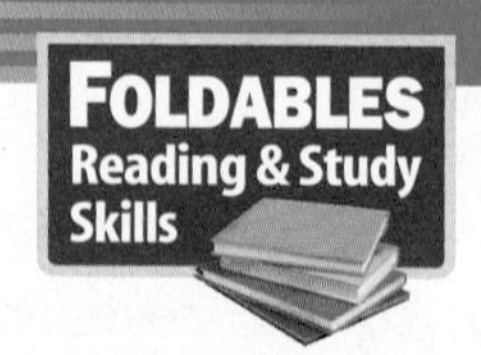

Under the tabs of your Foldable, write and explain the characteristics of solids, liquids, and gases.

Visualizing Main Ideas

Complete the following concept map on states of matter.

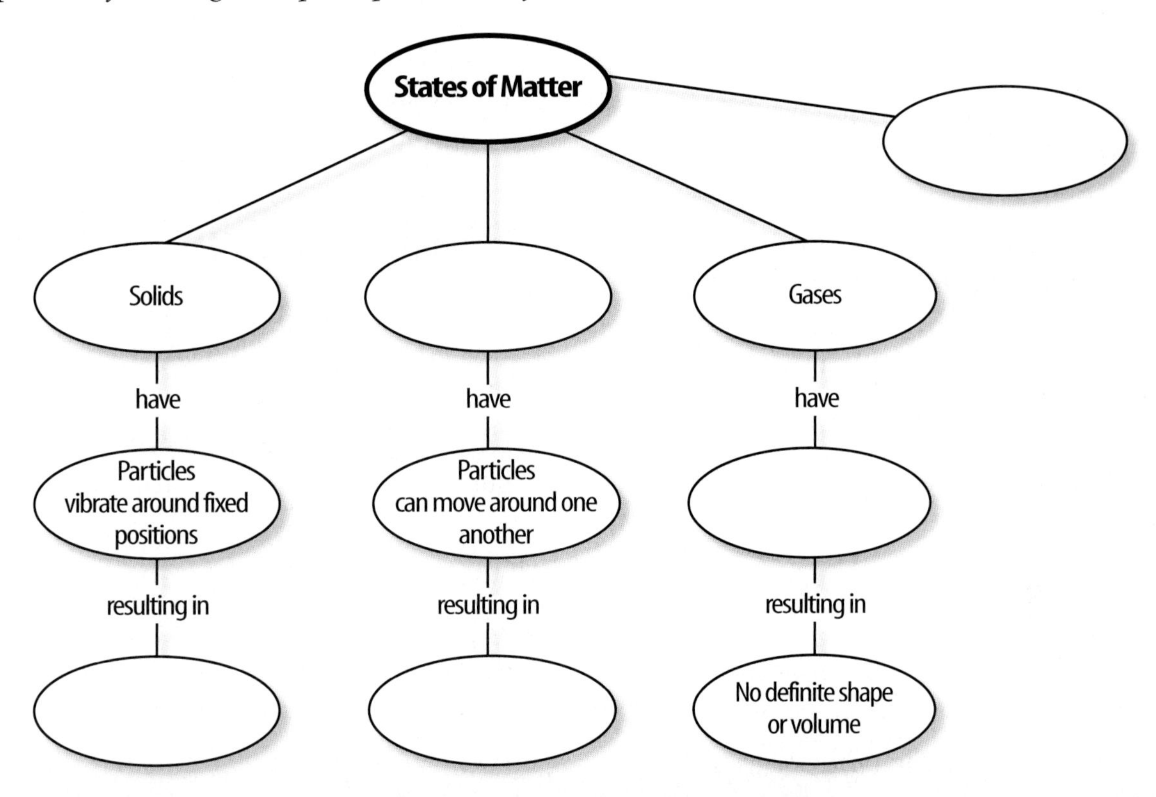

Vocabulary Review

Vocabulary Words

a. boiling point
b. buoyancy
c. diffusion
d. heat of fusion
e. heat of vaporization
f. kinetic theory
g. melting point
h. pascal
i. plasma
j. pressure
k. thermal expansion
l. viscosity

Study Tip

THE PRINCETON REVIEW

Make flashcards for new vocabulary words. Put the word on one side and the definition on the other. Use them to quiz yourself.

Using Vocabulary

Answer the following questions using complete sentences.

1. What is the property of a fluid that represents its resistance to flow?
2. What is the SI unit of pressure?
3. What term is used to describe the amount of force exerted per unit of area?
4. What is the temperature when a solid begins to liquefy?
5. What theory is used to explain the behavior of particles in matter?
6. What is the ability of a fluid to exert an upward force on an object?

Chapter 16 Assessment

Checking Concepts

Choose the word or phrase that best answers the question.

1. What is the temperature at which all particle motion of matter ceases?
A) absolute zero **C)** boiling point
B) melting point **D)** heat of fusion

2. What is the state of matter that has a definite volume and a definite shape?
A) solid **C)** gas
B) liquid **D)** plasma

3. What is the most common state of matter in the universe?
A) solid **C)** gas
B) liquid **D)** plasma

4. Which of the following would be used to measure pressure?
A) gram **C)** kilopascals
B) newtons **D)** kilograms

5. Which of the following uses Pascal's principle?
A) aerodynamics **C)** buoyancy
B) hydraulics **D)** changes of state

6. Which of the following uses Bernoulli's principle?
A) airplanes **C)** boats
B) pistons **D)** snowboards

7. The particles in which of the following are farthest from each other?
A) gas **C)** liquid
B) solid **D)** plasma

8. In which state would a material fill a room at the quickest rate?
A) solid **C)** water
B) liquid **D)** gas

9. What is the upward force in a liquid?
A) pressure **C)** buoyancy
B) kinetic theory **D)** diffusion

10. What is the amount of energy needed to change a solid to a liquid at its melting point called?
A) heat of fusion
B) heat of vaporization
C) temperature
D) absolute zero

Thinking Critically

11. Use the Temperature-Pressure law to explain why you should check your tire pressure when the temperature changes.

12. Describe the changes that occur inside a helium balloon as it rises from sea level.

13. Why do aerosol cans have a "do not incinerate" warning?

14. The Dead Sea is a solution that is so dense that you float on it easily. Explain why you are able to float easily, using the terms *density* and *buoyant force.*

Developing Skills

15. Making and Using Graphs A group of students heated ice until it turned to steam. They measured the temperature each minute. Their graph is provided below. Explain what is happening at each letter (a, b, c, d) in the graph.

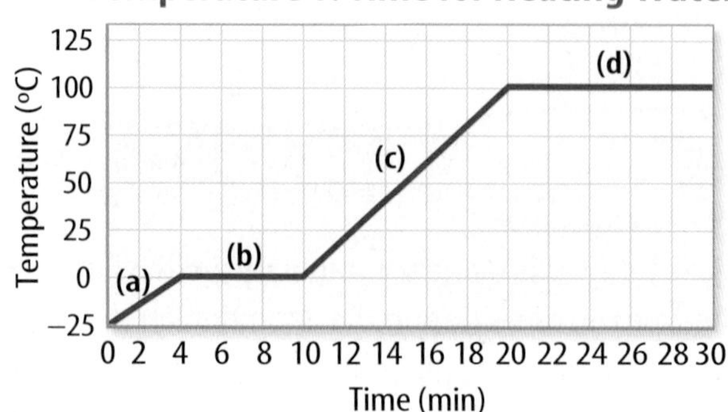

16. Concept Mapping Use a cycle map to show the changes in particles as cool water boils, changes to steam, and then changes back to cool water.

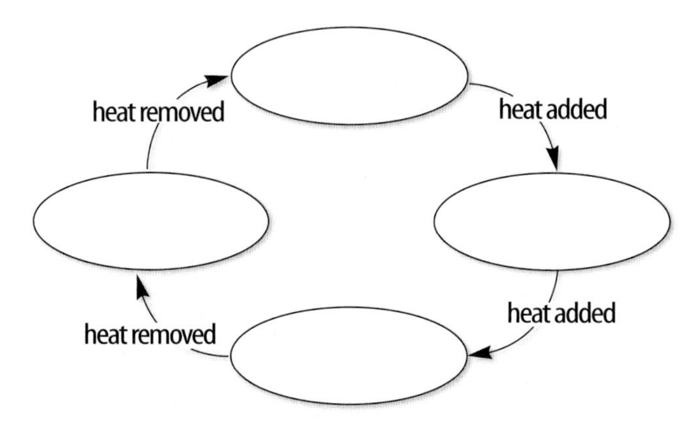

17. Interpreting Data As elevation increases, boiling point decreases. List each of the following locations as *at sea level, above sea level,* or *below sea level.* (Boiling point of water is given in parenthesis.)

Death Valley (100.3°C), Denver (94°C), Madison (99°C), Mt. Everest (76.5°C), Mt. McKinley (79°C), New York City (100°C), Salt Lake City (95.6°C)

Performance Assessment

18. Researching Research the effects of pressure changes on the human body and write a report. Include in your report any precautions that must be taken in dealing with pressure changes in space and when deep-sea diving.

Technology

Go to the Glencoe Science Web site at **science.glencoe.com** or use the **Glencoe Science CD-ROM** for additional chapter assessment.

Test Practice

A geologist was investigating the effects of the change in seasons on rock formations. One of his observations is shown in the diagram below.

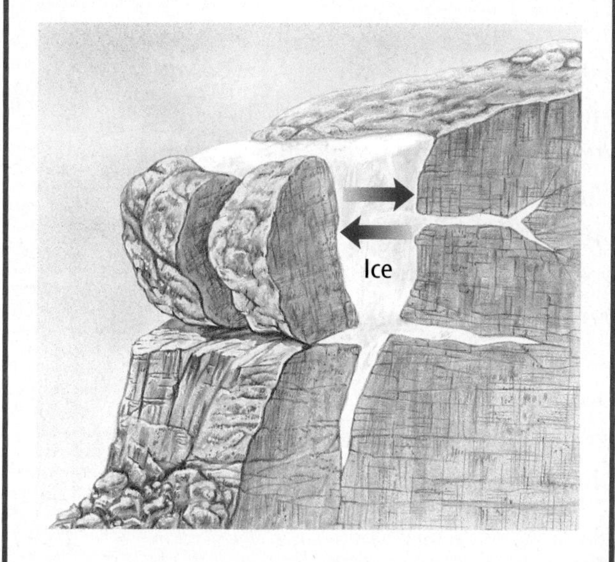

Study the picture and answer the following questions.

1. Which of these is the most likely cause of the rocks being wedged apart?

A) Water contains dissolved chemicals that eat away at rocks.

B) Cold water is more acidic than warm water.

C) Ice is more dense than water.

D) Water expands when it freezes.

2. The ice pushes outward in this manner in response to _____.

F) atmospheric pressure

G) temperature changes

H) glacial movement

J) gravity

CHAPTER 17 Classification of Matter

Paintings like this one are mixtures that combine many different chemical pigments. They can be mixed skillfully to achieve a well blended color or the artist can intentionally show the individual pigments within one brush stroke. All the materials in a painting can undergo physical and chemical changes with time. In this chapter, you will learn about mixtures and how to separate them. You also will learn about the physical and chemical properties of matter and you'll learn to distinguish between physical and chemical change.

What do you think?

Science Journal Look at the picture below with a classmate. Discuss what you think this might be or what is happening. Here's a hint: *Let the layers settle it.* Write your answer or your best guess in your Science Journal.

Imagine yourself marooned on an island without fresh water. What could you do to get clean drinking water? Could you purify seawater? One way you could do this is distillation. Discover how to purify water in this activity.

Demonstrate the distillation of water 

1. Place 75 mL of water in a 200-mL beaker and add 20 drops of red food coloring.
2. Place the beaker on a hot plate.
3. Add ice to an evaporating dish until the dish is half full. Place the evaporating dish on the beaker as shown in the photo.
4. Turn on the hot plate and slowly bring the water and food coloring solution to a boil.
5. After boiling the solution for five minutes, carefully remove the evaporating dish using tongs. Touch the drops of liquid on the bottom of the dish to a piece of white paper.
6. Observe the liquid on the paper.

Observe

In your Science Journal, discuss where the liquid came from. What was in the beaker that is not in the liquid on the paper?

Before You Read

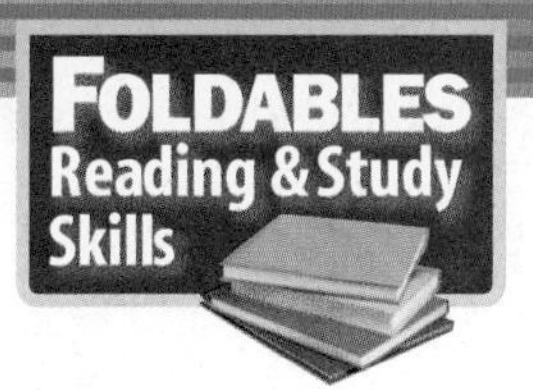

Making a Vocabulary Study Fold **Make the following Foldable to ensure you have understood the content by defining the vocabulary terms from this chapter.**

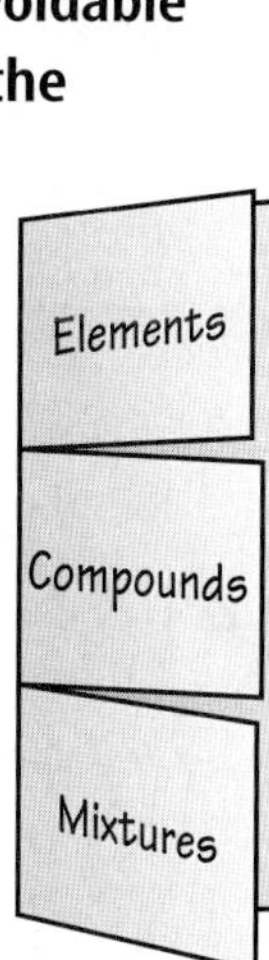

1. Place a sheet of paper in front of you so the short side is at the top. Fold the paper in half from the left side to the right side.
2. Fold the top and bottom in. Unfold the paper so three sections show.
3. Through the top thickness of the paper, cut along each of the fold lines to the fold on the left, forming three tabs. Label *Elements, Compounds,* and *Mixtures* across the front of the tabs as shown.
4. As you read the chapter, define each term and list examples of each under the tabs.

SECTION 1

Composition of Matter

As You Read

What You'll Learn

- **Define** substances and mixtures.
- **Identify** elements and compounds.
- **Compare and contrast** solutions, colloids, and suspensions.

Vocabulary

substance
element
compound
heterogeneous mixture
homogeneous mixture
solution
colloid
Tyndall effect
suspension

Why It's Important

You can form a better picture of your world when you understand the concepts of elements and compounds.

Pure Substances

Have you ever seen a picture hanging on a wall that looked just like a real painting? Did you have to touch it to find out? If so, the rough or smooth surface told you which it was. Each material has its own properties. The properties of materials can be used to classify them into general categories.

Materials are made of a pure substance or a mixture of substances. A pure **substance,** or simply a substance, is either an element or a compound. Substances cannot be broken down into simpler components and still maintain the properties of the original substance. Some substances you might recognize are helium, aluminum, water, and salt.

Elements All substances are built from atoms. If all the atoms in a substance are alike, that substance is an **element.** The graphite in your pencil point and copper coating of most pennies are examples of elements. In graphite all the atoms are carbon atoms, and in a copper sample, all the atoms are copper atoms. The metal substance beneath the copper in the penny is another element—zinc. There are 90 elements found in nature. More than 20 others have been made in laboratories, but most are unstable and exist only for short periods of time. Some elements you might recognize are shown in **Figure 1.** Some less common elements and their properties are shown in **Figure 2.**

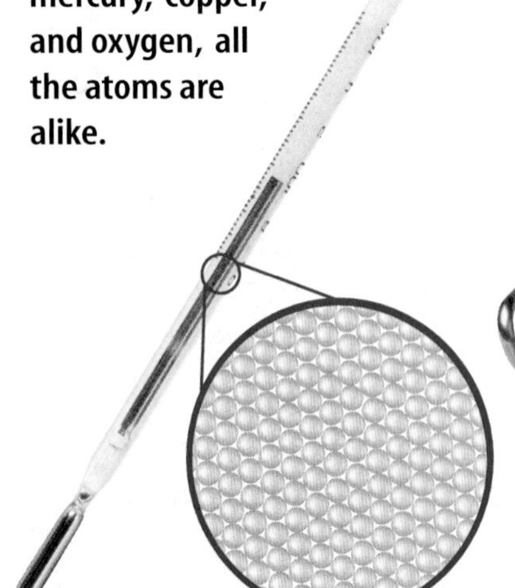

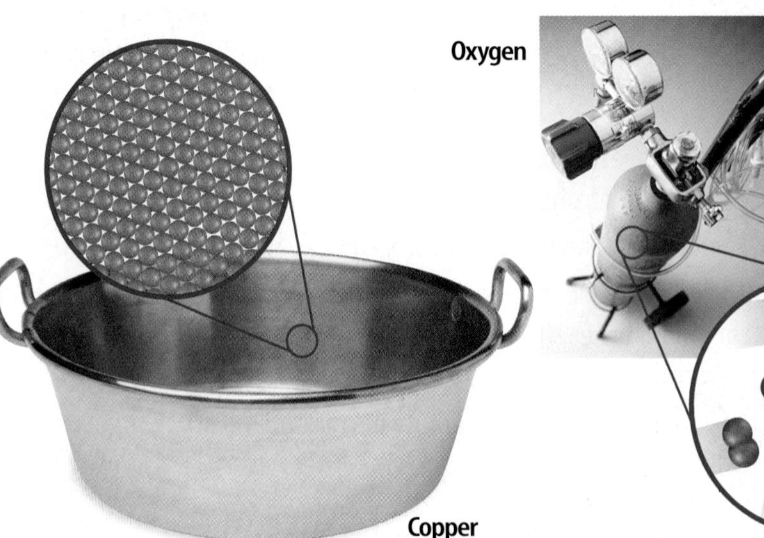

Figure 1
In elements such as mercury, copper, and oxygen, all the atoms are alike.

NATIONAL GEOGRAPHIC VISUALIZING ELEMENTS

Figure 2

Most of us think of gold as a shiny yellow metal used to make jewelry. However, it is an element that is also used in more unexpected ways, such as in spacecraft parts. On the other hand, some less common elements, such as americium (am-uh-REE-see-um), are used in everyday objects. Some elements and their uses are shown here.

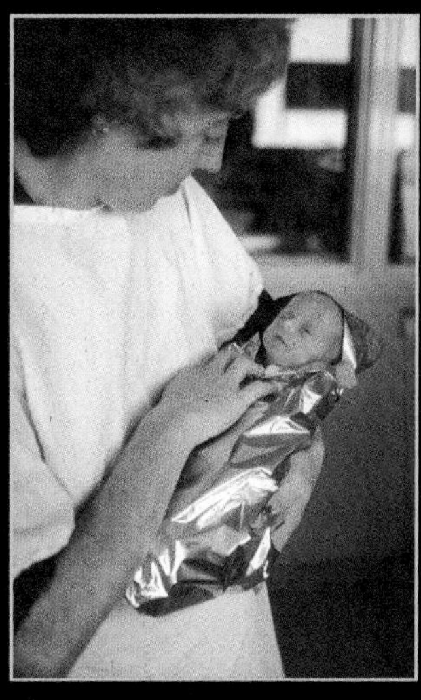

▲ ALUMINUM Aluminum is an excellent reflector of heat. Here, an aluminum plastic laminate is used to retain the body heat of a newborn baby.

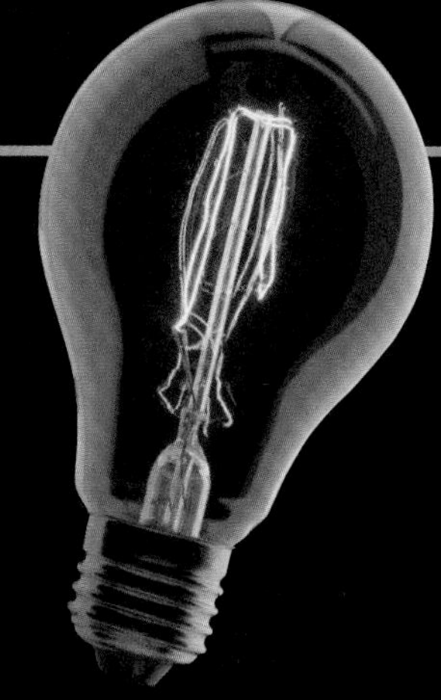

▲ TUNGSTEN Although tungsten can be combined with steel to form a very durable metal, in its pure form it is soft enough to be stretched to form the filament of a lightbulb. Tungsten has the highest melting point of any metal.

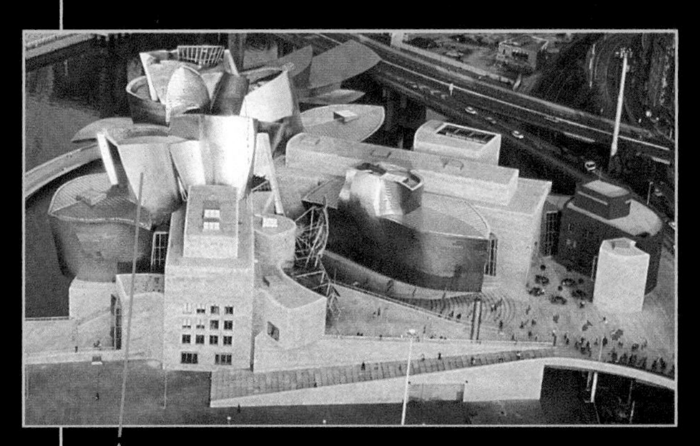

▲ TITANIUM (Tie-TAY-nee-um) Parts of the exterior of the Guggenheim Museum in Bilbao, Spain, are made of titanium panels. Strong and lightweight, titanium is also used for body implants.

▲ GOLD Gold's resistance to corrosion and its ability to reflect infrared radiation make it an excellent coating for space vehicles. The electronic box on the six-wheel Sojourner Rover, above, part of NASA's Pathfinder 1997 mission to Mars, is coated with gold.

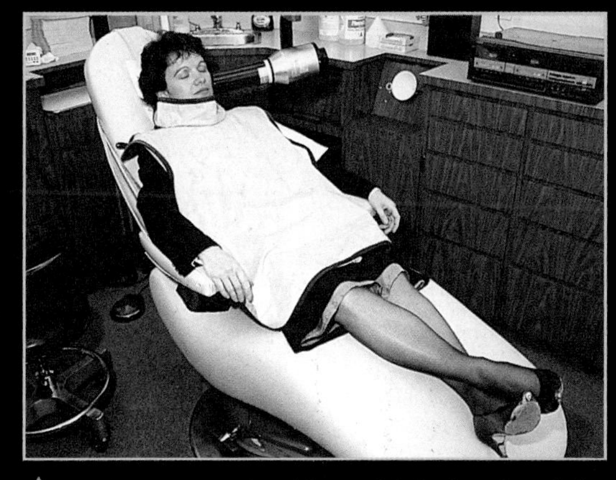

▲ LEAD Because lead has a high density, it is a good barrier to radiation. Dentists drape lead aprons on patients before taking x-rays of the patient's teeth to reduce radiation exposure.

◀ AMERICIUM Named after America, where it was first produced, americium is a component of this smoke detector. It is a radioactive metal that must be handled with care to avoid contact.

Compounds Two or more elements can combine to form substances called compounds. A **compound** is a substance in which the atoms of two or more elements are combined in a fixed proportion. For example, water is a compound in which two atoms of the element hydrogen combine with one atom of the element oxygen. Chalk contains calcium, carbon and oxygen in the proportion of one atom of calcium and carbon to three atoms of oxygen.

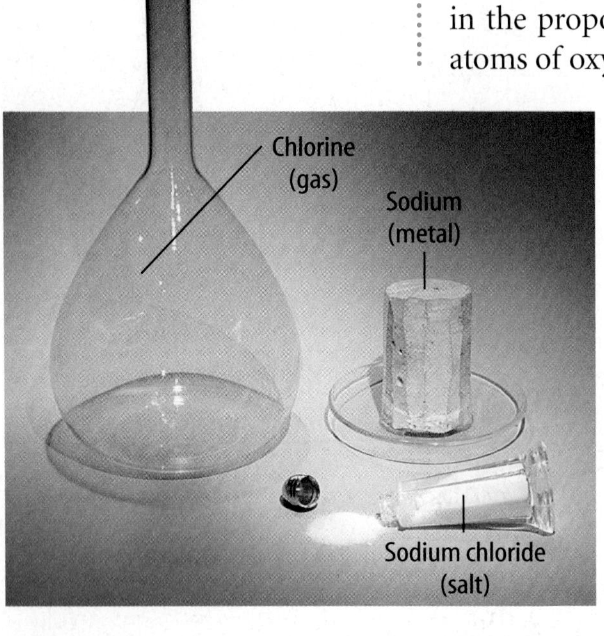

Figure 3
Chlorine gas and sodium metal combine dramatically in the ratio of one to one to form sodium chloride.

Reading Check *How are elements and compounds related?*

Can you imagine yourself putting something made from a silvery metal and a greenish-yellow, poisonous gas on your food? You might have shaken some on your food today—table salt is a chemical compound that fits this description. Even though it looks like white crystals and adds flavor to food, its components—sodium and chlorine—are neither white nor salty, as shown in **Figure 3.** Like salt, compounds usually look different from the elements in them.

Mixtures

Are pizza and a soft drink one of your favorite lunches? If so, you enjoy two foods that are classified as mixtures—but two different kinds of mixtures. A mixture, such as the pizza or soft drink shown in **Figure 4,** is a material made up of two or more substances that can be easily separated by physical means.

Figure 4
Pizza and soft drinks, like most foods, are mixtures.

Heterogeneous Mixtures Unlike compounds, mixtures do not always contain the same proportions of the substances that make them up—the pizza chef doesn't measure precisely how much of each topping is sprinkled on. You easily can see most of the toppings on a pizza. A mixture in which different materials can be distinguished easily is called a **heterogeneous** (het uh ruh JEE nee us) **mixture.** Granite, concrete, and dry soup mixes are other heterogeneous mixtures you can recognize.

You might be wearing another heterogeneous mixture—clothing made of permanent-press fabric like that seen in **Figure 5A.** Such fabric contains fibers of two materials—polyester and cotton. The amounts of polyester and cotton can vary from one article of clothing to another, as shown by the label. Though you might not be able to distinguish the two fibers just by looking at them with your naked eye, you probably could tell using a microscope, as shown in **Figure 5B.** Therefore, a permanent-press fabric is also a heterogeneous mixture.

Most of the substances you come in contact with every day are heterogeneous mixtures. Some components are easy to see, like the ingredients in pizza, but others are not. In fact, the component you see can be a mixture itself. For example, the cheese in pizza is also a mixture, but you cannot see the individual components. Cheese contains many compounds, such as milk proteins, butterfat, colorings, and other food additives.

TRY AT HOME Mini LAB

Separating Mixtures

Procedure

1. Put equal amounts of **soil, clay, sand, gravel,** and **pebbles** in a **clear-plastic container.** Add **water** until the container is almost full. Wash your hands well after handling the materials.
2. Stir or shake the mixture thoroughly. Predict the order in which the materials will settle.
3. Observe what happens and compare your observations to your predictions.

Analysis

1. In what order did the materials settle?
2. Explain why the materials settled in the order they did.

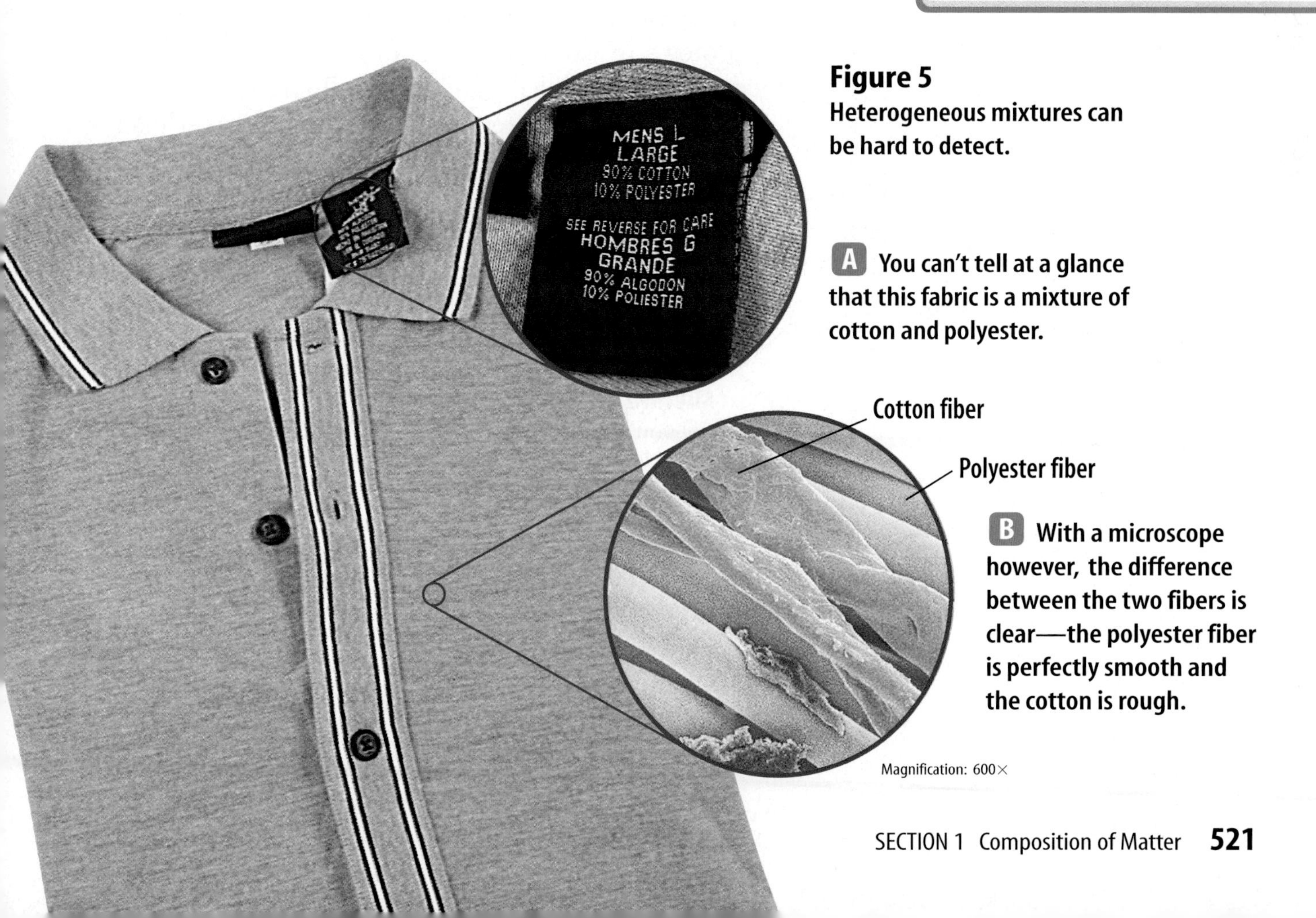

Figure 5
Heterogeneous mixtures can be hard to detect.

A You can't tell at a glance that this fabric is a mixture of cotton and polyester.

B With a microscope however, the difference between the two fibers is clear—the polyester fiber is perfectly smooth and the cotton is rough.

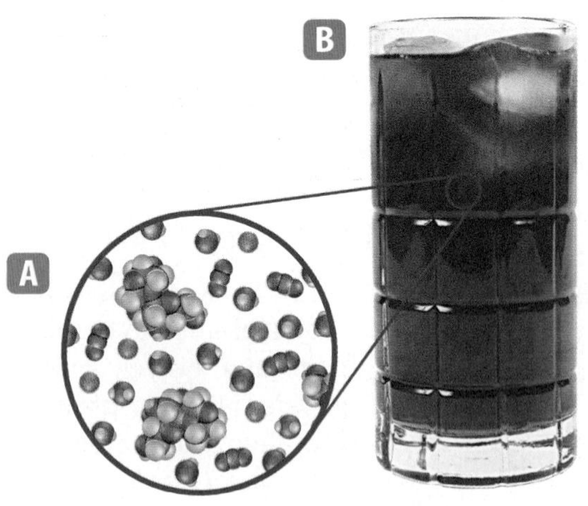

Figure 6
A soft drink can be either heterogeneous or homogeneous.
A As carbon dioxide fizzes out it is a heterogeneous mixture.
B The resulting flat soft drink is a homogeneous mixture of water, sugar, flavor, color and some remaining carbon dioxide.

Homogeneous Mixtures Remember that soft drink you had with your pizza? Regular and diet soft drinks look alike but taste different and contain different amounts of calories.

Cold soft drinks in sealed bottles are examples of homogeneous mixtures. A **homogeneous** (hoh muh JEE nee us) **mixture** contains two or more gaseous, liquid, or solid substances blended evenly throughout. Soft drinks contain water, sugar, flavoring, coloring, and carbon dioxide gas. **Figure 6** will help you to visualize these particles in a liquid soft drink.

Vinegar is another homogeneous mixture. It appears clear even though it is made up of particles of acetic acid mixed with water. Another name for homogeneous mixtures like vinegar and a cold soft drink is solution. A **solution** is a homogeneous mixture of particles so small that they cannot be seen with a microscope and will never settle to the bottom of their container. Solutions remain constantly and uniformly mixed. The differences between substances and mixtures are summarized in **Figure 7.**

✔ Reading Check *What kind of mixture is a solution?*

Colloids Milk is an example of a specific kind of mixture called a colloid. Like a heterogeneous mixture, it contains water, fats, and proteins in varying proportions. Like a solution, its components won't settle if left standing. A **colloid** (KAH loyd) is a type of mixture that never settles. Its particles are larger than those in solutions but not heavy enough to settle. The word *colloid* comes from a Greek word for glue. The first colloids studied were in gelatin, a source of some types of glue.

Paint is an example of a liquid with suspended colloid particles. Gases and solids can contain colloidal particles, too. For example, fog consists of particles of liquid water suspended in air, and smoke contains solids suspended in air.

Figure 7
All matter can be divided into substances and mixtures.

Matter
Has mass and takes up space

Substance
Composition definite

Mixture
Composition variable

Compound
Two or more kinds of atoms

Heterogeneous
Unevenly mixed

Element
One kind of atom

Homogeneous
Evenly mixed; a solution

9999 FINE GOLD 19154 88

Figure 8
Fog is a colloid composed of water droplets suspended in air.

A The light from the headlights is scattered by fog.

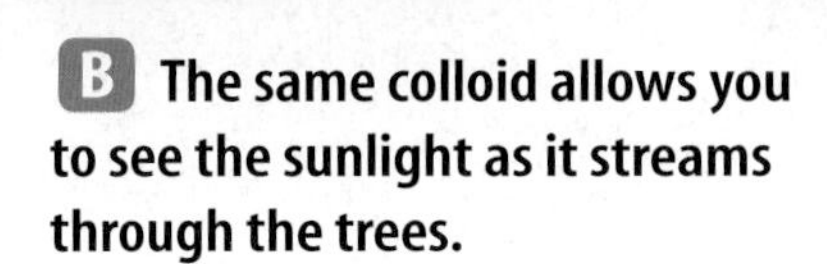

B The same colloid allows you to see the sunlight as it streams through the trees.

Detecting Colloids One way to distinguish a colloid from a solution is by its appearance. Fog appears white because its particles are large enough to scatter light as shown in **Figure 8.** Sometimes it is not so obvious that a liquid is a colloid. For example, some shampoos and gelatins are colloids called gels that appear almost clear. You can tell for certain if a liquid is a colloid by passing a beam of light through it, as shown in **Figure 9.** A light beam is invisible as it passes through a solution, but can be seen readily as it passes through a colloid. This occurs because the particles in the colloid are large enough to scatter light, but those in the solution are not. This scattering of light by colloidal particles is called the **Tyndall effect.**

✔ Reading Check *How can you distinguish a colloid from a solution?*

Figure 9
Because of the Tyndall effect, a light beam is scattered by the colloid suspension on the right, but passes invisibly through the solution on the left.

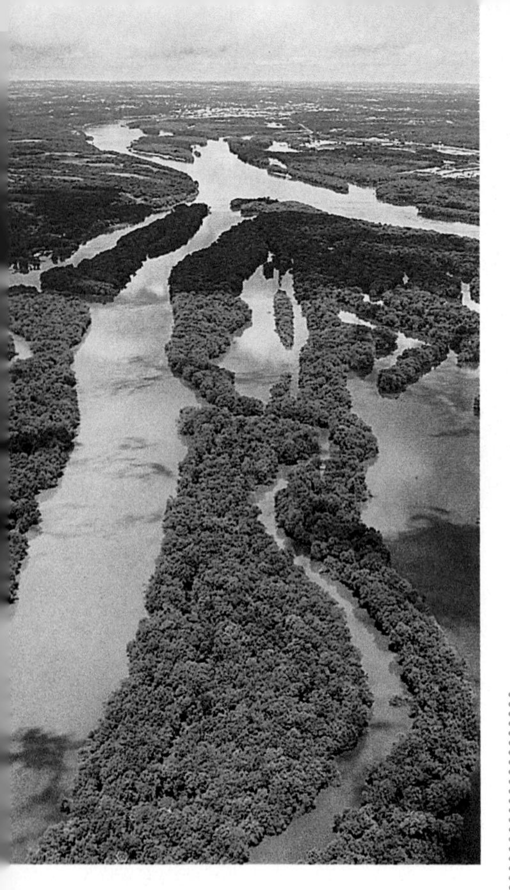

Figure 10
Layers of mud build up until they can be thousands of feet thick. The mud deposited by the Mississippi River is said to be more than 10,000 m thick.

Table 1 Comparing Solutions, Colloids, and Suspensions

Description	Solutions	Colloids	Suspensions
Settle upon standing?	no	no	yes
Separate using filter paper?	no	no	yes
Particle Size	0.1–1 nm	1–100 nm	> 100 nm
Scatter Light	no	yes	yes

Suspensions Some mixtures are neither solutions nor colloids. One example is muddy pond water. If pond water stands long enough, some mud particles will fall to the bottom, and the water clears. Pond water is a **suspension,** which is a heterogeneous mixture containing a liquid in which visible particles settle. **Table 1** summarizes the properties of different types of mixtures.

River deltas are a large scale example of how a suspension settles. Rivers flow swiftly through narrow channels, picking up soil and debris along the way. As the river widens, it flows more slowly. Suspended particles settle forming deltas at the mouth, as shown in **Figure 10.**

Section 1 Assessment

1. How is a compound similar to a homogeneous mixture? How is it different?
2. Distinguish between a substance and a mixture. Give examples.
3. Describe the differences between colloids and suspensions.
4. Why is vinegar considered a solution?
5. **Think Critically** Why do the words "Shake well before using" on a bottle of fruit juice indicate that the fruit juice is a suspension?

Skill Builder Activities

6. **Comparing and Contrasting** In terms of suspensions and colloids, compare and contrast a glass of milk and a glass of fresh-squeezed orange juice. **For more help, refer to the Science Skill Handbook.**
7. **Communicating** In your Science Journal, make a list of the liquid products you find in your home. Classify each as a solution, a colloid, or a suspension. **For more help, refer to the Science Skill Handbook.**

Activity

Elements, Compounds, and Mixtures

Elements, compounds, and mixtures all contain atoms. In elements, the atoms all have the same identity. In compounds, two or more elements have been combined in a fixed ratio. In a mixture, the ratio of substances can vary.

What You'll Investigate

What are some differences among elements, compounds, and mixtures?

Materials

plastic freezer bag containing the following labeled items:

copper wire
small package of salt
pencil
aluminum foil
chalk (calcium carbonate)
piece of granite
sugar water in a vial

Goals

- **Determine** whether several materials are elements, compounds, or mixtures.

Safety Precautions   

Procedure

1. Copy the data table into your Science Journal and use it to record your observations.
2. Obtain a bag of objects. Identify each object and classify it as an element, compound, heterogeneous mixture, or homogeneous mixture. The elements appear in the periodic table. Compounds are named as examples in Section 1.

Conclude and Apply

1. If you know the name of a substance, how can you find out whether or not it is an element?
2. **Examine** the contents of your refrigerator at home. Classify what you find as elements, compounds, or mixtures.
3. Then, identify whether the mixtures are homogeneous or heterogeneous, and whether they are colloids or suspensions.

Classification of Objects

Object	Identity	Classification
1		
2		
3		
4		
5		
6		
7		

Communicating Your Data

Enter your data in the data table and compare your findings with those of your classmates. **For more help, refer to the Science Skill Handbook.**

Properties of Matter

As You Read

What You'll Learn

- **Identify** substances using physical properties.
- **Compare and contrast** physical and chemical changes.
- **Compare and contrast** chemical and physical properties.
- **Determine** how the law of conservation of mass applies to chemical changes.

Vocabulary

physical property
physical change
distillation
chemical property
chemical change
law of conservation of mass

Why It's Important

Understanding chemical and physical properties can help you use materials properly.

Physical Properties

You can stretch a rubber band, but you can't stretch a piece of string very much, if at all. You can bend a piece of wire, but you can't easily bend a matchstick. In each case, the materials change shape, but the identity of the substances—rubber, string, wire, wood—does not change. The abilities to stretch and bend are physical properties. Any characteristic of a material that you can observe or attempt to observe without changing the identity of the substances that make up the material is a **physical property.** Examples of other physical properties are color, shape, size, melting point, and boiling point. What physical properties can you use to describe the items in **Figure 11?**

Appearances How would you describe a tennis ball? You could begin by describing its shape, color, and state of matter. For example, you might describe the tennis ball as a brightly colored hollow sphere. You can measure some physical properties, too. For instance, you could measure the diameter of the ball. What physical property of the ball is measured with a balance?

To describe a soft drink in a cup, you could start by calling it a liquid with a brown color and sweet taste. You could measure its volume and temperature. Each of these characteristics is a physical property of that soft drink.

Figure 11
Appearance is the most obvious physical property. *How would you describe the appearance of these items?*

Figure 12
The best way to separate substances depends on their physical properties.

A Size is the property that helps separate poppy seeds from sunflower seeds.

Behavior Some physical properties describe the behavior of a material or a substance. As you might know, objects that contain iron, such as a safety pin, are attracted by a magnet. Attraction to a magnet is a physical property of the substance iron. Every substance has a specific combination of physical properties that make it useful for certain tasks. Some metals, such as copper, can be drawn out into wires. Others, such as gold, can be pounded into sheets as thin as 0.1 micrometers (μm), about 4 millionths of an inch. This property of gold makes it useful for decorating picture frames and other objects. Gold that has been beaten or flattened in this way is called gold leaf.

Think again about your soft drink. If you knock over the cup, the drink will spread out over the table or floor. If you knock over a jar of molasses however, it does not flow as easily. The ability to flow is a physical property of liquids.

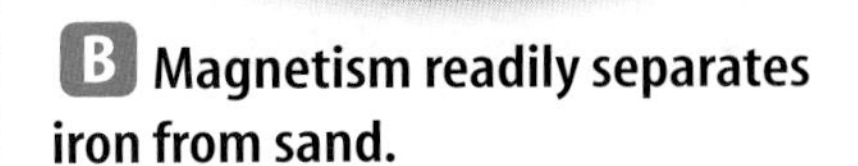

B Magnetism readily separates iron from sand.

Using Physical Properties to Separate Do you lick the icing from the middle of a sandwich cookie before eating the cookie? If you do, you are using physical properties to identify the icing and separate it from the rest of the cookie. **Figure 12A** shows a mixture of poppy seeds and sunflower seeds. You can identify the two kinds of seeds by differences in color, shape, and size. By sifting the mixture, you can separate the poppy seeds from the sunflower seeds quickly because their sizes differ.

Now look at the mixture of iron filings and sand shown in **Figure 12B.** You probably won't be able to sift out the iron filings because they are similar in size to the sand particles. What you can do is pass a magnet through the mixture. The magnet attracts only the iron filings and pulls them from the sand. This is an example of how a physical property, such as magnetic attraction, can be used to separate substances in a mixture. Something like this is done to separate iron for recycling.

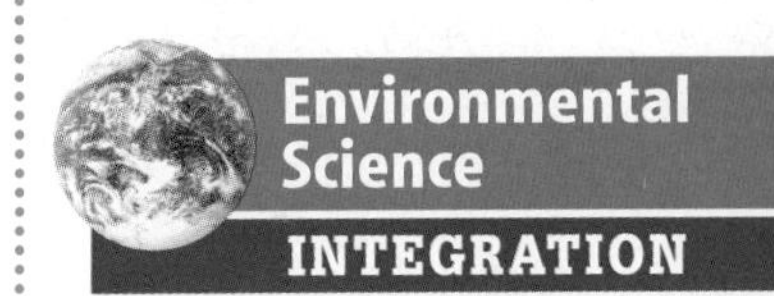

Recycling conserves natural resources. In some large recycling projects, it is difficult to separate aluminum metal from scrap iron. What physical properties of the two metals would help separate them?

Identifying Changes

Procedure

WARNING: *Clean up any spills promptly. Potassium permanganate can stain clothing.*

1. Add **water** to a **250-mL beaker** until it is half-full.
2. Add a crystal of **potassium permanganate** to the water and observe what happens.
3. Add 1 g of **sodium hydrogen sulfite** to the solution and stir it until the solution becomes colorless.

Analysis

1. Is dissolving a chemical or a physical change?
2. What evidence of a chemical change did you see?

Physical Change

If you break a piece of chewing gum, you change some of its physical properties—its size and shape. However, you have not changed the identity of the materials that make up the gum. Each piece still tastes and chews the same.

The Identity Remains the Same The changes in state that you have studied are all examples of physical changes. When a substance freezes, boils, evaporates, or condenses, it undergoes physical changes. A change in size, shape, or state of matter is called a **physical change.** These changes might involve energy changes, but the kind of substance—the identity of the element or compound—does not change.

✔ Reading Check ***Does a change in state mean that a new substance has formed? Explain.***

Iron is a substance that can change states if it absorbs or releases enough energy—at high temperatures, it melts. However, in both the solid and liquid state, iron has physical properties that identify it as iron. Color changes can accompany a physical change, too. For example, when iron is heated it first glows red. Then, if it is heated to a higher temperature, it turns white, as shown in **Figure 13.**

Using Physical Change to Separate A cool drink of water is something most people take for granted, but in some parts of the world, drinkable water is scarce. Not enough drinkable water can be obtained from wells. Many such areas that lie close to the sea obtain drinking water by using physical properties of water to separate it from the salt. One of these methods, which uses the property of boiling point, is a type of distillation.

Figure 13
Heating iron raises its energy level and it changes color. These energy changes are physical changes because it is still iron.

Distillation **Distillation** is a process for separating substances in a mixture by evaporating a liquid and recondensing its vapor. It usually is done in the laboratory using an apparatus similar to that shown in **Figure 14.** As you can see, the liquid vaporizes and condenses, leaving the solid material behind.

Two liquids having different boiling points can be separated in a similar way. The mixture is heated slowly until it begins to boil. Vapors of the liquid with the lowest boiling point form first and are condensed and collected. Then, the temperature is increased until the second liquid boils, condenses and is collected. Distillation is used often in industry. For instance, natural oils such as mint are distilled.

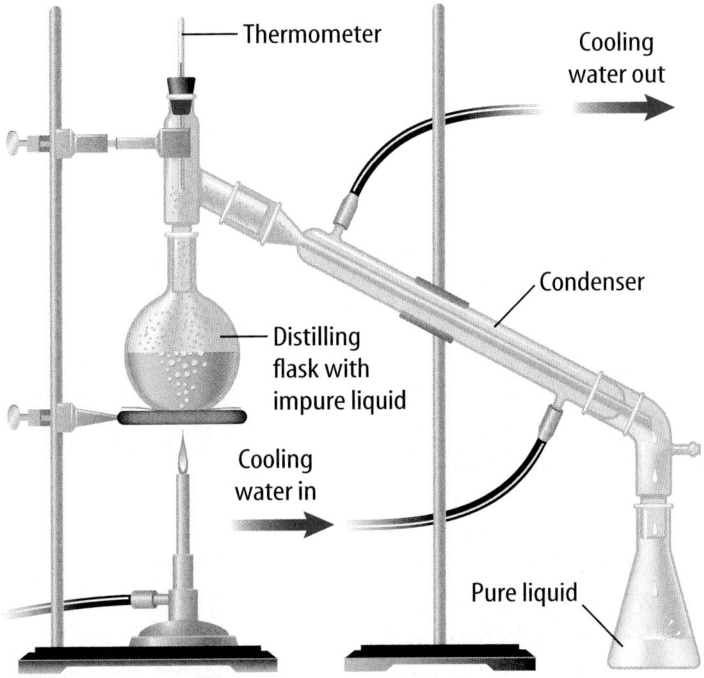

Thermometer
Cooling water out
Condenser
Distilling flask with impure liquid
Cooling water in
Pure liquid

Figure 14
Distillation can easily separate liquids from solids dissolved in them. The liquid is heated until it vaporizes and moves up the column. Then, as it touches the water-cooled surface of the condenser, it becomes liquid again.

Chemical Properties and Changes

You probably have seen warnings on cans of paint thinners and lighter fluids for charcoal grills that say these liquids are flammable (FLA muh buhl). The tendency of a substance to burn, or its flammability, is an example of a chemical property because burning produces new substances during a chemical change. A **chemical property** is a characteristic of a substance that indicates whether it can undergo a certain chemical change. Many substances used around the home, such as lighter fluids, are flammable. Knowing which ones are flammable helps you to use them safely.

A less dramatic chemical change can affect some medicines. Look at **Figure 15.** You probably have seen bottles like this in a pharmacy. Many medicines are stored in dark bottles because they contain compounds that share chemical properties; that is, they can change chemically if they are exposed to light.

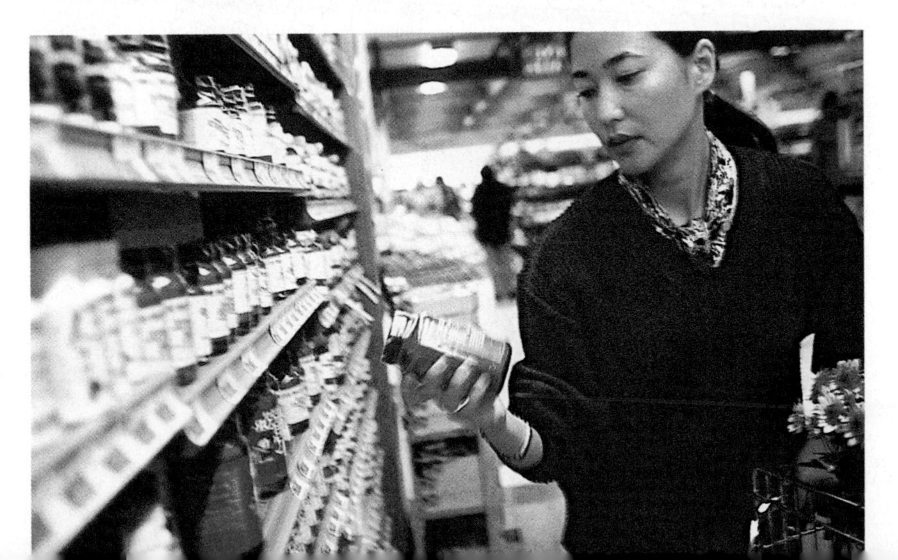

Figure 15
The brown color of these bottles tells you that these vitamins may react to light. Reaction to light is a chemical property.

Detecting Chemical Change

If you leave a pan of chili cooking unattended on the stove for too long, your nose soon tells you that something is wrong. Instead of a spicy aroma, you detect an unpleasant smell that alerts you that something is burning. This burnt odor is a clue telling you that a new substance has formed.

Earth Science INTEGRATION

In a thunderstorm, light and sound tell you that changes have taken place. The pungent smell of ozone indicates that a chemical reaction also took place. Lightning converts oxygen gas, O_2, into ozone, O_3. Ozone is unstable and soon breaks up forming oxygen again.

The Identity Changes The smell of rotten eggs and the formation of rust on bikes or car fenders, are signs that a chemical change has taken place. A change of one substance to another is a **chemical change.** The foaming of an antacid tablet in a glass of water and the smell in the air after a thunderstorm are other signs of new substances being produced. In some chemical changes, a rapid release of energy—detected as heat, light, and sound—is a clue that changes are occurring.

✔ Reading Check *What is a chemical change?*

Clues, such as heat, cooling, or the formation of bubbles or solids in a liquid, are helpful indicators that a reaction is taking place. However, the only sure proof is that a new substance is produced. Consider the following example. The heat, light, and sound produced when hydrogen gas combines with oxygen in a rocket engine are clear evidence that a chemical reaction has taken place. But no clues announce the reaction that takes place when iron combines with oxygen to form rust because the reaction takes place so slowly. The only clue that iron has changed into a new substance is the presence of rust. Burning and rusting are chemical changes because new substances form. You sometimes can follow the progress of a chemical reaction visually. For example, you can see silver tarnish being removed in **Figure 16.**

Figure 16
Some reactions are visible only after they take place.

A Tarnish mars the surface of this silver pitcher.

B You can remove the tarnish using another chemical reaction with aluminum foil and baking soda.

C The tarnish is gone and no silver is lost.

Using Chemical Change to Separate One case where you might separate substances using a chemical change is in cleaning tarnished silver. Tarnish is silver sulfide formed from sulfur compounds in the air. It can be changed back into silver using a chemical reaction. To do this you place the tarnished item in a pot of water containing baking soda and some crumpled aluminum foil. Then you heat the pot. The procedure and its results are shown in **Figure 16.**

You don't usually separate substances using chemical changes in the home. In industry and chemical laboratories, however, this kind of separation is common. For example, many metals are separated from their ores and then purified using chemical changes.

Math Skills Activity

Calculations with the Law of Conservation of Mass

When a chemical reaction takes place, the total mass of the reactants equals the total mass of the products. The total number of atoms of reactants also equals the total number of atoms of products.

Example Problem

In the following reaction, 18 g of hydrogen react completely with 633 g of chlorine. How many grams of HCl are formed? $H_2 + Cl_2 \rightarrow 2HCl$

Solution

1. *This is what you know.* mass $H_2 = 18$ g; mass $Cl_2 = 633$ g
2. *This is what you need to find:* mass of HCl
3. *This is the equation you need to use:* mass reactants = mass products
4. *Solve for the mass of HCl:* $(g\ H_2 + g\ Cl_2) = (g\ HCl)$

   Substitute the known values: $(18\ g + 633\ g) = 651\ g\ HCl$

   Check your answer by subtracting the mass of H_2 from the mass of HCl. Do you obtain the mass of the Cl_2?

Practice Problems

1. In the following reaction, 24 g of CH_4 react with 96 g of O_2 to form 66 g of CO_2. How many grams of H_2O are formed? $CH_4 + 2O_2 \rightarrow CO_2 + 2H_2O$
2. In the following equation, 54.0 g of Al react with 409.2 g of $ZnCl_2$ to form 196.2 g of Zn metal. How many grams of $AlCl_3$ are formed? $2Al + 3ZnCl_2 \rightarrow 3Zn + 2AlCl_3$

For more help, refer to the Math Skill Handbook.

A Flowing water shaped and smoothed these rocks in a physical process.

B Both chemical and physical changes shaped the famous White Cliffs of Dover lining the English Channel.

Figure 17
Weathering can involve physical or chemical change.

Weathering—Chemical or Physical Change?

The forces of nature continuously shape Earth's surface. Rocks split, deep canyons are carved out, sand dunes shift, and curious limestone formations decorate caves. Do you think these changes, often referred to as weathering, are physical or chemical? The answer is both. Geologists, who use the same criteria that you have learned in this chapter, say that some weathering changes are physical and some are chemical.

Physical Large rocks can split when water seeps into small cracks, freezes, and expands. However, the smaller pieces of newly exposed rock still have the same properties as the original sample. This is a physical change. Streams can cut through softer rock, forming canyons, and can smooth and sculpt harder rock, as shown in **Figure 17A.** In each case, the stream carries rock particles far downstream before depositing them. Because the particles are unchanged, the change is a physical one.

Chemical In other cases, the change is chemical. For example, solid calcium carbonate, a compound found in limestone, does not dissolve easily in water. However, when the water is even slightly acidic, as it is when it contains some dissolved carbon dioxide, calcium carbonate reacts. It changes into a new substance, calcium hydrogen carbonate, which does dissolve in water. This change in limestone is a chemical change because the identity of the calcium carbonate changes. The White Cliffs of Dover, shown in **Figure 17B,** are made of limestone and undergo such chemical changes, as well as physical changes. A similar chemical change produces caves and the icicle-shaped rock formations that often are found in them.

Research Visit the Glencoe Science Web site at **science.glencoe.com** to find out more about cave formations. Communicate to your class what you learn.

The Conservation of Mass

Figure 18
This reaction appears to be destroying these logs. When it is over, only ashes will remain. Yet you know that no mass is lost in a chemical reaction. *How can you explain this?*

Wood is combustible, or burnable. As you just learned this is a chemical property. Suppose you burn a large log in the fireplace, as shown in **Figure 18,** until nothing is left but a small pile of ashes. Smoke, heat, and light are given off and the changes in the appearance of the log confirm that a chemical change took place. At first, you might think that matter was lost during this change because the pile of ashes looks much smaller than the log did. In fact, the mass of the ashes is less than that of the log. However, suppose that you could collect all the oxygen in the air that was combined with the log during the burning and all the smoke and gases that escaped from the burning log and measure their masses, too. Then you would find that no mass was lost after all.

Not only is no mass lost during burning, mass is not gained or lost during any chemical change. In other words, matter is neither created nor destroyed during a chemical change. According to the **law of conservation of mass,** the mass of all substances that are present before a chemical change equals the mass of all the substances that remain after the change.

Reading Check *Explain what is meant by the law of conservation of mass.*

Section 2 Assessment

1. In terms of substances, explain why evaporation of water is a physical change and not a chemical change.
2. Name four physical properties you could use to describe a liquid.
3. Why is flammability a chemical property rather than a physical property?
4. How does the law of conservation of mass apply to chemical changes?
5. **Think Critically** The law of conservation of mass applies to physical changes as well as to chemical changes. How might you demonstrate this law for melting ice and distillation of water?

Skill Builder Activities

6. **Drawing Conclusions** What evidence tells you that chemical and physical changes take place in a candle as it burns? **For more help, refer to the Science Skill Handbook.**
7. **Solving One-Step Equations** Two chemicals with a combined mass of 25.48 g react in a flask that has a mass of 142.05 g. A gas is produced that totally escapes into a flask that has an empty mass of 141.65 g. After the reaction, the first flask and its contents have a mass of 167.16 g. Calculate the total mass of the second flask and gas. **For more help, refer to the Math Skill Handbook.**

Activity

Design Your Own Experiment

Checking Out Chemical Changes

Mixing materials together does not always produce a chemical change. You must find evidence of a new substance with new properties being produced before you can conclude that a chemical change has taken place. Try this activity and use your observation skills to deduce what kind of change has occurred.

Recognize the Problem

What evidence indicates a chemical change?

Form a Hypothesis

Think about what happens when small pieces of limestone are mixed with sand. What happens when limestone is mixed with an acid? Based on these thoughts, form a hypothesis about how to determine when mixing substances together produces a chemical change.

Goals

- **Observe** the results of adding dilute hydrochloric acid to baking soda.
- **Infer** that the production of new substances indicates that a chemical change has occurred.
- **Design** an experiment that allows you to compare the activity of baking soda with that of a product formed when baking soda reacts.

Possible Materials

baking soda
small evaporating dish
hand lens
1*M* hydrochloric acid (HCl)
10-mL graduated cylinder
electric hot plate

Safety Precautions

Limestone

Sand

Test Your Hypothesis

Plan

1. As a group, agree upon a hypothesis and decide how to test it. Write the hypothesis statement.
2. To test your hypothesis, devise a plan to compare two different mixtures. The first mixture consists of 3 mL of hydrochloric acid and 0.5 g of baking soda. The second mixture is 3 mL of hydrochloric acid and the solid product of the first mixture. Describe exactly what you will do at each step.
3. Make a list of the materials needed to complete your experiment.
4. **Design** a table for data and observations in your Science Journal so that it is ready to use as your group observes what happens.

Do

1. Make sure your teacher approves your plan before you start.
2. Read over your entire experiment to make sure that all steps are in logical order.
3. **Identify** any constants and the variables of the experiment.
4. Should you run any test more than once? How will observations be summarized?
5. Assemble your materials and carry out the experiment according to your plan. Be sure to record your results as you work.

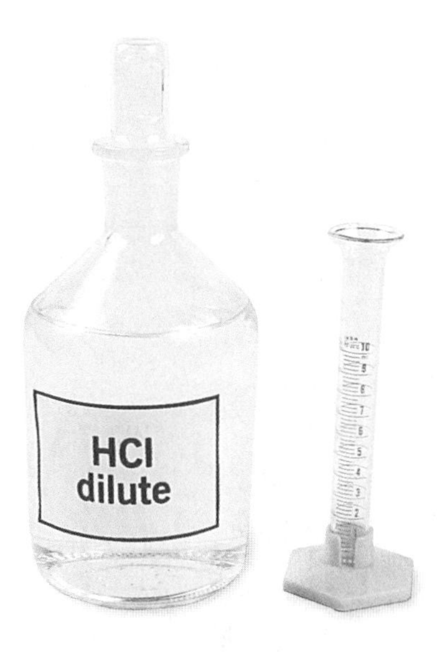

Analyze Your Data

1. What happened to the baking soda? Did anything happen to the product formed from the first mixture? Explain why this occurred.
2. What different properties of any new substances did you observe after adding hydrochloric acid to the baking soda?

Draw Conclusions

1. Did the results support your hypothesis? Explain.
2. If you had used vinegar, which contains acetic acid as the acid, do you think a new susbstance would have formed? How could you test this?

Write a description of your observations in your Science Journal. **Compare** your results with those of other groups. **Discuss** your conclusions.

Science Stats

Intriguing Elements

Did you know...

...Silver-white cobalt, which usually is combined with other elements in nature, is used to create rich paint pigments. It can be used to form powerful magnets, treat cancer patients, build jet engines, and prevent disease in sheep.

Few scientists have seen the rare and elusive element astatine. Earth's supply is lean—probably only about 28 g of astatine total. Chemically, it's similar to iodine.

...You have something in common with diamonds—carbon. Diamonds form from carbon under extremely high pressures and temperatures deep inside Earth. Carbon is an essential element in living organisms, making up about 18 percent of the human body.

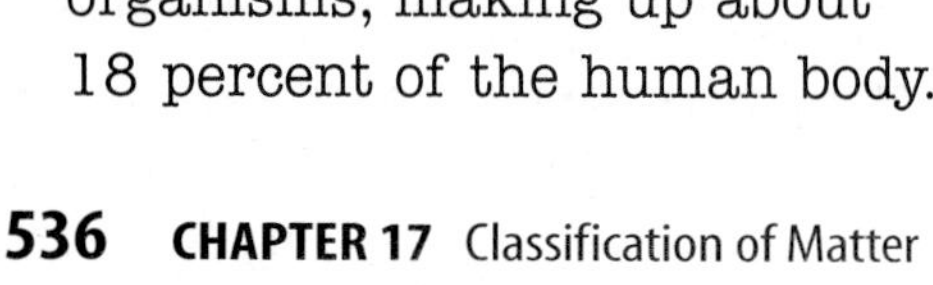

...Gold is the most ductile (stretchable) of all the elements. Just 29 g of gold—about ten wedding bands—can be pulled into a wire 100 km long. That's long enough to stretch from Toledo, Ohio, to Detroit, Michigan, and beyond.

...Zinc makes chewing gum taste better. Up to 0.3 mg of zinc acetate can be added per 1,000 mg of chewing gum to provide a tart, zingy flavor.

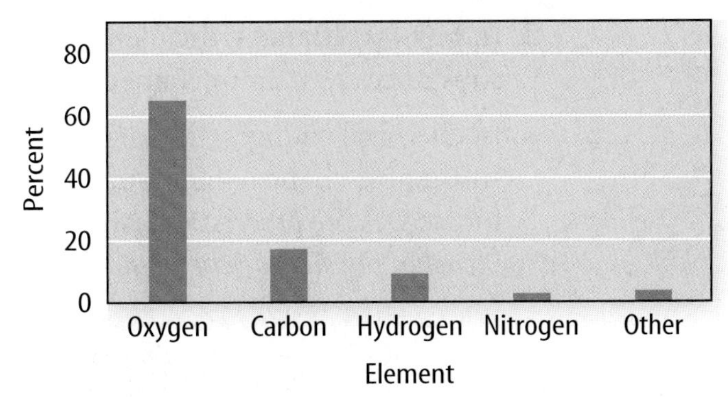

Do the Math

1. For kids, the recommended daily allowance (RDA) of zinc is 10 mg. If chewing gum were your only source of zinc, how many milligrams of gum with the maximum amount of zinc would you have to chew to get your RDA of zinc?
2. If you made the thinnest possible gold wire with 100 g of gold, how long would your wire be?
3. If you had 50 atoms of hydrogen, how many atoms each of oxygen, carbon, and nitrogen would you need to have the elements in the same proportion as they are in your body?

Go Further

What is the element most recently discovered by scientists? Go to the Glencoe Science Web site at **science.glencoe.com** to find out.

Chapter 17 Study Guide

Reviewing Main Ideas

Section 1 Composition of Matter

1. Elements and compounds are substances. A mixture is composed of two or more substances.
2. You can distinguish the different materials in a heterogeneous mixture either using your naked eye or using a microscope.
3. Colloids and suspensions are two types of heterogeneous mixtures. The particles in a suspension will settle eventually. Particles of a colloid will not. *What simple observation tells you that the substance shown here is a colloid?*

4. In a homogeneous mixture, the particles are distributed evenly and are not visible, even when using a microscope. Homogeneous mixtures can be composed of solids, liquids, or gases.
5. A solution is another name for a homogeneous mixture that remains constantly and uniformly mixed. *Is the substance in the container below a solution? How do you know?*

Section 2 Properties of Matter

1. Physical properties are characteristics of materials that you can observe without changing the identity of the substance.
2. Chemical properties indicate what chemical changes substances can undergo. *What type of chemical change might these bottles prevent?*
3. In physical changes, the identities of substances remain unchanged.
4. In chemical changes, the identities of substances change—new substances are formed. *Is the process of cleaning rust with bleach a physical or chemical change?*

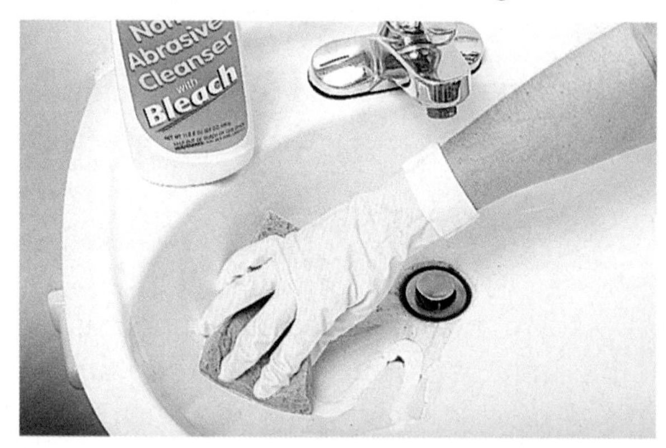

5. The law of conservation of mass states that during any chemical change, matter is neither created nor destroyed.

After You Read

FOLDABLES Reading & Study Skills

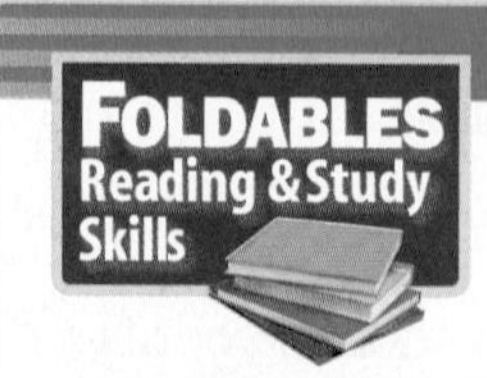

To help you review the classifications of matter, use the Vocabulary Study Fold you made at the beginning of the chapter.

Chapter 17 Study Guide

Visualizing Main Ideas

Complete the concept map below about matter.

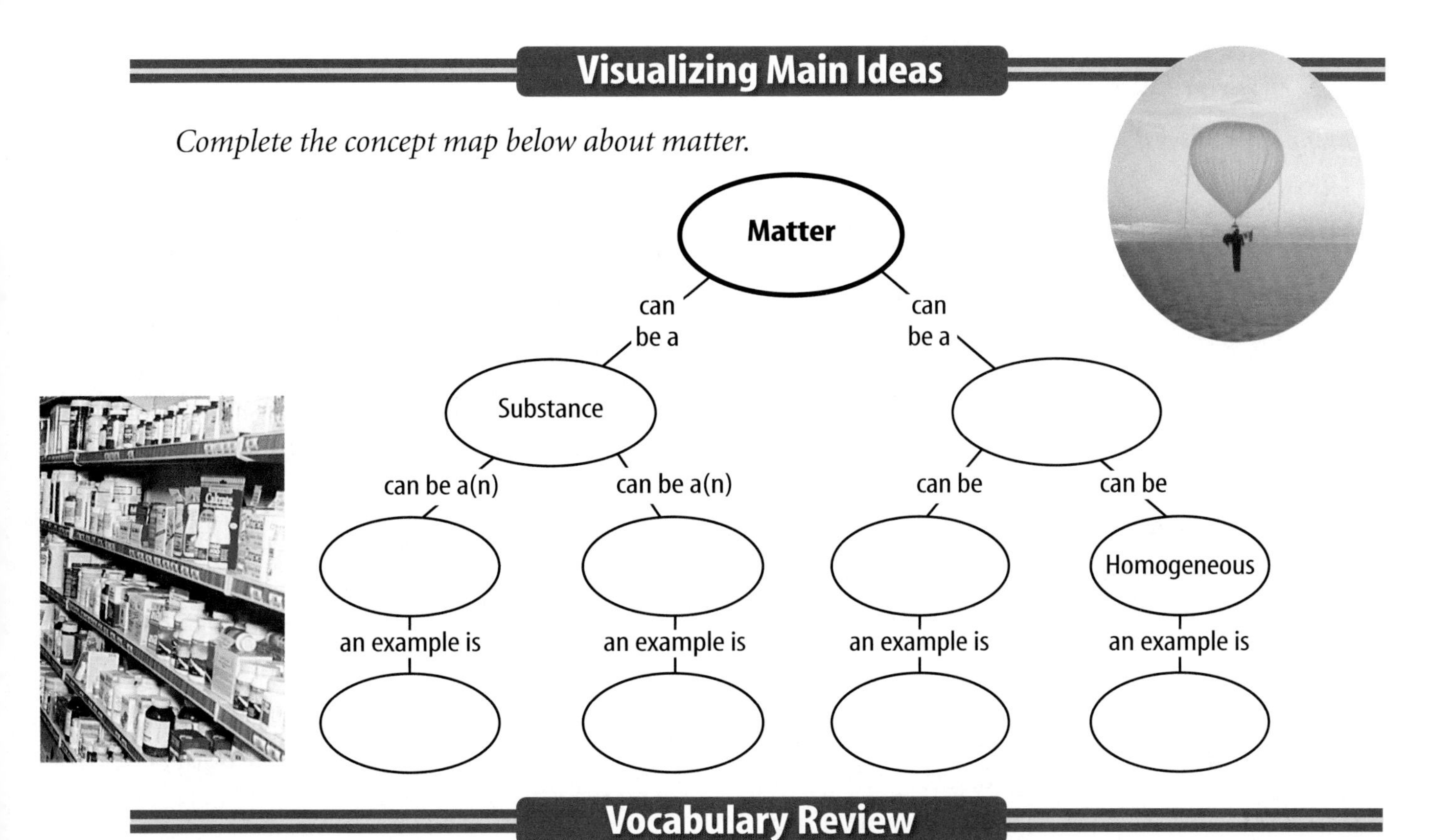

Vocabulary Review

Vocabulary Words

a. chemical change
b. chemical property
c. colloid
d. compound
e. distillation
f. element
g. heterogeneous mixture
h. homogeneous mixture
i. law of conservation of mass
j. physical change
k. physical property
l. solution
m. substance
n. suspension
o. Tyndall effect

Study Tip

Make sure to read over your class notes after each lesson. Reading them will help you better understand what you've learned, as well as prepare you for the next day's lesson.

Using Vocabulary

Replace the underlined words with the correct vocabulary words.

1. Substances formed from atoms of two or more elements are called <u>mixtures</u>.
2. A <u>colloid</u> is a heterogeneous mixture in which visible particles settle.
3. Freezing, boiling, and evaporation are all examples of <u>chemical change</u>.
4. According to the <u>Tyndall effect</u>, matter is neither created nor destroyed during a chemical change.
5. A mixture in which different materials are easily identified is a <u>homogeneous mixture</u>.
6. Compounds are made from the atoms of two or more <u>colloids</u>.
7. Distillation is a process that can separate two liquids using <u>chemical change</u>.

Chapter 17 Assessment

Checking Concepts

Choose the word or phrase that best answers the question.

1. Bending a copper wire is an example of what type of property?
   A) chemical **C)** conservation
   B) physical **D)** element
2. Which of the following is NOT an element?
   A) water **C)** oxygen
   B) carbon **D)** hydrogen
3. Which of the following is an example of a chemical change?
   A) boiling **C)** evaporation
   B) burning **D)** melting
4. What type of substance is gelatin?
   A) colloid **C)** substance
   B) compound **D)** suspension
5. A visible sunbeam is an example of which of the following?
   A) an element **C)** a suspension
   B) a solution **D)** the Tyndall effect
6. You start to eat some potato chips from an open bag you found in your locker and notice that they taste unpleasant. What do you think might cause this unpleasant taste?
   A) combustion **C)** physical change
   B) chemical change **D)** melting
7. How would you classify the color of a rose?
   A) chemical change **C)** physical change
   B) chemical property **D)** physical property
8. How would you describe the process of evaporating water from seawater?
   A) chemical change **C)** physical change
   B) chemical property **D)** physical property
9. Which of these warnings refers to a chemical property of the material?
   A) Fragile **C)** Handle with Care
   B) Flammable **D)** Shake Well
10. Which of the following is a substance?
    A) colloid **C)** mixture
    B) element **D)** solution

Thinking Critically

11. Describe the contents of a carton of milk using at least four physical properties.
12. Black carbon and the colorless gases hydrogen and oxygen combine to form sugar. How do you know sugar is a compound?
13. The word *colloid* means "gluelike." Why was this term chosen to name certain mixtures?
14. Use a nail rusting in air to explain the law of conservation of mass.
15. Mai says that ocean water is a solution. Tom says that it's a suspension. Can they both be correct? Explain.

Developing Skills

16. **Making and Using Tables** Different colloids can involve different states. For example, gelatin is formed from solid particles in a liquid. Complete this table using these colloids: *smoke, marshmallow, fog,* and *paint.*

Common Colloids

Colloid	Example
Solid in a liquid	Gelatin
Solid in a gas	
Gas in a solid	
Solid in a liquid	
Liquid in a gas	

17. **Comparing and Contrasting** Give examples of solutions, suspensions, and colloids from your daily life and compare and contrast their properties.

18. Using Variables, Constants, and Controls Marcos took a 100-cm^3 sample of a suspension, shook it well, and poured equal amounts into four different test tubes. He placed one test tube in a rack, one in hot water, one in warm water, and the fourth in ice water. He then observed the time it took for each suspension to settle. What was the variable in the experiment? What was one constant?

19. Interpreting Data Hannah started with a 25-mL sample of pond water. Without shaking the sample, she poured 5 mL through a piece of filter paper. She repeated this with four more pieces of filter paper. She dried each piece of filter paper and measured the mass of the sediment. Why did the last sample have a higher mass than did the first sample?

20. Concept Mapping Make a network tree to show types of liquid mixtures. Include these terms: *homogeneous mixtures, heterogeneous mixtures, solutions, colloids,* and *suspensions.*

Performance Assessment

21. Design an Experiment Assume that some sugar was put into some rice by mistake. Design an experiment to separate the mixture. In your Science Journal, list your hypothesis and your experimental steps. Then carry out the experiment, and report the results.

Technology

Go to the Glencoe Science Web site at **science.glencoe.com** or use the **Glencoe Science CD-ROM** for additional chapter assessment.

Test Practice

A student did some research about which elements are found in the human body. The information is shown below.

Elements in the Human Body

Element	Percent
Oxygen	65%
Calcium	2.0%
Carbon	18.0%
Hydrogen	10.0%
Phosphorus	1.0%
Other elements	4.0%

Study the table and answer the following questions.

1. Which element makes up 1.0 percent of the human body?
A) calcium
B) hydrogen
C) phosphorus
D) carbon

2. About how much greater is the percentage of carbon in the human body than hydrogen?
F) 3 percent
G) 8 percent
H) 10 percent
J) 15 percent

3. Which element, together with phosphorus, makes up 3 percent of the human body?
A) hydrogen
B) oxygen
C) calcium
D) nitrogen

4. Which two elements make up three fourths of the human body?
F) carbon and calcium
G) oxygen and hydrogen
H) oxygen and nitrogen
J) phosphorus and carbon

CHAPTER

18 Properties of Atoms and the Periodic Table

It might surprise you to know that these clouds of Jupiter, your pencil or pen, and you have something in common. Much of the universe is made up of particles so small they cannot even be seen, called atoms. But there are particles even smaller than atoms. What are these tiny pieces of the universe? In this chapter you will learn about atoms and their components—protons, neutrons, and electrons.

What do you think?

Science Journal Look at the picture below with a classmate. Discuss what you think this might be. Here's a hint: *It's a winning combination you couldn't live without.* Write your answer or your best guess in your Science Journal.

How do detectives solve a crime when no witnesses saw it happen? How do scientists study atoms when they cannot see them? In situations such as these, techniques must be developed to find clues to answer the question. Do the activity below to see how clues might be gathered.

Inferring what you can't observe 

1. Take an envelope from your teacher.
2. Place an assortment of dried beans in the envelope and seal it. **WARNING:** *Do not eat any lab materials.*
3. Trade envelopes with another group.
4. Without opening the envelope, try to figure out the types and number of beans that are in the envelope. Record a hypothesis about the contents of the envelope in your Science Journal.
5. After you record your hypothesis, open the envelope and see what is inside.

Observe

How many of each kind of bean did you find? Was your hypothesis correct?

Before You Read

Making a Know-Want-Learn Study Fold **Make the following Foldable to help you identify what you already know and what you want to know about properties of atoms.**

1. Stack two sheets of paper in front of you so the short side of both sheets is at the top.
2. Slide the top sheet up so that about 4 cm of the bottom sheet shows.
3. Fold both sheets top to bottom to form four tabs and staple along the fold, as shown.
4. Label the top flap *Atoms.* Then label the other flaps *Know, Want,* and *Learned* as shown. Before you read the chapter, write what you know about atoms on the *Know* tab and what you want to know on the *Want* tab.
5. As you read the chapter, list the things you learn about atoms on the *Learned* tab.

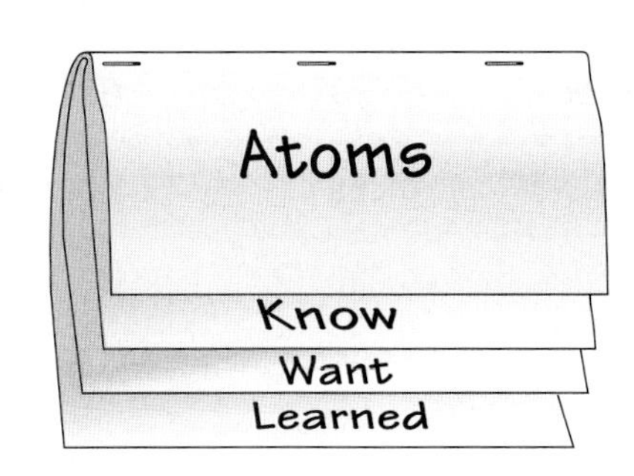

SECTION

Structure of the Atom

As You Read

What You'll Learn

- **Identify** the names and symbols of common elements.
- **Identify** quarks as subatomic particles of matter.
- **Describe** the electron cloud model of the atom.
- **Explain** how electrons are arranged in an atom.

Vocabulary

atom
nucleus
electron
proton
neutron
quark
electron cloud

Why It's Important

Everything you see, touch, and breathe is composed of tiny atoms.

Scientific Shorthand

Do you have a nickname? Do you use abbreviations for long words or the names of states? Scientists also do this. In fact, scientists have developed their own shorthand for dealing with long, complicated names.

Do the letters C, Al, Ne, and Ag mean anything to you? Each letter or pair of letters is a chemical symbol, which is a short or abbreviated way to write the name of an element. Chemical symbols, such as those in **Table 1,** consist of one capital letter or a capital letter plus one or two small letters. For some elements, the symbol is the first letter of the element's name. For other elements, the symbol is the first letter of the name plus another letter from its name. Some symbols are derived from Latin. For instance, *Argentum* is Latin for "silver." Elements have been named in a variety of ways. Some elements are named to honor scientists, for places, or for their properties. Other elements are named using rules established by an international committee. Regardless of the origin of the name, scientists derived this international system for convenience. It is much easier to write H for hydrogen, O for oxygen, and H_2O for dihydrogen oxide (water). Because scientists worldwide use this system, everyone understands what the symbols mean.

Table 1 Symbols of Some Elements

Element	Symbol	Element	Symbol	Element	Symbol
Aluminum	Al	**Gold**	Au	**Mercury**	Hg
Calcium	Ca	**Helium**	He	**Nitrogen**	N
Carbon	C	**Hydrogen**	H	**Oxygen**	O
Chlorine	Cl	**Iodine**	I	**Potassium**	K
Copper	Cu	**Iron**	Fe	**Silver**	Ag
Fluorine	F	**Magnesium**	Mg	**Sodium**	Na

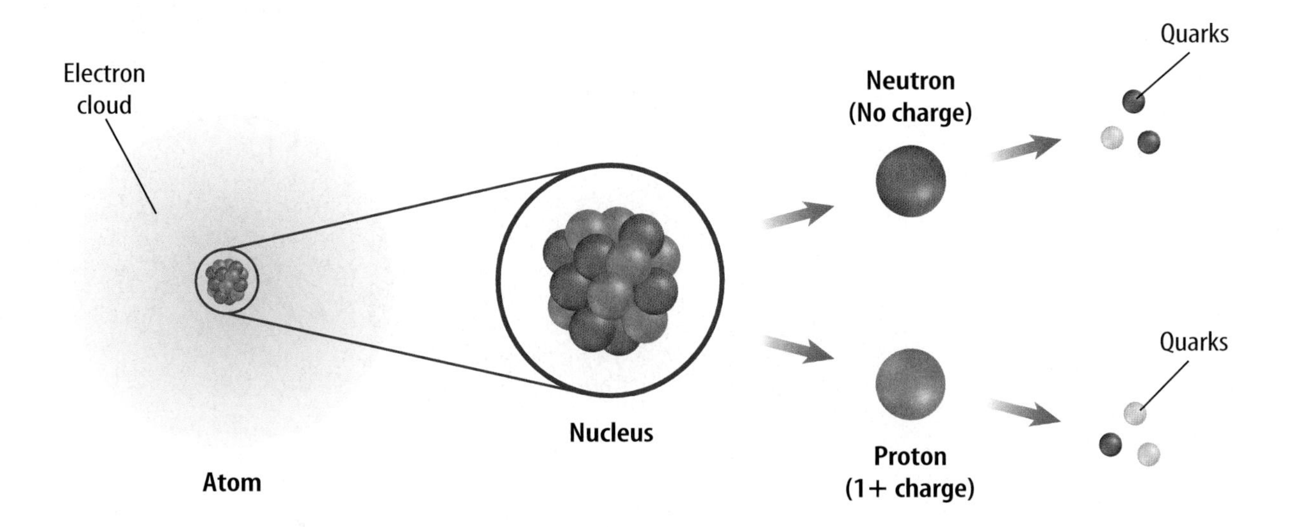

Figure 1
The nucleus of the atom contains protons and neutrons that are composed of quarks. The proton has a positive charge and the neutron has no charge. A cloud of negatively charged electrons surrounds the nucleus of the atom.

Atomic Components

An element is matter that is composed of one type of **atom,** which is the smallest piece of matter that still retains the property of the element. For example, the element silver is composed of only silver atoms and the element hydrogen is composed of only hydrogen atoms. Atoms are composed of particles called protons, neutrons, and electrons, as shown in **Figure 1.** Protons and neutrons are found in a small, positively-charged center of the atom called the **nucleus** that is surrounded by a cloud containing electrons. **Protons** are particles with an electrical charge of 1+. **Neutrons** are neutral particles that do not have an electrical charge. **Electrons** are particles with an electrical charge of 1−. Atoms of different elements differ in the number of protons they contain.

What are the particles that make up the atom and where are they located?

Quarks—Even Smaller Particles

Are the protons, electrons, and neutrons that make up atoms the smallest particles that exist? Scientists hypothesize that electrons are not composed of smaller particles and are one of the most basic types of particles. Protons and neutrons, however, are made up of smaller particles called **quarks.** So far, scientists have confirmed the existence of six uniquely different quarks. Scientists theorize that an arrangement of three quarks held together with the strong nuclear force produces a proton. Another arrangement of three quarks produces a neutron. The search for the composition of protons and neutrons is an ongoing effort.

Research Visit the Glencoe Science Web site at **science.glencoe.com** for more information about ongoing research at the Fermi National Accelerator Laboratory. Communicate to your class what you learn.

Figure 2
The Tevatron is a huge machine. A This aerial photograph of Fermi National Accelerator Laboratory shows the circular outline of the Tevatron particle accelerator. B This close-up photograph of the Tevatron gives you a better view of the tunnel. *Why is such a long tunnel needed?*

A

Finding Quarks To study quarks, scientists accelerate charged particles to tremendous speeds and then force them to collide with—or smash into—protons. This collision causes the proton to break apart. The Fermi National Accelerator Laboratory, a research laboratory in Batavia, Illinois, houses a machine that can generate the forces that are required to collide protons. This machine, the Tevatron, shown in **Figure 2,** is approximately 6.4 km in circumference. Electric and magnetic fields are used to accelerate, focus and collide the fast-moving particles.

The particles that result from the collision are detected in a device called a bubble chamber in which the particles leave tracks. Just as police investigators can reconstruct traffic accidents from tire marks and other clues at the scene, scientists are able to examine and gather information from the tracks left after proton collisions, as shown in **Figure 3.** Studying the tracks reveals information about the inner structure of the atom. Scientists then use inference to identify the subatomic particles.

Figure 3
Bubble chambers are used by scientists to study the tracks left by subatomic particles.

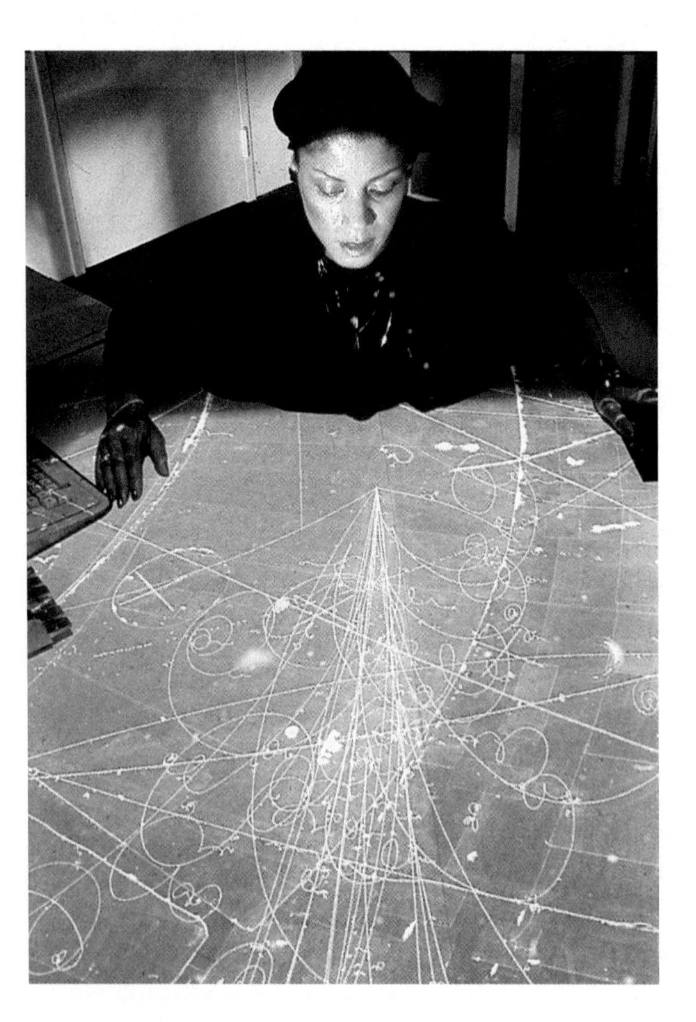

The Sixth Quark Finding evidence for the existence of the quarks was not an easy task. Scientists found five quarks and hypothesized that a sixth quark existed. However, it took a team of nearly 450 scientists from around the world several years to find the sixth quark. The tracks of the sixth quark were hard to detect because only about one billionth of a percent of the proton collisions performed showed the presence of a sixth quark—typically referred to as the *top* quark.

Models—Tools for Scientists

Scientists and engineers use models to represent things that are difficult to visualize—or picture in your mind. You might have seen models of buildings, the solar system, or airplanes. These are scaled-down models. Scaled-down models allow you to see either something too large to see all at once, or something that has not been built yet. Scaled-up models are often used to visualize things that are too small to see. Atoms are very small. To give you an idea of how small the atom is, it would take about 24,400 atoms stacked one on top of the other to equal the thickness of a sheet of aluminum foil. To study the atom, scientists have developed scaled-up models that they can use to visualize how the atom is constructed. For the model to be useful, it must support all of the information that is known about matter and the behavior of atoms. As more information about the atom is collected, scientists change their models to include the new information.

Reading Check *Why do scientists use models?*

The Changing Atomic Model You know now that all matter is composed of atoms, but this was not always known. Around 400 B.C., Democritus proposed the idea that atoms make up all substances. However, a famous Greek philosopher, Aristotle, disputed Democritus's theory and proposed that matter was uniform throughout and was not composed of smaller particles. Aristotle's incorrect theory was accepted for about 2,000 years. In the 1800s, John Dalton, an English scientist, was able to offer proof that atoms exist.

Dalton's model of the atom, a solid sphere shown in **Figure 4,** was an early model of the atom. As you can see in **Figure 5,** the model has changed somewhat over time. As each scientist performed experiments and learned a little bit more about the structure of the atom, the model was modified. The model in use today is the accumulated knowledge of almost two hundred years.

Modeling an Aluminum Atom

Procedure

1. Arrange 13 3-cm circles cut from **orange paper** and 14 3-cm circles cut from **blue paper** on a **flat surface** to represent the nucleus of an atom. Each orange circle represents one proton, and each blue circle represents one neutron.
2. Position 2 holes punched from **red paper** about 20 cm from your nucleus.
3. Position 8 punched holes about 40 cm from your nucleus.
4. Position 3 punched holes about 60 cm from your nucleus.

Analysis

1. How many protons, neutrons, and electrons does an aluminum atom have?
2. Explain how your circles model an aluminum atom.
3. Explain why your model does not accurately represent the true size and distances in an aluminum atom.

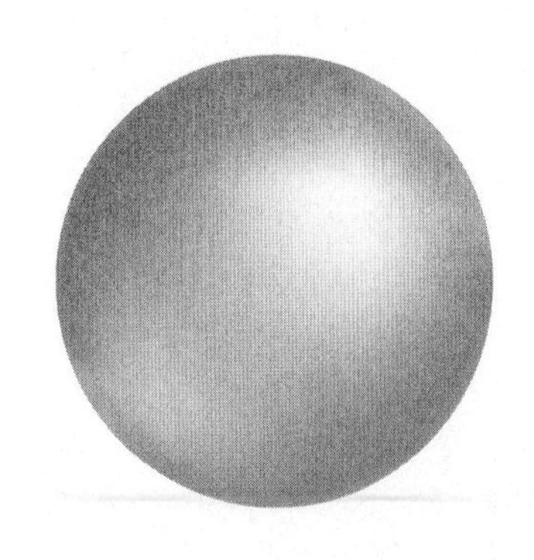

Figure 4
John Dalton's atomic model was a simple sphere.

VISUALIZING THE ATOMIC MODEL

Figure 5

The ancient Greek philosopher Democritus proposed that elements consisted of tiny, solid particles that could not be subdivided (A). He called these particles *atomos,* meaning "uncuttable." This concept of the atom's structure remained largely unchallenged until the 1900s, when researchers began to discover through experiments that atoms were composed of still smaller particles. In the early 1900s, a number of models for atomic structure were proposed (B-D). The currently accepted model (E) evolved from these ideas and the work of many other scientists.

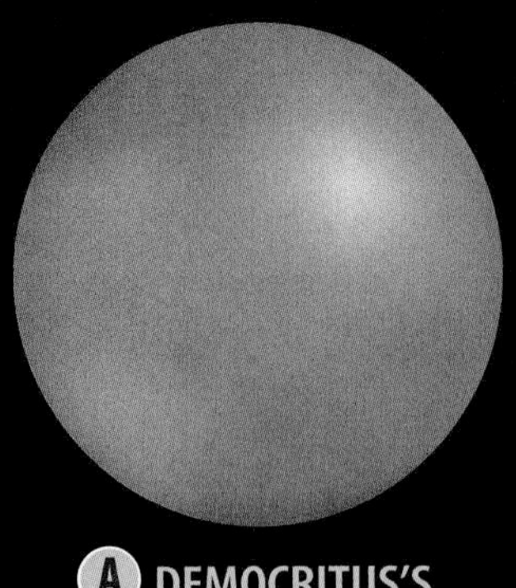

A DEMOCRITUS'S UNCUTTABLE ATOM

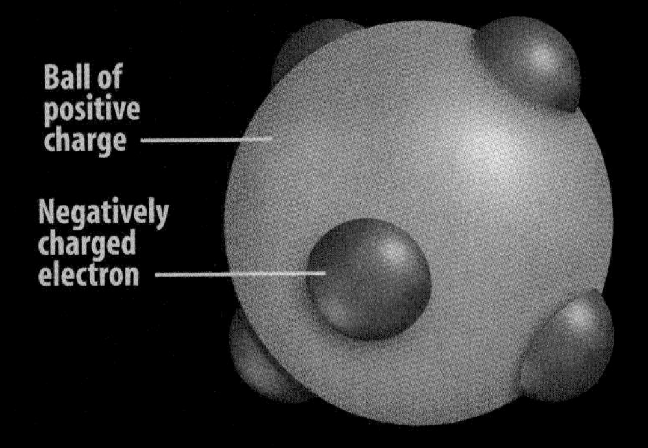

B THOMSON MODEL, 1904 English physicist Joseph John Thomson inferred from his experiments that atoms contained small, negatively charged particles. He thought these "electrons" (in red) were evenly embedded throughout a positively charged sphere, much like chocolate chips in a ball of cookie dough.

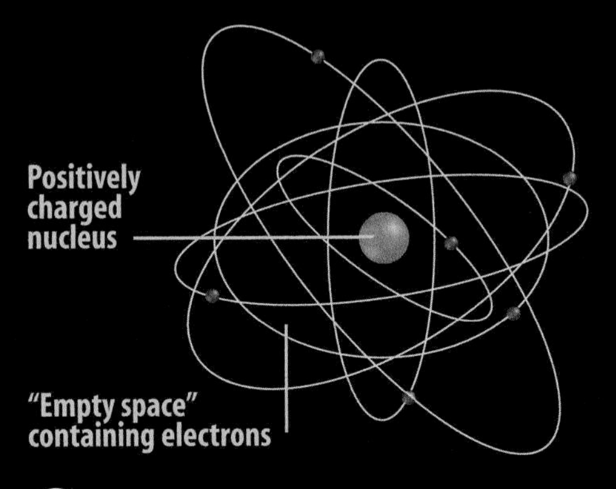

C RUTHERFORD MODEL, 1911 Another British physicist, Ernest Rutherford, proposed that almost all the mass of an atom—and all its positive charges—were concentrated in a central atomic nucleus surrounded by electrons.

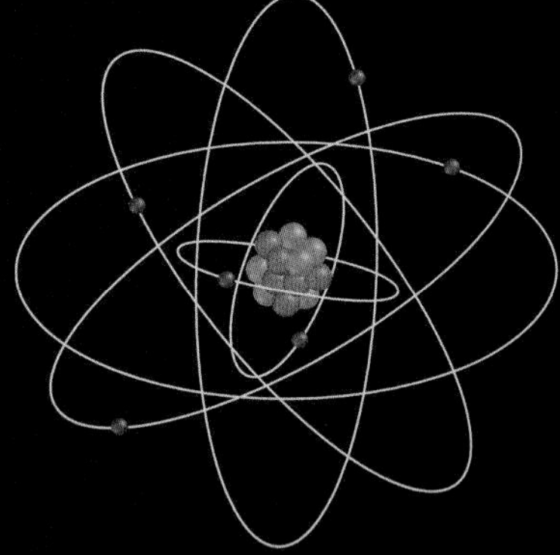

D BOHR MODEL, 1913 Danish physicist Niels Bohr hypothesized that electrons traveled in fixed orbits around the atom's nucleus. James Chadwick, a student of Rutherford, concluded that the nucleus contained positive protons and neutral neutrons.

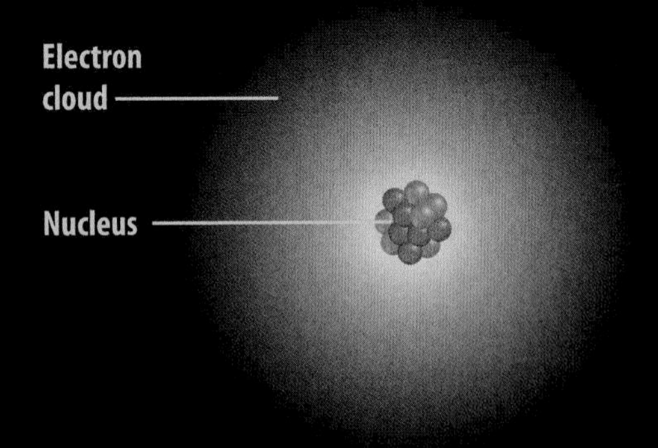

E ELECTRON CLOUD MODEL, CURRENT According to the currently accepted model of atomic structure, electrons do not follow fixed orbits but tend to occur more frequently in certain areas around the nucleus at any given time.

The Electron Cloud Model By 1926, scientists had developed the electron cloud model of the atom that is in use today. An **electron cloud** is the area around the nucleus of an atom where its electrons are most likely found. The electron cloud is 100,000 times larger than the diameter of the nucleus. In contrast, each electron in the cloud is much smaller than a single proton.

Because an electron's mass is small and the electron is moving so quickly around the nucleus, it is impossible to describe its exact location in an atom. Picture the spokes on a moving bicycle wheel. They are moving so quickly that you can't pinpoint any single spoke. All you see is a blur that contains all of the spokes somewhere within it. In the same way, an electron cloud is a blur containing all of the electrons of the atom somewhere within it. **Figure 6** illustrates what the electron cloud might look like.

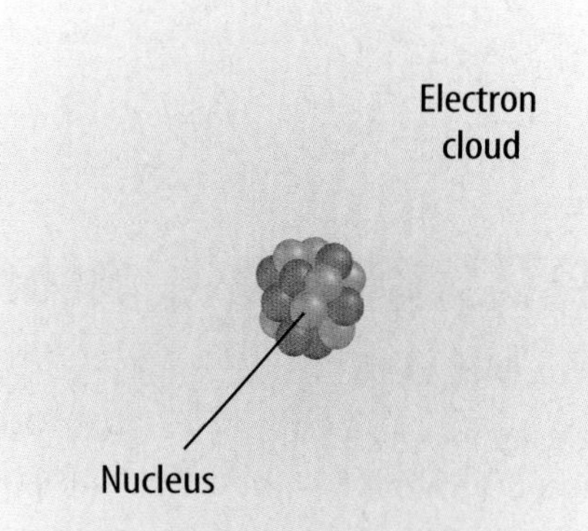

Figure 6
The electrons are located in an electron cloud surrounding the nucleus of the atom.

Scientists have determined that the electron cloud is more than just a blur. Each electron travels at an average distance from the nucleus, depending on its energy. These average distances are referred to as energy levels. Energy levels are areas of the cloud where the electrons are more likely to be found.

Section  Assessment

1. Write the chemical symbols for the elements carbon, aluminum, hydrogen, oxygen, and sodium.
2. What are the names, charges, and locations of three kinds of particles that make up an atom?
3. What is the smallest particle of matter? How were they discovered?
4. Describe the electron cloud model of the atom.
5. **Think Critically** Explain how a rotating electric fan might be used to model the atom. Explain how the rotating fan is unlike an atom.

Skill Builder Activities

6. **Concept Mapping** Make a concept map for the parts of an atom. Include the following terms: *electron cloud, nucleus, electrons, protons, quarks,* and *neutrons.* Provide the location of each particle in the atom. **For more help, refer to the Science Skill Handbook.**
7. **Solving One-Step Equations** The mass of a proton is estimated to be 1.6726×10^{-24} g and the mass of an electron is estimated to be 9.1093×10^{-28} g. How many times larger is the mass of a proton compared to the mass of an electron? **For more help, refer to the Math Skill Handbook.**

SECTION 2

Masses of Atoms

As You Read

What You'll Learn

- **Compute** the atomic mass and mass number of an atom.
- **Identify** isotopes of common elements.
- **Interpret** the average atomic mass of an element.

Vocabulary

atomic number
mass number
isotope
average atomic mass

Why It's Important

Most elements exist in more than one form. Some are radioactive, and others are not.

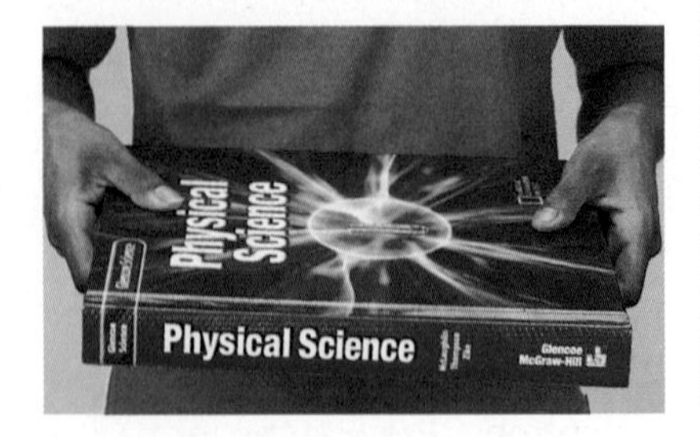

Figure 7
If you held a textbook and placed a paper clip on it, you wouldn't notice the added mass because the mass of a paper clip is small compared to the mass of the book. In a similar way, the masses of an atom's electrons are negligible compared to an atom's mass.

Atomic Mass

The nucleus contains most of the mass of the atom because protons and neutrons are far more massive than electrons. The mass of a proton is about the same as that of a neutron—approximately 1.6726×10^{-24} g, as shown in **Table 2.** The mass of each is approximately 1,836 times greater than the mass of the electron. The electron's mass is so small that it is considered negligible when finding the mass of an atom, as shown in **Figure 7.**

If you were asked to estimate the height of your school building, you probably wouldn't give an answer in kilometers. The number would be too cumbersome to use. Considering the scale of the building, you would more likely give the height in a smaller unit, meters. When thinking about the small masses of atoms, scientists found that even grams were not small enough to use for measurement. Scientists need a unit that results in more manageable numbers. The unit of measurement used for atomic particles is the atomic mass unit (amu). The mass of a proton or a neutron is almost equal to 1 amu. This is not coincidence—the unit was defined that way. The atomic mass unit is defined as one-twelfth the mass of a carbon atom containing six protons and six neutrons. Remember that the mass of the carbon atom is contained almost entirely in the mass of the protons and neutrons that are located in the nucleus. Therefore, each of the 12 particles in the nucleus must have a mass nearly equal to one.

Reading Check ***Where is the majority of the mass of an atom located?***

Table 2 Comparison of Particles in an Atom

Particle	Mass (g)	Charge	Location in Atom
Proton	1.6726×10^{-24}	1+	Nucleus
Neutron	1.6749×10^{-24}	0	Nucleus
Electron	9.1093×10^{-28}	1−	Cloud surrounding nucleus

Table 3 Mass Numbers of Some Atoms

Element	Symbol	Atomic Number	Protons	Neutrons	Mass Number	Average Atomic Mass*
Boron	B	5	5	6	11	10.81
Carbon	C	6	6	6	12	12.01
Oxygen	O	8	8	8	16	16.00
Sodium	Na	11	11	12	23	22.99
Copper	Cu	29	29	34	63	63.55

* The atomic mass units are rounded to two decimal places.

Protons Identify the Element You learned earlier that atoms of different elements are different because they have different numbers of protons. In fact, the number of protons tells you what type of atom you have and vice versa. For example, every carbon atom has six protons. Also, all atoms with six protons are carbon atoms. Atoms with eight protons are oxygen atoms. The number of protons in an atom is equal to a number called the **atomic number.** The atomic number of carbon is six. Therefore, if you are given any one of the following—the name of the element, the number of protons in the element, or the atomic number of the element, you can determine the other two.

✔ Reading Check *Which element is an atom with six protons in the nucleus?*

Mass Number The **mass number** of an atom is the sum of the number of protons and the number of neutrons in the nucleus of an atom. Look at **Table 3** and see if this is true.

If you know the mass number and the atomic number of an atom, you can calculate the number of neutrons. The number of neutrons is equal to the atomic number subtracted from the mass number.

$$\text{number of neutrons} = \text{mass number} - \text{atomic number}$$

Atoms of the same element with different numbers of neutrons can have different properties. For example, carbon with a mass number equal to 12 or carbon-12 is the most common form of carbon. Carbon-14 is present on Earth in much smaller quantities. Carbon-14 is radioactive and carbon-12 is not.

✔ Reading Check *How is the mass number calculated?*

Living organisms on Earth contain carbon. Carbon-12 makes up 99 percent of this carbon. Carbon-13 and carbon-14 make up the other one percent. Which isotopes are archaeologists most interested in when they determine the age of carbon-containing remains? Explain your answer in your Science Journal.

Isotopes

Not all the atoms of an element have the same number of neutrons. Atoms of the same element that have different numbers of neutrons are called **isotopes.** Suppose you have a sample of the element boron. Naturally occurring atoms of boron have mass numbers of 10 or 11. How many neutrons are in a boron atom? It depends upon the isotope of boron to which you are referring. Obtain the number of protons in boron from the periodic table. Then use the formula on the previous page to calculate the number of neutrons in each boron isotope. You can determine that boron can have five or six neutrons.

Uranium-238 has 92 protons. How many neutrons does it have?

Problem-Solving Activity

Radioactive Isotopes Help Tell Time

Atoms can be used to measure the age of bones or rock formations that are millions of years old. The time it takes for half of the radioactive atoms in a piece of rock or bone to change into another element is called its half-life. Scientists use the half-lives of radioactive isotopes to measure geologic time.

Half-lives of Radioactive Isotopes

Radioactive Element	Changes to this Radioactive Element	Half-Life
uranium-238	lead-206	4,460 million years
potassium-40	argon-40, calcium-40	1,260 million years
rubidium-87	strontium-87	48,800 million years
carbon-14	nitrogen-14	5,715 years

Identifying the Problem

The table above lists the half-lives of a sample of radioactive isotopes and into which elements they change. For example, it would take 5,715 years for half of the carbon-14 atoms in a rock to change into atoms of nitrogen-14. After another 5,715 years, half of the remaining carbon-14 atoms will change, and so on. You can use these radioactive clocks to measure different periods of time.

Solving the Problem

1. How many years would it take half of the rubidium-87 atoms in a piece of rock to change into strontium-87? How many years would it take for 75% of the atoms to change?
2. After a long period, only 25% of the atoms in a rock remained uranium-238. How many years old would you predict the rock to be? The other 75% of the atoms are now which radioactive element?

Identifying Isotopes Models of two isotopes of boron are shown in **Figure 8.** Because the numbers of neutrons in the isotopes are different, the mass numbers are also different. You use the name of the element followed by the mass number of the isotope to identify each isotope: boron-10 and boron-11. Because most elements have more than one isotope, each element has an average atomic mass. The **average atomic mass** of an element is the weighted-average mass of the mixture of its isotopes. For example, four out of five atoms of boron are boron-11, and one out of five is boron-10. To find the weighted-average or the average atomic mass of boron, you would solve the following equation:

$$\frac{4}{5}(11 \text{ amu}) + \frac{1}{5}(10 \text{ amu}) = 10.8 \text{ amu}$$

The average atomic mass of the element boron is 10.8 amu. Note that the average atomic mass of boron is close to the mass of its most abundant isotope, boron-11.

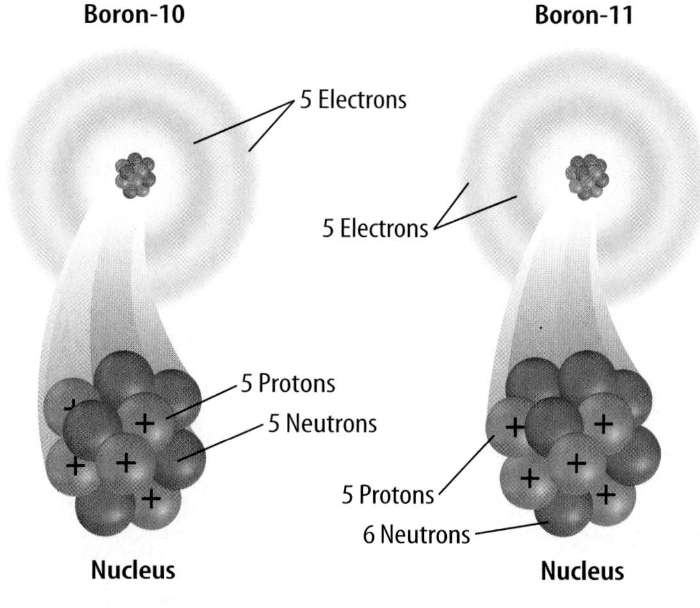

Figure 8
Boron-10 and boron-11 are two isotopes of boron. These two isotopes differ by one neutron.

Section  Assessment

1. A chlorine atom has 17 protons and 18 neutrons. What is its mass number? What is its atomic number?
2. How are the isotopes of an element alike and how are they different?
3. Why is the atomic mass of an element an average mass?
4. How would you calculate the number of neutrons in potassium-40?
5. **Think Critically** Chlorine has an average atomic mass of 35.45 amu. The two naturally occurring isotopes of chlorine are chlorine-35 and chlorine-37. Why does this indicate that most chlorine atoms contain 18 neutrons?

Skill Builder Activities

6. **Comparing and Contrasting** How does the average atomic mass relate to the mass number of an atom? Use carbon as an example in your explanation. **For more help, refer to the Science Skill Handbook.**
7. **Using an Electronic Spreadsheet** Use a spreadsheet to construct a table organizing information about the atomic numbers; atomic mass numbers; and the number of protons, neutrons, and electrons in atoms of oxygen-16, oxygen-17, carbon-12, carbon-14, chlorine-35, and chlorine-36. Record your calculations in your Science Journal. **For more help, refer to the Technology Skill Handbook.**

SECTION

The Periodic Table

As You Read

What You'll Learn

- **Explain** the composition of the periodic table.
- **Use** the periodic table to obtain information.
- **Explain** what the terms *metal, nonmetal,* and *metalloid* mean.

Vocabulary

periodic table
group
electron dot diagram
period

Why It's Important

The periodic table is an organized list of the elements that compose all living and nonliving things that are known to exist in the universe.

Organizing the Elements

On a clear evening, you can see one of the various phases of the Moon. Each month, the Moon seems to grow larger, then smaller, in a repeating pattern. This type of change is periodic. *Periodic* means "repeated in a pattern." Look at a calendar. The days of the week are periodic because they repeat themselves every seven days. Months repeat every 12 months. The calendar is a periodic table of days and months. Calendars are used to organize your schedule into a convenient format.

In the late 1800s, Dmitri Mendeleev, a Russian chemist, searched for a way to organize the elements. When he arranged all the elements known at that time in order of increasing atomic masses, he discovered a pattern. **Figure 9** shows Mendeleev's early periodic chart. Chemical properties found in lighter elements could be shown to repeat in heavier elements. Because the pattern repeated, it was considered to be periodic. Today, this arrangement is called a periodic table of elements. In the **periodic table,** the elements are arranged by increasing atomic number and by changes in physical and chemical properties.

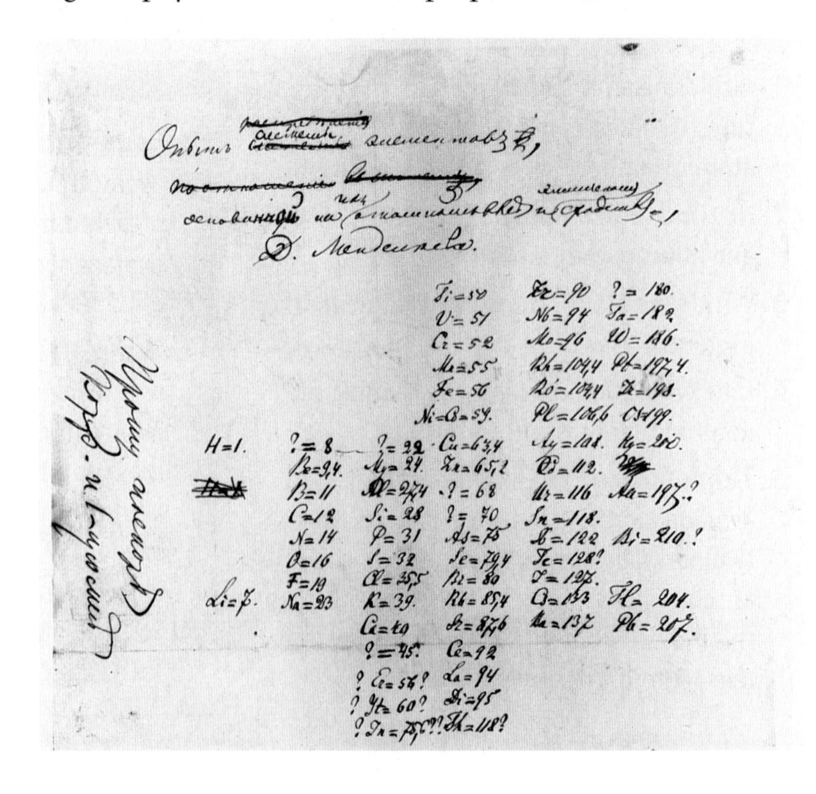

Figure 9
Mendeleev discovered that the elements had a periodic pattern in their chemical properties. Notice the question marks in his chart. These were elements that had not been discovered at that time.

Table 4 Mendeleev's Predictions

Predicted Properties of Ekasilicon (Es)	Actual Properties of Germanium (Ge)
Predicted Properties	*Actual Properties*
Atomic mass = 72	Atomic mass = 72.61
High melting point	Melting point = 938°C
Density = 5.5 g/cm^3	Density = 5.323 g/cm^3
Dark gray metal	Gray metal
Density of EsO_2 = 4.7 g/cm^3	Density of GeO_2 = 4.23 g/cm^3

Mendeleev's Predictions Mendeleev had to leave blank spaces in his periodic table to keep the elements properly lined up according to their chemical properties. He looked at the properties and atomic masses of the elements surrounding these blank spaces. From this information, he was able to predict the properties and the mass numbers of new elements that had not yet been discovered. **Table 4** shows Mendeleev's predicted properties for germanium, which he called ekasilicon. His predictions proved to be accurate. Scientists later discovered these missing elements and found that their properties were extremely close to what Mendeleev had predicted.

Reading Check *How did Mendeleev organize his periodic chart?*

Improving the Periodic Table Although Mendeleev's arrangement of elements was successful, it did need some changes. On Mendeleev's table, the atomic mass gradually increased from left to right. If you look at the modern periodic table, shown in **Table 5,** you will see several examples, such as cobalt and nickel, where the mass decreases from left to right. You also might notice that the atomic number always increases from left to right. In 1913, the work of Henry G.J. Moseley, a young English scientist, led to the arrangement of elements based on their increasing atomic numbers instead of an arrangement based on atomic masses. This new arrangement seemed to correct the problems that had occurred in the old table. The current periodic table uses Moseley's arrangement of the elements.

Reading Check *How is the modern periodic table arranged?*

Organizing a Personal Periodic Table

Procedure

1. Collect as many of the following items as you can find: **feather, penny, container of water, pencil, dime, strand of hair, container of milk, container of orange juice, square of cotton cloth, nickel, crayon, quarter, container of soda, golf ball, sheet of paper, baseball, marble, leaf,** and **paper clip.**
2. Organize these items into several columns based on their similarities to create your own periodic table.

Analysis

1. Explain the system you used to group your items.
2. Were there any items on the list that did not fit into any of your columns?
3. Infer how your activity modeled Mendeleev's work in developing the Periodic Table of the Elements.

PERIODIC TABLE OF THE ELEMENTS

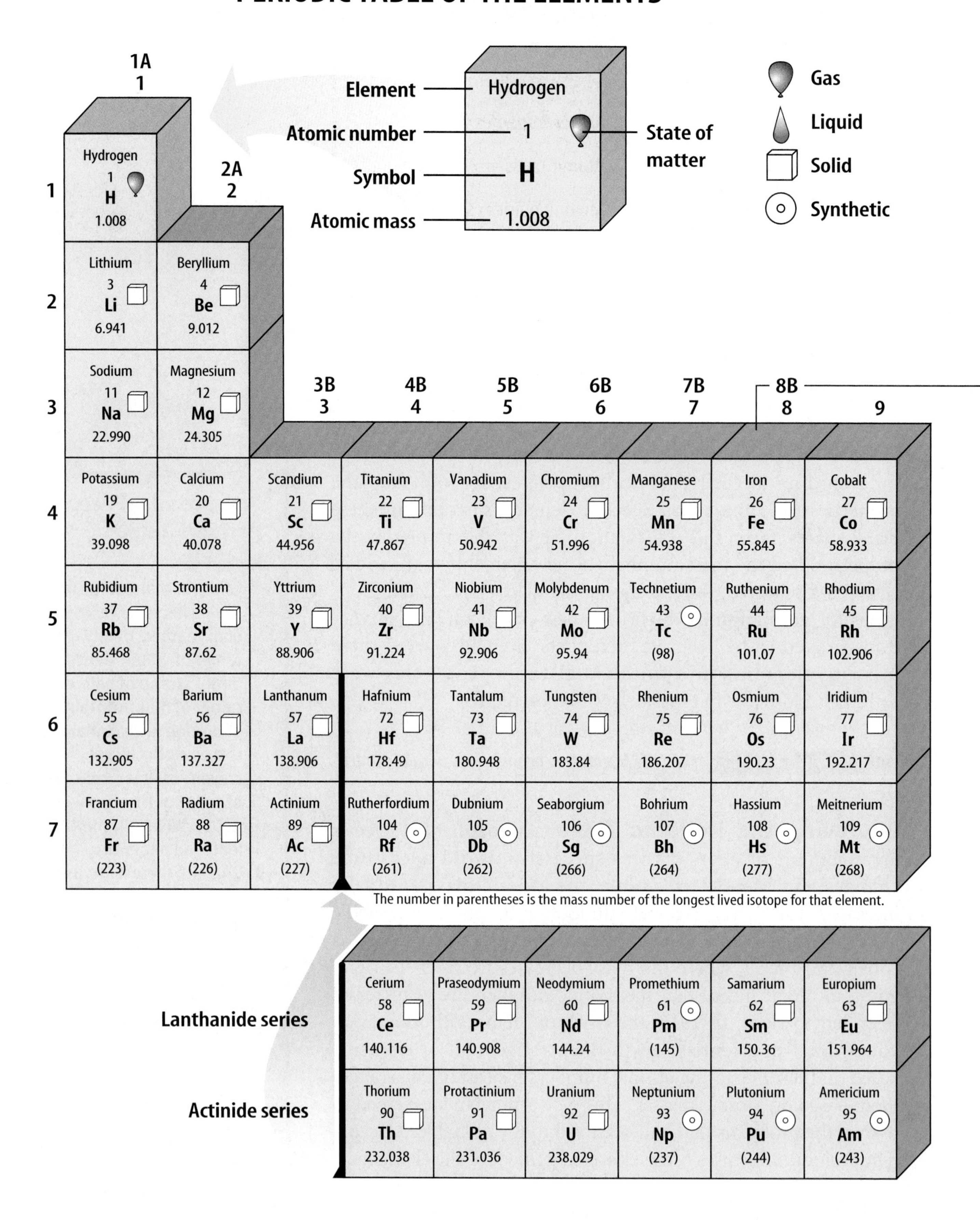

Metal
Metalloid
Nonmetal
Recently discovered

	10	1B 11	2B 12	3A 13	4A 14	5A 15	6A 16	7A 17	8A 18
									Helium 2 He 4.003
				Boron 5 B 10.811	Carbon 6 C 12.011	Nitrogen 7 N 14.007	Oxygen 8 O 15.999	Fluorine 9 F 18.998	Neon 10 Ne 20.180
				Aluminum 13 Al 26.982	Silicon 14 Si 28.086	Phosphorus 15 P 30.974	Sulfur 16 S 32.065	Chlorine 17 Cl 35.453	Argon 18 Ar 39.948
	Nickel 28 Ni 58.693	Copper 29 Cu 63.546	Zinc 30 Zn 65.39	Gallium 31 Ga 69.723	Germanium 32 Ge 72.64	Arsenic 33 As 74.922	Selenium 34 Se 78.96	Bromine 35 Br 79.904	Krypton 36 Kr 83.80
	Palladium 46 Pd 106.42	Silver 47 Ag 107.868	Cadmium 48 Cd 112.411	Indium 49 In 114.818	Tin 50 Sn 118.710	Antimony 51 Sb 121.760	Tellurium 52 Te 127.60	Iodine 53 I 126.904	Xenon 54 Xe 131.293
	Platinum 78 Pt 195.078	Gold 79 Au 196.967	Mercury 80 Hg 200.59	Thallium 81 Tl 204.383	Lead 82 Pb 207.2	Bismuth 83 Bi 208.980	Polonium 84 Po (209)	Astatine 85 At (210)	Radon 86 Rn (222)
	Ununnilium * 110 Uun (281)	Unununium * 111 Uuu (272)	Ununbium * 112 Uub (285)		Ununquadium * 114 Uuq (289)		Ununhexium * 116 Uuh (289)		Ununoctium * 118 Uuo (293)

* Names not officially assigned. Discovery of elements 114, 116, and 118 recently reported. Further information not yet available.

Gadolinium 64 Gd 157.25	Terbium 65 Tb 158.925	Dysprosium 66 Dy 162.50	Holmium 67 Ho 164.930	Erbium 68 Er 167.259	Thulium 69 Tm 168.934	Ytterbium 70 Yb 173.04	Lutetium 71 Lu 174.967
Curium 96 Cm (247)	Berkelium 97 Bk (247)	Californium 98 Cf (251)	Einsteinium 99 Es (252)	Fermium 100 Fm (257)	Mendelevium 101 Md (258)	Nobelium 102 No (259)	Lawrencium 103 Lr (262)

The Atom and the Periodic Table

Objects often are sorted or grouped according to the properties they have in common. This also is done in the periodic table. The vertical columns in the periodic table are called **groups,** or families, and are numbered 1 through 18. Elements in each group have similar properties. For example, in Group 11, copper, silver, and gold have similar properties. Each is a shiny metal and a good conductor of electricity and heat. What is responsible for the similar properties? To answer this question, look at the structure of the atom.

Physics INTEGRATION

When a glass rod is rubbed with silk, the rod becomes positively charged. Infer what type of particle in the atoms in the rod has been removed. Explain your answer in your Science Journal.

Electron Cloud Structure You have learned about the number and location of protons and neutrons in an atom. But where are the electrons located? How many are there? In a neutral atom, the number of electrons is equal to the number of protons. Therefore, a carbon atom, with an atomic number of six, has six protons and six electrons. These electrons are located in the electron cloud surrounding the nucleus.

Scientists have found that electrons within the electron cloud have different amounts of energy. Scientists model the energy differences of the electrons by placing the electrons in energy levels, as in **Figure 10.** Energy levels nearer the nucleus have lower energy than those levels that are farther away. These energy levels fill with electrons from inner levels—closer to the nucleus, to outer levels—farther from the nucleus.

Elements that are in the same group have the same number of electrons in their outer energy level. It is the number of electrons in the outer energy level that determines the chemical properties of the element. It is important to understand the link between the location on the periodic table, chemical properties, and the structure of the atom.

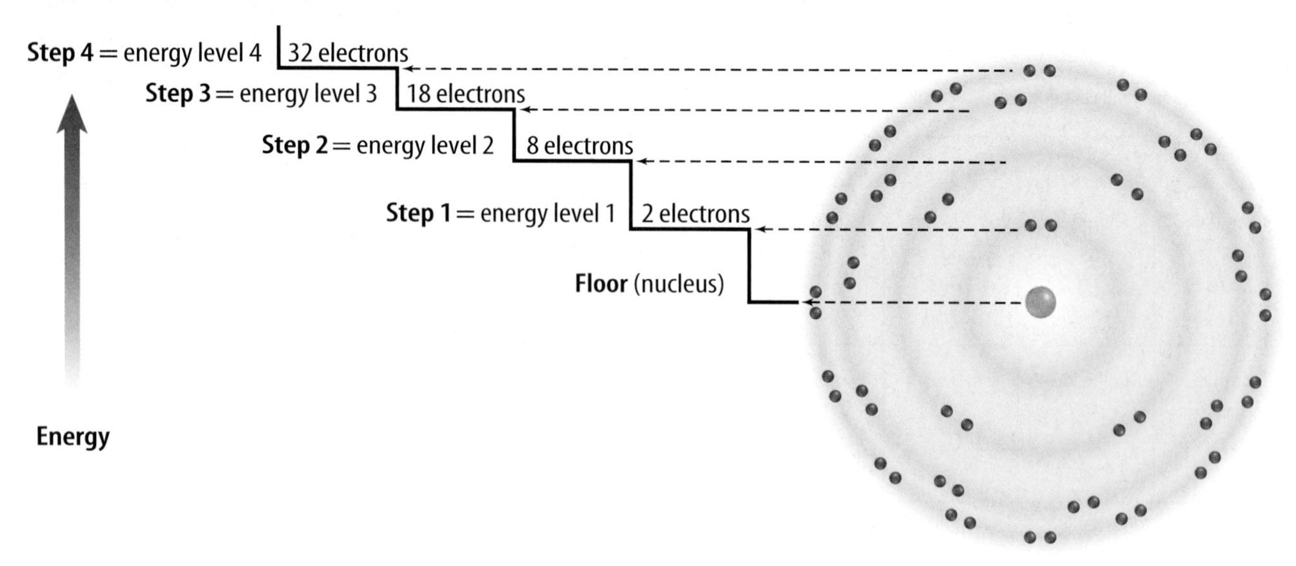

Figure 10
Energy levels in atoms can be represented by a flight of stairs. Each stair step away from the nucleus represents an increase in the amount of energy within the electrons. The higher energy levels contain more electrons.

Energy Levels These energy levels are named using numbers one to seven. The maximum number of electrons that can be contained in each of the first four levels is shown in **Figure 10.** For example, energy level one can contain a maximum of two electrons. Energy level two can contain a maximum of eight electrons. Notice that energy levels three and four contain several electrons. A complete and stable outer energy level will contain eight electrons. In elements in periods three and higher, additional electrons can be added to inner energy levels although the outer energy level contains only eight electrons.

Rows on the Table Remember that the atomic number found on the periodic table is equal to the number of electrons in an atom. Look at **Figure 11.** The first row has hydrogen with one electron and helium with two electrons both in energy level one. Because energy level one is the outermost level containing an electron, hydrogen has one outer electron. Helium has two outer electrons. Recall from **Figure 10** that energy level one can hold only two electrons. Therefore, helium has a full or complete outer energy level. The second row begins with lithium, which has three electrons—two in energy level one and one in energy level two. Lithium has one outer electron. Lithium is followed by beryllium with two outer electrons, boron with three, and so on until you reach neon with eight outer electrons. Again, looking at **Figure 10,** energy level two can only hold eight electrons. Therefore, neon has a complete outer energy level. Do you notice how the row in the periodic table ends when an outer energy level is filled? In the third row of elements, the electrons begin filling energy level three. The row ends with argon, which has a full outer energy level of eight electrons.

Research Visit the Glencoe Science Web site at **science.glencoe.com** for more information about the structure of atomic energy levels. Communicate to your class what you learned.

Figure 11
One proton and one electron are added to each element as you go across a period in the periodic table.

Hydrogen 1 **H**							Helium 2 **He**
Lithium 3 **Li**	Beryllium 4 **Be**	Boron 5 **B**	Carbon 6 **C**	Nitrogen 7 **N**	Oxygen 8 **O**	Fluorine 9 **F**	Neon 10 **Ne**
Sodium 11 **Na**	Magnesium 12 **Mg**	Aluminum 13 **Al**	Silicon 14 **Si**	Phosphorus 15 **P**	Sulfur 16 **S**	Chlorine 17 **Cl**	Argon 18 **Ar**

Figure 12
The elements in Group 1 have one electron in their outer energy level. This electron dot diagram represents that one electron.

H·

Li·

Na·

K·

Rb·

Cs·

Fr·

Electron Dot Diagrams Did you notice that hydrogen, lithium, and sodium have one electron in their outer energy level? Elements that are in the same group have the same number of electrons in their outer energy level. These outer electrons are so important in determining the chemical properties of an element that a special way to represent them has been developed. An **electron dot diagram** uses the symbol of the element and dots to represent the electrons in the outer energy level. **Figure 12** shows the electron dot diagram for Group 1 elements. Electron dot diagrams are used also to show how the electrons in the outer energy level are bonded when elements combine to form compounds.

Same Group—Similar Properties The elements in Group 17, the halogens, have electron dot diagrams similar to chlorine, shown in **Figure 13.** All halogens have seven electrons in their outer energy levels. Since all of the members of a group on the periodic table have the same number of electrons in their outer energy level, group members will undergo chemical reactions in similar ways.

A common property of the halogens is the ability to form compounds readily with elements in Group 1. Group 1 elements have only one electron in their outer energy level. **Figure 13** shows an example of a compound formed by one such reaction. The Group 1 element, sodium, reacts easily with the Group 17 element, chlorine. The result is the compound sodium chloride, or NaCl—ordinary table salt.

Not all elements will combine readily with other elements. The elements in Group 18 have complete outer energy levels. This special configuration makes Group 18 elements relatively unreactive. You will learn more about why and how bonds form between elements in the later chapters.

Reading Check *Why do elements in a Group undergo similar chemical reactions?*

Figure 13
Electron dot diagrams show the electrons in an element's outer energy level.

·Cl:

A The electron dot diagram for Group 17 consists of three sets of paired dots and one single dot.

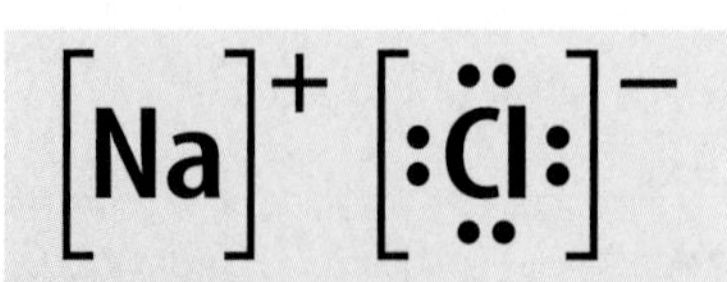

B Sodium combines with chlorine to give each element a complete outer energy level in the resulting compound.

C Neon, a member of Group 18, has a full outer energy level. Neon has eight electrons in its outer energy level, making it unreactive.

Regions on the Periodic Table

The periodic table has several regions with specific names. The horizontal rows of elements on the periodic table are called **periods.** The elements increase by one proton and one electron as you go from left to right in a period.

All of the elements in the blue squares in **Figure 14** are metals. Iron, zinc, and copper are examples of metals. Most metals exist as solids at room temperature. They are shiny, can be drawn into wires, can be pounded into sheets, and are good conductors of heat and electricity.

Those elements on the right side of the periodic table, in yellow, are classified as nonmetals. Oxygen, bromine, and carbon are examples of nonmetals. Most nonmetals are gases, are brittle, and are poor conductors of heat and electricity at room temperature. The elements in green are metalloids or semimetals. They have some properties of both metals and nonmetals. Boron and silicon are examples of metalloids.

What are the properties of the elements located on the left side of the periodic table?

Research Visit the Glencoe Science Web site at **science.glencoe.com** for more information about newly discovered elements. Communicate to your class what you learn.

A Growing Family Scientists around the world are continuing their research into the synthesis of elements. In 1994, scientists at the Heavy-Ion Research Laboratory in Darmstadt, Germany, discovered element 111. As of 1998, only one isotope of element 111 has been found. This isotope had a life span of 0.002 s. In 1996, element 112 was discovered at the same laboratory. As of 1998, only one isotope of element 112 has been found. The life span of this isotope was 0.00048 s. Both of these elements are produced in the laboratory by joining smaller atoms into a single atom. The search for elements with higher atomic numbers continues. Scientists believe they have synthesized elements 114, 116, and 118, too. However, the discovery of these elements has not been confirmed.

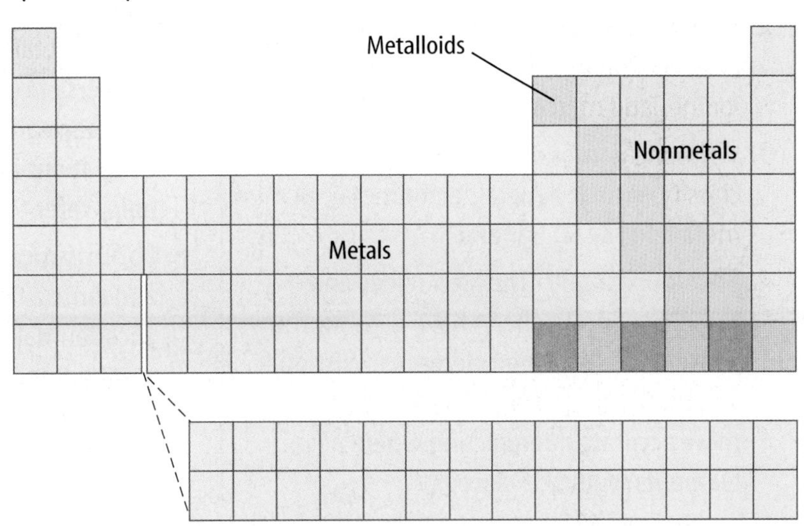

Figure 14
Metalloids are located along the green stair-step line. Metals are located to the left of the metalloids. Nonmetals are located to the right of the metalloids.

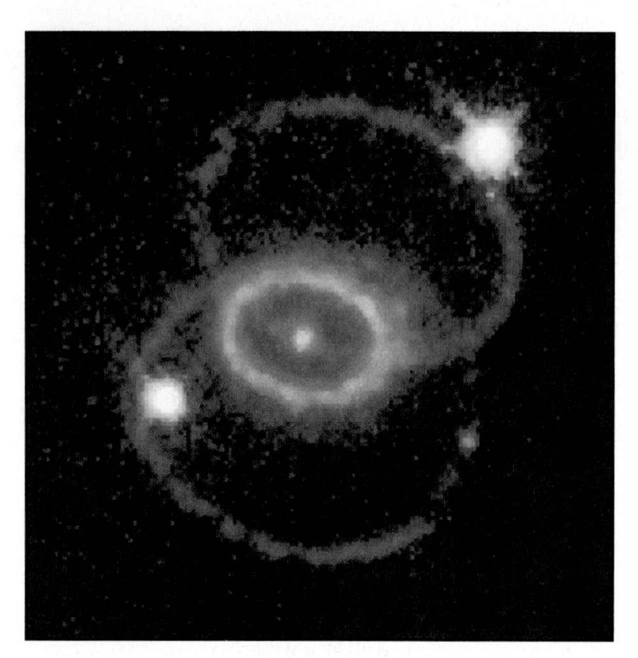

Figure 15
Scientists believe that naturally occurring elements are manufactured within stars.

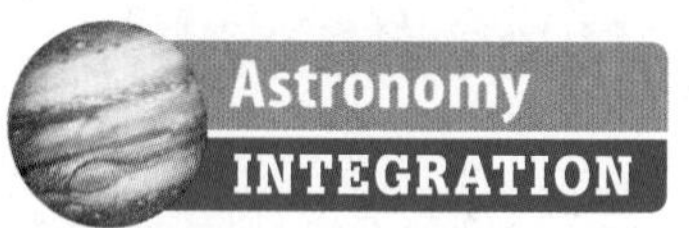

Elements in the Universe

Are all of the elements throughout the universe the same? Scientists have made giant strides in answering this question. Using the technology that is available today, scientists are finding the same elements throughout the universe. They have been able to study only a small portion of the universe, though, because it is so vast. Many scientists believe that hydrogen and helium are the building blocks of the other naturally occurring elements. Atoms join together within stars to produce elements with atomic numbers greater than 1 or 2—the atomic numbers of hydrogen and helium. Exploding stars, or supernovas, shown in **Figure 15,** give scientists evidence to support this theory. When stars go supernova a mixture of elements, including the heavy elements such as iron, are flung into the galaxy. Many scientists believe that supernovas have spread naturally occurring elements that are found throughout the universe. It is important to note that all of the superheavy elements—those with an atomic number above 92—are human-made and found only in laboratories. Technetium and promethium also do not occur naturally.

Section  Assessment

1. Use the periodic table to find the name, atomic number, and average atomic mass of the following elements: N, Ca, Kr, and W.
2. Give the period and group in which each of these elements is found: nitrogen, sodium, iodine, and mercury.
3. Write the names of these elements and classify each as a metal, a nonmetal, or a metalloid: K, Si, Ba, and S.
4. **Think Critically** The Mendeleev and Moseley periodic charts have gaps for the as-then-undiscovered elements. Why do you think the chart used by Moseley was more accurate at predicting where new elements would be placed?

Skill Builder Activities

5. **Making and Using Graphs** Construct a circle graph showing the percentage of elements classified as metals, metalloids, and nonmetals. Use markers or colored pencils to distinguish clearly between each section on the graph. Record your calculations in your Science Journal. **For more help, refer to the** Science Skill Handbook.
6. **Communicating** Choose a synthetic element and write a brief paragraph about it in your Science Journal. Include information about the element's name, location of the synthesis research, and the people responsible for its discovery. **For more help, refer to the** Science Skill Handbook.

Activity

A Periodic Table of Foods

Use your favorite foods to create a periodic table of foods.

What You'll Investigate

How can you create a periodic table to organize your favorite foods?

Materials

11 × 17 paper
12- or 18-inch ruler
colored pencils or markers

Goals

- **Organize** 20 of your favorite foods into a periodic table of foods.
- **Analyze** and **evaluate** your periodic table for similar characteristics among groups or family members on your table.
- **Infer** where new foods added to your table would be placed.

Procedure

1. **List** 20 of your favorite foods and drinks.
2. **Describe** basic characteristics of each of your food and drink items. For example, you might describe the primary ingredient, nutritional value, taste, and color of each item. You also could identify the food group of each item such as fruits/vegetables, grains, dairy products, meat, and sweets.
3. **Create** a data table to organize the information that you collect.
4. Using your data table, construct a periodic table of foods on your 11 × 17 sheet of paper. Determine which characteristics you will use to group your items. Create families (columns) of food and drink items that share similar characteristics on your table.

For example, potato chips, pretzels, and cheese-flavored crackers could be combined into a family of salty tasting foods. Create as many groups as you need, and you do not need to have the same number of items in every family.

Conclude and Apply

1. **Evaluate** the characteristics you used to make the groups on your periodic table. Do the characteristics of each group adequately describe all the family members? Do the characteristics of each group distinguish its family members from the family members of the other groups?
2. Analyze the reasons why some items did not fit easily into a group.
3. Infer why chemists have not created a periodic table of compounds.

Communicating Your Data

Construct a bulletin board of the periodic table of foods created by the class. **For more information, refer to the Science Skill Handbook.**

Activity Use the Internet

What's in a Name?

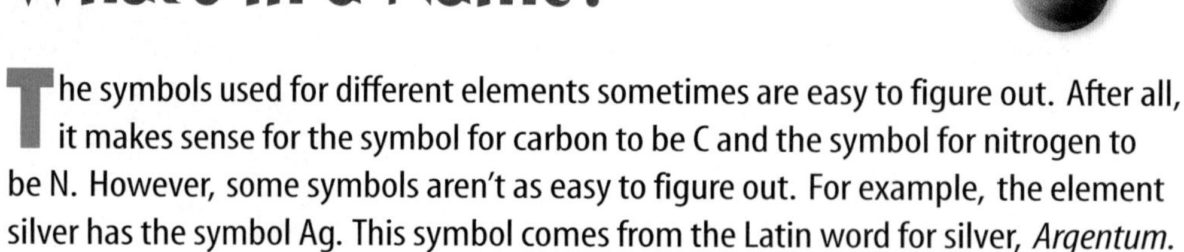

The symbols used for different elements sometimes are easy to figure out. After all, it makes sense for the symbol for carbon to be C and the symbol for nitrogen to be N. However, some symbols aren't as easy to figure out. For example, the element silver has the symbol Ag. This symbol comes from the Latin word for silver, *Argentum.*

Recognize the Problem

How are symbols and names chosen for elements?

Form a Hypothesis

How are elements named? Are newly discovered elements named the same way they were hundreds of years ago? How are the symbols for these elements determined? Form a hypothesis about why certain elements have their names and symbols.

Goals

- **Research** the names and symbols of various elements.
- **Study** the methods that are used to name elements and how they have changed through time.
- **Organize** your data by making your own periodic table.
- **Study** the history of certain elements and their discoveries.
- **Create** a table of your findings and communicate them to other students.

Data Source

SCIENCE*Online* Go to the Glencoe Science Web site at **science.glencoe.com** for more information on naming elements, elements' symbols, and the discovery of new elements, and for data from other students.

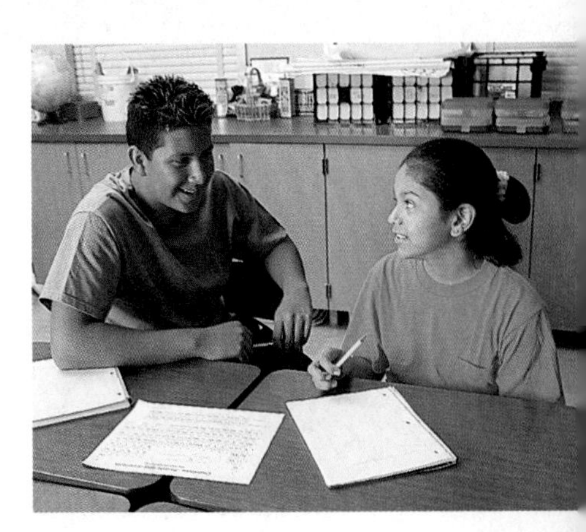

Test Your Hypothesis

Plan

1. Make a list of particular elements you wish to study.
2. **Compare and contrast** these elements' names to their symbols.
3. **Research** the discovery of these elements. Do their names match their symbols? Were they named after a property of the element, a person, their place of discovery, or a system of nomenclature? What was that system?

Do

1. Make sure your teacher approves your plan before you start.
2. Visit the Glencoe Science Web site for links to different sites about elements, their history, and how they were named.
3. **Research** these elements.
4. Carefully record your data in your Science Journal.

Analyze Your Data

1. **Record** in your Science Journal how the symbols for your elements were chosen. What were your elements named after?
2. Make a periodic table that includes the research information on your elements that you found.
3. Make a chart of your class's findings. Sort the chart by year of discovery for each element.
4. How are the names and symbols for newly discovered elements chosen? Make a chart that shows how the newly discovered elements will be named.

Draw Conclusions

1. **Compare** your findings to those of your classmates. Did anyone's data differ for the same element? Were all the elements in the periodic table covered?
2. What system is used to name the newly discovered elements today?
3. Some elements were assigned symbols based on their name in another language. Do these examples occur for elements discovered today or long ago?

Communicating Your Data

SCIENCE Online Find this *Use the Internet* activity on the Glencoe Science Web site at **science.glencoe.com** and post your data in the table provided. **Compare** your data to those of other students. Combine your data with those of other students to complete your periodic table with all of the elements.

TIME **SCIENCE AND HISTORY**

SCIENCE CAN CHANGE THE COURSE OF HISTORY!

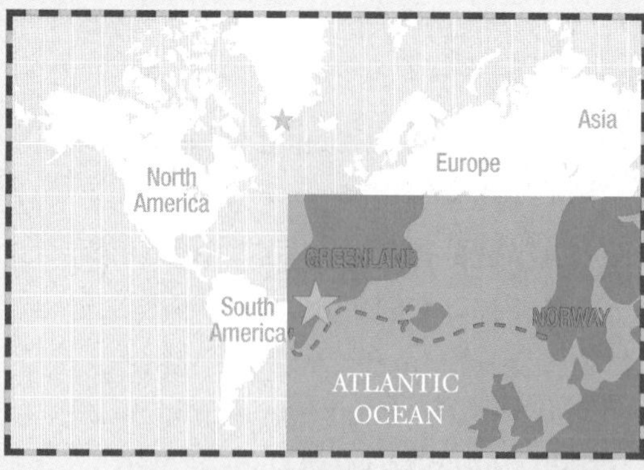

The orange line shows the route the Norse sailed from Norway to Greenland.

A group of people vanish without a trace. Scientists turn to ice for the answer.

A Chilling

A scientist inspects an ice core sample from the Greenland Ice Sheet. The samples are stored in a freezer at −36°C.

Drilling to the bottom of the Greenland Ice Sheet takes place in a deep trench dug in the snow surface, and uses an electronically controlled drill.

Story

Air bubbles and dirt trapped in ice provide clues to Earth's past climate.

Picture this: It's 1361. A ship from Norway arrives at a Norwegian settlement in Greenland. The ship's crew hopes to trade its cargo with the people living there. The crew gets off the ship. They look around. The settlement is deserted. More than 1,000 people had vanished!

New evidence has shed some light on the mysterious disappearance of the Norse settlers. The evidence came from a place on the Greenland Ice Sheet over 600 km away from the settlement. This part of Greenland is so cold that snow never melts. As new snow falls, the existing snow is buried and turns to ice. By drilling deep into this ice, scientists can recover an ice core. The core is made up of ice formed from snowfalls going way, way back in time.

By measuring the ratio of oxygen isotopes in the ice core, scientists can estimate Greenland's past air temperatures. The cores provide a detailed climate history going back over 80,000 years. Individual ice layers can be dated much like tree rings to determine their age, and the air bubbles trapped within each layer are used to learn about climate variations. Dust and pollen trapped in the ice also yield clues to ancient climates.

A Little Ice Age

Based on their analysis, scientists think the Norse moved to Greenland during an unusually warm period. Then in the 1300s, the climate started to cool and a period known as the Little Ice Age began. The ways the Norse hunted and farmed were inadequate for survival in this long chill. Since they couldn't adapt to their colder surroundings, the settlers died out.

Examining ice cores fascinates scientists. It gives them an idea of what Earth's climate was like long ago. The ice cores also may help scientists better understand why global temperatures have been rising since the end of the Little Ice Age.

Though life in Greenland ended in a chilling way for the Norse, that hasn't been the story for other people living there. The native Inuit have flourished by finding ways to adapt their lifestyles to their environment—no matter what the weather!

CONNECTIONS Research Report Evidence seems to show that Earth is warming. Rising temperatures could affect our lives. Research global warming to find out how Earth may change. Report to the class.

SCIENCE *Online*
For more information, visit science.glencoe.com

Chapter 18 Study Guide

Reviewing Main Ideas

Section 1 Structure of the Atom

1. A chemical symbol is a shorthand way of writing the name of an element.
2. An atom consists of a nucleus made of protons and neutrons surrounded by an electron cloud. *In the figure to the right, how many protons and neutrons are in the nucleus?*

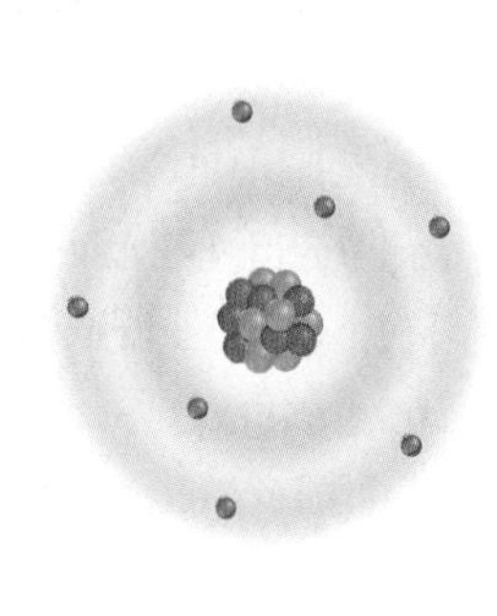

3. Quarks are particles of matter that make up protons and neutrons.
4. The model of the atom changes over time. As new information is discovered, scientists incorporate it into the model.

Section 2 Masses of Atoms

1. The number of neutrons in an atom can be computed by subtracting the atomic number from the mass number.
2. The isotopes of an element are atoms of that same element that have different numbers of neutrons. *What is the mass number of the isotopes below?*

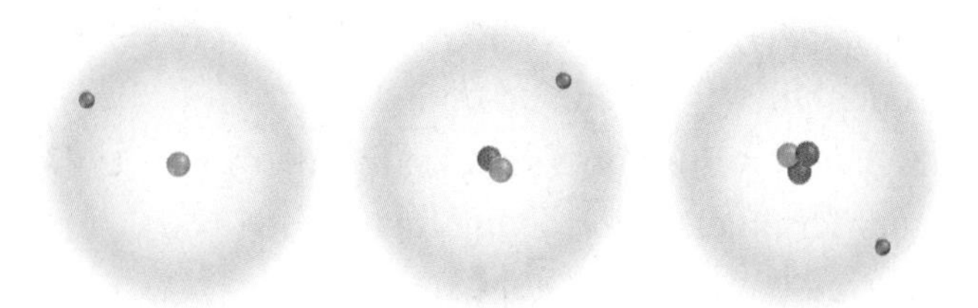

3. The average atomic mass of an element is the weighted-average mass of the mixture of its isotopes. Isotopes are named by using the element name, followed by a dash, and its mass number.

Section 3 The Periodic Table

1. In the periodic table, the elements are arranged by increasing atomic number resulting in periodic changes in properties. Knowing that the number of protons, electrons, and atomic number are equal gives you partial composition of the atom.
2. In the periodic table, the elements are arranged in 18 vertical columns, or groups, and seven horizontal rows, or periods.
3. Metals are found at the left of the periodic table, nonmetals at the right, and metalloids along the line that separates the metals from the nonmetals. *In the diagram below, what is the name of the group of elements that is highlighted in blue?*

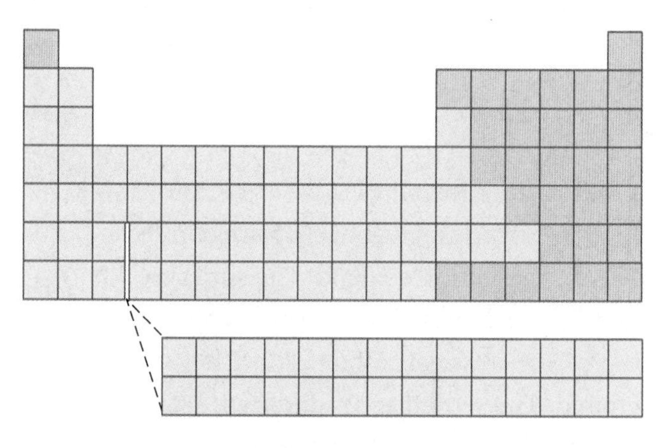

4. Elements are placed on the periodic table in order of increasing atomic number. A new row on the periodic table begins when the outer energy level of the element is filled.

After You Read

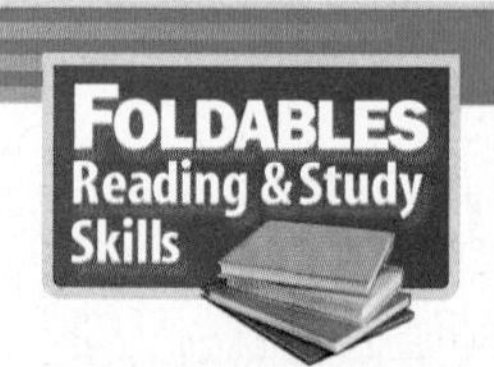

Without looking at the chapter or at your Foldable, write what you learned about atoms on the *Learned* fold of your Foldable.

Visualizing Main Ideas

Complete the following concept map on the parts of the atom.

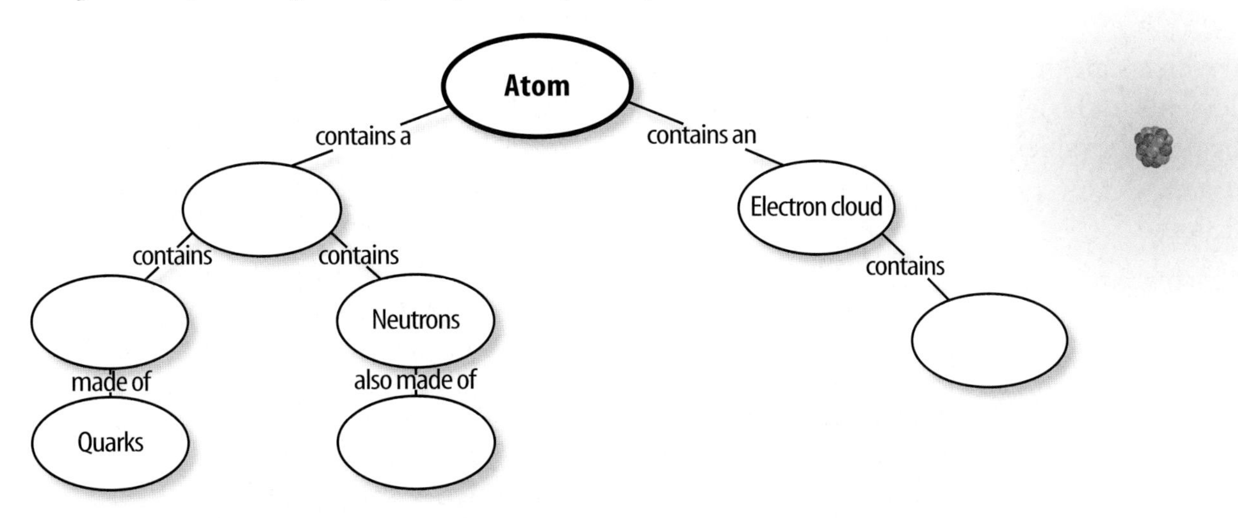

Vocabulary Review

Vocabulary Words

a. atom
b. atomic number
c. average atomic mass
d. electron
e. electron cloud
f. electron dot diagram
g. group
h. isotope
i. mass number
j. neutron
k. nucleus
l. period
m. periodic table
n. proton
o. quark

Study Tip

Don't memorize definitions. Write complete sentences using new vocabulary words to be certain you understand what they mean.

Using Vocabulary

Answer the following questions using complete sentences.

1. What is the name of the organized table of elements that was designed by Mendeleev?

2. What are two elements with the same number of protons but a different number of neutrons called?

3. What is the weighted-average mass of all the known isotopes for an element called?

4. What is the name of the positively charged center of an atom?

5. What are the particles that make up protons and neutrons?

6. What is the name of a horizontal row in the periodic table called?

7. What is the sum of the number of protons and neutrons called?

8. In the current model of the atom, where are the electrons located?

Chapter 18 Assessment

Checking Concepts

Choose the word or phrase that best answers the question.

1. In which state of matter are most of the elements to the left of the stair-step line in the periodic table?
   A) gas
   B) liquid
   C) plasma
   D) solid

2. Which is a term for a pattern that repeats?
   A) isotopic
   B) metallic
   C) periodic
   D) transition

3. Which of the following is an element that would have similar properties to those of neon?
   A) aluminum
   B) argon
   C) arsenic
   D) silver

4. Which of the following terms describes boron?
   A) metal
   B) metalloid
   C) noble gas
   D) nonmetal

5. How many outer level electrons do lithium and potassium have?
   A) 1
   B) 2
   C) 3
   D) 4

6. Which of the following is NOT found in the nucleus of an atom?
   A) proton
   B) neutron
   C) electron
   D) quark

7. The halogens are located in which group?
   A) 1
   B) 11
   C) 15
   D) 17

8. In which of the following states is nitrogen found at room temperature?
   A) gas
   B) metalloid
   C) metal
   D) liquid

9. Which of the elements below is a shiny element that conducts electricity and heat?
   A) chlorine
   B) sulfur
   C) hydrogen
   D) magnesium

10. The atomic number of Re is 75. The atomic mass of one of its isotopes is 186. How many neutrons are in an atom of this isotope?
    A) 75
    B) 111
    C) 186
    D) 261

Thinking Critically

11. Lead and mercury are two pollutants in the environment. From information about them in the periodic table, determine why they are called heavy metals.

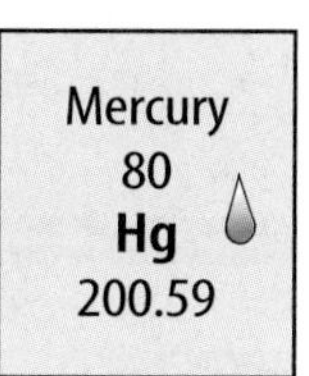

12. Ge and Si are used in making semiconductors. Are these two elements in the same group or the same period?

13. Using the periodic table, predict how many outer level electrons will be in elements 114, 116, and 118. Explain your answer.

14. Ca is used by the body to make bones and teeth. Radioactive Sr is in nuclear waste. Yet one is safe for people and the other is hazardous. Why is Sr hazardous to people?

Developing Skills

15. **Making and Using Tables** Use the periodic table to list a metal, a metalloid, and a nonmetal each with five outer-level electrons.

16. **Comparing and Contrasting** From the information found in the periodic table and reference books, compare and contrast the properties of chlorine and bromine.

17. Interpreting Data If scientists have determined that a neutral atom of rubidium has an atomic number of 37 and a mass number of 85, how many protons, neutrons, and electrons does the atom have?

18. Concept Mapping As a star dies, it becomes more dense. Its temperature rises to a point where He nuclei are combined with other nuclei. When this happens, the atomic numbers of the other nuclei are increased by 2 because each gains the two protons contained in the He nucleus. For example, Cr fused with He becomes Fe. Complete the concept map showing the first four steps in He fusion.

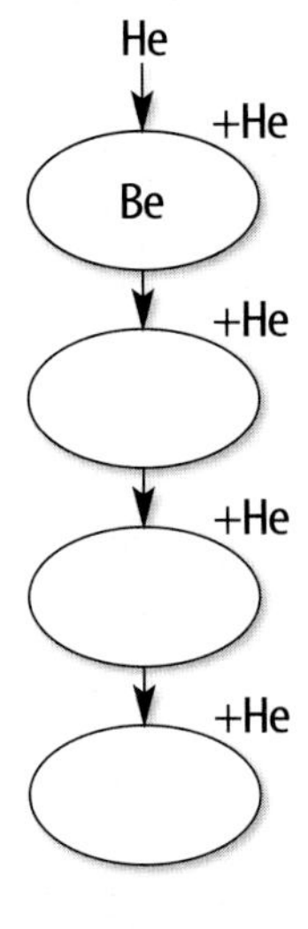

Performance Assessment

19. Interview Research the attempts made by Johann Döbereiner and John Newlands to classify the elements. Research the work of these scientists and write your findings in the form of an interview.

20. Display Make a display of pictures demonstrating the uses of several elements. List the name, symbol, atomic number, average atomic mass, and several other uses for each element used.

Technology

Go to the Glencoe Science Web site at **science.glencoe.com** or use the **Glencoe Science CD-ROM** for additional chapter assessment.

Test Practice

Larry researched elements that are nutrients in food. He found that all packaged foods have nutritional labels on them. This makes it very easy to determine the nutritional value of the foods you eat. As he was eating breakfast, he noticed the label on his cereal that is shown below.

Sweet Wheat Cereal Nutrition Facts

Serving Size	About 24 biscuits (59g/2.1 oz.)	
Servings per Container		About 12
Amount Per Serving	**Cereal**	**Cereal with 1/2 cup fat free milk**
Calories	200	240
Calories from Fat	10	10
	% Daily Value	
Total Fat 1 g	2%	2%
Saturated Fat 0 g	0%	0%
Monounsaturated Fat 0 g		
Polyunsaturated Fat 0.5 g		
Cholesterol 0 mg	0%	0%
Sodium 5 mg	0%	3%
Potassium 200 mg	6%	12%
Total Carbohydrate 48 g	16%	18%
Dietary Fiber 6 g	24%	24%
Sugars 12 g		
Other Carbohydrate 30 g		
Protein 6 g		

Study the chart and answer the following questions.

1. According to the chart, which nutrient below is provided only when the cereal is eaten with 1/2 cup of milk?

A) saturated fat
B) sodium
C) potassium
D) dietary fiber

2. If Larry were to eat two servings of this cereal each with 1/2 cup of milk, about how many grams of dietary fiber would he eat?

F) 48 g
G) 24 g
H) 12 g
J) 6 g

Chemical Bonds

Together, holding hands, these sky divers are a stable group instead of separately falling objects. A chemical bond between elements is similar. Atoms combine when the compound formed is more stable than the separate atoms. Like the circle of sky divers, the compounds formed have properties unlike those found in the separate elements. In this chapter, you will read about how chemical bonds form and learn how to write chemical formulas and equations.

What do you think?

Science Journal Look at the picture below with a classmate. Discuss what you think this might be or what is happening. Here's a hint: *Bonds are forming to make something that you can sprinkle on popcorn.* Write your answer or best guess in your Science Journal.

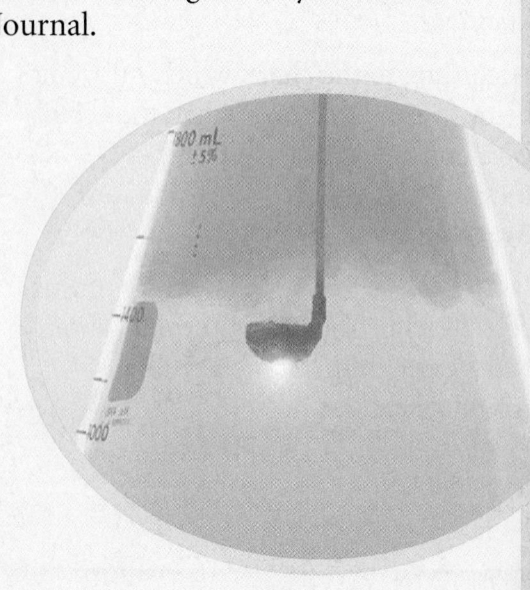

You have probably noticed how the liquids in oil and vinegar salad dressings will not stay mixed after the bottle is shaken. Why do some liquids such as water and rubbing alcohol mix together, but others like oil and water do not? The compounds that make up the liquids are different. Some, like water and rubbing alcohol, can attract each other. Others, like water and oil, do not. In this chapter, these different types of compounds will be discussed further.

Separating liquids

1. Pour 20 mL of water into a 100 mL graduated cylinder.
2. Pour 20 mL of vegetable oil into the same cylinder. Vigorously swirl the two liquids together, and observe for several minutes.
3. Add two drops of food dye and observe.
4. After several minutes, slowly pour 30 mL of rubbing alcohol into the cylinder.
5. Add two more drops of food dye and observe.

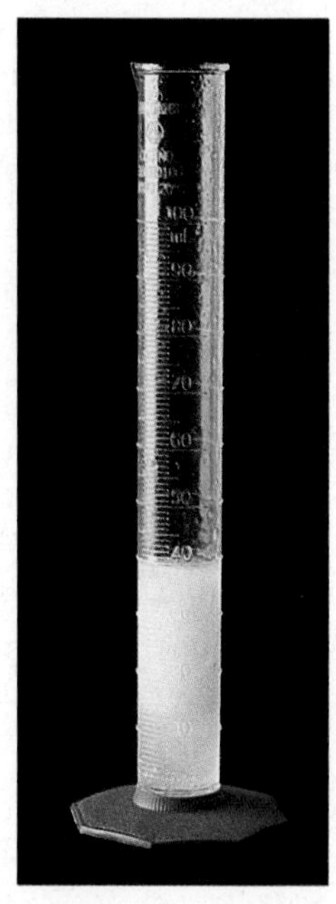

Observe

In your Science Journal, write a paragraph describing how the different liquids reacted to each other.

Before You Read

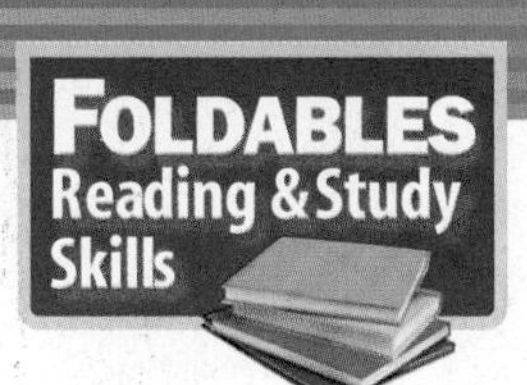

Making a Chemical Formula Study Fold

Make the following Foldable to help you identify chemical formulas from this chapter.

1. Place a sheet of notebook paper in front of you so the short side is at the top and the holes are on the right side. Fold the paper in half from the left side to the right side.
2. Through the top thickness of paper, cut along every third line from the outside edge to the centerfold, forming tabs as shown.
3. Before you read the chapter, collect ten chemical formulas and write them on the front of the tabs. As you read the chapter, write what compound the formula represents under the tabs.

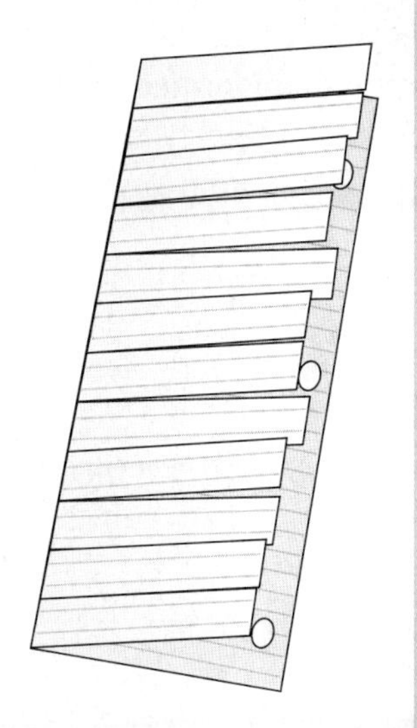

SECTION 1

Stability in Bonding

As You Read

What You'll Learn

- **Describe** how a compound differs from its component elements.
- **Explain** what a chemical formula represents.
- **State** a reason why chemical bonding occurs.

Vocabulary

chemical formula
chemically stable
chemical bond

Why It's Important

Understanding why atoms form compounds depends on knowing how an atom can become stable.

Combined Elements

Have you ever noticed the color of the Statue of Liberty? Why is it green? Did the sculptor purposely choose green? Why wasn't white, or tan, or even some other color like purple chosen? Was it painted that way? No, the Statue of Liberty was not painted. The Statue of Liberty is made of the metal copper, which is an element. Pennies, too, are made of copper. Wait a minute, you say. Copper isn't green—it's ... well, copper colored.

You are right. Uncombined, elemental copper is a bright, shiny copper color. So again the question arises: Why is the Statue of Liberty green?

Some of the matter around you is in the form of uncombined elements such as copper, sulfur, and oxygen. But, like many other sets of elements, these three elements unite chemically to form a compound when the conditions are right. The green coating on the Statue of Liberty and some old pennies is a result of this chemical change. One compound in this coating, seen in contrast with elemental copper in **Figure 1,** is a new compound called copper sulfate. Copper sulfate isn't shiny and copper colored like elemental copper. Nor is it a pale-yellow solid like sulfur or a colorless, odorless gas like oxygen. It has its own unique properties.

Figure 1
The difference between the copper metal and the copper compound formed on the Statue of Liberty is striking. A new compound has formed.

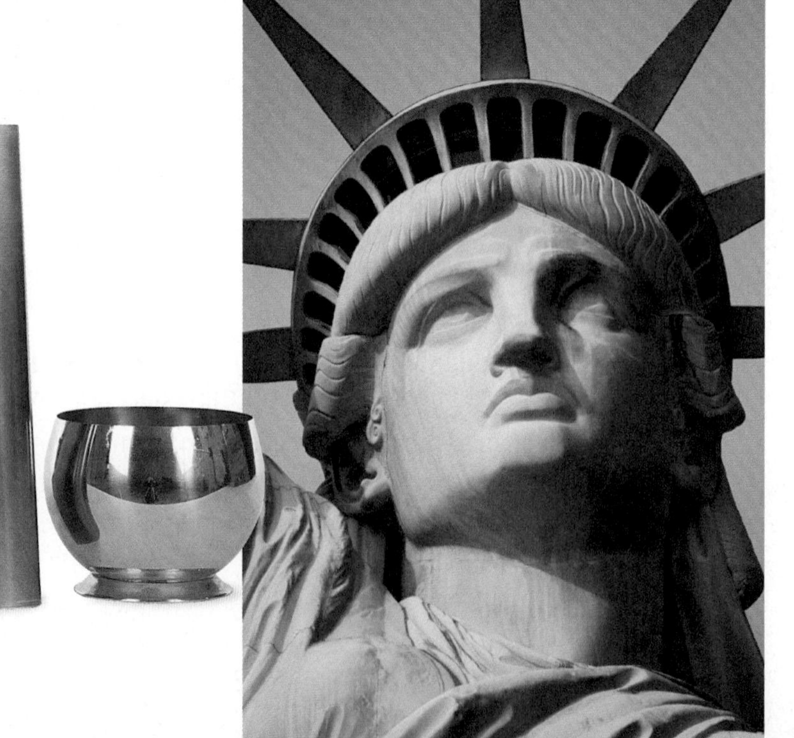

New Properties One interesting observation you will make is that the compound formed when elements combine often has properties that aren't anything like those of the individual elements. Sodium chloride, for example, shown in **Figure 2,** is a compound made from the elements sodium and chlorine. Sodium is a shiny, soft, silvery metal that reacts violently with water. Chlorine is a poisonous greenish-yellow gas. Would you have guessed that these elements combine to make ordinary table salt?

Figure 2
Sodium is a soft, silvery metal that combines with chlorine, a greenish-yellow gas, to form sodium chloride, which is a white crystalline solid. *How are the properties of table salt different from those of sodium and chlorine?*

Formulas

The chemical symbols Na and Cl represent the elements sodium and chlorine. When written as NaCl, the symbols make up a formula, or chemical shorthand, for the compound sodium chloride. A **chemical formula** tells what elements a compound contains and the exact number of the atoms of each element in a unit of that compound. The compound that you are probably most familiar with is H_2O, more commonly known as water. This formula contains the symbols H for the element hydrogen and O for the element oxygen. Notice the subscript number 2 written after the H for hydrogen. *Subscript* means "written below." A subscript written after a symbol tells how many atoms of that element are in a unit of the compound. If a symbol has no subscript, the unit contains only one atom of that element. A unit of H_2O contains two hydrogen atoms and one oxygen atom.

Look at the formulas for each compound listed in **Table 1.** What elements combine to form each compound? How many atoms of each element are required to form each of the compounds?

Table 1 Some Familiar Compounds

Familiar Name	Chemical Name	Formula
Sand	Silicon dioxide	SiO_2
Milk of magnesia	Magnesium hydroxide	$Mg(OH)_2$
Cane sugar	Sucrose	$C_{12}H_{22}O_{11}$
Lime	Calcium oxide	CaO
Vinegar	Acetic acid	$HC_2H_3O_2$
Laughing gas	Dinitrogen oxide	N_2O
Grain alcohol	Ethanol	C_2H_5OH
Battery acid	Sulfuric acid	H_2SO_4
Stomach acid	Hydrochloric acid	HCl

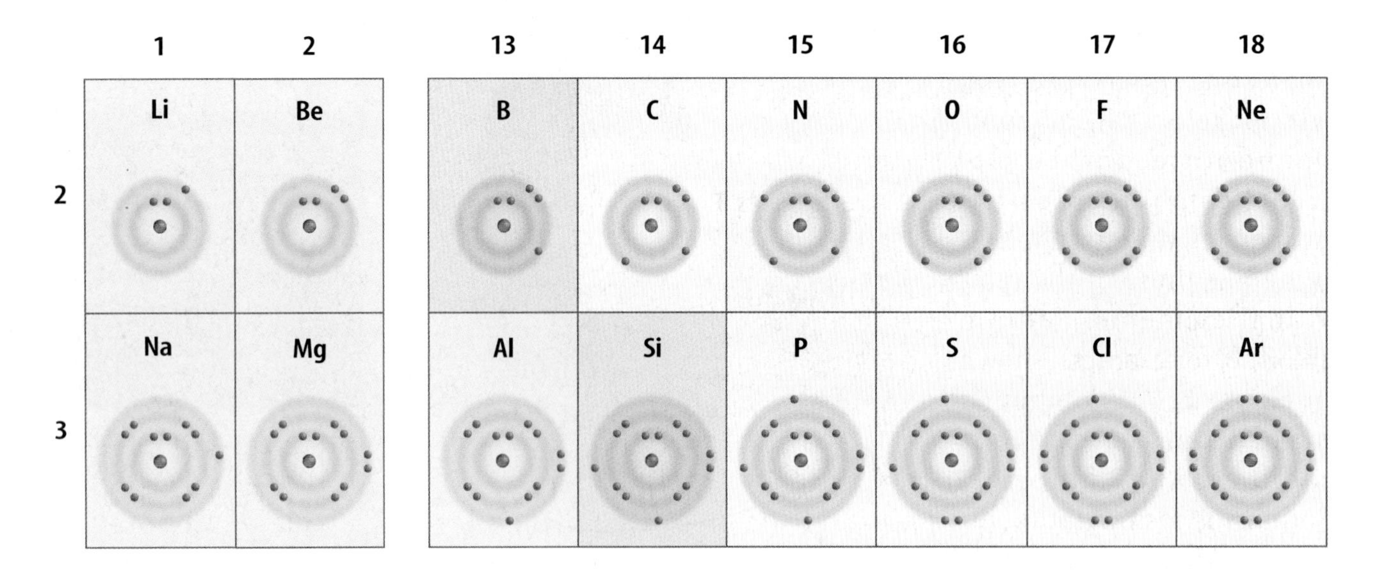

Figure 3
The number of electrons in each group's outer level increases across the table until the noble gases in Group 18 have a complete outer energy level.

Atomic Stability

Why do atoms form compounds? Atoms combine when the compound formed is more stable than the separate atoms. The periodic table on the inside back cover of your book lists 115 elements, most of which can combine with other elements. However, the six noble gases in Group 18 seldom form compounds. Why is this so? Atoms of noble gases are unusually stable. Compounds of these atoms rarely form because they are almost always less stable than the original atoms.

The Unique Noble Gases To understand the stability of the noble gases, you must look at electron dot diagrams. Electron dot diagrams show only the electrons in the outer energy level of an atom. They contain the chemical symbol for the element surrounded by dots representing its outer electrons. How do you know how many dots to make? For Groups 1 and 2 and 13 through 18, you can use a periodic table or the portion of it shown in **Figure 3.** Look at the outer ring of each of the elements. Group 1 has one outer electron. Group 2 has two. Group 13 has three, Group 14, four, and so on to Group 18, the noble gases, which have eight.

What makes the noble gases more stable than other elements? An atom is **chemically stable** when its outer energy level is complete. Recall that the outer energy levels of helium and hydrogen are stable with two electrons. The outer energy levels of all the other elements are stable when they contain eight electrons. The noble gases are stable because they each have a complete outer energy level. **Figure 4** shows electron dot diagrams of some of the noble gases. Notice that eight dots surround Kr, Ne, Xe, Ar, and Rn, and two dots surround He.

Figure 4
Electron dot diagrams of noble gases show that they all have a stable, filled outer energy level.

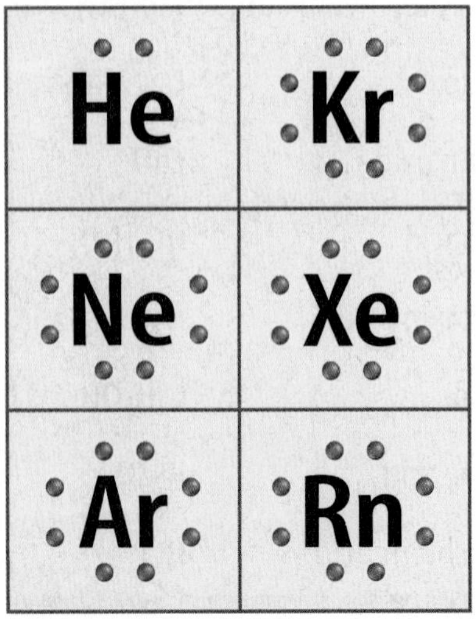

He	Kr
Ne	Xe
Ar	Rn

Energy Levels and Other Elements How do the dot diagrams represent other elements, and how does that relate to their ability to make compounds? Hydrogen and helium, the elements in row one of the periodic table, can hold a maximum of two electrons in their outer energy levels. Hydrogen contains one electron in its lone energy level. A dot diagram for hydrogen has a single dot next to its symbol. This means that hydrogen's outer energy level is not full. Therefore, it is more stable when it is part of a compound. This is why so many hydrogen-containing compounds, including water, exist on Earth.

In contrast, helium's outer energy level contains two electrons. Its dot diagram has two dots—a pair of electrons—next to its symbol. Helium already has a full outer energy level by itself and is chemically stable. Helium rarely forms compounds but, by itself, the element is a commonly used gas.

When you look at the elements in Groups 13 through 17, you see that each of them falls short of having a stable energy level. Each group contains too few electrons for a stable level of eight electrons. The elements are not stable because of it and commonly form compounds.

Research Visit the Glencoe Science Web site at **science.glencoe.com** for more information about using dot diagrams to represent outer energy level electrons. Communicate to your class what you learn.

Outer Levels—Getting Their Fill As you just learned, hydrogen is an element that does not have a full outer energy level. How does hydrogen, or any other element find or get rid of extra electrons? Atoms with partially stable outer energy levels can lose, gain, or share electrons to obtain a stable outer energy level. They do this by combining with other atoms that also have partially complete outer energy levels. As a result, each achieves stability. **Figure 5** shows electron dot diagrams for sodium and chlorine. When they combine, sodium loses one electron and chlorine gains one electron. You can see from the electron dot diagram that chlorine now has a stable outer energy level similar to a noble gas. But what about sodium?

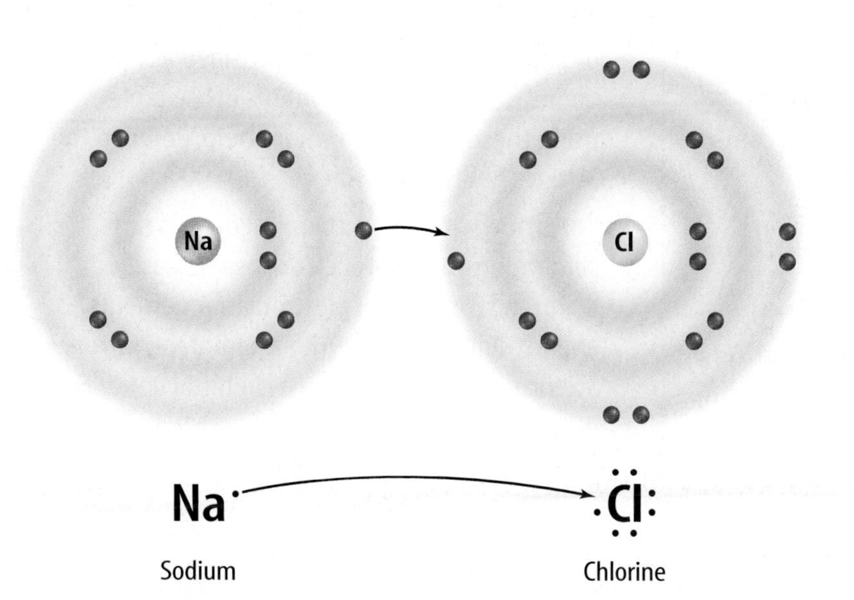

Figure 5
Each of these atoms has the potential of having a stable outer energy level by just adding or taking away one electron.

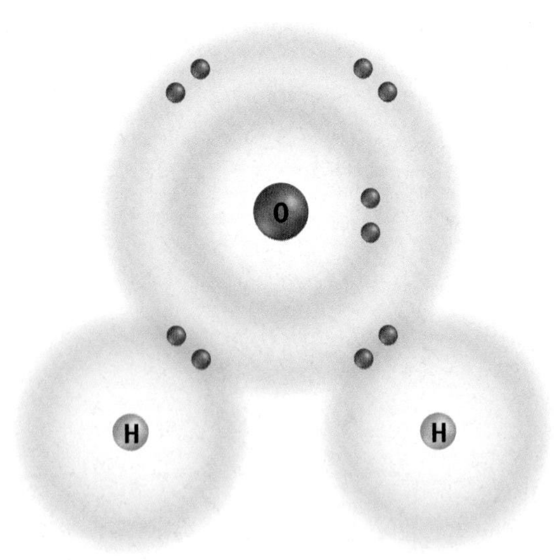

Figure 6
In water, hydrogen contributes one electron and oxygen contributes the other to each hydrogen-oxygen bond. The atoms share those electrons instead of giving them up.

Stability Is Reached Sodium had only one electron in its outer energy level, which it lost to combine with chlorine in sodium chloride. However, look back to the next, outermost energy level of sodium. This is now the new outer energy level, and it is stable with eight electrons. When the outer electron of sodium is removed, a complete inner energy level is revealed and now becomes the new outer energy level. Sodium and chlorine are stable now because of the exchange of an electron.

In the compound water, each hydrogen atom needs one electron to fill its outer energy level. The oxygen atom needs two electrons for its outer level to be stable with eight electrons. Hydrogen and oxygen become stable and form bonds in a different way than sodium and chlorine. Instead of gaining or losing electrons, they share them. **Figure 6** shows how hydrogen and oxygen share electrons to achieve a more stable arrangement of electrons.

When atoms gain, lose, or share electrons, an attraction forms between the atoms, pulling them together to form a compound. This attraction is called a chemical bond. A **chemical bond** is the force that holds atoms together in a compound. In Section 2 you will learn how these chemical bonds are formed.

Reading Check *What is a chemical bond?*

Section 1 Assessment

1. Describe what happens to the properties of elements when atoms form compounds.
2. What does the formula BaF_2 tell you about this compound?
3. Why are some elements stable on their own while others are more stable in compounds?
4. In what ways can a chemical bond form?
5. **Think Critically** The label on a box of cleanser states that it contains $HC_2H_3O_2$. What elements are in this compound? How many atoms of each element can be found in a unit of $HC_2H_3O_2$?

Skill Builder Activities

6. **Making and Using Tables** The compounds in **Table 1** that contain carbon are classified as organic, and the others are classified as inorganic. Reorganize the contents of the table using these groups. **For more help, refer to the Science Skill Handbook.**
7. **Communicating** A chemical bond is not something you can touch or observe easily. Using the concepts of electron arrangements, energy, and stability, write a paragraph in your Science Journal describing a chemical bond. **For more help, refer to the Science Skill Handbook.**

Activity

Atomic Trading Cards

Perhaps you have seen or collected trading cards for famous athletes. Usually each card has a picture of the athlete on one side with important statistics related to the sport on the back. Atoms can also be identified by their properties and statistics.

What You'll Investigate

How can a visible model show how energy levels fill when atoms combine?

Materials

4 × 6 inch index cards
periodic table

Goals

- **Display** the electrons of elements according to their energy levels.
- **Compare and classify** elements according to their outer energy levels.

Procedure

1. Get an assigned element from the teacher. Write the following information for your element on your index card: name, symbol, Group, atomic number, atomic mass, metal/nonmetal/metalloid.
2. On the other side of your index card show the number of protons and neutrons in the nucleus (e.g. 6p for six protons and 6n for six neutrons for carbon.)
3. Draw circles around the nucleus to represent the energy levels of your element. The number of circles you will need is the same as the row the element is in on the periodic table.
4. Draw dots on each circle to represent the electrons in each energy level. Remember that level one can hold two electrons and levels two and three can hold eight electrons.
5. Look at the picture side only of four or five of your classmates' cards. Determine which element they have and to which group it belongs.

Conclude and Apply

1. As you classify the elements according to their Group number, what pattern do you see in the number of electrons in the outer energy level?
2. Atoms that give up electrons combine with atoms that gain electrons in order to form compounds. In your Science Journal, predict some pairs of elements that would combine in this way.

Communicating Your Data

Make a graph that relates the groups to the number of electrons in their outer energy level. **For more help, refer to the Science Skill Handbook.**

SECTION 2

Types of Bonds

As You Read

What You'll Learn

- **Describe** ionic bonds and covalent bonds.
- **Identify** the particles produced by ionic bonding and by covalent bonding.
- **Distinguish** between a nonpolar covalent bond and a polar covalent bond.

Vocabulary

ion
ionic bond
molecule
polar molecule
nonpolar molecule
covalent bond

Why It's Important

Bond type determines other properties of the compound.

Gain or Loss of Electrons

When you participate in a sport you might talk about gaining or losing an advantage. To gain an advantage, you want to have a better time than your opponent. It is important that you keep practicing because you don't want to lose that advantage. Gaining or losing an advantage happens as you try to meet a standard for your sport.

Atoms, too, lose or gain to meet a standard—a stable energy level. They do not lose or gain an advantage. Instead, they lose or gain electrons. An atom that has lost or gained electrons is called an ion. An **ion** is a charged particle because it now has either more or fewer electrons than protons. The positive and negative charges are not balanced.

Some of the most common compounds are made by the loss and gain of just one electron. They include an element from Group 1 on the periodic table and an element from Group 17. Some examples are sodium chloride, commonly known as table salt; sodium fluoride, an anticavity ingredient in some toothpastes; and potassium iodide, an ingredient in iodized salt.

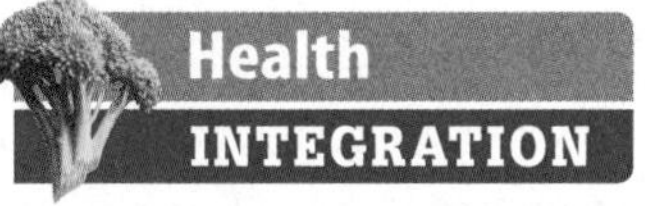

Why do people need iodine? A lack of iodine causes a wide range of problems in the human body. The most obvious is an enlarged thyroid gland shown in **Figure 7,** but the problems can include mental retardation, neurological disorders, and physical problems. In infants the problems can be irreversible and can cause death.

Adding iodine to salt is as easy as spraying it with potassium iodide. However, even though the solution is relatively simple, more countries have iodine deficiency problems than do not. Much progress has been made in the movement to solve this problem, and the work continues.

Figure 7
Goiter, a condition that causes an enlargement of the thyroid gland in the neck, is caused by iodine deficiency.

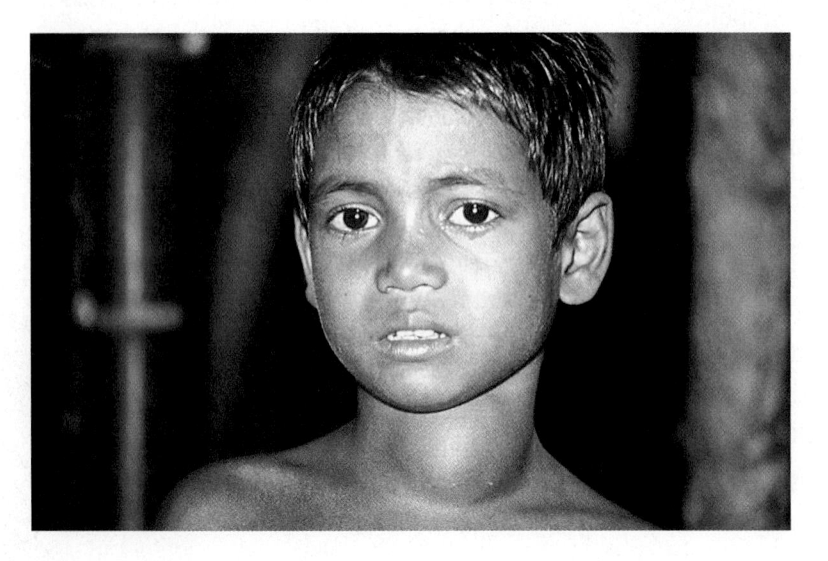

A Bond Forms What happens when potassium and iodine atoms come together? A neutral atom of potassium has one electron in its outer level. This is not a stable outer energy level. When potassium forms a compound with iodine, potassium loses one electron from its fourth level, and the third level becomes a complete outer level. However, the atom is no longer neutral. The potassium atom has become an ion. When a potassium atom loses an electron, the atom becomes positively charged because there is one electron less in the atom than there are protons in the nucleus. The 1+ charge is shown as a superscript written after the element's symbol, K^+, to indicate its charge. *Superscript* means "written above."

The iodine atom in this reaction undergoes change, as well. An iodine atom has seven electrons in its outer energy level. Recall that a stable outer energy level contains eight electrons. During the reaction with potassium, the iodine atom gains an electron, leaving its outer energy level with eight electrons. This atom is no longer neutral because it gained an extra negative particle. It now has a charge of 1– and is called an iodine ion, written as I^-. The compound formed between potassium and iodine is called potassium iodide. The dot diagrams for the process are shown in **Figure 8.**

Reading Check *What part of an ion's symbol indicates its charge?*

Another way to look at the electron in the outer shell of a potassium atom is as an advertisement to other atoms saying, "Available: One electron to lend." The iodine atom would have the message, "Wanted: One electron to borrow." When the two atoms get together, each becomes a stable ion. Notice that the resulting compound has a neutral charge because the positive and negative charges of the ions cancel each other.

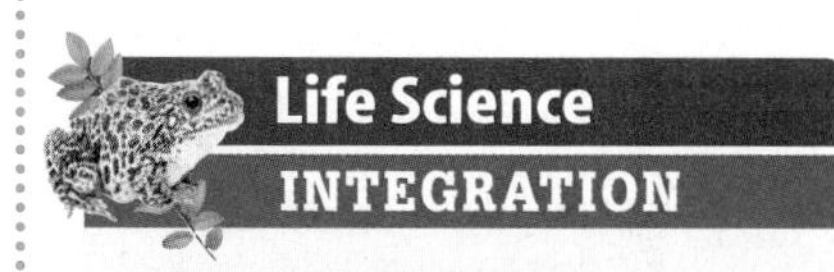

Ions are important in many processes in your body. The movement of muscles is just one of these processes. Muscle movement would be impossible without the movement of ions and out of nerve cells.

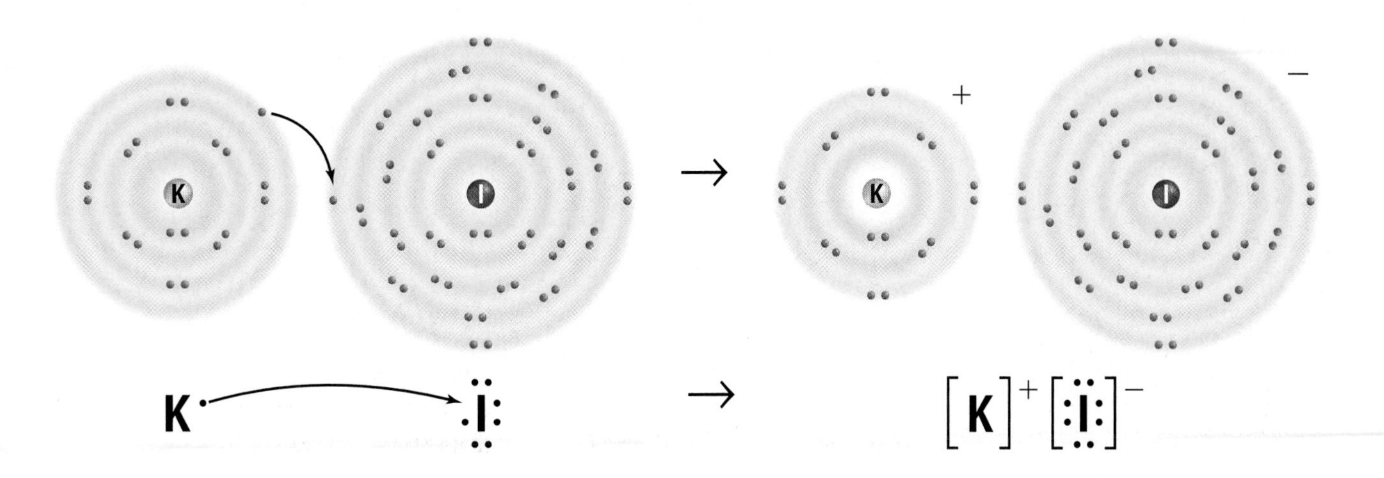

Figure 8
Potassium and iodine must perform a transfer of one electron. Potassium and iodine end up with stable outer energy levels.

The Ionic Bond

When ions cooperate in this way, a bond is formed. An **ionic bond** is the force of attraction between the opposite charges of the ions in an ionic compound. In an ionic bond, a transfer of electrons takes place. If an element loses electrons, one or more elements must gain an equal number of electrons to maintain the neutral charge of the compound.

Now that you have seen how an ionic bond forms when one electron is involved, see how it works when more than one is involved. The formation of magnesium chloride, $MgCl_2$, is another example of ionic bonding. When magnesium reacts with chlorine, a magnesium atom loses two electrons and becomes a positively charged ion, Mg^{2+}. At the same time, two chlorine atoms gain one electron each and become negatively charged chloride ions, Cl^{-}. In this case, a magnesium atom has two electrons to lend, but a single chlorine atom needs to borrow only one electron. Therefore, it takes two chlorine atoms, as shown in **Figure 9,** to take the two electrons from the magnesium ion.

SCIENCE Online

Research Visit the Glencoe Science Web site at **science.glencoe.com** for more information about ionic bonding. Communicate to your class what you learn.

Zero Net Charge The result of this bond is a neutral compound. The compound as a whole is neutral because the sum of the charges on the ions is zero. The positive charge of the magnesium ion is exactly equal to the negative charge of the two chloride ions. In other words, when atoms form an ionic compound, their electrons are shifted to other atoms, but the overall number of protons and electrons of the combined atoms remains equal and unchanged. Therefore, the compound is neutral.

Ionic bonds usually are formed by bonding between metals and nonmetals. Looking at the periodic table, you will see that the elements that bond ionically are often across the table from each other. Ionic compounds are often crystalline solids with high melting points.

Figure 9
A magnesium atom gives an electron to each of two chlorine atoms to form $MgCl_2$.

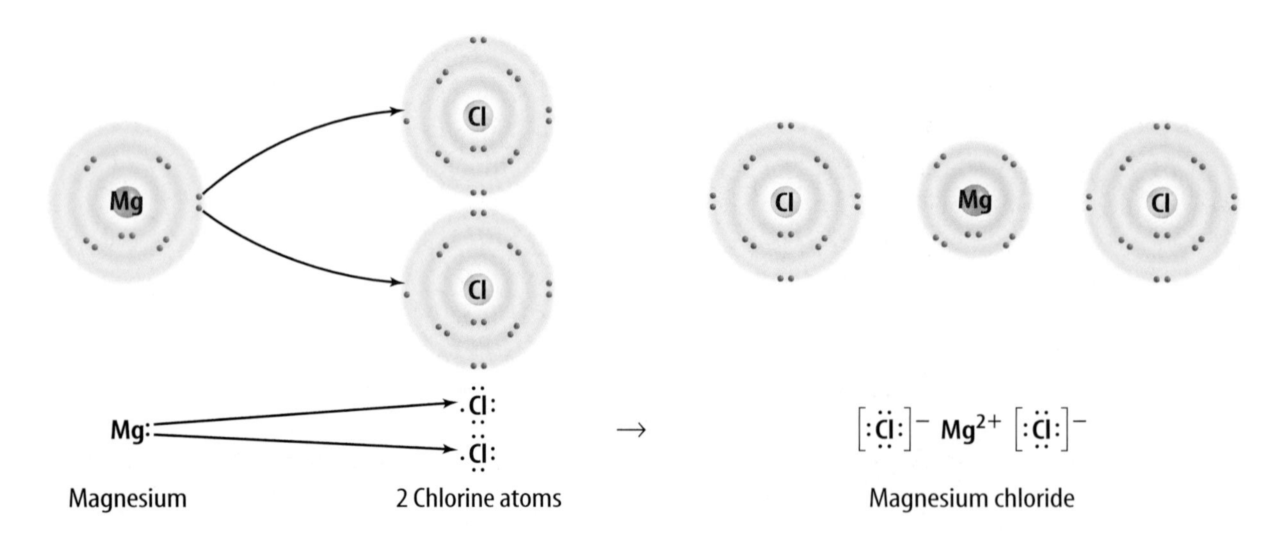

Sharing Electrons

Some atoms of nonmetals are unlikely to lose or gain electrons. For example, the elements in Group 4 of the periodic table have four electrons in their outer level. They would have to either gain or lose four electrons in order to have a stable outer level. The loss of this many electrons takes a great deal of energy. Each time an electron is removed, the nucleus holds the remaining electrons even more tightly. Therefore, these atoms become more chemically stable by sharing electrons, rather than by losing or gaining electrons.

The attraction that forms between atoms when they share electrons is known as a **covalent bond.** A neutral particle that forms as a result of electron sharing is called a **molecule.**

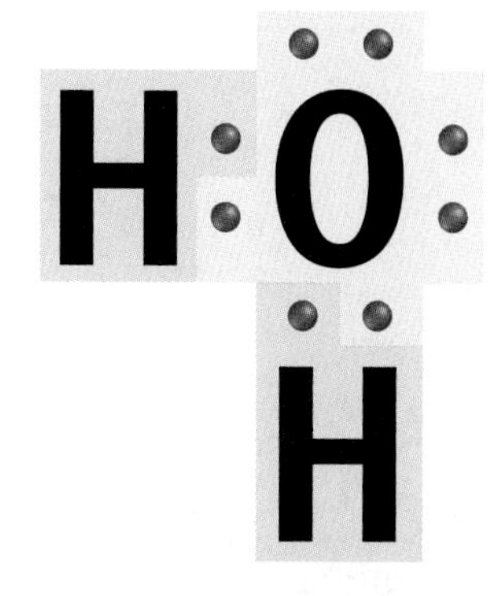

Figure 10
Each of the pairs of electrons between the two hydrogens and the oxygen is shared as each atom contributes one electron to the pair to make the bond.

Single Covalent Bonds A single covalent bond is made up of two shared electrons. Usually, one of the shared electrons comes from one atom in the bond and the other comes from the other atom in the bond. A water molecule contains two single bonds. In each bond, a hydrogen atom contributes one electron to the bond and the oxygen atom contributes the other. The two electrons are shared, forming a single bond. The result of this type of bonding is a stable outer energy level for each atom in the molecule. Each hydrogen atom is stable with two electrons, and the oxygen atom is stable with eight outer energy level electrons.

Multiple Bonds A covalent bond also can contain more than one pair of electrons. An example of this is the bond in nitrogen (N_2), shown in **Figure 11.** A nitrogen atom has five electrons in its outer energy level and needs to gain three electrons to become stable. It does this by sharing its three electrons with another nitrogen atom. The other nitrogen atom also shares its three electrons. When each atom contributes three electrons to the bond, the bond contains six electrons, or three pairs of electrons. Each pair of electrons represents a bond. Therefore, three pairs of electrons represent three bonds, or a triple bond. Each nitrogen atom is stable with eight electrons in its outer energy level. In a similar way, a bond that contains two shared pairs of electrons is a double bond.

Covalent bonds form between nonmetallic elements. These elements are close together in the upper right-hand corner of the periodic table. Many covalent compounds are liquids or gases at room temperature.

N + N → N N

Figure 11
The dot diagram shows that the two nitrogen atoms in nitrogen gas share six electrons.

Observing Bond Type

Procedure

1. Turn on the faucet to produce a thin **stream of water.**
2. Rub an inflated **balloon** with **wool or fur.**
3. Bring the balloon near the stream of water, and describe what you see.

Analysis

1. Explain your observations.
2. Relate the attraction between the balloon and the water to the attraction between the north and south poles of two magnets. Why might water act like a magnet?

Unequal Sharing Electrons are not always shared equally between atoms in a covalent bond. The strength of the attraction of each atom to its electrons is related to the size of the atom, the charge of the nucleus, and the total number of electrons the atom contains. Part of the strength of attraction has to do with how far away from the nucleus the electron being shared is. For example, a magnet has a stronger pull when it is right next to a piece of metal rather than several centimeters away. The other part of the strength of attraction has to do with the size of the positive charge in the nucleus. Using a magnet as an example again, a strong magnet will hold the metal more firmly than a weak magnet.

One example of this unequal sharing is found in a molecule of hydrogen chloride, HCl. In water, HCl is hydrochloric acid, which is used in laboratories, in industry to clean metal, and is found in your stomach where it digests food. Chlorine atoms have a stronger attraction for electrons than hydrogen atoms do. As a result, the electrons shared in hydrogen chloride will spend more time near the chlorine atom than near the hydrogen atom, as shown in **Figure 12.** The chlorine atom has a partial negative charge represented by a lower case Greek symbol delta followed by a negative superscript, δ^-. The hydrogen atom has a partial positive charge represented by a δ^+.

Tug-of-War You might think of the bond as the rope in a tug-of-war, and the shared electrons as the knot in the center of the rope. **Figure 13** illustrates this concept. Each atom in the molecule attracts the electrons that they share. However, sometimes the atoms aren't the same size. The same thing happens in tug-of-war. Sometimes one team is larger or has stronger participants than the other.

When this is true, the knot in the middle of the rope ends up closer to the stronger team. Similarly, the electrons being shared in a molecule are held more closely to the atoms with the stronger pull or larger nucleus.

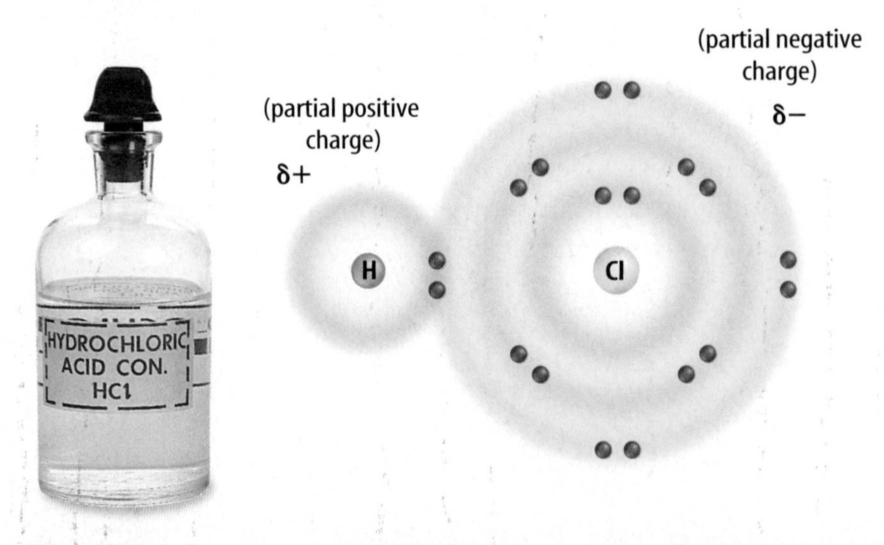

Figure 12
The chlorine atom exerts the greater pull on the electron in hydrogen chloride which forms hydrochloric acid in water.

Figure 13

When playing tug-of-war, if there are more—or stronger—team members on one end of the rope than the other, there is an unequal balance of power. The stronger team can pull harder on the rope and has the advantage. A similar situation exists in polar molecules, in which electrons are attracted more strongly by one type of atom in the molecule than another. Because of this unequal sharing of electrons, polar molecules have a slightly negative end and a slightly positive end, as shown below.

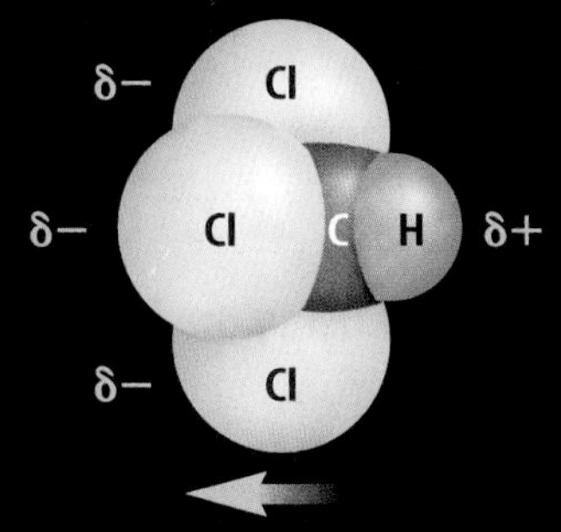

CHLOROFORM In a molecule of chloroform ($CHCl_3$), or trichloromethane (tri klor oh ME thayn), the three chlorine atoms attract electrons more strongly than the hydrogen atom does, creating a partial negative charge on the chlorine end of the molecule and a partial positive charge on the hydrogen end. This polar molecule is a clear, sweet-smelling liquid once widely used as an anesthetic in human and veterinary surgery.

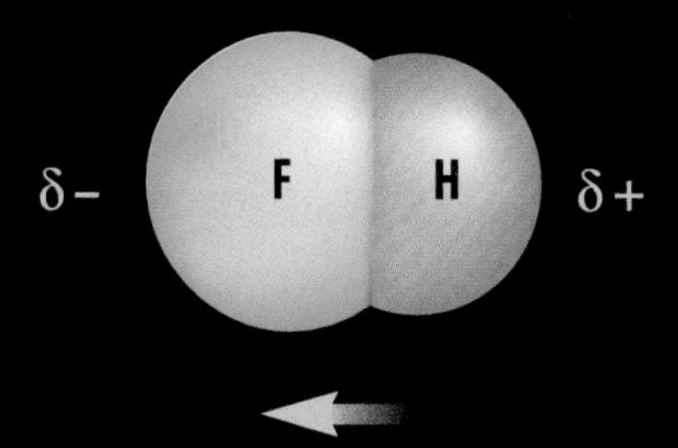

HYDROGEN FLUORIDE Hydrogen and fluorine react to form hydrogen fluoride (HF). In an HF molecule, the two atoms are bound together by a pair of electrons, one contributed by each atom. But the electrons are not shared equally because the fluorine atom attracts them more strongly than the hydrogen atom does. The result is a polar molecule with a slightly positive charge near the hydrogen end and a slightly negative charge near the fluorine end.

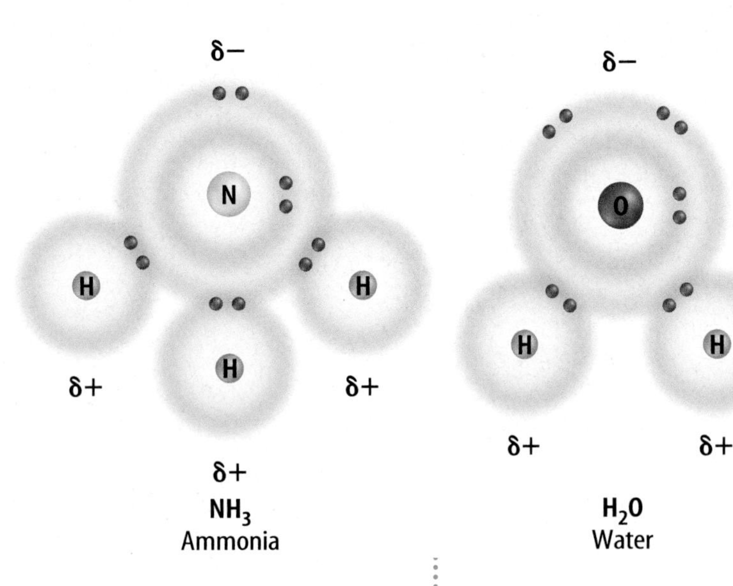

Figure 14
A model of the ammonia molecule illustrates the charge difference as the nitrogen holds the electron more closely. The polarity of water is responsible for many of its unique properties.

Polar or Nonpolar? For the molecule involved in this electron tug-of-war, there is another consequence. Again, look at the molecule of hydrogen chloride. This unequal sharing of electrons gives each chlorine atom a slight negative charge and each hydrogen atom a slight positive charge. The atom holding the electron more closely always will have a slightly negative charge. The charge is balanced but not equally distributed. This type of molecule is called polar. The term *polar* means "having opposite ends." A **polar molecule** is one that has a slightly positive end and a slightly negative end although the overall molecule is neutral. Ammonia and water are examples of a polar molecule, as shown in **Figure 14.**

Reading Check *What is a polar molecule?*

Two atoms that are exactly alike can share their electrons equally, forming a nonpolar molecule. A **nonpolar molecule** is one in which electrons are shared equally in bonds. Such a molecule does not have oppositely charged ends. This is true of molecules made from two identical atoms or molecules that are symmetric, such as CCl_4.

Section 2 Assessment

1. Why does an atom make an ionic bond only with certain other atoms?
2. Compare the possession of electrons in ionic and covalent bonds.
3. What types of particles are formed by covalent bonds?
4. What is the difference between polar and nonpolar molecules?
5. **Think Critically** From the following list of symbols, choose two elements that are likely to form an ionic bond: O, Ne, S, Ca, K. Next, select two elements that would likely form a covalent bond. Explain.

Skill Builder Activities

6. **Concept Mapping** Using the following terms, make a network tree concept map of chemical bonding: *ionic, covalent, ions, positive ions, negative ions, molecules, polar,* and *nonpolar.* **For more help, refer to the Science Skill Handbook.**
7. **Solving One-Step Equations** Aluminum oxide, Al_2O_3, can be produced during space shuttle launches. Show that the sum of the positive and negative charges in a unit of Al_2O_3 equals zero. **For more help, refer to the Math Skill Handbook.**

SECTION

Writing Formulas and Naming Compounds

Symbols and Shorthand

Does the table in **Figure 15** look like it has anything to do with chemistry? It is an early table of the elements made by alchemists—scientists who tried to make gold from other elements. The alchemist used symbols like these to write the formulas of substances like silver tarnish. The modern chemist uses symbols from the modern periodic table and writes the formula Ag_2S. When you get to the end of this section, you, too, will name compounds and write their formulas.

As You Read

What **You'll Learn**

- **Explain** how to determine oxidation numbers.
- **Write** formulas and names for ionic compounds.
- **Describe** hydrates and write their formulas.
- **Write** formulas and names for covalent compounds.

Vocabulary

binary compound, polyatomic ion, oxidation number, hydrate

Why **It's Important**

The name and formula of a compound convey information about the compound.

Binary Ionic Compounds

The first formulas of compounds you will write are for binary ionic compounds. A **binary compound** is one that is composed of two elements. Potassium iodide, the salt additive discussed in Section 2, is a binary ionic compound. However, before you can write a formula, you must have all the needed information at your fingertips. What will you need to know?

Are electrons gained or lost? You need to know which elements are involved and what number of electrons they lose, gain, or share in order to become stable. How can you determine this? Section 1 discussed the relationship between an element's position on the periodic table and the number of electrons it gains or loses. This is called the **oxidation number** of an element. An oxidation number tells you how many electrons an atom has gained, lost, or shared to become stable.

For ionic compounds the oxidation number is the same as the charge on the ion. For example, a sodium ion has a charge of 1+ and an oxidation number of 1+. A chlorine ion has a charge of 1− and an oxidation number of 1−.

Figure 15
This old chart of the elements used pictorial symbols to represent elements.

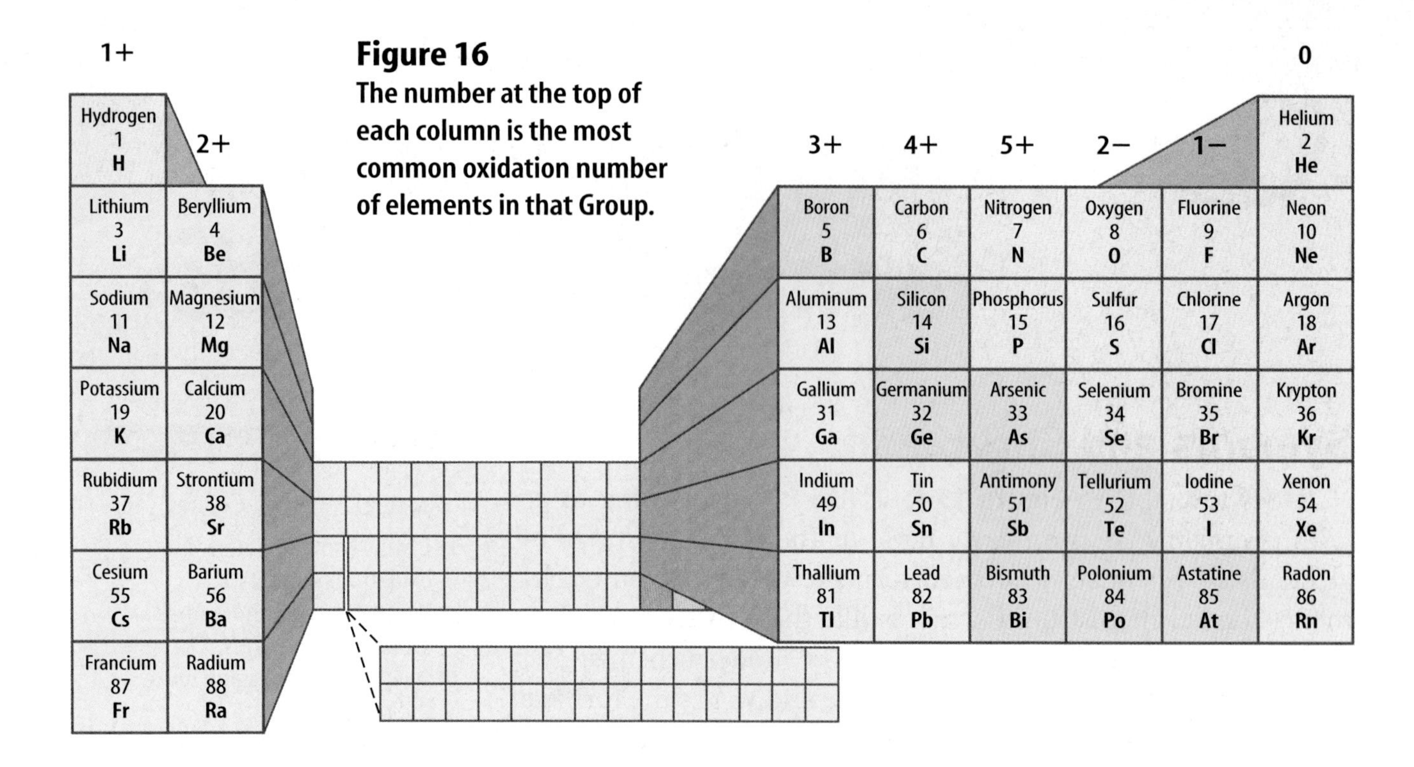

Figure 16
The number at the top of each column is the most common oxidation number of elements in that Group.

Oxidation Numbers The numbers with positive or negative signs in **Figure 16** are the oxidation numbers for these elements. Notice how they fit with the periodic table groupings.

The elements in **Table 2** can have more than one oxidation number. When naming these compounds, the oxidation number is expressed in the name with a roman numeral. For example, the oxidation number of iron in iron(III) oxide is 3+.

Table 2 Special Ions

Name	Oxidation Number
Copper (I)	1+
Copper (II)	2+
Iron (II)	2+
Iron (III)	3+
Chromium (II)	2+
Chromium (III)	3+
Lead (II)	2+
Lead (IV)	4+

Compounds are Neutral When writing formulas it is important to remember that although the individual ions in a compound carry charges, the compound itself is neutral. A formula must have the right number of positive ions and the right number of negative ions so the charges balance. For example, sodium chloride is made up of a sodium ion with a 1+ charge and a chlorine ion with a 1− charge. One of each ion put together makes a neutral compound with the formula NaCl.

However, what if you have a compound like calcium fluoride? A calcium ion has a charge of 2+ and a fluoride ion has a charge of 1−. In this case you need to have two fluoride ions for every calcium ion in order for the charges to cancel and the compound to be neutral with the formula CaF_2.

Some compounds require more figuring. Aluminum oxide contains an ion with a 3+ charge and an ion with a 2− charge. You must find the least common multiple of 3 and 2 in order to determine how many of each ion you need. You need two aluminum ions and three oxygen ions in order to have a 6+ charge and a 6− charge and therefore, the neutral compound Al_2O_3.

Writing Formulas After you've learned how to find the oxidation numbers and their least common multiple, you can write formulas for ionic compounds by using the following rules in this order.

1. Write the symbol of the element or polyatomic ion (ions containing more than one atom) that has the positive oxidation number or charge. Hydrogen, the ammonium ion (NH_4^+) and all metals have positive oxidation numbers.
2. Write the symbol of the element or polyatomic ion with the negative oxidation number. Nonmetals other than hydrogen and polyatomic ions other than NH_4^+ have negative oxidation numbers.
3. Use subscripts next to each ion so that the sum of the charges of all the ions in the formula is zero.

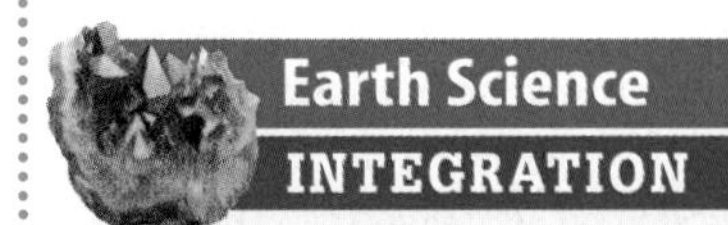

Farmers must sometimes add lime, which is calcium oxide, to soil in their fields. What is the formula of calcium oxide?

Math Skills Activity

Writing Formulas

Example Problem

What is the formula for lithium nitride?

Solution

1 *This is what you know:*

Write the symbol and oxidation number of the positive element:

Lithium (Li) = 1+

Write the symbol and oxidation number of the negative element:

Nitrogen (N) = 3−

Least common multiple = 3

2 *This is what you need to do:*

Add subscripts so that the sum of the oxidation numbers is zero

3 Li = (3) (1+) = 3+

1 N = 3−

Complete the formula: Li_3N

Check your answer by multiplying each subscript by the oxidation number and compare.

Li_3 gives (3)(1+) = 3+.

N gives (1)(3−) = 3− and (3+) + (3−) = 0.

Practice Problem

What is the formula for lead(IV) phosphide?

For more help, refer to the Math Skill Handbook.

Writing Names You can name a binary ionic compound from its formula by using these rules.

1. Write the name of the positive ion.
2. Using **Table 2,** check to see if the positive ion is capable of forming more than one oxidation number. If it is, determine the oxidation number of the ion from the formula of the compound. To do this, keep in mind that the overall charge of the compound is zero and the negative ion has only one possible charge. Write the charge of the positive ion using roman numerals in parentheses after the ion's name. If the ion has only one possible oxidation number, proceed to step 3.
3. Write the root name of the negative ion. The root is the first part of the element's name. For chlorine the root is *chlor-*. For oxygen it is *ox-*.
4. Add the ending *-ide* to the root. **Table 3** lists several elements and their *-ide* counterparts. For example, BaF_2 is named barium fluoride.

Notice that the subscripts of the positive and negative ion are not part of the name of the compound except when the positive ion has more than one possible charge.

Table 3 Elements in Binary Compounds

Element	-ide Name
Oxygen	oxide
Phosphorus	phosphide
Nitrogen	nitride
Sulfur	sulfide

Problem-Solving Activity

Can you name binary ionic compounds?

What would a chemist name the compound CuCl?

Identifying the problem

There are four simple steps in naming binary ionic compounds.

1. Write the name of the positive ion in the compound. In CuCl, the name of the positive ion is copper.
2. Check **Table 2** to determine if copper is one of the elements that can have more than one oxidation number. Looking at **Table 2,** you can see that copper can have a 1+ or a 2+ oxidation number. You need to determine which to use. Looking at the compound, you see that there is one copper atom and one chlorine atom. You know that the overall charge of the compound is zero and that chlorine only forms a 1− ion. For the charge of the compound to be zero, the charge of the copper ion must be 1+. Write this charge using roman numerals in parentheses after the element's name, copper (I).
3. Write the root name of the negative ion. The negative ion is chlorine and its root is *chlor-*.
4. Add the ending *-ide* to the root, chloride.
5. The full name of the compound CuCl is copper (I) chloride.

Solving the Problem

1. What is the name of CuO?
2. What is the name of $AlCl_3$?

Compounds with Complex Ions

Not all compounds are binary. Baking soda—used in cooking, as a medicine, and for brushing your teeth—has the formula $NaHCO_3$. This is an example of an ionic compound that is not binary. Which four elements does it contain? Some compounds, including baking soda, are composed of more than two elements. They contain polyatomic ions. The prefix *poly-* means "many," so the term *polyatomic* means "having many atoms." A **polyatomic ion** is a positively or negatively charged, covalently bonded group of atoms. So the compound as a whole contains three or more elements. The polyatomic ion in baking soda is the bicarbonate or hydrogen carbonate ion, HCO_3^-.

Table 4 Polyatomic Ions

Charge	Name	Formula
1+	ammonium	NH_4^+
1–	acetate	$C_2H_3O_2^-$
	chlorate	ClO_3^-
	hydroxide	OH^-
	nitrate	NO_3^-
2–	carbonate	CO_3^{2-}
	sulfate	SO_4^{2-}
3–	phosphate	PO_4^{3-}

Writing Names **Table 4** lists several polyatomic ions. To name a compound that contains one of these ions, first write the name of the positive ion. Use **Table 4** to find the name of a polyatomic ion. Then write the name of the negative ion. For example, K_2SO_4 is potassium sulfate. What is the name of $Sr(OH)_2$? Begin by writing the name of the positive ion, strontium. Then find the name of the polyatomic ion, OH^-. Table 4 lists it as hydroxide. Thus the name is strontium hydroxide.

Writing Formulas To write formulas for these compounds, follow the rules for binary compounds, with one addition. When more than one polyatomic ion is needed, write parentheses around the polyatomic ion before adding the subscript. How would you write the formula of barium chlorate?

First, identify the symbol of the positive ion. Barium has a symbol of Ba and forms a 2+ ion, Ba^{2+}. Next, identify the negative chlorate ion. **Table 4** shows that it is ClO_3^-. Finally, you need to balance the charges of the ions to make the compound neutral. It will take two chlorate ions with a 1– charge to balance the 2+ charge of the barium ion. Since the chlorate ion is polyatomic, you use parentheses before adding the subscript. Therefore, the formula is $Ba(ClO_3)_2$.

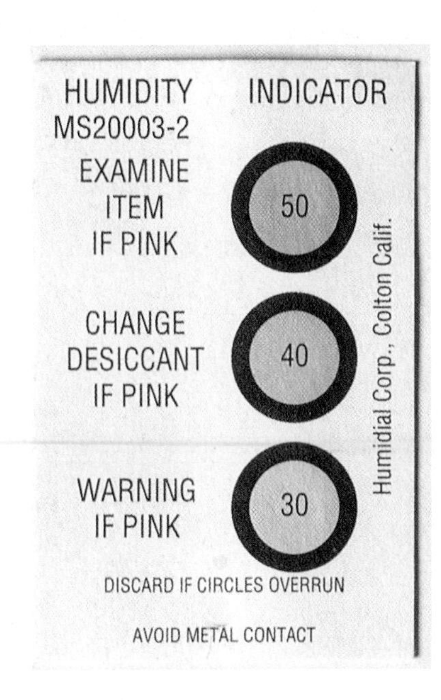

Figure 17
This humidity predictor uses cobalt chloride.

Making a Hydrate

Procedure

1. Mix 150 mL of **plaster of paris** with 75 mL of water in a small **bowl.**
2. Let the plaster dry overnight and take the hardened plaster out of the bowl.
3. Lightly tap the plaster with a **rubber hammer.**
4. Heat the plaster with a **hair dryer** on the hottest setting and observe.
5. Lightly tap the plaster with the hammer after heating it.

Analysis

1. What happened to the plaster when you tapped it before and after heating it?
2. What did you observe happening to the plaster as you heated it? Explain.

Compounds with Added Water

Have you seen weather predictors made from blue paper that turns pink in humid air? **Figure 17** shows an example of these humidity detectors. What properties allow this change? Some ionic compounds have water molecules as part of their structure. These compounds are called hydrates. A **hydrate** is a compound that has water chemically attached to its ions.

Common Hydrates The term *hydrate* comes from a word that means "water." When a solution of cobalt chloride evaporates, pink crystals that contain six water molecules for each unit of cobalt chloride are formed. The formula for this compound is $CoCl_2 \cdot 6H_2O$ and is called cobalt chloride hexahydrate.

You can remove water from these crystals by heating them. The resulting blue compound is called anhydrous, which means "without water." If you apply this anhydrous compound to paper, the paper will gain water molecules easily. How will this blue paper react to the presence of water vapor?

The plaster of paris shown in **Figure 18** also forms a hydrate when water is added. It becomes calcium sulfate dihydrate, which is also known as gypsum. The water that was added to the powder became a part of the compound.

Naming Binary Covalent Compounds

Covalent compounds are those formed between elements that are nonmetals. Some pairs of nonmetals can form more than one compound with each other. For example, nitrogen and oxygen can form N_2O, NO, NO_2, and N_2O_5. In the system you have learned so far, each of these compounds would be called nitrogen oxide. You would not know from that name what the composition of the compound is.

Figure 18
The presence of water changes this powder into a medium that can be used to create art.

Using Prefixes Scientists use the Greek prefixes in **Table 5** to indicate how many atoms of each element are in a binary covalent compound. The nitrogen and oxygen compounds N_2O, NO, NO_2, and N_2O_5 would be named dinitrogen oxide, nitrogen oxide, nitrogen dioxide, and dinitrogen pentoxide. Notice that the last vowel of the prefix is dropped when the second element begins with a vowel as in pentoxide. Often the prefix *mono-* is omitted, although it is used for emphasis in some cases. Carbon monoxide is one example.

✔ Reading Check ***What prefix would be used for seven atoms of one element in a covalent compound?***

These same prefixes are used when naming the hydrates previously discussed. The main ionic compound is named the regular way, but the number of water molecules in the hydrate is indicated by the Greek prefix.

You have learned how to write formulas of binary ionic compounds and of compounds containing polyatomic ions. Using oxidation numbers to write formulas, you can predict the ratio in which atoms of elements might combine to form compounds. You also have seen how hydrates have water molecules as part of their structures and formulas. Finally, you saw how to use prefixes in naming binary covalent compounds. As you continue to study, you will see many uses of formulas.

Table 5 Prefixes for Covalent Compounds

Number of Atoms	Prefix
1	*mono-*
2	*di-*
3	*tri-*
4	*tetra-*
5	*penta-*
6	*hexa-*
7	*hepta-*
8	*octa-*

Section  Assessment

1. What compounds can be formed from element *X*, with oxidation numbers 3+ and 5+, and element *Z*, with oxidation numbers 2− and 3−? Write their formulas.
2. Write formulas for the following compounds: *potassium iodide, magnesium hydroxide, aluminum sulfate,* and *chlorine heptoxide.*
3. Write the names of these compounds: KCl, Cr_2O_3, $Ba(ClO_3)_2$, NH_4Cl, and PCl_3.
4. Name $Mg_3(PO_4)_2 \cdot 4H_2O$, and write the formula for calcium nitrate trihydrate.
5. **Think Critically** Explain why sodium and potassium will or will not form a bond.

Skill Builder Activities

6. **Testing a Hypothesis** Design an experiment to distinguish between crystals that are hydrates and those that are not. Include crystals of iron(II) chloride, crystals of copper(II) nitrate, and crystals of sucrose. **For more help, refer to the Science Skill Handbook.**
7. **Solving One-Step Equations** The overall charge on the polyatomic sulfate ion, found in some acids, is 2−. Its formula is SO_4^{2-}. If the oxygen ion has a 2− oxidation number, determine the oxidation number of sulfur in this polyatomic ion. **For more help, refer to the Math Skill Handbook.**

Activity

Design Your Own Experiment

Become a Bond Breaker

The basic structural units of ionic compounds are ions. For covalent substances, molecules make up the basic units. By using controlled heat to melt substances, you can test various compounds to rate the attractive forces between their basic units. Would a substance that is difficult to melt have strong forces or weak forces holding its basic units together?

Recognize the Problem

How do the attractive forces between ions compare to the attractive forces between molecules?

Form a Hypothesis

Based on what you know about ions and molecules, state a hypothesis about which generally would have stronger attractions between their structural units.

Goals

- **Observe** the effect of heat on melting points of selected substances.
- **Design** an experiment that allows you to make some inferences that relate ease of melting and forces of attraction between particles of a substance.

Possible Materials

small samples of crushed ice, table salt, and sugar
wire test-tube holder
test tubes
laboratory burner
stopwatch

Safety Precautions

Keep a safe distance from the open flame of the lab burner. Wear proper eye protection. Do not continue heating beyond 5 min.

Test Your Hypothesis

Plan

1. As a group, agree upon and write a hypothesis statement.
2. As a group, write a detailed list of steps that are needed to test your hypothesis. Determine what your control will be.
3. As you heat materials in a test tube, what variables are held constant?
4. How will you time the heating of the individual substances?
5. Will you run any tests more than one time?
6. Make a list of materials that you will need to complete your experiment.
7. **Design** a data table in your Science Journal to record your observations.
8. Make sure your teacher approves your plan before you start.

Do

1. Carry out the experiment exactly as planned.
2. While you are observing the heating of each substance, think about the movement of the particles. Which particles are held together by ionic bonds? Which are made up of covalent molecules? How does that affect their movement?
3. Be sure to write down exactly how long it takes to melt each tested substance.

Analyze Your Data

1. **Compare** your results with those of other groups in the class.
2. **Classify** your tested substances as more likely ionic or covalent.
3. Which substances are generally more difficult to melt?
4. Did you have a control in this experiment? Variables?

Draw Conclusions

1. How did the results of your experiment support or disprove your hypothesis?
2. Sugar is known as a polar covalent compound. Knowing this, infer from your results how polarity affects melting point.

Make a chart showing your results and pointing out ways to distinguish between the different kinds of bonds.

Oops! Accidents in SCIENCE

SOMETIMES GREAT DISCOVERIES HAPPEN BY ACCIDENT!

A strong adhesive glue was a lucky accident

A Sticky Subject

In 1942, a research team was working on creating a new kind of glass. The group was working with some cyanoacrylate monomers (si uh noh A kruh layt • MAH nuh muhrz) which showed promise, but there was a problem that kept coming up. Everything the monomers touched stuck to everything else!

Cyanoacrylate is the chemical name for instant, super-type glues. The researcher was so focused on finding a different type of glass that at the time nobody recognized an important new adhesive. Not until a few years later.

Super-type glues make it possible to perfectly repair broken objects.

In 1952, a member of the research team, working on new materials for jet plane canopies, made a similar complaint. The ethyl cyanoacrylate they were working with again made everything stick together. This time, the insight stuck to the scientists like, well, like GLUE! "I began gluing everything I could lay my hands on—glass plates, rubber stoppers, metal spatulas, wood, paper, plastic. Everything stuck to everything, almost instantly, and with bonds I could not break apart," recalls the head of the research group.

Stick to It

Most adhesives, commonly called glues, are long chains of bonded molecules called polymers. Cyanoacrylate, however, exists as monomers—single molecules with double bonds. And it stays that way until it hits anything with moisture in it—like air. Yes, even the small amount of moisture in air and on the surfaces of most materials is enough to dissolve the double bonds in the monomers of cyanoacrylate, making them join together in long chains. The chains bond to surfaces as they polymerize.

The discovery of cyanoacrylates had an immediate impact on the automobile and airplane industries. And it soon "held" a spot in almost every household toolbox. Since the 1990s, however, cyanoacrylate glues are also finding a place in the doctor's office. A doctor can apply a thin layer of instant glue instead of putting stitches in a cut. This specially made medical glue was approved by the U.S. Food and Drug Administration in 1998. Cyanoacrylates are also used in dental and eye surgery and to stop bleeding in internal organs.

The repaired mug is ready for use.

CONNECTIONS **Take Note** Visit a store and make a table of different kinds of glues. List their common names, their chemical names, what they are made of, how long it takes them to set, and the types of surfaces for which they are recommended. Note any safety precautions.

SCIENCE Online
For more information, visit science.glencoe.com

Chapter 19 Study Guide

Reviewing Main Ideas

Section 1 Stability in Bonding

1. The properties of compounds are generally different from the properties of the elements they contain.
2. A chemical formula for a compound indicates the composition of a unit of the compound. *What can you infer about the composition of water from the model pictured?*

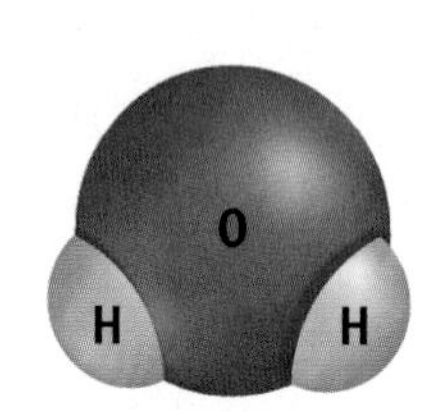

3. Chemical bonding occurs because atoms of most elements become more stable by gaining, losing, or sharing electrons in order to obtain a stable outer energy level.

Section 2 Types of Bonds

1. Ionic bonds between atoms are formed by the attraction between ions. Covalent bonds are formed by the sharing of electrons. *Which kind of bond is shown in the dot diagram?*

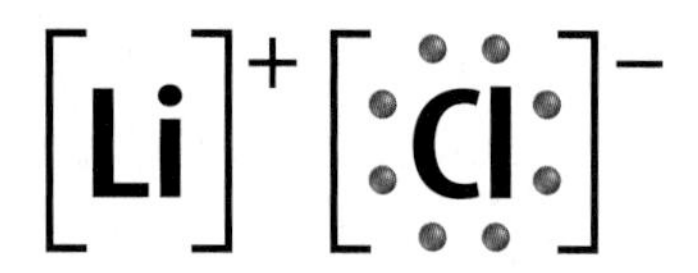

2. Ionic bonding occurs between charged particles called ions and produces ionic compounds. Covalent bonding produces units called molecules and occurs between nonmetallic elements.
3. The unequal sharing of electrons produces compounds that contain polar bonds, and the equal sharing of electrons produces nonpolar compounds.

Section 3 Writing Formulas and Naming Compounds

1. An oxidation number indicates how many electrons an atom has gained, lost, or shared when bonding with other atoms.
2. In the formula of an ionic compound, the element or ion with the positive oxidation number is written first, followed by the one with the negative oxidation number.
3. The name of a binary compound is derived from the names of the two elements that compose the compound. *What is the name of chalk, shown below? It contains the CO_3^{2-} ion and the Ca^{2+} ion.*

4. A hydrate is a compound that has water chemically attached to its ions and written into its formula.
5. Greek prefixes are used in the names of covalent compounds. These indicate the number of each atom present.

After You Read

FOLDABLES Reading & Study Skills

Use the chemical formulas on your Foldable to test your knowledge. See how many you can name and identify as common objects.

Visualizing Main Ideas

Complete the following concept map.

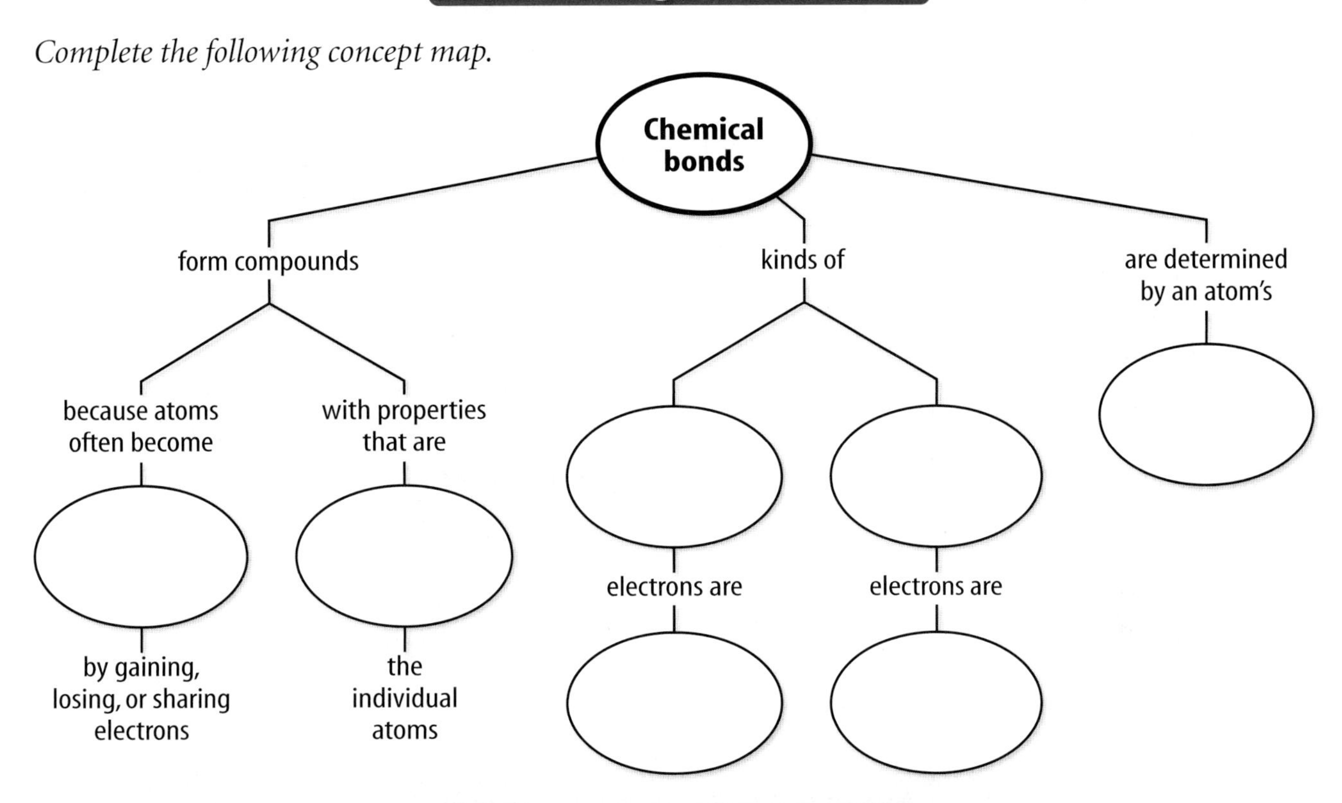

Vocabulary Review

Vocabulary Words

a. binary compound
b. chemical bond
c. chemical formula
d. chemically stable
e. covalent bond
f. hydrate
g. ion
h. ionic bond
i. molecule
j. nonpolar molecule
k. oxidation number
l. polar molecule
m. polyatomic ion

Study Tip

Read the chapters before you go over them in class. Being familiar with the material before your teacher explains it gives you a better understanding and provides you with a good opportunity to ask questions.

Using Vocabulary

Match each phrase with a vocabulary word.

1. a charged group of atoms
2. a compound composed of two elements
3. a molecule with partially charged ends
4. a positively or negatively charged atom
5. a chemical bond between oppositely charged ions
6. a bond formed from shared electrons
7. crystalline substance that contains water
8. outer energy level is filled with electrons
9. shows an element's combining ability
10. tells which elements are in a compound and their ratios

Chapter 19 Assessment

Checking Concepts

Choose the word or phrase that best answers the question.

1. Which elements are least likely to react with other elements?
   A) metals **C)** nonmetals
   B) noble gases **D)** transition elements

2. What is the oxidation number of Fe in the compound Fe_2S_3?
   A) 1^+ **C)** 3^+
   B) 2^+ **D)** 4^+

3. What is the name of CuO?
   A) copper oxide
   B) copper(I) oxide
   C) copper(II) oxide
   D) copper(III) oxide

4. What is the formula for copper(II) chlorate?
   A) $CuClO_3$ **C)** $Cu(ClO_3)_2$
   B) CuCl **D)** $CuCl_2$

5. Which of the following formulas represents a nonpolar molecule?
   A) N_2 **C)** NaCl
   B) H_2O **D)** HCl

6. How many electrons are in the outer energy level of Group 17 elements?
   A) 1 **C)** 17
   B) 2 **D)** 7

7. Which is a binary ionic compound?
   A) O_2 **C)** H_2SO_4
   B) NaF **D)** $Cu(NO_3)_2$

8. Which of these is an example of an anhydrous compound?
   A) H_2O **C)** $CuSO_4 \cdot 5H_2O$
   B) $CaSO_4$ **D)** $CaSO_4 \cdot 2H_2O$

9. Which of the following is an atom that has gained an electron?
   A) negative ion **C)** polar molecule
   B) positive ion **D)** nonpolar molecule

10. Which of these is an example of a covalent compound?
    A) sodium chloride **C)** calcium chloride
    B) calcium fluoride **D)** sulfur dioxide

Thinking Critically

11. Anhydrous magnesium chloride is used to make wood fireproof. Draw a dot diagram of magnesium chloride.

12. Baking soda, which is sodium hydrogen carbonate, and vinegar, which contains hydrogen acetate, can be used as household cleaners. Write the chemical formulas for these two compounds.

13. Artificial diamonds are made using thallium carbonate. If thallium has an oxidation number of 1+, what is the formula for the compound?

14. The formula for a compound that composes kidney stones is $Ca_3(PO_4)_2$. What is the chemical name of this compound?

15. Ammonium sulfate is used as a fertilizer. What is its chemical formula?

Developing Skills

16. **Comparing and Contrasting** Compare and contrast polar and nonpolar molecules.

17. **Interpreting Scientific Illustrations** Write the name and formula for the compound illustrated to the right.

18. **Drawing Conclusions** Ammonia gas and water react to form household ammonia, which contains NH_4^+ and OH^- ions. If the formula for water is H_2O, what is the formula for ammonia gas?

19. **Predicting** Elements from one family (vertical row) of the periodic table generally combine with elements from another family and polyatomic ions in the same ratio. For example, one calcium atom combines with two chlorine atoms to give $CaCl_2$ (calcium chloride) as it does with two fluorine atoms to give CaF_2 (calcium fluoride). For each formula in the table below, using a periodic table as a guide, predict which of the two compounds on the right side of the table is more likely to exist.

Which compounds exist?

Formula	Possible Compounds
SF_6	AlF_6 or TeF_6
K_2SO_4	Na_2SO_4 or Ba_2SO_4
CO_2	CCl_2 or CS_2
$CaCO_3$	OCO_3 or $BaCO_3$

Performance Assessment

20. **Drawing Conclusions** The name of a compound called copper (II) sulfate is written on a bottle. What is the charge of the copper ion? What is the charge of the sulfate ion?

21. **Model** One common form of phosphorus, white phosphorus, has the formula P_4 and is formed by four covalently bonded phosphorus atoms. Make a model of this molecule, showing that all four atoms are now chemically stable.

TECHNOLOGY

Go to the Glencoe Science Web site at **science.glencoe.com** or use the **Glencoe Science CD-ROM** for additional chapter assessment.

Test Practice

Cindy constructed a graph of ionization energy, which is the energy required to remove one electron and also the number of outer electrons an atom has in relation to its atomic number. Her finished graphs are shown in the diagram below.

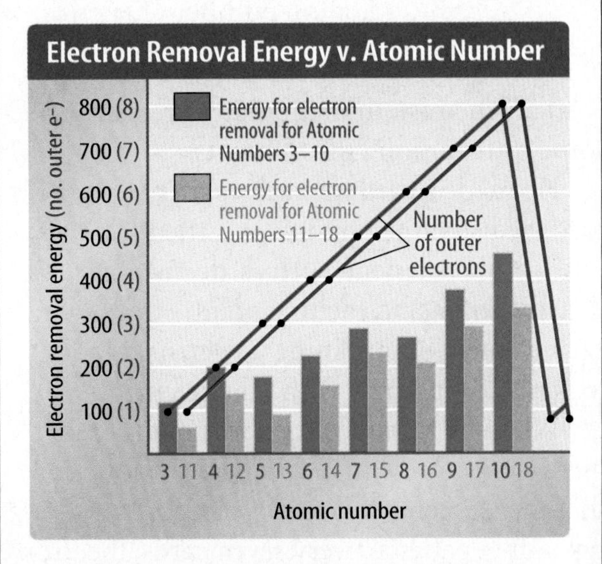

Study the diagram and answer the following questions.

1. A reasonable hypothesis based on these data is that as the number of outer electrons increases, the _____.
   A) energy required to remove one electron decreases
   B) energy required to remove one electron increases
   C) atomic number decreases
   D) potential to form ions increases

2. The noble gases have full outer energy levels. Which atomic numbers on the graph represent noble gases?
   F) 3 and 11
   G) 7 and 15
   H) 9 and 17
   J) 10 and 18

Standardized Test Practice

Reading Comprehension

Read the passage. Then read each question that follows the passage. Decide which is the best answer to each question.

Mendeleev and the Periodic Table

By 1860, scientists had discovered a total of 63 chemical elements. Dmitri Mendeleev, a Russian chemist, thought that there had to be some order among the elements.

He made a card for each element. On the card, he listed the physical and chemical properties of the element, such as atomic mass, density, color, and melting point. He also wrote each element's combining power, or its ability to form compounds with other elements.

When he arranged the cards in order of increasing atomic mass, Mendeleev noticed that the elements followed a periodic, or repeating, pattern. Every seven cards, the properties repeated. He placed each group of seven cards in rows, one under another so that the elements in a column, or group, had similar chemical and physical properties.

In a few places, Mendeleev had to move his cards one space to the left or right to follow the pattern. This left a few empty spaces. He predicted that these spaces would be filled with elements that were unknown. He even predicted the properties of the unknown elements. Fifteen years later, three new elements were discovered and placed in the empty spaces of the Periodic Table. Their physical and chemical properties agreed with Mendeleev's predictions.

Today there are more than 100 known elements. An extra column has been added for the noble gases, a group of elements not known to exist in Mendeleev's time. Members of this group almost never combine with other elements. As new elements are discovered or are made artificially, scientists can place them in their proper place on the Periodic Table thanks to Mendeleev.

Test-Taking Tip To answer questions about sequence of events, make a time line of what happened in each paragraph of the passage.

All elements can be organized according to their physical and chemical properties.

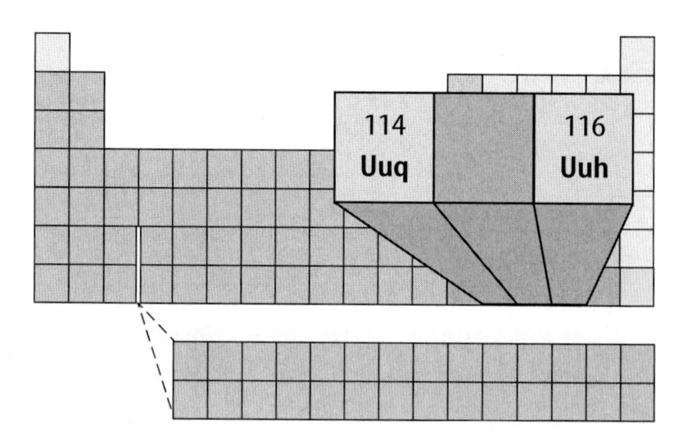

1. Which of the following occurred FIRST in the passage?
   A) Three new elements were discovered fifteen years after Mendeleev developed the Periodic Table.
   B) The noble gases were discovered and placed in the Periodic Table.
   C) Mendeleev predicted properties of the unknown elements.
   D) New elements were made in the laboratory and then placed in the Periodic Table.

2. The word artificially in this passage means _____.
   F) unnaturally
   G) artistically
   H) atomically
   J) radioactively

Standardized Test Practice

Reasoning and Skills

Read each question and choose the best answer.

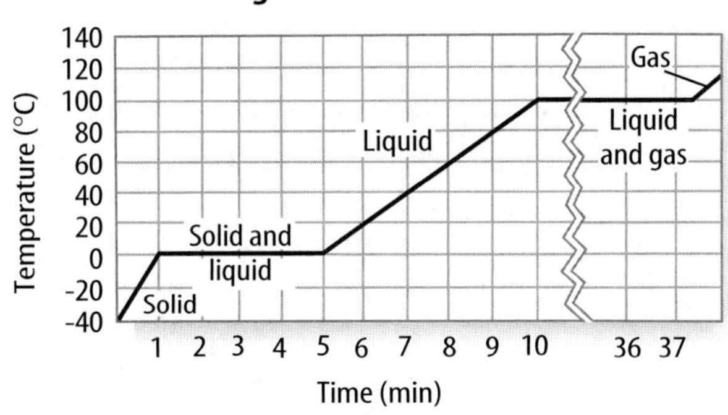

1. The graph shows the change in temperature as ice is heated until it changes completely to gas. How much higher than the starting temperature is the boiling point?

A) 40°C
B) 100°C
C) 140°C
D) 180°C

Test-Taking Tip Boiling point is the flat section of the graph where liquid changes to gas.

2. What is being measured in the illustration?

F) boiling point
G) melting point
H) density
J) flammability

Test-Taking Tip Think about what you might be measuring when you use a thermometer in a liquid that you are heating.

Selected Properties of Selected Pure Substances

Substance	Melting Point (°C)	Boiling Point (°C)	Color
Aluminum	660.4	2,519	silver metallic
Argon	-189.2	-185.7	colorless
Mercury	-38.8	356.6	silver metallic
Water	0	100	colorless

3. Room temperature is about 20°C. In the table, which substance is a solid at room temperature?

A) aluminum
B) argon
C) mercury
D) water

Test-Taking Tip Remember that negative temperatures are below zero.

Read this question carefully before writing your answer on a separate sheet of paper.

4. The density of pure water is 1.00 g/cm³. Since ice floats on water, we know that the density of ice is *less* than that of water. Design an experiment to determine the density of an ice cube. List all the necessary steps.
(Volume = Length × Width × Height;
Density = Mass / Volume)

Test-Taking Tip Make sure to consider all the information provided in the question.

UNIT 5 Diversity of Matter

How Are Billiards & Bottles Connected?

Billiards, a popular table game of the 1800s, used balls carved from ivory. In the 1860s, an ivory shortage prompted one billiard-ball manufacturer to offer a reward of $10,000 to anyone who could come up with a suitable substitute. In an attempt to win the prize, an inventor combined certain organic compounds, put them into a mold, and subjected them to heat and pressure. The result was a hard, shiny lump that sparked a major new industry—the plastics industry. By the mid-1900s, chemists had invented many different kinds of moldable plastic. Today, plastic is made into countless products—everything from car parts to soda bottles.

SCIENCE CONNECTION

POLYMERS Plastics are polymers—long molecules made up of smaller units called monomers. By using different monomers, it's possible to create plastics with different characteristics. Explore your home or school and make a list of plastic items. How do the physical characteristics of different plastics—hardness, flexibility, and so forth—relate to their function? Using your list as a reference, write a description of one part of your day, identifying all the plastics you use.

CHAPTER

Elements and Their Properties

It takes many different elements to build an airplane like this one. Some of those elements are metals, some are nonmetals, and some are metalloids. Each element has distinct properties. Knowing the properties of elements allows you to use them in a variety of practical ways—from treating diseases to building an airplane. This chapter will describe groups of elements and help you learn how they are related.

What do you think?

Science Journal Look at the picture below with a classmate. Discuss what you think this might be. Here's a hint: *The gas in the tubes stands apart from the crowd.* Write your best guess in your Science Journal.

It may be surprising to you that every known physical object is made from one or more of 115 known elements. When they are not bonded to atoms of other elements, the atoms of an element have specific, identifiable properties. In this activity, you'll observe how heated atoms of elements absorb energy and then in a short time release the absorbed energy, which you see as colored light.

Observe colorful clues

1. Using tongs, carefully hold a clean paper clip in the hottest part of a lab burner flame for 45 seconds.
2. Dip the hot paper clip into a solution of copper(II) sulfate.
3. Using the tongs with the same paper clip, repeat step 1, observing any color change.
4. Repeat all three steps using solutions of strontium chloride and sodium chloride with clean tongs.

Observe

1. Which element—chlorine or strontium—was responsible for the color observed when strontium chloride was placed in the flame? How do you know?
2. In your Science Journal devise a plan to determine whether copper or sulfate was responsible for the color in step 2.

Before You Read

FOLDABLES Reading & Study Skills

Making a Classify Study Fold Make the following Foldable to help you classify and organize objects into groups based on their common features.

Metals | Nonmetals

1. Place a sheet of paper in front of you so the long side is at the top. Fold the paper in half from the left side to the right side. Fold top to bottom and crease. Then unfold.
2. Through the top thickness of paper, cut along the middle fold line to form two tabs as shown.
3. Label the tabs *Metals* and *Nonmetals* as shown.
4. Before you read the chapter, list all the metal and nonmetal elements you know on the front of the tabs. As you read the chapter, check your list and make changes as needed.

Metals

As You Read

***What* You'll Learn**

- **Describe** the properties of a typical metal.
- **Identify** the alkali metals and alkaline earth metals.
- **Differentiate** among three groups of transition elements.

Vocabulary

metal
malleable
ductile
metallic bonding
radioactive element
transition element

***Why* It's Important**

Metals are a part of your everyday life—from electric cords to the cars you ride in.

Properties of Metals

The first metal used about 6,000 years ago was gold. The use of copper and silver followed a few thousand years later. Then came tin and iron. Aluminum wasn't refined until the 1800s because it must go through a much more complicated refining process that earlier civilizations had not yet developed.

In the periodic table, metals are elements found to the left of the stair-step line. In the table on the inside back cover of your book, the metal element blocks are colored blue. **Metals** usually have common properties—they are good conductors of heat and electricity, and all but one are solid at room temperature. Mercury is the only metal that is not a solid at room temperature. Metals also reflect light. This is a property called luster. Metals are **malleable** (MAH lee uh bul), which means they can be hammered or rolled into sheets, as shown in **Figure 1A.** Metals are also **ductile,** which means they can be drawn into wires like the ones shown in **Figure 1B.** These properties make metals suitable for use in objects ranging from eyeglass frames to computers to building structures.

Figure 1
The various properties of metals make them useful.

A Metals, like the one shown, can be hammered into thin sheets. *What is one use for a sheet of metal?*

B Metals can be drawn into wires, like the wire that is being used here. *What is this property of metals called?*

Ionic Bonding in Metals The atoms of metals generally have one to three electrons in their outer energy levels. In chemical reactions, metals tend to give up electrons easily because they are close to having an empty outer energy level. When metals combine with nonmetals, the atoms of the metals tend to lose electrons to the atoms of nonmetals, forming ionic bonds, as shown in **Figure 2.** Both metals and nonmetals become more chemically stable when they form ions. They take on the electron structure of the nearest noble gas.

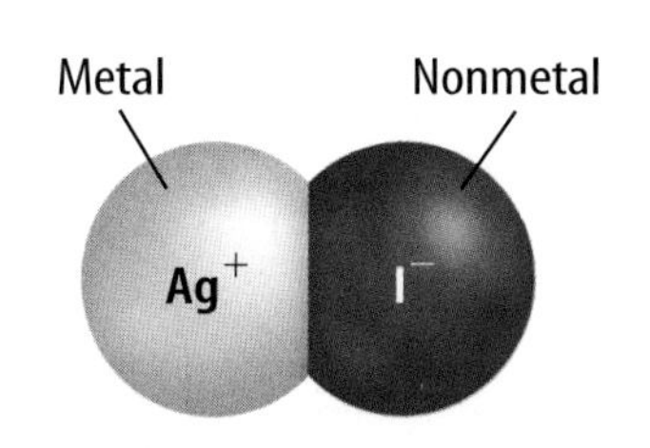

Figure 2
Metals can form ionic bonds with nonmetals.

Metallic Bonding Another type of bonding, neither ionic nor covalent, occurs among the atoms in a metal. In **metallic bonding,** positively charged metallic ions are surrounded by a cloud of electrons. Outer-level electrons are not held tightly to the nucleus of an atom. Rather, the electrons move freely among many positively charged ions. As shown in **Figure 3,** the electrons form a cloud around the ions of the metal.

The idea of metallic bonding explains many of the properties of metals. For example, when a metal is hammered into a sheet or drawn into a wire, it does not break because the ions are in layers that slide past one another without losing their attraction to the electron cloud. Metals are also good conductors of electricity because the outer-level electrons are weakly held.

Reading Check *Why do metals conduct electricity?*

Look at the periodic table inside the back cover of your book. How many of the elements in the table are classified as metals? All of the blue-shaded boxes represent metals. Except for hydrogen, all the elements in Groups 1 through 12 are metals, as well as the elements under the stair-step line in Groups 13 through 15. You will learn more about metals in some of these groups throughout this chapter.

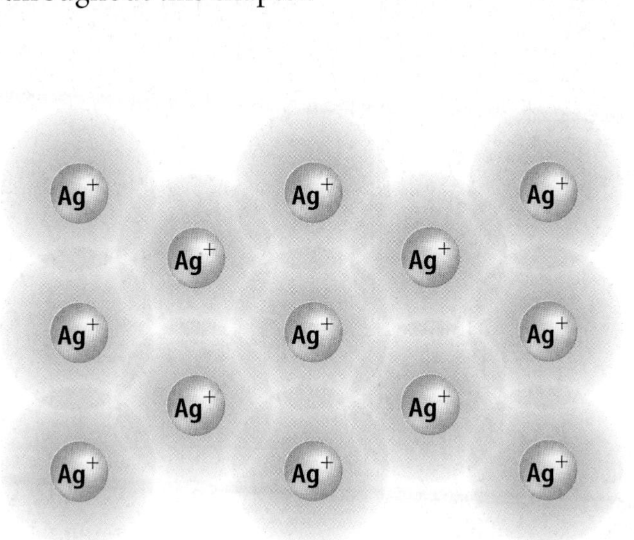

Figure 3
In metallic bonding, the electrons represented by the cloud are not attached to any one silver ion. This allows them to move and conduct electricity.

The Alkali Metals

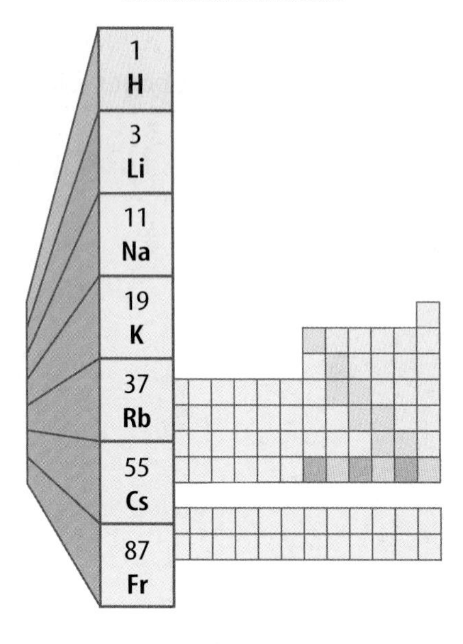

The elements in Group 1 of the periodic table are the alkali (AL kuh li) metals. Like other metals, Group 1 metals are shiny, malleable, and ductile. They are also good conductors of heat and electricity. However, they are softer than most other metals. The alkali metals are the most reactive of all the metals. They react rapidly—sometimes violently—with oxygen and water, as shown in **Figure 4.** Because they combine so readily with other elements, alkali metals don't occur in nature in their elemental form and are stored in substances that are unreactive, such as an oil.

Each atom of an alkali metal has one electron in its outer energy level. This electron is given up when an alkali metal combines with another atom. As a result, the alkali metal becomes a positively charged ion in a compound such as sodium chloride, NaCl, or potassium bromide, KBr.

Alkali metals and their compounds have many uses. You and other living things need potassium and sodium compounds to stay healthy. Doctors use lithium compounds to treat bipolar depression. The lithium keeps chemical levels that are important to mental health within a narrow range. The operation of some photocells depends upon rubidium or cesium compounds. Francium, the last element in Group 1, is extremely rare and radioactive. A **radioactive element** is one in which the nucleus breaks down and gives off particles and energy. Francium can be found in uranium minerals, but only 25 g to 30 g of francium are in all of Earth's crust at one time.

Figure 4
Alkali metals are very reactive.

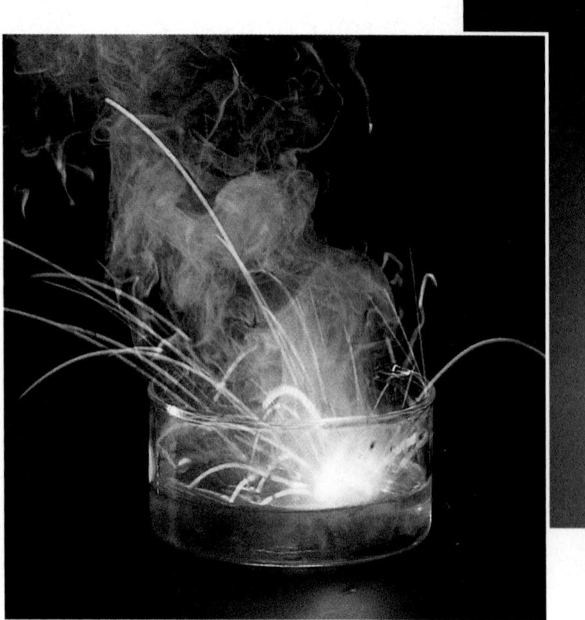

A Potassium reacts strongly in water.

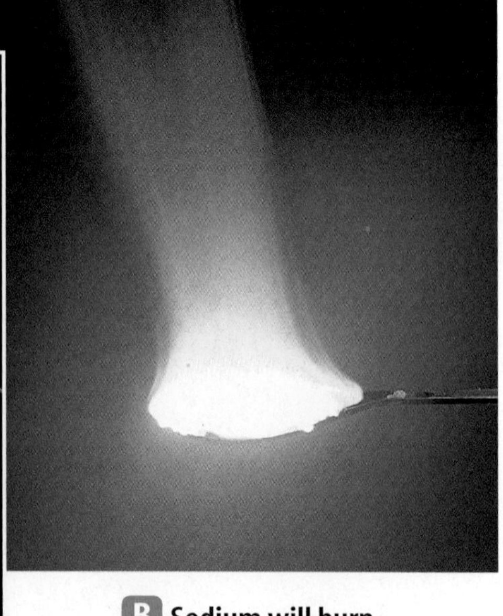

B Sodium will burn in air if it is heated.

The Alkaline Earth Metals

The alkaline earth metals make up Group 2 of the periodic table. Like most metals, these metals are shiny, malleable, and ductile. They are also similar to alkali metals in that they combine so readily with other elements that they are not found as free elements in nature. Each atom of an alkaline earth metal has two electrons in its outer energy level. These electrons are given up when an alkaline earth metal combines with a nonmetal. As a result, the alkaline earth metal becomes a positively charged ion in a compound such as calcium fluoride, CaF_2.

Fireworks and Other Uses Magnesium metal is one of the metals used to produce the brilliant white color in fireworks like the ones in **Figure 5.** Compounds of strontium produce the bright red flashes. Magnesium's lightness and strength account for its use in cars, planes, and spacecraft. Magnesium also is used in compounds to make such things as household ladders and baseball and softball bats. Most life on Earth depends upon chlorophyll, a magnesium compound that enables plants to make food. Marble statues and some countertops are made of the calcium compound calcium carbonate.

The Alkaline Earth Metals and Your Body Calcium is seldom used as a free metal, but its compounds are needed for life. You may take a vitamin with calcium. Calcium phosphate in your bones helps make them strong.

The Alkaline Earth Metals

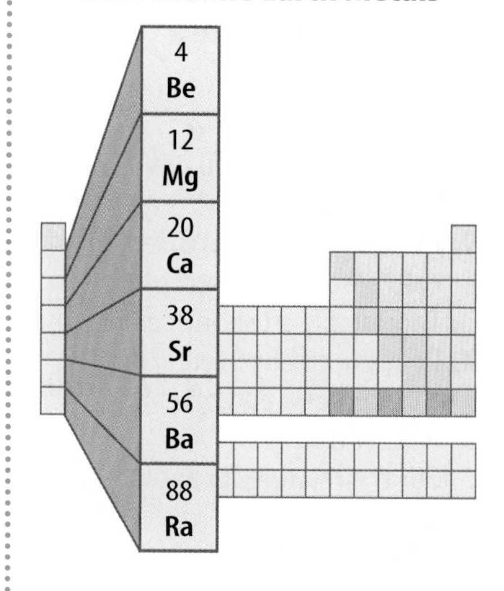

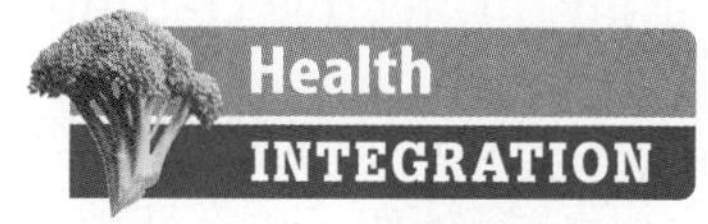

The barium compound $BaSO_4$ is used to diagnose some digestive disorders because it absorbs X-ray radiation well. First, the patient swallows a barium compound. Next, an X ray is taken while the barium compound is going through the digestive tract. A doctor can then see where the barium is in the body. In this way, doctors can diagnose internal abnormalities in the body.

Radium, the last element in Group 2, is radioactive and is found associated with uranium. It was once used to treat cancers. Today, other radioactive elements that are more readily available are replacing radium in cancer therapy.

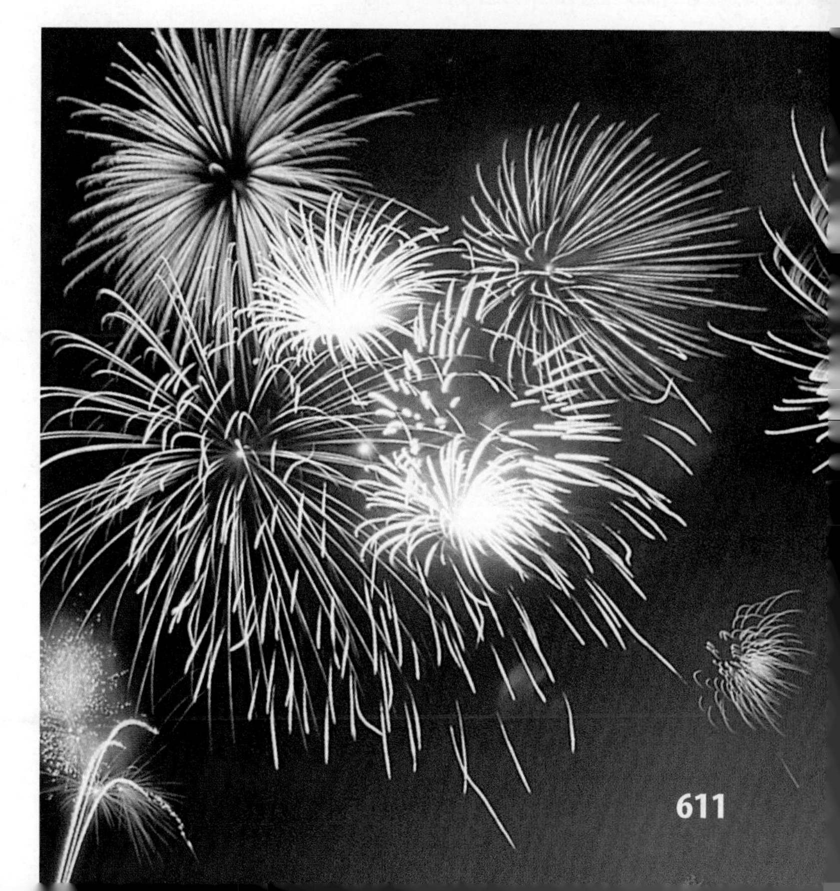

Figure 5
Alkaline earth metals make spectacular fireworks.

Discovering What's in Cereal

Procedure

1. **Tape** a small, strong **magnet** to a **pencil** at the eraser end.
2. Place some **dry, fortified, cold cereal** in a **plastic bag.**
3. Thoroughly crush the cereal.
4. Pour the crushed cereal into a **deep bowl** and cover it with **water.**
5. Stir the mixture for about 10 min with your pencil/magnet. Stir slowly for the last minute.
6. Remove the magnet and examine it carefully. Record your observations.

Analysis

1. What common element is attracted to your magnet?
2. Why is this element added to the cereal?

Transition Elements

A titanium bike frame and a glowing tungsten lightbulb filament are examples of objects made from transition elements. **Transition elements** are those elements in Groups 3 through 12 in the periodic table. They are called transition elements because they are considered to be elements in transition between Groups 1 and 2 and Groups 13 through 18. Look at the periodic table inside the back cover of your book. Which elements do you think of as being typical metals? Transition elements are the most familiar because they often occur in nature as uncombined elements, unlike Group 1 and Group 2 metals which are less stable.

Transition elements often form colored compounds. The gems in **Figure 6** show brightly colored compounds containing chromium. Cadmium yellow and cobalt blue paints are made from compounds of transition elements. However, cadmium and cobalt paints are so toxic that their use is limited.

Iron, Cobalt, and Nickel The first elements in Groups 8, 9, and 10—iron, cobalt, and nickel—form a unique cluster of transition elements. These three sometimes are called the iron triad. All three elements are used in the process to create steel and other metal mixtures.

Iron—the main component of steel—is the most widely used of all metals. It is the second most abundant metallic element in Earth's crust after aluminum. Other metals are added to steel to give it various characteristics. Some steels contain cobalt or nickel. Nickel is added to some metals to give them strength. Also, nickel is used to give a shiny, protective coating to other metals.

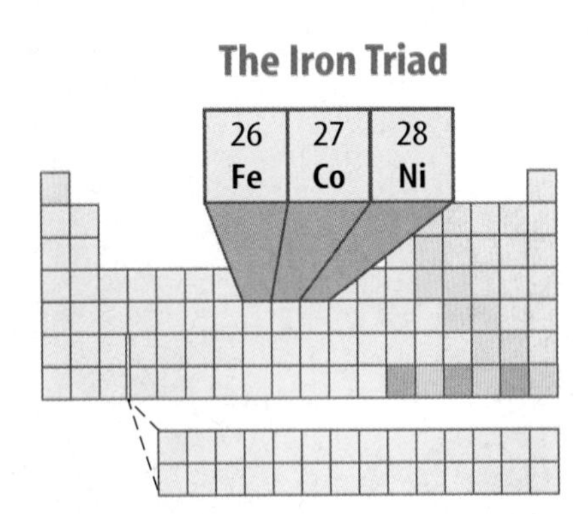

Figure 6
The colors of the ruby and emerald are due to the transition element chromium.

Figure 7
The coinage metals have many uses.

A **Because gold and silver are so expensive, copper is more common in coins.**

B **Silver is used in compounds to make photographic materials.**

C **Gold frequently is used in jewelry.**

Copper, Silver, and Gold The main metals in the objects in **Figure 7** are copper, silver, and gold—the three elements in Group 11. Because they are so stable and malleable and can be found as free elements in nature, these metals were once used widely to make coins. For this reason, they are known as the coinage metals. However, because they are so expensive, silver and gold rarely are used in coins anymore. The United States stopped using gold in the production of its coins in 1933 and silver in 1964. Most coins now are made of nickel and copper.

Copper often is used in electrical wiring because of its superior ability to conduct electricity and its relatively low cost. Can you imagine a world without photographs and movies? Silver iodide and silver bromide break down when exposed to light, producing an image on paper. Consequently, these compounds are used to make photographic film and paper. Silver and gold are used in jewelry because of their attractive color, relative softness, resistance to corrosion, and rarity.

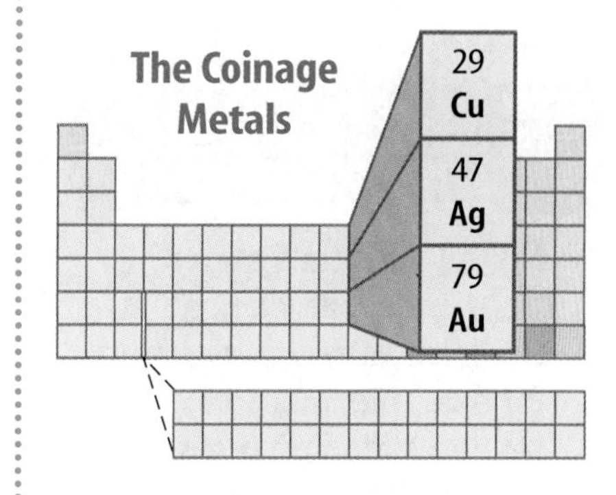

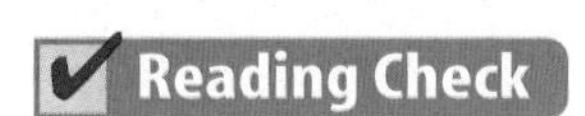

Why does gold's relative softness make it a good choice for jewelry?

Zinc, Cadmium, and Mercury Zinc, cadmium, and mercury are found in Group 12 of the periodic table. Zinc combines with oxygen in the air to form a thin, protective coating of zinc oxide on its surface. Zinc and cadmium often are used to coat, or plate, other metals such as iron because of this protective quality. Cadmium is used also in rechargeable batteries.

Mercury is a silvery, liquid metal—the only metal that is a liquid at room temperature. It is used in thermometers, thermostats, switches, and batteries. Mercury is poisonous and can accumulate in the body. People have died of mercury poisoning after eating fish that lived in mercury-contaminated water.

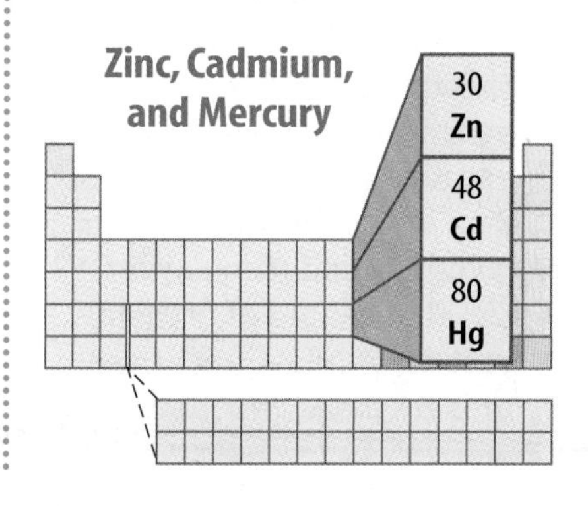

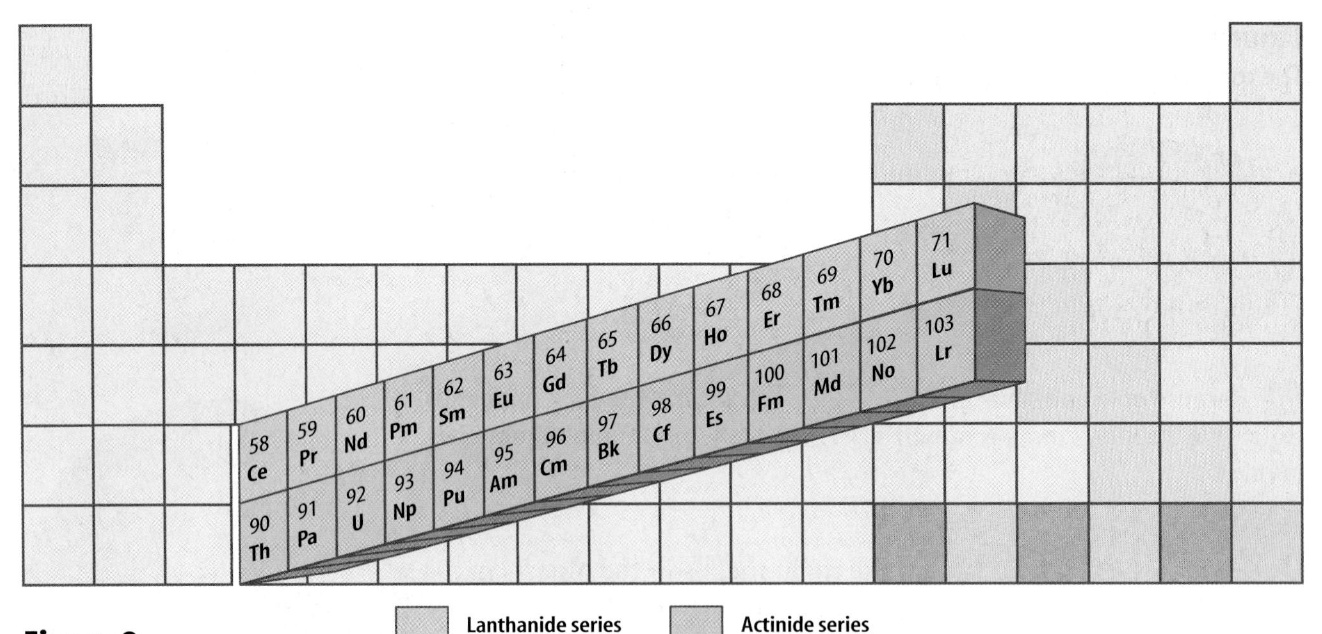

Figure 8
To save space, the periodic table usually isn't shown with the inner transition elements positioned where they should be.
Where are the lanthanide and actinide series?

The Inner Transition Metals

The two rows of elements that seem to be disconnected from the rest on the periodic table are called the inner transition elements. They are called this because like the transition elements, they fit in the periodic table between Groups 3 and 4 in periods 6 and 7, as shown in **Figure 8.** To save room, they are listed below the table.

The Lanthanides The first row includes a series of elements with atomic numbers of 58 to 71. These elements are called the lanthanide series because they follow the element lanthanum.

Lanthanum, cerium, praseodymium, and samarium are used with carbon to make a compound that is used extensively by the motion picture industry. Europium, gadolinium, and terbium are used to produce the colors you see on your TV screen.

The Actinides The second row of inner transition metals includes elements with atomic numbers ranging from 90 to 103. These elements are called the actinide series because they follow the element actinium. All of the actinides are radioactive and unstable. Their unstable nature makes researching them difficult. Thorium and uranium are the actinides found in the Earth's crust in usable quantities. Thorium is used in making the glass for high-quality camera lenses because it bends light without much distortion. Uranium is best known for its use in nuclear reactors and in weapons applications, but one of its compounds has been used as photographic toner, as well. More will be said about these elements at the end of Section 3.

Earth Science INTEGRATION

Not all materials mined from Earth can be used in the form they come in. Iron, for example, must be refined from iron ore in large blast furnaces. In these furnaces, charcoal is burned to create carbon monoxide gas that combines with the oxygen in the iron ore and carries it away. Other ways to extract materials from their ores include mixing them in solutions and using electricity. Find examples of metals that are purified from their ores by these two methods.

Metals in the Crust

Earth's hardened outer layer, called the crust, contains many compounds and a few uncombined metals such as gold and copper. Metals must be mined and separated from their ores, as shown in **Figure 9.**

Some metals are more abundant in one place than in another. Most of the world's platinum is found in South Africa. Chromium is important because it is used to harden steel, to manufacture stainless steel, and to form other alloys. The United States imports most of its chromium from South Africa, the Philippines, and Turkey.

Figure 9
Copper is mined in the United States at the Bingham Canyon Copper Mine in Utah.

Ores: Minerals and Mixtures Metals in Earth's crust that combined with other elements are found as ores. Most ores consist of a metal compound, or mineral, within a mixture of clay or rock. After an ore is mined from Earth's crust, the rock is separated from the mineral. Then the mineral often is converted to another physical form. This step usually involves heat and is called roasting. Finally, the metal is refined into a pure form. Later it can be alloyed with other metals.

Removing the waste rock can be expensive. If the cost of removing the waste rock becomes greater than the value of the desired material, the mineral no longer is classified as an ore.

Section  Assessment

1. How would you test a piece of palladium to see whether it is a metal?
2. How does the arrangement of the iron triad differ from the arrangements of coinage metals? Of the zinc group?
3. What characteristics are shared by alkali metals and alkaline earth metals?
4. How do metallic bonds differ from ionic and covalent bonds?
5. **Think Critically** If X stands for a metal, how can you tell from the following formulas—XCl and XCl_2—which compound contains an alkali metal and which contains an alkaline earth metal?

Skill Builder Activities

6. **Interpreting Scientific Illustrations** Draw dot diagrams to show the similarity among the chlorides of three alkali metals: *lithium chloride, sodium chloride,* and *potassium chloride.* **For more help, refer to the Science Skill Handbook.**
7. **Using Percentages** Pennies used to be made of 95 percent copper and 5 percent zinc, and they each weighed 3.11 g. Today, pennies are made of copper-plated zinc, and each weighs 2.5 g. A new penny weighs what percent of an old penny? **For more help, refer to the Math Skill Handbook.**

Nonmetals

SECTION 2

As You Read

What You'll Learn

- **Recognize** hydrogen as a nonmetal.
- **Compare and contrast** properties of the halogens.
- **Describe** properties and uses of the noble gases.

Vocabulary

nonmetal
diatomic molecule

Why It's Important

Nonmetals are not only all around you, they are an essential part of your body.

Properties of Nonmetals

Most of your body's mass is made of oxygen, carbon, hydrogen, and nitrogen, as shown in **Figure 10.** Calcium, a metal, and other elements make up the remaining four percent of your body's mass. Phosphorus, sulfur, and chlorine are among these other elements found in your body. These elements are classified as nonmetals. **Nonmetals** are elements that usually are gases or brittle solids at room temperature. Because solid nonmetals are brittle or powdery, they are not malleable or ductile. Most nonmetals do not conduct heat or electricity well, and generally they are not shiny.

In the periodic table, all nonmetals except hydrogen are found at the right of the stair-step line. On the table in the inside back cover of your book, the nonmetal element blocks are colored yellow. The noble gases, Group 18, make up the only group of elements that are all nonmetals. Group 17 elements, except for astatine, are also nonmetals. Other nonmetals, found in Groups 13 through 16, will be discussed later.

Figure 10
As a percentage of mass, humans are made up of mostly nonmetals.

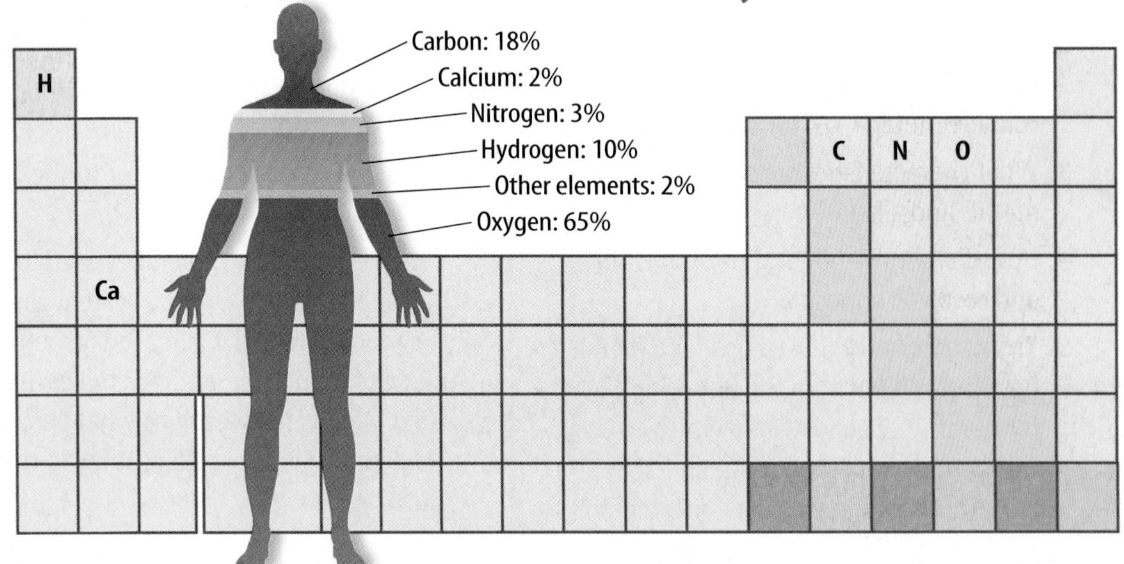

A Lead and sulfur bond ionically to form lead sulfide, PbS, also known as galena.

B Carbon and oxygen can bond covalently to form carbon dioxide, CO_2.

Figure 11
Nonmetals form ionic bonds with metals and covalent bonds with other nonmetals.

Bonding in Nonmetals The electrons in most nonmetals are strongly attracted to the nucleus of the atom. So, as a group, nonmetals are poor conductors of heat and electricity.

Most nonmetals can form ionic and covalent compounds. Examples of these two kinds of compounds are shown in **Figure 11.**

When nonmetals gain electrons from metals, the nonmetals become negative ions in ionic compounds. An example of such an ionic compound is potassium iodide, KI, which often is added to table salt. KI is formed from the nonmetal iodine and the metal potassium. On the other hand, when bonded with other nonmetals, atoms of nonmetals usually share electrons to form covalent compounds. An example is ammonia, NH_3, the strong, unpleasant-smelling compound you notice when you open a bottle of some household cleaners.

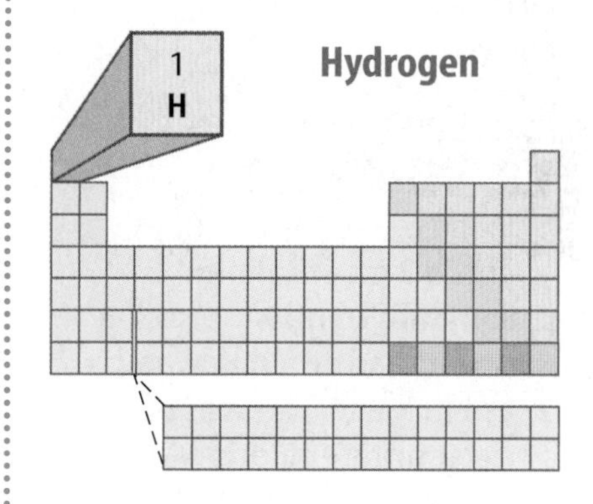

Hydrogen

If you could count all the atoms in the universe, you would find that about 90 percent of them are hydrogen. Most hydrogen on Earth is found in the compound water. The word *hydrogen* is derived from the Greek term for "water forming." When water is broken down into its elements, hydrogen becomes a gas made up of diatomic molecules. A **diatomic molecule** consists of two atoms of the same element in a covalent bond.

Hydrogen is highly reactive. A hydrogen atom has a single electron, which the atom shares when it combines with other nonmetals. For example, hydrogen burns in oxygen to form water, H_2O, in which hydrogen shares electrons with oxygen.

Hydrogen can gain an electron when it combines with alkali and alkaline earth metals. The compounds formed are hydrides, such as sodium hydride, NaH.

What is a diatomic molecule?

The Halogens

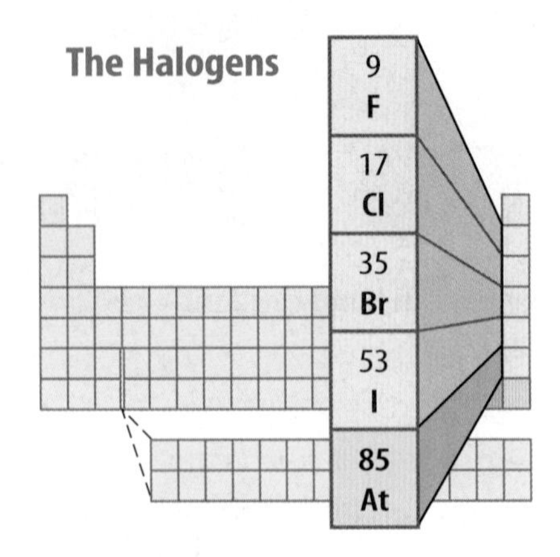

Halogen lights contain small amounts of bromine or iodine. These elements, as well as fluorine, chlorine, and astatine, are called halogens and are in Group 17. They are very reactive in their elemental form, and their compounds have many uses. As shown in **Figure 12,** fluorides are added to toothpastes and to city water systems to prevent tooth decay, and chlorine compounds are added to water to disinfect it.

Because an atom of a halogen has seven electrons in its outer energy level, only one electron is needed to complete this energy level. If a halogen gains an electron from a metal, an ionic compound, called a salt, is formed. In the gaseous state, the halogens form reactive diatomic covalent molecules and can be identified by their distinctive colors. Chlorine is greenish yellow, bromine is reddish orange, and iodine is violet.

Fluorine is the most chemically active of all elements. Hydrofluoric acid, a mixture of hydrogen fluoride and water, is used to etch glass and to frost the inner surfaces of lightbulbs.

Figure 12
The halogens have many uses.

A **Chlorine compounds are used in pools to disinfect the water.**

B **Fluoride compounds are used in toothpaste to prevent tooth decay.**

Mini LAB

Identifying Chlorine Compounds in Your Water

Procedure

1. In three labeled **test tubes,** obtain 2 mL of **chlorine standard solution, distilled water,** and **drinking water.**
2. Carefully add five drops of **silver nitrate solution** to each and stir. **WARNING:** *Avoid contact with the silver nitrate solution.*

Analysis

1. Which solution will definitely show a presence of chlorine? How did this result compare to the result with distilled water?
2. Which result most resembled your drinking water?

Figure 13
This ocean-salt recovery site uses evaporation to separate the halogen compounds from the water so the salts can be refined further.

Uses of Halogens The odor you sometimes smell near a swimming pool is chlorine. Chlorine compounds are used to disinfect water. Chlorine, the most abundant halogen, is obtained from seawater at ocean-salt recovery sites like the one in **Figure 13.** Household and industrial bleaches used to whiten flour, clothing, and paper also contain chlorine compounds.

Bromine, the only nonmetal that is a liquid at room temperature, also is extracted from compounds in seawater. Other bromine compounds are used as dyes in cosmetics.

Iodine, a shiny purple-gray solid at room temperature, is another halogen obtained from seawater. When heated, iodine changes directly to a purple vapor. The process of a solid changing directly to a vapor without forming a liquid is called sublimation, as shown in **Figure 14.** Iodine is essential in your diet for the production of the hormone thyroxin and to prevent goiter, an enlarging of the thyroid gland in the neck.

Reading Check *What is sublimation?*

Astatine is the last member of Group 17. It is radioactive and rare, but has many properties similar to those of the other halogens. There are no known uses due to its rarity.

Environmental Science INTEGRATION

Compounds called chlorofluorocarbons are used in refrigeration systems. If released, these compounds destroy ozone in the atmosphere. The ozone protects you from some of the harmful rays from the Sun. Find the advantages and disadvantages of these compounds. Write your answer in your Science Journal.

Figure 14
Frozen carbon dioxide, or dry ice, is used to make inexpensive, visible gas for theatrical productions. The carbon dioxide is brought out as a solid, then it sublimes as shown here.

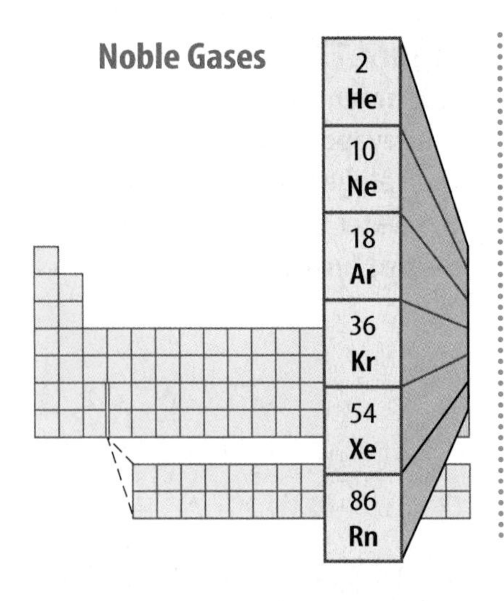

The Noble Gases

The noble gases exist as isolated atoms. They are stable because their outermost energy levels are full. No naturally occurring noble gas compounds are known, but several compounds of xenon and krypton with fluorine have been created in a laboratory.

The stability of noble gases is what makes them useful. In addition, the light weight of helium makes it useful in lighter-than-air blimps and balloons. Neon and argon are used in "neon lights" for advertising. Argon and krypton are used in electric lightbulbs to produce light in lasers, as seen in **Figure 15.**

Figure 15
Noble gases are used to produce spectacular laser light shows.

Section  Assessment

1. What are two ways in which hydrogen combines with other elements?
2. Rank the following nonmetals from lowest number of electrons in the outer level to highest: Cl^-, H^+, *He,* and *H.* What property of noble gases makes them useful?
3. How are solid nonmetals different from solid metals?
4. How can you tell that a gas is a halogen?
5. **Think Critically** What is the process of a solid changing directly into a vapor? Which element undergoes this process?

Skill Builder Activities

6. **Interpreting Data** Within the following compounds, identify the nonmetal and list its oxidation number: *MgO, NaH,* $AlBr_3$, and *FeS.* **For more help, refer to the Science Skill Handbook.**
7. **Communicating** In your Science Journal, write a paragraph explaining why nonmetals form ionic and covalent compounds, but not metallic bonds. Give some examples of ionic and covalent bonds. **For more help, refer to the Science Skill Handbook.**

Activity

What type is it?

Suppose you want an element for a certain use. You might be able to use a metal but not a nonmetal. In this activity, you will test several metals and nonmetals and compare their properties.

Observing Properties				
Element	Appearance	Malleable or Brittle	Electrical Conductivity	Shiny or Dull
Carbon				
Magnesium				
Aluminum				
Sulfur				
Tin				

What You'll Investigate

How can you use properties to distinguish metals from nonmetals?

Materials

samples of C, Mg, Al, S, and Sn
dishes for the samples
conductivity tester
spatula
small hammer

Goals

- **Observe** physical properties.
- **Test** the malleability of the materials.
- **Identify** electrical conductivity in the given materials.

Safety Precautions

Procedure

1. **Prepare** a table in your Science Journal like the one shown.
2. **Observe** and record the appearance of each element sample. Include its physical state, color, and whether it is shiny or dull.
3. Remove a small sample of one of the elements. Place it on a hard surface chosen by your teacher. Gently tap the sample with a hammer. The sample is malleable if it flattens when tapped and brittle if it shatters. Record your results.
4. Repeat step 3 for each sample.
5. Test the conductivity of each element by touching the electrodes of the conductivity tester to a sample. If the bulb lights, the element conducts electricity. Record your results.

Conclude and Apply

1. Locate each element you used on the periodic table. Compare your results with what you would expect from an element in that location.
2. Locate palladium, Pd, on the periodic table. Use the results you obtained during the activity to predict some of the properties of palladium.
3. **Infer** why some elements might show properties of metals as well as properties of nonmetals.

Compare your results with those of other students. **For more help, refer to the Science Skill Handbook.**

SECTION

Mixed Groups

As You Read

What You'll Learn

- **Distinguish** among metals, nonmetals, and metalloids.
- **Describe** the nature of allotropes.
- **Recognize** the significance of differences in crystal structure in carbon.
- **Understand** the importance of synthetic elements.

Vocabulary

metalloid
allotrope
semiconductor
transuranium element

Why It's Important

The elements in mixed groups affect your life every day, because they are in everything from the computer you use to the air you breathe.

Properties of Metalloids

Can an element be a metal and a nonmetal? In a sense, some elements called metalloids are. Metalloids share unusual characteristics. **Metalloids** can form ionic and covalent bonds with other elements and can have metallic and nonmetallic properties. Some metalloids can conduct electricity better than most nonmetals, but not as well as some metals, giving them the name semiconductor. With the exception of aluminum, the metalloids are the elements in the periodic table that are located along the stair-step line. The mixed groups—13, 14, 15, 16, and 17—contain metals, nonmetals, and metalloids. Group 17, the halogens, were discussed in Section 2.

The Boron Group

Boron, a metalloid, is the first element in Group 13. If you look around your home, you might find two compounds of boron. One of these is borax, which is used in some laundry products to soften water. The other is boric acid, a mild antiseptic. Boron also is used as a grinding material and as boranes, which are compounds used for jet and rocket fuel.

Aluminum, a metal in Group 13, is the most abundant metal in Earth's crust. It is used in soft-drink cans, foil wrap, cooking pans, and as siding. Aluminum is strong and light and is used in the construction of airplanes such as the one in **Figure 16.**

The Boron Group

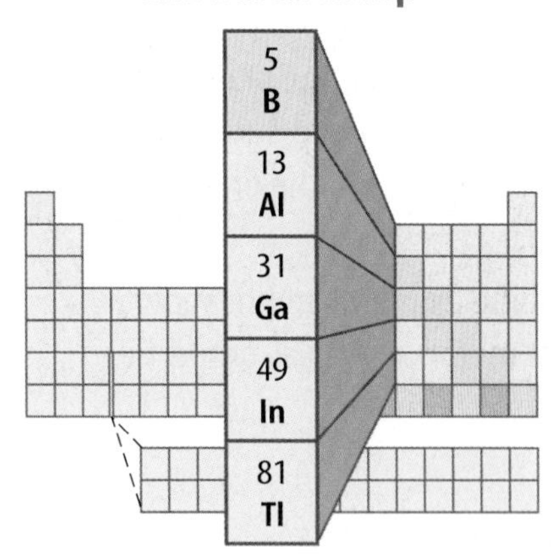

Figure 16
Aluminum is used frequently in the construction of airplanes because it is light and strong.

Figure 17
Elements in Group 14 have many uses.

A Silicon is used to make the chips that allow this computer to run.

B These tin cans are made of steel with a tin coating.

The Carbon Group

Each element in Group 14, the carbon family, has four electrons in its outer energy level, but this is where much of the similarity ends. Carbon is a nonmetal, silicon and germanium are metalloids, and tin and lead are metals. Carbon occurs as an element in coal and as a compound in oil, natural gas, and foods. Carbon in these materials can combine with oxygen to produce carbon dioxide, CO_2. In the presence of sunlight, plants utilize CO_2 to make food. Carbon compounds, many of which are essential to life, can be found in you and all around you. All organic compounds contain carbon, but not all carbon compounds are organic.

Silicon is second only to oxygen in abundance in Earth's crust. Most silicon is found in sand, SiO_2, and almost all rocks and soil. The crystal structure of silicon dioxide is similar to the structure of diamond. Silicon occurs as two allotropes. **Allotropes,** which are different forms of the same element, have different molecular structures. One allotrope of silicon is a hard, gray substance, and the other is a brown powder.

Reading Check *What are allotropes?*

Silicon is the main component in **semiconductors**—elements that conduct an electric current under certain conditions. Many of the electronics that you use every day, like the computer in **Figure 17A,** need semiconductors to run. Germanium, the other metalloid in the carbon group, is used along with silicon in making semiconductors. Tin is used to coat other metals to prevent corrosion, like the tin cans in **Figure 17B.** Tin also is combined with other metals to produce bronze and pewter. Lead was used widely in paint at one time, but because it is toxic, lead no longer is used.

The Carbon Group

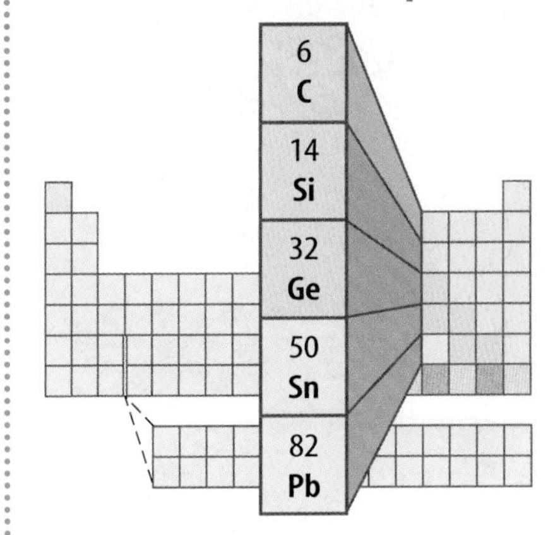

Allotropes of Carbon What do the diamond in a diamond ring and the graphite in your pencil have in common? They are both carbon. Diamond, graphite, and buckminsterfullerene, shown in **Figure 18,** are allotropes of an element.

A diamond is clear and extremely hard. In a diamond, each carbon atom is bonded to four other carbon atoms at the vertices, or corner points, of a tetrahedron. In turn, many tetrahedrons join together to form a giant molecule in which the atoms are held tightly in a strong crystalline structure. This structure accounts for the hardness of diamond.

Research Visit the Glencoe Science Web site at **science.glencoe.com** for more information about buckminsterfullerene. Communicate to your class what you learn.

Graphite is a black powder that consists of hexagonal layers of carbon atoms. In the hexagons, each carbon atom is bonded to three other carbon atoms. The fourth electron of each atom is bonded weakly to the layer next to it. This structure allows the layers to slide easily past one another, making graphite an excellent lubricant. In the mid 1980s, a new allotrope of carbon called buckminsterfullerene was discovered. This soccer-ball-shaped molecule, informally called a buckyball, was named after the architect-engineer R. Buckminster Fuller, who designed structures with similar shapes.

In 1991, scientists were able to use the buckyballs to synthesize extremely thin, graphitelike tubes. These tubes, called nanotubes, are about 1 billionth of a meter in diameter. That means you could stack tens of thousands of nanotubes just to get the thickness of one piece of paper. Nanotubes might be used someday to make computers that are smaller and faster and to make strong building materials.

Figure 18
Three allotropes of carbon are depicted here. *What geometric shapes make up each allotrope?*

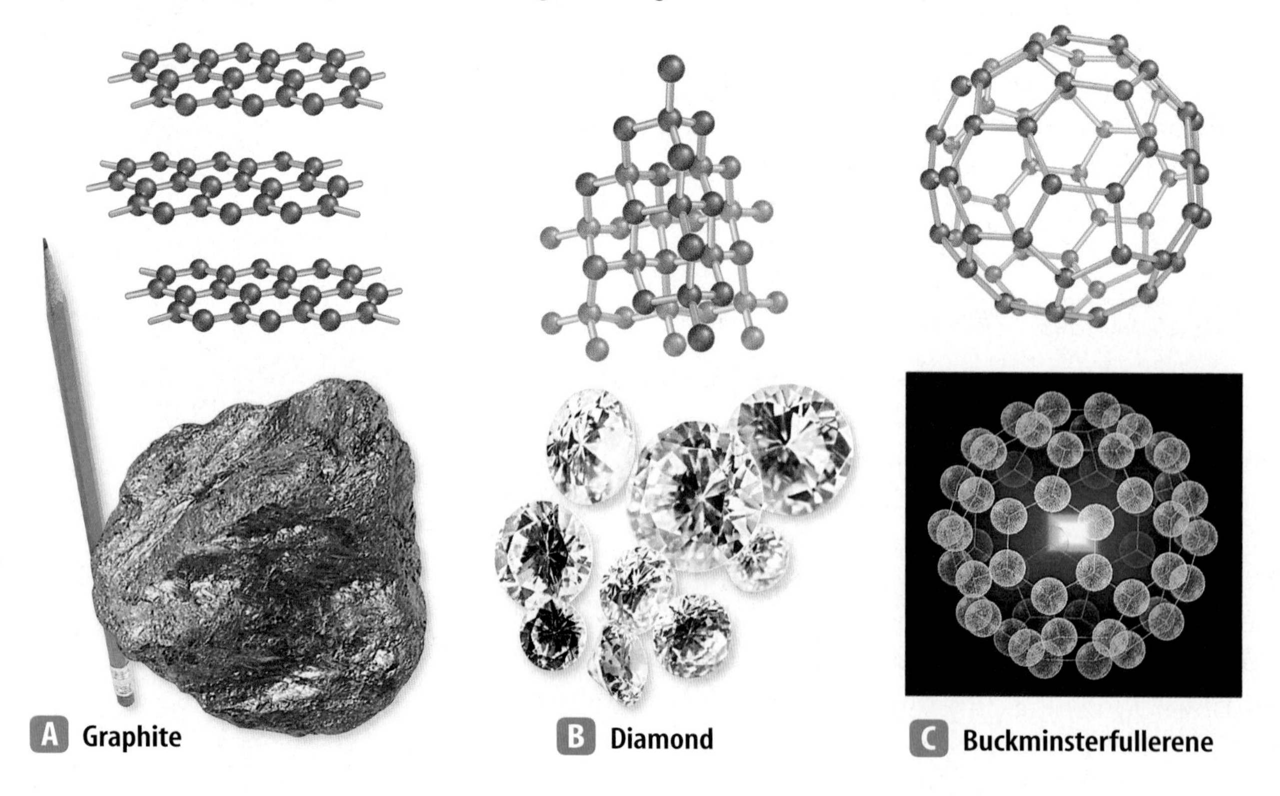

The Nitrogen Group

The nitrogen family makes up Group 15. Each element has five electrons in its outer energy level. These elements tend to share electrons and to form covalent compounds with other elements. Nitrogen often is used to make nitrates and ammonia, NH_3, both of which are used in fertilizers. Nitrogen is the fourth most abundant element in your body. Each breath you take is about 80 percent gaseous nitrogen in the form of diatomic molecules, N_2. Yet you and other animals and plants can't use nitrogen in its diatomic form. The nitrogen must be combined into compounds, such as nitrates—compounds that contain the nitrate ion, NO_3^-.

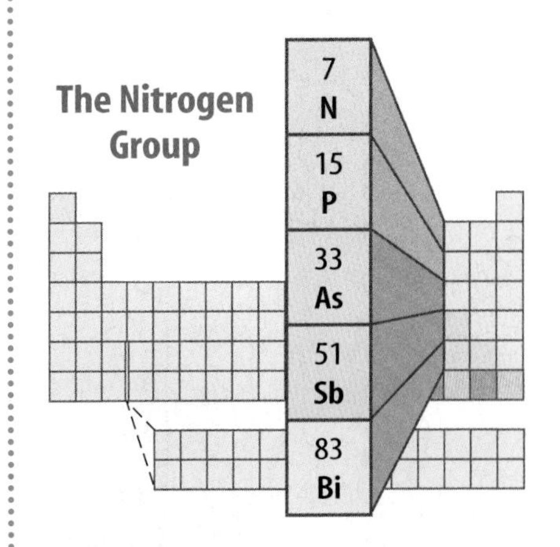

Math Skills Activity

Using Circle Graphs to Illustrate Data

Example Problem

Oxygen, the predominant element in Earth's crust, makes up approximately 46.6 percent of the crust. If you were to show this information on a circle graph, how many degrees would be used to represent oxygen?

Solution

1. *This is what you know:* oxygen = 46.6%; circle contains 360°
2. *This is what you want to find:* the part of 46.6 to 360°
3. *This is the equation you need to use:* $\frac{46.6}{100} = \frac{x}{360°}$
4. *Solve the equation for x:* $x = \frac{46.6 \times 360°}{100} = 167.76°$ or $168°$

Check your answer by dividing your answer by 360, then multiplying by 100.

Practice Problems

1. The percentages of remaining elements in Earth's crust are: silicon, 27.7; aluminum, 8.1; iron, 5.0; calcium, 3.6; sodium, 2.8; potassium, 2.6; magnesium, 2.1; and other elements, 1.5. Illustrate in a circle graph.
2. The approximate percentages of the elements in the body are: oxygen, 65; carbon, 18; hydrogen, 10; nitrogen, 3; calcium, 2; and other elements which account for 2. Illustrate these percentages in a circle graph.

For more help, refer to the Math Skill Handbook.

Uses of the Nitrogen Group Phosphorus is a nonmetal that has three allotropes. Phosphorous compounds can be used for many things from water softeners to fertilizers, match heads, and even in fine china. Antimony is a metalloid, and bismuth is a metal. Both elements are used with other metals to lower their melting points. It is because of this property that the metal in automatic fire-sprinkler heads contains bismuth.

Reading Check *Why is bismuth used in fire-sprinkler heads?*

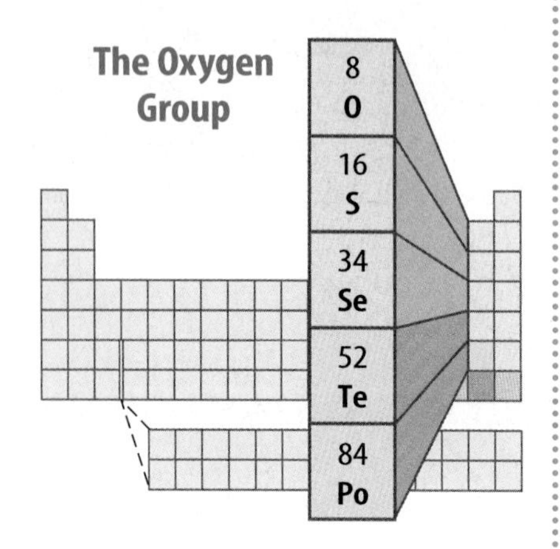

The Oxygen Group

Group 16 on the periodic table is the oxygen group. You can live for only a short time without oxygen, which makes up about 21 percent of air. Oxygen, a nonmetal, exists in the air as diatomic molecules, O_2. During electrical storms, some oxygen molecules, O_2, change into ozone molecules, O_3. Oxygen also has several uses in compound form, including the one shown in **Figure 19A.**

Nearly all living things on Earth need O_2 for respiration. Living things also depend on a layer of O_3 around Earth for protection from some of the Sun's radiation.

The second element in the oxygen group is sulfur. Sulfur is a nonmetal that exists in several allotropic forms. It exists as different-shaped crystals and as a noncrystalline solid. Sulfur combines with metals to form sulfides of such distinctive colors that they are used as pigments in paints.

The nonmetal selenium and two metalloids—tellurium and polonium—are the other Group 16 elements. Selenium is the most common of these three. This element is one of several that you need in trace amounts in your diet. Many multivitamins contain this nonmetal as an ingredient. But selenium is toxic if too much of it gets into your system. Selenium also is used in photocopiers like the one in **Figure 19B.**

Figure 19
Group 16 compounds have a variety of uses.

A **Solutions of hydrogen peroxide, H_2O_2, are used to clean minor wounds.**

B **Selenium is used in xerography to make photocopies.**

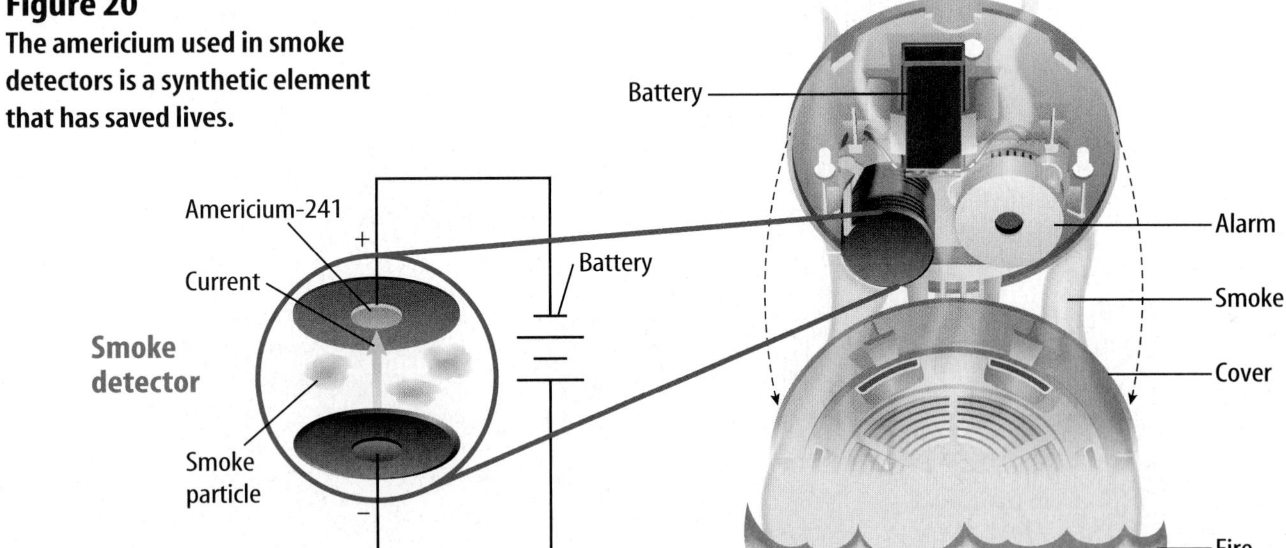

Figure 20
The americium used in smoke detectors is a synthetic element that has saved lives.

Synthetic Elements

If you made something that always fell apart, you might think you were not successful. However, nuclear scientists are learning to do just that. By smashing existing elements with particles accelerated in a heavy ion accelerator, they have been successful in creating elements not typically found on Earth. Except for technetium 43 and promethium 61, each synthetic element has more than 92 protons.

Bombarding uranium with protons can make neptunium, element 93. Half of the synthesized atoms of neptunium disintegrate in about two days. This may not sound useful, but when neptunium atoms disintegrate, they form plutonium. This highly toxic element has been produced in control rods of nuclear reactors and is used in bombs. Plutonium also can be changed to americium, element 95. This element is used in home smoke detectors such as the one in **Figure 20.** In smoke detectors, a small amount of americium emits harmless, charged particles. An electric plate in the smoke detector attracts some of these charged particles. When a lot of smoke is in the air, it interferes with the electric current, which immediately sets off the alarm in the smoke detector.

Transuranium Elements Elements having more than 92 protons, the atomic number of uranium, are called **transuranium elements.** These elements do not belong exclusively to the metal, nonmetal, or metalloid group. These are the elements toward the bottom of the periodic table. Some are in the actinide series, and some are on the bottom row of the main periodic table. All of the transuranium elements are synthetic and unstable, and many of them disintegrate quickly.

The Transuranium Elements

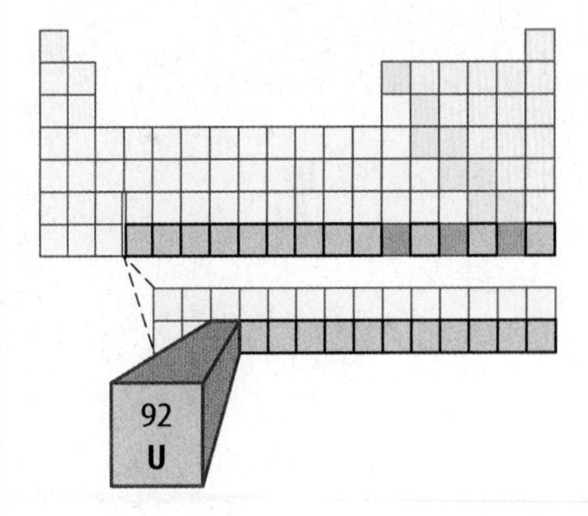

Figure 21

Some elements, such as gold, silver, tin, carbon, copper, and lead, have been known and used for thousands of years. Most others were discovered much more recently. Even at the time of the American Revolution in 1776, only 24 elements were known. The timeline below shows the dates of discovery of selected elements, ancient and modern.

Au - GOLD
Prized since the Stone Age

Ag - SILVER
Found in tombs dating to 4000 B.C.

A.D. 1774
Cl - CHLORINE
Pale green, toxic gas

10,000 B.C. — A.D. 1700

1817
Cd - CADMIUM
Used to color yellow and red paint

1825
Al - ALUMINUM
Most abundant element in Earth's crust

1868
He - HELIUM
Lighter-than-air gas used to fill balloons

1898
Ne - NEON
Glows when electricity flows through it

1898
Po - POLONIUM and Ra - RADIUM
Radioactive elements discovered by Marie and Pierre Curie

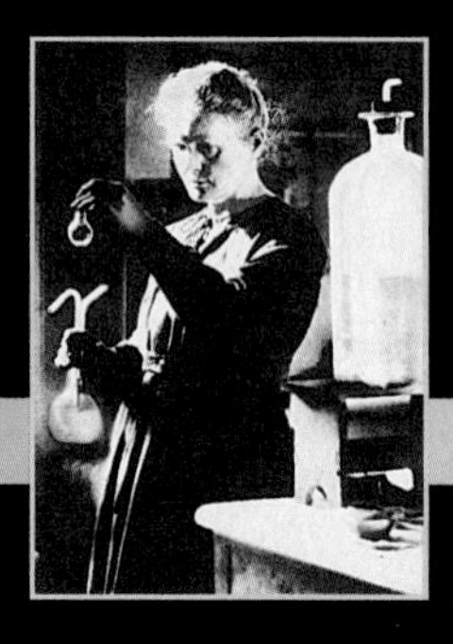

A.D. 1800

1900
Rn - RADON
Radioactive gas that may cause cancer

1952
Es - EINSTEINIUM
Radioactive gas named after Albert Einstein

1981–1996
Bh- BOHRIUM, Uun - UNUNNILIUM
Elements isolated by a heavy ion accelerator such as the UNILAC, below

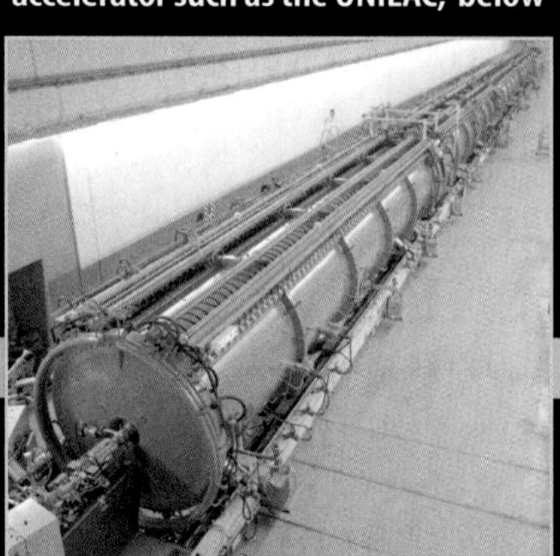

A.D. 1900 — A.D. 2000

Why make elements? **Figure 21** shows when some of the elements were discovered throughout history. The processes used to discover these elements have varied widely. The most recently discovered elements are synthetic. By studying how the synthesized elements form and disintegrate, you can gain an understanding of the forces holding the nucleus together. When these atoms disintegrate, they are said to be radioactive.

Radioactive elements can be useful. For example, technetium's radioactivity makes it ideal for many medical applications. At this time, many of the synthetic elements last only small fractions of seconds after they are constructed and can be made only in small amounts. However, the value of applications that might be discovered easily could offset their costs.

Seeking Stability Element 114, discovered in 1999, appears to be much more stable than most synthetic elements of its size. It lasted for 30 s before it broke apart. This may not seem like long, but it lasts 100,000 times longer than an atom of element 112. Perhaps this special combination of 114 protons and 175 neutrons allows the nucleus to hold together despite the enormous repulsion between the protons.

In the 1960s, scientists theorized that stable synthetic elements exist. Finding one might help scientists understand how the forces inside the atom work. Perhaps someday you'll read about some of the everyday uses this discovery has brought.

Data Update For an online update about synthetic elements, visit the Glencoe Science Web site at **science.glencoe.com** and select the appropriate chapter.

Section  Assessment

1. Why are Groups 14 and 15 better representatives of mixed groups than Groups 13 and Group 16?
2. Describe how the allotropes of silicon differ in appearance.
3. How is an element classified as a transuranium element?
4. What type of structure does a diamond have? Describe how you would build a model of this.
5. **Think Critically** Graphite and a diamond are both made of the element carbon. Why is graphite a lubricant and diamond the hardest gem known?

Skill Builder Activities

6. **Concept Mapping** Make a concept map for allotropes of carbon using the terms *graphite, diamond, buckminsterfullerene, sphere, hexagon,* and *tetrahedron.* **For more help, refer to the Science Skill Handbook.**
7. **Using a Computerized Card Catalog** Some people are concerned about the use of radioactive elements, such as americium, in smoke detectors and other products. Research smoke detectors and radioactive elements using a computerized card catalog. Summarize what you learn in a table. **For more help, refer to the Technology Skill Handbook.**

Activity Design Your Own Experiment

Slippery Carbon

Often, a lubricant is needed when two metals touch each other. For example, a sticky lock sometimes works better with the addition of a small amount of graphite. What gives this allotrope of carbon the slippery property of a lubricant?

Recognize the Problem

Why do certain arrangements of atoms in a material cause the material to feel slippery?

Form a Hypothesis

Based on your understanding of how carbon atoms bond, form a hypothesis about the relationship of graphite's molecular structure to its physical properties.

Possible Materials

thin spaghetti
small gumdrops
thin polystyrene sheets
flat cardboard
scissors

Goals

- **Make a model** that will demonstrate the molecular structure of graphite.
- **Compare and contrast** the strength of the different bonds in graphite.
- **Infer** the relationship between bonding and physical properties.

Safety Precautions

Use care when working with scissors and uncooked spaghetti.

Test Your Hypothesis

Plan

1. As a group, agree upon a logical hypothesis statement.
2. As a group, sequence and list the steps you need to take to test your hypothesis. Be specific, describing exactly what you will do at each step to make a model of the types of bonding present in graphite.
3. Remember from **Figure 18** that graphite consists of rings of six carbons bonded in a flat hexagon. These rings are bonded to each other. In addition, the flat rings in one layer are weakly attached to other flat layers.
4. List possible materials you plan to use.
5. Read over the experiment to make sure that all steps are in logical order.
6. Will your model be constructed with materials that show weak and strong attractions?

Do

1. Make sure your teacher approves your plan before you start.
2. Have you selected materials to use in your model that demonstrate weak and strong attractions? Carry out the experiment as planned.
3. Once your model has been constructed, list any observations that you make and include a sketch in your Science Journal.

Analyze Your Data

1. **Compare** your model with designs and results of other groups.
2. How does your model illustrate two types of attractions found in the graphite structure?
3. How does the bonding of graphite that you explored in the lab explain graphite's lubricating properties? Write your answer in your Science Journal.

Draw Conclusions

1. Did the results you obtained from your experiment support your hypothesis? Explain.
2. **Describe** why graphite makes a good lubricant.
3. What kinds of bonds do you think a diamond has?

Explain to a friend why graphite makes a good lubricant and how the two types of bonds make a difference. **For more help, refer to the Science Skill Handbook.**

The GAS that glows

Neon has made the world a more colorful place

"Nothing in the world gave a glow such as we had seen." With these words, two British chemists recorded their discovery of neon in 1898. Neon is a noble gas that emits a spectacular red-orange glow when an electric current is passed through it. It also makes up a tiny portion of the air we breathe.

But neon's presence remained undetected until a technology called spectroscopy allowed the chemists to view that "blaze of crimson light" in their lab—a light that soon lit up the world in fantastical ways.

A neon coffee cup lights up the Seattle night.

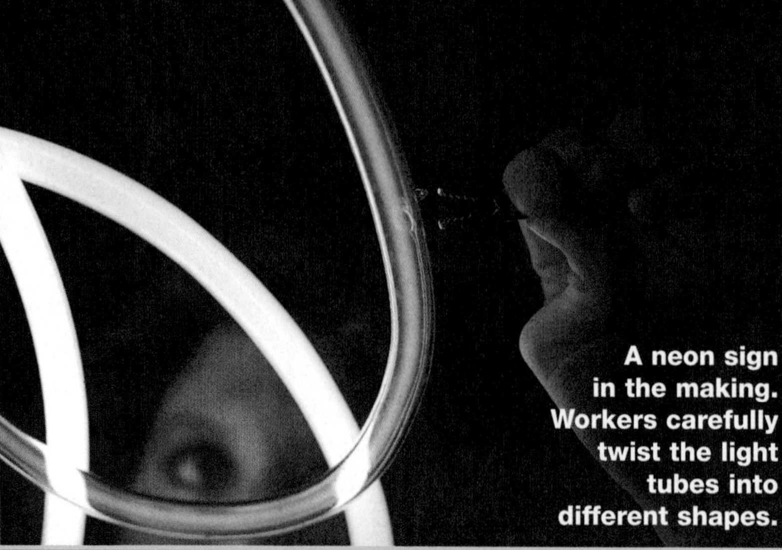

A neon sign in the making. Workers carefully twist the light tubes into different shapes.

Signs of Change

Pink flamingos, cowboys on bucking broncos, deep-sea fish afloat in the air—neon signs make any building or billboard come alive in a kaleidoscope of colors. Barely a decade after neon was discovered, a chemist developed the first neon sign. The chemist took the air out of a glass tube and replaced it with neon gas. When the gas was jolted with electricity, it glowed like a fiery sunset. The chemist sold the light to a barber, who hung it over his storefront. By the 1920s, neon lights were used to advertise everything from cars to diners.

When a touch of mercury is added to neon, it glows a tropical blue. The other colors seen in "neon lights" actually come from other noble gases. Krypton, for instance, glows yellow. Xenon shines like a bluish-white star.

Elemental Snobs

The noble gases include neon, helium, argon, krypton, xenon, and radon. Scientists called these gases "noble" because that word sometimes implies a snobbish attitude—the perfect description for gases that rarely react with other elements. Like the other noble gases, neon is stable.

For an element to be chemically stable, it needs eight electrons in its outer energy level. Neon has eight electrons in its outer shell—so it doesn't react with other elements because it already has a complete outer energy level. Although scientists have discovered that some noble gases do interact with other elements, thus far, neon remains aloof.

Other Uses for Neon

The vivid light emitted by neon can penetrate the densest fog, making it a natural choice for airplane beacons. Neon also is used to manufacture lasers and television tubes. Neon definitely helps light up our lives!

Neon lights up a Hong Kong, China night.

CONNECTIONS Design a Sign As a group, brainstorm a new product or business, then design a neon sign to advertise your idea. See if other groups can correctly guess what your sign represents.

SCIENCE Online
For more information, visit science.glencoe.com

Chapter 20 Study Guide

Reviewing Main Ideas

Section 1 Metals

1. A typical metal is a hard, shiny solid that, due to metallic bonding, is malleable, ductile, and a good conductor.
2. Groups 1 and 2 are the alkali and alkaline earth metals, which have some similar and some contrasting properties. *Why aren't Group 1 elements typically found in their elemental form?*

3. The iron triad, the coinage metals, and the elements in Group 12 are examples of transition elements.
4. The lanthanides and actinides have atomic numbers 58 through 71 and 90 through 103, respectively.

Section 2 Nonmetals

1. Nonmetals are typically brittle and dull. They are also poor conductors of electricity.
2. As a typical nonmetal, hydrogen is a gas that forms compounds by sharing electrons with other nonmetals and by forming ionic bonds with metals.
3. All the halogens, Group 17, have seven outer electrons and form covalent and ionic compounds, but each halogen has some properties that are unlike each of the others in the group.
4. The noble gases, Group 18, are elements whose properties and uses are related to their chemical stability.

Section 3 Mixed Groups

1. Groups 13 through 16 include metals, nonmetals, and metalloids. *Which objects below appear to be metallic and which appear to be nonmetallic?*

2. Allotropes are forms of the same element having different molecular structures.
3. The properties of three forms of carbon—graphite, diamond, and buckminsterfullerene—depend upon the differences in their crystal structures.
4. All synthetic elements are short-lived. With two exceptions, they have atomic numbers greater than 92 and are referred to as transuranium elements.

After You Read

FOLDABLES Reading & Study Skills

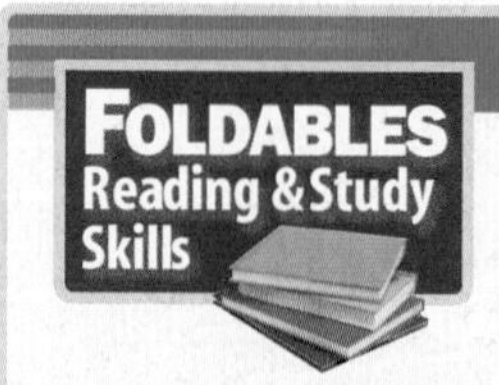

Under the tabs of the Foldable you created, write the common characteristics of metal elements and of nonmetal elements.

Visualizing Main Ideas

Complete the concept map using the following: Transition elements, hydrogen, Metals, Gas or brittle solid at room temperature, Inner transition metals, Metalloids, Noble gases.

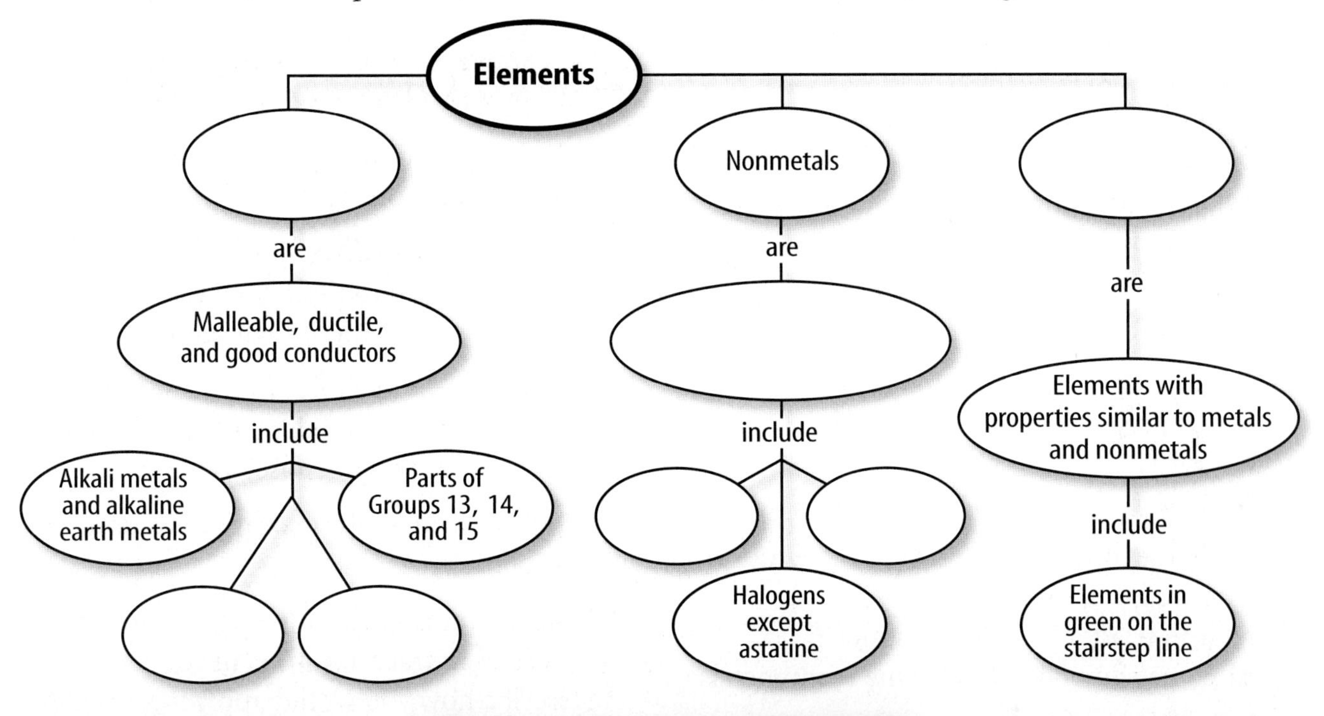

Vocabulary Review

Vocabulary Words

a. allotrope
b. diatomic molecule
c. ductile
d. malleable
e. metal
f. metallic bonding
g. metalloid
h. nonmetal
i. radioactive element
j. semiconductor
k. transition element
l. transuranium element

Study Tip

Outline the chapter to make sure that you're understanding the key ideas. Writing down the main points of the chapter will help you to remember the important details and understand the larger themes.

Using Vocabulary

Each of the following sentences is false. Make the sentence true by replacing the underlined word with a word from the list.

1. The <u>radioactive elements</u> are located to the left of the stair-step line on the periodic table.
2. Different structural forms of the same element are called <u>diatomic molecules</u>.
3. Positively charged ions are surrounded by freely moving electrons in a <u>nonmetal</u> element.
4. An <u>allotrope</u> is a molecule comprised of two atoms.
5. The <u>nonmetals</u> are in Groups 3 through 12 on the periodic table.

Chapter 20 Assessment

Checking Concepts

Choose the word or phrase that best answers the question.

1. When magnesium and fluorine react, what type of bond is formed?
A) metallic **C)** covalent
B) ionic **D)** diatomic

2. What type of bond is found in a piece of pure gold?
A) metallic **C)** covalent
B) ionic **D)** diatomic

3. Because electrons move freely in metals, which property describes metals?
A) brittle **C)** dull
B) hard **D)** conductors

4. Which set of elements makes up the most reactive group of all metals?
A) iron triad **C)** alkali metals
B) coinage metals **D)** alkaline earth metals

5. Which element is the most reactive of all nonmetals?
A) fluorine **C)** hydrogen
B) uranium **D)** oxygen

6. Which element is always found in nature combined with other elements?
A) copper **C)** magnesium
B) gold **D)** silver

7. Which elements are least reactive?
A) metals **C)** noble gases
B) halogens **D)** actinides

8. What element is formed when neptunium disintegrates?
A) ytterbium **C)** americium
B) promethium **D)** plutonium

9. Which element is an example of a radioactive element?
A) astatine **C)** chlorine
B) bromine **D)** fluorine

10. Which is the only group that is completely nonmetallic?
A) 1 **C)** 17
B) 2 **D)** 18

Thinking Critically

11. Why is mercury rarely used in thermometers that take body temperature?

12. The density of hydrogen is lower than air and can be used to fill balloons. Why is helium used instead of hydrogen?

13. Copper is a good choice for use in electrical wiring. What type of elements would not work well for this purpose? Why?

14. Why are various silver compounds used in photography?

15. Like selenium, chromium is poisonous but is needed in trace amounts in your diet. Describe how you would apply this information in order to use vitamin and mineral pills safely.

Developing Skills

16. Making and Using Tables Use the periodic table to classify each of the following as a lanthanide or actinide: californium, europium, cerium, nobelium, terbium, and uranium.

17. Comparing and Contrasting Aluminum is close to carbon on the periodic table. Explain why aluminum is a metal and carbon is not.

18. Drawing Conclusions You are shown two samples of phosphorus. One is white and burns if exposed to air. The other is red and burns if lit. Infer why the properties of two samples of the same element differ.

19. Recognizing Cause and Effect Plants need nitrogen compounds. Nitrogen fixing changes free nitrogen into nitrates. Lightning and legumes are possible examples of nitrogen fixing. What are the cause and effect of nitrogen fixing in this example?

20. Concept Mapping Complete the concept map using the following: *Na, Fe, Actinides, Hg, Ba, Alkali,* and *Inner transition.*

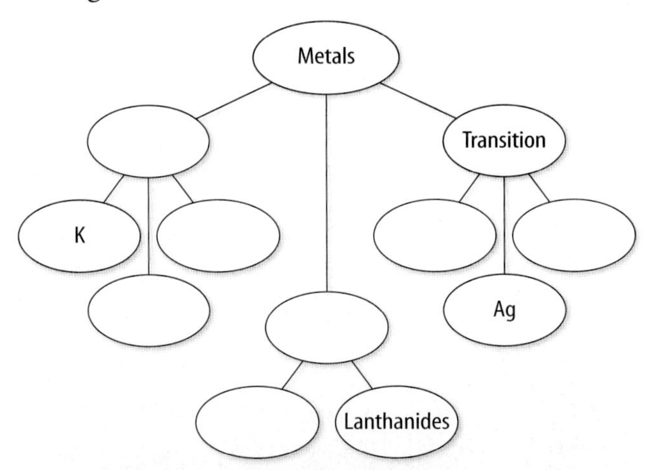

Performance Assessment

21. Analyze the Data The model made in the two-page activity of this chapter helps explain the lubricating ability of graphite. Use that same model to explain how the graphite in a pencil is able to leave a mark on a piece of paper.

22. Investigate a Controversial Issue Research the pros and cons of using nuclear energy to produce electricity. Prepare a report that includes data as well as your informed opinion on the subject.

Technology

Go to the Glencoe Science Web site at **science.glencoe.com** or use the **Glencoe Science CD-ROM** for additional chapter assessment.

Test Practice

All of the elements that make up matter can be divided into groups by looking at their different characteristics. Examples of two such groups of elements are listed in Tables A and B below.

Table A
Iron
Silver
Copper

Table B
Helium
Carbon
Sulfur

Study the tables and answer the following questions.

1. The elements in Table A are different from the elements in Table B because only the elements in Table A ____.

A) are shiny and malleable
B) are gases
C) dissolve instantly in water
D) are yellow

2. Which of these belongs with Table A?

F) nitrogen gas
G) chlorine gas
H) aluminum foil
J) water

3. The title of Table A is likely to be ____.

A) metals
B) nonmetals
C) metalloids
D) elements

4. Which of the following elements does not belong in Table B?

F) oxygen
G) fluorine
H) sodium
J) bromine

CHAPTER 21 Organic Compounds

You may not see it, but carbon is everywhere. It is in your clothes, your backpack, your books, and your food. It is even in you. Carbon is an amazing element. In this chapter, you will learn how carbon can form more than 16 million known compounds. Also, you will see how black, gooey petroleum is transformed into useful substances. You will be introduced to some carbon compounds essential to life itself—among them proteins and DNA.

What do you think?

Science Journal Look at the picture below with a classmate. Discuss what this might be. Here's a hint: *This pattern offers an infinite variety of forms.* Write your answer or best guess in your Science Journal.

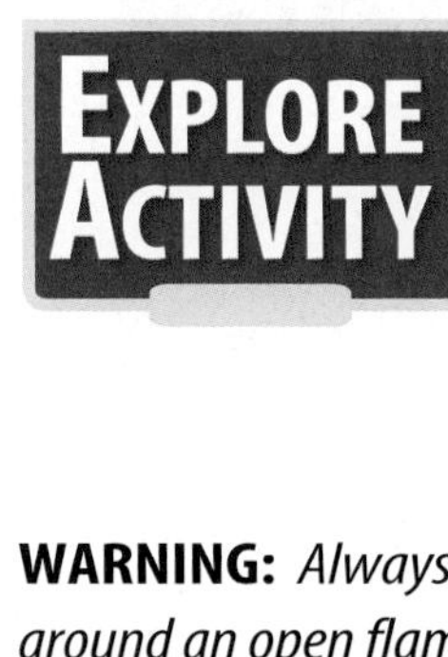

The element carbon exists in three very different forms. It exists as dull, black charcoal; slippery, gray graphite; and bright, sparkling diamond. However, this is nothing compared with the millions of different compounds that carbon can form. In this activity, you will seek out the carbon hidden in two common substances.

WARNING: *Always use extreme caution around an open flame. Point test tubes away from yourself and others.*

Identify a common element

1. Place a small piece of bread in a test tube.
2. Using a test-tube holder, hold the tube over the flame of a laboratory burner until you observe changes in the bread.
3. Using a clean test tube and a small amount of paper instead of bread, repeat step 2.

Observe

1. What happened when you heated the tubes? Did you see anything leave the two samples?
2. What remained in each of the test tubes? Describe it in your Science Journal. What do you think it is?

Before You Read

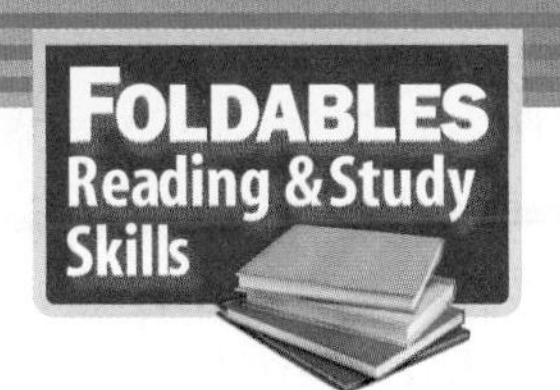

Making a Vocabulary Study Fold **Make the following vocabulary Foldable to ensure you have understood the content by defining the vocabulary terms from this chapter.**

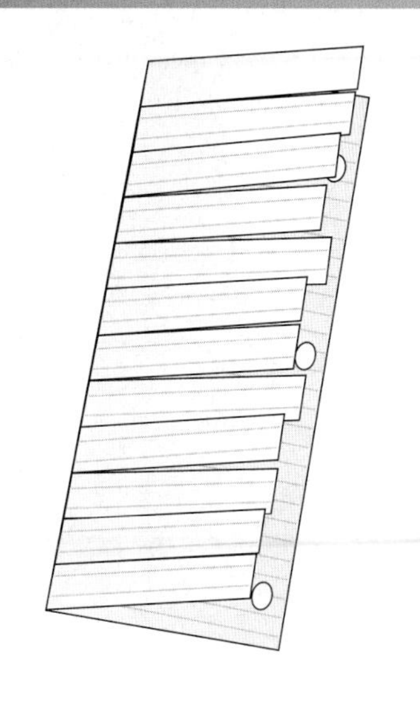

1. Place a sheet of notebook paper in front of you so the short side is at the top. Fold the paper in half from the left side to the right side.
2. Through the top thickness of paper, cut along every third line from the outside edge to the centerfold, forming ten tabs, as shown.
3. On the front of each tab, write a vocabulary word listed on the first page of each section in this chapter. On the back of each tab, define the word.

SECTION 1

Simple Organic Compounds

As You Read

What You'll Learn

- **Identify** the difference between organic and inorganic carbon compounds.
- **Examine** the structures of some organic compounds.
- **Differentiate** between saturated and unsaturated hydrocarbons.
- **Identify** isomers of organic compounds.

Vocabulary

organic compound
hydrocarbon
saturated hydrocarbon
isomer
unsaturated hydrocarbon

Why It's Important

Carbon compounds surround you—they're in your food, your body, and most materials you use every day.

Organic Compounds

What do you have in common with your athletic shoes, sunglasses, and backpack? All the items shown in **Figure 1** contain compounds of the element carbon—and so do you. Most compounds containing the element carbon are **organic compounds.**

At one time, scientists thought that only living organisms could make organic compounds, which is how they got their name. Did you notice the similarity between the terms *organic* and *organism?* By 1830, scientists could make organic compounds in laboratories, but they continued to call them organic.

Of the millions of carbon compounds known today, more than 90 percent of them are considered organic. The other ten percent, including carbon dioxide and the carbonates, are considered inorganic compounds because they are found in nonliving things, such as air, rocks, and minerals.

Bonding You may wonder why carbon can form so many organic compounds. The main reason is that a carbon atom has four electrons in its outer energy level. This means that each carbon atom can form four covalent bonds with atoms of carbon or with other elements. As you have learned, a covalent bond is formed when two atoms share a pair of electrons. Four is a large number of bonds compared to the number of bonds that atoms of other elements can form. This allows carbon to form many types of compounds ranging from small compounds used as fuel, to complex compounds found in medicines and dyes, and the polymers used in plastics and textile fibers.

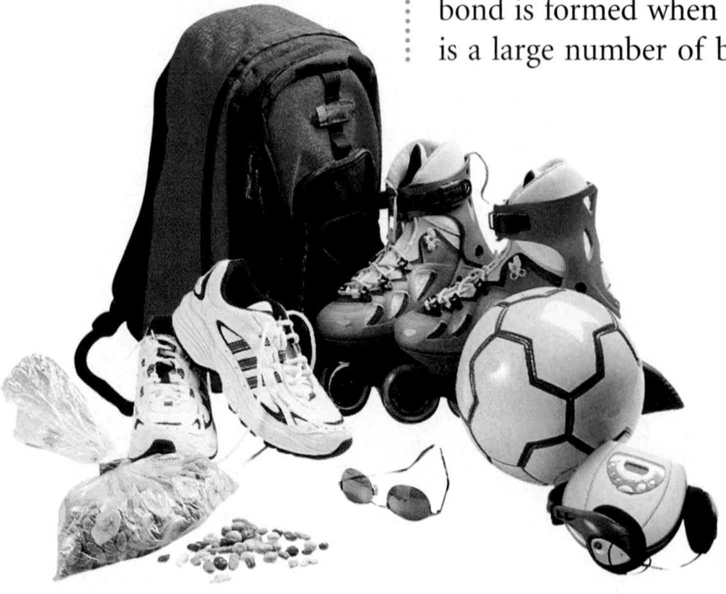

Figure 1
Most items used every day contain carbon.

Arrangement Another reason carbon can form so many compounds is that carbon can link together with other carbon atoms in many different arrangements—chains, branched chains, and even rings. It also can form double and triple bonds as well as single bonds. In addition, carbon can bond with atoms of many other elements, such as hydrogen and oxygen. **Figure 2** shows some possible arrangements for carbon compounds.

Hydrocarbons

Carbon forms an enormous number of compounds with hydrogen alone. A compound made up of only carbon and hydrogen atoms is called a **hydrocarbon.** Does the furnace, stove, or water heater in your home burn natural gas? A main component of the natural gas used for these purposes is the hydrocarbon methane. The chemical formula of methane is CH_4.

Methane can be represented in two other ways, as shown in **Figure 3A.** The structural formula shows that four hydrogen atoms are bonded to one carbon atom in a methane molecule. Each line between atoms represents a single covalent bond. The space-filling model in **Figure 3A** shows a more realistic picture of the size and arrangement of the atoms in the molecule. Chemists might refer to space saving models occasionally, but use chemical and structural formulas to write about reactions.

Another hydrocarbon used as fuel is propane. Some stoves, most outdoor grills, and the heaters in hot-air balloons burn this hydrocarbon, which is found in bottled gas. Propane's structural formula and space-filling model are shown in **Figure 3B.**

Methane and other hydrocarbons produce more than 90 percent of the energy humans use. Carbon compounds also are important in medicines, foods, and clothing. To understand how carbon can play so many roles, you must understand how it forms bonds.

Figure 2
A Carbon atoms bond to form chains, as in heptane found in gasoline. **B** The chains can be branched, as in isoprene, found in natural rubber. **C** The chains also can be rings, as in vanillin, found in vanilla flavor.

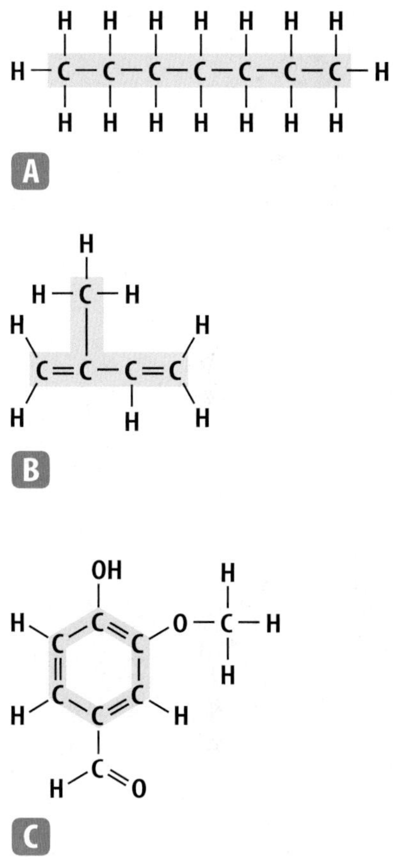

Figure 3
A Natural gas is mostly methane, CH_4. **B** Bottled gas is mostly propane, C_3H_8.

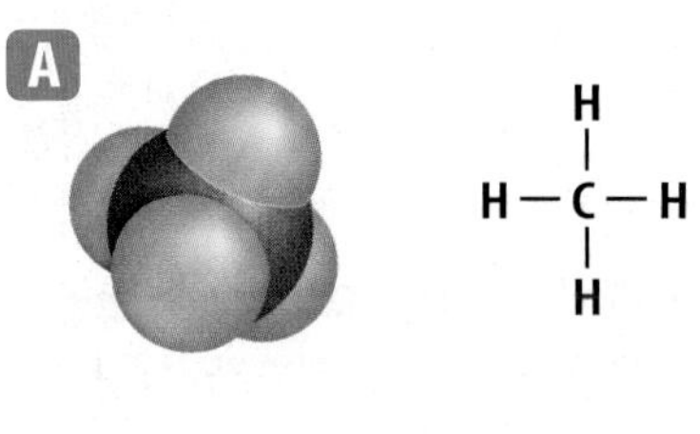

Methane
CH_4

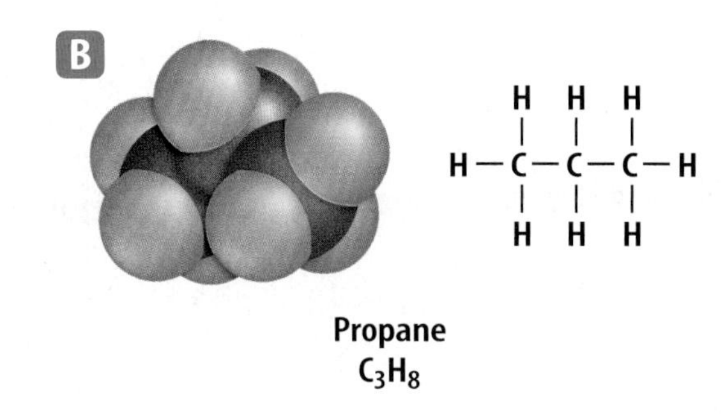

Propane
C_3H_8

Table 1 Some Hydrocarbons

Name	Chemical Formula	Structural Formula
Methane	CH_4	H | H—C—H | H
Ethane	C_2H_6	H H | | H—C—C—H | | H H
Propane	C_3H_8	H H H | | | H—C—C—C—H | | | H H H
Butane	C_4H_{10}	H H H H | | | | H—C—C—C—C—H | | | | H H H H

Single Bonds

In some hydrocarbons, the carbon atoms are joined by single covalent bonds. Hydrocarbons containing only single-bonded carbon atoms are called **saturated hydrocarbons.** Saturated means that a compound holds as many hydrogen atoms as possible—it is saturated with hydrogen atoms.

What are saturated hydrocarbons?

Table 1 lists four saturated hydrocarbons. Notice how each carbon atom appears to be a link in a chain connected by single covalent bonds. **Figure 4** shows a graph of the boiling points of some hydrocarbons. Notice the relationship between boiling points and the addition of carbon atoms.

Structural Isomers Perhaps you have seen or know about butane, which is a gas that sometimes is burned in camping stoves and lighters. The chemical formula of butane is C_4H_{10}. Another hydrocarbon called isobutane has exactly the same chemical formula. How can this be? The answer lies in the arrangement of the four carbon atoms. Look at **Figure 5.** In a molecule of butane, the carbon atoms form a continuous chain. The carbon chain of isobutane is branched. The arrangement of carbon atoms in each compound changes the shape of the - molecule. Isobutane and butane are isomers.

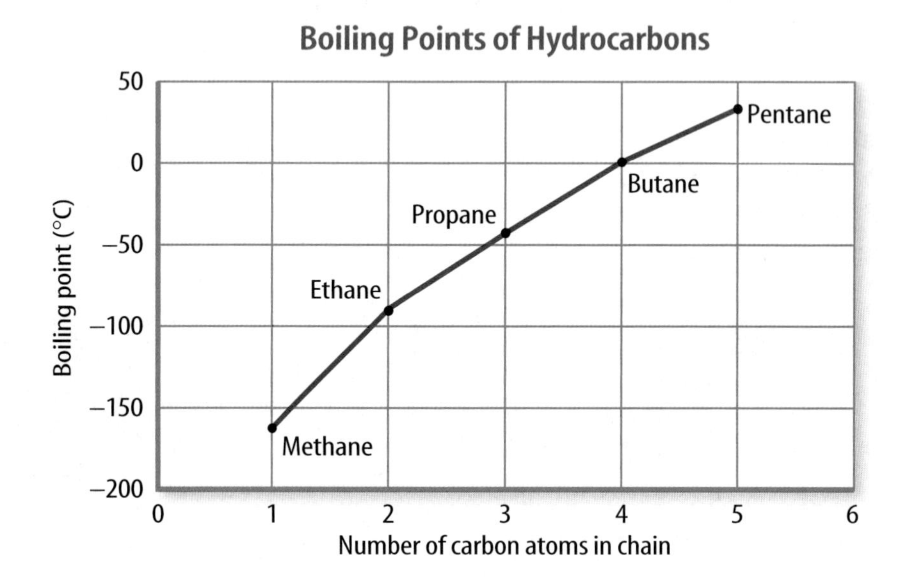

Figure 4
Boiling points of hydrocarbons increase as the number of carbon atoms in the chain increases.

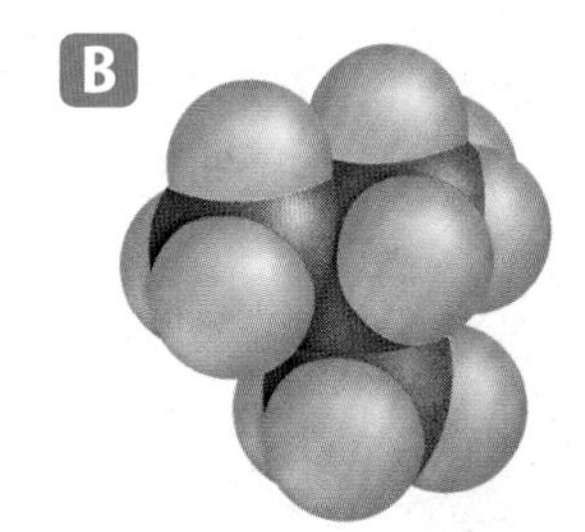

```
  H   H   H   H
  |   |   |   |
H—C—C—C—C—H
  |   |   |   |
  H   H   H   H
```

Butane
C_4H_{10}

```
  H   H   H
  |   |   |
H—C—C—C—H
  |   |   |
  H   |   H
      |
    H—C—H
      |
      H
```

Isobutane
C_4H_{10}

Figure 5
Butane has two isomers. A One isomer has a straight chain. B The other isomer has a branched chain.

Isomers are compounds that have identical chemical formulas but different molecular structures and shapes. Thousands of isomers exist among the hydrocarbons. Generally, melting points and boiling points decrease as the amount of branching in an isomer increases. You can see this pattern in **Table 2,** which lists properties of butane and isobutane.

Sometimes properties of isomers can vary amazingly. For example, the isomer of octane having all eight carbons in a straight chain melts at −56.8°C, but the most branched octane melts at 100.7°C. In this case, the high melting point results from the symmetry of the molecule and its globular shape. Look for this isomer when you do the Try at Home MiniLAB.

Other Isomers There are many other kinds of isomers in organic and inorganic chemistry. Some isomers differ only slightly in how their atoms are arranged in space. Such isomers form what is often called right- and left-handed molecules, like mirror images. Two such isomers may have nearly identical physical and chemical properties.

Table 2 Properties of Butane Isomers

Property	Butane	Isobutane
Description	Colorless gas	Colorless gas
Density	0.60 kg/L	0.603 kg/L
Melting Point	−135°C	−145°C
Boiling Point	−0.5°C	−10.2°C

Modeling Structures of Octane

Procedure

1. To model octane, C_8H_{18}, a hydrocarbon found in gasoline, use soft **gumdrops** to represent carbon atoms.
2. Use **raisins** to represent hydrogen atoms.
3. Use **toothpicks** for chemical bonds.
   WARNING: *NEVER eat any food in the laboratory.*

Analysis

1. How do you distinguish one structure from another?
2. What was the total number of different molecules found in your class?

Multiple Bonds

Peaches are among the many fruits that can form small quantities of ethylene gas, which aids in ripening. Ethylene is another name for the hydrocarbon ethene, C_2H_4. This contains one double bond in which two carbon atoms share two pairs of electrons. The hydrocarbon ethyne contains a triple bond in which three pairs of electrons are shared. Hydrocarbons, such as ethene and ethyne, that contain at least one double or triple bond are called **unsaturated hydrocarbons.** They are shown in **Figure 6.**

Reading Check *What is another name for ethene?*

Remember that saturated hydrocarbons contain only single bonds and have as many hydrogen atoms as possible. Hydrocarbons having more than one double or triple bond are called polyunsaturated, because the prefix *poly* means many.

Figure 6
Hydrocarbons can contain double or triple bonds between carbon atoms. A Ethyne, also called acetylene, is used in torches for welding. B Ethene or ethylene gas ripens fruit.

Section 1 Assessment

1. Explain why carbon can form so many organic compounds?
2. Compare and contrast *ethane*, *ethene*, and *ethyne*.
3. How is an unsaturated hydrocarbon different from a saturated hydrocarbon?
4. How did organic compounds get this name?
5. **Think Critically** Cyclopropane is a saturated hydrocarbon containing three carbon atoms. In this compound, each carbon atom is bonded to two other carbon atoms. Draw its structural formula. Are cyclopropane and propane isomers? Explain. Carbon can form single, double, and triple bonds easily. How many electrons are shared between two carbon atoms joined in a triple bond?

Skill Builder Activities

6. **Making and Using Graphs** Make a graph of **Table 1.** For each compound, plot the number of carbon atoms on one axis and the number of hydrogen atoms on the other axis. Use the graph to predict the formula of hexane, which has six carbon atoms. **For more help, refer to the Science Skill Handbook.**
7. **Solving One-Step Equations** Compare the formulas of three saturated hydrocarbons: C_2H_6, C_3H_8, and C_4H_{10}. What mathematical relationship do you see between the number of carbon atoms and the number of hydrogen atoms in each? Compare the number of carbon and the number of hydrogen atoms in the unsaturated hydrocarbons C_2H_4, C_3H_6, and C_4H_8. **For more help, refer to the Math Skill Handbook.**

Other Organic Compounds

Aromatic Compounds

Chewing flavored gum or dissolving a candy mint in your mouth releases pleasant flavors and aromas. Many chemical compounds produce pleasant odors but others have less pleasant flavors and smells. For example, aspirin, which has an unpleasant, sour taste, is shown in **Figure 7A.** The compound that produces the fresh fragrance of wintergreen is shown in **Figure 7B.** Both of these compounds are considered aromatic compounds. In addition to the fragrances mentioned here, aromatic compounds contribute to the smell of cloves, cinnamon, anise, and vanilla.

You might assume that aromatic compounds are so named because they are smelly—and most of them are. However, smell is not what makes a compound aromatic in the chemical sense. To a chemist, an **aromatic compound** is one that contains a benzene structure having a ring with six carbons.

Reading Check ***What structure is found in all aromatic compounds?***

As You Read

What You'll Learn

- **Define** aromatic compounds.
- **Identify** the nature of alcohols and acids.
- **Identify** organic compounds you use in daily life.

Vocabulary

aromatic compound
substituted hydrocarbon
alcohol

Why It's Important

Aromatic compounds are building blocks of thousands of useful compounds, such as flavorings and medicines.

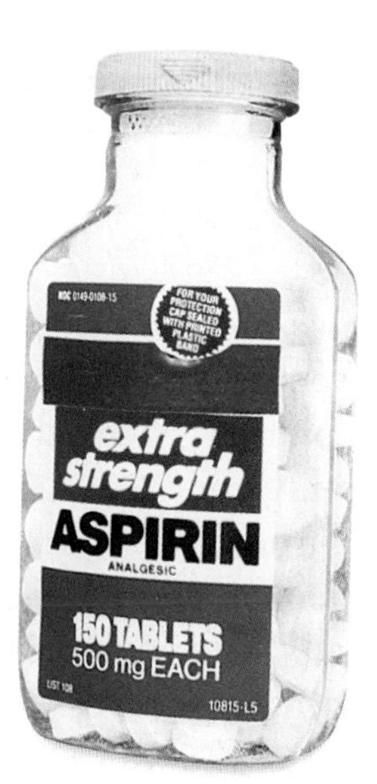

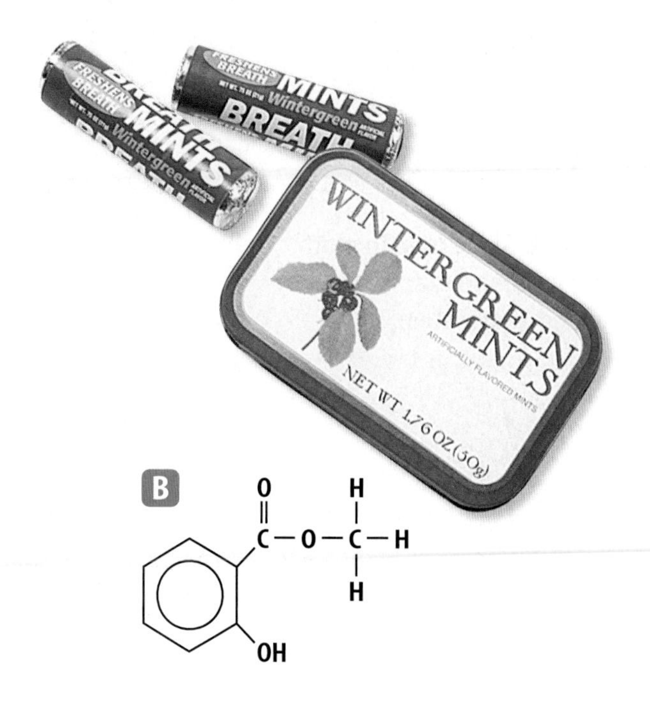

Figure 7
You can see the six-carbon benzene ring in these aromatic compounds. A Aspirin is acetyl salicylic acid. B Wintergreen is methyl salicylate.

A (benzene ring with $-C(=O)-OH$ and $-O-C(=O)-CH_3$)

B (benzene ring with $-C(=O)-O-CH_3$ and $-OH$)

Figure 8
Benzene, C_6H_6, can be represented by a A space-filling model, B a structural formula, or C the benzene symbol.

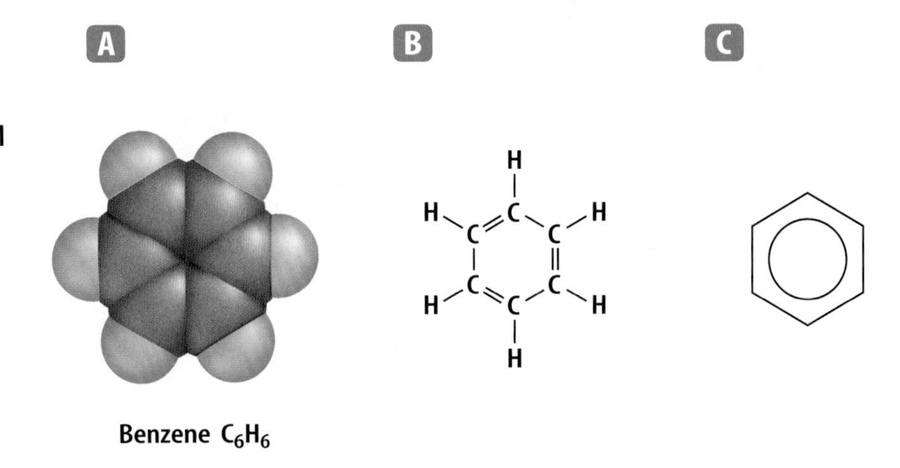

Benzene Look at a model of benzene, C_6H_6, and its structural formula in **Figure 8.** As you can see, the benzene molecule has six carbon atoms bonded into a ring. The electrons shown as alternating double and single bonds that form the ring are shared by all six carbon atoms in the ring. This equal sharing of electrons is represented by the special symbol shown in **Figure 8C.** The sharing of these electrons causes the benzene molecule to be very stable because all six carbon atoms are bound in a rigid, flat structure. Many compounds contain this stable ring structure. The stable ring acts as a framework upon which new molecules can be built.

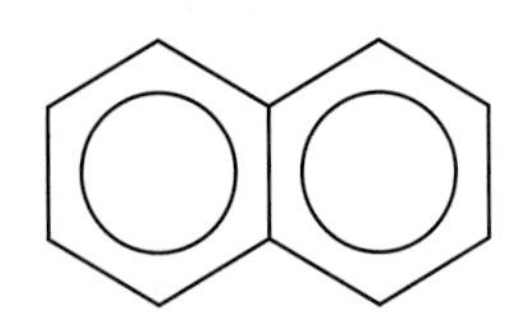

Figure 9
Naphthalene used in moth crystals is an example of a fused-ring system.

Reading Check *What is responsible for the stability of the benzene ring?*

Fused Rings Moth crystals have a distinct odor. One type of moth crystal is made of naphthalene (NAF thuh leen). This is a different type of aromatic compound that is made up of two ring structures fused together, as shown in **Figure 9.** Many known compounds contain three or more rings fused together. Tetracycline (te truh SI kleen) antibiotics are based on a fused ring system containing four fused rings.

Substituted Hydrocarbons

Usually a cheeseburger is a hamburger covered with melted American cheese and served on a bun. However, you can make a cheeseburger with Swiss cheese and serve it on toast. Such substitutions would affect the taste of this cheeseburger.

In a similar way chemists change hydrocarbons into other compounds having different physical and chemical properties. They may include a double or triple bond or add different atoms or groups of atoms to compounds. These changed compounds are called substituted hydrocarbons.

A **substituted hydrocarbon** has one or more of its hydrogen atoms replaced by atoms or groups of other elements. Depending on what properties are needed, chemists decide what to add. Examples of substituted hydrocarbons are shown in **Figure 10.**

Astronomy INTEGRATION

Carbon compounds also are found in space. About five percent of meteorites contain water and carbon compounds. Carbon compounds, such as formic acid and a form of acetylene, have been detected in outer space using radio telescopes. The areas where they are found are thought to be regions of space where new stars are forming.

Alcohols and Acids Rubbing alcohol gets its name from the fact that it was used for rubbing on aching muscles. Rubbing alcohol is a substituted hydrocarbon. Alcohols are an important group of organic compounds. They serve often as solvents and disinfectants, and more importantly can be used as pieces to assemble larger molecules. An **alcohol** is formed when –OH groups replace one or more hydrogen atoms in a hydrocarbon. **Figure 10A** shows ethanol, an alcohol produced by the fermentation of sugar in grains and fruit.

✔ Reading Check ***Why are alcohols considered substituted hydrocarbons?***

Organic acids form when a carboxyl group, –COOH, is substituted for one of the hydrogen atoms attached to a carbon atom. Look at **Figure 10.** The structures of ethane, ethanol, and acetic acid are similar. Do you see that acetic acid, found in vinegar, is a substituted hydrocarbon? You know some other organic acids, too—citric acid found in citrus fruits, such as oranges and lemons, and lactic acid found in sour milk.

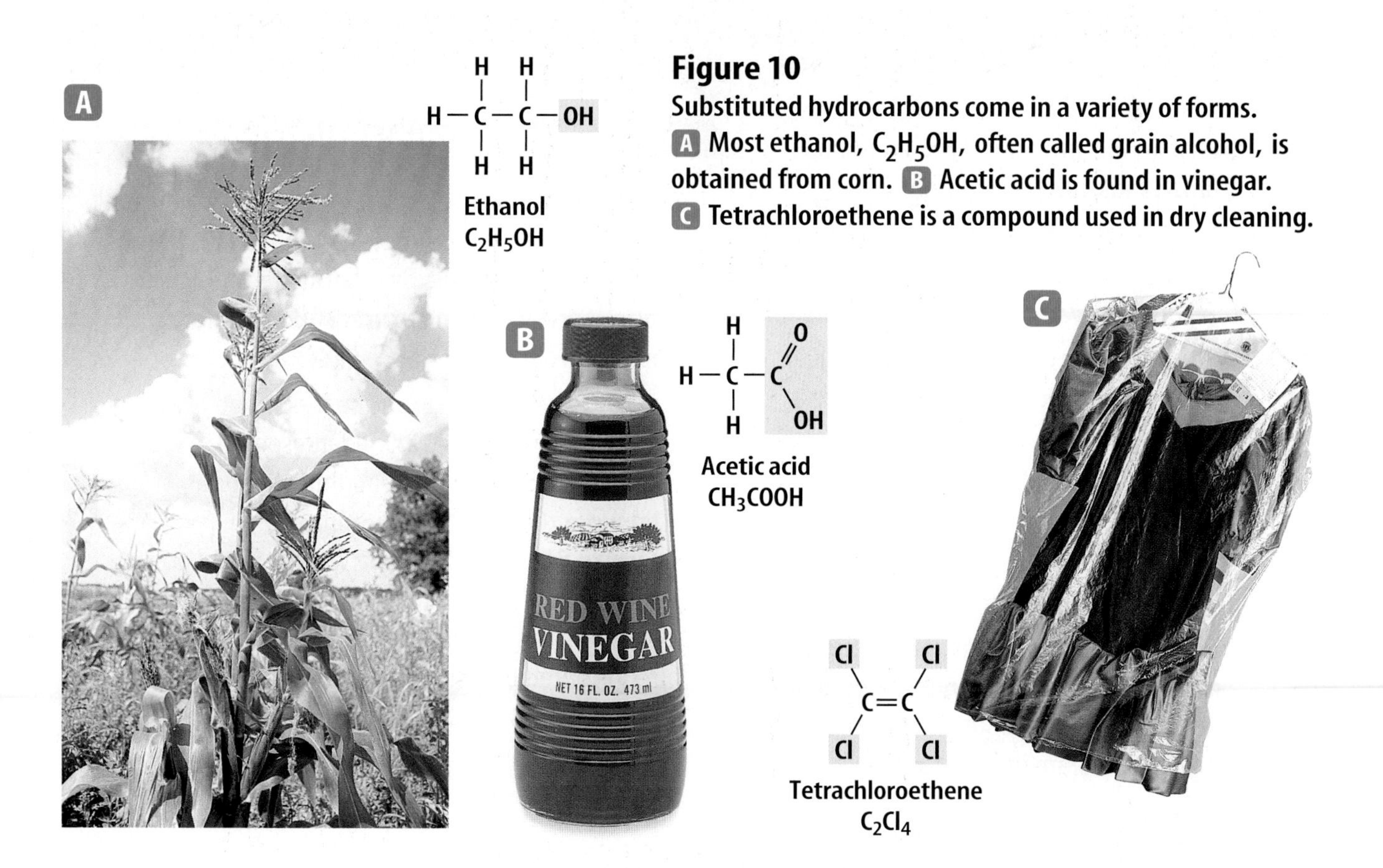

Figure 10
Substituted hydrocarbons come in a variety of forms.
A Most ethanol, C_2H_5OH, often called grain alcohol, is obtained from corn. B Acetic acid is found in vinegar. C Tetrachloroethene is a compound used in dry cleaning.

Figure 11
Strangely, small concentrations of foul-smelling compounds are often found in pleasant-smelling substances. For example, the mercaptan in skunk spray is among the 834 components of coffee aroma.

Substituting Other Elements Other atoms besides hydrogen and oxygen can be added to hydrocarbons. One is chlorine. When four chlorine atoms replace four hydrogen atoms in ethylene, the result is tetrachloroethene (teh truh klor uh eth EEN), a solvent used in dry cleaning. It is shown in **Figure 10C.** Adding four fluorine atoms to ethylene makes a compound that can be transformed into a black, shiny material used for nonstick surfaces in cookware. Among other possible substituted hydrocarbons are molecules containing nitrogen, bromine, and sulfur.

When sulfur replaces oxygen in the –OH group of an alcohol, the resulting compound is called a thiol, or more commonly a mercaptan. Most mercaptans have unpleasant odors. This can be useful to animals like the skunk shown in **Figure 11.**

Mercaptan odors are not only unpleasant, they are also powerful. You can smell skunk spray even in concentrations as low as 0.5 parts per million. Though you might not think so, such a powerful stink can be an asset, and not just for skunks. In fact, smelly mercaptans can save lives. As you know natural gas has no odor of its own so it is impossible to smell a gas leak. For this reason, gas companies add small amounts of a bad-smelling mercaptan to the gas to make people aware of any leaks before gas can accumulate to dangerous levels.

Section  Assessment

1. What do the structures of all aromatic compounds have in common?
2. How is each of the following a substituted hydrocarbon: *tetrachloroethene, ethanol,* and *acetic acid?*
3. What elements other than oxygen could be added to a hydrocarbon to produce a substituted hydrocarbon?
4. Explain why chemists might want to prepare substituted hydrocarbons. Give two examples of possible substitutions.
5. **Think Critically** Chloroethane, C_2H_5Cl, can be used as a spray-on anesthetic for localized injuries. How does chloroethane fit the definition of a substituted hydrocarbon? Diagram its structure.

Skill Builder Activities

6. **Interpreting Scientific Illustrations** Formic acid, HCOOH, is the simplest organic acid. Draw its structural formula by referring to the structure of acetic acid in **Figure 10B. For more help, refer to the Science Skill Handbook.**
7. **Communicating** Examine the different ways a benzene molecule can be represented, as shown in **Figure 8**. Do you think the double bonds in the structural formula shown in **Figure 8B** must always be written in the positions shown? In your Science Journal, explain your answer and discuss why benzene can be represented by a six-sided figure containing a circle. **For more help, refer to the Science Skill Handbook.**

Activity

Alcohols and Organic Acids

Have you ever wondered how chemists change one substance into another? You have learned that changing the bonding among atoms holds the key to that process.

What You'll Investigate

How can an alcohol change into an acid?

Materials

large test tube and stopper
0.01*M* potassium permanganate solution (1 mL)
6*M* sodium hydroxide solution (1 mL)
ethanol (3 drops)
10-mL graduated cylinder

Safety Precautions

Always handle these chemicals with care; immediately flush any spill with water.

Goals

- **Control** the immediate environment of a reaction to produce a specific compound.
- Gather evidence to form conclusions about the identity of a new compound formed from a chemical reaction.

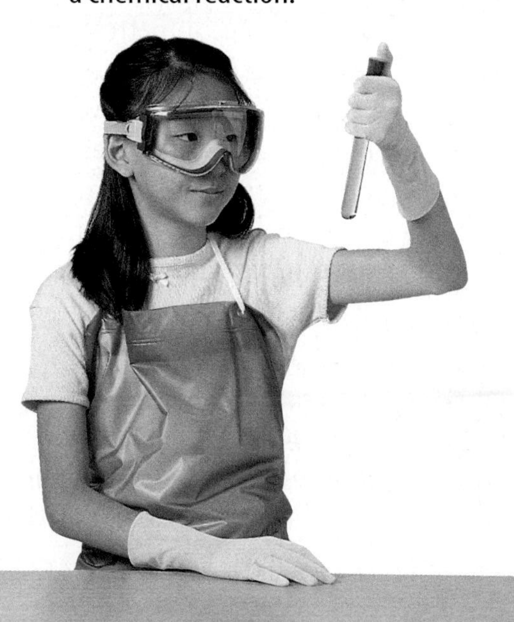

Procedure

1. Pour 1 mL of 0.01*M* potassium permanganate solution and 1 mL of 6*M* sodium hydroxide solution into a test tube.
2. Add 3 drops of ethanol to the test tube.
3. Stopper the test tube. Gently shake it for 1 minute. Observe and record any changes in the solution for 5 minutes.

Conclude and Apply

1. What is the structural formula for ethanol?
2. What part of a molecule identifies a compound as an alcohol?
3. What part of a molecule identifies a compound as an organic acid?
4. **Explain** how you know that a chemical change took place in the test tube.
5. In the presence of potassium permanganate, an alcohol may undergo a chemical change into an acid. If ethanol is used, predict the formula of the acid produced.
6. The acid from ethanol is found in a common household product—vinegar. What is the acid's chemical name?

Communicating Your Data

Design a table and record what changes take place in the color of the solution. Compare your observations with those of other students in your class. **For more help, refer to the Science Skill Handbook.**

SECTION 3

Petroleum—A Source of Carbon Compounds

As You Read

What You'll Learn

- **Explain** how carbon compounds are obtained from petroleum.
- **Compare and contrast** differences among the various petroleum-based fuels.
- **Determine** how carbon compounds can form long chains.
- **Identify** some of the ways petroleum enriches your world.

Vocabulary

polymer
monomer
polyethylene

Why It's Important

Petroleum gives us fuels, plastics, clothing, and many other products.

What is petroleum?

Do you carry a comb in your pocket or purse? What is it made from? If you answer plastic, you are probably right, but do you know where that plastic came from? Chances are it came from petroleum—a dark, flammable liquid, often called crude oil, that is found deep within Earth. Like coal and natural gas, this dark, foul-smelling substance is formed from the remains of fossilized material. For this reason these substances often are called fossil fuels.

How can a thick, dark liquid like petroleum be transformed into a hard, brightly colored, useful object like a comb? The answer lies in the nature of petroleum. Petroleum is a mixture of thousands of carbon compounds. To make items such as combs, the first step is to extract the crude oil from its underground source, as shown in **Figure 12.** Then, chemists and engineers separate it into its individual compounds using a physical property that you have learned about. This property is the boiling point. Even though petroleum contains a large number of compounds, each compound has its own boiling point.

Separating Petroleum Compounds The separation process is known as fractional distillation. It takes place in petroleum refineries. If you have ever driven past a refinery, you may have seen these big, metal towers called fractionating towers. They often rise as high as 35 m and can be 18 m wide and have pipes and metal scaffolding attached to the outside.

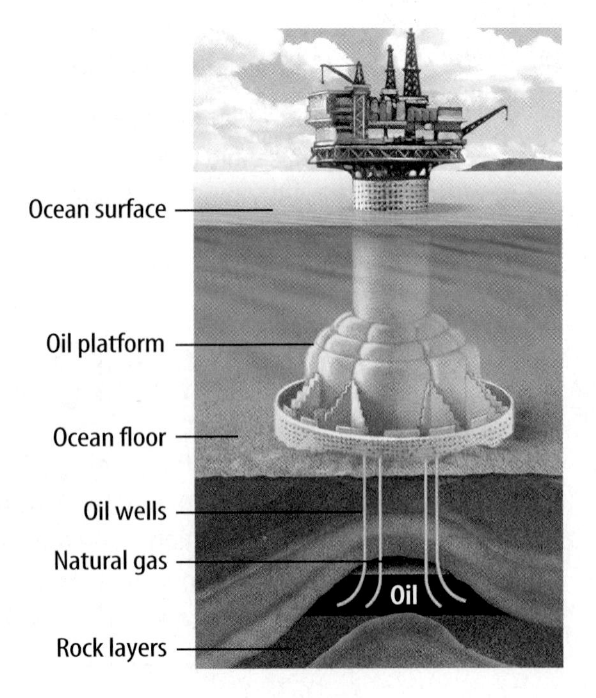

Figure 12
Drilling for petroleum beneath the ocean floor requires huge platforms.

The Tower Inside the tower is a series of metal plates arranged like the floors of a building. These plates have small holes so that vapors can pass through. On the outside you can see a maze of pipes at various levels. The tower separates crude oil into fractions containing compounds having a range of boiling points. Within a fraction, boiling points may range more than 100°C.

How It Happens The crude petroleum at the base of the tower is heated to more than 350°C. At this temperature most hydrocarbons in the mixture become vapor and start to rise. The higher boiling fractions reach only the lower plates before they condense, forming shallow pools that drain off through pipes on the sides of the tower and are collected.

Fractions with lower boiling points may climb higher to the middle plates before condensing. Finally, those with the lowest boiling points condense on the topmost plates or never condense at all and are collected as gases at the top of the tower. **Figure 13** shows some typical fractions and how they are used.

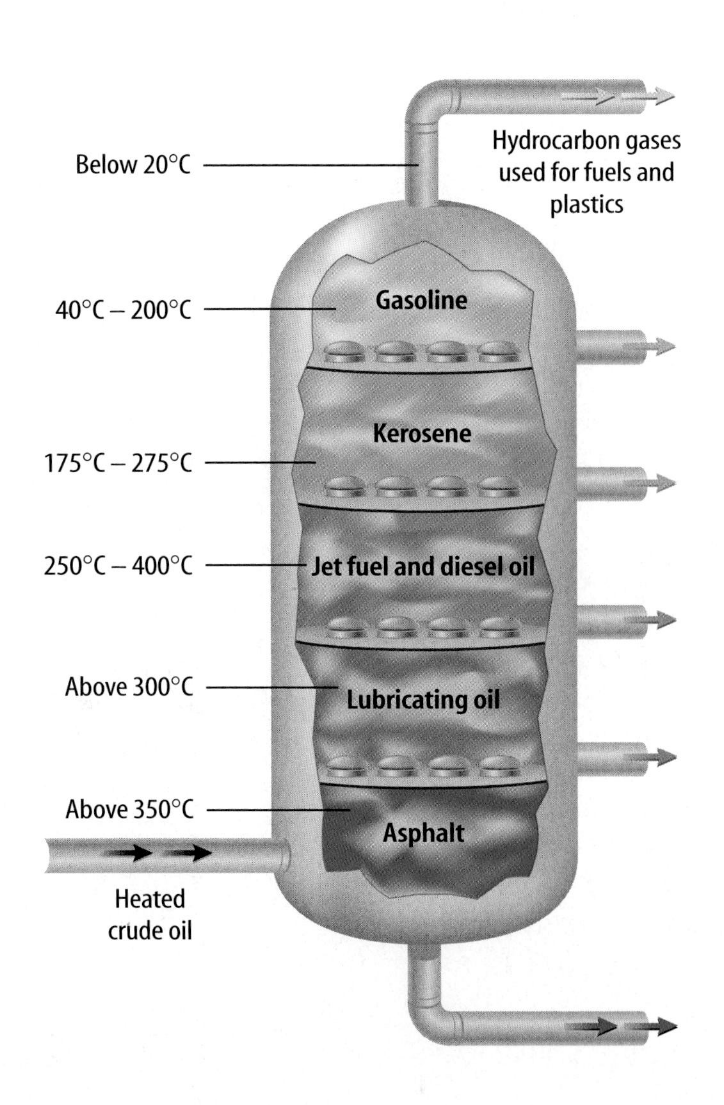

Figure 13
Typical fractions are separated in a fractionating tower by their boiling points.

Why don't the condensed liquids fall back through the holes? The reason is that pressure from the rising vapors prevents this. In fact, the separation of the fractions is improved by the interaction of rising vapors with condensed liquid. The processes involved vary. For example, some towers add steam at the bottom to aid vaporization. The design and process used depend on the type of crude oil and on the fractions desired.

Uses for Petroleum Compounds

Some fractions are used directly for fuel—the lightest fractions from the top of the tower include butane and propane. The fractions that condense on the upper plates and contain from five to ten carbons are used for gasoline and solvents. Below these are fractions with 12 to 18 carbons that are used for kerosene and jet fuel. The bottom fractions go into lubricating oil, and the residue is used for paving asphalt. **Figure 14** shows the variety of useful products that can be obtained from petroleum, in addition to its use as a fuel.

NATIONAL GEOGRAPHIC

VISUALIZING PETROLEUM PRODUCTS

Figure 14

Petroleum, or crude oil, provides the raw material for a huge number of products that have become essential to modern life. After it has been refined, petroleum can be used to make various types of fuel, plastics, and synthetic fibers, as well as paint, dyes, and medicines.

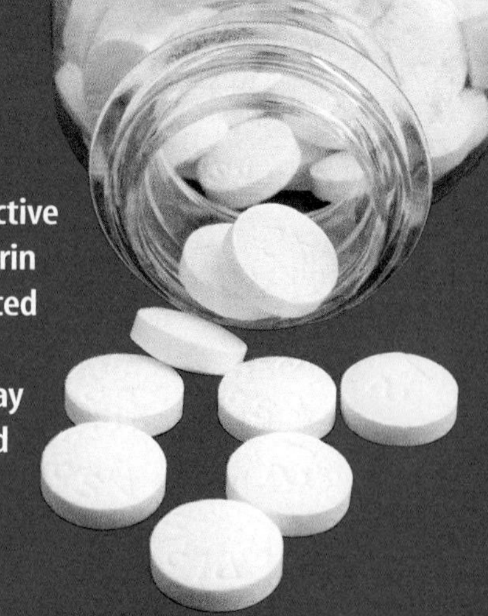

MEDICINES The active ingredient in aspirin used to be extracted from the bark of willow trees. Today it is manufactured from petroleum.

FABRICS Like the fleece used to make these gloves, many modern fabrics are made from synthetic, rather than natural, fibers. Some of the most popular synthetic fibers—polyester and nylon are petroleum-based.

PRINTING INK The ink used in newspapers is made from carbon black, another product from petroleum.

FUELS This commuter jet is being refueled at an airport. Most of the world's petroleum is still used in the form of fuel.

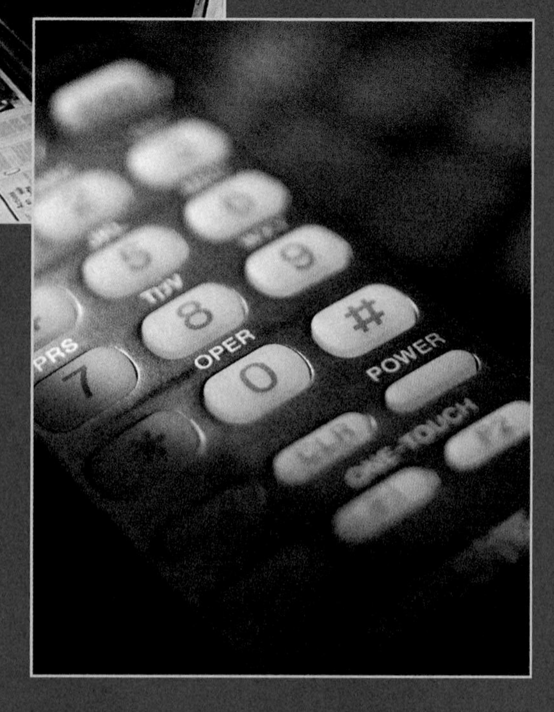

PLASTICS The durability of hard plastic makes it the ideal material for a cell phone keypad.

Polymers

Did you ever loop together strips of paper to make paper chains for decorations, or have you ever strung paper clips together? A paper chain can represent the structure of a polymer as shown in **Figure 15.** Some of the smaller molecules from petroleum can act like links in a chain. When these links are hooked together, they make new, extremely large molecules known as **polymers.** The small molecule, which forms a link in the polymer chain, is called a **monomer.** *Mono* means one.

✔ Reading Check *How are polymers similar to paper chains or linked paper clips?*

Common Polymers One common polymer or plastic is made from the monomer ethene or ethylene. Under standard room-temperature conditions, this small hydrocarbon is a gas. However, when ethylene combines with itself repeatedly, it forms a polymer called **polyethylene.** Polyethylene (pah lee EH thuh leen) is used widely in shopping bags and plastic bottles. Another common polymer is polypropylene (pah lee PRO puh leen) used to make glues and carpets. Often two or more different monomers, known as copolymers, combine to make one polymer molecule.

Polymers can be made light and flexible or so strong that they can be used to make plastic pipes, boats, and even some auto bodies. In many cases, they have replaced natural building materials, such as wood and metal. Because so many things used today are made of synthetic polymers, some people call this "The Age of Plastics."

Visualizing Polymers

Procedure

1. Use **paper clips** to represent monomers in a synthetic polymer. Hook about 20 together to make a chain.
2. Cut 20 **strips of colored paper** and mark each with a different letter of the alphabet from A to T.
3. Assemble these strips in random order to make a paper chain.

Analysis

1. Imagine both chains extended to contain 10,000 or more units. Compare them in terms of ease of construction and degree of complexity.
2. Compare the paper chains made by your class. How many different combinations of letters are there?

Figure 15
Imagine this paper chain extended by 10,000 units. Then imagine each link as a monomer. Now you have an idea of what a typical polymer used to make plastic looks like.

Figure 16
Processing can modify a polymer's properties. A Polystyrene used in CD cases is clear, hard, and brittle. B Polystyrene used in cups is opaque, light-weight, and foamy.

Designing Polymers The properties of polymers depend mostly on which monomers are used to make them. Also, like hydrocarbons, polymers can have branches in their chains. The amount of branching and the shape of the polymer greatly affect its properties.

Polymer materials can be shaped in many ways. Some are molded to make containers or other rigid materials. Sometimes the same polymer can take two completely different forms. For example, polystyrene (pah lee STI reen) that is made from styrene, shown in **Figure 16,** forms brittle, transparent cases for CDs and lightweight, opaque, foam cups and packing materials. To make this transformation, a gas such as carbon dioxide is blown into melted polystyrene as it is molded. Bubbles remain within the polymer when it cools, making polystyrene foam an efficient insulator.

Other polymers can be spun into threads, which are used to make clothing or items such as suitcases and backpacks. Fibers can be made strong and durable for products that receive wear and tear. Others can resist strong impacts. For example, bullet-proof vests are made of a tightly woven, synthetic polymer. Polymer fibers also can be made stretchy and resilient for fabric products like exercise garments. Some polymers remain rigid when heated, but others become soft and pliable when heated and harden again when cooled.

SCIENCE Online
Research Visit the Glencoe Science Web site at **science.glencoe.com** to learn more about polymers. Communicate to your class what you learn.

✔ Reading Check *Name some applications of polymer fibers.*

Other Petroleum Products

Other fractions obtained from fractional distillation of petroleum may be purified further by different techniques to isolate individual compounds. After these are separated, they can be converted into substituted hydrocarbons, as you learned in the last section. Chemists use these to make products ranging from medicines such as aspirin to dyes, insecticides, printers' ink, and flavorings.

Many of the products that come from petroleum are aromatic compounds. For example, the sweetener saccharin is related to toluene, a substituted benzene molecule. It has from 200 to 700 times the sweetening power of sugar, but no calories. In fact, it passes unchanged through the digestive tract. Although other sweeteners are available today, saccharin still is used widely in many diet foods and to sweeten mouthwash, as shown in **Figure 17.**

Also, aromatic dyes have replaced natural dyes, such as indigo and alizarin, almost completely. The first synthetic dye was a bright purple called mauve that was discovered accidentally in coal tar compounds. Today, most dyes come from petroleum.

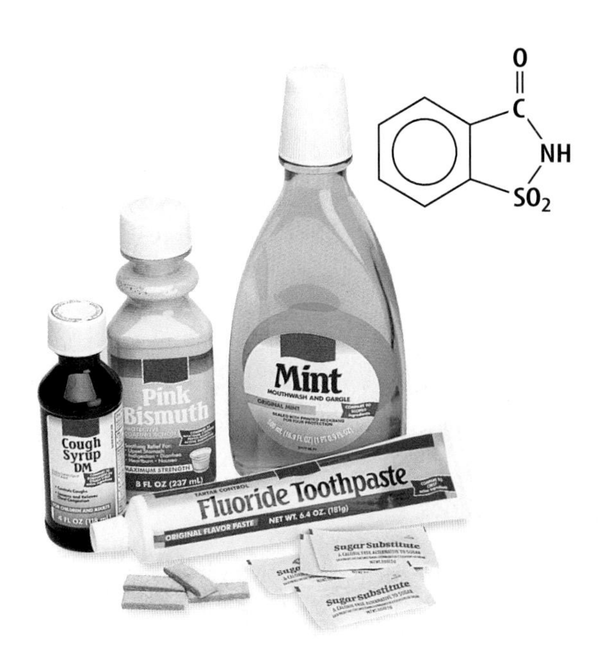

Figure 17
Saccharin is used as a sweetener because it doesn't damage teeth like sugar can.

Section  Assessment

1. What is petroleum and where does it come from?
2. Why are some fuels referred to as fossil fuels?
3. Which fractions of petroleum are used directly for fuel? How are they obtained from petroleum?
4. What is the name for substances made from thousands of small molecules that are linked together? What are the small molecules called?
5. **Think Critically** Petroleum fractions obtained by fractional distillation contain many individual compounds. How might these mixtures be separated further? What physical property could be used?

Skill Builder Activities

6. **Predicting** Based on the names of the polymers and monomers in this section, what do you think polymers made from the monomers terpene and urethane would be called? **For more help, refer to the Science Skill Handbook.**
7. **Communicating** Research at least two synthetic polymers other than those mentioned in this section. Write in your Science Journal *which monomers they include, the approximate length of their chains, and how they are used.* Describe their properties and, if possible, how they are related to polymer structure. Report your findings to your class. **For more help, refer to the Science Skill Handbook.**

SECTION

Biological Compounds

As You Read

What **You'll Learn**

- **Compare and contrast** proteins, nucleic acids, carbohydrates, and lipids.
- **Identify** the structure of polymers found in basic food groups.
- **Identify** the structure of large biological polymers.

Vocabulary

protein
nucleic acid
deoxyribonucleic acid (DNA)
carbohydrate
lipid

Why **It's Important**

All life processes depend on large biological compounds.

Biological Polymers

Like the polymers that are used to make the plastics and fibers, biological polymers are huge molecules. Also, they are made of many smaller monomers that are linked together. The monomers of biological polymers are usually larger and more complex in structure. Still, you can picture a biological monomer as one link in a very long chain.

Many of the important biological compounds in your body are polymers. Among them are the proteins, which often contain hundreds of units.

Proteins

Proteins are large organic polymers formed from organic monomers called amino acids. Even though only 20 amino acids are commonly found in nature, they can be arranged in so many ways that millions of different proteins exist. Proteins come in numerous forms and make up many of the tissues in your body, such as muscles and tendons, as well as your hair and fingernails. In fact, proteins account for 15 percent of your total body weight.

Each amino acid contains an amine ($-NH_2$) group.

Each amino acid also contains a carboxylic acid ($-COOH$) group.

A Glycine

Cysteine

B

Water is formed from the $-OH$ of a carboxylic group and the $-H$ of an amine group.

Peptide bonds link molecules of amino acids.

Figure 18
In a protein polymer, peptide bonds link together molecules of amino acids.

Figure 19
Four peptide chains coil around each other in the protein polymer hemoglobin. Each chain has an atom of iron, which carries oxygen.

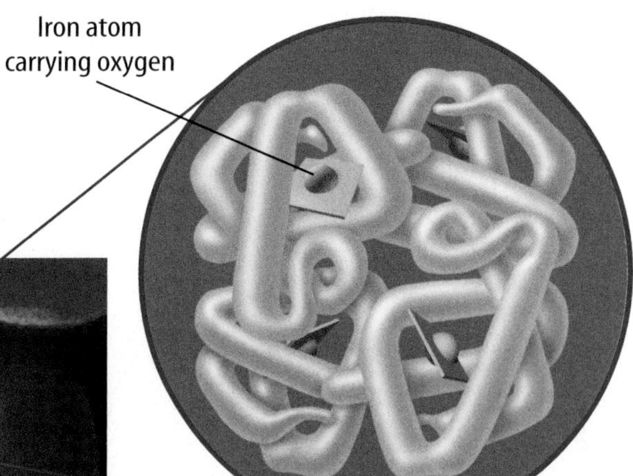

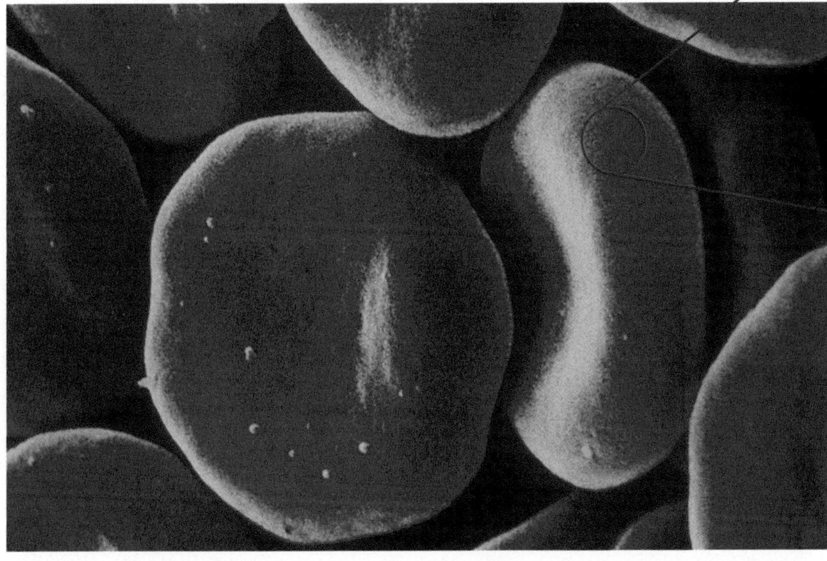

Protein Monomers Amino acids are the monomers that combine to form proteins. Two amino acids are shown in **Figure 18A.** The $-NH_2$ group is the amine group and the $-COOH$ group is the carboxylic acid group. Both groups appear in every amino acid.

Amine groups of one amino acid can combine with the carboxylic acid group of another amino acid, linking them together to form a compound called a peptide as shown in **Figure 18B.** The bond joining them is known as a peptide bond. When a peptide contains a large number of amino acids—about 50 or more—the molecule is called a protein.

Approximately how many amino acid units does a protein contain?

Protein Structure Long protein molecules tend to twist and coil in a manner unique to each protein. For example, hemoglobin, which carries oxygen in your blood, has four chains that coil around each other as shown in **Figure 19.** Each chain contains an iron atom that carries the oxygen. If you look closely, you can see all four iron atoms in hemoglobin.

When you eat foods that contain proteins, such as meat, dairy products, and some vegetables, your body breaks down the proteins into their amino acid monomers. Then your body uses these amino acids to make new proteins that form muscles, blood, and other body tissues.

Nucleic Acids

The **nucleic acids** are another important group of organic polymers that are essential for life. They control the activities and reproduction of cells. One kind of nucleic acid, called **deoxyribonucleic** (dee AHK sih ri boh noo klay ihk) **acid** or DNA, is found in the nuclei of cells where it codes and stores genetic information. This is known as the genetic code.

Data Update Visit the Glencoe Science Web site at **science.glencoe.com** for recent news of magazine articles about DNA finger-printing. Communicate to your class what you learn.

Nucleic Acid Monomers The monomers that make up DNA are called nucleotides. Nucleotides are complex molecules containing an organic base, a sugar, and a phosphoric acid unit. In DNA two nucleotide chains twist around each other forming what resembles a twisted ladder or what is called the double helix. Human DNA contains only four different organic bases, but they can form millions of combinations. The bases on one side of the ladder pair with bases on the other side, as shown in **Figure 20.** The genetic code gives instructions for making other nucleotides and proteins needed by your body.

Problem-Solving Activity

Selecting a Balanced Diet

What do you like to eat? You probably choose your foods by how good they taste. A better way might be to look at their nutritional value. Your body needs nutrients like proteins, carbohydrates, and fats to give it energy and help it build cells. Almost every food has some of these nutrients in it. The trick is to pick your foods so you don't get too much of one thing and not enough of another.

Identifying the Problem

The table on the right lists some basic nutrients for a variety of foods. The amount of the protein, carbohydrate, and fat is recorded as the number of grams in 100 g of the food. By examining these data, can you select the foods that best provide each nutrient?

Solving the Problem

1. Using the table, list the foods that supply the most protein and carbohydrates. What might be the problem with eating too many potato chips?
2. In countries where meat and dairy products are hard to get, people eat a lot of food made from soybeans. Can you think of reasons why people might wish to substitute meat and dairy products with soybean based products?

Nutritional Values for Some Common Foods

Food (100 g)	Protein (g)	Carbohydrate (g)	Fat (g)
Cheddar cheese	25	1	33
Hamburger	17	23	17
Soybeans	13	11	7
Wheat	15	68	2
Potato chips	7	53	35

DNA Fingerprinting Human DNA contains more than 5 billion base pairs. The DNA of each person differs in some way from that of everyone else, except for identical twins, who share the same DNA sequence. The unique nature of DNA offers crime investigators a way to identify criminals from hair or fluids left at a crime scene. DNA from bloodstains or cells in saliva found on a cigarette can be extracted in the laboratory. Then, chemists can break up the DNA into its nucleotide components and use radioactive and X-ray methods to obtain a picture of the nucleotide pattern. Comparing this pattern to one made from the DNA of a suspect can link that suspect to the crime scene.

Carbohydrates

If you hear the word *carbohydrate,* you may think of bread, cookies, or pasta. Have you heard of carbohydrate loading by athletes? Runners, for example, often prepare for a long-distance race by eating, or loading up on, carbohydrates in foods such as vegetables and pasta. **Carbohydrates** are compounds containing carbon, hydrogen, and oxygen, that have twice as many hydrogen atoms as oxygen atoms. Carbohydrates include the sugars and starches.

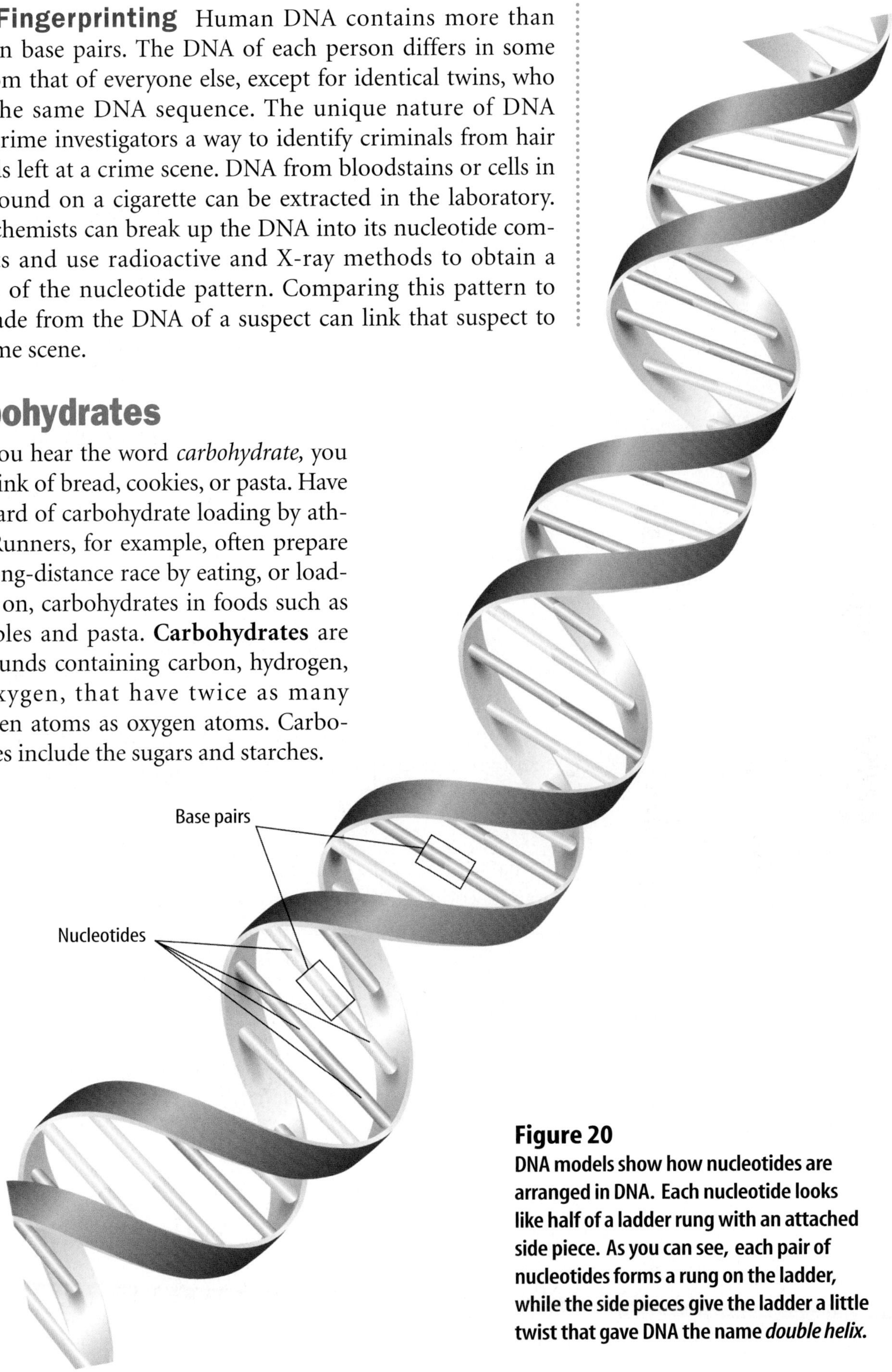

Figure 20
DNA models show how nucleotides are arranged in DNA. Each nucleotide looks like half of a ladder rung with an attached side piece. As you can see, each pair of nucleotides forms a rung on the ladder, while the side pieces give the ladder a little twist that gave DNA the name *double helix.*

Sucrose $C_{12}H_{22}O_{11}$

Glucose $C_6H_{12}O_6$

Sugars Sugars are a major group of carbohydrates, as shown in **Figure 21.** The sugar glucose is found in your blood and also in many sweet foods such as grapes and honey. Common table sugar, known as sucrose, is broken down by digestion into two simpler sugars—fructose, often called fruit sugar, and glucose. Unlike starches, sugars provide quick energy soon after eating.

Figure 21
Sucrose and glucose are sugars found in foods. Fruits contain glucose and another simple sugar called fructose. *Why are sugars carbohydrates?*

Starches Starch, shown in **Figure 22,** is a carbohydrate that is also a polymer. It is made of units or monomers of the sugar glucose. During digestion, the starch is broken down into smaller molecules of glucose and other similar sugars, which release energy in your body cells.

Athletes, especially long-distance runners, use starches to provide high-energy, long-lasting fuel for the body. The energy from starches can be stored in liver and muscle cells in the form of a compound called glycogen. During a long race, this stored energy is released, giving the athlete a fresh burst of power.

Figure 22
Starch is the major component of pasta.

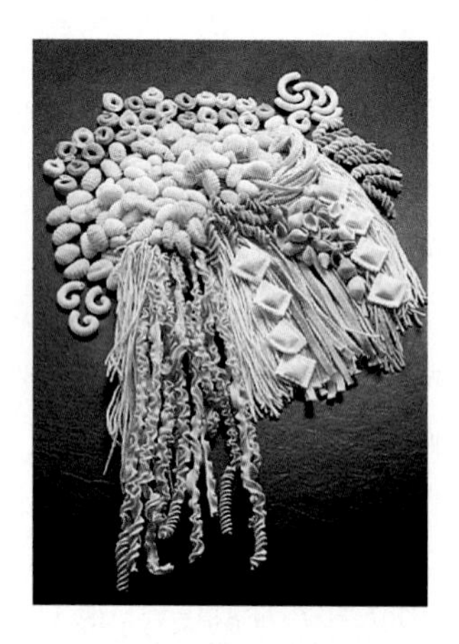

Lipids

Fats, oils, and related compounds make up a group of organic compounds known as **lipids.** Lipids include animal fats such as butter, and vegetable oils such as corn oil. Lipids contain the same elements as carbohydrates but in different proportions. For example, lipids have fewer oxygen atoms and contain carboxylic acid groups.

Reading Check *How are lipids like carbohydrates? How are they different?*

Fats and Oils These substances are similar in structure to hydrocarbons. They can be classified as saturated or unsaturated, according to the types of bonds in their carbon chains. Saturated fats contain only single bonds between carbon atoms. Unsaturated fats having one double bond are called monounsaturated, and those having two or more double bonds are called polyunsaturated. Animal lipids or fats tend to be saturated and are solids at room temperature. Plant lipids called oils are unsaturated and are usually liquids, as shown in **Figure 23.** Sometimes hydrogen is added to vegetable oils to form more saturated solid compounds known as hydrogenated vegetable shortenings.

Have you heard that eating too much fat can be unhealthy? Evidence shows that too much saturated fat and cholesterol in the diet may contribute to some heart disease and that unsaturated fats may help to prevent heart disease. It appears that saturated fats are more likely to be converted to substances that can block the arteries leading to the heart. A balanced diet includes some fats, just as it includes proteins and carbohydrates.

Figure 23
At room temperature, fats are normally solids, and oils are usually liquids.

Cholesterol Another lipid that is often in the news is cholesterol, which is found in meats, eggs, butter, cheese, and fish. Even if you never eat foods containing cholesterol, your body makes its own. Some cholesterol is needed by the body to build cell membranes. It is also found in bile, a digestive fluid. Too much cholesterol may cause serious damage to heart and blood vessels, similar to the damage caused by saturated fats.

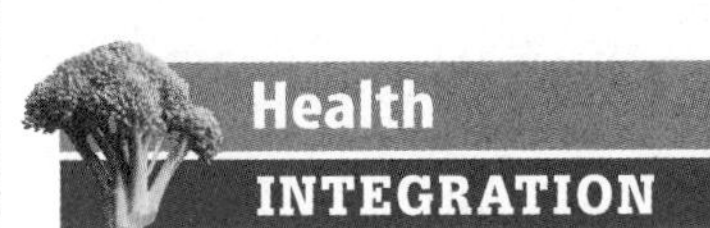

Check the label on any available container of milk. What percentage of fat is in the milk? Infer any advantages of drinking low-fat milk.

Section 4 Assessment

1. What is a polymer? Why are polymers important organic compounds? Give some specific examples.
2. Compare and contrast proteins and nucleic acids in terms of their structures and their functions in your body.
3. Where does your body get the amino acids it needs to build proteins?
4. Explain the difference between saturated and unsaturated fats and oils. Which type is considered healthier in foods?
5. **Think Critically** Is ethanol a carbohydrate? Explain.

Skill Builder Activities

6. **Comparing and Contrasting** In terms of DNA fingerprinting, compare and contrast identical twins and two people who are not identical twins. **For more help, refer to the Science Skill Handbook.**
7. **Solving One-Step Equations** Changing just one amino acid in a chain changes the function of the entire protein. If the letters G, L, C, and I represent four amino acids, how many different peptides containing four amino acids can be made? **For more help, refer to the Math Skill Handbook.**

Activity

Preparing an Ester

Are esters aromatic compounds? Organic compounds known as acids and alcohols react to form another type of organic compound called an ester. Esters frequently produce a recognizable and often pleasant fragrance, even though they are not aromatic in the chemical sense—they might not contain a benzene ring. Esters are responsible for many fruit flavors, such as apple, pineapple, pear, and banana.

What You'll Investigate

How do an acid and an alcohol combine to produce a compound with different characteristics? Can the presence of the new compound formed be detected by its odor?

Goals

- Prepare an ester from an alcohol and an acid.
- Detect the results of the reaction by the odor of the product.

Materials

medium-size test tube
test-tube holder
250-mL beaker
10-mL graduated cylinder
water
hot plate
ring stand
thermometer
salicylic acid (0.2 g)
methyl alcohol (2 mL)
concentrated sulfuric acid (2 or 3 drops to be added by teacher)

Safety Precautions

WARNING: *Sulfuric acid is caustic. Avoid all contact. Mix all the contents together using a glass stirring rod. Do not use the thermometer as a stirring rod.*

Procedure

WARNING: *Any compound you can smell has entered your body, and unknown compounds can be toxic or caustic. To detect an aroma safely, hold the container about 10 cm in front of your face and wave your hand over the opening to direct air currents to your nose.* See the illustration below for the proper way to detect odors in the laboratory.

1. Add about 150 mL of water to the beaker and heat it on the hot plate to 70°C.
2. Place approximately 0.2 g of salicylic acid in a test tube. Does this material have an odor?
3. Add 2 mL of methyl alcohol to the test tube. Before adding it, check to see if this compound has an odor. If so, try to remember what it smells like.
4. Ask your teacher to add carefully three to five drops of concentrated sulfuric acid.
5. Place the test tube in the hot water and leave it untouched for about 12 to 15 minutes.
6. Remove the tube from the hot water using a test-tube holder and allow it to cool. Check to see if you can detect a new aroma.

Conclude and Apply

1. What did you smell in step 6?
2. Look closely at the surface of the liquid in the test tube. Do you see any small droplets of an oily substance? What do you think it is?
3. Look at the equation for the reaction below. One product is given. What do you think is the second product formed in this reaction?

WARNING: *When checking an odor, waft the vapor toward your face gently.*

Communicating Your Data

Write a description of your experiment in your Science Journal. Suggest how you might modify the experiment to produce a different ester. **For more help, refer to the Science Skill Handbook.**

Oops! Accidents in SCIENCE

SOMETIMES GREAT DISCOVERIES HAPPEN BY ACCIDENT!

A SPILL for a Spill

If you ever spill something on a chair and it doesn't stain, you can thank Patsy Sherman

"How many great discoveries would never have occurred were it not for accidents?" asks Sherman.

In 1953, American chemist Patsy Sherman invented a way to protect fabrics from accidental spills. Strangely enough, this discovery came about because of an accidental spill in her lab.

"We were trying to develop a new kind of rubber for jet aircraft fuel lines," Sherman once explained, "when one of the lab assistants accidentally dropped a glass bottle that contained a batch of synthetic latex I had made. Some of the latex mixture splashed on the assistant's canvas tennis shoe and the result was remarkable."

The latex mixture didn't stain the shoe or change it in any way. But it simply would not come off.

Neither soap nor alcohol nor any other cleaning material could remove the stubborn mixture from the shoe. In fact, the mixture actually made water bead and run off the shoe in much the same way that water runs off a duck's back.

Perfecting the Product

Although the lab assistant was frustrated by the mixture's staying power, Sherman was inspired. She realized that it could be used to protect fabrics from oil, water, and dirt. She spent three years working with another chemist to perfect the product, which came on the market in 1956. The substituted hydrocarbon compound that Sherman developed makes fabrics more durable as well as stain resistant. It bonds to the fibers in the fabric and protects them like an invisible shield. Today, the carpet in your home, the fabric that covers the couch in your living room, and some of the clothes that you wear are likely treated with the fabric protector invented by Sherman. Her product is especially useful on uniforms for people who spend a lot of time outdoors in difficult, dirty, or messy situations, such as firefighters, police officers, and military personnel.

Encouraging Young Inventors

Sherman had a successful career as a chemist and inventor before she retired in 1992. She often speaks to students about the life of an inventor and encourages them to pursue their dreams. Sherman stresses that a creative mind is a scientist's best tool. "Anyone can become an inventor," she insists, "as long as they keep an open and inquiring mind and never overlook the possible significance of an accident or apparent failure."

Patsy Sherman and her team have made mud on the rug easier to deal with.

CONNECTIONS **Experiment** **Pour a small amount of water on a piece of cloth that has been treated with fabric protector. Do the same to a piece of untreated cloth. What happened to the water in both cases? What happened to the pieces of cloth?**

SCIENCE Online
For more information, visit science.glencoe.com

Chapter 21 Study Guide

Reviewing Main Ideas

Section 1 Simple Organic Compounds

1. Carbon is an element with a structure that enables it to form a large number of compounds, known as organic compounds.
2. Saturated hydrocarbons contain only single bonds between carbon atoms. Unsaturated hydrocarbons contain double or triple bonds.
3. Many camp stoves burn butane. *How many carbon atoms are in a butane molecule?*

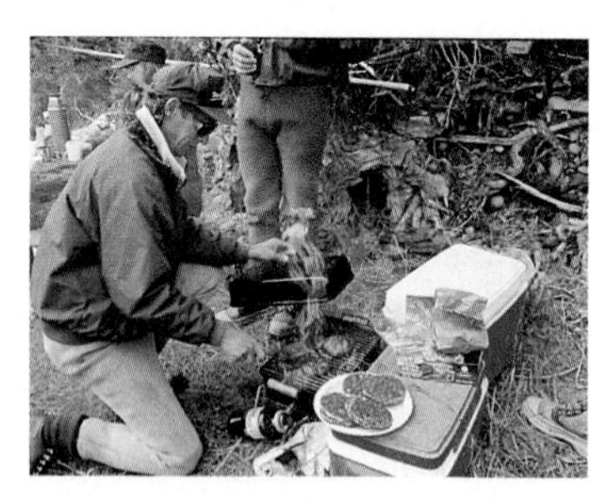

4. Isomers of organic compounds have identical formulas but different molecular shapes.

Section 2 Other Organic Compounds

1. Aromatic compounds, many of which have odors, contain the benzene ring structure.
2. Cookware often has a nonstick coating. This coating is a hydrocarbon polymer in which fluorine replaces some hydrogen atoms. *What are such hydrocarbons called?*

3. Benzene rings are stable because all six carbon atoms are bound tightly on one plane.
4. Aromatic compounds include those having two or more rings fused together.

Section 3 Petroleum—A Source of Carbon Compounds

1. Petroleum is a mixture of thousands of carbon compounds.
2. A fractionating tower separates petroleum into groups of compounds or fractions based on their boiling points.
3. Small hydrocarbons obtained from petroleum can be combined to make long chains called polymers, which are used for plastics.
4. Polymers can be spun into fibers designed to have specific properties. *What property is important in spandex fibers?*

Section 4 Biological Compounds

1. Proteins, nucleic acids, carbohydrates, and lipids are major groups of biological organic compounds.
2. Many important biological compounds are polymers, huge organic molecules made of smaller units, or monomers.
3. The pain-producing components of wasp venom are peptides. *What are the monomers of peptides?*

After You Read

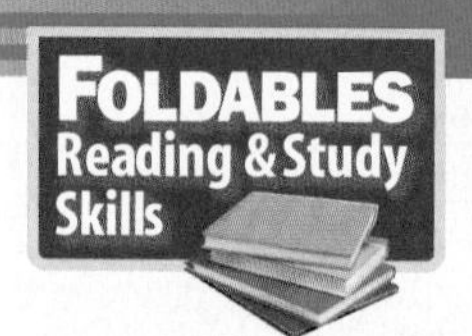

Use each of the vocabulary words on your Foldable in a complete sentence that explains something about the organic compounds discussed in this chapter.

Chapter 21 Study Guide

Visualizing Main Ideas

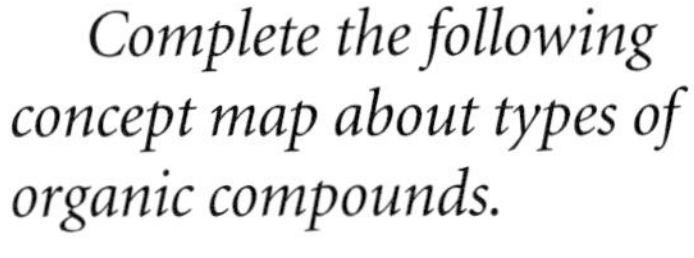

Complete the following concept map about types of organic compounds.

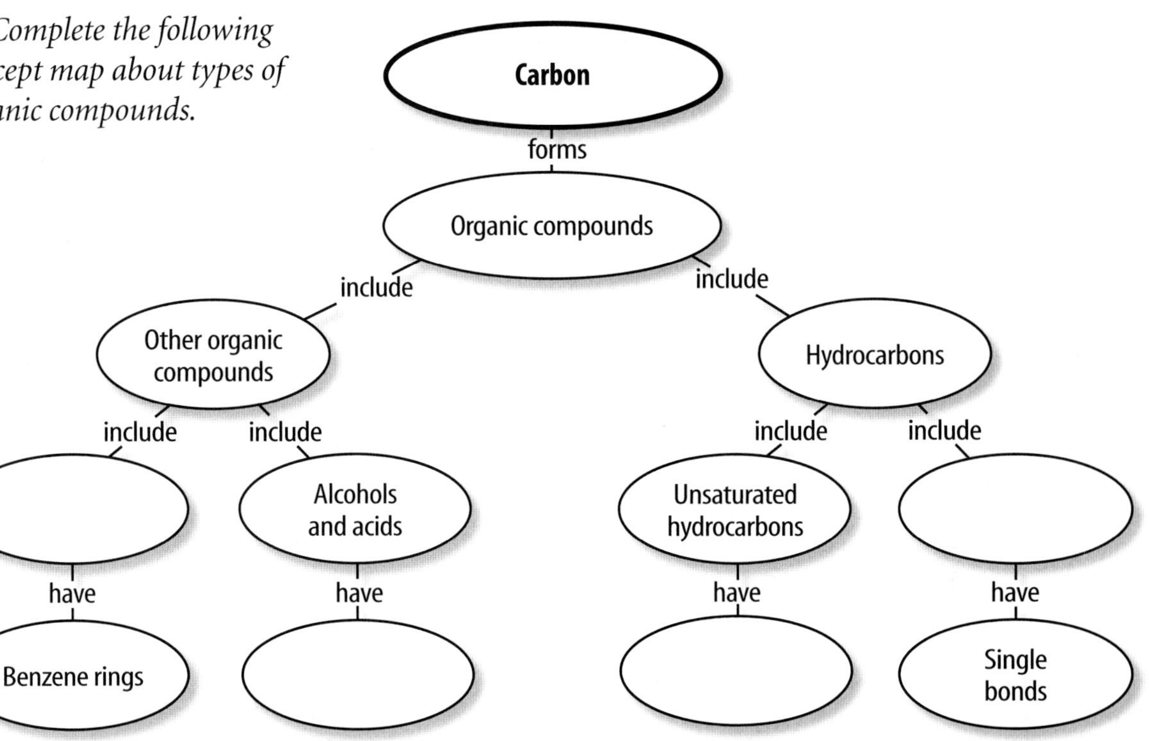

Vocabulary Review

Vocabulary Words

a. alcohol
b. aromatic compound
c. carbohydrate
d. deoxyribonucleic acid (DNA)
e. hydrocarbon
f. isomer
g. lipid
h. monomer
i. nucleic acid
j. organic compound
k. polyethylene
l. polymer
m. protein
n. saturated hydrocarbon
o. substituted hydrocarbon
p. unsaturated hydrocarbon

Study Tip

Think of other ways that you might design an experiment to prove scientific principles. Consider controls and variables.

Using Vocabulary

Replace the underlined words with the correct vocabulary words.

1. Isomers are compounds that contain the element carbon.
2. An alcohol forms a link in a polymer chain.
3. Protein is the nucleic acid that contains your genetic information.
4. An unsaturated hydrocarbon contains the benzene-ring structure.
5. Organic compounds such as sugars and starches are called proteins.
6. Organic compounds such as fats and oils are called isomers.
7. Lipids are compounds with identical chemical formulas but different structures.

Chapter 21 Assessment

Checking Concepts

Choose the word or phrase that best answers the question.

1. How would you describe a benzene ring?
   A) rare **C)** unstable
   B) stable **D)** saturated
2. How would you classify hydrocarbons, such as alcohols and organic acids?
   A) aromatic **C)** substituted
   B) saturated **D)** unsaturated
3. What are the small units that make up polymers called?
   A) monomers **C)** plastics
   B) isomers **D)** carbohydrates
4. What type of compound is hemoglobin found in red blood cells?
   A) carbohydrate **C)** nucleic acid
   B) lipid **D)** protein
5. RNA tells your body how to make proteins. What does DNA code and store?
   A) lipids **C)** protein
   B) nucleic acids **D)** genetic information
6. What type of compounds form the DNA molecule?
   A) amino acids **C)** polymers
   B) nucleotides **D)** carbohydrates
7. Glucose and fructose both have the formula $C_6H_{12}O_6$. What are such compounds called?
   A) amino acids **C)** isomers
   B) alcohols **D)** polymers
8. If a carbohydrate has 16 oxygen atoms, how many hydrogen atoms does it have?
   A) 4 **C)** 16
   B) 8 **D)** 32
9. What type of compound is cholesterol?
   A) sugar **C)** protein
   B) starch **D)** lipid
10. Which petroleum fractions are collected at the top of a fractionating tower?
   A) highest boiling **C)** lowest boiling
   B) liquid **D)** polymer

Thinking Critically

11. Too much saturated or unsaturated fat in your diet is unhealthful. How do they differ in composition?
12. Propyl alcohol and isopropyl alcohol have the formula C_3H_80. How might their structures differ?
13. Draw a diagram to explain single, double, and triple bonds in hydrocarbons. Draw a chain of carbon atoms that shows each type of bond.
14. Explain the term *fraction* when used to describe the products produced during petroleum refining. Describe the process by which these products are obtained.

Developing Skills

15. **Making and Using Graphs** Using the following table, plot the number of carbon atoms on one axis and the boiling point on the other axis on a graph. Use the graph to predict the boiling points of butane, octane, and dodecane ($C_{12}H_{26}$).

Hydrocarbons

Name	Formula	Boiling Point (°C)
Methane	CH_4	−162
Ethane	C_2H_6	−89
Propane	C_3H_8	−42

16. **Recognizing Cause and Effect** Anthracene is a compound containing three fused rings similar to the benzene ring. How would you explain the stability of this compound?

17. Interpreting Scientific Illustrations Which of the following terms apply to the illustration below: *alcohol, aromatic, carbohydrate, hydrocarbon, lipid, organic compound, polymer, saturated,* and *substituted hydrocarbon.*

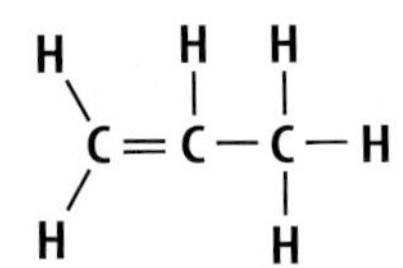

18. Concept Mapping Make a network tree to describe types of fats. Include the terms *saturated fats, unsaturated fats, single bonds,* and *double bonds.*

19. Hypothesizing A weight-reduction diet allows no food other than lettuce and fresh fruit for three days a week. What is missing from the diet on these days?

Performance Assessment

20. Scientific Drawing Research three hydrocarbons containing five carbon atoms. Draw diagrams of their structures, name them, and tell their uses.

21. Surveying and Graphing Record the fiber content of your clothing, noting whether it is synthetic or natural. Make a circle graph comparing the percentages of natural and synthetic fibers.

TECHNOLOGY

Go to the Glencoe Science Web site at **science.glencoe.com** or use the **Glencoe Science CD-ROM** for additional chapter assessment.

Test Practice

The common alcohol methanol is represented by the following chemical symbol:

CH_3OH

Methanol

Study the chemical formula above and answer the following questions.

1. According to this chemical formula, all of the following elements are found in methanol EXCEPT ________.
   A) carbon
   B) nitrogen
   C) hydrogen
   D) oxygen

2. How many carbon atoms are there in the compound methanol?
   F) one
   G) two
   H) three
   J) four

3. How many electrons are shared in each of the bonds linking the carbon atom in methanol to the three hydrogen atoms?
   A) one
   B) two
   C) three
   D) four

CHAPTER

22

New Materials Through Chemistry

This *X-33* test aircraft was designed to test hydrogen holding tanks made from a carbon reinforced plasticlike material and a new heat protection system made from nickel. If tests like these are successful, a future aircraft will carry payloads to the *International Space Station* using this technology. In this chapter, you will learn about various types of materials—old, new, and improved.

What do you think?

Science Journal Look at the picture below with a classmate. Discuss what this might be. Here's a hint: *It is a strong and lightweight material.* Write your answer or best guess in your Science Journal.

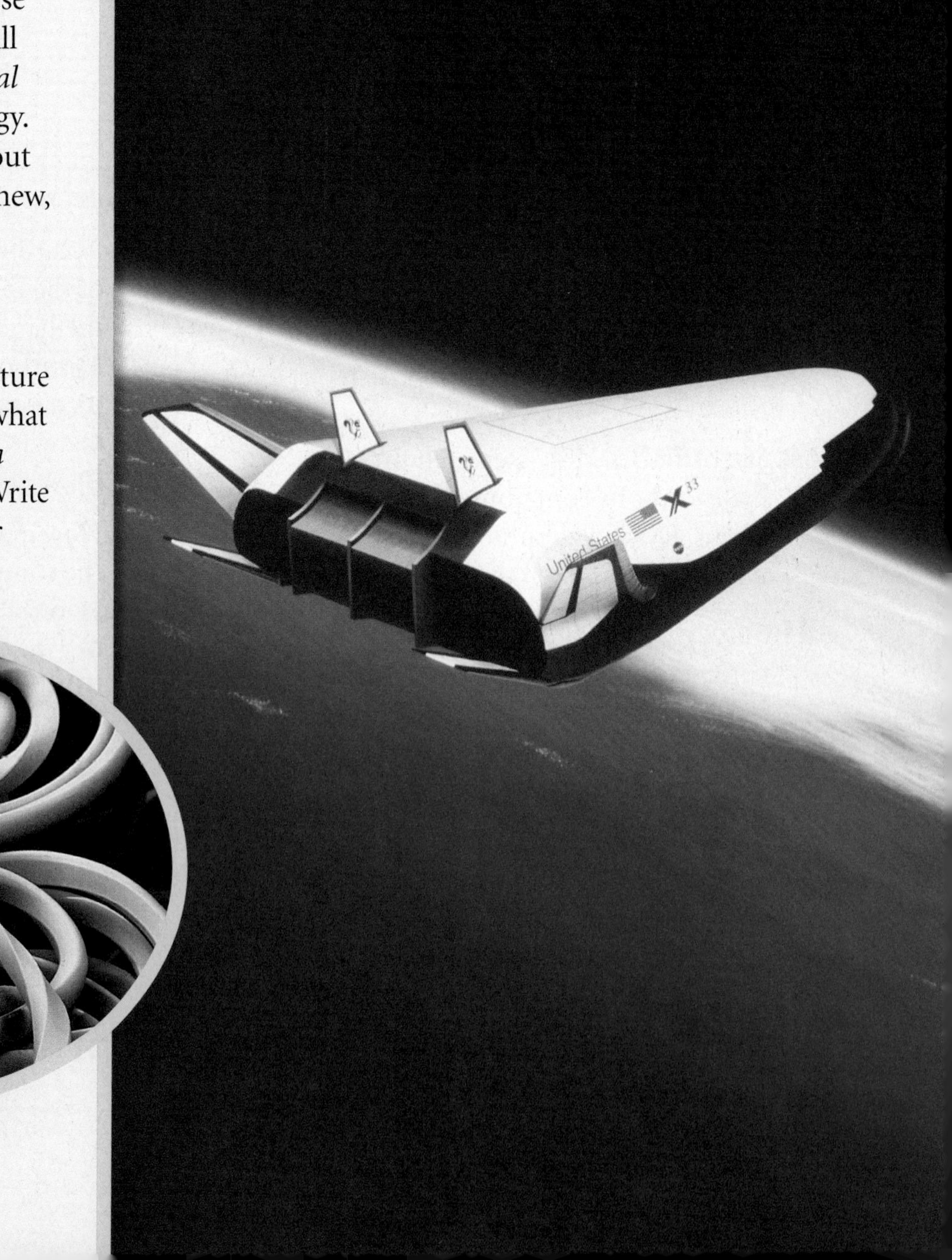

When an engineer designs a vehicle, bridge, or building, the materials used for construction must be selected. Can the manufacturing process affect a material's performance?

Safety Precautions:

WARNING: *Use proper protection when handling hot objects or near an open flame. Tie back hair; roll up sleeves.*

Compare Properties

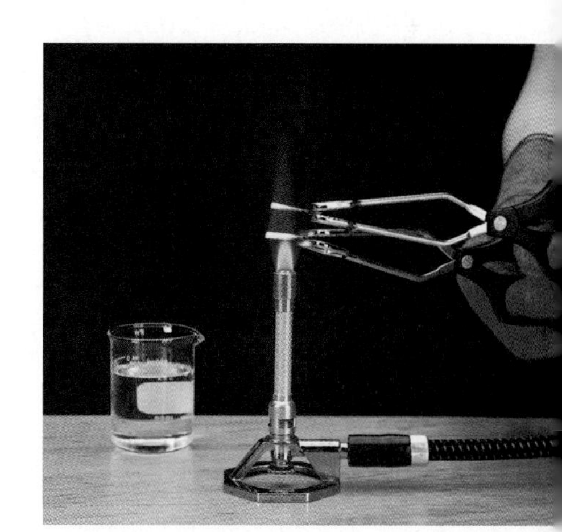

1. Using tongs, hold a 5-cm piece of steel wire in a lab burner flame until the wire glows red-hot for 30 seconds.
2. Quickly drop the hot wire into a beaker of cold water.
3. Repeat step 1 with another 5-cm piece of steel wire, but place this hot wire on a heat-proof surface to cool instead of in water.
4. After both pieces of wire are cool, compare the flexibility of the wires.

Observe

In your Science Journal, write what you observe about the flexibility of the two wires. Suggest reasons.

Before You Read

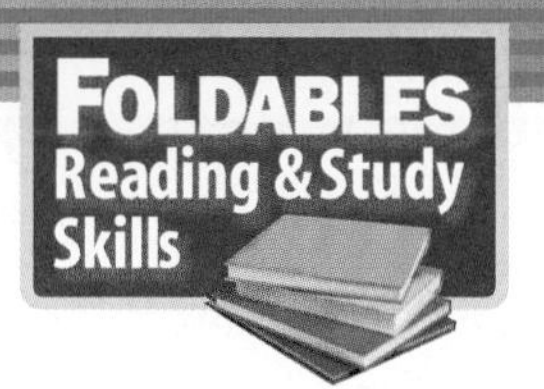

Making a Classify Study Fold **Make the following Foldable to help you organize objects into groups based on their common features.**

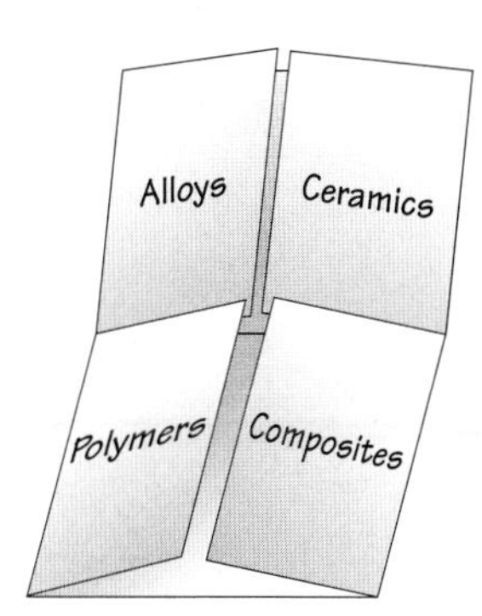

1. Place a sheet of paper in front of you so the long side is at the top. Fold the paper in half from the left side to the right side and then unfold.
2. Fold each side in to the centerfold line to divide the paper into fourths. Fold the paper in half from top to bottom and unfold.
3. Through the top thickness of paper, cut along both of the middle fold lines to form four tabs as shown. Label the tabs *Alloys*, *Ceramics*, *Polymers*, and *Composites*.
4. As you read the chapter, list three or more examples of common materials around you under the tab for each group.

Materials with a Past

As You Read

What You'll Learn

- **Identify** how different alloys are used.
- **Explain** how the properties of alloys determine their use.

Vocabulary

alloy
luster
ductility
malleability
conductivity

Why It's Important

Alloys make modern cities, space travel, and many other things possible.

Alloys

For ages, people have searched for better materials to use to make their lives more comfortable and their tasks easier. Ancient cultures used stone tools until methods for processing metals became known. Today, advances in metal processing are still occurring as scientists continue to improve the art of blending metals, or making alloys, to make better metal products. An **alloy** is a mixture of a metal with one or more other elements where the mixture retains the properties of the metal. Alloys are produced to obtain a material with improved properties such as greater hardness, strength, lightness, or durability. This idea of creating better materials to make better tools is not new, as you will soon learn.

Alloys Through Time In about 3500 B.C., historians believe that ancient Sumerians in the Tigris-Euphrates Valley (now Iraq) accidentally discovered bronze. They believe that Sumerians used rocks rich in copper and tin ore to make fire rings to keep their campfires from spreading. The hot campfire melted the copper and tin ores within the rocks, creating bronze. This first known mixture of metals became so popular and widely used that a 2,000-year span of history is known as the Bronze Age.

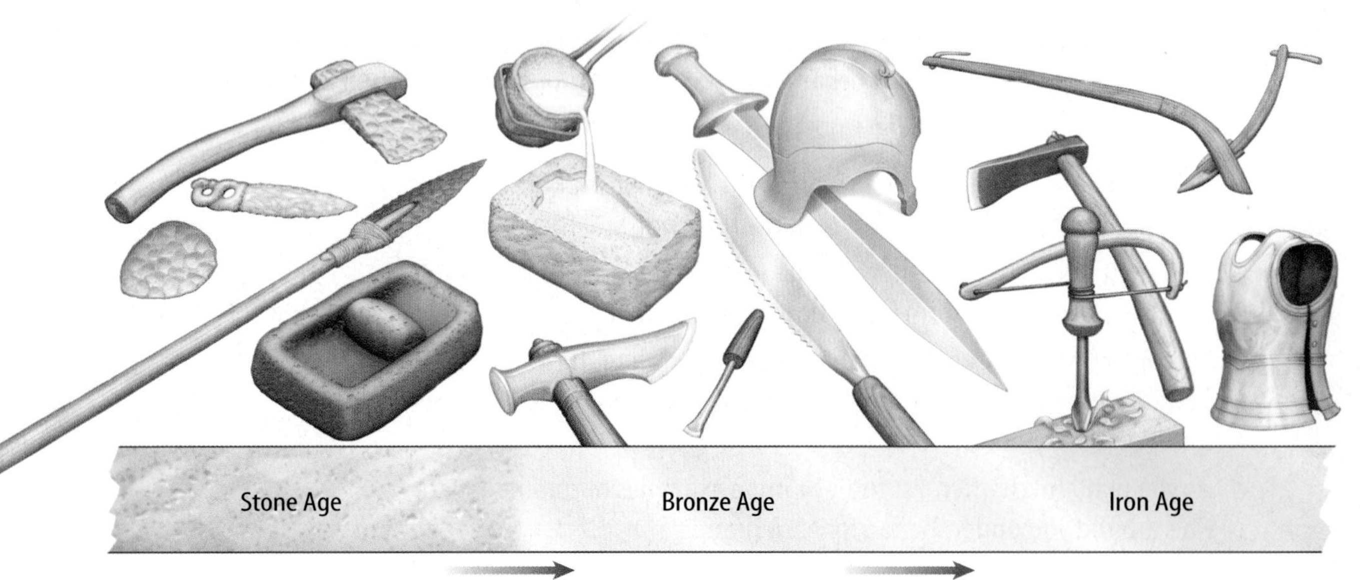

Figure 1
This historic time line shows the Stone Age, Bronze Age, and Iron Age. Some artifacts common from those times are shown above the time line.

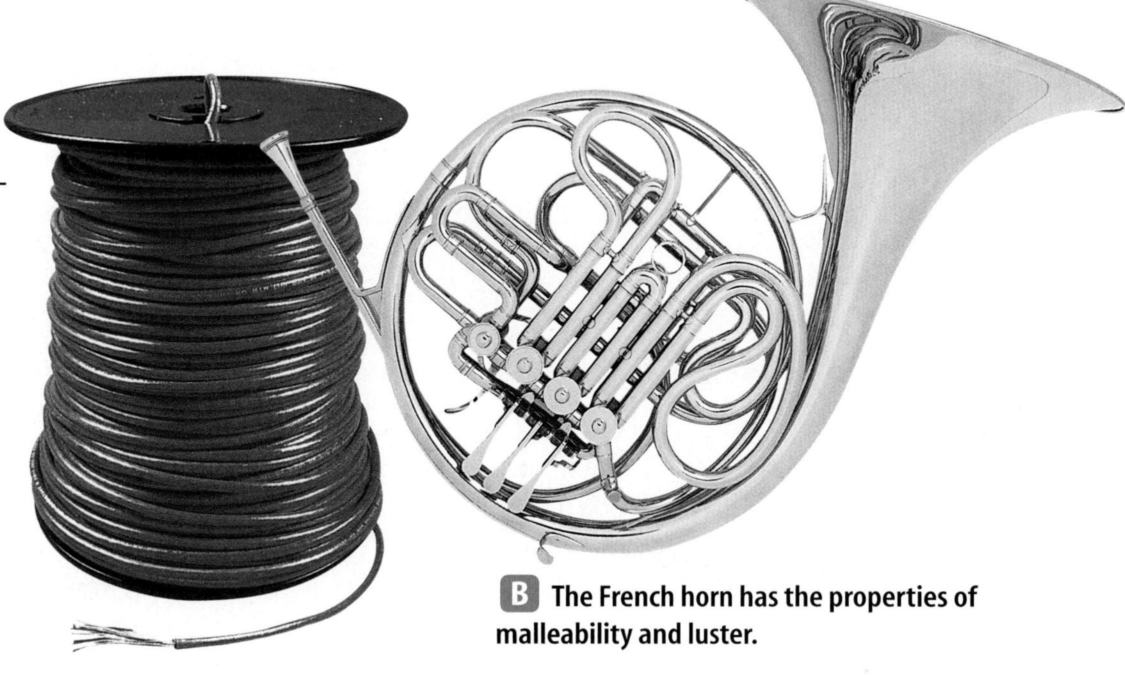

Figure 2

A This roll of copper wire reflects the metallic properties of ductility and conductivity.

B The French horn has the properties of malleability and luster.

Materials Change Typical objects from the Bronze Age and the following Iron Age are shown in **Figure 1.** Bronze and iron are still used today, but it is doubtful that these ancient people would recognize them. The methods of processing these alloys have undergone many changes. Other alloys also have been developed through the ages giving people a large selection of materials to choose from today.

Properties of Metals and Alloys Alloys retain the metallic properties of metals, some of which are shown in **Figure 2,** but what are the properties of metals? Metals have **luster,** which means they reflect light or have a shiny appearance. The shiny appearance of aluminum foil and copper wire demonstrates the property of luster. **Ductility** (duk TIH luh tee) means the metal can be pulled into wires. The copper electrical wire in your home demonstrates the ductility of metals and alloys. **Malleability** (ma lee uh BIH luh tee) is the property that allows metals and alloys to be hammered or rolled into thin sheets. Aluminum foil that is used in food preparation and food storage demonstrates the malleability of aluminum. The French horn above demonstrates the luster and malleability of brass. **Conductivity** (kahn duk TIH vuh tee) means that heat or electrical charges can move easily through the material. Metals and alloys have high conductivity because some of their electrons are not tightly held by their atoms. Metals and alloys usually are good conductors of heat and electricity because of these loosely bound electrons. Copper is used to carry electricity because it is conductive and ductile.

Reading Check *What are five other examples of items that you know of that have metallic properties?*

Mini LAB

Observing Properties of Alloys

Safety Precautions:

Procedure

1. Observe a small sheet of **aluminum foil.**
2. Using a **conductivity tester,** test the following items for their ability to conduct electric current: aluminum foil, **paper, pencil, ink pen,** and **paper clip.**

Analysis

1. What metallic properties of the foil do you observe?
2. Explain why each item was or was not able to conduct electric current.

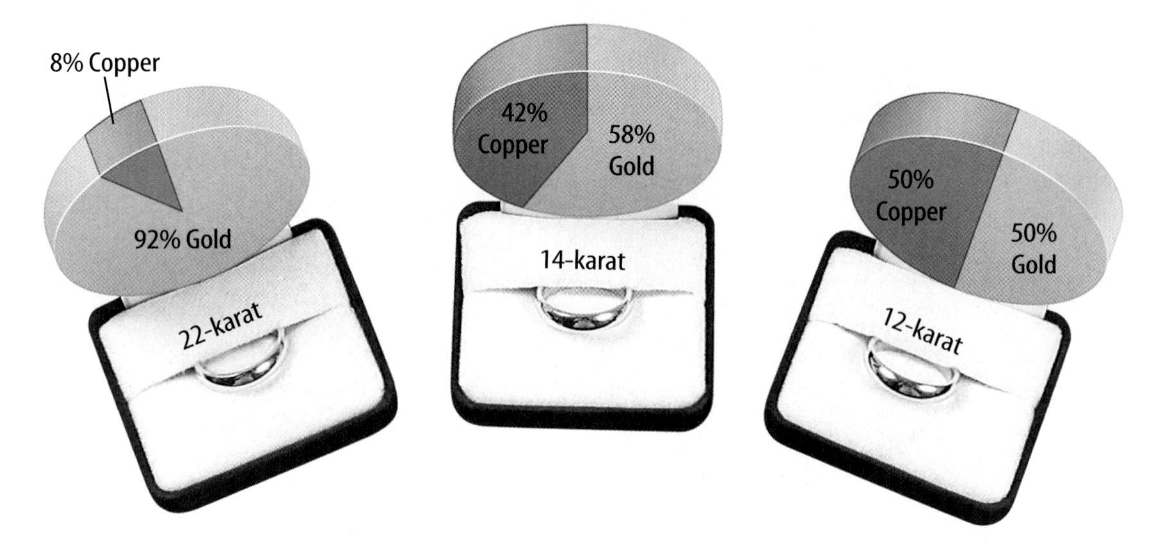

Figure 3
These gold rings appear to look alike, but they vary in the amount of copper that has been added to the gold. The composition of the alloy that was used to make the ring will determine its properties.

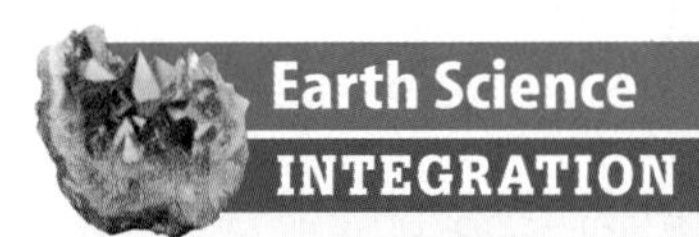

During the California gold rush, prospectors obtained gold by panning. Panning is a process in which dirt and gravel are washed away with water, leaving gold in the bottom of the pan. In your Science Journal describe the properties of gold that were important in making this technique useful.

Choosing an Alloy What properties of an alloy are most important? The answer depends upon how the alloy will be used and which characteristics are the most desirable. Look at the characteristics of familiar objects made from alloys such as the gold jewelry shown in **Figure 3.** The rings appear to be made of pure gold, but they are made from alloys.

Gold is a bright, expensive metal that is soft and bends easily. Copper, on the other hand, is an inexpensive metal that is harder than gold. When gold and copper are melted, mixed, and allowed to cool, an alloy forms. The properties will vary depending upon the amount of each metal that is added. A ring made with a higher percentage of gold will bend easily due to gold's softness. This ring will be more valuable because it contains a higher percentage of gold. A ring with a higher percentage of copper will not bend as easily because copper is harder than gold. This ring will be less valuable because it contains more copper, a less-expensive metal.

Which properties are needed? Choosing the alloy to make a piece of jewelry and deciding which alloy to use to manufacture a drill bit do not appear to have much in common. However, the characteristics of the final product must be considered in both situations before the product is constructed.

How hard does the alloy have to be to prevent the object from breaking when it is used? Will the object be exposed to chemicals that will react with the alloy and cause the alloy to fail? These questions relate to the properties of the alloy and its intended use. This represents only two of the many possible questions that must be answered while a product is being designed.

Uses of Alloys Alloys are used in a variety of products, as shown in **Table 1.** If you see an object that looks metallic, it is most likely an alloy. Alloys that are exceptionally strong are used to manufacture industrial machinery, construction beams, and railroad cars and rails. Automobile and aircraft bodies that require strong materials are constructed of alloys that are corrosion resistant and lightweight but able to carry heavy loads. Other types of alloys are used in products such as food cans, carving knives, and roller skates. **Figure 4** shows some additional examples of alloys.

Table 1 Common Alloys

Name	Composition	Use
Bronze	copper, tin	jewelry, marine hardware
Brass	copper, zinc	hardware, musical instruments
Sterling Silver	silver, copper	tableware
Pewter	tin, copper, antimony	tableware
Solder	lead, tin	plumbing
Wrought Iron	iron, carbon	porch railings, fences, sculpture

If you have a tooth filling, your dentist might have used a silver and mercury alloy to fill it, preventing further tooth decay. Other alloys that are resistant to tissue rejection can be used inside the human body. Special pins and screws made from alloys are used by surgeons to connect broken bones, as shown in **Figure 5.** Alloys also are used as metal plates to repair damage to the skull. These plates protect the brain from injury, and are safe to use inside the body.

✔ **Reading Check** *What are several uses for alloys?*

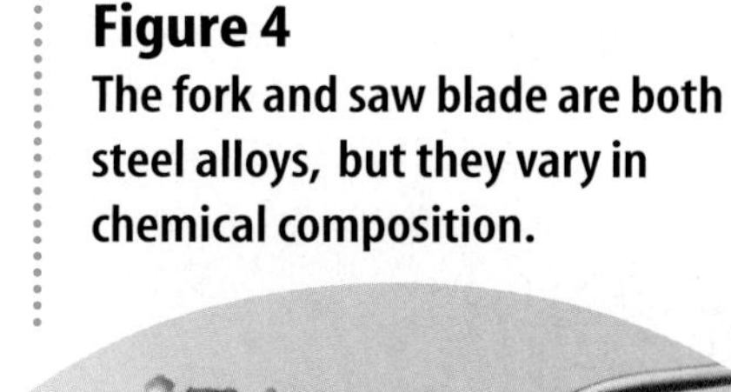

Figure 4
The fork and saw blade are both steel alloys, but they vary in chemical composition.

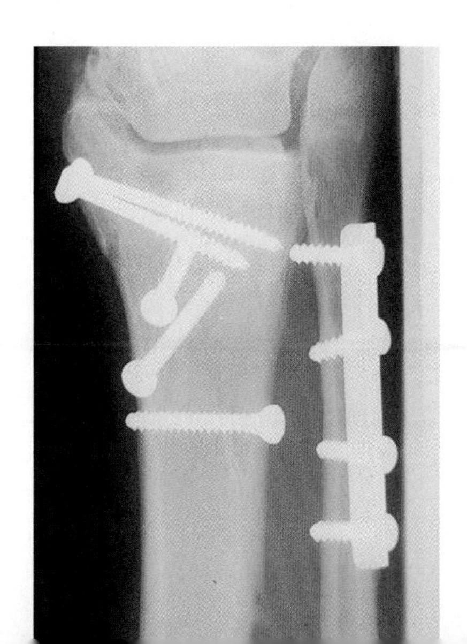

Figure 5
Surgical steel can be used to join bones. *What properties of this type of steel are most important?*

Steel—An Important Alloy There are various classes of steel. They are classified by the amount of carbon and other elements present, as well as by the manufacturing process that is used to refine the iron ore. The classes of steel have different properties and therefore different uses. Steel is a strong alloy and is used often if a great deal of strength is required. Office buildings have steel beams to support the weight of the structure. Bridges, overpasses, and streets also are reinforced with steel. Ship hulls, bedsprings, and automobile gears and axles are made from steel. Another class of steel, called stainless steel, is used in surgical instruments, cooking utensils, and large vessels where food products are prepared.

SCIENCE Online

Research Visit the Glencoe Science Web site at **science.glencoe.com** for recent news on research for new materials that can be used for space travel. Communicate what you learn to your class.

Reading Check *Why is steel an important alloy?*

New Aluminum Alloys Steel is not the only common type of alloy. Aluminum is familiar because it is used to make soda cans and cooking foil. Did you know that engineers also are using new aluminum and titanium alloys to build large commercial aircraft? The aircraft shown in **Figure 6** shows how extensively alloys are used in new aircraft construction. The new alloys are strong, lightweight and last longer than alloys used in the past. Also, a lighter plane is less expensive to fly.

Space Age Alloys Titanium alloy panels, developed for the space shuttle heat shield, might be used on future reusable launch vehicles that are designed to carry payloads to the *International Space Station.* The *X-33* experimental aircraft, shown at the beginning of the chapter, was a test vehicle for this technology.

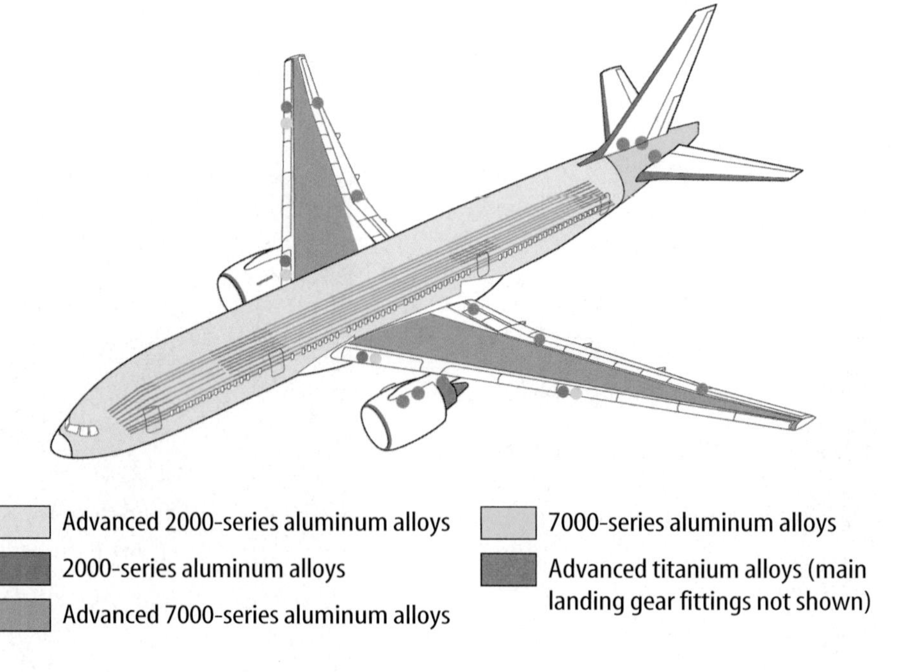

Figure 6
This commercial aircraft uses new alloys in its construction. Notice that the aircraft skin is mostly alloy construction.

New Titanium Alloy Heat Shield The original heat shield on the space shuttle, shown in **Figure 7A,** uses ceramic tiles that are prone to cracking as a result of the high temperature and stress they experience during reentry into Earth's atmosphere. Each broken ceramic tile must be removed carefully and a new one glued into place.

The new titanium alloy panels, **7B,** are much larger and easier to attach to the heat shield than the ceramic tiles are. A lower maintenance cost for the heat shield is expected by using the new alloy. Scientists and engineers also predict that the new alloy will protect the space shuttle as well as the old ceramic tiles did.

Figure 7
New alloys are being tested for use on the space shuttle. A The ceramic heat shield is constructed of small tiles, which are glued in place. They are more difficult to replace if damage occurs. B The shell of this test aircraft is made up almost entirely of titanium alloy panels.

Reading Check *Why is the new heat shield desirable?*

Section  1 Assessment

1. Give two medical uses of alloys.
2. What are the properties of metals and alloys?
3. How are steels classified?
4. Why must the property of a material be considered before it is used to make a product?
5. **Think Critically** If you were designing a skyscraper in an earthquake zone, what properties would the structural materials need?

Skill Builder Activities

6. **Comparing and Contrasting** Compare and contrast ceramic tiles and titanium alloys used on the space shuttle heat shield. **For more help, refer to the Science Skill Handbook.**
7. **Solving One-Step Equations** Use the information from **Figure 3** to calculate the actual amount of gold in a 65-g, 14-karat gold necklace. How many grams of copper are in the necklace? **For more help, refer to the Math Skill Handbook.**

SECTION

Versatile Materials

As You Read

What You'll Learn

- **Examine** the versatile properties of ceramics.
- **Identify** how ceramic materials are used.
- **Explain** what a semiconductor is.

Vocabulary

ceramics
doping
integrated circuit
semiconductor

Why It's Important

Ceramics and semiconductors are classes of materials with versatile properties that are used in many modern devices.

Ceramics

Do you think of floor tiles or pottery when you see the word *ceramic?* **Ceramics** are materials made from dried clay or clay-like mixtures. Ceramics have been around for centuries—in fact, pieces of clay pottery from 10,000 B.C. have been found. The first walled town, Jericho, was built about 8,000 B.C. The wall surrounding Jericho, as well as the homes inside the walls were constructed of bricks made from mud and straw that were baked in the Sun. Around 1,500 B.C., the first glass vessels were made and kilns were used to fire and glaze pottery. By 50 B.C. the Romans developed concrete and used it as a building material. Some of the structures built by the Romans still stand today. About the same time that Romans were developing concrete, the Syrians were developing glass-blowing techniques to make glass vessels. Pottery, bricks, glass, and concrete are examples of ceramics.

How are ceramics made? Traditional ceramics are made from easily obtainable raw materials—clay, silica (sand), and feldspar (crystalline rocks). These raw materials were used by ancient civilizations to make ceramic materials and still are used today. However, some of the more recent ceramics are made from compounds of metallic elements and carbon, nitrogen, or sulfur.

Reading Check *What raw materials are used to make traditional ceramic objects?*

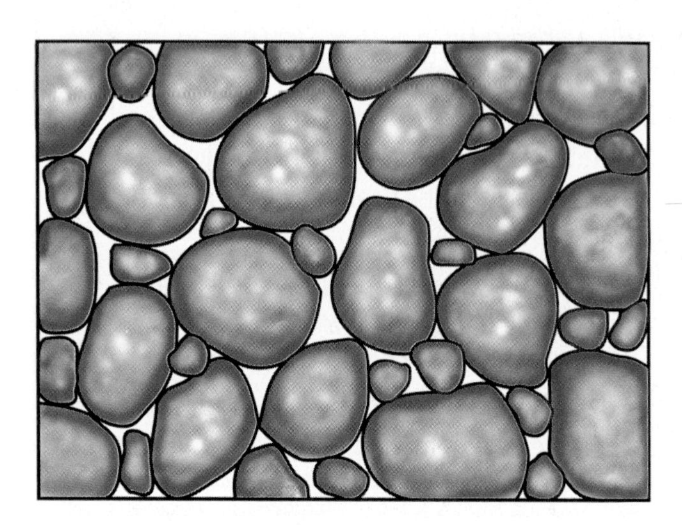

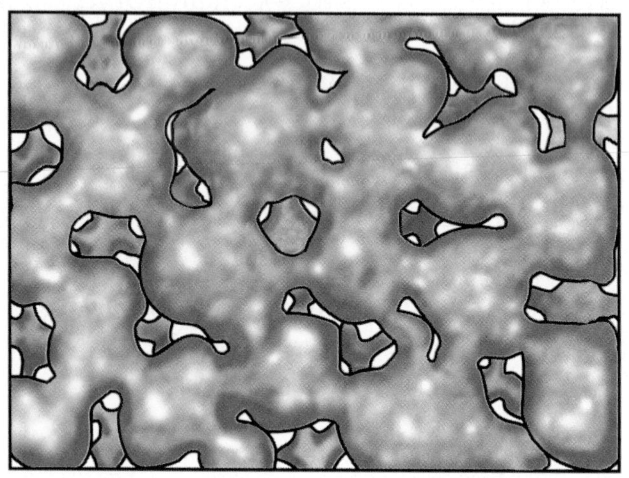

Figure 8
Ceramics are molded and then heated to high temperatures to force the particles to merge. The object shrinks and the structure becomes more dense.

Figure 9
Ceramic materials are used for a wide range of products. Glass, pottery, bricks, and tile are all made from ceramics.

Firing Ceramics After the raw materials are processed, ceramics are usually made by molding the ceramic into the desired shape, then heating it to temperatures between 1,000°C and 1,700°C. The heating process, called firing, causes the spaces between the particles to shrink, as shown in **Figure 8.** The entire object shrinks as the spaces become smaller. This extremely dense internal structure gives ceramics their strength. This is demonstrated by the use of ceramics on the space shuttle heat shield. They are able to withstand the high temperatures and stress of reentry into Earth's atmosphere. However, these same ceramics also are fragile and will break if they are dropped or if the temperature changes too quickly.

Traditional Ceramics Ceramics are known also for their chemical resistance to oxygen, water, acids, bases, salts, and strong solvents. These qualities make ceramics useful for applications where they may encounter these substances. For instance, ceramics are used for tableware because your foods contain acids, water, and salts. Ceramic tableware is not damaged by contact with foods containing these substances.

Traditional ceramics are used also as insulators because they do not conduct heat or electricity. You may have seen electric wires attached to poles or posts with ceramic insulators. These insulators keep the current flowing through the wire instead of into the ground.

The properties of ceramics can be customized which makes them useful for a wide variety of applications as shown in **Figure 9.** Changing the composition of the raw materials or the manufacturing process changes the properties of the ceramic. Manufacturing ceramics is similar to manufacturing alloys because scientists determine which properties are required and then attempt to create the ceramic material.

TRY AT HOME Mini LAB

Modeling a Composite Material

Safety Precautions

WARNING: *Wear goggles and an apron while doing this lab. Wash your hands before leaving the lab.*

Procedure
1. Mix four tablespoons of **sand,** four tablespoons of **aquarium gravel or small pebbles,** and six tablespoons of **white glue** in a paper cup.
2. Add enough water to thoroughly mix the ingredients.
3. Stir the mixture until it is smooth.
4. Allow the mixture to sit for several days and observe.
5. Dispose of the cup as instructed by your teacher.

Analysis
1. Describe what happened to your mixture after several days. Is it a ceramic?
2. How is your mixture similar to concrete?
3. What are some of the properties of your product?

Figure 10
This ceramic hip socket is used to replace damaged ones in the human body.

Modern Ceramics Ceramics can be customized to have nontraditional properties, too. For instance, ceramics traditionally are used as insulators, but there are exceptions. For instance, chromium dioxide conducts electricity as well as most metals and some copper-based ceramics have superconductive properties. One application of nontraditional ceramics uses a transparent, electrically conductive ceramic in aircraft windshields to keep them free of ice and snow.

Ceramics have medical uses too. **Figure 10** shows a ceramic replacement hip socket for use in the human body. Ceramics can be used in the body because they are strong and resistant to body fluids, which can damage other materials. In the medical field, surgeons use ceramics for the repair and replacement of joints such as hips, knees, shoulders, elbows, fingers, and wrists. Dentists use ceramics for tooth replacements, repair, and braces.

Problem-Solving Activity

Can you choose the right material?

Scientists are learning about atoms and how they bond at a rapid pace. With this new knowledge, technology also is advancing quickly. Chemists today are able to create substances with a wide range of properties. This is especially evident in the production of specialized ceramics.

Ceramic Properties

Material	Wear Resistant	Conducts Electricity	Reacts with Chemicals	Melting Point
ceramic A	highly resistant	no	no	3,000°C
ceramic B	wears easily	no	yes	100°C
ceramic C	moderately resistant	yes	no	1,500°C
ceramic D	resistant	yes	no	500°C

Identifying the Problem

As an engineer working on the design of a new car, you need to select the right ceramic materials to build parts of the car's engine and its onboard computer. The table above shows the materials you have to choose from. Using the properties given in the table, decide which materials should be used for the engine parts and the onboard computer. Be prepared to explain your answer.

Solving the Problem

1. Which of the above materials would you use when you build the engine? Explain the factors that you considered to make your decision.
2. Which of the above materials would you select when building the onboard computer? Explain your selection.
3. If you had to choose a material for building the car's bumper, what factors would you consider? Do you think that a ceramic material would be the best choice? Explain your answer.

Semiconductors

Another class of versatile materials is semiconductors. **Semiconductors** are the materials that make computers and other electronic devices possible.

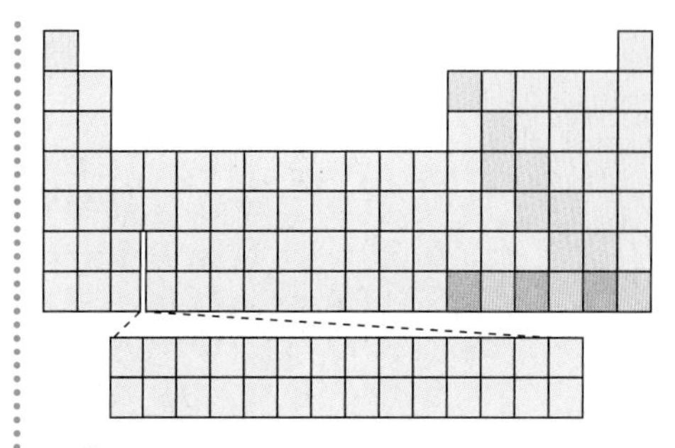

Figure 11
This outline of the periodic table clearly shows the metalloids, which appear in green.

The Periodic Table What are semiconductors? To answer this question, think about the periodic table. The elements on the left side and in the center of the table are metals. Metals are good conductors of electricity. Nonmetals, on the right side of the table, are poor conductors of electricity and are electrical insulators. The small number of elements found along the staircase-shaped border shown in **Figure 11** between the metals and nonmetals are metalloids. Some metalloids, such as silicon and germanium, are semiconductors. Semiconductors are poorer conductors of electricity than metals but better conductors than nonmetals, and their electrical conductivity can be controlled. It is this property that makes semiconductor devices useful and versatile.

Controlling Conductivity Adding impurities to some metalloids will alter their conductive properties. For example, silicon is used to make semiconductor devices. Its electrical conductivity can be increased by introducing impurities, such as atoms of arsenic or gallium, into its crystal structure, as shown in **Figure 12.** Adding even a single atom of one of these elements to a million silicon atoms significantly changes the conductivity. By controlling the type and amount of elements that are added, the conductivity of silicon can be made to vary over a wide range.

Figure 12
Pure silicon is a poor conductor of electricity. Adding an impurity to the crystal changes the conductivity.

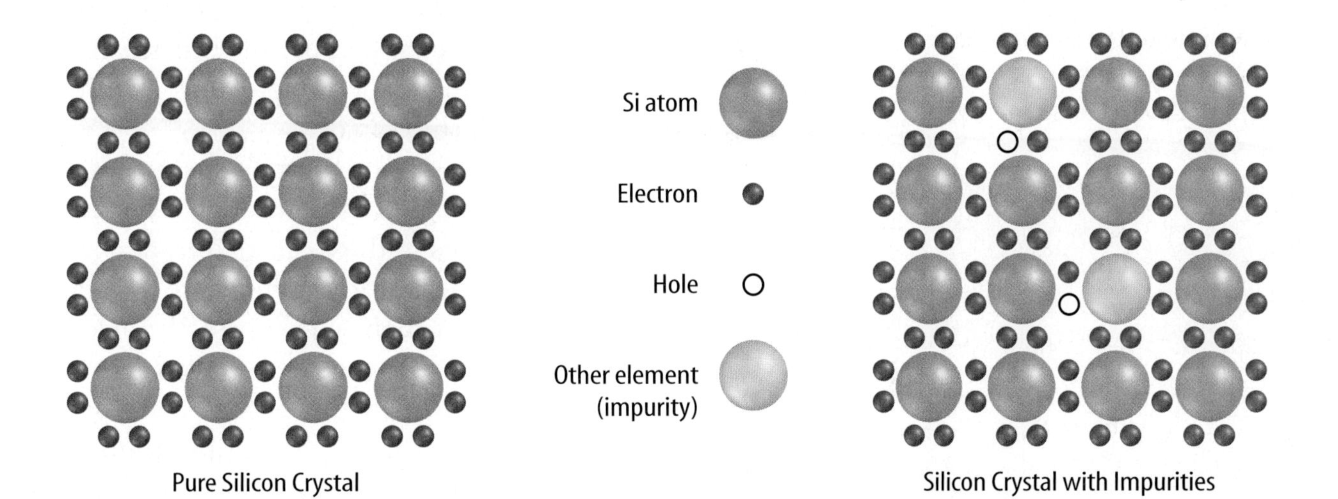

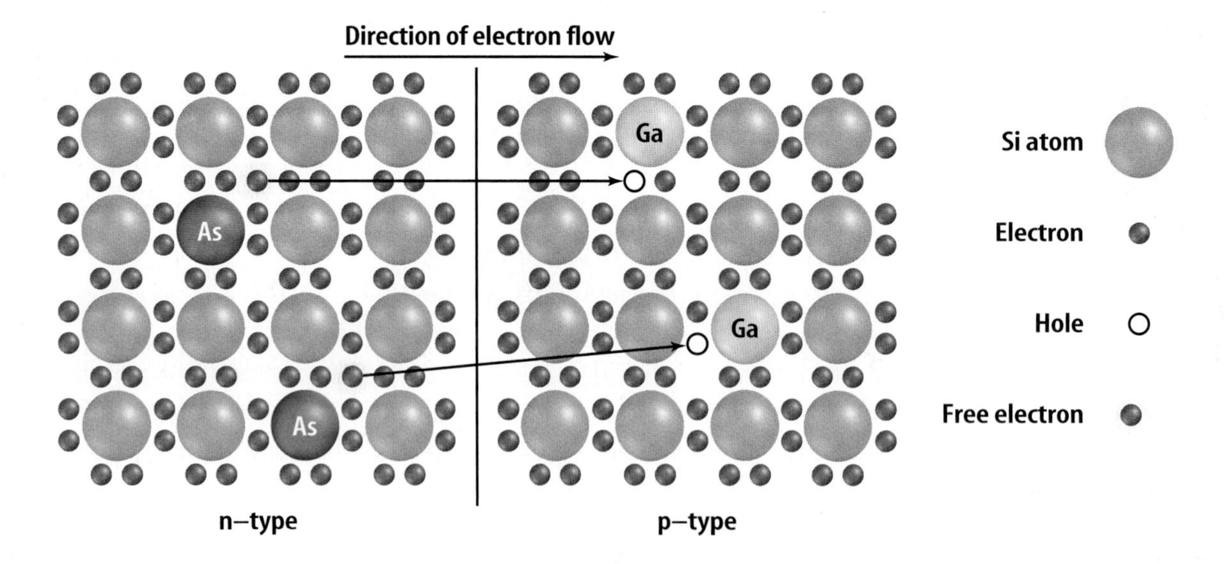

Figure 13
The electrons flow from the arsenic doped silicon to the germanium doped silicon, filling the available holes. The electron flow is controlled by the sequence of n-type or p-type semiconductors.

Doping The process of adding impurities or other elements to a semiconductor to increase the conductivity is called **doping.** Depending on the element added, the overall number of electrons in the semiconductor is increased or decreased. If the impurity causes the overall number of electrons to increase, the semiconductor is called an *n-type* semiconductor. If doping reduces the overall number of electrons, the semiconductor is called a *p-type* semiconductor.

Integrated Circuits By placing n-type and p-type semiconductors together, semiconductor devices such as transistors and diodes can be made. These devices are used to control the flow of electrons in electrical circuits, as shown in **Figure 13.** During the 1960s, methods were developed for making these components extremely small. At the same time the integrated circuit was developed.

An **integrated circuit** contains many semiconducting devices. Integrated circuits as small as 1 cm on a side can contain millions of semiconducting devices. Because of their small size, integrated circuits are sometimes called microchips. **Figure 14** shows how small an integrated chip can be.

Being able to pack so many circuit components onto a tiny integrated circuit was a technological breakthrough. This makes it possible for today's televisions, radios, calculators, and other devices to be smaller in size, cheaper to manufacture and capable of more advanced functions than older ones. Also because the circuit components are so close together, it takes less time for electric current to travel through the circuit. This enables electronic signals to be processed more rapidly by computers, cell phones, and other electronic appliances. **Figure 15** demonstrates how integrated circuits have given us faster, smaller, and more capable computers since the 1940s.

Figure 14
A single integrated circuit is tiny. This allows computers and other electronic devices to be compact.

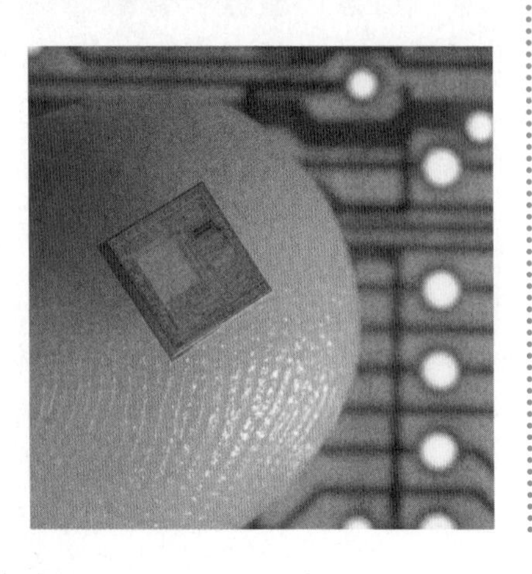

Figure 15

The earliest, room-size computers relied on vacuum tubes to store data. Today's computers use microchips, tiny flakes of silicon engraved with millions of circuit components. A selection of computers is shown here, beginning with the Electronic Numerical Integrator and Computer (ENIAC), developed by the Army in 1946.

Ⓐ A technician programs the ENIAC, the first electronic computer. Some of the 18,000 vacuum tubes that ran the ENIAC are shown at right.

Ⓑ A young woman operates a 1960s-era computer. The inset photo shows an integrated circuit from such a computer.

Ⓒ Teenagers surf the Internet on a modern personal computer. The microchips that store computer programs are now smaller than a fingernail.

Figure 16
Desktop computers use semiconductors to perform their tasks.

Research Visit the Glencoe Science Web site at **science.glencoe.com** to learn more about semiconductors and computers. Communicate what you learn to your class.

Semiconductors and Computers Semiconductors make our computers possible. A desktop computer is an example of a device that uses semiconductors. A computer has three main jobs. First, it must be able to receive and store the information that is needed to solve a problem. Next, it must be able to follow instructions to perform tasks in a logical way. Finally, a computer must communicate information to the outside world. All three jobs can be done with a combination of hardware and software components.

Computer hardware refers to the major permanent components of a computer, such as the keyboard, monitor, and central processing unit, CPU. These components are shown in **Figure 16.** Software refers to the instructions that tell the computer what to do. When a computer system is functioning properly, the hardware and software work together to perform tasks.

Section 2 Assessment

1. Describe how ceramic materials are made.
2. List five uses of ceramic materials.
3. Describe electrical conductivity of ceramics.
4. Explain what semiconductors are and where they are used.
5. **Think Critically** Computers and software have changed the way in which businesses operate. If you operated a distribution center for a manufacturer, how would you use computers to assist you?

Skill Builder Activities

6. **Drawing Conclusions** What properties of ceramics are useful in coffee mugs? **For more help, refer to the Science Skill Handbook.**
7. **Communicating** In your Science Journal, explain how computers aid scientists and engineers in their profession. Include information about problem solving and communicating in your explanation. **For more help, refer to the Science Skill Handbook.**

Polymers and Composites

Polymers

Polymers are similar to alloys and ceramics because they represent another class of materials. **Polymers** are a class of natural or manufactured substances that are composed of molecules arranged in large chains with small, simple, repeating units called monomers. Each link in the chain is a monomer. A **monomer** is one specific molecule that is repeated in the polymer chain. Polypropylene, for example, might have 50,000 to 200,000 monomers in its chain. Several examples of manufactured polymers are shown in **Table 2.** Not all polymers are manufactured. Some polymers occur naturally. Proteins, cellulose, and nucleic acids are polymers found in living things. In this section, the focus will be on manufactured polymers or synthetic polymers. **Synthetic** means that the polymer does not occur naturally, but it was manufactured in a laboratory or chemical plant.

Reading Check *What are some similarities and differences between monomers and polymers?*

As You Read

***What* You'll Learn**

- **Identify** what a polymer is and the variety of polymers around us.
- **Explain** what a composite material is and why composites are used.

Vocabulary

polymer
monomer
synthetic
composite

***Why* It's Important**

Polymers and composite materials can replace natural materials such as metal, wood, and paper.

Table 2 Common Polymers

Polymer	Monomer	Uses
Polyethylene	$-[CH_2-CH_2]-$	bottles, garment bags
Polyvinyl Chloride (PVC)	$-[CH_2-CHCl]-$	pipe, bottles, compact discs, computer housing
Polypropylene	$-[CH_2-CH(CH_3)]-$	rope, luggage, carpet, film
Polystyrene	$-[CH(C_6H_5)-CH_2]-$	toys, packaging, egg cartons, flotation devices
Polytetrafluoroethylene	$-[CF_2-CF_2]-$	nonstick cookware, gaskets, bearings

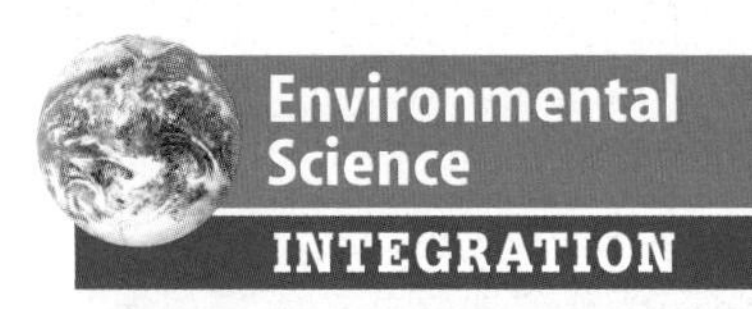

Synthetic polymers are used to make many disposable items such as plates, diapers, trash bags, and utensils. These items are used once, then thrown away. Most synthetic polymers do not decompose in landfills. In your Science Journal, infer the problems that this might cause and suggest solutions to these problems.

History of Synthetic Polymers Humankind has used natural polymers for centuries. The ancient Egyptians soaked their burial wrappings in natural resins to help preserve their dead. Animal horns and turtle shells, which contain natural resins, were used to make combs and buttons for many years. In the 1800s, scientists began developing processes to improve natural polymers and to create new ones in the laboratory.

In 1839, Charles Goodyear, an American inventor, found that heating sulfur and natural rubber together improved the qualities of natural rubber. By treating the rubber with sulfur, the natural rubber was no longer brittle when it became cold or soft when it became hot. In the late 1860s, John Hyatt developed celluloid as a replacement for ivory in billiard balls. Celluloid was used in other products such as umbrella handles and toys. These early polymers had many drawbacks, but they were the beginning of the development of a huge class of materials now referred to as polymers. Today, so many types of synthetic polymers exist that they tend to be divided into groups such as plastics, synthetic fibers, adhesives, surface coatings, and natural and synthetic rubbers. **Figure 17** shows a time line of when some of these materials were created.

Hydrocarbons Today, synthetic polymers usually are made from fossil fuels such as oil, coal, or natural gas. Fossil fuels are composed primarily of carbon and hydrogen and are referred to as hydrocarbons. Because synthetic polymers are made from hydrocarbons, carbon and hydrogen are the primary components of most synthetic polymers.

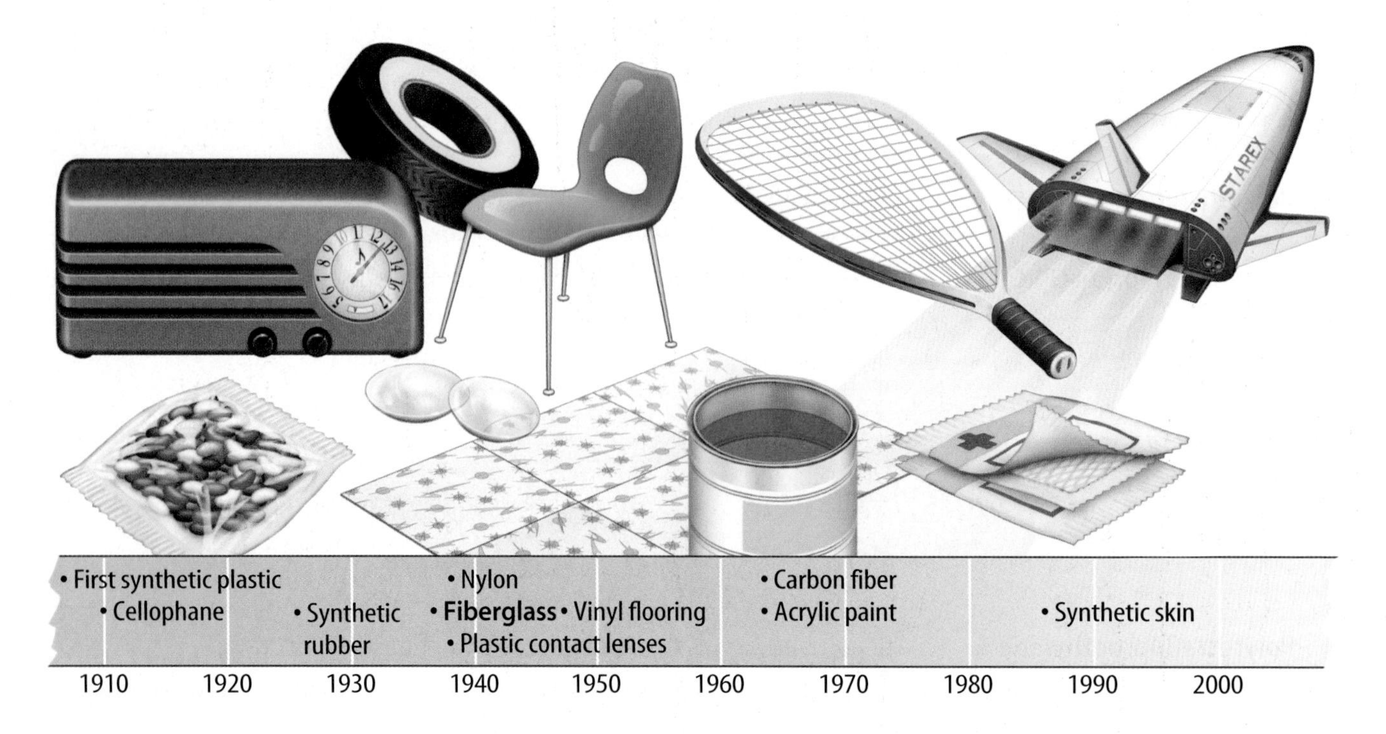

Figure 17
Polymers were developed in the late 1800s, but they did not become widely used until after World War II in 1945.

Changing Properties Polymers are a class of materials with a wide range of uses. The reason that polymers can be used for so many applications is directly related to the ease with which their properties can be modified. Polymers are long chains of monomers. If the composition or arrangement of monomers is changed, then the properties of the material will change.

Figure 18 shows how the monomer ethylene can be modified to produce a polymer with different properties and uses. Ethylene has only two carbon atoms and six bonding sites. The number of carbon atoms in the polymer can be high, and each bonding site represents a possibility of a change in properties. Polyethylene can be high density or low density depending upon how the molecules are attached to the monomer. One of the substances on the monomer can be replaced by another substance or a group of substances and the properties will change, too. The possibilities for creating new materials are almost limitless.

SCIENCE Online

Research Visit the Glencoe Science Web site at **science.glencoe.com** to learn more about how changing the monomer changes the properties of the polymer. Communicate what you learn to your class.

The Plastics Group Plastics are widely used for many products because they have desirable properties. Plastics are usually lightweight, strong, impact resistant, waterproof, moldable, chemical resistant, and inexpensive. Examples of plastics are easy to find. They are used to make toys, computer housing, telephones, containers, plates, and so on. The properties of plastics vary widely within this group. Some plastics are clear, some melt at high temperature, and some are flexible. Transparency, melting temperature, and flexibility are properties of plastics that relate to the composition of the polymer.

Figure 18
The arrangement of the branches along the chain can affect the properties of the polymer.

A **Low-Density Polyethylene, LDPE, is flexible, tough, and chemical resistant. The chain has a great deal of side-branching, which causes low density.**

B **High-Density Polyethylene, HDPE, is firmer, stronger, and less translucent than LDPE. This chain has little side-branching, which allows the chain to pack closer together, thus giving it a higher density and different properties.**

C **Polyvinyl chloride (PVC) is used in building materials. The substitution of chlorine for a hydrogen in the polyethylene chain makes the polymer harder and more heat resistant.**

Figure 19
Some synthetic polymers are used in hazardous conditions. A This aramid is a fiber that is fire proof. Clothing made from this fabric is used by firefighters and flight crews. B Survival cells are manufactured from another aramid for use in Indy race cars. C A survival cell in an Indy car provides protection for the driver during a crash.

A

B

Synthetic Fibers Nylon, polyester, acrylic, and polypropylene are examples of polymers that can be manufactured as fibers. Most synthetic fibers are composed of carbon chains because they are produced from petroleum or natural gas. Synthetic fibers can be mass-produced to almost any set of desired properties. Nylon is often used in wind and water-resistant clothing such as lightweight jackets. Polyester and polyester blended with natural fibers such as cotton often are used in clothing. Polyester fiber also is used to fill pillows and quilts. Polyurethane is the foam used in mattresses and pillows.

Special fibers called aramids are a family of nylons with special properties. **Figure 19** shows some uses for these materials. Aramids are used to make fireproof clothing. Firefighters, military pilots, and race car drivers are examples of professionals that make use of this special fabric. Another aramid is used to make bulletproof vests, race car survival cells, puncture-resistant gloves, and motorcycle clothing. Although they are light, these aramids are five times stronger than steel.

Adhesives Synthetic polymers are used to make adhesives that can be modified to provide the best properties for a particular application. Contact cements are used in the manufacture of automobile parts, furniture, leather goods, and decorative laminates. They bond instantly with the bond getting stronger after it dries. Structural adhesives are used in construction projects. One structural adhesive, silicone, is used to seal windows and doors to prevent heat loss in homes and other buildings. Ultraviolet-cured adhesives are used by orthodontists to adhere brace brackets to teeth. These adhesives bond after exposure to ultraviolet light. Other types of adhesives are hot-melt and transparent, pressure-sensitive tape.

✔ Reading Check *What are five uses of adhesives?*

C

Figure 20
Composite materials are used to make some cars and boats. A The glass fiber embedded in the plastic reinforces the plastic structure. B This composite material is strong, lightweight, and corrosion resistant.

Surface Coatings and Elastic Polymers Many surface coatings use synthetic polymers. Polyurethane is a popular polymer that is used to protect and enhance wood surfaces. Many paints use synthetic polymers in their composition, too.

Synthetic rubber is a synthetic elastic polymer. It is used to manufacture tires, gaskets, belts, and hoses. The soles of some shoes also are made from this rubber.

Taking a Cue from Nature Spinning long fibers into threads and fabrics is not an original idea. Spiders spun fibers for their webs long before humans copied the idea and began spinning fibers themselves. Nylon fiber is another idea borrowed from nature. The silkworm produces a highly desirable fiber that is woven into fabric for items such as blouses and stockings. Can you imagine how long it would take a silkworm to produce enough silk for one blouse? Nylon was produced in the laboratory as a possible substitute for silk. Why do you think natural silk fabric is more expensive than nylon fabric?

Research Visit the Glencoe Science Web site at **science.glencoe.com** for recent news on newly created polymers and new uses for older ones. Write a short passage in your Science Journal about the information that you find.

Composites

The properties of a synthetic polymer can be altered by using more than one material. A **composite** is a mixture of two or more materials—one embedded or layered in the other. Composite materials of plastic are used to construct boat and car bodies, as shown in **Figure 20.** These bodies are made of a glass-fiber composite that is a mixture of small threads or fibers of glass embedded in a plastic. The structure of the fiberglass reinforces the plastic, making a strong, lightweight composite. If a substance is light but brittle, such as some plastics, embedding flexible fibers into it can alter the brittleness property. After the substance has the flexible fibers embedded, the product is less brittle and can withstand greater forces before it breaks. Glass fibers are used often to reinforce plastics because glass is inexpensive, but other materials can be used, as well.

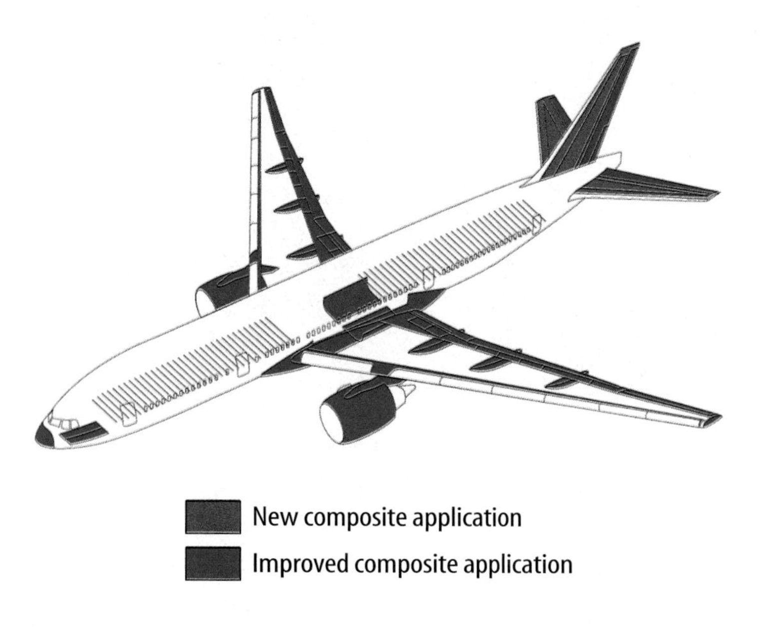

Figure 21
Commercial aircraft use composite materials in some locations. Composites provide corrosion resistance and are simple to repair.

Composites in Flight Composite materials are used in the construction of satellites. Lighter-weight satellites are less expensive to launch into orbit, yet the structure is able to withstand the stress of the launch. Carbon fibers are used to strengthen the plastic body, creating a material that is four times more firm and 40 percent stronger than aluminum. Satellites made of graphite composites are about 13 percent lighter than satellites made of aluminum. The composite material is stronger and lighter in weight than aluminum therefore, it is less expensive to launch and can endure the stress of the launch better.

Commercial aircraft use composite materials in their construction, as shown in **Figure 21.** Aircraft made of composites also benefit from the strong yet lightweight properties of composite materials. The weight of this aircraft was reduced by more than 2,600 kg by using advanced alloys and composite materials. The lower weight results in cost savings by reducing the amount of fuel required to operate the aircraft.

Reading Check *Why are composites used in aircraft?*

Section 3 Assessment

1. Explain what a polymer is and give three examples of items that are made from polymers.
2. What are the raw materials that are used to make most synthetic polymers?
3. Explain what a composite material is and give three examples of items that are made from composites.
4. Synthetic polymers can be separated into groups based upon their uses. What are some of these groups?
5. **Thinking Critically** How are synthetic polymers creating waste-disposal problems? Discuss possible solutions to this problem.

Skill Builder Activities

6. **Classifying** Make a list of 20 items in your home that are made of polymers. Group the items based on the following properties: elastic, heat resistant, and shatterproof. Did any of your items have all three properties? Why or why not? **For more help, refer to the Science Skill Handbook.**
7. **Communicating** You've created a new synthetic polymer. In your Science Journal, describe the properties of your polymer and suggest uses for it. **For more help, refer to the Science Skill Handbook.**

Activity

What can you do with this stuff?

This substance is fun to play with. But how do you describe its properties?

What You'll Investigate

What are the properties of this new material and what can it be used for?

Materials

white glue
borax laundry soap
warm water
250-mL beaker or cup
100-mL beaker or cup
graduated cylinder
craft stick for mixing

Goals

- **Predict** the properties of this material.
- **Determine** possible uses for the material.

Safety Precautions

WARNING: *Never eat lab materials.*

Procedure

1. Prepare a data table to record your observations of the following: stretched slowly, stretched quickly, rolled into a ball and left alone, pressed onto newspaper ink, dropped on a hard surface.
2. Put about 100 mL of warm water in the larger beaker and add borax laundry soap until soap no longer dissolves.
3. Put 5 mL of water and 10 mL of white glue into the smaller beaker and mix completely.
4. Add 5 mL of the borax solution to the glue solution and continue mixing for a couple of minutes.

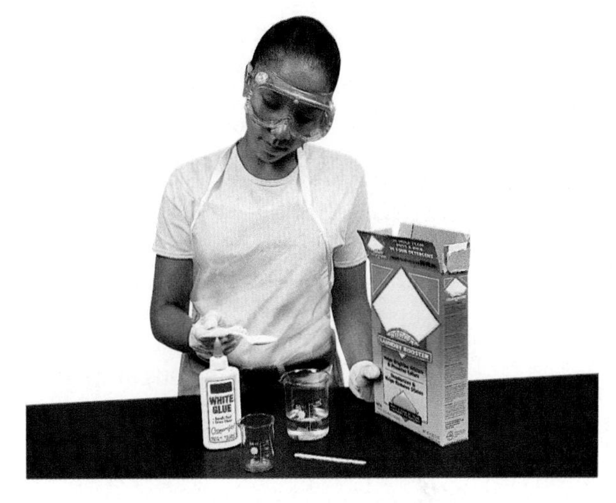

5. When the substance firms up, remove it from the container and continue to mix it by pressing with your fingers until it is like soft clay.
6. **Examine** the properties of this material and record them in your data table.

Conclude and Apply

1. What were the properties of this material?
2. Get together with other students and brainstorm. What could this material be used for, and which of its properties would make it useful for that purpose?
3. You're in charge of marketing this product. Prepare an advertisement with text and graphics on a sheet of notebook paper. Which magazine would you place this ad in and why?

Compare your conclusions with those of other students in your class. **For more help, refer to the Science Skill Handbook.**

Activity

Why are composite materials used instead of wood or metal in high performance applications? Scientists and engineers test many materials before selecting the best one for a specific use. Composites are used in aircraft parts, sports equipment, and space vehicles because of their strength and low weight. What other factors might be important? How do you measure performance?

Recognize the Problem

How can you compare the performance of a composite with other materials?

Thinking Critically

How do you choose the best material for an application?

Goals

- **Model** appropriate equipment to test wood, steel, and fiberglass composite rods.
- **Measure** the force required to flex the test rods.
- **Calculate** the relative flexibility of each rod.
- **Estimate** the performance of each material.

Possible Materials

meterstick
spring scale, 0–12-kg range and 0–2-kg range
wood, steel, and fiberglass composite rods, 6.35 mm in diameter by 50 cm long
supports to hold the test rods
graph paper

Safety Precautions

WARNING: *Wear safety goggles at all times during this lab.*

Material	Distance Flexed (cm)	Force Required (kg)
composite	1	
composite	2	
composite	3	
steel	1	
steel	2	
steel	3	
wood	1	
wood	2	
wood	3	

Planning the Model

1. To test the model, hook the spring scale to the center of the fiberglass rod. Have a team member pull down on the spring scale until the top of the test rod moves down 1 cm from the zero point. Record the scale reading on the data table. Pull down on the spring scale until the rod flexes 2 cm, then 3 cm. Record both of the spring scale readings on the data chart.
2. Repeat the tests on the steel and wood rods. Record the data in your table.

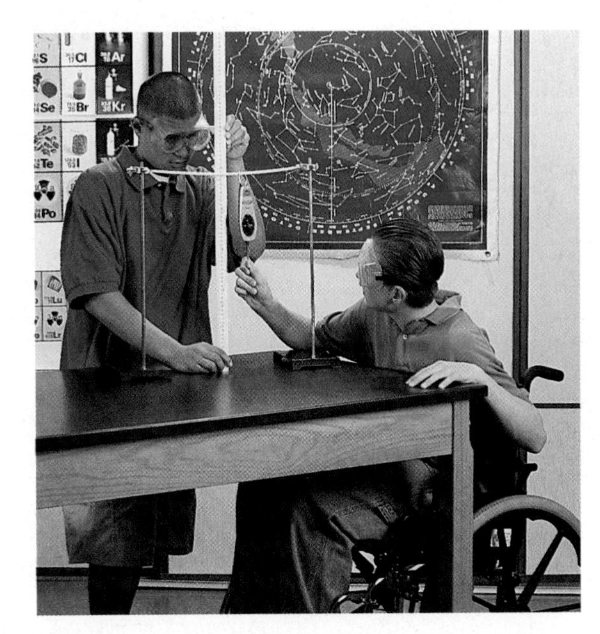

Analyzing and Applying Results

1. For each of the rods, graph the force measured on the *y*-axis and the distances on the *x*-axis. Calculate the slope of each line in kilograms per centimeter. The slope is a relative measure of the flexibility of the samples.
2. Divide the slope of each line by the density of the corresponding material to get a specific performance number, which may be used to compare different materials. The densities are: composite = 1.2 g/cm^3, steel = 7.9 g/cm^3, and wood = 0.5 g/cm^3.
3. **Determine** which rod had the highest specific performance number. What is meant by the statement that a polymer composite is twice as strong as steel?
4. **Analyze** which variables could affect the flexibility measurement.

Making the Model

1. Using the data that you have gathered, create a model exhibit showing possible construction uses for each of these materials.
2. Indicate the reason the specific material was chosen.

Give an oral presentation on choosing the best material for a specific application to another class of students using your model.

Wonder Fiber

In 1964, Stephanie Kwolek was a chemist working at a research laboratory. Her assignment? Create a new type of tough, light-weight fiber. Kwolek's routine at the lab was about the same each day. She combined different substances in test tubes. She stirred them. She heated them. Then she would have any new substance spun into fibers. The idea was to get a fiber that was both strong and lightweight. But coming up with a new superfiber was difficult.

This police dog can thank Stephanie Kwolek for its bulletproof vest!

A close-up view of fiber-optic cables wrapped in Kwolek's discovery.

Stephanie Kwolek

A Shocking Discovery

At one point, Kwolek was working with two polymers. She wanted to use heat to combine them, but they would not melt. So she decided to use a solvent to dissolve them. But when she poured the solvent onto one of the polymers, she got something unlike anything she had ever seen. Not only did it look different, it behaved differently when she stirred it. It separated into two distinct layers.

Kwolek thought this strange liquid might be something special. She asked one of her coworkers to spin it into fibers using a machine called a spinneret. The other chemist refused at first, saying the liquid wouldn't form fibers. And besides, it would probably gum up the equipment. But Kwolek had a hunch about this liquid. So she persisted. Finally, after several days, the other chemist agreed to try to spin the liquid into fibers.

A New Type of Fiber

What they found shocked Kwolek and everyone else in the lab. The fibers that formed in the spinneret were very lightweight but also extremely stiff and strong. Kwolek had accidentally discovered a new type of synthetic fiber—a fiber made from a new substance called a liquid-crystal solution.

This new fiber was five times stronger than steel, and over the decades since its discovery, it has been put to many uses. It is found in a number of products where a material that is light but very strong is needed. These products include bulletproof vests, boat hulls, fiber-optic cables, cut-resistant gloves, airplane parts, skis, tennis rackets, and parts of spacecraft.

The discovery was a huge accomplishment for Kwolek, but an even more important one for the world. The fiber she discovered 40 years ago has benefited many people and saved many lives, from the creation of bulletproof vests used by police officers to the tough clothing used by firefighters.

CONNECTIONS Research Use resources in your school's media center or the Glencoe Science Web site to find out more about the superfiber Kwolek discovered. List as many of the possible uses as you can. Compare what you uncover with what others in the class find.

SCIENCE *Online*
For more information, visit science.glencoe.com

Chapter 22 Study Guide

Reviewing Main Ideas

Section 1 Materials with a Past

1. People have been making and using alloys for thousands of years. Some common alloys include bronze, brass, and various alloys of iron.

2. An alloy is a mixture of a metal with one or more other elements. Metals and alloys have the properties of luster, ductility, malleability, and conductivity. *Which property or properties does the figure above demonstrate?*

Section 2 Versatile Materials

1. Ceramics are used in a wide range of products. This is due to the ability of scientists to customize the properties of ceramics. *What properties does this ceramic material in the windshield have?*

2. Semiconductors are made from silicon doped with other elements.

3. Ceramic materials are made by molding the object, and then heating the object to high temperatures. This process increases the density of the material.

4. Integrated circuits contain n-type and p-type semiconducting devices.

Section 3 Polymers and Composites

1. Polymers are a class of natural or human-made substances that are composed of molecules that are in large chains with simple repeating units called monomers.

2. Synthetic polymers can be produced in many forms, ranging from thin films to thick slabs or blocks. Synthetic fibers are produced in thin strands that can be woven into fabrics.

3. A composite is a mixture of two materials, one embedded in the other. Reinforced concrete and fiberglass are examples of composites. The skateboard in the figure to the right is constructed of a fiberglass composite. *What advantages would composite construction have over wood construction?*

After You Read

FOLDABLES Reading & Study Skills

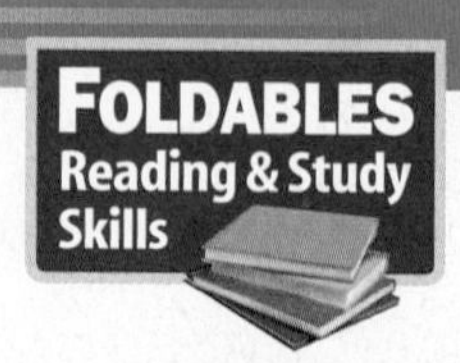

Number the tabs on your Foldable in the order the materials were discovered. Predict new materials of the future.

Visualizing Main Ideas

Complete the following concept map on materials.

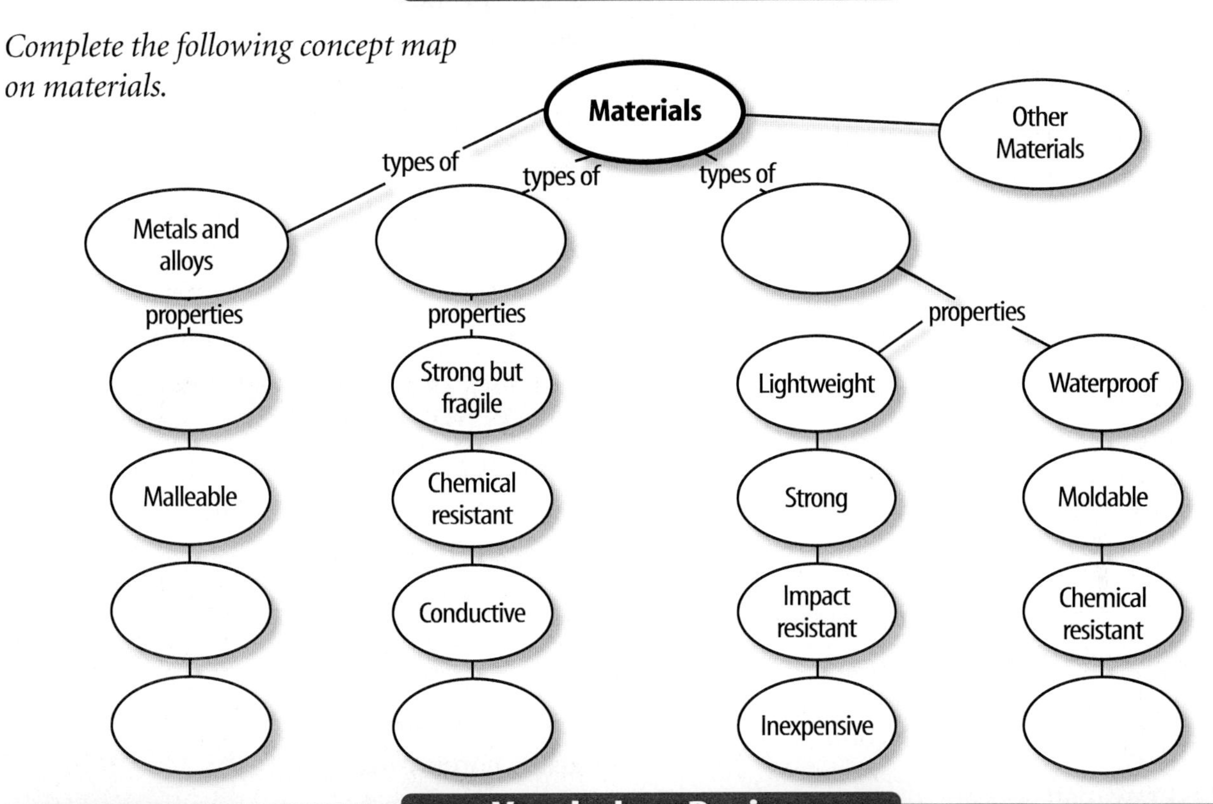

Vocabulary Review

Vocabulary Words

a. alloy
b. ceramics
c. composite
d. conductivity
e. doping
f. ductility
g. integrated circuit
h. luster
i. malleability
j. monomer
k. polymer
l. semiconductor
m. synthetic

Study Tip

After each day's lesson, make a practice quiz for yourself. Later, when you're studying for the test, take the practice quizzes that you created.

Using Vocabulary

Replace the underlined words with the correct vocabulary word(s).

1. The ability to be hammered or rolled into thin sheets is a property of metals and alloys.
2. Compounds made from clay, silica, and feldspar are used on the space shuttle heat shield and might be replaced by an alloy.
3. Fiberglass is a human-made material that is used to make boats and skateboards.
4. Fiberglass is a mixture of two materials.
5. A chrome bumper has a reflective surface.
6. A molecule that is formed by a long chain of repeating units is used to make food storage containers.

Chapter 22 Assessment

Checking Concepts

Choose the word or phrase that best answers the question.

1. Which metal replaces bronze as a widely used metal?
A) copper
B) tin
C) zinc
D) iron

2. Why are metals and alloys good conductors of heat and electricity?
A) They have loosely bound electrons within the atom.
B) They have luster and malleability.
C) They are composed of mixtures.
D) They have a shiny appearance.

3. An alloy of steel will contain iron and what element?
A) mercury
B) tin
C) zinc
D) carbon

4. What raw materials are many synthetic polymers made from?
A) hydrocarbons
B) iron ore
C) fiberglass
D) ceramics

5. What type of fibers are used to reinforce polymers in some automobile bodies?
A) ceramic
B) metal alloy
C) hydrocarbon
D) glass

6. Which element below is found in both brass and bronze?
A) mercury
B) copper
C) tin
D) zinc

7. Which of the following is a natural fiber?
A) nylon
B) polyester
C) silk
D) acrylic

8. Which group of materials below is not classified as synthetic?
A) ceramics
B) alloys
C) composites
D) metal ores

9. Customizing properties is NOT likely in which of the following?
A) alloys
B) synthetic polymers
C) ceramics
D) pure metals

10. Which of the following elements is used to dope silicon crystals?
A) carbon
B) zinc
C) copper
D) gallium

Thinking Critically

11. A lower karat gold has less gold in it than a higher karat gold. Why might you prefer a ring that is 10-karat gold over a ring that is 20-karat gold?

12. Explain the advantages and disadvantages of using composites in the world of sports.

13. A synthetic fiber might be preferred over a natural fiber for use outdoors because it will not rot. Explain how this could negatively affect the environment.

Developing Skills

14. Interpreting Scientific Illustrations Look at the polymer below. Draw the monomer upon which the polymer is based.

```
[ H   H   H   H   H   H  ]
[ |   |   |   |   |   |  ]
[ C — C — C — C — C — C  ]
[ |   |   |   |   |   |  ]
[ H   CN  H   CN  H   CN ]
```

15. Measuring in SI A bronze trophy has a mass of 952 g. If the bronze is 85 percent copper, how many grams of tin are contained in the trophy?

16. **Comparing and Contrasting** Compare and contrast alloys and ceramics.

17. **Recognizing Cause and Effect** A student performing the two-page lab did not see any difference in the flexing of the rods. What are some possible causes of this result?

18. **Concept Mapping** Draw a network tree about polymers, moving from the most general term to the most specific. Use the terms *composites, hydrocarbons, polymers, adhesives, plastics, synthetic fibers,* and *surface coatings.*

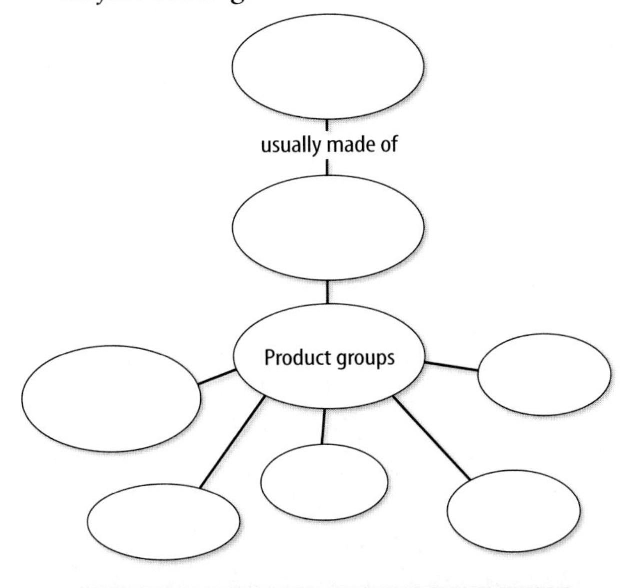

Performance Assessment

19. **Project** Research the production of silk fabric and compare it with production of synthetic fabrics such as nylon. Display samples of each.

Technology

Go to the Glencoe Science Web site at **science.glencoe.com** or use the **Glencoe Science CD-ROM** for additional chapter assessment.

Test Practice

An engineer studied various composite materials and created the table below.

Thermal Properties of Composite Ceramics

Material	Melting Temperature (°C)	Thermal Conductivity (W/m•K)
Alumina	2,050	30.00
Fused Silica	1,650	2.00
Soda-lime Glass	700	1.70
Polyethylene	120	0.38

Use this table and answer the following questions.

1. According to the table, what is the thermal conductivity of Fused Silica?
   A) 30 W/m•K
   B) 2.0 W/m•K
   C) 1.7 W/m•K
   D) 0.38 W/m•K

2. According to this table, which composite ceramic will melt first if heated?
   F) Alumina
   G) Fused Silica
   H) Soda-Lime Glass
   J) Polyethylene

Reading Comprehension

Read the passage. Then read each question that follows the passage. Decide which is the best answer to each question.

Medicine and Synthetic Compounds

In the constant battle to develop new treatments for fatal illnesses such as cancer and AIDS, scientists have studied the possibilities provided by synthetic compounds.

Synthetic compounds are human-made, and as a result, can be adapted to suit multiple needs and purposes. For example, a doctor discovered that a rare metalloid, germanium, improves patients' stamina and endurance. This could increase the quality of life for both cancer and AIDS patients. But because it is so rare, medications would need to be made from synthesized germanium.

Other research has shown that medicine containing synthetic opiates might be able to limit the growth of the HIV virus as well as limit the development of secondary infections.

Having learned about the effectiveness of synthetic polyanionic compounds in treating cancer and developing antibiotics, a research team is now working to use synthetic polyanionic compounds to treat AIDS.

In addition to these cases, many other studies are being conducted to learn about the medicinal benefits of synthetics. From plastics in computers to non-stick surfaces on our pots and pans, synthetic compounds are assisting our quality of life.

Test-Taking Tip Consider how the title of the passage relates to the main idea of the passage.

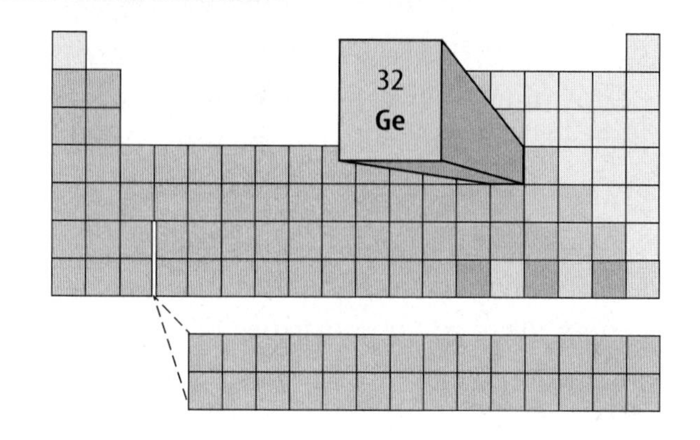

Scientists are studying the medicinal benefits of synthetic germanium.

1. You can tell from the passage that synthetic means _____.
- **A)** natural
- **B)** rare
- **C)** helpful
- **D)** human-made

2. Based on the information in the passage, the reader can conclude that _____.
- **F)** synthetics are only being studied for their possible medicinal purposes
- **G)** medicines made out of plastic could help cancer and AIDS patients
- **H)** synthetic medicines could improve the quality of life for patients
- **J)** germanium is a common element found in Earth's crust

3. According to the passage, why would medications need to be made from synthesized germanium?
- **A)** because natural germanium is very dangerous
- **B)** because natural germanium is not as effective as synthesized germanium
- **C)** because natural germanium is rare
- **D)** because synthesized germanium is very hard to produce

Standardized Test Practice

Reasoning and Skills

Read each question and choose the best answer.

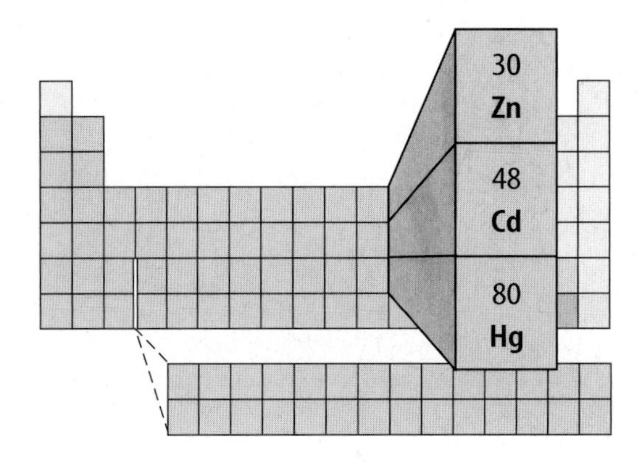

1. Zinc, cadmium and mercury are examples of _____.

A) synthetic elements
B) metals
C) nonmetals
D) metalloids

Test-Taking Tip Review the information about groups in the periodic table.

Nutrients Found in Certain Foods	
Food	**Nutrient**
Fruits	Sugars
Pasta, Bread Cookies	Starches
Vegetables	Vitamins & Minerals
Meat & Fish	Protein

2. According to the chart, spaghetti would be a good source of _____.

F) sugars
G) starches
H) vitamins and minerals
J) protein

Test-Taking Tip Review information about carbohydrates and lipids.

3. Ceramics are made from clay, sand, and/or crystalline rocks. According to this information, which of the following is a ceramic?

A)

B)

C)

D)

Test-Taking Tip Carefully consider the information in both the graphic and the question before answering the question.

Consider this question carefully before writing your answer on a separate sheet of paper.

4. Plastic and nylon are examples of synthetic polymers. Discuss the benefits of developing synthetic polymers.

Test-Taking Tip Review the history and specialized uses of synthetic polymers.

UNIT 6 Interactions of Matter

How Are Algae & Photography Connected?

In the mid-1800s, scientists experimented with light-sensitive chemicals. They found that when paper was treated with such chemicals and then exposed to light, the resulting reaction changed the paper's color. If an object blocked some of the light, a silhouette of the object was created. One set of chemicals produced prints—called cyanotypes—of white images on a blue background. A botanist named Anna Atkins saw the potential of this process. Until that time, the only way to create pictures of plants had been to draw them. Atkins used cyanotypes to create impressions of the plants. In 1843, she published a book of cyanotype images of algae, including the two seen at lower right. It was the first book ever to be illustrated by photography. Since Atkins' time, photography has gone through many changes. But it is still a powerful tool for making images of the natural world—which includes this giant jellyfish, whose image is being captured by an underwater photographer.

SCIENCE CONNECTION

CHEMICAL REACTIONS The paper used in the cyanotype process is treated with two chemicals—ferric ammonium citrate and potassium ferricyanide—that react when exposed to light. Research the chemicals and reactions involved in modern film photography. Create an illustrated poster that compares and contrasts the two processes. If commercially prepared cyanotype paper is available, create your own cyanotypes (also called "sunprints") to help illustrate the poster.

CHAPTER 23

Solutions

What do lemonade, seawater, and suntan lotion have in common? Not only are they all liquids, they are all solutions. In this chapter, you will learn about solutions—how they form and why some form faster than others. You'll learn why something like lemon juice forms a solution with water while something like suntan lotion does not. In addition, you'll learn about the properties of some solutions and how you apply them to your everyday life.

What do you think?

Science Journal Look at the picture below with a classmate. Discuss what you think this might be or what is happening. Here's a hint: *It's not snow, but it's closer to snow than you realize.* Write your best guess in your Science Journal.

What do you like to drink when you're thirsty? Do you prefer water from the faucet, bottled water, or a sports drink that contains substances added to replace those lost during sweating? What do these thirst quenchers contain? Try the following activity to find out.

Identify the solution by solvent subtraction

1. Obtain three solution samples from your teacher and place equal amounts of each in separate, marked, 100-mL beakers.
2. Carefully, boil each solution on a hot plate. As soon as the liquid is gone, remove each beaker to a heat-proof surface using a thermal mitt. Let cool.
3. Examine the inside of your cooled beakers. What do you see? Guess the identity of each solution.

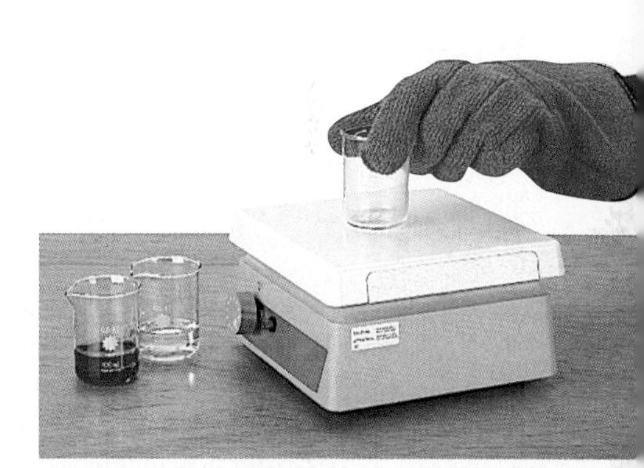

Observe

Describe in your Science Journal what remained in each of the three containers and explain how solutions may look alike but contain different substances.

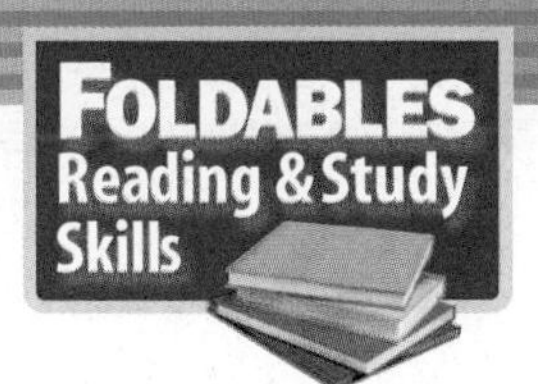

Before You Read

Making a Venn Diagram Study Fold Make the following Foldable to compare and contrast the characteristics of solvents and solutes.

Solvents | Both | Solutes

1. Place a sheet of paper in front of you so the short side is at the top. Fold the paper in half from top to bottom.
2. Fold both sides in to divide the paper into thirds. Unfold the paper so three sections show.
3. Through the top thickness of paper, cut along each of the fold lines to the topfold, forming three tabs. Label the tabs *Solvents, Both,* and *Solutes* as shown.
4. Draw circles across the front of the paper as shown. Before you read the chapter, define the terms and list characteristics of each under their tabs.

SECTION

How Solutions Form

As You Read

What You'll Learn

- **Identify** three types of solutions.
- **Determine** how things dissolve.
- **Examine** the factors that affect the rates at which solids and gases dissolve in liquids.

Vocabulary
solution
solute
solvent
alloy
polar

Why It's Important
Many chemical reactions take place in solution—the food you eat is digested, or chemically changed, by the solution that is in your stomach.

What is a solution?

Hummingbirds are fascinating creatures. They can hover for long periods while they sip nectar from flowers through their long beaks. To attract hummingbirds, many people use feeder bottles containing a red liquid. The liquid is a solution of sugar and red food coloring in water.

Suppose you are making some hummingbird food. When you add sugar to water and stir, the sugar crystals disappear. When you add a few drops of red food coloring and stir, the color spreads evenly throughout the sugar water. Why does this happen?

Hummingbird food is one of many solutions. A **solution** is a mixture that has the same composition, color, density, and even taste throughout. The reason you no longer see the sugar crystals and the reason the red dye spreads out evenly is that they have formed a completely homogeneous mixture. The sugar crystals broke up into sugar molecules, the red dye into its molecules, and both mixed evenly among the water molecules.

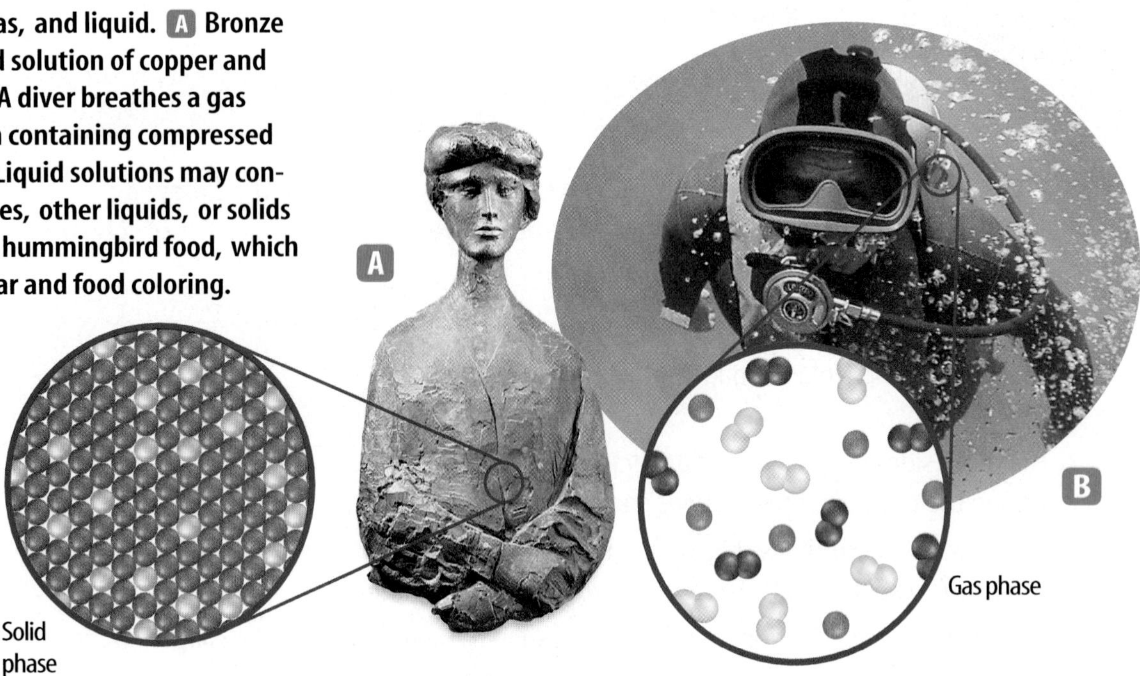

Figure 1
Three phases of solutions exist—solid, gas, and liquid. A Bronze is a solid solution of copper and tin. B A diver breathes a gas solution containing compressed air. C Liquid solutions may contain gases, other liquids, or solids like this hummingbird food, which has sugar and food coloring.

Solutes and Solvents

To describe a solution, you may say that one substance is dissolved in another. The substance being dissolved is the **solute,** and the substance doing the dissolving is the **solvent.** When a solid dissolves in a liquid, the solid is the solute and the liquid is the solvent. Thus, in salt water, salt is the solute and water is the solvent. In carbonated soft drinks, carbon dioxide gas is one of the solutes and water is the solvent. When a liquid dissolves in another liquid, the substance present in the larger amount is usually called the solvent.

Reading Check ***How do you know which substance is the solute in a solution?***

Solutions can also be gaseous or even solid. Examples of all three solution phases are shown in **Figure 1.** Did you know that the air you breathe is a solution? In fact, all mixtures of gases are solutions. Air is a solution of 78 percent nitrogen, 20 percent oxygen, and small amounts of other gases such as argon, carbon dioxide and hydrogen. The sterling silver and brass used in musical instruments is an example of a solid solution. The sterling silver contains 92.5 percent silver and 7.5 percent copper. The brass is a solution of copper and zinc metals. Solid solutions are known as **alloys.** They are made by melting the metal solute and solvent together. Most coins, as shown in **Figure 2,** are alloys.

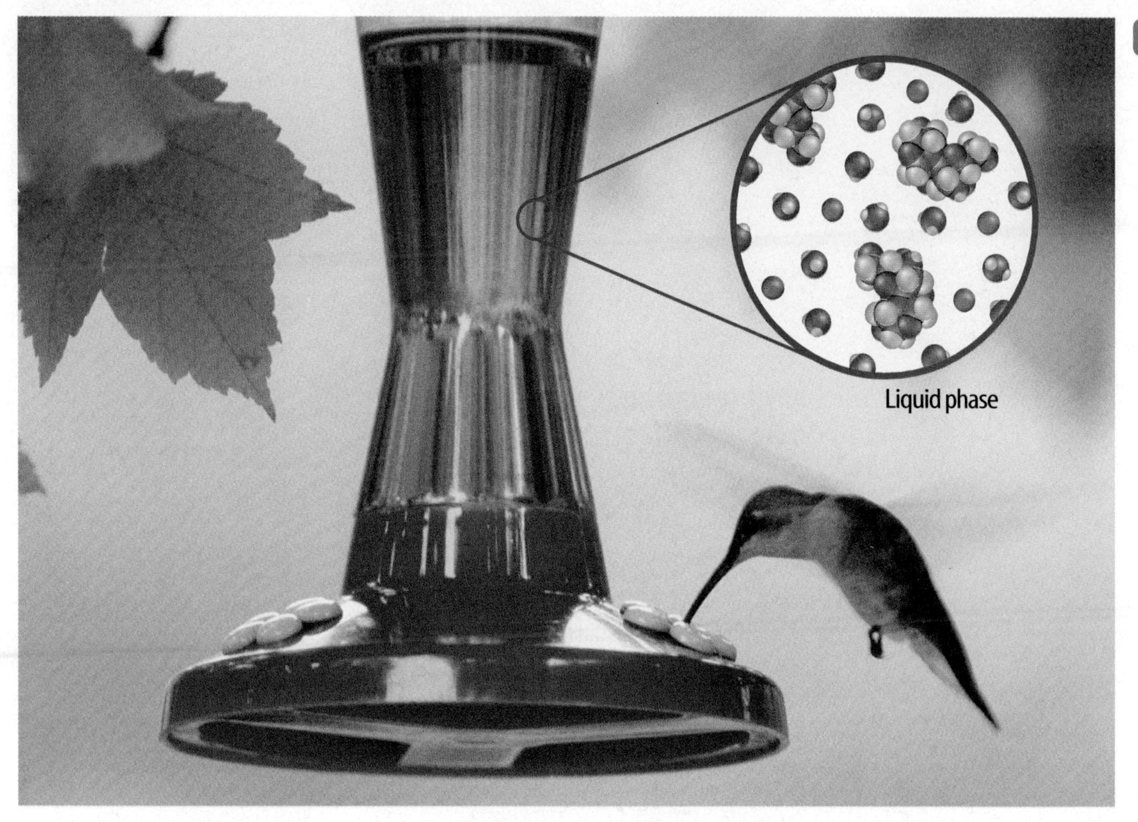

VISUALIZING METAL ALLOYS

Figure 2

Have you ever accidentally put a non-United States coin into a vending machine? Of course, the vending machine didn't accept it. If a vending machine is that selective, how can it be fooled by two coins that look and feel very different? This is exactly the case with the silver Susan B. Anthony dollar and the new golden Sacagawea dollar. Vending machines can't tell them apart.

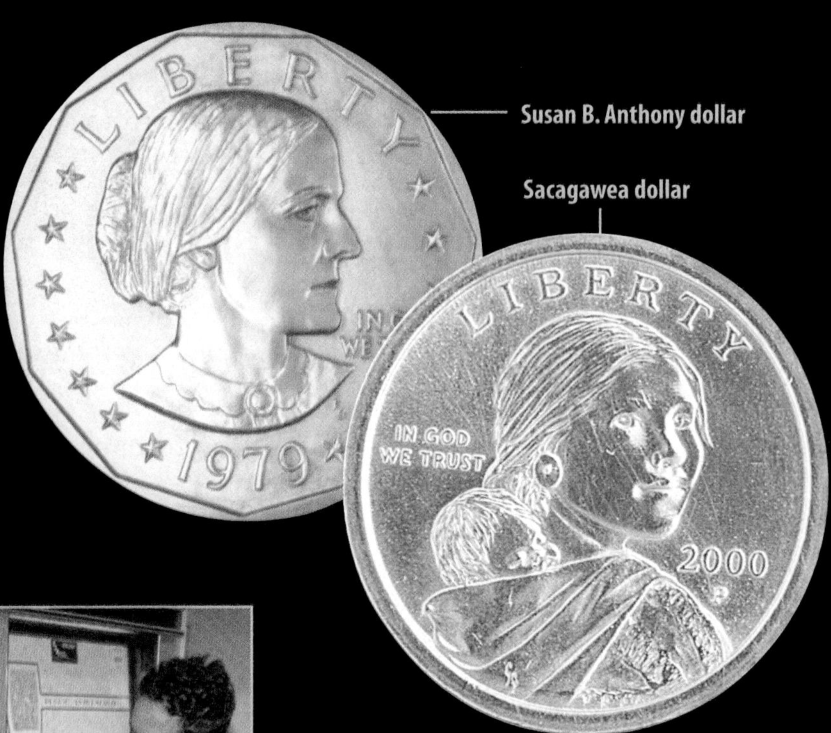

◀ Vending machines recognize coins by size, weight, and electrical conductivity. The size and weight of the Susan B. Anthony coin were easy to copy. Copying the coin's electrical conductivity was more difficult.

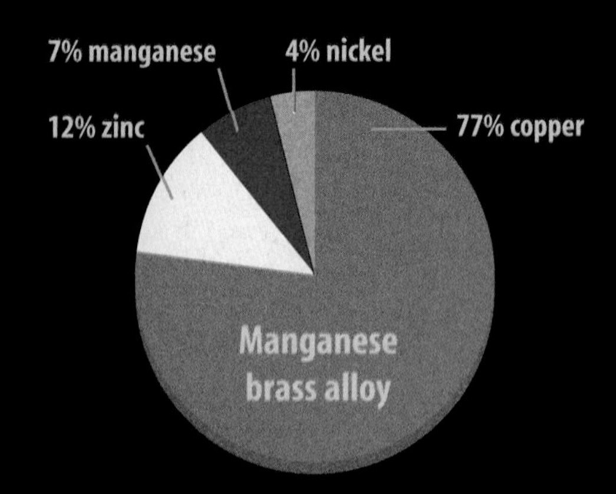

▲ Over 30,000 samples of coin coatings were tested to find an alloy and thickness that would copy the conductivity of the Susan B. Anthony dollar. The final composition of the alloy is shown in the graph above. The key ingredient? Manganese.

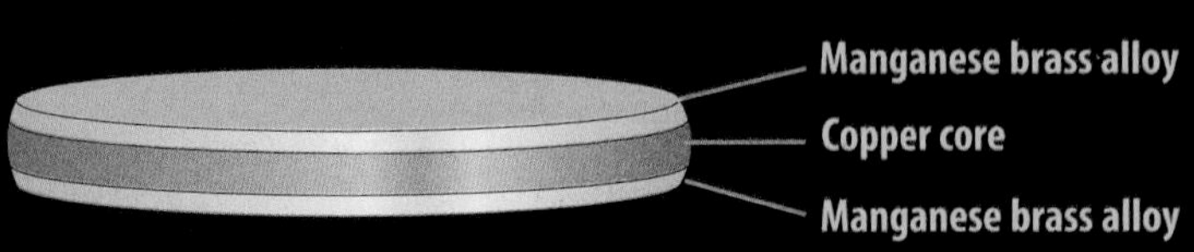

▲ The dollar's copper core is half the coin's thickness. It is sandwiched between two layers of manganese brass alloy.

Dissolving

Fruit drinks and sports drinks are examples of solutions made by dissolving solids in liquids. Like hummingbird food, both contain sugar as well as other substances that add color and flavor. How do solids such as sugar dissolve in water?

The dissolving of a solid in a liquid occurs at the surface of the solid. To understand how water solutions form, keep in mind two things you have learned about water. Like the particles of any substance, water molecules are constantly moving. Also, water molecules are **polar,** which means they have a positive area and a negative area. Molecules of sugar are also polar.

How It Happens **Figure 3** shows molecules of sugar dissolving in water. First, in **Figure 3A,** water molecules cluster around sugar molecules with their negative ends attracted to the positive ends of the sugar. Then, in **Figure 3B,** the water molecules pull the sugar molecules into solution. Finally, in **Figure 3C,** the water molecules and the sugar molecules mix evenly, forming a solution.

Reading Check *How do water molecules help sugar molecules dissolve?*

The process described in **Figure 3** repeats as layer after layer of sugar molecules moves away from the crystal, until all the molecules are evenly spread out. The same three steps occur for any solid solute dissolving in a liquid solvent.

Dissolving Liquids and Gases A similar but more complex process takes place when a gas dissolves in a liquid. Particles of liquids and gases move much more freely than do particles of solids. When gases dissolve in gases or when liquids dissolve in liquids, this movement spreads solutes evenly throughout the solvent.

Dissolving Solids in Solids How can you mix solids to make alloys? Although solid particles do move a little, this movement is not enough to spread them evenly throughout the mixture. The solid metals are first melted and then mixed together. In this liquid state, the metal atoms can spread out evenly and will remain mixed when cooled.

Figure 3
The dissolving of sugar in water can be thought of as a three-step process.

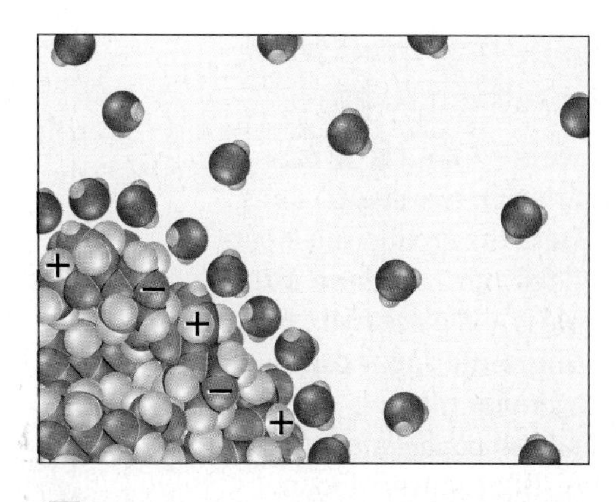

A **The moving water molecules cluster around the sugar molecules as their negative ends are attracted to the positive ends of the sugar molecules.**

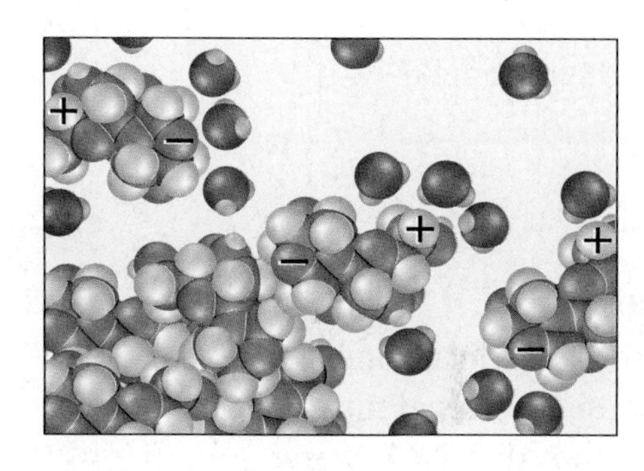

B **The water molecules pull the sugar molecules into solution.**

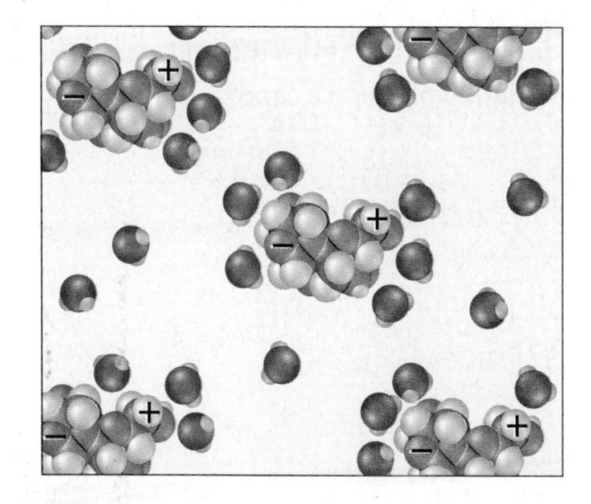

C **The water molecules and sugar molecules spread out to form a homogeneous mixture.**

Mini LAB

Observing the Effect of Surface Area

Procedure

1. Grind up two **sugar cubes.**
2. Place the ground sugar particles into a **medium-sized glass** and place two **unground sugar cubes** into a **similar glass.**
3. Add an equal amount of **distilled water** at room temperature to each glass.

Analysis

1. Compare the times required to dissolve each.
2. What do you conclude about the dissolving rate and surface area?

Rate of Dissolving

How can you speed up the dissolving process? Think about how you make a drink from a powdered mix. After you add the mix to water, you stir it. Stirring a solution speeds up dissolving because it brings more fresh solvent into contact with more solute. The fresh solvent attracts the particles of solute, causing the solid solute to dissolve faster.

Crystal Size A second way to speed the dissolving of a solid in a liquid is to grind large crystals into smaller ones. Suppose you have to use a 5-g crystal of rock candy to sweeten your lemonade. If you put the whole crystal into a glass of lemonade, it might take several minutes to dissolve, even with stirring. However, if you first grind the crystal of rock candy into a powder, it will dissolve in the same amount of lemonade in a few seconds.

Why does breaking up a solid cause it to dissolve faster? Breaking the solid into smaller pieces greatly increases its surface area, as you can see in **Figure 4.** Because dissolving takes place at the surface of the solid, increasing the surface area allows more solvent to come into contact with more solid solute. Therefore, the speed of the dissolving process increases.

Temperature A third way to increase the rate at which most solids dissolve is to increase the temperature of the solvent. Think about making hot chocolate from a mix. You can make the sugar in the chocolate mix dissolve faster by putting it in hot water instead of cold water. Increasing the temperature of a solvent speeds up the movement of its particles. This increase causes more solvent particles to bump into the solute. As a result, solute particles break loose and dissolve faster.

Figure 4
Crystal size affects solubility. A Large crystals of sugar called rock candy dissolve in water slowly because the surface area is limited. B Crushing the crystal increases its surface area. C Further crushing produces a surface area many times that of the original piece rock candy.

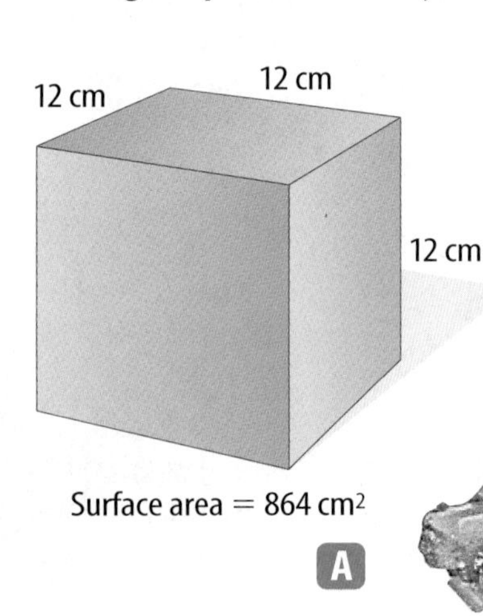

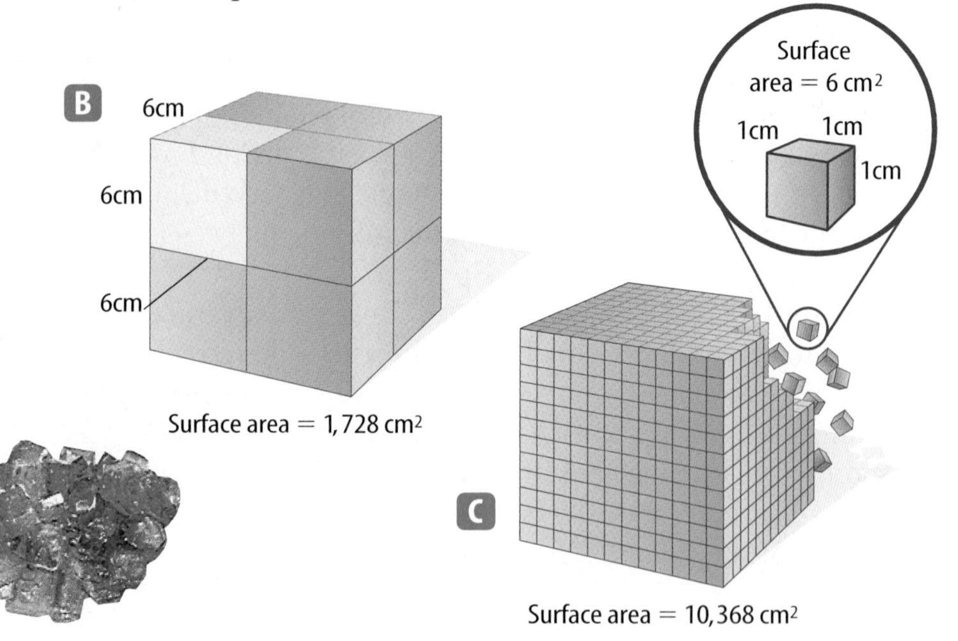

Math Skills Activity

Calculating Surface Area

Example Problem

The length, height, and width of a cube are each 1 cm. If the cube is cut in half to form two rectangular solids, what is the total surface area of the new pieces?

Solution

1 *This is what you know:*
The cube has dimensions of $l = h = w = 1$ cm.
The rectangular solid has a width $w = 0.5$ cm.
The rectangular solid has length and height $l = h = 1$ cm each.

2 *This is what you want to find:* The total surface area of the two rectangular solids

3 *Here is how you get the equation you need:* The cube and the rectangular solid have six faces.

Front and back $= (h \times w)$

Left and right $= (h \times l)$

Top and bottom $= (w \times l)$

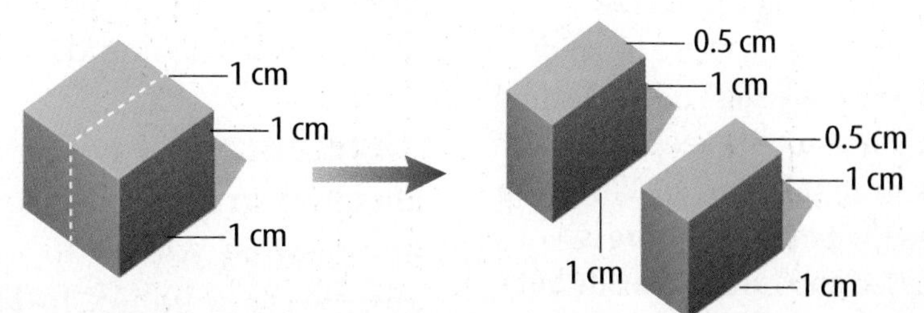

The total surface area of the cube or the rectangular solid is the sum of these areas, or
$2(h \times w) + 2(h \times l) + 2(w \times l) =$ total surface area.

4 *Solve the equation by substituting the appropriate numbers.*

Total surface area of the cube is:
$2(1 \text{ cm} \times 1 \text{ cm}) + 2(1 \text{ cm} \times 1 \text{ cm}) + 2(1 \text{ cm} \times 1 \text{ cm}) = 6 \text{ cm}^2$

Total surface area of the rectangular solid is:
$2(1 \text{ cm} \times 0.5 \text{ cm}) + 2(1 \text{ cm} \times 1 \text{ cm}) + 2(0.5 \text{ cm} \times 1 \text{ cm}) = 4 \text{ cm}^2$

Because there are two rectangular solids, their combined surface area is:
$4 \text{ cm}^2 + 4 \text{ cm}^2 = 8 \text{ cm}^2$

To find out how much new surface area has been created, compare the two results:
$8 \text{ cm}^2 - 6 \text{ cm}^2 = 2 \text{ cm}^2$

Check your answer by considering what you've done by splitting the cube in two. You have created two new faces, each with a surface area of 1 cm². Can you think of a way to simplify the equation for the total surface area of the cube?

Practice Problem

A cube of salt with a length, height, and width of 5 cm is attached along a face to another cube of salt with the same dimensions. What is the combined surface area of the new rectangular solid? How much surface area has been lost?

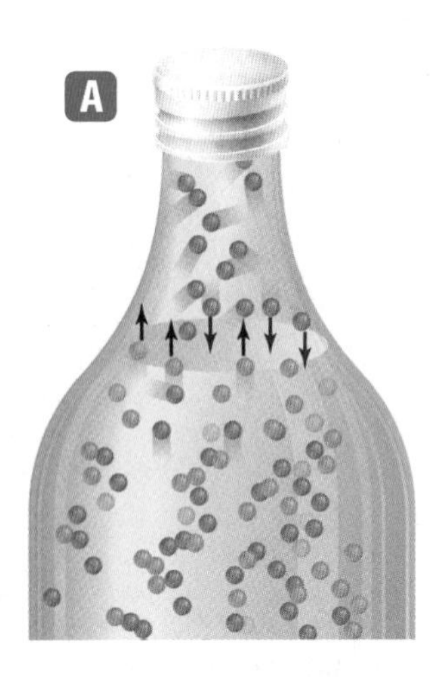

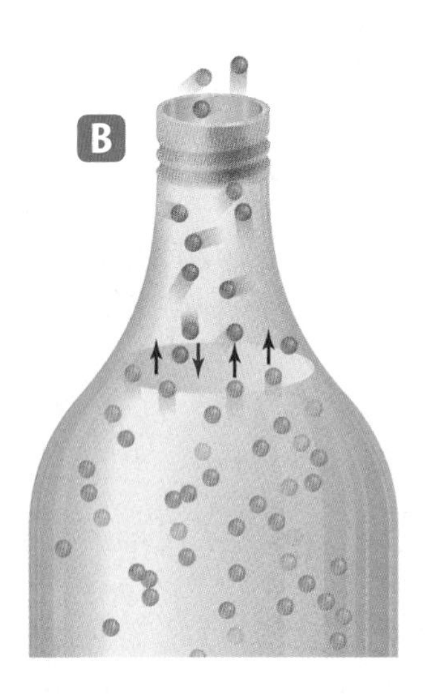

Figure 5
Solutions of gases behave differently from those of solids or liquids. A Soda pop is bottled under pressure to keep carbon dioxide in solution. B When you open the bottle, pressure is released and carbon dioxide bubbles out of solution.

Gases in Solution

When you shake an opened bottle of soda, it bubbles up and may squirt out. Shaking or pouring a solution of a gas in a liquid causes gas to come out of solution. Agitating the solution exposes more gas molecules to the surface, where they escape from the liquid.

Pressure What might you do if you want to dissolve more gas in a liquid? One thing you can do is increase the pressure of that gas over the liquid. Soft drinks are bottled under increased pressure. This increases the amount of carbon dioxide that dissolves in the liquid.

Temperature Another way to increase the amount of gas that dissolves in a liquid is to cool the liquid. This is just the opposite of what you do to increase the speed at which most solids dissolve in a liquid. In **Figure 5,** you can see what happens to the carbon dioxide when a bottle of soft drink is opened. Even more carbon dioxide will bubble out of a soft drink as it gets warmer.

Section  Assessment

1. Name three phases of solutions. Give an example of each type.
2. What three methods can increase the rate of dissolving a solid in a liquid?
3. How can you keep a bottle of soda from going "flat"?
4. How are the metals in an alloy mixed so that they are evenly distributed?
5. **Think Critically** Amalgams, which are sometimes used in tooth fillings, are alloys of mercury with other metals. Is an amalgam a solution? Explain.

Skill Builder Activities

6. **Comparing and Contrasting** Compare and contrast the effects on the rate of dissolving (1) a solid in a liquid and (2) a gas in a liquid, when the solution (a) is cooled, (b) is stirred, and (c) the pressure on it is lowered. **For more help, refer to the Science Skill Handbook.**
7. **Communicating** Write a paragraph in your Science Journal explaining why many fish aquariums have a device that bubbles air into the water. **For more help, refer to the Science Skill Handbook.**

Dissolving Without Water

When Water Won't Work

Water often is referred to as the universal solvent because it can dissolve so many things. However, there are some things it can't do. For example, it can't dissolve a lipstick stain on a linen napkin or clean an oil-soaked seabird. To explain why water can't dissolve some materials, you must know what these materials are and how they differ from water.

Water can't dissolve some solutes because of its polarity. As you learned in the first section, water has positive and negative areas that allow it to attract polar solutes. However, **nonpolar** materials have no separated positive and negative areas. Because of this, they do not attract polar materials, which means they do not attract water molecules. Nonpolar materials do not dissolve in water except to a small extent, if at all.

As You Read

What **You'll Learn**

- **Identify** the kinds of solutes that do not dissolve well in water.
- **Explain** how solvents work.
- **Determine** how to choose the right solvent for the job.

Vocabulary
nonpolar

Why **It's Important**

Many solutes do not dissolve in water.

Nonpolar Solutes An example of a nonpolar substance that does not dissolve in water can be seen on many dinner tables. The vinegar-and-oil salad dressing shown in **Figure 6** has two distinct layers—the bottom layer is vinegar, which is a solution of acetic acid in water, and the top layer is salad oil.

Most salad oils contain large molecules made of carbon and hydrogen atoms, which are called hydrocarbons. In hydrocarbons, carbon and hydrogen atoms share electrons in a nearly equal manner.

This equal distribution of electrons means that the molecule has no separate positive and negative areas. Because it has no charge separation, the nonpolar oil molecule is not attracted to the polar water molecules in the vinegar solution. That's why you must shake this kind of dressing to mix it just before you pour it on your salad.

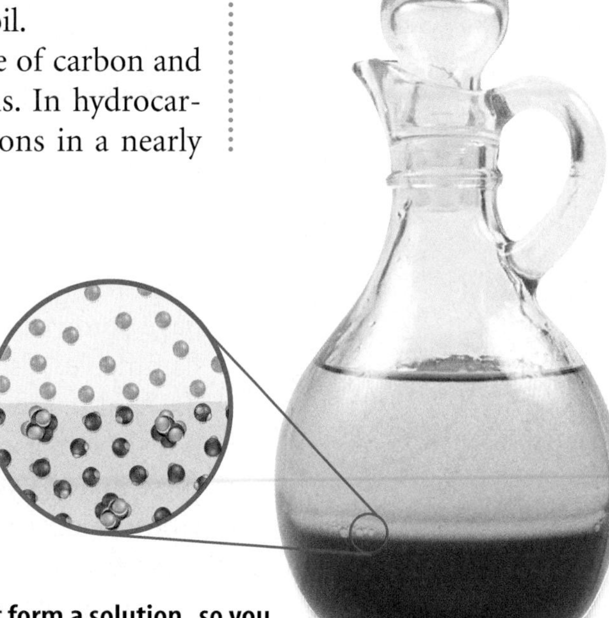

Figure 6
Oil and vinegar do not form a solution, so you must shake this salad dressing before using it.

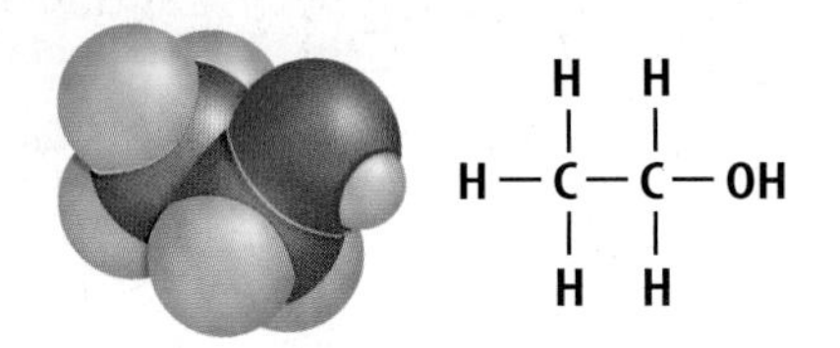

Figure 7
Ethanol, C_2H_5OH, has a polar –OH group at one end but the $–C_2H_5$ section is nonpolar.

Versatile Alcohol Some substances form solutions with polar as well as nonpolar solutes because their molecules have a polar and a nonpolar end. Ethanol, shown in **Figure 7,** is such a molecule. The polar end dissolves polar substances, and the nonpolar end dissolves nonpolar substances. For example, ethanol dissolves iodine, which is nonpolar, as well as water, which is polar.

Reading Check *How can alcohol dissolve both polar and nonpolar substances?*

Useful Nonpolar Molecules

Some materials around your house may be useful as nonpolar solvents. For example, mineral oil may be used as a solvent to remove candle wax from glass or metal candleholders. Both the mineral oil and the candle wax are nonpolar materials. Mineral oil can also aid in removing bubble gum from some surfaces for the same reason. Oil-based paints contain pigments that are dissolved in oils. In order to thin or remove such paints, a nonpolar solvent must be used. The gasoline you use in your car and lawnmower is a solution of hydrocarbons, which are nonpolar substances.

Dry cleaners use nonpolar solvents when removing oily stains. The word *dry* refers to the fact that no water is used in the process. Molecules of a nonpolar solute can slip easily among molecules of a nonpolar solvent. That is why dry cleaning can remove stains of grease and oil that you cannot clean easily yourself. A general statement that describes which substance dissolves which is the phrase "like dissolves like."

Many nonpolar solvents are connected with specific jobs. People who paint pictures using oil-based paints probably used the solvent turpentine. It comes from the sap of a pine tree. **Figure 8** shows how well turpentine dissolves nonpolar paint.

Figure 8
With no polarity to interfere, paint molecules slide smoothly among molecules of turpentine.

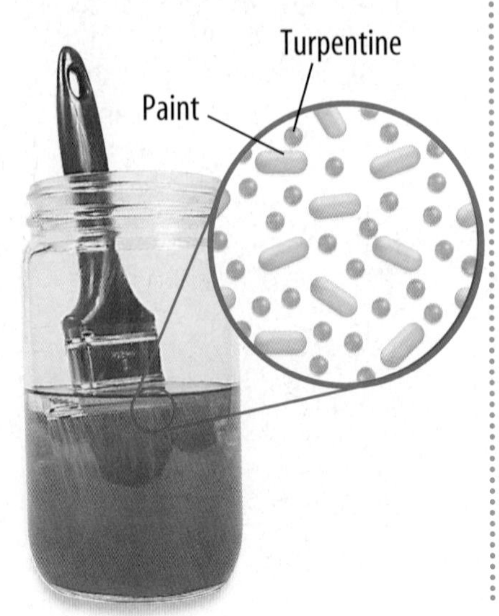

Toxicity Although nonpolar solvents have many uses, they have some drawbacks, too. First, many nonpolar solvents are flammable. Also, some are toxic, which means that contact with the skin or inhaling vapors of these solvents can be dangerous. For these reasons, you must always be careful when handling these materials and never use them in a closed area. Good ventilation is critical, because nonpolar solvents tend to evaporate more readily than water, and even small amounts of a nonpolar liquid can produce high concentrations of harmful vapor in the air.

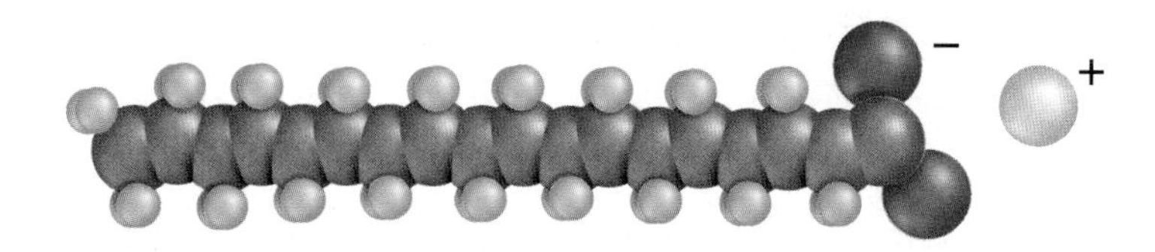

Figure 9
The long hydrocarbon tail of sodium stearate is nonpolar. The head is ionic.

How Soap Works The oils on human skin and hair keep them from drying out, but the oils can also attract and hold dirt. The oily dirt is a nonpolar mixture, so washing with water alone won't clean away the dirt. This is where soap comes in. Soaps, you might say, have a split personality. They are substances that have polar and nonpolar properties. Soaps are salts of fatty acids, which are long hydrocarbon molecules with a carboxylic acid group –COOH at one end. When a soap is made, the hydrogen atom of the acid group is removed, leaving a negative charge behind, and a positive ion of sodium or potassium is attached. This is shown in **Figure 9.**

Reading Check *Why doesn't water alone clean oily dirt?*

Thus, soap has an ionic end that will dissolve in water and a long hydrocarbon portion that will dissolve in oily dirt. In this way, the dirt is removed from your skin, hair, or a fabric, suspended in the wash water and washed away, as shown in **Figure 10.**

Clinging Molecules

Procedure

1. Lay **two clean pennies** side by side and heads up on a **paper towel.**
2. Slowly place drops of **water** from a **dropper** onto the head of one penny. Count each drop and continue until the accumulated water spills off the edge of the penny.
3. With adult supervision, repeat step 2 using **rubbing alcohol,** which is approximately 30 percent isopropyl alcohol, and the other penny.

Analysis

1. Which penny held the most drops before liquid spilled over the edge of the penny?
2. Isopropyl alcohol has the formula C_3H_7OH. How polar do you think it is?
3. How do the results of the experiment support the concept of polarity and molecules sticking to each other?

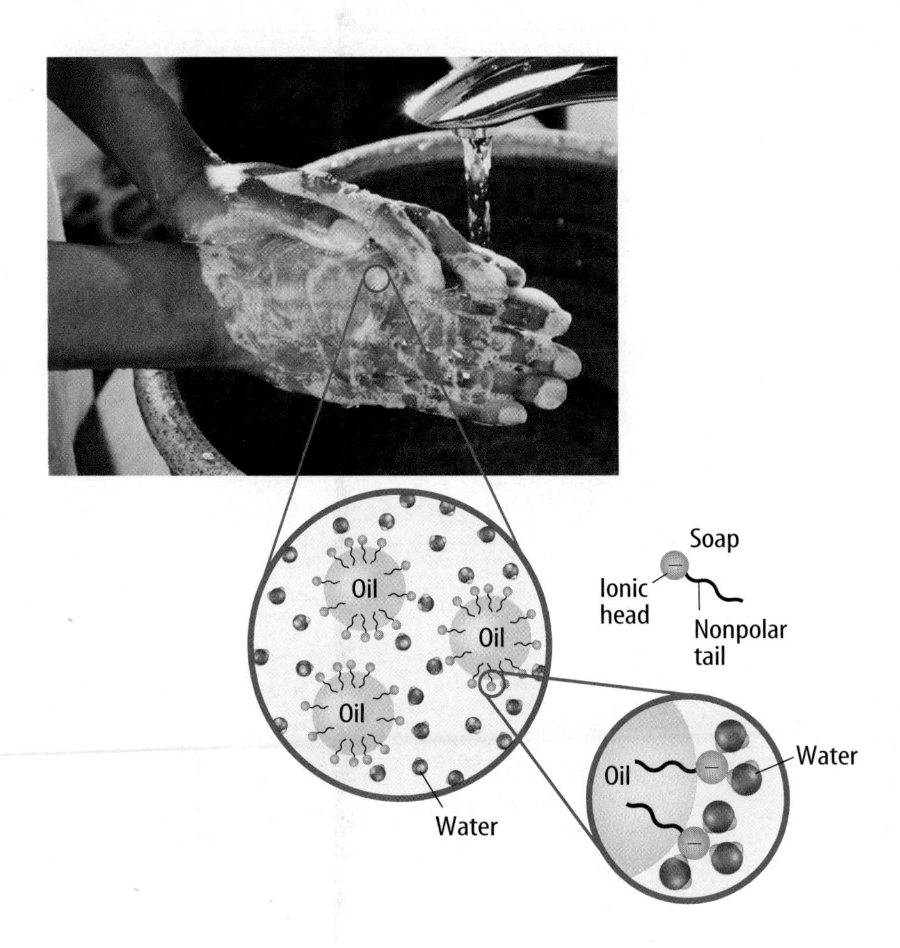

Figure 10
Soap cleans because its nonpolar hydrocarbon part dissolves in oily dirt and its ionic part interacts strongly with water. The oil and water mix and the dirt is washed away.

Polarity and Vitamins

Having the right kinds and amounts of vitamins is important for your health. The B vitamins and vitamin C are polar compounds. They dissolve readily in the water present in cells throughout your body. Because water is passing constantly through your system, these dissolved vitamins can be flushed out before they are used. For this reason, you must replace these vitamins by eating enough of the foods that contain them or by taking vitamin supplements. Look at **Table 1** for some good sources of vitamin C.

Research Visit the Glencoe Science Web site at **science.glencoe.com** for more information about vitamins. Prepare a report on your research.

When you look at the structure of vitamin C, shown in **Figure 11,** you will see that it has several carbon-to-carbon bonds. This might make you think that it is nonpolar. But, if you look again, you will see that it also has several oxygen-to-hydrogen bonds that resemble those found in water. This makes it polar. That is why water dissolves vitamin C.

Fat is stored in various specialized cells of your body. Fat molecules consist of long chains of carbon atoms bonded to each other by nonpolar bonds. This makes them nonpolar. Therefore, fat molecules do not dissolve in the water around them.

Table 1 Sources of Vitamin C

Food	Amount	mg
Orange juice, fresh	1 cup	124
Green peppers, raw	1/2 cup	96
Broccoli, raw	1/2 cup	70
Cantaloupe	1/4 melon	70
Strawberries	1/2 cup	42

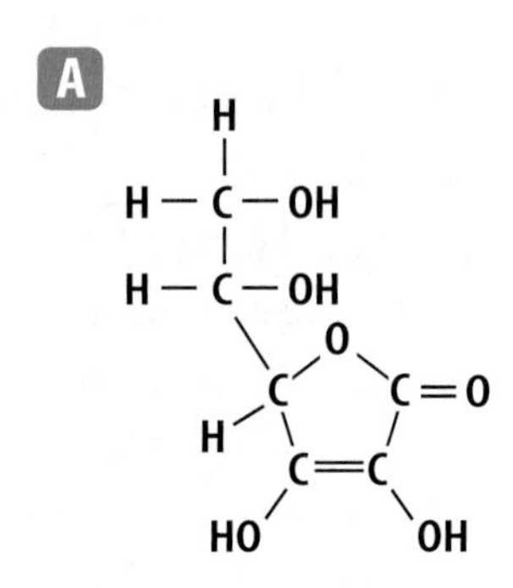

Figure 11
Vitamin C helps heal wounds and helps the body absorb iron.
A Although it has carbon-to-carbon bonds, it also has polar groups and is water soluble. B These foods are good sources of vitamin C as you can see in Table 1.

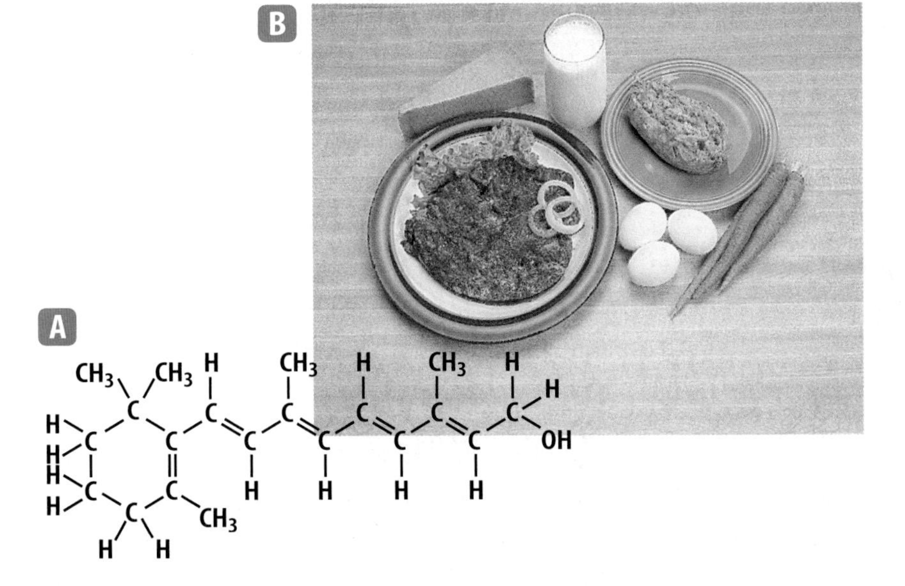

Figure 12
Vitamin A is a nonpolar, fat-soluble vitamin. A You can see the long hydrocarbon chain that makes it nonpolar in this structural formula. B These foods are good sources of vitamin A.

Vitamin Dosage Some of the vitamins you need, such as vitamin A shown in **Figure 12,** are also nonpolar and can dissolve in fat. Because they do not wash away with water, fat-soluble vitamins are not easily eliminated and they can accumulate in our tissues. Some of these vitamins are toxic in high concentrations, so taking large doses can be dangerous. In general, the best way to stay healthy is to eat a variety of healthy foods. Such a diet will supply all the vitamins you need with no risk of overdoses. Vitamins D, E, and K are also fat-soluble.

✔ Reading Check ***Why do you need to replace some of the vitamins used in your body?***

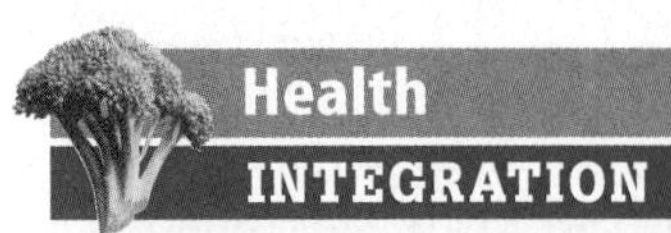

Drinking too much carrot juice, which is rich in beta carotene, a substance related to vitamin A, can cause the palms of your hands and soles of your feet to turn orange. Though this condition is not serious and the color fades in time, taking too much of several vitamins can be dangerous. *Research what might result from taking too much vitamin B-6, vitamin D, and niacin.*

Section  Assessment

1. Explain how a solute can dissolve in polar and nonpolar solvents.
2. Explain the phrase "like dissolves like" and give an example of two "like" substances.
3. Explain how soap cleans greasy dirt from your hands.
4. Some small engines require a mixture of oil and gasoline. Gasoline evaporates easily. What conclusion can be drawn about the polarity of the engine oil?
5. **Think Critically** What might happen to your skin if you washed with soap too often?

Skill Builder Activities

6. **Drawing Conclusions** Some people believe in taking large doses of vitamin C to prevent colds. How does the polarity of vitamin C aid in preventing accumulation of this vitamin? Explain your reasoning. **For more help, refer to the Science Skill Handbook.**
7. **Communicating** Make a survey of your home and list in your Science Journal the nonpolar solvents you find. Read the labels. What precautions should be observed while using them? **For more help, refer to the Science Skill Handbook.**

SECTION 3

Solubility and Concentration

As You Read

What You'll Learn

- **Determine** how temperature affects solubility.
- **Identify** how to express the concentration of solutions.
- **Compare and contrast** saturated, unsaturated, and supersaturated solutions.

Vocabulary

solubility
saturated solution
unsaturated solution
supersaturated solution

Why It's Important

Many products that we use and foods that we eat are in the form of solutions.

How much can dissolve?

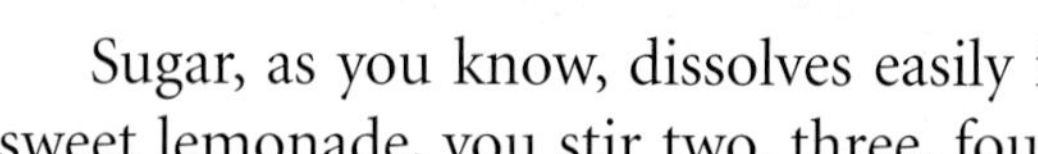

Sugar, as you know, dissolves easily in water. To make extra sweet lemonade, you stir two, three, four, or more teaspoons of sugar into a glass of lemon and water. It all dissolves. You try adding more sugar, but eventually, no more sugar dissolves, and the excess granules sink to the bottom of the glass. You now see how soluble sugar is in water. **Solubility** (sol yuh BIHL ih tee) is the maximum amount of a solute that can be dissolved in a given amount of solvent at a given temperature.

Reading Check *What is solubility?*

Comparing Solubilities The amount of a substance that can dissolve in a solvent depends on the nature of these substances. To determine the solubility of two substances, A and B, you stir 1 g of substance A into a beaker containing 100 mL of water. All of substance A dissolves. But, when you try to add more of substance A to that beaker, it doesn't dissolve. Then you stir 1 g of substance B into another beaker containing 100 mL of water. Substance B dissolves completely. You add more of substance B, and find that you can add two more grams before no more will dissolve. If the temperature of the water is the same in both beakers, you can conclude that substance B is more soluble than substance A as shown in **Figure 13.** Another way to state this is that substance B has a higher solubility in water than substance A does.

Table 2 shows how the solubility of several substances varies at 20°C. For solutes that are gases, the pressure also must be given.

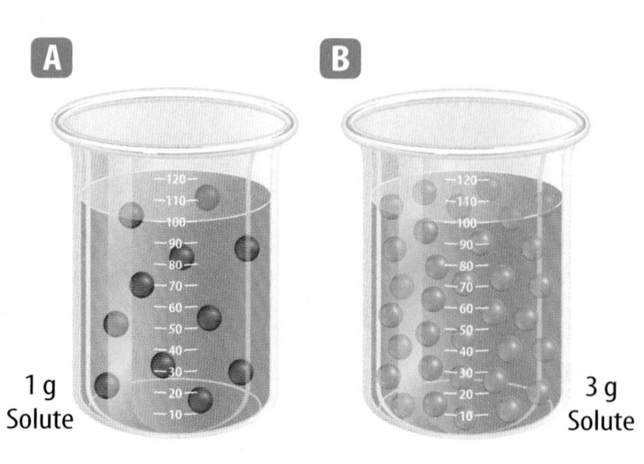

Figure 13
Only 1 g of A dissolves in 100 mL of water, but more than 3 g of B dissolves in the same amount of water at the same temperature. Substance B is more soluble in water than substance A.

Concentration

Suppose you add one teaspoon of lemon juice to a glass of water to make lemonade. Your friend adds four teaspoons of lemon juice to another glass of water the same size. You could say that your glass of lemonade is dilute and your friend's lemonade is concentrated, because your friend's drink now has more lemon flavor than yours. A concentrated solution is one in which a large amount of solute is dissolved in the solvent. A dilute solution is one that has a small amount of solute in the solvent.

Precise Concentrations How much real fruit juice is there in one of those boxed fruit drinks? You can read the label to find out. *Concentrated* and *dilute* are not precise terms. However, concentrations of solutions can be described precisely. One way is to state the percentage by volume of the solute. The percentage by volume of the juice in the drink shown in **Figure 14** is ten percent. Adding 10 mL of juice to 90 mL of water makes 100 mL of this drink. Commonly, fruit-flavored drinks can contain from ten percent to 100 percent fruit juice. Generally, if two or more liquids are being mixed, the concentration is given in percentage by volume, stated in number of milliliters of solute plus enough solvent to make 100 mL of solution.

Table 2 Solubility of Substances in Water at 20°C

Substance	Solubility in g/100 g of Water
Solid Substances	
Salt (sodium chloride)	35.9
Baking soda (sodium bicarbonate)	9.6
Washing soda (sodium carbonate)	21.4
Lye (sodium hydroxide)	109.0
Sugar (sucrose)	203.9
Gaseous Substances*	
Hydrogen	0.00017
Oxygen	0.005
Carbon dioxide	0.16

*at normal atmospheric pressure

Figure 14
The concentrations of fruit juices often are given in percent by volume like these. Concentrations can range from ten percent to 100 percent juice.

Types of Solutions

How much solute can dissolve in a given amount of solvent? That depends on a number of factors, including the solubility of the solute. Here you will examine the types of solutions based on the amount of a solute dissolved.

Saturated Solutions If you add 35 g of copper(II) sulfate, $CuSO_4$, to 100 g of water at 20°C, only 32 g will dissolve. You have a saturated solution because no more copper(II) sulfate can dissolve. A **saturated solution** is a solution that contains all the solute it can hold at a given temperature. However, if you heat the mixture to a higher temperature, more copper(II) sulfate can dissolve. Generally, as the temperature of a liquid solvent increases, the amount of solid solute that can dissolve in it also increases. **Table 3** shows the amounts of a few solutes that can dissolve in 100 g of water at different temperatures, forming saturated solutions. Some of these data also are shown on the accompanying graph.

Table 3 Solubility of Compounds in g/100 g of Water at Various Temperatures

Compound	0°C	20°C	60°C	100°C
Ammonium chloride	29.4	37.2	55.3	77.3
Copper(II) sulfate	23.1	32.0	61.8	114
Lead(II) chloride	0.67	1.0	1.94	3.2
Potassium bromide	53.6	65.3	85.5	104
Potassium chloride	28.0	34.0	45.8	56.3
Potassium nitrate	13.9	31.6	109	245
Sodium acetate	36.2	46.4	139	170.15
Sodium chlorate	79.6	95.9	137	204
Sodium chloride	35.7	35.9	37.1	39.2
Sodium nitrate	73.0	87.6	122	180
Sucrose (sugar)	179.2	203.9	287.3	487.2

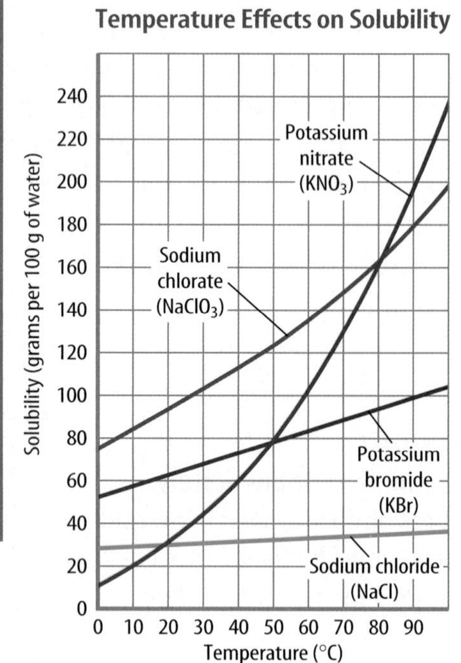

Solubility Curves Each line on the graph is called a solubility curve for a particular substance. You can use a curve to figure out how much solute will dissolve at any temperature given on the graph. For example, about 78 g of KBr (potassium bromide) will form a saturated solution in 100 g of water at 47°C. How much NaCl will form a saturated solution with 100 g of water at the same temperature?

Unsaturated Solutions An **unsaturated solution** is any solution that can dissolve more solute at a given temperature. Each time a saturated solution is heated to a higher temperature, it becomes unsaturated. The term *unsaturated* isn't precise. If you look at **Table 3,** you'll see that at 20°C, 37.2 g of NH_4Cl (ammonium chloride) forms a saturated solution in 100 g of water. However, an unsaturated solution of NH_4Cl could be any amount less than 37.2 g in 100 g of water at 20°C.

What happens to a saturated solution if it is heated?

Supersaturated Solutions If you make a saturated solution of potassium nitrate at 100°C and then let it cool to 20°C, part of the solute comes out of solution. This is because, at the lower temperature, the solvent cannot hold as much solute. Most other saturated solutions behave in a similar way when cooled. However, if you cool a saturated solution of sodium acetate from 100°C to 20°C without disturbing it, no solute comes out. At this point, the solution is supersaturated. A **supersaturated solution** is one that contains more solute than a saturated one at the same temperature. If a seed crystal of sodium acetate is dropped into the supersaturated solution, excess sodium acetate crystallizes out, as in **Figure 15.**

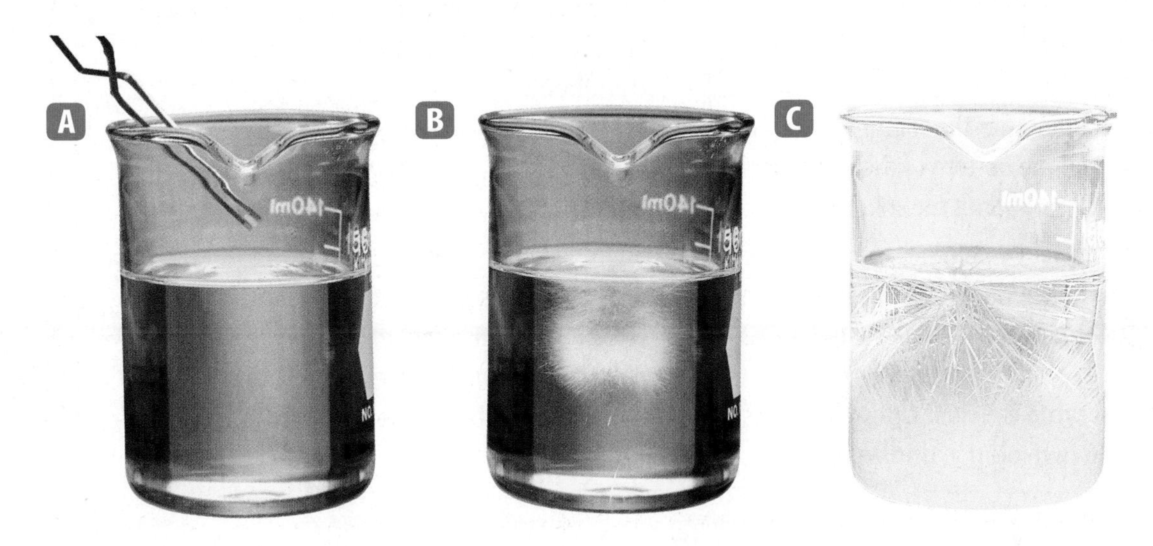

Figure 15
A supersaturated solution is unstable. A When a seed crystal of sodium acetate is added to a supersaturated solution of sodium acetate, B C excess solute immediately crystallizes from solution.

Figure 16
In this cold pack there are sealed bags of water and ammonium nitrate. When the pack is squeezed, the inner bags break and the water and ammonium nitrate mix. As the ammonium nitrate dissolves, the pack cools.

Solution Energy As the supersaturated solution of sodium acetate crystallizes, the solution becomes hot. Energy is given off as new bonds form between the ions and the water molecules. Some portable heat packs use crystallization from supersaturated solutions to produce heat. After crystallization, the heat pack can be reused by heating it to again dissolve all the solute.

You may have seen a "cold pack," like the one shown in **Figure 16,** applied to a sport's injury to reduce swelling. Some substances, such as ammonium nitrate, *need* energy to dissolve. The energy needed is taken from the surroundings, so the temperature of the water will drop several degrees.

Section  Assessment

1. Do all solutes dissolve to the same extent in the same solvent? How do you know?
2. Using **Table 3** state the following: the mass of $NaNO_3$ (sodium nitrate) that would have to be dissolved in 100 g of water to form a saturated, an unsaturated, and a supersaturated solution of $NaNO_3$ at 20°C.
3. Suppose you add a solute crystal to a solution of the solute. After the solution is stirred, the crystal is on the bottom of the container. What do you know about the solution?
4. **Think Critically** By volume, orange drink is ten percent each of orange juice and corn syrup. A 1.5-L can of the drink costs $0.95. A 1.5-L can of orange juice is $1.49, and 1.5 L of corn syrup is $1.69. Does it cost less to make your own orange drink or buy it ?

Skill Builder Activities

5. **Making and Using Graphs** Using **Table 3,** make a graph of solubility versus temperature for $CuSO_4$ (copper(II) sulfate) and $NaNO_3$ (sodium nitrate). How would you make a saturated solution of each substance at 80°C? **For more help, refer to the Math Skill Handbook.**
6. **Using an Electronic Spreadsheet** Use **Table 3** to prepare a spreadsheet showing the number of grams of solute needed to make 10 mL, 50 mL, 100 mL, 500 mL, and 1,000 mL of a saturated solution at 20°C of the following compounds: *ammonium chloride, lead(II) chloride, sodium chlorate,* and *sucrose.* **For more help, refer to the Technology Skill Handbook.**

SECTION

Particles in Solution

Particles With a Charge

Did you know that there are charged particles in your body that conduct electricity? In fact, you could not live without them. Some help nerve cells transmit messages. Each time you blink your eyes or wave your hand nerves control how muscles will respond. These charged particles, called **ions,** are in the fluids that are in and around all the cells in your body. The compounds that produce solutions of ions that conduct electricity in water are known as **electrolytes.** Some substances, like sodium chloride, are strong electrolytes and conduct a strong current. Strong electrolytes exist completely in the form of ions in solution. Other substances, like the acetic acid in vinegar, remain mainly in the form of molecules when they dissolve in water. They produce few ions and conduct current only weakly. They are called weak electrolytes. Substances that form no ions in water and cannot conduct electricity are called **nonelectrolytes.** Among these are organic molecules like ethyl alcohol and sucrose.

Reading Check *What is a nonelectrolyte?*

How Ionic Solutions Form

Ionic solutions form in two ways. Electrolytes, such as hydrogen chloride, are molecules made up of neutral atoms. To form ions, the molecules must be broken apart in such a way that the atoms take on a charge. This process of forming ions is called **ionization.** The process is shown in **Figure 17,** using hydrogen chloride as a model. Both hydrogen chloride and water are polar molecules. Water surrounds the hydrogen chloride molecules and pulls them apart, forming positive hydrogen ions and negative chloride ions. The hydrogen chloride is a strong electrolyte because it exists completely as ions in water.

As You Read

What You'll Learn

- **Examine** how some solutes break apart in water solutions to form positively and negatively charged particles.
- **Determine** how some solutions conduct electricity.
- **Describe** how antifreeze works.

Vocabulary

ion
electrolyte
ionization
nonelectrolyte
dissociation

Why It's Important

We use the properties of solutes every day.

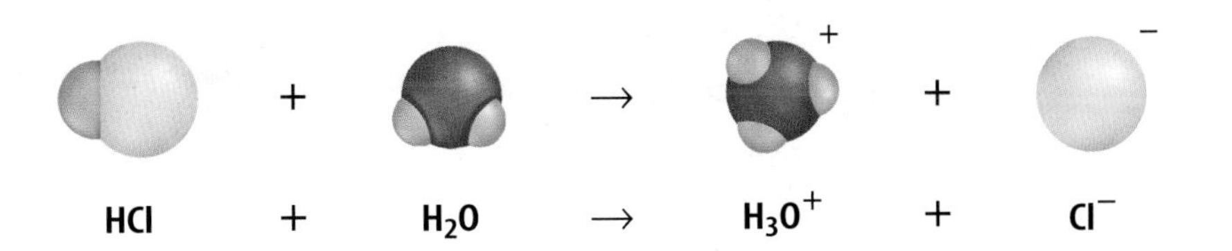

Figure 17
When hydrogen chloride dissolves in water, H_2O molecules surround and pull apart the HCl molecules, forming chloride and hydrogen ions, which are often shown as H_3O^+ to emphasize the role water plays in ionization.

Figure 18
This is a model of a sodium chloride crystal. Each chloride atom is surrounded by six sodium atoms and vice versa.

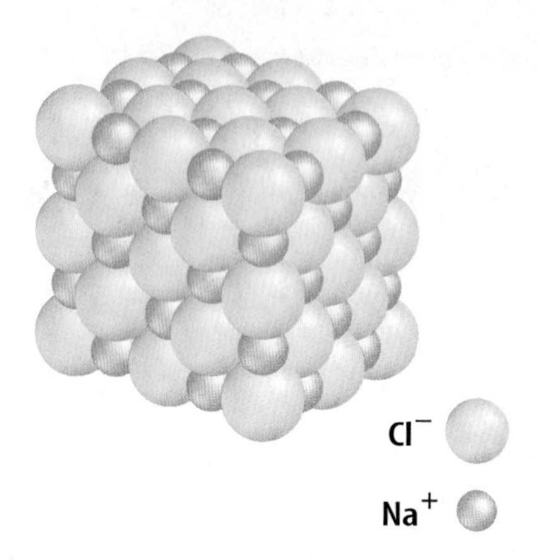

Dissociation The second way that ionic solutions form is by the dissociation of ionic compounds. The ions already exist in the ionic compound and are attracted into the solution by the surrounding polar water molecules. **Dissociation** is the process in which an ionic solid, such as sodium chloride, separates into its positive and negative ions. A model of a sodium chloride crystal is shown in **Figure 18.** In the crystal, each positive sodium ion is attracted to six negative chloride ions. Each of the negative chloride ions is attracted to six sodium ions, a pattern that exists throughout the crystal.

When placed in water, the crystal begins to break apart under the influence of water molecules. Remember that water is polar, which means that the positive areas of the water molecules are attracted to the negative chloride ions. Likewise the negative oxygen part of the water molecules is attracted to the sodium ions.

In **Figure 19,** water molecules are approaching the sodium and chloride ions in the crystal. The water molecules surround the sodium and chloride ions, having pulled them away from the crystal and into solution. The sodium and chloride ions have dissociated from one another. The solution now consists of sodium and chloride ions mixed with water. The ions move freely through the solution and are capable of conducting an electric current.

Reading Check ***What are the differences and similarities between dissociation and ionization?***

Figure 19
Sodium chloride dissociates as water molecules attract and pull the sodium and chloride ions from the crystal. Water molecules then surround and separate the Na^+ and Cl^- ions.

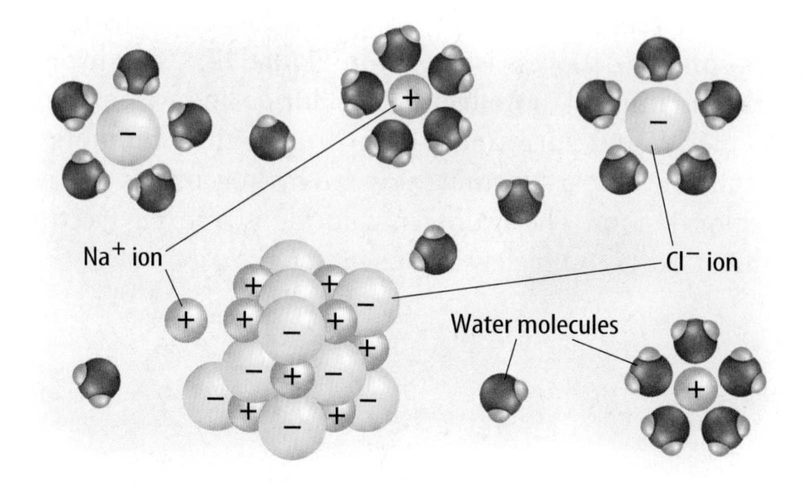

Effects of Solute Particles

All solute particles—polar and nonpolar, electrolyte and nonelectrolyte—affect the physical properties of the solvent, such as its freezing point and its boiling point. These effects can be useful. For example, adding antifreeze to water in a car radiator lowers the freezing point of the radiator fluid. Sugar and salt also would do the same thing. The effect that a solute has on the freezing point or boiling point of a solvent depends on the *number* of solute particles in solution, not on the chemical nature of the particles.

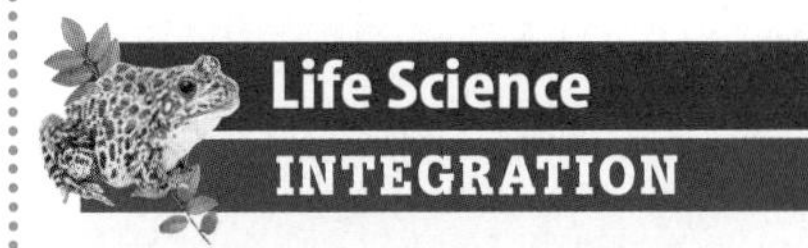

Certain animals that live in extremely cold climates have their own kind of antifreeze. Caribou, for example, contain substances in the lower section of their legs that prevent freezing in subzero temperatures. The caribou can stand for long periods of time in snow and ice with no harm to their legs.

Lowering Freezing Point Adding a solute such as antifreeze to a solvent lowers the freezing point of the solvent. How much the freezing point goes down depends upon how many solute particles you add. How does this work?

As a substance freezes, its particles arrange themselves in an orderly pattern. The added solute particles interfere with the formation of this pattern, making it harder for the solvent to freeze as shown in **Figure 20.** To overcome this interference, a lower temperature is needed to freeze the solvent.

Raising Boiling Point Surprisingly, antifreeze also raises the boiling point of the water. How can it do this? The amount the boiling point is raised depends upon the number of solute molecules present. Solute particles interfere with the evaporation of solvent particles. Thus, more energy is needed for the solvent particles to escape from the liquid surface, and so the boiling point of the solution will be higher than the boiling point of solvent alone.

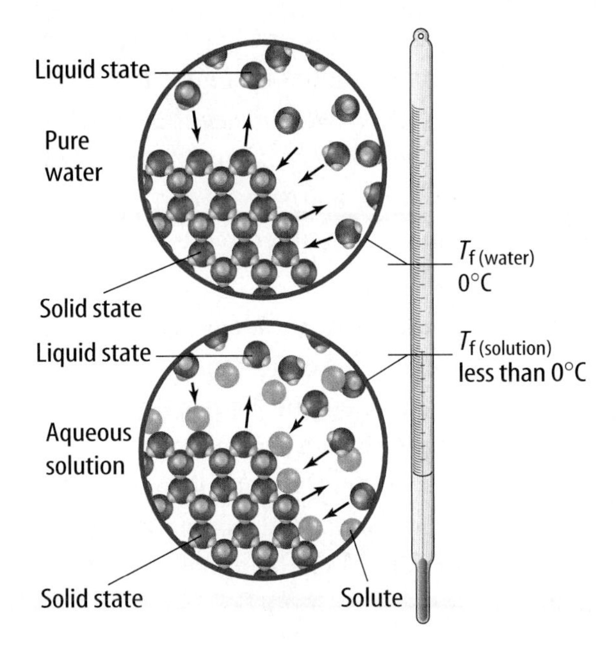

Figure 20
Solute molecules interfere with the freezing process by blocking molecules of solvent as they try to join the growing crystal lattice. For example, antifreeze molecules added to water block the formation of ice crystals.

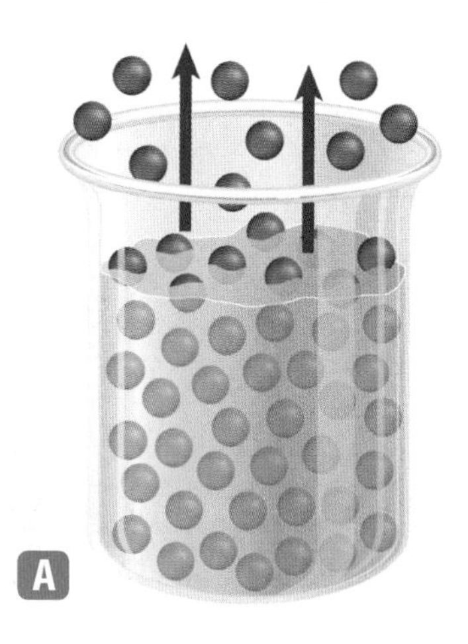

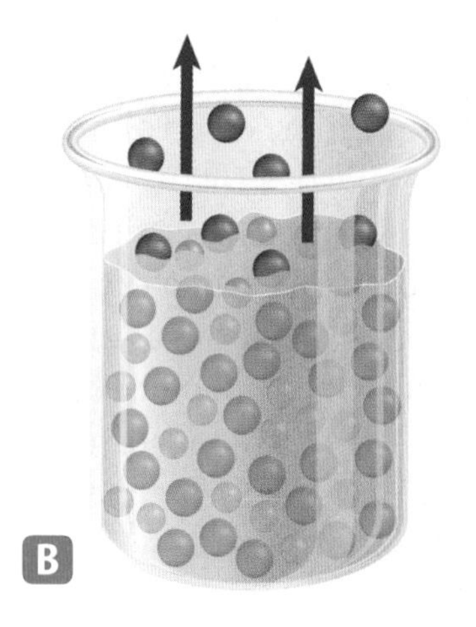

Figure 21
Solute particles raise the boiling point of a solution. A Solvent particles vaporize freely from the surface. B Solute particles block part of the surface, making it more difficult for solvent to vaporize.

Car Radiators The beaker in **Figure 21A** represents a car radiator when it contains water molecules only—no antifreeze. Some of those molecules on the surface will vaporize, and the number of molecules that do vaporize depends upon the temperature of the solvent. As temperature increases, water molecules move faster, and more molecules vaporize. Finally, when the pressure of the water vapor equals atmospheric pressure, the water boils. Have you ever seen a vehicle at the side of the road with vapors rising from the radiator?

The result of adding antifreeze is shown in **Figure 21B.** Particles of solute are distributed evenly throughout the solution, including the surface area. Now fewer water molecules can reach the surface and evaporate, making the vapor pressure of the solution lower than that of the solvent. This means that it will take a higher temperature to make the car's radiator boil over.

Section  Assessment

1. Explain (a) how ionization is different from dissociation and (b) how the two processes are similar.
2. What kinds of solute particles are present in water solutions of electrolytes and nonelectrolytes?
3. Describe how an ionic substance dissociates in water.
4. How does the concentration of a solution influence its boiling point?
5. **Think Critically** In cold weather, people often put salt on ice that forms on sidewalks and driveways. The salt helps melt the ice, forming a saltwater solution. Explain why this solution may not refreeze.

Skill Builder Activities

6. **Concept Mapping** Draw a concept map to show the relationship among the following terms: *electrolytes, nonelectrolytes, dissociation, ionization, ionic compounds, certain polar compounds,* and *other polar compounds.* **For more help, refer to the Science Skill Handbook.**
7. **Communicating** Many insect eggs can survive extremely cold temperatures. What can you conclude about the fluids in these eggs? Research chicken eggs. What temperatures can they withstand? Summarize your conclusions in your Science Journal. **For more help, refer to the Science Skill Handbook.**

Activity

Boiling Points of Solutions

Adding small amounts of salt to water that is being boiled to cook pasta and adding antifreeze to a radiator have a common result—increasing the boiling point.

What You'll Investigate

How much can the boiling point of a solution be changed?

Materials

distilled water (400 mL)
Celsius thermometer,
table salt, NaCl (72 g)
ring stand
hot plate
250-mL beaker

Safety Precautions

Goals

- **Determine** how adding salt affects the boiling point of water.

Procedure

1. Bring 100 mL of distilled water to a gentle boil in a 250-mL beaker. Record the temperature. Do not touch the hot plate surface.
2. Dissolve 12 g of NaCl in 100 mL of distilled water. Bring this solution to a gentle boil and record its boiling point.
3. Repeat step 2, using 24 g of NaCl, then 36 g.
4. Make a graph of your results. Put boiling point on the *x*-axis and grams of NaCl on the *y*-axis.

The Effects of Solute on Boiling Point

Grams of NaCl Solute	Boiling Point (°C)
0	
12	
24	
36	

Conclude and Apply

1. **Explain** the difference between the boiling points of pure water and a water solution.
2. Instead of doubling the amount of NaCl in step 3, what would have been the effect of doubling the amount of water?
3. **Predict** what would happen if you continued to add more salt. Would your graph continue in the same pattern or eventually level off? Explain your prediction.

Communicating Your Data

Compare your results with those of other groups and discuss any differences in the results obtained. **For more help, refer to the Science Skill Handbook.**

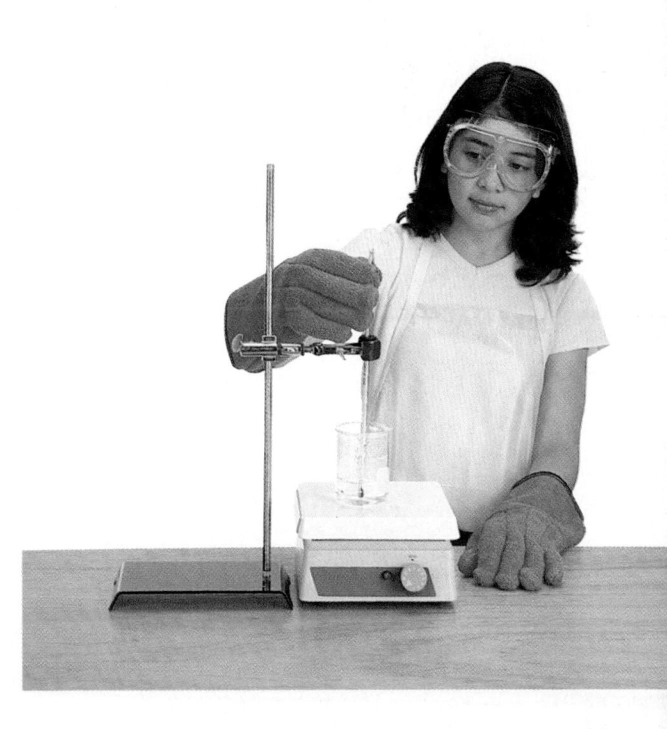

Activity

Saturated Solutions

Two major factors to consider when you are dissolving a solute in water are temperature and the ratio of solute to solvent. What happens to a solution as the temperature changes? To be able to draw conclusions about the effect of temperature, you must keep other variables constant. For example, you must be sure to stir each solution in a similar manner.

What You'll Investigate

How does solubility change as temperature is increased?

Materials

distilled water at room temperature
large test tubes
Celsius thermometer
table sugar
copper wire stirrer, bent into a spiral as shown on the next page
test-tube holder
graduated cylinder (10 mL)
beaker (250 mL) with 150 mL of water
electric hot plate
test-tube rack
ring stand

Goals

- **Observe** the effects of temperature on the amount of solute that dissolves.

Safety Precautions

WARNING: *Do NOT touch the test tubes or hot plate surface when hot plate is turned on or cooling down. When heating a solution in a test tube, keep it pointed away from yourself and others. Do NOT remove goggles until clean up including washing hands is completed.*

Procedure

1. Place 20 mL of distilled water in a test tube.
2. Add 30 g of sugar.
3. Stir. Does this dissolve?
4. If it dissolves completely, add another 5 g of sugar to the test tube. Does it dissolve?
5. Continue adding 5-g amounts of sugar until no more sugar dissolves.
6. Now place the beaker of water on the hot plate and hang the thermometer from the ring stand so that the bulb is immersed about halfway into the beaker, making sure it does not touch the sides or bottom. Record the starting temperature.
7. Using a test-tube holder, place the test tube into the water.
8. Gradually increase the temperature of the hot plate, while stirring the solution in the tube, until all the sugar dissolves.
9. Note the temperature at which this happens.
10. Add another 5 g of sugar and continue. Note the temperature at which this additional sugar dissolves.
11. Continue in this manner until you have at least four data points. Note the total amount of sugar that has dissolved. Record your data on the data table.

Dissolving Sugar in Water

Temperature	Total Grams of Sugar Dissolved

Conclude and Apply

1. How did the saturation change as the temperature was increased?
2. **Graph** your results using a line graph. Place grams of solute per 100 g of water (multiply the number of grams by five because you used only 20 mL of water) on the *y*-axis and place temperature on the *x*-axis.
3. Using your graph, estimate the solubility of sugar at 100°C and at 0°C, the boiling and freezing point of water, respectively.
4. **Compare** your results with those given in **Table 3.**

Compare your results with those of other groups and discuss any differences noted. Why might these differences have occurred? **For more help, refer to the Science Skill Handbook.**

Weird Solutions

Did you know...

. . . The "brightest" solutions can glow like a streetlight. Glowing rods called glow or light sticks are an example. Each rod contains two liquids that are kept apart using glass or plastic containers. When you flex the rod, the containers break, and the solutions mix and react to produce luminescence, or glowing light. A similar process called bioluminescence allows some living organisms, like fireflies and squid, to glow.

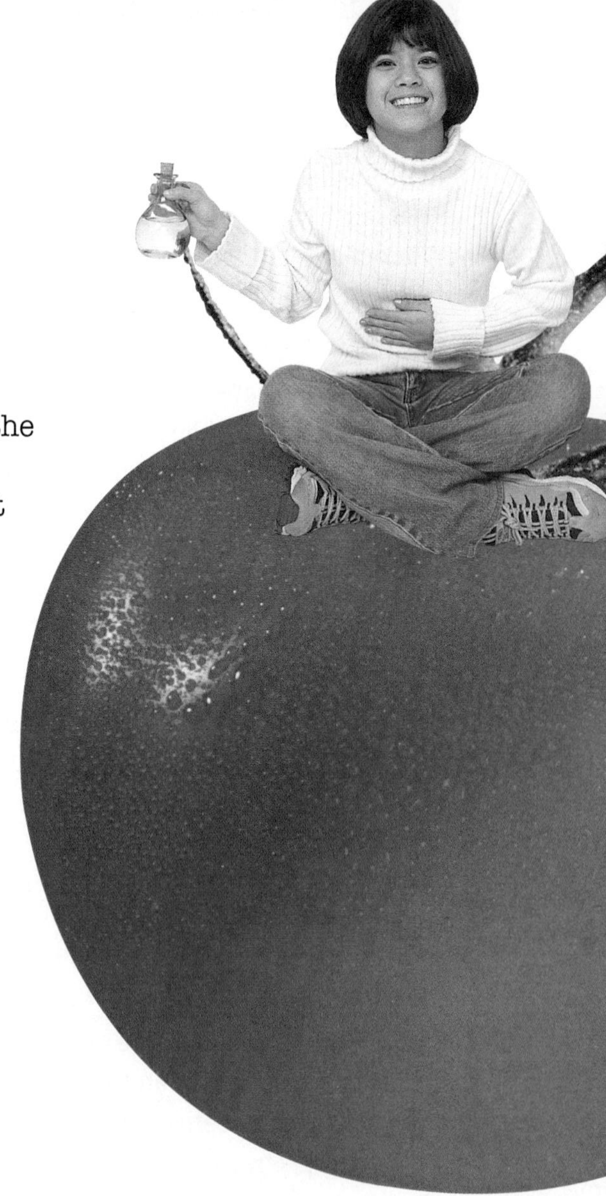

. . . One of the hardest solutions is steel, a solid solution of iron, carbon, and other elements. When you add some chromium and nickel to the mix, you get stainless steel, which is a tough, rust-resistant solution. In 1998, the United States produced nearly 100 million metric tons of raw steel—enough to make more than 1,800 Empire State Buildings.

. . . One of the strongest acids around is the solution found in your stomach. Stomach acid contains hydrochloric acid and is about 10 times stronger than vinegar. Vinegar is 10 times more acidic than fresh tomatoes. Fortunately, chemical buffers in the stomach neutralize the hydrochloric acid when it touches the stomach walls.

Connecting To Math

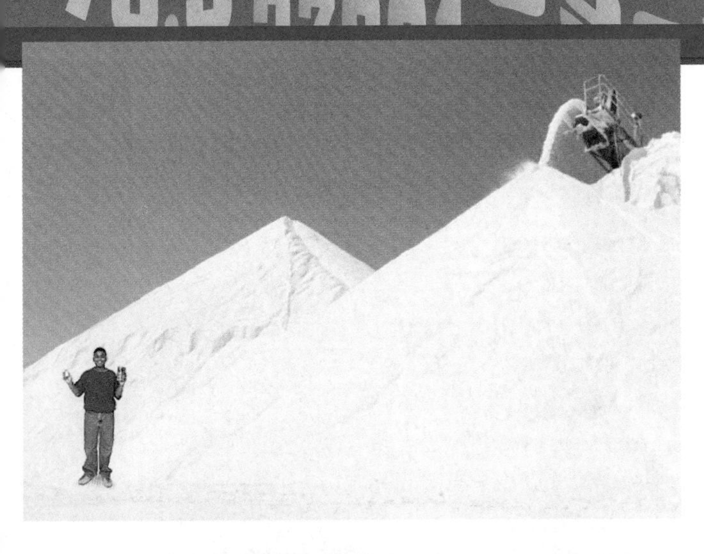

. . . The saltiest and largest body of solution in the western hemisphere is the Great Salt Lake in Utah. If all the salt in the lake dried out and hardened, the result would be a rock with a mass of about $4\frac{1}{2}$ trillion kg—as heavy as 300 million large trucks.

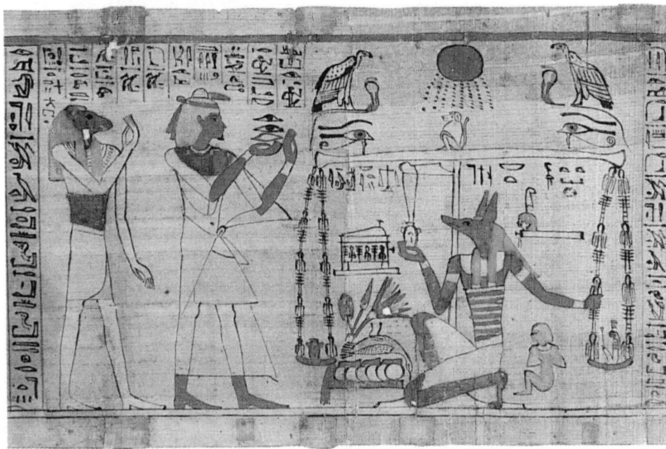

. . . Some of the stickiest solutions are glues. Gluing is nothing new. Papyrus—ancient Egyptian paper—was held together with glues. Natural adhesives come from animal proteins and plant carbohydrates. During the twentieth century, chemists developed many synthetic adhesives.

Do the Math

1. Great Salt Lake's salinity is 5 percent when the water is highest and 30 percent when the water is lowest. What is the lake's salinity when the water level is halfway between its highest and lowest levels?
2. Glow sticks shine for about 10 h. If you kept a glow stick glowing continuously in your window for seven days, how many sticks would you need?
3. One formula for making stainless steel is 1 percent carbon, 54 percent iron, and twice as much chromium as nickel. Calculate the percentage of chromium used.

Go Further

The human body contains many solutions. On **science.glencoe.com** find out how many liters of solutions are in an average adult body.

Chapter 23 Study Guide

Reviewing Main Ideas

Section 1 How Solutions Form

1. A solution is a mixture that has the same composition, color, density, and taste throughout.
2. The substance being dissolved is called a solute, and the substance that does the dissolving is called a solvent.
3. The rate of dissolving can be increased by stirring or increasing temperature.
4. Under similar conditions, small particles of solute dissolve faster than large particles.

Section 2 Dissolving Without Water

1. Water cannot dissolve all solutes.
2. Nonpolar solvents are needed to dissolve nonpolar solutes.
3. Some vitamins are nonpolar and dissolve in the fat contained in some body cells.
4. Nonpolar solvents can be dangerous as well as helpful. *What precautions are needed when using substances like these?*

Section 3 Solubility and Concentration

1. Some compounds are more soluble than others, and this can be measured.
2. *Concentrated* and *dilute* are not precise terms used to describe concentration of solutions.
3. Concentrations can be expressed as percent by volume.
4. An unsaturated solution can dissolve more solute, and a saturated solution cannot. A supersaturated solution is made by raising the temperature of a saturated solution and adding more solute. If it is cooled carefully, the supersaturated solution will retain the dissolved solute. *Is this tea unsaturated, saturated, or supersaturated with sugar? Explain.*

Section 4 Particles in Solution

1. Substances that dissolve in water to produce solutions that conduct electricity are called electrolytes.
2. When water pulls apart the molecules of a polar substance, forming ions, the process is called ionization.
3. When ionic solids dissolve in water, the process is called dissociation, because the ions are already present in the solid.

After You Read

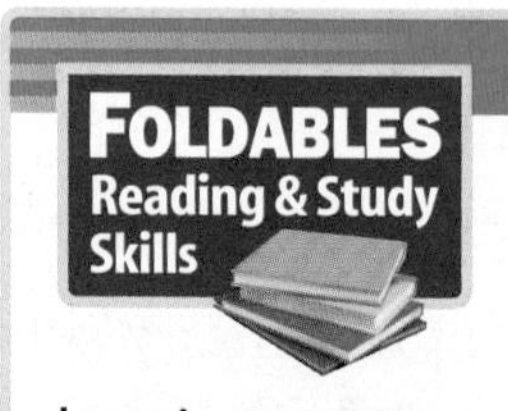

Use your Foldable to determine what characteristics solvents and solutes have in common.

Visualizing Main Ideas

Complete the following concept map on solutions.

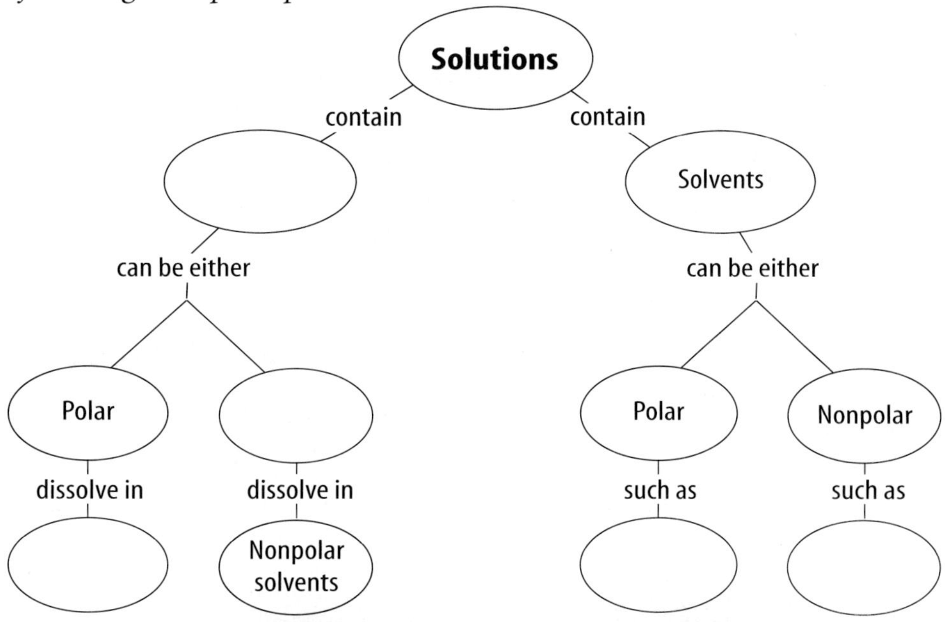

Vocabulary Review

Vocabulary words

a. alloy
b. dissociation
c. electrolyte
d. ion
e. ionization
f. nonelectrolyte
g. nonpolar
h. polar
i. saturated solution
j. solubility
k. solution
l. solute
m. solvent
n. supersaturated solution
o. unsaturated solution

Study Tip

Pay attention to the chapter's illustrations. Try to figure out exactly what point the picture is trying to stress.

Using Vocabulary

The sentences below include terms that are used incorrectly. Change the incorrect terms so that the sentence reads correctly. Underline your change.

1. In lemonade, sugar is the solvent and water is the solution.
2. During ionization, particles in an ionic solid are separated and drawn into solution.
3. If more of substance B dissolves in water than substance A, then substance B has a higher dissociation than substance A.
4. Adding a seed crystal may cause solute to crystallize from an unsaturated solution.
5. More solute can be added to a saturated solution.
6. A substance that does not form ions in water is a solute.

Chapter 23 Assessment

Checking Concepts

Choose the word or phrase that best answers the question.

1. Which of the following is NOT a solution?
A) glass of soda
B) air in a SCUBA tank
C) bronze alloy
D) mud in water

2. What term is NOT appropriate to use when describing solutions?
A) heterogeneous
B) gaseous
C) liquid
D) solid

3. When iodine is dissolved in alcohol, what term is used to describe the alcohol?
A) alloy
B) solvent
C) solution
D) solute

4. What word is used to describe a mixture that is 85 percent copper and 15 percent tin?
A) alloy
B) solvent
C) saturated
D) solute

5. Solvents, such as paint thinner and gasoline evaporate more readily than water because they are what type of compounds?
A) ionic
B) nonpolar
C) dilute
D) polar

6. What can a polar solvent dissolve?
A) any solute
B) a polar solute
C) a nonpolar solute
D) no solute

7. If a water solution conducts electricity, what must the solute be?
A) gas
B) electrolyte
C) liquid
D) nonelectrolyte

8. In forming a water solution, what process does an ionic compound undergo?
A) dissociation
B) electrolysis
C) ionization
D) no change

9. What can you increase to make a gas more soluble in a liquid?
A) particle size
B) pressure
C) stirring
D) temperature

10. If a solute crystallizes out of a solution when a seed crystal is added, what kind of solution is it?
A) unsaturated
B) saturated
C) supersaturated
D) dilute

Thinking Critically

11. Why might potatoes cook more quickly in salted water than in unsalted water?

12. Explain what happens when an ionic compound such as copper(II) sulfate, $CuSO_4$, dissolves in water.

13. Why is the term *dilute* not precise?

14. Explain how you would make a 25 percent solution by volume of apple juice.

15. Why is the statement, "Water is the solvent in a solution," not always true?

Developing Skills

16. Making and Using Tables Using the data in **Table 3,** fill in the following table. Use the terms *saturated, unsaturated,* and *supersaturated* to describe the type of solution.

Limits of Solubility

Compound	Type of Solution	Solubility in 100 g Water at 20°C
$CuSO_4$		32.0 g
KCl		45.8 g
KNO_3		31.6 g
$NaClO_3$		79.6 g

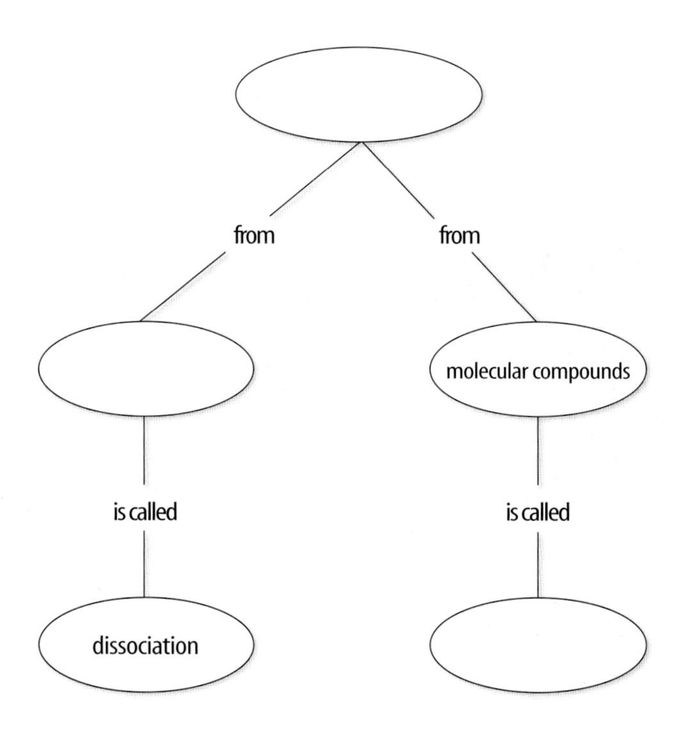

17. Concept Mapping Compare and contrast the processes of ionization and dissociation in the concept map above.

18. Measuring in SI 153 g of potassium nitrate have been dissolved in enough water to make 1 L of this solution. You use a graduated cylinder to measure 80 mL of solution. What mass of potassium nitrate is in the 80-mL sample?

Performance Assessment

19. Wise consumer Visit a grocery store and identify 20 examples of solutions. Classify them according to the type of solution—gas, liquid, or solid. Present your results to your class.

TECHNOLOGY

Go to the Glencoe Science Web site at **science.glencoe.com** or use the **Glencoe Science CD-ROM** for additional chapter assessment.

Test Practice

Pictured below are a scoop of cherry drink mix, a glass of water, and a glass of cherry drink made by adding the mix to the water.

1. Which of the following best describes what happens when the cherry drink mix is added to the water?

A) The size and number of cherry drink mix particles increases.

B) The particles reject the cherry drink mix particles.

C) The cherry drink mix particles become evenly mixed with the water particles.

D) Each cherry drink particle enters a water particle.

2. Which of the following best represents a solution that contains a high level of cherry drink mix?

F) 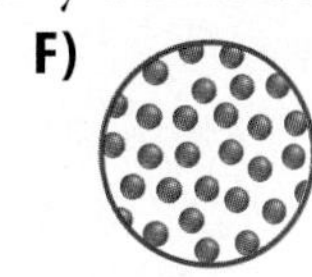 **H)** 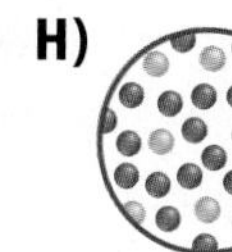

G) 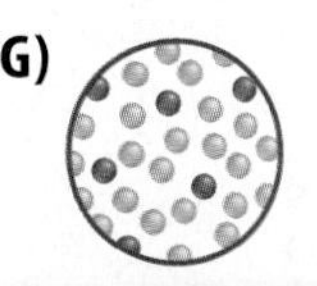 **J)** 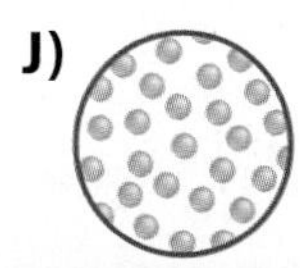

CHAPTER

24 Chemical Reactions

Ohhhh! Ahhhh! The crowd gasps as fireworks explode overhead into patterns of brightly colored sparks. Deafening booms split the air and children squeal with delight. All these effects are produced by chemical reactions with oxygen. The colors come from small amounts of metal ions—blue from copper, red from lithium, gold from sodium, and green from barium. In this chapter, you will learn about chemical reactions—what they are, how to describe them, and how they release or absorb energy.

What do you think?

Science Journal Look at the picture below with a classmate. Discuss what you think this might be or what is happening. Here's a hint: *These bubbles are lighter than air.* Write down your best guess in your Science Journal.

Like exploding fireworks, rusting is a chemical reaction in which iron metal combines with oxygen. Other metals combine with oxygen, too—some more readily than others. In this activity, you will compare how iron and aluminum react with oxygen.

Safety Precautions

Observe which metal reacts

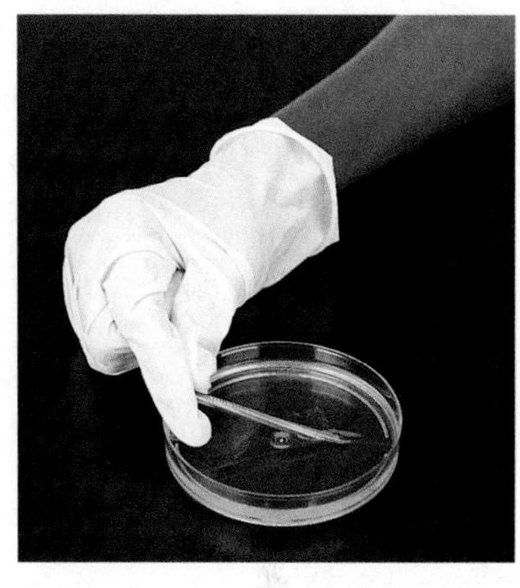

1. Place a clean, iron or steel nail in a dish prepared by your teacher.
2. Place a clean, aluminum nail in a second dish. These dishes contain agar gel and an indicator that detects a reaction with oxygen.
3. Observe both nails after one hour. Record any changes around the nails in your Science Journal.
4. Carefully examine both of the dishes the next day. Record any differences between the two dishes in your Science Journal.

Observe

Did a reaction occur in either dish? How could you tell? What might have caused any differences you observed between the two nails?

Before You Read

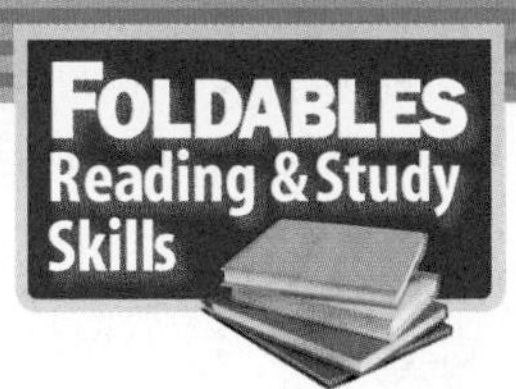

Making a Cause and Effect Study Fold **Make the following Foldable to help you understand the cause and effect relationship of chemical reactions.**

1. Place a sheet of paper in front of you so the short side is at the top. Fold the paper in half from the left side to the right side.
2. Now fold the paper in half from top to bottom. Then fold it in half again top to bottom. Unfold the last two folds you did.
3. Through one thickness of paper, cut along each of the fold lines to form four tabs as shown.
4. Label the four sections "R" for reactants, "⟶" for produce, and "P" for products as shown. As you read the chapter, record examples of chemical reactions under the tabs.

Chemical Changes

As You Read

What You'll Learn

- **Identify** the reactants and products in a chemical reaction.
- **Determine** how a chemical reaction satisfies the law of conservation of mass.
- **Determine** how chemists express chemical changes using equations.

Vocabulary

chemical reaction
chemical equation
coefficient
reactant
product

Why It's Important

Chemical reactions cook our food, warm our homes, and provide energy to our bodies.

Describing Chemical Reactions

Dark mysterious mixtures react, gas bubbles up and expands, and powerful aromas waft through the air. Where are you? Are you in a chemical laboratory carrying out a crucial experiment? No. You are in the kitchen baking a chocolate cake. Nowhere in the house do so many chemical reactions take place as in the kitchen.

Actually, chemical reactions are taking place all around you and even within you. A **chemical reaction** is a change in which one or more substances are converted into new substances. The substances that react are called **reactants.** The new substances produced are called **products.** This relationship can be written as follows:

$$\text{reactants} \xrightarrow{\text{produce}} \text{products}$$

Conservation of Mass

By the 1770s, chemistry was changing from the art of alchemy to a true science. Chemical reactions were studied carefully. Through such study, the French chemist Antoine Lavoisier established that the mass of the products always equals the mass of the reactants. This principle is demonstrated in **Figure 1.**

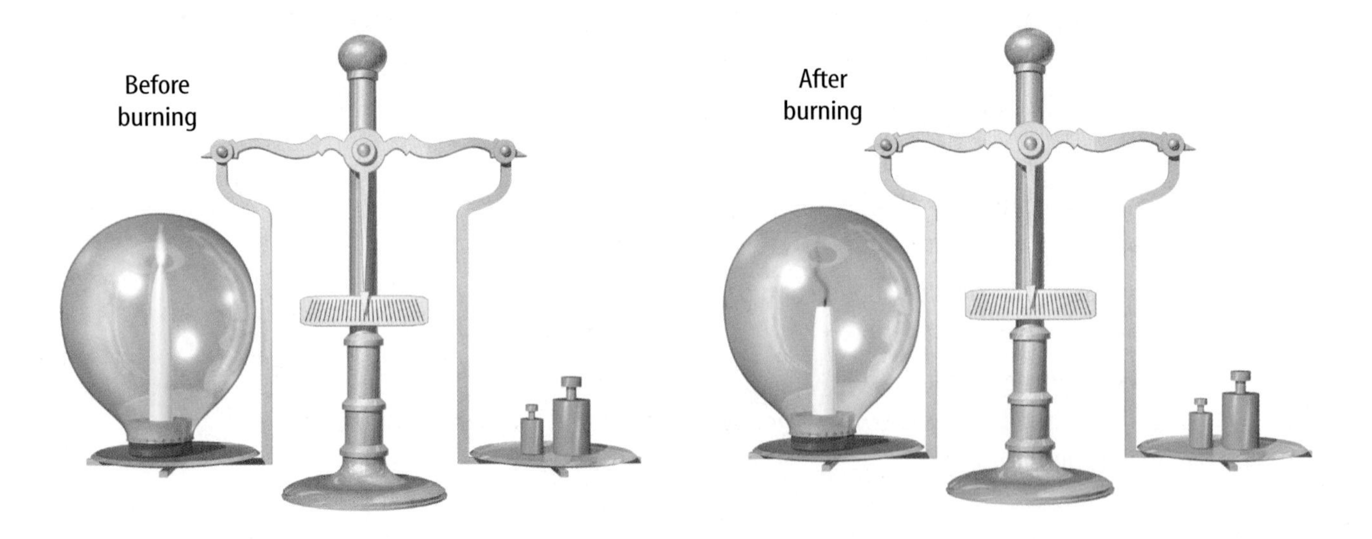

Figure 1
The mass of the candles and oxygen before burning is exactly equal to the mass of the remaining candle and gaseous products.

Figure 2
Antoine Lavoisier's wife, Marie-Anne, drew this view of Lavoisier in his laboratory performing studies on oxygen. She depicted herself at the right taking notes.

Lavoisier's Contribution

In one experiment, Lavoisier placed a carefully measured mass of solid mercury(II) oxide, which he knew as mercury calx, into a sealed container. When he heated this container, he noted a dramatic change. The red powder had been transformed into a silvery liquid that he recognized as mercury metal, and a gas was produced. When he determined the mass of the liquid mercury and gas, their combined masses were exactly the same as the mass of the red powder he had started with.

mercury(II) oxide		oxygen	plus	mercury
10.0 g	=	0.7 g	+	9.3 g

Lavoisier also established that the gas produced by heating mercury oxide, which we call oxygen, was a component of air. He did this by heating mercury metal with air and saw that a portion of the air combined to give red mercury oxide. He studied the effect of this gas on living animals, including himself. Hundreds of experiments carried out in his laboratory, shown in **Figure 2,** confirmed that in a chemical reaction, matter is not created or destroyed, but is conserved. This principle became known as the law of conservation of mass. This means that the starting mass of the reactants equals the final mass of the products.

Reading Check ***What does the law of conservation of mass state?***

This principle is the basis for modern chemistry. It made it possible for chemists to see clearly what happens in chemical reactions. More importantly, the concept of balance between reactants and products led to a valuable tool of chemistry—chemical equations.

Research Visit the Glencoe Science Web site at **science.glencoe.com** for more information about Antoine Lavoisier and his contributions to chemistry. Communicate to your class what you learn.

Writing Equations

If you wanted to describe the chemical reaction shown in **Figure 3,** you might write something like this:

nickel(II) chloride, dissolved in water, plus sodium hydroxide, dissolved in water, produces solid nickel(II) hydroxide plus sodium chloride, dissolved in water

This series of words is rather cumbersome, but all of the information is important. The same is true of descriptions of most chemical reactions. Many words are needed to state all the important information. As a result, scientists have developed a shorthand method to describe chemical reactions. A **chemical equation** is a way to describe a chemical reaction using chemical formulas and other symbols. Some of the symbols used in chemical equations are listed in **Table 1.**

Table 1 Symbols Used in Chemical Equations

Symbol	Meaning
$\rightarrow$	produces or forms
$+$	plus
(s)	solid
(l)	liquid
(g)	gas
(aq)	aqueous, a substance is dissolved in water
$\xrightarrow{\text{heat}}$	the reactants are heated
$\xrightarrow{\text{light}}$	the reactants are exposed to light
$\xrightarrow{\text{elec.}}$	an electric current is applied to the reactants

The chemical equation for the reaction described above in words and shown in **Figure 3** looks like this:

$$NiCl_2(aq) + 2NaOH(aq) \rightarrow Ni(OH)_2(s) + 2NaCl(aq)$$

It is much easier to tell what is happening by writing the information in this form. Later, you will learn how chemical equations make it easier to calculate the quantities of reactants that are needed and the quantities of products that are formed.

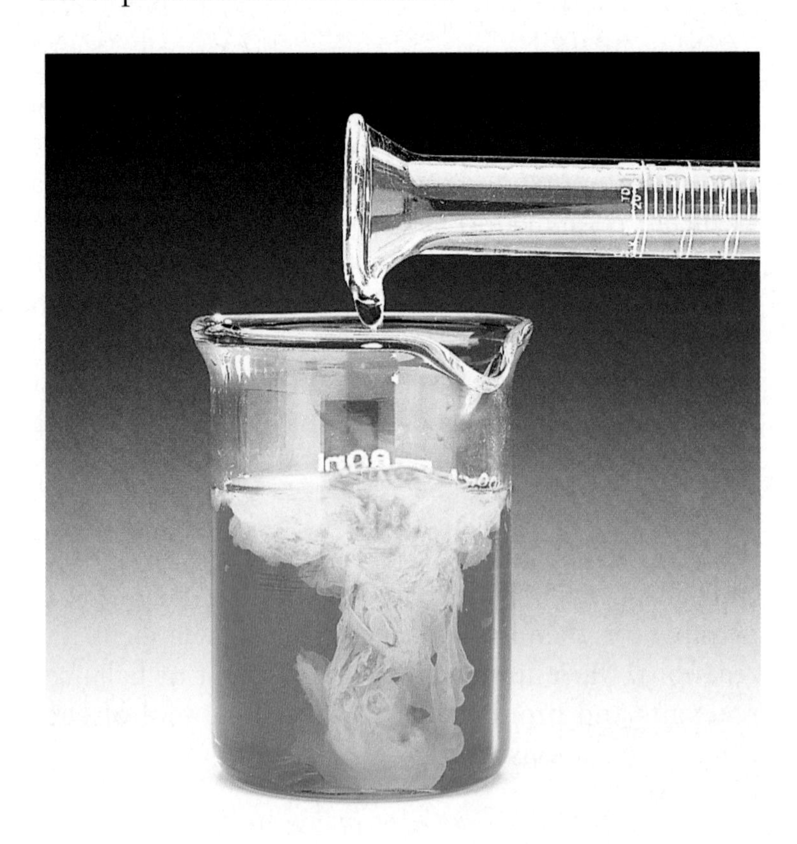

Figure 3
A white precipitate of nickel(II) hydroxide forms when sodium hydroxide is added to a green solution of nickel(II) chloride. Sodium chloride, the other product formed, is in the solution.

Unit Managers

What do the numbers to the left of the formulas for reactants and products mean? Remember that according to the law of conservation of mass, matter is neither made nor lost during chemical reactions. Atoms are rearranged but never lost or destroyed. These numbers, called **coefficients,** represent the number of units of each substance taking part in a reaction. Coefficients can be thought of as unit managers.

Reading Check *What is the function of coefficients in a chemical equation?*

Imagine that you are responsible for making sandwiches for a picnic. You have been told to make a certain number of three kinds of sandwiches, and that no substitutions can be made. You would have to figure out exactly how much food to buy so that you had enough without any food left over. You might need two loaves of bread, four packages of turkey, four packages of cheese, two heads of lettuce, and ten tomatoes. With these supplies you could make exactly the right number of each kind of sandwich.

In a way, your sandwich-making effort is like a chemical reaction. The reactants are your bread, turkey, cheese, lettuce, and tomatoes. The number of units of each ingredient are like the coefficients of the reactants in an equation. The sandwiches are like the products, and the numbers of each kind of sandwich are like coefficients, also.

Knowing the number of units of reactants enables chemists to add the correct amounts of reactants to a reaction. Also, these units, or coefficients, tell them exactly how much product will form. An example of this is the reaction of one unit of $NiCl_2$ with two units of NaOH to produce one unit of $Ni(OH)_2$ and two units of NaCl. You can see these units in **Figure 4.**

TRY AT HOME Mini LAB

Designing a Team Equation

Procedure

1. Obtain **15 index cards** and mark each as follows: five with *guard*, five with *forward*, and five with *center*.
2. Group the cards to form as many complete basketball teams as possible. Each team needs two guards, two forwards, and one center.

Analysis

1. Write the formula for a team. Write the formation of a team as an equation. Use coefficients in front of each type of player needed for a team.
2. How is this equation like a chemical equation? Why can't you use the remaining cards?

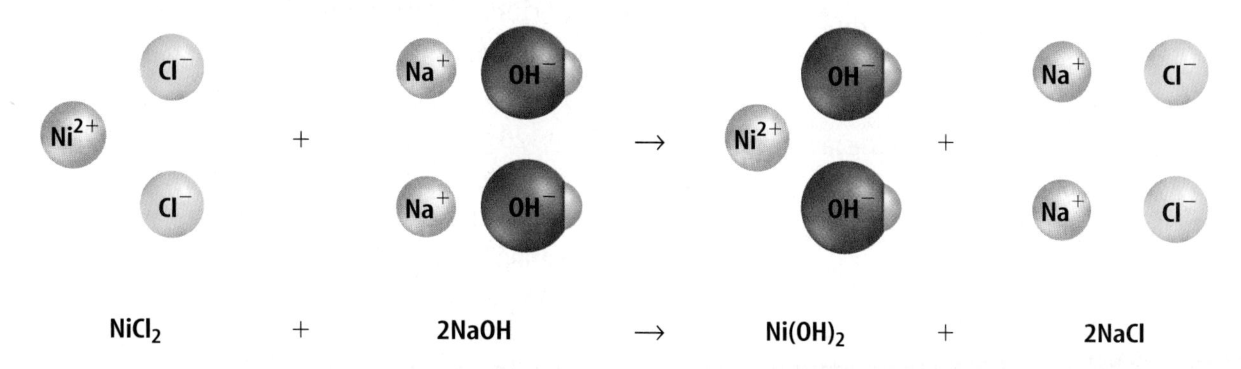

Figure 4
Each coefficient in the equation represents the number of units of each type in this reaction.

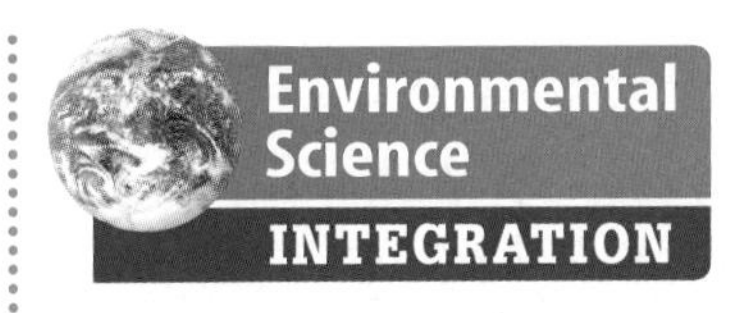

Metals and the Atmosphere

When iron is exposed to air and moisture, it corrodes or rusts, forming hydrated iron(III) oxide. Rust can seriously damage iron structures because it crumbles and exposes more iron to the air. This leads to more oxidation and eventually can destroy the structure. However, not all reactions of metals with the atmosphere are damaging like rust. Some are helpful.

Aluminum also reacts with oxygen in the air to form aluminum oxide. Unlike rust, aluminum oxide adheres to the aluminum surface, forming an extremely thin layer that protects the aluminum from further attack. You can see this thin layer of aluminum oxide on aluminum outdoor furniture. It makes the once shiny aluminum look dull.

Copper is another metal that corrodes when it is exposed to air forming a blue-green coating called a patina. You can see this type of corrosion on many public monuments and also on the Statue of Liberty, shown in **Figure 5.**

Figure 5
The blue-green patina that coats the Statue of Liberty contains copper(II) sulfate among other copper corrosion products.

Section 1 Assessment

1. Identify the reactants and the products in the following chemical equation.
   $Cd(NO_3)_2(aq) + H_2S(g) \rightarrow CdS(s) + 2HNO_3(aq)$
2. What is the state of matter of each substance in the following reaction?
   $Zn(s) + 2HCl(aq) \rightarrow H_2(g) + ZnCl_2(aq)$
3. Why is the reaction of oxygen with iron a problem, but the reaction of oxygen with aluminum is not?
4. What is the importance of the law of conservation of mass?
5. **Think Critically** Why do you think the copper patina was kept when the Statue of Liberty was restored?

Skill Builder Activities

6. **Recognizing Cause and Effect** Lavoisier heated mercury(II) oxide in a sealed flask. Explain the effect on Lavoisier's conclusions about the law of conservation of mass if he had used an open container. **For more help, refer to the Science Skill Handbook.**
7. **Solving One-Step Equations** When making soap, if 890 g of a specific fat react completely with 120 g of sodium hydroxide, the products formed are soap and 92 g of glycerin. Calculate the mass of soap formed to satisfy the law of conservation of mass. **For more help, refer to the Math Skill Handbook.**

Chemical Equations

Balanced Equations

Lavoisier heated mercury(II) oxide—HgO—and obtained mercury and oxygen. This reaction is shown in **Figure 6.** When written as a chemical equation, the reaction is:

$$HgO(s) \xrightarrow{\text{heat}} Hg(l) + O_2\ (g)$$

Notice that the number of mercury atoms is the same on both sides of the equation but that the number of oxygen atoms is not the same. One oxygen atom appears on the reactant side of the equation and two appear on the product side

Atoms	**HgO**	$\rightarrow$	**Hg**	+	$\mathbf{O_2}$
Hg	1		1		
O	1				2

Now wait a minute, you say. The law of conservation of mass states that atoms are neither created nor destroyed in a chemical reaction. One oxygen atom can't just become two. Are the formulas wrong? Can you fix this problem simply by adding the subscript 2 and writing HgO_2 instead of HgO?

Think for a moment about water, H_2O, and hydrogen peroxide, H_2O_2. Although their formulas are similar, they are very different compounds. Similarly, the formulas HgO_2 and HgO do not represent the same compound. In fact, HgO_2 does not exist. In Lavoisier's experiment, he heated HgO, not HgO_2. The formulas in a chemical equation must accurately represent the compounds that react. They can't be changed to something different.

✔ **Reading Check** ***Why can't you change the subscripts in a chemical equation?***

If you can't change the subscripts in an equation, how can you fix Lavoisier's equation so that the number of oxygen atoms is the same on both sides? Fixing this equation requires a process called balancing. Balancing an equation doesn't change what happens in a reaction—it simply changes the way the reaction is represented. The balancing process involves changing coefficients in a reaction to achieve a **balanced chemical equation,** which has the same number of atoms of each element on both sides of the equation.

As You Read

What **You'll Learn**

- **Identify** what is meant by a balanced chemical equation.
- **Determine** how to write balanced chemical equations.

Vocabulary

balanced chemical equation

Why **It's Important**

Chemical equations are the language used to describe chemical change.

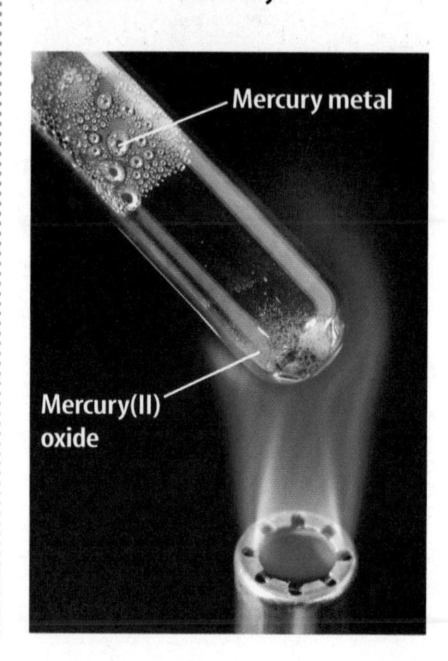

Figure 6
Mercury metal forms when mercury oxide is heated. Because mercury is poisonous, this reaction is never performed in a classroom laboratory.

Research Visit the Glencoe Science Web site at **science.glencoe.com** for more information on balancing chemical equations. Communicate to your class what you learned.

Choosing Coefficients Finding out which coefficients to use to balance an equation is often a trial-and-error process, but with practice the process becomes easier. In the equation for Lavoisier's experiment, the number of mercury atoms is balanced, but one oxygen atom is on the left and two are on the right. Therefore, this equation isn't balanced. If you put a coefficient of 2 before the mercury(II) oxide on the left, the oxygen atoms will be balanced. However, you have two mercury atoms on the left and only one on the right. To correct this, put a 2 in front of mercury on the right. The equation is now balanced.

Atoms	$2HgO$	$\rightarrow$	$2Hg$	$+$	O_2
Hg	2		2		
O	2				2

Try Your Balancing Act Magnesium burns with such a brilliant, white light that it is often used in emergency flares as shown in **Figure 7.** Burning leaves a white powder called magnesium oxide. To write a balanced chemical equation for this and most other reactions, follow these four steps.

Step 1 Write a chemical equation for the reaction using formulas and symbols. Review how to write formulas for compounds. Recall that oxygen gas is a diatomic molecule.

$$Mg(s) + O_2(g) \rightarrow MgO(s)$$

Step 2 Check the equation for atom balance.

Atoms	Mg	$+$	O_2	$\rightarrow$	MgO
Mg	1				1
O			2		1

The magnesium atoms are balanced, but the oxygen atoms are not. Therefore, this equation isn't balanced.

Step 3 Choose coefficients that balance the equation. Remember, never change subscripts of a correct formula to balance an equation. Try putting a coefficient of 2 before MgO.

$$Mg(s) + O_2(g) \rightarrow 2MgO(s)$$

Step 4 Recheck the numbers of each atom on each side of the equation and adjust coefficients again if necessary.

Now two Mg atoms are on the right side and only one is on the left side. So a coefficient of 2 is needed for Mg also.

$$2Mg(s) + O_2(g) \rightarrow 2MgO(s)$$

The equation is now balanced.

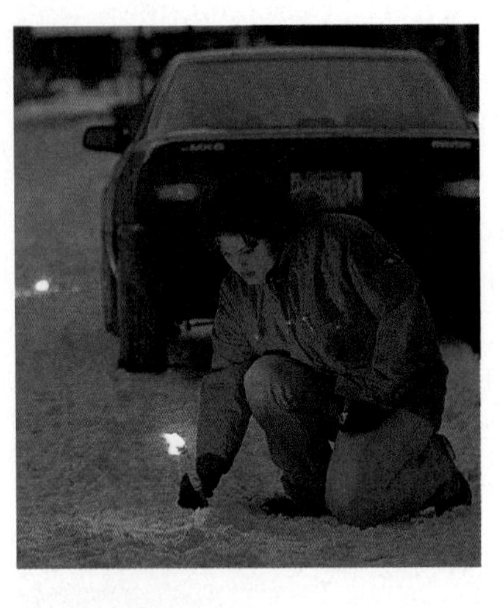

Figure 7
Magnesium combines with oxygen giving an intense white light.

Polish Your Skill When lithium metal is treated with water, hydrogen gas and lithium hydroxide are produced, as shown in **Figure 8.**

Step 1 Write the chemical equation.

$$Li(s) + H_2O \rightarrow H_2(g) + LiOH(aq)$$

Step 2 Check for balance by counting the atoms.

Atoms	Li	+	H_2O	$\rightarrow$	LiOH	+	H_2
Li	1				1		
H			2		1		2
O			1		1		

This equation is not balanced. There are three hydrogen atoms on the right and only two on the left. Complete steps 3 and 4 to balance the equation. After each step, count the atoms of each element. When equal numbers of atoms of each element are on both sides, the equation is balanced.

Just as in the first example, this balancing does not change the reaction that took place between lithium and water. It just states what happened more accurately. This accurate statement tells chemists how much lithium metal to use to produce a certain amount of hydrogen gas.

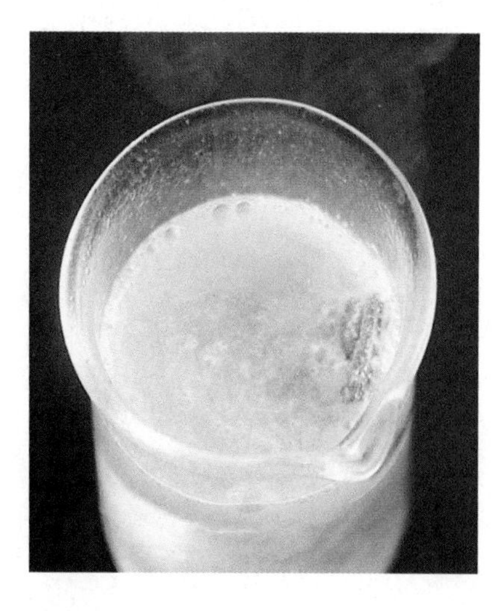

Figure 8
When lithium metal is added to water, it reacts, producing a solution of lithium hydroxide and bubbles of hydrogen gas.

Section  Assessment

1. Give two reasons for balancing equations for chemical reactions.
2. Write a balanced chemical equation for each of the following reactions.
   a. Iron metal plus oxygen produces iron(II) oxide.
   b. Sodium metal plus water produces sodium hydroxide plus hydrogen gas.
3. Explain why oxygen gas must always be written as O_2 in a chemical equation.
4. What is understood if no coefficient is written before a formula in a chemical equation?
5. **Think Critically** Explain why the sum of the coefficients on the reactant side of a balanced equation does not have to equal the sum of the coefficients on the product side of the equation.

Skill Builder Activities

6. **Predicting** Silver tarnish, Ag_2S, forms from sulfur compounds in the air and in some foods, such as eggs. Polishing removes this tarnish. Why might people not wish to polish their silverware very frequently? How might they prevent tarnish from forming and thereby reduce how often they need to polish? **For more help, refer to the Science Skill Handbook.**
7. **Using an Electronic Spreadsheet** Use a spreadsheet and design a table for the reaction of $CaCO_3$ with HCl to form CO_2, $CaCl_2$, and water. Remember that equal numbers of each element must be on both sides of a balanced equation. **For more help, refer to the Technology Skill Handbook.**

SECTION 3 Classifying Chemical Reactions

As You Read

What You'll Learn

- **Identify** the four general types of chemical reactions.
- **Predict** which metals will replace other metals in compounds.

Vocabulary

synthesis reaction
decomposition reaction
single-displacement reaction
double-displacement reaction
precipitate

Why It's Important

Classifying reactions helps you understand what is happening and predict the outcome of reactions.

Types of Reactions

You might have begun to notice that there are all sorts of chemical reactions. You could just memorize them, but that would be difficult. Fortunately there is a better way.

Imagine if the public library shelved its books without any order. Every time you wanted a particular book you would have to search through all the books or remember where you had seen it last. Fortunately libraries classify books into groups, such as biography, history, and fiction. This way you know exactly where to look.

You can do the same thing with chemical reactions. Also, once a chemical reaction is placed into a certain type, you can learn a great deal by comparing it to others of that type. One system is based upon the way the compounds interact. Most reactions can be divided into four main types—synthesis, decomposition, single displacement, and double displacement.

Synthesis Reactions One of the easiest reaction types to recognize is a synthesis reaction. In a **synthesis reaction,** two or more substances combine to form another substance. The generalized formula for this reaction type is as follows.

$$A + B \rightarrow AB$$

The reaction in which hydrogen burns in oxygen to form water is an example of a synthesis reaction.

$$2H_2(g) + O_2(g) \rightarrow 2H_2O(g)$$

This reaction is used to power some types of rockets. Another synthesis reaction is the combination of oxygen with iron in the presence of water to form hydrated iron(II) oxide or rust. This reaction is shown in **Figure 9.**

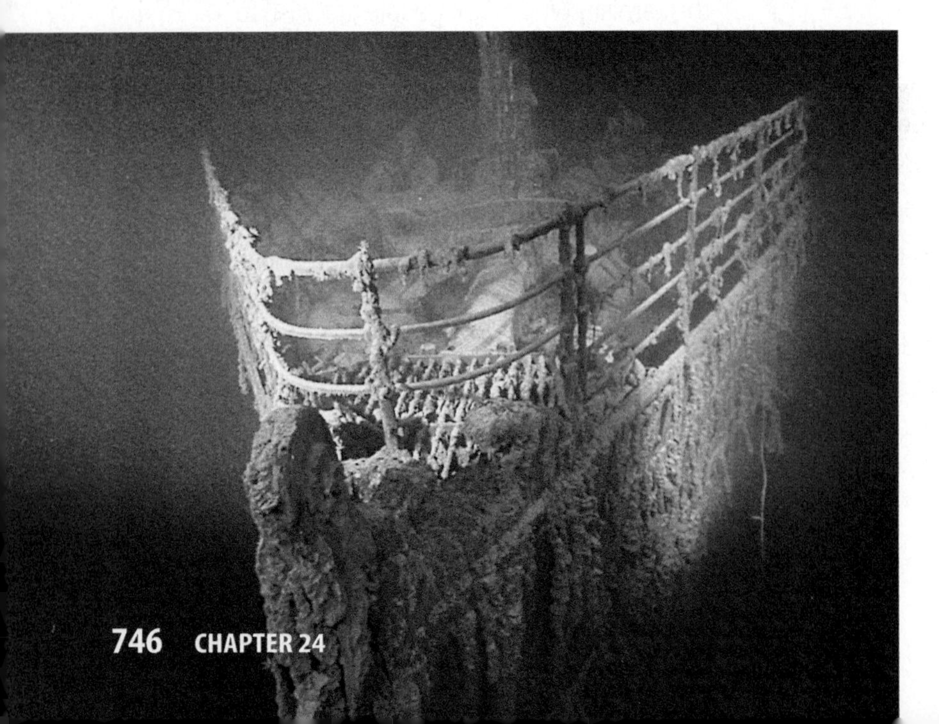

Figure 9
Rust has accumulated on the *Titanic* since it sank in 1912.

Decomposition Reactions A decomposition reaction is just the opposite of a synthesis. Instead of two substances coming together to form a third, a **decomposition reaction** occurs when one substance breaks down, or decomposes, into two or more substances. The general formula for this type of reaction can be expressed as follows:

$$AB \rightarrow A + B$$

Most decomposition reactions require the use of heat, light, or electricity. For example, an electric current passed through water produces hydrogen and oxygen as shown in **Figure 10.**

$$2H_2O(l) \xrightarrow{\text{elec.}} 2H_2(g) + O_2(g)$$

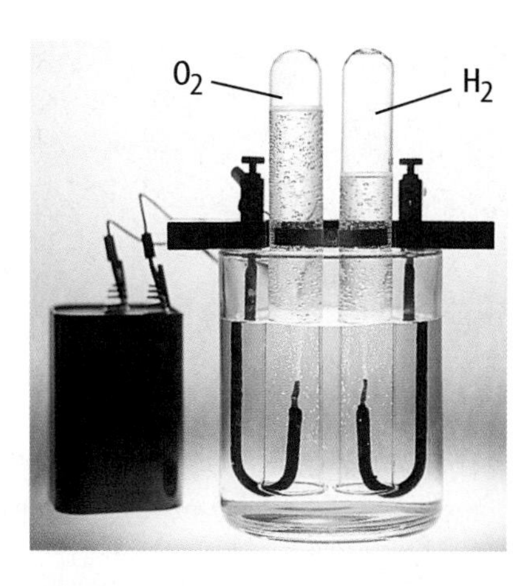

Figure 10
Water decomposes into hydrogen and oxygen when an electric current is passed through it. A small amount of sulfuric acid is added to increase conductivity. Notice the proportions of the gases collected. ***How is this related to the coefficients of the products in the equation?***

Single Displacement When one element replaces another element in a compound, it is called a **single-displacement reaction.** There are two generalized types of this reaction.

In the first case, A replaces B as follows:

$$A + BC \rightarrow AC + B$$

In the second case, D replaces C:

$$D + BC \rightarrow BD + C$$

The first case is illustrated in **Figure 11,** where a copper wire is put into a solution of silver nitrate. Because copper is a more active metal than silver, it replaces the silver, forming a blue copper(II) nitrate solution. The silver, which is not soluble, forms on the wire.

$$Cu(s) + 2AgNO_3(aq) \rightarrow Cu(NO_3)_2\ (aq) + 2Ag(s)$$

Reading Check ***Describe a single-displacement reaction.***

Sometimes single-displacement reactions can cause problems. For example, if iron-containing vegetables such as spinach are cooked in aluminum pans, aluminum can displace iron from the vegetable. This causes a black deposit of iron to form on the sides of the pan. For this reason, it is better to use stainless steel or enamel cookware when cooking spinach.

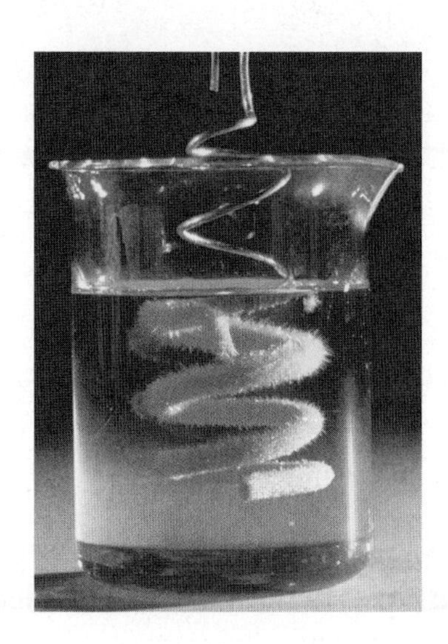

Figure 11
Copper in a wire replaces silver in silver nitrate, forming a blue solution of copper(II) nitrate.

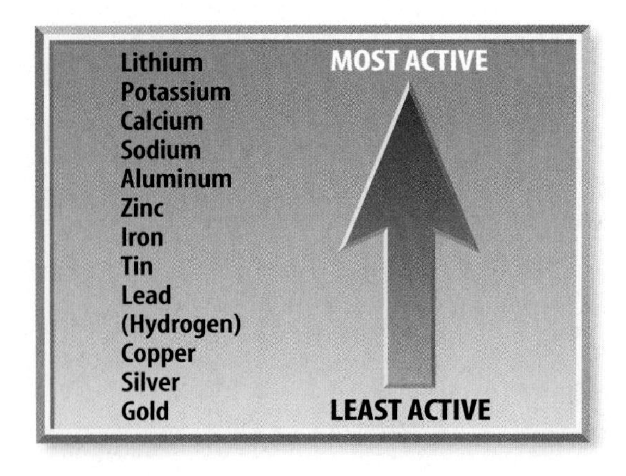

Figure 12
This figure shows the activity series of metals. A metal will replace any other metal that is less active.

The Activity Series We can predict which metal will replace another using the diagram shown in **Figure 12,** which lists metals according to how active they are. A metal will replace any less active metal. Notice that copper, silver, and gold are the least active metals on the list. That is why these elements often occur as deposits of the relatively pure element. For example, gold is sometimes found as veins in quartz rock, and copper is found in pure lumps known as native copper. Other metals occur as compounds.

Math Skills Activity

Using Coefficients to Balance Chemical Equations

Example Problem

A sample of barium sulfate is placed on a piece of paper, which is then ignited. Barium sulfate reacts with the carbon from the burned paper producing barium sulfide and carbon monoxide. Write a balanced chemical equation for this reaction.

Solution

1. *Write a chemical equation using formulas and symbols:*
   $BaSO_4(s) + C(s) \rightarrow BaS(s) + CO(g)$
2. *Check the equation for atom balance by setting up a table:*

Kind of Atom	Number of Atoms Before Reaction	Number of Atoms After Reaction
Ba	1	1
S	1	1
O	4	1
C	1	1

The oxygen atoms are not equal—this equation is not balanced.

3. *Choose coefficients that balance the equation. Try putting a 4 in front of CO. Now you have 4 oxygens, but 4 carbons also, so add another 4 in front of the C in the reactants to balance the remaining elements.*
   $BaSO_4(s) + 4C(s) \rightarrow BaS(s) + 4CO(g)$

Practice Problem

1. HCl is slowly added to aqueous Na_2CO_3 forming NaCl, H_2O, and CO_2. Follow the steps above to write a balanced equation for this reaction.

For more help with solving equations, refer to the Math Skill Handbook.

Double Displacement In a **double-displacement reaction,** the positive ion of one compound replaces the positive ion of the other to form two new compounds. A double-displacement reaction takes place if a precipitate, water, or a gas forms when two ionic compounds in solution are combined. A **precipitate** is an insoluble compound that comes out of solution during this type of reaction. The generalized formula for this type of reaction is as follows.

$$AB + CD \rightarrow AD + CB$$

The reaction of barium nitrate with potassium sulfate is an example of this type of reaction. A precipitate—barium sulfate—forms, as shown in **Figure 13.** The chemical equation is as follows:

$$Ba(NO_3)_2(aq) + K_2SO_4(aq) \rightarrow BaSO_4(s) + 2KNO_3(aq)$$

These are a few examples of chemical reactions classified into types. Many more reactions of each type occur around you.

Reading Check *What type of reaction produces a precipitate?*

Figure 13
A precipitate of barium sulfate settles to the bottom of a test tube containing potassium nitrate. Because it is opaque to X rays, barium sulfate is used to take X-ray photographs of the intestinal tract.

Section  Assessment

1. Classify each of the following reactions:
   a. $CaO(s) + H_2O \rightarrow Ca(OH)_2\ (aq)$
   b. $Fe(s) + CuSO_4(aq) \rightarrow FeSO_4(aq) + Cu(s)$
   c. $NH_4NO_3(s) \rightarrow N_2O(g) + 2H_2O(g)$
2. Sulfur trioxide, (SO_3), a pollutant released by coal-burning plants, can react with water in the atmosphere to produce sulfuric acid, H_2SO_4. Write a balanced equation for this reaction.
3. Explain the difference between synthesis and decomposition reactions.
4. Why can't you change the subscripts of reactants and products when balancing a chemical equation?
5. **Think Critically** Balance the following equation and identify the reaction type.
   $Au(CN)_2^-(aq) + Zn(s) \rightarrow Au(s) + Zn(CN)_4^{2-}(aq)$

Skill Builder Activities

6. **Forming Hypotheses** Group 1 metals replace hydrogen in water in reactions that are often violent. This sample equation shows what happens when potassium reacts with water.
   $2K(s) + 2HOH(l) \rightarrow 2KOH(aq) + H_2(g)$
   Use **Figure 12** to help you hypothesize why these metals replace hydrogen in water. **For more help, refer to the Science Skill Handbook.**
7. **Using Proportions** The following equation, showing the formation of iron oxide, is balanced, but the coefficients used are larger than necessary. Rewrite this equation using the smallest coefficients that give a balanced equation. **For more help, refer to the Math Skill Handbook.**
   $9Fe(s) + 12H_2O(g) \rightarrow 3Fe_3O_4(s) + 12H_2(g)$

SECTION

Chemical Reactions and Energy

As You Read

What You'll Learn

- **Identify** the source of energy changes in chemical reactions.
- **Compare and contrast** exergonic and endergonic reactions.
- **Examine** the effects of catalysts and inhibitors on the speed of chemical reactions.

Vocabulary

exergonic reaction
exothermic reaction
endergonic reaction
endothermic reaction
catalyst
inhibitor

Why It's Important

Chemical reactions provide energy to cook your food, keep you warm, and transform the food you eat into substances you need to live and grow.

Chemical Reactions—Energy Exchanges

Often a crowd gathers to watch a building being demolished using dynamite. In a few breathtaking seconds, tremendous structures of steel and cement that took a year or more to build are reduced to rubble and a large cloud of dust. A dynamite explosion, as shown in **Figure 14,** is an example of a rapid chemical reaction.

Most chemical reactions proceed more slowly, but all chemical reactions release or absorb energy. This energy can take many forms, such as heat, light, sound and electricity. The heat produced by a wood fire and the light emitted by a candle are two examples of reactions that release energy.

Chemical bonds are the source of this energy. When most chemical reactions take place, some chemical bonds in the reactants must be broken, and breaking these bonds takes energy. In order for products to be produced, new bonds must form. Bond formation releases energy. Reactions such as dynamite combustion require much less energy to break chemical bonds than the energy released when new bonds are formed. The result is a release of energy and sometimes a loud explosion. Another release of energy is used to power rockets, as shown in **Figure 15.**

Figure 14
When its usefulness is over, a building is sometimes demolished using dynamite. Dynamite charges must be placed carefully so that the building collapses inward, where it cannot harm people or property.

Figure 15

Rockets burn fuel to provide the thrust necessary to propel them upward. In 1926, engineer Robert Goddard used gasoline and liquid oxygen to propel the first ever liquid-fueled rocket. Although many people at the time ridiculed Goddard's space travel theories, his rockets eventually served as models for those that have gone to the Moon and beyond. A selection of rockets—including Goddard's—is shown here. The number below each craft indicates the amount of thrust—expressed in newtons (N)—produced during launch.

SPACE SHUTTLE The main engines produce enormous amounts of energy by combining liquid hydrogen and oxygen. Coupled with solid rocket boosters, they produce over 32.5 million newtons (N) of thrust to lift the system's 2 million kg off the ground.

JUPITER C This rocket launched the first United States satellite in 1958. It used a fuel called hydyne plus liquid oxygen.

GODDARD'S MODEL ROCKET Although his first rocket rose only 12.6 m, Goddard successfully launched 35 rockets in his lifetime. The highest reached an altitude of 2.7 km.

LUNAR MODULE Smaller rocket engines, like those used by the Lunar Module to leave the Moon, use hydrazine-peroxide fuels. The number shown below indicates the fixed thrust from one of the module's two engines; the other engine's thrust was adjustable.

400 N

369,350 N

32,500,000 N

15,920 N

Mini LAB

Colorful Chemical Reaction

Procedure

1. Pour 5 mL of **water** into a **test tube.**
2. Sprinkle a few crystals of **copper(II) bromide** into the test tube and observe the color change of the crystals.
3. Slowly add more water and observe what happens.

Analysis

1. What color were the copper(II) bromide crystals after you added them to the test tube of water?
2. What color were they when you added more water?
3. What caused this color change?

More Energy Out

You have probably seen many reactions that release energy. Chemical reactions that release energy are called **exergonic** (ek sur GAH nihk) **reactions.** In these reactions less energy is required to break the original bonds than is released when new bonds form. As a result, some form of energy, such as light or heat, is given off by the reaction. The familiar glow from the reaction inside a glow stick, shown in **Figure 16,** is an example of an exergonic reaction, which produces visible light. In other reactions however, the energy given off can produce heat. This is the case with some heat packs that are used to treat muscle aches and other problems that require heat.

When the energy given off in a reaction is primarily in the form of heat, the reaction is called an **exothermic reaction.** The burning of wood and the explosion of dynamite are exothermic reactions. Iron rusting is also exothermic, but the reaction proceeds so slowly that it's difficult to detect any temperature change.

Reading Check *Why is a log fire considered to be an exothermic reaction?*

Exothermic reactions provide most of the power used in homes and industries. Fossil fuels that contain carbon, such as coal, petroleum, and natural gas, combine with oxygen to yield carbon dioxide gas and energy. Unfortunately impurities in these fuels, such as sulfur, burn as well, producing pollutants such as sulfur dioxide. Sulfur dioxide combines with water in the atmosphere, producing acid rain.

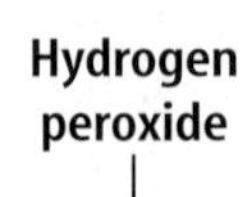

Figure 16
Glow sticks contain three different chemicals—an ester and a dye in the outer section and hydrogen peroxide in a center glass tube. Bending the stick breaks the tube and mixes the three components.

More Energy In

Sometimes a chemical reaction requires more energy to break bonds than is released when new ones are formed. These reactions are called **endergonic reactions.** The energy absorbed can be in the form of light, heat, or electricity.

Electricity is often used to supply energy to endergonic reactions. For example, electroplating deposits a coating of metal onto a surface, as shown in **Figure 17.** Also, aluminum metal is obtained from its ore using the following endergonic reaction.

$$2Al_2O_3(l) \xrightarrow{\text{elec.}} 4Al(l) + 3O_2(g)$$

In this case, electrical energy provides the energy needed to keep the reaction going.

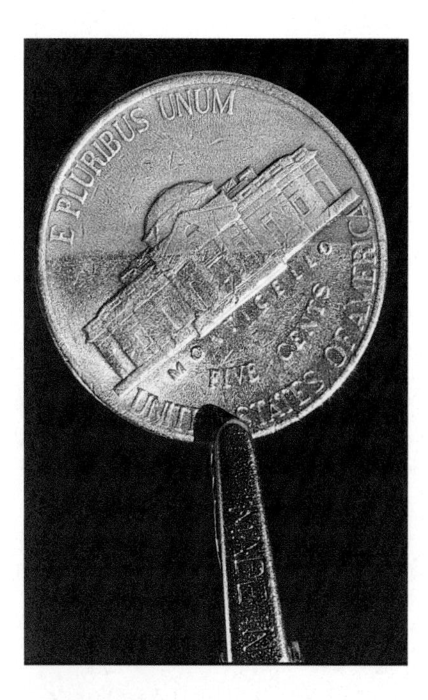

Figure 17
Electroplating of a metal is an endergonic reaction that requires electricity. A coating of copper was plated onto this coin.

Heat Absorption When the energy needed is in the form of heat, the reaction is called an **endothermic reaction.** The term *endothermic* is not just related to chemical reactions. It also can describe physical changes. The process of dissolving a salt in water is a physical change. If you ever had to soak a swollen ankle in an Epsom salt solution, you probably noticed that when you mixed the Epsom salt in water, the solution became cold. The dissolving of Epsom salt absorbs heat. Thus it is a physical change that is endothermic.

Cold packs also use endothermic processes, as shown in **Figure 18.** They contain ammonium nitrate crystals and a water container. When the water container is ruptured, water mixes with the ammonium nitrate. Almost immediately the pack cools dramatically.

B

Figure 18
Cold packs are often used in sports. A Because they cool so quickly, it's like carrying an instant bag of ice. They are handy for treating athletic injuries on the field. B They start to work only when you press the pack, breaking a container of water that mixes with ammonium nitrate crystals in a process that absorbs heat.

Metals, such as platinum and palladium, are used as catalysts in the exhaust systems of automobiles. What reactions do you think they catalyze?

Catalysts and Inhibitors Some reactions proceed too slowly to be useful. To speed them up, a catalyst can be added. A **catalyst** is a substance that speeds up a chemical reaction without being permanently changed itself. When you add a catalyst to a reaction, the mass of the product that is formed remains the same, but it will form more rapidly. The catalyst remains unchanged and often is recovered and reused. Catalysts are used to speed many reactions in industry, such as polymerization to make plastics and fibers.

✔ Reading Check ***Why would a catalyst be needed for a chemical reaction?***

At times, it is worthwhile to prevent certain reactions from occurring. Substances called **inhibitors** are used to combine with one of the reactants. This ties up the reactant and prevents it from undergoing the original reaction. The food preservatives BHT and BHA are inhibitors that prevent spoilage of certain foods, such as cereals and crackers.

Sometimes you might want to inhibit a catalyst. For example, the cut surfaces of fruits darken rapidly due to an enzyme-catalyzed reaction with air. Brushing or spraying the surfaces with lemon juice prevents the catalyst from acting.

One thing to remember when thinking about catalysts and inhibitors is that they do not change the amount of product produced. They only change the rate of production. Catalysts increase the rate and inhibitors decrease the rate. Therefore, foods containing preservatives will spoil eventually, but preservatives are inhibitors that often add months or even years to the shelf time.

Section  Assessment

1. Discuss the difference between exergonic and endergonic reactions.
2. What happens to a catalyst in a reaction?
3. Crackers containing BHT stay fresh longer than those without it. Explain why.
4. Classify the reaction that makes a firefly glow in terms of energy input or output.
5. **Think Critically** To develop a product that warms people's hands, would you choose an exothermic or endothermic reaction to use? Why?

Skill Builder Activities

6. **Concept Mapping** Construct a concept map to show the relationship between energy and bond formation and bond breakage. **For more help, refer to the Science Skill Handbook.**
7. **Communicating** In your Science Journal, compare the profit or loss in an investment or a business deal with the gain or loss of energy in a chemical reaction. **For more help, refer to the Science Skill Handbook.**

Activity

Catalyzed Reaction

A balanced chemical equation tells nothing about the rate of a reaction. One way to affect the rate is to use a catalyst.

What You'll Investigate

How does the presence of a catalyst affect the rate of a chemical reaction?

Materials

test tubes (3)
test-tube stand
3% hydrogen peroxide, H_2O_2 (15 mL)
10-mL graduated cylinder
small plastic teaspoon
sand (1/4 teaspoon)
hot plate
wooden splint
beaker of hot water
manganese dioxide, MnO_2 (1/4 teaspoon)

Goals

- **Observe** a catalyst on the rate of reaction.
- **Conclude** based on your observations whether the catalyst remained unchanged.

Safety Precautions

WARNING: *Hydrogen peroxide can irritate skin and eyes. Wipe up spills promptly. Point test tubes away from other students.*

Procedure

1. Set three test tubes in a test-tube stand. Pour 5 mL of hydrogen peroxide into each tube.
2. Place about 1/4 teaspoon of sand in tube 2 and the same amount of MnO_2 in tube 3.
3. In the presence of a catalyst, H_2O_2 decomposes rapidly producing oxygen gas, O_2. Test the gas produced as follows: Light a wooden splint, blow out the flame, and insert the glowing splint into the tube. The splint will relight if oxygen is present.

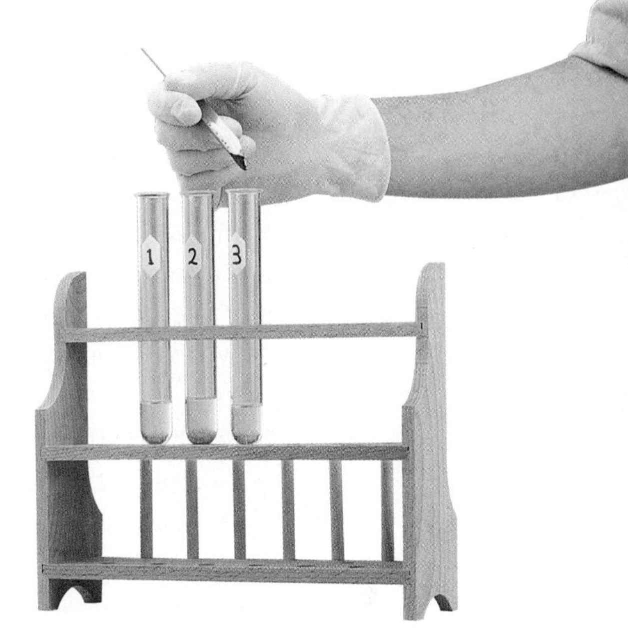

4. Place all three tubes in a beaker of hot water. Heat on a hot plate until all of the remaining H_2O_2 is driven away and no liquid remains.

Conclude and Apply

1. What changes did you observe when the solids were added to the tubes?
2. In which tube was oxygen produced rapidly, and how do you know?
3. Which substance, sand or MnO_2, caused the rapid production of gas from the H_2O_2?
4. What remained in each tube after the H_2O_2 was driven away?

Communicating Your Data

Compare your results with those of your classmates and discuss any differences observed. **For more help refer to the Science Skill Handbook.**

Activity **Use the Internet**

Fossil Fuels and Greenhouse Gases

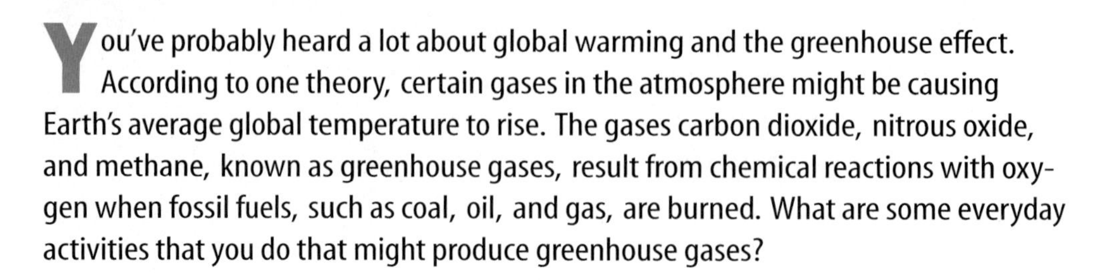

You've probably heard a lot about global warming and the greenhouse effect. According to one theory, certain gases in the atmosphere might be causing Earth's average global temperature to rise. The gases carbon dioxide, nitrous oxide, and methane, known as greenhouse gases, result from chemical reactions with oxygen when fossil fuels, such as coal, oil, and gas, are burned. What are some everyday activities that you do that might produce greenhouse gases?

Recognize the Problem

What do you do to produce greenhouse gases?

Form a Hypothesis

Think of the different ways you use fossil fuels every day. Why are fossil fuels important? Form a hypothesis about how certain activities add greenhouse gases to our atmosphere.

Goals

- **Observe** how you use fossil fuels in your daily life.
- **Gather data** on the process of burning fossil fuels and how greenhouse gases are released.
- **Research** the chemical reactions that produce greenhouse gases.
- **Identify** the importance of fossil fuels and their effect on the environment.
- **Communicate** your findings to other students.

Data Source

SCIENCE *Online* Go to the Glencoe Science Web site at **science.glencoe.com** for more information about fossil fuels, the chemical reactions that produce greenhouse gases, uses of fossil fuels, their effects on the environment, and data from other students.

Test Your Hypothesis

Plan

1. **Observe** the activities of your daily life. How do you use energy from fossil fuels each day?
2. **Develop** a way to categorize the different chemical reactions and the greenhouse gases they produce.
3. **Search** reference sources to learn what chemical reactions produce greenhouse gases.
4. What are some of the most common uses of fossil fuels? Is it possible to never use fossil fuels?

Do

1. Make sure your teacher approves your plan before you start.

2. **Research** the different greenhouse gases and the chemical reactions that produce them.
3. **Compare** the different reactions and their products.
4. **Record** your data in your Science Journal.

Analyze Your Data

1. **Record** in your Science Journal what activities contribute to the greatest amount of greenhouse gases in our atmosphere.
2. **Analyze** the types of chemical reactions that produce greenhouse gases. What types of reactions are they?
3. **Compare** your results with other students. What greenhouse gas is most often released?
4. Make a table of your data.

Draw Conclusions

1. How do you think your data would be affected if you had performed this experiment 100 years ago?
2. What processes in nature might also contribute to the release of greenhouse gases? Compare their impact to that made by fossil fuels.

Communicating Your Data

Find this *Use the Internet* activity on the Glencoe Science Web site at **science.glencoe.com.** Post your data in the table provided. Compare your data to that of other students. Combine your data with that of other students and write an entry in your Science Journal that explains how the production of greenhouse gases could be reduced.

Oops! Accidents in SCIENCE

SOMETIMES GREAT DISCOVERIES HAPPEN BY ACCIDENT!

A Clumsy Move Pays Off

Hilaire de Chardonnet

Clumsy (KLUM zee)—lacking physical coordination
Procrastinate (proh KRAS tuh nayt)—to put off doing something until later

Silkworms at work—spinning their cocoons

Most people might not think that a person who had the two qualities described above would make a great scientific discovery. But they never met a chemist named Hilaire de Chardonnet (hee LAYR • duh • shar doh NAY). In 1878, Chardonnet accidentally knocked over some nitrate chemicals. He put off cleaning up the mess—and ended up inventing artificial silk.

Silk is produced naturally by silkworms. In the mid-1800s, though, silkworms were dying from disease and the silk industry was suffering. Businesses were going under and people were put out of work. Many scientists were working to develop a solution to this problem.

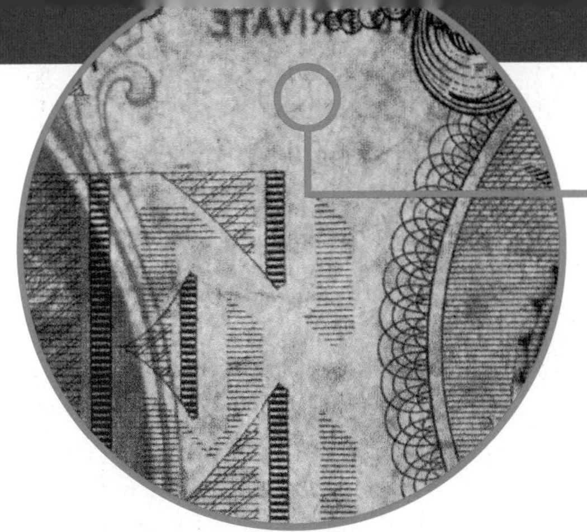

To help prevent counterfeiting, dollars are printed on paper which contains red and blue rayon fibers. If you can scratch off the red or blue, that means it's ink—and your bill is counterfeit. If you can pick out the red or blue fiber with a needle, it's a real bill.

Chardonnet had been searching for a silk substitute for years—he just didn't plan to find it by knocking it over!

A Messy Discovery

Chardonnet was in his darkroom developing photographs when the accidental spill took place. He decided to clean up the spill later—and finish what he was working on. By the time he returned to wipe up the spill, the chemical solution had turned into a thick, gooey mess. When he pulled the cleaning cloth away, the goop formed long, thin strands of fiber that stuck to the cloth. The chemicals had reacted with the cellulose in the wooden table and liquefied it. The strands of fiber looked just like the raw silk made by silkworms.

Within six years, Chardonnet had developed a way to make the fibers into an artificial silk. Other scientists extended his work, developing a fiber called rayon—"ray" for its shiny appearance, and "on" to remind people of the word "cotton." The rayon we wear on our backs today, called viscose rayon, is made with sodium hydroxide. The sodium hydroxide is mixed with wood fibers. The wood turns into a liquid that can be squeezed through tiny holes to make strands or fibers. The fibers are then woven together to make cloth.

Polymers

Rayon is a polymer. A polymer is a big molecule made up of smaller molecules that are linked together to form a straight chain, a twisting ladder, or a branching tree, depending on how the smaller molecules link together. The cellulose in plant cells is a natural polymer, as is the DNA in your body. To make artificial polymers, such as rubber or plastics, scientists studied natural polymers and then tried to mimic their structure in the lab.

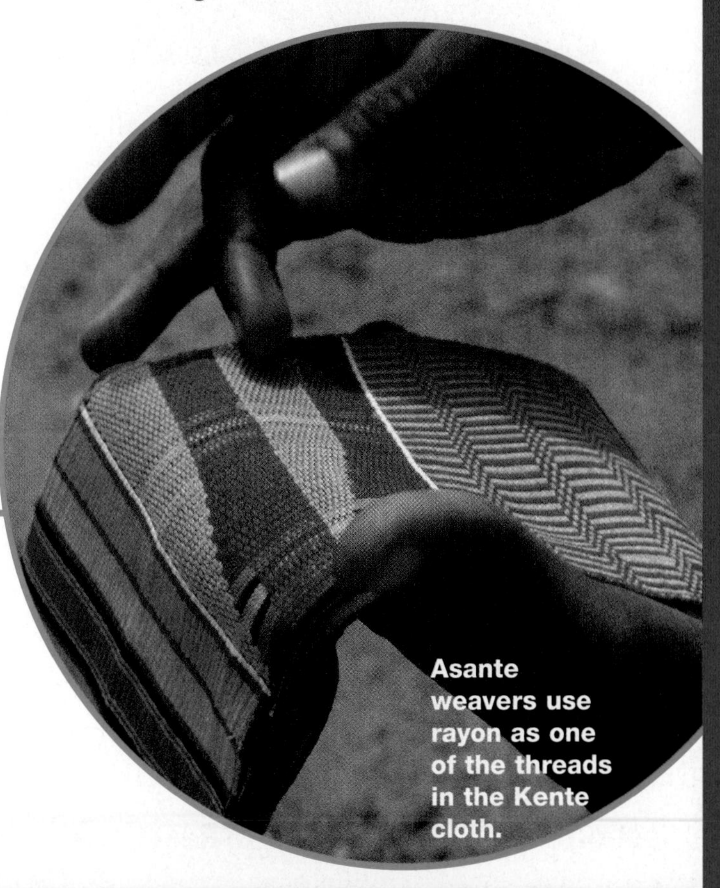

Asante weavers use rayon as one of the threads in the Kente cloth.

CONNECTIONS **Create** Make a data table. Work with a partner to examine the labels on the inside collars of your clothes. These labels list the different materials found in the clothes. Research the materials, then make a data table that identifies their characteristics.

SCIENCE Online
For more information, visit science.glencoe.com

Chapter 24 Study Guide

Reviewing Main Ideas

Section 1 Chemical Changes

1. In a chemical reaction, one or more substances are changed to new substances.
2. The substances that react are called reactants, and the new substances formed are called products. Charcoal, the reactant shown below, is almost pure carbon. *What is a major product formed?*

3. The law of conservation of mass states that in chemical reactions, matter is neither created nor destroyed, just rearranged.
4. Chemical equations efficiently describe what happens in chemical reactions.

Section 2 Chemical Equations

1. A balanced chemical equation has the same number of atoms of each element on both sides of the equation.
2. A balanced equation tells the ratio of products and reactants.
3. When balancing equations, change only the coefficients of the formulas, never the subscripts.

Section 3 Classifying Chemical Reactions

1. In synthesis reactions, two or more substances combine to form another substance.
2. Bleach—sodium hypochlorite ($NaClO$)—decomposes in two ways. In one way, it forms oxygen and another product. *What common household substance is the other product?*
3. In single-displacement reactions, one element replaces another in a compound.
4. In double-displacement reactions, ions in two compounds switch places, often forming a gas or insoluble compound.

Section 4 Chemical Reactions and Energy

1. Energy in the form of light, heat, sound or electricity is released from some chemical reactions known as exergonic reactions. *What types of energy are produced in the reaction shown here?*

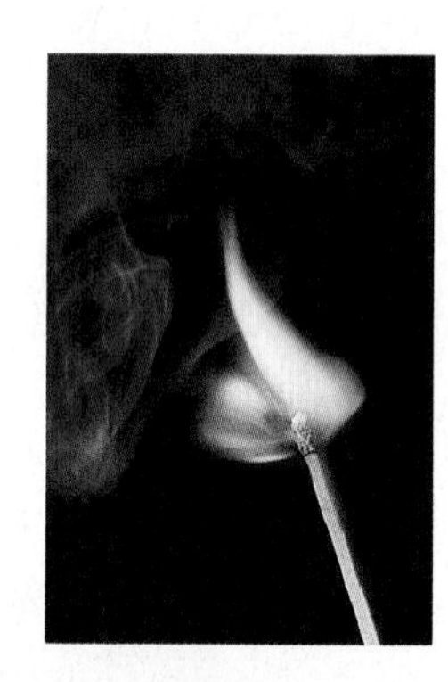

2. Reactions that need energy to proceed are called endergonic reactions.
3. Reactions may be sped up by adding catalysts and slowed down by adding inhibitors.
4. When energy is released in the form of heat, the reaction is exothermic.

After You Read

Under the tabs of your Foldable write the chemical reactions you recorded using a sentence or chemical equation.

Visualizing Main Ideas

Complete the spider map about chemical reactions.

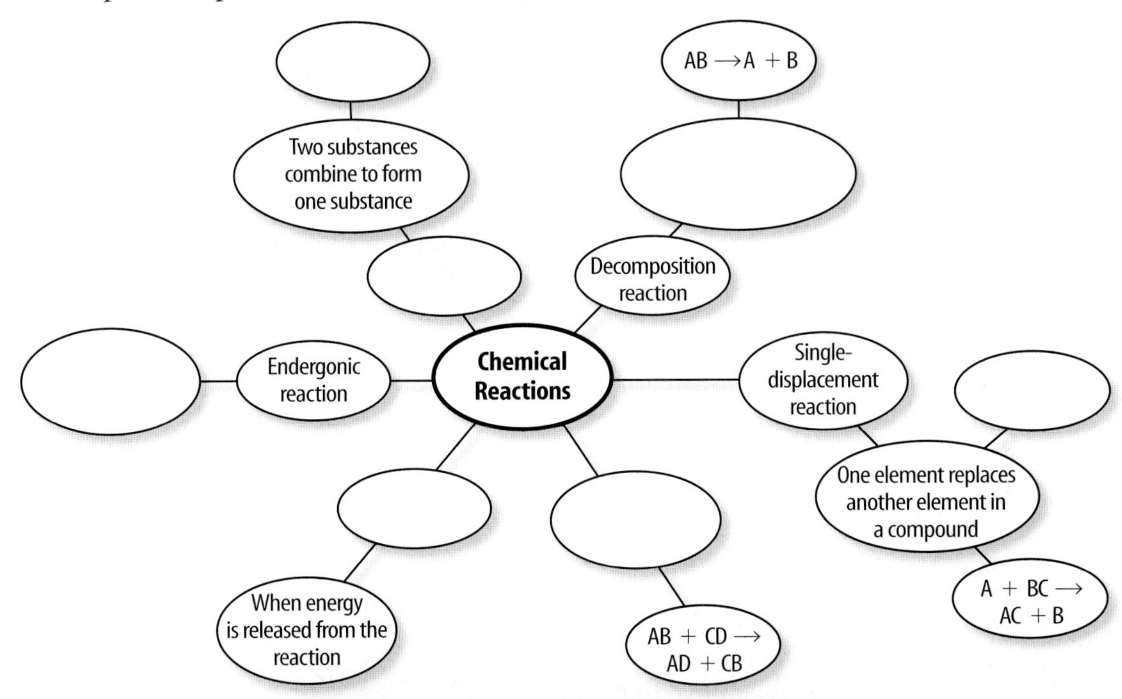

Vocabulary Review

Vocabulary Words

a. balanced chemical equation
b. chemical equation
c. catalyst
d. chemical reaction
e. coefficient
f. decomposition reaction
g. double-displacement reaction
h. endergonic reaction
i. endothermic reaction
j. exergonic reaction
k. exothermic reaction
l. inhibitor
m. precipitate
n. product
o. reactant
p. single-displacement reaction
q. synthesis reaction

Study Tip

Use acronyms to help you remember important facts. For example, the acronym SDSD will help you remember the types of chemical reactions—Synthesis, Decomposition, Single Displacement, and Double Displacement.

Using Vocabulary

For each set of vocabulary words below, explain the relationship that exists.

1. coefficient, balanced chemical equation
2. synthesis reaction, decomposition reaction
3. reactant, product
4. catalyst, inhibitor
5. exothermic reaction, endothermic reaction
6. chemical reaction, product
7. endergonic reaction, exergonic reaction
8. single-displacement reaction, double-displacement reaction
9. chemical reaction, synthesis reaction

Chapter 24 Assessment

Checking Concepts

Choose the word or phrase that best answers the question.

1. Oxygen gas is always written as O_2 in chemical equations. What term is used to describe the *two* in this formula?
A) product **C)** catalyst
B) coefficient **D)** subscript

2. What law is based on the experiments of Lavoisier?
A) chemical reaction **C)** coefficients
B) conservation of mass **D)** gravity

3. What must an element be in order to replace another element in a compound?
A) more active **C)** more inhibiting
B) a catalyst **D)** more soluble

4. How do you indicate that a substance in an equation is a solid?
A) *(l)* **C)** *(s)*
B) *(g)* **D)** *(aq)*

5. What term is used to describe the 4 in the expression 4 $Ca(NO_3)_2$?
A) coefficient **C)** subscript
B) formula **D)** symbol

6. What type of compound is the food additive BHA?
A) catalyst **C)** inhibitor
B) formula **D)** CFC

7. How do you show that a substance is dissolved in water when writing an equation?
A) *(aq)* **C)** *(g)*
B) *(s)* **D)** *(l)*

8. What word would you use to describe HgO in the reaction that Lavoisier used to show conservation of mass?
A) catalyst **C)** product
B) inhibitor **D)** reactant

9. When hydrogen burns, what is oxygen's role?
A) catalyst **C)** product
B) inhibitor **D)** reactant

10. Give an example of a chemical reaction.
A) bending **C)** melting
B) evaporation **D)** photosynthesis

Thinking Critically

11. Chromium is produced by reacting its oxide with aluminum. If 76 g of Cr_2O_3 and 27 g of Al completely react to form 51 g of Al_2O_3, how many grams of Cr are formed?

12. Propane, $C_3H_8(g)$, burns in oxygen to form carbon dioxide and water vapor. Write a balanced equation for burning propane.

13. Use the balanced chemical equation from question 12 to explain the law of conservation of mass.

14. Zn is placed in a solution of $Cu(NO_3)_2$ and Cu is placed in a $Zn(NO_3)_2$ solution. In which of these will a reaction occur?

15. If lye, NaOH*(s)*, is put in water, the solution gets hot. What kind of energy process is this?

Developing Skills

16. Recognizing Cause and Effect Sucrose, or table sugar, is a disaccharide. This means that sucrose is composed of two simple sugars chemically bonded together. Sucrose can be separated into its components by heating it in an aqueous sulfuric acid solution. Research what products are formed by breaking up sucrose. What role does the acid play?

17. Classifying Make an outline with the general heading "Chemical Reactions." Include the four types of reactions, with a description and example of each.

18. Concept Mapping The arrow in a chemical equation tells the reaction direction. Some reactions are reversible. Sometimes, the bond formed is weak, and a product breaks apart as it's formed. Double arrows are used in the equation to indicate this reaction. Fill in the concept map, using the words *product(s)* and *reactant(s)*. In the blank in the center, fill in the formulas for the substances appearing in the reversible reaction.

$$H_2(g) + I_2(g) \rightleftarrows 2HI(g)$$

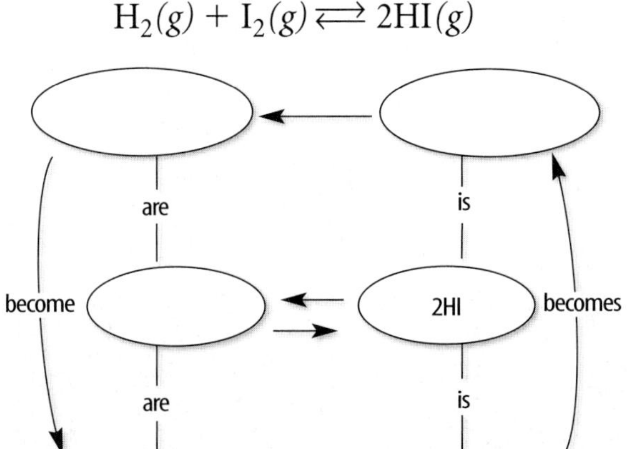

19. Interpreting Data When 46 g of sodium were exposed to dry air, 62 g of sodium oxide formed. How many grams of oxygen from the air were used?

Performance Assessment

20. Portfolio List five chemical reactions you have observed. Illustrate them using diagrams or photos. Include a brief description of the compounds involved.

Technology

Go to the Glencoe Science Web site at **science.glencoe.com** or use the **Glencoe Science CD-ROM** for additional chapter assessment.

Test Practice

Students were studying how to balance chemical equations. The table below shows some equations they wrote for the reaction of sulfuric acid with potassium hydroxide.

1	$H_2SO_4 + KOH \rightarrow K_2SO_4 + H_2O$
2	$H_2SO_4 + 2KOH \rightarrow K_2SO_4 + H_2O$
3	$H_2SO_4 + 2KOH \rightarrow K_2SO_4 + 2H_2O$
4	$2H_2SO_4 + 4KOH \rightarrow K_2SO_4 + H_2O$

Study the table above and answer the following questions:

1. Which of the equations in the table is correctly balanced?

A) one
B) two
C) three
D) four

2. All of the chemical reactions in the table show _____.

F) reactants combining with products to produce new substances
G) reactants not being changed by the chemical reaction
H) reactants combining to produce new products
J) products combining to produce new reactants

CHAPTER

Acids, Bases, and Salts

Salt is essential for most animals including humans. Farmers often provide salt to livestock, and many wild animals get salt from natural salt licks. Butterflies like those shown here suck salt from moist ground—a behavior known as puddling. In this chapter you'll learn about this specific salt—sodium chloride—among other salts and about the acids and bases that react to form them.

What do you think?

Science Journal Look at the picture below with a classmate. Discuss what you think this might be. Here's a hint: *These and many more like them keep the oceans from being big lakes.* Write your answer in your Science Journal.

Many limestone caves and rock formations are shaped by water containing carbon dioxide. Higher levels of carbon dioxide in acid rain can damage marble structures. Observe this reaction using soda water to represent acid rain and chalk, which like limestone and marble is calcium carbonate.

Determine how acid rain affects limestone and marble

1. Measure approximately 5 g of classroom chalk.
2. Crush it slightly and place it in a 100-mL beaker.
3. Add 50 mL of plain, bottled carbonated water to the beaker.
4. After several minutes, stir the mixture.
5. When the mixture stops reacting, filter it using a paper filter in a glass funnel.
6. Dry the residue overnight and determine its mass.

Observe

1. How did the mass change?
2. In your Science Journal, write your conclusions about the effect that acid rain can have on marble buildings and monuments.

Before You Read

Making a Venn Diagram Study Fold **A Venn Diagram can be used to compare and contrast the characteristics of acids, bases, and salts.**

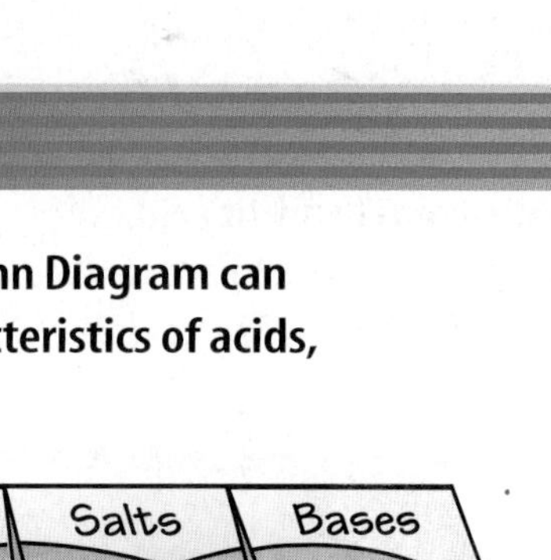

1. Place a sheet of paper in front of you so the short side is at the top. Fold the paper in half from top to bottom.
2. Draw and label *Acids, Salts,* and *Bases* across the front of the paper, as shown. Fold both sides in.
3. Unfold the paper so that three columns show.
4. Through the top thickness of paper, cut along each of the fold lines to the topfold, forming three tabs.
5. As you read the chapter, collect information about each and write it under its tab.

SECTION

Acids and Bases

As You Read

What You'll Learn

- **Compare** and **contrast** acids and bases and identify the characteristics they have.
- **Examine** some formulas and uses of acids and bases you encounter every day.
- **Determine** how the processes of ionization and dissociation apply to acids and bases.

Vocabulary

acid
hydronium ion
indicator
base

Why It's Important

Acids and bases are found almost everywhere—from fruit juice and gastric juice to soaps.

Acids

What comes to mind when you hear the word *acid?* Do you think of a substance that can burn your skin or even burn a hole through a piece of metal? Do you think about sour foods like those shown in **Figure 1?** Does your mouth water when you think about biting into a sour pickle or sinking your teeth into a lemon? Although some acids can burn and are dangerous to handle, most acids in foods are safe to eat. What all acids have in common, however, is that they contain at least one hydrogen atom that can be removed when the acid is dissolved in water.

Properties of Acids

When an acid dissolves in water, some of the hydrogen is released as hydrogen ions, H^+. An **acid** is a substance that produces hydrogen ions in a water solution. It is the ability to produce these ions that gives acids their characteristic properties. When an acid dissolves in water, H^+ ions interact with water molecules to form H_3O^+ ions, which are called **hydronium ions** (hi DROH nee um • I ahnz).

Acids have several common properties. For one thing, all acids taste sour. The familiar sour taste of many foods is due to the presence of acids. However, taste never should be used to test for the presence of acids. Some acids can damage tissue by producing painful burns.

Acids are corrosive. Some acids react strongly with certain metals, seeming to eat away the metals as metallic compounds and hydrogen gas form. Acids also react with indicators to produce predictable changes in color. An **indicator** is an organic compound that changes color in acid and base. For example, the indicator litmus paper turns red in acid.

Figure 1
Pickles contain acetic acid.
Can you guess which of these foods contain lactic and citric acids?

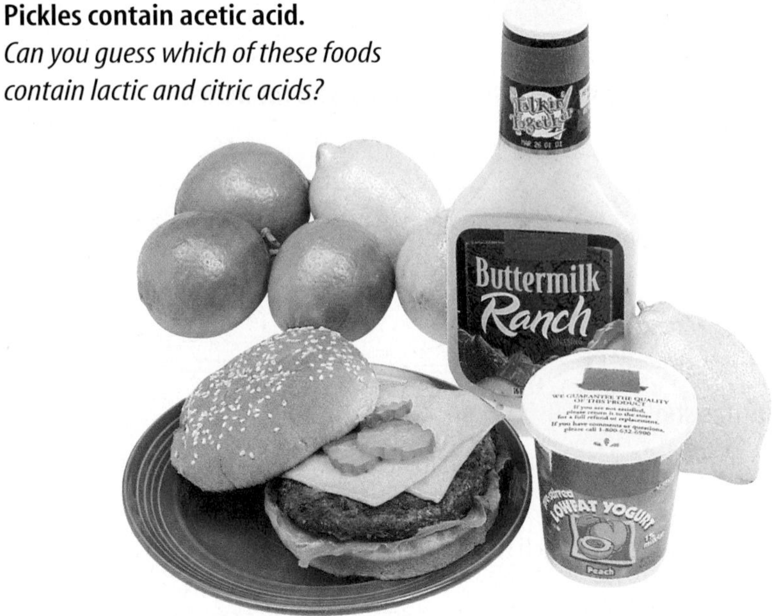

Common Acids

Many foods contain acids. In addition to citric acid in citrus fruits, lactic acid is found in yogurt and buttermilk, and any pickled food contains vinegar, also known as acetic acid. Your stomach uses hydrochloric acid to help digest your food. Four acids are vital to industry—sulfuric, phosphoric, nitric and hydrochloric.

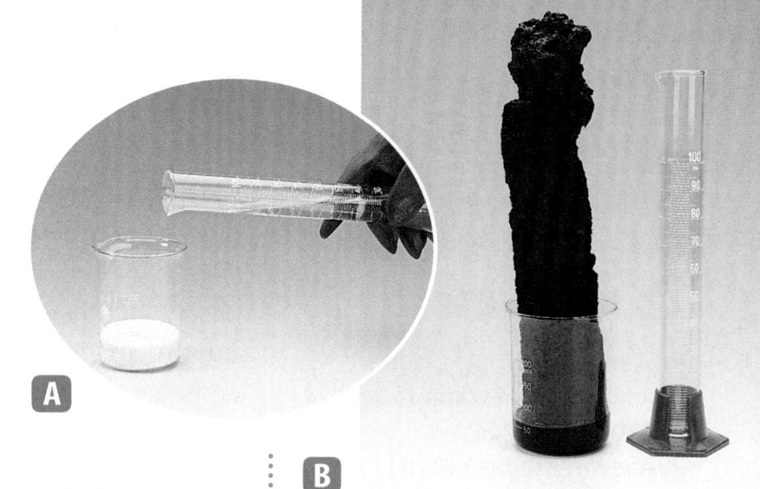

Reading Check *Which four acids are important for industry?*

Table 1 lists the names and formulas of a few acids, their uses, and some properties. Three acids are used to make fertilizers—most nitric and sulfuric acid and 90 percent of phosphoric acid are used for this purpose. Many acids can burn, but sulfuric acid can burn by removing water from your skin as easily as it takes water from sugar, as shown in **Figure 2.**

Figure 2
A When sulfuric acid is added to sugar B the mixture foams, removing hydrogen and oxygen atoms as water and leaving air-filled carbon.

Table 1 Common Acids and Their Uses

Name, Formula	Use	Other Information
Acetic Acid, CH_3COOH	Food preservation and preparation	When in solution with water, it is known as vinegar.
Acetylsalicylic Acid, $HOOC-C_6H_4-OOCCH_3$	Pain relief, fever relief, to reduce inflammation	Known as aspirin
Ascorbic Acid, $H_2C_6H_6O_6$	Antioxidant, vitamin	Called vitamin C
Carbonic Acid, H_2CO_3	Carbonated drinks	Involved in cave, stalactite, and stalagmite formation and acid rain
Hydrochloric Acid, HCl	Digestion as gastric juice in stomach, to clean steel in a process called pickling	Commonly called muriatic acid
Nitric Acid, HNO_3	To make fertilizers	Colorless, yet yellows when exposed to light
Phosphoric Acid, H_3PO_3	To make detergents, fertilizers, and soft drinks	Slightly sour but pleasant taste, detergents containing phosphates cause water pollution
Sulfuric Acid, H_2SO_4	Car batteries, to manufacture fertilizers and other chemicals	Dehydrating agent, causes burns by removing water from body cells

Observing Acid Relief

Procedure
1. Add 150 mL of **water** to a **250-mL beaker.**
2. Add three drops **1 M HCl** and 12 drops of **universal indicator.**
3. Observe the color of the solution.
4. Add an **antacid tablet** and observe for 15 minutes.

Analysis
1. Describe any changes that took place in the solution.
2. Explain why these changes occurred.

Bases

You might not be as familiar with bases as you are with acids. Although you can eat some foods that contain acids, you don't consume many bases. Some foods, such as egg whites, are slightly basic. Other basic materials are baking powder and weak bases, such as amines found in some foods. Medicines, such as milk of magnesia and antacids are basic, too. Still, you come in contact with many bases every day. For example, each time you wash your hands using soap, you are using a base. One characteristic of bases is that they feel slippery, like soapy water. Bases are important in many types of cleaning materials, as shown in **Figure 3.** Bases are important in industry, also. For example, sodium hydroxide is used in the paper industry to separate fibers of cellulose from wood pulp. The freed cellulose fibers are made into paper.

Bases can be defined in two ways. Any substance that forms hydroxide ions, OH^-, in a water solution is a **base.** In addition, a base is any substance that accepts H^+ from acids. The definitions are related, because the OH^- ions produced by some bases do accept H^+ ions.

Properties of Bases

One way to think about bases is as the complements, or opposites, of acids. Although acids and bases share some common features, the bases have their own characteristic properties.

In the pure, undissolved state, many bases are crystalline solids. In solution, bases feel slippery and have a bitter taste. Like strong acids, strong bases are corrosive, and contact with skin can result in severe burns. Therefore, taste and touch never should be used to test for the presence of a base. Finally, like acids, bases react with indicators to produce changes in color. The indicator litmus turns blue in bases.

Figure 3
Bases are commonly found in many cleaning products used around the home.
What property of bases is evident in soaps?

Figure 4
Two applications of bases are shown here.

A **Aluminum hydroxide is a base used in water-treatment plants. Its sticky surface collects impurities, making them easier to filter.**

B **Some drain cleaners contain NaOH, which dissolves grease, and small pieces of aluminum. The aluminum reacts with NaOH, producing hydrogen and dislodging solids, such as hair.**

Common Bases

You probably are familiar with many common bases because they are found in cleaning products used in the home. These and some other bases are shown in **Table 2,** which also includes their uses and some information about them. **Figure 4** shows two uses of bases that you might not be familiar with.

Table 2 Common Bases and Their Uses

Name, Formula	Use	Other Information
Aluminum Hydroxide, $Al(OH)_3$	Color-fast fabrics, antacid, water purification as shown in **Figure 4A**	Sticky gel that collects suspended clay and dirt particles on its surface
Calcium Hydroxide, $Ca(OH)_2$	Leather-making, mortar and plaster, lessen acidity of soil	Called caustic lime
Magnesium Hydroxide, $Mg(OH)_2$	Laxative, antacid	Called milk of magnesia
Sodium Hydroxide, NaOH	To make soap, oven cleaner, drain cleaner, textiles, and paper	Called lye and caustic soda; generates heat (exothermic) when combined with water, reacts with metals to form hydrogen
Ammonia, NH_3	Cleaners, fertilizer, to make rayon and nylon	Irritating odor that is damaging to nasal passages and lungs

Solutions of Acids and Bases

You have learned that substances such as HCl, HNO_3, and H_2SO_4 are acids because of their ability to produce H^+ ions. This ionization process is shown in **Figure 5A,** using HCl as an example. When a polar molecule, such as HCl, is dissolved in water, the negatively charged area of nearby water molecules attracts the positively charged area of the polar molecule. The H^+ is removed from the polar molecule, and a hydronium ion (H_3O^+) is formed. Thus, acid describes any compound that can be ionized in water to form hydronium ions. When an organic acid ionizes, the H at the end of the −COOH group separates as H^+ and combines with water to form the hydronium ion as shown in **Figure 5B.**

Life Science INTEGRATION

Some ants add sting to their bite by injecting a solution of formic acid. In fact, formic acid was named for ants, which make up the genus *Formica*. Still ants are considered tasty treats by many animals. For example, one woodpecker called a flicker has saliva that is basic enough to take the sting out of ants.

Bases Compounds that can form hydroxide ions (OH^-) in solution are classified as bases. If you look at **Table 2,** you will find that most of the substances that are listed contain OH in their formulas. Except for ammonia, inorganic compounds that produce bases in aqueous solution are ionic compounds—that is, they already are made up of ions. As the following equation shows, when such a compound dissolves in water, the ions dissociate and exist as individual ions in solution.

$$NaOH(s) \xrightarrow{H_2O} Na^+(aq) + OH^-(aq)$$

These substances release OH^- ions in water. If a solution contains more OH^- ions than H_3O^+ ions, it is referred to as a basic solution.

Figure 5
Both inorganic and organic acids form hydronium ions in water.

A **Hydrogen chloride gas dissolves in water and ionizes to produce hydronium ions and chloride ions.**

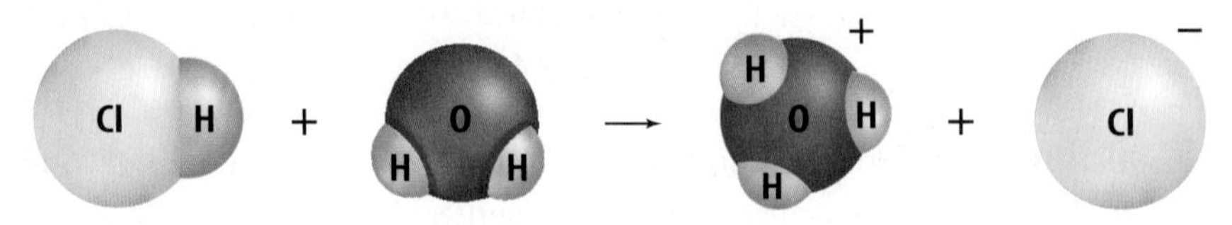

B **Organic acids also form hydronium ions when they dissolve in water. Here, formic acid forms hydronium ions and formate ions.**

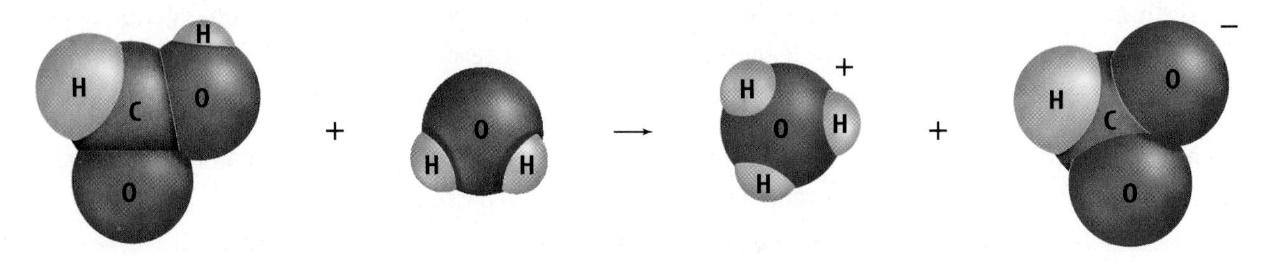

Figure 6
Even though ammonia is not a hydroxide, it reacts with water to produce some hydroxide ions. Therefore, it is a base.

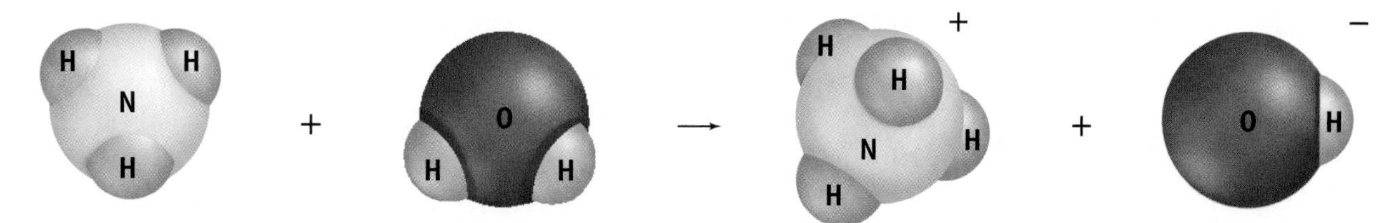

Ammonia Ammonia is a polar compound. In a water solution, ionization takes place when the ammonia molecule attracts a hydrogen ion from a water molecule, forming an ammonium ion. This leaves a hydroxide ion, as shown in **Figure 6.**

Reading Check *How does ammonia react in a water solution?*

Ammonia is a popular household cleaner. However, products containing ammonia never should be used with other cleaners that contain chlorine (sodium hypochlorite), such as some bathroom bowl cleaners and bleach. A reaction between sodium hypochlorite and ammonia produces toxic vapors of hydrazine and chloramine. Breathing these vapors can severely damage lung tissues and cause death.

Solutions of both acids and bases produce some ions that are capable of carrying electric current to some extent. Thus, they are said to be conductors.

SCIENCE Online

Research Visit the Glencoe Science Web site at **science.glencoe.com** for more information about the dangers of mixing ammonia cleaners with chlorine or hydrochloric acid cleaners. Communicate to your class what you learn.

Section 1 Assessment

1. Name three important acids and three important bases and describe their uses.
2. What is an indicator?
3. What metallic compound would you predict forms when sulfuric acid reacts with magnesium metal?
4. If an acid donates H^+ and a base produces OH^-, what compound is likely to be produced when acids react with bases?
5. **Think Critically** Vinegar contains acetic acid, CH_3COOH. Is acetic acid organic or inorganic? How do you know? Write an equation showing how acetic acid ionizes.

Skill Builder Activities

6. **Predicting** Remember that solutions of acids and bases produce ions that can conduct electric current. Name two substances found in your home that could conduct electricity. Explain why. **For more help, refer to the Science Skill Handbook.**
7. **Using a Database** Make a database that compares acids and bases. Include the following: *home and commercial uses, properties, where they are found,* and *what ions they form in solution.* **For more help, refer to the Technology Skill Handbook.**

SECTION 2

Strength of Acids and Bases

As You Read

What You'll Learn

- **Determine** what is responsible for the strength of an acid or a base.
- **Compare and contrast** strength and concentration of acids and bases.
- **Determine** the meaning of pH.
- **Examine** the relationship between pH and acid or base strength.

Vocabulary

strong acid
weak acid
strong base
weak base
pH
buffer

Why It's Important

Understanding the strength of acids and bases helps you use them safely.

Strong and Weak Acids and Bases

Some acids must be handled with great care. For example, sulfuric acid found in car batteries can burn your finger, yet you drink acids such as citric acid in orange juice and carbonic acid in soft drinks. Obviously, some acids are stronger than others. One measure of acid strength is the ability to ionize in solution.

The strength of an acid or base depends on how completely a compound separates into ions when dissolved in water. An acid that ionizes almost completely in solution is a **strong acid.** HCl, HNO_3, and H_2SO_4 are examples of strong acids. Conversely, an acid that only partly ionizes in solution is a **weak acid.** Acetic acid and carbonic acid are examples of weak acids.

Ions in solution can carry electric current. **Figure 7** shows how strengths of acids can be compared using conductivity. When more ions are present, more electricity can be conducted.

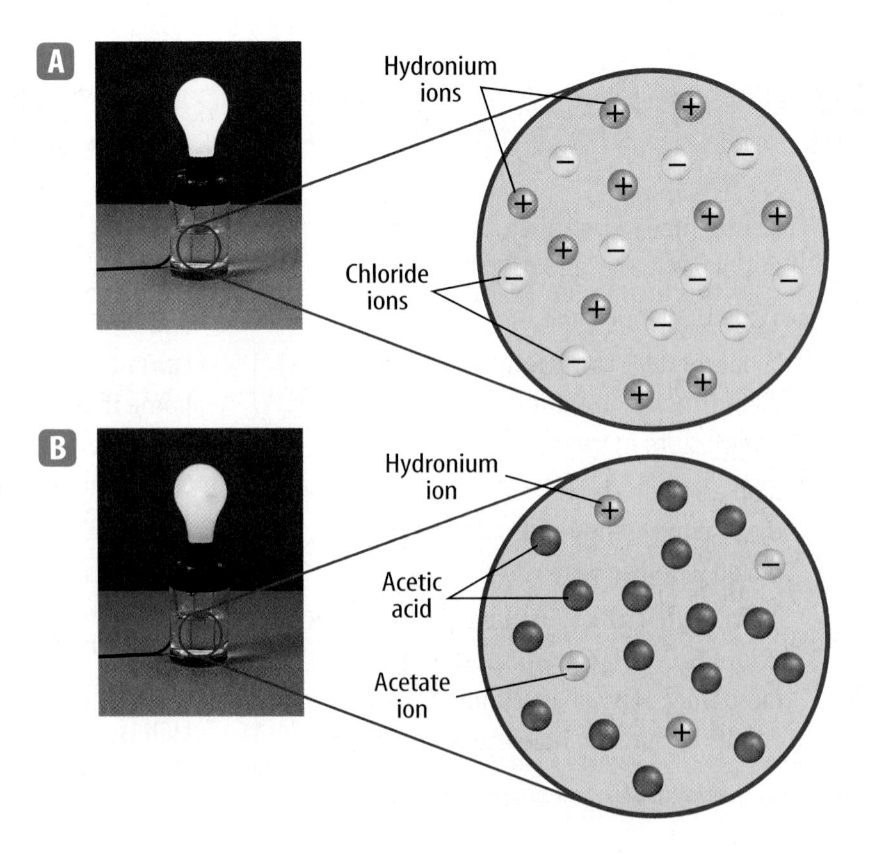

Figure 7
This test compares the conductivity of two acids that are prepared in the same concentration. A HCl, a strong acid, ionizes completely, so the bulb burns brightly. B Acetic acid, a weak acid, produces fewer ions to carry the current so the bulb is dimmer.

Acids Equations describing ionization can be written in two ways. In strong acids, such as HCl, nearly all the acid ionizes. This is shown by writing the equation using a single arrow pointing toward the ions that are formed.

$$HCl(g) + H_2O(l) \longrightarrow H_3O^+(aq) + Cl^-(aq)$$

Almost 100 percent of the particles in solution are H_3O^+ and Cl^- ions, and only a negligible number of HCl molecules are present. Equations describing the ionization of weak acids, such as acetic acid, are written using double arrows pointing in opposite directions. This means that only some of the CH_3COOH ionizes. Thus, the reaction is incomplete.

$$CH_3COOH(l) + H_2O(l) \rightleftarrows H_3O^+(aq) + CH_3COO^-(aq)$$

Double arrows usually indicate that the reaction proceeds in both directions and that the forward reaction does not go to completion. In an acetic acid solution, most of the particles are CH_3COOH molecules, and only a few CH_3COO^- and H^+ ions are in solution.

Figure 8
You can have a dilute solution of a strong acid and a concentrated solution of a weak acid.

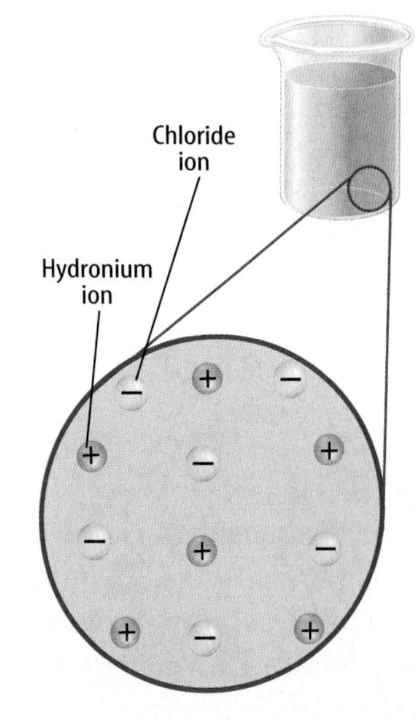

A **This is a dilute solution of HCl.**

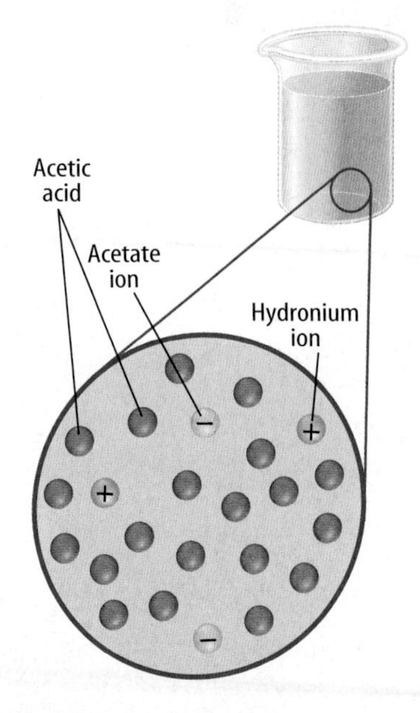

B **This is a concentrated solution of acetic acid.**

Bases Remember that many bases are ionic compounds that dissociate to produce ions when they dissolve. A **strong base** dissociates completely in solution. The following equation shows the dissociation of sodium hydroxide, a strong base.

$$NaOH(s) + H_2O \longrightarrow Na^+(aq) + OH^-(aq)$$

The ionization of ammonia, which is a weak base, is shown using double arrows to indicate that not all the ammonia ionizes. A **weak base** is one that does not ionize completely.

$$NH_3(aq) + H_2O(l) \rightleftarrows NH_4^+(aq) + OH^-(aq)$$

Because ammonia produces only a few ions and most of the ammonia remains in the form of NH_3, ammonia is a weak base.

Strength and Concentration Sometimes, when talking about acids and bases, the terms *strength* and *concentration* can be confused. The terms *strong* and *weak* are used to classify acids and bases. The terms refer to the ease with which an acid or base dissociates in solution. *Strong* acids and bases dissociate completely; *weak* acids and bases dissociate only partially. In contrast, the terms *dilute* and *concentrated* are used to indicate the concentration of a solution, which is the amount of acid or base dissolved in the solution. It is possible to have dilute concentrations of strong acids and bases and concentrated solutions of weak acids and bases, as shown in **Figure 8.**

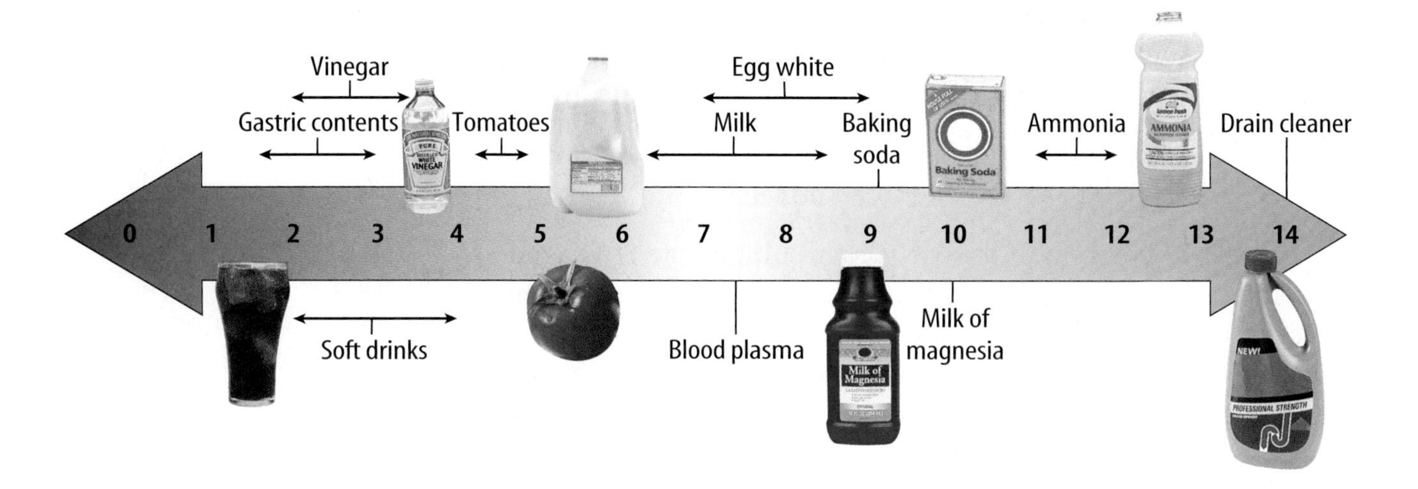

Figure 9
The pH scale helps you classify solutions as acidic or basic.

pH of a Solution

If you have a swimming pool or keep tropical fish, you know that the pH of the water must be controlled. Also, many products such as shampoos claim to control pH so it suits your type of hair. The **pH** of a solution is a measure of the concentration of H^+ ions in it. The greater the H^+ concentration is, the lower the pH is and the more acidic the solution is. The pH measures how acidic or basic a solution is. To indicate pH, a scale ranging from 0 to 14 has been devised, as shown in **Figure 9.**

As the scale shows, solutions with a pH lower than 7 are described as acidic, and the lower the value is, the more acidic the solution is. Solutions with a pH greater than 7 are basic, and the higher the pH is, the more basic the solution is. A solution with a pH of exactly 7 indicates that the concentrations of H^+ ions and OH^- ions are equal. These solutions are considered neutral. Pure water at 25°C has a pH of 7.

One way to determine pH is by using a universal indicator paper. This undergoes a color change in the presence of H_3O^+ ions and OH^- ions in solution. The final color of the pH paper is matched with colors in a chart to find the pH, as shown in **Figure 10A.**

An instrument called a pH meter can be used to determine the pH of a solution more precisely. This meter is operated by immersing the electrodes in the solution to be tested and reading the dial. Small, battery-operated pH meters with digital readouts are convenient for use outside the laboratory when testing the pH of soils and streams, as shown in **Figure 10B.**

Figure 10
The pH of a sample can be measured in several ways. A Indicator paper gives an approximate value quickly. B A pH meter is quick and more precise.

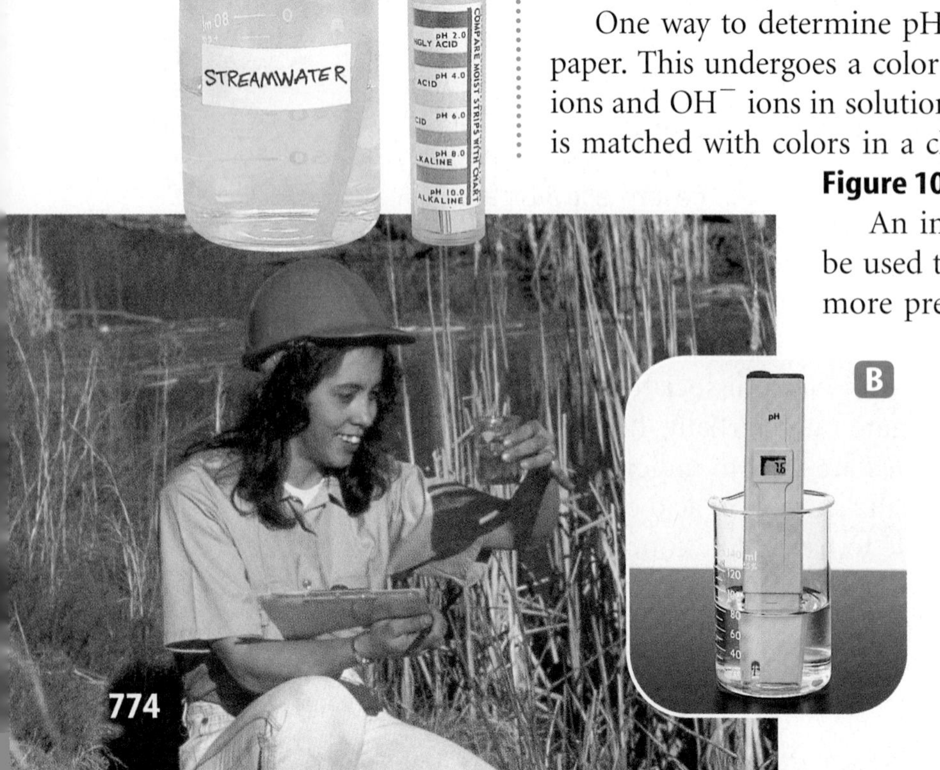

Blood pH Your blood circulates throughout your body carrying oxygen, removing carbon dioxide, and absorbing nutrients from food that you have eaten. In order to carry out its many functions properly, the pH of blood must remain between 7.0 and 7.8. The main reason for this is that enzymes, the protein molecules that act as catalysts for many reactions in the body, cannot work outside this pH range. Yet you can eat foods that are acidic without changing the pH of your blood. How can this be? The answer is that your blood contains compounds called buffers that enable small amounts of acids or bases to be absorbed without harmful effects.

Buffers are solutions containing ions that react with additional acids or bases to minimize their effects on pH. One buffer system in blood involves bicarbonate ions, HCO_3^-. Because of these buffer systems, adding even a small amount of concentrated acid will not change pH much, as shown in **Figure 11.** Buffers help keep your blood close to a nearly constant pH of 7.4.

Figure 11
This experiment shows how well blood plasma acts as a buffer. A Adding 1 mL of concentrated HCl to 1 L of salt water changes the pH from 7.4 to 2.0. B Adding the same amount of concentrated HCl to 1 L of blood plasma changes the pH from 7.4 to 7.2.

✔ Reading Check *What are buffers and how are they important for health?*

Section  Assessment

1. What determines the strength of an acid? A base?
2. How can you make a dilute solution of a strong acid? Include an example.
3. Explain the meaning of pH.
4. Describe pH values of 9.1, 1.2, and 5.7 as basic, acidic, or very acidic.
5. **Think Critically** The proper pH range for a swimming pool is between 7.2 and 7.8. Most pools use two substances, Na_2CO_3 and HCl, to maintain this range. How would you adjust the pH if you found it was 8.2? What would you do if it was 6.9?

Skill Builder Activities

6. **Concept Mapping** Make a concept map of pH values. Start with three boxes labeled *Acidic, Neutral,* and *Basic* and indicate the pH range of each box. Below each box, give some examples of solutions that belong in each pH range. **For more help, refer to the Science Skill Handbook.**
7. **Communicating** In your Science Journal, explain why a weak acid in solution has a higher pH than a strong acid of the same concentration. Extend your explanation to include strong and weak bases. **For more help, refer to the Science Skill Handbook.**

Activity

Strong and Weak Acids

When you read about the behavior of things as small as molecules, it may be difficult to visualize what causes the behavior. This activity will help you do just that.

What You'll Investigate

How can you make a model that represents the degree of ionization of weak and strong acids?

Materials

typing paper marked with rectangles measuring about 5 cm × 7 cm
thick cardboard marked with rectangles measuring about 5 cm × 7 cm
scissors
test tubes (2)
droppers (2)
dilute hydrochloric acid (HCl)
dilute acetic acid (CH_3COOH)
pH paper
timer with a second hand

Goals

- **Classify** acids as weak or strong on the basis of this model.

Safety Precautions

WARNING: *Handle acids with care. Avoid direct contact and report any spills immediately.*

Procedure

1. Mark your paper and cardboard with equal numbers of rectangles. Mark every other rectangle with a plus sign (+) to represent hydronium ions. Then mark the blank rectangles with a minus sign (−) to represent negative ions.
2. Cut the typing paper into as many rectangles as possible in 60 s. Use care handling scissors.
3. Repeat step 2 with the piece of cardboard.
4. Obtain 2 mL each of dilute HCl and dilute CH_3COOH in test tubes.
5. Using a different dropper for each acid, add two drops of each to separate strips of pH paper. Determine their respective pH values and record them in the data table.

Comparing Cutting Difficulty to Ease of Ionization

Number of Ions Removed		pH of Solution	
from typing paper		pH of HCl solution	
from cardboard		pH of acetic acid solution	

Conclude and Apply

1. Was the typing paper or the cardboard easier to cut into pieces?
2. Which acid formed ions more easily? How do you know?
3. Which acid is stronger—HCl or CH_3COOH? How do you know?
4. Does the typing paper or the cardboard represent the stronger acid? Explain.

Communicating Your Data

Compare your results with those of other students in your class. **For more help, refer to the Science Skill Handbook.**

Salts

Neutralization

Advertisements for antacids describe how effectively these products neutralize excess stomach acid that causes indigestion. What does this mean? Normally, gastric juice in your stomach contains a dilute solution of hydrochloric acid. Too much of the acid can produce discomfort. Antacids contain bases or other compounds containing sodium, potassium, calcium, magnesium, or aluminum, that react with acids and lower acid concentration. **Figure 12** shows what happens when you take an antacid tablet containing sodium bicarbonate—$NaHCO_3$. The equation is:

$$HCl(aq) + NaHCO_3(s) \longrightarrow NaCl(aq) + CO_2(g) + H_2O(l)$$

Too many H^+ ions can cause indigestion. Therefore, the important part of this reaction for a person suffering from indigestion is the cancelling or neutralizing of these ions.

Neutralization is a chemical reaction between an acid and a base that takes place in a water solution. For example, when HCl is neutralized by NaOH, hydronium ions from the acid combine with hydroxide ions from the base to produce neutral water.

$$H_3O^+(aq) + OH^- \longrightarrow 2H_2O(l)$$

Salt Formation The equation above accounts for only half of the ions present in the solution. What happens to the remaining ions? They react to form a salt. A **salt** is a compound formed when the negative ions from an acid combine with the positive ions from a base. In the reaction above the salt formed in water solution is sodium chloride.

$$Na^+(aq) + Cl^-(aq) \longrightarrow NaCl(aq)$$

Acid-base reactions in water can be represented by the general equation:

$$\text{acid} + \text{base} \longrightarrow \text{salt} + \text{water}$$

Another neutralization reaction occurs between HCl, an acid and $Ca(OH)_2$, a base.

$$2HCl(aq) + Ca(OH)_2(aq) \longrightarrow CaCl_2(aq) + 2H_2O(l)$$

The products are water and the salt $CaCl_2$.

As You Read

What You'll Learn

- **Identify** a neutralization reaction.
- **Determine** what a salt is and how salts form.
- **Compare and contrast** soaps and detergents.
- **Examine** how esters are made and what they are used for.

Vocabulary

neutralization, salt, titration, soap

Why It's Important

You need salt to live and soaps and detergents to keep yourself and your clothing clean.

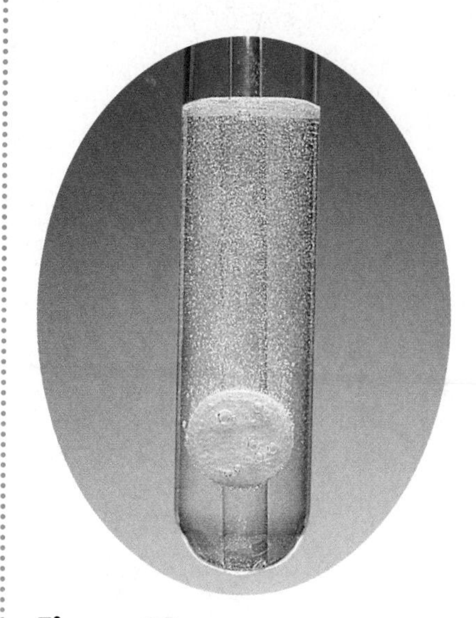

Figure 12
An antacid tablet reacts in your stomach much as it does in this dilute HCl. Usually, people chew antacid tablets before swallowing them. *How would this affect the rate of the reaction?*

Figure 13
Like many animals, elephants get salt from natural deposits. Salt helps to maintain body processes.

Salts

Salt is essential for many animals large and small. Some animals find it at natural deposits, as shown in **Figure 13.** Even insects, such as butterflies, need salt and often are found clustered on moist ground as you saw earlier. You need salt too, especially because you lose salt in perspiration. How humans obtain one salt—sodium chloride—is shown in **Figure 14.**

There are many other salts, however, a few of which are shown in **Table 3.** Most salts are composed of a positive metal ion and an ion with a negative charge, such as Cl^- or CO_3^{2-}. Ammonium salts contain the ammonium ion, NH_4^+, rather than a metal.

Salts also form when acids react with metals. These are single-displacement reactions in which a metal displaces hydrogen from the acid. For example, zinc displaces hydrogen from sulfuric acid forming hydrogen gas and zinc sulfate.

$$H_2SO_4(aq) + Zn(s) \longrightarrow ZnSO_4(aq) + H_2(g)$$

Table 3 Some Common Salts and Their Uses

Name, Formula	Common Name	Uses
Sodium Chloride, $NaCl$	Salt	Food preparation, manufacture of chemicals
Sodium Hydrogen Carbonate, $NaHCO_3$	Sodium bicarbonate Baking soda	Food preparation, antacids
Calcium Carbonate, $CaCO_3$	Calcite, chalk	Manufacture of paint and rubber tires
Potassium Nitrate, KNO_3	Saltpeter	Fertilizers
Potassium Carbonate, K_2CO_3	Potash	Manufacture of soap and glass
Sodium Phosphate, Na_3PO_4	TSP	Detergents
Ammonium Chloride, NH_4Cl	Sal ammoniac	Dry-cell batteries

Figure 14

The salt you use every day comes from both the land and the sea. Some salt can be mined from the ground in much the same way as coal, or salt can be obtained by the process of evaporation in crystallizing ponds.

▲ MINING SALT Underground salt deposits are found where there was once a sea. Salt mines can be located deep underground or near Earth's surface in salt domes. Salt domes, such as the one above on Avery Island, Louisiana, form when pressure from Earth pushes buried salt deposits close to the surface, where they are easily mined.

◀ EVAPORATION PROCESS Workers fill evaporation ponds, like these near San Francisco Bay, California, with salt water, or brine. They move the brine from pond to pond as it becomes saltier through evaporation. (Red-tinted ponds have a higher salt content.) The saltiest water is then pumped from evaporation ponds into crystallizing ponds, where the remaining water is drained off. In the five years it takes to produce a crop of salt, brine may move through as many as 23 different ponds.

Unit cell of sodium chloride (NaCl)

◀ TABLE SALT Raw sodium chloride is washed in chemicals and water to remove impurities before it appears on your dining-room table as salt. Iodine is added to table salt to ensure against iodine deficiency in the diet.

▼ SALT MOUNDS When the crystallizing ponds are drained, the result is huge piles of salt, like these on the Caribbean island of Bonaire.

Titration

Sometimes you need to know the concentration of an acidic or basic solution; for example, to determine the purity of a commercial product. This can be done using a process called **titration** (ti TRAY shun), in which a solution of known concentration is used to determine the concentration of another solution. **Figure 15** shows a titration experiment.

A titration uses a solution of known concentration, called the standard solution. This is added slowly and carefully to a solution of unknown concentration to which an acid/base indicator has been added. If the solution of unknown concentration is a base, a standard acid solution is used. If the unknown is an acid, a standard base solution is used.

Research Visit the Glencoe Science Web site at **science.glencoe.com** for more information about acids, bases, and indicators. Communicate to your class what you learn.

The Endpoint Has a Color The titration shown in **Figure 15** shows how you could find the concentration of an acid solution. First, you would add a few drops of an indicator, such as phenolphthalein (fee nul THAY leen), to a carefully measured amount of the solution of unknown concentration. Phenolphthalein is colorless in an acid but turns bright pink in the presence of a base.

Then, you would slowly and carefully add a base solution of known concentration to this acid-and-indicator mixture. Toward the end of the titration you must add base drop by drop until one last drop of the base turns the solution pink and the color persists. The point at which the color persists is known as the end point, the point at which the acid is completely neutralized by the base. When you know what volume of base was used, you use that value and the known concentration of the base to calculate the concentration of the acid solution.

Figure 15
In this titration, a base of known concentration is being added to an acid of unknown concentration. The swirl of pink color shows that the endpoint is near.

Figure 16
Natural indicators include red cabbage, radishes, and roses.

Many natural substances are acid-base indicators. In fact, the indicator litmus comes from a lichen—a combination of a fungus and an algae or a cyanobacterium. Flowers that are indicators include hydrangeas, which produce blue blossoms when the pH of the soil is acidic and pink blossoms when the soil is basic. This is just the opposite of litmus.

Other natural indicators possess a range of color. For example, the color of red cabbage varies from deep red at pH 1 to lavender at pH 7 and yellowish green at pH 10. Grape juice is also an indicator, as you can find out by doing the Try at Home Minilab.

TRY AT HOME Mini LAB

Testing a Grape Juice Indicator

Procedure

1. Add one-half cup of **water** to each of **two small glasses.**
2. Add 1 tablespoon of **purple grape juice** to each glass.
3. To one glass, add 1 teaspoon of **baking soda.** Stir.
4. To the other glass add 1 teaspoon of **white vinegar.**
5. Note the color after each addition in steps 2, 3, and 4.

Analysis

1. Did the color change when you added baking soda? Why?
2. Did the color change when you added vinegar? Did your grape juice contain any citric or ascorbic acid? How would this affect your experiment?

Problem-Solving Activity

How can you handle an upsetting situation?

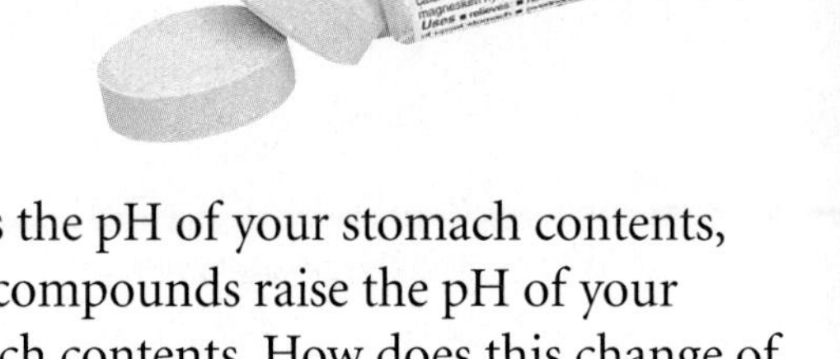

Most of us have, at some time, experienced an upset stomach. Often, the cause is the excess acid within our stomachs. For digestive purposes, our stomachs contain dilute hydrochloric acid with a pH between 1.6 and 3.0. A doctor might recommend an antacid treatment for an upset stomach. What type of compound is "anti acid"?

Identifying the Problem

You have learned that neutralization reactions change acids and bases into salts. Antacids typically contain small amounts of $Ca(OH)_2$, $Al(OH)_3$, or $NaHCO_3$, which are bases. Whereas having an excess of acid lowers the pH of your stomach contents, these compounds raise the pH of your stomach contents. How does this change of pH make you feel better?

Solving the Problem

1. What compounds are produced from a reaction of HCl and $Mg(OH)_2$?
2. Why is it important to have some acid in your stomach?
3. How could you compare how well antacid products neutralize acid? Describe the procedure you would use.

Figure 17
Soaps that contain sodium like this one made from stearic acid are solids, those that contain potassium are liquids.

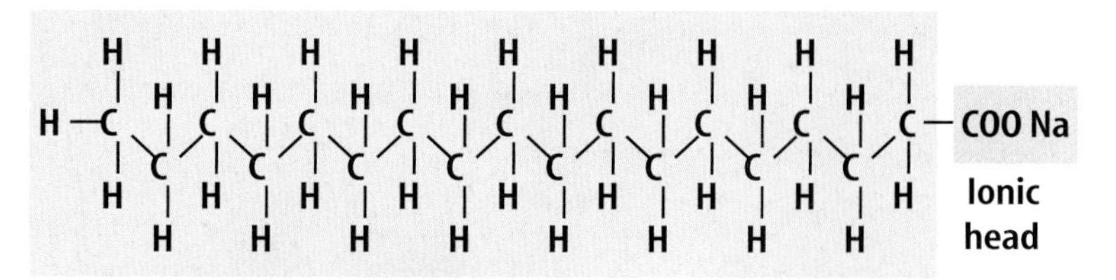

Soaps and Detergents

The next time you are in a supermarket, go to the aisle with soaps and detergents. You'll see all kinds of products—solid soaps, liquid soaps, and detergents for washing clothes and dishes. What are all these products? Do they differ from one another? Yes, they do differ slightly in how they are made and in the ingredients included for color and aroma. Still, all these products are classified into two types—soaps and detergents.

Soaps The reason soaps clean so well is explained by polar and nonpolar molecules. **Soaps** are organic salts. They have a nonpolar organic chain of carbon atoms on one end and either a sodium or potassium salt of a carboxylic acid (kar bahk SIHL ihk), $-COOH$, group at the other end. Look at **Figure 17.** The nonpolar, hydrocarbon end interacts with oils and dirt so that they can be removed readily, and the ionic end, COONa or COOK, helps them dissolve in water.

To make an effective soap, the acid must contain ten to 18 carbon atoms. If it contains fewer than ten atoms, it will not be able to mix well with and clean oily dirt. If it has too many carbon atoms, its sodium or potassium salt will not be soluble in water. **Figure 18** shows how soap interacts with dirt particles to clean your hands.

Figure 18
This is how soaps clean. A The long hydrocarbon tail of a soap molecule mixes well with oily dirt while the ionic head attracts water molecules. B Dirt now linked with the soap rinses away as water flows over it.

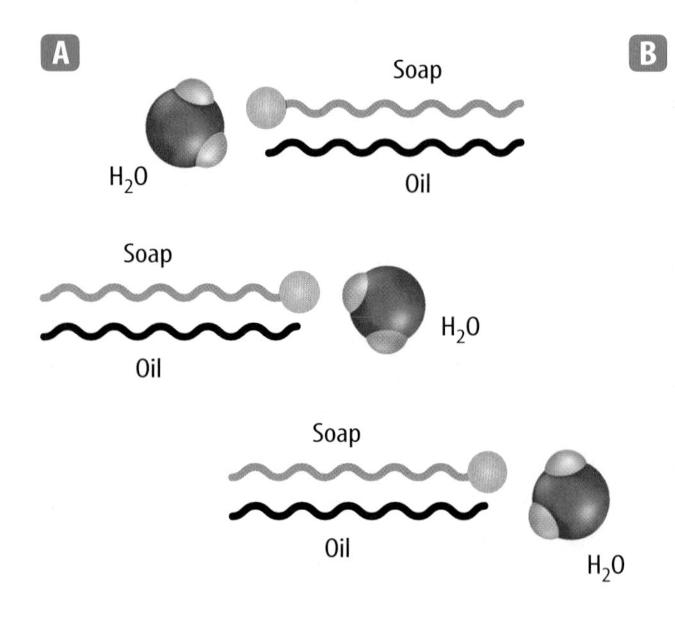

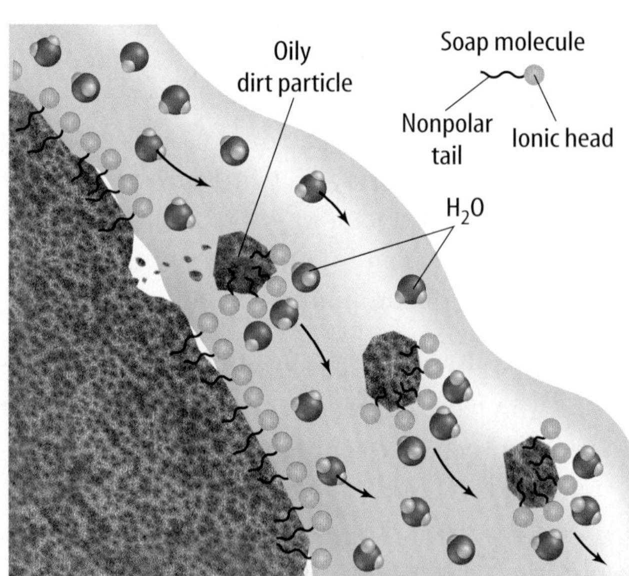

Commercial Soaps A simple soap like the one shown in **Figure 17** can be made by reacting a long-chain fatty acid with sodium or potassium hydroxide. The fatty acids used to make commercial soaps come from natural sources, such as cottonseed, palm, and coconut oils. One problem with all soaps, however, is that the sodium and potassium ions can be replaced by ions of calcium, magnesium, and iron found in some water known as hard water. When this happens, the salts formed are insoluble. They precipitate out of solution in the form of soap scum. Detergents were developed to avoid this problem.

Reading Check *How are simple soaps made?*

Environmental Science INTEGRATION

When phosphates are added to detergents and washed into streams, they can act like a strong fertilizer. This can cause algae and water plants to grow uncontrollably. Research to find if this is a problem in your area. Communicate your findings to your class.

Detergents Like soaps, detergents have long hydrocarbon chains, but instead of a carboxylic acid group (−COOH) at the end, they contain either a sulfonic acid or phosphoric acid group. These acids form more soluble salts with the ions in hard water and thereby lessen the problem of soap scum. They can also be used in cold water. Most detergents contain other substances called builders to further improve cleaning in hard water.

Detergents introduce other problems, however. For example, phosphates act as fertilizers that cause overgrowth of vegetation in streams. For this reason, many states limit the amount of phosphates that can be used in detergents. Also, one form of sulfonic acid detergent causes excessive foaming in water-treatment plants and streams, as shown in **Figure 19B.** These detergents are not degraded or broken down easily by bacteria. They stay in the environment over long periods. Most detergents now in use resemble the one shown in **Figure 19A.**

Figure 19
Some detergents can have adverse effects on the environment.

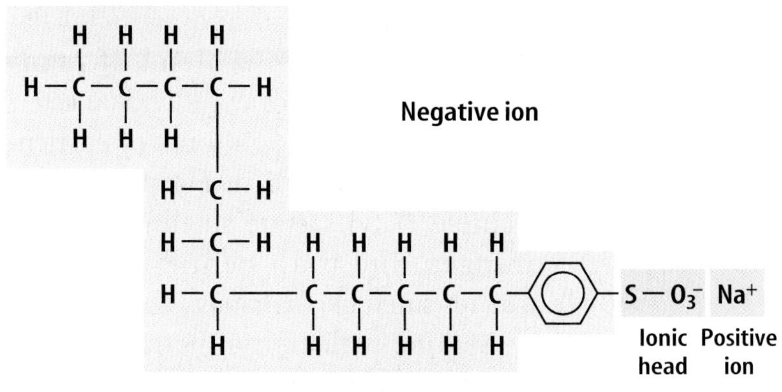

A **Newer, biodegradable detergents like the one shown here do not produce excess foam.**

B **Foam from nonbiodegradable detergents can build up in waterways.**

$$\text{Butyric acid} + \text{Ethyl alcohol} \rightarrow \text{Ethyl butyrate} + H_2O$$

Butyric acid | Ethyl alcohol | Ethyl butyrate | Water

Figure 20
This structural equation shows the formation of the ester ethyl butyrate, an ester that tastes and smells like pineapple. *Which alcohol would you use to prepare butyl butyrate?*

Versatile Esters

In a way esters can be thought of as the organic counterparts of salts. Like salts, esters are made from acids, and water is formed in the reaction used to prepare them. The difference is that salts are made from bases and esters come from alcohols that are not bases but have a hydroxyl group.

Esters have many different applications. Esters of the alcohol glycerine are used commercially to make soaps. Other esters are used widely in flavors and perfumes, and still others can be transformed into fibers to make clothing.

Reading Check *What are three types of products that are made from esters?*

Esters for Flavor Many fruit-flavored soft drinks and desserts taste like the real fruit. If you look at the label though, you might be surprised to find that no fruit was used—only artificial flavor. Most likely this artificial flavor contains some esters.

The reaction to prepare esters involves removing a molecule of water from an acid and an alcohol. Often concentrated sulfuric acid is added to aid this reaction. **Figure 20** shows the reaction of butyric (byew TIHR ihk) acid and ethyl alcohol to produce water and the ester, ethyl butyrate, which is a component in pineapple flavor.

Although natural and artificial flavors often contain a blend of many esters, the odor of some individual esters immediately makes you think of particular fruits, as shown in **Figure 21.** For example, octyl acetate smells much like oranges, and both pentyl and butyl acetates smell like bananas.

Making realistic synthetic flavors is an art, in which chemists vary the composition to achieve the desired taste. Strawberry flavor, for example, may contain several esters.

Figure 21
These esters have strong fruity aromas.

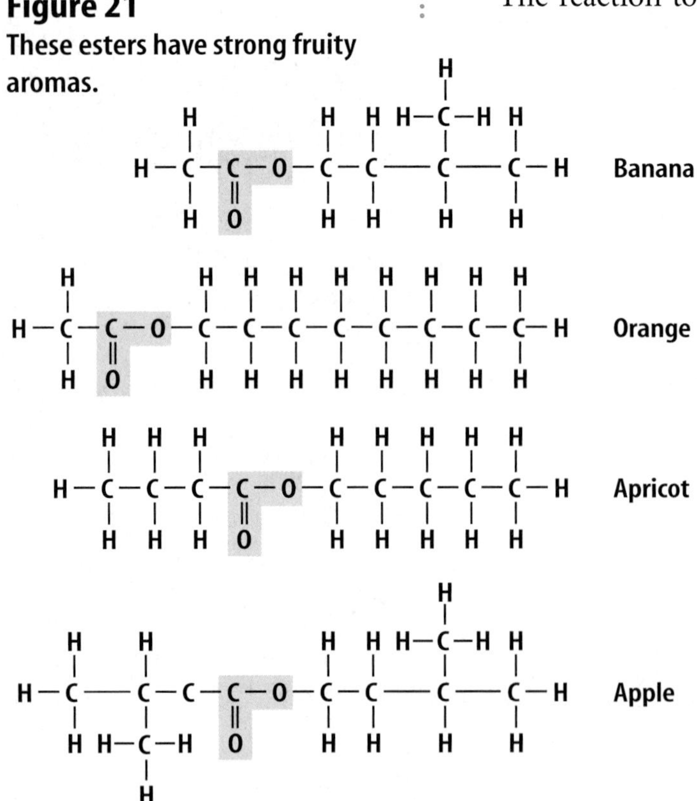

Figure 22
Polyesters and nylons are polymers most often used for clothing fibers.

$$HO-\overset{O}{\overset{\|}{C}}-C_6H_4-\overset{O}{\overset{\|}{C}}-OH + HO-\underset{H}{\overset{H}{C}}-\underset{H}{\overset{H}{C}}-OH \rightarrow$$

Organic acid   Alcohol

$$\sim\left[\overset{O}{\overset{\|}{C}}-C_6H_4-\overset{O}{\overset{\|}{C}}-O-\underset{H}{\overset{H}{C}}-\underset{H}{\overset{H}{C}}-O\right]\sim + 2H_2O$$

Polymer (1 unit)   Water

A Polyesters contain long, nonpolar chains that tend to repel water.

Polyesters Synthetic fibers known as polyesters are polymers; that is, they are chains containing many or *poly* esters. They are made from an organic acid that has two $-COOH$ groups and an alcohol that has two $-OH$ groups, as shown in **Figure 22A.** The two compounds form long chains that are closely packed together. This adds strength to the polymer fiber. Many varieties of polyesters are made, depending on what alcohols and acids are used. They can be woven or knitted into fabrics that are durable, colorfast, and do not wrinkle easily. Because of their low moisture content however, they tend to build up a static electric charge that causes them to cling. Polyesters often are combined with natural fibers, as shown in **Figure 22B.**

B Blends of polyester and cotton fibers make comfortable activewear.

Section  Assessment

1. What is a neutralization reaction? What are the products of such reactions?
2. In a titration experiment, what is the purpose of an indicator?
3. How does the composition of detergents differ from that of soaps?
4. What small molecule is produced in the reaction between an alcohol and an acid to form an ester?
5. **Think Critically** Give the names and formulas of the salts formed in these neutralizations: sulfuric acid and calcium hydroxide, nitric acid and potassium hydroxide, and carbonic acid and aluminum hydroxide.

Skill Builder Activities

6. **Interpreting Data** Three salts—calcium sulfate, sodium chloride, and potassium nitrate—were obtained in reactions of the acid-base pairs shown here. Match each salt with the acid-base pair that produced it. **For more help, refer to the Science Skill Handbook.**

   $HCl + NaOH$; $HNO_3 + KOH$; $H_2SO_4 + Ca(OH)_2$

7. **Calculating Ratios** In the following reaction:

   $2HCl(aq) + Ca(OH)_2(aq) \longrightarrow CaCl_2(aq) + 2H_2O(l)$

   acid reacts with base in what ratio? How many molecules of HCl are needed to produce four molecules of H_2O? **For more help, refer to the Math Skill Handbook.**

Activity
Design Your Own Experiment

Be a Soda Scientist

Carbonated soft drinks contain carbonic acid and sometimes phosphoric acid. You have learned that bases can neutralize acids. Using a proper indicator and a base solution, how could you compare the acidity levels in soft drinks?

Recognize the Problem

How can the acidity level of soft drinks be compared effectively?

Form a Hypothesis

Based on observations about acids and bases, develop a hypothesis about how neutralization reactions can be used to rank the acidity of soft drinks.

Possible Materials

different colorless soft drinks (3)
test tubes (3)
25-mL graduated cylinder
droppers (2)
1% phenolphthalein indicator solution
dilute NaOH solution

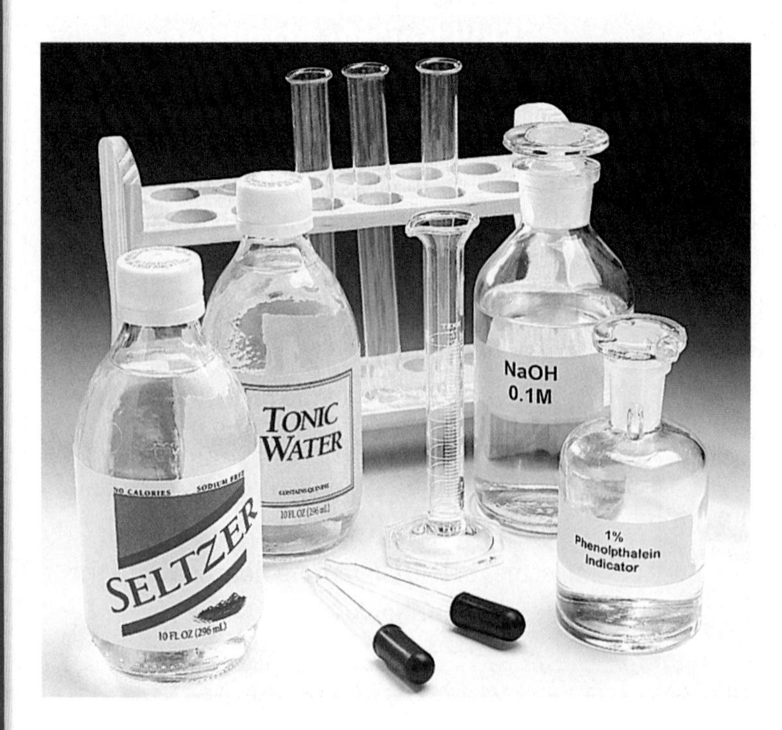

Goals

- **Observe** evidence of a neutralization reaction using an indicator.
- **Compare** the acidity levels in three different soft drinks.
- **Design** an experiment that uses the independent variable of acid content of soft drinks and the dependent variable of amount of base added to the soft drinks to determine the acidity of the drinks.

Safety Precautions

WARNING: *Sodium hydroxide is caustic. Wear eye protection and avoid any skin contact with the solution. Flush thoroughly under a stream of water if any of the NaOH touches your skin. Keep your hands away from your face.*

Test Your Hypothesis

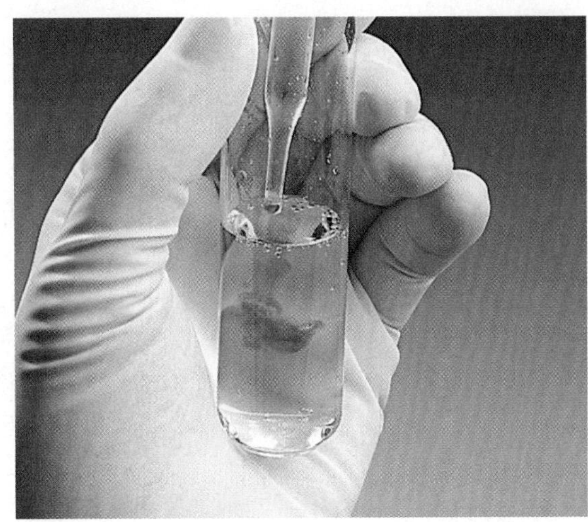

Plan

1. As a group, agree upon and write the hypothesis statement.
2. In a logical manner, list the specific steps that you will use to test your hypothesis.
3. **List** all of the materials that you will need to test your hypothesis.
4. **Design** a data table in your Science Journal that will allow you to record the amount of NaOH that was required to neutralize each soda sample.
5. **Decide** the amount of soda to be tested in each trial as a control. Decide also how many times to repeat each trial.
6. **Predict** whether you can test only colorless solutions with this procedure and explain why.

Do

1. Make sure your teacher approves your plan before you start.
2. What color change does the indicator phenolphthalein undergo in a solution that changes from an acidic pH to a basic pH?
3. While doing the experiment, write your observations and complete the data table in your Science Journal.

Analyze Your Data

1. **Classify** the sodas you tested based on their acidities. Rank them in the order of most acidic to least acidic.
2. Can your acidity values be compared with those of other groups if they used different amounts of soda?

Draw Conclusions

1. Did the results support your hypothesis? Explain why or why not.
2. At warmer temperatures less gas dissolves in a liquid. How would this affect the results of an experiment comparing two sodas stored at different temperatures?

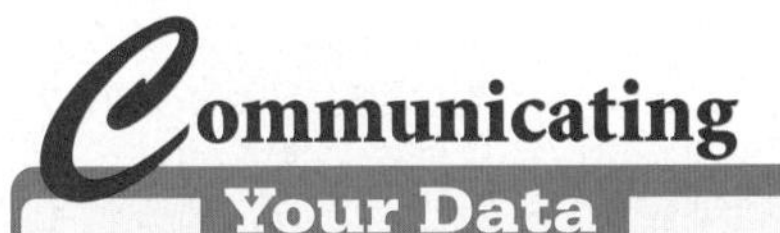

Compare your soda rankings with those of other class groups. **Discuss** possible reasons for any differences observed.

Acid Rain

Protecting Earth from the damaging effects of chemically-loaded precipitation

In a forest in eastern Canada, the trees grow slower than normal. It's the height of summer, but their leaves are brown and spotted. In a nearby lake, dead fish rise to the surface of the water. The trees and the fish are victims of a form of pollution known as acid rain.

Acid rain is rain, snow, or sleet that is more acidic than unpolluted precipitation. It's caused by the burning of fossil fuels, such as coal, oil, and natural gas. In the United States, most gasoline and electricity come from fossil fuels. People burn fossil fuels each time they drive a car, heat a building, or turn on a light.

Normally, raindrops pick up particles and natural chemicals in the air. When rain falls, it mixes with the carbon dioxide in the atmosphere, giving clean rain a slightly acidic pH of 5.6. Then natural chemicals found in the air and soil balance out the acidity, giving most lakes and streams a pH between six and eight. But when pollutants are introduced, this delicate balance is upset. This imbalance can affect entire ecosystems.

The burning of fossil fuels releases sulfur dioxide and nitrogen oxides into the air. In the clouds these chemicals combine with moisture to form mild solutions of sulfuric acid and nitric acid.

The natural chemicals, or bases, that balance clean rain's acidity are not strong enough to neutralize these solutions. Wind can carry this acidic moisture for hundreds of miles before it falls to Earth as acid rain. That's why even an isolated forest in Canada, lying far from the nearest power plant, can be devastated by acid rain. And that's why acid rain is a worldwide problem.

Eating Away at History

Like all acids, acid rain can corrode, or eat away at, substances. Many historical monuments, such as the Mayan temples in Mexico, and the Parthenon in Greece, have been slowly but steadily damaged by acid rain. The Environmental Protection Agency (EPA) notes that this kind of damage can be fixed, though it costs billions of dollars. But if ancient monuments and buildings, like the Colosseum in Rome, are destroyed, they can never be replaced.

Some Solutions

In some countries, high acid levels in lakes and streams have been lowered by adding lime to the water. Lime, a natural base, balances out the damaging chemicals. In the United States, all new cars must have catalytic converters, which help reduce the amount of exhaust pollution that vehicles give off. Also, coal-burning power plants can use scrubbers that wash sulfur out of the coal during or after burning. This, too, helps reduce air pollution.

Ride a bike! It saves fuel, is nonpolluting, and helps preserve the environment.

You also can make a difference. Turning off the lights when you are not using them, means a power plant does not have to produce as much electricity. By carpooling, using public transportation, and walking, there is less pollution from cars. The results of all these individual actions can make a huge difference in preserving our environment.

CONNECTIONS List Go to a local park or forest. List any effects of acid rain that you see. Make a list of the things you do that use energy or cause pollution. Did you turn lights out? How do you go to school? Think about what your family can do to reduce pollution and save energy. Share your list with an adult.

Chapter 25 Study Guide

Reviewing Main Ideas

Section 1 Acids and Bases

1. An acid is a substance that produces hydrogen ions, H^+, in solution. A base produces hydroxide ions, OH^-, in solution.
2. Properties of acids and bases are due, in part, to the presence of the H^+ and OH^- ions. *Which of the foods shown here are acidic?*

3. Common acids include hydrochloric acid, sulfuric acid, nitric acid, and phosphoric acid. Common bases include sodium hydroxide, calcium hydroxide, and ammonia.
4. Acidic solutions form when certain polar compounds ionize as they dissolve in water. Except for ammonia, basic solutions form when certain ionic compounds dissociate upon dissolving in water.

Section 2 Strength of Acids and Bases

1. The strength of an acid or base is determined by how completely it forms ions when it is in solution.
2. Strength and concentration are not the same thing. Concentration involves the relative amounts of solvent and solute in a solution, whereas strength is related to the extent to which a substance dissociates.
3. pH measures the concentration of hydronium ions in water solution using a scale ranging from 0 to 14.
4. For acidic solutions of equal concentration, the stronger the acid is, the lower its pH is. For basic solutions of equal concentration, the stronger the base is, the higher its pH is.

Section 3 Salts

1. In a neutralization reaction, the H_3O^+ ions from an acid react with the OH^- ions from a base to produce water molecules. The products of a neutralization reaction are a salt and water.
2. Salts form when negative ions from an acid combine with positive ions from a base.
3. Soaps and detergents are organic salts. Unlike soaps, detergents do not react with compounds in hard water to form soap scum. *Which type of product was used to wash materials in this sink?*

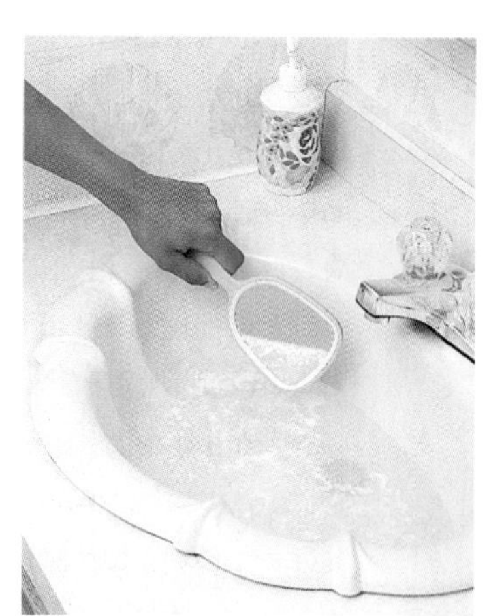

4. Esters are organic compounds formed by the reaction of an organic acid and an alcohol.

After You Read

Foldables Reading & Study Skills

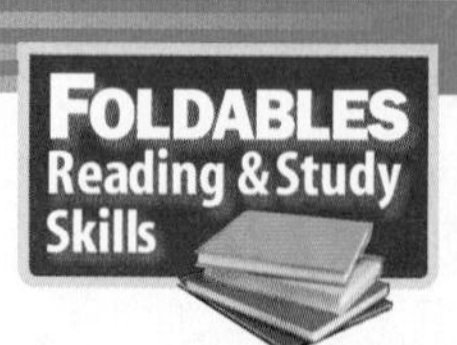

List examples of acids, bases, and salts on the front of your Venn Diagram study fold. Contrast the properties of acids and bases. Explain how salts are formed from acids and bases.

Visualizing Main Ideas

Complete the concept map using the following: salt, slippery feel, pH less than 7, tastes sour, forms hydroxide ions in solution, water, forms hydrogen ions in solution, pH higher than 7, corrosive, *and* bitter taste.

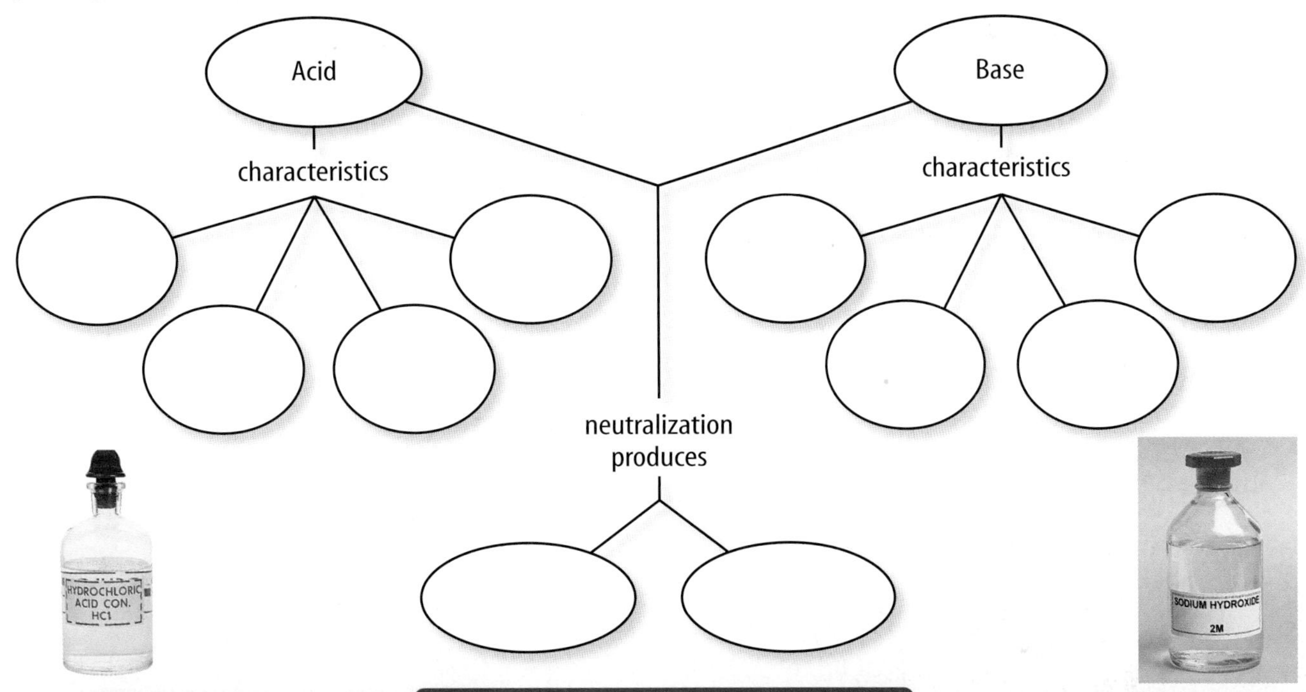

Vocabulary Review

Vocabulary Words

a. acid
b. base
c. buffer
d. hydronium ion
e. indicator
f. neutralization
g. pH
h. salt
i. soap
j. strong acid
k. strong base
l. titration
m. weak acid
n. weak base

THE PRINCETON REVIEW Study Tip

Take good notes, even during lab. Lab experiments reinforce key concepts, and looking back on these notes can help you better understand what happened and why.

Using Vocabulary

Explain the differences between each set of vocabulary words given below. Then explain how the words are related.

1. acid, base
2. acid, salt
3. salt, soap
4. base, soap
5. neutralization, salt
6. strong acid, pH
7. hydronium ion, acid
8. indicator, titration
9. pH, buffer
10. weak base, strong base

Chapter 25 Assessment

Checking Concepts

Choose the word or phrase that best answers the question.

1. What best describes solutions of equal concentrations of HCl and CH_3COOH?
A) do not have the same pH
B) will react the same with metals
C) will make the same salts
D) have the same amount of ionization

2. What is hydrochloric acid also known as?
A) battery acid **C)** stomach acid
B) citric acid **D)** vinegar

3. What is 90 percent of phosphoric acid used to produce?
A) batteries **C)** petroleum products
B) fertilizer **D)** plastics

4. Which of the following acids ionizes only partially in water?
A) HCl **C)** HNO_3
B) H_2SO_4 **D)** CH_3COOH

5. Which of the following is another name for sodium hydroxide (NaOH)?
A) ammonia **C)** lye
B) caustic lime **D)** milk of magnesia

6. Carrots have a pH of 5.0, so how would you describe them?
A) acidic **C)** neutral
B) basic **D)** an indicator

7. What is the pH of pure water at 25°C?
A) 0 **C)** 7
B) 5.2 **D)** 14

8. A change of what property permits certain materials to act as indicators?
A) acidity **C)** concentration
B) color **D)** taste

9. What type of substance is KBr?
A) acid **C)** indicator
B) base **D)** salt

10. Which of the following might you use to titrate an oxalic acid solution?
A) HBr **C)** NaOH
B) $Ca(NO_3)_2$ **D)** NH_4Cl

Thinking Critically

11. When hydrogen chloride, HCl, is dissolved in water to form hydrochloric acid, what happens to the HCl?

12. Explain how the hydroxide ion in NaOH differs from the −OH group in an alcohol.

13. Why is ammonia considered a base, even though it contains no hydroxide ions? Is it a strong or weak base?

14. Explain why a concentrated acid is not necessarily a strong acid.

15. Ashes from a wood fire are basic. Explain how this helped early settlers to make soap.

Developing Skills

16. Comparing and Contrasting How would the pH of a dilute solution of HCl compare with the pH of a concentrated solution of the same acid?

17. Recognizing Cause and Effect Ramon often saw his mother cleaning white deposits from inside her teakettle using vinegar. When she added vinegar, bubbles formed. When she finished, all the white deposits were gone. What do you think these white deposits might be? Do you think dish detergent would have worked as well?

18. Interpreting Data A soil test indicates that the pH of the soil in a field is 4.8. To neutralize this soil, would you add a substance that contained H_3PO_4 or a substance that contained $Ca(OH)_2$? Explain.

19. **Comparing and Contrasting** Compare and contrast the reactions that would occur when a marshmallow is burned and when acid in paper causes damage over time.

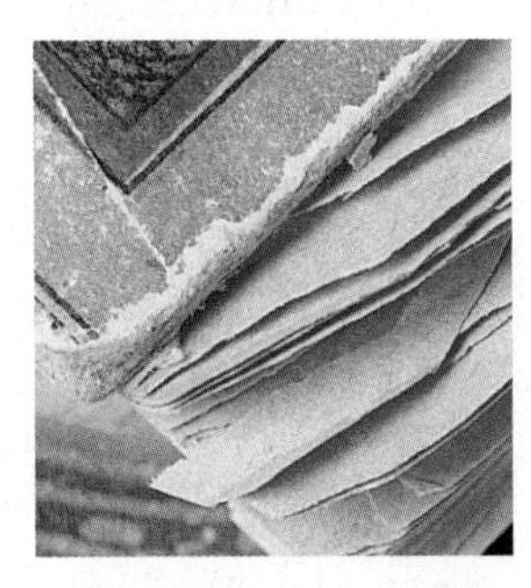

20. **Drawing Conclusions** You have equal amounts of three colorless liquids: A, B, and C. You add several drops of phenolphthalein to each liquid. A and B remain colorless, but C turns pink. Next, you add some C to A and the pink color disappears. Then, you add the rest of C to B and the mixture remains pink. What can you infer about each of these liquids? Which original liquid could have had a pH of 7?

Performance Assessment

21. **Display** Make a display and write a report about common household acids and bases, their uses, and cautions to be taken.

22. **Lab Report** Do research to find out how acids and bases are used in making fertilizer. Name several acids and one base, tell why each is needed, and explain how their proportions can vary.

TECHNOLOGY

Go to the Glencoe Science Web site at **science.glencoe.com** or use the **Glencoe Science CD-ROM** for additional chapter assessment.

Test Practice

A chemist is measuring the pH levels of several different substances. The table below lists some common substances and their average pH values.

pH Readings

Substance	pH
Battery acid	1.5
Lemon juice	2.5
Apple	3
Milk	6.7
Seawater	8.5
Ammonia	12

Study the table and answer the following questions.

1. The chemist measures a pH of 8.4 for an unknown substance. What is the substance likely to be?
   A) Apples **C)** Lemon juice
   B) Sea water **D)** Milk

2. According to this information, what is an unknown substance with a pH of 3.1 likely to be?
   F) ammonia **H)** an apple
   G) lemon juice **J)** battery acid

3. Among the substances listed in the table, which one would be most effective for neutralizing battery acid?
   A) seawater **C)** lemon juice
   B) ammonia **D)** milk

4. Which of the substances listed in the table has a pH value that is closest to being neutral?
   F) apple **H)** milk
   G) seawater **J)** ammonia

Standardized Test Practice

Reading Comprehension

Read the passage. Then read each question that follows the passage. Decide which is the best answer to each question.

Test-Taking Tip Remember that clues and information can always be found in the passage.

The Eyes Have It

If you've ever seen ancient Egyptian paintings, then you were probably struck by the dramatic appearance of the dark-rimmed eyes of their subjects. Anthropologists know from the records of ancient scribes that the Egyptians achieved this effect in real life by using cosmetics. However, until recently, they were unaware of the sophisticated chemistry that went into the ancient Egyptians' creation of these early cosmetics.

The Louvre museum in Paris has an extensive collection of ancient Egyptian makeup containers. Amazingly, many of them still retain remnants of their original contents. When the technology to analyze the remains of the Egyptians' creams and powders became available, French scientists went to work. They found that the cosmetics contained several lead-based chemicals. Two of them, lead sulfide (PbS) and lead carbonate ($PbCO_3$), are common in nature. However, two of the others found in the makeup are not. PbOHCl and $Pb_2Cl_2CO_3$ are formed only when lead minerals oxidize in a combination of chlorinated and carbonated water. Scientists believe that the Egyptians created these compounds artificially, using a sophisticated process called "wet chemistry" to synthesize their molecules.

Scientists had already found written records indicating that the ancient Greeks and Romans used wet chemistry. However, the Egyptian discovery pushes the date of the first known use of this technique back to 2000 B.C., almost 2,000 years before the birth of Cleopatra.

This painting reflects regular use of cosmetics by ancient Egyptians.

1. Scientists believe that Egyptians created two of the compounds found in the ancient cosmetics artificially because _____.

A) scientists had never seen these compounds before
B) they are not often found in nature
C) the process used to make them was simple
D) only artificially created compounds could survive such a long time

2. Historians knew that ancient Greeks and Romans used "wet chemistry" because _____.

F) they had found samples of cosmetics for testing
G) they saw evidence of the use of cosmetics in their artwork
H) surviving text from the period indicates the use of "wet chemisty"
J) the Greek and Roman civilizations followed the Egyptian model

Standardized Test Practice

THE PRINCETON REVIEW

Reasoning and Skills

Read each question and choose the best answer.

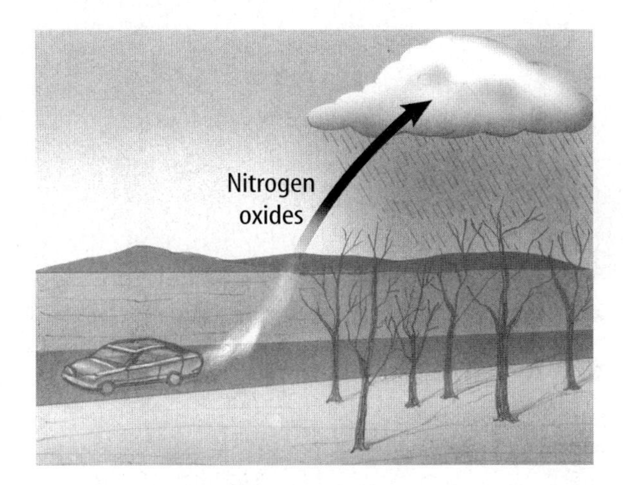

1. This picture indicates that the trees are probably losing their leaves because of _____.
   A) soil erosion
   B) acid rain
   C) nutrient-poor soil
   D) over-watering

Test-Taking Tip The best answer is always supported by any information in a chart or graphic.

2. What is the source of the nitrogen oxides?
   F) the soil
   G) the air
   H) the car exhaust
   J) the cloud

Common Bases and Applications	
Bases	**Product**
Sodium hydroxide	Drain cleaner
Magnesium hydroxide	Laxative
Ammonium hydroxide	Window cleaner
Aluminum hydroxide	Antacid

3. According to the information on the chart, ammonium hydroxide is used in the manufacturing of _____.
   A) window cleaner
   B) laxative
   C) drain cleaner
   D) antacid

Test-Taking Tip Always consider every answer choice as you work through a question.

4. A balanced chemical equation has the same number of atoms of each element on both sides of the equation. Which of the following is a balanced equation?
   F) $2Al + 3Ag_2S \rightarrow Al_2S_3 + Ag$
   G) $2H_2CO_3 \rightarrow H_2O + CO_2$
   H) $Fe + O_2 \rightarrow Fe_2O_3$
   J) $2Ag + H_2S \rightarrow Ag_2S + H_2$

Test-Taking Tip Double-check that you have identified the correct answer by making sure that the other answer choices are not balanced equations.

Consider this question carefully before writing your answer on a separate sheet of paper.

5. Touch and litmus tests can be used to help determine if a substance is a base. Discuss how each of these tests indicates the presence of a base. Also, discuss the possible benefits or drawbacks of using that type of test.

Test-Taking Tip Consider the different types of experiments that you have done in class or have read about in your textbooks as you answer this question.

Student Resources

CONTENTS

Organizing Information

As you study science, you will make many observations and conduct investigations and experiments. You will also research information that is available from many sources. These activities will involve organizing and recording data. The quality of the data you collect and the way you organize it will determine how well others can understand and use it. In **Figure 1,** the student is obtaining and recording information using a thermometer.

Putting your observations in writing is an important way of communicating to others the information you have found and the results of your investigations and experiments.

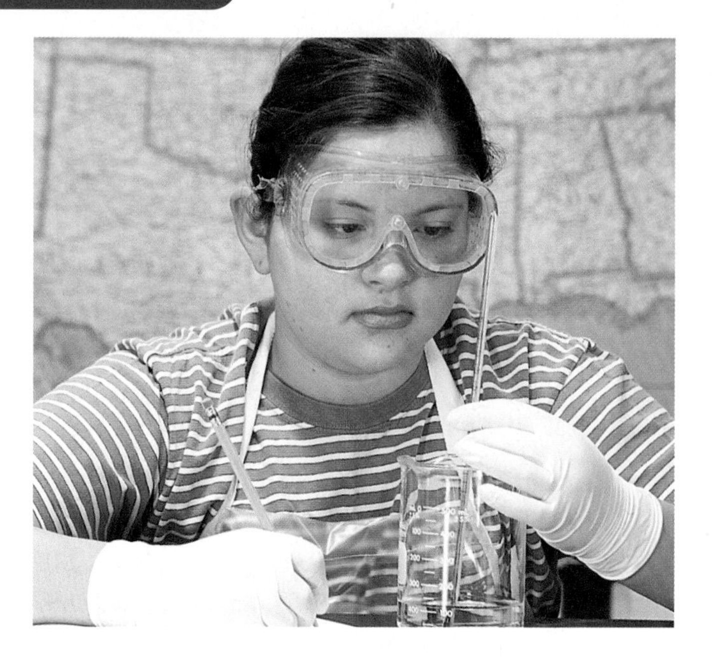

Figure 1
Making an observation is one way to gather information directly.

Researching Information

Scientists work to build on and add to human knowledge of the world. Before moving in a new direction, it is important to gather the information that already is known about a subject. You will look for such information in various reference sources. Follow these steps to research information on a scientific subject:

Step 1 Determine exactly what you need to know about the subject. For instance, you might want to find out about one of the elements in the periodic table.

Step 2 Make a list of questions, such as: Who discovered the element? When was it discovered? What makes the element useful or interesting?

Step 3 Use multiple sources such as textbooks, encyclopedias, government documents, professional journals, science magazines, and the Internet.

Step 4 List where you found the sources. Make sure the sources you use are reliable and the most current available.

Evaluating Print and Nonprint Sources

Not all sources of information are reliable. Evaluate the sources you use for information, and use only those you know to be dependable. For example, suppose you want to find ways to make your home more energy efficient. You might find two Web sites on how to save energy in your home. One Web site contains "Energy-Saving Tips" written by a company that sells a new type of weatherproofing material you put around your door frames. The other is a Web page on "Conserving Energy in Your Home" written by the U.S. Department of Energy. You would choose the second Web site as the more reliable source of information.

In science, information can change rapidly. Always consult the most current sources. A 1985 source about saving energy would not reflect the most recent research or findings.

Interpreting Scientific Illustrations

As you research a science topic, you will see drawings, diagrams, and photographs. Illustrations help you understand what you read. Some illustrations are included to help you understand an idea that you can't see easily by yourself. For instance, you can't see the tiny particles in an atom, but you can look at a diagram of an atom as labeled in **Figure 2** that helps you understand something about it. Visualizing a drawing helps many people remember details more easily. Illustrations also provide examples that clarify difficult concepts or give additional information about the topic you are studying.

Most illustrations have a label or caption that identifies the illustration or provides additional information to better explain it. Can you find the caption or labels in **Figure 2?**

Figure 2
This drawing shows an atom of carbon with its six protons, six neutrons, and six electrons.

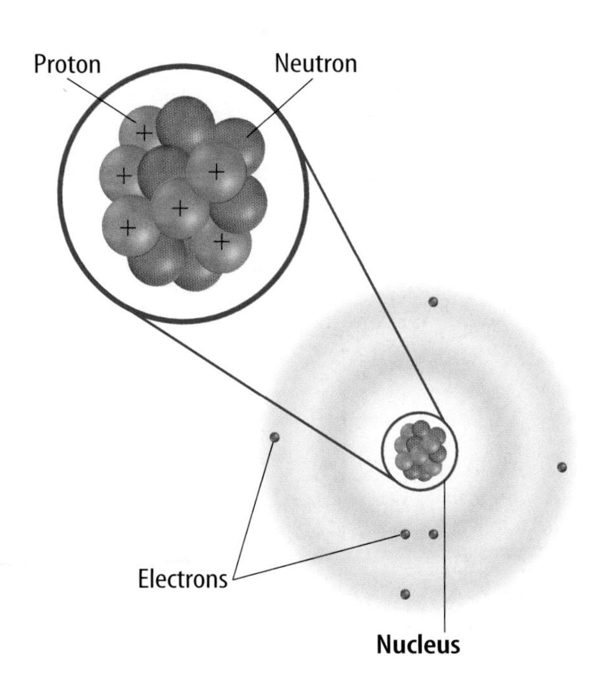

Venn Diagram

Although it is not a concept map, a Venn diagram illustrates how two subjects compare and contrast. In other words, you can see the characteristics that the subjects have in common and those that they do not.

The Venn diagram in **Figure 3** shows the relationship between two different substances made from the element carbon. However, due to the way their atoms are arranged, one substance is the gemstone diamond, and the other is the graphite found in pencils.

Concept Mapping

If you were taking a car trip, you might take some sort of road map. By using a map, you begin to learn where you are in relation to other places on the map.

A concept map is similar to a road map, but a concept map shows relationships among ideas (or concepts) rather than places. It is a diagram that visually shows how concepts are related. Because a concept map shows relationships among ideas, it can make the meanings of ideas and terms clear and help you understand what you are studying.

Overall, concept maps are useful for breaking large concepts down into smaller parts, making learning easier.

Figure 3
A Venn diagram shows how objects or concepts are alike and how they are different.

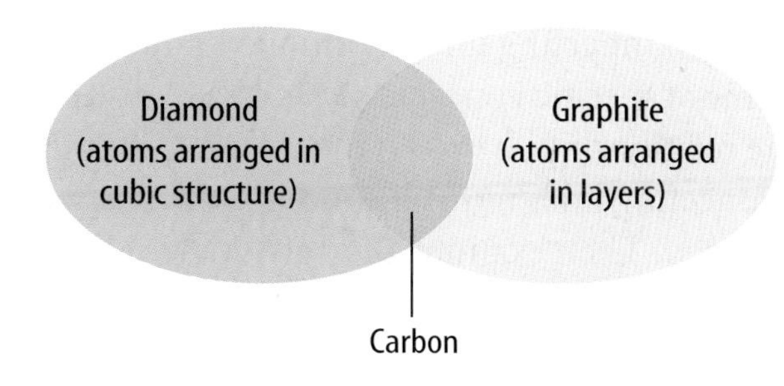

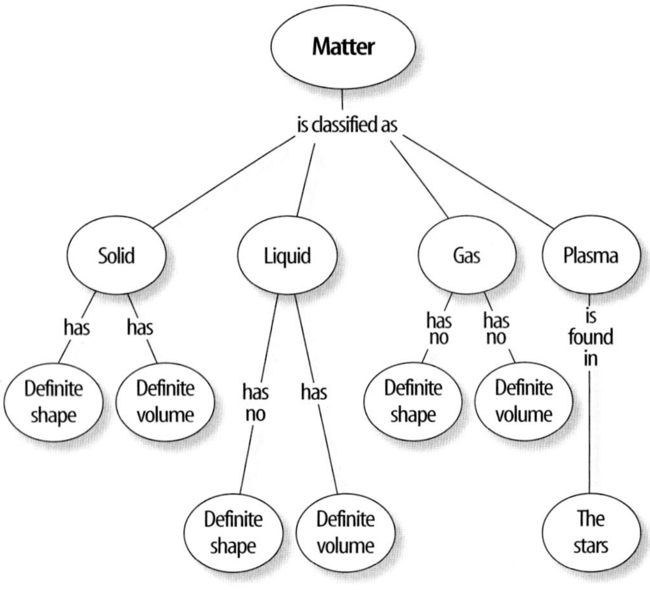

Figure 4
A network tree shows how concepts or objects are related.

Network Tree Look at the network tree in **Figure 4,** that describes the different types of matter. A network tree is a type of concept map. Notice how some words are in ovals while others are written across connecting lines. The words inside the ovals are science terms or concepts. The words written on the connecting lines describe the relationships between the concepts.

When constructing a network tree, write the topic on a note card or piece of paper. Write the major concepts related to that topic on separate note cards or pieces of paper. Then arrange them in order from general to specific. Branch the related concepts from the major concept and describe the relationships on the connecting lines. Continue branching to more specific concepts. If necessary, write the relationships between the concepts on the connecting lines until all concepts are mapped. Then examine the network tree for relationships that cross branches and add them to the network tree.

Events Chain An events chain is another type of concept map. It models the order, or sequence, of items. In science, an events chain can be used to describe a sequence of events, the steps in a procedure, or the stages of a process.

When making an events chain, first find the one event that starts the chain. This event is called the initiating event. Then, find the next event in the chain and continue until you reach an outcome. Suppose you are asked to describe why and how a sound might make an echo. You might draw an events chain such as the one in **Figure 5.** Notice that connecting words are not necessary in an events chain.

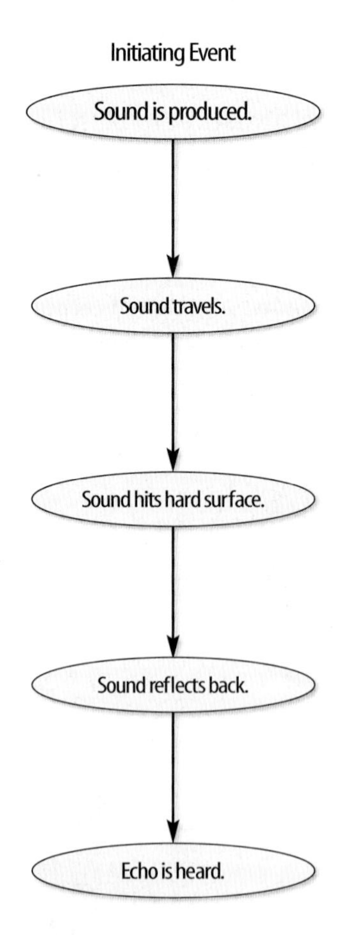

Figure 5
Events chains show the order of steps in a process or event.

Cycle Map A cycle concept map is a specific type of events chain map. In a cycle concept map, the series of events does not produce a final outcome. Instead, the last event in the chain relates back to the beginning event.

You first decide what event will be used as the beginning event. Once that is decided, you list events in order that occur after it. Words are written between events that describe what happens from one event to the next. The last event in a cycle concept map relates back to the beginning event. The number of events in a cycle concept varies, but is usually three or more. Look at the cycle map, as shown in **Figure 6.**

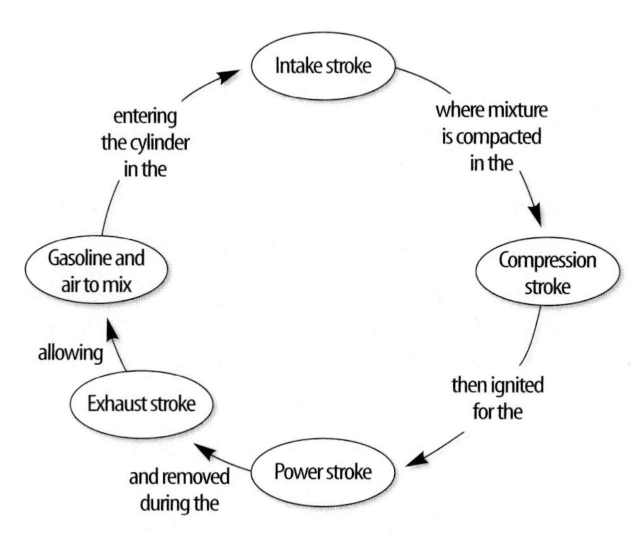

Figure 6
A cycle map shows events that occur in a cycle.

Spider Map A type of concept map that you can use for brainstorming is the spider map. When you have a central idea, you might find you have a jumble of ideas that relate to it but are not necessarily clearly related to each other. The spider map on sound in **Figure 7** shows that if you write these ideas outside the main concept, then you can begin to separate and group unrelated terms so they become more useful.

Figure 7
A spider map allows you to list ideas that relate to a central topic but not necessarily to one another.

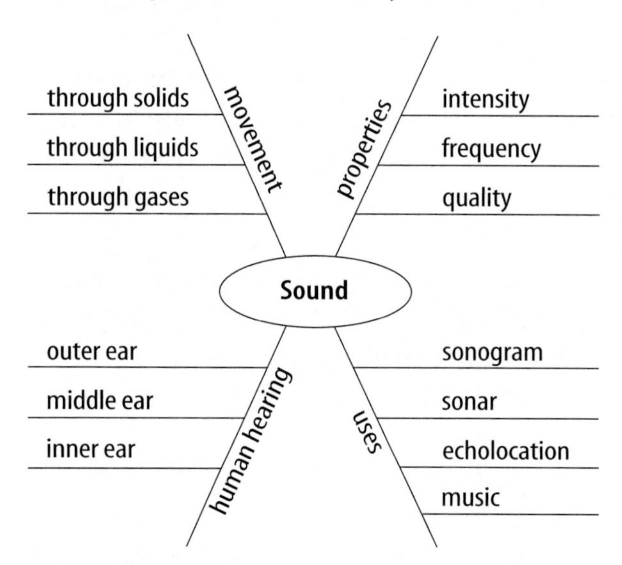

Writing a Paper

You will write papers often when researching science topics or reporting the results of investigations or experiments. Scientists frequently write papers to share their data and conclusions with other scientists and the public. When writing a paper, use these steps.

Step 1 Assemble your data by using graphs, tables, or a concept map. Create an outline.

Step 2 Start with an introduction that contains a clear statement of purpose and what you intend to discuss or prove.

Step 3 Organize the body into paragraphs. Each paragraph should start with a topic sentence, and the remaining sentences in that paragraph should support your point.

Step 4 Position data to help support your points.

Step 5 Summarize the main points and finish with a conclusion statement.

Step 6 Use tables, graphs, charts, and illustrations whenever possible.

Investigating and Experimenting

You might say the work of a scientist is to solve problems. When you decide to find out why your neighbor's hydrangeas produce blue flowers while yours are pink, you are problem solving, too. You might also observe that your neighbor's azaleas are healthier than yours are and decide to see whether differences in the soil explain the differences in these plants.

Scientists use orderly approaches to solve problems. The methods scientists use include identifying a question, making observations, forming a hypothesis, testing a hypothesis, analyzing results, and drawing conclusions.

Scientific investigations involve careful observation under controlled conditions. Such observation of an object or a process can suggest new and interesting questions about it. These questions sometimes lead to the formation of a hypothesis. Scientific investigations are designed to test a hypothesis.

Identifying a Question

The first step in a scientific investigation or experiment is to identify a question to be answered or a problem to be solved. You might be interested in knowing how beams of laser light like the ones in **Figure 8** look the way they do.

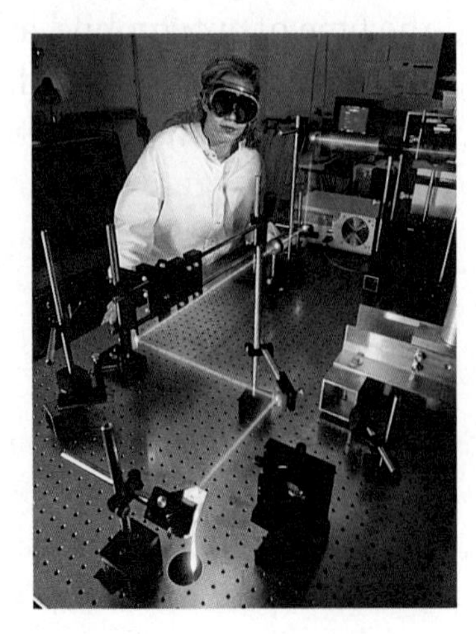

Figure 8
When you see lasers being used for scientific research, you might ask yourself, "Are these lasers different from those that are used for surgery?"

Forming Hypotheses

Hypotheses are based on observations that have been made. A hypothesis is a possible explanation based on previous knowledge and observations.

Perhaps a scientist has observed that certain substances dissolve faster in warm water than in cold. Based on these observations, the scientist can make a statement that he or she can test. The statement is a hypothesis. The hypothesis could be: *A substance dissolves in warm water faster.* A hypothesis has to be something you can test by using an investigation. A testable hypothesis is a valid hypothesis.

Predicting

When you apply a hypothesis to a specific situation, you predict something about that situation. First, you must identify which hypothesis fits the situation you are considering. People use predictions to make everyday decisions. Based on previous observations and experiences, you might form a prediction that if substances dissolve in warm water faster, then heating the water will shorten mixing time for powdered fruit drinks. Someone could use this prediction to save time in preparing a fruit punch for a party.

Testing a Hypothesis

To test a hypothesis, you need a procedure. A procedure is the plan you follow in your experiment. A procedure tells you what materials to use, as well as how and in what order to use them. When you follow a procedure, data are generated that support or do not support the original hypothesis statement.

For example, premium gasoline costs more than regular gasoline. Does premium gasoline increase the efficiency or fuel mileage of your family car? You decide to test the hypothesis: "If premium gasoline is more efficient, then it should increase the fuel mileage of my family's car." Then you write the procedure shown in **Figure 9** for your experiment and generate the data presented in the table below.

Figure 9
A procedure tells you what to do step by step.

Procedure

1. Use regular gasoline for two weeks.
2. Record the number of kilometers between fill-ups and the amount of gasoline used.
3. Switch to premium gasoline for two weeks.
4. Record the number of kilometers between fill-ups and the amount of gasoline used.

Gasoline Data

Type of Gasoline	Kilometers Traveled	Liters Used	Liters per Kilometer
Regular	762	45.34	0.059
Premium	661	42.30	0.064

These data show that premium gasoline is less efficient than regular gasoline in one particular car. It took more gasoline to travel 1 km (0.064) using premium gasoline than it did to travel 1 km using regular gasoline (0.059). This conclusion does not support the hypothesis.

Are all investigations alike? Keep in mind as you perform investigations in science that a hypothesis can be tested in many ways. Not every investigation makes use of all the ways that are described on these pages, and not all hypotheses are tested by investigations. Scientists encounter many variations in the methods that are used when they perform experiments. The skills in this handbook are here for you to use and practice.

Identifying and Manipulating Variables and Controls

In any experiment, it is important to keep everything the same except for the item you are testing. The one factor you change is called the independent variable. The factor that changes as a result of the independent variable is called the dependent variable. Always make sure you have only one independent variable. If you allow more than one, you will not know what causes the changes you observe in the dependent variable. Many experiments also have controls—individual instances or experimental subjects for which the independent variable is not changed. You can then compare the test results to the control results.

For example, in the fuel-mileage experiment, you made everything the same except the type of gasoline that was used. The driver, the type of automobile, and the type of driving did not change. In this way, you could be sure that any mileage differences were caused by the type of fuel—the independent variable. The amount of fuel per kilometer was the dependent variable.

If you could repeat the experiment using several automobiles of the same type on a standard driving track with the same driver, you could make one automobile a control by using regular gasoline over the four-week period.

Collecting Data

Whether you are carrying out an investigation or a short observational experiment, you will collect data, or information. Scientists collect data accurately as numbers and descriptions and organize it in specific ways.

Observing Scientists observe items and events, then record what they see. When they use only words to describe an observation, it is called qualitative data. For example, a scientist might describe the color, texture, or odor of a substance produced in a chemical reaction. Scientists' observations also can describe how much there is of something. These observations use numbers, as well as words, in the description and are called quantitative data. For example, if a sample of the element gold is described as being "shiny and very dense," the data are clearly qualitative. Quantitative data on this sample of gold might include "a mass of 30 g and a density of 19.3 g/cm^3." Quantitative data often are organized into tables. Then, from information in the table, a graph can be drawn. Graphs can reveal relationships that exist in experimental data.

When you make observations in science, you should examine the entire object or situation first, then look carefully for details. If you're looking at an element sample, for instance, check the general color and pattern of the sample before using a hand lens to examine its surface for any smaller details or characteristics. Remember to record accurately everything you see.

Scientists try to make careful and accurate observations. When possible, they use instruments such as microscopes, metric rulers, graduated cylinders, thermometers, and balances. Measurements provide numerical data that can be repeated and checked.

Sampling When working with large numbers of objects or a large population, scientists usually cannot observe or study every one of them. Instead, they use a sample or a portion of the total number. To *sample* is to take a small, representative portion of the objects or organisms of a population for research. By making careful observations or manipulating variables within a portion of a group, information is discovered and conclusions are drawn that might apply to the whole population.

Estimating Scientific work also involves estimating. To *estimate* is to make a judgment about the amount of something or the number of something without measuring every part of an object or counting every member of a population. Scientists first measure or count the amount or number in a small sample. A geologist, for example, might remove a 10-g sample from a large rock that is rich in copper ore, as in **Figure 10.** Then a chemist would determine the percentage of copper by mass and multiply that percentage by the total mass of the rock to estimate the total mass of copper in that large rock.

Figure 10
Determining the percentage of copper by mass in a piece from a large rock that is rich in copper ore, can help estimate the total mass of copper ore in the large rock.

Measuring in SI

The metric system of measurement was developed in 1795. A modern form of the metric system, called the International System, or SI, was adopted in 1960. SI provides standard measurements that all scientists around the world can understand.

The metric system is convenient because unit sizes vary by multiples of 10. When changing from smaller units to larger units, divide by a multiple of 10. When changing from larger units to smaller, multiply by a multiple of 10. To convert millimeters to centimeters, divide the millimeters by 10. To convert 30 mm to centimeters, divide 30 by 10 (30 mm equal 3 cm).

Prefixes are used to name units. Look at the table below for some common metric prefixes and their meanings. Do you see how the prefix *kilo-* attached to the unit *gram* is *kilogram*, or 1,000 g?

Metric Prefixes

Prefix	Symbol	Meaning	
kilo-	k	1,000	thousand
hecto-	h	100	hundred
deka-	da	10	ten
deci-	d	0.1	tenth
centi-	c	0.01	hundredth
milli-	m	0.001	thousandth

Now look at the metric ruler shown in **Figure 11.** The centimeter lines are the long, numbered lines, and the shorter lines are millimeter lines.

When using a metric ruler, line up the 0-cm mark with the end of the object being measured, and read the number of the unit where the object ends, in this instance it would be 4.5 cm.

Figure 11
This metric ruler has centimeter and millimeter divisions.

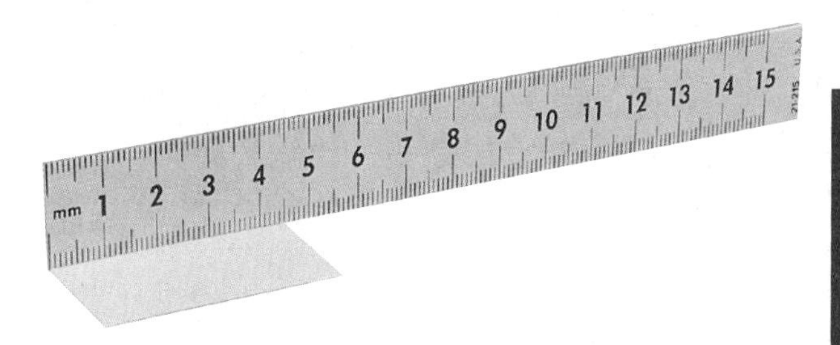

Liquid Volume In some science activities, you will measure liquids. The unit that is used to measure liquids is the liter. A liter has the volume of 1,000 cm^3. The prefix *milli-* means "thousandth (0.001)." A milliliter is one thousandth of 1 L, and 1 L has the volume of 1,000 mL. One milliliter of liquid completely fills a cube measuring 1 cm on each side. Therefore, 1 mL equals 1 cm^3.

You will use beakers and graduated cylinders to measure liquid volume. A graduated cylinder, as illustrated in **Figure 12,** is marked from bottom to top in milliliters. This one contains 79 mL of a liquid.

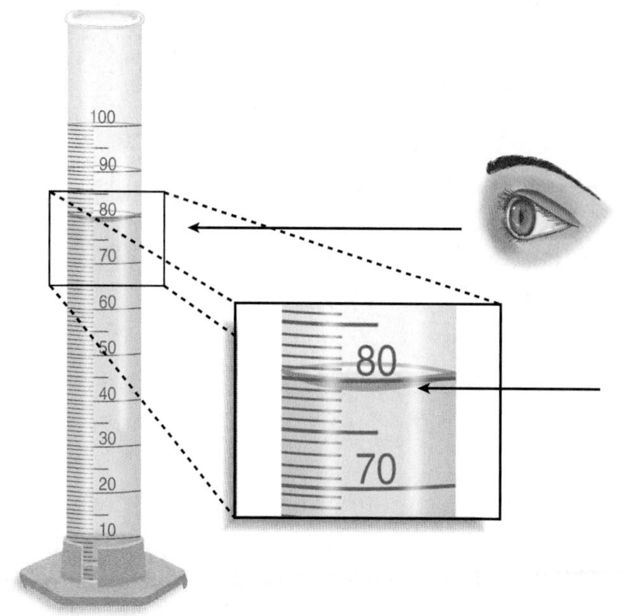

Figure 12
Graduated cylinders measure liquid volume.

Mass Scientists measure mass in grams. You might use a beam balance similar to the one shown in **Figure 13.** The balance has a pan on one side and a set of beams on the other side. Each beam has a rider that slides on the beam.

Before you find the mass of an object, slide all the riders back to the zero point. Check the pointer on the right to make sure it swings an equal distance above and below the zero point. If the swing is unequal, find and turn the adjusting screw until you have an equal swing.

Place an object on the pan. Slide the largest rider along its beam until the pointer drops below zero. Then move it back one notch. Repeat the process on each beam until the pointer swings an equal distance above and below the zero point. Sum the masses on each beam to find the mass of the object. Move all riders back to zero when finished.

Figure 13
A triple beam balance is used to determine the mass of an object.

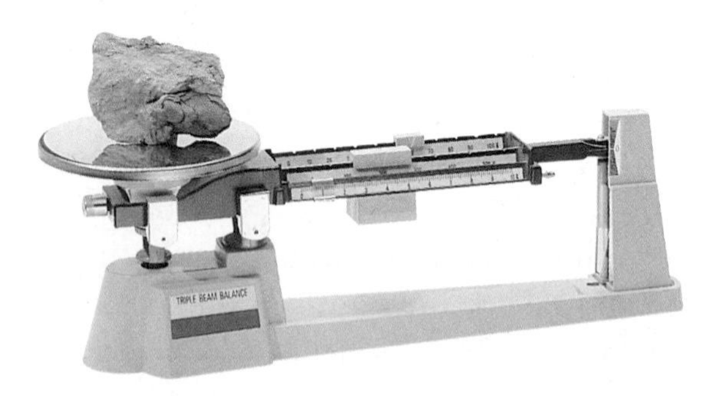

You should never place a hot object on the pan or pour chemicals directly onto the pan. Instead, find the mass of a clean container. Remove the container from the pan, then place the chemicals in the container. Find the mass of the container with the chemicals in it. To find the mass of the chemicals, subtract the mass of the empty container from the mass of the filled container.

Making and Using Tables

Browse through your textbook and you will see tables in the text and in the activities. In a table, data, or information, are arranged so that they are easier to understand. Activity tables help organize the data you collect during an activity so results can be interpreted.

Making Tables To make a table, list the items to be compared in the first column and the characteristics to be compared in the first row. The title should clearly indicate the content of the table, and the column or row heads should tell the reader what information is found in there. The table below lists materials collected for recycling on three weekly pick-up days. The inclusion of kilograms in parentheses also identifies for the reader that the figures are mass units.

Recyclable Materials Collected During Week

Day of Week	Paper (kg)	Aluminum (kg)	Glass (kg)
Monday	5.0	4.0	12.0
Wednesday	4.0	1.0	10.0
Friday	2.5	2.0	10.0

Using Tables How much paper, in kilograms, is being recycled on Wednesday? Locate the column labeled "Paper (kg)" and the row "Wednesday." The information in the box where the column and row intersect is the answer. Did you answer "4.0"? How much aluminum, in kilograms, is being recycled on Friday? If you answered "2.0," you understand how to read the table. How much glass is collected for recycling each week? Locate the column labeled "Glass (kg)" and add the figures for all three rows. If you answered "32.0," then you know how to locate and use the data provided in the table.

Recording Data

To be useful, the data you collect must be recorded carefully. Accuracy is key. A well-thought-out experiment includes a way to record procedures, observations, and results accurately. Data tables are one way to organize and record results. Set up the tables you will need ahead of time so you can record the data right away.

Record information properly and neatly. Never put unidentified data on scraps of paper. Instead, data should be written in a notebook like the one in **Figure 14.** Write in pencil so information isn't lost if your data get wet. At each point in the experiment, record your data and label it. That way, your information will be accurate and you will not have to determine what the figures mean when you look at your notes later.

Figure 14
Record data neatly and clearly so they are easy to understand.

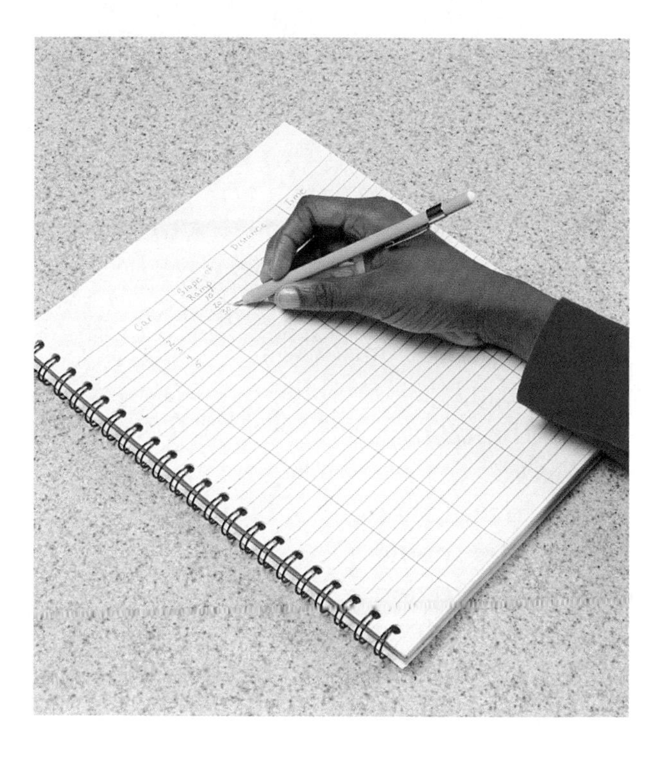

Recording Observations

It is important to record observations accurately and completely. That is why you always should record observations in your notes immediately as you make them. It is easy to miss details or make mistakes when recording results from memory. Do not include your personal thoughts when you record your data. Record only what you observe to eliminate bias. For example, when you record the time required for five students to climb the same set of stairs, you would note which student took the longest time. However, you would not refer to that student's time as "the worst time of all the students in the group."

Making Models

You can organize the observations and other data you collect and record in many ways. Making models is one way to help you better understand the parts of a structure you have been observing or the way a process for which you have been taking various measurements works.

Models often show things that are too large or too small for normal viewing. For example, you normally won't see the inside of an atom. However, you can understand the structure of the atom better by making a three-dimensional model of an atom. The relative sizes, the positions, and the movements of protons, neutrons, and electrons can be explained in words. An atomic model made of a plastic-ball nucleus and pipe-cleaner electron shells can help you visualize how the parts of the atom relate to each other.

Other models can be devised on a computer. Some models, such as those that illustrate the chemical combinations of different elements, are mathematical and are represented by equations.

Making and Using Graphs

After scientists organize data in tables, they might display the data in a graph that shows the relationship of one variable to another. A graph makes interpretation and analysis of data easier. Three types of graphs are the line graph, the bar graph, and the circle graph.

Line Graphs A line graph like in **Figure 15** is used to show the relationship between two variables. The variables being compared go on two axes of the graph. For data from an experiment, the independent variable always goes on the horizontal axis, called the *x*-axis. The dependent variable always goes on the vertical axis, called the *y*-axis. After drawing your axes, label each with a scale. Next, plot the data points.

A data point is the intersection of the recorded value of the dependent variable for each tested value of the independent variable. After all the points are plotted, connect them.

Distance v. Time

Figure 15
This line graph shows the relationship between distance and time during a bicycle ride lasting several hours.

Bar Graphs Bar graphs compare data that do not change continuously. Vertical bars show the relationships among data.

To make a bar graph, set up the *y*-axis as you did for the line graph. Draw vertical bars of equal size from the *x*-axis up to the point on the *y*-axis that represents the value of *x*.

Figure 16
The amount of aluminum collected for recycling during one week can be shown as a bar graph or circle graph.

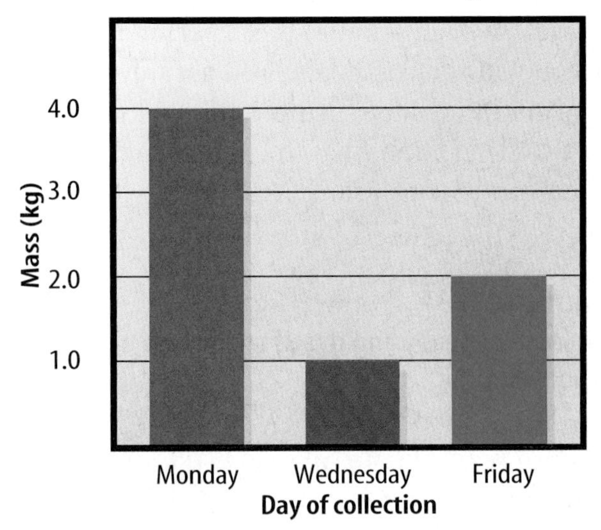

Circle Graphs A circle graph uses a circle divided into sections to display data as parts (fractions or percentages) of a whole. The size of each section corresponds to the fraction or percentage of the data that the section represents. So, the entire circle represents 100 percent, one-half represents 50 percent, one-fifth represents 20 percent, and so on.

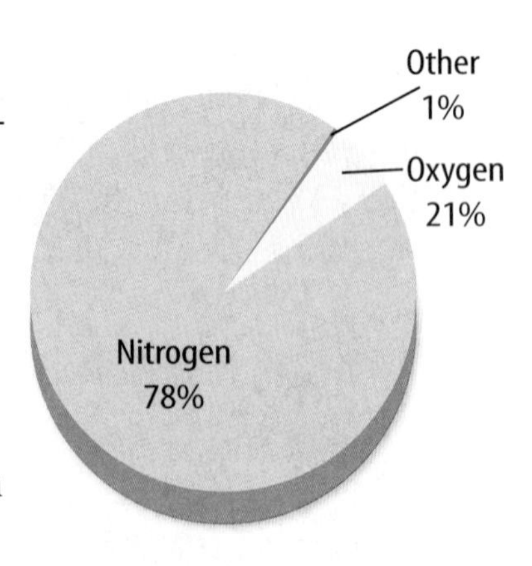

Analyzing and Applying Results

Analyzing Results

To determine the meaning of your observations and investigation results, you will need to look for patterns in the data. You can organize your information in several of the ways that are discussed in this handbook. Then you must think critically to determine what the data mean. Scientists use several approaches when they analyze the data they have collected and recorded. Each approach is useful for identifying specific patterns in the data.

Forming Operational Definitions

An operational definition defines an object by showing how it functions, works, or behaves. Such definitions are written in terms of how an object works or how it can be used; that is, they describe its job or purpose.

For example, a ruler can be defined as a tool that measures the length of an object (how it can be used). A ruler also can be defined as something that contains a series of marks that can be used as a standard when measuring (how it works).

Classifying

Classifying is the process of sorting objects or events into groups based on common features. When classifying, first observe the objects or events to be classified. Then select one feature that is shared by some members in the group but not by all. Place those members that share that feature into a subgroup. You can classify members into smaller and smaller subgroups based on characteristics.

How might you classify a group of chemicals? You might first classify them by state of matter, putting solids, liquids, and gases into separate groups. Within each group, you could then look for another common feature by which to further classify members of the group, such as color or how reactive they are.

Remember that when you classify, you are grouping objects or events for a purpose. For example, classifying chemicals can be the first step in organizing them for storage. Both at home and at school, poisonous or highly reactive chemicals should all be stored in a safe location where they are not easily accessible to small children or animals. Solids, liquids, and gases each have specific storage requirements that may include waterproof, airtight, or pressurized containers. Are the dangerous chemicals in your home stored in the right place? Keep your purpose in mind as you select the features to form groups and subgroups.

Figure 17
Color is one of many characteristics that are used to classify chemicals.

Comparing and Contrasting

Observations can be analyzed by noting the similarities and differences between two or more objects or events that you observe. When you look at objects or events to see how they are similar, you are comparing them. Contrasting is looking for differences in objects or events. The table below compares and contrasts the characteristics of two elements.

Elemental Characteristics		
Element	Aluminum	Gold
Color	silver	gold
Classification	metal	metal
Density (g/cm^3)	2.7	19.3
Melting Point (°C)	660	1064

Recognizing Cause and Effect

Have you ever heard a loud pop right before the power went out and then suggested that an electric transformer probably blew out? If so, you have observed an effect and inferred a cause. The event is the effect, and the reason for the event is the cause.

When scientists are unsure of the cause of a certain event, they design controlled experiments to determine what caused it.

Interpreting Data

The word *interpret* means "to explain the meaning of something." Look at the problem originally being explored in an experiment and figure out what the data show. Identify the control group and the test group so you can see whether or not changes in the independent variable have had an effect. Look for differences in the dependent variable between the control and test groups.

These differences you observe can be qualitative or quantitative. You would be able to describe a qualitative difference using only words, whereas you would measure a quantitative difference and describe it using numbers. If there are differences, the independent variable that is being tested could have had an effect. If no differences are found between the control and test groups, the variable that is being tested apparently had no effect.

For example, suppose that three beakers each contain 100 mL of water. The beakers are placed on hot plates, and two of the hot plates are turned on, but the third is left off for a period of 5 min. Suppose you are then asked to describe any differences in the water in the three beakers. A qualitative difference might be the appearance of bubbles rising to the top in the water that is being heated but no rising bubbles in the unheated water. A quantitative difference might be a difference in the amount of water that is present in the beakers.

Inferring Scientists often make inferences based on their observations. An inference is an attempt to explain, or interpret, observations or to indicate what caused what you observed. An inference is a type of conclusion.

When making an inference, be certain to use accurate data and accurately described observations. Analyze all of the data that you've collected. Then, based on everything you know, explain or interpret what you've observed.

Drawing Conclusions

When scientists have analyzed the data they collected, they proceed to draw conclusions about what the data mean. These conclusions are sometimes stated using words similar to those found in the hypothesis formed earlier in the process.

Conclusions To analyze your data, you must review all of the observations and measurements that you made and recorded. Recheck all data for accuracy. After your data are rechecked and organized, you are almost ready to draw a conclusion such as "salt water boils at a higher temperature than freshwater."

Before you can draw a conclusion, however, you must determine whether the data allow you to come to a conclusion that supports a hypothesis. Sometimes that will be the case, other times it will not.

If your data do not support a hypothesis, it does not mean that the hypothesis is wrong. It means only that the results of the investigation did not support the hypothesis. Maybe the experiment needs to be redesigned, but very likely, some of the initial observations on which the hypothesis was based were incomplete or biased. Perhaps more observation or research is needed to refine the hypothesis.

Avoiding Bias Sometimes drawing a conclusion involves making judgments. When you make a judgment, you form an opinion about what your data mean. It is important to be honest and to avoid reaching a conclusion if no supporting evidence for it exists or if it was based on a small sample. It also is important not to allow any expectations of results to bias your judgments. If possible, it is a good idea to collect additional data. Scientists do this all the time.

For example, the *Hubble Space Telescope* was sent into space in April, 1990, to provide scientists with clearer views of the universe. *Hubble* is the size of a school bus and has a 2.4-m-diameter mirror. *Hubble* helped scientists answer questions about the planet Pluto.

For many years, scientists had only been able to hypothesize about the surface of the planet Pluto. *Hubble* has now provided pictures of Pluto's surface that show a rough texture with light and dark regions on it. This might be the best information about Pluto scientists will have until they are able to send a space probe to it.

Evaluating Others' Data and Conclusions

Sometimes scientists have to use data that they did not collect themselves, or they have to rely on observations and conclusions drawn by other researchers. In cases such as these, the data must be evaluated carefully.

How were the data obtained? How was the investigation done? Was it carried out properly? Has it been duplicated by other researchers? Were they able to follow the exact procedure? Did they come up with the same results? Look at the conclusion, as well. Would you reach the same conclusion from these results? Only when you have confidence in the data of others can you believe it is true and feel comfortable using it.

Communicating

The communication of ideas is an important part of the work of scientists. A discovery that is not reported will not advance the scientific community's understanding or knowledge. Communication among scientists also is important as a way of improving their investigations.

Scientists communicate in many ways, from writing articles in journals and magazines that explain their investigations and experiments, to announcing important discoveries on television and radio, to sharing ideas with colleagues on the Internet or presenting them as lectures.

People who study science rely on computers to record and store data and to analyze results from investigations. Whether you work in a laboratory or just need to write a lab report with tables, good computer skills are a necessity.

Using a Word Processor

Suppose your teacher has assigned a written report. After you've completed your research and decided how you want to write the information, you need to put all that information on paper. The easiest way to do this is with a word processing application on a computer.

A computer application that allows you to type your information, change it as many times as you need to, and then print it out so that it looks neat and clean is called a word processing application. You also can use this type of application to create tables and columns, add bullets or cartoon art to your page, include page numbers, and check your spelling.

Helpful Hints

- If you aren't sure how to do something using your word processing program, look in the help menu. You will find a list of topics there to click on for help. After you locate the help topic you need, just follow the step-by-step instructions you see on your screen.
- Just because you've spell checked your report doesn't mean that the spelling is perfect. The spell check feature can't catch misspelled words that look like other words. If you've accidentally typed *cold* instead of *gold*, the spell checker won't know the difference. Always reread your report to make sure you didn't miss any mistakes.

Figure 18
You can use computer programs to make graphs and tables.

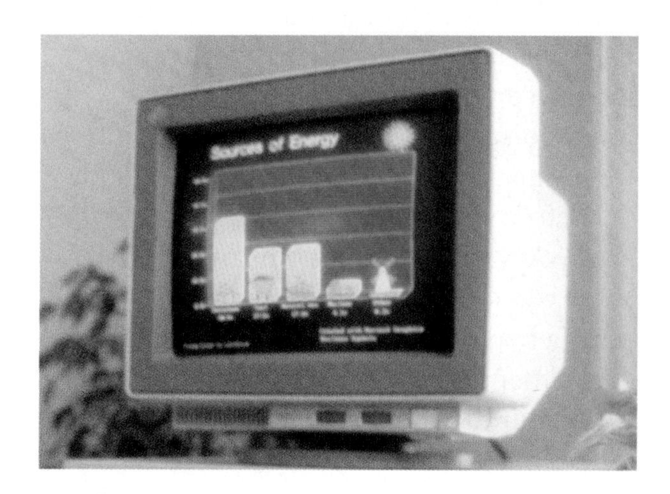

Using a Database

Imagine you're in the middle of a research project, busily gathering facts and information. You soon realize that it's becoming more difficult to organize and keep track of all the information. The tool to use to solve information overload is a database. Just as a file cabinet organizes paper records, a database organizes computer records. However, a database is more powerful than a simple file cabinet because at the click of a mouse, the contents can be reshuffled and reorganized. At computer-quick speeds, databases can sort information by any characteristics and filter data into multiple categories.

Helpful Hints

- Before setting up a database, take some time to learn the features of your database software by practicing with established database software.
- Periodically save your database as you enter data. That way, if something happens such as your computer malfunctions or the power goes off, you won't lose all of your work.

Doing a Database Search

When searching for information in a database, use the following search strategies to get the best results. These are the same search methods used for searching internet databases.

- Place the word *and* between two words in your search if you want the database to look for any entries that have both the words. For example, "gold *and* silver" would give you information that mentions both gold and silver.
- Place the word *or* between two words if you want the database to show entries that have at least one of the words. For example "gold *or* silver" would show you information that mentions either gold or silver.
- Place the word *not* between two words if you want the database to look for entries that have the first word but do not have the second word. For example, "gold *not* jewelry" would show you information that mentions gold but does not mention jewelry.

In summary, databases can be used to store large amounts of information about a particular subject. Databases allow biologists, Earth scientists, and physical scientists to search for information quickly and accurately.

Using an Electronic Spreadsheet

Your science fair experiment has produced lots of numbers. How do you keep track of all the data, and how can you easily work out all the calculations needed? You can use a computer program called a spreadsheet to record data that involve numbers. A spreadsheet is an electronic mathematical worksheet.

Type your data in rows and columns, just as they would look in a data table on a sheet of paper. A spreadsheet uses simple math to do data calculations. For example, you could add, subtract, divide, or multiply any of the values in the spreadsheet by another number. You also could set up a series of math steps you want to apply to the data. If you want to add 12 to all the numbers and then multiply all the numbers by 10, the computer does all the calculations for you in the spreadsheet. Below is an example of a spreadsheet that records test car data.

Helpful Hints

- Before you set up the spreadsheet, identify how you want to organize the data. Include any formulas you will need to use.
- Make sure you have entered the correct data into the correct rows and columns.
- You also can display your results in a graph. Pick the style of graph that best represents the data with which you are working.

Figure 19
A spreadsheet allows you to display large amounts of data and do calculations automatically.

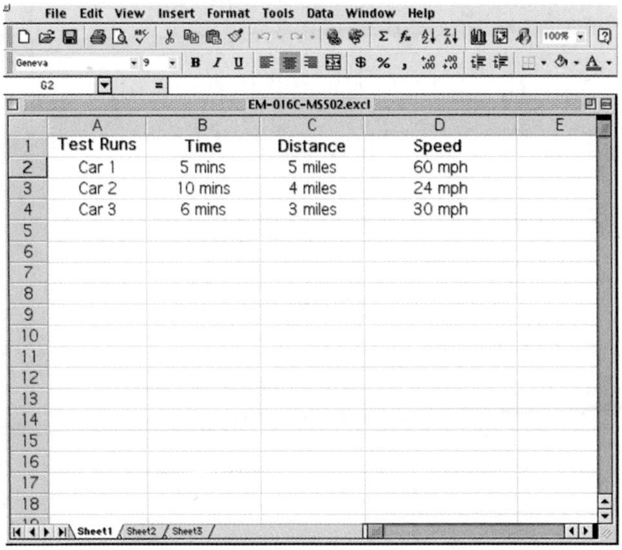

File Edit View Insert Format Tools Data Window Help

EM-016C-MSS02.excl

	A	B	C	D	E
1	Test Runs	Time	Distance	Speed	
2	Car 1	5 mins	5 miles	60 mph	
3	Car 2	10 mins	4 miles	24 mph	
4	Car 3	6 mins	3 miles	30 mph	

Sheet1 / Sheet2 / Sheet3

Using a Computerized Card Catalog

When you have a report or paper to research, you probably go to the library. To find the information you need in the library, you might have to use a computerized card catalog. This type of card catalog allows you to search for information by subject, by title, or by author. The computer then will display all the holdings the library has on the subject, title, or author requested.

A library's holdings can include books, magazines, databases, videos, and audio materials. When you have chosen something from this list, the computer will show whether an item is available and where in the library to find it.

Helpful Hints

- Remember that you can use the computer to search by subject, author, or title. If you know a book's author but not the title, you can search for all the books the library has by that author.
- When searching by subject, it's often most helpful to narrow your search by using specific search terms, such as *and, or,* and *not.* If you don't find enough sources this way, you can broaden your search.
- Pay attention to the type of materials found in your search. If you need a book, you can eliminate any videos or other resources that come up in your search.
- Knowing how your library is arranged can save you a lot of time. If you need help, the librarian will show you where certain types of materials are kept and how to find specific holdings.

Using Graphics Software

Are you having trouble finding that exact piece of art you're looking for? Do you have a picture in your mind of what you want but can't seem to find the right graphic to represent your ideas? To solve these problems, you can use graphics software. Graphics software allows you to create and change images and diagrams in almost unlimited ways. Typical uses for graphics software include arranging clip art, changing scanned images, and constructing pictures from scratch. Most graphics software applications work in similar ways. They use the same basic tools and functions. Once you master one graphics application, you can use other graphics applications.

Figure 20
Graphics software can use your data to draw bar graphs.

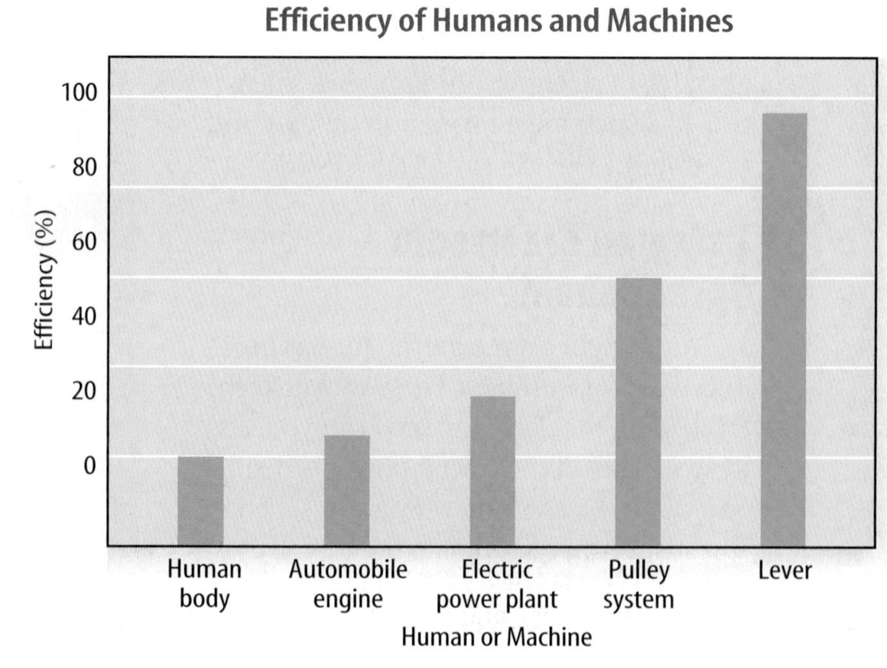

Figure 21
Graphics software can use your data to draw circle graphs.

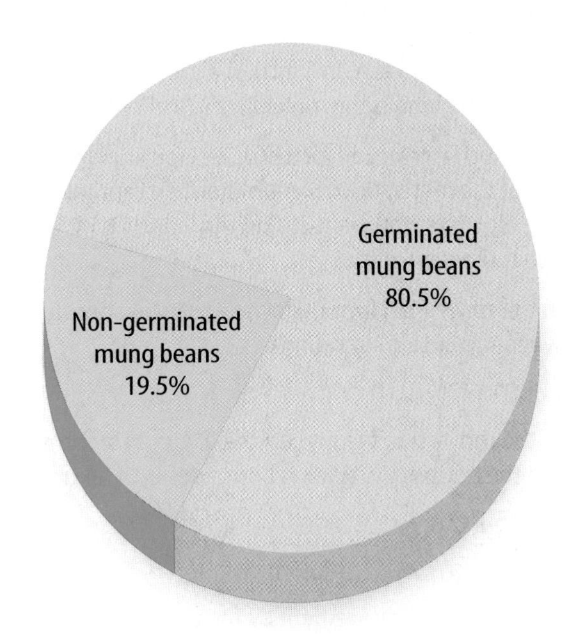

Helpful Hints

- As with any method of drawing, the more you practice using the graphics software, the better your results will be.
- Start by using the software to manipulate existing drawings. Once you master this, making your own illustrations will be easier.
- Clip art is available on CD-ROMs and the Internet. With these resources, finding a piece of clip art to suit your purposes is simple.
- As you work on a drawing, save it often.

Developing Multimedia Presentations

It's your turn—you have to present your science report to the entire class. How do you do it? You can use many different sources of information to get the class excited about your presentation. Posters, videos, photographs, sound, computers, and the Internet can help show your ideas.

First, determine what important points you want to make in your presentation. Then, write an outline of what materials and types of media would best illustrate those points. Maybe you could start with an outline on an overhead projector, then show a video, followed by something from the Internet or a slide show accompanied by music or recorded voices. You might choose to use a presentation builder computer application that can combine all these elements into one presentation. Make sure the presentation is well constructed to make the most impact on the audience.

Figure 22
Multimedia presentations use many types of print and electronic materials.

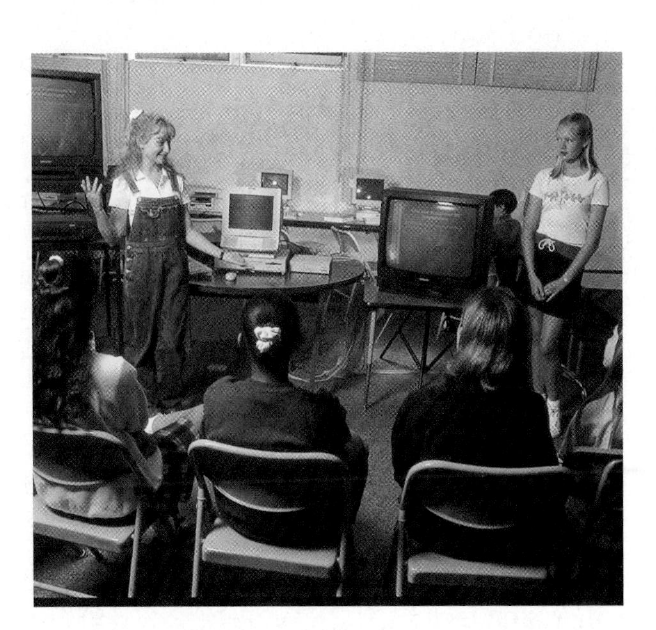

Helpful Hints

- Carefully consider what media will best communicate the point you are trying to make.
- Make sure you know how to use any equipment you will be using in your presentation.
- Practice the presentation several times.
- If possible, set up all of the equipment ahead of time. Make sure everything is working correctly.

Use this Math Skill Handbook to help solve problems you are given in this text. You might find it useful to review topics in this Math Skill Handbook first.

Converting Units

In science, quantities such as length, mass, and time sometimes are measured using different units. Suppose you want to know how many miles are in 12.7 km.

Conversion factors are used to change from one unit of measure to another. A conversion factor is a ratio that is equal to one. For example, there are 1,000 mL in 1 L, so 1,000 mL equals 1 L, or:

$$1{,}000 \text{ mL} = 1 \text{ L}$$

If both sides are divided by 1 L, this equation becomes:

$$\frac{1{,}000 \text{ mL}}{1 \text{ L}} = 1$$

The **ratio** on the left side of this equation is equal to one and is a conversion factor. You can make another conversion factor by dividing both sides of the top equation by 1,000 mL:

$$1 = \frac{1 \text{ L}}{1{,}000 \text{ mL}}$$

To **convert units,** you multiply by the appropriate conversion factor. For example, how many milliliters are in 1.255 L? To convert 1.255 L to milliliters, multiply 1.255 L by a conversion factor.

Use the **conversion factor** with new units (mL) in the numerator and the old units (L) in the denominator.

$$1.255 \cancel{\text{L}} \times \frac{1{,}000 \text{ mL}}{1 \cancel{\text{L}}} = 1{,}255 \text{ mL}$$

The unit L divides in this equation, just as if it were a number.

Example 1 There are 2.54 cm in 1 inch. If a meterstick has a length of 100 cm, how long is the meterstick in inches?

Step 1 Decide which conversion factor to use. You know the length of the meterstick in centimeters, so centimeters are the old units. You want to find the length in inches, so inch is the new unit.

Step 2 Form the conversion factor. Start with the relationship between the old and new units.

$$2.54 \text{ cm} = 1 \text{ inch}$$

Step 3 Form the conversion factor with the old unit (centimeter) on the bottom by dividing both sides by 2.54 cm.

$$1 = \frac{2.54 \text{ cm}}{2.54 \text{ cm}} = \frac{1 \text{ inch}}{2.54 \text{ cm}}$$

Step 4 Multiply the old measurement by the conversion factor.

$$100 \cancel{\text{cm}} \times \frac{1 \text{ inch}}{2.54 \cancel{\text{cm}}} = 39.37 \text{ inches}$$

The meterstick is 39.37 inches long.

Example 2 There are 365 days in one year. If a person is 14 years old, what is his or her age in days? (Ignore leap years)

Step 1 Decide which conversion factor to use. You want to convert years to days.

Step 2 Form the conversion factor. Start with the relation between the old and new units.

$$1 \text{ year} = 365 \text{ days}$$

Step 3 Form the conversion factor with the old unit (year) on the bottom by dividing both sides by 1 year.

$$1 = \frac{1 \text{ year}}{1 \text{ year}} = \frac{365 \text{ days}}{1 \text{ year}}$$

Step 4 Multiply the old measurement by the conversion factor:

$$14 \cancel{\text{years}} \times \frac{365 \text{ days}}{1 \cancel{\text{year}}} = 5{,}110 \text{ days}$$

The person's age is 5,110 days.

Practice Problem A book has a mass of 2.31 kg. If there are 1,000 g in 1 kg, what is the mass of the book in grams?

Using Fractions

A **fraction** is a number that compares a part to the whole. For example, in the fraction $\frac{2}{3}$, the 2 represents the part and the 3 represents the whole. In the fraction $\frac{2}{3}$, the top number, 2, is called the numerator. The bottom number, 3, is called the denominator.

Sometimes fractions are not written in their simplest form. To determine a fraction's **simplest form,** you must find the greatest common factor (GCF) of the numerator and denominator. The greatest common factor is the largest common factor of all the factors the two numbers have in common.

For example, because the number 3 divides into 12 and 30 evenly, it is a common factor of 12 and 30. However, because the number 6 is the largest number that evenly divides into 12 and 30, it is the **greatest common factor.**

After you find the greatest common factor, you can write a fraction in its simplest form. Divide both the numerator and the denominator by the greatest common factor. The number that results is the fraction in its **simplest form.**

Example Twelve of the 20 chemicals used in the science lab are in powder form. What fraction of the chemicals used in the lab are in powder form?

Step 1 Write the fraction.

$$\frac{\text{part}}{\text{whole}} = \frac{12}{20}$$

Step 2 To find the GCF of the numerator and denominator, list all of the factors of each number.

Factors of 12: 1, 2, 3, 4, 6, 12 (the numbers that divide evenly into 12)

Factors of 20: 1, 2, 4, 5, 10, 20 (the numbers that divide evenly into 20)

Step 3 List the common factors.

1, 2, 4.

Step 4 Choose the greatest factor in the list of common factors.

The GCF of 12 and 20 is 4.

Step 5 Divide the numerator and denominator by the GCF.

$$\frac{12 \div 4}{20 \div 4} = \frac{3}{5}$$

In the lab, $\frac{3}{5}$ of the chemicals are in powder form.

Practice Problem There are 90 rides at an amusement park. Of those rides, 66 have a height restriction. What fraction of the rides has a height restriction? Write the fraction in simplest form.

Calculating Ratios

A **ratio** is a comparison of two numbers by division.

Ratios can be written 3 to 5 or 3:5. Ratios also can be written as fractions, such as $\frac{3}{5}$. Ratios, like fractions, can be written in simplest form. Recall that a fraction is in **simplest form** when the greatest common factor (GCF) of the numerator and denominator is 1.

Example A chemical solution contains 40 g of salt and 64 g of baking soda. What is the ratio of salt to baking soda as a fraction in simplest form?

Step 1 Write the ratio as a fraction. $\frac{\text{salt}}{\text{baking soda}} = \frac{40}{64}$

Step 2 Express the fraction in simplest form. The GCF of 40 and 64 is 8.

$$\frac{40}{64} = \frac{40 \div 8}{64 \div 8} = \frac{5}{8}$$

The ratio of salt to baking soda in the solution is $\frac{5}{8}$.

Practice Problem Two metal rods measure 100 cm and 144 cm in length. What is the ratio of their lengths in simplest fraction form? 25/36

Using Decimals

A **decimal** is a fraction with a denominator of 10, 100, 1,000, or another power of 10. For example, 0.854 is the same as the fraction $\frac{854}{1,000}$.

In a decimal, the decimal point separates the ones place and the tenths place. For example, 0.27 means twenty-seven hundredths, or $\frac{27}{100}$, where 27 is the **number of units** out of 100 units. Any fraction can be written as a decimal using division.

Example Write $\frac{5}{8}$ as a decimal.

Step 1 Write a division problem with the numerator, 5, as the dividend and the denominator, 8, as the divisor. Write 5 as 5.000.

Step 2 Solve the problem.

$$\begin{array}{r} 0.625 \\ 8\overline{)5.000} \\ \underline{48} \\ 20 \\ \underline{16} \\ 40 \\ \underline{40} \\ 0 \end{array}$$

Therefore, $\frac{5}{8} = 0.625$.

Practice Problem Write $\frac{19}{25}$ as a decimal. 0.76

Using Percentages

The word *percent* means "out of one hundred." A **percent** is a ratio that compares a number to 100. Suppose you read that 77 percent of Earth's surface is covered by water. That is the same as reading that the fraction of Earth's surface covered by water is $\frac{77}{100}$. To express a fraction as a percent, first find an equivalent decimal for the fraction. Then, multiply the decimal by 100 and add the percent symbol. For example, $\frac{1}{2} = 1 \div 2 = 0.5$. Then $0.5 \times 100 = 50 = 50\%$.

Example Express $\frac{13}{20}$ as a percent.

Step 1 Find the equivalent decimal for the fraction.

$$\begin{array}{r} 0.65 \\ 20\overline{)13.00} \\ \underline{120} \\ 100 \\ \underline{100} \\ 0 \end{array}$$

Step 2 Rewrite the fraction $\frac{13}{20}$ as 0.65.

Step 3 Multiply 0.65 by 100 and add the % sign.

$$0.65 \cdot 100 = 65 = 65\%$$

So, $\frac{13}{20} = 65\%$.

Practice Problem In one year, 73 of 365 days were rainy in one city. What percent of the days in that city were rainy? 20%

Using Precision and Significant Digits

When you make a **measurement,** the value you record depends on the precision of the measuring instrument. When adding or subtracting numbers with different precision, the answer is rounded to the smallest number of decimal places of any number in the sum or difference. When multiplying or dividing, the answer is rounded to the smallest number of significant figures of any number being multiplied or divided. When counting the number of **significant figures,** all digits are counted except zeros at the end of a number with no decimal such as 2,500, and zeros at the beginning of a decimal such as 0.03020.

Example The lengths 5.28 and 5.2 are measured in meters. Find the sum of these lengths and report the sum using the least precise measurement.

Step 1 Find the sum.

5.28 m	2 digits after the decimal
+ 5.2 m	1 digit after the decimal
10.48 m	

Step 2 Round to one digit after the decimal because the least number of digits after the decimal of the numbers being added is 1.

The sum is 10.5 m.

Practice Problem Multiply the numbers in the example using the rule for multiplying and dividing. Report the answer with the correct number of significant figures. 27.5 m^2

Skill Handbooks

Solving One-Step Equations

An **equation** is a statement that two things are equal. For example, $A = B$ is an equation that states that A is equal to B.

Sometimes one side of the equation will contain a **variable** whose value is not known. In the equation $3x = 12$, the variable is x.

The equation is solved when the variable is replaced with a value that makes both sides of the equation equal to each other. For example, the solution of the equation $3x = 12$ is $x = 4$. If the x is replaced with 4, then the equation becomes $3 \cdot 4 = 12$, or $12 = 12$.

To solve an equation such as $8x = 40$, divide both sides of the equation by the number that multiplies the variable.

$$8x = 40$$
$$\frac{8x}{8} = \frac{40}{8}$$
$$x = 5$$

You can check your answer by replacing the variable with your solution and seeing if both sides of the equation are the same.

$$8x = 8 \cdot 5 = 40$$

The left and right sides of the equation are the same, so $x = 5$ is the solution.

Sometimes an equation is written in this way: $a = bc$. This also is called a **formula.** The letters can be replaced by numbers, but the numbers must still make both sides of the equation the same.

Example 1 Solve the equation $10x = 35$.

Step 1 Find the solution by dividing each side of the equation by 10.

$$10x = 35 \qquad \frac{10x}{10} = \frac{35}{10} \qquad x = 3.5$$

Step 2 Check the solution.

$$10x = 35 \qquad 10 \times 3.5 = 35 \qquad 35 = 35$$

Both sides of the equation are equal, so $x = 3.5$ is the solution to the equation.

Example 2 In the formula $a = bc$, find the value of c if $a = 20$ and $b = 2$.

Step 1 Rearrange the formula so the unknown value is by itself on one side of the equation by dividing both sides by b.

$$a = bc$$
$$\frac{a}{b} = \frac{bc}{b}$$
$$\frac{a}{b} = c$$

Step 2 Replace the variables a and b with the values that are given.

$$\frac{a}{b} = c$$
$$\frac{20}{2} = c$$
$$10 = c$$

Step 3 Check the solution.

$$a = bc$$
$$20 = 2 \times 10$$
$$20 = 20$$

Both sides of the equation are equal, so $c = 10$ is the solution when $a = 20$ and $b = 2$.

Practice Problem In the formula $h = gd$, find the value of d if $g = 12.3$ and $h = 17.4$.

Using Proportions

A **proportion** is an equation that shows that two ratios are equivalent. The ratios $\frac{2}{4}$ and $\frac{5}{10}$ are equivalent, so they can be written as $\frac{2}{4} = \frac{5}{10}$. This equation is an example of a proportion.

When two ratios form a proportion, the **cross products** are equal. To find the cross products in the proportion $\frac{2}{4} = \frac{5}{10}$, multiply the 2 and the 10, and the 4 and the 5. Therefore **$2 \cdot 10 = 4 \cdot 5$, or $20 = 20$.**

Because you know that both proportions are equal, you can use cross products to find a missing term in a proportion. This is known as **solving the proportion.** Solving a proportion is similar to solving an equation.

Example The heights of a tree and a pole are proportional to the lengths of their shadows. The tree casts a shadow of 24 m at the same time that a 6-m pole casts a shadow of 4 m. What is the height of the tree?

Step 1 Write a proportion.

$$\frac{\text{height of tree}}{\text{height of pole}} = \frac{\text{length of tree's shadow}}{\text{length of pole's shadow}}$$

Step 2 Substitute the known values into the proportion. Let h represent the unknown value, the height of the tree.

$$\frac{h}{6} = \frac{24}{4}$$

Step 3 Find the cross products.

$$h \cdot 4 = 6 \cdot 24$$

Step 4 Simplify the equation.

$$4h = 144$$

Step 5 Divide each side by 4.

$$\frac{4h}{4} = \frac{144}{4}$$

$$h = 36$$

The height of the tree is 36 m.

Practice Problem The ratios of the weights of two objects on the Moon and on Earth are in proportion. A rock weighing 3 N on the Moon weighs 18 N on Earth. How much would a rock that weighs 5 N on the Moon weigh on Earth?

Using Statistics

Statistics is the branch of mathematics that deals with collecting, analyzing, and presenting data. In statistics, there are three common ways to summarize the data with a single number—the mean, the median, and the mode.

The **mean** of a set of data is the arithmetic average. It is found by adding the numbers in the data set and dividing by the number of items in the set.

The **median** is the middle number in a set of data when the data are arranged in numerical order. If there were an even number of data points, the median would be the mean of the two middle numbers.

The **mode** of a set of data is the number or item that appears most often.

Another number that often is used to describe a set of data is the range. The **range** is the difference between the largest number and the smallest number in a set of data.

A **frequency table** shows how many times each piece of data occurs, usually in a survey. The frequency table below shows the results of a student survey on favorite color.

Color	Tally	Frequency
red	||||	4
blue	~~||||~~	5
black	||	2
green	|||	3
purple	~~||||~~ ||	7
yellow	~~||||~~ |	6

Based on the frequency table data, which color is the favorite?

Example The speeds (in m/s) for a race car during five different time trials are 39, 37, 44, 36, and 44.

To find the mean:

Step 1 Find the sum of the numbers.

$$39 + 37 + 44 + 36 + 44 = 200$$

Step 2 Divide the sum by the number of items, which is 5.

$$200 \div 5 = 40$$

The mean measure is 40 m/s.

To find the median:

Step 1 Arrange the measures from least to greatest.

36, 37, <u>39</u>, 44, 44

Step 2 Determine the middle measure.

The median measure is 39 m/s.

To find the mode:

Step 1 Group the numbers that are the same together.

44, 44, 36, 37, 39

Step 2 Determine the number that occurs most in the set.

<u>44, 44</u>, 36, 37, 39

The mode measure is 44 m/s.

To find the range:

Step 1 Arrange the measures from largest to smallest.

44, 44, 39, 37, 36

Step 2 Determine the largest and smallest measures in the set.

<u>44</u>, 44, 39, 37, <u>36</u>

Step 3 Find the difference between the largest and smallest measures.

$$44 - 36 = 8$$

The range is 8 m/s.

Practice Problem Find the mean, median, mode, and range for the data set 8, 4, 12, 8, 11, 14, 16.
mean, 10; median, 11; mode, 8; range, 12

List of Formulas

Chapter 2 Motion

$$\text{Speed (m/s)} = \frac{\text{distance (m)}}{\text{time (s)}}$$

$$s = \frac{d}{t}$$

$$\text{Acceleration (m/s}^2\text{)} = \frac{\text{change in speed (m/s)}}{\text{time (s)}}$$

$$a = \frac{\text{final speed} - \text{initial speed}}{\text{time}}$$

$$a = \frac{v_f - v_i}{t}$$

Chapter 3 Forces

$$\text{Acceleration (m/s}^2\text{)} = \frac{\text{net force (N)}}{\text{mass (kg)}}$$

$$a = \frac{f}{m}$$

$$\text{Weight (N)} = \text{mass (kg)} \times \text{acceleration due to gravity (m/s}^2\text{)}$$

$$W = m \times g$$

$$\text{Momentum (kg m/s)} = \text{mass (kg)} \times \text{speed (m/s)}$$

$$p = m \times v$$

Chapter 4 Energy

$$\text{Kinetic energy (J)} = \frac{1}{2}\text{mass (kg)} \times \text{velocity (m/s)} \times \text{velocity (m/s)}$$

$$KE = \frac{1}{2} \times m \times v^2$$

$$\text{Gravitational potential energy (J)} = \text{weight (N)} \times \text{height (m)}$$

$$GPE = W \times h$$

$$\text{Gravitational potential energy (J)} = \text{mass (kg)} \times \text{gravitational acceleration (m/s}^2\text{)} \times \text{height (m)}$$

$$GPE = m \times g \times h$$

Chapter 5 Work and Machines

$$\text{Work (J)} = \text{force (N)} \times \text{distance (m)}$$

$$W = F \times d$$

$$\text{Power (W)} = \frac{\text{work (J)}}{\text{time (s)}}$$

$$P = \frac{W}{t}$$

Mechanical advantage = resistance force (N)/effort force (N)

$$MA = \frac{F_r}{F_e}$$

$$\text{efficiency} = 100\% \times \frac{\text{output work (J)}}{\text{input work (J)}}$$

$$= \frac{W_{out}}{W_{in}} \times 100\%$$

Chapter 6 Thermal Energy

Change in thermal energy (J) =
mass (kg) × change in temperature (K) × specific heat (J/kg · K)

$$Q = m \times (T_{final} - T_{initial}) \times C$$

Chapter 7 Solids, Liquids, and Gases

$$\text{Pressure (N/m}^2\text{)} = \frac{\text{Force (N)}}{\text{area (m}^2\text{)}}$$

$$P = \frac{F}{A}$$

Chapter 17 Electricity

Voltage (V) = current (A) × resistance (ohms)

$$V = I \times R$$

Electrical power (W) = current (A) × Voltage (V)

$$P = I \times V$$

Electrical energy used (kWh) = electric power(kW) × time (h)

$$E = P \times t$$

Chapter 21 Waves

Wave speed (m/s) = frequency (1/s) × wavelength (m)

$$V = f \times \lambda$$

Practice Problems

Chapter 2

1. It took you 6.5 h to drive 550 km. What was your speed?
2. You are riding in a train that is traveling at a speed of 120 km/h. How long will it take to travel 950 km?
3. A car goes from rest to a speed of 90 km/h in 10 s. What is the car's acceleration in m/s^2?
4. A cart rolling at a speed of 10 m/s comes to a stop in 2 s. What is the cart's acceleration?

Chapter 3

1. A book with a mass of 1 kg is sliding on a table. If the frictional force on the book is 5 N, calculate the book's acceleration. Is it speeding up or slowing down?
2. What is the weight of a person with a mass of 80 kg?
3. A car with a mass of 1,200 kg has a speed of 30 m/s. What is the car's momentum?

Chapter 4

1. A car with a mass of 900 kg is traveling at a speed of 25 m/s. What is the kinetic energy of the car in joules?
2. What is the gravitational potential energy of a diver with a mass of 60 kg who is 10 m above the water?
3. If your weight is 500 N, and you are standing on a floor that is 20 m above the ground, what is your gravitational potential energy?

Chapter 5

1. By applying a force of 50 N, a pulley system can lift a box with a mass of 20 kg. What is the mechanical advantage of the pulley system?
2. A person pushes a box up a ramp that is 3 m long, and 1 m high. If the box has a mass of 20 kg, and the person pushes with a force of 80 N, what is the efficiency of the ramp?
3. A first-class lever has a mechanical advantage of 5. How large would a force need to be to lift a rock with a mass of 100 kg?
4. If a person has a mass of 50 kg, how much power is used to climb a flight of stairs in 5 s if the stairs are 4 m high?

Chapter 6

1. How much heat is needed to raise the temperature of 100 g of water by 50 K, if the specific heat of water is 4,184 J/kg·K?
2. A sample of an unknown metal has a mass of 0.5 kg. Adding 1,985 J of heat to the metal raises its temperature by 10 K. What is the specific heat of the metal?

Chapter 7

1. A skater has a weight of 500 N. The skate blades are in contact with the ice over an area of 0.001 m^2. What is the pressure exerted on the ice by the skater?
2. A piston applies a pressure of 5,000 N/m^2. If the piston has a surface area of 0.1 m^2, how much force can the piston apply?

Chapter 17

1. What is the resistance of a lightbulb that draws 0.5 amp of current when plugged into an outlet of voltage 120 V?
2. How much current flows through a 100-W lightbulb that is plugged into a 120-V outlet?
3. 8 amps of current flow through a hair dryer connected to a 120-V outlet. How much electrical power does the hair dryer use?
4. Compare the electrical energy that is used by a 100-W lightbulb that burns for 10 h, and a 1,200-W hair dryer that is used for 15 min.

Chapter 21

1. A sound wave has a wavelength of 50 m and a frequency of 22 cycles per second. What is the speed of the sound wave?
2. A tidal wave, or tsunami, travels across the ocean at a speed of 500 km/h. If the distance between wave crests is 200 km, what is the frequency of the waves?

Safety in the Science Classroom

1. Always obtain your teacher's permission to begin an investigation.
2. Study the procedure. If you have questions, ask your teacher. Be sure you understand any safety symbols shown on the page.
3. Use the safety equipment provided for you. Goggles and a safety apron should be worn during most investigations.
4. Always slant test tubes away from yourself and others when heating them or adding substances to them.
5. Never eat or drink in the lab, and never use lab glassware as food or drink containers. Never inhale chemicals. Do not taste any substances or draw any material into a tube with your mouth.
6. Report any spill, accident, or injury, no matter how small, immediately to your teacher; then follow his or her instructions.
7. Know the location and proper use of the fire extinguisher, safety shower, fire blanket, first aid kit, and fire alarm.
8. Keep all materials away from open flames. Tie back long hair and tie down loose clothing.
9. If your clothing should catch fire, smother it with the fire blanket, or get under a safety shower. NEVER RUN.
10. If a fire should occur, turn off the gas; then leave the room according to established procedures.

Follow these procedures as you clean up your work area

1. Turn off the water and gas. Disconnect electrical devices.
2. Clean all pieces of equipment and return all materials to their proper places.
3. Dispose of chemicals and other materials as directed by your teacher. Place broken glass and solid substances in the proper containers. Make sure never to discard materials in the sink.
4. Clean your work area. Wash your hands thoroughly after working in the laboratory.

First Aid

Injury	Safe Response ALWAYS NOTIFY YOUR TEACHER IMMEDIATELY
Burns	Apply cold water.
Cuts and Bruises	Stop any bleeding by applying direct pressure. Cover cuts with a clean dressing. Apply ice packs or cold compresses to bruises.
Fainting	Leave the person lying down. Loosen any tight clothing and keep crowds away.
Foreign Matter in Eye	Flush with plenty of water. Use eyewash bottle or fountain.
Poisoning	Note the suspected poisoning agent.
Any Spills on Skin	Flush with large amounts of water or use safety shower.

Reference Handbook B

SI—Metric/English, English/Metric Conversions

	When you want to convert:	To:	Multiply by:
Length	inches	centimeters	2.54
	centimeters	inches	0.39
	yards	meters	0.91
	meters	yards	1.09
	miles	kilometers	1.61
	kilometers	miles	0.62
Mass and Weight*	ounces	grams	28.35
	grams	ounces	0.04
	pounds	kilograms	0.45
	kilograms	pounds	2.2
	tons (short)	tonnes (metric tons)	0.91
	tonnes (metric tons)	tons (short)	1.10
	pounds	newtons	4.45
	newtons	pounds	0.22
Volume	cubic inches	cubic centimeters	16.39
	cubic centimeters	cubic inches	0.06
	liters	quarts	1.06
	quarts	liters	0.95
	gallons	liters	3.78
Area	square inches	square centimeters	6.45
	square centimeters	square inches	0.16
	square yards	square meters	0.83
	square meters	square yards	1.19
	square miles	square kilometers	2.59
	square kilometers	square miles	0.39
	hectares	acres	2.47
	acres	hectares	0.40
Temperature	To convert °Celsius to °Fahrenheit	°C × 9/5 + 32	
	To convert °Fahrenheit to °Celsius	5/9 (°F − 32)	

*Weight is measured in standard Earth gravity.

PERIODIC TABLE OF THE ELEMENTS

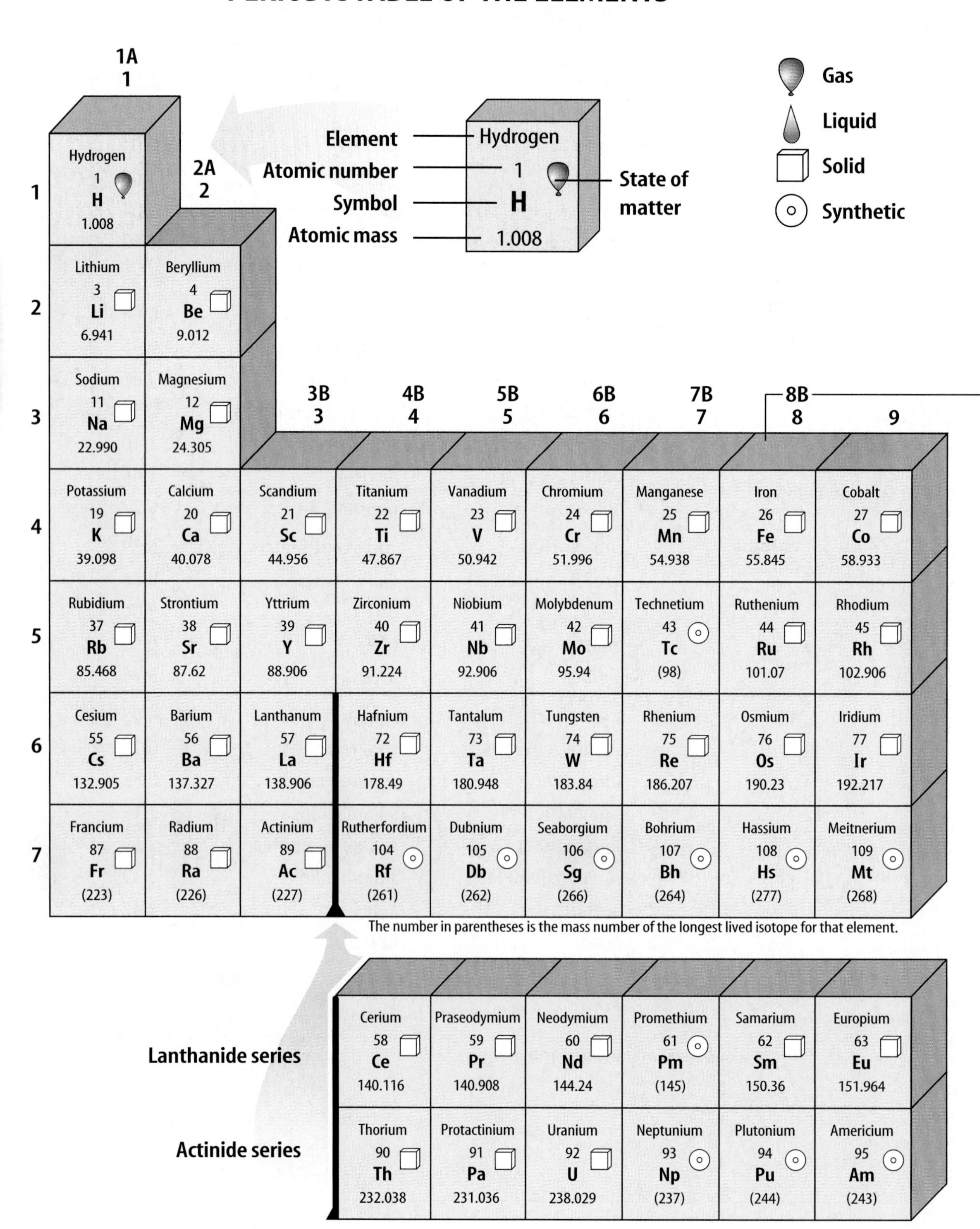

	1A 1	2A 2	3B 3	4B 4	5B 5	6B 6	7B 7	8	9
1	Hydrogen 1 H 1.008								
2	Lithium 3 Li 6.941	Beryllium 4 Be 9.012							
3	Sodium 11 Na 22.990	Magnesium 12 Mg 24.305							
4	Potassium 19 K 39.098	Calcium 20 Ca 40.078	Scandium 21 Sc 44.956	Titanium 22 Ti 47.867	Vanadium 23 V 50.942	Chromium 24 Cr 51.996	Manganese 25 Mn 54.938	Iron 26 Fe 55.845	Cobalt 27 Co 58.933
5	Rubidium 37 Rb 85.468	Strontium 38 Sr 87.62	Yttrium 39 Y 88.906	Zirconium 40 Zr 91.224	Niobium 41 Nb 92.906	Molybdenum 42 Mo 95.94	Technetium 43 Tc (98)	Ruthenium 44 Ru 101.07	Rhodium 45 Rh 102.906
6	Cesium 55 Cs 132.905	Barium 56 Ba 137.327	Lanthanum 57 La 138.906	Hafnium 72 Hf 178.49	Tantalum 73 Ta 180.948	Tungsten 74 W 183.84	Rhenium 75 Re 186.207	Osmium 76 Os 190.23	Iridium 77 Ir 192.217
7	Francium 87 Fr (223)	Radium 88 Ra (226)	Actinium 89 Ac (227)	Rutherfordium 104 Rf (261)	Dubnium 105 Db (262)	Seaborgium 106 Sg (266)	Bohrium 107 Bh (264)	Hassium 108 Hs (277)	Meitnerium 109 Mt (268)

The number in parentheses is the mass number of the longest lived isotope for that element.

Lanthanide series	Cerium 58 Ce 140.116	Praseodymium 59 Pr 140.908	Neodymium 60 Nd 144.24	Promethium 61 Pm (145)	Samarium 62 Sm 150.36	Europium 63 Eu 151.964
Actinide series	Thorium 90 Th 232.038	Protactinium 91 Pa 231.036	Uranium 92 U 238.029	Neptunium 93 Np (237)	Plutonium 94 Pu (244)	Americium 95 Am (243)

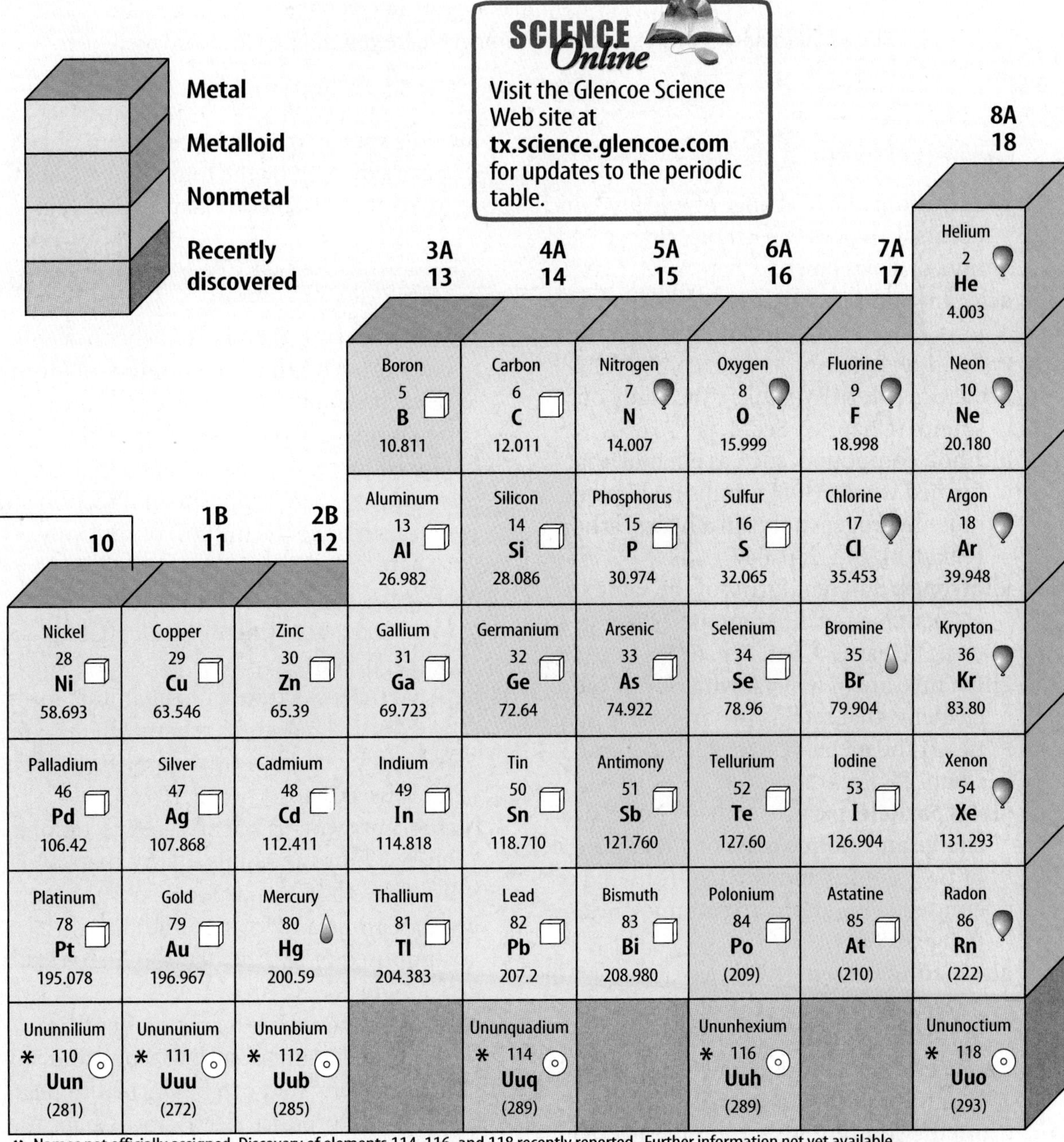

10	1B 11	2B 12	3A 13	4A 14	5A 15	6A 16	7A 17	8A 18
								Helium 2 He 4.003
			Boron 5 B 10.811	Carbon 6 C 12.011	Nitrogen 7 N 14.007	Oxygen 8 O 15.999	Fluorine 9 F 18.998	Neon 10 Ne 20.180
			Aluminum 13 Al 26.982	Silicon 14 Si 28.086	Phosphorus 15 P 30.974	Sulfur 16 S 32.065	Chlorine 17 Cl 35.453	Argon 18 Ar 39.948
Nickel 28 Ni 58.693	Copper 29 Cu 63.546	Zinc 30 Zn 65.39	Gallium 31 Ga 69.723	Germanium 32 Ge 72.64	Arsenic 33 As 74.922	Selenium 34 Se 78.96	Bromine 35 Br 79.904	Krypton 36 Kr 83.80
Palladium 46 Pd 106.42	Silver 47 Ag 107.868	Cadmium 48 Cd 112.411	Indium 49 In 114.818	Tin 50 Sn 118.710	Antimony 51 Sb 121.760	Tellurium 52 Te 127.60	Iodine 53 I 126.904	Xenon 54 Xe 131.293
Platinum 78 Pt 195.078	Gold 79 Au 196.967	Mercury 80 Hg 200.59	Thallium 81 Tl 204.383	Lead 82 Pb 207.2	Bismuth 83 Bi 208.980	Polonium 84 Po (209)	Astatine 85 At (210)	Radon 86 Rn (222)
Ununnilium * 110 Uun (281)	Unununium * 111 Uuu (272)	Ununbium * 112 Uub (285)		Ununquadium * 114 Uuq (289)		Ununhexium * 116 Uuh (289)		Ununoctium * 118 Uuo (293)

* Names not officially assigned. Discovery of elements 114, 116, and 118 recently reported. Further information not yet available.

Gadolinium 64 Gd 157.25	Terbium 65 Tb 158.925	Dysprosium 66 Dy 162.50	Holmium 67 Ho 164.930	Erbium 68 Er 167.259	Thulium 69 Tm 168.934	Ytterbium 70 Yb 173.04	Lutetium 71 Lu 174.967
Curium 96 Cm (247)	Berkelium 97 Bk (247)	Californium 98 Cf (251)	Einsteinium 99 Es (252)	Fermium 100 Fm (257)	Mendelevium 101 Md (258)	Nobelium 102 No (259)	Lawrencium 103 Lr (262)

This glossary defines each key term that appears in bold type in the text. It also shows the chapter, section, and page number where you can find the word used.

A

acceleration: rate of change of velocity, which occurs if an object speeds up, changes direction, or slows down. (Chap. 2, Sec. 2, p. 47)

acid: any substance that produces hydrogen ions, H+, in a water solution. (Chap. 25, Sec. 1, p. 766)

acoustics (uh KEW stihks): the study of sound. (Chap. 12, Sec. 4, p. 375)

alcohol: compound, such as ethanol, that is formed when −OH groups replace one or more hydrogen atoms in a hydrocarbon. (Chap. 21, Sec. 2, p. 647)

allotropes: different forms of the same element having different molecular structures. (Chap. 20, Sec. 3, p. 623)

alloy: mixture of a metal with one or more elements that retains the original properties of the metal. (Chap. 22, Sec. 1, p. 672; Chap. 23, Sec. 1, p. 707)

alpha particle: particle with a +2 electrical charge and atomic mass of 4 that is made of two protons and two neutrons and is emitted from a decaying atomic nucleus. (Chap. 9, Sec. 2, p. 263)

alternating current (AC): electric current that reverses its direction regularly and is used to run appliances. (Chap. 8, Sec. 3, p. 244)

amplitude: a measure of the energy carried by a wave. (Chap. 11, Sec. 2, p. 336)

aromatic compound: compound that contains the benzene ring structure and can have a pleasant or unpleasant odor and flavor. (Chap. 21, Sec. 2, p. 645)

atom: the smallest piece of matter that still retains the property of the element. (Chap. 18, Sec. 1, p. 545)

atomic number: number of protons in an atom's nucleus. (Chap. 18, Sec. 2, p. 551)

average atomic mass: weighted-average mass of the mixture of an element's isotopes. (Chap. 18, Sec. 2, p. 553)

average speed: total distance that an object travels divided by the total time it takes to travel that distance. (Chap. 2, Sec. 1, p. 41)

B

balanced chemical equation: chemical equation with the same number of atoms of each element on both sides of the equation. (Chap. 24, Sec. 2, p. 743)

balanced forces forces on a body that are equal in size and opposite in direction and do not change the motion of the body. (Chap. 2, Sec. 3, p. 53)

base: any substance that forms hydroxide ions, OH^-, in a water solution. (Chap. 25, Sec. 1, p. 768)

beta particle: electron with an atomic mass of 0 and a –1 electrical charge that is emitted at high speed from a decaying atomic nucleus. (Chap. 9, Sec. 2, p. 265)

bias: occurs when a scientist's expectations change how the results of an experiment are viewed. (Chap. 1, Sec. 1, p. 10)

binary compound: compound that is composed of two elements. (Chap. 19, Sec. 3, p. 587)

biomass: renewable organic matter from plants and animals, such as wood and animal manure, that can be burned to heat water and produce electricity. (Chap. 10, Sec. 3, p. 310)

boiling point: temperature at which the pressure of the atmosphere is equal to the pressure of a liquid's vapor, and gas molecules can escape the attractive force between the molecules. (Chap. 16, Sec. 1, p. 491)

bubble chamber: a type of tool that uses superheated liquid to detect and monitor the paths of nuclear particles. (Chap. 9, Sec. 3, p. 269)

buffer: solution containing ions that react with added acids or bases and minimize their effects on pH. (Chap. 25, Sec. 2, p. 775)

buoyancy: ability of a fluid to exert an upward force on an object immersed in the fluid. (Chap. 16, Sec. 2, p. 497)

C

carbohydrates: group of biological compounds, such as sugars and starches, with twice as many hydrogen atoms as oxygen atoms. (Chap. 21, Sec. 4, p. 659)

carrier wave: specific electromagnetic wave frequency that a radio station is assigned. (Chap. 13, Sec. 3, p. 403)

catalyst: substance that speeds up a chemical reaction without being permanently changed itself. (Chap. 24, Sec. 4, p. 754)

cathode-ray tube: sealed vacuum tube that uses one or more beams of electrons that strike tiny colored rectangles on a screen and produce full-color images on a TV. (Chap. 13, Sec. 3, p. 406)

centripetal acceleration: acceleration of an object toward the center of a curved or circular path. (Chap. 3, Sec. 2, p. 81)

centripetal force: a force directed toward the center of the circle for an object moving in circular motion. (Chap. 3, Sec. 2, p. 81)

ceramics: versatile materials with customizable properties; produced by a process in which an object is molded and then heated to high temperatures, increasing its density. (Chap. 22, Sec. 2, p. 678)

chain reaction: ongoing series of fission reactions. (Chap. 9, Sec. 4, p. 274)

charging by contact: process of transferring charge between objects by touching or rubbing. (Chap. 7, Sec. 1, p. 197)

charging by induction: process of transferring charge between objects by bringing a charged object near a neutral object. (Chap. 7, Sec. 1, p. 198)

chemical bond: force that holds atoms together in a compound. (Chap. 19, Sec. 1, p. 578)

chemical change: change of one substance into a new substance. (Chap. 17, Sec. 2, p. 530)

chemical equation: shorthand method to describe chemical reactions using chemical formulas and other symbols. (Chap. 24, Sec. 1, p. 740)

chemical formula: chemical shorthand that uses symbols to tell what elements are in a compound and their ratios. (Chap. 19, Sec. 1, p. 575)

chemically stable: describes an atom whose outer energy level is completely filled with all the electrons allowed in that level. (Chap. 19, Sec. 1, p. 576)

chemical potential energy: energy stored in chemical bonds. (Chap. 4, Sec. 1, p. 103)

chemical property: any characteristic of a substance, such as flammability, that indicates whether it can undergo a certain chemical change. (Chap. 17, Sec. 2, p. 529)

chemical reaction: process in which one or more substances are changed into new substances. (Chap. 24, Sec. 1, p. 738)

circuit: closed conducting loop through which an electric current can flow. (Chap. 7, Sec. 2, p. 203)

cloud chamber: a type of tool that uses water or ethanol vapor to detect alpha and beta particle radiation. (Chap. 9, Sec. 3, p. 268)

cochlea (KAHK lee uh): spiral-shaped, fluid-filled structure in the inner ear that converts sounds waves to nerve impulses. (Chap. 12, Sec. 1, p. 362)

coefficient: number in a chemical equation that represents the number of units of each substance taking part in a chemical reaction. (Chap. 24, Sec. 1, p. 741)

coherent light: light of a single wavelength that travels with its crests and troughs aligned in the same direction. (Chap. 14, Sec. 3, p. 434)

colloid (KAHL oyd): heterogeneous mixture whose particles never settle. (Chap. 17, Sec. 1, p. 522)

composite: mixture of two materials, one of which is embedded in the other. (Chap. 22, Sec. 3, p. 689)

compound: substance formed from two or more elements in which the exact combination and proportion of elements is always the same. (Chap. 17, Sec. 1, p. 520)

compound machine: combination of two or more simple machines. (Chap. 5, Sec. 3, p. 146)

compressional wave: a type of wave where the matter in the medium moves back and forth in the same direction that the wave travels; has compressions and rarefactions. (Chap. 11, Sec. 1, p. 328)

concave lens: diverges light rays to form reduced, upright, virtual images; is thinner in the middle and thicker at the edges and is usually used in combination with other lenses. (Chap. 15, Sec. 2, p. 462)

concave mirror: reflects light from a surface that curves inward and can magnify objects or create beams of light. (Chap. 15, Sec. 1, p. 454)

conduction: transfer of energy through matter by colliding particles; takes place because the particles are in constant motion. (Chap. 6, Sec. 2, p. 164)

conductivity (kahn duk TIHV ut ee): property of metals and alloys that allows them to be good conductors of heat and electricity. (Chap. 22, Sec. 1, p. 673)

conductor: material, such as copper wire, through which an excess of electrons can move easily. (Chap. 7, Sec. 1, p. 196)

constant: in an experiment, a variable that does not change when other variables change. (Chap. 1, Sec. 1, p. 9)

control: standard used for comparison of test results in an experiment. (Chap. 1, Sec. 1, p. 9)

convection: transfer of energy by the motion of heated particles in a fluid. (Chap. 6, Sec. 2, p. 165)

convex lens: converges light rays, is thicker in the middle than at the edges, and can form real or virtual images. (Chap. 15, Sec. 2, p. 460)

convex mirror: reflects light from a surface that curves outward and forms a reduced, upright, virtual image. (Chap. 15, Sec. 1, p. 457)

cornea: transparent covering on the eyeball through which light enters. (Chap. 15, Sec. 2, p. 463)

covalent bond: attraction formed between atoms when they share electrons. (Chap. 19, Sec. 2, p. 583)

crest: highest point of a transverse wave. (Chap. 11, Sec. 2, p. 332)

critical mass: amount of fissionable material required so that each fission reaction produces approximately one more fission reaction. (Chap. 9, Sec. 4, p. 274)

D

decibel (DES uh bel): unit on the scale for sound intensity; abbreviated dB. (Chap. 12, Sec. 2, p. 365)

decomposition reaction: chemical reaction in which one substance breaks down into two or more substances. (Chap. 24, Sec. 3, p. 747)

density: mass per unit volume of a material. (Chap. 1, Sec. 2, p. 19)

deoxyribonucleic (dee AHK sih ri boh noo klay ihk) **acid:** a type of essential biological compound that codes and stores genetic information, controls the production of RNA, and is found in the nuclei of cells. (Chap. 21, Sec. 4, p. 658)

dependent variable: factor that changes as the independent variable changes. (Chap. 1, Sec. 1, p. 9)

derived unit: unit, such as cubic meter, m^3, that is obtained by combining different SI units. (Chap. 1, Sec. 2, p. 19)

diatomic molecule: a molecule that consists of two atoms of the same element. (Chap. 20, Sec. 2, p. 617)

diffraction: describes the bending of waves around a barrier; can also occur when waves pass through a narrow opening. (Chap. 11, Sec. 3, p. 342)

diffusion: spreading of particles throughout a given volume until they are uniformly distributed. (Chap. 16, Sec. 1, p. 491)

direct current (DC): electric current that flows through a wire in only one direction. (Chap. 8, Sec. 3, p. 244)

displacement: distance and direction of an object's change in position from the starting point. (Chap. 2, Sec. 1, p. 39)

dissociation: process in which ions are separated from a crystal and drawn into solution. (Chap. 23, Sec. 4, p. 724)

distance: how far an object moves. (Chap. 2, Sec. 1, p. 39)

distillation: process than can separate two substances in a mixture by evaporating a liquid and recondensing its vapor. (Chap. 17, Sec. 2, p. 529)

doping: process of adding impurities to a semiconductor to increase its conductivity. (Chap. 22, Sec. 2, p. 682)

Doppler effect: change in pitch or frequency that occurs when the source of a sound wave is moving relative to an observer. (Chap. 12, Sec. 2, p. 367)

double-displacement reaction: chemical reaction that produces a precipitate, water, or a gas when two ionic compounds in solution are combined. (Chap. 24, Sec. 3, p. 749)

ductile: ability of metals to be drawn into wires. (Chap. 20, Sec. 1, p. 608)

ductility (duk TIHL uh tee): ability of metals or alloys to be pulled into wires. (Chap. 22, Sec. 1, p. 673)

E

eardrum: tough membrane in the outer ear that is about 0.1 mm thick and passes sound vibrations into the middle ear. (Chap. 12, Sec. 1, p. 361)

echolocation: process in which objects are located by emitting sounds and interpreting sound waves that are reflected back. (Chap. 12, Sec. 4, p. 375)

efficiency: measure of how much of the work put into a machine is changed into work done by a machine. (Chap. 5, Sec. 2, p. 136)

effort force: force exerted on a machine that is used to do work. (Chap. 5, Sec. 2, p. 134)

elastic potential energy: energy stored by things that stretch or twist or compress. (Chap. 4, Sec. 1, p. 103)

electric current: flow of electric charge through a wire or any conductor; measured in amperes, A, in a circuit. (Chap. 7, Sec. 2, p. 203)

electric motor: device that converts electrical energy to mechanical energy to do work; basically has a power supply, a permanent magnet, and an electromagnet that can rotate. (Chap. 8, Sec. 2, p. 237)

electrical power: rate at which electrical energy is converted to another form of energy; expressed in watts, W. (Chap. 7, Sec. 3, p. 212)

electrolyte: compound that breaks apart in water, forming charged particles (ions) that can conduct electricity. (Chap. 23, Sec. 4, p. 723)

electromagnet (ih lek troh MAG nut): temporary magnet made by placing an iron core inside a current-carrying coil of wire and whose strength can be increased by adding more turns to the wire loop or by increasing the amount of current passing through the wire. (Chap. 8, Sec. 2, p. 234)

electromagnetic induction (ih lek troh mag NET ihk • ihn DUK shun): process in which electric current is produced by moving a wire loop through a magnetic field or by moving a magnet through a loop of wire. (Chap. 8, Sec. 3, p. 240)

electromagnetic waves: waves created by vibrating electric charges; can travel through a vacuum or through matter and can have a wide variety of frequencies and wavelengths. (Chap. 13, Sec. 1, p. 390)

electron: negatively charged particles surrounding the center of an atom. (Chap. 18, Sec. 1, p. 545)

electron cloud: area around the nucleus of an atom where the atom's electrons are most likely to be found. (Chap. 18, Sec. 1, p. 549)

electron dot diagram: uses the symbol for an element and dots representing the number of electrons in the element's outer energy level. (Chap. 18, Sec. 3, p. 560)

element: substance with atoms that are all alike. (Chap. 17, Sec. 1, p. 518)

endergonic reaction: chemical reaction; requires energy to proceed. (Chap. 24, Sec. 4, p. 753)

endothermic reaction: chemical reaction that requires heat energy to proceed. (Chap. 24, Sec. 4, p. 753)

exergonic reaction: chemical reaction that releases some form of energy, such as light or heat. (Chap. 24, Sec. 4, p. 752)

exothermic reaction: chemical reaction in which energy is primarily given off in the form of heat. (Chap. 24, Sec. 4, p. 752)

experiment: organized procedure for testing a hypothesis that tests the effect of one thing on another under controlled conditions. (Chap. 1, Sec. 1, p. 8)

F

fluorescent light: light that results when ultraviolet radiation produced inside a fluorescent bulb causes the phosphor coating inside the bulb to glow. (Chap 14, Sec. 3, p. 431)

focal length: distance from the center of a lens or mirror to the focal point. (Chap. 15, Sec. 1, p. 454)

focal point: a single point on the optical axis of a concave mirror or convex lens where light rays pass through. (Chap. 15, Sec. 1, p. 454)

force: push or pull that one body exerts on another. (Chap. 2, Sec. 3, p. 52)

fossil fuels: nonrenewable energy resources, such as oil, natural gas, and coal, that are formed from the decayed remains of ancient plants and animals; can be burned to supply energy and generate electricity. (Chap. 10, Sec. 1, p. 291)

frequency: measures how many wavelengths pass a fixed point each second and is expressed in hertz, Hz. (Chap. 11, Sec. 2, p. 333; Chap. 13, Sec. 1, p. 394)

friction: force that opposes motion between two touching surfaces. (Chap. 3, Sec. 1, p. 70)

G

galvanometer (gal vuh NAHM ut ur): instrument that uses an electromagnet to measure electric current. (Chap. 8, Sec. 2, p. 236)

gamma ray: electromagnetic wave with no mass and no charge that travels at the speed of light and is usually emitted with alpha or beta particles. (Chap. 9, Sec. 2, p. 265; Chap. 13, Sec. 2, p. 401)

Geiger counter: device that measures radioactivity by producing an electric current when radiation is present; emits a clicking sound or flashing light. (Chap. 9, Sec. 3, p. 270)

generator: device that converts mechanical energy to electrical energy by rotating a wire coil through the magnetic field of a permanent magnet. (Chap. 8, Sec. 3, p. 240)

geothermal energy: thermal energy contained in hot magma; can be converted by a power plant into electrical energy. (Chap. 10, Sec. 3, p. 309)

Global Positioning System (GPS): uses satellites, ground monitoring systems, and receivers to help people determine their exact location at or above Earth's surface. (Chap. 13, Sec. 3, p. 409)

graph: visual display of information or data that can provide a quick way to communicate a lot of information clearly in a small amount of space. (Chap. 1, Sec. 3, p. 22)

gravitational potential energy: energy stored by things attracted to each other by the force of gravity. (Chap. 4, Sec. 1, p. 104)

group: vertical column in the periodic table. (Chap. 18, Sec. 3, p. 558)

H

half-life: amount of time it takes for half the nuclei in a sample of a radioactive isotope to decay. (Chap. 9, Sec. 2, p. 266)

heat: thermal energy that flows from a warmer material to a cooler material; is measured in joules, J. (Chap. 6, Sec. 1, p. 160)

heat engine: device that converts thermal energy produced by combustion into mechanical energy. (Chap. 6, Sec. 3, p. 176)

heat mover: device, such as a refrigerator, that removes thermal energy from one place and transfers it to another place at a different temperature. (Chap. 6, Sec. 3, p. 177)

heat of fusion: amount of energy required to change a substance from the solid phase to the liquid phase. (Chap. 16, Sec. 1, p. 490)

heat of vaporization: amount of energy required for liquid particles to escape the attractive forces within the liquid or the energy required to change from a liquid to a gas. (Chap. 16, Sec. 1, p. 490)

heterogeneous (het uh ruh JEE nee us) **mixture:** mixture, such as concrete or a dry soup mix, in which different materials are unevenly distributed and are easily identified. (Chap. 17, Sec. 1, p. 521)

holography: technique that produces a complete photographic image of a three-dimensional object. (Chap. 14, Sec. 4, p. 437)

homogeneous (hoh moh JEE nee us) **mixture:** solid, liquid, or gas that contains two or more substances blended evenly throughout. (Chap. 17, Sec. 1, p. 522)

hydrate: compound that has water chemically attached to its ions and written into its chemical formula. (Chap. 19, Sec. 3, p. 592)

hydrocarbon: saturated or unsaturated compound containing only carbon and hydrogen atoms. (Chap. 21, Sec. 1, p. 641)

hydroelectricity: electricity produced from the energy of moving water. (Chap. 10, Sec. 3, p. 307)

hydronium ions (hi DROH nee um • I ahnz): H_3O^+ ions, which form when an acid dissolves in water and H^+ ions interact with water. (Chap. 25, Sec. 1, p. 766)

hypothesis: educated guess using what you know and what you observe. (Chap. 1, Sec. 1, p. 8)

I

incandescent light: light produced by heating a piece of metal, usually tungsten, until it glows. (Chap. 14, Sec. 3, p. 430)

inclined plane: simple machine that consists of a sloping surface, such as a ramp, that reduces the amount of force needed to lift something by increasing the distance over which the force is applied. (Chap. 5, Sec. 3, p. 144)

incoherent light: light that contains more than one wavelength and does not travel with its crests and troughs aligned in the same direction. (Chap. 14, Sec. 3, p. 434)

independent variable: factor that, as it changes, affects the measure of another variable. (Chap. 1, Sec. 1, p. 9)

index of refraction: property of a material indicating how much light slows down when traveling in the material. (Chap. 14, Sec. 1, p. 422)

indicator: organic compound that changes color in acids and bases. (Chap. 25, Sec. 1, p. 766)

inertia (ihn UR shuh)**:** resistance of an object to a change in its motion. (Chap. 2, Sec. 3, p. 54)

infrared wave: electromagnetic wave that has a frequency slightly lower than visible light, transmits thermal energy, and is used in television remote controls and bar-code scanners. (Chap. 13, Sec. 2, p. 398)

inhibitor: substance that slows down a chemical reaction or prevents it from occurring by combining with a reactant. (Chap. 24, Sec. 4, p. 754)

instantaneous speed: speed of an object at a given point in time, which is constant for an object moving with constant speed and is different at each point in time for an object that is slowing down or speeding up. (Chap. 2, Sec. 1, p. 41)

insulator: material that doesn't allow electrons to move through it easily. (Chap. 7, Sec. 1, p. 197)

insulators: materials, such as fleece or fiberglass, that do not allow heat to move easily through them. (Chap. 6, Sec. 2, p. 169)

integrated circuit: tiny chip of semiconductor material that can contain millions of transistors, diodes, and other components. (Chap. 22, Sec. 2, p. 682)

intensity: amount of energy a sound wave carries, which can be measured in decibels, dB. (Chap. 12, Sec. 2, p. 364)

interference: occurs when two or more waves overlap and combine to form a new wave; can form a larger wave (constructive interference) or a smaller wave (destructive interference). (Chap. 11, Sec. 3, p. 344)

internal combustion engine: a type of heat engine that burns fuel inside the engine. (Chap. 6, Sec. 3, p. 176)

ion: charged particle that has more or fewer electrons than protons. (Chap. 19, Sec. 2, p. 580) (Chap. 23, Sec. 4, p. 723)

ionic bond: attraction formed between oppositely charged ions in an ionic compound. (Chap. 19, Sec. 2, p. 582)

ionization: process in which electrolytes dissolve in water and separate into charged particles. (Chap. 23, Sec. 4, p. 723)

isomers: compounds with identical chemical formulas but different molecular structures and shapes, which can affect physical properties such as melting points. (Chap. 21, Sec. 1, p. 643)

isotopes: atoms of the same element that have different numbers of neutrons. (Chap. 18, Sec. 2, p. 552)

J

joule (JEWL)**:** SI unit of energy. (Chap. 4, Sec. 1, p. 102)

K

kilowatt-hour: unit of electrical energy, which is 1000 W of power used for 1 h. (Chap. 7, Sec. 3, p. 215)

kinetic energy: energy in the form of motion; depends on the mass and velocity of the object. (Chap. 4, Sec. 1, p. 102)

kinetic theory: explanation of the behavior of molecules in matter; states that all matter is made of constantly moving particles that collide without losing energy. (Chap. 16, Sec. 1, p. 488)

L

law of conservation of charge: states that charge can be transferred from one object to another but cannot be created or destroyed. (Chap. 7, Sec. 1, p. 195)

law of conservation of energy: states that energy can never be created or destroyed. (Chap. 4, Sec. 2, p. 111)

law of conservation of mass: states that the mass of all substances that are present before a chemical change equals the mass of all the substances that are remaining after the change. (Chap. 17, Sec. 2, p. 533)

law of gravitation: states that any two masses exert an attractive force on each other, the amount of which depends on the mass of the two objects and the distance between them. (Chap. 3, Sec. 2, p. 75)

lever: simple machine made from a bar that is free to pivot about a fixed point. (Chap. 5, Sec. 3, p. 138)

lipids: group of biological compounds that contains the same elements as carbohydrates but in different arrangements and combinations, and includes saturated and unsaturated fats and oils. (Chap. 21, Sec. 4, p. 660)

loudness: describes how humans perceive sound intensity. (Chap. 12, Sec. 2, p. 365)

luster: property of metals and alloys that describes having a shiny appearance or reflecting light. (Chap. 22, Sec. 1, p. 673)

M

machine: device that makes doing work easier by increasing the force applied to an object, by changing the direction of an applied force, or by increasing the distance over which a force can be applied. (Chap. 5, Sec. 2, p. 132)

magnetic domain: group of atoms having aligned magnetic poles. (Chap. 8, Sec. 1, p. 231)

magnetic pole: region where a magnet's magnetic force is strongest; like poles repel and opposite poles attract. (Chap. 8, Sec. 1, p. 227)

magnetism: the properties and interactions of magnets. (Chap. 8, Sec. 1, p. 226)

malleability (mal yuh BIHL yt ee)**:** ability of metals and alloys to be rolled into thin sheets or hammered. (Chap. 22, Sec. 1, p. 673)

malleable (MAH lee uh bul): ability of metals to be hammered or rolled into thin sheets. (Chap. 20, Sec. 1, p. 608)

mass: amount of matter in an object. (Chap. 1, Sec. 2, p. 19)

mass number: sum of the number of protons and neutrons in an atom's nucleus. (Chap. 18, Sec. 2, p. 551)

mechanical advantage (MA): number of times a machine multiplies the effort force applied to it. (Chap. 5, Sec. 2, p. 136)

mechanical energy: sum of potential and kinetic energy in a system. (Chap. 4, Sec. 2, p. 108)

medium: any material—a solid, liquid, gas, or combination of these—that a wave transfers energy through. (Chap. 11, Sec. 1, p. 327)

melting point: temperature at which a solid begins to liquefy. (Chap. 16, Sec. 1, p. 490)

metal: element that typically is a hard, shiny solid, is malleable, and is a good conductor of heat and electricity. (Chap. 20, Sec. 1, p. 608)

metallic bonding: occurs because electrons move freely among a metal's positively charged ions and explains properties such as ductility and the ability to conduct electricity. (Chap. 20, Sec. 1, p. 609)

metalloid: element that shares some properties with metals and some with nonmetals. (Chap. 20, Sec. 3, p. 622)

microscope: uses convex lenses to magnify small, close objects. (Chap. 15, Sec. 3, p. 471)

microwave: radio wave with a wavelength of less than 1 m. (Chap. 13, Sec. 2, p. 397)

mirage: image of a distant object that results when air at ground level is much warmer or cooler than the air layers above it, which makes the image refract and appear at a different location from where it actually is. (Chap. 14, Sec. 1, p. 424)

model: can be used to represent an idea, object, or event that is too big, too small, too complex, or too dangerous to observe or test directly. (Chap. 1, Sec. 1, p. 11)

momentum: property that a moving object has because of its mass and velocity. (Chap. 3, Sec. 3, p. 86)

monomer: small molecule that forms a link in the polymer chain and can be made to combine with itself repeatedly. (Chap. 21, Sec. 3, p. 653; Chap. 22, Sec. 3, p. 685)

music: sound created using specific pitches and sound qualities used in a set pattern. (Chap. 12, Sec. 3, p. 369)

N

net force: sum of the forces that are acting on an object. (Chap. 2, Sec. 3, p. 53)

neutralization: chemical reaction that occurs when the H_3O^+ ions from an acid react with the OH^- ions from a base to produce water molecules. (Chap. 25, Sec. 3, p. 777)

neutron: neutral particle, composed of quarks, inside the nucleus of an atom. (Chap. 18, Sec. 1, p. 545)

Newton's second law of motion: states that a net force acting on an object causes the object to accelerate in the direction of the net force. (Chap. 3, Sec. 1, p. 69)

Newton's third law of motion: describes action-reaction pairs—to every action force there is an equal and opposite reaction force. (Chap. 3, Sec. 3, p. 83)

nonelectrolyte: substance that does not ionize in water and cannot conduct electricity. (Chap. 23, Sec. 4, p. 723)

nonmetal: element that usually is a gas or brittle solid at room temperature, is not malleable or ductile, is a poor conductor of heat and electricity, and typically is not shiny. (Chap. 20, Sec. 2, p. 616)

nonpolar: material that has no separated positive and negative areas, does not attract water molecules, and does not dissolve easily in water. (Chap. 23, Sec. 2, p. 713)

nonpolar molecule: molecule that shares electrons equally and does not have oppositely charged ends. (Chap. 19, Sec. 2, p. 586)

nonrenewable resources: any natural resources, such as all fossil fuels, that cannot be replaced by natural processes as quickly as they are used. (Chap. 10, Sec. 1, p. 297)

nuclear fission: process of splitting a large atomic nucleus into two nuclei with smaller masses. (Chap. 9, Sec. 4, p. 273)

nuclear fusion: process of fusing together two atomic nuclei with low masses to form one nucleus with a larger mass. (Chap. 9, Sec. 4, p. 275)

nuclear reactor: uses energy from a controlled nuclear chain reaction to generate electricity. (Chap. 10, Sec. 2, p. 298)

nuclear waste: any radioactive by-product that results when radioactive materials are used; must be stored safely away from human activity. (Chap. 10, Sec. 2, p. 302)

nucleic acids: essential organic polymers that control the activities and reproduction of cells. (Chap. 21, Sec. 4, p. 658)

nucleus: positively charged center of an atom that contains protons and neutrons and is surrounded by a cloud of electrons. (Chap. 18, Sec. 1, p. 545)

O

Ohm's law: states that the current in a circuit equals the voltage difference divided by the resistance. (Chap. 7, Sec. 2, p. 207)

opaque (oh PAYK): material that absorbs or reflects all light. (Chap. 14, Sec. 1, p. 420)

optical axis: imaginary straight line on a concave mirror or convex lens that contains the focal point. (Chap. 15, Sec. 1, p. 454)

organic compounds: large number of compounds containing the element carbon. (Chap. 21, Sec. 1, p. 640)

overtone: vibration whose frequency is a multiple of the fundamental frequency, or main tone. (Chap. 12, Sec. 3, p. 370)

oxidation number: positive or negative number that indicates how many electrons an atom has gained, lost, or shared to become stable. (Chap. 19, Sec. 3, p. 587)

P

parallel circuit: circuit in which electric current has more than one path to follow. (Chap. 7, Sec. 3, p. 210)

pascal: SI unit of pressure. (Chap. 16, Sec. 3, p. 502)

period: horizontal row in the periodic table. (Chap. 18, Sec. 3, p. 561)

periodic table: organized list of all known elements that are arranged according to repeated changes in properties. (Chap. 18, Sec. 3, p. 554)

petroleum: crude oil—a highly flammable fossil fuel formed from decayed remains of ancient organisms; contains hydrocarbons and can be refined into fuels and used to make plastics. (Chap. 10, Sec. 1, p. 293)

pH: a measure of the concentration of hydronium ions in a solution using a scale ranging from 0 to 14, with 0 being the most acidic and 14 being the most basic. (Chap. 25, Sec. 2, p. 774)

photon: a type of particle that can describe light. (Chap. 13, Sec. 1, p. 395)

photovoltaic cell: device that converts solar energy into electricity; also called a solar cell. (Chap. 10, Sec. 3, p. 305)

physical change: any change in size, shape, or state of matter in which the identity of the substance remains the same. (Chap. 17, Sec. 2, p. 528)

physical property: any characteristic of a material, such as size or shape, that you can observe or attempt to observe without changing the identity of the material. (Chap. 17, Sec. 2, p. 526)

pigment: colored material that absorbs some colors and reflects others. (Chap. 14, Sec. 2, p. 428)

pitch: how high or low a sound appears to be, which is related to the frequency of the sound waves. (Chap. 12, Sec. 2, p. 366)

plane mirror: flat, smooth mirror that reflects light to form upright, life-sized, virtual images. (Chap. 15, Sec. 1, p. 453)

plasma: high-temperature gas with an overall neutral charge that is the most common state of matter in the universe. (Chap. 16, Sec. 1, p. 492)

polar molecule: molecule with a slightly positive end and a slightly negative end as a result of electrons being shared unequally. (Chap. 19, Sec. 2, p. 586)

polarized light: light in which transverse waves vibrate in only one direction. (Chap. 14, Sec. 4, p. 436)

polyatomic ion: positively or negatively charged, covalently bonded group of atoms. (Chap. 19, Sec. 3, p. 591)

polyethylene: polymer formed from a chain containing many ethylene units; often used in plastic bags and plastic bottles. (Chap. 21, Sec. 3, p. 653)

polymer: new, huge molecule made up of many monomers linked together to form a long chain that can be spun into fibers and used for plastics. (Chap. 21, Sec. 3, p. 653; Chap. 22, Sec. 3, p. 685)

potential energy: stored energy due to position; can be converted to kinetic energy when something acts to release it. (Chap. 4, Sec. 1, p. 103)

power: amount of work done, or the amount of energy transferred, in a certain amount of time; measured in watts, W. (Chap. 5, Sec. 1, p. 129)

precipitate: insoluble compound that comes out of solution during a double-displacement reaction. (Chap. 24, Sec. 3, p. 749)

pressure: amount of force exerted per unit area; SI unit is the pascal, Pa. (Chap. 16, Sec. 3, p. 498)

product: in a chemical reaction, the new substance that is formed. (Chap. 24, Sec. 1, p. 738)

proteins: large, complex, biological polymers formed from amino acids; make up many body tissues such as muscles, tendons, hair, and fingernails. (Chap. 21, Sec. 4, p. 656)

proton: positively charged particle, composed of quarks, inside the nucleus of an atom. (Chap. 18, Sec. 1, p. 545)

pulley: simple machine that consists of a grooved wheel with a rope, chain, or cable that runs along a groove, changes the direction of the effort force, and can be fixed or movable. (Chap. 5, Sec. 3, p. 141)

Q

quality: describes the differences among sounds having the same pitch and loudness. (Chap. 12, Sec. 3, p. 370)
quarks: particles of matter that make up protons and neutrons. (Chap. 18, Sec. 1, p. 545)

R

radiant energy: energy carried by an electromagnetic wave. (Chap. 13, Sec. 1, p. 393)
radiation: transfer of energy in the form of electromagnetic waves. (Chap. 6, Sec. 2, p. 166)
radio wave: low-frequency, electromagnetic wave with a wavelength from less than 1 cm to about 1,000 m that can carry the signal from a radio station to a radio. (Chap. 13, Sec. 2, p. 396)
radioactive element: element, such as radium, whose nucleus breaks down and emits particles and energy. (Chap. 20, Sec. 1, p. 610)
radioactivity: process of nuclear decay that takes place when the strong force is not able to hold unstable nuclei together permanently. (Chap. 9, Sec. 1, p. 260)
rarefaction (rar uh FAK shun): the least dense region of a compressional wave. (Chap. 11, Sec. 2, p. 332)
reactant: in a chemical reaction, the substance that reacts. (Chap. 24, Sec. 1, p. 738)
real image: upside-down, optical image formed when light rays pass through an object. (Chap. 15, Sec. 1, p. 455)
reflecting telescope: uses a concave mirror, a plane mirror, and a convex lens to collect and focus light from distant objects. (Chap. 15, Sec. 3, p. 469)
refracting telescope: uses two convex lenses to gather and focus light from distant objects. (Chap. 15, Sec. 3, p. 469)
refraction: bending of a wave as it changes speed, moving from one medium to another. (Chap. 11, Sec. 3, p. 340)
renewable resource: energy source, such as solar energy, that is replaced almost as quickly as it is used. (Chap. 10, Sec. 3, p. 305)
resistance: tendency for a material to oppose electron flow and change electrical energy into thermal energy and light; measured in ohms. (Chap. 7, Sec. 2, p. 205)
resistance force: force applied by a machine to overcome resistance. (Chap. 5, Sec. 2, p. 134)
resonance (RE zun unts)**:** ability of an object to vibrate by absorbing energy at its natural frequency. (Chap. 11, Sec. 3, p. 347)
resonator (RE zen ay tur)**:** hollow, air-filled chamber of a stringed instrument that amplifies sound when its air vibrates. (Chap. 12, Sec. 3, p. 371)
retina: inner lining of the eye that has cells that convert light images into electrical signals. (Chap. 15, Sec. 2, p. 463)

S

salt: compound formed when negative ions from an acid combine with positive ions from a base. (Chap. 25, Sec. 3, p. 777)
saturated hydrocarbon: compound, such as propane or methane, that contains only single bonds between carbon atoms. (Chap. 21, Sec. 1, p. 642)
saturated solution: any solution that contains all the solute it can hold at a given temperature. (Chap. 23, Sec. 3, p. 720)
scientific law: statement about how things work in nature that seems to be true all the time. (Chap. 1, Sec. 1, p. 12)
scientific method: organized set of investigation procedures that can include stating a problem, forming a hypotheses, researching and gathering information, testing a hypothesis, analyzing data, and drawing conclusions. (Chap. 1, Sec. 1, p. 7)

screw: simple machine that consists of an inclined plane wrapped in a spiral around a cylindrical post. (Chap. 5, Sec. 3, p. 145)

semiconductors: elements (silicon and germanium) that will conduct an electric current under certain conditions. (Chap. 20, Sec. 3, p. 623)

series circuit: circuit in which electric current has only one path to follow. (Chap. 7, Sec. 3, p. 209)

SI: International System of Units—the improved, universally accepted version of the metric system that is based on multiples of ten and includes the meter (m), liter (L), and kilogram (kg). (Chap. 1, Sec. 2, p. 15)

simple machine: machine that does work with only one movement; includes the lever, pulley, wheel and axle, inclined plane, screw, and wedge. (Chap. 5, Sec. 3, p. 138)

single-displacement reaction: chemical reaction in which one element replaces another element in a compound. (Chap. 24, Sec. 3, p. 747)

soaps: organic salts with nonpolar, hydrocarbon ends that interact with oils and dirt and polar ends that helps them dissolve in water. (Chap. 25, Sec. 3, p. 782)

solar collector: solar heating system device that absorbs radiant energy from the Sun. (Chap. 6, Sec. 3, p. 175)

solar energy: energy from the Sun that is free, renewable, and can be used to heat homes and buildings. (Chap. 6, Sec. 3, p. 174)

solubility: maximum amount of a solute that can be dissolved in a given amount of solvent at a given temperature. (Chap. 23, Sec. 3, p. 718)

solute: in a solution, the substance being dissolved. (Chap. 23, Sec. 1, p. 707)

solution: mixture that appears to have the same composition, color, density, and taste throughout and is mixed at the atomic or molecular level. (Chap. 17, Sec. 1, p. 522; Chap. 23, Sec. 1, p. 706)

solvent: in a solution, the substance in which the solute is dissolved. (Chap. 23, Sec. 1, p. 707)

sonar: system that uses the reflection of sound waves to find objects that are underwater. (Chap. 12, Sec. 4, p. 377)

specific heat: amount of energy needed to raise the temperature of 1 kg of a material 1 K (Chap. 6, Sec. 1, p. 161)

speed: distance an object travels per unit of time, which can be calculated by $v = d/t$, where v = speed, d = distance, and t equals time. (Chap. 2, Sec. 1, p. 39)

standard: exact, agreed-upon quantity used for comparison. (Chap. 1, Sec. 2, p. 14)

standing wave: a type of wave that forms when waves of equal wavelength and amplitude, but traveling in opposite directions, continuously interfere with each other; has points called nodes that do not move. (Chap. 11, Sec. 3, p. 346)

static electricity: electricity generated when more of one type of charge is on an object. (Chap. 7, Sec. 1, p. 194)

strong acid: any acid that ionizes almost completely in solution. (Chap. 25, Sec. 2, p. 772)

strong base: any base that ionizes completely in solution. (Chap. 25, Sec. 2, p. 773)

strong force: force that acts between protons and neutrons in an atomic nucleus and keeps them together. (Chap. 9, Sec. 1, p. 259)

substance: element or compound that cannot be broken down into simpler components and maintain the properties of the original substance. (Chap. 17, Sec. 1, p. 518)

substituted hydrocarbon: hydrocarbon with one or more of its hydrogen atoms replaced by atoms or groups of other elements. (Chap. 21, Sec. 2, p. 647)

supersaturated solution: any solution that contains more solute than a saturated solution at the same temperature. (Chap. 23, Sec. 3, p. 721)

suspension: heterogeneous mixture containing a liquid in which visible particles settle. (Chap. 17, Sec. 1, p. 524)

synthesis reaction: chemical reaction in which two or more substances combine to form a different substance. (Chap. 24, Sec. 3, p. 746)

synthetic: describes polymers, such as plastics, adhesives, and surface coatings, that are made from hydrocarbons. (Chap. 22, Sec. 3, p. 685)

T

technology: application of science to help people. (Chap. 1, Sec. 1, p. 13)

temperature: a measure of the average kinetic energy of all the particles in an object. (Chap. 6, Sec. 1, p. 159)

theory: explanation of things or events that is based on knowledge gained from many observations and experiments. (Chap. 1, Sec. 1, p. 12)

thermal energy: total energy, including kinetic and potential energy, of the particles that make up a material; is transferred by conduction, convection, and radiation. (Chap. 6, Sec. 1, p. 159)

thermal expansion: increase in the size of a substance that results from the separation of its molecules when the temperature is increased. (Chap. 16, Sec. 1, p. 493)

titration (ti TRAY shun)**:** process in which a solution of known concentration is used to determine the concentration of another solution. (Chap. 25, Sec. 3, p. 780)

total internal reflection: occurs when light strikes a surface between two materials and completely reflects back into the first material. (Chap. 14, Sec. 4, p. 438)

tracer: a radioisotope that is used to find or keep track of molecules in an organism. (Chap. 9, Sec. 4, p. 276)

transceiver: transmits one radio signal and receives another radio signal from a base unit, allowing a cordless phone user to talk and listen at the same time. (Chap. 13, Sec. 3, p. 407)

transformer: device that increases or decreases alternating current generated by a power plant so that it can enter homes safely. (Chap. 8, Sec. 3, p. 244)

transition elements: elements in Groups 3 through 12 of the periodic table, occur in nature as uncombined elements, and include the iron triad and coinage metals. (Chap. 20, Sec. 1, p. 612)

translucent (trans LEWS unt)**:** material that allows some light to pass through but not enough to see objects clearly through it. (Chap. 14, Sec. 1, p. 420)

transmutation: process of changing one element to another through radioactive decay. (Chap. 9, Sec. 2, p. 264)

transparent: material that transmits almost all the light striking it so that objects can be seen clearly through it. (Chap. 14, Sec. 1, p. 420)

transuranium element: elements having more than 92 protons. (Chap. 20, Sec. 3, p. 627)

transverse wave: a type of wave, such as a water wave, where the matter in the medium moves back and forth at right angles to the direction that the wave travels; has crests and troughs. (Chap. 11, Sec. 1, p. 328)

trough: lowest point of a transverse wave. (Chap. 11, Sec. 2, p. 332)

turbine: large wheel that rotates when pushed by steam, wind, or water and that provides mechanical energy to a generator. (Chap. 8, Sec. 3, p. 242)

Tyndall effect: scattering of a light beam as it passes through a colloid. (Chap. 17, Sec. 1, p. 523)

U

ultrasonic: sound waves with frequencies above 20,000 Hz. (Chap. 12, Sec. 2, p. 366)

ultraviolet wave: electromagnetic wave with frequencies slightly higher than visible light that can have useful and harmful effects. (Chap. 13, Sec. 2, p. 399)

unsaturated hydrocarbon: compound, such as ethene or ethyne, that contains at least one double or triple bond between carbon atoms. (Chap. 21, Sec. 1, p. 644)

unsaturated solution: any solution that can dissolve more solute at a given temperature. (Chap. 23, Sec. 3, p. 721)

V

variable: factor that can cause a change in the results of an experiment. (Chap. 1, Sec. 1, p. 9)

velocity: describes the speed and direction of a moving object. (Chap. 2, Sec. 1, p. 44)

virtual image: upright, reflected image that is perceived by your brain even though no light rays pass through the image. (Chap. 15, Sec. 1, p. 454)

viscosity: a fluid's resistance to flow. (Chap. 16, Sec. 2, p. 501)

visible light: electromagnetic waves with wavelengths 390 billionths to 770 billionths of a meter that can be seen by human eyes. (Chap. 13, Sec. 2, p. 399)

voltage difference: push that causes electrical charges to flow through a conductor; measured in volts, V. (Chap. 7, Sec. 2, p. 202)

volume: amount of space occupied by an object. (Chap. 1, Sec. 2, p. 18)

W

wave: a rhythmic disturbance that transfers energy through matter or space; exists only as long as it has energy to carry. (Chap. 11, Sec. 1, p. 326)

wavelength: distance between one point on a wave and the nearest point just like it on the following wave; as frequency increases, wavelength always decreases. (Chap. 11, Sec. 2, p. 333)

weak acid: any acid that only partly ionizes in solution. (Chap. 25, Sec. 2, p. 772)

weak base: any base that does not ionize completely in solution. (Chap. 25, Sec. 2, p. 773)

wedge: simple machine that consists of an inclined plane with one or two sloping sides. (Chap. 5, Sec. 3, p. 145)

weight: gravitational force exerted on an object by Earth. (Chap. 3, Sec. 2, p. 77)

wheel and axle: simple machine that consists of two different-sized wheels that rotate together. (Chap. 5, Sec. 3, p. 143)

work: transfer of energy that occurs when a force makes an object move; measured in joules. (Chap. 5, Sec. 1, p. 126)

X

X ray: ultrahigh-frequency, energetic electromagnetic wave that can travel through matter and is often used for medical imaging. (Chap. 13, Sec. 2, p. 401)

Spanish Glossary

Este glosario define cada término clave que aparece en negrillas en el texto. También muestra el capítulo y el número de página en donde se usa dicho término.

A

acceleration / aceleración: tasa de cambio de la velocidad, la cual ocurre si un cuerpo acelera, cambia de dirección o decelera. (Cap. 2, Sec. 2, pág. 47)

acid / ácido: toda sustancia que produce iones hidrógeno, H+, en una solución acuosa. (Cap. 25, Sec. 1, pág. 766)

acoustics / acústica: estudio del sonido. (Cap. 12, Sec. 4, pág. 375)

alcohol / alcohol: compuesto como el etanol, que se forma cuando los grupos –OH reemplazan uno o más átomos de hidrógeno en un hidrocarburo. (Cap. 21, Sec. 2, pág. 647)

allotropes / formas alotrópicas: formas diferentes del mismo elemento que poseen distintas estructuras moleculares. (Cap. 20, Sec. 3, pág. 623)

alloy / aleación: mezcla de un metal con uno o más elementos, la cual retiene las propiedades originales del metal. (Cap. 22, Sec. 1, pág. 672; Cap. 23, Sec. 1, pág. 707)

alpha particle / partícula alfa: partícula con una carga eléctrica de +2 y una masa atómica de 4 compuesta por dos protones y dos neutrones y que se emite desde un núcleo atómico en desintegración. (Cap. 9, Sec. 2, pág. 263)

alternating current (AC) / corriente alterna (CA): corriente eléctrica que invierte su dirección regularmente y que se usa en el funcionamiento de electrodomésticos. (Cap. 8, Sec. 3, pág. 244)

amplitude / amplitud: medida de la energía que transporta una onda. (Cap. 11, Sec. 2, pág. 336)

aromatic compound / compuesto aromático: compuesto que contiene la estructura de anillo bencénico y que puede tener un olor y sabor agradables o desagradables. (Cap. 21, Sec. 2, pág. 645)

atom / átomo: el trozo más pequeño de materia que todavía retiene las propiedades del elemento. (Cap. 18, Sec. 1, pág. 545)

atomic number / número atómico: número de protones en el núcleo de un átomo. (Cap. 18, Sec. 2, pág. 551)

average atomic mass / masa atómica promedio: masa media ponderada de la mezcla de los isótopos de un elemento. (Cap. 18, Sec. 2, pág. 553)

average speed / rapidez promedio: distancia total que viaja un objeto dividida entre el tiempo total que le lleva viajar tal distancia. (Cap. 2, Sec. 1, pág. 41)

B

balanced chemical equation / ecuación química equilibrada: ecuación química que posee el mismo número de átomos de cada elemento en ambos lados de la ecuación. (Cap. 24, Sec. 2, pág. 743)

balanced forces / fuerzas equilibradas: fuerzas sobre un cuerpo que son iguales en tamaño, pero opuestas en dirección y que no alteran el movimiento del cuerpo. (Cap. 2, Sec. 3, pág. 53)

base / base: toda sustancia que forma iones hidróxido, OH^-, en una solución acuosa. (Cap. 25, Sec. 1, pág. (768)

beta particle / partícula beta: electrón con una masa atómica de 0 y una carga eléctrica de –1 que se emite a gran velocidad desde un núcleo atómico en desintegración. (Cap. 9, Sec. 2, pág. 265)

bias / sesgo: se presenta cuando las expectativas de un científico cambian el modo en que se enfocan los resultados de un experimento. (Cap. 1, Sec. 1, pág. 10)

Spanish Glossary

binary compound / compuesto binario: compuesto formado por dos elementos. (Cap. 19, Sec. 3, pág. 587)

biomass / biomasa: materia orgánica renovable proveniente de plantas y animales, como por ejemplo, el abono vegetal y animal, que se puede quemar para calentar agua y producir electricidad. (Cap. 10, Sec. 3, pág. 310)

boiling point / punto de ebullición: temperatura a la cual la presión de la atmósfera es igual a la presión del vapor de un líquido, y las moléculas gaseosas pueden escapar de la fuerza de atracción entre las moléculas. (Cap. 16, Sec. 1, pág. 491)

bubble chamber / cámara de burbujas: tipo de instrumento que utiliza un líquido supercalentado con el fin de detectar y monitorear las trayectorias de partículas nucleares. (Cap. 9, Sec. 3, pág. 269)

buffer / tampón: solución que contiene iones que reaccionan con ácidos o bases que se le añaden y minimiza sus efectos sobre el pH. (Cap. 25, Sec. 2, pág. 775)

buoyancy / flotabilidad: capacidad que tiene un líquido de ejercer una fuerza ascendente sobre un cuerpo inmerso en tal líquido. (Cap. 16, Sec. 2, pág. 497)

C

carbohydrates / carbohidratos: grupo de compuestos biológicos que poseen el doble de átomos de hidrógeno que de oxígeno, por ejemplo, los azúcares y almidones. (Cap. 21, Sec. 4, pág. 659)

carrier wave / onda portadora: frecuencia de onda electromagnética específica que se le asigna a una radioemisora. (Cap. 13, Sec. 3, pág. 403)

catalyst / catalizador: sustancia que acelera una reacción química sin cambiar ella misma permanentemente. (Cap. 24, Sec. 4, pág. 754)

cathode-ray tube / tubo de rayos catódicos: tubo sellado al vacío que utiliza uno o más haces de electrones que dirigen e impactan diminutos rectángulos coloreados sobre una pantalla y producen imágenes a todo color en un televisor. (Cap. 13, Sec. 3, pág. 406)

centripetal acceleration / aceleración centrípeta: aceleración de un cuerpo hacia el centro de una trayectoria curva o circular. (Cap. 3 Sec. 2, pág. 81)

centripetal force / fuerza centrípeta: una fuerza dirigida hacia el centro de un círculo para un objeto que se mueve en movimiento circular. (Cap. 3 Sec. 2, pág. 81)

ceramics / cerámica: materiales versátiles cuyas propiedades son adaptables; se producen mediante un proceso en el cual un objeto se moldea y luego se calienta a altas temperaturas, aumentando su densidad. (Cap. 22, Sec. 2, pág. 678)

chain reaction / reacción en cadena: serie progresiva de reacciones de fisión nuclear. (Cap. 9, Sec. 4, pág. 274)

charging by contact / cargar por contacto: proceso de transferir cargas entre objetos al tocarlos o frotarlos. (Cap. 7, Sec. 1, pág. 197)

charging by induction / cargar por inducción: proceso de transferir cargas entre objetos al acercar un objeto con carga a un objeto neutro. (Cap. 7, Sec. 1, pág. 198)

chemical bond / enlace químico: fuerza que mantiene unidos a los átomos de un compuesto. (Cap. 19, Sec. 1, pág. 578)

chemical change / cambio químico: cambio de una sustancia a una nueva sustancia. (Cap. 17, Sec. 2, pág. 530)

chemical equation / ecuación química: método abreviado de describir reacciones químicas usando fórmulas químicas y otros símbolos. (Cap. 24, Sec. 1, pág. 740)

chemical formula / fórmula química: abreviación química que usa símbolos para indicar los elementos presentes en un compuesto y sus razones. (Cap. 19, Sec. 1, pág. 575)

chemically stable / químicamente estable: describe un átomo cuyo nivel de energía externo está completamente lleno con todos los electrones que se permiten a ese nivel. (Cap. 19, Sec. 1, pág. 576)

chemical potential energy / energía potencial química: energía almacenada en enlaces químicos. (Cap. 4, Sec. 1, pág. 103)

chemical property / propiedad química: cualquier característica de una sustancia, como la combustibilidad, que indica si puede sufrir cierto cambio químico. (Cap. 17, Sec. 2, pág. 529)

chemical reaction / reacción química: proceso en el cual una o más sustancias se transforman en nuevas sustancias. (Cap. 24, Sec. 1, pág. 738)

circuit / circuito: bucle conductor cerrado a través del cual puede fluir una corriente eléctrica. (Cap. 7, Sec. 2, pág. 203)

cloud chamber / cámara de niebla: tipo de instrumento que utiliza agua o vapor de etanol con el fin de detectar la radiación de partículas alfa y beta. (Cap. 9, Sec. 3, pág. 268)

cochlea / cóclea: estructura en forma de espiral y llena de líquido ubicada en el oído interno, la cual convierte las ondas sonoras en impulsos nerviosos. (Cap. 12, Sec. 1, pág. 362)

coefficient / coeficiente: número en una ecuación química que representa el número de unidades de cada sustancia que participa en una reacción química. (Cap. 24, Sec. 1, pág. 741)

coherent light / luz coherente: luz de una longitud de onda individual que viaja con sus crestas y senos alineados en la misma dirección. (Cap. 14, Sec. 3, pág. 434)

colloid / coloide: mezcla heterogénea cuyas partículas nunca se asientan. (Cap. 17, Sec. 1, pág. 522)

composite / compuesto: mezcla de dos materiales, uno de los cuales está incrustado en el otro. (Cap. 22, Sec. 3, pág. 689)

compound / compuesto: sustancia que se forma a partir de dos o más elementos, en la cual la combinación y proporción exactas de elementos es siempre la misma. (Cap. 17, Sec. 1, pág. 520)

compound machine / máquina compuesta: combinación de dos o más máquinas simples. (Cap. 5, Sec. 3, pág. 146)

compressional wave / onda de compresión: tipo de onda donde la materia del medio poseen un movimiento de vaivén en la misma dirección en que viaja la onda; tiene compresiones y rarefacciones. (Cap. 11, Sec. 1, pág. 328)

concave lens / lente cóncava: desvía los rayos luminosos para formar imágenes reducidas, verticales y virtuales; es más delgada en el centro y más gruesa en los bordes y, por lo general, se utiliza conjuntamente con otras lentes. (Cap. 15, Sec. 2, pág. 462)

concave mirror / espejo cóncavo: refleja la luz desde una superficie que se curva hacia adentro y puede aumentar objetos o crear haces de luz. (Cap. 15, Sec. 1, pág. 454)

conduction / conducción: transferencia de energía a través de la materia debido a partículas que chocan; se lleva a cabo porque las partículas se encuentran en constante movimiento. (Cap. 6, Sec. 2, pág. 164)

conductivity / conductividad: dícese de la propiedad de los metales y aleaciones que les permite ser buenos conductores de calor y electricidad. (Cap. 22, Sec. 1, pág. 673)

conductor / conductor: material, como el alambre de cobre, a través del cual puede moverse fácilmente un exceso de electrones. (Cap. 7, Sec. 1, pág. 196)

constant / constante: en un experimento, una variable que no cambia cuando cambian otras variables. (Cap. 1, Sec. 1, pág. 9)

control / control: estándar que se usa para efectos de comparar los resultados de las pruebas en un experimento. (Cap. 1, Sec. 1, pág. 9)

convection / convección: transferencia de energía debido al movimiento de partículas calentadas en un líquido. (Cap. 6, Sec. 2, pág. 165)

convex lens / lente convexa: converge rayos luminosos, es más gruesa en el centro que en los bordes y puede formar imágenes reales o virtuales. (Cap. 15, Sec. 2, pág. 460)

convex mirror / espejo convexo: refleja la luz desde una superficie que se curva hacia afuera y forma una imagen reducida, vertical y virtual. (Cap. 15, Sec. 1, pág. 457)

cornea / córnea: cubierta transparente en el globo ocular por la cual entra la luz. (Cap. 15, Sec. 2, pág. 463)

covalent bond / enlace covalente: atracción que se presenta entre los átomos cuando comparten electrones. (Cap. 19, Sec. 2, pág. 583)

crest / cresta: el punto más alto de una onda transversal. (Cap. 11, Sec. 2, pág. 332)

critical mass / masa crítica: cantidad de materia fisible que se requiere para que cada reacción nuclear produzca aproximadamente una reacción nuclear más. (Cap. 9, Sec. 4, pág. 274)

D

decibel / decibel: unidad de la escala de intensidad sonora, abreviada dB. (Cap. 12, Sec. 2, pág. 365)

decomposition reaction / reacción de descomposición: reacción química en la cual una sustancia se descompone en dos o más sustancias. (Cap. 24, Sec. 3, pág. 747)

density / densidad: masa por unidad de volumen de un material. (Cap. 1, Sec. 2, pág. 19)

deoxyribonucleic acid / ácido desoxirribonucleico: tipo de compuesto biológico esencial que codifica y almacena información genética, controla la producción de RNA y se encuentra en el núcleo de las células. (Cap. 21, Sec. 4, pág. 658)

dependent variable / variable dependiente: factor que cambia conforme cambia la variable independiente. (Cap. 1, Sec. 1, pág. 9)

derived unit / unidad derivada: unidad, como el metro cúbico (m^3), que se obtiene combinando diferentes unidades SI. (Cap. 1, Sec. 2, pág. 19)

diatomic molecule / molécula diatómica: molécula formada por dos átomos del mismo elemento. (Cap. 20, Sec. 2, pág. 617)

diffraction / difracción: describe la flexión de las ondas alrededor de un obstáculo; también puede ocurrir cuando las ondas atraviesan una abertura estrecha. (Cap. 11, Sec. 3, pág. 342)

diffusion / difusión: propagación de partículas a través de un volumen dado hasta que su distribución es uniforme. (Cap. 16, Sec. 1, pág. 491)

direct current (DC) / corriente directa (CD): corriente eléctrica que fluye a través de un alambre en una sola dirección. (Cap. 8, Sec. 3, pág. 244)

displacement / desplazamiento: distancia y dirección del cambio de posición de un cuerpo desde el punto de partida. (Cap. 2, Sec. 1, pág. 39)

dissociation / disociación: proceso en el cual se separan los iones de un cristal para llevarlos a una solución. (Cap. 23, Sec. 4, pág. 724)

distance / distancia: el recorrido o trayecto que se mueve un cuerpo. (Cap. 2, Sec. 1, pág. 39)

distillation / destilación: proceso que puede separar dos sustancias en una mezcla al evaporar un líquido y recondensar su vapor. (Cap. 17, Sec. 2, pág. 529)

doping / doping o adulteración: proceso de añadir impurificaciones a un semiconductor para aumentar su conductividad. (Cap. 22, Sec. 2, pág. 682)

Doppler effect / efecto Doppler: cambio de tono o frecuencia que ocurre cuando la fuente de una onda sonora se mueve en relación con el observador. (Cap. 12, Sec. 2, pág. 367)

double-displacement reaction / reacción de desplazamiento doble: reacción química que produce un precipitado, agua o un gas cuando se combinan dos compuestos iónicos en solución. (Cap. 24, Sec. 3, pág. 749)

ductile / dúctil: propiedad de los metales de extenderse en alambres. (Cap. 20, Sec. 1, pág. 608)

ductility / ductilidad: capacidad de los metales o aleaciones de extenderse en alambres. (Cap. 22, Sec. 1, pág. 673)

E

eardrum / tímpano: membrana resistente del oído externo de aproximadamente 0.1 mm de grosor y que transmite las vibraciones sonoras al oído medio. (Cap. 12, Sec. 1, pág. 361)

echolocation / ecolocalización: proceso en el cual se localizan objetos emitiendo sonidos e interpretando las ondas sonoras que regresan y son reflejadas de nuevo. (Cap. 12, Sec. 4, pág. 375)

efficiency / rendimiento: mide la cantidad de trabajo realizado por una máquina, en comparación con el trabajo que se le pone a la máquina. (Cap. 5, Sec. 2, pág. 136)

effort force / fuerza de esfuerzo: fuerza ejercida sobre una máquina que se utiliza para hacer trabajo. (Cap. 5, Sec. 2, pág. 134)

elastic potential energy / energía potencial elástica: energía almacenada por cuerpos que se estiran, torsionan o comprimen. (Cap. 4, Sec. 1, pág. 103)

electric current / corriente eléctrica: flujo de carga eléctrica a través de un alambre o cualquier otro conductor, medido en amperios (A), en un circuito. (Cap. 7, Sec. 2, pág. 203)

electric motor / motor eléctrico: dispositivo que convierte la energía eléctrica en energía mecánica para hacer trabajo; básicamente consta de una fuente de potencia, un imán permanente y un electroimán que puede girar. (Cap. 8, Sec. 2, pág. 237)

electrical power / potencia eléctrica: tasa a la cual la energía eléctrica se convierte en otra forma de energía; se expresa en vatios (W). (Cap. 7, Sec. 3, pág. 212)

electrolyte / electrólito: compuesto que se descompone en el agua formando partículas con carga eléctrica (iones) que pueden conducir electricidad. (Cap. 23, Sec. 4, pág. 723)

electromagnet / electroimán: imán temporal que se fabrica colocando un núcleo de hierro dentro de un rollo de alambre que transporta corriente y cuya fuerza se puede aumentar añadiendo más vueltas de alambre o aumentando la cantidad de corriente que pasa por el alambre. (Cap. 8, Sec. 2, pág. 234)

electromagnetic induction / inducción electromagnética: proceso de producción de corriente eléctrica mediante el movimiento de un bucle de alambre a través de un campo magnético o al mover un imán a través de un bucle de alambre. (Cap. 8, Sec. 3, pág. 240)

electromagnetic waves / ondas electromagnéticas: ondas creadas por la vibración de cargas eléctricas; pueden viajar a través del vacío o de la materia y pueden tener una amplia variedad de frecuencias y longitudes de onda. (Cap. 13, Sec. 1, pág. 390)

electron / electrón: partícula cargada negativamente que rodea el centro de un átomo. (Cap. 18, Sec. 1, pág. 545)

electron cloud / nube electrónica: área alrededor del núcleo de un átomo donde es más probable que se hallen los electrones del átomo. (Cap. 18, Sec. 1, pág. 549)

electron dot diagram / diagrama de puntos electrónicos: utiliza el símbolo para un elemento y los puntos representan el número de electrones que se hallan en el nivel de energía externo del elemento. (Cap. 18, Sec. 3, pág. 560)

element / elemento: sustancia cuyos átomos son todos semejantes. (Cap. 17, Sec. 1, pág. 518)

endergonic reaction / reacción endergónica: reacción química que requiere energía con el fin de seguir su curso. (Cap. 24, Sec. 4, pág. 753)

endothermic reaction / reacción endotérmica: reacción química que requiere energía térmica con el fin de seguir su curso. (Cap. 24, Sec. 4, pág. 753)

exergonic reaction / reacción exergónica: reacción química que libera algún tipo de energía, como por ejemplo, energía luminosa o energía calórica. (Cap. 24, Sec. 4, pág. 752)

exothermic reaction / reacción exotérmica: reacción química en la cual la energía es emitida primordialmente en forma de calor. (Cap. 24, Sec. 4, pág. 752)

experiment / experimento: procedimiento organizado para probar una hipótesis que prueba el efecto de un fenómeno sobre otro bajo condiciones controladas. (Cap. 1, Sec. 1, pág. 8)

F

fluorescent light / luz fluorescente: luz que resulta cuando la radiación ultravioleta producida dentro de un bombillo fluorescente hace brillar el revestimiento fosforescente dentro del bombillo. (Cap. 14, Sec. 3, pág. 431)

focal length / distancia focal: distancia desde el centro de una lente o espejo hasta el punto focal. (Cap. 15, Sec. 1, pág. 454)

focal point / punto focal: punto individual en el eje óptico de un espejo cóncavo o de una lente convexa por donde pasan los rayos luminosos. (Cap. 15, Sec. 1, pág. 454)

force / fuerza: empuje o halón que un cuerpo ejerce sobre otro. (Cap. 2, Sec. 3, pág. 52)

fossil fuel / combustible fósil: recursos energéticos no renovables, como el petróleo, el gas natural y el carbón, que se forman a partir de los restos descompuestos de plantas y animales antiguos; se pueden quemar para suministrar energía y generar electricidad. (Cap. 10, Sec. 1, pág. 291)

frequency / frecuencia: mide el número de longitudes de onda que pasan por un punto fijo cada segundo y se expresa en hertz (Hz). (Cap. 11, Sec. 2, pág. 333; Cap. 13, Sec. 1, pág. 394)

friction / fricción: fuerza que resiste el movimiento entre dos superficies en contacto. (Cap. 3 Sec. 1, pág. 70)

G

galvanometer / galvanómetro: instrumento que usa un electroimán para medir la corriente eléctrica. (Cap. 8, Sec. 2, pág. 236)

gamma ray / rayo gamma: onda electromagnética que no posee masa ni carga, la cual puede viajar a la velocidad de la luz y que por lo general es emitida por partículas alfa o beta. (Cap. 9, Sec. 2, pág. 265; Cap. 13, Sec. 2, pág. 401)

Geiger counter / contador Geiger: dispositivo que mide la radiactividad al producir una corriente eléctrica cuando detecta radiación; emite un clic o una luz centelleante. (Cap. 9, Sec. 3, pág. 270)

generator / generador: dispositivo que convierte la energía mecánica en energía eléctrica al rotar un rollo de alambre a través del campo magnético de un imán permanente. (Cap. 8, Sec. 3, pág. 240)

geothermal energy / energía geotérmica: energía térmica presente en el magma caliente; una planta eléctrica puede convertirla en energía eléctrica. (Cap. 10, Sec. 3, pág. 309)

Global Positioning System (GPS) / Sistema global de posicionamiento: utiliza satélites, sistemas de monitoreo desde tierra y receptores para ayudar a la gente a determinar su ubicación o posición exacta ya sea en o encima de la superficie terrestre. (Cap. 13, Sec. 3, pág. 409)

graph / gráfica: exhibición visual de información o de datos que puede ofrecer una manera rápida de comunicar, de manera clara, una gran cantidad de información, en una cantidad pequeña de espacio. (Cap. 1, Sec. 3, pág. 22)

gravitational potential energy / energía potencial gravitatoria: energía almacenada por cuerpos que se atraen entre sí debido a la fuerza de gravedad. (Cap. 4, Sec. 1, pág. 104)

group / grupo: columna vertical en la tabla periódica. (Cap. 18, Sec. 3, pág. 558)

H

half-life / media vida: cantidad de tiempo que tarda en desintegrarse la mitad de los núcleos en una muestra de un isótopo radioactivo. (Cap. 9, Sec. 2, pág. 266)

heat / calor: energía térmica que fluye de un cuerpo con mayor temperatura a otro con menor temperatura; se mide en julios (J). (Cap. 6, Sec. 1, pág. 160)

heat engine / motor térmico: dispositivo que convierte, en energía mecánica, la energía térmica producida por combustión. (Cap. 6, Sec. 3, pág. 176)

heat mover / máquina térmica: dispositivo, como el refrigerador, que extrae energía térmica de un lugar y la transfiere a otro a una temperatura diferente. (Cap. 6, Sec. 3, pág. 177)

heat of fusion / calor de fusión: cantidad de energía que se requiere para cambiar una sustancia de la fase sólida a la fase líquida. (Cap. 16, sec. 1, pág. 490)

heat of vaporization / calor de vaporización: cantidad de energía que se requiere para que las partículas de un líquido escapen de las fuerzas de atracción dentro del líquido o la energía que se requiere para cambiar de líquido a gas. (Cap. 16, Sec. 1, pág. 490)

heterogeneous mixture / mezcla heterogénea: mezcla, como el concreto o una mezcla de sopa seca, en la cual diferentes materiales se hallan distribuidos desigualmente y se pueden identificar con facilidad. (Cap. 17, Sec. 1, pág. 521)

holography / holografía: técnica que produce una imagen fotográfica completa de un objeto tridimensional. (Cap. 14, Sec. 4, pág. 437)

homogeneous mixture / mezcla homogénea: sólido, líquido o gas que contiene dos o más sustancias combinadas uniformemente. (Cap. 17, Sec. 1, pág. 522)

hydrate / hidrato: compuesto que contiene agua añadida químicamente a sus iones y que se indica en su fórmula química. (Cap. 19, Sec. 3, pág. 592)

hydrocarbon / hidrocarburo: compuesto saturado o no saturado que contiene sólo átomos de carbono e hidrógeno. (Cap. 21, Sec. 1, pág. 641)

hydroelectricity / hidroelectricidad: electricidad producida por la energía del agua en movimiento. (Cap. 10, Sec. 3, pág. 307)

hydronium ions / iones hidronio: iones H_3O^+, que se forman cuando un ácido se disuelve en agua y los iones H^+ interactúan con el agua. (Cap. 25, Sec. 1, pág. 766)

hypothesis / hipótesis: estimación razonada o bien fundada que usa el conocimiento y la observación. (Cap. 1, Sec. 1, pág. 8)

I

incandescent light/ luz incandescente: luz producida al calentar un trozo de metal, por lo general tungsteno, hasta que brille. (Cap. 14, Sec. 3, pág. 430)

inclined plane / plano inclinado: máquina simple que consta de una superficie inclinada, como una rampa, que reduce la cantidad de fuerza que se necesita para levantar algo al aumentar la distancia sobre la cual se aplica la fuerza. (Cap. 5, Sec. 3, pág. 144)

incoherent light / luz incoherente: luz que contiene más de una longitud de onda y que no viaja con sus crestas y senos alineados en la misma dirección. (Cap. 14, Sec. 3, pág. 434)

independent variable / variable independiente: factor que a medida que cambia, afecta la medida de otra variable. (Cap. 1, Sec. 1, pág. 9)

index of refraction / índice de refracción: propiedad de un material que indica cuánto aminora la velocidad de la luz cuando viaja por el material. (Cap. 14, Sec. 1, pág. 422)

indicator / indicador: compuesto orgánico que cambia de color en ácidos y bases. (Cap. 25, Sec. 1, pág. 766)

inertia / inercia: resistencia de un cuerpo a un cambio en su movimiento. (Cap. 2, Sec. 3, pág. 54)

infrared wave / onda infrarroja: onda electromagnética con una frecuencia ligeramente menor que la frecuencia de la luz visible, transmite energía térmica y se usa en los controles remotos de los televisores y en lectores o escáneres de código de barras. (Cap. 13, Sec. 2, pág. 398)

inhibitor / inhibidor: sustancia que retarda una reacción química o evita que ocurra al combinarse con un reactivo. (Cap. 24, Sec. 4, pág. 754)

instantaneous speed / rapidez instantánea: rapidez de un cuerpo en un punto dado en el tiempo, la cual es constante para un cuerpo que se mueve a velocidad constante y diferente en cada punto del tiempo, para un cuerpo que decelera o acelera. (Cap. 2, Sec. 1, pág. 41)

insulator / aislante: material que no permite que los electrones se muevan fácilmente a través de él. (Cap. 7, Sec. 1, pág. 197)

insulators / aisladores: materiales, como por ejemplo, el vellocino o la fibra de vidrio, que no permiten el movimiento fácil del calor a través de ellos. (Cap. 6, Sec. 2, pág. 169)

integrated circuit / circuito integrado: microplaqueta de material semiconductor que puede contener millones de transistores, diodos y otros componentes. (Cap. 22, Sec. 2, pág. 682)

intensity / intensidad: cantidad de energía que transporta una onda sonora, la cual se puede medir en decibeles (dB). (Cap. 12, Sec. 2, pág. 634)

interference / interferencia: ocurre cuando dos o más ondas se traslapan y se combinan para formar una nueva onda; puede formar una onda más grande (interferencia constructiva) o una onda más pequeña (interferencia destructiva). (Cap. 11, Sec. 3, pág. 344)

internal combustion engine / motor de combustión interna: tipo de motor térmico que quema el combustible dentro del motor. (Cap. 6, Sec. 3, pág. 176)

ion / ion: partícula cargada que posee o más o menos electrones que protones. (Cap. 19, Sec. 2, pág. 580) (Cap. 23, Sec. 4, pág. 723)

ionic bond / enlace iónico: atracción que se presenta entre iones de carga opuesta en un compuesto iónico. (Cap. 19, Sec. 2, pág. 582)

ionization / ionización: proceso en el cual los electrólitos se disuelven en agua y se separan en partículas con carga. (Cap. 23, Sec. 4, pág. 723)

isomers / isómeros: compuestos con fórmulas químicas idénticas, pero diferentes estructuras y formas moleculares, los cuales pueden afectar las propiedades físicas como el punto de fusión. (Cap. 21, Sec. 1, pág. 643)

isotopes / isótopos: átomos de un mismo elemento que poseen diferentes números de neutrones. (Cap. 18, Sec. 2, pág. 552)

J

joule / julio: unidad SI de energía. (Cap. 4, Sec. 1, pág. 102)

K

kilowatt-hour / kilovatio-hora: unidad de energía eléctrica, que corresponde a 1000 vatios de potencia usados durante una hora. (Cap. 7, Sec. 3, pág. 215)

kinetic energy / energía cinética: energía en forma de movimiento; depende de la masa y velocidad del cuerpo. (Cap. 4, Sec. 1, pág. 102)

kinetic theory / teoría cinética: explicación del comportamiento de las moléculas en la materia; establece que toda la materia se compone de partículas en constante movimiento que chocan sin perder energía. (Cap. 16, Sec. 1, pág. 488)

L

law of conservation of charge / ley de conservación de cargas: establece que las cargas se pueden transferir de un objeto a otro, pero que no se puede crear o destruir. (Cap. 7, Sec. 1, pág. 195)

law of conservation of energy / ley de conservación de la energía: establece que la energía no se crea ni se destruye. (Cap. 4, Sec. 2, pág. 111)

law of conservation of mass / ley de conservación de la masa: establece que la masa de todas las sustancias, presente antes de un cambio químico, equivale a la masa de todas las sustancias que quedan después del cambio. (Cap. 17, Sec. 2, pág. 533)

law of gravitation / ley de gravitación: establece que cualquier par de masas ejercen una fuerza de atracción entre sí, la cantidad de la cual depende de la masa de los dos cuerpos y la distancia entre ellos. (Cap. 3 Sec. 2, pág. 75)

lever / palanca: máquina simple hecha de una barra libre para girar sobre un punto fijo. (Cap. 5, Sec. 3, pág. 138)

lipids / lípidos: grupo de compuestos biológicos que contiene los mismos elementos que los carbohidratos, pero en arreglos y combinaciones diferentes; incluye las grasas y los aceites saturados e insaturados. (Cap. 21, Sec. 4, pág. 660)

loudness / volumen: describe cómo los seres humanos perciben la intensidad sonora. (Cap. 12, Sec. 2, pág. 365)

luster / lustre: propiedad de metales y aleaciones que describe su apariencia brillante o la propiedad de reflejar la luz. (Cap. 22, Sec. 1, pág. 673)

M

machine / máquina: dispositivo que facilita el trabajo aumentando la fuerza aplicada a un cuerpo, cambiando la dirección de una fuerza aplicada o aumentando la distancia sobre la cual se puede aplicar una fuerza. (Cap. 5, Sec. 2, pág. 132)

magnetic domain / dominio magnético: grupo de átomos cuyos polos magnéticos están alineados. (Cap. 8, Sec. 1, pág. 231)

magnetic pole / polo magnético: región donde la fuerza magnética de un imán es más fuerte; los polos iguales se repelen y los polos opuestos se atraen. (Cap. 8, Sec. 1, pág. 227)

magnetism / magnetismo: las propiedades e interacciones de los imanes. (Cap. 8, Sec. 1, pág. 226)

malleability / maleabilidad: propiedad de los metales y aleaciones de extenderse en láminas delgadas o de poder martillarse. (Cap. 22, Sec. 1, pág. 673)

malleable / maleable: propiedad de los metales de ser martillados o extenderse en láminas delgadas. (Cap. 20, Sec. 1, pág. 608)

mass / masa: cantidad de materia que posee un cuerpo. (Cap. 1, Sec. 2, pág. 19)

mass number / número de masa: suma del número de protones y neutrones en el núcleo de un átomo. (Cap. 18, Sec. 2, pág. 551)

mechanical advantage (MA) / ventaja mecánica (VM): número de veces que una máquina multiplica la fuerza de esfuerzo que se le aplica. (Cap. 5, Sec. 2, pág. 136)

mechanical energy / energía mecánica: suma de la energía potencial y la cinética de un sistema. (Cap. 4, Sec. 2, pág. 108)

[illegible] medio: cualquier material, ya sea [illegible] líquido, gas o una combinación de [illegible], a través del cual una onda [illegible] e energía. (Cap. 11, Sec. 1, [illegible])

[illegible] point / punto de fusión: tempera[illegible] cual un sólido comienza a vol[illegible] uido. (Cap. 16, Sec. 1, pág. 490)

m[illegible] tal: elemento que típicamente es [illegible] o duro, brillante y maleable y que [illegible] en conductor del calor y la elec[illegible]. (Cap. 20, Sec. 1, pág. 608)

m[illegible] nding / enlace metálico: se pre[illegible] bido al movimiento libre de los [illegible] es entre iones de un metal carga[illegible] tivamente y explica las propie[illegible] mo la ductibilidad y la capacidad [illegible] ucir electricidad. (Cap. 20, Sec. 1, [illegible])

m[illegible] / metaloide: elemento que com[illegible] unas propiedades con los meta[illegible] nas con los no metales. (Cap. 20, [illegible] g. 622)

m[illegible] / microscopio: utiliza lentes [illegible] para aumentar objetos [illegible] s cercanos. (Cap. 15, Sec. 3, [illegible])

m[illegible] / microondas: onda radial con [illegible] gitud de onda menor que 1 m. [illegible], Sec. 2, pág. 397)

m[illegible] spejismo: imagen de un objeto [illegible] ue resulta cuando el aire a nivel [illegible] o es más caliente o más frío que [illegible] s de aire por encima, lo cual hace [illegible] magen se refracte y aparezca en [illegible] cación diferente de la cual se [illegible] tra en realidad. (Cap. 14, Sec. 1, [illegible] 4)

mo[illegible] modelo: se puede utilizar para re[illegible] sentar una idea, un objeto o un evento que es demasiado grande, demasiado pequeño, demasiado complejo o demasiado peligroso para ser observado o probado directamente. (Cap. 1, Sec. 1, pág. 11)

momentum / momento: propiedad que posee un cuerpo en movimiento debido a su masa y velocidad. (Cap. 3 Sec. 3, pág. 86)

monomer / monómero: molécula pequeña que forma un enlace en una cadena polímera y que se puede combinar consigo misma repetidamente. (Cap. 21, Sec. 3, pág. 653; Cap. 22, Sec. 3, pág. 685)

music / música: sonido creado al usar tonos específicos y cualidades sonoras que se utilizan en un patrón fijo. (Cap. 12, Sec. 3, pág. 369)

N

net force / fuerza neta: suma de las fuerzas que actúan sobre un cuerpo. (Cap. 2, Sec. 3, pág. 53)

neutralization / neutralización: reacción química que ocurre cuando los iones H_3O^+ de un ácido reaccionan con los iones OH^- de una base para producir moléculas de agua. (Cap. 25, Sec. 3, pág. 777)

neutron / neutrón: partícula neutra, compuesta de quarks, en el interior del núcleo de un átomo. (Cap. 18, Sec. 1, pág. 545)

Newton's second law of motion / segunda ley del movimiento de Newton: establece que una fuerza neta que actúa sobre un cuerpo hace que éste acelere en la dirección de la fuerza neta. (Cap. 3 Sec. 1, pág. 69)

Newton's third law of motion / tercera ley del movimiento de Newton: describe pares de acción y reacción: por cada fuerza de acción existe una fuerza de reacción igual y opuesta. (Cap. 3 Sec. 3, pág. 83)

nonelectrolyte / no electrólito: sustancia que no se ioniza en el agua y que no puede conducir electricidad. (Cap. 23, Sec. 4, pág. 723)

nonmetal / no metal: elemento que por lo general es un gas o un sólido frágil a temperatura ambiente, no es maleable o dúctil, es mal conductor de calor y electricidad y típicamente no es brillante. (Cap. 20, Sec. 2, pág. 616)

nonpolar / no polar: material que no posee áreas positivas y negativas separadas, no atrae las moléculas de agua y no se disuelve con facilidad en el agua. (Cap. 23, Sec. 2, pág. 713)

nonpolar molecule / molécula no polar: molécula que comparte electrones equitativamente y que no posee extremos con carga opuesta. (Cap. 19, Sec. 2, pág. 586)

nonrenewable resources / recursos no renovables: todos los recursos naturales, como los combustibles fósiles, que los procesos naturales no pueden reemplazar con la misma rapidez con que se utilizan. (Cap. 10, Sec. 1, pág. 297)

nuclear fission / fisión nuclear: proceso que separa un núcleo atómico grande en dos núcleos de menor masa. (Cap. 9, Sec. 4, pág. 273)

nuclear fusion / fusión nuclear: proceso que une dos núcleos atómicos de menor masa para formar un núcleo con una masa mayor. (Cap. 9, Sec. 4, pág. 275)

nuclear reactor / reactor nuclear: utiliza energía de una reacción nuclear en cadena controlada con el fin de generar electricidad. (Cap. 10, Sec. 2, pág. 298)

nuclear waste / desecho nuclear: todo subproducto radiactivo que resulta cuando se usan materiales radiactivos; se deben almacenar de manera segura lejos de la actividad humana. (Cap. 10, Sec. 2, pág. 302)

nucleic acids / ácidos nucleicos: polímeros orgánicos esenciales que controlan las actividades y reproducción de las células. (Cap. 21, Sec. 4, pág. 658)

nucleus / núcleo: centro de un átomo, cargado positivamente, que contiene protones y neutrones y que se halla rodeado de una nube electrónica. (Cap. 18, Sec. 1, pág. 545)

O

Ohm's law / ley de Ohm: establece que la corriente de un circuito equivale a la diferencia de potencial dividida entre la resistencia. (Cap. 7. Sec. 2, pág. 207)

opaque / opaco: material que absorbe o refleja la luz en su totalidad. (Cap. 14, Sec. 1, pág. 420)

optical axis / eje óptico: recta imaginaria en un espejo cóncavo o una lente convexa que contiene el punto focal. (Cap. 15, Sec. 1, pág. 454)

organic compounds / compuestos orgánicos: extenso número de compuestos que contienen el elemento carbono. (Cap. 21, Sec. 1, pág. 640)

overtone / armónico: vibración cuya frecuencia es un múltiplo de la frecuencia fundamental o tono principal. (Cap. 12, Sec. 3, pág. 370)

oxidation number / número de oxidación: número positivo o negativo que indica la cantidad de electrones que un átomo ha ganado, perdido o compartido para poder estabilizarse. (Cap. 19, Sec. 3, pág. 587)

P

parallel circuit / circuito paralelo: circuito en el cual la corriente eléctrica tiene más de una trayectoria que puede seguir. (Cap. 7, Sec. 3, pág. 210)

pascal / pascal: unidad SI de presión. (Cap. 16, Sec. 3, pág. 502)

period / período: hilera horizontal en la tabla periódica. (Cap. 18, Sec. 3, pág. 561)

periodic table / tabla periódica: lista organizada de todos los elementos conocidos que están ordenados según cambios repetidos de propiedades. (Cap. 18, Sec. 3, pág. 554)

petroleum / petróleo bruto: petróleo crudo, un combustible fósil altamente inflamable que se formó a partir de los restos de organismos antiguos; contiene hidrocarburos, se puede refinar para producir combustibles y se utiliza en la elaboración de plásticos. (Cap. 10, Sec. 1, pág. 293)

pH / pH: medida de la concentración de iones hidronio en una solución que usa una escala que va de 0 a 14, donde 0 denota la mayor acidez y 14 denota la mayor basicidad. (Cap. 25, Sec. 2, pág. 774)

photon / fotón: un tipo de partícula que puede describir la luz. (Cap. 13, Sec. 1, pág. 395)

photovoltaic cell / célula fotovoltaica: dispositivo que convierte la energía solar en electricidad; también recibe el nombre de célula solar. (Cap. 10, Sec. 3, pág. 305)

physical change / cambio físico: todo cambio de tamaño, forma o estado de la materia en el cual no cambia la identidad de la sustancia. (Cap. 17, Sec. 2, pág. 528)

physical property / propiedad física: cualquier característica de un material, como el tamaño o la forma, que se puede observar o intentar observar sin alterar la identidad del material. (Cap. 17, Sec. 2, pág. 526)

pigment / pigmento: material coloreado que absorbe algunos colores y refleja otros. (Cap. 14, Sec. 2, pág. 428)

pitch / tono: agudeza o gravedad de un sonido, el cual se relaciona con la frecuencia de las ondas sonoras. (Cap. 12, Sec. 2, pág. 366)

plane mirror / espejo plano: espejo liso que refleja la luz para formar imágenes verticales, virtuales y de tamaño real. (Cap. 15, Sec. 1, pág. 453)

plasma / plasma: gas de alta temperatura con una carga neutra total que es el estado de materia más común en el universo. (Cap. 16, Sec. 1, pág. 492)

polar molecule / molécula polar: molécula que tiene un extremo ligeramente positivo y un extremo ligeramente negativo, como resultado de compartir electrones desigualmente. (Cap. 19, Sec. 2, pág. 586)

polarized light / luz polarizada: luz en la cual las ondas transversales vibran en una sola dirección. (Cap. 14, Sec. 4, pág. 436)

polyatomic ion / ion poliatómico: grupo de átomos cargados positiva o negativamente y que están enlazados covalentemente. (Cap. 19, Sec. 3, pág. 591)

polyethylene / polietileno: polímero formado a partir de una cadena que contiene muchas unidades de etileno; se usa a menudo en la fabricación de bolsas y botellas plásticas. (Cap. 21, Sec. 3, pág. 653)

polymer / polímero: nueva molécula enorme compuesta de muchos monómeros enlazados entre sí para formar una cadena larga que puede hilarse en fibras y ser usada para los plásticos. (Cap. 21, Sec. 3, pág. 653; Cap. 22, Sec. 3, pág. 685)

potential energy / energía potencial: energía almacenada debido a la posición de los cuerpos; se puede convertir en energía cinética cuando algo actúa para liberarla. (Cap. 4, Sec. 1, pág. 103)

power / potencia: cantidad de trabajo realizado, o la cantidad de energía transferida, en cierta cantidad de tiempo; se mide en vatios (W). (Cap. 5, Sec. 1, pág. 129)

precipitate / precipitado: compuesto insoluble que proviene de una solución durante una reacción de desplazamiento doble. (Cap. 24, Sec. 3, pág. 749)

pressure / presión: cantidad de fuerza ejercida por área unitaria o unidad de superficie; la unidad SI es el pascal (Pa). (Cap. 16, Sec. 3, pág. 502)

product / producto: en una reacción química, la nueva sustancia que se forma. (Cap. 24, Sec. 1, pág. 738)

proteins / proteínas: polímeros biológicos complejos y grandes formados a partir de aminoácidos; las proteínas son los constituyentes de muchos tejidos corporales, como los músculos, los tendones, el cabello y las uñas. (Cap. 21, Sec. 4, pág. 656)

proton / protón: partícula cargada positivamente, compuesta de quarks, que se halla en el interior de un átomo. (Cap. 18, Sec. 1, pág. 545)

pulley / polea: máquina simple compuesta de una rueda acanalada con una cuerda, cadena, o cable que corre a lo largo de una ranura; una polea cambia la dirección de la fuerza de esfuerzo y puede ser fija o movible. (Cap. 5, Sec. 3, pág. 141)

Q

quality / calidad: describe las diferencias entre sonidos que poseen el mismo tono y volumen sonoro. (Cap. 12, Sec. 3, pág. 370)

quarks / quarks: partículas de materia que son los constituyentes de los protones y neutrones. (Cap. 18, Sec. 1, pág. 545)

R

radiant energy / energía radiante: energía transmitida por una onda electromagnética. (Cap. 13, Sec. 1, pág. 393)

radiation / radiación: transferencia de energía en forma de ondas electromagnéticas. (Cap. 6, Sec. 2, pág. 166)

radio wave / onda radial: onda electromagnética de baja frecuencia cuya longitud de onda varía entre menos de 1 cm hasta unos 1,000 m y que puede portar la señal desde una radioemisora a un radiorreceptor. (Cap. 13, Sec. 2, pág. 396)

radioactive element / elemento radiactivo: elemento, como el radio, cuyo núcleo se desintegra y emite partículas y energía. (Cap. 20, Sec. 1, pág. 610)

radioactivity / radiactividad: proceso de desintegración nuclear que ocurre cuando la fuerza fuerte no es capaz de mantener juntos los núcleos inestables de manera permanente. (Cap. 9, Sec. 1, pág. 260)

rarefaction / rarefacción: la región menos densa de una onda de compresión. (Cap. 11, Sec. 2, pág. 332)

reactant / reactante: en una reacción química, la sustancia que reacciona. (Cap. 24, Sec. 1, pág. 738)

real image / imagen real: imagen óptica invertida que se forma cuando los rayos luminosos atraviesan un cuerpo. (Cap. 15, Sec. 1, pág. 455)

reflecting telescope / telescopio reflector: utiliza un espejo cóncavo, un espejo plano y una lente convexa para recoger y enfocar la luz de cuerpos distantes. (Cap. 15, Sec. 3, pág. 469)

refracting telescope / telescopio refractor: utiliza dos lentes convexas para recoger y enfocar la luz de cuerpos distantes. (Cap. 15, Sec. 3, pág. 469)

refraction / refracción: flexión de una onda a medida que cambia de rapidez, al moverse de un medio a otro. (Cap. 11, Sec. 3, pág. 340)

renewable resource / recurso renovable: fuente energética, como la energía solar, que se recupera o se reproduce casi al mismo ritmo con que se utiliza. (Cap. 10, Sec. 3, pág. 305)

resistance / resistencia: la tendencia de un material a oponerse al flujo de electrones y a transformar la energía eléctrica en energía térmica y luminosa; se mide en ohmios. (Cap. 7, Sec. 2, pág. 205)

resistance force / fuerza de resistencia: fuerza que aplica una máquina para vencer una resistencia. (Cap. 5, Sec. 2, pág. 134)

resonance / resonancia: capacidad que tiene un objeto de vibrar al absorber energía a su frecuencia natural. (Cap. 11, Sec. 3, pág. 347)

resonator / resonador: cámara hueca y llena de aire de un instrumento de cuerda que amplifica el sonido cuando su aire vibra. (Cap. 12, Sec. 3, pág. 371)

retina / retina: revestimiento interno del ojo que tiene células que convierten las imágenes luminosas en señales eléctricas. (Cap. 15, Sec. 2, pág. 463)

S

salt / sal: compuesto que se forma cuando los iones negativos de un ácido se combinan con los iones positivos de una base. (Cap. 25, Sec. 3, pág. 777)

saturated hydrocarbon / hidrocarburo saturado: compuesto que contiene sólo enlaces sencillos entre los átomos de carbono, como el propano o el metano. (Cap. 21, Sec. 1, pág. 642)

saturated solution / solución saturada: solución que contiene todo el soluto que puede sostener a una temperatura dada. (Cap. 23, Sec. 3, pág. 720)

scientific law / ley científica: enunciado sobre cómo funcionan los elementos en la naturaleza, el cual parece ser siempre cierto. (Cap. 1, Sec. 1, pág. 12)

scientific method / método científico: conjunto organizado de procedimientos de investigación que puede incluir la enunciación de un problema, la formulación de una hipótesis, la investigación y la recopilación de información, el someter a prueba una hipótesis, el análisis de datos y el sacar conclusiones. (Cap. 1, Sec. 1, pág. 7)

screw / tornillo: máquina simple que consta de un plano inclinado enrollado en espiral alrededor de un poste cilíndrico. (Cap. 5, Sec. 3, pág. 145)

semiconductors / semiconductores: elementos (silicio y germanio) que pueden conducir una corriente eléctrica bajo ciertas condiciones. (Cap. 20, Sec. 3, pág. 623)

series circuit / circuito en serie: circuito en el cual la corriente eléctrica tiene solamente una trayectoria que puede seguir. (Cap. 7, Sec. 3, pág. 209)

SI International System of Units / SI (Sistema internacional de unidades): la versión mejorada y universalmente aceptada del sistema métrico que se basa en múltiplos de diez e incluye el metro (m), el litro (L) y el kilogramo (kg). (Cap. 1, Sec. 2, pág. 15)

simple machine / máquina simple: máquina que realiza trabajo con un movimiento solamente; incluye la palanca, la polea, la rueda y eje, el plano inclinado, el tornillo y la cuña. (Cap. 5, Sec. 3, pág. 138)

single-displacement reaction / reacción de desplazamiento simple: reacción química en la cual un elemento reemplaza a otro elemento en un compuesto. (Cap. 24, Sec. 3, pág. 747)

soaps / jabones: sales orgánicas con extremos no polares hidrocarburos que interactúan con los aceites y la suciedad y extremos polares que les ayudan a disolverse en el agua. (Cap. 25, Sec. 3, pág. 782)

solar collector / colector solar: dispositivo de sistemas de calefacción solar que absorbe la energía radiante del Sol. (Cap. 6, Sec. 3, pág. 175)

solar energy / energía solar: energía proveniente del Sol que es gratis, renovable y se puede utilizar para calentar hogares y otros inmuebles. (Cap. 6, Sec. 3, pág. 174)

solubility / solubilidad: cantidad máxima de un soluto que se puede disolver en una cantidad dada de disolvente a una temperatura dada. (Cap. 23, Sec. 3, pág. 718)

solute / soluto: en una solución, la sustancia que se disuelve. (Cap. 23, Sec. 1, pág. 707)

solution / solución: mezcla que parece tener la misma composición, color, densidad y sabor y que se mezcla a nivel molecular o atómico. (Cap. 17, Sec. 1, pág. 522; Cap. 23, Sec. 1, pág. 706)

solvent / disolvente: en una solución, la sustancia en la cual se disuelve el soluto. (Cap. 23, Sec. 1, pág. 707)

sonar / sonar: sistema que utiliza el reflejo de las ondas sonoras para hallar objetos debajo del agua. (Cap. 12, Sec. 4, pág. 377)

specific heat / calor específico: cantidad de energía que se necesita para elevar en 1 K la temperatura de 1 kg de material. (Cap. 6, Sec. 1, pág. 161)

speed / rapidez: distancia que viaja un objeto por unidad de tiempo, la cual se puede calcular por $v = d/t$, donde v = rapidez, d = distancia y t es igual a tiempo. (Cap. 2, Sec. 1, pág. 39)

standard / estándar: cantidad exacta que se ha acordado y que se usa para efectos de comparación. (Cap. 1, Sec. 2, pág. 14)

standing wave / onda estacionaria: tipo de onda que se forma cuando las ondas con longitud de onda y amplitud idénticas, pero que viajan en direcciones opuestas, interfieren continuamente entre sí; posee puntos llamados nodos que no se mueven. (Cap. 11, Sec. 3, pág. 346)

static electricity / electricidad estática: electricidad que se genera cuando hay más de un tipo de carga en un objeto. (Cap. 7, Sec. 1, pág. 194)

strong acid / ácido fuerte: todo ácido que se ioniza casi totalmente en una solución. (Cap. 25, Sec. 2, pág. 772)

strong base / base fuerte: toda base que se ioniza totalmente en una solución. (Cap. 25, Sec. 2, pág. 773)

strong force / fuerza fuerte: fuerza que actúa entre protones y neutrones en un núcleo atómico para mantenerlos juntos. (Cap. 9, Sec. 1, pág. 259)

substance / sustancia: elemento o compuesto que no se puede descomponer en componentes más simples y mantener las propiedades de la sustancia original. (Cap. 17, Sec. 1, pág. 518)

substituted hydrocarbon / hidrocarburo de sustitución: hidrocarburo en que uno o más de sus átomos de hidrógeno son reemplazados por átomos o grupos de otros elementos. (Cap. 21, Sec. 2, pág. 647)

supersaturated solution / solución supersaturada: toda solución que contiene más soluto que una solución saturada a igual temperatura. (Cap. 23, Sec. 3, pág. 721)

suspension / suspensión: mezcla heterogénea que contiene un líquido en el cual se asientan partículas visibles. (Cap. 17, Sec. 1, pág. 524)

synthesis reaction / reacción de síntesis: reacción química en la cual dos o más sustancias se combinan para formar una sustancia diferente. (Cap. 24, Sec. 3, pág. 746)

synthetic / sintético: describe a los polímeros, como los plásticos, los adhesivos y los revestimientos de superficies, que están hechos de hidrocarburos. (Cap. 22, Sec. 3, pág. 685)

T

technology / tecnología: aplicación de la ciencia para ayudar a la gente. (Cap. 1, Sec. 1, pág. 13)

temperature / temperatura: medida de la energía cinética promedio de todas las partículas de un cuerpo. (Cap. 6, Sec. 1, pág. 159)

theory / teoría: explicación de fenómenos o eventos que se basa en el conocimiento adquirido a través de la observación y la experimentación. (Cap. 1, Sec. 1, pág. 12)

thermal energy / energía térmica: energía total, que incluye la energía cinética y la energía potencial, de las partículas que forman un material; se transfiere por conducción, convección y radiación. (Cap. 6, Sec. 1, pág. 159)

thermal expansion / expansión térmica: aumento en el tamaño de una sustancia como resultado de la separación de sus moléculas, al aumentar la temperatura. (Cap. 16, Sec. 1, pág. 493)

titration / titulación o análisis volumétrico: proceso en el cual una solución de concentración conocida se usa para determinar la concentración de otra solución. (Cap. 25, Sec. 3, pág. 780)

total internal reflection / reflexión interna total: se produce cuando la luz choca contra una superficie entre dos materiales y se vuelve a reflejar completamente en el primer material. (Cap. 14, Sec. 4, pág. 438)

tracer / indicador radiactivo: radioisótopo que se utiliza para encontrar o rastrear moléculas en un organismo. (Cap. 9, Sec. 4, pág. 276)

transceiver / transceptor: transmite una señal de radio y recibe otra señal de radio desde una unidad base, lo cual le permite al usuario de teléfonos inalámbricos hablar y escuchar al mismo tiempo. (Cap. 13, Sec. 3, pág. 407)

transformer / transformador: dispositivo que aumenta o disminuye la corriente alterna generada por una planta eléctrica, de modo que pueda entrar en las casas sin presentar peligros. (Cap. 8, Sec. 3, pág. 244)

transition elements / elementos de transición: elementos de los Grupos 3 al 12 en la tabla periódica; se presentan en la naturaleza como elementos libres e incluyen la tríada del hierro y los metales de acuñación. (Cap. 20, Sec. 1, pág. 612)

translucent / traslúcido: material a través del cual pasa la luz pero no en suficiente cantidad como para ver cuerpos claramente a través de él. (Cap. 14, Sec. 1, pág. 420)

transmutation / transmutación: proceso que transforma un elemento en otro por medio de la desintegración radiactiva. (Cap. 9, Sec. 2, pág. 264)

transparent / transparente: material que transmite casi toda la luz que choca contra él y a través del cual pueden verse de forma clara los cuerpos. (Cap. 14, Sec. 1, pág. 420)

transuranium element / elemento transuránico: elementos con más de 92 protones. (Cap. 20, Sec. 3, pág. 627)

transverse wave / onda transversal: tipo de onda, como la ola, en que la materia del medio se mueve de un lado a otro formando ángulos rectos a la dirección en que viaja la onda; posee crestas y senos. (Cap. 11, Sec. 1, pág. 328)

trough / seno: el punto más bajo de una onda transversal. (Cap. 11, Sec. 2, pág. 332)

turbine / turbina: rueda grande que gira cuando es empujada por el vapor, por el viento o por el agua y que provee energía mecánica a un generador. (Cap. 8, Sec. 3, pág. 242)

Tyndall effect / efecto de Tyndall: dispersión de un haz luminoso conforme atraviesa un coloide. (Cap. 17, Sec. 1, pág. 523)

U

ultrasonic / ultrasónico: ondas sonoras con frecuencias por encima de 20,000 Hz. (Cap. 12, Sec. 2, pág. 366)

ultraviolet wave / onda ultravioleta: onda electromagnética con frecuencias ligeramente más altas que la luz visible y la cual puede presentar efectos tanto útiles como perjudiciales. (Cap. 13, Sec. 2, pág. 399)

unsaturated hydrocarbon / hidrocarburo insaturado: compuesto, como el eteno o el etino, que contiene por lo menos un enlace doble o triple entre los átomos de carbono. (Cap. 21, Sec. 1, pág. 644)

unsaturated solution / solución no saturada: toda solución que puede disolver más soluto a una temperatura dada. (Cap. 23, Sec. 3, pág. 721)

V

variable / variable: factor que puede causar un cambio en los resultados de un experimento. (Cap. 1, Sec. 1, pág. 9)

velocity / velocidad: describe tanto la rapidez como la dirección de un cuerpo en movimiento. (Cap. 2, Sec. 1, pág. 44)

virtual image / imagen virtual: imagen vertical y reflejada que percibe el encéfalo aunque ningún rayo luminoso atraviese la imagen. (Cap. 15, Sec. 1, pág. 454)

viscosity / viscosidad: resistencia de un líquido a fluir. (Cap. 16, Sec. 2, pág. 501)

visible light / luz visible: ondas electromagnéticas con longitudes de onda de 390 a 770 billonésimas de metro que el ojo humano puede observar. (Cap. 13, Sec. 2, pág. 399)

voltage difference / diferencia de voltaje: empuje que hace que las cargas eléctricas fluyan por un conductor; se mide en voltios (V). (Cap. 7, Sec. 2, pág. 202)

volume / volumen: cantidad de espacio que ocupa un cuerpo. (Cap. 1, Sec. 2, pág. 18)

W

wave / onda: perturbación rítmica que transfiere energía a través de la materia o del espacio; existe solamente mientras posea energía para transportar. (Cap. 11, Sec. 1, pág. 326)

wavelength / longitud de onda: distancia entre un punto en una onda y el punto más cercano semejante en la onda siguiente; a medida que aumenta la frecuencia, la longitud de onda siempre disminuye. (Cap. 11, Sec. 2, pág. 333)

weak acid / ácido débil: todo ácido que sólo se ioniza parcialmente en una solución. (Cap. 25, Sec. 2, pág. 772)

weak base / base débil: toda base que no se ioniza totalmente en una solución. (Cap. 25, Sec. 2, pág. 773)

wedge / cuña: máquina simple que comprende un plano inclinado con uno o dos lados inclinados. (Cap. 5, Sec. 3, pág. 145)

weight / peso: fuerza gravitatoria que la Tierra ejerce sobre un cuerpo. (Cap. 3 Sec. 2, pág. 77)

wheel and axle / rueda y eje: máquina simple que consta de dos ruedas de diferente tamaño que giran juntas. (Cap. 5, Sec. 3, pág. 143)

work / trabajo: transferencia de energía que ocurre cuando una fuerza causa el movimiento de un cuerpo; se mide en julios. (Cap. 5, Sec. 1, pág. 126)

X

X ray / rayos X: onda electromagnética de hiperfrecuencia que puede viajar a través de la materia y que se utiliza a menudo en la formación de imágenes médicas. (Cap. 13, Sec. 2, pág. 401)

Index

The index for *Glencoe Physical Science* will help you locate major topics in the book quickly and easily. Each entry in the index is followed by the number of the pages on which the entry is discussed. A page number given in boldfaced type indicates the page on which that entry is defined. A page number given in italic type indicates a page on which the entry is used in an illustration or photograph. The abbreviation *act.* indicates a page on which the entry is used in an activity.

A

Index

D

E

Index

Index

F

G

H

M

N

Index

Q

R

Index

T

Art Credits

Glencoe would like to acknowledge the artists and agencies who participated in illustrating this program: Absolute Science Illustration; Andrew Evansen; Argosy; Articulate Graphics; Craig Attebery represented by Frank & Jeff Lavaty; CHK America; Gagliano Graphics; Pedro Julio Gonzalez represented by Melissa Turk & The Artist Network; Robert Hynes represented by Mendola Ltd.; Morgan Cain & Associates; JTH Illustration; Laurie O'Keefe; Matthew Pippin represented by Beranbaum Artist's Representative; Precision Graphics; Publisher's Art; Rolin Graphics, Inc.; Wendy Smith represented by Melissa Turk & The Artist Network; Kevin Torline represented by Berendsen and Associates, Inc.; WILDlife ART; Phil Wilson represented by Cliff Knecht Artist Representative; Zoo Botanica.

Photo Credits

Abbreviation Key: AH=Aaron Haupt; AP=Amanita Pictures; CB=CORBIS; BD=Bob Daemmrich; DM=Doug Martin; FI=First Image; FP=Fundamental Photographs; GB=Geoff Butler; GH=Grant Heilman Photography; IS=Index Stock; KS=KS Studios; LA=Liaison Agency; MM=Matt Meadows; PA=Peter Arnold, Inc.; PD=PhotoDisc; PE=PhotoEdit; PQ=PictureQuest; PR=Photo Researchers; PT=PhotoTake NYC; RM=Richard Megna; SB=Stock Boston; SPL=Science Photo Library; SS=SuperStock; TIB=The Image Bank; TIW=The Image Works; TSA=Tom Stack & Associates; TSM=The Stock Market; VU=Visuals Unlimited.

Cover Simon Terry/SPL/PR; **x** John S. Shelton; **xiv** Hank Morgan/Science Source/PR; **xvi** MM; **xvii** Dale Sloat/PT/PQ; **xviii** Walter H. Hodge/PA; **xix** MM; **xx** Mark E. Gibson/VU; **xxii** Ray Ellis/PR; **1** Tony Walker/PE; **2-3** John Terence Turner/FPG; **3** (t)Artville, (b)Charles L. Perrin; **4** GH; **4-5** AFP/CB; **5** FI; **6** Shuttle Mission Imagery/LA; **7** Will McIntyre/PR; **9** James L. Amos/CB; **10** David Young-Wolff/PE; **11** (l)Roger Ressmeyer/CB, (r)Douglas Mesney/TSM; **12** J. Marshall/TIW; **13** (t)Jonathan Nourok/PE, (b)Tony Freeman/PE; **14** FI; **15** Courtesy Bureau International Des Poids et Mesures; **16** MM; **17** (t)AP, (bl)Runk/Schoenberger from GH, (br)CB; **20** Stephen R. Wagner; **24** FI; **28** (t) AP, (b) BD; **29** Icon Images; **30** (l)Rosalie Winard, (r)courtesy Temple Grandin; **31** NASA; **32** (l)TSADO/NCDC/NOAA/TSA, (r)David Ball/TSM; **34** Clement Mok/PQ; **35** AP: **36** Alan Towse/ Ecoscene/CB; **36-37** Robert Mathena/FP; **37** Frank Lane/ Parfitt/Stone; **38** Icon Images; **41** Paul Silverman/FP; **42** A. & J. Verkaik/Skyart/TSM; **43** Terje Rakke/TIB; **44** (t)SS, (b)Robert Homes/CB; **49** (t)RDF/VU, (c)Ron Kimball, (b)RM/FP; **50** (l)The Image Finders, (r)Richard Hutchings; **51** Bill Aaron/PE; **52** Holway & Lobel Globus; **53** BD; **54** Rob Nelson/SB/PQ; **55** (t)Paul Kennedy/LA, (b)Courtesy Insurance Institute for Highway Safety; **56** Donald Johnston/ Stone; **57** FI; **58** (t)Rod Joslin, (b)Icon Images; **59** Icon Images; **60** Sylvain Grandadam/Stone; **61** Andreas Fechner; **62** (t)Tony Freeman/PE, (c)Peter Newton/Stone, (b)Tony Freeman/PE; **63** (l)Don C. Nieman, (r)Jim Steinberg/ PR; **64** Duomo; **66** NASA; **66-67** Tim Wright/CB; **67** Dominic Oldershaw; **68** David Young-Wolff/PE; **70** (t)Pictor, (c)M.W. Davidson/PR, (b)PD; **71** BD; **72** (t)BD, (b)Peter Fownes/ Stock South/PQ; **73** (t)Michael Newman/PE, (b)James Sugar/Black Star/PQ; **74** Keith Kent/PA; **76** CB; **77** Peticolas-Megna/FP; **78** NASA; **80** (tl)KS, (tr)David Young-Wolff/PE, (b)RM/FP; **81** Pictor; **83** Steven Sutton/Duomo; **84** Philip Bailey/TSM; **85** (t)North American Rockwell, (cl)NASA, (cr)Teledyne Ryan Aeronautical, (b)North American Rockwell; **86** StockTrek/CB; **87** (t)Tony Freeman/PE, (bl)Jeff Smith/Fotosmith, (br)RM/FP; **88** AP; **90** MM; **91** John Evans; **92** Bettmann/CB; **92-93** CB; **93** (l)CB, (r)George Hall/CB; **94** (t)Mark Burnett, (c)Robert Brenner/PE, (b)Dominic Oldershaw; **95** (l)Dominic Oldershaw, (r)Gunter Marx/CB; **96** Mary Ann McDonald/CB; **98** Alfred Pasieka/PA; **98-99** SS; **99** GB; **100** (l)Runk/Schoenberger from GH, (c)Michael Newman/PE, (r)Richard Clintsman/Stone; **101** (tl)Tony Walker/PE, (tr)D. Boone/CB, (b)FI; **105** KS; **109** Walter H. Hodge/PA; **110** RFD/VU; **114** Rudi Von Briel; **117** MM; **118** Maurites Escher; **119** (l)Brompton Studios, (r)Hank Morgan/PR; **120** (t)SS, (c)Telegraph Colour Library/FPG, (b)Jana R. Jirak/VU; **122** Courtesy Six Flags Amusement Parks; **124** Tom Pantages; **124-125** Eduardo Di Baia/AP/ Wide World Photos; **125** Timothy Fuller; **126 127** Michael Newman/PE; **129** Michelle Bridwell/PE; **131** Jules Frazier/PD; **132** Michael Newman/PE; **133** (t)Mark Burnett, (b)Tony Freeman/PE; **134** Joseph P. Sinnot/FP; **135** RM/FP; **138** Yoav Levy/PT; **139** (tl)Tom Pantages, (tr)Mark Burnett, (b)Tony Freeman/PE; **140** (t)Richard T. Nowitz, (c)David Madison/Stone, (b)John Henley/TSM; **141** A.J. Copley/VU; **143** Mark Burnett; **144** Tom Pantages; **145** (tl tc)file photo, (tr)Mark Burnett, (b)AP; **146** SS; **147** Mark Burnett; **148** Bruno P. Zehnder/PA; **149** Hickson-Bender; **150** Joe Lertola; **151** (t)Sandia National Laboratory, (b)courtesy D. Carr & H. Craighead, Cornell University; **152** (t)Ed Lallo/LA, (bl)file photo, (br)C Squared Studios/PD; **153** (t bl)Siede Preis/PD, (br)Glenn Mitsui/PD; **156** Stone; **156-157** Charles E. Rotkin/CB; **157** MM; **160** AH; **165** Peter Ardito/IS; **166** (t)Earth Imaging/Stone, (bl)K&K Arman/ Bruce Coleman, Inc./PQ, (br)Charles & Josette Lenars/CB; **168** (l)Doug Cheeseman/PA, (c)Tim Davis/TSM, (r)Ed Reschke/PA; **169** David Madison; **171** Leonard Lee Rue III/Bruce Coleman, Inc./PQ; **173** Steve Callahan/VU; **175** Stu Rosner/SB; **180** MM; **182** (t)NASA, (c)Roy Johnson/TSA, (b)NASA; **182-183** StockTrek/PD; **183** (t)Michael Newman/PE, (b)NASA; **184** (t)Dean Conger/CB, (c)Myrleen Ferguson Cate/PE, (b)MM; **185** Sami Sarkis/PD; **188** Tony Freeman/PE; **190-191** K. Yamashita/Panoramic Images; **191** Digital Stock; **192** Kip Peticolas/FP; **192-193** Richard Pasley/ SB/PQ; **193** Michael Newman/PE; **195** GB; **197 198** KS; **199** T. Wiewandt/DRK Photo; **200** Dale Sloat/PT/PQ; **205** Ray Ellis/PR; **206** Thomas Veneklasen; **208 209 210** GB; **212** (l)GB, (r)Tony Freeman/PE; **213** AH; **216** GB; **217** GB; **218** B. Gotfryd/Woodfin Camp & Associates; **219** Courtesy Carnegie Mellon University, School of Computer Science; **220** (t)file photo, (bl)Lester Lefkowitz/TSM, (br)Harold Stucker/Black Star; **224** Yoav Levy/PT; **224-225** Steele Hill/NASA; **225** Thomas Veneklasen; **227 228** RM/FP; **231** (l)Stephen Frisch/SB, (c)Mark Burnett, (r)Stephen Frisch/SB; **232** Mark Burnett; **235** Icon Images; **242** (t)Tom Campbell/FPG, (b)Russell D. Curtis/PR; **243** (l, t to b)Clement Mok/PQ, Tim Ridley/DK Images, Brian Gordon Green, (r, t to b) Joseph Palmieri/Pictor, Robin Adshead/

Military Picture Library/CB, Tim Ridley/DK Images; **244** Mark Burnett; **247 248 249** KS; **250** Laurence Dutton/Stone; **251** (l)Brian Blauser/Stock Shop, (r)Mehau Kulyk/PR; **252** (t)Mark Burnett, (c)Jeff Greenberg/PA, (b)Michael Newman/PE; **256** Timothy Fuller; **256-257** NASA/LA; **257** Dominic Oldershaw; **258** (l)FI, (r)David Frazier/TIW; **262** American Institute of Physics; **264** AP; **268** (l)courtesy Supersaturated Environments, (r)SPL/PR; **269** (t)CERN, P. Loiez/SPL/PR, (b)RM/FP; **270** Glen Allison/PD; **271** Hank Morgan/PR; **276** Oliver Meckes/Nicole Ottawa/PR; **277** (t)Davis Meltzer, (b)UCLA School of Medicine; **278** J.L. Carson/Custom Medical Stock Photo; **279** Dominic Oldershaw; **280** (t)Fermilab/VU, (b)MM; **281** MM; **282** Hulton-Getty/LA; **283** (t)Nicholas Veasey/Stone, (b)Rob Maass/Sipa Press; **284** (t)Photri/TSM, (b)Martha Cooper/PA; **285** Richard Green/PR; **286** Oliver Meckes/Nicole Ottawa/PR; **288** Joseph Sohm/ChromoSohm, Inc./CB; **288-289** Andy Sacks/Stone; **289** MM; **290** (l)Jan Halaska/PR, (c)James Sugar/Black Star, (r)Bill Heinsohn/Stone; **292** Robert E. Pratt; **293** AP; **294** (t)Digital Vision/PQ, (b)DM; **298** Robert Essel/TSM; **299** PR; **302** AFP/CB; **303** Tim Wright/CB; **305** AP; **307** S.K. Patrick/TSA; **308** (t)John Elk III/SB, (b)Lester Lefkowitz/TSM; **310** Martin Bond/SPL/PR; **312** Timothy Fuller; **313** Dominic Oldershaw; **314** (l)Sally A. Morgan/Ecoscene/CB, (r)Lester Lefkowitz/TSM; **315** (t)Ahmet Sel/Sipa Press, (b)CB Sygma; **316** (t)AP, (c)Kathy Ferguson/PE, (b)US Department of Energy/SPL/PR; **317** (l)Paul A. Souders/CB, (r)Mark Newham/Eye Ubiquitous/CB; **318** Lester Lefkowitz/TSM; **320** Robert Holmes/CB; **322** John Eastcott & Yva Momatiuk/Woodfin Camp & Associates; **322-323** Nonstock Inc.; **324** Francois Gohier/PR; **324-325** Will Dickey, Florida Times Union/AP/Wide World; **325** MM; **326** James H. Karales/PA; **328** SS; **330** (t)Galen Rowell, (b)Stephen R. Wagner; **338** DM; **339** Richard Hutchings/PR; **340** (t)RM/FP, (b)SIU/VU; **342** (t)John S. Shelton, (b)RM/FP; **343** RM/FP; **346** Kodansha; **348 349** DM; **350** U.S. Naval Institute; **351** Ralph White/CB; **352** (tl)Stan Osolinski/TSM, (tr)Bruce Heinemann/PD, (b)Kevin Fleming/CB; **356** TSM; **356-357** Jonathan Blair/CB; **357** Mark Burnett; **360** AH; **362** Prof. P.L. Motta/Dept. of Anatomy/University la Sapienza, Rome/SPL/PR; **364** Michael Newman/PE; **365** (tl)Ian O'Leary/Stone, (tr)David Young-Wolff, (bl)Mark A. Schneider/VU, (bc)Rafael Macia/PR, (br)SS; **368** NOAA/OAR/ERL/National Severe Storms Laboratory (NSSL); **369** (t)Jean Miele/TSM, (b)FI; **371** (t)Charles Gupton/Stone, (b)Will Hart/PE; **372** (t)Nick Rowe/PD, (b)Derick A. Thomas, Dat's Jazz/CB; **374** DM; **375** Araldo de Luca/CB; **376** (tl)Gary W. Carter/CB, (tr)Merlin D. Tuttle, (b)Stephen R. Wagner; **378** DM; **380** (t)Hisham F. Ibrahim/PD, (b)John Wang/PD; **381** David Young-Wolff/PE; **382** (t)Zig Leszczynski/Earth Scenes, (cl)Australian Picture Library/CB, (c)PD, (cr)CB, (bl)David Sacks, (br)Ethan Miller/CB; **384** (tl)Planet Earth Pictures/FPG, (tr)CMCD/PD, (b)Robert Brenner/PE; **385** (l)Steve Cole/PD, (c)C Squared Studios/PD, (r)CMCD/PD; **386** KS; **388** Vince Streano/CB; **388-389** VCG/FPG; **389** Spencer Grant/PE; **393** Matt Jacob/TIW; **398** (t)Pete Saloutos/TSM, (bl)NASA/GSFC/JPL, MISR and AirMISR Teams, (br)NASA/GSFC/MITI/ERSDAC/JAROS, and USA/Japan ASTER Science Team; **399** AP/Wide World Photos; **401** Telegraph Colour Library/FPG; **403** StudiOhio; **404** (t bl)FI, (br)PD; **405** (t)Mark Richards/PE/PQ, (c)Stephen Frisch/SB/PQ, (bl)CB, (bc)Dorling Kindersley, (br)Michael Newman/PE/PQ; **406** Nino Mascardi/TIB; **407** Photolink/PD; **408** VCG/FPG; **410** Len Delessio/IS; **411** Jim Wark/IS; **412-413** It Stock International/IS; **414** (tl)Martin Dohrn/SPL/PR, (tr)Mickey Pfleger/Photo 20-20, (b)Joe McDonald/TSA; **415** (t)Susan Leavises/PR, (c)Alain Evrard/LA, (b)Myrleen Ferguson; **416** Shaun Egan/Stone; **418** Mickey Pfleger/Photo 20-20/PQ; **418-419** Richard During/Stone; **419** Dominic Oldershaw; **420** FI; **422** (tl)Bruce Iverson, (tr)Myrleen Cate/IS/PQ, (b)Stephen Frisch/SB/PQ; **423** David Parker/SPL/PR; **424** Charles O'Rear/CB; **426** FI; **427** (t)Ralph C. Eagle, Jr./PR, (b)Diane Hirsch/FP; **428** MM; **431** Ginger Chih/PA; **432** FI; **433** (t)Will & Deni McIntyre/ PR, (others)Marvin J. Fryer; **435** (t)Paul Silverman/FP, (b)Andrew Syred/SPL/PR; **437** Larry Mulvehill/PR; **439** Paul Silverman/FP; **440** David Young-Wolff/PE; **441** DM; **442** (t)Timothy Fuller, (b)Dominic Oldershaw; **443** Dominic Oldershaw; **444** (t)Archivo Iconografico, S.A./CB, (b)Ashmolean Museum, Oxford, UK/Bridgeman Art Library, London/NY; **445** Courtesy Maria Martinez-Cañas and Fredric Snitzer Gallery; **446** (l)Dick Poe/VU; (r)Timothy Fuller; **448** AH; **450** Bob Rowan/Progressive Image/CB; **450-451** John Elk III/SB/PQ; **451** Mark Burnett; **452** CB; **453** Jeff Greenberg/VU; **456** George B. Diebold/TSM; **457** Paul A. Souders/CB; **459** MM; **460** David Parker/SPL/ PR; **466** (tr)Stephen R. Wagner, (bc br)courtesy Optobionics Corp., (others) National Eye Institute, National Institutes of Health; **470** NASA/Consolidated News Pictures/Archive Photos; **472 473** Breck P. Kent/Earth Scenes; **474** (t)NASA, (b)Dominic Oldershaw; **475** Dominic Oldershaw; **476** Hemsey/LA; **477** (tl)Max Aquilera-Hellweg/Timepix, (tr)AFP/CB, (b)Louis Psihoyos/Matrix; **478** (t)David Young-Wolff/PE, (c)John Durham/SPL/PR, (b)Mark Burnett; **479** (l)DM, (r)Croan Studio; **484-485** (foreground) National Geographic Society, (background)Granger Collection, NY; **486** Alfred Pasieka/PA; **486-487** TSM; **487** MM; **488** Icon Images; **491** DM; **492** JISAS/Lockheed/SPL/PR; **493** (t)CB, (b)Mark E. Gibson/VU; **495** AP; **496** MM; **500** BD; **502** file photo; **504** Mark Burnett/PR; **507** MM; **508** John Evans; **510** (t)Geoff Tomkinson/SPL/PR, (c)BD/SB, (b)Michael Dwyer/SB; **510-511** StockTrek/PD; **511** (l)StockTrek/PD, (r)Jules Frazier/PD; **512** (t)Breck P. Kent/Animals Animals, (c)AH, (b)Icon Images; **516** MM; **516-517** Michael Newman/PE; **517** MM; **518** (l)Siede Pries/PD, (c)Ryan McVay/PD, (r)PD; **519** (tl)Yann Arthus-Bertrand/CB, (tc)James L. Amos/CB, (tr)Stockbyte/PQ, (c)CalTech, (bl)Robert Fried/SB/PQ, (br)Dorling Kindersley; **520** (t)RM/FP, (b)Icon Images; **521** (t)Icon Images, (b)Andrew Syred/SPL/PR; **522** (t)Icon Images, (bl)PD, (br)C Squared Studios/PD; **523** (tl)Chuck Keeler, Jr./TSM, (tr)Kaj R. Svensson/SPL/PR, (b)Kip and Pat Peticolas/FP; **524** Larry Mayer/LA; **525** Icon Images; **526** MM; **527** (t)AP, (b)DM; **528** RM/FP; **529** Jed & Kaoru Share/Stone; **530** Tim Courlas; **532** (l)M. Romesser, (r)SS; **533** Chris Rogers/TSM; **534** (t)MM, (b)Icon Images; **535** Icon Images; **536** (tl)Dominic Oldershaw, (tr)George Hall/CB, (b)The Denver Post/AP/Wide World Photos; **536-537** (br)Peter Johnson/CB; **537** (l)DM, (r)Frank Siteman/SB/PQ; **538** (tl)Icon Images, (tr)Will Ryan/TSM, (c)Icon Images, (b)Tony Freeman/PE; **539** (l)Mary Kate Denny/PE, (r)Dean Conger/CB; **542** PR; **542-543** CB; **543** Icon Images; **546** (tl)SPL/PR, (tr)Hank Morgan/Science Source/PR, (b)PR; **547** Sheila Terry/SPL/PR; **548** file photo; **550** MM; **554** (l)Bettmann/CB, (r)Science Museum/Science & Society Picture Library; **562** CB; **563** TIB; **564** Tom Pantages; **565** PE; **566** Roger Ressmeyer/CB; **567** (t)John Bolzan, (b)Maria Stenzel/National Geographic Image Collection; **572** Yoav Levy/PT/PQ; **572-573** Aerial Focus/IS; **573** AP; **574** (l)John Evans, (r)CB; **575** (l)Runk/Schoenberger from GH, (c)RM/FP, (r)TIW; **579** FI; **580** John Paul Kay/PA; **584** John Evans; **585** (t)Patricia Lanza, (bl br)file photo; **587** Bettmann/CB; **591** Courtesy Süd-Chemie Performance Packaging; **592** AP; **594** (t)AP, (b)Richard Hutchings; **595** Richard Hutchings; **596 597** Daniel Belknap; **598** AP; **603** Yoav Levy/PT; **604-605** Jon Feingersh/TSM; **605** Dorling Kindersley; **606** Dan McCoy from Rainbow/PQ; **606-607** Georg Gerster/PR; **607** KS; **608** (l)Henry Groskinsky/TimePix, (r)Nubar Alexanian/CB; **610** (l)Stephen Frisch/SB, (r)Jerry Mason/SPL/PR; **611** Firefly Productions/TSM; **612** Mark Burnett/PR; **613** (tl)MM, (tr)Icon Images, (b)Jeff J. Daly/FP; **615** Melvyn P. Lawes/Papilio/CB; **617** (l to r)Russ Lappa/ Science Source/PR, Gregory G. Dimijian/PR, Stephen Frisch/SB, Andrew J. Martinez/PR, DM, Charles Gupton/ SB; **618** (t)Spencer Grant/PE/PQ, (b)Dick Frank/TSM; **619** (t)Norris Blake/VU, (b)Steven Senne/AP/Wide World Photos; **620** Gene J. Puskar/AP/Wide World Photos; **622** George Hall/CB; **623** (l)Icon Images, (r)Charles D. Winters/PR; **624** (l)AH, (c)Rick Gayle Studio/TSM, (r)Alfred Pasieka/ SPL/PR; **626** (l)RMIP/Richard Haynes, (r)PD; **628** (tl)Gianni Dagli Orti/CB, (tc)Jonathan Blair/CB, (tr)Patricia Lanza, (cl)CB, (ccl)Tony Freeman/CB, (c)PD, (ccr)Paul A. Souders/CB, (cr)Bettmann/CB, (bl)Steve Cole/CB, (bc)Artville, (br)Achim Zschau; **630** (t)Charles D. Winters/PR, (b)Richard Hutchings; **631** Richard Hutchings; **632** (t)Ken Reid/FPG, (b)Lynn Johnson/Aurora; **633** Flip Chalfant/TIB; **634** (tl)Charles D. Winters/PR, (tcl)Stephen Frisch/SB, (tcr)Charles E. Zirkle, (tr)DM, (bl)Russ Lappa/Science Source/PR, (br)Karl Hartmann/Sachs/PT; **638** Ken Eward/Science Source/PR; **638-639** KS; **639 640** KS; **642** (tl)courtesy Land's End, Inc., (tr)CB, (c)Bernard Roussel/TIB, (bl)George Hall/Check Six/PQ, (br)PD; **644** (l)Jens Jorgen Jensen/TSM, (r)PD; **645** (l)AH, (r)FI; **646** Ken Frick; **647** (l)AH, (c r)Timothy Fuller; **648** Lynn M. Stone/IS; **649** KS; **654** (l)GB, (r)Charles D. Winters/PR; **655** AP; **657** David M. Phillips/VU; **660** (t)Bob Mullenix, (b)Diana Calder/TSM; **661** Ken Frick; **662** MM; **664 665** Courtesy 3M; **666** (tl)Francis & Donna Caldwell/VU, (tr)Richard Hutchings/PR, (bl)PD, (br)IS; **670** Meckes/Ottawa/PR; **670-671** NASA/LA; **671** MM; **673** (l)Icon Images, (r)IS; **674** Icon Images; **675** (t)Mark Thomas/FoodPix, (bl)SIU/VU, (br)Jonathan Blair/CB; **676** Courtesy Boeing; **677** (l)James L. Amos/PA, (r)NASA; **679** KS; **680** Courtesy Zimmer, Inc.; **682** Michael Philip Manheim/Photo Network/PQ; **683** (t)Bettmann/CB, (cl)L. Mulvehill/PR, (c)courtesy Texas Instruments, (cr)Bettmann/CB, (bl)David Young-Wolff/PE, (br)Charles O'Rear/CB; **684** AP; **687** (l)Leonard Lessin/PA, (c)Myrleen Ferguson/PE, (r)Robert Rathe/SB; **688** (t)Doug Wilson/CB, (bl)courtesy Dallara, (br)Jon Eisberg/FPG; **689** (l)Herb Charles Ohlemeyer/Fran Heyl Associates, (r)Cindy Lewis; **690** Courtesy Boeing; **691** AH; **692** Icon Images; **693** Timothy Fuller; **694** (l)Steve Taylor/Stone, (r)Steve Labadessa/Timepix; **694-695** Bernard Roussel/TIB; **695** Courtesy 3M; **696** (t)Tony Freeman/PE, (bl)Mark Wagner/Stone, (br)Flexdex Skateboards www.flexdex.com. Photo by John Durant Photography; **698** KS; **701** (t)Elaine Shay, (c)Ken Frick, (bl)Phillip A. Harrington/PA, (br)Jules Frazier/PD; **702-703** Norbert Wu; **703** Spencer Collection, NY Public Library, Astor, Lenox & Tilden; **704** RM/FP; **704-705** Michael Pogany/Columbus Zoo; **705** MM; **706** (l)Christie's Images, (r)Flip Schulke/Black Star; **707** Annie Grifiths Belt/CB; **708** (tl)Bettmann/CB, (tr)Brian Gordon Green, (b)Michael Newman/PE/PQ; **710 713 714** Icon Images; **715** Steve Mason/PD; **716 717** KS; **719** Mark Thayer; **721** RM/FP; **722** David Young-Wolff/PE; **727 728 729** MM; **730** (t)Tony Freeman/PE, (bl)AISI/VU, (br)John Evans; **730-731** CB; **731** (t)Standard-Examiner, Kort Duce/AP/Wide World Photos, (b)Alinari/Art Resource, NY; **732** Icon Images; **736** RM/FP; **736-737** Transglobe/IS; **737** MM; **739** Mansell Collection/TimePix; **740** RM/FP; **742** CB; **743** RM/FP; **744** DM/PR; **745** RM/FP; **746** Emory Kristof/National Geographic Image Collection; **747** (t)Charles D. Winters/PR, (b)Stephen Frisch/SB/PQ; **749** RM/FP; **750** Tim Matsui/LA; **751** (t)Roger Ressmeyer/NASA/CB, (c)Bettmann/CB, (bl)Bettmann/CB, (br)NASA/CB; **752** MM; **753** (t)Jeff J. Daly/FP, (b)Spencer Grant/PE; **755** Timothy Fuller; **756** UNEP-Topham/TIW; **757** Michael Newman/PE; **758** (t)Bibliotheque du Museum D'Histoire Naturelle, (b)Paul Chesley/Stone; **759** (t)Brad Maushart/ Graphistock, (b)Owen Franken/CB; **760** (t)Icon Images, (c)Charles D. Winters/PR, (b)Joseph P. Sinnot/FP; **764** Michael Dalton/FP; **764-765** Luiz C. Marigo/PA; **765** Bjorn Bolstad/PA; **766** KS; **767** MM; **768** GB; **769** Rick Poley/IS; **772** MM; **774** (vinegar)Elaine Shay, (milk)Brent Turner/BLT Productions, (baking soda)MM, (ammonia) Elaine Shay, (cola, tomato)CB, (milk of magnesia)Icon Images, (drain cleaner)StudiOhio, (c)Mark Burnett/SB/PQ, (bl)Dominic Oldershaw, (br)MM; **777** Charles D. Winters/ PR; **778** The Purcell Team/CB; **779** (t)Wolkmar Wentzel, (cl)Kevin Schafer/CB, (cr)CB, (b)Kevin Schafer/CB; **780** RM/FP; **781** (l)MM, (r)AP; **783** Chinch Gryniewicz/ Ecoscene/CB; **784** AP; **785** PD; **786** (t)Bill Lyons/LA, (b)RM/FP; **787** RM/FP; **790** (l)KS, (r)MM; **791** (l)John Evans, (r)Martyn F. Chillmaid/SPL/PR; **793** (l)Mike & Carol Werner/SB/PQ, (r)Paul Silverman/FP; **794** Egyptian National Museum, Cairo, Egypt/SS; **796-797** PD; **798** Timothy Fuller; **802** Roger Ball/TSM; **804** (l)GB, (r)Coco McCoy from Rainbow/PQ; **805** Dominic Oldershaw; **806** StudiOhio; **807** FI; **809** MM; **812** Paul Barton/TSM; **815** Davis Barber/PE.

Acknowledgments

Two haiku from A Haiku Garden, copyright © 1996 by Stephen Addis. Reprinted by permission of Weatherhill, Inc.; From *A Brave and Startling Truth,* by Maya Angelou, copyright © 1995 by Maya Angelou. Used by permission of Random House, Inc.; From *The Invisible Man* by Ralph Ellison. Copyright © 1952 by Ralph Ellison. Reprinted by permission of Random House, Inc.; From *Thinking in Pictures,* by Temple Grandin, copyright © 1995 by Temple Grandin. Used by permission of Doubleday, a division of Random House, Inc.

Handbook for Texas Students & Parents

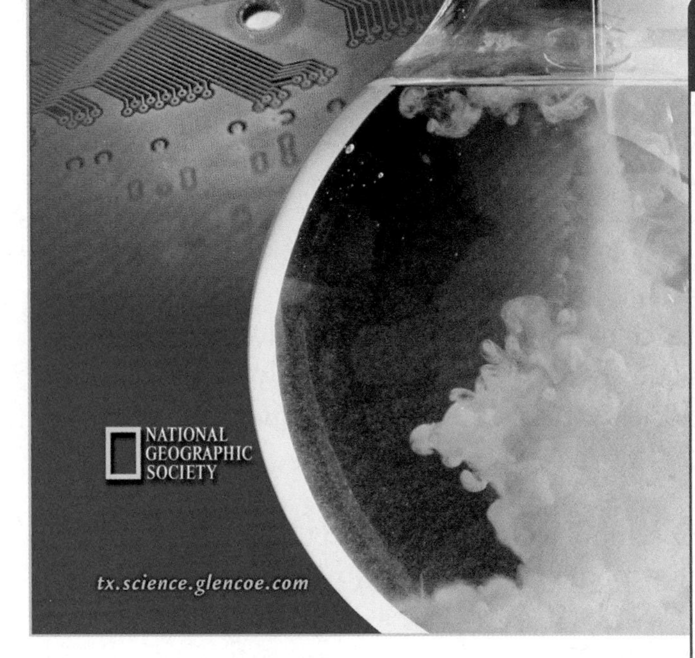

Visit us at tx.science.glencoe.com for additional TAKS practice.

Table of Contents

New York, New York Columbus, Ohio Woodland Hills, California Peoria, Illinois

To Parents and Students

Preparing to Take the TAKS Test

The following pages have been provided to help you prepare for the TAKS examination that each Texas student must take at the end of tenth and eleventh grade. Your involvement can help your student have a positive attitude about the test and help you focus your student's study efforts on the content that will be tested. The TAKS test may cover selected TEKS objectives from Grade 8 Science, from Integrated Physics and Chemistry, and from Biology.

In these pages, you will find the following:

- **A review of the Grade 8 Science TEKS for physics and chemistry concepts that your student learned in Grade 8 Science**
  Some of the Grade 8 Science TEKS may be tested on TAKS. You can review the objectives tested on TAKS by visiting the Glencoe Science Web site at **tx.science.glencoe.com**.

- **A sample of the types of questions that your student will see on the TAKS test**
  These questions cover each of the six TAKS test objectives. For complete sample tests, as well as additional questions that review specific objectives, visit the Glencoe Science Web site at **tx.science.glencoe.com**.

TAKS Periodically, this symbol will appear within the list of TEKS Student Expectations. The symbol indicates that this TEKS objective is on TAKS.

For complete TAKS practice tests, as well as additional practice for each objective, visit **tx.science.glencoe.com**.

TEKS, Integrated Physics and Chemistry

TAKS *Highlighted TEKS are tested on TAKS.*

TEKS

(1) Scientific processes. The student, for at least 40% of instructional time, conducts field and laboratory investigations using safe, environmentally appropriate, and ethical practices. The student is expected to:
(A) demonstrate safe practices during field and laboratory investigations; and
(B) make wise choices in the use and conservation of resources and the disposal or recycling of materials.

(2) Scientific processes. The student uses scientific methods during field and laboratory investigations. The student is expected to:
(A) plan and implement investigative procedures including asking questions, formulating testable hypotheses, and selecting equipment and technology;
(B) collect data and make measurements with precision;
(C) organize, analyze, evaluate, make inferences, and predict trends from data; and
(D) communicate valid conclusions.

(3) Scientific processes. The student uses critical thinking and scientific problem solving to make informed decisions. The student is expected to:
TAKS *(A)* analyze, review, and critique scientific explanations, including hypotheses and theories, as to their strengths and weaknesses using scientific evidence and information;
TAKS *(B)* draw inferences based on data related to promotional materials for products and services;
(C) evaluate the impact of research on scientific thought, society, and the environment;
(D) describe connections between physics and chemistry, and future careers; and
(E) research and describe the history of physics, chemistry, and contributions of scientists.

(4) Science concepts. The student knows concepts of force and motion evident in everyday life. The student is expected to:
TAKS *(A)* calculate speed, momentum, acceleration, work, and power in systems such as in the human body, moving toys, and machines;
TAKS *(B)* investigate and describe applications of Newton's laws such as in vehicle restraints, sports activities, geological processes, and satellite orbits;
(C) analyze the effects caused by changing force or distance in simple machines as demonstrated in household devices, the human body, and vehicles; and
TAKS *(D)* investigate and demonstrate mechanical advantage and efficiency of various machines such as levers, motors, wheels and axles, pulleys, and ramps.

(5) Science concepts. The student knows the effects of waves on everyday life. The student is expected to:
TAKS *(A)* demonstrate wave types and their characteristics through a variety of activities such as modeling with ropes and coils, activating tuning forks, and interpreting data on seismic waves;
(B) demonstrate wave interactions including interference, polarization, reflection, refraction, and resonance within various materials;
(C) identify uses of electromagnetic waves in various technological applications such as fiber optics, optical scanners, and microwaves; and
(D) demonstrate the application of acoustic principles such as in echolocation, musical instruments, noise pollution, and sonograms.

(6) Science concepts. The student knows the impact of energy transformations in everyday life. The student is expected to:

TAKS *(A)* describe the law of conservation of energy;
TAKS *(B)* investigate and demonstrate the movement of heat through solids, liquids, and gases by convection, conduction, and radiation;
(C) analyze the efficiency of energy conversions that are responsible for the production of electricity such as from radiant, nuclear, and geothermal sources, fossil fuels such as coal, gas, oil, and the movement of water or wind;
TAKS *(D)* investigate and compare economic and environmental impacts of using various energy sources such as rechargeable or disposable batteries and solar cells;
(E) measure the thermal and electrical conductivity of various materials and explain results;
TAKS *(F)* investigate and compare series and parallel circuits;
(G) analyze the relationship between an electric current and the strength of its magnetic field using simple electromagnets; and
(H) analyze the effects of heating and cooling processes in systems such as weather, living, and mechanical.

(7) Science concepts. The student knows relationships exist between properties of matter and its components. The student is expected to:
TAKS *(A)* investigate and identify properties of fluids including density, viscosity, and buoyancy;
(B) research and describe the historical development of the atomic theory;
(C) identify constituents of various materials or objects such as metal salts, light sources, fireworks displays, and stars using spectral-analysis techniques;
TAKS *(D)* relate the chemical behavior of an element including bonding, to its placement on the periodic table; and
TAKS *(E)* classify samples of matter from everyday life as being elements, compounds, or or mixtures.

(8) Science concepts. The student knows that changes in matter affect everyday life. The student is expected to:
TAKS *(A)* distinguish between physical and chemical changes in matter such as oxidation, digestion, changes in states, and stages in the rock cycle;
(B) analyze energy changes that accompany chemical reactions such as those occurring in heat packs, cold packs, and glow sticks to classify them as endergonic or exergonic reactions;
TAKS *(C)* investigate and identify the law of conservation of mass;
(D) describe types of nuclear reactions such as fission and fusion and their roles in applications such as medicine and energy production; and
(E) research and describe the environmental and economic impact of the end-products of chemical reactions.

(9) Science concepts. The student knows how solution chemistry is a part of everyday life. The student is expected to:
(A) relate the structure of water to its function as the universal solvent;
TAKS *(B)* relate the concentration of ions in a solution to physical and chemical properties such as pH, electrolytic behavior, and reactivity;
(C) simulate the effects of acid rain on soil, buildings, statues, or microorganisms;
TAKS *(D)* demonstrate how various factors influence solubility including temperature, pressure, and nature of the solute and solvent; and
(E) demonstrate how factors such as particle size, influence the rate of dissolving.

Grade 8 TEKS Review

On the following pages, you will find a review of the major physics and chemistry concepts that you learned in Grade 8 Science. The review is based upon the Grade 8 TEKS. Use this review as you study the physics and chemistry science concepts required by the TEKS for Integrated Physics and Chemistry.

Most Grade 8 Science TEKS contain topics that are not called for in the Integrated Physics and Chemistry TEKS. Therefore, the Grade 8 Science TEKS reviewed here are only those that support content concepts covered in the Integrated Physics and Chemistry TEKS that you will study this year.

CONTENTS

There Is a Relationship Between Force and Motion

TEKS 8.7

There is a relationship between force and motion.

8.7 A demonstrate how unbalanced forces cause changes in the speed or direction of an object's motion

8.7 B recognize that waves are generated and can travel through different media

Directions *Review these concepts from Grade 8. They will help prepare you to study Chapters 2, 3, 11, 12, and 13 on the relationship between force and motion, a topic required by the Integrated Physics and Chemistry TEKS.*

For 8.7 A, recall that:

A force is a push or a pull. Usually, two or more forces act on an object at the same time. The sum of the forces acting on an object is called the net force. When two or more forces act on an object, the forces are balanced or unbalanced.

Balanced forces are two or more forces that push or pull equally in opposite directions and cause no change in the speed or direction of an object's motion. Balanced forces have a net force of zero. A can of soda resting on a table is pulled down by the force of gravity and pushed up by the table. These are balanced forces, so the motion of the can does not change.

When unbalanced forces act on an object, the object changes speed or direction of motion. You might assume that any time an object is moving, it is being acted upon by unbalanced forces. However, only when an object is slowing down, speeding up, or changing directions are the acting forces unbalanced. Dragging your foot on the ground while on your bike, as shown in **Figure 1,** changes the speed of the bike. While your bike is slowing down, the forces acting on it are unbalanced. Unbalanced forces produce a net force that is not equal to zero.

Unbalanced forces also cause changes in the direction of movement. If you throw a baseball, it does not continue forever in a straight line. Instead, the force of gravity pulls the ball downward so it curves toward the ground.

Figure 1
When this bike rider exerts a force to slow the speed of the bike, the forces acting on the bike are unbalanced.

For 8.7 B, recall that:

A wave carries energy, but it does not carry matter. Waves are classified by whether or not they require matter to travel through. Both types of waves are an important part of your everyday life.

TEKS 8.7 A Assessment

1. What is the value of the net force when two balanced forces act on an object?
2. What are three indications that unbalanced forces are acting on a moving object?
3. **Thinking Critically** Describe your trip to school this morning. During which parts of the trip were unbalanced forces acting on you or on a vehicle?

Mechanical Waves

Any wave that can travel only through matter, called a medium, is a mechanical wave. Mechanical waves can be further classified as either compressional or transverse.

Compressional waves cause the matter that makes up the medium to move in the same direction as the wave. Compressional waves can travel through media that are liquid, solid, or gaseous. Sound waves, like the ones generated by the harp in **Figure 2,** are compressional waves.

Transverse waves cause matter in the medium to move at right angles to the direction the wave travels. Transverse mechanical waves usually travel through solid media.

Electromagnetic Waves

Electromagnetic waves can travel through space that contains no matter, as well as through solids, liquids, and gases. Radio waves, microwaves, and light are examples of electromagnetic waves. Electromagnetic waves are a type of transverse waves.

Figure 2
The sound waves generated by this harp are mechanical, compressional waves. The medium these waves travel through is air.

TEKS 8.7 B Assessment

1. Give an example of a compressional wave and a transverse wave traveling through a liquid medium.
2. Does a wave carry matter only, energy only, or both matter and energy?
3. **Thinking Critically** What type or types of waves are involved when you enjoy listening to your favorite song on the radio?

TEKS Review

Matter is Composed of Atoms

TEKS 8.8

Matter is composed of atoms.

8.8 A describe the structure and parts of an atom

8.8 B identify the properties of an atom including mass and electrical charge

Directions *Review these concepts from Grade 8. They will help prepare you to study Chapters 16, 17, 18, and 19 on matter and atoms, a topic required by the Integrated Physics and Chemistry TEKS.*

For 8.8 A, recall that:

Much matter is composed of atoms, which are tiny particles, which, under usual conditions, can't be further divided. Atoms are so small that about 2.5 million of them could fit on the head of a pin. The work of many scientists has contributed to today's understanding of the structure of the atom.

Atoms have three main components. Protons are particles with a positive charge, neutrons are neutrally charged particles, and electrons have a negative charge. The model currently used to describe the structure of the atom is the electron cloud model, as shown in **Figure 1**. Protons and neutrons are located in the center of the atom, called the nucleus. The electrons move in a cloud that surrounds the nucleus. While the electrons in the cloud are most likely to be located near the nucleus, no electron has a set position. Equal numbers of protons and electrons are found in a neutral atom, but other combinations are possible.

Elements are composed of atoms. Each atom of a particular element always has the same number of protons in the nucleus.

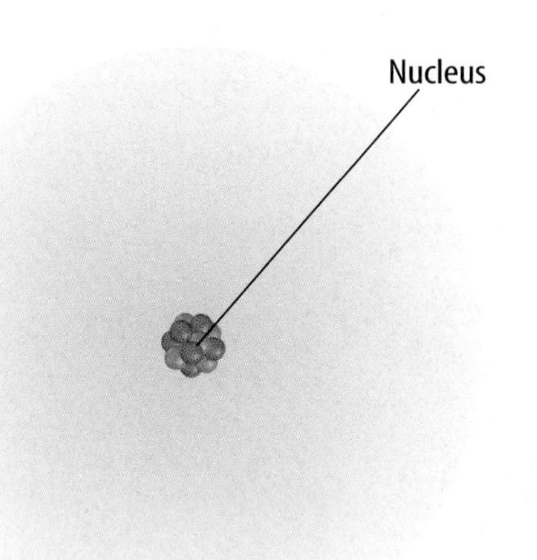

Figure 1
The atom's electrons are located in the electron cloud surrounding the nucleus.

TEKS 8.8 A Assessment

1. What components of an atom are found in the nucleus?
2. What model do scientists currently use to describe the structure of the atom?
3. **Thinking Critically** Why are the electrons in the electron cloud most likely to be located near the nucleus?

Table 1 Isotopes of Carbon

Isotope	Atomic Number	Mass Number	Number of Protons	Number of Neutrons	Number of Electrons
carbon-12	6	12	6	6	6
carbon-13	6	13	6	7	6
carbon-14	6	14	6	8	6

For 8.8 B recall that:

Atoms can be identified by several different properties. The mass number of an atom is determined by adding the number of neutrons and the number of protons in its nucleus. Recall that all atoms of the same element have the same number of protons in the nucleus. However, there can be variation in the number of neutrons, as shown in **Table 1.** This table shows how mass number, atomic number, and the number of subatomic particles are related for three forms of the element carbon. These different forms of an element are called isotopes. They each have the same number of protons but different mass numbers.

The atomic number of an atom is the number of protons in the nucleus. Atoms of different elements always will have different atomic numbers. Isotopes, because they are different forms of the same element, always have the same atomic number.

Neutral atoms have no net electrical charge. The equal numbers of positively charged protons and negatively charged electrons cause the net charge to be zero. However, atoms can gain or lose electrons. This can happen in a collision with another atom or in a chemical reaction. When this occurs the atom is called an ion, and it has a positive or negative charge.

TEKS 8.8 B Assessment

1. What determines the mass number of an atom?
2. What is an atom with an electrical charge called?
3. **Thinking Critically** An atom of carbon has 12 protons and no net electrical charge. How many electrons must this atom have? Why?

TEKS Review

Substances Have Chemical and Physical Properties

TEKS 8.9

Substances have chemical and physical properties.

8.9 A demonstrate that substances may react chemically to form new substances

8.9 B interpret information on the periodic table to understand that physical properties are used to group elements

8.9 C recognize the importance of formulas and equations to express what happens in a chemical reaction

8.9 D identify that physical and chemical properties influence the development and application of everyday materials such as cooking surfaces, insulation, adhesives, and plastics

Directions *Review these concepts from Grade 8. They will help prepare you to study Chapters 18, 19, 20, 24, and 25 on the chemical and physical properties of substances, a topic required by the Integrated Physics and Chemistry TEKS.*

For 8.9 A, recall that:

Substances can undergo physical and chemical changes. Physical changes do not change the identity of a substance. Only its form changes. Chemical changes are changes that result in a new substance or substances, with properties that differ from the starting material.

Chemical reaction is the term used for the process of chemical change. Chemical reactions can be described using a chemical equation, which shows the beginning substances, called the reactants, and the ending substances, called the products. Chemical reactions occur at many different rates.

Endothermic chemical reactions require added energy to occur. Photosynthesis is an endothermic chemical reaction. Exothermic chemical reactions give off energy. Burning, as shown in **Figure 1,** is an exothermic reaction.

TEKS 8.9 A Assessment

1. What is the term used to describe a chemical reaction that gives off energy?
2. What are the substances formed in a chemical reaction called?
3. Would respiration be classified as an endothermic or an exothermic chemical reaction? Why?

Figure 1
Burning is an exothermic chemical reaction.

For 8.9 B, recall that:

An element is a substance that can't be broken down into simpler components. The 115 known elements are organized on a chart called the periodic table of the elements. The elements are arranged in rows, called periods, in order by atomic number. The periodic table of the elements classifies elements by physical properties. The location of an element on the periodic table can give clues about the characteristics of the element.

On the periodic table, the symbol for each element is shown in a box, called an element key. The element key for the element hydrogen is shown in **Figure 2.** The element key contains the name of the element, the abbreviation used for the element, the atomic number, the atomic mass, and the state of matter that the element usually exists in.

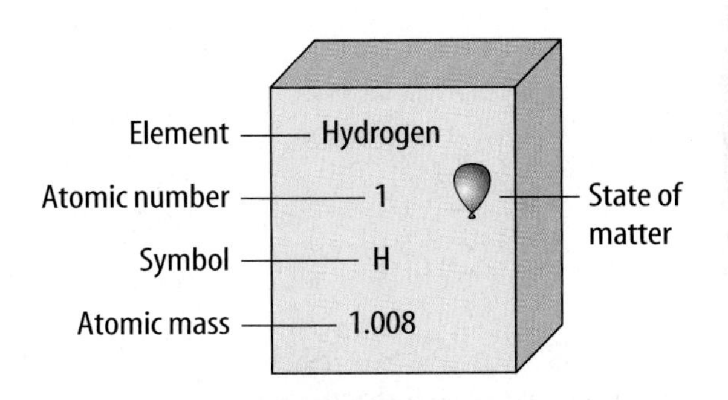

Figure 2
This element key gives information about hydrogen.

The elements in each column on the periodic table are called a group, or a family. The elements in each group have similar physical and chemical characteristics. For example, the elements in the first column, with the exception of hydrogen, are all silvery solids.

Elements on the periodic table also can be classified into three types: metals, nonmetals, and metalloids, by physical properties. Metals are ductile, malleable, lustrous, and conduct electricity and heat. Nonmetals are poor conductors of heat and electricity and do not have luster. Metalloids have some properties of metals and some properties of nonmetals.

TEKS 8.9 B Assessment

1. What is a term for the elements in a column of the periodic table?
2. What are three characteristics of metals?
3. **Thinking Critically** Why are metallic elements used more often than nonmetallic elements for making pots and pans?

TEKS Review

For 8.9 C, recall that:

A chemical reaction occurs when a substance is converted into another substance, or substances, with properties different from the starting material. The original substances are called reactants. The resulting substances are called products.

Chemical formulas are shortened ways to show the number of individual elements present. For example, the chemical formula for water, H_2O, indicates that one molecule of water contains two atoms of hydrogen and one atom of oxygen.

A chemical equation is a way to describe a chemical reaction.

Figure 3
The chemical reaction between vinegar and baking soda can be expressed using a chemical equation. A Chemical formulas for the reactants are shown to the left of the arrow, and chemical formulas of the products are shown to the right of the arrow. B A balanced chemical equation has the same number of atoms of each element on both sides of the equation.

Reactants — **Products**

A $HC_2H_3O_2 + NaHCO_3 \rightarrow NaC_2H_3O_2 + H_2O + CO_2$

B HCCHHHOO NaHCOOO — NaCCHHHOO HHO COO

vinegar, baking soda — sodium acetate, water, carbon dioxide

A chemical equation, shown in **Figure 3A,** includes the chemical formulas and the relative amounts of all of the substances in the reaction. An arrow indicates the direction of the reaction. The chemical formulas of the reactants are to the left of the arrow and the products are to the right of the arrow. The arrow means "produces." In this equation, the reactants are vinegar and baking soda. The products of this chemical reaction are sodium acetate, water, and carbon dioxide.

Atoms cannot be created or destroyed during a chemical reaction, so the number of atoms of each element are the same on both sides of the equation, as in **Figure 3B.** In the reaction above, the number of atoms of oxygen, hydrogen, carbon, and sodium in the reactants are equal to those in the products. This is an example of a balanced equation.

TEKS 8.9 C Assessment

1. What abbreviation do chemists use to describe a molecule?
2. Why must chemical equations be balanced?
3. **Thinking Critically** Why are some chemical equations written with arrows pointing both directions?

For 8.9 D, recall that:

New materials are developed with chemical and physical properties that are useful in everyday life.

Ceramics The properties that make ceramics so important are strength, hardness, and resistance to many chemicals. For hundreds of years, ceramics have been used decoratively and as materials for flooring. Modern ceramics are often made of metallic elements combined with oxygen, carbon, nitrogen, or sulfur. Ceramics can be custom made with the properties required to perform a certain function. Certain ceramics are used as insulators.

Alloys An alloy is a mixture of a metal with another element or several elements. Some alloys, such as bronze, have been used for centuries for tools and weapons. Steel is shown in **Figure 4.**

Figure 4
The properties of steel make it useful in construction. Steel is used as the framework for many buildings. Other alloys are used to make food cans and fillings for teeth.

Synthetic Polymers A polymer is a substance made of molecules in long chains. The chain is constructed with small units called monomers.

Plastics are made from synthetic polymers. These are substances that do not occur naturally but must be manufactured. Plastics can be customized to have the properties required for a particular function. Synthetic polymers are used for nonstick surfaces on cooking utensils. Other synthetic polymers, such as acrylics, are developed and used for their adhesive properties.

TEKS 8.9 D Assessment

1. What is an alloy made up of?
2. What are the smaller units that compose a polymer called?
3. **Thinking Critically** Synthetic polymers often are used to make the plumbing that carries water throughout homes and apartments. What is one property that should be considered when designing a polymer used for plumbing?

TEKS 8.10 Review

Complex Interactions Between Matter and Energy

What You'll Review

TEKS 8.10

Complex interactions occur between matter and energy.

8.10 A illustrate interactions between matter and energy including specific heat

8.10 B describe interactions among solar, weather, and ocean systems

8.10 C identify and demonstrate that loss or gain of heat energy occurs during exothermic and endothermic chemical reactions

TEKS Review

Directions *Review these concepts from Grade 8. They will help prepare you to study Chapters 4 and 24 on interactions between matter and energy, a topic required by the Integrated Physics and Chemistry TEKS.*

For 8.10 A, recall that:

Most matter on Earth is composed of atoms and molecules. The specific types and numbers of atoms and molecules present in a substance determine the unique qualities of the substance.

The energy of the moving particles in a substance is called kinetic energy. A substance with faster-moving particles will have a higher temperature, as shown in **Figure 1.** Specific heat is the amount of energy required to increase the temperature of one kilogram of a substance one degree Celsius. Every substance has its own specific heat.

Energy can be transferred between substances. Thermal energy can be transferred from a warm object to a cooler object in three ways. Radiation is the transfer of energy by electromagnetic waves. Radiation does not require matter to travel through. Heat from the Sun is transferred to Earth by the process of radiation. Conduction is the transfer of energy between substances that are in direct contact with one another. A pan on a stove burner is heated through the process of conduction. Convection is a transfer of energy due to a flow of heated material. Matter that is liquid or gaseous can transfer energy through convection.

Figure 1
The particles in the hot tea move faster than those in iced tea. The particles in the hot tea have more kinetic energy.

TEKS 8.10 A Assessment

1. What are three ways that thermal energy can be transferred?
2. What is the term for the energy contained in the moving particles of a substance?
3. **Thinking Critically** Explain what types of energy transfer take place if you burn your hand on a playground slide on a hot day.

For 8.10 B, recall that:

Wind, air pressure, water, and temperature determine weather. Weather is affected by the Sun and the oceans. Oceans are the main source of water for precipitation. As shown in **Figure 2,** oceans also absorb much of the thermal energy radiated by the Sun that strikes Earth, moderating Earth's temperatures.

The thermal energy that determines temperature is energy transferred by radiation from the Sun. Energy that heats the ocean also is transferred from place to place by convection in water currents. Thermal energy is also transferred from the oceans to the atmosphere, affecting the formation of storms and other weather systems.

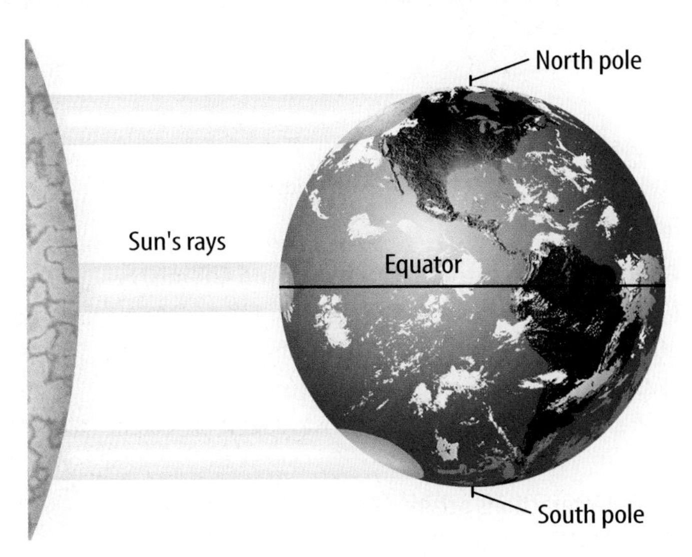

Figure 2
The Sun's rays strike different parts of Earth at different angles. The area near the equator receives the most direct sunlight.

TEKS 8.10 B Assessment

1. How is thermal energy from the Sun transferred to Earth?
2. What is one way that Earth's oceans affect weather?
3. **Thinking Critically** What are two reasons that the weather on Mercury is different from the weather on Earth?

For 8.10 C, recall that:

In all chemical reactions, energy must be added to break bonds between atoms. When atoms form bonds, energy is released, often as heat or light.

Figure 3
The chemical process that takes place when a cold compress is activated is an endothermic reaction.

Endothermic Reactions Endothermic reactions require added energy to occur. They absorb energy. One example of an endothermic reaction is shown in **Figure 3.**

Exothermic Reactions Chemical reactions that give off energy as heat are called exothermic reactions.

TEKS 8.10 C Assessment

1. What type of chemical reaction gives off energy?
2. What chemical reaction needs more energy to break bonds in the reactants than is given off during formation of bonds in products?
3. **Thinking Critically** An endothermic reaction occurs in a beaker on a desk. Where does the energy for the reaction come from?

TEKS Review

TAKS Objectives

This list of TAKS will be corrected if the TEA makes changes.

Objective 1

The student will demonstrate an understanding of the nature of science.

Biology (1) and Integrated Physics and Chemistry (1) Scientific Processes. The student, for at least 40% of instructional time, conducts field and laboratory investigations using safe, environmentally appropriate, and ethical practices. The student is expected to

(A) demonstrate safe practices during field and laboratory investigations.

Biology (2) and Integrated Physics and Chemistry (2) Scientific Processes. The student uses scientific methods during field and laboratory investigations. The student is expected to

(A) plan and implement investigative procedures including asking questions, formulating testable hypotheses, and selecting equipment and technology;

(B) collect data and make measurements with precision;

(C) organize, analyze, evaluate, make inferences, and predict trends from data; and

(D) communicate valid conclusions.

Integrated Physics and Chemistry (3) Scientific Processes. The student uses critical thinking and scientific problem solving to make informed decisions. The student is expected to

(A) analyze, review, and critique scientific explanations, including hypotheses and theories, as to their strengths and weaknesses using scientific evidence and information; and

(B) draw inferences based on data related to promotional materials for products and services.

Objective 2

The student will demonstrate an understanding of the organization of living systems.

Biology (4) Science Concepts. The student knows that cells are the basic structures of all living things and have specialized parts that perform specific functions and that viruses are different from cells and have different properties and functions. The student is expected to

(B) investigate and identify cellular processes including homeostasis, permeability, energy production, transportation of molecules, disposal of wastes, function of cellular parts, and synthesis of new molecules.

Biology (6) Science Concepts. The student knows the structures and functions of nucleic acids in the mechanisms of genetics. The student is expected to

(A) describe components of deoxyribonucleic acid (DNA), and illustrate how information for specifying the traits of an organism is carried in the DNA;

(C) identify and illustrate how changes in DNA cause mutations and evaluate the significance of these changes; and

(D) compare genetic variations observed in plants and animals.

Biology (8) Science Concepts. The student knows applications of taxonomy and can identify its limitations. The student is expected to

(C) identify characteristics of kingdoms including monerans, protists, fungi, plants, and animals.

Biology (10) Science Concepts. The student knows that at all levels of nature living systems are found within other living systems, each with its own boundary and limits. The student is expected to

(A) interpret the functions of systems in organisms including circulatory, digestive, nervous, endocrine, reproductive, integumentary, skeletal, respiratory, muscular, excretory, and immune

Objective 3

The student will demonstrate an understanding of the interdependence of organisms and the environment.

Biology (4) Science Concepts. The student knows that cells are the basic structures of all living things and have specialized parts that perform specific functions, and that viruses are different from cells and have different properties and functions. The student is expected to

(C) compare the structures and functions of viruses to cells and describe the role of viruses in causing diseases and conditions such as acquired immune deficiency syndrome, common colds, smallpox, influenza, and warts.

(D) identify and describe the role of bacteria in maintaining health such as in digestion and in causing diseases such as in streptococcus infections and diphtheria.

Biology (7) Science Concepts. The student knows the theory of biological evolution. The student is expected to
(B) illustrate the results of natural selection in speciation, diversity, phylogeny, adaptation, behavior, and extinction.

Biology (12) Science Concepts. The student knows that interdependence and interactions occur within an ecosystem. The student is expected to
(B) interpret interactions among organisms exhibiting predation, parasitism, commensalism, and mutualism; and
(E) investigate and explain the interactions in an ecosystem including food chains, food webs, and food pyramids.

Biology (13) Science Concepts. The student knows the significance of plants in the environment. The student is expected to
(A) evaluate the significance of structural and physiological adaptations of plants to their environments.

Objective 4

The students will demonstrate an understanding of the structures and properties of matter.

Integrated Physics and Chemistry (7) Science Concepts. The student knows relationships exist between properties of matter and its components. The student is expected to
(A) investigate and identify properties of fluids including density, viscosity, and buoyancy;
(E) classify samples of matter from everyday life as being elements, compounds, or mixtures.

Integrated Physics and Chemistry (8) Science Concepts. The student knows that changes in matter affect everyday life. The student is expected to
(A) distinguish between physical and chemical changes in matter such as oxidation, digestion, changes in states, and stages in the rock cycle; and
(C) investigate and identify the law of conservation of mass.

Integrated Physics and Chemistry (9) Science Concepts. The student knows how solution chemistry is a part of everyday life. The student is expected to
(A) relate the structure of water to its function [as the universal solvent]; and
(D) demonstrate how various factors influence solubility including temperature, pressure, and nature of the solute and solvent.

Objective 5

The student will demonstrate an understanding of motion, forces, and energy.

Integrated Physics and Chemistry (4) Science Concepts. The student knows concepts of force and motion evident in everyday life. The student is expected to
(A) calculate speed, momentum, acceleration, work, and power in systems such as in the human body, moving toys, and machines;
(B) investigate and describe applications of Newton's laws such as in vehicle restraints, sports activities, geological processes, and satellite orbits

Integrated Physics and Chemistry (5) Science Concepts. The student knows the effects of waves on everyday life. The student is expected to
(A) demonstrate wave types and their characteristics through a variety of activities such as modeling with ropes and coils, activating tuning forks, and interpreting data on seismic waves.

Integrated Physics and Chemistry (6) Science Concepts. The student knows the impact of energy transformations in everyday life. The student is expected to
(A) describe the law of conservation of energy;
(B) investigate and demonstrate the movement of heat through solids, liquids, and gases by convection, conduction, and radiation;
(F) investigate and compare series and parallel circuits.

TAKS Test Practice

How to Use This Practice Test

The following questions will help you practice and review for the TAKS exam that you will take at the end of tenth grade. The exam includes questions on physics, chemistry, biology, and Earth science. The TAKS objectives that cover physics and chemistry are reviewed here and are labeled following each question.

SCIENCE Online For complete TAKS practice tests, as well as additional TAKS practice for each objective, visit **tx.science.glencoe.com**.

1.

3 Li
11 Na
19 K
37 Rb
55 Cs
87 Fr

The alkali metals are in Group 1A in the periodic table. Which of the following statements about the alkali metals is FALSE?
Objective 4, IPC 7D

A) They react violently with cold water.
B) They are metallic solids at room temperature.
C) They are not found in nature in an uncombined state.
D) They are relatively unreactive.

2.

Animal	Blood proteins
A	QWWYXYYXXZZQZ
B	QWWYVYYVVZZQZ
C	RWWYVYYVVZZQZ
D	QWWSXYYXXZZQZ
E	PWWSXYYXXZZQZ
F	RWWTVTYVVZZQZ
G	PWWYXYYXXZZQZ
H	QWWTVYYVVZZQZ

Each series of letters in the chart represents blood proteins of different animals. If the animals were divided into two groups of closely related animals based on evolutionary changes, which of the following groupings would be correct?
Objective 3, Biology 7A

F) Group 1: A, B, D, H
Group 2: C, E, F, G
G) Group 1: A, C, E, F
Group 2: B, D, G, H
H) Group 1: A, D, E, H
Group 2: B, C, F, G
J) Group 1: A, D, E, G
Group 2: B, C, F, H

Table A	Table B	Table C
Element	Compound	Mixture
carbon	water	salt water
hydrogen	sugar	air
oxygen		soil

3. An element is a form of matter that contains only one kind of atom. A compound is composed of two or more combined elements that bond chemically. A mixture is composed of two or more elements or compounds that are not bonded. Which of the following items would best fill in the blank in **Table B?**
Objective 4, IPC 4.7 E

A) helium
B) salt
C) muddy water
D) copper

4. Bacteria cause tuberculosis, strep throat, lyme diseases, and food poisoning. In the human intestines, bacteria produce the vitamin necessary for blood clotting. Bacteria also decompose dead organisms, enrich soil, and are used to produce foods and medicines. Which of the following statements best summarizes the role of bacteria in human health?
Objective 3, Biology 4D

F) All bacteria are harmful to human health.
G) All bacteria are beneficial to human health.
H) Some bacteria are harmful while other bacteria are helpful.
J) Bacteria have no effect on human health.

Wave motion blocked

5. Which of the following wave interactions does the demonstration shown in the picture above model above?
Objective 5, IPC 5B

A) refraction
B) reflection
C) polarization
D) interference

6. Symmetry refers to the arrangement of parts of an object or organism. All animals have some type of symmetry. Butterflies, for example, have bilateral symmetry, which means they can be divided down their length into two halves that are mirror images of one another. Which of the animals does NOT have bilateral symmetry?
Objective 2, Biology 8 C

F)

G)

H)

J)

7. When the skier moves down the hill, her potential energy changes to kinetic energy. Which of the following statements describes what happens during this energy change?
Objective 5, IPC 6A

A) The total amount of energy is increased.
B) A small amount of energy is destroyed.
C) The total amount of energy is not changed.
D) The total amount of energy decreases.

	R	r
R	RR	Rr
r	Rr	rr

8. The Punnett square shows the possible offspring of two heterozygous plants. Dominant alleles are represented by capital letters and recessive alleles are represented by small letters. What are the chances that the offspring will show the dominant trait?
Objective 2, Biology 6 D

F) 1 out of 4
G) 2 out of 4
H) 3 out of 4
J) 4 out of 4

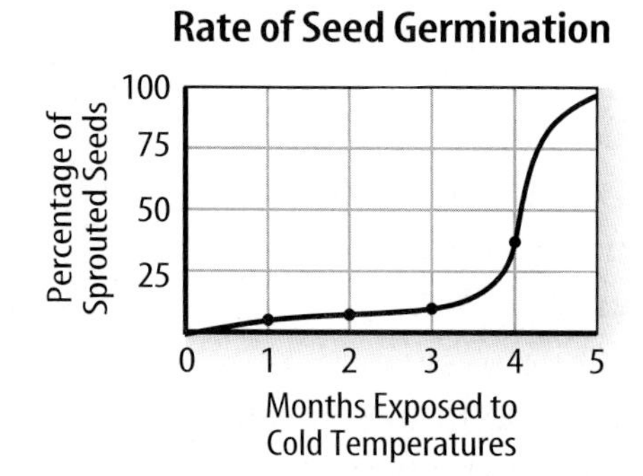

9. The seeds of different plant species germinate, or sprout into new plants, under different environmental conditions. Based on the graph, what can you infer about the conditions needed for the seeds of this particular species to sprout?
Objective 3, Biology 13 A

A) The seeds require a warm environment.
B) The seeds require a dry environment.
C) The seeds require a moist environment.
D) The seeds require a cold environment.

---

10. Osmosis is the diffusion of water through a selectively permeable membrane. The arrows in the diagrams show the direction of water movement into and out of cells. Which diagram represents a cell in equilibrium?
Objective 2, Biology 4 B

F)
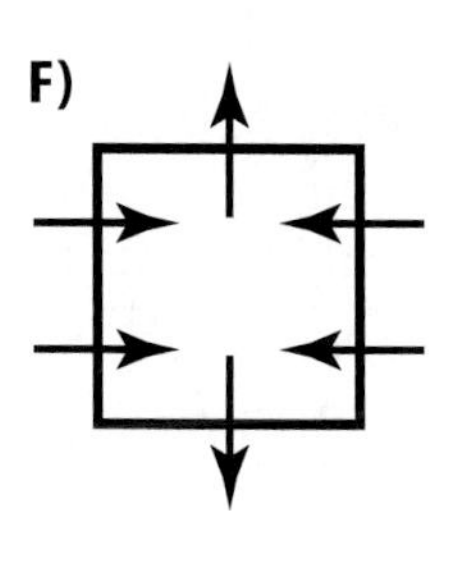

H)
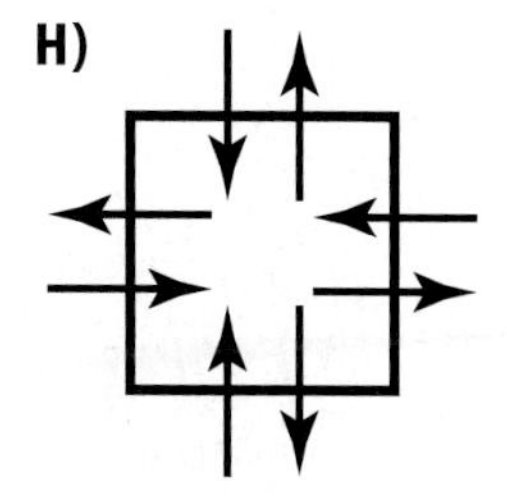

G)
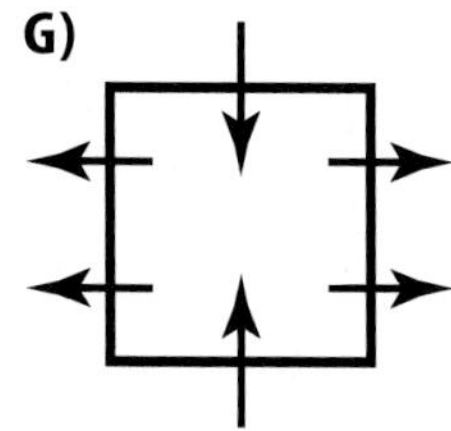

J)
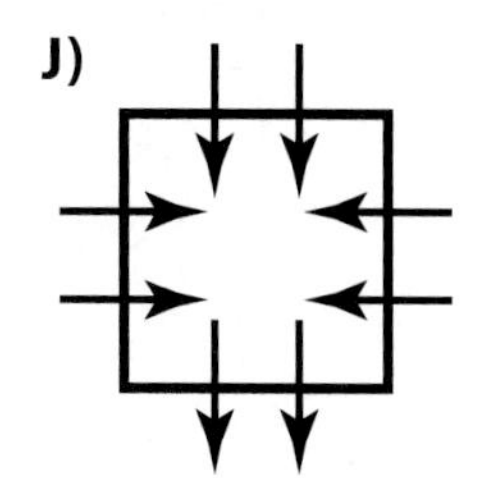

11. Substances that dissolve most readily in water include ionic compounds and polar covalent compounds. Which of the following compounds would NOT dissolve in water?
Objective 4, IPC 9A

A) NaCl (sodium chloride)
B) CH_4 (methane)
C) CO_2 (carbon dioxide)
D) $MgCl_2$ (magnesium chloride)

Human Disease Caused by Viruses

Disease	Affected System	Vaccine
AIDS	Immune system	No
Common cold	Respiratory system	No
Measles	Skin	Yes
Polio	Nervous system	Yes
Rabies	Nervous system	Yes

12. A virus is a nonliving particle that infects and multiplies inside a living cell. The table shows some human diseases caused by viruses. Which statement best represents the data in the table?
Objective 3, Biology 4 C

F) Rabies affects the immune system.
G) Polio affects the nervous system.
H) Measles cannot be prevented by a vaccine.
J) AIDS can be prevented by a vaccine.

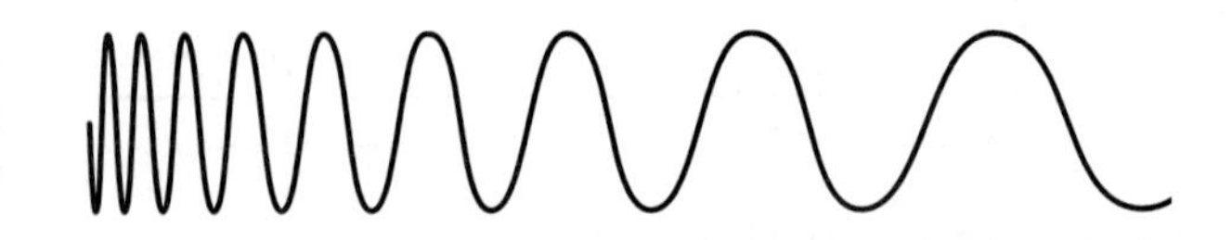

13. Examine the diagram of a transverse wave. What happens to wavelength as frequency decreases?
Objective 5, IPC 5 A

A) It increases.
B) It decreases.
C) It stays the same.
D) It disappears.

14. Which of the following items is not used for safety protection in the laboratory?
Objective 1, IPC 1 A

F)
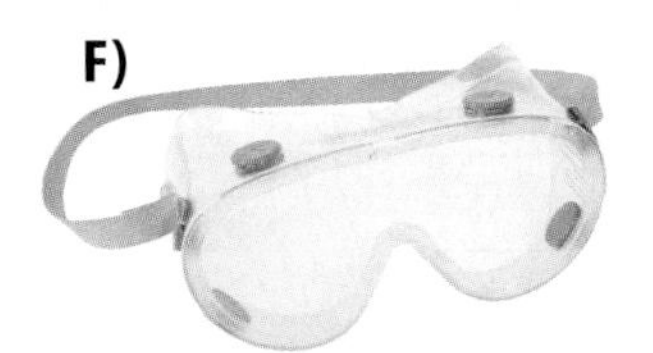

G)
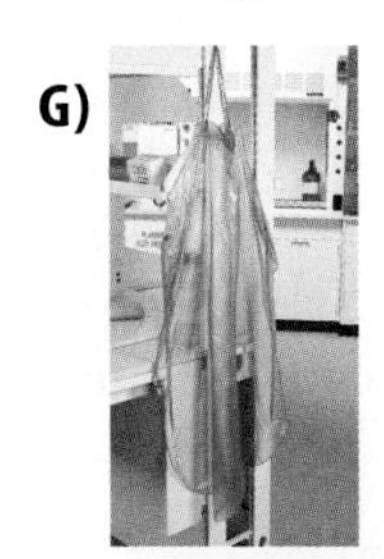

H)
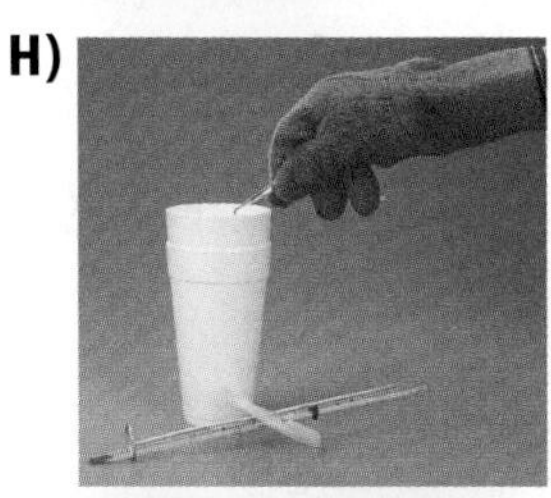

J)
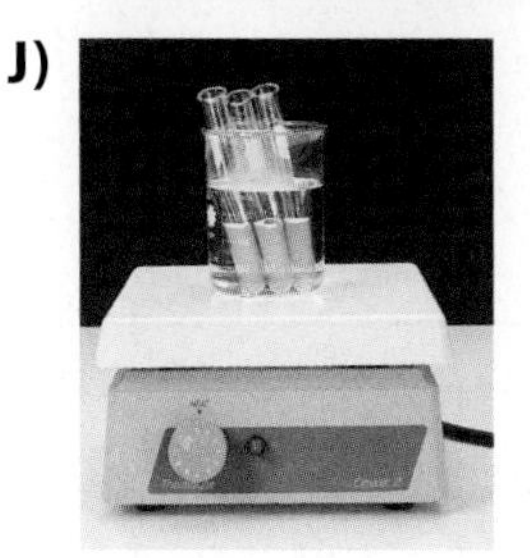

15. Company A is advertising a new detergent, that claims to whiten with no bleach. They say that it has been tested in 50 homes on normal wash loads. Company B is also advertising a new detergent claiming identical results. Company B published their results showing test results from 500 homes on normal wash loads. As a student of scientific methods, which product would you try first, and why?
Objective 1, IPC 3 B

A) Company A because they claim to have tested their product.
B) Company B because their test results were the same as Company A's results.
C) Company A because they wouldn't advertise unless their claim were true.
D) Company B because their research results come from a larger sample of homes.

16. A parallel circuit provides more than one path for an electric current to follow. A simple circuit, however, provides only one path for the current. Which of the following diagrams shows a simple circuit?
Objective 5, IPC 6 F

F)
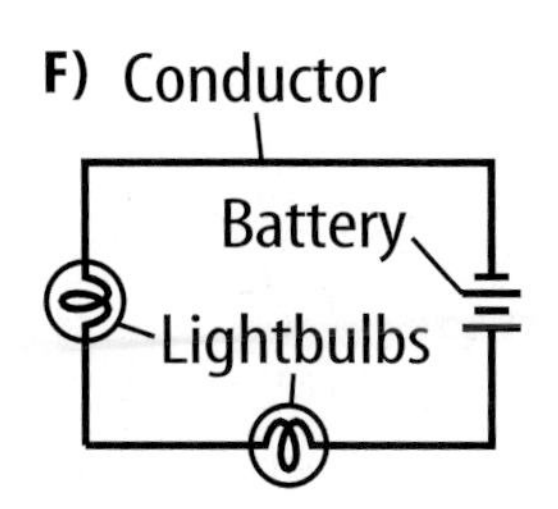

G)
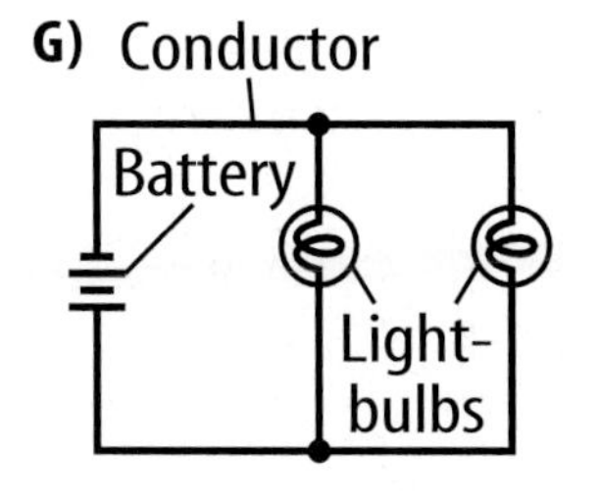

H)
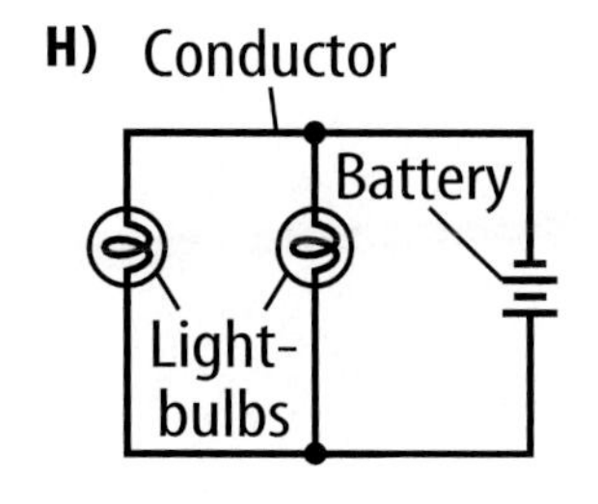

J)
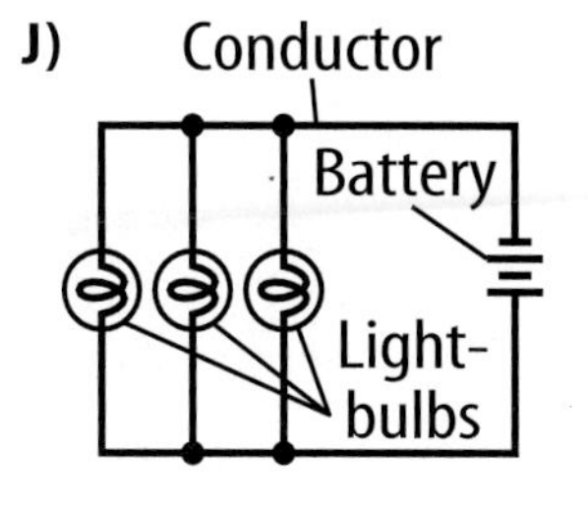

TAKS Test Practice

Hypothesis: Wet slopes are less stable than dry slopes	
Step	**Experiment Procedure**
1	Add water to sediment, then make three model slopes of differing sizes.
2	Use rocks to test the stability of the model slopes.
3	Record observations and analyze results.

17. A student designed an experiment to test the effect of water on the stability of steep slopes. The table shows the steps he planned to test his hypothesis. Before performing the experiment, he gave the plan to his teacher to evaluate. The teacher told him that his procedure was flawed. What statement below best reflects the teacher's concern?
Objective 1, IPC 3 A

A) The student did not form a hypothesis.
B) The student did not use a control.
C) The student did not analyze results.
D) The student did not list all steps.

18. The illustration shows different species of birds that are thought to have evolved from a common ancestor. What factor most distinguishes the birds from one another?
Objective 3, Biology 7 B

F) the shape of their heads
G) the color of their features
H) the placement of their eyes
J) the shape of their beaks

19. When must you wear safety goggles in a science laboratory?
Objective 1, IPC 1 A

A) at all times
B) only when working with hot things
C) whenever you also have to wear an apron
D) when disposing of materials used in the lab

20. The theory of the structure of the atom today is different from what it was 200 years ago. One reason for this is:
Objective 1, IPC 3 A

F) Scientists are smarter now than scientists then.
G) Scientists can see the atom better now.
H) A theory can change with new information.
J) A theory is the work of only one scientist.

21. What is the most likely reason that scientists have not been able to develop a successful long-term vaccination for influenza?
Objective 2, Biology 6C

A) The virus that causes influenza has not been isolated and identified.
B) The development and production of a vaccine is too expensive.
C) The genetic code for the viral surface protein mutates often.
D) Influenza is caused by an autoimmune response.

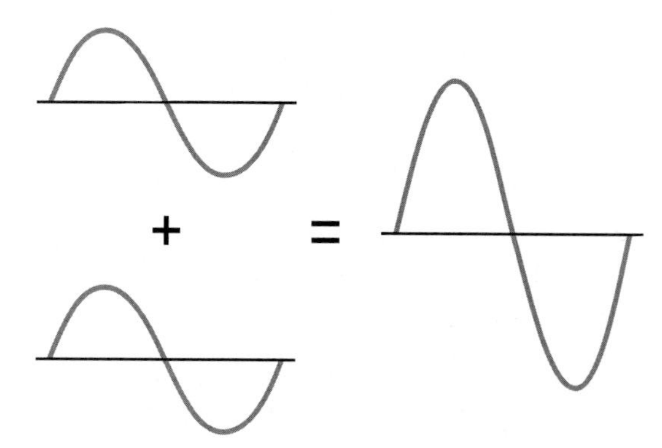

22. Which of the following statements describes what happens when the crest of the two identical waves shown overlap?
Objective 5, IPC 5B

F) A wave with half the amplitude results.
G) A wave with twice the amplitude results.
H) No wave results because of wave interference.
J) The waves have no effect on each other.

23. A lack of leguminous plants in a farmer's field might result in which of the following?
Objective 6, 8.12 C

A) decreased oxygen fixation
B) increased photosynthesis
C) decreased nitrogen fixation
D) increased cellular respiration

24. According to Newton's first law, if you are a passenger in a vehicle which stops suddenly, how will your body behave if you are not wearing a safety belt?
Objective 5, IPC 4B

F) It will stop with the vehicle.
G) It will continue to move straight ahead.
H) It will move suddenly backward.
J) It will tend to move to one side.

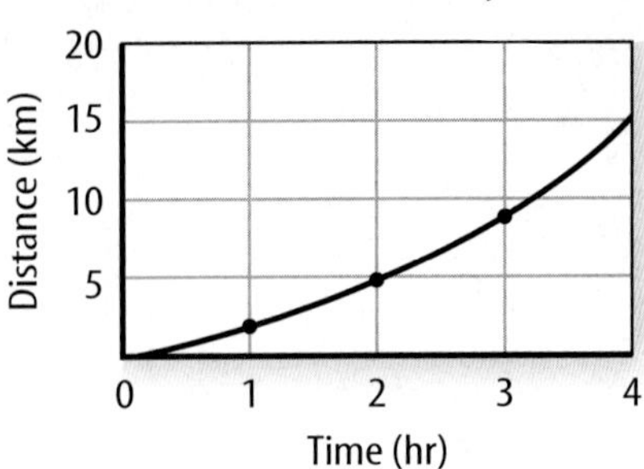

25. The graph shows the distance traveled by a hiker over the course of four hours. What can you infer about the hiker's speed?
Objective 5, IPC 4 A

A) She started out fast, then slowed down.
B) She did not change her rate of speed.
C) She started out slowly, then quickened her pace.
D) Her speed cannot be determined from the graph.

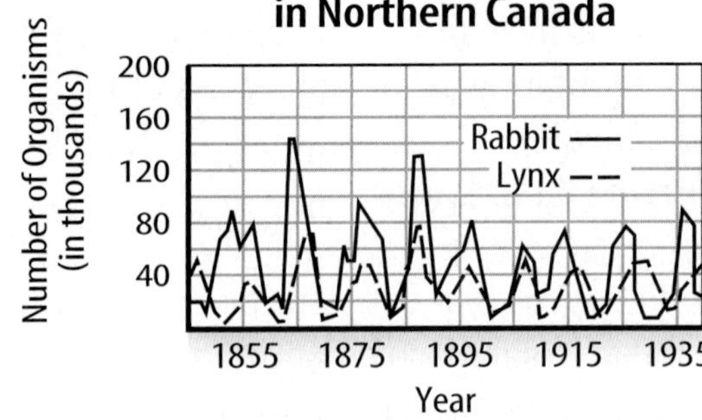

26. Lynx are predators that feed on other animals, such as hares. The graph shows how populations of lynxes and hares in northern Canada have changed over time. Which statement best summarizes the data in the graph?
Objective 3, Biology 12 B

F) As prey population increased, predator population decreased.
G) As prey population increased, predator population increased.
H) As prey population decreased, predator population increased.
J) There is no measureable relationship between prey and predator populations.

27. A hypothesis is a statement about a problem that must be testable. At the end of an investigation, which of the following may happen?
Objective 1, IPC 2 A

A) The hypothesis is accepted because it is testable.
B) The hypothesis is accepted because it is supported by data.
C) The hypothesis is rejected because it is wrong.
D) The hypothesis is rejected because it is supported by data.

28. Isaac Newton was known for his thorough investigation of problems. He planned his experiments precisely. However, he often did not publish his findings for many years. What aspect of scientific research did Newton avoid?
Objective 1, IPC 2 D

F) collecting data
G) analyzing data
H) communicating conclusions
J) asking questions

29. Energy is transferred by the processes of radiation, convection, and conduction. Radiation is the transfer of energy by electromagnetic waves. Convection is the transfer of energy by the flow of a heated material. Conduction is the transfer of energy through matter by direct contact of particles. Which illustration shows the transfer of energy by convection?
Objective 5, IPC 6 B

A) 

B) 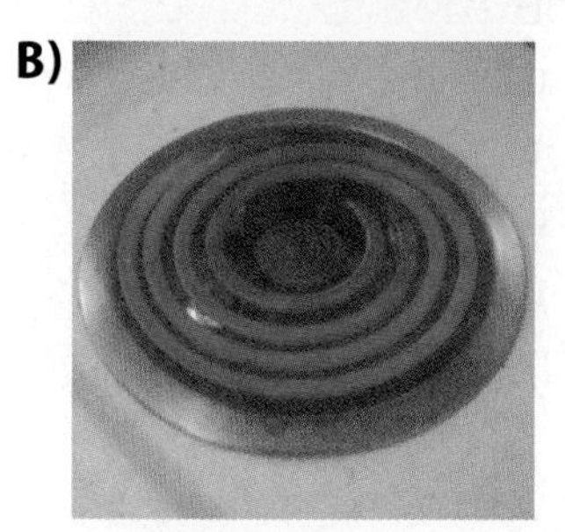

C) 

D) 

30. DNA is the master copy of an organism's information code. It is composed of four bases: adenine (A), guanine (G), cytosine (C), and thymine (T). These bases always occur in certain pairs. Which diagram represents a correct pair of bases?
Objective 2, Biology 6 A

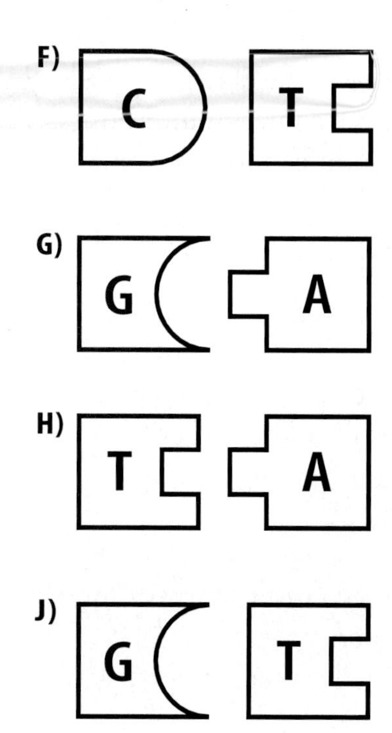

Effect of Cooling on Crystal Size

Trials	Crystal Size		
Rate of Cooling	Fine-grained	Medium-grained	Large-grained
Trial 1			
Fast	✓		
Slow			✓
Moderate		✓	
Trial 2			
Fast	✓		
Slow			✓
Moderate		✓	

31. A student designed an experiment to test the effect of cooling rates on crystal size. She performed several trials. The table shows the results of her experiment. What pattern do you observe?
Objective 1, IPC 2 C

A) Fine-grained crystals form at moderate cooling rates.
B) Fine-grained crystals form at slow cooling rates.
C) Large-grained crystals form at fast cooling rates.
D) Large-grained crystals form at slow cooling rates.

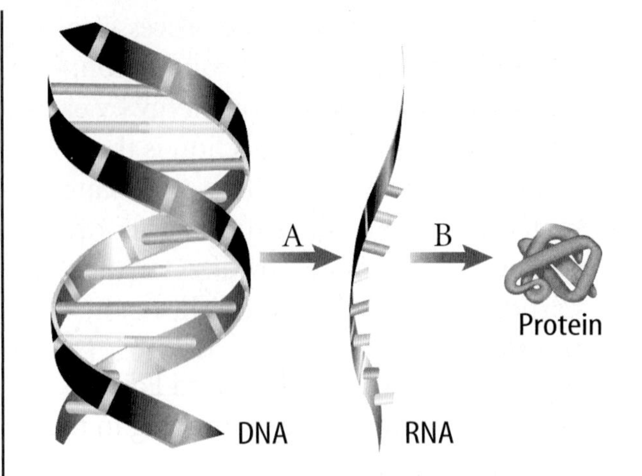

32. In the transfer of genetic information shown above, which stage is shown by arrow A?
Objective 2, Biology 6B

F) translation
G) transcription
H) repression
J) replication

33. Of the systems in the human body, which one of the following directly interacts with all of the other systems?
Objective 2, Biology 10 A

A) digestive system
B) reproductive system
C) skeletal system
D) circulatory system

34. A dancer lifts a 400-N ballerina 1.4 m off the ground and holds her there for five seconds. Using the formula $W = f \times d$, how much work did he do?
Objective 5, IPC 4 A

F) 258 J
G) 560 J
H) 285 J
J) 600 J

35. If the force of Earth's gravity suddenly stopped pulling on the Moon, what would happen according to Newton's laws?
Objective 5, IPC 4B

A) The Moon would move off into space in a straight line.
B) The Moon would continue to orbit Earth.
C) The Moon would fall toward Earth.
D) The Moon would begin moving in the opposite direction.

36. Gymnosperms and angiosperms have many common adaptations that allow them to survive on land. Which of the following is an adaptation that is unique to angiosperms?
Objective 3, Biology 13A

F) Each of their seeds contains a young plant and stored food.
G) Seeds are produced enclosed in fruits.
H) Seeds can survive long periods of time in unfavorable environments.
J) Pollen grains do not require moisture for carrying sperm to the egg.

Densities of Common Materials

Material	Density (mL)
Oxygen	0.0015
Water	1.0
Sugar	1.6
Copper	

37. The table shows the densities of some common materials. If a sample of copper has a mass of 178 g and a volume of 20 mL, what is its density?
Objective 4, IPC 7 A

A) 8.9 g/mL
B) 3560 g/mL
C) 198 g/mL
D) 0.11 g/mL

38. A food web is a model of how many different overlapping food chains exist in an area. In the food web above, which organisms do not compete for the same food?
Objective 3, Biology 12 E

F) mice and deer
G) hawks and snakes
H) deer and mountain lions
J) rabbits and mice

39. The bar graph below shows the approximate number of invertebrates and vertebrates currently identified by scientists. Which category includes humans?
Objective 2, Biology 8 C

A) insects
B) other invertebrates
C) vertebrates
D) Category is not listed in table.

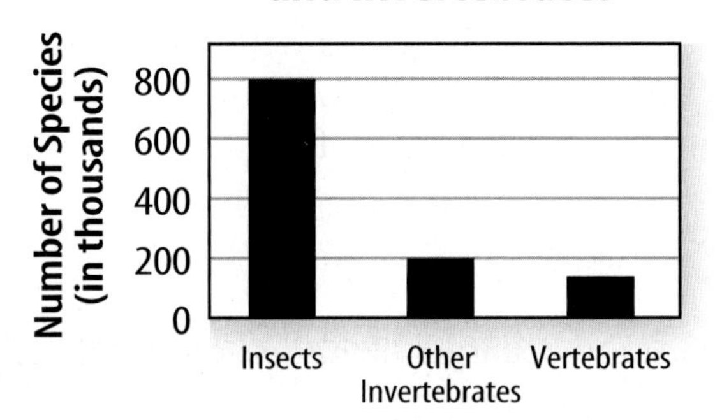

40. After a warm and rainy summer, you go out to the garage and find that your family's bikes are rusted badly. What chemical reaction have the bikes undergone?
Objective 4, IPC 8 A

F) respiration
G) decomposition
H) digestion
J) oxidation

41. Dissolve a large sugar cube in a container of water marked *A*. Then dissolve eight smaller cubes into another container, *B* which is the same temperature and contains the same amount of water as *A*. What would you expect to happen?
Objective 4, IPC 9 D

A) The sugar will dissolve at the same rate in both containers.
B) The large cube will dissolve faster than the smaller cubes.
C) The smaller cubes will dissolve faster even though there are more of them.
D) None of the sugar cubes will dissolve unless the water in each container is heated.